GLOSSARY OF SYMBOLS: Roman Letters

Symbol	Meaning	Chapter(s)
m_i	Number of elements in cluster *i*	19
$\bar{m}$	Average cluster size	19
M_2	Average squared deviation	2
M_3	Average cubed deviation	2
M_4	Average of fourth-power deviation	2
n	Sample size	all
n	Number of trials, binomial and hypergeometric distribution	5
N	Population size	all
n_E	Number of occurrences of event E in repeated trials	3
N_E	Number of possible outcomes in event E	3
N_F	Population number of failures	5
N_S	Population number of successes	5
o_i	*i*th outcome	3
$P_{XY}(x,y)$	Joint probability distribution of discrete random variables *X* and *Y*	4
$P_{Y\|X}(y\|x)$	Conditional distribution of discrete random variable *Y*, given $X = x$	4
${}_kP_r$	Number of sequences (permutations) of *k* things out of *r* things	5
R^2	Coefficient of determination	13–15
R.R.	Rejection region	9ff
r_s	Rank correlation	13
r_{yx}	Correlation between variables *Y* and *X*	13–15
$\binom{r}{k}$	Number of sets (combinations) of *k* things out of *r* things	5
$R_{y \cdot x_1 \ldots x_k}$	Multiple correlation of variable *Y* with the best-predicting combination of $X_1, \ldots, X_k$	14, 15
s	Sample standard deviation	all
s^2	Sample variance	all
s_ϵ	Residual standard deviation	13–15
s_p	Pooled standard deviation	10
s_θ	Estimated standard error of the statistic $\hat{\theta}$	8–15

STATISTICAL THINKING FOR MANAGERS

STATISTICAL THINKING FOR MANAGERS

SECOND EDITION

DAVID K. HILDEBRAND
UNIVERSITY OF PENNSYLVANIA

LYMAN OTT
MERRELL DOW RESEARCH INSTITUTE

DUXBURY PRESS
BOSTON

PWS PUBLISHERS

Prindle, Weber & Schmidt • Duxbury Press • PWS Engineering • Breton Publishers •
20 Park Plaza • Boston, Massachusetts 02116

PWS Publishers is a division of Wadsworth, Inc.

Library of Congress Cataloging-in-Publication Data

Hildebrand, David K.
Statistical thinking for managers.

Ott's name appears first on earlier edition.
Bibliography: p.
Includes index.
1. Industrial management—Statistical methods. 2. Industrial management—Statistical methods—Data processing. 3. Economics—Statistical methods. 4. Statistics—Data processing. I. Ott, Lyman. II. Title.
HD30.215.H54 1987 519.5'024658 86-29222
ISBN 0-87150-036-1
Printed in the United States of America

87 88 89 90 91—10 9 8 7 6 5 4 3 2 1

Sponsoring Editor: Michael Payne
Production Coordinator: Ellie Connolly
Production: Stacey Sawyer, Complete Editorial Production Services
Interior and Cover Design: Ellie Connolly
Copy Editor: John Thomas
Cover Art: "Sedona II" by Richard Helmick; a screenprint used with permission of the artist.
Typesetting: Syntax International
Cover Printing: John P. Pow Company, Inc.
Printing and Binding: R. R. Donnelley & Sons Company

To Pat, Marty, and Jeff
Sally, Curtis, and Kathy
who all helped

PREFACE

Events since the publication of the first edition have confirmed our judgment that statistics texts must reflect the wide availability of computers. Now that there are literally dozens of statistical packages available for mainframe and personal computers, the emphasis in managerial statistics courses obviously must shift away from computation and toward thoughtful selection of methods and careful interpretation of results.

We have made several substantial changes to *Statistical Thinking for Managers* to increase its usefulness as a computer-era text.

- Review exercises have been added following several sets of chapters. These exercises allow the student to practice selecting appropriate methods without having the artificial clue of exercise placement within a chapter.
- Computer simulations of the performance of several standard methods, under many different conditions, have been incorporated in the text and exercises. These simulations provide concrete and often striking evidence of the effects of choosing a poor method or of violating key assumptions. (Transparency masters of these and many other simulations are available from the publisher.)
- Many additional exercises have been added to chapters. Some of these exercises are straightforward drills; many ask for thoughtful, critical responses.
- In recognition of the role of experiments in modern quality control, a discussion of quality control and multifactor experiments has been added. Of course, the computations are done by computer packages.
- Normal plots and box plots (including Tukey's "fence" check for outliers) of data are discussed. These plots are readily obtained by all good computer packages and are useful in choosing appropriate analyses of the data.
- A loosely structured "case exercise" has been added to the chapter on regression modeling to give the reader the opportunity to try hands-on regression modeling. It is no more likely than most regression studies to yield good results to simplistic models.

- In addition, we have rearranged some material for greater clarity and accessibility. The three basic forms of statistical inference—point estimation, interval estimation, and hypothesis testing—are discussed in separate chapters. The concepts of covariance and correlation of random variables are discussed much earlier than they were in the first edition, in the context of variance of a random variable. All methods for grouped data have been brought together in a single section. In recognition of the limited class time available, we have omitted the first edition's Chapter 3 (descriptive measures of statistical relation); where needed, the material has been added to the corresponding chapters on inference methods.
- We have added a chapter describing some of the data management chores necessary before an analysis, as well as guidelines for a statistical analysis and report.
- The calculus content of the text has been upgraded, particularly in the discussions of probability distributions. We have added many examples and exercises plus a new section on joint probability densities. However, knowledge of calculus is not a prerequisite for use of this text. Sections and exercises involving differential and integral calculus have been marked by derivative and integral signs, respectively, for the convenience of those who wish to skip calculus-based material.
- Geometric, negative binomial, uniform, and exponential distributions are discussed, along with the binomial, Poisson, and normal distributions. These additions should give the reader a better sense of the enormous variety of applications of probability theory.
- The material on analysis of variance has been expanded to include a discussion of randomized block designs.

These major changes, along with numerous other less substantive changes, have been made in response to constructive comments from our reviewers and users.

We gratefully acknowledge the contributions of many people to the making of this edition. Dr. Patricia J. Hildebrand ran all the computer simulations, calculated many of the tables, and prepared many of the answers. Editor Michael Payne put up with the authors with great fortitude. Ellie Connolly was in charge of converting a manuscript into a text. Professor Martin Puterman of the University of British Columbia provided extremely thoughtful and perceptive reviews of the work in progress. Other very useful reviews were done by Bruce L. Bowerman, Miami University; Jonathan D. Cryer, University of Iowa; Jerome F. Heavey, Lafayette College; Steven Hillmer, The University of Kansas; James E. Holstein, University of Missouri; Raj Jagannathan, University of Iowa; Frank Kelly, University of New Mexico; Paul Nelson, Kansas State University; Jacqueline F. Redder, Virginia Polytechnic and State University; George Sadler, University of Southern Colorado; Charles J. Stone, University of California at Berkeley; and Paul Thompson, Ohio State University. Of course, we have not always agreed with their suggestions; even those ideas that we have rejected have forced us to rethink our positions. Our thanks to them all.

D. Hildebrand
L. Ott

ABOUT THE AUTHORS

David K. Hildebrand is Professor and Chair of Statistics at the Wharton School, University of Pennsylvania. Since arriving at Wharton, he has taught full-time undergraduate, part-time undergraduate, full-time MBA, part-time MBA, and Ph.D. students of business (not all at the same time). Besides numerous articles in various statistical journals, he has published two other books, *Prediction Analysis of Cross Classifications* (New York: John Wiley and Sons, 1977), with J. D. Laing and H. Rosenthal, and *Statistical Thinking for Behavioral Scientists* (Boston: Duxbury, 1985), and a monograph *Analysis of Ordinal Data* (Santa Monica: Sage University Papers, 1977), also with J. D. Laing and H. Rosenthal. Among his other activities on behalf of the American Statistical Association, he has been an associate editor of the journal and has presented a short course before the national meetings in 1979. Dr. Hildebrand holds a Ph.D. in statistics from Carnegie-Mellon University and a bachelor's degree in mathematics from Carleton College. He was responsible for most of the initial drafting of this book.

Lyman Ott is Director of Biomedical Information Systems at Merrell Dow Research Institution, and an Adjunct Professor in the Division of Biostatistics at the University of Cincinnati. At Merrell Dow he is responsible for Biostatistics, Computer Services, Clinical Data Processing, Laboratory Automation, and Scientific Information Services. Prior to his career in the pharmaceutical industry, he was a faculty member in the Department of Statistics at the University of Florida, where he taught service courses as well as courses for majors at both the undergraduate and graduate levels. Dr. Ott has published research articles in various statistical journals and several textbooks, including *An Introduction to Statistical Methods and Data Analysis,* 2nd Edition, (Duxbury Press, 1984), *Understanding Statistics,* 4th Edition (Duxbury Press, 1985) coauthored with W. Mendenhall, and *Elementary Survey Sampling,* 3rd Edition, (Duxbury Press, 1986) coauthored with R. Scheaffer and W. Mendenhall. He is a Fellow of the American Statistical Association (ASA) and has served on the ASA Board of Directors. He holds a Ph.D. in Statistics from Virginia Polytechnic Institute and an undergraduate degree in Mathematics and Education from Bucknell University.

CONTENTS

1 STATISTICS: MAKING SENSE OF DATA 1

1.1 Gathering Data 1
1.2 Summarizing Data 2
1.3 The Role of Probability 3
1.4 Making Inferences from Sample Data 3
1.5 The Role of the Computer 4
1.6 Using Data in Making Decisions 5
Summary 6

2 SUMMARIZING DATA ABOUT ONE VARIABLE 7

2.1 The Distribution of Values of a Variable 7
2.2 On the Average: Typical Values 15
2.3 Measuring Variability 19
2.4 Other Summary Measures of Data 26
2.5 Calculators and Computer Software Systems 30
2.6 Summarizing Grouped Data 31
Summary 34
Appendix: Summation Notation 38

3

A FIRST LOOK AT PROBABILITY 41

3.1 Interpretations of Probability 41
3.2 Basic Concepts and Axioms of Probability Theory 44
3.3 Probability Laws 47
3.4 Statistical Independence 55
3.5 Probability Tables and Probability Trees 58
Summary 66

Review Exercises Chapters 2–3 70

4

RANDOM VARIABLES AND PROBABILITY DISTRIBUTIONS 74

4.1 Random Variable: The Basic Idea 74
4.2 Probability Distributions for Discrete Random Variables 77
4.3 Probability Distributions for Continuous Random Variables (∂, ∫) 82
4.4 Expected Value, Variance, and Standard Deviation: Discrete Random Variables 89
4.5 Expected Value, Variance, and Standard Deviation: Continuous Random Variables (∫) 94
4.6 Joint Probability Distributions and Independence 97
4.7 Covariance and Correlation of Random Variables 101
4.8 Joint Probability Densities for Continuous Random Variables (∫) 105
Summary 110
Appendix 4A: Properties of Expected Values and Variances 113
Appendix 4B: Some Reminders About Calculus 116

5

SOME SPECIAL PROBABILITY DISTRIBUTIONS 121

5.1 Counting Possible Outcomes 121
5.2 Bernoulli Trials and the Binomial Distribution 124
5.3 Hypergeometric Distribution 130
5.4 Geometric and Negative Binomial Distributions 133
5.5 Poisson Distribution 136
5.6 Uniform Distribution 139
5.7 Exponential Distribution (∫) 141
5.8 Normal Distribution 144

5.9 Normal Approximations to Binomial and Poisson Probabilities 150
Summary 154

6

RANDOM SAMPLING AND SAMPLING DISTRIBUTIONS 162

6.1 Random Sampling 162
6.2 Sample Statistics and Sampling Distributions 166
6.3 Expected Values and Standard Errors of Sample Sums and Sample Means 170
6.4 Sampling Distributions for Means and Sums 173
6.5 Uses and Misuses of the Central Limit Theorem 180
6.6 Computer Simulations 181
Summary 194
Appendix: Standard Error of a Mean 197

Review Exercises Chapters 4–6 198

7

POINT ESTIMATION 202

7.1 Point Estimators 203
7.2 Sampling with and without Replacement 211
7.3 Maximum Likelihood Estimation (∂) 213
Summary 219

8

INTERVAL ESTIMATION 226

8.1 Interval Estimation of a Population Mean with Known Standard Deviation 227
8.2 Confidence Intervals for a Proportion 230
8.3 How Large a Sample Is Needed? 232
8.4 The *t* Distribution 236
8.5 Confidence Intervals Using the *t* Distribution 241
8.6 Assumptions for Interval Estimation 244
8.7 Confidence Intervals for a Median 246
Summary 250
Appendix: The Chi-square and *t* Distributions 254

9

HYPOTHESIS TESTING 257

9.1 A Test for a Binomial Proportion 257
9.2 Type II Error, β Probability, and the Power of a Test 262
9.3 A Test for a Population Mean with Known Standard Deviation 265
9.4 The β Probability for z Tests 272
9.5 The p-value for a Hypothesis Test 276
9.6 Hypothesis Testing with the t Distribution 281
9.7 The Effect of Population Nonnormality 286
9.8 Tests About a Population Median 290
9.9 Testing a Population Proportion Using a Normal Approximation 295
9.10 The Relationship Between Hypothesis Tests and Confidence Intervals 296
9.11 Hypothesis Testing as a Decision Method 298
Summary 300

Review Exercises Chapters 7–9 302

10

COMPARING TWO SAMPLES 306

10.1 Comparing the Means of Two Populations with Known Standard Deviations 306
10.2 Comparing the Means of Two Populations with Unknown Standard Deviations 312
10.3 A Nonparametric Alternative: The Wilcoxon Rank Sum Test 323
10.4 Paired-sample Methods 332
10.5 Two-sample Procedures for Proportions 342
Summary 346
Appendix: The Mathematics of Pooled-variance t Methods 351

11

CHI-SQUARE AND F METHODS 352

11.1 Chi-square Methods for a Variance 352
11.2 Comparing Variances or Standard Deviations for Two Independent Samples 357
11.3 Tests for Several Proportions 361
11.4 Chi-square Tests of Independence 365
11.5 Measuring Strength of Relation 371
Summary 375

12

INTRODUCTION TO THE ANALYSIS OF VARIANCE 380

12.1 Testing the Equality of Several Population Means 380
12.2 Comparing Several Distributions by a Rank Test 387
12.3 Specific Comparisons Among Means 391
12.4 Two-factor ANOVA: Basic Ideas 396
12.5 Two-factor ANOVA: Methods 399
12.6 Randomized Block Designs 411
12.7 More Complex Experiments 416
Summary 420
Appendix: Sums of Squares 429

Review Exercises Chapters 10–12 430

13

LINEAR REGRESSION AND CORRELATION METHODS 436

13.1 Assumptions in Linear Regression Problems 437
13.2 Estimating Model Parameters 440
13.3 Inferences About Parameters 449
13.4 Predicting New *y* Values Using Regression 454
13.5 Correlation 459
13.6 Rank Correlation 465
Summary 468
Appendix: The Mathematics of Least Squares 474

14

MULTIPLE REGRESSION METHODS 477

14.1 The Multiple Regression Model 477
14.2 Estimating Multiple Regression Coefficients 483
14.3 Inferences in Multiple Regression 492
14.4 Inferences Based on the Coefficient of Determination 497
14.5 Forecasting Using Multiple Regression 502
14.6 Some Multiple Regression Theory 506
Summary 512

15

CONSTRUCTING A MULTIPLE REGRESSION MODEL 521

15.1 Selecting Candidate Independent Variables (Step 1) 522
15.2 Lagged Predictor Variables (Step 1) 526
15.3 Nonlinear Regression Models (Step 2) 530
15.4 Choosing Among Regression Models (Step 3) 543
15.5 Residuals Analysis: Nonnormality and Nonconstant Variance (Step 4) 549
15.6 Residuals Analysis: Autocorrelation (Step 4) 557
15.7 Model Validation 569
Summary 572

Review Exercises Chapters 13–15 596

16

TIME SERIES ANALYSIS 615

16.1 Index Numbers 616
16.2 The Classical Trend, Cycle, and Seasonal Approach 620
16.3 Smoothing Methods 630
16.4 The Box-Jenkins Approach 637
Summary 650

17

SOME IDEAS OF DECISION THEORY 657

17.1 The Components of Decision Theory 657
17.2 Decision Making Using Expected Values 662
17.3 The Element of Risk 667
17.4 The Basic Ideas of Expected Utility Theory 675
Summary 679
Appendix: More Expected Utility Theory 680

18

USING SAMPLING INFORMATION IN MAKING DECISIONS 686

18.1 Joint Probabilities and Bayes' Theorem 686
18.2 Buying and Using Information 692
18.3 Decision Trees 700
18.4 Decisions Based on a Mean or a Proportion (ʃ) 708
18.5 Bayesian Decision Theory and Hypothesis Testing 716
Summary 717

19

SOME ALTERNATIVE SAMPLING METHODS 722

19.1 Taking a Simple Random Sample 723
19.2 Stratified Random Sampling 728
19.3 Cluster Sampling 734
19.4 Selecting the Sample Size 740
19.5 Other Sampling Techniques 745
Summary 747

20

DATA MANAGEMENT AND REPORT PREPARATION 755

20.1 Preparing Data for Statistical Analysis 755
20.2 Guidelines for a Statistical Analysis and Report 759
20.3 Documentation and Storage of Results 760
Summary 760

Appendix 762

References 791

Answers 793

Index 884

1

STATISTICS: MAKING SENSE OF DATA

Statistics, as a subject, is the study of making sense of data. Almost every manager—corporate president, cabinet member, hospital director, third assistant to the associate vice comptroller—must deal with data. For statistical purposes, the word *data* means any collection of **numerical** values, along with an explicit or implicit definition of how these values were measured. This book is an introduction to statistical methods: methods for gathering data, for summarizing data in a coherent way, for making predictions and forecasts from data, and for making decisions based on data.

numerical

1.1 GATHERING DATA

Part of the business of statistics is to indicate good and bad ways to gather data. If a manager wants to conduct market research on a new product, evaluate the occupancy rate of hospital beds, experiment with the effects of different compensation plans, audit the accounts receivable of a chain store, or survey the opinions of businesspeople about a proposed set of government regulations, the first requirement is intelligent data gathering.

Statistical sampling theory and the theory of experimental design provide useful guides to good methods of data collection. Usually, statistical sampling is more or less passive; the aim is to gather (or survey) data on existing conditions, attitudes, or behavior. For example, a manager of a large manufacturing plant might be interested in determining the attitudes of production supervisors toward an incentive (bonus) plan. To do this the manager might survey their opinions. Experimental studies tend to be more active; the person conducting such a study deliberately varies certain conditions, such as the noise level in a manufacturing plant, and observes the results, gain or loss in productivity. Often studies combine elements of sampling and experimentation, as in a market research study in which individuals indicate their purchasing habits and also indicate preferences among several formulations of a product.

Gathering useful data is a major part of statistical thinking. Unless someone has noted which supplier produced a certain part, the source of a rash of defects in that part is likely to go unidentified. Unless a market researcher obtains a good cross-section of a target group, that group's likes and dislikes may go unheeded. Unless an auditor takes a careful sample of a company's receivables, serious errors may go undetected. The concept of a random, probabilistic sample is an integral part of statistical thinking. Some of the basic ideas of random sampling are discussed in Chapter 6; more elaborate sampling schemes are the subject of Chapter 19.

Since most managers are not closely involved in the data-gathering process, we assume in the early sections of the text that the data have been gathered in an intelligent manner. This is not done to negate the importance of the data-gathering process; rather, it allows us to focus initially on the important topics of statistics related to summarizing data, analyzing data for meaningful trends, and analyzing data to develop predictions (or forecasts) of future events. Some of the material in the latter sections of the text, particularly Chapter 19 on statistical sampling methods, deals directly with data-gathering methods. Other ideas for collecting data are scattered throughout the book.

1.2 SUMMARIZING DATA

Once the data are gathered, they need to be summarized before any meaningful interpretations can be made. Imagine facing a computer printout of the current balances of every MasterCard holder, or of the occupancy status of every bed in a large hospital for the last five years, and being asked to describe the data. It's almost impossible to describe such detailed data; instead, the raw data must be summarized to be understandable. **The first step in trying to summarize data is to graph the data.** Draw some intelligent pictures and try to make some sense of the data.

Next focus on the average or typical value in the data set and some measure of the range or spread of the data. For example, in trying to summarize data on starting salaries for college graduates with a chemical engineering degree, it would be important to know that the starting salaries ranged from a low of $10,500 to a high of $35,000 and that the average starting salary was $21,200. This focus on the typical or average is necessary and probably inevitable in making sense of data.

In some situations, though, the average or typical values may be irrelevant. For example, we would prefer not to live next to a flood-control system that was designed to handle the average rainfall in a given place. Here we would look for protection against large values, which are greatly different from the average or typical value. One of the more important aspects of summarizing data is to discover if there are wildly extreme cases, ones that differ greatly from the average or typical value.

Measuring the average value of a variable is not nearly enough for good statistical thinking; **measuring variability is absolutely crucial**. If a change in the monthly growth rate of the gross national product is well within the usual random variation of that figure, it does not make much sense to devise elaborate interpretations of the reasons for the change. A large part of this book deals with issues of variability—how to measure it, how to reduce it, and how to make decisions despite it.

In addition to average values and measures of variability, other summary statistics are useful in dealing intelligently with data. Data should be checked for **skewness**—asymmetry around the average value—and for **outliers**—wild values far from the bulk of the data. Both skewness and outliers can distort the usual summary statistics badly.

Basic statistical methods for summarizing data are discussed in Chapter 2.

1.3 THE ROLE OF PROBABILITY

population
sample

We define a **population** of measurements as the complete set of measurements of interest to the manager, and a **sample** of measurements as the subset of data that is available for analysis and interpretation. For example, in an audit of accounts receivable for a large, nationwide chain of stores, the credit balance for every account receivable could represent the population of data of interest to the auditor. However, due to the time and expense involved in verifying each account, it may be feasible to examine only a subset of credit balances from the complete set of all credit balances in accounts receivable for the chain. More important, managers are faced with making sense out of existing data in the face of less than complete information. Necessarily, then, a bit of uncertainty is associated with any prediction or forecast based on the available sample data.

Probability theory is the language of uncertainty. The concepts and theorems of probability allow us to specify probable errors when making inferences and forecasts. Probability calculations can be used to process the available data in a coherent, rational way. **In all, probability theory is one of the most useful branches of mathematics for managers and provides the basis for measuring the uncertainty of all statistical inferences (predictions, forecasts, decisions).**

In this book, probability ideas are developed as they are needed. There is no one section or chapter that contains everything you need to know about probability. The basic language is defined in Chapters 3 and 4, some basic probability theorems are developed in Chapters 5 and 6, and the use of probability theory in evaluating information and risk is discussed in Chapter 18. Other probability ideas are developed in the context of various statistical procedures, so most of the book can be regarded as an illustration of the uses of probability theory.

1.4 MAKING INFERENCES FROM SAMPLE DATA

The fact that sample data are usually incomplete forces a manager to make inferences about the population from which the sample was drawn. A market research study reaches only a few of the potential buyers of a product; the probable reaction of all consumers must be inferred from the reactions of those included in the study (sample). Certainly the individuals included in the sample will not be exactly, perfectly representative of the whole population; but if the study has been designed carefully, the results should be close to those for the whole population of interest.

Statistics makes use of probability to develop the probable error associated with an inference. For instance, if sample data are used to compute the percentage of consumers who react favorably to a proposed new product, this percentage is an **estimate** of the percentage of all consumers in the population who would react favorably to the product. Because the sample data are incomplete, the estimate will not be exactly equal to the corresponding percentage of favorable responses in the population. Hence we need to know the **probable error** for the estimate.

estimate

probable error

The concept of probable error is illustrated as follows. If an auditor samples 2000 accounts and finds 84 (4.2%) to be in error, then we could say with a high degree of certainty (to be defined later) that the sample percentage 4.2% is within plus or minus .9% of the actual unknown percentage of accounts in error for the population. This plus-or-minus factor is the probable error of the sample estimate, 4.2%. The concept of probable error is a fundamental notion in statistics. The smaller the probable error is for a given problem, the more information we have about the unknown characteristic in the population.

Application of statistics also allows a manager to assess whether an apparent change can reasonably be attributed to random fluctuation, rather than to a real change. For example, suppose that the historical error rate for receivables has been 3.5%. If a sample of 2000 accounts gives a 4.2% rate, it may appear that the error rate has increased seriously. However, as the .9% plus-or-minus figure indicates, it is possible that the discrepancy between the observed 4.2% and the historical 3.5% is simply the luck of the draw—a fluke of statistical sampling. Thus an auditor should be cautious in assuming that the error rate has really increased.

The basic ideas of statistical inference are introduced in Chapters 7, 8, and 9. Applications and methods of statistical inference are the fundamental topics of much of the rest of the book.

1.5 THE ROLE OF THE COMPUTER

Making sense of data has become vastly easier with the advent of the modern computer. Computer programs can be and have been written to perform the most tedious, arcane, nasty computations described in this book, as well as other, even more elaborate computations. Equally useful, a good computer program plots the data, so one can see typical values, variability, skewness, and outliers. Properly used, a computer can be an immense help in analyzing data. Yet the mere fact that data analysis has been done by computer doesn't guarantee that it has been done intelligently. A human must still choose the method of analysis, check the assumptions, and interpret the results. Until artificial intelligence improves greatly, real intelligence will still be needed. In this book we present some ideas for interpreting computer output sensibly.

Computers are useful in another way in analyzing data. Every statistical inference method is based on some assumptions. The inevitable question is, What happens if the assumptions are violated? Using a computer, one can estimate the effect of any particular violation of assumptions by having the computer perform the statistical procedure repeatedly using artificial data generated with the specified violation of assumptions. This computer-simulation method can be used to demonstrate results, without the use of difficult mathematical arguments. In this book we use computer simulations repeatedly to evaluate the effect of violating various assumptions.

Finally, computers help managers use data more efficiently. With the availability of computer power for data plots, analyses, and simulations, managers can spend less time *doing* the necessary calculations and more time *interpreting* the results. Thus the manager's role changes from that of a statistical analyst who does all the required calculations to that of a statistical critic who assesses the appropriateness and meaning of the results.

There are many packages of computer programs available to analyze data. The most commonly used systems are BMDP, Minitab, SAS, and SPSS^X. Each is available in a version for a mainframe computer and in a version for a (large) personal computer. In addition, a large number of packages have been developed specifically for use on a personal computer. It is not necessary to know computer programming to use any of these packages. Most people find that they can use any of these packages easily, though they may be frustrated at first by minor errors. The utility of these packages in analyzing large amounts of data and drawing many data plots more than repays the initial investment.

Generally, it's not necessary to learn everything about a software package to use it. You need to learn only the parts of the package that do what you need. The steps of a computerized analysis typically include describing and entering the data, manipulating the data into a desired format, possibly selecting subsets of the data, and calling up the desired parts of the package to do the analysis and plotting. The results may be displayed on a terminal or relayed to a printer.

This is not a text on computer usage. Therefore we don't spend time on the mechanics of using a package, which vary from package to package and from installation to installation. Our main interest is in interpreting the output from these programs. Often the designers of a package try to include in the output everything that a user could conceivably want to know. As a result, in any particular situation, some of the output is irrelevant. **When reading computer output, look for the values you need and don't worry about the rest.** As you learn more, you'll understand more of the output. In the meantime, look for what you need and disregard the rest.

1.6 USING DATA IN MAKING DECISIONS

A manager does not go to the trouble and expense of gathering data just for fun. In many cases, the purpose is simply understanding; a manager wants to know what is going on! In other cases, a specific decision must be made. For example, a market research study usually leads to a decision as to which form of a new product is to be introduced, or indeed if the product is to be introduced at all. Statistical decision theory provides a framework for using data in making decisions.

Decision theory has two important components: the assessment of risk and the evaluation of imperfect information. One of the most difficult problems managers face is uncertainty; the results of a decision may be less favorable or more favorable than expected. Even if they gather huge amounts of information beforehand, they must make decisions in the face of this element of risk. Decision theory provides a language and some methods for dealing intelligently with risk. Once again the language is based on probability theory. The basic language and concepts are discussed in Chapter 17.

Information obtained from data can reduce the element of risk. A manager can combine data with personal beliefs or hunches to make better decisions. But it's expensive and time-consuming to gather data and to extract information from the data. Decision theory can guide a manager in using information, both in modifying original opinions and in deciding whether it is worth the trouble and expense to gather the data at all. These issues are considered in Chapter 18.

SUMMARY

This chapter sets the agenda for the book. **Statistics as a subject is the study of making sense of data**—summarizing data for coherence, making inferences and forecasts from data, and using data in making decisions. Keep this in mind as you study the material in the book, and at the end of each chapter try to visualize how the material of that chapter fits into the overall objective of statistics—making sense of data.

2

SUMMARIZING DATA ABOUT ONE VARIABLE

An unorganized mass of numbers is virtually useless in understanding anything. The first task in making sense out of a data set is to summarize it. In this chapter we focus on three aspects of the data-summarization problem: finding and displaying the frequencies of various data values, calculating a typical value, and indicating the degree of variability around that typical value. Most statistical problems encountered in practice involve several key variables. But an important first step in making sense of such data is to summarize the variables one by one; methods for doing so are described in this chapter.

2.1 THE DISTRIBUTION OF VALUES OF A VARIABLE

frequency table

One of the first steps in summarizing data from a single variable is to find the frequencies with which various values occur and to display them in a **frequency table** so that patterns can be seen.

EXAMPLE 2.1 A small firm has a total of 20 salespeople working in four offices. The offices are numbered 1 through 4. The firm's records show that the respective offices of the salespeople (listed in alphabetical order) are

1 4 1 3 3 2 1 1 1 3 4 4 2 2 1 1 2 4 4 1

Summarize the data by displaying the frequencies associated with the values 1, 2, 3, and 4 in a frequency table.

Solution The easiest way to obtain the frequencies is to list the values and count the frequencies. Then the frequency table is as follows:

Value (office number)	1	2	3	4
Frequency (number of salespeople)	8	4	3	5

□

relative frequency

It's helpful, particularly with large amounts of data, to convert the counts to percentages or proportions. In statistical jargon, the **relative frequency** of a value is the proportion of all observations that have that value. Relative frequency is calculated by dividing the frequency (number of occurrences) of the value by the total number of observations.

EXAMPLE 2.2 Convert the frequencies of Example 2.1 to relative frequencies.

Solution In Example 2.1, the total number of observations was 20. Therefore,

Value	1	2	3	4
Relative frequency	$\frac{8}{20} = .40$	.20	.15	.25

To convert relative frequencies to percentages, simply multiply them by 100. □

grouped data

When there are many possible values, it's usually better to combine values into groups. The resulting data are referred to as **grouped data**. Suppose you had a list of the amounts paid by 850 customers at a supermarket's checkout counters in one day. If you listed the amounts to the penny and counted frequencies, you would probably list over 800 values with a frequency of 1 and a few values with frequencies of 2 or 3. That wouldn't be a very useful summary! In such a case you might want to round off the amounts to the nearest dollar or the nearest 5 dollars. This rounding-off process creates groups of similar checkout amounts.

classes

The choice of groups (often called **classes** for summarizing data from a single variable is somewhat arbitrary, but the selection of classes should conform to the requirement that each measurement falls into one and only one class. In addition, it is desirable to choose the classes so that (1) no gaps appear between the classes, and (2) the classes have a common width. There are many rules of thumb for determining the number of classes and the class width; here are a few suggestions:

1. **Choose a sufficient number of classes** so that the data are not all lumped into two or three groups but not so many that the frequency table becomes unwieldy. Approximately 5 to 20 classes seems about right for most purposes, with a small number of classes (near 5) for data sets with a small number of observations and more (near 20) in larger data sets. This practice diminishes the possibility of obtaining many classes with a frequency of 0 or 1.
2. **If possible, the classes should have equal width** so that they are comparable. If necessary, an open-ended class such as "over $50" can be used at the high (or low) end of the values. For most purposes, classes such as $0.01 to $10.00, $10.01 to $20.00, $20.01 to $30.00, $30.01 to $40.00, $40.01 and up are preferable to $0.01 to $5.00, $5.01 to $15.00, $15.01 to $40.00, $40.01 and up.

3. **Choose convenient midpoints for the classes.** A frequency chart with midpoints $5, $15, $25 is easier to understand than one with midpoints $7.20, $14.40, $21.60.
4. **Make sure that the class boundaries are not ambiguous.** In suggestion 2 we stated the intervals as $0.01 to $10.00, $10.01 to $20.00, . . . , rather than as $0 to $10, $10 to $20, . . . , to avoid the problem that would arise with a purchase of exactly $10.00.

Some statistics textbooks give long, detailed instructions for grouping data and point out some minor, but irritating, pitfalls. The suggestions given here, plus a bit of common sense, should be adequate for simple summaries of data.

EXAMPLE 2.3 Suppose that the 20 salespeople in Example 2.1 had the following commission incomes (excluding salaries) in a certain month:

$850	$1265	$895	$575	$2410	$470	$660	$1820	$1510	$1100
$620	$425	$751	$965	$840	$1505	$1375	$695	$1125	$1475

Use these data to construct a frequency table with suitable class intervals.

Solution With only 20 observations, we want a small number of classes, such as 5. The incomes range from $425 to $2410; one convenient choice of classes takes incomes to the nearest $500. The results of such a grouping of data are displayed below in a frequency table.

Class	$250 to $749	$750 to $1249	$1250 to $1749	$1750 to $2249	$2250 to $2749
Frequency	6	7	5	1	1
Relative frequency	.30	.35	.25	.05	.05

Note that each measurement (recorded in dollars) falls into exactly one class, that there are no gaps between the intervals, and that the classes have a common interval width of 499.

Some would argue that the common interval width is 500 since, theoretically, the class intervals are 250–750, 750–1250, ... Others would argue that the class intervals are really 249.5–749.5, 749.5–1249.5, . . . , with interval width of 500. We will not quibble with the fine points. Since, for a set of classes, all measurements fall into one and only one class, and since the classes are of equal width and have no gaps, these different groups satisfy the major requirements for the selection of classes.

Many other groupings would be reasonable for the same data. We might redefine the groups as $252 to $751, $752 to $1251, . . . (which changes the group for the individual who earned exactly $751). We might choose to center the groups at $600, $1000, $1400, $1800, and $2200. While arbitrary choices such as these affect details of the frequency table, they usually do not affect the broad pattern of the data. □

Frequency tables are helpful in summarizing data; pictures or graphs derived from these tables are even better, because they allow you to see the basic data pattern easily. There are many kinds of graphical methods for data summarization. We present a few standard ones in this section.

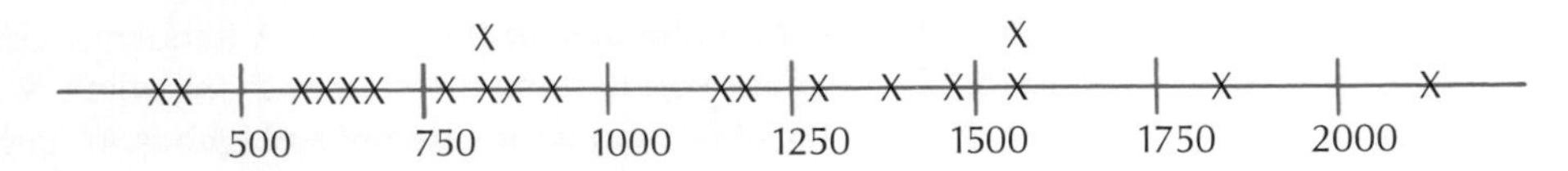

FIGURE 2.1 Line Plot of Commission Income Data

line plot

When there are relatively few observations in the data (no more than 40 or so), one quick and easy method is to mark each observation by a check or X along a line. If two or more observations are equal or close to each other, the marks may be piled up. Such a graph might be called a **line plot**.

EXAMPLE 2.4 Draw a line plot for the data of Example 2.3.

Solution The commission income data of Example 2.3 yield the line plot shown in Figure 2.1. The stacked Xs represent the $840 and $850 values and the $1505 and $1510 values. Note the roughly even spread over most of the range and the two large values. □

A line plot is a very easy, rough, "back of an envelope" method that is fine for getting a quick look at the data. For more careful or more formal presentations, bar charts or histograms (among other methods) are preferable.

bar chart

Bar charts and histograms use rectangles to portray the data. The base of each rectangle indicates a value or a group (class) of values. The height of each rectangle represents the frequency or the relative frequency of each value or class. To create a **bar chart**, label the horizontal axis with values or categories of the variable, label the vertical axis with frequencies, and construct separate rectangles for each value with height corresponding to the frequency or relative frequency of that value. Figure 2.2 shows a bar chart of the data from Example 2.1.

histograms

Histograms are constructed in the same way as bar charts. The only difference is that the rectangles in a bar chart are separated by some space, whereas the rectangles in a histogram are directly adjacent. A histogram is often used with grouped data. Figure 2.3 shows a histogram of the commission income data grouped as in Example 2.3.

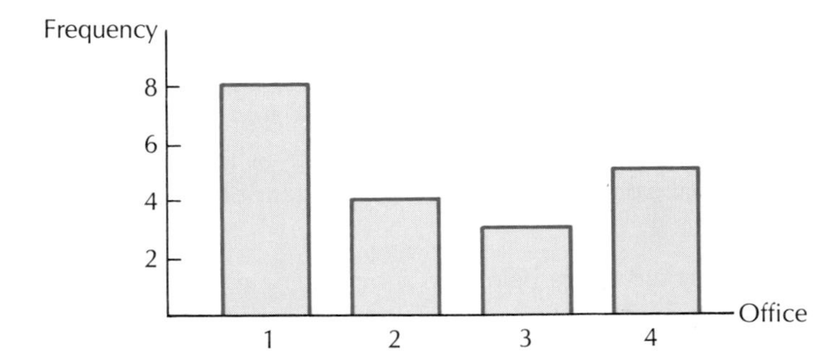

FIGURE 2.2 Bar Chart of Sales Office Data

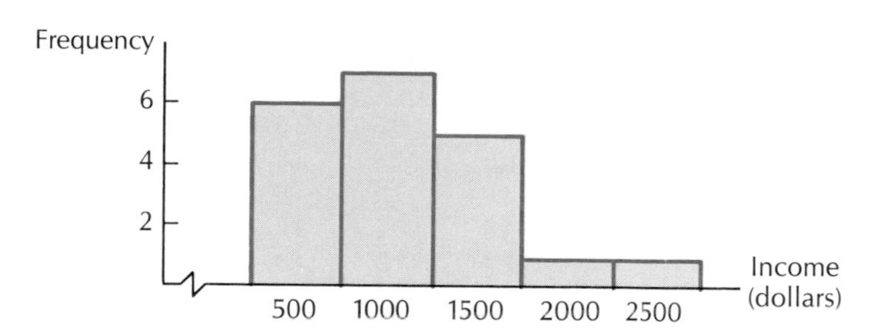

FIGURE 2.3 Histogram of Grouped Commission Income Data

There are some minor variations in the presentation of histograms. The horizontal axis can be labeled with either class midpoints, as in Figure 2.3, or class endpoints. Thus in Figure 2.3 we could have labeled the edges of the rectangles as \$250, \$750, . . . (or equivalently as \$249.5, \$749.5, . . .). The vertical axis of a histogram is labeled with frequencies or with **relative frequencies**. Because relative frequencies are proportional to frequencies, the frequency histogram and the relative frequency histogram have the same shape. The specific choice is mostly a matter of taste.

relative frequencies

The distinction between bar charts and histograms is based on the distinction between *qualitative* and *quantitative* variables. The values of **qualitative** variables vary in kind but not degree and hence are not measurements. For example, "office number" in Example 2.1 is a qualitative variable; although the offices were numbered 1, 2, 3, and 4, they could equally well have been labeled New York, Atlanta, Chicago, and San Francisco. Numerical values can be assigned to qualitative variables, as was done in Example 2.1, but these values are only convenient codes. In contrast, values of a **quantitative** variable result from an actual measurement with some sort of "yardstick." For example, "commission income" in Example 2.3 is a quantitative variable. A good test for whether a variable should be regarded as qualitative or quantitative is whether a unit of measure can be identified. Commission income is measured in dollars; office number has no unit of measurement. Quantitative variables usually have units; qualitative variables don't.*

qualitative

quantitative

Bar charts are used with qualitative variables; the separation of rectangles suggests that each value of the variable represents a distinct category. Histograms are used with quantitative variables. The fact that rectangles in a histogram are adjacent (for intervals with nonzero frequencies) suggests that the variable's values are measured along a scale.

EXAMPLE 2.5 Monthly values of the increase in the consumer price index (CPI) converted to a yearly basis for the period from 1975 to 1978 are listed on the top of the next page. Use these data to construct a frequency histogram.

* Variables such as sales (in units of dozens) or hotel rooms (simply in units of numbers) can be regarded as quantitative. Sometimes variables such as college grade points ($A = 4$, $B = 3$, and so on) are treated as quantitative, even though there is no obvious unit.

			(1975–1978) CPI				
6.0	10.8	6.0	4.8	9.6	3.6	9.6	6.0
7.2	3.6	2.4	6.0	12.0	4.8	7.2	7.2
4.8	6.0	2.4	4.8	7.2	4.8	9.6	9.6
6.0	7.2	6.0	4.8	9.6	3.6	10.8	9.6
6.0	6.0	7.2	2.4	7.2	4.8	10.8	6.0
9.6	6.0	4.8	4.8	6.0	4.8	10.8	7.2

Solution A frequency histogram is shown in the accompanying computer output.

```
MIDDLE OF INTERVAL            NUMBER OF OBSERVATIONS
        2.                            3   ***
        3.                            0
        4.                            3   ***
        5.                           10   **********
        6.                           12   ************
        7.                            8   ********
        8.                            0
        9.                            0
       10.                            7   *******
       11.                            4   ****
       12.                            1   *
```

□

stem-and-leaf diagram

A clever device that constructs a histogram-like picture is the **stem-and-leaf diagram**. It is best explained by an example.

EXAMPLE 2.6 Suppose that the grades of 40 job applicants on an aptitude test are as follows. Construct a stem-and-leaf diagram for these data.

42	21	46	69	87	29	34	59	81	97
64	60	87	81	69	77	75	47	73	82
91	74	70	65	86	87	67	69	49	57
55	68	74	66	81	90	75	82	37	94

Solution The scores range from 21 to 97. The first digits, 2 through 9, are placed in a column on the left of the diagram. The respective second digits are recorded in the appropriate row. The first three scores, 42, 21, and 46, would be represented as

```
2 | 1
3 |
4 | 2 6
```

The full stem-and-leaf diagram is shown in Figure 2.4. □

2	1 9
3	4 7
4	2 6 7 9
5	9 7 5
6	9 4 0 9 5 7 9 8 6
7	7 5 3 4 0 4 5
8	7 1 7 1 2 6 7 1 2
9	7 1 0 4

FIGURE 2.4 **Stem-and-Leaf Diagram**

The diagram can be made a bit neater by first ordering the data within a row, from lowest to highest score, but this process is time-consuming if done by hand. The end result of a stem-and-leaf diagram looks much like a histogram turned sideways. The advantage of such a diagram is that it not only reflects frequencies but also contains the actual values. No information is lost.

There are many possible variations. A display of the data from Example 2.6 with more classes is shown in the stem-and-leaf diagram of Figure 2.5. This Minitab stem-and-leaf diagram divides each interval in half by placing observations from the 90–94 range, for example, in row 9*, and observations in the 95–99 range in row 9. Similar subdivisions are made in the other rows. Note also the ordering of observations within a row. The LO

```
STEM-AND-LEAF DISPLAY OF GRADES
LEAF DIGIT UNIT = 1.0000
1  2 REPRESENTS 12

LO    21

2.    9
3*    4
3.    7
4*    2
4.    679
5*
5.    579
6*    04
6.    5678999
7*    0344
7.    557
8*    11122
8.    6777
9*    014
9.    7
```

FIGURE 2.5 **Computer Stem-and-Leaf Diagram**

21 entry is a Minitab suggestion that 21 may be an outlier—a wild, extreme value. We would not agree.

The stem-and-leaf diagram, and many other data-summarization ideas, are discussed in Tukey (1977).

SECTION 2.1 EXERCISES

2.1 An automobile manufacturer routinely keeps records on the number of finished (passing all inspections) cars produced per 8-hour shift. The data for the last 28 shifts are

366	390	324	385	380	375	384	383	375	339	360	386	387	384
379	386	374	366	377	385	381	359	363	371	379	385	367	364

a. Construct a histogram using convenient class intervals. Can you think of an explanation for the apparent shape of the histogram?

b. Construct a stem-and-leaf diagram of the data and compare it to the histogram. The left-hand "stem" should have an initial 3 in each value.

2.2 A city manager receives 17 bids for supplying new electric typewriters. The dollar costs per typewriter are

847	849	838	841	852	846	812	838	850	836	871	849
824	846	864	843	839							

a. Construct a histogram using an appropriate number of classes.

b. Construct a stem-and-leaf diagram for these data. The left-hand "stem" should start with 81 and end with 87.

2.3 A purchasing agent tests samples of 10 batteries for hand calculators from each of two manufacturers. Each battery is tested in a calculator that is programmed to do a continuous "loop" of typical calculations; the time in hours to failure of each battery is recorded in the table below:

Manufacturer	Time				
E	11.80	11.91	11.95	12.00	12.02
	12.03	12.04	12.07	12.13	12.20
S	12.06	12.14	12.18	12.19	12.20
	12.20	12.21	12.23	12.27	12.33

a. Construct a combined histogram of all 20 times; use 6 or 7 classes.

b. Construct separate histograms with the classes of part (a) for each manufacturer.

c. How do the separate histograms help to explain the appearance of the combined histogram?

2.4 Price–earnings ratios for a sample of 24 publicly owned companies involved in the sale of computer software used in manufacturing operations are given below:

7.7	8.5	9.6	10.3	13.6	14.5	19.5	10.1	9.7	11.4	17.8	15.9
14.2	13.7	20.7	22.1	25.9	29.1	32.6	36.7	32.4	35.9	40.1	45.9

a. Construct a relative frequency histogram.

b. Use the same data to construct a stem-and-leaf plot. How can you handle the problem that most of the price–earnings ratios have three digits rather than two? Compare the histogram to the stem-and-leaf plot; which one do you find more informative?

2.2 ON THE AVERAGE: TYPICAL VALUES

Frequency tables, bar charts, histograms, and stem-and-leaf displays all give a general sense of the pattern or distribution of values in a data set. They do not indicate a typical, middle, or average value explicitly. In this section we define some standard measures of the typical or average value.

The word *average* has at least three meanings. It can mean the most common value, the mode; it can mean the middle value, the median; or it can mean the arithmetic average, the mean. This section contains more careful definitions of these three basic concepts of typical values.

Mode

The mode of a variable is the value or category with the highest frequency in the data. It is most commonly used with qualitative data but can also be used with quantitative data.

In Example 2.1, the modal category for the qualitative variable "office number" is 1, because that office has the largest number of salespeople.

EXAMPLE 2.7 The data shown below represent the percentage of eligible voters for a sample of 10 voting districts:

63 56 32 48 48 39 45 45 45 41

Determine the modal value.

Solution It is clear that the modal value is 45%, because the value 45 occurs most frequently in the sample. □

Because we do not usually have the original measurements that have been used to form a frequency table, the mode for grouped data is defined as the midpoint of the class with the highest frequency. This value approximates the mode of the actual (ungrouped) data.

In Example 2.3 the modal commission income is $1000, the midpoint of the class with endpoints of $750 to $1250. Unfortunately, the modal value is very sensitive to small changes in data values or class definitions. Had we defined the classes as $252 to $751, $752 to $1251, and so on in Example 2.3, the frequency table would change slightly, as shown here:

Class	Frequency
252–751	7
752–1251	6
1252–1751	5
1752–2251	1
2252–2751	1

The mode for these grouped data would be 501, the midpoint of the first interval, rather than 1000 as we found previously.

As was illustrated above, a mode calculated from a small amount of data or based on arbitrary class definitions is not too reliable and shouldn't be taken too seriously. One final word about modes: There may be data sets with more than one mode, for example, sets for which there are two or more values (classes) with the highest frequency; these data sets are referred to as bimodal, trimodal, and so on.

Median

The median of a set of data is the middle value when the data are arranged from lowest to highest. If n, the sample size, is odd, the median is the $(n + 1)/2$th value; if n is even, the median is the average of the $n/2$th and $(n + 2)/2$th values.

EXAMPLE 2.8 Suppose that the number of units per day of whole blood used in transfusions at a hospital over the previous 11 days is

25 16 61 12 18 15 20 24 17 19 28

Solution Arranged in increasing order, the values are

12 15 16 17 18 19 20 24 25 28 61

The median value is the sixth value, 19. □

If n, the number of measurements, is odd, there is no problem in finding the middle value. However, where n is even, there is no value exactly in the middle; for this situation, the median is conventionally defined as the average of the middle two values when the data are ordered from smallest to largest. If in Example 2.8 there had been a twelfth value of 72, the median would be the average of the sixth and seventh values, $(19 + 20)/2 = 19.5$. Whether n is odd or even, there are equal numbers of observations above and below the median.

Mean

The mean of a variable is the sum of the measurements taken on that variable divided by the number of measurements. It is meaningful only for quantitative data.

In Example 2.3 the mean commission income was (\$850 + \$1265 + . . . + \$1475)/20, or \$1066.55. We use two different symbols for the mean, depending on whether we want to regard the data as the entire population of interest or as a sample of measurements from the population of interest.

As we indicated in Chapter 1, a population of measurements is the complete set of measurements of interest to the manager; a sample of measurements is a subset of measurements selected from the population of interest. If we let $y_1, y_2, \ldots, y_n$ represent a sample of n measurements selected from a population, the sample mean is denoted by the symbol $\bar{y}$:

$$\bar{y} = \frac{\sum_i y_i}{n}$$

The corresponding mean of the population from which the sample was drawn is denoted by the Greek letter μ.

In most situations we do not know the population mean; the sample mean is used to make inferences about the corresponding unknown population mean. This is discussed in greater detail beginning in Chapter 7.

The mean is the most useful and convenient measure of the average value; however, any user of statistics should be alert to the possibility of distortions due to skewness or outliers. An extreme value can pull the mean in its direction and cause the mean to appear atypical of the values in the set. For this reason, statisticians have developed the idea of **trimmed means**. To find a 40% trimmed mean, for instance, drop the highest 20% and lowest 20% of the values and average the remaining values. This trimming process eliminates the effect of **outliers**, data values that lie far above or below the preponderance of the data. Trimmed means are particularly useful in estimation of a population mean based on sample data, because they reduce the impact of possible "freak" values.

trimmed means

outliers

The mode, median, mean, and trimmed means are called **measures of location** or **measures of central tendency**; they indicate the center or general location of data values. The relation among these measures depends on the skewness of the data. If the distribution is mound shaped—symmetric around a single peak—the mode, median, mean, and trimmed means are all equal. For a skewed distribution, one having a long tail in one direction and a single peak, the mean is pulled out toward the long tail; usually the median falls between the mode and the mean, and a trimmed mean between the median and the mean (see Figure 2.6).

measures of location

measures of central tendency

The mean is the most widely used measure of central tendency; one reason for this popularity is its utility in statistical inference problems. Another reason is that it is possible to combine subgroup means into an overall mean. This property does not hold for the other measures of central tendency. For example, if the means for three equal-sized subgroups are 23, 25, and 30, the mean for the combined group must be $(23 + 25 + 30)/3 = 26$. However, for the same data, if we know that the subgroup medians are 23, 25, and 30, we know only that the combined group median is somewhere between 23 and 30. Subgroup modes are even less informative. In general, the overall mean is a **weighted average** of the subgroup means, with weights equal to the subgroup sizes; that is, if $\bar{y}_j$

weighted average

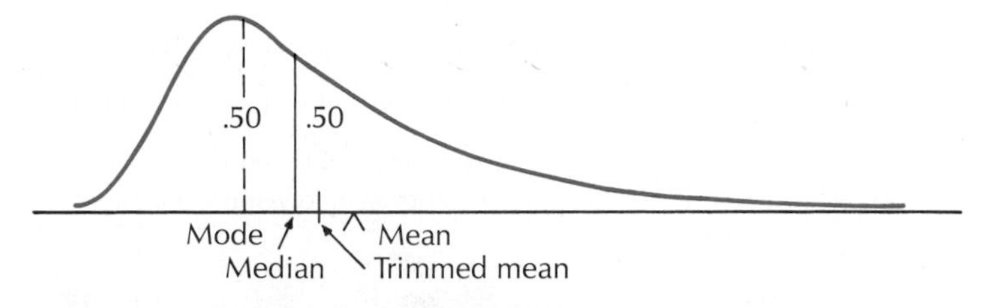

FIGURE 2.6 **Measures of Location for Skewed Distribution**

denotes the mean in subgroup j, $j = 1, \ldots, k$, and n_j is the number in subgroup j, then

$$\bar{y} = \frac{\sum_j n_j \bar{y}_j}{\sum_j n_j}$$

EXAMPLE 2.9 Find $\bar{y}$ for the following 10 measurements based on the three subgroup means.

Subgroup	Values	Total	$\bar{y}_j$	n_j
1	21, 21, 23, 24, 26	115	23	5
2	23, 26, 26	75	25	3
3	29, 31	60	30	2
		Grand total = 250		$n = 10$

Solution Substituting into the formula for $\bar{y}$ using subgroup means, we obtain

$$\bar{y} = \frac{5(23) + 3(25) + 2(30)}{5 + 3 + 2} = \frac{250}{10} = 25.0$$

Note that the larger weight given to $\bar{y}_1$ causes the overall mean to fall below the simple average of the three means $(23 + 25 + 30)/3 = 26$. □

Statistical thinking focuses very heavily on typical, average values. In many situations, inferences related to average or typical values will suffice, but there is a limit to such thinking. Sometimes the average value isn't important. A single, dramatically effective drug can propel a small pharmaceutical firm's ledger from the red to the black very quickly; a single disaster can propel an insurance firm into bankruptcy court. But it would be foolish to ignore statistical thinking on these grounds. Most products will not be overwhelming successes; most accidents will not be bankruptcy-causing disasters. Managers who assume that their results will always be better than average are in grave danger of becoming ex-managers. Measures of variability provide another dimension to our thinking; they are discussed in the next section.

SECTION 2.2 EXERCISES

2.5 a. Find the mean and median for the car production data of Exercise 2.1.

b. Find the modal class for the histogram and for the stem-and-leaf diagram constructed in that exercise.

c. Arrange the mode, median, and mean for this data set from lowest to highest.

2.6 Compute the mean, median, and mode for the following data:

11 17 18 10 22 23 15 17 14 13 10 12 18 18 11 14

2.7 The Insurance Institute for Highway Safety published data on the total damage suffered by compact automobiles in a series of controlled, low-speed collisions. The data, in dollars, with brand names removed, are

361	393	430	543	566	610	763	851	886	887	976	1039
1124	1267	1328	1415	1425	1444	1476	1542	1544	2048	2197	

a. Draw a histogram of the data, using 6 or 7 categories.
b. On the basis of the histogram, what would you guess the mean to be?
c. Calculate the median and mean.
d. What does the relation between the mean and median indicate about the shape of the data?

2.8 Production records for an automobile manufacturer show the following figures for production per shift (maximum production is 720 cars per shift).

688	711	625	701	688	667	694	630	547	703	688	697	703
656	677	700	702	688	691	664	688	679	708	699	667	703

a. Would the mode be a useful summary statistic for these data?
b. Find the median.
c. Find the mean.
d. What does the relation between the mean and median indicate about the shape of the data?

2.9 Draw a stem-and-leaf plot of the data in Exercise 2.8. The "stem" should include (from highest to lowest) 71, 70, 69, ... Does the shape of the stem-and-leaf display confirm your judgment in part (d) of Exercise 2.8?

2.10 Data are collected on the weekly expenditures of a sample of urban households on food (including restaurant expenditures). The data, obtained from diaries kept by each household, are grouped by number of members of the household. The expenditures were as follows:

1 member:	67	62	168	128	131	118	80	53	99	68		
	76	55	84	77	70	140	84	65	67	183		
2 members:	129	116	122	70	141	102	120	75	114	81	106	95
	94	98	85	81	67	69	119	105	94	94	92	
3 members:	79	99	171	145	86	100	116	125				
	82	142	82	94	85	191	100	116				
4 members:	139	251	93	155	158	114	108					
	111	106	99	132	62	129	91					
5+ members:	121	128	129	140	206	111	104	109	135	136		

a. Calculate the mean expenditure separately for each number of members.
b. Calculate the median expenditure separately for each number of members.

2.11 Answer the following for the data in Exercise 2.10:
a. Calculate the mean of the combined data, using the raw data.
b. Can the combined mean be calculated from the means for each number of members?
c. Calculate the median of the combined data, using the raw data.
d. Can the combined median be calculated from the medians for each number of members?

2.3 MEASURING VARIABILITY

The average value in a set of measurements is only one important summary figure. It is equally important to summarize the extent to which values differ among themselves or around a central value. Suppose that the average daily number of beds occupied in a hospital for a given year is 70 (out of an available total of 100). If the actual daily number of beds occupied throughout the year is always 69, 70, or 71, there is very low variability and the hospital can reasonably reduce the total number of beds available. At the other extreme, if the actual daily occupancy ranges all the way from 20 to 100, there is a great deal of variability, and reducing the number of available beds may be very risky. In this section we discuss several statistical measures of such variability.

range

The simplest but least useful of the measures of variability is the **range**, which is defined as the difference between the largest and smallest measurements. Because it depends only on the most extreme data values, the range does not reflect the general pattern of the data and is very sensitive to flukes (outliers). Therefore we do not use the range as a measure of variability.

deviations from the mean

More useful measures of variability are based on **deviations from the mean**. The deviation of a sample measurement y_i from its mean $\bar{y}$ is defined as $(y_i - \bar{y})$. Some of these deviations are positive, others negative; their algebraic sum is always zero:

$$\sum_i (y_i - \bar{y}) = (y_1 - \bar{y}) + (y_2 - \bar{y}) + \ldots + (y_n - \bar{y})$$

$$= y_1 + y_2 + \ldots + y_n - n\bar{y} = n\bar{y} - n\bar{y} = 0$$

Thus the positive and negative deviations exactly balance. To measure variability, we must look at the magnitudes (ignoring the positive or negative sign) of the deviations; large positive and negative deviations indicate large variability.

average absolute deviation

The simplest measure of the magnitudes of the deviations is the **average absolute deviation** (AD), defined as the average of the absolute values of the deviations:

$$\text{AD} = \frac{\sum_i |y_i - \bar{y}|}{n}$$

EXAMPLE 2.10 Regard the data values 11, 12, 13, 14, and 30 as a sample of five measurements selected from a population of interest. Compute the average absolute deviation for the sample data.

Solution It is easy to show that $\bar{y} = 80/5 = 16$, so

$$\text{AD} = \frac{|11 - 16| + |12 - 16| + |13 - 16| + |14 - 16| + |30 - 16|}{5}$$

$$= \frac{5 + 4 + 3 + 2 + 14}{5} = 5.6$$

Thus, on the average, an individual measurement deviates 5.6 units from the mean. □

The average absolute deviation is easy to compute and to interpret. However, it is difficult to deal mathematically with absolute values. Most statistical methods deal with squared deviations instead.

The most useful statistical measures of variability are the *variance* and the *standard deviation*. The variance of a sample of n measurements $y_1, y_2, \ldots, y_n$ is defined as the sum of the squared deviations divided by $(n - 1)$. We denote the sample variance by s^2. The sample standard deviation s of the measurements is the (positive) square root of the variance. The corresponding population variance and standard deviation are denoted by σ^2 and σ, respectively.

$$s^2 = \frac{\sum_i (y_i - \bar{y})^2}{n - 1}$$

$$s = \sqrt{s^2}$$

EXAMPLE 2.11 Compute the sample variance and sample standard deviation for the data of Example 2.10.

Solution With $\bar{y} = 16$, we can substitute into the formula for s^2 to obtain

$$s^2 = \frac{(11 - 16)^2 + (12 - 16)^2 + (13 - 16)^2 + (14 - 16)^2 + (30 - 16)^2}{5 - 1}$$

$$= \frac{250}{4} = 62.5$$

and thus

$$s = \sqrt{62.5} = 7.906$$ □

The use of $(n - 1)$ as the denominator of s^2 is not arbitrary. This definition of the sample variance makes it an "unbiased estimator" of the population variance σ^2. This roughly means that if we were to draw a very large number of samples each of size n from the population and if we computed s^2 for each sample, the average sample variance would equal the population variance σ^2. Had we divided by n in the definition of s^2, the average sample variance would be less than the population variance, and hence s^2 would tend to underestimate σ^2.

The definitions of variance and standard deviation depend on whether the data are regarded as a population or as a sample. In practice, the available data are usually only a sample. Unless specifically indicated otherwise, we regard all data sets as samples and consider population means and variances as conceptual quantities to be estimated from the samples.*

Computation of s^2 and s is often simplified by the use of the algebraic identity

$$\sum_i (y_i - \bar{y})^2 = \sum_i y_i^2 - \frac{\left(\sum_i y_i\right)^2}{n}$$

This identity yields a shortcut formula for s^2 and therefore for s.

Shortcut Formula for s^2 and s

$$s^2 = \frac{1}{n - 1}\left[\sum_i y_i^2 - \frac{\left(\sum_i y_i\right)^2}{n}\right]$$

$$s = \sqrt{s^2}$$

* When a set of data is regarded as a population, the population variance is defined as the sum of squared deviations from the population mean μ, all divided by N, the population size—not by $(N - 1)$.

We have defined the variance and standard deviation but have not yet indicated how to interpret them. What does a standard deviation of \$552.51 mean? There are two possible interpretations, one a rough but reasonably accurate approximation, the other a mathematically guaranteed (but not so accurate) inequality. The approximation assumes that the measurements have roughly a mound-shaped histogram—that is, a symmetric, single-peaked histogram that tapers off reasonably smoothly from the peak toward the tails. We call the approximation the **Empirical Rule**.

Empirical Rule

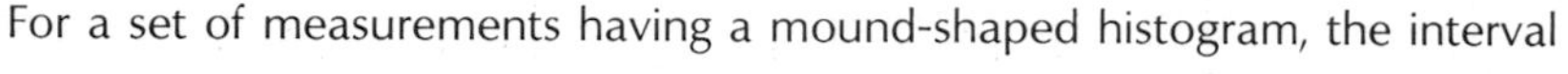

For a set of measurements having a mound-shaped histogram, the interval

$\bar{y} \pm 1s$ contains approximately 68% of the measurements,
$\bar{y} \pm 2s$ contains approximately 95% of the measurements,
$\bar{y} \pm 3s$ contains approximately all of the measurements.

EXAMPLE 2.12 A sample of 20 days throughout the previous year indicates that the average wholesale price per pound for steers at a particular stockyard was \$.61 and that the standard deviation was \$.07. If the histogram for the measurements is mound-shaped, describe the variability of the data using the Empirical Rule.

Solution Applying the Empirical Rule, we conclude that

the interval $.61 \pm .07$, or \$.54 to \$.68, contains approximately 68% of the measurements;
the interval $.61 \pm .14$, or \$.47 to \$.75, contains approximately 95% of the measurements;
the interval $.61 \pm .21$, or \$.40 to \$.82, contains approximately all of the measurements. □

The mathematically guaranteed bound is known as Chebyshev's Inequality.

Chebyshev's Inequality

For any set of measurements and a constant $c > 1$, the interval

$\bar{y} \pm cs$ contains **at least** $1 - 1/c^2$ of the measurements.

Thus for $c = 2$ and $c = 3$,

$\bar{y} \pm 2s$ contains **at least** 3/4 of the measurements,

and

$\bar{y} \pm 3s$ contains **at least** 8/9 of the measurements.

Although Chebyshev's Inequality is always mathematically correct, the Empirical Rule is often a better approximation. If the assumption of a mound-shaped histogram is exactly correct (specifically, if the data follow the normal distribution defined in Chapter 5), the Empirical Rule is exactly correct. Even if the data don't follow an exactly mound-shaped distribution, the Empirical Rule is still fairly accurate.

EXAMPLE 2.13 The following data represent the percentages of family income allocated to groceries for a sample of 30 shoppers:

26	28	30	37	33	30
29	39	49	31	38	36
33	24	34	40	29	41
40	29	35	44	32	45
35	26	42	36	37	35

For these data, $\sum_i y_i = 1043$ and $\sum_i y_i^2 = 37{,}331$.

a. Compute the mean, variance, and standard deviation of the percentage of income spent on food.
b. Verify that Chebyshev's Inequality holds with $c = 2$. Which is closer to the actual proportion, this inequality or the Empirical Rule?

Solution The sample mean is

$$\bar{y} = \frac{\sum_i y_i}{30} = \frac{1043}{30} = 34.77$$

The corresponding sample variance and standard deviation are

$$s^2 = \frac{1}{n-1}\left[\sum_i y_i^2 - \frac{\left(\sum_i y_i\right)^2}{n}\right]$$

$$= \frac{1}{29}[37{,}331 - 36{,}261.63] = \frac{1069.37}{29} = 36.87$$

$$s = \sqrt{36.87} = 6.07$$

Chebyshev's Inequality indicates that at least .75 of the data should fall in the range 34.77 − 2(6.07) to 34.77 + 2(6.07), or 22.63 to 46.91. In fact, 29 of the 30 values (a proportion of .967) fall within this range; the proportion is much closer to the Empirical Rule value. □

An alternative way to approach variability is the interquartile range (IQR). The quartiles of a data distribution are the 25th and 75th percentiles—the values that mark the bottom one-fourth and top one-fourth of the data. Tukey (1977) calls these values "hinges" and notes that they can be found by taking medians of each half of the data. The median of the bottom half of the data is the 25th percentile, simply because half of one-half is one-fourth; similarly, the median of the top half of the data is the 75th percentile. The interquartile range is the difference between the two quartiles.

Interquartile Range

To find the quartiles of a set of data on a variable.

1. Sort the data and find the median.
2. Divide the data into top and bottom halves (above and below the median). If the sample size n is odd, arbitrarily include the median in both halves.
3. Find the medians of both halves. These are the 25th and 75th percentiles, or "hinges."
4. IQR = 75th percentile − 25th percentile.

EXAMPLE 2.14 Find the interquartile range for the data of Example 2.13.

Solution First we sort the data.

24	26	26	28	29	29	29	30	30	31	32	33	33	34	35
35	35	36	36	37	37	38	39	40	40	41	42	44	45	49

Because $n = 30$, the median is the average of the fifteenth and sixteenth values (both of which are 35), namely 35. The bottom half of the data is the lowest 15 values; the top half is the highest 15 values. The median of the bottom half (that is, the 25th percentile) is the eighth value, 30. The 75th percentile is the eighth value from the top, 39. The IQR is $39 - 30 = 9$. □

The IQR is extremely useful in checking for outliers. Tukey (1977) defines the inner fences as follows:

lower inner fence = 25th percentile − 1.5 IQR
upper inner fence = 75th percentile + 1.5 IQR

Any value outside the inner fences is a outlier candidate. The choice of the number 1.5 is arbitrary, but it seems to work decently. Similarly, the outer fences are defined as follows:

lower outer fence = 25th percentile − 3.0 IQR
upper outer fence = 75th percentile + 3.0 IQR

Any data value outside the outer fences is a serious outlier.

EXAMPLE 2.15 Calculate the various fences for the data of Example 2.13. Does the fence test identify any outliers? Do there appear to be outliers on visual inspection?

Solution A plot of the data, or merely a scan by eye, does not indicate that any value should be regarded as an outlier. The closest thing to an outlier is the value 49, but it is close to several other values. The 25th percentile and 75 percentile were found in Example 2.14 to be 30 and 39, so the IQR is 9. The inner fences are 30 − 1.5(9) = 16.5 and 39 + 1.5(9) = 52.5. No data value even comes close to the inner fences, so the fence test identifies no outliers. □

The interquartile range is the basis for still another Tukey (1977) idea, the box plot, sometimes called the "box-and-whiskers plot."

Box Plot

1. Draw the edges of a box at the 25th and 75th percentile. Draw a vertical bar through the box at the median.
2. Draw lines ("whiskers") from the edges of the box to the adjacent values—the smallest and largest non-outliers.
3. Plot each outlier candidate separately, conventionally using a * symbol. Plot each serious outlier, conventionally using a 0 symbol.

EXAMPLE 2.16 Suppose that the return on investments for 21 companies in a certain industry for a certain year is

−24.6	−2.6	2.4	2.7	3.8	5.6	5.9	6.7	7.0	7.2	7.5
8.0	8.2	8.5	8.6	8.8	9.0	9.2	9.7	10.0	20.5	

Draw a box plot of these data.

Solution With $n = 21$, the median is the eleventh score, 7.5. The 25th percentile is the median of the bottom 11 scores, namely 5.6. (Note that, because n is an odd number, we include the median in both halves of the data. Thus the bottom half of the data comprises 11 scores, not 10.) The 75th percentile is the median of the top 11 scores, namely 8.8. Thus IQR = 8.8 − 5.6 = 3.2. The fences are

$$\begin{aligned}
\text{lower outer fence} &= 5.6 - 3.0(3.2) = -4.0\\
\text{lower inner fence} &= 5.6 - 1.5(3.2) = 0.8\\
\text{upper inner fence} &= 8.8 + 1.5(3.2) = 13.6\\
\text{upper outer fence} &= 8.8 + 3.0(3.2) = 18.4
\end{aligned}$$

The fence test identifies two serious outliers, −24.6 and 20.5, and one outlier candidate, −2.6. (A plot of the data indicates that the serious outliers are obviously extreme and that the outlier candidate is debatably out.) The adjacent values are the smallest and largest non-outliers, 2.4 and 10.0. The resulting box plot is shown in Figure 2.7. □

FIGURE 2.7 Box Plot for Example 2.16

SECTION 2.3 EXERCISES

2.12 Consider the typewriter bid data of Exercise 2.2.

a. Compute the mean and standard deviation.

b. Compute the average absolute deviation and range.

c. How many observations actually fall within two standard deviations of the mean? Compare the actual result to the Empirical Rule and to Chebyshev's Inequality approximations.

2.13 Suppose that the data from two samples are

Sample 1:	15	19	21	25									
Sample 2:	14	17	18	19	19	20	20	20	21	21	22	23	26

a. Find the range for each sample.

b. Find the standard deviation for each sample.

c. Which sample shows more variability? Construct a histogram or line plot to support your opinion.

2.14 Compute the sample variance and sample standard deviation for the data given in Exercise 2.6.

2.15 Consider the data in Exercise 2.7.

a. Find the standard deviation of dollar damages.

b. Of the 23 data values, how many fall within two standard deviations of the mean? Is the fraction approximately what would be expected according to the Empirical Rule?

c. Does the histogram found in part (a) of Exercise 2.7 suggest that the Empirical Rule should work fairly well?

2.16 Again consider the data in Exercise 2.7.

a. Find the IQR.

b. Calculate the inner and outer outlier fences for the data. Are there any outliers according to this calculation?

c. Construct a box plot of the data.

2.17 Data for car production per shift given in Exercise 2.8 are reproduced here:

688	711	625	701	688	667	694	630	547	703	688	697	703
656	677	700	702	688	691	664	688	679	708	699	667	703

a. Find the mean and standard deviation.

b. How well does the Empirical Rule work for the fraction of data falling within one standard deviation of the mean?

2.18 a. Find the median and IQR for the data in Exercise 2.17.

b. Find the inner and outer fences. Are there outliers?

c. Draw a box plot of the data.

2.19 Draw box plots of each subgroup (each number of members of the household) for the data in Exercise 2.10. Include a check for outliers.

2.20 Calculate standard deviations separately for each subgroup of data in Exercise 2.10.

2.4 OTHER SUMMARY MEASURES OF DATA

The mean and standard deviation are by far the most important and widely used summary statistics for a set of data. In this section we define some other statistics that are occasionally useful in measuring the skewness and "tail-heaviness" of data.

In Section 2.2 we said that when the data are skewed to the right, the mean is larger than the median; for data that are skewed to the left, the mean is smaller than the median. One simple measure of skewness is

$$\frac{\text{mean} - \text{median}}{\text{standard deviation}}$$

A positive value of this measure indicates that the data are right-skewed; a negative value, left-skewed. This measure is zero if the data are perfectly symmetric (because the mean and median both equal the point of symmetry). It is also possible that the measure equals zero for certain patterns of nonsymmetric data.

EXAMPLE 2.17 A random sample of 36 intermediate-level executives from health-related corporations are interviewed to determine their estimated increases in expenditures for research and development over the next 10 years. These data (recorded as percentages of sales) are shown here:

2.10	2.47	1.75	2.94	1.69	2.75	2.82	2.52	2.77	1.98	2.70	2.43
2.17	2.80	2.82	2.38	2.68	2.39	1.99	2.10	2.67	2.65	2.06	2.55
2.22	2.92	3.05	2.77	2.36	1.80	3.09	2.20	2.93	1.85	2.28	1.96

A Minitab output is shown below (some steps are omitted).

```
MTB > HISTOGRAM OF '$R-AND-D'

$R-AND-D

MIDDLE OF INTERVAL          NUMBER OF OBSERVATIONS
        1.6                          1   *
        1.8                          3   ***
        2.0                          4   ****
        2.2                          6   ******
        2.4                          5   *****
        2.6                          5   *****
        2.8                          7   *******
        3.0                          5   *****

MTB >
  Describe '$R-and-D'
  $R-AND-D   N = 36     MEAN = 2.4336     ST.DEV. = .397

MTB >
  Median of '$R-and-D', put in k1
  MEDIAN = 2.4500

MTB >
  Print k5 = coef of skewness, k6 = kurtosis
  K5     -.188922
  K6     1.79995
```

a. Use these data to compute $(\bar{y} - \text{median})/s$, a measure of skewness.
b. Do the data suggest a skewness?

Solution a. The mean is 2.4336, the median is 2.4500, and the standard deviation is 0.397. So $(\bar{y} - \text{median})/s = (2.4336 - 2.45)/0.397 = -0.041$, a very small number.

b. The histogram for the sample data is shown in the output. There is no blatant skewness; if anything, there is a slight left-skewness, because the mean is slightly below the median. □

Another measure of skewness involves raising deviations from the mean to the third power. This measure, called the coefficient of skewness, is defined as

$$\frac{M_3}{M_2^{3/2}}$$

where

$$M_3 = \sum (y - \bar{y})^3/n$$
$$M_2 = \sum (y - \bar{y})^2/n$$

This measure also is positive for right-skewed data, negative for left-skewed data, and 0 for symmetric data.

EXAMPLE 2.18 Refer to the data and Minitab output of Example 2.17. Locate the coefficient of skewness.

Solution The coefficient of skewness $M_3/M_2^{3/2}$ is shown as K5 in the output. The value K5 = −.188922 indicates left-skewness. □

This measure works as a measure of skewness because the cube (third power) of a large number is vastly bigger than the cube of a smaller number. In right-skewed data there are a few relatively large positive deviations, resulting in a positive value for this skewness measure. Unfortunately, this cubing process makes the measure very sensitive to extreme values in the data; for this reason, we prefer to use (mean − median)/s as a measure of skewness.

kurtosis

A measure of the heaviness of the tails of a distribution of data is the **kurtosis**. This involves fourth powers of deviations and can be defined as

$$\frac{M_4}{M_2^2}$$

where

$$M_4 = \sum (y - \bar{y})^4/n$$
$$M_2 = \sum (y - \bar{y})^2/n$$

If the data follow a bell-shaped distribution, as shown below, the kurtosis is approximately 3. A symmetric distribution having heavier tails than a bell-shaped distribution has a kurtosis larger than 3. A heavy-tailed distribution is indicated in Figure 2.8. A light-tailed distribution has a kurtosis less than 3.

The difficulty with the kurtosis measure is that, because it uses fourth powers, it is very sensitive to extreme values. Modest changes in extreme values can lead to large

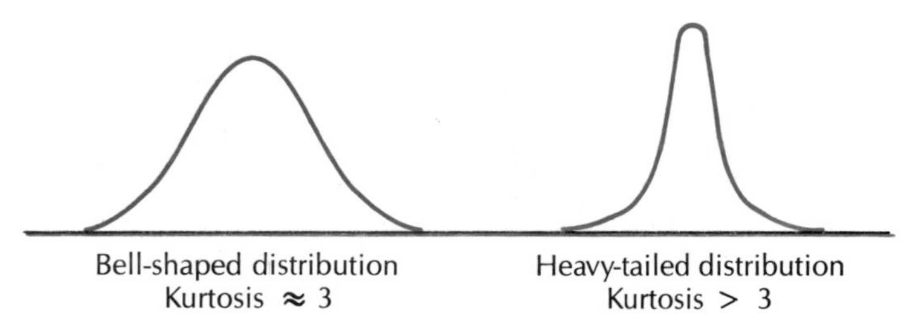

FIGURE 2.8 **Bell-shaped and Heavy-tailed Distributions**

changes in the kurtosis measure. However, we know of no other measure of heavy-tailness that is more satisfactory than the kurtosis. If this measure is to be used, it is important to plot the data to see if one or two very extreme values are present; such values can cause a misleading value of the kurtosis.

EXAMPLE 2.19 Find the coefficient of kurtosis for the data of Example 2.17.

Solution The Minitab output from Example 2.17 also displays the coefficient of kurtosis, 1.79995, indicating a light-tailed distribution. The histogram for the sample data indicates no extreme outliers. □

SECTION 2.4 EXERCISES

2.21 Use the SAS computer output shown here to answer (a)–(c).

a. Locate the measures of skewness and kurtosis.

b. Plot the data in convenient form. Does the skewness coefficient indicate the same sign (positive or negative) as does the plot?

c. Verify the computation of the skewness coefficient.

OBS	1	2	3	4	5	6	7	8	9	10	11	12	13	14	15
CROWD	42.6	12.1	10.6	12.7	14.5	19.2	16.7	20.1	15.5	9.9	14.6	16.1	19.4	23.3	29.7

	16	17	18	19	20	21	22	23	24	25	26	27	28	29	30
	22.6	15.7	18.9	12.8	19.1	33.5	24.7	18.6	18.2	19.9	15.5	21.2	20.2	16.5	18.6

ATTENDANCE DATA FOR 30 BASEBALL GAMES

VARIABLE	N	MEAN	STANDARD DEVIATION	MINIMUM VALUE	MAXIMUM VALUE	VARIANCE
CROWD	30	19.10000000	6.76930190	9.90000000	42.60000000	45.82344828

SKEWNESS	KURTOSIS
1.76647488	4.37158198

2.22 Refer to the data of Exercise 2.2 (p. 14) and the histogram constructed in part (a) of that exercise.

a. Calculate the kurtosis coefficient.

b. Does the coefficient indicate the kind of tailness suggested by the histogram?

2.23 Indicate the skewness pattern you would expect to find in each of the following data situations:

a. sale prices of existing houses in southern California,

b. scores on a fairly easy 100-point classroom examination;

c. lumber yield in a stand of widely spaced, equal-age trees.

2.24 As part of a facilities-expansion study, the monthly precipitation (inches) at a major metropolitan airport is obtained for a two-year period. The data are as follows:

Year 1:	0.1	0.7	0.5	4.5	6.7	1.1	0.7	3.2	4.8	1.0	0.9	0.8
Year 2:	7.9	1.8	1.4	0.2	0.7	1.5	2.6	5.5	0.6	0.5	1.7	3.3

a. Compute $\bar{y}$ and s for each year.

b. Compute two measures of skewness for each year. Do both measures indicate skewness?

2.25 Combine the data in Exercise 2.24 for both years.

a. Determine the average monthly precipitation over the two-year period. Can it be found from the yearly averages, or must you go back to the original data?

b. Use the monthly data to compute the standard deviation of the combined data. Is the combined standard deviation equal to the average of the yearly standard deviations?

2.26 The automobile manufacturer in Exercise 2.8 can produce a maximum of 720 cars per shift. Normally the plant can come close to that maximum, but occasional downtime reduces production.

a. From this description, what skewness should be observed in the data?

b. Calculate $(\bar{y} - \text{median})/s$ for the data. Does this measure reflect the expected type of skewness?

2.27 The data in Exercise 2.7 for the amount of damage suffered by automobiles in low-speed crashes are reproduced here:

361	393	430	543	566	610	763	851	886	887	976	1039
1124	1267	1328	1415	1425	1444	1476	1542	1544	2048	2197	

a. Does a plot of the data reveal extreme skewness?

b. Does the value $(\bar{y} - \text{median})/s$ indicate extreme skewness?

2.28 Calculate the kurtosis for the data of Exercise 2.27. If a computer package is available to you, this is the spot to use it. What does the value of the kurtosis indicate about the shape of the data?

2.5 CALCULATORS AND COMPUTER SOFTWARE SYSTEMS

Electronic hand calculators can be great aids in performing some of the calculations mentioned in this chapter, especially for small data sets. For example, many calculators have keys for obtaining the sample mean and standard deviation directly after the data are entered. Others can be used with the shortcut formulas of the previous sections to obtain $\bar{y}$, s^2, and s. For larger data sets, even hand calculators are of little use because of the time required for entering data and the inability to update an erroneous entry without reentering the entire data set. In these situations, a computer can be of help. Specific programs or more general software systems can be used to perform statistical analyses almost instantaneously, even for very large data sets, after the data are entered into the computer from a terminal, magnetic tape, or disk storage. It is not necessary to have knowledge of computer programming in order to make use of specific programs or software systems for planned analyses. Most have user's manuals that give detailed directions for their use. Others, developed for use at a terminal, provide program prompts that lead the user through the analysis of choice.

If you have access to a computer and are interested in using it, inquire about how one obtains an account, what programs and software systems are available for doing statistical analyses, and where one obtains instruction on data entry for these programs and software systems.

There are dangers in using statistical software packages carelessly. A computer is a mindless beast, and it does anything asked of it, no matter how absurd the result may be. For instance, suppose that the data include age, sex (1 = female, 2 = male), religion (1 = Catholic, 2 = Jewish, 3 = Protestant, 4 = other or none), and monthly income of a group of people. If we ask the computer to calculate means, we get means for the quantitative variables, age and income, and also (absurdly) means for the qualitative variables, sex and religion. Furthermore, it is very unlikely that a standard program will warn a user that extreme skewness is distorting a mean value or that the data contain a gross outlier. Used intelligently, these packages are convenient, powerful, and useful; but the most important acronym of computer technology is GIGO: Garbage In, Garbage Out.

Computers can be of great assistance in data description as well as in statistical analyses. With the advent of more sophisticated computer software and hardware, it is now possible, not only to generate simple line plots, histograms, and stem-and-leaf diagrams (such as those displayed earlier in the chapter) on a computer, but also to produce high-quality computer graphs that can be photographed for scientific, marketing, and management presentations.

Software designed specifically for statistical applications will be illustrated throughout the remainder of the text to display data and the results of statistical analyses. It is hoped that through applications in the text, you as a manager will see the tremendous potential and utility of this rapidly expanding technology in computer packages.

2.6 SUMMARIZING GROUPED DATA

Occasionally, the original data aren't available and only grouped data (frequencies in each interval) can be used. With the advent of computerized data bases, grouped-data situations are becoming less common. Much of federal and state government data are still presented in grouped form to protect the privacy of individuals. With grouped data, it isn't possible to compute the mean, median, and standard deviation exactly, but one can often obtain good approximations.

It is possible to approximate the median on the assumption that the actual data values are evenly spread over the class interval. The approximation formula is

$$\text{median} \approx L + \frac{w}{f_m}(.5n - cf_b)$$

where

L = lower limit of the interval containing the median
w = width of that interval
f_m = frequency in that interval
cf_b = cumulative frequency (total count) below that interval
n = total number of values

EXAMPLE 2.20 The commission income data grouped as in Example 2.3 can be tabulated as follows:

Class Interval	Frequency f_i	Cumulative Frequency cf_i
$250–749	6	6
750–1249	7	13
1250–1749	5	18
1750–2249	1	19
2250–2749	1	20
	$n = 20$	

Obtain an approximate value for the median.

Solution The median is in the interval 750–1249; it is approximated as $750 + (500/7)\,[.5(20) - 6] \approx 1036$. The actual median, calculated from the ungrouped values, is $(895 + 965)/2 = 930$. The error in the approximation occurs because the data happen to be concentrated in the lower part of the interval 750–1249. It is obviously better to use the actual data rather than the grouped-data approximation whenever possible. □

As in the case of the median, it is possible to approximate the mean for grouped data. It turns out that it doesn't matter whether we assume that the values are evenly spread through a class interval or concentrated at the midpoint; the second assumption makes the arithmetic easier. If we denote the ith class midpoint as m_i and the ith class frequency as f_i, the approximation formula for the sample mean is

$$\bar{y} \approx \frac{\sum_i f_i m_i}{n}$$

EXAMPLE 2.21 The grouped commission income data is tabulated in Example 2.20. Compute the approximate sample mean.

Solution The midpoints may be taken as 500, 1000, 1500, 2000, and 2500, so the sample mean can be approximated as

$$\bar{y} \approx \frac{6(500) + 7(1000) + 5(1500) + 1(2000) + 1(2500)}{20} = 1100$$

For these data we have both the sample measurements and the frequency table. The actual sample mean is \$1066.55, whereas the approximate value computed from the frequency table is \$1100. Whenever possible, use the sample measurements to compute the sample mean. □

It is possible to approximate the sample variance if only grouped data are available. The conventional formula for computing s^2 from grouped data assumes that all data values occur at the class midpoint rather than being spread out through the class interval. This assumption minimizes the variability within intervals and therefore tends to yield a slight underestimate of the variance. It leads to the following formula for a grouped-data approximation to the sample variance:

$$s^2 \approx \frac{1}{n-1}\left[\sum_i f_i m_i^2 - \frac{\left(\sum_i f_i m_i\right)^2}{n}\right]$$

where

f_i = frequency for class i
m_i = midpoint for class i
n = sample size (the total frequency)

Again $s = \sqrt{s^2}$.

EXAMPLE 2.22 Compute the sample variance and sample standard deviation for the grouped data of Example 2.20.

Solution Before we compute s^2, we need the quantities $\sum f_i m_i^2$ and $\sum f_i m_i$.

$$\sum f_i m_i^2 = 6(500)^2 + 7(1000)^2 + \ldots + 1(2500)^2 = 30{,}000{,}000$$

and

$$\sum f_i m_i = 22{,}000$$

Then

$$s^2 \approx \frac{1}{19}\left[30{,}000{,}000 - \frac{(22{,}000)^2}{20}\right] = 305{,}263.16$$

$$s \approx 552.51$$

□

SECTION 2.6 EXERCISES

2.29 In a market research study, 740 consumers are given samples of a standard brand of frozen dinner and a newly developed competitor. Each consumer is asked to rate the new product in comparison to the established brand on a scale of 0 to 100. The comparison points indicated to the consumer are

0 = New product is inedible; wouldn't eat it if it were free.
20 = New product is much worse than competitor; would eat it but not buy it.
40 = New product is somewhat worse; would consider buying it on sale.
60 = Established product is somewhat worse than new product.
80 = Established product is much worse.
100 = Established product is vastly worse than new; would never consider using established product.

The 740 responses were grouped as follows:

5–14.9	15–24.9	25–34.9	35–44.9	45–54.9	55–64.9	65–74.9	75–84.9	85–94.9
0	6	15	68	192	266	147	45	1

a. Find the approximate median response.
b. Find the approximate mean response.

2.30 Starting salaries for 40 recent MBA graduates from a major university are shown here (thousand dollars).

Range	Frequency
24.9 to 29.9	6
29.9 to 34.9	10
34.9 to 39.9	15
39.9 to 44.9	7
44.9 to 49.9	2

a. Construct a frequency histogram.
b. Using the histogram in part (a), guess the mean starting salary.
c. Compute the mean starting salary. Why must this mean be regarded as an approximation?

2.31 Find an approximate value for s using the grouped data of Exercise 2.29.

2.32 Compute approximate values for the variance and standard deviation using the grouped MBA salary data of Exercise 2.30.

SUMMARY

Chapter 2 presented the important concepts that enable us to organize and summarize data. Data description for values from a single variable were discussed here. Extensions to more than one variable can be derived from what we learned about a single variable.

We began by introducing a frequency table and graphs of the frequencies (or relative frequencies) associated with the values (or classes) from the frequency table. These graphs, called *histograms*, provide a simple visual description of the distribution of measurements for a variable. Bar graphs, which are a form of histogram for qualitative variables, were also discussed.

We then introduced measures of central tendency, which indicate "typical" values in a data set. The measures of typical or average values discussed were the mode, median, and mean. We discussed the calculation of these quantities for both grouped and ungrouped sample data.

Measures of variability presented in Chapter 2 were the range, average absolute deviation, variance, standard deviation, and interquartile range. The Empirical Rule and Chebyshev's Inequality provide means for interpreting the variability of sample data using the sample mean and sample standard deviation.

Finally, we discussed measures of skewness and kurtosis.

Having considered data description for values from a single variable, we turn now to ways of summarizing relations among two or more variables.

KEY FORMULAS: Summarizing Data from One Variable

1. Sample mean: $\bar{y} = \dfrac{\sum_i y_i}{n}$

2. Sample variance: $s^2 = \dfrac{1}{n-1}\left[\sum_i y_i^2 - \dfrac{\left(\sum_i y_i\right)^2}{n}\right]$

3. Sample standard deviation: $s = \sqrt{s^2}$

4. Median, grouped data: median $\approx L + \dfrac{w}{f_m}(.5n - cf_b)$

5. Sample mean, grouped data: $\bar{y} \approx \dfrac{\sum_i f_i m_i}{n}$

6. Sample variance, grouped data: $s^2 \approx \dfrac{1}{n-1}\left[\sum_i f_i m_i^2 - \dfrac{\left(\sum_i f_i m_i\right)^2}{n}\right]$

CHAPTER 2 EXERCISES

2.33 A study of sick-leave days over one year for a sample of 20 workers in a company yields the following numbers, arranged in increasing order:

0 0 0 0 0 0 1 1 1 1 2 2 2 3 3 4 6 9 14 31

a. Verify that the sample mean is 4.0 and the sample standard deviation is 7.27.
b. What fraction of the observations actually fall within one standard deviation of the mean? What aspect of the data explains the discrepancy between this fraction and the Empirical Rule approximation?
c. Calculate the skewness index $(\bar{y} - \text{median})/s$. Does its sign make sense?

2.34 If a computer package is available to you, use it to find the mean, median, standard deviation, and skewness for the data of Exercise 2.33. What skewness coefficient is calculated? Is the standard deviation calculated with a denominator of $(n - 1)$ or n?

2.35 Consider the following artificial data:

8	9	10	10	10	10	10	10	11	12	19	20	20
20	21	28	29	30	30	30	30	30	30	31	32	

a. Plot the data using 5 classes. What pattern of skewness and modality do you see?
b. The mean of these values is 20. How can this result be obtained without computation? How about the median?
c. Calculate the standard deviation. What must the skewness coefficient equal?
d. What fraction of the values fall within one standard deviation of the mean? How does this compare with the Empirical Rule fraction? Explain the discrepancy.
e. What fraction of the values fall within two standard deviations of the mean? How does this fraction compare with the Chebyshev Inequality fraction?

2.36 Here is another artificial data set:

−36	−1	−1	−1	−1	0	0	1	1	2	3	4	6	9	14

a. Verify that the mean is 0 and that the standard deviation is 10.84.
b. Calculate the skewness coefficient $(\bar{y} - \text{median})/s$ and $M_3/(M_2)^{3/2}$.
c. Plot the data. Can you identify an obvious skewness?
d. What causes the discrepancy between the two skewness statistics?

2.37 A firm's transportation department studies the performance of four trucking companies. For a sample of 40 loads of goods shipped in one month they ascertain the promised and actual arrival dates, as well as the reported condition of each load. The relevant variables are

SHIPCODE: a code identifying the trucker
LOTCODE: 1 for full truckload, 2 for less than a truckload
PROMISED: the promised day of arrival
ACTUAL: the actual day of arrival
DIFF: number of days late; a negative value occurs if the load arrives early
CONDN: a code for the reported condition ranging from 0 (no damage) to 10 (unusable)

The person performing the study requests the mean, kurtosis, and skewness for each of the variables. The results are shown in the following output:

a. For which variables are the means and standard deviations meaningful?
b. Locate these values in the output.
c. What would you request from a computer program to make more informative summaries of the data?

LOAD	SHIPCODE	LOTCODE	PROMISED	ACTUAL	DIFF	CONDN
1	1.	2.	2.	4.	2.	1.
2	4.	1.	3.	3.	0.	0.
3	2.	2.	4.	4.	0.	0.
4	3.	1.	4.	7.	3.	3.
5	4.	2.	4.	4.	0.	0.
6	1.	2.	7.	7.	0.	0.
7	1.	1.	7.	7.	0.	0.
8	2.	1.	7.	9.	2.	0.
9	3.	2.	7.	7.	0.	0.
10	3.	2.	8.	10.	2.	0.
11	1.	2.	8.	8.	0.	0.
12	4.	1.	9.	9.	0.	1.
13	3.	1.	9.	9.	0.	0.
14	2.	1.	10.	12.	2.	0.
15	2.	2.	10.	10.	0.	0.
16	3.	1.	10.	14.	4.	5.
17	1.	2.	11.	11.	0.	0.
18	3.	2.	11.	14.	3.	2.
19	2.	2.	11.	11.	0.	0.
20	4.	2.	14.	16.	2.	0.
21	3.	1.	14.	14.	0.	0.
22	4.	1.	14.	14.	0.	0.
23	3.	2.	14.	15.	1.	0.
24	2.	2.	15.	15.	0.	0.
25	1.	1.	16.	15.	−1.	0.
26	4.	1.	16.	16.	0.	0.
27	1.	2.	17.	18.	1.	1.
28	4.	2.	17.	17.	0.	0.
29	2.	2.	18.	18.	0.	0.
30	3.	2.	18.	21.	3.	1.
31	2.	1.	21.	21.	0.	0.
32	2.	1.	21.	21.	0.	0.
33	1.	1.	22.	21.	−1.	0.
34	4.	2.	23.	24.	1.	0.
35	4.	1.	23.	23.	0.	0.
36	2.	2.	24.	24.	0.	0.
37	3.	1.	25.	28.	3.	2.
38	1.	1.	28.	28.	0.	0.
39	3.	2.	28.	31.	3.	0.
40	2.	2.	28.	29.	1.	1.

```
VARIABLE SHIPCODE

MEAN            2.500      KURTOSIS     -1.257    SKEWNESS     -0.000
MINIMUM         1.000      MAXIMUM       4.000

VALID OBSERVATIONS   40
----------------------------------------------------------------------
VARIABLE LOTCODE

MEAN            1.550      KURTOSIS     -2.062    SKEWNESS     -0.209
MINIMUM         1.000      MAXIMUM       2.000

VALID OBSERVATIONS   40
----------------------------------------------------------------------
VARIABLE PROMISED

MEAN           13.950      KURTOSIS     -0.849    SKEWNESS      0.353
MINIMUM         2.000      MAXIMUM      28.000

VALID OBSERVATIONS   40
----------------------------------------------------------------------
VARIABLE ACTUAL

MEAN           14.725      KURTOSIS     -0.611    SKEWNESS      0.433
MINIMUM         3.000      MAXIMUM      31.000

VALID OBSERVATIONS   40
----------------------------------------------------------------------
VARIABLE DIFF

MEAN            0.775      KURTOSIS     -0.147    SKEWNESS      0.999
MINIMUM        -1.000      MAXIMUM       4.000

VALID OBSERVATIONS   40
----------------------------------------------------------------------
VARIABLE CONDN

MEAN            0.425      KURTOSIS     11.014    SKEWNESS      3.124
MINIMUM         0.0        MAXIMUM       5.000

VALID OBSERVATIONS   40
```

APPENDIX: SUMMATION NOTATION

Most readers of this book are familiar with summation notation; however, a brief indication of the way it is used here might be useful. We denote numerical values (observations) by y or x; when there are many observations, we use an indexing subscript such as i or j to differentiate them. For example, if our observations are 11, 30, 12, 14, 13, we let $y_1 = 11$, $y_2 = 30$, . . . , $y_5 = 13$.

The Greek symbol $\sum$ (uppercase sigma) is an instruction to do something—namely, to add up whatever is indicated next. Thus we have $\sum_i y_i = y_1 + \ldots + y_5 = 11 + 30 + 12 + 14 + 13 = 80$. A more elaborate version of the notation indicates which values of the index i are used; in the example, this notation would be $\sum_{i=1}^{5} y_i = y_1 + \ldots + y_5 = 80$. Since we always want to sum over all possible values of the index, we do not use the elaborate form.

EXAMPLE 2.23 Let $y_1 = 2$, $y_2 = 4$, $y_3 = 10$, $y_4 = 1$, $y_5 = 20$, $y_6 = 7$, and $y_7 = 3$. Determine $\sum_i y_i$.

Solution

$$\begin{aligned}\textstyle\sum_i y_i &= y_1 + y_2 + \ldots + y_7 \\ &= 2 + 4 + \ldots + 3 \\ &= 47\end{aligned}$$

□

Most complicated formulas may also be understood by translating $\sum$ as "the sum of." For instance, $\sum_i y_i^2$ means "the sum of the squared observations." In the example, $\sum_i y_i^2 = (11)^2 + \ldots + (13)^2 = 1530$. Note that we square first, then sum. In contrast, the square of the sum of the observations is indicated by $(\sum_i y_i)^2$. In our example, $(\sum_i y_i)^2 = (80)^2 = 6400$. Similarly, translate $\sum_i (y_i - \bar{y})^2$ as "the sum of the squared deviations." For this example, $\sum_i (y_i - \bar{y})^2 = 250$.

EXAMPLE 2.24 Use the data in Example 2.23 to compute $\sum y_i^2$. Also calculate $(\sum_i y_i)^2$.

Solution $\sum_i y_i^2 = 2^2 + 4^2 + \ldots + 3^2 = 579$. From the solution to Example 2.23, we know immediately that $(\sum_i y_i)^2 = (47)^2 = 2209$. □

Basic properties of sums may be demonstrated simply by writing out the sums. We will use the following results in the rest of the book:

$$\textstyle\sum_i (x_i + y_i) = \sum_i x_i + \sum_i y_i$$
$$\textstyle\sum_i cy_i = c\sum_i y_i, \text{ where } c \text{ is a constant}$$
$$\textstyle\sum_i c = nc, \text{ where } c \text{ is a constant and } n \text{ is the number of values of } i$$

EXAMPLE 2.25 Find $\sum_i 3y_i$ and $\sum (y_i + 3)$ using the values for y_i in Example 2.23.

Solution The values are 2, 4, 10, 1, 20, 7, and 9. Multiplying each value by 3 gives 6, 12, 30, 3, 60, 21, and 27. Therefore, $\sum_i 3y_i = 6 + 12 + \ldots + 27 = 141$. Alternatively, we found in Example 2.23 that $\sum_i y_i = 47$, so $\sum_i 3y_i = 3(47) = 141$.

Adding 3 to each y_i value yields 5, 7, 13, 4, 23, 10, and 12. So $\sum_i (y_i + 3) = 5 + 7 + \ldots + 12 = 68$. Alternatively, $\sum_i (y_i + 3) = \sum_i y_i + \sum_i 3 = 47 + 7(3) = 68$. □

In the next chapter we deal with several observations from each of several groups, so we need two indexing subscripts; i, the first subscript, indicates the group, and j, the second subscript, indexes individual observations within the group. Thus $y_{2,5}$ indicates the fifth observation in the second group. The comma is often omitted to save typesetters a second level of subscript, so y_{25} should be read as y-two-five, not y-twenty-five. Then $\sum_{i,j} y_{ij}$ should be read as "the sum of the individual values over all individuals and all groups," while $\sum_j y_{ij}$ should be read as "the sum of the individual values over all individuals in group i." For the latter case, there are as many sums as there are groups.

EXAMPLE 2.26 The following y values have been obtained:

Group 1: 13 8 2 6
Group 2: 5 7 10
Group 3: 6 1 8 7 9

What are y_{12} and y_{21}? Find $\sum_j y_{ij}$ and $\sum_{i,j} y_{ij}$.

Solution y_{12} is the second score in the first group—namely 8; y_{21} is the first score in the second group—namely 5. Adding separately in each group, we get the following:

i	$\sum_j y_{ij}$
1	29
2	22
3	31

Therefore, $\sum_{i,j} = 29 + 22 + 31 = 82$. □

Summation notation is merely a device to save words and space. When in doubt, translate the expression into English or write out the sum.

3

A FIRST LOOK AT PROBABILITY

Probability theory is the basis of statistical inference; it is also fundamental in analyzing decision-making problems involving risk or incomplete information. In this chapter we introduce the basic concepts and principles of probability theory. Before treating the technical aspects of the theory, we sketch some alternative interpretations of probability statements. Then we introduce the concepts of sample space, outcome, and event, the mathematical axioms of probability theory, the basic mathematical principles that underlie the more complex computations, and the important idea of statistical independence. Finally, we describe some techniques that can be used to combine the basic mathematical principles in the solution of more complicated problems.

This is an important foundation chapter for our later discussion of statistical inference. The illustrations and examples that we use are relatively simple—too simple to be realistic examples of managerial applications. They do, however, suggest a wide variety of applications.

3.1 INTERPRETATIONS OF PROBABILITY

The theory of probability forms the basis for statistical inference and for decision theory. If 20% of the work force at a textile company have signed union election cards, and if 8 randomly chosen workers are to be fired, then the probability of choosing 8 workers who have signed the cards is very small. Thus, if it turns out that all 8 workers fired by the company had signed union election cards, then we can make the inference that the firings were related to union activity. The probability that a new grocery product will receive supermarket shelf space is obviously a crucial factor in the decision whether or not to introduce the product. The first step in the study of probability is understanding the possible interpretations of probability statements.

The earliest mathematics, and the first interpretation, of probability theory arose from various games of chance. "The probability that a flip of a balanced coin will show heads is 1/2" and "the probability that a card selected at random from a standard deck of 52 cards will be a king is 4/52" are typical examples of this kind of probability statement. The numerical probability values arise from the physical nature of the experiment. A coin flip has only two possible outcomes: heads or tails; the probability that a head occurs should therefore be 1 out of 2. In a standard deck of 52 cards, 4 are kings, so the probability of drawing a king should be 4 out of 52.

classical interpretation

These probability calculations are based on the **classical interpretation** of probability. In this interpretation each distinct possible result of an experiment is called an **outcome**; an **event** is identified with certain of these outcomes. In the card-drawing illustration there are 52 possible outcomes, 4 of which are identified with the event, drawing a king. According to the classical interpretation, the probability of an event E is taken to be the ratio of the number of outcomes favorable to an event N_E to the total number of possible outcomes N, or symbolically,

$$P(\text{Event E}) = \frac{N_E}{N}$$

The usefulness of this interpretation depends completely on the assumption that all possible outcomes are **equally likely**. If that assumption is false—if the coin is loaded or the deck marked, for instance—the classical interpretation does not apply.

EXAMPLE 3.1 An ordinary thumbtack is dropped on a hard surface. It can come to rest point up or on its side. Are the two outcomes equally likely?

Solution There is no reason to assume that the two possible outcomes are equally likely. □

random sample

The classical interpretation has some use, even outside the gambling casino. A **random sample**, by definition, is taken in such a way that any possible sample (of a specific size) has the same probability as any other of being selected. Therefore the outcomes (possible samples) are equally likely, and probabilities can be found by counting favorable outcomes. We use this idea extensively throughout this book.

long-run relative frequency

Situations that do not readily allow a classical interpretation sometimes can be given a **long-run relative frequency** interpretation. If an experiment has been repeated over a huge number of trials, and if 24% of these trials have resulted in a particular event E, then the probability of the event E should be .24, at least to a very good approximation. **Symbolically, if an experiment is repeated over n trials and the event E occurs in n_E trials, the probability of event E is approximately n_E divided by n:**

$$P(\text{event E}) \approx \frac{n_E}{n}$$

EXAMPLE 3.2 Suppose that in the thumbtack-tossing experiment of Example 3.1, it is claimed that the probability of the tack landing point up is .70. Give a long-run relative frequency interpretation that would justify the claim.

Solution The claim could be justified on grounds of long-run relative frequency if a tack had been tossed many times with 70% of the results being point up. □

The long-run relative frequency interpretation of probability is based on observations of a large number of trials; just how large must the number of trials be? As we see in Chapter 8, it is possible to approximate the true probability of an event in a finite number of trials and to assess how good the approximation should be. The relative frequency interpretation is often convenient. We use it whenever it seems sensible to imagine a large number of repeated trials of an experiment.

There are many applications of probability, particularly in management problems, that seem to be "one-shot" situations, those in which it's hard to imagine repeated trials. A product manager who estimates the probability that a new item will receive adequate shelf space is not imagining a long series of trials with this item; it can be introduced as a new product only once. The director of a state welfare agency who estimates the probability that a proposed revision in eligibility rules will pass the state legislature is not imagining a long series of identical proposals; the proposal will be considered only once in the particular legislative session. What then is the meaning of, say, a .6 probability of adequate shelf space or a .3 probability of legislative approval? Such probabilities are **subjective** or **personal** probabilities. One interpretation of such probabilities is that they represent willingness to make certain bets. If someone states that the probability of a certain event is .5, that person regards an even-money bet on the occurrence of that event as a fair bet (and is willing to take either side of the gamble). A subjective probability of .6 for an event means that "lose $6 if the event doesn't occur, win $4 if it does" is regarded as a fair bet. This subjective, betting interpretation of probability is natural in describing the risk-taking questions that confront most managers.

subjective/ personal interpretation

Subjective probability estimates are matters of opinion. Two people might well assign very different probabilities to the same event, even if they have the same information. Such diversity of opinion is seen in economic projections for the next calendar year by various economic and investment advisors. Although many of these individuals work from the same data, they formulate diverse opinions about the most probable economic conditions. Such projections are inherently subjective.

The mathematical laws of probability can help one make logically consistent probability estimates, but they cannot guarantee that those estimates are correct. A good manager will take better risks than a bad one and should do better in the long run. But there is no way to guarantee that one assessment of a particular subjective probability is correct and another incorrect. That, as they say, is what makes horse races!

EXAMPLE 3.3 Give a subjective probability interpretation of the statement "the probability that a thumbtack will land point up is .5."

Solution If you made such a statement, you would be saying that you would take either side of a bet of $1 that the tack would land point up against $1 that it would not. As we suggested in Example 3.1, we would not agree with you. We believe that the tack is more likely to land point up and would prefer that side of the "even-money" bet. □

There are many philosophical arguments about which probability interpretation is most appropriate. We do not need to make a hard and fast choice in this book. The mathematics of probability theory is valid regardless of the interpretation chosen. If a particular statistical procedure is most naturally developed under a particular interpretation, we follow it. Otherwise, you're free to impose whichever interpretation seems most sensible.

SECTION 3.1 EXERCISES

3.1 For each of the following situations, indicate which interpretation of the probability statement seems most appropriate. (In many situations, it is arguable which is the best interpretation.)

a. A new statistics textbook for managers is about to be published. The editor states that the probability that at least enough copies will sell to break even is .8.
b. A small manufacturing firm produces a certain kind of dial for various electrical devices. A critical component of the dial assembly is a certain gear. The probability that a particular gear fails to satisfy tolerances is .002.
c. A random sample of 100 employees is to be taken from the 13,000 employees of a firm. It is known that 55% of the employees are men. As a check on the sampling process, the number of men in the sample will be counted. The probability that there will be 42 or fewer men in the sample is .0061.
d. The probability that the West German inflation rate next year will exceed 4% is .3.
e. The probability that on a given day the demand for coronary-care beds at a local hospital exceeds the normal capacity is .004.

3.2 Give your own subjective probability for each of the following statements. If an entire class does this problem, it might be interesting to tabulate the various probabilities.

a. The Soviet Union will purchase wheat from the United States next year.
b. The next elected president of the United States will be a Democrat.
c. The increase in tuition costs for the major state university in your state will exceed 7% next year.
d. It will rain next week.

3.2 BASIC CONCEPTS AND AXIOMS OF PROBABILITY THEORY

In the previous section we discussed several different interpretations of probability statements. Now we formalize some of the basic definitions and assumptions that enable us to calculate the probability of an event.

Probabilities are defined for a specified **experiment**. The word *experiment* is used in an extremely broad sense to mean any situation that has more than one possible result;

in this usage, experiments are not necessarily conducted under controlled laboratory conditions. An experiment could be

1. recording the total number of hours each of 2200 families spend watching television during a particular week;
2. measuring the sales volume of 194 supermarkets over a one-year period;
3. recording the daily production of an automobile assembly plant for each of 240 working days.

An experiment can be defined by stating all the possible results that might occur. In very small experiments, it is possible to list the possibilities; in most cases, it is necessary to describe but not list the possibilities.

The words *outcomes* and *event* have differing technical meanings in probability theory; they are not synonymous. An outcome is exactly one specific result of an experiment (such as "king of hearts drawn"), whereas an event may include several possibilities (such as "king drawn"). These concepts are formally defined in terms of set theory.

Sample Space *S*

A **sample space *S*** is the set of all distinct possible results of an experiment. An **outcome** is one element of *S*. An **event** is any collection of outcomes, which is a subset of *S*.

An outcome is sometimes called a *simple event* or a *nondecomposable event,* and an event is sometimes called a *compound event.*

EXAMPLE 3.4 A coin is to be flipped three times. Define the sample space by listing all outcomes of the form (result on flip #1, result on flip #2, result on flip #3).

Solution Because on any toss of the coin we can observe either a head (H) or tail (T), the sample space *S* consists of the 8 possible outcomes

$$S = \{(HHH); (HHT); (HTH); (THH); (HTT); (THT); (TTH); (TTT)\}$$

Note that none of the outcomes can be decomposed. □

This same problem can be phrased in more managerial terms as follows. Suppose that intensive audits are performed on the service records of three new car dealers chosen at random from a geographic area of the country. For each company audited, we mark an "H" if there are no unresolved service complaints of more than a two-month duration, and we mark a "T" otherwise. The sample space of Example 3.4 describes the possible outcomes.

EXAMPLE 3.5 In the experiment of Example 3.4, identify these events: A, observe exactly one head; B, observe an odd number of heads; and C, observe no heads.

Solution

$$A = \{(HTT), (THT), (TTH)\}$$
$$B = \{(HTT), (THT), (TTH), (HHH)\}$$
$$C = \{(TTT)\}$$

□

One way to calculate probabilities of events is to assign probabilities first to each outcome. In many cases, the individual outcomes can be assumed to be equally likely; techniques described later in this chapter can also be used to make the assignment. One of the basic principles of probability is that the probability of any event is the sum of the probabilities of all outcomes in the event.

Addition of Outcome Probabilities

If an event A consists of the outcomes $0_1, \ldots, 0_k$, then

$$P(A) = P(0_1) + \ldots + P(0_k)$$

EXAMPLE 3.6 Assume that each of the outcomes in Example 3.4 has probability 1/8. Find the probabilities of the events A, B, and C defined in Example 3.5.

Solution

$$\begin{aligned} P(A) &= P(\text{exactly one H}) \\ &= P(HTT) + P(THT) + P(TTH) = 1/8 + 1/8 + 1/8 = 3/8 \\ P(B) &= P(\text{odd number of H's}) = P(HTT) + P(THT) + P(TTH) + P(HHH) \\ &= 1/8 + 1/8 + 1/8 + 1/8 = 4/8 \\ P(C) &= P(\text{no heads}) = P(TTT) = 1/8 \end{aligned}$$

Note that we could also have taken each probability as

$$\frac{\text{number of outcomes included in the event}}{8}$$

Such classical-interpretation probabilities require equally likely outcomes. □

This approach to calculating probabilities is not always feasible. When there are many possible outcomes with unequal probabilities, it may be excessively difficult to assign probabilities to every outcome. In the rest of this chapter we provide some other methods for calculating probabilities. Regardless of how probabilities are assigned to events, probabilities must satisfy certain mathematical requirements, or axioms.

Probability Axioms

1. For any event A, $0 \leq P(A) \leq 1$.
2. $P(S) = 1$.
3. If the events A and B have no outcomes in common,

$$P(\text{either A or B occurs}) = P(A) + P(B)$$

The first two axioms merely say that probabilities are conventionally taken between 0 and 1, and that a probability of 1 is assigned to the sample space event, which by definition is certain to occur. The third axiom is a generalization of the idea of adding outcome probabilities; as long as events A and B have no outcomes in common, the probability that one or the other occurs is the sum of the separate event probabilities. We build on these ideas in Section 3.3, where we introduce some other ideas for probability calculation.

SECTION 3.2 EXERCISES

3.3 A corporation consists of three divisions, each headed by an executive vice president. Within each division are two groups, each headed by a group vice president. Final decisions about year-end bonuses are made by a committee consisting of one executive vice president and two group vice presidents; membership on the committee each year is determined by a lottery. Define a sample space by listing all possible committees. Designate the executive vice presidents as A, B, and C, and the group vice presidents as 1, 2, 3, 4, 5, and 6.

3.4 In Exercise 3.3, what is the probability that all three committee members belong to the same division? to three different divisions?

3.5 An audit is performed on the receivables of a department store. One hundred accounts are selected at random and verified. Each account is coded 0 (correct) or 1 (erroneous). Describe a typical outcome of the sample space. Should all outcomes be considered equally likely?

3.6 A sample space consists of 6 outcomes with the following probabilities:

Outcome:	1	2	3	4	5	6
Probability:	.25	.20	.20	.15	.15	.05

Event A consists of outcomes 1, 2, 3, and 4; event B consists of outcomes 3, 4, and 5. Find *P*(A), *P*(B), *P*(A and B both occur), and *P*(either A or B, or both, occurs).

3.7 In Exercise 3.6, should *P*(A or B) = *P*(A) + *P*(B)? Why?

3.3 PROBABILITY LAWS

Not all probability problems can be solved by the sample-space approach in which we first list all outcomes, assign reasonable probabilities to those outcomes such that $0 \leq P(0_i) \leq 1$ and $P(S) = 1$, and then compute the probability of any event A by summing the probabilities of the outcomes in A. Sometimes merely listing the outcomes becomes a burdensome chore. For example, suppose a judge has ordered the formation of a panel of arbitrators to hear salary demands by the fire and police union of a large metropolitan area. One individual is to be selected from the list of 14 submitted by the fire and police union, one from the list of 29 submitted by the city, and a third arbitrator from another list of 11 neutrals. If we assume that no arbitrator appears on more than one list, there are a total of 14(29)(11) = 4466 possible outcomes consisting of three arbitrators, one from each list.

In situations such as this, short of listing the entire set of outcomes, we must rely on certain event relationships and probability laws in order to calculate the probability of an event. These concepts are usually stated in the language of set theory, so some preliminary definitions are needed.

Complement, Union, and Intersection

The **complement** of an event A is the set of all outcomes in S that are not included in A; it is denoted as $\bar{A}$ and read as "not A." The **union** of the events A and B is the set of all outcomes that are included in A or in B (or in both); it is denoted as $A \cup B$ and read as "A union B" or "A or B." The **intersection** of the events A and B is the set of all outcomes that are included in both A and B; it is denoted as $A \cap B$ and read as "A intersection B" or "A and B."

These definitions formalize the simplest ideas of logic. The event $\bar{A}$ occurs whenever A does *not* occur, $A \cup B$ occurs whenever A *or* B occurs, and $A \cap B$ occurs whenever A *and* B occur. Often we simply say "not," "or," or "and" in place of the formal complement, union, or intersection.

A handy picture of these concepts is provided by Venn diagrams, as in Figure 3.1. Think of the probability of an event as its area; the whole rectangle, which represents S, has area 1. In Figure 3.1(a), event A is shaded; its complement is the entire white set. In Figure 3.1(b), $A \cup B$ is the entire shaded set, whereas $A \cap B$ is the heavily shaded set.

We have already used the idea of union in the third probability axiom, which stated that *if* events A and B have no outcomes in common the probability that either A or B occurs—that is, $P(A \cup B)$—is $P(A) + P(B)$. The condition that A and B have no outcomes in common is important enough to warrant a name.

Mutually Exclusive Events

Two events A and B are **mutually exclusive** (or disjoint, or logically incompatible) if they have no outcomes in common. For mutually exclusive events, $A \cap B$ contains no outcomes; the occurrence of one event automatically means that the other cannot occur. Events A, B, C, D, . . . are mutually exclusive if all possible pairs of events are mutually exclusive. If one such event occurs, none of the others can occur.

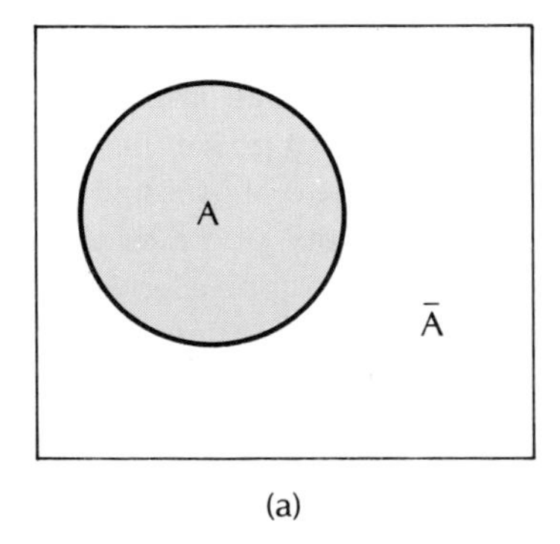

(a)

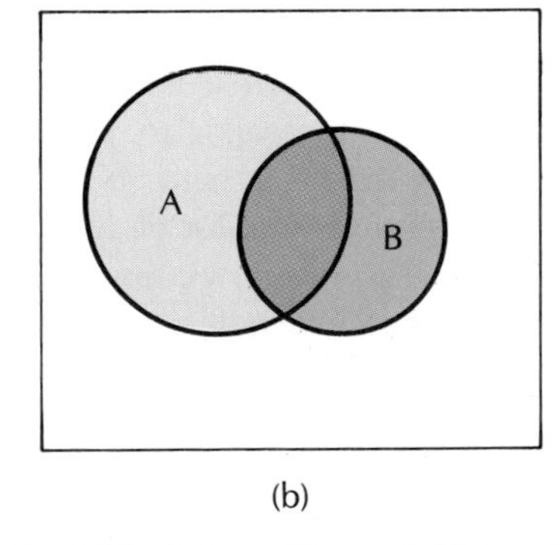

(b)

FIGURE 3.1 **Venn Diagrams Illustrating $P(A)$, $P(\bar{A})$, $P(A \cup B)$, and $P(A \cap B)$**

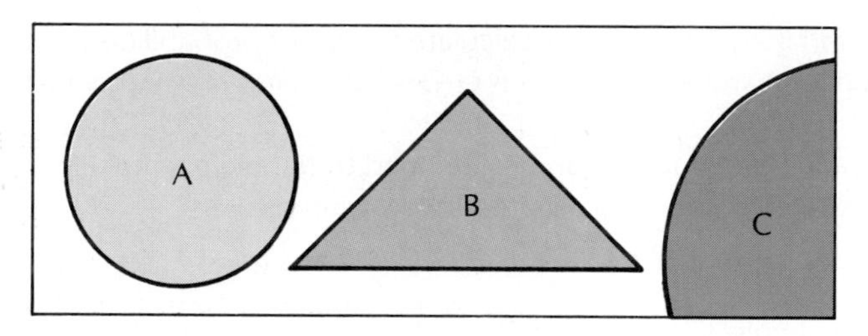

FIGURE 3.2 Three Mutually Exclusive Events

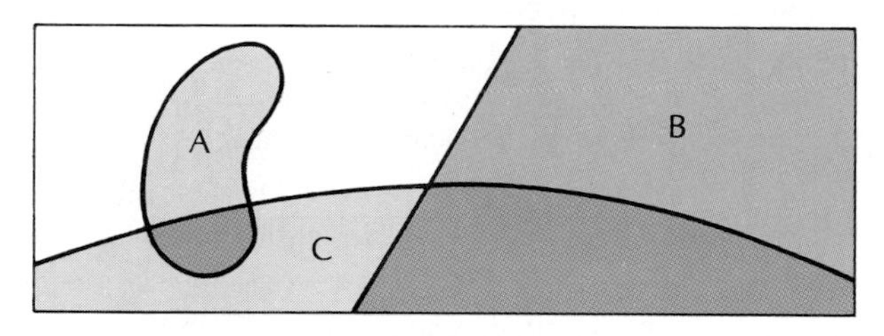

FIGURE 3.3 Venn Diagram for Example 3.7

The third axiom of mathematical probability theory can be translated into a basic law for calculating probabilities, using this language of set theory.

Addition Law for Mutually Exclusive Events

If the events A and B are mutually exclusive,

$$P(\text{A or B occurs}) = P(A \cup B) = P(A) + P(B)$$

The logical "or" corresponds to addition of probabilities, provided that the events have been defined to be mutually exclusive. Of course this idea is not restricted to two events; it applies to any finite or infinite number of events.*

In Figure 3.2, no event intersects any other one, so all three events are mutually exclusive. The area of the shaded event, $P(A \cup B \cup C)$, is obviously just the sum of the three separate areas, $P(A) + P(B) + P(C)$.

EXAMPLE 3.7 Events A, B, and C have respective probabilities .2, .5, and .4. Events A and B are mutually exclusive, but A and C are not mutually exclusive, nor are B and C. Which of the probabilities $P(A \cup B)$, $P(A \cup C)$, $P(B \cup C)$, $P(A \cup B \cup C)$ can be calculated with the given information?

Solution A suitable Venn diagram is shown in Figure 3.3. Because A and B do not have any overlap, $P(A \cup B) = P(A) + P(B) = .2 + .5 = .7$. But since we have no information about the areas

* For mathematical purists: The third axiom can be extended by induction to cover any finite number of events, but it must be restated to cover an infinite number of events.

of $A \cap C$ and $B \cap C$, we cannot calculate the other probabilities. If we erroneously added probabilities, we would get $P(A \cup B \cup C) = .2 + .5 + .4 = 1.1$, which is impossible. □

The addition principle can be extended to handle events that aren't mutually exclusive. The extension can be illustrated by an example.

EXAMPLE 3.8 Assume in Example 3.7 that $P(A \cap C) = .05$ and $P(B \cap C) = .18$. Find $P(A \cup C)$ and $P(B \cup C)$.

Solution When we add $P(A)$ to $P(C)$ we count the area of the intersection, $P(A \cap C)$, twice. To correct for this, we can subtract the intersection area (once). So $P(A \cup C) = .2 + .4 - .05 = .55$. Similarly, $P(B \cup C) = .5 + .4 - .18 = .72$. □

The general addition law follows by the same reasoning.

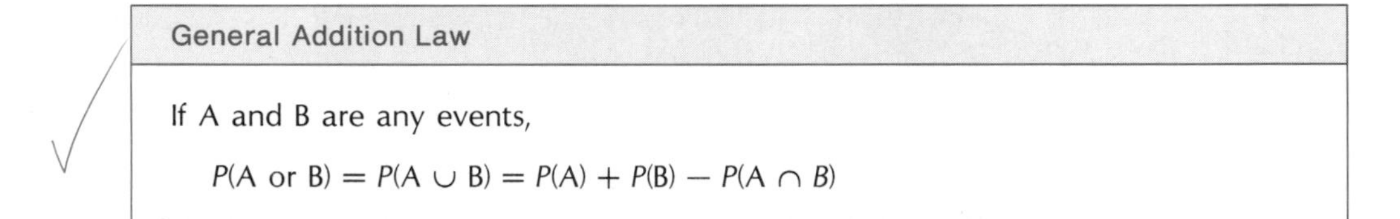

General Addition Law

If A and B are any events,

$$P(A \text{ or } B) = P(A \cup B) = P(A) + P(B) - P(A \cap B)$$

When A and B are mutually exclusive, $A \cap B$ contains no outcomes, so $P(A \cap B) = 0$. In this case, the general addition law reduces to simple addition of probabilities. This law can be extended to the case of many events, but accounting for double-counts, triple-counts, and so on becomes very awkward. Usually, it's better strategy to break up an event into mutually exclusive components, so that simple addition of probabilities can be used.

EXAMPLE 3.9 In Example 3.8, find $P(A \cup B \cup C)$ and P(exactly one of the events A, B, or C occurs).

Solution From the information given in Example 3.8, we can deduce the probabilities shown in Figure 3.4. For $P(A \cup B \cup C)$ we add all the probabilities corresponding to the occurrence of one or more of the events:

$$P(A \cup B \cup C) = .15 + .05 + .17 + .18 + .32 = .87$$

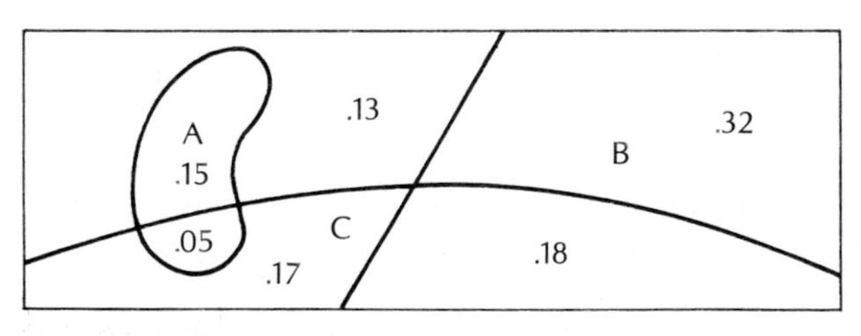

FIGURE 3.4 **Venn Diagram for Example 3.9**

For P(exactly one of A, B, or C), we do not add the intersection probabilities .05 and .18, because the intersection represents the occurrence of two events. So P(exactly one of A, B, or C occurs) = .15 + .17 + .32 = .64. □

A useful probability law follows directly from the addition principle. By definition, an event A and its complement $\bar{A}$ are mutually exclusive; $A \cup \bar{A}$ includes the entire sample space and has probability 1. Therefore,

$$P(A \cup \bar{A}) = P(A) + P(\bar{A}) = 1$$

Complement Law

$$P(A) = 1 - P(\bar{A}) = 1 - P(\text{not } A)$$

It is often easier to find the probability that an event does *not* happen. Then, by the complement law, $P(A)$ is simply $1 - P(\bar{A})$.

EXAMPLE 3.10 In putting together a new microcomputer system, it is observed that 16% of the newly assembled systems have exactly one defect, 4% have exactly two defects, and 1% have three or more defects. What is the probability that a randomly chosen system has no defects?

Solution The event "0 defects" is the complement of the event "1 or more defects": P(0 defects) = 1 − P(1 or more defects). In turn, by the addition principle for mutually exclusive events,

$$\begin{aligned} P(\text{1 or more defects}) &= P(\text{exactly 1}) + P(\text{exactly 2}) + P(\text{3 or more}) \\ &= .16 + .04 + .01 = .21 \end{aligned}$$

Thus P(0 defects) = 1 − .21 = .79. □

One other concept is crucial in developing basic probability laws: the conditional probability of an event B given that another event A has occurred. For example, it may be that 10% of the purchasers of a luxury automobile order special high-performance audio speakers, so P(special speakers) = .1. But if it is known that the customer has ordered the top-of-the-line stereo receiver for the car, the probability that the customer will also order the high-performance speakers is presumably higher. Thus, a condition, in this case that the customer has ordered the special receiver, may change the probability of an event, that the customer will order the high-performance speakers. For another example, suppose that auditors take a random sample from a list of 216 accounts receivable, of which 24 contain errors. The probability that the second account chosen contains an error must be 24/216, if there is no information about the status of the first account. But if it is given that the first account contains an error, then there are only 23 erroneous accounts among the remaining 215; given that the first account contains an error, the probability that the second account also contains an error changes to 23/215. We use the definition of conditional probability in terms of unconditional probabilities heavily in the following chapters.

Definition of Conditional Probability

The conditional probability of event B given that event A occurs denoted by $P(B|A)$ and read as "the probability of B given A," is

$$P(B|A) = \frac{P(A \cap B)}{P(A)}$$

Similarly,

$$P(A|B) = \frac{P(A \cap B)}{P(B)}$$

A conditional probability is formally defined in terms of unconditional probabilities. The definition can be understood in terms of long-run relative frequencies. Imagine a very large number n of trials. We take as given that event A occurs; there will be n_A such trials. The conditional probability is that fraction of the n_A trials in which both A and B occur. There are n_{AB} trials in which A and B both occur, so the conditional probability is

$$\frac{n_{AB}}{n_A}$$

Now divide numerator and denominator of this fraction by n, the total number of trials, to get

$$P(B|A) = \frac{\frac{n_{AB}}{n}}{\frac{n_A}{n}}$$

But the numerator is the unconditional probability $P(A \cap B)$ and the denominator is $P(A)$. Therefore, we get back to the definition,

$$P(B|A) = \frac{P(A \cap B)}{P(A)}$$

EXAMPLE 3.11 Refer to Example 3.4. If event A is "observe 1 or more heads in 3 tosses of a coin and event B is "observe exactly 1 head," compute $P(B|A)$ using the definition of conditional probability.

Solution

A: composed of all outcomes in S except (TTT); hence $P(A) = 7/8$

B: composed of the outcomes (HTT), (THT), and (TTH)

$A \cap B$: composed of the outcomes (HTT), (THT), and (TTH); $P(A \cap B) = 3/8$

The conditional probability of event B given that event A occurs is

$$P(B|A) = \frac{P(A \cap B)}{P(A)} = \frac{3/8}{7/8} = 3/7$$

That is, there are 3 chances in 7 of B occurring given that event A has occurred. □

EXAMPLE 3.12 Suppose that, in the receivables-auditing illustration discussed on p. 51, 16 of the 80 large (over $10,000) accounts contain errors, and that a single account is chosen at random. What is the probability that the account selected is in error, given that it is large?

Solution The answer ought to be 16/80 = .2. According to the definition

$$P(\text{error} \mid \text{large}) = \frac{P(\text{error} \cap \text{large})}{P(\text{large})}$$

There are 216 accounts, 80 of which are large and 16 of which are large and in error. Therefore,

$$P(\text{error} \mid \text{large}) = \frac{16/216}{80/216} = 16/80$$

as it should be. □

The definition of conditional probability leads directly to the multiplication law of probabilities.

Multiplication Law

If A and B are any events,

$$P(\text{A and B}) = P(\text{A} \cap \text{B}) = P(\text{A})P(\text{B} \mid \text{A})$$

Note: By reversing the roles of A and B, $P(\text{A} \cap \text{B}) = P(\text{B})P(\text{A} \mid \text{B})$.

joint probability
marginal probability

The only difference between the multiplication law and the definition of conditional probability lies in which probabilities are assumed and which are to be calculated. When the so-called **joint probability** $P(\text{A} \cap \text{B})$ and the **marginal probability** $P(\text{A})$ are assumed to be known, the conditional probability $P(\text{B} \mid \text{A})$ can be calculated from the definition. When $P(\text{A})$ and $P(\text{B} \mid \text{A})$ are assumed to be known, $P(\text{A} \cap \text{B})$ can be calculated by the multiplication law.

EXAMPLE 3.13 An evaluation team of two people is to be selected from a group of eligible men and women consisting of 10 males and 6 females. If each group of two people has the same chance of being selected, find the probability of selecting an evaluation team composed of two females.

Solution Let A be the event that the first person selected is female and let B be the event that the second person selected is female. Then we want $P(\text{A} \cap \text{B})$. By the multiplication law,

$$P(\text{A} \cap \text{B}) = P(\text{A})P(\text{B} \mid \text{A})$$

It should be clear that $P(\text{A}) = 6/16$ and $P(\text{B} \mid \text{A}) = 5/15$. Substituting, we get

$$P(\text{A} \cap \text{B}) = \left(\frac{6}{16}\right)\left(\frac{5}{15}\right) = .125$$

□

The multiplication law can be extended to deal with the joint probability of three events, or even more:

$$P(A \cap B \cap C) = P(A)P(B|A)P(C|A \cap B)$$

$$P(A \cap B \cap C \cap D) = P(A)P(B|A)P(C|A \cap B)P(D|A \cap B \cap C)$$

and so on.

EXAMPLE 3.14 Suppose that three people are to be selected in Example 3.13. What is the probability that all three are female?

Solution Define A as the first person selected is female, B as the second person selected is female, and C as the third person selected is female. $P(C|A \cap B)$ must be 4/14, so

$$P(A \cap B \cap C) = P(A)P(B|A)P(C|A \cap B)$$

$$= \left(\frac{6}{16}\right)\left(\frac{5}{15}\right)\left(\frac{4}{14}\right) = .036$$

□

SECTION 3.3 EXERCISES

3.8 A manufacturing firm has two emergency generators, either of which can supply sufficient power for basic operations. Each generator is subject to failure. Let A be the event that generator 1 works properly and B the event that generator 2 works properly. Describe each of the following events verbally: $\bar{A}$; $A \cup B$; $A \cap B$; $\bar{A} \cap \bar{B}$. What is the complement of $A \cup B$?

3.9 Assume in Exercise 3.8 that $P(A) = .96$, $P(B) = .94$, and $P(A \cap B) = .93$. Draw a Venn diagram. Find $P(A \cap \bar{B})$, $P(\bar{A} \cap B)$, and $P(\bar{A} \cap \bar{B})$.

3.10 Use the probabilities in Exercise 3.9 to find $P(B|A)$, $P(\bar{B}|A)$, $P(B|\bar{A})$, and $P(\bar{B}|\bar{A})$. Give a verbal statement of each of these probabilities. Does $P(B|A) + P(\bar{B}|A) = 1$? Does $P(B|A) + P(B|\bar{A}) = 1$?

3.11 Are the events A and B in Exercises 3.8 and 3.9 mutually exclusive? Find $P(A \text{ or } B)$.

3.12 The director of a federal agency responsible for housing grants to small communities finds that 14.2% of all applications were past deadline, 8.7% were incomplete, and 15.9% were ineligible. Define A as "past deadline," B as "incomplete," and C as "ineligible." Which pairs, if any, of events are logically mutually exclusive? Interpret the event $A \cap B \cap C$.

3.13 For the situation of Exercise 3.12, suppose $P(A \cap B) = .046$, $P(A \cap C) = .092$, $P(B \cap C) = .035$, and $P(A \cap B \cap C) = .016$. Construct a Venn diagram and deduce probabilities for mutually exclusive events. For instance, since $P(A \cap B) = .046$ and $P(A \cap B \cap C) = .016$, it must be true that $P(A \cap B \cap \bar{C}) = .030$.

3.14 Use the Venn diagram constructed in Exercise 3.13 to find $P(\bar{A} \cap \bar{B} \cap \bar{C})$, $P(A \cup B \cup C)$, and $P(\bar{A} \cap \bar{B})$. Give verbal statements of these events.

3.15 In a manufacturing process, a hole must be drilled in a block of metal to rather precise specifications. A defect in drilling can render the block useless, a condition that can be discovered only after assembly. Experience indicates that 90% of the holes are drilled to specification. Each hole is measured by an inspector. If the hole is not drilled to specification, there is a 90% chance that the inspector detects the defect.

a. What is the probability that a block is defectively drilled and the inspector detects the defect?

b. What is the probability that a block is defectively drilled and the inspector does not detect the defect?

3.16 Suppose in Exercise 3.15 that, with a reallocation of effort, the rate of drilling to specification can be increased to 99%. With this new strategy, however, less time is available for inspection, so that only 80% of defectively drilled blocks are detected by the inspector.

a. What is the probability that a block is defectively drilled and the inspector does not detect the defect?

b. Compare your answer to part (a) with the answer to part (b) of Exercise 3.15. In which case is the probability of an undetected defect lower? Would it matter to your answer if the rate of defect detection was only 40%, rather than 80%?

3.17 Refer again to Exercise 3.15. Suppose that a defect in a drilled block that is not detected by the inspector is detected during assembly with probability .80. What is the probability that a randomly chosen block is defectively drilled and the defect is not detected either by the inspector or at assembly?

3.4 STATISTICAL INDEPENDENCE

statistical independence

The concept of **statistical independence** is fundamental in probability theory and particularly in its statistical applications. Suppose that, in the receivables-auditing example of the previous section, 36 of the 216 accounts are "foreign" and 4 of the 36 foreign accounts are in error. Does the probability of error given a foreign account differ from the overall (unconditional) probability of error? Recall that there were 24 erroneous accounts in the group of 216, so $P(\text{error}) = 24/216 = 1/9$. The conditional probability of error given a foreign account, $P(\text{error}|\text{foreign}) = P(\text{foreign and error})/P(\text{foreign}) = (4/216)/(36/216) = 4/36 = 1/9$, also. Because the conditional probability of error is exactly the same as the unconditional probability, the events, observing a foreign account and observing an account in error, are said to be **statistically independent**.

statistically independent

The idea of statistical independence is that the occurrence of event A does not change the probability that event B occurs. In other words, the conditional probability $P(B|A)$ is the same as the unconditional probability $P(B)$.

Definition of Independent Events

Events A and B are statistically independent if and only if $P(B|A) = P(B)$. Otherwise they are dependent. Note: If A and B are independent, it also follows that $P(A|B) = P(A)$.

Hereafter, we simply say *independent events* and omit the word *statistically*.

EXAMPLE 3.15 Refer to the receivables-auditing illustration. Determine whether the events "first account erroneous" and "second account erroneous" are independent.

Solution Because the receivables sampling is done without replacement, the occurrence of an erroneous account on the first draw (slightly) reduces the probability of an erroneous account on the second. Therefore the events are not independent. We know that P(second is erroneous | first is erroneous) = 23/215 = .107 and that the unconditional probability P(second is erroneous) = 24/216 = .111. The numerical difference in probabilities is very small, so the events are nearly, but not quite, independent. □

EXAMPLE 3.16 Suppose that in a university computing center 192 of 960 jobs are high-priority jobs: 128 of these are submitted by students and 64 by faculty. Of all jobs, 640 are from students and 320 from faculty. If one job is selected at random, are the events "high-priority job" and "student job" independent?

Solution Let A be the event that the job is submitted by a student and B the event that the job is a high-priority job. In order for events A and B to be independent, we must show

$$P(A|B) = P(A)$$

or

$$P(B|A) = P(B)$$

For this example we can compute $P(B|A)$ using the definition of conditional probability:

$$P(B|A) = \frac{P(A \cap B)}{P(A)}$$

$$= \frac{128/960}{640/960} = \frac{128}{640} = .200$$

Also, $P(B) = 192/960 = .200$, so the events A and B are independent. □

The definition of independent events leads to a special case of the multiplication law that applies to independent events.

Multiplication Law for Independent Events

If events A and B are independent,

$$P(A \cap B) = P(A)P(B)$$

This result is an alternate definition of independence.

EXAMPLE 3.17 Use the multiplication law for independent events to verify that the two events of Example 3.16 are independent.

Solution We showed previously that $P(A \cap B) = 128/960 = .133$. Similarly, $P(A)P(B) = (640/960) \times (192/960) = .133$. Because $P(A \cap B) = P(A)P(B)$, events A and B are independent. □

The definition of independence suggests that we should find $P(A \cap B)$, $P(A)$, and $P(B)$ and then check to determine if the events are independent. Sometimes this is in fact the procedure. More often, independence is a natural assumption. For example, sampling with replacement leads naturally to assumed independence. The multiplication principle is then used to calculate joint probabilities such as $P(A \cap B)$.

EXAMPLE 3.18 Suppose that 70% of the teachers in a school district are rated as satisfactory, that 59% are age 40 or more, and that rating and age are assumed to be independent. What is the probability that a randomly chosen teacher is (a) rated satisfactory and over 40; (b) not rated satisfactory and not over 40; (c) not rated satisfactory, given under 40?

Solution a. Because the events "rated satisfactory" and "over 40 years of age" are independent, it follows that

$$\begin{aligned} P(\text{satisfactory} \cap \text{over 40}) &= P(\text{satisfactory})P(\text{over 40}) \\ &= (.70)(.59) = .413 \end{aligned}$$

b. Because the events "rated satisfactory" and "over 40 years of age" are independent, their complements ("not rated satisfactory" and "not over 40") are also independent. Hence,

$$\begin{aligned} P(\text{not satisfactory} \cap \text{not over 40}) &= (1 - .70)(1 - .59) \\ &= .123 \end{aligned}$$

c. $$P(\text{not satisfactory} \mid \text{not over 40}) = \frac{P(\text{not satisfactory} \cap \text{not over 40})}{P(\text{not over 40})}$$

$$= \frac{.123}{.41} = .30$$

This is exactly the probability of the event not "rated satisfactory." □

The multiplication law can be extended to more than two independent events, but to do so, we need the idea of independent processes. Processes (essentially separate sample spaces) are independent if any event from one process is independent of events from all other processes. If, for instance, there are four independent processes and events A, B, C, and D, then the probability of their intersection is

$$P(A \cap B \cap C \cap D) = P(A)P(B)P(C)P(D)$$

EXAMPLE 3.19 Assume that the probability that a buyer of a new automobile orders factory-installed air conditioning is .6 and that the various buyers' decisions are independent processes. What is the probability that the next five buyers all order factory air conditioning?

Solution Let A_1, A_2, A_3, A_4, A_5 be the events that buyers 1, 2, 3, 4, 5 order factory air conditioning. Then

$$\begin{aligned} P(\text{all five order factory air conditioning}) &= P(A_1 \cap A_2 \cap A_3 \cap A_4 \cap A_5) \\ &= P(A_1)P(A_2)P(A_3)P(A_4)P(A_5) \\ &= (.6)(.6)(.6)(.6)(.6) \\ &= .07776 \end{aligned}$$

□

SECTION 3.4 EXERCISES

3.18 A personnel officer for a firm that employs many part-time salespeople tries out a sales-aptitude test on several hundred applicants. Because the test is unproven, results are not used in hiring. Forty percent of applicants show high aptitude on the test and 12% of those hired both show high aptitude and achieve good sales records. The firm's experience shows that 30% of all salespeople achieve good sales. Let A be the event "shows high aptitude" and let B be the event "achieves good sales."

a. Find $P(A)$, $P(A \cap B)$, and $P(B|A)$.

b. Are A and B independent?

c. How useful is the test in predicting good sales achievement?

3.19 Construct a Venn diagram for Exercise 3.18.

a. Find $P(A \cap \bar{B})$ and $P(\bar{B}|A)$.

b. Are A and $\bar{B}$ independent?

3.20 A survey of workers in two plants of a manufacturing firm includes the question "How effective is management in responding to legitimate grievances of workers?" In plant 1, 48 of 192 workers respond "poor"; in plant 2, 80 of 248 workers respond "poor". An employee of the manufacturing firm is to be selected randomly. Let A be the event "worker comes from plant 1" and let B be the event "response is poor."

a. Find $P(A)$, $P(B)$, and $P(A \cap B)$.

b. Are the events A and B independent?

c. Find $P(B|A)$ and $P(B|\bar{A})$. Are they equal?

3.21 Show that if A and B are independent, $P(B|A) = P(B|\bar{A})$. (Hint: B and $\bar{A}$ are also independent.)

3.22 A school district must staff two primary schools and one high school. On any particular day, the probability that no substitute for an absent teacher is needed at primary school 1 is .60; the same probability holds for primary school 2. At the high school, the probability that no substitute is needed is .50. Assume that absenteeism at the three schools defines three independent processes. Find the probability that no substitute is needed at any of the schools on a particular day.

3.23 Do you believe that the assumption of independent processes in Exercise 3.22 is realistic?

3.5 PROBABILITY TABLES AND PROBABILITY TREES

Many probability problems require the successive use of several of the basic principles to obtain a solution. These problems can be solved algebraically, but it is helpful to have some devices to keep the logic straight.

EXAMPLE 3.20 A firm has found that 46% of its junior executives have two-career marriages, 37% have single-career marriages, and 17% are unmarried. The firm estimates that 40% of the two-career marriage executives would refuse a transfer to another office, as would 15% of the single-career marriage executives and 10% of the unmarried executives. If a transfer offer is made to a randomly selected executive, what is the probability that it will be refused?

Solution First, the event "refused" can be thought of as "(refused $\cap$ two-career) $\cup$ (refused $\cap$ single-career) $\cup$ (refused $\cap$ unmarried)." The three possibilities are mutually exclusive, so the addition law yields

$$P(\text{refused}) = P(\text{refused} \cap \text{two-career}) + P(\text{refused} \cap \text{single-career}) + P(\text{refused} \cap \text{unmarried})$$

Second, each of the three joint probabilities can be evaluated by the multiplication law. For instance,

$$P(\text{refused} \cap \text{two-career}) = P(\text{two-career})P(\text{refused} \mid \text{two-career}) = (.46)(.40)$$

Putting the two ideas together, we have

$$\begin{aligned} P(\text{refused}) &= P(\text{two-career})P(\text{refused} \mid \text{two-career}) \\ &\quad + P(\text{single-career})P(\text{refused} \mid \text{single-career}) \\ &\quad + P(\text{unmarried})P(\text{refused} \mid \text{unmarried}) \\ &= (.46)(.40) + (.37)(.15) + (.17)(.10) = .2565 \end{aligned}$$

□

EXAMPLE 3.21 Investments of \$100 each are made in two projects. Project A is assumed to yield a net return of \$8, \$10, or \$12, with respective probabilities .2, .6, and .2. Project B is assumed to yield a net return of \$8, \$10, or \$12, with respective probabilities .3, .4, and .3. The returns from the two projects are assumed to be independent. What is the probability that the total of the two returns is exactly \$20?

Solution By the addition law,

$$\begin{aligned} P(\text{total} = \$20) &= P(\text{A yields } \$8 \cap \text{B yields } \$12) \\ &\quad + P(\text{A yields } \$10 \cap \text{B yields } \$10) \\ &\quad + P(\text{A yields } \$12 \cap \text{B yields } \$8) \end{aligned}$$

The multiplication law for independent events may be applied to each joint probability to obtain

$$\begin{aligned} P(\text{total} = \$20) &= P(\text{A yields } \$8)P(\text{B yields } \$12) \\ &\quad + P(\text{A yields } \$10)P(\text{B yields } \$10) \\ &\quad + P(\text{A yields } \$12)P(\text{B yields } \$8) \\ &= (.2)(.3) + (.6)(.4) + (.2)(.3) = .36 \end{aligned}$$

□

There are no new ideas involved in the solution of such problems, but it is sometimes tricky to find the right order in which to apply the basic principles. With larger, more complicated problems, the difficulty is increased. Several methods have been invented to help clarify the reasoning involved in solving a problem.

One useful approach is to construct a table of joint probabilities. The desired answer can sometimes be found by adding the appropriate table entries.

EXAMPLE 3.22 Construct a joint probability table for marital status versus action on transfer offers for the data of Example 3.20. Use it to find *P*(refused).

Solution First, put any known marginal probabilities on the appropriate margins of the table.

	Two-career	Single-career	Unmarried
Refused			
Accepted			
	.46	.37	.17

Now the body of the table can be filled in using the multiplication law. The remaining marginals can be found by addition.

	Two-career	Single-career	Unmarried	
Refused	(.46)(.40) = .1840	(.37)(.15) = .0555	(.17)(.10) = .0170	.2565
Accepted	(.46)(.60) = .2760	(.37)(.85) = .3145	(.17)(.90) = .1530	.7435
	.46	.37	.17	

P(Refused) is shown, in the right margin, to be .2565, as in Example 3.20. □

EXAMPLE 3.23 Construct a joint probability table and find *P*(total return = \$20) for the data of Example 3.21.

Solution In this case, both sets of marginal probabilities have been specified.

		Return from A \$8	Return from A \$10	Return from A \$12	
Return from B	\$8				.3
	\$10				.4
	\$12				.3
		.2	.6	.2	

The multiplication law for independent events can be used to fill in the body of the table.

		Return from A \$8	Return from A \$10	Return from A \$12	
Return from B	\$8	.06	.18	.06	.3
	\$10	.08	.24	.08	.4
	\$12	.06	.18	.06	.3
		.2	.6	.2	

The entries that correspond to a total return of \$20 are **brown**. The addition law yields

$$P(\text{total return} = \$20) = .06 + .24 + .06 = .36,$$

as in Example 3.21. □

Probability tables are a convenient, compact way of solving many problems. As a by-product, they often yield the solution to related problems as well. You should have no difficulty, for instance, in finding *P*(total return = $22) or *P*(total return = $16) in Example 3.23. For problems involving more than two categories of events, probability tables are at best awkward to use. If there had also been a project C in Example 3.21, some sort of three-dimensional table would have been necessary.

probability tree

Another device that often can be used is a **probability tree**. This method is hard to describe but easy to illustrate.

EXAMPLE 3.24 Use a probability tree to solve Example 3.20.

Solution First, construct branches for a set of events with known marginal probabilities:

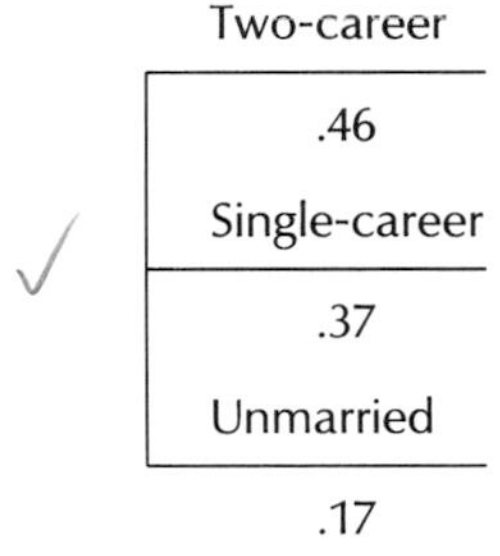

Then, at the tip of each of these branches, construct branches for another set of events, using conditional probabilities (given the appropriate first branch).

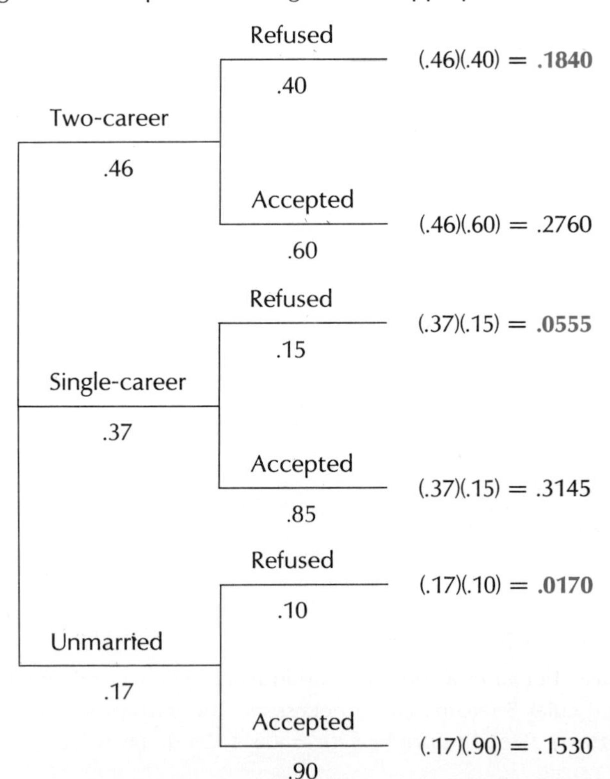

(Had there been another set of relevant events, we would have added another set of branches.) The probability for each specific path (sequence of branches) is found by multiplying the probabilities along that path, as shown. The probability for an event can be found by adding the probabilities of all paths that satisfy that event. The paths corresponding to "refused" are in **brown**: $P(\text{refused}) = .1840 + .0555 + .0170 = .2565$, once again. □

EXAMPLE 3.25 Use a probability tree to solve Example 3.21.

Solution Marginal probabilities are specified for both project A and project B, so we may use the returns to either project for the first set of branches. Out of sheer perversity, we begin by branching on project B.

B returns		A returns		
$8	.3	$8	.2	.06
		$10	.6	.18
		$12	.2	**.06**
$10	.4	$8	.2	.08
		$10	.6	**.24**
		$12	.2	.08
$12	.3	$8	.2	**.06**
		$10	.6	.18
		$12	.2	.06

In this case, because of the assumed independence, conditional probabilities of A returns given particular B returns are unnecessary. The path probabilities for a total return of $20 are in **brown**: $P(\text{total return} = \$20) = .06 + .24 + .06 = .36$. □

To give correct answers, a probability tree must be constructed according to the following rules:

Rules for Constructing a Probability Tree

1. Events forming the first set of branches must have known marginal probabilities, must be mutually exclusive, and must exhaust all possibilities (so that the sum of the branch probabilities is 1).
2. Events forming the second set of branches must be entered at the tip of each of the sets of first branches. Conditional probabilities, given the relevant first branch, must be entered, unless assumed independence allows the use of unconditional probabilities. Again, the branches must be mutually exclusive and exhaustive (so that the sum of the probabilities branching from any one tip is 1).
3. If there are further sets of branches, the probabilities must be conditional on all preceding events. As always, the branches must be mutually exclusive and exhaustive.
4. The sum of path probabilities must be taken over all paths included in the relevant event.

With a little practice, most people find probability trees quite easy to use. Trees and tables are both very useful in clarifying the logic of a solution. Both methods in effect construct appropriate sample spaces; a particular outcome corresponds to a path in a probability tree or an entry in a probability table. Trees can be used in a wider variety of problems. The only difficulty with using a tree for a large, complicated problem is that the tree can become impractically large. As long as one is willing to use a lot of paper, it is possible to solve some rather nasty problems surprisingly quickly.

EXAMPLE 3.26 Suppose that 40% of all theoretically plausible pharmaceutical drug concepts are biologically active. Of the active drugs, 70% show serious side effects. Of those drugs that prove to be inactive, 20% can be reformulated to be active, and among these reformulated drugs, 80% show serious side effects. All drugs that are to be marketed must be approved by a government agency. The probability that a drug will be approved, given that it is biologically active and shows no side effects, is .90. Of drugs that are biologically active but show side effects, 5% will be approved. If a drug is not biologically active, it will not be approved.

a. What is the probability that a new drug concept will result in an approved drug?
b. What is the probability that a new drug concept will lead to a drug with side effects?
c. If a drug is approved, what is the probability that it will lead to side effects?

Solution A probability tree can be constructed. In this case, as often happens, the natural sequence of branches is chronological. The natural first branch reflects whether or not the drug is active. Whether the drug can be reformulated, whether it shows side effects, and finally whether it is approved are then considered.

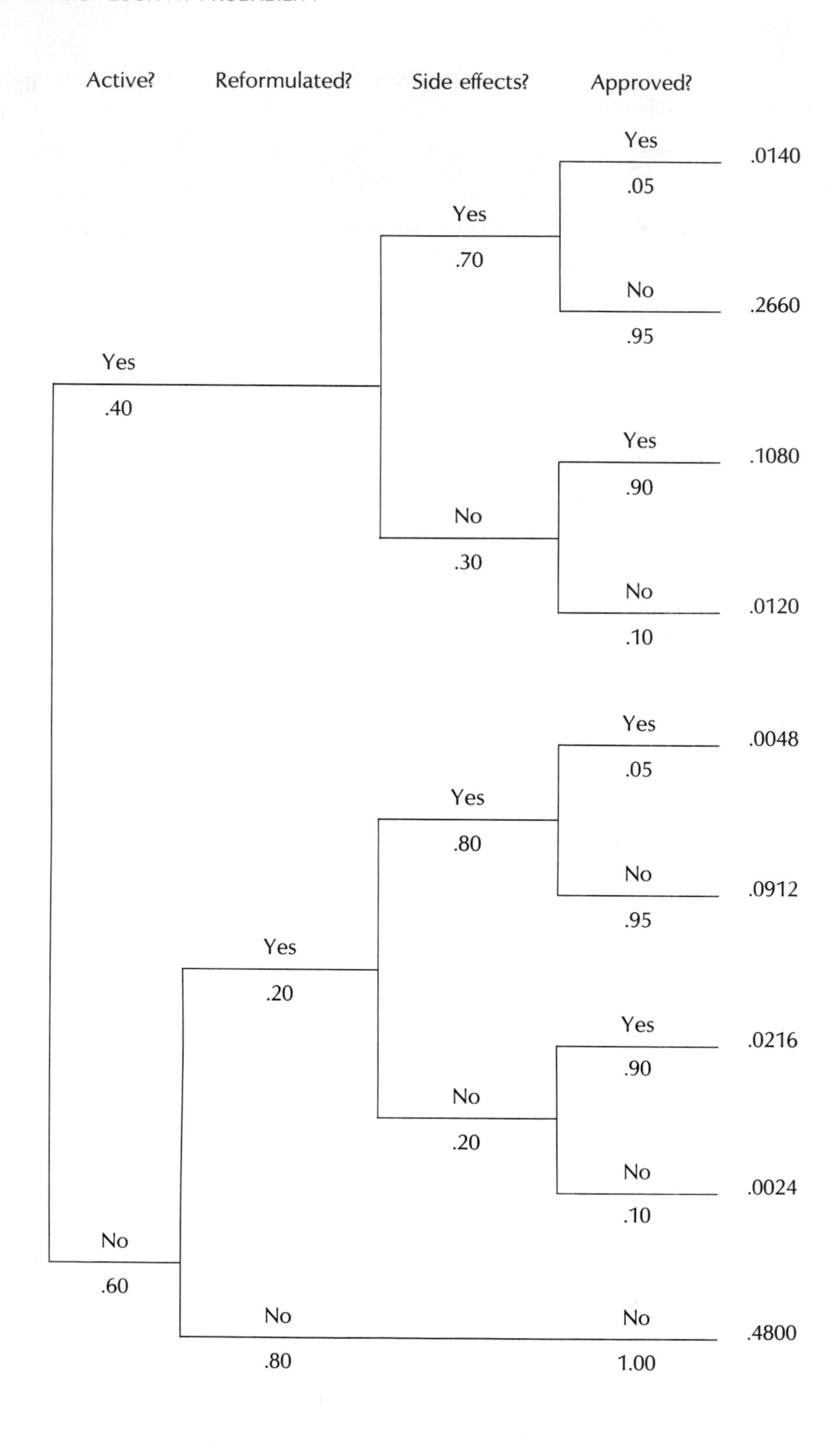
Active?
Reformulated?
Side effects?
Approved?
Yes
.40
No
.60
Yes
.20
No
.80
Yes
.70
No
.30
Yes
.80
No
.20
Yes
.05
No
.95
Yes
.90
No
.10
Yes
.05
No
.95
Yes
.90
No
.10
No
1.00
.0140
.2660
.1080
.0120
.0048
.0912
.0216
.0024
.4800

a. To find the probability that a drug is approved, simply add the probabilities for all paths corresponding to a "yes" branch on the "approved" question:

$$P(\text{approved}) = .0140 + .1080 + .0048 + .0216 = .1484$$

b. Again we must add the appropriate path probabilities. The paths corresponding to existence of side effects are the first, second, fifth, and sixth paths:

$$P(\text{side effects}) = .0140 + .2660 + .0048 + .0912 = .3760$$

Alternatively, we can draw the tree without the "approved" branches and obtain

$$P(\text{side effects}) = .40(.70) + .60(.20)(.80) = .280 + .096 = .376$$

once again.

c. To find a conditional probability like this one, we use the definition of conditional probability:

$$P(\text{side effects} \mid \text{approved}) = \frac{P(\text{side effects} \cap \text{approved})}{P(\text{approved})}$$

In part (a), we found $P(\text{approved}) = .1484$. Note that the first and fifth branches, with respective probabilities .0140 and .0048, are the only ones corresponding to "side effects $\cap$ approved." Thus

$$P(\text{side effects} \mid \text{approved}) = (.0140 + .0048)/(.1484)$$

□

SECTION 3.5 EXERCISES

3.24 The experience of a data-processing firm has shown that the first time a new program is tested there is a .6 probability of finding one or more major "bugs"—flaws in the program that cause the program to fail completely. There is a .3 chance of detecting minor bugs—flaws that allow the program to run but produce erroneous results in certain situations—and a .1 chance that no bugs are detected. In each case, an attempt is made to correct all detected programming errors. Then the program is retested on a more extensive basis. The likely results of the retest are summarized in the following table of conditional probabilities:

		Retest		
		Major	Minor	None
First Test	Major	.3	.5	.2
	Minor	.1	.3	.6
	None	0	.2	.8

a. Construct a table giving the joint probabilities of all the possible combinations of first test and retest results.

b. Find the probability that major bugs are still found at the retest.

c. Find the probabilities of minor bugs at retest and of no bugs.

3.25 Construct a probability tree to answer Exercise 3.21.

3.26 In the data-processing firm of Exercise 3.21, programs that still show major or minor bugs on retesting are sent through one more round of correction. Programs that had major bugs in retest have a .1 probability of retaining major bugs in the third test (regardless of the result of the initial test) and a .2 probability of showing minor bugs. Those that showed only minor bugs at retesting have essentially no chance of showing major bugs at third test but a .1 chance of showing minor bugs (again regardless of the result of the first test). It is assumed that programs showing no bugs at retesting need not go through a third round.

a. Construct a probability tree for this situation.
b. Find the probability that a program will show major bugs at all three tests.
c. Find the probability that a program will show major bugs at the third test. Why is this answer different from that of part (b)?
d. Find the probability that a program will achieve no-bug status (whether after two or three tests).

3.27 A purchasing unit for a state government has found that 60% of the winning bids for office-cleaning contracts come from regular bidders, 30% from occasional bidders, and 10% from first-time bidders. The services provided by successful bidders are rated satisfactory or unsatisfactory after one year on the job. Experience indicates that 90% of the jobs done by regular bidders are satisfactory, as are 80% of the jobs done by occasional bidders and 60% of the jobs done by first-time bidders.

a. What is the probability that a job will be done by a first-time bidder and will be satisfactory?
b. What is the probability that a job will be satisfactory?
c. Given that a job is satisfactory, what is the probability that it was done by a first-time bidder?

SUMMARY

In this chapter we developed the basic concepts of probability theory. Probability provides the basis for all statistical inference by allowing the experimenter to assess the chance or likelihood of various outcomes; based on these probabilities, an inference can be made. To see this, consider the following situation. A candidate for reelection to mayor of a community claims victory in the upcoming election. Not willing to take the candidate's word, we decide to conduct a small survey of registered voters. Of the 40 persons contacted, none favors the incumbent. If in fact the incumbent will win, we would expect at least half of the potential voters to favor the incumbent, so the probability on any given trial of selecting a person favoring the incumbent is .5 (or more). If the events are independent, that is, if one person's response does not affect how another person surveyed will respond, then the probability of none favoring the incumbent in a sample of 40 is approximately

$$P(\text{none}) = (.5)^{40} \approx .00000000000091$$

Notice we have used probability theory to calculate the probability of this event occurring assuming the incumbent wins. The statistical inference based on this probability is that the incumbent will not win. Although it's not impossible to observe 0 out of 40 favoring the incumbent assuming he or she wins, it is highly unlikely (improbable).

The details related to statistical inferences are discussed in later chapters. Chapter 3 helps provide the foundation we need in basic probability distributions.

Key topics and formulas from this chapter are presented next.

KEY TOPICS AND FORMULAS: A First Look at Probability

1. Interpretations of probability
   a. Classical interpretation. $P(\text{Event E}) = \dfrac{N_E}{N}$
   b. Long-run relative frequency: $P(\text{Event E}) = \dfrac{n_E}{n}$
   c. Subjective or personal probability
2. Basic concepts
   a. Experiment: any situation that has more than one possible result
   b. Sample space, S: the set of all possible results for an experiment
   c. Outcome: an element of the sample space
   d. Event: a collection of outcomes
3. Basic axioms of probability for any event A
   a. $0 \leq P(A) \leq 1$
   b. $P(S) = 1$
   c. If two events A and B have no outcomes in common, $P(\text{A or B}) = P(A) + P(B)$.
4. Event relations and probability laws
   a. Complement of an event A: the set of all outcomes in S that are not in A
   b. Union, $A \cup B$: the set of all outcomes either in A or B or in both
   c. Intersection, $A \cap B$: the set of outcomes in both A and B
   d. Mutually exclusive events: A and B are mutually exclusive if $A \cap B$ contains no outcomes.
   e. Addition law: $P(A \cup B) = P(A) + P(B) - P(A \cap B)$
   f. Complement law: $P(A) = 1 - P(\bar{A})$
   g. Multiplication law: $P(A \cap B) = P(A)P(B|A) = P(B)P(A|B)$
   h. Statistical independence: A and B are independent if $P(A|B) = P(A)$, or equivalently, if $P(B|A) = P(B)$, or if $P(A \cap B) = P(A)P(B)$.

CHAPTER 3 EXERCISES

3.28 Airlines often accept tickets bought for other airlines' flights to the same destination. Suppose that final accounting and settlement of all such tickets is made yearly, and that approximate monthly settlements are made on the basis of random samples of the month's accumulated tickets. Airline A draws a monthly sample of 60 tickets, which may have been bought from airlines B, C, or D. Indicate what a typical outcome of this experiment would be. Should all the outcomes be regarded as equally likely?

3.29 Suppose that in the packaged-cereals industry, 29% of all vice presidents hold MBA degrees, 24% hold undergraduate business degrees, and 8% hold both. A vice president is to be selected at random.

a. Construct a Venn diagram for this situation.

b. What is the probability that the vice president holds either an MBA or an undergraduate business degree (or both)?

c. What is the probability that the vice president holds neither degree?

3.30 In Exercise 3.29, what is the probability that the vice president holds one degree or the other, but not both?

3.31 Suppose that the records of an automobile maker show that, for a certain compact car model, 50% of all customers order air conditioning, 49% order power steering, and 26% order both. An order is selected randomly.

a. Draw a Venn diagram for this situation.

b. What is the probability that air conditioning is ordered but power steering is not?

c. What is the probability that neither option is ordered?

3.32 In Exercise 3.31, suppose that 68% of all customers order automatic transmissions, 19% order automatic transmissions and power steering without air conditioning, 13% order automatic transmissions and air conditioning without power steering, and 21% order all three options.

a. Construct a Venn diagram for this situation.

b. What is the probability that at least one of the options is ordered?

c. What is the probability that exactly one option is ordered?

3.33 Use the data of Exercises 3.31 and 3.32 to find $P(\text{automatic transmission} \cap \text{air conditioning})$. Are these events independent?

3.34 Proponents of the random walk theory of stock prices hold that predictions of whether a particular stock will do better or worse than the market in the short run (say, over a one-month period) are no better than what could be obtained by flipping a fair coin. Suppose that a securities analyst selects eight stocks that are predicted to beat the market in the next month.

a. What is the probability that all eight stocks do beat the market, assuming the validity of the random walk theory?

b. State the assumptions you made in answering part (a).

3.35 Refer to Exercise 3.34 and assume that the random walk theory of stock behavior is valid. Suppose that each of 100 different analysts selects eight stocks.

a. What is the probability that no analyst gets eight winners?

b. What is the probability that at least one analyst gets eight winners?

3.36 A paperback book seller estimates the following probabilities for the weekly sales of a particular historical romance:

Sales:	10	20	30	40
Probability:	.40	.30	.20	.10

Assume independence of sales from week to week.

a. Construct a probability table for the joint probabilities of various sales levels in week 1 and week 2.

b. Find the probability that the average sales level per week (over a two-week period) is 25.

3.37 Do you believe that the independence assumption made in Exercise 3.36 is reasonable?

3.38 A purchasing department finds that 75% of its special orders are received on time. Of those orders that are on time, 80% meet specifications completely; of those orders that are late, 60% meet specifications.

a. Find the probability that an order is on time and meets specifications.
b. Construct a probability table or tree for this situation.
c. Find the probability that an order meets specifications.

3.39 For the situation of Exercise 3.38, suppose that four orders are placed.
a. Find the probability that all four orders meet specifications.
b. State what assumptions you made in answering part (a).

3.40 A large credit card company finds that 50% of all cardholders pay a given monthly bill in full.
a. Suppose two cardholders are chosen at random. What is the probability that both pay a monthly bill in full? (The number of cardholders is so large that you need not worry about whether the choice is made with replacement or without.)
b. Suppose a cardholder is chosen at random. What is the probability that the holder pays both of two consecutive monthly bills in full?
c. What did you assume in answering parts (a) and (b)? Does the assumption seem unreasonable in either case?

3.41 A more detailed examination of the records of the credit card company in Exercise 3.40 shows that 90% of the customers who pay one monthly bill in full also pay the next monthly bill in full; only 10% of the customers who pay less than the full amount of one monthly bill pay the next monthly bill in full.
a. Find the probability that a randomly chosen customer pays two consecutive monthly bills in full.
b. Find the probability that a randomly chosen customer pays neither of two consecutive monthly bills in full.
c. Find the probability that a randomly chosen customer pays exactly one of two consecutive monthly bills in full.

3.42 In Exercise 3.41, if a randomly chosen customer pays the second monthly bill in full, what is the probability that that customer also pays the first monthly bill in full.

3.43 Records of a men's clothing shop show that alterations are required for 40% of the suit jackets that are bought and for 30% of the suit trousers. Alterations are required for both jacket and trousers in 22% of the purchases.
a. Find the probability that no alterations are required in a randomly chosen purchase. You may want to draw a Venn diagram.
b. Find the probability that alterations are required for either the jacket or the trousers, but not for both.

3.44 Are the events in Exercise 3.43, "alteration to jacket" and "alteration to trousers," independent?

3.45 Suppose that in Exercise 3.43 a customer purchases two suits made by different manufacturers.
a. What is the probability that both suit jackets require alterations?
b. What did you assume in answering part (a)? Is the assumption reasonable?

3.46 A newly hired security guard is given a ring with eight keys for offices in a particular suite. One of the keys is a master key that fits all locks; the remaining seven fit only one door apiece—the main suite door and the six office doors in the suite. The guard does not know the code on the keys and therefore tries out keys at random, discarding the ones that do not fit.
a. What is the probability that the first key the guard tries will open the main suite door?
b. What is the probability that the guard will open the main door with one of the first three keys tried?

3.47 If the guard in Exercise 3.46 opens the suite door on the second try, what is the probability that the key used is in fact the master key? You may find a probability tree helpful; if so, you will want to distinguish between choosing the master key and choosing the main suite key.

REVIEW EXERCISES CHAPTERS 2–3

These exercises are intended to help you check your understanding of the topics in chapters just completed. The problems are *not* in any particular order, so you can't tell how to do a problem by its location.

R1 Samples of car door locks from four different suppliers are tested to determine the number of times that the locks can be operated before they fail. The data, in thousands, are

Supplier	Operations before failure									
A	24.7	19.8	22.0	37.6	21.8	25.4	20.6	48.7	23.9	22.6
B	26.8	25.7	39.7	25.8	28.0	52.4	29.4	31.1	26.0	28.4
C	15.3	35.7	18.2	15.3	21.0	19.9	42.6	21.1	18.9	19.7
D	31.4	21.2	24.5	22.0	26.7	61.0	22.6	23.5	25.0	22.6

a. Summarize the data separately for each supplier. Be sure to discuss average, variability, and skewness.

b. Can the Empirical Rule be expected to work well for these data? Why or why not?

R2 Find the mean and variance of the combined data in Exercise R1. Can the mean be found directly from the supplier means in Exercise R1? Can the variance be found directly from the R1 variances?

R3 Prices posted on the shelves of a supermarket do not always match the correct current price of the item, because of errors in posting price changes. Suppose that over time 60% of the price changes are increases and 40% are decreases. Also suppose that 93% of the price increases are posted correctly, as are 98% of the price decreases. If a price change is not posted correctly, what is the probability that the change is a decrease?

R4 A study of small savings and loan associations yielded the following financial information:

Deposits ($000,000)	Capital ($000,000)	Reserves ($000,000)	Bad Debts (percent of portfolio)	Type of Bank (1 = savings, 2 = joint S & L, 3 = stock S & L)
3.68	1.14	0.97	1.62	2
11.64	4.03	3.28	0.97	1
31.62	10.63	9.22	2.00	3
2.62	0.85	0.53	3.97	2
1.97	0.61	0.79	0.75	1
15.21	5.21	3.77	1.11	3
3.88	0.65	1.10	1.77	2
5.01	1.00	1.15	0.32	1
7.53	1.16	3.02	4.31	3
3.67	0.89	0.92	1.12	2

a. Calculate means and standard deviations for all relevant variables.

b. Are there any outliers in any of the variables?

R5 A coal-burning electric generator occasionally is improperly stoked and emits unacceptable amounts of various gases. In the long run, this problem occurs in 1% of the generator's operating time. An air sample is taken and analyzed every hour. The analysis is not a perfect indicator of gas emissions. Calibration tests indicate that if the generator is emitting acceptable levels of the gases, the test shows excess emissions 4% of the time, borderline emissions 5% of the time, and acceptable emissions 91% of the time. If the generator is emitting excessive amounts of the gases, the test shows excess emissions 92% of the time, borderline emissions 5% of the time, and acceptable emissions 3% of the time. If the test indicates excess emissions, what is the probability that the generator is in fact emitting unacceptable amounts of the gases?

R6 Show that "borderline emissions in the test" and "unacceptable emissions by the generator" are independent events in Exercise R5.

R7 A supermarket chain does a study of the effectiveness of its own coupons in inducing additional sales in the meat department. The data from the preliminary pilot study were

X_1 Cents Off	X_2 Current Price (Cents)	X_3 Type of Meat	X_4 Normal Sales	X_5 Sales in Coupon Week
29	379	1	37,000	42,000
19	109	2	67,200	79,900
50	399	1	21,200	32,500
25	199	5	11,600	12,900
59	209	4	18,800	22,800
100	379	1	37,000	51,300
20	109	2	67,200	83,100
40	229	3	12,000	13,200
79	399	1	21,200	36,000
50	209	4	18,800	20,100
29	109	2	67,200	83,900
30	379	1	37,000	40,900
50	229	3	12,000	14,100

a. Calculate the mean and standard deviation of X_3.
b. What is the interpretation of the numbers determined in (a)?

R8 Calculate means, medians, and standard deviations of X_4, X_5, and $Y = X_5 - X_4$ in Exercise R7. Is there a simple relation among the means? Does the same relation hold for the medians? for the standard deviations?

R9 Calculate the skewness value of Y as defined in Exercise R8. Does the resulting number confirm your visual impression of the skewness of Y?

R10 Experience indicates that about 10% of new television shows place in the top third of all shows in audience ratings during the first year. About 40% place in the middle third and about 50% place in the bottom third. Of new shows placing in the top third, only 2% are cancelled; 40% of shows placing in the middle third are cancelled, as are 85% of shows placing in the bottom third.
a. Find the probability that a randomly selected new show will be cancelled.
b. If a show is cancelled, what is the probability that it was in the bottom third?

R11 In Exercise R10, are rating and cancellation assumed to be independent? What would independence mean in this context?

R12 Junior managers in a firm are rated by their bosses in terms of current performance and managerial potential. The current performance ratings are 18% excellent, 71% satisfactory, and 11% unsatisfactory. The managerial potential ratings are 24% definite, 40% possible, and 36% unlikely.

a. Find the probability that a randomly selected junior manager will be rated "excellent" on the performance scale and "definite" on the potential scale.

b. What did you assume in answering part (a)? Are the assumptions reasonable? If not, is the probability you calculated likely to be too low or too high?

R13 Records for a sample of employees in a large firm indicate the following distribution of claimed deductions on W-4 tax withholding forms.

Deductions:	0	1	2	3	4	5	6
	7	8	9	10	11	12	
Frequency:	201	287	364	332	151	97	52
	28	11	5	2	0	3	

a. Find the mean number of deductions.

b. Find the standard deviation. Does it make much difference if the data are regarded as a sample rather than a population?

c. How well does the Empirical Rule work for data within one standard deviation of the mean?

R14 A W-4 form is drawn at random from the data of Exercise R13.

a. What is the probability that it claims at least one deduction?

b. If the form claims at least one deduction, what is the probability that it claims at most three?

R15 Data are collected on the total compensation (salary plus bonuses) of samples of men and women junior managers in a firm. The data (in thousands of dollars per year) were

Men:	39.6	28.9	35.4	36.8	33.7	32.8	35.1	36.7	38.4	35.7	33.1
	31.6	34.7	33.8	36.2	34.9	35.7	40.2	36.5	37.4	35.2	36.6
Women:	34.2	31.8	32.7	27.6	33.0	38.1	33.0	31.5	29.8	31.8	44.7
	22.5	30.0	34.3	31.0	32.5						

a. From the looks of the data, should the mean and median compensation for men be similar? Calculate them.

b. Show that the smaller group has a larger range than the other. Explain what causes this phenomenon.

R16 Construct box plots for both sets of data in Exercise R15. Include a check for outliers.

R17 Calculate the mean and median compensation for the combined sample of managers in Exercise R15. How do these values relate to the means and medians of the two groups separately?

R18 Data from an automobile manufacturer indicate that, of all cars repaired under warranty, 57% require engine work, 47% require interior work, and 30% require exterior work. Also, 23% require both engine and interior work, 7% both engine and exterior work, and 13% both interior and exterior work; 5% require all three types of work. There are some cars that require other types of work.

a. Find the probability that a car repaired under warranty requires engine work but no interior or exterior work.

b. Find the probability that a car requires exactly one of the three types of work.

c. Are the events "engine work" and "interior work" independent?

R19 A cereal manufacturer collects samples of the time required for workers to clean out the manufacturing line when switching from production of one cereal to production of another.

The data, in actual worker-hours expended, were

Previous Flour Base	Time									
Corn	10.0	11.0	11.5	9.5	10.0	12.5	8.5	9.0	10.0	10.5
	11.5	13.0	9.5	16.5	14.5	11.0	10.5	10.0	11.0	15.0
Oats	13.5	11.0	10.0	11.5	12.0	10.5	11.0	16.5	13.0	19.0
	12.5	17.0	11.0	13.5	12.0	11.0	13.5	15.0		
Wheat	28.0	31.0	33.0	35.0	30.0	28.5	27.5	26.5	32.0	24.0
	30.5	32.0	31.5	40.5	31.0	33.0	30.5	33.0	28.5	47.5
	31.0	33.5	35.0	33.5	30.0	36.5	39.5	29.0	30.5	

a. Draw appropriate plots of the three sets of times. What is the general shape of the data?
b. Calculate the means and medians for the three sets of times. Does the relation between the resulting means and medians confirm your judgment about the shapes?

R20 Calculate the mean and median for the combined samples in Exercise R19. Is this a reasonable summary figure for a typical cleaning time?

4

RANDOM VARIABLES AND PROBABILITY DISTRIBUTIONS

The probability ideas of Chapter 3 can be specialized and extended to deal with quantitative outcomes of an experiment. In this chapter we introduce the concepts of (quantitative) random variable and probability distribution, which are fundamental in statistical inference and decision theory. With these ideas, we make connections with the definitions of mean, variance, and standard deviation of Chapter 2, and with the definition of independence of Chapter 3. Finally, in an appendix we develop some of the relevant mathematics for means, variances, and standard deviations.

4.1 RANDOM VARIABLE: THE BASIC IDEA

The basic language of probability developed in the preceding chapter deals with many different kinds of events. We are interested in calculating the probabilities associated with both quantitative and qualitative events. For example, we developed techniques that could be used to calculate the probability that a person selected at random for a Nielsen survey of television viewing habits would favor the ABC nightly news program (as opposed to that of CBS or NBC). These same techniques are also applicable to finding the probability that a person selected for the Nielsen survey watches television more than 30 hours per week.

These qualitative and quantitative events can be classified as events (or outcomes) associated with qualitative and quantitative **variables**. For example, in the Nielsen survey, responses to the question "Which evening television news program do you prefer: ABC, CBS, or NBC?" are observations on a qualitative variable, since the possible responses vary in kind but not in any numerical degree. Since we cannot predict with certainty what a particular person's response will be, the variable is classified as a **qualitative random variable**. Other examples of qualitative random variables that are commonly measured are political party affiliation, socioeconomic status, geographic location, and sex/race classification.

qualitative random variable

There are a finite (and typically quite small) number of possible outcomes associated with any qualitative variable. Using the methods of Chapter 3, it is possible to calculate the probabilities associated with these events.

Many times the events of interest in an experiment are quantitative outcomes associated with a **quantitative random variable**, since the possible responses vary in numerical magnitude. For example, in a Nielsen survey, responses to the question "How many hours a week do you watch television?" are observations on a quantitative random variable. Events of interest, such as viewing television more than 30 hours per week, are measured by this quantitative random variable. Other examples of quantitative random variables are the change in earnings per share over the next year, the increase in total sales over the next year, and the number of persons voting for the incumbent in an upcoming election. Again, the methods of Chapter 3 can be applied to calculate the probability associated with any particular event.

quantitative random variable

There are major advantages to dealing with quantitative random variables. The numerical yardstick underlying a quantitative variable makes the mean and standard deviation (for instance) sensible. With qualitative random variables, there isn't much more to be said than has already been said; the methods of Chapter 3 can be used to calculate the probabilities of various events, and that's about all. With quantitative random variables we can do much more: we can average the resulting quantities, find standard deviations, and assess probable errors, among other things. Hereafter, we use the phrase *random variable* to mean *quantitative random variable*; virtually all texts on probability theory use the phrase this way.

The first task is to define the idea of random variable more carefully. So far, we have suggested that a random variable is any numerical (quantitative) measurement associated with an experiment. The crucial information about a random variable is the specification of possible values and their respective probabilities. In order to carry out these ideas, we must apply the probability apparatus developed in Chapter 3. For this purpose, we give the following definition:

Definition of a Random Variable
Given a sample space S, a random variable is a rule (function) that assigns a numerical value to each outcome in S.

The probability associated with each value of a random variable is found by adding the probabilities for all outcomes that are assigned that value. According to this formal definition, we must specify a sample space to define a random variable. If $Y =$ number of heads in three flips of a fair coin, we have the following sample space:

Outcome	HHH	HHT	HTH	THH	HTT	THT	TTH	TTT
Probability	1/8	1/8	1/8	1/8	1/8	1/8	1/8	1/8
Value assigned by Y	3	2	2	2	1	1	1	0

Then, for instance,

$$P(Y = 2) = P(\text{HHT}) + P(\text{HTH}) + P(\text{THH}) = 1/8 + 1/8 + 1/8 = 3/8$$

In practice we don't need to follow the formal definition too rigidly. Stating the values and probabilities for a random variable implicitly defines a sample space; namely, the values themselves. A perfectly valid sample space for the coin-flipping situation is $S = \{0, 1, 2, 3\}$, with the same 1/8, 3/8, 3/8, 1/8 probabilities assumed. There's no logical need to do more than specify possible values and their probabilities unless it is convenient to list all the outcomes first.

The custom is to denote random variables by capital letters at the end of the alphabet; thus we might define $X =$ number of heads observed in three flips of a coin and $Y =$ number of Theater Guild subscribers in a random sample of 200 persons. Possible values of a random variable are usually denoted by the corresponding lower-case letter; we would say that x could be 0, 1, 2, or 3 and y could be 0, 1, 2, . . . , 200. The subtle distinction between Y, the random variable itself, and y, one of its possible values, is important; it becomes clear with practice.

EXAMPLE 4.1 Suppose that a random sample of two persons is to be selected from a large population consisting of 30% Theater Guild subscribers and 70% nonsubscribers.

a. List the outcomes that make up the sample space.
b. Assign probabilities.
c. Define the quantitative random variable Y as the number of Theater Guild subscribers in the sample. Specify the possible values that the random variable may assume and determine the probability of each.

Solution a. If we let S designate a subscriber and N a nonsubscriber, then the possible outcomes for the two persons sampled are

$$S = \{(S, S); (S, N); (N, S) \text{ and } (N, N)\}$$

b. From the statement of the problem we know that $P(S) = .3$ and $P(N) = .7$. Under the assumption that the outcomes for the two persons sampled are independent, we have the following probabilities associated with the four outcomes.

$$P(S, S) = (.3)^2 = .09$$
$$P(S, N) = (.3)(.7) = .21$$
$$P(N, S) = (.7)(.3) = .21$$
$$P(N, N) = (.7)^2 = .49$$
$$1.00$$

c. If the random variable Y is the number of subscribers in a sample of two from the populations of interest, then the possible values for Y are 0, 1, and 2. The probabilities associated with these values can be determined from probabilities for the outcomes that make up each numerical event.

Outcome	Probability	y	$P(y)$
(N, N)	.49	0	.49
(N, S)	.21	1	.42
(S, N)	.21	1	
(S, S)	.09	2	.09

□

discrete and continuous random variables

The random variables we have considered so far have been **discrete**: their possible values have been distinct and separate, like 0 or 1 or 2 or 3. Other random variables are most usefully considered to be **continuous**: their possible values form a whole interval (or range, or continuum). For instance, the one-year return per dollar invested in a common stock could range from 0 to something quite large. In practice, virtually all random variables assume a discrete set of values; the return per dollar of a million-dollar common-stock investment could be 1.06219423 or 1.06219424 or 1.06219425 or . . . . But, when there are many, many possible values for a random variable, it is sometimes mathematically useful to treat that random variable as continuous. In fact, one of the most important theoretical probability specifications—the bell-shaped, so-called normal distribution—formally applies only to continuous random variables. In Section 4.2 we define some language and notation for discrete random variables. In Section 4.3 we extend these ideas to continuous random variables.

4.2 PROBABILITY DISTRIBUTIONS FOR DISCRETE RANDOM VARIABLES

probability distribution

The **probability distribution** for a discrete random variable Y is a function $P_Y(y)$ that assigns a probability to each value y of the random variable Y. **The probability distribution for Y can be expressed as a formula, a graph, or a table.**

The properties of the probability distribution for a discrete random variable Y are listed here.

Properties of the Probability Distribution for a Discrete Random Variable Y

1. The probability $P_Y(y)$ associated with each value of Y must lie in the interval

   $0 \leq P_Y(y) \leq 1$

2. The sum of the probabilities for all values of Y equals 1.

   $\sum_{\text{all } y} P_Y(y) = 1$

3. Because different values of Y are mutually exclusive events, their probabilities are additive. Thus

   $P(Y = a \text{ or } Y = b) = P_Y(a) + P_Y(b)$

For the random variable Y = number of heads in three tosses of a fair coin, we might define $P_Y(y)$ by a table, as follows:

y	0	1	2	3
$P_Y(y)$	1/8	3/8	3/8	1/8

Or we might use the formula

$$P_Y(y) = \frac{3!}{y!(3 - y)!}\left(\frac{1}{8}\right)$$

where in general $k! = k(k - 1)(k - 2) \ldots (1)$ and $0! = 1$ by convention. Substituting $y = 0, 1, 2,$ and 3 into the formula yields the same probabilities as those listed in the previous table:

y	0	1	2	3
$P_Y(y)$	$\frac{3 \cdot 2 \cdot 1}{(1)(3 \cdot 2 \cdot 1)} \frac{1}{8} = \frac{1}{8}$	$\frac{3 \cdot 2 \cdot 1}{(1)(2 \cdot 1)} \frac{1}{8} = \frac{3}{8}$	$\frac{3 \cdot 2 \cdot 1}{(2 \cdot 1)(1)} \frac{1}{8} = \frac{3}{8}$	$\frac{3 \cdot 2 \cdot 1}{(3 \cdot 2 \cdot 1)(1)} \frac{1}{8} = \frac{1}{8}$

probability histogram

cumulative distribution function

A graph of this probability distribution, called a **probability histogram**, is shown in Figure 4.1. The discrete random variable Y is the number of heads in three tosses of a fair coin.

The **cumulative distribution function** is another function that is particularly appropriate when calculating probabilities and has applications in the simulation methods to be discussed in Section 6.6. In general, the cumulative distribution function F_Y for a discrete random variable Y is a function that specifies the probability that $Y \leq y$ for all values of y. Thus

$$F_Y(y) = P(Y \leq y) = P_Y(0) + P_Y(1) + \ldots + P_Y(y)$$

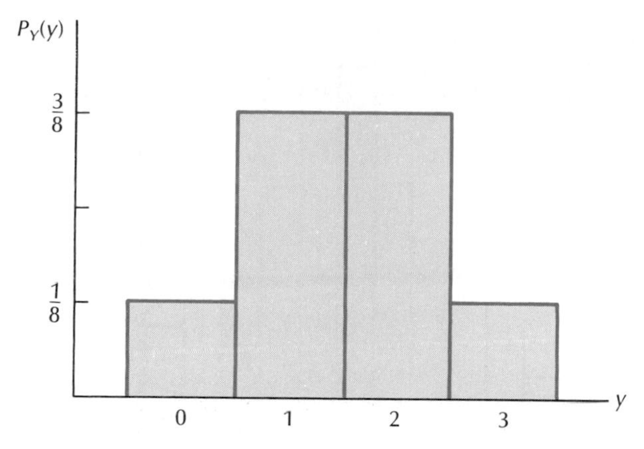

FIGURE 4.1 Graph of $P_Y(y)$ for the Coin-tossing Experiment

This can be illustrated for the coin-tossing example discussed previously:

y	0	1	2	3
$P_Y(y)$	1/8	3/8	3/8	1/8
$F_Y(y)$	1/8	4/8	7/8	8/8

As the name suggests and these data illustrate, the cumulative distribution function at a particular value y sums all probabilities for $Y \leq y$. For example,

$$F_Y(2) = P(Y \leq 2) = \frac{1}{8} + \frac{3}{8} + \frac{3}{8} = \frac{7}{8}$$

and

$$F_Y(3) = P(Y \leq 3) = 1$$

The cumulative distribution function (abbreviated cdf) is often used in constructing probability tables, so that the table user does not have to add up many table entries to find a certain probability. As an illustration, suppose that a large metropolitan teaching hospital has data on the number of acute coronary cases Y arriving at the hospital in a given day. The cdf is tabulated as follows:

y	0	1	2	3	4	5	6	7	8
$F_Y(y)$	.001	.003	.006	.011	.024	.061	.139	.224	.336

y	9	10	11	12	13	14	15	16	17
$F_Y(y)$	.510	.672	.782	.870	.925	.964	.988	.997	1.000

Suppose that the hospital has 14 coronary-care beds available at the beginning of a particular day. The probability that the number of new cases Y is less than or equal to 14 can be read directly from the table as .964. It's almost as easy to find the probability that

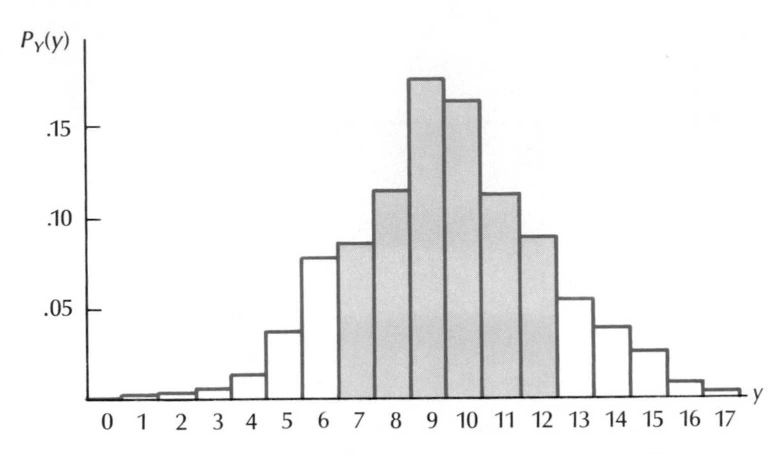

FIGURE 4.2 **Probability Histogram for the Coronary-care Illustration**

Y is 15 or more; $P(Y \geq 15) = 1 - P(Y \leq 14) = 1 - .964 = .036$. Had the table been stated in terms of individual probabilities $P(y)$, it would have been necessary to add up many entries to find these probabilities.

General use of cdf tables is easy enough if you draw a probability histogram. A probability histogram for the coronary-care illustration is shown in Figure 4.2; the probability $P_Y(y)$ of each particular value y is indicated by the height of the rectangle erected atop that y value.

For example, suppose we want $P(7 \leq Y \leq 12)$. We want the total area of the rectangles above $y = 7, 8, 9, 10, 11$, and 12, which are shaded in Figure 4.2. $F_Y(12)$ is the total area of all the rectangles above $y = 0, 1, \ldots, 12$; to find $P(7 \leq Y \leq 12)$ we must subtract the area of the rectangles above $y = 0, 1, 2, 3, 4, 5$, and 6, namely, $F_Y(6)$, from $F_Y(12)$:

$$P(7 \leq Y \leq 12) = F_Y(12) - F_Y(6) = .870 - .139 = .731$$

Generally, it's useful to draw a probability histogram whenever you want to use tables to calculate probabilities.

EXAMPLE 4.2 Suppose that a cosmetic company plans to market a new perfume. The product manager has assessed the following subjective probabilities for the first-year sales (denoted by X) in millions of bottles:

x	0	1	2	3	4	5	6	7	8
$F_X(x)$	.05	.20	.40	.60	.75	.85	.90	.95	1.00

Find the following probabilities, as assessed by the product manager:

a. $P(X \geq 5)$
b. $P(2 \leq X \leq 4)$
c. $P(X \leq 1)$

Solution A probability histogram for this example is shown in Figure 4.3. Areas relevant to each problem are indicated by a, b, or c.

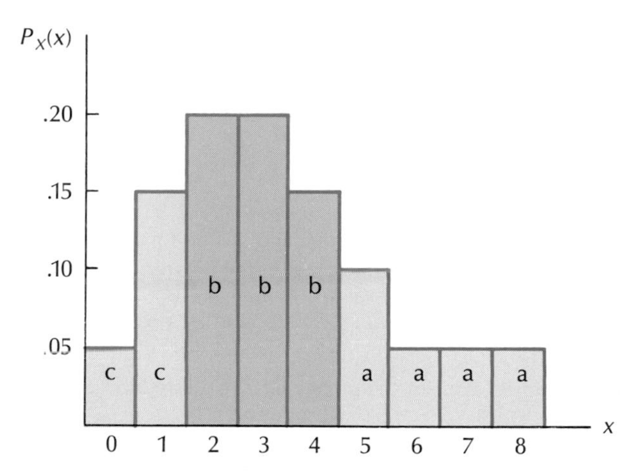

FIGURE 4.3 **Probability Histogram for Example 4.2**

a. $P(X \geq 5) = 1 - P(X \leq 4) = 1.00 - .75 = .25$

The area of all the rectangles is 1.00. We must subtract the areas of all rectangles through $x = 4$.

b. Subtract the areas for $x = 0, 1$ from the areas for $x = 0, 1, 2, 3, 4$ to get $P(2 \leq X \leq 4)$;

$$P(2 \leq X \leq 4) = F_X(4) - F_X(1) = .75 - .20 = .55$$

c. By definition, $P(X \leq 1) = F_X(1) = .20$; no subtraction is needed.

SECTIONS 4.1 AND 4.2 EXERCISES

4.1 The personnel department of a large firm employs five men and three women as college recruiters. For visits to certain large campuses, a team of two people is sent. Suppose that two of the eight recruiters are chosen at random. Let Y = numbeer of women selected.
a. Construct the sample space; call the recruiters A, B, . . . , H.
b. Find the value of Y for each outcome in the sample space.

4.2 In Exercise 4.1, find $P_Y(y)$ by counting. Construct a probability histogram.

4.3 Find the cdf of Y in Exercise 4.1. Plot $F_Y(y)$ against y.

4.4 An appliance store has the following probabilities for Y = number of major appliances sold on a given day:

y:	0	1	2	3	4	5	6	7	8	9	10
$P_Y(y)$:	.100	.150	.250	.140	.090	.080	.060	.050	.040	.025	.015

a. Construct a probability histogram.
b. Find $P(Y \leq 2)$.
c. Find $P(Y \geq 7)$.
d. Find $P(1 \leq Y \leq 5)$.

4.5 Calculate the cdf corresponding to $P_Y(y)$ in Exercise 4.4. Use this cdf to find $P(Y \leq 2)$, $P(Y \geq 7)$, and $P(1 \leq Y \leq 5)$.

4.6 The weekly demand X for copies of a popular word-processing program at a computer store has the probability distribution shown here.

x:	0	1	2	3	4	5	6	7	8	9	10
$P_X(x)$:	.06	.14	.16	.14	.12	.10	.08	.07	.06	.04	.03

a. What is the probability that three or more copies of the program will be demanded in a particular week?
b. What is the probability that the demand will be for at least two but no more than six?
c. The store policy is to have eight copies of the program available at the beginning of every week. What is the probability that the demand will exceed the supply in a given week?

4.7 a. Find the cumulative distribution function (cdf) $F_X(x)$ for the probability distribution shown in Exercise 4.6.
b. Use the cdf to recalculate the probabilities requested in Exercise 4.6.

4.3 PROBABILITY DISTRIBUTIONS FOR CONTINUOUS RANDOM VARIABLES (∂, $\int$)

In Section 4.2 we distinguished between discrete random variables, which can assume only distinct, separate values, and continuous random variables, which can assume (for all practical purposes) a complete range of values along some interval. In this section we develop the basic concepts and notation that apply to continuous random variables.

To illustrate, suppose that a U.S. resident is to be chosen at random according to that person's nine-digit Social Security number. Define $Y =$ the Social Security number chosen. Literally speaking, Y is a discrete random variable that can take on any one of the billion possible values 000–00–0000 to 999–99–9999. We are not overly eager to specify one billion different probabilities, so for practical reasons we regard Y as a continuous random variable that can assume all possible values between 0 and 1 billion.

It seems plausible to assume that Y probabilities are **uniform**; no one value is more likely than any other. Suppose we construct a probability histogram. First we consider only the first digit of the Social Security number drawn. Based on the assumption of

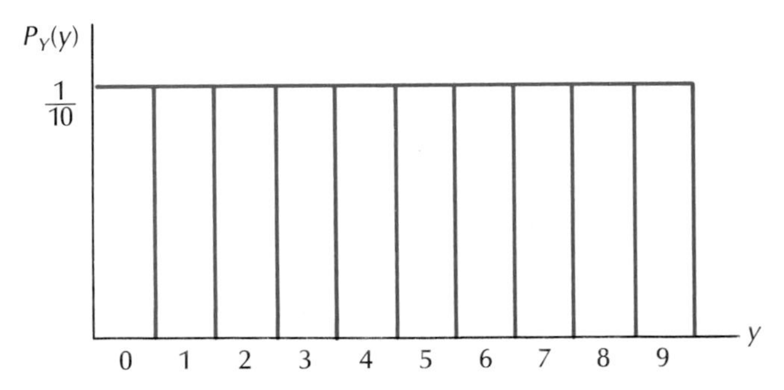

FIGURE 4.4 Uniform Probabilities: First Digit

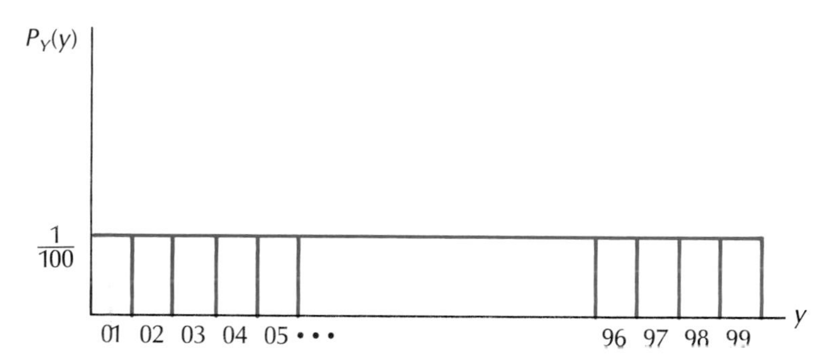

FIGURE 4.5 **Uniform Probabilities: First Two Digits**

uniform probability, the histogram should assign equal probabilities to all rectangles, as in Figure 4.4. If we had considered the first two digits, we would have a 100-rectangle histogram, as suggested in Figure 4.5.

As we refine this process—considering the first three digits, then the first four, and so on, we get more and more, thinner and thinner rectangles. Very soon (and in the limit, mathematically) the rectangles disappear into a continuous blur.

EXAMPLE 4.3 Suppose that a personnel manager measures Y, the actual weekly work time of supermarket employees. Construct histograms that indicate the probability distribution of Y when measurements are made

a. to the nearest hour;
b. to the nearest 10 minutes;
c. to the nearest second.

Solution a. Assume a nominal 40-hour week with modest overtime. The nearest-hour histogram might look like Figure 4.6a.

b. The nearest 10-minutes histogram might look like Figure 4.6b.

c. For all practical purposes, the nearest-second histogram would look like Figure 4.6c. □

Probability histograms were introduced in Section 4.2 when we defined the cumulative distribution function (cdf) F. The cdf idea works just as well with continuous random variables. For a continuous random variable Y, the **cumulative distribution function** is defined as before:

cumulative distribution function

$$F_Y(y) = P(Y \leq y)$$

For any particular example of a continuous random variable, the cdf is almost inevitably defined by a formula. For example, suppose that a file-transfer computer program sends lines of a program across a noisy transmission device. One important variable is $X =$ the proportion of lines correctly transmitted. Suppose that, as a model, the cdf is assumed

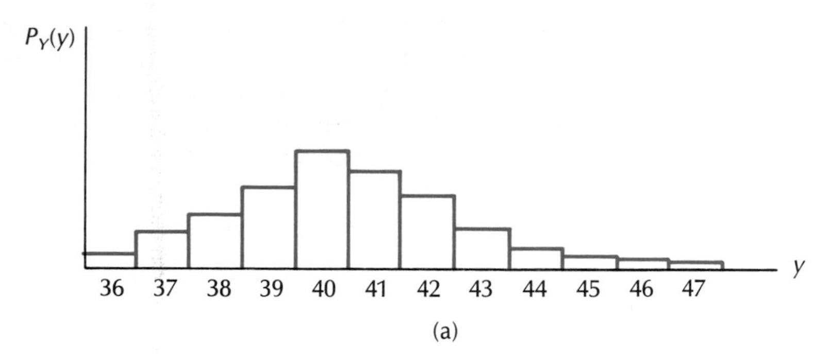

(a)

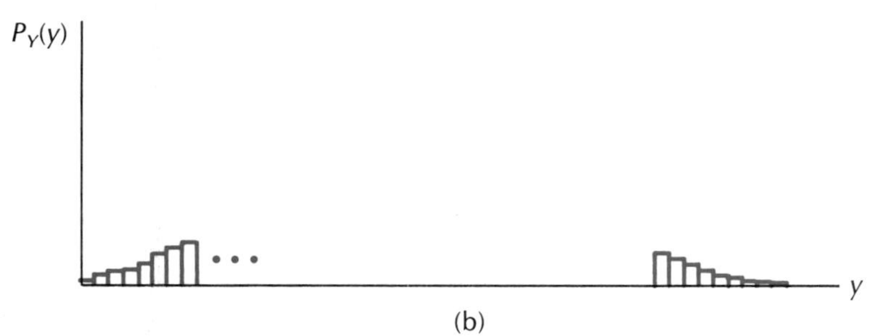

(b)

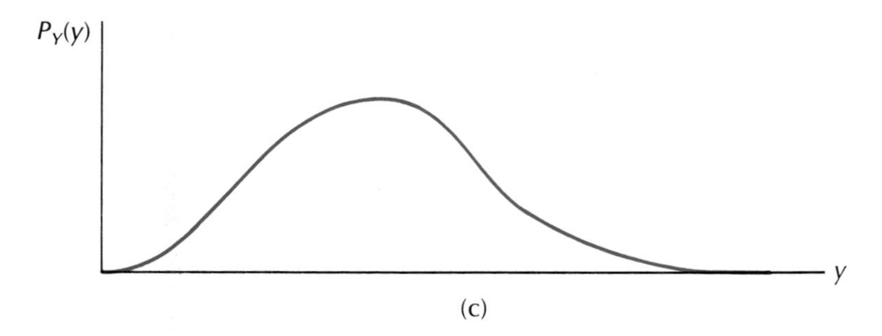

(c)

FIGURE 4.6

to be

$$F_X(x) = 21x^{20} - 20x^{21}, \qquad \text{for } 0 < x < 1$$

Then the probability that the proportion of correctly transmitted lines is less than .9 is

$$F_X(.9) = 21(.9)^{20} - 20(.9)^{21} = .3647$$

and the probability that the proportion is greater than .9 is

$$1 - F_X(.9) = 1 - .3647 = .6353$$

Furthermore, the probability that the proportion X is between .7 and .9 is

$$F_X(.9) - F_X(.7) = .3647 - .0056 = .3591$$

Recalling the calculations made for discrete random variables, you might expect that in computing $P(.7 \leq X \leq .9)$ we would subtract $F_X(.6)$ or perhaps $F_X(.69)$ rather than $F_X(.7)$. But the probability that the continuous random variable X equals exactly .7000 . . . is negligibly small; as a mathematical idealization, we may take the probability to be zero. Thus, in the continuous case, we may ignore the probability that the random variable is "right on the boundary."

EXAMPLE 4.4 Suppose that the reservations manager for a large airline assumes that the time T (measured in minutes) between successive phone calls to the reservation center is a continuous random variable with cdf

$$F_T(t) = 1 - e^{-2t}, \qquad \text{for } t \geq 0$$

where

$e \approx 2.7183$, the base of the natural logarithm

Find

a. $P(T \geq 5)$
b. $P(2 \leq T \leq 4)$
c. $P(T \leq 1)$

Solution The three parts of this example seem to be identical to those of Example 4.2. Because T is continuous, however, the solution procedure differs from that of Example 4.2.

a. $P(T \geq 5) = 1 - P(T < 5)$

But, because T is a continuous random variable, $P(T = 5.000000\ldots)$ is assumed to be zero and $P(T < 5) = P(T \leq 5)$.

$$\begin{aligned} P(T \geq 5) &= 1 - P(T \leq 5) = 1 - F_T(5) \\ &= 1 - (1 - e^{-2(5)}) = .0000454 \end{aligned}$$

(Values of e^x can be calculated by most hand-held calculators or obtained from tables.)

b. $P(2 \leq T \leq 4) = P(T \leq 4) - P(T < 2)$

The event $T = 2.000\ldots$ has probability zero, so

$$\begin{aligned} P(2 \leq T \leq 4) &= P(T \leq 4) - P(T \leq 2) \\ &= F_T(4) - F_T(2) \\ &= (1 - e^{-2(4)}) - (1 - e^{-2(2)}) \\ &= .0180 \end{aligned}$$

c. $P(T \leq 1) = F_T(1)$, by definition
$= 1 - e^{-2(1)} = .865$ □

The cumulative distribution function F means the same thing for discrete and continuous random variables. For any random variable Y, $F_Y(y) = P(Y \leq y)$. For continuous random variables, another function, the probability density function, is widely used. For a random variable Y, the probability density function is denoted $f_Y(y)$. It is roughly analogous to the probability distribution $P_X(x)$ defined for discrete random variables in that it measures how the probability is spread out—distributed—over the range of possible values of the random variable. But for a continuous random variable Y, the probability that Y exactly equals a particular number is zero. The probability density function does not yield probabilities directly. Instead this function defines a smooth curve; probability is calculated as area under the curve, using integral calculus. If both the cdf $F_Y(y)$ and the probability density function $f_Y(y)$ are known, we can compute the probability that Y is between numbers a and b in two ways.

$$P(a \leq Y \leq b) = F_Y(b) - F_Y(a)$$

or

$$P(a \leq Y \leq b) = \int_a^b f_Y(y)\,dy$$

In the example in which $X =$ the proportion of lines correctly transmitted, it can be shown that

$$f_X(x) = 21(20)x^{19}(1 - x), \qquad 0 < x < 1$$

The probability that X is larger than .9 can be computed by integrating the probability density over the region $.9 < x < 1$, because X cannot be larger than 1.

$$\begin{aligned} P(.9 < X) &= \int_{.9}^{1} 21(20)x^{19}(1 - x)\,dx \\ &= \int_{.9}^{1} 21(20)(x^{19} - x^{20})\,dx \\ &= (21x^{20} - 20x^{21})\Big|_{.9}^{1} \\ &= 1 - .3647 = .6353 \end{aligned}$$

as we found previously.

EXAMPLE 4.5 The probability density function for the random variable T of Example 4.4 can be shown to have probability density

$$f_T(t) = 2e^{-2t}, \qquad t \geq 0$$

Calculate the probability that T is between 2 and 4 using this probability density.

Solution To solve this problem, we need the elementary calculus result that the indefinite integral of ce^{-ct} is $-e^{-ct}$. Then

$$\begin{aligned} P(2 \leq T \leq 4) &= \int_2^4 2e^{-2t}\,dt = -e^{-2t}\Big|_2^4 \\ &= -.000335 - (-.018316) = .0180 \end{aligned}$$

as in Example 4.4. The calculation is shown in Figure 4.7. □

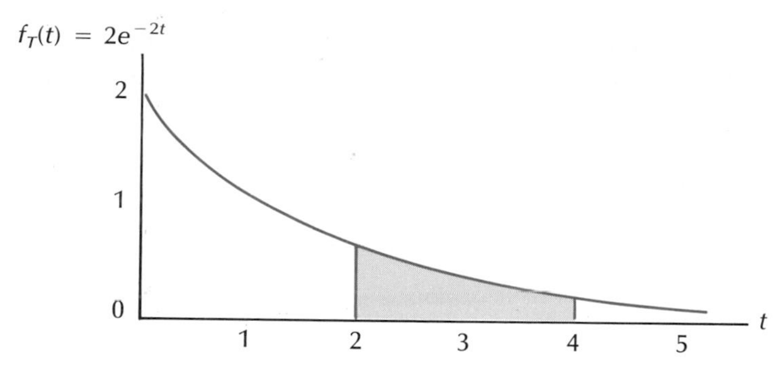

FIGURE 4.7 **Area (Probability) Found in Example 4.5**

The probability density function $f_Y(y)$ of a continuous random variable Y may be either specified directly or derived from the cdf $F_Y(y)$. Because the calculus operation of integration is the opposite of the operation of differentiation, it follows that

$$f_Y(y) = \frac{d}{dy} F_Y(y)$$

For example, we initially specified the cdf of X = the proportion of correctly transmitted lines as

$$F_X(x) = 21x^{20} - 20x^{21}, \qquad 0 < x < 1$$

Differentiation yields the expression that we used for the probability density function:

$$f_X(x) = \frac{d}{dx}(21x^{20} - 20x^{21}) = 21(20)x^{19} - 21(20)x^{20}$$

$$= 21(20)x^{19}(1 - x)$$

as previously indicated.

The process of finding the density function from the cdf may be reversed. Just as in the discrete case, where

$$F_X(x) = \sum_{x' \leq x} P_X(x')$$

in the continuous case

$$F_X(x) = \int_{-\infty}^{x} f_X(x')\, dx'$$

In performing the integration, one must be a bit careful to exclude regions where the random variable cannot occur and thus has a probability density of zero. If a random variable X is by definition nonnegative, the probability that it is less than zero (that is, $\int_{-\infty}^{0} f_X(x)\, dx$ is zero.

EXAMPLE 4.6 Show that the probability density specified in Example 4.5 yields the cdf of Example 4.4.

Solution We note first that T, as an elapsed time measure, cannot be negative. Therefore the probability density function of T must be zero for all values $t < 0$. The probability density of T is stated in Example 4.5 to be $f_T(t) = 2e^{-2t}$, for $t > 0$. Thus

$$F_T(t) = \int_0^t 2e^{-2t'}\,dt' = -e^{-2t'}\Big|_0^t = -e^{-2t} - (-1)$$
$$= 1 - e^{-2t}$$

as stated in Example 4.4. □

Generally speaking, every calculation of a discrete random variable involving a summation has an analogous calculation of a continuous random variable involving an integration. The technical difficulties in performing a particular summation or integration should not obscure the fact that the operations of summation and integration are direct analogues.

SECTION 4.3 EXERCISES

4.8 A certain charity is planning a direct mail campaign. The fraction Y of nonrespondents is taken to be a continuous random variable with the following cdf:

$$F_Y(y) = 5y^4 - 4y^5, \qquad 0 \leq y \leq 1$$

a. Calculate $F_Y(y)$ for various values of y between 0 and 1. Sketch the function $F_Y(y)$.
b. Use the graph of $F_Y(y)$ to calculate $P(Y \leq .8)$, $P(Y \geq .6)$, and $P(.5 \leq Y \leq .9)$.

4.9 A data-processing company owns a large computer. Access to the computer is gained via a large number of remote terminals. A reasonable probability model for the time Y (in minutes) between successive job submissions to the computer assumes that

$$F_Y(y) = 1 - e^{-.5y}, \qquad 0 \leq y < \infty$$

a. Calculate numerical values of $F_Y(y)$ for $y = 1.0, 2.0, \ldots,$ until $F_Y(y)$ exceeds .98 or so. Graph $F_Y(y)$ versus y.
b. Use the graph of $F_Y(y)$ to find $P(Y \leq .75)$, $P(Y \geq 4.0)$, and $P(2.0 \leq Y \leq 3.5)$.

∫ **4.10** a. Show that the probability density of Y in Exercise 4.8 is

$$f_Y(y) = 20(y^3 - y^4), \qquad 0 \leq y \leq 1$$

b. Use this density to calculate $P(.2 \leq Y \leq .7)$, $P(Y \leq .6)$, and $P(Y \geq .5)$.
c. Use the cdf $F_Y(y)$ to find the probabilities indicated in part (b).

∫ **4.11** A securities analyst summarizes some subjective probability estimates for the after-tax profit per share Y of a particular stock in the following continuous probability density:

$$f_Y(y) = \frac{4}{27}(9y - 6y^2 + y^3), \qquad 0 \leq y \leq 3$$

a. Calculate $f_Y(y)$ for several y values (such as 0, .25, .50, ...) and graph $f_Y(y)$ against y.
b. Find $P(Y \leq 1.50)$, $P(Y \geq 2.00)$, and $P(1 \leq Y \leq 2.50)$.
c. Find $F_Y(y)$ and sketch it.

∫ **4.12** The "back office" of a brokerage house owns a mainframe computer on which it processes all records of transactions. Excess computer time is sold to other companies. To determine how much time to sell, the firm has studied the distribution of Y = the computer time (in minutes) needed daily to process its own transactions. To a good approximation, the density of Y is

$$f_Y(y) = .0009375[40 - .1(y - 100)^2], \qquad \text{for } 80 < y < 120$$
$$= 0, \qquad \text{outside } 80 < y < 120$$

If the firm reserves 115 minutes of time daily, what is the probability that the actually required time will be longer?

$\int$ **4.13** Calculate the cdf $F_Y(y)$ for the density in Exercise 4.12. Use the cdf to recalculate the probability in that exercise.

$\int$ **4.14** The time in worker-hours required to assemble a complex manufactured item is random, with probability density

$$f_Y(y) = 3y^{-4}, \qquad \text{for } 1 < y < \infty$$

a. Find the probability that an item will require between two and four worker-hours to assemble. How important is it whether 2.00 and 4.00 hours are included or excluded from the interval?
b. Find the probability that an item will require between 0.5 and 1.5 worker hours. (Think before integrating.)

$\int$ **4.15** a. Find the cdf $F_Y(y)$ corresponding to the density in Exercise 4.14.
b. Find the 99th percentile of times; that is, solve $F_Y(y) = .99$.

4.4 EXPECTED VALUE, VARIANCE, AND STANDARD DEVIATION: DISCRETE RANDOM VARIABLES

In Sections 4.2 and 4.3 we introduced the language of random variables, distinguishing between discrete and continuous random variables. Now we further characterize the probability distribution for a random variable in terms of its mean (or expected value) and variance. We consider discrete random variables in this section, then extend the concepts to continuous random variables in the next section.

expected value of a discrete random variable

The **expected value of a discrete random variable** Y with probability distribution $P_Y(y)$ is the probability-weighted average of its possible values. The expected value is denoted by $E(Y)$ or μ_Y.

Definition of $E(Y)$

For a discrete random variable Y with probability distribution $P_Y(y)$, the **expected value** of Y is

$$E(Y) = \sum_{\text{all } y} yP_Y(y)$$

To find $E(Y)$, take each possible value y, multiply (weight) it by the associated probability $P_Y(y)$, and add the results.

EXAMPLE 4.7 A firm is considering two possible investments. As a rough approximation, the firm assigns (subjective) probabilities to losing 20% per dollar invested, losing 10%, breaking even, gaining 10%, and gaining 20%. Let Y be the return per dollar invested in the first project and Z the return per dollar invested in the second. The firm's probabilities are

y:	$-.20$	$-.10$	0	$+.10$	$+.20$
$P_Y(y)$:	.1	.2	.4	.2	.1

z:	$-.20$	$-.10$	0	$+.10$	$+.20$
$P_Z(z)$:	.01	.04	.10	.50	.35

Calculate expected returns per dollar invested in each project. Which project appears to be the more attractive investment?

Solution Project Y, by any reasonable standard, appears less attractive. It is thought to be as likely to lose 20% as to gain 20%, and as likely to lose 10% as to gain 10%. Project Z is thought to be very likely to gain 10% or 20% and relatively unlikely to lose.

Calculations:

y	$P_Y(y)$	$yP_Y(y)$	z	$P_Z(z)$	$zP_Z(z)$
$-.20$	.1	$-.02$	$-.20$	.01	$-.002$
$-.10$	.2	$-.02$	$-.10$	.04	$-.004$
0	.4	0	0	.10	0
$+.10$	.2	$+.02$	$+.10$	.50	$+.050$
$+.20$	.1	$+.02$	$+.20$	.35	$+.070$
		$E(Y) = 0$			$E(Z) = .114$

The expected Y return is (as anticipated) less than the expected Z return. □

interpretations of $E(Y)$

The expected value (mean) of a random variable Y can usefully be interpreted in several ways. First, it is simply a **probability-weighted average**, a summary figure that takes into account the relative probabilities of different values of Y. Second, it can be thought of as a **long-run average** of Y. For example, imagine that the firm of Example 4.7 could invest in a very large number of projects with the same return probabilities as Project Z. The average return per dollar invested would be .114, or 11.4%. (This fact follows because about 1 of 100 such projects would lose 20%, 4 of 100 would lose 10%, 10 of 100 would break even, 50 of 100 would earn 10%, and 35 of 100 would earn 20%.) Third, the expected value defines, in a certain sense*, the **fair value** of a gamble. Suppose that a casino gambling game pays 3 dollars with probability 12/38 and nothing with probability 26/38. If we let Y = the return on one play of the game, $E(Y) = 3(12/38) + 0(26/38) = 36/38$, or about .947. In a fair game, the casino ought to charge the player 36/38 of a dollar, or about 94.7 cents, to play. (Devotees of the American form of the game of roulette will recognize the nature of the game, the fact that the actual charge is 1 dollar, and the long-run effect of the 2/38 increment on casino profitability.) And finally, the expected value of Y represents a

* This interpretation ignores the risk factor, to which we turn in Chapter 17.

generalization of the concept of a population mean μ. If Y is a discrete random variable corresponding to a value drawn at random from a discrete population of values, then $E(Y) = \mu$, the mean of the population.

EXAMPLE 4.8 Suppose that a population consists of the following values and associated frequencies:

Value:	1000	2000	3000	4000
Frequency: ($N = 200$)	80	60	40	20

The population mean is 2000. Let Y denote a single value drawn at random from the population. Find $P_Y(y)$ and $E(Y)$.

Solution The possible values and their probabilities are

y:	1000	2000	3000	4000
$P_Y(y)$:	$80/200 = .4$	$60/200 = .3$	$40/200 = .2$	$20/200 = .1$

The expected value is

$$\begin{aligned} E(Y) &= 1000(.4) + 2000(.3) + 3000(.2) + 4000(.1) \\ &= 400 + 600 + 600 + 400 = 2000 \end{aligned}$$

$E(Y)$ is exactly equal to the population mean. □

variance of a discrete random variable

We have discussed the different interpretations associated with the expected value of a discrete random variable. **Equally important characteristics of a discrete random variable are the variance and standard deviation, which measure the probability dispersion or variability of a random variable.** The variance of a random variable Y, Var(Y), is the probability-weighted average of squared deviations from the mean (expected value).

Definition of Var(Y) and σ_Y

If Y is a discrete random variable,

$$\sigma_Y^2 = \text{Var}(Y) = \sum_{\text{all } y} (y - \mu_y)^2 P_Y(y), \qquad \text{where } \mu_Y = E(Y)$$

The standard deviation of Y, denoted σ_Y, is (as for other standard deviations) the positive square root of the variance:

$$\sigma_Y = \sqrt{\text{Var}(Y)}$$

To calculate Var(Y), take each value y, subtract the expected value $\mu_Y = E(Y)$, square the result, multiply by the probability $P_Y(y)$, and sum.

EXAMPLE 4.9 Find the variance and standard deviation for Y and Z in Example 4.7.

Solution In Example 4.7 we found $\mu_Y = E(Y) = 0$ and $\mu_Z = E(Z) = .114$. A worksheet shows the computations required.

y	$P_Y(y)$	$(y - \mu_Y)$	$(y - \mu_Y)^2$	$(y - \mu_Y)^2 P_Y(y)$
−.20	.1	−.20	.04	.004
−.10	.2	−.10	.01	.002
0	.4	0	0	0
.10	.2	.10	.01	.002
.20	.1	.20	.04	.004

$$\sigma_Y^2 = .012$$
$$\sigma_Y = \sqrt{.012} = .110$$

z	$P_Z(z)$	$(z - \mu_Z)$	$(z - \mu_Z)^2$	$(z - \mu_Z)^2 P_Z(z)$
−.20	.01	−.314	.098596	.00098596
−.10	.04	−.214	.045796	.00183184
0	.10	−.114	.012996	.00129960
.10	.50	−.014	.000196	.00009800
.20	.35	.086	.007396	.00258860

$$\sigma_Z^2 = .00680400$$
$$\sigma_Z = .082$$

The Y distribution has greater variability. The bulk of the Z distribution is concentrated on the larger values .10 and .20, while the Y probabilities are somewhat more spread out over all possible values. The variance for a return on investment is often taken as a measure of risk, with larger variances indicating greater risk. In this example, the Z investment has both a higher expected return and a lower risk. □

The computation of a variance can be clumsy, involving lots of many-digit numbers, as in Var(Z) in Example 4.9. There is a shortcut formula for variance computations that can be of help.

Shortcut Method for Var(Y)

If Y is a **discrete random variable**,

$$\text{Var}(Y) = \sum_{\text{all } y} y^2 P_Y(y) - \mu_Y^2, \qquad \text{where } \mu_Y = E(Y)$$

We can square the original values, weight by $P_Y(y)$, and add. At the end of that computation we subtract the square of the mean (expected value) to get the variance.

EXAMPLE 4.10 Use the shortcut formula to repeat the variance calculations of Example 4.9.

Solution For Y it is the same calculation, because $\mu_Y = E(Y) = 0$. It doesn't matter when we subtract 0, or even 0^2. For Z with $\mu_Z = E(Z) = .114$,

z	$P_Z(z)$	z^2	$z^2P_Z(z)$
−.20	.01	.04	.0004
−.10	.04	.01	.0004
0	.10	0	0
.10	.50	.01	.0050
.20	.35	.04	.0140
			.0198

So $\text{Var}(Z) = .0198 - (.114)^2 = .006804$, as in Example 4.9. □

Chebyshev's Inequality and the Empirical Rule, introduced for samples and populations in Chapter 2, apply to random variables as well.

Chebyshev's Inequality and the Empirical Rule for Random Variables

If a random variable Y has finite mean and variance,

$P(Y \text{ falls within } c\sigma_Y \text{ of its mean } \mu_Y) \geq 1 - 1/c^2$

If Y has roughly a mound-shaped probability histogram,

$P(Y \text{ falls within one } \sigma_Y \text{ of its mean } \mu_Y) \approx .68,$ and

$P(Y \text{ falls within two } \sigma_Y \text{ of its mean } \mu_Y) \approx .95$

For the random variable Y of Examples 4.7 and 4.9, $E(Y) = 0$ and $\sigma_Y = .110$. The actual probabilities are

$$\begin{aligned} P(Y \text{ falls within } \sigma_Y \text{ of its mean}) &= P(-.110 \leq Y \leq .110) \\ &= P(Y = -.10) + P(Y = 0) + P(Y = .10) \\ &= .80 \end{aligned}$$

and

$$P(Y \text{ falls within two } \sigma_Y \text{ of its mean}) = P(-.220 \leq Y \leq .220) = 1.00$$

Chebyshev's Inequality indicates that these probabilities must be at least $1 - 1/(1)^2 = 0$ and $1 - 1/(2)^2 = .75$, respectively; as usual, the inequalities are true by a large margin. The Empirical Rule approximation is mediocre in this case, in part because Y takes on a small number of values. Had the firm assessed subjective probabilities for returns of, say, $-.25, -.20, -.15, \ldots, +.15, +.20, +.25$, the Empirical Rule would most likely have been a somewhat better approximation, although the distribution might not be mound shaped.

Just as the mean of a random variable is a generalization of the idea of a population mean, so **the variance of a random variable Y is a generalization of a population variance**. If Y is a discrete random variable corresponding to a value drawn randomly from a population, $\sigma_Y^2 = \sigma^2$.

4.5 EXPECTED VALUE, VARIANCE, AND STANDARD DEVIATION: CONTINUOUS RANDOM VARIABLES ($\int$)

We have defined the expected value, variance, and standard deviation for discrete random variables. The mathematical definitions of their counterparts for continuous random variables necessarily involve calculus. The definition of expected value for a continuous random variable Y is as follows.

Expected Value for a Continuous Random Variable

$$E(Y) = \int_{-\infty}^{\infty} y f_Y(y)\,dy$$

Take each value y, multiply (weight) by the probability density $f_Y(y)$, and integrate (instead of adding). The technicalities should not obscure the fact that $E(Y)$ is a probability-weighted average, with the same long-run average and fair-gamble interpretations.

EXAMPLE 4.11 Find $E(T)$, where T is the time between calls of Example 4.4 and $f_T(t) = 2e^{-2t}$, $t \geq 0$. What is the interpretation of this figure?

Solution Implicitly, $f_T(t) = 0$ for $t < 0$, because $t < 0$ is impossible in this context. It is necessary to know that

$$\int_0^{\infty} te^{-ct}\,dt = \frac{1}{c^2}$$

Then it follows that

$$E(T) = \int_{-\infty}^{\infty} tf_T(t)\,dt$$

$$= 2\int_0^{\infty} te^{-2t}\,dt$$

(because $f_T(t) = 0$ for $t < 0$ and $f_T(t) = 2e^{-2t}$)

$$= 2\left(\frac{1}{2^2}\right) = \frac{1}{2}$$

Because T is the time in minutes between successive calls, $E(T) = 1/2$ means that, in the long run, calls come, on the average, half a minute apart. □

The definition and shortcut formula for Var(Y) were stated only for discrete random variables. When speaking of continuous random variables we substitute a probability density for a discrete probability distribution and an integral for a sum.

Definition of Var(Y)

If Y is a **continuous random variable**

$$\mathrm{Var}(Y) = \int_{-\infty}^{\infty} (y - \mu_Y)^2 f_Y(y)\,dy$$

and a shortcut formula is

$$\mathrm{Var}(Y) = \int_{-\infty}^{\infty} y^2 f_Y(y)\,dy - \mu_Y^2$$

EXAMPLE 4.12 Find Var(T) where T is defined in Examples 4.4 and 4.11. It is known that $\int_0^\infty t^2 e^{-ct}\,dt = 2/c^3$.

Solution Use the shortcut formula and notice that the integral may be taken from 0 to ∞, because $f_T(t) = 0$ for $t < 0$. In Example 4.11 we found $\mu_T = E(T) = 1/2$.

$$\begin{aligned}\mathrm{Var}(T) &= \int_0^\infty t^2 f_T(t)\,dt - \mu_T^2 \\ &= \left(\int_0^\infty t^2 2e^{-2t}\,dt\right) - \left(\frac{1}{2}\right)^2 \\ &= 2\int_0^\infty t^2 e^{-2t}\,dt - \frac{1}{4} \\ &= 2\left(\frac{2}{2^3}\right) - \frac{1}{4} = \frac{1}{2} - \frac{1}{4} = \frac{1}{4}\end{aligned}$$

□

SECTIONS 4.4 AND 4.5 EXERCISES

4.16 The product-development laboratory of a paint manufacturer is asked to develop a modified paint for automobiles. The director of the laboratory estimates the following probabilities for the required development time (in months):

y:	2	3	4	5	6	7	8	9	10	11	12
$P_Y(y)$:	.20	.30	.15	.10	.08	.06	.04	.03	.02	.01	.01

a. Construct a probability histogram.
b. Calculate the expected value of Y.
c. Mark $E(Y)$ on the histogram. How does the shape of the histogram affect $E(Y)$?

4.17 Refer to Exercise 4.16.

a. Calculate the standard deviation of Y. Use the definition.
b. Use the shortcut method to calculate σ_Y.

4.18 Refer to Exercise 4.16. What is the actual probability that Y differs from μ_Y by less than one standard deviation? Why does this probability differ from the Empirical Rule estimate?

4.19 An investment syndicate is trying to decide which of two $200,000 apartment houses to buy. An advisor estimates the following probabilities for the five-year net returns (in thousands of dollars):

Return:	−50	0	50	100	150	200	250
Probability for house 1:	.02	.03	.20	.50	.20	.03	.02
Probability for house 2:	.15	.10	.10	.10	.30	.20	.05

a. Calculate the expected net return for house 1 and for house 2.
b. Calculate the respective variances and standard deviations.

4.20 Refer to Exercise 4.19.
a. Is one investment better than the other in terms of both expected return and risk?
b. If you had a spare $200,000 to invest, which investment would you prefer?

4.21 In Exercise 4.6 we considered the probability distribution

x:	0	1	2	3	4	5	6	7	8	9	10
$P_X(x)$:	.06	.14	.16	.14	.12	.10	.08	.07	.06	.04	.03

a. Find the mean of X.
b. Use the definition to calculate the variance of X.
c. Use the shortcut method to recalculate the variance of X.

4.22 Calculate the probability that X in Exercise 4.21 is within two standard deviations of its mean. How does this probability compare to the theoretical values given by the Empirical Rule and by Chebyshev's Inequality?

∫ **4.23** a. Sketch the density $f_Y(y)$ of Exercise 4.12. What should the expected value of Y equal?
b. Calculate the expected value.
c. Calculate the variance of Y, using either the definition or the short cut method.

∫ **4.24** How well should the Empirical Rule work for the density $f_Y(y)$ in Exercise 4.12? Calculate the actual probability that Y will be within one standard deviation of the mean.

∫ **4.25** In Exercise 4.14, we defined

$$f_Y(y) = 3y^{-4}, \qquad \text{for } 1 < y < \infty$$

a. Calculate the mean. What is the interpretation of the resulting number?
b. Calculate the variance and standard deviation. The shortcut method may be easier.

∫ **4.26** a. Calculate values of the density in Exercise 4.25 for $y = 1.0, 1.5, 2.0, 2.5$, and 3.0. Sketch the density.
b. Should the Empirical Rule work well for this density? Find the probability that Y will be within one standard deviation of its mean. Remember that Y can't be less than 1.

∫ **4.27** Specifications call for a metal rod used in an assembly to have a diameter of 10 cm. Rods are inspected, and any rods with diameters less than 9.90 cm or greater than 10.10 cm are discarded. Careful measurements indicate that the density of $Y =$ the diameter of a randomly chosen rod (after inspection) is

$$\begin{aligned} f_Y(y) &= 100(y - 9.9) && \text{if } 9.9 < y < 10 \\ &= 100(10.1 - y) && \text{if } 10 < y < 10.1 \end{aligned}$$

The density is 0 for all other y values.
a. What must the expected value of Y equal?
b. Find the standard deviation of Y.

4.6 JOINT PROBABILITY DISTRIBUTIONS AND INDEPENDENCE

In Sections 4.2 and 4.3 we developed some basic language for dealing with one random variable. In this section we extend that language to deal with joint probability distributions for two random variables, X and Y. We define everything in terms of two discrete random variables. Those who are tolerably comfortable with calculus should be able to supply the analogues for continuous random variables.

When we deal with two random variables X and Y, it is convenient to work with joint probabilities. In Chapter 3 the **joint probability** of events A and B was the probability of the intersection, $P(A \cap B)$. Let A be the event $X = x$ and B, the event $Y = y$. Define the **joint probability distribution** $P_{XY}(x, y)$ to be a function that supplies the joint probability for each pair of values x and y.

joint probability distribution

EXAMPLE 4.13 Suppose that, in the emergency room of a small hospital, the most serious cases involve coronary attack and trauma (injury by violence or severe accident). Define X = number of coronary cases, and Y = number of trauma cases, arriving on a particular weekday night. It is assumed that

$$P_{XY}(x, y) = (x + 1)(y + 2)/84, \qquad x = 0, 1, 2,\ y = 0, 1, 2, 3$$

Calculate a numerical table of joint probabilities.

Solution Simply substitute the desired x and y values to get the joint probabilities: $P_{XY}(0, 3) = P(X = 0 \text{ and } Y = 3) = (0 + 1)(3 + 2)/84 = 5/84$, and so on. A tabular display of the joint probability distribution $P_{XY}(x, y)$ is shown here:

	y			
x	0	1	2	3
0	2/84	3/84	4/84	5/84
1	4/84	6/84	8/84	10/84
2	6/84	9/84	12/84	15/84

□

marginal probabilities

Once a joint probability distribution has been specified, **marginal probabilities** can be calculated by summation. In Chapter 3, when we dealt with joint probabilities like $P(A \cap B)$, $P(A \cap \bar{B})$, and so on, the term *marginal probability* referred to the probability of one event alone, like $P(A)$. We calculated marginal probabilities by the Additive Law. Because this section is different only notationally from Chapter 3, the same principle can be used here.

EXAMPLE 4.14 Find the marginal probability distribution of X and the marginal probability distribution of Y in Example 4.13.

Solution Sum across rows to get X probabilities and down columns to get Y probabilities.

x	y: 0	1	2	3	$P_X(x)$
0	2/84	3/84	4/84	5/84	14/84
1	4/84	6/84	8/84	10/84	28/84
2	6/84	9/84	12/84	15/84	42/84
$P_Y(y)$	12/84	18/84	24/84	30/84	

This idea can be expressed in a formula. To find the probability $P_X(x)$, add up the joint probabilities of that x value and each possible y value:

$$P_X(x) = \sum_{\text{all } y} P_{XY}(x, y)$$

In this example,

$$\begin{aligned} P(Y =) &= \sum_{\text{all } x} P_{XY}(x, 1) \\ &= P_{XY}(1, 0) + P_{XY}(1, 1) + P_{XY}(1, 2) + P_{XY}(1, 3) \\ &= \frac{4}{84} + \frac{6}{84} + \frac{8}{84} + \frac{10}{84} = \frac{28}{84} \end{aligned}$$

In the same way, the marginal probabilities for Y can be computed as

$$P_Y(y) = \sum_{\text{all } x} P_{XY}(x, y)$$

In this example,

$$\begin{aligned} P(Y = 1) &= \sum_{\text{all } x} P_{XY}(x, 1) \\ &= P_{XY}(0, 1) + P_{XY}(1, 1) + P_{XY}(2; 1) \\ &= \frac{3}{84} + \frac{6}{84} + \frac{9}{84} = \frac{18}{84} \end{aligned}$$

The idea is simply a translation of the addition principle.* □

We can extend basic probability notation to conditional probabilities. Just as we defined the conditional probability of B given A as

$$P(B|A) = \frac{P(A \cap B)}{P(A)}$$

* Now we can explain why we use the apparently redundant notation $P_X(x)$, $P_Y(y)$. If we merely wrote $P(x)$ or $P(y)$, we would not know whether $P(1)$ meant $P(X = 1) = P_X(1)$ or $P(Y = 1) = P_Y(1)$.

conditional distribution

We can define the **conditional distribution** of Y given $X = x$ as

$$P_{Y|X}(y|x)$$

Thus for any value of Y,

$$P(Y = y|X = x) = \frac{P(X = x \cap Y = y)}{P(X = x)}$$

$$= \frac{P_{XY}(x, y)}{P_X(x)}$$

The need for this notation arises from the idea of independence. Remember that we had two equivalent definitions of independence for events A and B:

$$P(B|A) = P(B)$$

$$P(A \cap B) = P(A)P(B)$$

equivalent definitions of statistical independence

We also have two **equivalent definitions of statistical independence** for random variables X and Y:

$$P_{Y|X}(y|x) = P_Y(y), \qquad \text{for all } x, y$$

$$P_{XY}(x, y) = P_X(x)P_Y(y), \qquad \text{for all } x, y$$

For mathematical ease, we usually use the second form of the independence definition in this text.

EXAMPLE 4.15 Show that X and Y of Examples 4.13 and 4.14 are independent.

Solution In Example 4.14 we found $P_X(x)$ and $P_Y(y)$. When we multiply appropriate $P_X(x)$ and $P_Y(y)$, we get the following table:

	y				
x	0	1	2	3	
0	(12/84)(14/84)	(18/84)(14/84)	(24/84)(14/84)	(30/84)(14/84)	14/84
1	(12/84)(28/84)	(18/84)(28/84)	(24/84)(28/84)	(30/84)(28/84)	28/84
2	(12/84)(42/84)	(18/84)(42/84)	(24/84)(42/84)	(30/84)(42/84)	42/84
	12/84	18/84	24/84	30/84	

When we reduce the fractions in this table, we find that every table entry equals the $P_{XY}(x, y)$ entry in Example 4.13. Therefore $P_{XY}(x, y) = P_X(x)P_Y(y)$ for all x and y; that is, X and Y are independent. □

The assumption of independence was built into the mathematical form of this particular $P_{XY}(x, y)$. In practice, we often assume that X and Y are independent; once we specify $P_X(x)$ and $P_Y(y)$, this assumption lets us calculate $P_{XY}(x, y)$ as the product $P_X(x)P_Y(y)$. Example 4.15 is one situation in which the independence assumption seems reasonable. The number of coronary cases arriving at an emergency room should have no relevance to predictions of the number of trauma cases.

SECTION 4.6 EXERCISES

4.28 A manufacturer of television sets sells two principal models. Define X = sales of model A next December (nearest 100,000) and Y = sales of model B next December. The marketing staff estimates that the joint probabilities $P_{XY}(x, y)$ are

	y			
x	1	2	3	4
1	.030	.055	.070	.075
2	.055	.070	.075	.070
3	.070	.075	.070	.055
4	.075	.070	.055	.030

a. Find $P(X = 1, Y = 2)$.
b. Find $P(X \leq 2, Y \leq 2)$.
c. Find $P_X(x)$ and $P_Y(y)$.
d. Are X and Y independent?

4.29 Show that the formula

$$P_{XY}(x, y) = .005(-10 + 10x + 10y - x^2 - y^2 - 2xy)$$

yields the joint probability table of Exercise 4.28. Can you find a formula for $P_X(x)$?

4.30 The owner of a small sound-system store determines the following probabilities for X = the number of amplifiers sold during a weekday and for Y = the number of speaker systems sold during the same day:

x:	0	1	2	3	4
$P_X(x)$:	.10	.40	.25	.20	.05

y:	0	1	2	3	4	5
$P_Y(y)$:	.10	.30	.25	.20	.10	.05

a. Assuming that X and Y are independent, calculate the joint probability distribution $P_{XY}(x, y)$.
b. Check your work by finding the marginal probabilities $P_X(x)$ and $P_Y(y)$.

4.31 Do you believe that independence is a reasonable assumption for Exercise 4.30?

4.32 A small management-consulting firm presents both written and oral proposals in an effort to get new consulting contracts. Records indicate that the probability distribution $P_{XY}(x, y)$ of X = number of oral proposals in a week and Y = number of written proposals in that week is given by the following table:

	y				
x	0	1	2	3	4
0	.010	.015	.030	.075	.050
1	.020	.030	.045	.060	.040
2	.030	.045	.100	.045	.030
3	.040	.060	.045	.030	.020
4	.050	.075	.030	.015	.010

a. Find the probability that there are two oral proposals and two written proposals in a particular week.
b. Find the probability that there are exactly two oral proposals and two or fewer written proposals in a particular week.
c. Find the probability that there are two or fewer oral proposals and two or fewer written proposals in a particular week.

4.33 a. Use the probability distribution of Exercise 4.32 to calculate the marginal probability distributions of X and of Y.
b. Assuming these probabilities, are X and Y independent?

4.34 Calculate the conditional distribution of Y given each possible value of X, using the probability distribution of Exercise 4.32. Do these conditional probability distributions indicate that X and Y are independent?

4.7 COVARIANCE AND CORRELATION OF RANDOM VARIABLES

In Section 4.6 we defined the *independence* of two random variables. Now we consider how to measure the degree of *dependence* between two random variables. There are many kinds of dependence that two variables might have and many measures of dependence that one might use. Two measures, covariance and correlation, are particularly important because they are closely related to the concept of variance of a random variable.

Again we begin with an example. A trust officer of a bank assumes the following (subjective) joint probabilities for the percentage returns (interest plus change in market value) of two utility bonds. The returns are labeled X and Y:

			Y			
X	8	9	10	11	12	$P_X(x)$
8	.03	.04	.03	.00	.00	.10
9	.04	.06	.06	.04	.00	.20
10	.02	.08	.20	.08	.02	.40
11	.00	.04	.06	.06	.04	.20
12	.00	.00	.03	.04	.03	.10
$P_Y(y)$	.09	.22	.38	.22	.09	

There is a relation between X and Y. For example, given $x = 8$, the Y probabilities are concentrated on the smaller values $y = 8$, 9, and 10. At the other extreme, given $x = 12$, the Y probabilities are concentrated on the larger values $y = 10$, 11, and 12. In general, there is a tendency for the X and Y outcomes to vary together.

Covariance of Random Variables X and Y

If X and Y are discrete random variables with respective expected values μ_X and μ_Y, and with joint probability distribution $P_{XY}(x, y)$, the **covariance** of X and Y, denoted by $\text{Cov}(X, Y)$ is defined as

$$\text{Cov}(X, Y) = \sum_x \sum_y (x - \mu_X)(y - \mu_Y)P_{XY}(x, y)$$

A shortcut method for computing the covariance is

$$\text{Cov}(X, Y) = \left[\sum_x \sum_y xyP_{XY}(x, y)\right] - \mu_X\mu_Y$$

EXAMPLE 4.16 Compute Cov(X, Y) for the joint distribution of bond yields given in the preceding discussion. Use the definition first and check to see that the shortcut method gives the same answer.

Solution From the marginal probabilities $P_X(x)$ and $P_Y(y)$, we get the expected values:

$$\mu_X = 8(.10) + 9(.20) + 10(.40) + 11(.20) + 12(.10) = 10$$
$$\mu_Y = 8(.09) + 9(.22) + 10(.38) + 11(.22) + 12(.09) = 10$$

The covariance can be computed using the definition as follows:

$$\begin{aligned}\text{Cov}(X, Y) &= \sum_x \sum_y (x - \mu_X)(y - \mu_Y)P_{XY}(x, y)\\ &= (8 - 10)(8 - 10)(.03) + (8 - 10)(9 - 10)(.04)\\ &\quad + (8 - 10)(10 - 10)(.03) + \ldots + (12 - 10)(12 - 10)(.03)\\ &= .60\end{aligned}$$

Similarly, using the shortcut method,

$$\begin{aligned}\text{Cov}(X, Y) &= \sum \sum xyP_{XY}(x, y) - \mu_X\mu_Y\\ &= 8(8)(.03) + 8(9)(.04) + 8(10)(.03) + \ldots + 12(12)(.03) - 10(10)\\ &= 100.60 - 100 = .60\end{aligned}$$

□

The covariance of two random variables is closely related to their correlation.

Correlation of Random Variables X and Y

If X and Y are discrete random variables with respective standard deviations σ_X and σ_Y, their **correlation** ρ_{XY} is defined as

$$\rho_{XY} = \frac{\text{Cov}(X, Y)}{\sigma_X\sigma_Y}$$

It follows that

$$\text{Cov}(X, Y) = \rho_{XY}\sigma_X\sigma_Y$$

EXAMPLE 4.17 Find ρ_{XY} for the bond-yield distribution discussed earlier in this section.

Solution In Example 4.16 we found Cov(X, Y) = .60. To get ρ_{XY} we need the standard deviations of X and Y, which can be computed from the respective marginal probabilities. The formulas in Sections 4.4 and 4.5 can be used to compute σ_X^2 and σ_Y^2.

$$\sigma_X^2 = \sum_{Y} x^2 P_X(x) - \mu_X^2$$

$$= 8^2(.10) + 9^2(.20) + 10^2(.40) + 11^2(.20) + 12^2(.10) - (10)^2$$
$$= 101.20 - 100 = 1.20$$

and hence $\sigma_X = \sqrt{1.20} = 1.095$.

Similarly, we have

$$\sigma_Y^2 = \sum_{y} y^2 P_Y(y) - \mu_Y^2 = 1.16$$

and $\sigma_Y = \sqrt{1.16} = 1.077$.

Substituting into the formula for ρ_{XY}, we find

$$\rho_{XY} = \frac{\text{Cov}(X, Y)}{\sigma_X \sigma_Y} = \frac{.60}{1.095(1.077)} = .509$$

□

The correlation between X and Y ranges between -1.00 and $+1.00$. A value of -1.00 or $+1.00$ indicates perfect linear prediction in the population, while a value of zero indicates no linear predictive value.

If the random variables X and Y are independent, there should be no relation (linear or otherwise) between them. Reasonably enough, when X and Y are independent, Cov(X, Y) = 0 and therefore $\rho_{XY} = 0$ also. This fact can be seen easily by using the shortcut method for covariance. Remember that discrete random variables X and Y are independent if $P_{XY}(x, y) = P_X(x)P_Y(y)$ for all possible x and y. In that case,

$$\text{Cov}(X, Y) = \sum_{x,y} xyP_{XY}(x, y) - \mu_X\mu_Y$$
$$= \sum_{x,y} xyP_X(x)P_Y(y) - \mu_X\mu_Y$$
$$= \sum_{x} xP_X(x)\left[\sum_{y} yP_Y(y)\right] - \mu_X\mu_Y$$
$$= \mu_X\mu_Y - \mu_X\mu_Y = 0$$

EXAMPLE 4.18 An assembly line can be stopped temporarily to adjust for either bad parts alignment or bad welds. Production records indicate the following joint distribution for X = number of stops in a production shift for bad alignment and Y = number of stops in a production shift for bad welds.

	Y					
X	0	1	2	3	4	
0	.03	.06	.12	.06	.03	.30
1	.04	.08	.16	.08	.04	.40
2	.03	.06	.12	.06	.03	.30
	.10	.20	.40	.20	.10	1.00

a. What should Cov(X, Y) equal for these probabilities?
b. Verify your answer numerically.

Solution a. In every case $P_{XY}(x, y) = P_X(x)P_Y(y)$. For example, $P_{XY}(2, 4) = .03$ and $P_X(2)P_Y(4) = (.30)(.10) = .03$ also. Therefore, X and Y are independent, and Cov(X, Y) should equal zero.

b. By looking at the marginal probabilities for X and for Y, it is easy to see that $\mu_X = 1$ and $\mu_Y = 2$. So

$$\begin{aligned}\text{Cov}(X, Y) &= [0(0)(.03) + 0(1)(.06) + \ldots + 2(4)(.03)] - 1(2) \\ &= 2.00 - 2 = 0\end{aligned}$$

as it should be. □

It is mathematically possible to have Cov(X, Y) = 0 even though X and Y are dependent. The reason is that covariance and correlation measure only the strength of *linear* relation. If there is a relation between X and Y, but that relation cannot be approximated by a linear relation, the covariance can be zero.

EXAMPLE 4.19 Suppose that in Example 4.18 the following probabilities are obtained:

	Y					
X	0	1	2	3	4	
0	.01	.05	.18	.05	.01	.30
1	.03	.10	.14	.10	.03	.40
2	.06	.05	.08	.05	.06	.30
	.10	.20	.40	.20	.10	1.00

Are X and Y independent? What is the covariance between X and Y?

Solution No, there is dependence. For example, $P_{XY}(0, 0) = .01$, but $P_X(0)P_Y(0) = (.10)(.30) = .03$. However,

$$\begin{aligned}\text{Cov}(X, Y) &= [0(0)(.01) + 0(1)(.05) + \ldots + 2(4)(.06)] - 1(2) \\ &= 2.00 - 2 = 0\end{aligned}$$

(Note that $\mu_X = 1$ and $\mu_Y = 2$, as in Example 4.18.) The reason that the covariance is zero is that there is no linear relation. Note that when y is either 0 or 4, the most likely x value is 2, when y is either 1 or 3, the most likely x value is 1, and when y is 2, the most likely x value is 2. Computation of expected X values given each value y also shows a completely nonlinear pattern. □

4.8 JOINT PROBABILITY DENSITIES FOR CONTINUOUS RANDOM VARIABLES (∫)

Our discussion of joint probabilities has, up until now, focused on discrete random variables. Now we turn briefly to continuous random variables. As we shall see once again, every discrete summation has a direct analogue in a continuous integration. The technical details of calculus are needed in this section, but they should not obscure the recurring analogy of continuous integration with discrete summation.

When discussing discrete random variables, we considered joint probabilities $P_{XY}(x, y) = P(X = x \text{ and } Y = y)$. Probabilities for X, Y, or both were obtained by appropriate sums. When we turn to continuous random variables X and Y, we consider joint density functions $f_{XY}(x, y)$. In the continuous case, probabilities are obtained by integration (rather than summation) of the joint probability density. For example, suppose that a study of the time T required to prepare a bid in worker-days and the size of the bid U in millions of dollars indicates that the joint probability density is

$$f_{TU}(t, u) = .02(t + 1)(t + 2)(10 - t)u^t(1 - u), \qquad 0 < t < 10,\ 0 < u < 1$$

In this case both T and U are continuous random variables, as indicated by the fact that they vary over continuous ranges $0 < t < 10$ and $0 < u < 1$. Thus instead of finding the probability that T and U lie in specified intervals by summing probabilities, we find such probabilities by integrating densities.

Joint Probabilities for Continuous Random Variables

If X and Y are continuous random variables with joint probability density $f_{XY}(x, y)$, then probabilities concerning X and Y are calculated as

$$P(a < X < b, c < Y < d) = \int_{x=a}^{x=b} \int_{y=c}^{y=d} f_{XY}(x, y)\, dy\, dx$$

(Note for those not familiar with double integrals: The integrations are performed "from the inside out." Thus, in the expression above, the first integration is performed with respect to y, with x being regarded as a constant. Once the y variable has been integrated out, the single integration with respect to x is carried out.)

For example, suppose that continuous random variables X and Y have joint density

$$f_{XY}(x, y) = 1.2[2 - (x + y)^2], \qquad 0 < x < 1,\ 0 < y < 1$$

Find the probability that both X and Y are less than .5.

We note first that both X and Y can't be negative, so that we're finding $P(0 < X < .5$ and $0 < Y < .5)$. Thus

$$
\begin{aligned}
P(0 < X < .5 \text{ and } 0 < Y < .5) &= 1.2 \int_0^{.5} \int_0^{.5} [2 - (x + y)^2]\, dy\, dx \\
&= 1.2 \int_0^{.5} \left[2y - \frac{(x + y)^3}{3} \right] \Bigg|_{y=0}^{y=.5} dx \\
&= 1.2 \int_0^{.5} \left\{ \left[1 - \frac{(x + .5)^3}{3} \right] + \frac{x^3}{3} \right\} dx \\
&= 1.2 \left[x - \frac{(x + 5)^4}{12} + \frac{x^4}{12} \right] \Bigg|_{x=0}^{x=.5} \\
&= 1.2 \left[.5 - \left(\frac{1}{12} - \frac{(.5)^4}{12} \right) + \frac{(.5)^4}{12} \right] = .5125
\end{aligned}
$$

EXAMPLE 4.20 Histograms of past data on X = time required to cut bolts of cloth to a pattern and Y = time required to sew the same bolts of cloth for military uniforms (both measured in worker hours) indicate that

$$f_{XY}(x, y) = 72x^2(1 - x)y(1 - y), \qquad \text{for } 0 < x < 1 \qquad \text{and} \qquad 0 < y < 1$$

Find the probability that Y will be less than X and X will be between 0 and .5.

Solution This problem presents some technical difficulties in that the limits for integrating Y depend on the specified x value. We must have $y < x$ and $0 < x < .5$. Thus the region of integration of $f_{XY}(x, y)$ is $0 < x < .5$, $0 < y < x$ (because y, by definition, cannot be less than 0). So

$$
\begin{aligned}
P(0 < Y < X < .5) &= \int_{x=0}^{x=.5} \int_{y=0}^{y=x} 72x^2(1 - x)y(1 - y)\, dy\, dx \\
&= \int_{x=0}^{x=.5} 72x^2(1 - x) \int_{y=0}^{y=x} y(1 - y)\, dy\, dx \\
&= \int_{x=0}^{x=.5} 72x^2(1 - x) \left[\frac{y^2}{2} - \frac{y^3}{3} \right] \Bigg|_{y=0}^{y=x} dx \\
&= \int_{x=0}^{x=.5} 72x^2(1 - x) \left[\left(\frac{x^2}{2} - \frac{x^3}{3} \right) - (0 - 0) \right] dx \\
&= 36 \int_{x=0}^{x=.5} x^4(1 - x)\, dx - 24 \int_{x=0}^{x=.5} x^5(1 - x)\, dx \\
&= 36 \left(\frac{x^5}{5} - \frac{x^6}{6} \right) \Bigg|_{x=0}^{x=.5} - 24 \left(\frac{x^6}{6} - \frac{x^7}{7} \right) \Bigg|_{x=0}^{x=.5} \\
&= 36 \left[\frac{(.5)^5}{5} - \frac{(.5)^6}{6} \right] - 24 \left[\frac{(.5)^6}{6} - \frac{(.5)^7}{7} \right] \\
&= 0.0955, \text{ after some arithmetic}
\end{aligned}
$$

□

In Section 4.6, we showed that we could find the marginal probability distribution of X by summing over y in the joint probability distribution $P_{XY}(x, y)$. You may not be overwhelmingly surprised to hear that, when dealing with continuous random variables, we substitute an integral over y for a sum over y. In the continuous case,

$$f_X(x) = \int_{\text{all } y} f_{XY}(x, y)\, dy$$

For T = time spent in preparing a bid and U = size of the bid, with joint density

$$f_{TU}(t, u) = .02(t + 1)(t + 2)(10 - t)u^t(1 - u),\ 0 < t < 10,\ 0 < u < 1$$

$$f_T(t) = \int_{u=0}^{u=1} .02(t + 1)(t + 2)(10 - t)u^t(1 - u)\, du$$

$$= .02(t + 1)(t + 2)(10 - t)\left[\frac{u^{t+1}}{(t + 1)} - \frac{u^{t+2}}{(t + 2)}\right]\Bigg|_{u=0}^{u=1}$$

$$= .02(t + 1)(t + 2)(10 - t)\left[\frac{1}{(t + 1)} - \frac{1}{(t + 2)}\right] = .02(10 - t)$$

EXAMPLE 4.21 Find the marginal density of X in Example 4.20.

Solution

$$f_X(x) = \int_{y=0}^{y=1} 72x^2(1 - x)y(1 - y)\, dy$$

$$= 72x^2(1 - x)\int_{y=0}^{y=1} y(1 - y)dy$$

$$= 72x^2(1 - x)\left[\frac{1}{2} - \frac{1}{3}\right] = 12x^2(1 - x)$$

□

We may extend the definition of conditional probability distribution to the idea of the conditional density of Y given X. Just as we define the conditional probability distribution in the discrete case as the ratio of the joint probability $P_{XY}(x, y)$ to the marginal probability $P_X(x)$, we can define the conditional density as

$$f_{Y|X}(y|x) = \frac{f_{XY}(x, y)}{f_X(x)}$$

In particular, we can extend the definition of independence by saying that a continuous random variable Y is independent of another continuous random variable X if

$$f_{Y|X}(y|x) = f_Y(y)$$

or, equivalently, if $f_{XY}(x, y) = f_X(x)f_Y(y)$ for every x and y. If

$$f_{TU}(t, u) = .02(t + 1)(t + 2)(10 - t)u^t(1 - u), \qquad 0 < t < 10,\ 0 < u < 1$$

we have shown that $f_T(t) = .02(10 - t)$, so

$$f_{U|T}(u|t) = \frac{f_{TU}(t, u)}{f_T(t)} = \frac{.02(t + 1)(t + 2)(10 - t)u^t(1 - u)}{.02(10 - t)}$$

$$= (t + 1)(t + 2)u^t(1 - u)$$

Because the conditional density of U given $T = t$ is a function of t as well as u, U is not independent of T.

EXAMPLE 4.22 Refer to Example 4.20. Are X and Y independent in this case?

Solution We defined

$$f_{XY}(x, y) = 72x^2(1 - x)y(1 - y), \qquad 0 < x < 1, 0 < y < 1$$

In Example 4.21 we showed that

$$f_X(x) = 12x^2(1 - x)$$

so

$$f_{YX}(y, x) = \frac{72x^2(1 - x)y(1 - y)}{12x^2(1 - x)} = 6y(1 - y)$$

is a function only of y. (We must also be careful to notice that the range of definition of the formula for y is independent of x, as it is here.) Thus X and Y are independent.

Alternatively, we may calculate the marginal probability density of Y as $f_Y(y) = 6y(1 - y)$, for $0 < y < 1$. Thus

$$\begin{aligned} f_X(x)f_Y(y) &= 12x^2(1 - x)6y(1 - y), \qquad 0 < x < 1, 0 < y < 1 \\ &= f_{XY}(x, y) \end{aligned}$$

so once again X and Y are independent. □

SECTIONS 4.7 AND 4.8 EXERCISES

4.35 In Exercise 4.32 we considered the following joint distribution $P_{XY}(x, y)$ of X = number of oral proposals in a week and Y = number of written proposals in that week as given by the following table:

	y					
x	0	1	2	3	4	Total
0	.010	.015	.030	.075	.050	.180
1	.020	.030	.045	.060	.040	.195
2	.030	.045	.100	.045	.030	.250
3	.040	.060	.045	.030	.020	.195
4	.050	.075	.030	.015	.010	.180
Total	.150	.225	.250	.225	.150	

a. What are the means of X and Y? (Think, don't calculate.)
b. Calculate the standard deviations of X and Y.

4.36 a. Find the covariance of X and Y in Exercise 4.35.

b. Find the correlation of X and Y in Exercise 4.35. What does it indicate about the relation between X and Y? In particular, could X and Y be independent?

4.37 Find the conditional expection of Y, given $X = x$, for the probability distribution of Exercise 4.35. Does the conditional expectation change with X?

4.38 Define $T = X + Y$ to be the total number of proposals made by the firm in Exercises 4.32 and 4.35 in a particular week.

a. Calculate the probability distribution of T.

b. Calculate the expected value and variance of T directly from the probability distribution.

c. Use the results in the appendix to this chapter to recalculate the mean and variance of T.

4.39 In Exercises 3.15 and 3.17 we considered a manufacturing process in which holes are drilled in blocks. The probability that a hole is defectively drilled is .10. Let X = number of defects in a sample of two blocks (there is only one hole per block).

a. Find the probability distribution of X. You may wish to draw a tree.

b. Find the expected value and variance of X.

c. What have you assumed in answering parts (a) and (b)? Under what conditions might the assumption be unreasonable?

4.40 In Exercise 3.17 we assumed that an inspector fails to detect a defect with probability .10; implicitly we assumed that the inspector does not "detect" defects when in fact there are none. Let Y = number of detected defects. Use a probability tree to derive the joint distribution of X (from Exercise 4.39) and Y. Note that Y cannot be larger than X.

4.41 a. Find the mean and standard deviation of Y in Exercise 4.40.

b. Use the joint distribution of X and Y found in Exercise 4.40 to find the correlation of X and Y.

c. Explain why the correlation should naturally be positive.

4.42 A new-car dealer offers three packages of optional equipment for a particular model. There is an automatic-transmission package, with a profit of \$200 to the dealer, an air-conditioning package, with a profit of \$150, and an interior-decor package, with a profit of \$100. Data indicate that 80% of customers order the automatic-transmission package; 60% of these and 50% of those who don't order automatic transmissions also order the air-conditioning package. Of those who order both of these packages, 40% order the interior-decor package, as do 30% of those who order exactly one of the transmission and air-conditioning packages and 20% of those who order neither of the other packages. Let Y = the number of packages ordered on a randomly chosen new car.

a. Find the probability distribution of Y.

b. Find $P(Y \geq 2)$.

c. Find the cumulative distribution function of Y; use it to recalculate $P(Y \geq 2)$.

4.43 Find the mean and standard deviation of Y in Exercise 4.42.

4.44 Let X = the profit from sales of optional packages for the dealer in Exercise 4.42. Note that X is not directly a function of Y, because the profit depends not only on how many but also on which packages are sold.

a. Find the probability distribution of X.

b. Find the mean and standard deviation of X.

4.45 Refer to Exercise 4.44. Let T be the total profit from sales of optional packages to 18 randomly chosen customers.

a. Use the results from the appendix to this chapter to find the expected value and variance of T.

b. What did you assume in answering part (a)? Does any assumption appear clearly unreasonable?

∫ **4.46** A daily newspaper in a small city keeps records of the column-inches of classified ads in a given weekday's paper. Saturday and Sunday editions have different patterns of ads and are excluded. The probabilities for Y = number of column-inches (in thousands) on a randomly chosen day are approximated by the density function

$$f_Y(y) = 30y^4(1 - y), \qquad 0 < y < 1$$

Note that $f_Y(y) = 0$ for y outside the range $0 < y < 1$.

a. Calculate the density for $y = .1, .2, \ldots, .9$ and draw a sketch of the density.

b. Find the probability that between 700 and 900 column-inches are used (that is, that $.7 < Y < .9$) in a randomly chosen day.

c. Find the probability that $Y > .8$.

∫ **4.47** Find the mean and standard deviation of the number of column-inches used Exercise 4.46 according to the probability density.

∫ **4.48** The newspaper in Exercise 4.46 also keeps track of X = number of column-inches of commercial ads (in thousands). The distribution of X appears to be

$$f_X(x) = (6/125)x(5 - x), \qquad 0 < x < 5$$

and zero for all other x.

a. Find the cdf of X.

b. Find the probability that X is at least 3.

c. Find the mean and standard deviation of X.

4.49 The random variables X and Y in Exercises 4.48 and 4.46 appear (according to the records of the newspaper) to be independent of each other. If this is assumed, what is the correlation of X and Y?

∫ **4.50** A company offering overseas telephone service believes that the key variable cost factors for a call are X = number of seconds of computer time used in placing the call and Y = number of minutes of operator time used in placing the call. The probability structure can be represented by the following joint density:

$$f_{XY}(x, y) = (.0625)xe^{-.5y-x/y}, \qquad 0 < x < \infty, 0 < y < \infty$$

a. Find $f_Y(y)$. From calculus, it is known that $\int_0^\infty xe^{-kx} = 1/k^2$.

b. Find the conditional density of X, given $Y = y$.

c. In practice, should X and Y be independent? In this joint density, are they?

∫ **4.51** Find the conditional expected value of X, given $Y = y$, for the density given in Exercise 4.50.

∫ **4.52** Find the covariance of X and Y for the density of Exercise 4.50.

SUMMARY

This chapter combined the concepts of probability from Chapter 3 and the concepts of descriptive statistics from Chapter 2. The vehicle for this combination is the notion of a random variable, which leads to the related notion of a probability distribution. The mathematics for discrete and continuous random variables (and probability distributions) are parallel. A key idea in this chapter is the generalization of the concepts of mean, variance, and standard deviation for random variables. We have also considered joint probability distributions and independence of random variables. Covariance and correlation are measures of the amount of linear relation of random variables.

KEY TOPICS AND FORMULAS: Random Variables and Probability Distributions

1. Properties of the probability distribution $P_Y(y)$ for a discrete random variable Y
   a. $0 \leq P_Y(y) \leq 1$, for all y
   b. $\sum_{\text{all } y} P_Y(y) = 1$
   c. Since values of Y are mutually exclusive events, the probabilities are additive.
2. The probability distribution $P_Y(y)$ for a discrete random variable Y can be displayed by a table, a formula, or a graph (called a *probability histogram*).
3. ∫ The probability density function $f_Y(y)$ for a continuous random variable Y allows probabilities to be found by integration:

$$P(a \leq Y \leq b) = \int_a^b f_Y(y)\,dy$$

4. The expected value, variance, and standard deviation for a discrete random variable*

$$\mu_Y = E(Y) = \sum_{\text{all } y} yP_Y(y)$$

$$\sigma_Y^2 = \text{Var}(Y) = \sum_{\text{all } y} (y - \mu_Y)^2 P_Y(y)$$

$$\sigma_Y = \sqrt{\text{Var}(Y)}$$

5. Joint probability distribution, $P_{XY}(x, y)$
   a. Marginal distributions, $P_X(x)$ and $P_Y(y)$
   b. Conditional distribution, $P_{Y|X}(y|x)$
6. The cumulative distribution function (cdf) $F_Y(y) = P(Y \leq y)$ is defined for both discrete and continuous random variables Y.
7. The covariance of random variables X and Y,

$$\text{Cov}(X, Y) = \sum_x \sum_y (x - \mu_X)(y - \mu_Y)P_{XY}(x, y) = \sum_x \sum_y xyP_{XY}(x, y) - \mu_X\mu_Y$$

   (For continuous random variables, replace summation by integration.)
8. The correlation of X and Y

$$\text{Corr}(X, Y) = \rho_{XY} = \frac{\text{Cov}(X, Y)}{\sigma_X\sigma_Y}$$

9. X and Y are independent if $P_{XY}(x, y) = P_X(x)P_Y(y)$ for all x and y. (For continuous random variables, replace P by f.) If X and Y are independent, $\rho_{XY} = 0$.

* For continuous random variables, summation signs are replaced by integration signs and the probability density function replaces $P_Y(y)$.

CHAPTER 4 EXERCISES

4.53 The sales force of a small firm consists of four field engineers (three of whom are over 40 years old) and six sales representatives (two of whom are over 40 years old). One field engineer and two sales representatives are chosen, supposedly at random, to receive special training.

a. Construct the sample space for this experiment. Number the field engineers 1, . . . , 4 and the sales representatives 5, . . . , 10.

b. Let $Y =$ number of persons selected who are over 40 years old. Find $P_Y(y)$ and $F_Y(y)$ by counting.

4.54 In Exercise 4.53 find $E(Y)$ and σ_Y.

4.55 A state public health agency investigates reported unhealthful practices in restaurants, food stores, and the like. The number of cases varies from week to week. The data indicate the following:

Number of cases/week:	0	1	2	3	4	5	6
Probability:	.02	.13	.20	.30	.19	.15	.01

a. For $Y =$ number of cases in a specified week, find $F_Y(y)$.

b. Find $E(Y)$ and σ_Y.

c. Find $P(\mu_Y - \sigma_Y \leq Y \leq \mu_Y + \sigma_Y)$. Compare to the Empirical Rule approximation.

$\int$ **4.56** Consider the probability density function $f_Y(y) = 20(y^3 - y^4)$, $0 \leq y \leq 1$:

a. Find $E(Y)$ and σ_Y. Use the shortcut formula

$$\sigma_Y^2 = \int_{\text{all } y} y^2 f_Y(y)\,dy - \mu_Y^2$$

b. By finding an area under the $f_Y(y)$ curve, find

$$P(\mu_Y - 2\sigma_Y \leq Y \leq \mu_Y + 2\sigma_Y)$$

4.57 The fraction Y of column-inches devoted to display advertising in a certain newspaper on any particular Tuesday can be regarded as a continuous random variable with cdf

$$F_Y(y) = 10y^3 - 15y^4 + 6y^5, \qquad 0 \leq y \leq 1$$

a. Plot $F_Y(y)$.

b. Find $P(Y \leq .5)$, $P(.4 \leq Y \leq .6)$, and $P(Y \geq .7)$.

$(\partial, \int)$ **4.58** a. Show that the probability density in Exercise 4.57 is

$$f_Y(y) = 30y^2 - 60y^3 + 30y^3$$

b. Sketch $f_Y(y)$.

c. Find $E(Y)$ and $\text{Var}(Y)$.

d. Find $P(\mu_Y - \sigma_Y \leq Y \leq \mu_Y + \sigma_Y)$. Compare to the Empirical Rule approximation.

4.59 The records of a small auto-body repair shop indicate the following relative frequencies for the number of customers per day:

Number of customers:	0	1	2	3	4	5	6
Relative frequency:	.21	.38	.20	.11	.06	.03	.01

Let $Y =$ number of customers on one particular day.

a. Calculate $F_Y(y)$.

b. Find $E(Y)$ and σ_Y.

4.60 Assume that the numbers of customers on successive days in Exercise 4.59 are independent. Let Y_1 and Y_2 be the respective numbers of customers on two consecutive days.

a. Construct a table for $P_{Y_1Y_2}(y_1, y_2)$.

b. Define $S = Y_1 + Y_2$, the two-day total number of customers. Find $P_S(s)$.
c. Calculate $E(S)$ and σ_S.

∫ **4.61** Users of a computerized data base have established that X = the number of thousands of lines of instructions and Y = the time in minutes required to run the program have joint density

$$f_{XY}(x, y) = (3/320)(16 - 4x^2 - y^2 + 4xy), \qquad 0 < x < 2, 0 < y < 4$$

a. Find the probability that both X and Y will be less than 0.5.
b. Find the probability that Y will be larger than 1. (X could take any value.)

∫ **4.62** a. Calculate the marginal density of X, for the joint density given in Exercise 4.61.
b. Find $f_{Y|X}(y|x)$.

∫ **4.63** Considering their nature, should the random variables X and Y in Exercise 4.61 be independent? According to Exercise 4.62, are they?

∫ **4.64** An insurance firm receives records semiannually from independent agents. From past data, a model for the joint density of X = the proportion of records requiring a coverage update and Y = the proportion of records requiring an address change is

$$f_{XY}(x, y) = 240xy(1 - x)^2(1 - y)^3, \qquad \text{for } 0 < x < 1, 0 < y < 1$$

a. Find the probability that both X and Y are greater than 0.5.
b. Find the probability that Y is between 0.1 and 0.3.

∫ **4.65** a. Find the marginal densities of X and Y in Exercise 4.64.
b. What does the covariance of X and Y equal?

APPENDIX 4.A PROPERTIES OF EXPECTED VALUES AND VARIANCES

In this section we present some simple mathematical results about expected values and variances. The results are stated in the language of random variables. Because the notions of expected value and variance of a random variable are generalizations of the corresponding population concepts, the same results apply to populations.

The first results deal with the effect of adding or subtracting a constant. In analyzing the probable return on an investment, there must be a close relation between the gross return (which does not consider the original investment outlay) and net return (which subtracts that outlay).

> **Effect of Adding a Constant**
>
> If a is any constant and Y any random variable,
>
> $$E(Y + a) = E(Y) + a$$
> $$\text{Var}(Y + a) = \text{Var}(Y); \qquad \sigma_{Y+a} = \sigma_Y$$

If Y is the gross return and I the investment outlay, then the expected net return $E(Y - I)$ is, reasonably enough, the expected gross return less the investment, $E(Y) - I$. The variances of gross and net return are equal and therefore the standard deviations are also equal. The effect of subtracting I is to shift the whole probability histogram to the left by I units; since that shifting doesn't alter the spread of the histogram, the variance is unchanged.

It is fairly easy to prove these two results:

$$\begin{aligned} E(Y+a) &= \sum_{\text{all } y} (y+a)P_Y(y) \\ &= \sum_{\text{all } y} yP_Y(y) + \sum_{\text{all } y} aP_Y(y) \\ &= E(Y) + a\sum_{\text{all } y} P_Y(y) \\ &= E(Y) + a, \qquad \text{because } \sum_{\text{all } y} P_Y(y) = 1 \end{aligned}$$

$$\begin{aligned} \text{Var}(Y+a) &= \sum_{\text{all } y} [(y+a) - E(Y+a)]^2 P_Y(y) \\ &= \sum_{\text{all } y} [(y+a) - (E(Y)+a)]^2 P_Y(y) \\ &= \sum_{\text{all } y} [y - E(Y)]^2 P_Y(y) \\ &= \text{Var}(Y) \end{aligned}$$

Another set of results deals with multiplying or dividing by a constant. This mathematical operation is simply a change in the scale of measurement. For instance, multiplying a dollar amount by 100 changes to units of cents.

Effect of Multiplying by a Constant

If c is any constant and Y any random variable,

$$E(cY) = cE(Y)$$

$$\text{Var}(cY) = c^2\text{Var}(Y); \qquad \sigma_{cY} = |c|\sigma_Y$$

If $c = 100$ and Y is cost in dollars, then cY is cost in cents. The expected cost in cents $E(cY)$ is 100 times the expected cost in dollars, $cE(Y)$. The variance is multiplied by $10{,}000 = (100)^2$ because variance is average squared error; once we take square roots, the standard deviation of the cost in cents σ_{cY} is 100 times the standard deviation of cost in dollars, $c\sigma_Y$.* The proof is a matter of writing down the definitions of expected value and variance, then factoring out c and c^2, respectively.

EXAMPLE 4.23 A U.S. firm has an investment opportunity in France. The initial outlay is 5,000,000 francs. The firm estimates that the gross return Y has an expected value of 6,200,000 francs and a standard deviation of 500,000 francs. Find the expected value and standard deviation of the net return in dollars, assuming an exchange rate of 5 francs to the dollar.

Solution One way to proceed is to work first with net return in francs, then convert to dollars. The net return is $Y - 5{,}000{,}000$, so it has expected value $E(Y) - 5{,}000{,}000$ or 1,200,000

* The absolute value in the standard deviation formula takes care of multiplying by a negative number. Notice that $\sqrt{(-5)^2}$ is $+5$.

francs, and the standard deviation of 500,000 francs is unchanged. To convert to dollars, divide both the expected value and standard deviation of Y by 5. The expected net return is \$240,000 and the standard deviation is \$100,000. □

The last mathematical results we present involve adding two random variables. This operation arises in a wide variety of situations—the total return on two different investments, the total of cardiac and trauma cases in an emergency room, the total daily output from two automobile assembly lines. Of course there's nothing magic about adding *two* random variables; the results extend immediately to any number of random variables.

Mean and Variance for Sums of Random Variables

For any random variables X and Y,

$$E(X + Y) = E(X) + E(Y)$$

If X and Y are independent, then

$$\text{Var}(X + Y) = \text{Var}(X) + \text{Var}(Y); \qquad \sigma_{X+Y} = \sqrt{\text{Var}(X) + \text{Var}(Y)}$$

The proof of these results is a bit harder than the other proofs.

$$E(X + Y) = \sum_{\text{all } x} \sum_{\text{all } y} (x + y)P_{XY}(x, y)$$

$$= \sum_{\text{all } x} \sum_{\text{all } y} xP_{XY}(x, y) + \sum_{\text{all } x} \sum_{\text{all } y} yP_{XY}(x, y)$$

In the first double sum, think of summing over y first:

$$\sum_{\text{all } y} xP_{XY}(x, y) = x \sum_{\text{all } y} P_{XY}(x, y) = xP_X(x)$$

by definition of the marginal distribution $P_X(x)$. So the first double sum reduces to $\sum_{\text{all } x} xP_X(x) = E(X)$. A similar argument shows that the second double sum is $E(Y)$, which proves the expected value result.

The variance result assumes independence, $P_{XY}(x, y) = P_X(x)P_Y(y)$, and proceeds by expanding the square. Recall that $(a + b)^2 = a^2 + 2ab + b^2$.

$$\text{Var}(X + Y) = \sum_{\text{all } x} \sum_{\text{all } y} (x + y - \mu_{X+Y})^2 P_X(x)P_Y(y)$$

$$= \sum_{\text{all } x} \sum_{\text{all } y} (x - \mu_X + y - \mu_Y)^2 P_X(x)P_Y(y)$$

because we just proved that $\mu_{X+Y} = \mu_X + \mu_Y$. Now expand the square with $a = x - \mu_X$ and $b = y - \mu_Y$:

$$\text{Var}(X + Y) = \sum_{\text{all } x} \sum_{\text{all } y} (x - \mu_X)^2 P_X(x)P_Y(y)$$

$$+ 2 \sum_{\text{all } x} \sum_{\text{all } y} (x - \mu_X)(y - \mu_Y)P_X(x)P_Y(y)$$

$$+ \sum_{\text{all } x} \sum_{\text{all } y} (y - \mu_Y)^2 P_X(x)P_Y(y)$$

The first double sum is

$$\sum_{\text{all } x} (x - \mu_X)^2 P_X(x) \left[\sum_{\text{all } y} P_Y(y) \right] = \sum_{\text{all } x} (x - \mu_X)^2 P_X(x)[1] = \text{Var}(X)$$

The same procedure shows that the third double sum is Var(*Y*). The second double sum is 2 Cov(*X*, *Y*), by the definition of covariance given in Section 4.7. Therefore, in general,

$$\text{Var}(X + Y) = \text{Var}(X) + 2\ \text{Cov}(X, Y) + \text{Var}(Y)$$

In Section 4.7 we showed that Cov(*X*, *Y*) = 0 when *X* and *Y* are independent (as well as in some other cases). Therefore, if *X* and *Y* are independent, the covariance term in Var(*X* + *Y*) drops out, and we have Var(*X* + *Y*) = Var(*X*) + Var(*Y*).

APPENDIX 4.B SOME REMINDERS ABOUT CALCULUS

Calculus methods are not critical to understanding the essential ideas of this text, but there are a few occasions when it is covenient to use some basic ideas from calculus. This appendix gives a quick refresher in basic calculus methods; it is not intended as an introduction to calculus.

The first key concept is **function**. Informally, a function assigns an "output" number, say, *w*, to an "input" number, say, *x*, according to a specified rule. Because we want to reserve the letters *f* and *F* for other uses, we use *g* or *G* to indicate a function; we write $w = g(x)$.

The **derivative** of a function *g* at a point $x = a$ is informally defined as the slope of *g* when $x = a$. We may think of a line tangent to the curve $w = g(x)$ at $x = a$, as in Figure 4.8. The derivative is the slope of the tangent line. The derivative is denoted

$$\frac{d}{dx} g(x)$$

or $g'(x)$. We have no need to emphasize the particular value $x = a$ that is being considered.

A very brief table of derivatives is given in Table 4.1.

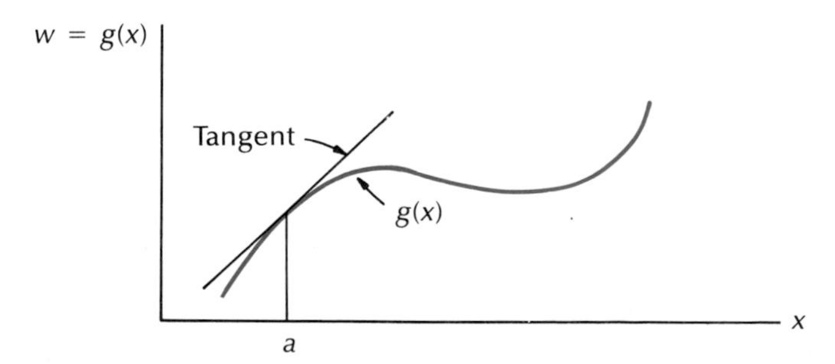

FIGURE 4.8 Tangent line at $x = a$

TABLE 4.1 **Elementary Derivatives**

$g(x)$	$\frac{d}{dx} g(x)$
c, a constant	0
x^n	nx^{n-1}
e^x	e^x
e^{cx}	ce^{cx}
$\log_e(x)$	$\frac{1}{x}$
$ag_1(x) + bg_2(x)$	$a \frac{d}{dx} g_1(x) + b \frac{d}{dx} g_2(x)$
$g_1(x)g_2(x)$	$g_1(x)\left[\frac{d}{dx} g_2(x)\right] + g_2(x)\left[\frac{d}{dx} g_1(x)\right]$

We sometimes use the **chain rule** for composite functions—functions that are defined in stages; that is, some functions may be thought of as taking an input x, transforming it to an intermediate value $w = g_1(x)$, then transforming w to a final value $v = g_2(w)$. We write such a two-stage function as

$$v = g_2[g_1(x)]$$

For example,

$$v = e^{x^2}$$

can be thought of as a two-stage function. Transform x to $w = x^2$; then transform w to

$$v = e^w = e^{x^2}$$

Alternatively, we write $w = g_1(x) = x^2$ and $v = g_2(w) = e^w$; so $v = g_1[g_2(x)]$. For such "stage-wise" functions, the derivative also goes in stages. First, find the derivative of $g_2(w)$, evaluated at $w = g_1(x)$; next multiply by the derivative of $g_1(x)$. Together, the chain rule asserts that

$$\frac{d}{dx} g_2[g_1(x)] = \left\{\frac{d}{dw} g_2[w = g_1(x)]\right\}\left[\frac{d}{dx} g_1(x)\right]$$

For $g(x) = e^{x^2}$, take $w = g_1(x) = x^2$ and $v = g_2(w) = e^w$. Then

$$\frac{d}{dx} g(x) = \left[\frac{d}{dw} e^w\right]\left[\frac{d}{dx} x^2\right]$$

$$= [e^w][2x] = 2xe^{x^2}$$

from elementary derivatives.

One of the important uses of derivatives is in finding relative maxima and minima. In Figure 4.9, notice that at both the peaks and the valleys of the function $g(x)$ the slope (derivative) of $g(x)$ is zero.

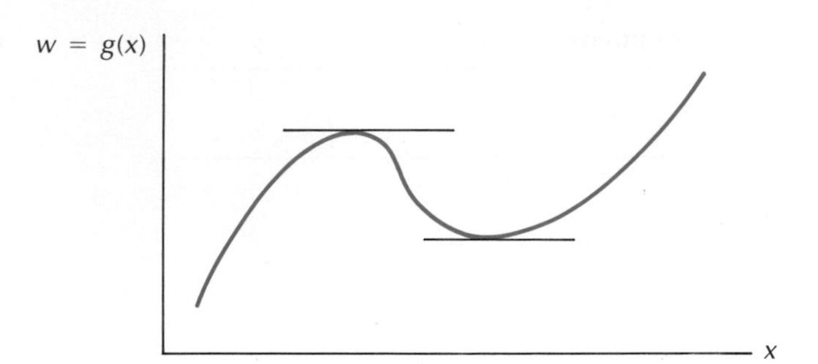

FIGURE 4.9 Derivative Is Zero at Minima and Maxima

Thus to locate a maximum or minimum of $g(x)$, we must solve the equation $(d/dx)g(x) = 0$ for x. In our problems it is usually obvious whether a particular solution is a minimum or a maximum; more sophisticated analyses (such as second-derivative tests) are not needed.

The ideas about derivatives discussed so far may be extended to functions of several variables, such as

$$w = g(x_1, x_2, x_3) = x_1^2 e^{3x_2} + \log_e(x_3)$$

We may take **partial derivatives** with respect to each variable by (temporarily) treating the other variables as constants. Partial derivative notation is, for example,

$$\frac{\partial}{\partial x_2}[x_1^2 e^{3x_2} + \log_e(x_3)]$$

In this example, x_1 (and thus x_1^2) and x_3 [and thus $\log_e(x_3)$] should be taken as constants, say, c_1 and c_3. Thus

$$\begin{aligned}\frac{\partial}{\partial x_2}(c_1 e^{3x_2} + c_3) &= c_1 \frac{\partial}{\partial x_2}(e^{3x_2}) + c_3 \\ &= c_1(e^{3x_2})(3) + 0 = 3x_1^2 e^{3x_2}\end{aligned}$$

where we have applied the chain rule to get

$$\frac{\partial}{\partial x_2} e^{3x_2} = (e^{3x_2}) \frac{\partial}{\partial x_2}(3x_2) = (e^{3x_2}(3)$$

To find a maximum or minimum of a function of several variables, we must equate *all* partial derivatives to zero and solve the resulting set of equations for the values of all the variables. In our problems, it is obvious whether the solution is a minimum or a maximum. To find a minimum or maximum of

$$w = g(x_1, x_2) = (x_1 - 4)^2 + (2x_1 + x_2 - 4)^2$$

we must solve the two equations

$$\frac{\partial}{\partial x_1}[(x_1 - 4)^2 + (2x_1 + x_2 - 4)^2] = 2(x_1 - 4) + 24(2x_1 + x_2 - 4) = 0$$

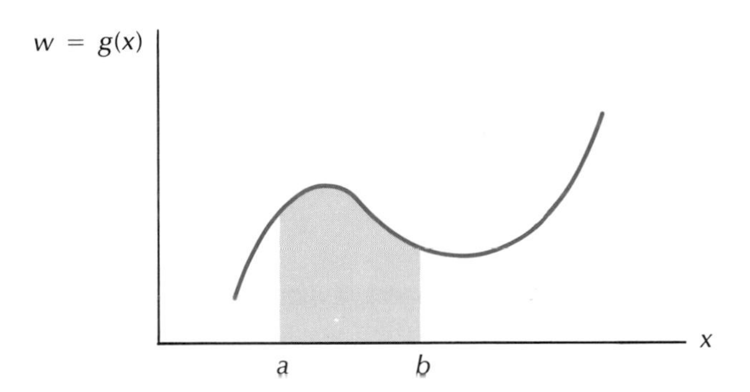

FIGURE 4.10 **Integral as Area**

and

$$\frac{\partial}{\partial x_2}[(x_1 - 4)^2 + (2x_1 + x_2 - 4)^2] = 2(2x_1 + x_2 - 4) = 0$$

and obtain $x_1 = 4$ and $x_2 = -4$. We note that $g(x_1, x_2)$ is never negative and that $g(4, -4) = (4 - 4)^2 + [2(4) + (-4) - 4]^2 = 0$. Therefore, we have clearly found the only minimum of g; because the given solution is the only one, there is no finite maximum of g.

We have some use for integral calculus as well as the previous differential (derivative) calculus. In general, we need to evaluate **definite integrals** of the form

$$\int_a^b g(x)\,dx$$

Recall that a definite integral is the area under the curve defined by $g(x)$ between the points a and b, as shown in Figure 4.10.

The standard way to evaluate definite integrals is to appeal to the Fundamental Theorem of Calculus. According to this theorem we should

1. find a function $G(x)$ having derivative $g(x)$—called the **antiderivative** of $g(x)$.
2. Evaluate $\int_a^b g(x)\,dx = G(x)|_a^b = G(b) - G(a)$

A few of the most useful antiderivatives are shown in Table 4.2.

TABLE 4.2 **Antiderivatives for Integration**

Function, $g(x)$	Antiderivative, $G(x)$
$(x + c)^n, \quad n \neq -1$	$(x + c)^{n+1}/(n + 1)$
x^{-1}	$\log_e(x)$
e^{cx}	e^{cx}/c
xe^{cx}	$xe^{cx}/c - e^{cx}/c^2$
$c_1 g_1(x) + c_2 g_2(x)$	$c_1 G_1(x) + c_2 G_2(x)$
$c_n x^n + c_{n-1}x^{n-1} + \ldots + c_0$	$c_n x^{n+1}/(n + 1)$ $+ c_{n-1}x^n/n + \ldots + c_0 x$

For example, to evaluate $\int_1^4 2e^{-2x}\,dx$, we find the antiderivative of $g(x) = e^{-2x}$ to be $G(x) = e^{-2x}/(-2)$ and note that the antiderivative of $2e^{-2x}$ is $2e^{-2x}/(-2) = -e^{-2x}$. It follows that

$$\int_1^4 2e^{-2x}\,dx = -e^{-2x}\Big|_1^4 = [-e^{-2(4)}] - [-e^{-2(1)}]$$
$$= e^{-2} - e^{-8}$$

Occasionally, we need to evaluate integrals with infinite endpoints, such as

$$\int_0^\infty g(x)\,dx = \lim_{M\to\infty} \int_0^M g(x)\,dx$$

assuming that the limit exists. The evaluation of the limit is usually clear, but it can sometimes be tricky. Consider the integral (see Table 4.2)

$$\int_0^\infty xe^{-x}\,dx = \int_0^\infty xe^{(-1)x}\,dx = [xe^{-x}/(-1) - e^{-x}/(-1)^2]\Big|_0^\infty$$

The problem is in evaluating $-xe^{-x}\big|_0^\infty$. As $x \to \infty$, $e^{-x} \to 0$ and $x \to \infty$, so that the limit of xe^{-x} (as $x \to \infty$) is "$0 \cdot \infty$." An application of a theorem of calculus called L'Hôpital's Rule indicates that $e^{-x} \to 0$ faster than $x \to \infty$, so

$$\lim_{x\to\infty} xe^{-x} = 0$$

Thus

$$\int_0^\infty xe^{-x}\,dx = (0 - 0) - [0 - (-1)] = 1$$

Finally, we need to evaluate a very few double integrals $\int_a^b \int_c^d g(x_1, x_2)\,dx_1\,dx_2$. Under certain technical conditions (which are always met in our situations), double integrals may be evaluated in two steps. First, regard x_2 as a constant c_2 and perform the integral with respect to x_1. Then integrate the result with respect to x_2. The integral may also be done in the reverse order, with respect to x_2 first, then x_1. For example, we may use the fact that the antiderivative of $(x + c)^n$ is $(x + c)^{n+1}/(n + 1)$ to evaluate

$$\int_0^1 \int_0^2 (x_1 + 2x_2)^2\,dx_1\,dx_2 = \int_0^1 \left[\int_0^2 (x_1 + 2x_2)^2\,dx_1\right] dx_2$$
$$= \int_0^1 \frac{(x_1 + 2x_2)^3}{3}\Bigg|_0^2 dx_2$$
$$= \int_0^1 \left[\frac{(2 + 2x_2)^3}{3} - \frac{(0 + 2x_2)^3}{3}\right] dx_2$$
$$= \left[\frac{8(1 + x_2)^4}{12} - \frac{8(x_2)^4}{12}\right]\Bigg|_0^1$$
$$= \left[\frac{8(2)^4}{12} - \frac{8(1)^4}{12}\right] - \left[\frac{8(1)^4}{12} - \frac{8(0)^4}{12}\right]$$
$$= 9.3333$$

One may also integrate with respect to x_2 first, then with respect to x_1; the result is once again 9.3333.

5

SOME SPECIAL PROBABILITY DISTRIBUTIONS

The ideas, notations, and results of the previous chapter apply to any random variables and any probability distributions. Now we identify some particular probability distributions that arise very often in practice. Although we present formulas for each of these distributions, the most important part of each section is the description of the kind of situation that indicates the use of each distribution.

5.1 COUNTING POSSIBLE OUTCOMES

This chapter contains a discussion of the probability distributions that apply to several commonly occurring situations. Among the most common is taking a random sample. The probability distributions that apply to the simplest sampling situations are discussed in Sections 5.2 and 5.3. As we suggested when discussing the classical interpretation of probability, we have considerable use for the idea that

$$P(\text{event}) = \frac{\text{number of outcomes favoring event}}{\text{total number of outcomes}}$$

To use this idea, we need a method for counting possible outcomes without the labor of actually listing the outcomes. This section contains a brief discussion of counting formulas; these formulas are critical in the development of the probability distributions of the next two sections; they arise as answers to the following two questions:

sequences and subsets

1. How many **sequences** of k symbols can be formed from a set of r distinct symbols, using each symbol no more than once?
2. How many **subsets** of k symbols can be formed from a set of r distinct symbols, using each symbol no more than once?

The only difference between a sequence and a subset is that order matters for sequences and not for subsets. The sequence ABC is not the same as the sequence CAB, but the subset {A, B, C} is the same as the subset {C, A, B}. As an example, consider sequences and subsets consisting of three of the first five letters. There are 60 sequences but only 10 subsets (Table 5.1).

First we derive a formula for the number of sequences. In the example, we could choose any of five letters to be first, then any of the four remaining letters to form 5×4 or 20 two-letter sequences. Each of these can be combined with any of the remaining three letters to form $(5 \times 4)(3) = 60$ three-letter sequences. In general, we have r choices for the first symbol, $r - 1$ for the second symbol, and so on. When we come to choose the kth and last symbol, we have already used up $k - 1$ symbols and have $r - (k - 1) = r - k + 1$ symbols remaining. Therefore,

$$\text{number of sequences} = r(r-1)\ldots(r-k+1)$$

of k symbols from r distinct symbols. This formula looks like a factorial ($r!$) except that it is truncated at $r - k + 1$ instead of continuing down to 1. The number of sequences is often called the number of **permutations** of r symbols taken k at a time and is denoted as ${}_kP_r$ or r_k. It can be expressed via factorials as

permutations

$${}_kP_r = \frac{r!}{(r-k)!} = r(r-1)\ldots(r-k+1)$$

The factors $r - k, r - k - 1, \ldots, 3, 2, 1$ in the denominator cancel the corresponding factors in the numerator, leaving only the factors $r, r - 1, \ldots, r - k + 1$.

The number of subsets is called the number of **combinations** of r symbols taken k at a time and is denoted as ${}_kC_4$ or $\binom{r}{k}$. Table 5.1 suggests an indirect way of finding $\binom{r}{k}$. Each row of the table of sequences corresponds to a particular subset. The six columns

TABLE 5.1 **Subsets and Sequences of the Five Letters A, B, C, D, and E**

Subsets	Sequences					
{A, B, C}	ABC	ACB	BAC	CAB	BCA	CBA
{A, B, D}	ABD	ADB	BAD	DAB	BDA	DBA
{A, B, E}	ABE	AEB	BAE	EAB	BEA	EBA
{A, C, D}	ACD	ADC	CAD	DAC	CDA	DCA
{A, C, E}	ACE	AEC	CAE	EAC	CEA	ECA
{A, D, E}	ADE	AED	DAE	EAD	DEA	EDA
{B, C, D}	BCD	BDC	CBD	DBC	CDB	DCB
{B, C, E}	BCE	BEC	CBE	EBC	CEB	ECB
{B, D, E}	BDE	BED	DBE	EBD	DEB	EDB
{C, D, E}	CDE	CED	DCE	ECD	DEC	EDC

correspond to all possible orderings of a given set of three letters. In general, for k symbols, there are $k(k-1)\ldots 3\cdot 2\cdot 1 = k!$ columns, because any of the k symbols may be first, any of the remaining $k-1$ symbols second, and so on. All k symbols are used, so the factorial is not truncated. Therefore,

$$\text{number of sequences} = (\text{number of rows})(\text{number of columns})$$

$$\frac{r!}{(r-k)!} = (\text{number of subsets})k!$$

Now solve for $\binom{r}{k}$, the number of subsets, and get

$$\binom{r}{k} = \frac{r!}{k!(r-k)!}$$

The symbol $\binom{r}{k}$ is read "r choose k," suggesting a choice of a subset of k things from a set of r things.

The combinations formula is particularly useful in random sampling, because choosing a sample of size k without replacement from a population of size r is exactly the same as choosing a subset of k things from a set of r things. We do not typically care about the ordering of items during sampling, so the permutation formula is somewhat less relevant.

EXAMPLE 5.1 In auditing the 87 accounts payable of a small firm, a sample of 10 account balances are checked. How many possible samples are there? Assuming that 13 of the accounts contain errors, how many samples contain exactly two erroneous accounts?

Solution There is no need to consider the sequence (order) in which the 10 accounts are drawn, because all 10 are checked. Therefore we can count the number of combinations. There are $\binom{87}{10} = 87!/10!77! \approx 4{,}000{,}000{,}000{,}000$ possible samples. To obtain all samples with two erroneous accounts, we can combine any of the $\binom{13}{2}$ choices of two from the 13 erroneous accounts with any of the $\binom{74}{8}$ choices of eight from the 74 correct accounts. Since any choice of two erroneous accounts can be matched with any choice of eight correct ones, there are $\binom{13}{2}\binom{74}{8} \approx 1{,}200{,}000{,}000{,}000$ samples with two erroneous and eight correct accounts. □

EXAMPLE 5.2 In a sales contest, the 10 top performers out of 612 salespeople receive prizes, ranging from a free vacation for the overall winner to $50 for the tenth-place finisher. How many different prize lists are possible?

Solution Here the ordering is certainly relevant, so the permutation formula applies. There are ${}_{10}P_{612} = 612!/602! \approx 6{,}800{,}000{,}000{,}000{,}000{,}000{,}000{,}000{,}000$ possibilities. □

SECTION 5.1 EXERCISES

5.1 In a certain state, an appeals court consists of seven judges. For a routine case, three judges are chosen at random as a panel to hear a case and render a decision. How many distinct panels can be formed?

5.2 Suppose that five of the seven judges on the appeals court in Exercise 5.1 are considered potentially sympathetic to a particular legal argument. How many panels can be formed having exactly two potentially sympathetic judges? How many panels have at least two such judges?

5.3 A grocery chain wants to taste-test a private-label cola drink. A tester is given eight unmarked glasses, four containing the private-label drink and four containing a nationally advertised cola. The tester is asked to identify the four glasses containing the private-label drink. How many different choices of four glasses can the taster make?

5.4 How many of the choices in Exercise 5.3 include three correct glasses and one incorrect glass?

5.2 BERNOULLI TRIALS AND THE BINOMIAL DISTRIBUTION

The simplest data-gathering process is counting the number of times a certain event occurs. When taking a random sample of registered voters, we can count the number who prefer the incumbent to the challenger. When sampling pistons for an auto engine assembly, we can count the number that fail to meet tolerances. When examining hiring practices, we can count the number of minority workers hired by a firm. When examining credit policies, we can count the number of bad-debt accounts. An almost endless variety of situations reduce to this simple counting process.

These examples, and many others, share certain common features. First, the overall process can be thought of as a series of **trials**, each trial yielding exactly one of two possible outcomes. In sampling registered voters, each person constitutes a trial. The incumbent is either preferred or not preferred. In sampling pistons, each trial yields a defective piston or a piston within tolerances. Each person hired is or is not a minority member, and each credit account is or is not a bad debt. The standard language is to call one outcome "success" and the other "failure." Which outcome is called success does not matter; a bad debt account could be called a success.

Second, in each of these situations, it is reasonable to assume that the probability of success π is constant over trials. The probability of finding a registered voter who favors the incumbent does not change in mid-sample (unless the sample is conducted over an extended period of time), nor does the probability of a defective piston, nor does the probability of a bad debt. If relative unemployment rates and the firm's hiring practices do not change, the probability that a given new employee is a minority worker does not change.

And finally, in each situation, the results of the various trials can be assumed to be independent. The preference of one voter for the incumbent should not affect the preference of another voter; at least that shouldn't occur in a carefully designed study. If one account happens to be a bad debt, that fact doesn't change the likelihood that the next account sampled will be good.

These three assumptions—each trial results in either a success or failure, constant probability π of success, and independence of trials—define a series of **Bernoulli trials**. The assumptions are assumptions; not every counting process can reasonably be modeled as Bernoulli trials. Whether these assumptions are reasonable depends on the situation.

Yes–no trials are not naturally independent and identical, but in many cases these assumptions hold to a good approximation, making Bernoulli trials a useful model.

EXAMPLE 5.3 Discuss whether or not a series of Bernoulli trials provides a reasonable model for each of the following situations.

a. A telephone researcher involved in a television viewing survey calls different homes (selected at random), one each 15 minutes between 5:30 P.M. and 10:00 P.M. Each person contacted is asked if anyone in the household is watching the ABC network program. A trial consists of contacting a household to determine whether or not someone in the house is watching an ABC network program.

b. A trust officer examines a sample of stock listings from those on the New York Stock Exchange to determine whether or not each stock has risen in price during the past week. Here a trial consists of selecting a stock and determining whether or not the price has risen during the past week.

c. Each of 50 newly hired management trainees is rated outstanding, acceptable, or unsatisfactory at the conclusion of a training program. Determining the rating for a newly hired management trainee constitutes a trial.

Solution a. The assumption of constant probability from trial to trial is not plausible in this survey, since the level of television watching in general is relatively lower early in the evening. Hence the probability of finding someone watching the ABC network program may vary depending on the time of the call.

b. The independence assumption is very dubious. During any particular time period, there's a moderately strong tendency for stock prices to move up or down together, because of interest rate changes, political news, or the herd instinct of investors. So for the stocks listed in the sample, the outcome on any one trial would depend heavily on price changes for the other stocks.

c. For this problem there are three possible outcomes on each trial, not two. However, if we define a success to be a rating of outstanding and a failure to be the complement (not rated outstanding), Bernoulli trials may be a good model. The key question is whether the trial outcomes are independent. If there is an effective ceiling or quota for the number (or proportion) of outstanding ratings (for instance, a restriction that the supervisor can rate no more than 10% of the group as outstanding), then the independence assumption is violated. But if each trainee is rated according to established, reasonably objective criteria, independence of trials (ratings) should be a reasonable assumption. □

There is one additional feature common to all these situations. We are counting the number of successes that occur in a fixed number n of trials, without regard to the particular order in which successes and failures occur. This would not be true if, for instance, a telephone interviewer called homes at random until 24 television-watching homes had been obtained. In this situation n is not fixed and the order of successes and failures *is* relevant; the last trial (call) is guaranteed to be a success.

binomial experiment

A collection of a fixed number n of Bernoulli trials in which the researcher is interested in the total number of successes defines a **binomial experiment**. The properties of a binomial experiment are listed here.

Properties of a Binomial Experiment

1. There are n Bernoulli trials; each one results in either a success (S) or a failure (F).
2. The probability of a success, $\pi = P(S)$, remains constant over trials $[P(F) = 1 - \pi]$.
3. The trials are independent. (Assumptions 1–3 define Bernoulli trials.)
4. The random variable of interest is Y, the number of successes in n trials (the ordering of successes is not important).

binomial random variable

binomial probability distribution

The random variable Y in a binomial experiment is called a **binomial random variable**. It is a discrete random variable that can assume any one of the values 0, 1, 2, . . . , n. The **binomial probability distribution** $P_Y(y)$, which assigns probabilities to each value of Y, is best understood by considering a simple example.

Suppose we take a random sample of three individuals from a population with a proportion π of successes. Figure 5.1 shows a probability tree for calculating the distri-

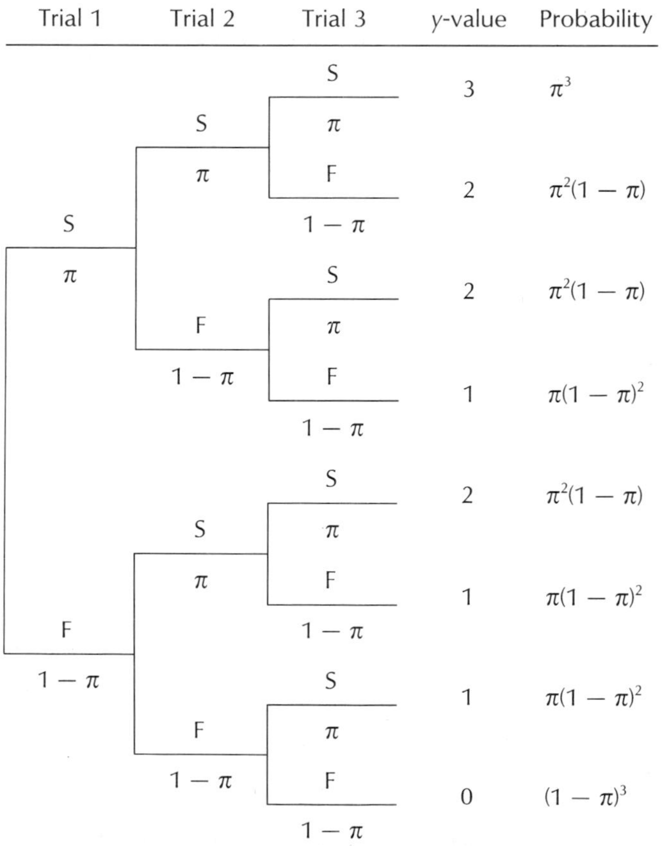

Trial 1	Trial 2	Trial 3	y-value	Probability
S (π)	S (π)	S (π)	3	π^3
S (π)	S (π)	F ($1 - \pi$)	2	$\pi^2(1 - \pi)$
S (π)	F ($1 - \pi$)	S (π)	2	$\pi^2(1 - \pi)$
S (π)	F ($1 - \pi$)	F ($1 - \pi$)	1	$\pi(1 - \pi)^2$
F ($1 - \pi$)	S (π)	S (π)	2	$\pi^2(1 - \pi)$
F ($1 - \pi$)	S (π)	F ($1 - \pi$)	1	$\pi(1 - \pi)^2$
F ($1 - \pi$)	F ($1 - \pi$)	S (π)	1	$\pi(1 - \pi)^2$
F ($1 - \pi$)	F ($1 - \pi$)	F ($1 - \pi$)	0	$(1 - \pi)^3$

FIGURE 5.1 Probability Tree for the Binomial Probability Distribution with $n = 3$

bution of Y. By adding up the probabilities of appropriate paths, we can find the binomial probability distribution for $n = 3$. For instance, the second, third, and fifth paths (counting from the top) give $y = 2$; each of these paths has probability $\pi^2(1 - \pi)$. We add the path probabilities to get $P(Y = 2)$; $P_Y(2) = \pi^2(1 - \pi) + \pi^2(1 - \pi) + \pi^2(1 - \pi) = 3\pi^2(1 - \pi)$. The complete probability distribution is

y:	0	1	2	3
$P_Y(y)$:	$(1 - \pi)^3$	$3\pi(1 - \pi)^2$	$3\pi^2(1 - \pi)$	π^3

EXAMPLE 5.4 Find the binomial distribution for $n = 4$.

Solution To save space, we have listed the paths instead of drawing the tree. You may wish to construct the probability tree that gives rise to these paths.

Path Number	Path Sequence	y	Probability
1	SSSS	4	π^4
2	SSSF	3	$\pi^3(1 - \pi)$
3	SSFS	3	$\pi^3(1 - \pi)$
4	SSFF	2	$\pi^2(1 - \pi)^2$
5	SFSS	3	$\pi^3(1 - \pi)$
6	SFSF	2	$\pi^2(1 - \pi)^2$
7	SFFS	2	$\pi^2(1 - \pi)^2$
8	SFFF	1	$\pi(1 - \pi)^3$
9	FSSS	3	$\pi^3(1 - \pi)$
10	FSSF	2	$\pi^2(1 - \pi)^2$
11	FSFS	2	$\pi^2(1 - \pi)^2$
12	FSFF	1	$\pi(1 - \pi)^3$
13	FFSS	2	$\pi^2(1 - \pi)^2$
14	FFSF	1	$\pi(1 - \pi)^3$
15	FFFS	1	$\pi(1 - \pi)^3$
16	FFFF	0	$(1 - \pi)^4$

All the paths corresponding to a particular y value have the same probability; for instance, each of the six paths that yield $y = 2$ has probability $\pi^2(1 - \pi)^2$. So adding up the path probabilities for a particular y value amounts to multiplying the number of paths by the appropriate probability.

y:	0	1	2	3	4
$P_Y(y)$:	$(1 - \pi)^4$	$4\pi(1 - \pi)^3$	$6\pi^2(1 - \pi)^2$	$4\pi^3(1 - \pi)$	π^4

□

A formula is needed to save the labor of actually counting paths. The methods of Section 5.1 can be used. One way to specify a path in a binomial experiment is to state the trials on which a success occurs. For example, if $n = 5$, specifying that successes occur only at trials 1 and 4—for short, S at (1, 4)—specifies the path SFFSF. The ordering of the trial numbers is irrelevant; S at (4, 1) also specifies the path SFFSF. Therefore, in n trials the number of paths containing y successes is the same as the number of subsets of size

y out of the first n integers. From Section 5.1, this number is

$$\binom{n}{y} = \frac{(n!)}{y!(n-y)!}$$

Using this expression for the relevant number of paths in a binomial probability tree, we obtain a general expression for the binomial probability distribution.

Binomial Probability Distribution

$$P_Y(y) = \frac{n!}{y!(n-y)!}\pi^y(1-\pi)^{n-y}, \qquad \text{for } y = 0, 1, \ldots, n$$

Appendix Table 1 (at the end of the book) contains numerical values of binomial probabilities. Each value of n determines a block of probabilities. For values of π below .5, values of π are read at the top of the block and values of y are read on the left. For values of π above .5, values of π are read at the bottom, and values of y on the right.

EXAMPLE 5.5 In Appendix Table 1 of binomial probabilities, find the probability distribution of a binomial random variable for $n = 5$ and (a) $\pi = .2$, (b) $\pi = .5$, and (c) $\pi = .7$.

Solution We look in the $n = 5$ block of Appendix Table 1. For $\pi = .2$ we read down the .20 column; for $\pi = .5$ we use the $\pi = .50$ column; for $\pi = .7$ we read *up* the $\pi = .70$ column. The resulting distribution is

y	0	1	2	3	4	5
$P_Y(y)$ for $\pi = .20$	.3277	.4096	.2048	.0512	.0064	.0003
$P_Y(y)$ for $\pi = .50$	.0313	.1563	.3125	.3125	.1563	.0313
$P_Y(y)$ for $\pi = .70$	.0024	.0284	.1323	.3087	.3602	.1681

□

EXAMPLE 5.6 In the long run, 20% of all management trainees are rated outstanding, 50% acceptable, and 30% unsatisfactory. In a sample of 20 randomly selected trainees, find the following probabilities:

a. Exactly 4 trainees are rated outstanding:
b. At least 4 trainees are rated outstanding;
c. Exactly 15 trainees are rated outstanding or acceptable;
d. At least 15 trainees are rated outstanding or acceptable.

Assume that we have a set of Bernoulli trials.

Solution a. Find the entry for $n = 20$, $\pi = .20$ (on top of the block), and $y = 4$ (to the left). The probability is .2182.

b. Add the entries for $n = 20$, $\pi = .20$, $y = 4, 5, 6, \ldots, 20$, and get .5886.

c. The probability that a rating is outstanding or acceptable is $.20 + .50 = .70$. Find the entry for $n = 20$, $\pi = .70$ (below the block), and $y = 15$ (to the right): .1789. Alternatively, this probability must equal the probability of exactly five unsatisfactory ratings, which has $\pi = .30$ and $y = 5$. This reasoning yields the same table entry: .1789.

d. Add the entries for $n = 20$, $\pi = .70$, $y = 15, 16, \ldots, 20$ to get .4163. Or add the entries for $n = 20$, $\pi = .30$, $y = 5, 4, \ldots, 0$ to get the equivalent probability of 5 or fewer unsatisfactory ratings. □

The expected value and variance of a binomial random variable Y depend, of course, on the values of n and π.

Mean and Variance of a Binomial Random Variable

$$E(Y) = n\pi$$

$$\text{Var}(Y) = n\pi(1 - \pi); \qquad \sigma_Y = \sqrt{n\pi(1 - \pi)}$$

The resulting expected value for a binomial random variable seems intuitively reasonable. If, on average, 30% of all trainees are rated outstanding, then in a sample of 20 trainees we would expect to find $20(.3) = 6$ who are rated outstanding.

SECTION 5.2 EXERCISES

5.5 Let Y be a binomial random variable. Compute $P_Y(y)$ for each of the following situations.

a. $n = 10$, $\pi = .2$, $y = 3$
b. $n = 4$, $\pi = .4$, $y = 2$
c. $n = 16$, $\pi = .7$, $y = 12$

5.6 It is given that Y has a binomial probability distribution with $n = 6$ and $\pi = .25$.

a. Calculate $P_Y(y)$ by hand for $y = 1$, 2, and 3. Compare your results to those listed in Appendix Table 1.
b. Draw a probability histogram of $P_Y(y)$.
c. Find the mean and standard deviation of Y.

5.7 Let $Y =$ the number of successes in 20 independent trials, where the probability of success on any one trial is .4. Find

a. $P(Y \geq 4)$
b. $P(Y > 4)$
c. $P(Y \leq 10)$
d. $P(Y > 16)$

5.8 Let Y be a binomial random variable with $n = 20$ and $\pi = .6$. Find $P(Y \leq 16)$ and $P(Y < 16)$. Compare these probabilities to the ones found in parts (a) and (b) of Exercise 5.7.

5.9 A chain of motels has adopted a policy of giving a 3% discount to customers who pay in cash rather than by credit cards. Its experience is that 30% of all customers take the discount. Let $Y =$ the number of discount takers among the next 20 customers.

a. Do you think the binomial assumptions are reasonable in this situation?
b. Assuming that binomial probabilities apply, find the probability that exactly 5 of the next 20 customers take the discount.
c. Find P(5 or fewer customers take the discount).
d. What is the most probable number of discount takers in the next 20 customers?

5.10 Find the expected value and standard deviation of the number of discount takers in Exercise 5.9.

5.11 Use the Empirical Rule to approximate the probability that Y in Exercise 5.9 falls within one standard deviation of its expected value. Use binomial tables to find the exact probability. How good is the Empirical Rule approximation?

5.12 A small company uses a parcel service to ship packages of special cheeses ordered as gifts. The company has found that 90% of all orders are delivered on time. A batch of 100 packages is sent out. Let Y = number of packages delivered on time.

a. Do the binomial assumptions seem reasonable in this situation?

b. Assuming that binomial probabilities apply, find $P(Y \geq 85)$.

5.13 Find $E(Y)$ and σ_Y in Exercise 5.12, assuming binomial probabilities.

5.14 A prescription drug manufacturer claims that only 10% of all new drugs that are shown to be effective in animal tests ever pass through all the additional testing required to be marketed. The manufacturer currently has 8 new drugs that have been shown to be effective in animal tests and await further testing and approval.

a. Find the probability that none of the 8 drugs is marketed.

b. Find the probability that at least 2 are marketed.

c. Find the expected number of marketed drugs among the 8.

5.15 Plot a probability histogram of $P_Y(y)$ in Exercise 5.14.

5.3 HYPERGEOMETRIC DISTRIBUTION

hypergeometric probability distribution

The counting formulas of Section 5.1 can be used to define the **hypergeometric probability distribution**. In this section we state a formula for this distribution and relate it to binomial distribution.

The situation that leads to the hypergeometric distribution is easily described. There must be a population consisting of some number N_S of successes and some number N_F of failures. The total population size is $N = N_S + N_F$.* A sample of size n is taken from the population without replacement. The relevant random variable is Y = observed number of successes in the sample. Example 5.1 illustrates one such situation: there are 13 successes (erroneous accounts) and 74 failures in the population, and the sample size $n = 10$. That example indicates the basic principle of the hypergeometric distribution, which is straight out of the classical interpretation of probability:

$$P(\text{event}) = \frac{\text{number of outcomes favoring the event}}{\text{total number of outcomes}}$$

In this context, outcome means sample. There are $\binom{N}{n}$ possible samples of size n that can be drawn from a population of size N. The samples that favor the event are those that have exactly y successes and exactly $n - y$ failures. As indicated in Example 5.1, there

* These Ns are not random variables, though we usually use capital letters to denote random variables. The Ns are constants.

are $\binom{N_S}{y}\binom{N_F}{n-y}$ such samples. This leads to the following hypergeometric distribution:

Hypergeometric Probability Distribution

Y = number of S's in a random sample of size n (taken without replacement) from a population consisting of N_S S's and N_F F's.

$$P_Y(y) = \frac{\binom{N_S}{y}\binom{N_F}{n-y}}{\binom{N}{n}}, \qquad y = 0, 1, \ldots, n$$

(If $y > N_S$, take $\binom{N_S}{y}$ and $P_Y(y)$ to be 0; if $n - y > N_F$, take $\binom{N_F}{n-y}$ and $P_Y(y)$ to be 0.)

EXAMPLE 5.7 For the situation of Example 5.1, find P(2 erroneous accounts in a sample of 10).

Solution We found in Example 5.1 that there were $\binom{87}{10} \approx 4{,}000{,}000{,}000{,}000$ possible samples. Of these, $\binom{13}{2}\binom{74}{8} \approx 1{,}200{,}000{,}000{,}000$ contained exactly two erroneous accounts. Therefore the probability of getting exactly two accounts in a sample size 10 (selected without replacement) is

$$P_Y(2) = \frac{\binom{13}{2}\binom{74}{8}}{\binom{87}{10}} \approx \frac{1{,}200{,}000{,}000{,}000}{4{,}000{,}000{,}000{,}000} = .3$$

□

Although we could set out many other illustrations of hypergeometric situations, we want to emphasize the close relation of hypergeometric and binomial probabilities. **If the population size N is large (relative to the sample size n), the distinction between the binomial and the hypergeometric is negligible.** If a random sample of 100 is taken from a population of 100,000,000, it does not matter in any serious way whether hypergeometric or binomial probabilities are used.

EXAMPLE 5.8 Find P(2 erroneous accounts) in Example 5.7 using a binomial probability distribution.

Solution We take $n = 10$ and $\pi = 13/87 \approx .149$. Thus $P(2 \text{ erroneous accounts}) = \binom{10}{2}(.149)^2(.851)^8 \approx .275$. This probability is calculated using the binomial probability distribution and is approximately equal to the probability of two erroneous accounts, .30, calculated using the hypergeometric probability distribution. As N gets larger, these probabilities become closer.

□

The close relation between hypergeometric and binomial probabilities extends to expected values and variances. We don't prove it, but the mean and variance for a hypergeometric probability distribution are as follows:

Expected Value and Variance for a Hypergeometric Random Variable Y

$$E(Y) = n\frac{N_S}{N}$$

$$\text{Var}(Y) = n\frac{N_S}{N}\left(1 - \frac{N_S}{N}\right)\frac{N-n}{N-1}$$

The ratio N_S/N is exactly π, the probability of success on a single trial; in Example 5.8 we took $\pi = 13/87$. Therefore the expected value of Y for the hypergeometric is $E(Y) = n\pi$, just like the binomial. By substituting $\pi = N_S/N$, the hypergeometric variance reduces to $n\pi(1-\pi)[(N-n)/(N-1)]$, as compared to the binomial variance $n\pi(1-\pi)$. There is an extra factor $(N-n)/(N-1)$, called the **finite population correction**. This factor is exactly equal to 1 when $n = 1$; otherwise, it is less than 1. For most practical situations in which the sample is a small fraction of the population (n is much smaller than N), the factor $(N-n)/(N-1)$ is nearly 1. For example, if $n = 100$ and $N = 100{,}000{,}000$, then $(N-n)/(N-1) \approx .999999$. Therefore we do not worry about the distinction between the hypergeometric and the binomial probability distributions in most situations; the distinction makes little numerical difference.

finite population correction

SECTION 5.3 EXERCISES

5.16 Let Y be a hypergeometric random variable with $N_S = 3$, $N_F = 4$, and $n = 3$.

a. Compute $P_Y(y)$ for $y = 0, 1, 2, 3$.

b. Graph this probability distribution.

5.17 Find the mean and standard deviation for the random variable Y defined in Exercise 5.16.

5.18 Compute $P_Y(2)$ for a hypergeometric random variable Y in each of the following situations:

a. $N_S = 2$, $N_F = 3$, $n = 3$

b. $N_S = 4$, $N_F = 4$, $n = 5$

c. $N_S = 5$, $N_F = 1$, $n = 3$

5.19 Compute the probability that $Y = 0$ in both of the following situations:

a. Y is binomial with $n = 5$ and $\pi = .40$.

b. Y is hypergeometric with $N_S = 2$, $N_F = 3$, and $n = 5$.

5.20 Refer to Exercises 5.3 and 5.4.

a. Find the probability that the tester selects the 4 correct glasses, assuming random selection.

b. Find $P_Y(y)$, where Y = number of correct choices.

5.21 Assume that in the 2500 business accounts of a bank, 125 have been fraudulently altered. The alterations are sufficiently subtle that only a detailed audit can uncover them. Fifty business accounts are chosen at random for detailed auditing. What is the probability that at least one of the alterations is discovered?

5.22 Find the expected value and variance of the number of altered accounts discovered during audit in Exercise 5.21.

5.23 Use a binomial approximation to answer Exercises 5.21 and 5.22. How close are the numerical answers?

5.24 Find the probability distribution of the number of potentially sympathetic judges selected in Exercise 5.2.

5.4 GEOMETRIC AND NEGATIVE BINOMIAL DISTRIBUTIONS

Bernoulli trials, yielding success or failure on each trial, with constant probabilities and independence from trial to trial, were discussed in Section 5.2. There we were concerned with the case in which the number of trials was fixed and the number of successes was random. In a number of cases, the situation is reversed: the number of successes is fixed and the number of trials is random. In this section, we deal with this situation, which leads to the geometric and negative binomial probability distributions.

Many banks supplement their usual teller services with card-operated automatic teller machines. There is some risk of unauthorized use of bank cards at these machines. Suppose that one of every thousand attempted automatic-teller transactions is based on an unauthorized use of a bank card. Regarding each transaction as a trial (and ignoring the possibility of repeated transactions using the same card), we can assume that transactions are a series of Bernoulli trials. The binomial distribution of Section 5.2 would apply to problems such as finding the probability that there are more than 20 unauthorized uses within the next 10,000 transactions. In such a problem, the number of trials (transactions) would be regarded as fixed and the number of successes (unauthorized uses) as random. Alternatively, we could ask for the number of transactions that occur before the next unauthorized use, or before the tenth unauthorized use. These questions lead to the geometric and negative binomial probability distributions.

The geometric distribution arises when we consider $Y =$ number of trials required to obtain the next success. The probability tree for a geometric random variable is very simple. To require y trials to obtain a success is to require that there be $y - 1$ consecutive failures followed by a success.

Geometric Probability Distribution

In a Bernoulli trials situation, define $Y =$ number of trials required to obtain a success. Then

$$P_Y(y) = \pi(1 - \pi)^{y-1}, \qquad y = 1, 2, 3, \ldots$$

where π is the probability of success on any trial.

These probabilities form a geometric series. If $\pi = .2$, the probabilities are .2, .2(.8), $.2(.8)^2$, ...

EXAMPLE 5.9 The labels on bottles of medication are examined with an optical scanner to see that they are properly affixed to the bottles. Assume that the probability of detecting an improperly affixed label is $\pi = .0001$ and compute the probability that the process will detect an improper label on the very first trial. Also compute the probability that the process will first detect an improper label on exactly the 10,000th bottle.

Solution The event "improper label at trial 1" is the same as the event "$Y = 1$," where Y = number of trials to find the first improper label. Assuming Bernoulli trials with $P(\text{Success}) = \pi = .0001$, the geometric distribution applies. $P(Y = 1) = P_Y(1) = (.0001)(.9999)^{1-1} = .0001$. The event "first improper label at bottle 10,000" is the same as the event "$Y = 10{,}000$" and has probability $P_Y(10{,}000) = (.0001)(.9999)^{10{,}000-1} = .0000368$. Note that, even though we expect one improper label every 10,000 bottles, there is a higher probability that the next bad label will occur at the very next bottle than that it will occur at precisely the next 10,000th bottle. □

The mean (expected value) and variance of a geometric random variable may be computed by another convenient shortcut formula.

Mean and Variance for a Geometric Random Variable

$$E(Y) = \frac{1}{\pi}$$

$$\text{Var}(Y) = \frac{1 - \pi}{\pi^2}$$

where π is the probability of success on any given trial.

EXAMPLE 5.10 Find the expected value and variance of the number of labels examined until the next improper label is found, using the assumptions of Example 5.9.

Solution We have $\pi = .0001$, so $E(Y) = 1/(.0001) = 10{,}000$. It is reasonable that, if one out of every 10,000 labels is improper, we will wait an average of 10,000 bottles to find the next improper label. The variance is $(1 - .0001)/(.0001)^2 = 99{,}990{,}000$; therefore, the standard deviation of Y is $\sqrt{99{,}990{,}000} = 9999.5$. □

The idea of counting the number of trials to the next success may be extended to counting the number of trials to the kth success. For example, a market research firm that needs to obtain $k = 100$ women who have full-time jobs and also watch a certain local television newscast has to interview a random number of potential candidates. Each interview is a trial; the most relevant random variable is Y = number of interviews needed to obtain 100 qualifying women. If the assumptions of Bernoulli trials (success or failure trials, constant probability of success, independent trials) hold, the probability distribution of Y = number of trials required to obtain k successes is negative binomial.

Negative Binomial Distribution

If Y = number of trials to obtain k successes, then

$$P_Y(y) = \frac{(y-1)!}{(k-1)!(y-k)!}\pi^k(1-\pi)^{y-k}, \qquad y = k, k+1, \ldots$$

The reason that $y - 1$ and $k - 1$ occur in the expression for the negative binomial distribution is that there must be $k - 1$ successes in the first $y - 1$ trials, followed by one success (at trial y).

EXAMPLE 5.11 In Example 5.9 we assumed that the probability of an improperly affixed label was .0001. Suppose that 50 improperly affixed labels are needed to study the cause of improper label fixing. Write an expression for the probability that 100,000 or more bottles are needed to obtain 50 improper labels.

Solution We may regard the number of successes (improperly affixed labels) as fixed and find the probability that $Y =$ number of bottles required is at least 100,000:

$$P(Y \geq 100{,}000) = \sum_{100{,}000}^{\infty} \frac{(y-1)!}{(50-1)!(y-50)!} (.0001)^{50}(.9999)^{y-50}$$ □

Because the negative binomial distribution is simply the extension of the geometric distribution to $k > 1$ successes, it is not surprising that expressions for the mean and variance of the negative binomial distribution are extensions of those for the geometric distribution.

Mean and Variance of the Negative Binomial Distribution

If $Y =$ number of trials required to obtain k successes,

$$E(Y) = \frac{k}{\pi}$$

$$\text{Var}(Y) = \frac{k(1-\pi)}{\pi^2}$$

EXAMPLE 5.12 Find the expected value and standard deviation of the number of bottles required to find 50 improperly affixed labels, assuming that the probability of an improperly affixed label is .0001.

Solution $E(Y) = 50/.0001 = 500{,}000$. $\text{Var}(Y) = 50(.9999)/(.0001)^2 = 4{,}999{,}500{,}000$. The standard deviation is $\sqrt{4{,}999{,}500{,}000} = 70{,}707$. □

Not all Bernoulli trials situations can be solved by binomial or negative binomial methods. If $Y =$ number of trials until two consecutive successes occur, then neither the number of trials nor the number of successes is fixed, so neither binomial nor negative binomial probabilities apply. In such cases, one must go back to basic principles to find the relevant probabilities.

5.5 POISSON DISTRIBUTION

A different sort of probability situation occurs when a succession of events seems to happen at random over time. An electrical utility faces occasional thunderstorms that down power lines or damage transformers. While the long-run probability of occurrence of such storms can be determined quite accurately, the timing of the next storm is rather unpredictable. A company that insures oil tankers cannot predict the time of the next sinking. The manager of a university computer center faces random variation in the timing of job submissions. It's important to be able to protect against probable variation in such situations.

Poisson probability distribution

The **Poisson probability distribution*** is the simplest and most widely used model of events occurring randomly in time. This distribution is the mathematical result of certain assumptions. If the assumptions are not correct, at least approximately, for a particular situation, then the Poisson distribution may be a bad model in that situation. The two crucial assumptions can be translated (without doing much violence to the mathematical niceties) as follows:

1. Events occur one at a time. Two or more events do not occur at precisely the same time.
2. The occurrence of the event of interest in a given period is independent of the occurrence of the event in a nonoverlapping period; that is, the occurrence (or nonoccurrence) of an event during one period does not change the probability of an event occurring in some later period.

In many discussions on this topic, a third assumption is added: that the expected number of events in a period of specified length stays constant, so that the expected number of events during any one period is the same as during any other period. This third assumption makes the math easier, but it has been proved to be essentially irrelevant. As long as the first two assumptions hold, the Poisson distribution results.

There are two approaches to assessing whether or not a Poisson distribution is a reasonable model in a given situation. One is to see if the assumptions seem reasonable in a given context, the other is to see if the actual data histogram looks like a Poisson probability histogram. Of course the ideal is to have both.

EXAMPLE 5.13 In the three situations described at the beginning of this section, should the Poisson assumptions hold?

Solution We would expect that the assumption of independence would be shaky for the electrical utility example. It seems to us that if lightning from one storm knocks out some equipment, it is quite likely that lightning from the same storm or another in the vicinity will knock out other equipment. For the oil tanker example, one could argue that, since one large tanker might collide with another, sinking both, the assumption that events happen one at a time doesn't hold. While this is certainly possible, we would guess that such flukes

* Named for Simeon Poisson, the mathematician who first derived it.

are sufficiently rare that the Poisson distribution is a decent model for the probability of a tanker sinking in a given period. In the computer center, much depends on the situation. If there are only a few terminals, which are tied up during the processing of a job, then the submission of a job now reduces the probability of submission of another job (from the same terminal) a bit later, violating the assumption of independence. But if there are many terminals or if a terminal is not tied up during processing, the Poisson assumptions look good to us. We would like to see some data! □

Poisson Probability Distribution

$$P_Y(y) = \frac{e^{-\mu}\mu^y}{y!}, \qquad y = 0, 1, 2, \ldots$$

where μ is the expected number of events occurring in a given period and $e = 2.71828 \ldots$.

A Poisson random variable Y is the number of random events that occur in a fixed period; in principle, there's no upper limit to the values y. In practice, very large values of y are extremely unlikely. Probabilities for the Poisson probability distribution are shown in Appendix Table 2.

EXAMPLE 5.14 On Saturday mornings, customers enter a boutique at a suburban shopping mall at an average rate of .50 per minute. Let Y = number of customers arriving in a specified 10-minute interval of time. Find the following probabilities:

a. $P(Y = 3)$
b. $P(Y \leq 3)$
c. $P(Y \geq 4)$
d. $P(4 \leq Y \leq 10)$

Solution The Poisson assumptions seem fairly reasonable in this context. We assume that customers don't arrive in groups (or else count the entire group as one arrival) and that the arrival of one customer neither decreases nor increases the probability of other arrivals.

To obtain μ, we note that, at an average rate of .50 per minute over a 10-minute time span, we would expect $\mu = (.50)(10) = 5.0$ arrivals. To find the probabilities, we consult Appendix Table 2.

a. $P(Y = 3)$ is read directly from Appendix Table 2 with $\mu = 5$ and $y = 3$: $P(Y = 3) = .1403$.
b. $P(Y \leq 3) = P(Y = 0) + P(Y = 1) + P(Y = 2) + P(Y = 3)$
$= .0067 + .0337 + .0843 + .1403 = .2650.$
c. $P(Y \geq 4) = 1 - P(Y \leq 3) = 1 - .2650 = .7350$
d. $P(4 \leq Y \leq 10) = P(Y = 4) + P(Y = 5) + \ldots + P(Y = 10)$
$= .1755 + .1755 + \ldots + .0181 = .7213$ □

As indicated in the definition of the Poisson probability distribution, the expected value is $E(Y) = \mu$. Coincidentally, the variance of a Poisson random variable is also μ.

Mean and Variance for a Poisson Random Variable
If Y has a Poisson distribution, then $E(Y) = \mu$ $\text{Var}(Y) = \mu$

EXAMPLE 5.15 Find the standard deviation of Y in Example 5.14.

Solution We noted in Example 5.14 that $\mu = 5.0$. Thus

$$\sigma_Y = \sqrt{\text{Var}(Y)} = \sqrt{5.0} = 2.24$$ □

Poisson approximation to binomial distributions

The Poisson distribution provides a good approximation to the binomial probability distribution when π is small and n is large but $n\pi$ is less than 5. The Poisson expected value μ is equated to the binomial expected value $n\pi$ for this approximation.

EXAMPLE 5.16 A sample of 1000 patients are treated with a new drug product during a large clinical trial. Compute the probability that none of the patients experiences a particular side effect (such as nausea) if we assume $\pi = .001$.

Solution The mean of the binomial distribution is $\mu = n\pi = 1000(.001) = 1$. Substituting into the Poisson probability distribution, with $\mu = 1$, we have

$$P_Y(0) = \frac{(1)^0 e^{-1}}{0!} = e^{-1} = .3679$$

The corresponding probability computed using the binomial probability distribution is $\binom{1000}{0}(.001)^0(.999)^{1000} \approx .3677$. □

SECTION 5.5 EXERCISES

5.25 Let Y denote a random variable with a Poisson distribution. Use Appendix Table 2 to calculate
a. $P_Y(1)$ for $\mu = .4$, $\mu = .7$, and $\mu = 4.8$;
b. $P(Y \leq 3)$ for $\mu = 1.6$ and $\mu = 7.0$;
c. $P(Y \leq 10)$ for $\mu = 2.1$ and $\mu = 10.0$.

5.26 Calculate and graph the Poisson probability distribution for $\mu = .5$. Is the distribution roughly symmetric?

5.27 A firm that insures homes against fire assumes that claims arise according to a Poisson distribution at an average rate of 2.25 per week. Let Y be the number of claims arising in a four-week period. Find (a) $P(Y \leq 10)$, (b) $P(Y \geq 7)$, and (c) $P(7 \leq Y \leq 11)$.

5.28 Find the expected value and standard deviation of Y in Exercise 5.27.

5.29 Can you think of insurance situations that would make the Poisson assumption in Exercise 5.27 unreasonable?

5.30 Logging trucks have a particular problem with tire failures due to blowouts, cuts, and large punctures; these trucks are driven fast over very rough, temporary roads. Assume that such failures occur according to a Poisson distribution at a mean rate of 4.0 per 10,000 miles.

a. If a truck drives 1000 miles in a given week, what is the probability that it does not have any tire failures?

b. What is the probability that it has at least two failures?

5.31 What are the expected value and standard deviation of the number of tire failures per 1000 miles driven in Exercise 5.30?

5.32 The Poisson distribution also applies to events occurring randomly over an area or in a volume. Chocolate chips spread through well-mixed cookie dough tend to follow a Poisson distribution. A commercial baker produces cookies with an average of 8 chips per cookie.

a. What is the probability of (horrors!) a chipless cookie?

b. A cookie is considered acceptable only if it has at least 5 chips. What fraction of the cookies are acceptable?

5.6 UNIFORM DISTRIBUTION

The simplest continuous distribution is uniform distribution. If Y has a uniform distribution, its probability density is spread out evenly over some range a to b. Uniform density is shown in Figure 5.2.

Uniform density arises naturally in random number picking. If $Y =$ a number chosen randomly between 0 and 1, then the probability density of Y is flat over the 0 to 1 range; no one number has higher probability (density) than any other.

The basic formulas—probability density, expected value, and variance—for a uniform random variable are very simple.

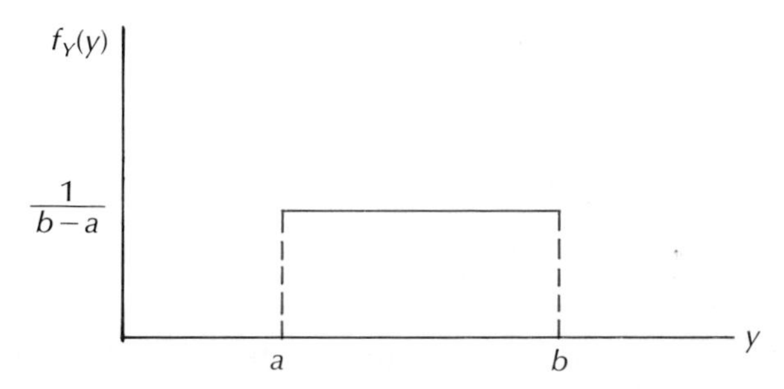

FIGURE 5.2 Uniform Probability Density

Probability Density, Mean, and Variance for a Uniform Random Variable

$$f_Y(y) = \begin{cases} \dfrac{1}{b-a}, & \text{if } a < y < b \\ 0, & \text{otherwise} \end{cases}$$

$$E(Y) = \frac{a+b}{2}$$

$$\text{Var}(Y) = \frac{(b-a)^2}{12}$$

Probabilities for uniform random variables may be found by simple geometry. For example, suppose that Y is uniformly distributed between 0 and 50; what is the probability that Y is between 10 and 40? Figure 5.3 shows the situation. The desired probability is the area of a rectangle—the base times the height. Therefore $P(10 < Y < 40) = 30(1/50) = .6$.

Alternatively, we may find uniform probabilities using elementary calculus. If Y is uniformly distributed between 0 and 50,

$$P(10 < Y < 40) = \int_{10}^{40} \left(\frac{1}{50}\right) dy = y\Big|_{10}^{40}$$

$$= \frac{(40-10)}{50} = .6$$

EXAMPLE 5.17 A mail order company specializing in software programs for microcomputers has found that between 5 and 15% of all orders in a day require special shipping. Suppose that, as a first approximation, we take the distribution of Y = percentage of orders on a randomly chosen day to be uniform over the range 5 to 15. Find the mean and standard deviation of Y and the probability that Y is between 9 and 12.

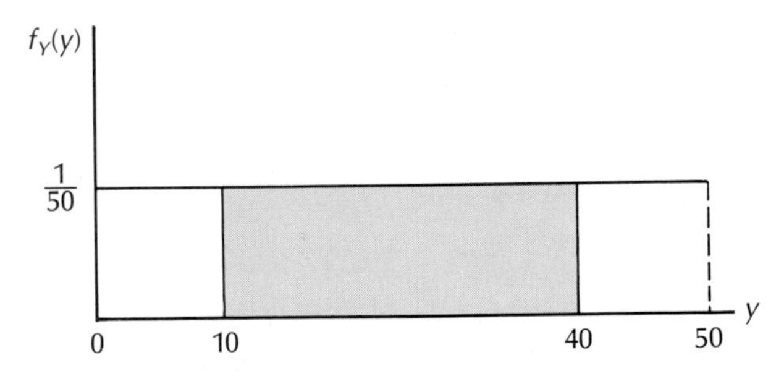

FIGURE 5.3 **Probabilities from a Uniform Distribution**

Solution We have $a = 5$ and $b = 15$. Therefore

$$E(Y) = \frac{(5 + 15)}{2} = 10$$

$$\text{Var}(Y) = \frac{(15 - 5)^2}{12} = 8.3333$$

$$\sigma_Y = \sqrt{8.3333} = 2.89$$

The probability that Y is between 9 and 12 is the area of a rectangle with base $12 - 9 = 3$ and height $1/(15 - 5) = .1$; therefore

$$P(9 < Y < 12) = 3(.1) = .3$$

□

SECTION 5.6 EXERCISES

5.33 Suppose Y is a uniformly distributed random variable on the interval $10 < y < 120$. Graph the density of Y and find the probability that Y lies in the interval $60 < y < 85$.

5.34 Find the expected value and standard deviation of Y in Exercise 5.33.

5.35 Calculate the following probabilities for a uniform random variable over the interval $0 < y < 200$:

a. $P(10 < Y < 50)$
b. $P(Y > 50)$
c. $P(Y \leq 120)$

5.36 A random telephone-dialing machine picks the last four digits of telephone numbers randomly between 0000 and 9999 (both included). Treat the random variable Y = number selected as continuous (even though there are 10,000 discrete possibilities) and uniformly distributed.

a. Find $P(0300 < Y \leq 1300)$.
b. Find the variance of Y.

5.37 On summer days, Y = time that a suburban commuter train is late can be modeled as uniformly distributed between 0 and 20 minutes.

a. Find the probability that the train is at least 8 minutes late.
b. Find the standard deviation of the amount of time the train is late.

5.7 EXPONENTIAL DISTRIBUTION (∫)

The Poisson distribution discussed in Section 5.5 applies to events occurring randomly over time. Specifically, it applies to Y = number of events occurring in a fixed period. If events occur randomly in time, we may also ask about W = waiting time until the next occurrence of an event. Given appropriate assumptions, the probability distribution of W follows an exponential distribution. In contrast to the discrete Poisson random variable, an exponential random variable is continuous.

Recall that the assumptions for a Poisson distribution are that events occur separately and that the occurrence of an event in one period does not change the probability of occurrence in another period. In a waiting-time problem, we must also assume that the expected rate of occurrence is constant over the period. Thus, if there is a rush-hour–off-hour situation in which events occur frequently and then infrequently, exponential probabilities do not apply.

One of the many uses of exponential probabilities is in reliability problems. If a component of a system fails only because of random occurrences (as opposed to wearing out), it's reasonable to assume that nonfailure in previous periods does not change the probability of failure in the next period and also that the rate of failure is constant over time. Of course there can be no clumping of multiple failures of a single component in a single period.

Given these assumptions, the probability density W = time to the next event is exponential.

Exponential Density

Assume that events occur randomly over time, with the expected time between events denoted μ. If W = time to the next event, then

$$f_W(w) = \left(\frac{1}{\mu}\right)e^{-w/\mu}, \qquad w > 0$$

$$E(W) = \mu$$

$$\text{Var}(W) = \mu^2$$

The exponential density is often called a "waiting-time" distribution because it is so often used as a model for the length of time one must wait to obtain the next event.

Probabilities involving the exponential density may be found using elementary integral calculus. It is not hard to show that

$$\int_a^b \left(\frac{1}{\mu}\right)e^{-w/\mu}\,dw = e^{-a/\mu} - e^{-b/\mu}$$

Most calculators and many computer programs evaluate e^x automatically.

EXAMPLE 5.18 The average length of time between submissions of jobs to a mainframe computer during the work day is 2.5 minutes. Assume that jobs arrive randomly over time at a constant expected rate.

a. What is the probability that the waiting time between jobs is between 2 and 5 minutes?
b. Find the mean and standard deviation of the waiting time between jobs.

Solution a. It is specified that $\mu = 2.5$ minutes,

$$P(2 < W < 5) = \int_2^5 \left(\frac{1}{2.5}\right)e^{-w/2.5}\,dw$$
$$= e^{-2/2.5} - e^{-5/2.5} = .4493 - .1353 = .3140$$

b. The mean of W was given as 2.5. The variance is $\mu^2 = (2.5)^2 = 6.25$. The standard deviation is, as always, the square root of the variance, $\sqrt{6.25} = 2.5$. Curiously, the standard deviation equals the mean in this distribution. □

The exponential density is only a model. Accurate use of the model requires that the underlying assumptions hold. The assumption of independence over time is particularly critical for the exponential density to apply. The assumption of a constant average rate is also important.

EXAMPLE 5.19 We assumed in Example 5.18 that jobs are submitted to the computer at a rate of 2.5 minutes per job. Suppose that

a. there is a tendency for one job submission to be followed immediately by another, related submission;
b. there is a tendency for jobs to be submitted "on the hour," when employees arrive from or leave for meetings.

What assumptions are called into question by each tendency?

Solution a. Here there is a dependence. If a job is submitted at one period, there is a higher probability that another job will be submitted shortly thereafter.

b. Here there is a nonconstant average rate of occurrence. The rate is higher "on the hour." □

SECTION 5.7 EXERCISES

5.38 Use a calculator that finds e^x to compute the value of the exponential density function $f_Y(y)$ for $\mu = 2.5$ and $y = 0, .5, 1.0, 1.5,$ and 2.0. Sketch the density function.

5.39 Compute the following probabilities for an exponential random variable with $\mu = 2$.
a. $P(Y > 2)$
b. $P(Y > 1)$
c. $P(1 < Y < 2)$
d. $P(1 \leq Y \leq 2)$
(Hint: For part (d), use your head, not your calculator.)

5.40 The time between arrivals at a rural emergency treatment center follows an exponential distribution with an average time between arrivals of 1.25 hours. Find the probability that the time between arrivals is more than 1 hour. Find the probability that the time between arrivals is more than 2 hours.

5.41 Rather than focus on the time between arrivals at the treatment center in Exercise 5.40, focus on the arrivals in a given period. Note that the assumptions for the exponential and Poisson distributions are identical; note also that an average time between arrivals of 1.25 hours indicates an average of $1/1.25 = .80$ arrivals per hour.
a. Using Poisson probabilities, find the probability that there are no arrivals in 1 hour.
b. Find the probability that there are no arrivals in 2 hours.
c. Compare your answers to this exercise with those of Exercise 5.40. What is the explanation?

5.42 The service times for unticketed passengers at an airline ticket counter follow an exponential distribution with a mean of 5 minutes.

a. Find the probability of a service time less than 2.5 minutes.
b. Find the probability of a service time longer than 10 minutes.

5.43 Consider the passenger service situation of Exercise 5.42.
a. What is the expected number of passengers served per minute?
b. Find the probability that at least one passenger is served within 2.5 minutes.
c. Find the probability that no passenger is served within 10 minutes.

5.44 "Unusual events"—minor operating problems—occur randomly over time at a nuclear power station. The average time between events is 40 days.
a. What is the probability that the time to the next "unusual event" is between 20 and 60 days?
b. Find the standard deviation of the time to the next "unusual event."

5.45 Examination of the records of the power station in Exercise 5.44 shows that "unusual events" occur much more frequently on weekends than on weekdays. What assumption underlying your answers to Exercise 5.44 is called into question?

5.46 A major league baseball team sells individual-game tickets at a downtown ticket office during normal working hours. Ticket buyers arrive at the office at the average rate of 12 per hour. Buyers arrive individually and randomly; the average rate stays essentially constant during the day.
a. Find the probability that there are more than 5 arrivals in a 10-minute (1/6 hour) period.
b. Find the probability that the next buyer arrives within 3 minutes. Note that the average time between arrivals is 1/12 hour, or 5 minutes.

5.47 Find a number k such that the probability of k or more arrivals in 1/4 hour in Exercise 5.46 is close to .10.

5.48 The time between "crashes" of a certain mainframe computer appears to have an exponential probability distribution. The average time is 5 days.
a. What is the probability that the time to the next crash will be at least one (7-day) week?
b. What is the probability that there will be a two-week period with no crashes?

5.49 What is the probability that there will be 4 or more crashes of the computer in Exercise 5.48 within a specified 7-day week?

5.8 NORMAL DISTRIBUTION

Now we turn to the most fundamental distribution used in statistical theory, normal distribution. Many standard statistical procedures discussed in later chapters are based on the formal mathematical assumption that the underlying population has a normal distribution. Many methods that are widely used in economics, finance, and marketing are based on an assumption of a normal population. This section is therefore important in understanding many later sections of the text.

A normally distributed random variable is continuous. Therefore it has a probability density, as shown below.

Normal Probability Density

$$f_Y(y) = \frac{1}{\sqrt{2\pi}\,\sigma} e^{-.5(y-\mu)^2/\sigma^2}$$

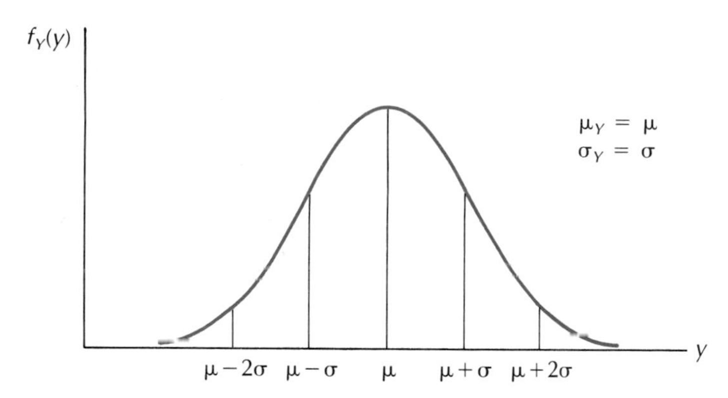

FIGURE 5.4 Normal Probability Distribution

The values μ and σ in the normal density function are in fact the mean and standard deviation of Y (though we don't prove it). A probability histogram, called the **normal curve**, for a normal random variable is bell shaped and symmetric around the mean μ, as shown in Figure 5.4.

normal curve

Tables of normal curve areas (probabilities) are always given for the **standard normal distribution**, which has mean 0 and standard deviation 1. Appendix Table 3 gives areas between 0 and a positive number z. For instance, the entry for $z = 1.00$ is .3413; if Z is the standard normal random variable, then $P(0 \leq Z \leq 1.00) = .3413$, as in Figure 5.5.

standard normal distribution

EXAMPLE 5.20 Let Z be a standard normal random variable. Find

a. $P(0 \leq Z \leq 1.96)$
b. $P(Z > 1.96)$
c. $P(-1.96 \leq Z \leq 1.96)$
d. $P(-1.00 \leq Z \leq 1.96)$

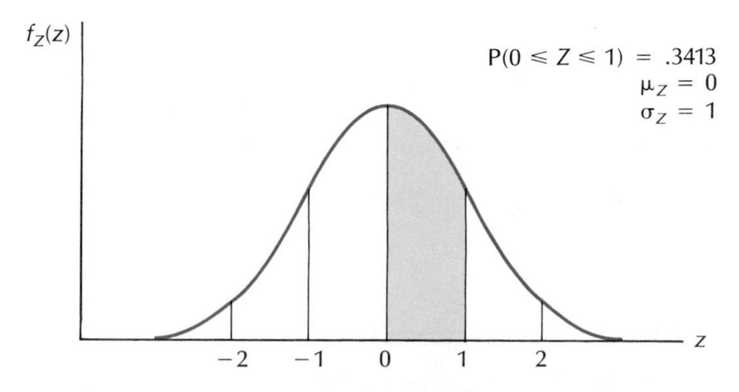

FIGURE 5.5 Standard Normal Probability Distribution

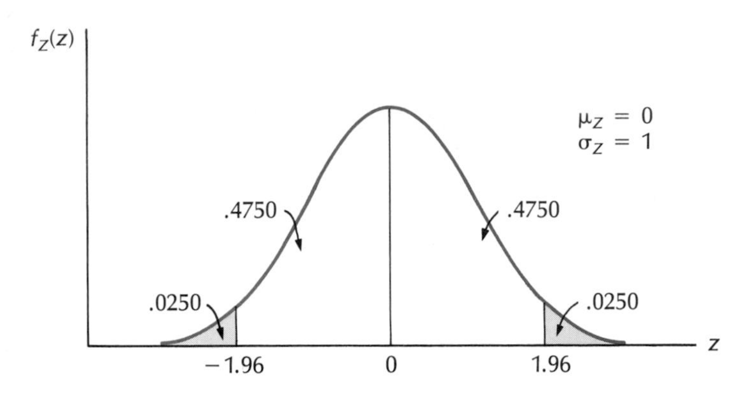

FIGURE 5.6 **Solution to Example 5.20**

Solution An illustration like Figure 5.5 makes it much easier to use normal tables. The entry for $z = 1.96$ (found by looking in the 1.9 row and .06 column) is .4750. Figure 5.6 is useful.

a. $P(0 \leq Z \leq 1.96) = .4750$.

b. Because the area to the right of 0 must be .5000 (the normal curve is symmetric and the total area beneath the curve is 1), $P(Z > 1.96) = .5000 - .4750 = .0250$.

c. By symmetry, the area between -1.96 and 0 must also be .4750. So $P(-1.96 \leq Z \leq 1.96) = .4750 + .4750 = .9500$.

d. $P(-1.00 \leq Z \leq 1.96) = .3413 + .4750 = .8163$. (Draw a picture.) □

EXAMPLE 5.21 Find k_1 such that $P(0 \leq Z \leq k_1) = .40$ and k_2 such that $P(-k_2 \leq Z \leq k_2) = .60$.

Solution This problem is in a sense the opposite of Example 5.20. In that problem, values are given and probabilities have to be found. Here probabilities are given and values have to be found. Again, a picture is helpful (see Figure 5.7).

a. Looking through Appendix Table 3 for an area of .40, we find that the closest z value is 1.28. Therefore $P(0 \leq Z \leq 1.28) = .40$; that is, $k_1 = 1.28$.

b. An area of .30 (half the desired probability as shown in Figure 5.7) corresponds to $z \approx .84$, so $P(-.84 \leq Z \leq .84) = .60$; that is, $k_2 = .84$. □

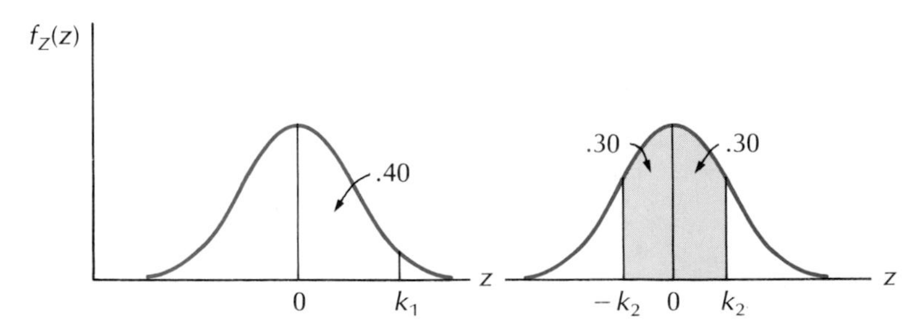

FIGURE 5.7 **Solution to Example 5.21**

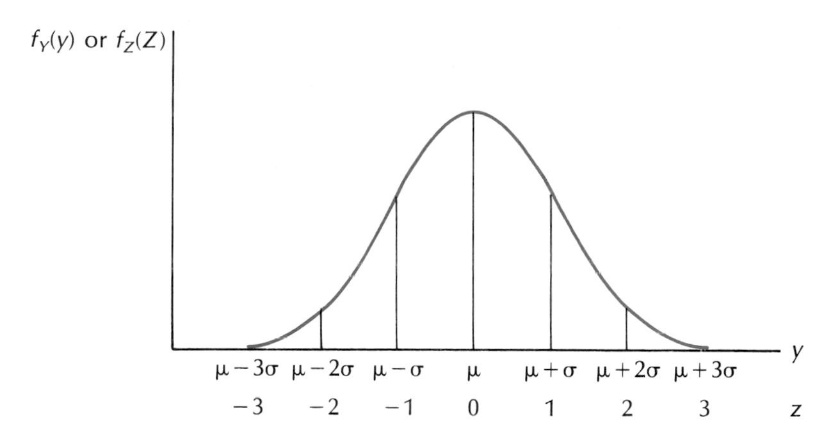

FIGURE 5.8 **Relation between Specific Values of Y and z-Scores**

Any normal random variable Y can be transformed to a standard normal random variable Z by subtracting the expected value μ and dividing the result by the standard deviation σ.

$$Z = \frac{Y - \mu}{\sigma}$$

***z*-score** For a given value of y, the corresponding value of z, sometimes called a ***z*-score**, is the number of standard deviations that y lies away from μ. If $\mu = 100$ and $\sigma = 20$, a y-value of 130 is 1.5 standard deviations above (to the right of) the mean μ and the corresponding z-score is $z = (130 - 100)/20 = 1.50$. A y-value of 85 is .75 standard deviations below (to the left of) the mean μ and

$$z = \frac{85 - 100}{20} = -.75$$

The relation between specific values of a normal random variable Y and the corresponding z-scores is shown in Figure 5.8. [Note that $z = (y - \mu)/\sigma$.]

EXAMPLE 5.22 Annual incomes for career service employees at a large university are approximately normally distributed with a mean of \$6200 and a standard deviation of \$900. Find the probability that an employee chosen at random has an annual income less than \$5000; an income greater than \$7000.

Solution First we draw a figure showing the areas in question (Figure 5.9). Now we must determine the area between 5000 and 6200.

$$z = \frac{y - \mu}{\sigma} = \frac{5000 - 6200}{900} = \frac{-1200}{900} = -1.33$$

The area between the mean of a normal distribution and a value 1.33 standard deviations to the left of the mean, from Appendix Table 3, is .4082. Hence the probability of observing

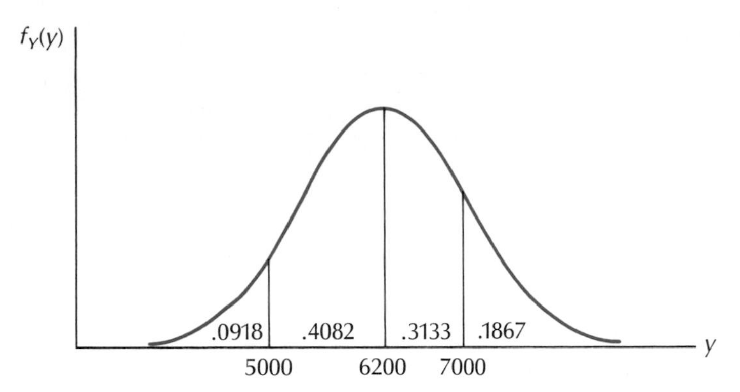

FIGURE 5.9 **Areas Greater Than 7000 and Smaller Than 5000 for $\mu = 6200$ and $\sigma = 900$; Example 5.22**

an annual income less than \$5000 is

$$.5 - .4082 = .0918$$

Similarly, to compute the probability of observing a salary under \$7000 we determine the area between 6200 and 7000:

$$z = \frac{y - \mu}{\sigma} = \frac{7000 - 6200}{900} = .89$$

The area corresponding to $z = .89$ is .3133. Hence the desired probability is

$$.5 - .3133 = .1867$$

□

EXAMPLE 5.23 If Y has a normal distribution with mean 500 and standard deviation 100, find

a. $P(500 \leq Y \leq 696)$;
b. $P(Y \geq 696)$;
c. $P(304 \leq Y \leq 696)$;
d. k such that $P(500 - k \leq Y \leq 500 + k) = .60$.

Solution a. A y-value of 696 is 1.96 standard deviations above the mean; $z = (696 - 500)/100 = 1.96$. Of course 500 is zero standard deviations above the mean, so $z = (500 - 500)/100 = 0.00$. Thus $P(500 \leq Y \leq 696) = P(0 \leq Z \leq 1.96) = .4750$.

b. $P(Y \geq 696) = P(Z \geq 1.96) = .0250$.

c. $P(304 \leq Y \leq 696) = P(-1.96 \leq Z \leq 1.96) = .9500$, because 304 corresponds to a z of $(304 - 500)/100 = -1.96$.

d. As in Example 5.21, $P(-.84 \leq Z \leq .84) = .60$, so we want a range for Y from .84 standard deviation below the mean $\mu = 500$ to .84 standard deviation above the mean: $P[500 - .84(100) \leq Y \leq 500 + .84(100)] = P(416 \leq Y \leq 584) = .60$ (see Figure 5.10). □

A little practice with such problems and a habit of drawing pictures makes normal probability calculations fairly easy.

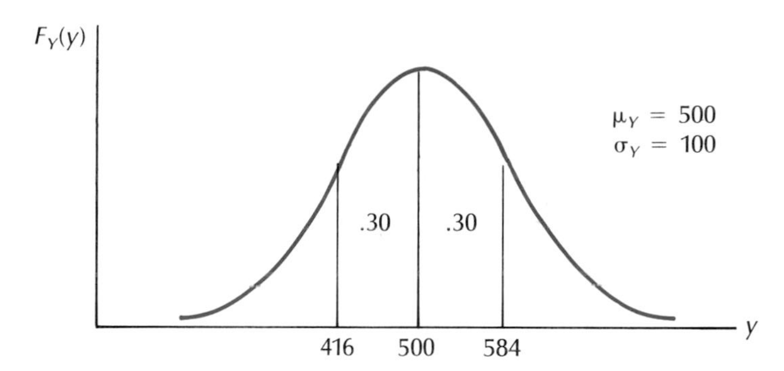

FIGURE 5.10 Solution to Example 5.23

As we've indicated previously, the concept of a continuous random variable is really an abstraction, since most variables of interest do have a finite number of possible values. But for many situations it is convenient to assume that the random variable of interest has a continuous distribution. In the same way, the normal random variable is an abstraction, since in theory any numerical value, negative as well as positive, is possible and the probability histogram is a smooth, symmetric, bell-shaped curve. In practice, negative values or positive values such as $612.3142769 may be impossible. Such issues often don't really matter. If a random variable Y is assumed normal with mean 500 and standard deviation 100, the probability that $Y < 0$ is by assumption $P(Z < -5)$, which is effectively zero. Whether or not Y can actually assume negative values hardly matters. Similarly, the errors incurred by rounding off $612.3142769 to $612.31 or to $612 are tiny. If a population histogram for a random variable is generally bell shaped, the normal probability distribution usually provides an excellent model for the actual probability distribution.

SECTION 5.8 EXERCISES

5.50 Suppose that Z represents a standard (tabled) normal random variable. Find the following probabilities.

a. $P(0 \leq Z \leq 1.00)$
b. $P(0 \leq Z \leq 1.65)$
c. $P(-1.00 \leq Z \leq 0)$
d. $P(-1.28 \leq Z \leq 0)$
e. $P(-1.65 \leq Z \leq 1.65)$
f. $P(-1.28 \leq Z \leq 1.28)$
g. $P(-1.07 \leq Z \leq 2.33)$
h. $P(Z \geq 2.65)$
i. $P(Z \leq -2.42)$
j. $P(Z \geq 1.39 \text{ or } Z \leq -1.39)$

Draw pictures.

5.51 For the standard normal random variable Z, solve the following equations for k:

a. $P(Z \geq k) = .01$
b. $P(-k \leq Z \leq k) = .98$
c. $P(Z \leq -k) = .01$
d. $P(-k \leq Z \leq k) = .6826$
e. $P(-k \leq Z \leq k) = .9544$
f. $P(Z \geq k) = .95$

Again, draw pictures.

5.52 Refer to the answers to Exercise 5.51, parts (d) and (e). How do these answers relate to the Empirical Rule?

5.53 Suppose that Y represents a normally distributed random variable with expected value (mean) equal to 100 and standard deviation 15.

a. Show that the event $(Y \leq 130)$ is equivalent to $(Z \leq 2)$.

b. Convert the event $(Y \geq 82.5)$ to z-score form.

c. Find $P(Y \leq 130)$ and $P(Y \geq 82.5)$.

d. Find $P(Y > 106)$, $P(Y < 94)$, and $P(94 \leq Y \leq 106)$.

e. Find $P(Y \leq 70)$, $P(Y \geq 130)$, and $P(70 < Y < 130)$.

5.54 Consider the random variable Y of Exercise 5.53. Find the value of k satisfying

a. $P(100 \leq Y \leq 100 + k) = .45$;

b. $P(100 - k \leq Y \leq 100 + k) = .90$;

c. $P(Y \geq k) = .20$;

d. $P(Y \leq k) = .30$;

e. $P(Y \leq k) = .80$;

f. $P(Y \geq k) = .70$.

Draw appropriate pictures for each part.

5.55 A financial analyst states that the (subjective probability) price Y of a long-term \$1000 government bond one year later is normally distributed with expected value \$980 and standard deviation \$40.

a. Find $P(Y \geq 1000)$.

b. Find $P(Y \leq 940)$.

c. Find $P(960 \leq Y \leq 1060)$.

5.56 Refer to the random variable Y of Exercise 5.55.

a. Find the value of k satisfying $P(Y \geq k) = .90$.

b. Find the value k such that the probability that the price of the bond (one year later) exceeds k is .60.

5.57 Assume that the hourly wage rate earned by a worker in a clothing factory (based on a piecework pay system) is normally distributed with expected value \$5.10 and standard deviation \$0.40.

a. Find the probability that a worker's hourly rate exceeds \$5.40.

b. Find the probability that a worker's hourly rate is between \$4.70 and \$5.50.

c. Find the probability that a worker's hourly rate exceeds a contractual minimum of \$3.90.

5.9 NORMAL APPROXIMATIONS TO BINOMIAL AND POISSON PROBABILITIES

One of the many uses of the normal curve is as an approximation to other probability distributions, particularly the binomial and Poisson distributions. This section indicates how such approximations work and when they are reasonably accurate.

Probabilities associated with values of y can be computed for a binomial experiment for any values of n or π, but as you might imagine, the task becomes more difficult when n gets large. For example, suppose a sample of 1000 voters is polled to determine sentiment toward the consolidation of a city and county government. What is the probability of observing 460 or fewer favoring consolidation if we assume that 50% of the entire population favor the change? Here we have a binomial experiment with $n = 1000$ and π, the probability of selecting a person favoring consolidation, equal to .5. To determine the probability of observing 460 or fewer favoring consolidation in the random sample of 1000 voters, we could compute $P_Y(y)$ using the binomial formula for $y = 460, 459, \ldots, 0$. The desired probability would then be

$$P(Y = 460) + P(Y = 459) + \ldots + P(Y = 0)$$

There would be 461 probabilities to calculate, with each one being somewhat difficult due to the factorials. For example, the probability of observing 460 favoring consolidation is

$$P(Y = 460) = \frac{1000!}{460!540!}(.5)^{460}(.5)^{540}$$

normal approximation to binomial

The binomial distribution can be approximated by a normal distribution for certain values of n and π. This can be proved by a Central Limit Theorem, as discussed in the next chapter. We don't prove the result here; rather, we show how and when to use the approximation. The basic idea is to pretend that a binomial random variable Y has a normal distribution, using the binomial mean $\mu = n\pi$ and standard deviation $\sqrt{n\pi(1 - \pi)}$. For example, a binomial variable with $n = 400$ and $\pi = .20$ can be treated as approximately normally distributed with $\mu = 400(.20) = 80$ and $\sigma = \sqrt{400(.20)(.80)} = 8$. To approximate $P(Y > 96)$, use normal tables to get $P(Y > 96) = P[Z > (96 - 80)/8] = P(Z > 2) = .0228$, or about .02.

EXAMPLE 5.24 A life insurance agency has set as a target that 10% of all prospects contacted buy insurance. Assume that independence holds among prospects, so that binomial probabilities apply. What is the probability that, out of 600 prospects, 30 or fewer buy?

Solution The exact solution involves binomial probabilities with $n = 600$ and $\pi = .10$, if in fact the target success rate is being achieved. Because we have no tables for $n = 600$, we use a normal approximation with $\mu = n\pi = 600(.10) = 60$ and $\sigma = \sqrt{n\pi(1 - \pi)} = \sqrt{600(.10)(.90)} = 7.348$.

$$P(Y \leq 30) = P\left(Z \leq \frac{30 - 60}{7.348}\right) = P(Z \leq -4.08)$$

which is virtually zero. If an agent sold only 30 policies to the last 600 prospects, we would have to conclude that the agent was not on target; the result (30 successes of 600 trials) can't reasonably be explained as a fluke attributable to chance. □

The normal approximation to the binomial distribution can be pretty bad if $n\pi < 5$ or $n(1 - \pi) < 5$. If π, the probability of success, is small, and n, the sample size, is modest, the actual binomial distribution is seriously skewed to the right. In such a case, the symmetric normal curve gives a bad approximation. If π is near 1, so $n(1 - \pi) < 5$, the actual binomial is skewed to the left and again the normal approximation isn't very good. The normal approximation, as described, is quite good when $n\pi$ and $n(1 - \pi)$ exceed about 10. In the middle zone, $n\pi$ or $n(1 - \pi)$ between 5 and 10, modification called a **continuity correction** makes a fairly big difference to the quality of the approximation.

continuity correction

The reason for the continuity correction is that we are using the continuous normal curve to approximate a discrete binomial distribution. The situation is shown in Figure 5.11. The binomial probability that $Y \leq 5$ is the sum of the areas of the rectangles above 5, 4, 3, 2, 1, and 0. This probability (area) is approximated by the area under the superimposed normal curve to the left of 5. Thus the normal approximation ignores half of the

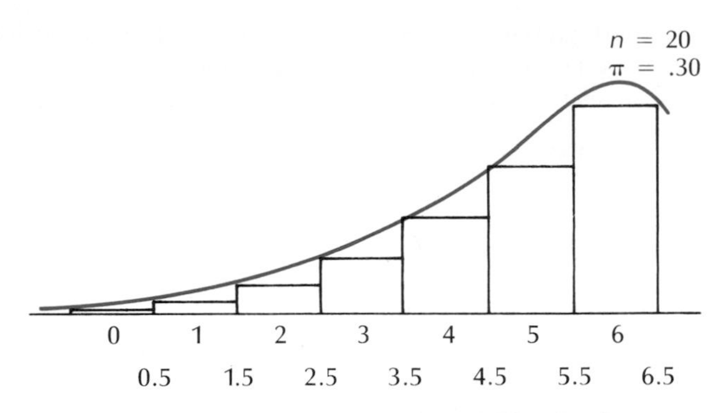

FIGURE 5.11 Normal Approximation to Binomial Distribution

rectangle above 5. The continuity correction simply includes the area between $y = 5$ and $y = 5.5$. For the binomial distribution with $n = 20$ and $\pi = .30$, the correction is to take $P(Y \leq 5)$ as $P(Y \leq 5.5)$. Instead of $P(Y \leq 5) = P\{Z \leq [5 - 20(.3)]/\sqrt{20(.3)(.7)}\} = P(Z \leq -.49) = .3121$, use $P(Y \leq 5.5) = P\{Z \leq [5.5 - 20(.3)]/(\sqrt{20(.3)(.7)}\} = P(Z \leq -.24) = .4052$. The actual binomial probability, from Appendix Table 1, is .4164. The general idea of the continuity correction is to add or subtract .5 from a binomial value before using normal probabilities. The best way to determine whether to add or subtract is to draw a picture like Figure 5.11.

Normal Approximation to the Binomial Probability Distribution

For large n and π not too near zero or one, a binomial random variable Y may be approximated by a normal distribution with $\mu = n\pi$ and $\sigma = \sqrt{n\pi(1 - \pi)}$. This approximation should be used only if $n\pi \geq 5$ and $n(1 - \pi) \geq 5$. A continuity correction improves the quality of the approximation in cases in which n is not overwhelmingly large.

EXAMPLE 5.25 A large drug company has 100 potential new prescription drugs under clinical test. About 20% of all drugs that reach this stage are eventually licensed for sale. What is the probability that at least 15 of the 100 drugs are eventually licensed? Assume that the binomial assumptions are satisfied, and use a normal approximation with continuity correction.

Solution The mean (expected value) of Y is $\mu = 100(.2) = 20$; the standard deviation is $\sigma = \sqrt{100(.2)(.8)} = 4.0$. The desired probability is that 15 or more drugs are approved. Because $Y = 15$ is included, the continuity correction is to take the event as $Y \geq 14.5$.

$$P(Y \geq 14.5) = P\left(Z \geq \frac{14.5 - 20}{4.0}\right) = P(Z \geq -1.375)$$

which is about .92. □

normal approximation to Poisson

The **normal approximation to the Poisson distribution** works the same way. If Y is a Poisson random variable with expected value μ, pretend that Y has a normal distribution; because the variance of a Poisson distribution equals μ, use mean μ and standard deviation $\sqrt{\mu}$.

EXAMPLE 5.26 Assume that Y, the number of jobs arriving at a computer center in a given half-hour period, has a Poisson distribution with mean .2 per minute. Use a normal approximation to find $P(Y \leq 10)$.

Solution $\mu = (.2/\text{minute})(30 \text{ minutes}) = 6$. Then from Appendix Table 3 we have

$$P(Y \leq 10) = P\left(Z \leq \frac{10 - 6}{\sqrt{6}}\right) = P(Z \leq 1.63) = .9484$$

The correct Poisson probability is .9574. □

The normal approximation is rather bad for $\mu < 5$ (corresponding to $n\pi < 5$ in the binomial case) and pretty good for $\mu > 10$. In the intermediate case, $5 \leq \mu \leq 10$, the same continuity correction makes the approximation accurate enough for most purposes.

EXAMPLE 5.27 Apply the continuity correction to Example 5.26.

Solution We still have $\mu = 6$ and $\sigma = \sqrt{6}$. Approximate $P(Y \leq 10)$ by $P(Y \leq 10.5)$. Draw a picture to see why.

$$P(Y \leq 10.5) = P\left(Z \leq \frac{10.5 - 6}{\sqrt{6}}\right) = P(Z \leq 1.84)$$

$$= .9671$$

The exact Poisson probability, again, is .9574. □

SECTION 5.9 EXERCISES

5.58 Suppose that Y has a binomial distribution with $n = 100$ and $\pi = .50$.
a. Use binomial tables to calculate $P(40 \leq Y \leq 60)$.
b. Use a normal approximation (without any continuity correction) to calculate the same probability. How good is the approximation?

5.59 Use a continuity-corrected normal approximation to give an answer to Exercise 5.58. Is this approximation much better than the one in Exercise 5.58?

5.60 Refer to Exercise 5.12. Use a normal approximation without continuity correction to find $P(Y \geq 85)$. How does the approximate answer compare to the exact binomial answer?

5.61 Use a continuity-corrected normal approximation to answer Exercise 5.60. Does the continuity correction improve the approximation?

5.62 Input errors to a computerized accounting system occur at a rate of 1.6 per 1000 entries. Under what conditions would you expect the Poisson assumptions to be a reasonable approximation?

5.63 Assume that in Exercise 5.62 Poisson probabilities apply. Let Y = number of errors in a set of 5000 entries.

a. Find $E(Y)$ and σ_Y.

b. Find $P(5 \leq Y \leq 11)$.

5.64 Use a normal approximation with continuity correction to answer Exercise 5.63, part (b).

SUMMARY

This chapter discusses several of the most important probability distributions. Each distribution occurs when specific assumptions hold. For each distribution, we specified the equation for the probability distribution or the probability density function, the mean, and the standard deviation.

The important discrete distributions are: binomial, for the number of successes in a fixed number of independent, constant-probability trials; hypergeometric, for sampling without replacement; geometric and negative binomial, for the number of trials required to obtain a specified number of successes; and Poisson, for the number of events (occurring randomly over time) in a fixed time period and also as an approximation to the binomial.

The most important continuous distributions are: uniform, for values spread out evenly over a range; exponential, for the waiting time to the next random event; and especially normal, the basis of many statistical procedures and probability models and also an approximation (preferably with a continuity correction) to binomial and Poisson probabilities.

KEY TOPICS AND FORMULAS: Some Special Probability Distributions

1. Counting rules
   a. The number of sequences (permutations) of r symbols taken k at a time is

$$_kP_r = \frac{r!}{(r-k)!}$$

   b. The number of subsets (combinations) of r symbols taken k at a time is

$$\binom{r}{k} = \frac{r!}{k!(r-k)!}$$

2. Binomial experiment
   a. There are n trials, each resulting in either success S or failure F.
   b. The probability of success $P(S) = \pi$ remains constant for all trials; $P(F) = 1 - \pi$.
   c. The trials are independent.
   d. The random variable of interest is Y, the number of successes in n trials.

3. Binomial random variable
   a. Binomial probability distribution

$$P_Y(y) = \frac{n!}{y!(n-y)!}\,\pi^y(1-\pi)^{n-y}$$

   b. Mean, variance, and standard deviation of Y

$$E(Y) = n\pi$$
$$\mathrm{Var}(Y) = n\pi(1-\pi)$$
$$\sigma_Y = \sqrt{n\pi(1-\pi)}$$

4. Hypergeometric random variable
   a. Hypergeometric probability distribution

$$P_Y(y) = \frac{\binom{N_S}{y}\binom{N_F}{n-y}}{\binom{N}{n}}$$

   b. Mean, variance, and standard deviation of Y

$$E(Y) = n\left(\frac{N_S}{N}\right)$$
$$\mathrm{Var}(Y) = n\left(\frac{N_S}{N}\right)\left(1-\frac{N_S}{N}\right)\frac{N-n}{N-1}$$
$$\sigma_Y = \sqrt{n\left(\frac{N_S}{N}\right)\left(1-\frac{N_S}{N}\right)\frac{N-n}{N-1}}$$

5. Geometric random variable
   a. Probability distribution

$$P_Y(y) = \pi(1-\pi)^{y-1}, \qquad y = 1, 2, \ldots$$

   b. Mean, variance, and standard deviation of Y

$$E(Y) = \frac{1}{\pi}$$
$$\mathrm{Var}(Y) = \frac{1-\pi}{\pi^2}$$
$$\sigma_Y = \frac{\sqrt{1-\pi}}{\pi}$$

6. Negative binomial random variable
   a. Probability distribution

$$P_Y(y) = \frac{(y-1)!}{(k-1)!(y-k)!}\,\pi^k(1-\pi)^{y-k}, \qquad y = k, k+1, \ldots$$

b. Mean, variance, and standard deviation of Y

$$E(Y) = \frac{k}{\pi}$$

$$\text{Var}(Y) = \frac{k(1 - \pi)}{\pi^2}$$

$$\sigma_Y = \frac{k\sqrt{(1 - \pi)}}{\pi}$$

7. Poisson random variable
   a. Poisson probability distribution

$$P_Y(y) = \frac{e^{-\mu}\mu^y}{y!}$$

   b. Mean, variance, and standard deviation of Y

$$E(Y) = \mu$$
$$\text{Var}(Y) = \mu$$
$$\sigma_Y = \sqrt{\mu}$$

8. Uniform random variable
   a. Probability density function

$$f_Y(y) = \frac{1}{b - a}, \qquad a < y < b$$

   b. Mean, variance, and standard deviation of Y

$$E(Y) = \frac{a + b}{2}$$

$$\text{Var}(Y) = \frac{b - a}{12}$$

$$\sigma_Y = \sqrt{\frac{b - a}{12}}$$

9. Exponential random variable
   a. Probability density function

$$f_Y(y) = \left(\frac{1}{\mu}\right)e^{-y/\mu}, \qquad y > 0$$

   b. Mean, variance, and standard deviation of Y

$$E(Y) = \mu$$
$$\text{Var}(Y) = \mu^2$$
$$\sigma_Y = \mu$$

10. The normal random variable
    a. Probability density function

$$f_Y(y) = \frac{1}{\sqrt{2\pi}\sigma} e^{-.5\left(\frac{y-\mu}{\sigma}\right)^2}$$

    b. Mean, variance, and standard deviation are μ, σ^2, and σ, respectively.
    c. Standard normal,

$$Z = \frac{Y - \mu}{\sigma}$$

    d. Areas under a normal curve are obtained by first computing a z-score and then referring to Appendix Table 3.

11. Normal approximations to binomial and Poisson probabilities
    a. For a normal approximation to the binomial distribution, use a normal distribution with mean $n\pi$ and standard deviation $\sqrt{n\pi(1 - \pi)}$, provided

$$n\pi \geq 5 \quad \text{and} \quad n(1 - \pi) \geq 5$$

    b. For a normal approximation to the Poisson distribution, use a normal distribution with mean μ and standard deviation $\sqrt{\mu}$, provided $\mu \geq 5$.

CHAPTER 5 EXERCISES

5.65 A telephone-sales firm is considering purchasing a machine that randomly selects and automatically dials telephone numbers. The firm would be using the machine to call residences during the evening; calls to business phones would be wasted. The manufacturer of the machine claims that its programming reduces the business-phone rate to 15%. As a test, 100 phone numbers are to be selected at random from a very large set of possible numbers.
a. Are the binomial assumptions satisfied in this situation?
b. Find the probability that at least 24 of the numbers belong to business phones.
c. If in fact 24 of the 100 numbers turn out to be business phones, does this cast serious doubt on the manufacturer's claim? Explain.

5.66 Refer to Exercise 5.65.
a. Find the expected value and variance of Y, the number of business phone numbers in the sample.
b. Use normal approximations (with and without continuity correction) to find $P(Y \geq 24)$. How close are the approximations?

5.67 It is estimated that 5% of all Medicaid claims in a particular city are fraudulent. A random sample of 50 claims is taken.
a. What is the probability that at most 1 claim in the sample is fraudulent?
b. What is the probability that at least 4 are fraudulent?

5.68 Refer to Exercise 5.67. Use a Poisson approximation to answer parts (a) and (b). How good is the approximation?

5.69 Of 30,000 bank credit cards in circulation in a particular city, 300 are subject to recall (because of theft or nonpayment). A merchant receives 100 different cards in one day.
a. How many different sets of 100 cards can the merchant receive? Don't work out a numerical answer unless you have an unnatural fondness for very large numbers.

b. How many sets of 100 cards contain no recallable cards? Again, avoid the arithmetic.

c. Write an expression for the probability of obtaining no recallable cards in the sample of 100.

d. Write an expression for the probability of obtaining 2 or fewer recallable cards in the sample.

5.70 Refer to Exercise 5.69, parts (c) and (d).

a. Write expressions for binomial approximations to these probabilities.

b. Use a Poisson approximation to get numerical answers to these probabilities.

5.71 It has been asserted that female managers tend to be placed in fringe areas, such as public relations or personnel management, as opposed to the central areas of production, marketing, and finance. Suppose that a firm has 24 male and 6 female managers at the assistant vice-president level. Of these positions, 14 are regarded as fringe positions.

a. In how many distinct ways can the 14 fringe managers be selected?

b. In how many ways can the fringe managers be selected such that 5 of the 6 women are included?

c. If the fringe managers had been randomly selected, what is the probability that at least 5 would be women?

5.72 Assume that lost-time industrial accidents occur in a plant according to a Poisson distribution with mean .12 per day. Let $Y =$ number of such accidents in a 10-day period.

a. Find $P(Y = 1)$ and $P(Y \leq 1)$.

b. Find $E(Y)$ and σ_Y.

5.73 Would a normal approximation in Exercise 5.72 be very accurate?

5.74 The weekly demand for 5-pound sacks of flour at a particular supermarket is assumed to be approximately normal with mean 72.0 cases and standard deviation 1.6 cases. Let $Y =$ demand in a particular week.

a. Find $P(Y \leq 72.8)$ and $P(71.2 \leq Y \leq 72.8)$.

b. Find $P(Y \geq 74.0)$.

c. The ordering policy of the market is that there be a 1% chance of stockout (demand exceeding supply) in any particular week. How much flour must be stocked to achieve this goal?

5.75 Refer to Exercise 5.74.

a. What is the probability that demand exceeds 73.0 cases in a particular week?

b. What is the probability that demand exceeds 73.0 cases in exactly three of four consecutive weeks? Assume independence from week to week.

5.76 A certain amount of material is wasted in cutting patterns for garments. A producer of army uniforms has found that the wastage is normally distributed with mean 4.1% and standard deviation .6%, from lot to lot.

a. In a particular lot, what is the probability that the wastage exceeds 5%?

b. If the actual amount of material required for a lot is 4700 yards, and 5000 yards of material are available, what is the probability that the supply of material is adequate?

5.77 Suppose that in Exercise 5.76 a particular cutter exceeds 5% wastage in 8 of 10 lots.

a. What is the probability of exceeding 5% in at least 8 of 10 lots?

b. Would such a result conclusively indicate that the cutter was inefficient?

5.78 A modem is an electronic device used in communication between computers. The specifications for a particular modem demand that the mean number of errors in transmitting through the device be 1 per 5000 words (or better). A particular modem is to be tested on a 25,000-word transmission. If 8 or more errors occur in transmission, the device is not accepted. Assume that Poisson probabilities apply and that the modem just meets the 1 per 5000 standard.

a. What is the probability that the device will be accepted?

b. Can you think of a reason why the Poisson assumptions may not hold?

5.79 Assume that the Poisson distribution applies in Exercise 5.78, but also that the modem has a mean error rate of 1 per 2500 words, thus not meeting specifications. What is the probability that the device will be accepted?

5.80 Executives at a soft drink company wish to test a new formulation of their chief product. The new formulation is tested in comparison to the current formulation. Each of 1000 potential customers is given a cup of the current formulation and a cup of the new formulation. The cups are labeled H and K to avoid bias. Each customer indicates a preference. Assume that, in fact, the customers can't detect a difference and are, in effect, guessing. Define Y to be the number (out of 1000) indicating preference for the new formulation.

a. What probability distribution should apply to Y? Do the assumptions underlying that distribution seem plausible in this context?

b. Find the mean and standard deviation of Y.

5.81 Find the approximate probability that the random variable Y in Exercise 5.80 is no larger than 460. Should the approximation be accurate?

5.82 A firm is considering using telemarketing techniques to supplement traditional marketing methods. It's estimated that 1 of every 100 calls results in a sale. Suppose that 250 calls are made in a single day.

a. Write an expression for the probability that there are 5 or fewer sales. Don't carry out any arithmetic.

b. What did you assume in answering part (a)? Are any of these assumptions grossly unreasonable?

5.83 a. Use a normal approximation to find a numerical value for the probability in Exercise 5.82, part (a).

b. Use a Poisson approximation to find a numerical value for the same probability.

c. Which approximation should be better? Why?

5.84 Refer again to Exercise 5.82. Let $Y =$ number of calls made up to and including the first sale.

a. Find the mean and standard deviation of Y.

b. Find $P(Y = 1)$ and $P(Y = 100)$. Which is larger?

5.85 In the telemarketing situation described in Exercise 5.82, define Y to be the number of calls made up to and including the fourth sale.

a. Write an expression for the probability that Y is at least 400. You need not carry out the arithmetic.

b. Find the variance of Y.

5.86 The chief executive officer (CEO) of a medium-size corporation must select 3 individuals to head the firm's annual drive for community charities. There are three divisions (A, B, and C) within the firm and 5, 6, and 4 individuals, respectively, within the divisions who could be chosen.

a. How many combinations of 3 individuals can be chosen such that 1 individual comes from each of the three divisions?

b. Suppose that the CEO chooses the individuals at random. What is the probability that at least 2 of them come from division A?

5.87 Referring to Exercise 5.87, part (b), let $Y =$ number of individuals chosen from division A. Find the expected value and variance of Y.

5.88 The computer that controls a bank's automatic teller machines "crashes" on occasion. The average time between crashes is 5.0 days. Define Y to be the waiting time until the next crash.

a. Find the probability that the system does not crash in a (7-day) week.

b. Find the probability that the time until the next crash is at least two weeks.

5.89 Find the expected value and standard deviation of the random variable Y of Exercise 5.88.

5.90 What assumption(s) did you make in Exercise 5.88? If it is known that one crash tends to be followed by another one in a relatively short time, what assumption would be violated?

5.91 An advertising campaign for a new product is targeted to make 20% of the adult population in a metropolitan area aware of the product. After the campaign, a random sample of 400 adults in the metropolitan area is obtained.

a. Find the approximate probability that 57 or fewer adults in the sample are aware of the product. Use a continuity correction.

b. Should the approximation be accurate?

c. The sample shows that 57 adults are aware of the product. The marketing manager argues that the low awareness rate is a random fluke of the particular sample. Based on your answer to part (a), do you agree?

5.92 Brand managers at a consumer-products company regard an introductory advertising campaign for a new product as successful if at least 20% of the target group are made aware of the product. After one such campaign, a market research study finds that 56 of 400 individuals sampled are aware of the product. The target group is all adults who possess driver's licenses in the United States.

a. Write an expression for the exact probability that 56 or fewer people in the sample are aware of the product, assuming that 20% of the target group are aware of the product. What probability distribution applies? What assumptions have you made?

b. Use a normal approximation to find a numerical value for this probability.

c. If you were the brand manager, would you believe that the advertising campaign had been successful?

5.93 A certain birth defect occurs with probability .0001; that is, 1 of every 10,000 babies has this defect. If 5000 babies are born at a particular hospital in a given year, what is the approximate probability that there is at least one baby with the defect? What approximation should be used?

5.94 Several states now have a Lotto lottery game. A player chooses 6 distinct integers in the range 1 to 40. If exactly those 6 numbers are selected as the winning numbers, the player receives a very large prize. What is the probability that a particular set of 6 numbers will be drawn? You may wish to think of the 6 numbers drawn as "success" numbers.

5.95 In the Lotto game described in Exercise 5.94, there are smaller prizes for selecting exactly 5 of the 6 winning numbers and even smaller prizes for selecting exactly 4 of the 6 winning numbers.

a. What is the probability of selecting exactly 4 of the 6 winning numbers?

b. What is the probability of selecting at least 4 of the 6 winning numbers?

5.96 Suppose that the Lotto game in Exercise 5.94 is changed such that 6 numbers were chosen in the range 1 to 42, rather than 1 to 40.

a. Without doing any arithmetic, determine if the probability of selecting all 6 winning numbers is larger or smaller than what it was in Exercise 5.94. Will the change be small or large?

b. Now compute the probability of selecting all 6 winning numbers, chosen from the numbers 1 to 42.

c. Compare your answer to part (b) with the answer to Exercise 5.94. Did the probability change as you expected in part (a) of this exercise?

5.97 Suppose that, in the Lotto game of Exercise 5.94, 1,000,000 players make independent choices of the 6 numbers.

a. What probability distribution applies to the random variable Y = number of players selecting all 6 numbers?

b. Find an expression for $P(Y = 0)$. Don't carry out the arithmetic.

c. Write an expression for $P(Y \geq 2)$.

5.98 a. Find the expected value and variance of the random variable Y in Exercise 5.97.

b. Use a Poisson approximation to find $P(Y = 0)$ and $P(Y \geq 2)$. How accurate should the approximation be?

c. If you have a suitable calculator, calculate a numerical value for the exact probability and compare it to the approximate probability found in part (b).

5.99 Should a normal approximation be used to approximate $P(Y = 0)$ and $P(Y \geq 2)$ in Exercise 5.97? Why?

5.100 If no one selects the correct 6 numbers in the Lotto game described in Exercise 5.97, the largest prize is not awarded; instead, the money is added to the pot for the next drawing. If there are no winners for several consecutive periods, the potential prize can be many millions of dollars. Suppose that, as in Exercise 5.97, 1,000,000 players independently select 6 numbers for each drawing. Define the random variable X = number of drawings required to obtain at least one winner.

a. What probability distribution applies to X? Defend your statement.

b. Find the mean and standard deviation of X.

c. Write an expression for $P(X = 3)$. In terms of what happens in the Lotto game, what does the event $X = 3$ mean?

5.101 There is an objection to the formulation in Exercise 5.100. In fact, if there is no winner in one or two consecutive drawings, the news media report the large potential prize, and many more people play the game. Show that this fact leads to a violation of one of the assumptions made in Exercise 5.100.

5.102 Fires in occupied homes in a particular city occur at the rate of 1 every 2 days.

a. What is the expected number of fires in homes over a 7-day week?

b. Find the probability that there are at least 4 fires in a particular week.

c. What are you assuming about the occurrence of fires in your answer to part (b)? Do any of the assumptions seem grossly unreasonable?

5.103 Refer to Exercise 5.102. Find the probability that at least 3 days elapse without any home fires. What is the expected time between home fires?

5.104 The operator of a mainframe computer system receives unscheduled requests to mount tapes. By policy, these requests must be answered as quickly as possible; therefore, they interrupt scheduled work flow. Data indicate that the rate of such requests during the 9 A.M.–5 P.M. shift is about 1.5 per hour. Let Y = number of requests received in a particular 9 A.M.–5 P.M. shift.

a. Find the mean and standard deviation of Y.

b. Find $P(Y > 8)$.

5.105 Refer to Exercise 5.104. Find the probability that the time between successive requests is at least two hours.

5.106 The computer system manager in Exercise 5.104 notes that the demand for unscheduled tape mounts varies during the normal work day. Between 9 A.M. and 1 P.M. there is an average of 1 request per hour; between 1 P.M. and 5 P.M. there is an average of 2 requests per hour.

a. Does this fact change your answers to part (a)?

b. Does this fact affect your answer to part (b)?

6

RANDOM SAMPLING AND SAMPLING DISTRIBUTIONS

This chapter contains the concepts and mathematical results that are most crucial for using probability theory in making statistical inferences. One basic notion is the idea of random sampling, which we discussed briefly in Chapter 5. Data collected from a random sample can be used to compute summary values called **statistics**. The most important concept in this chapter is that of a theoretical probability distribution (called a **sampling distribution**) for possible values of a sample statistic. In particular, we examine the sampling distributions of the sample mean and the sample sum. A Central Limit Theorem is used to justify the use of a normal distribution to approximate the sampling distributions of these sample statistics.

6.1 RANDOM SAMPLING

selection bias

Most statistics textbooks, including this one, tell you to use random sampling to collect data. A basic reason for using random sampling is to ensure that the inferences made from the sample data are not distorted by a **selection bias**. A selection bias exists whenever there is a systematic tendency to overrepresent or underrepresent some part of the population. For example, a telephone sample of households in a region, conducted entirely between the hours of 9:00 A.M. and 5:00 P.M., would be severely biased toward households with at least one nonworking number. Hence any inferences made from the sample data would be biased toward the attitudes or opinions of nonworking members and might not be truly representative of households in the region. Similarly, a sample of charge accounts taken by selecting a set of transactions would be biased toward active, many-transaction accounts, and away from inactive ones. Inferences from these data might not reflect the characteristics of the set of all accounts. A random sampling plan, by definition, avoids this kind of bias.

We indicated earlier that simple random sampling is a process whereby each possible sample of a given size has the same probability of being selected. Obtaining a truly, or even approximately, random sample requires some thought and effort. A random sample is not a casual or haphazard sample. The target population must be identified. In principle, a list of all elements (possible individual values) in the population ought to be constructed with elements to be included in the sample selected randomly from the list using a table of random numbers.

EXAMPLE 6.1 Suppose that the research staff of a Federal Reserve bank wishes to take a random sample of checks written on individual (nonbusiness) checking accounts to determine average amount, time to clearance, and insufficient-funds rates. How might they do so?

Solution First, the target population must be defined. Is it all individual checks written in a given period, or is it all checks processed by the Federal Reserve clearing house during that time? There's a difference, because a check that is cashed at the bank on which it was originally drawn never gets to the clearing house. Assume that the clearing-house definition is chosen. The next step is to establish a random sampling method. One could, in principle, put a numerical tag on every one of the 326,274 (or whatever number) checks processed by the center on a particular day. Then a random sample of 1000 could be drawn by selecting six-digit random numbers and the checks with corresponding tags (passing over 000000 and anything larger than 326274).

Obviously, this would be a very impractical and expensive way to obtain a random sample. Such a method only serves as an idealization against which a more practical method can be measured. Another possibility is to sample every 300th check processed. This method is not literally random sampling, because, for instance, two successive checks couldn't be included in the sample. No doubt one could dream up some situations for which sampling every 300th check would introduce some kind of selection bias; however, this process seems to yield a fairly good approximation to random sampling, at a manageable cost. □

The applicability of sampling methods is much broader than the familiar political polls and market research studies. Sampling should be considered whenever information is desired and the cost (in dollars, in labor, or in time) of obtaining complete information is excessive. For example, suppose that a processor of potato chips sells the product through 1943 retail outlets. A critical variable for the success of the product is the average amount of shelf space devoted to the product per outlet. It would be absurd for the processor to visit every outlet and measure that shelf space devoted to the product. Assuming that the potato-chip processor had a list of the retail outlets, it would be relatively easy to obtain a random sample of, say, 100 outlets and to measure the average shelf space in that sample.

Ideally, one has a list of all the elements of the target population. More often, one has a list that almost, but not quite, equals the target population. The almost-right list is called the **sampling frame**, to indicate that it is not exactly the same as the target population. A good sampling frame is sometimes fairly easy to obtain, as it would be in the case of the potato-chip processor, who most likely knows almost all, but not quite all, retail outlets. When sampling human populations, a good sampling frame is harder to develop. People move; a directory or a mailing list can become outdated rather quickly. Telephone directories are not a completely reliable source for developing a sampling frame; there are unlisted phone numbers and multiple phone numbers. Perhaps the most serious problem is that people without phones tend strongly to be poor people. This problem was a major cause of one of the most notorious failures in sampling, the *Literary Digest* poll before the 1936 U.S. presidential election. The *Literary Digest*, a popular magazine of the time, took a huge survey (2.4 million responses), based in large part on telephone books. In 1936, during the Great Depression, this procedure introduced a substantial bias. The magazine forecast that the Republican candidate would win the election; he won only two states. Much of the effort in conducting a good sampling should go into developing the sampling frame.

Given a decent sampling frame, random sampling can be done with a computer program that generates random numbers, or with a table of random numbers. Appendix Table 9 contains a small random number table. For all practical purposes, such a table can be entered at any point. (One way to enter such a table is to pull a dollar bill from one's wallet and use the first serial number digit to select the row and the second digit to indicate the column.) Suppose that we enter the table at row 3, column 1, go across the row, and use the first three digits. To obtain a random sample of 10 individuals from a population of 916 individuals, we number the individuals in the sampling frame from 000 to 915 and select the first 10 numbers, namely 24130 (using only 241), 483(60), 225(27), 972(75), 763(93), 648(09), 151(79), 248(30), 493(40), and 320(81). If we obtain any random numbers larger than 915, such as 972, they are ignored and replacement numbers drawn; we replace the 972 by 306(80), the next number in Appendix Table 9. Assuming that we wish to sample without replacement, we also ignore any repetitions of numbers. Of course, there's no need to depend on a table; virtually any decent computer system can generate a series of random numbers that serves equally well.

EXAMPLE 6.2 Suppose that a sample of size 4 is to be taken from a sampling frame of 1943 retail outlets. If we enter Appendix Table 9 of random numbers at row 3, column 1, and read vertically, down columns, the first 10 entries are 24130, 42167, 37570, 77921, 99562, 96301, 89579, 85475, 28918, 63553. Which stores should be sampled?

Solution One of the many ways that one can use these numbers is to ignore the last digit and select only those values between 0001 and 1943, both inclusive; thus ignore 2413, 4216, and so on. We continue down the column and pick stores 0942 (row 13), 1036 (row 14), 0711 (row 15), and 0236. Any repetitions are also ignored, assuming that we are sampling without replacement. (We could also have used a computer to select random samples in the range from 0001 to 1943.) □

Careful planning and a certain amount of ingenuity are required to have even a decent approximation to a random sample. This is especially true when the elements in the target population are people. People throw away mail questionnaires, are not home to answer telephone surveys, and modify answers to conform to social norms. Too often, surveys of people are done in a haphazard way, using a hastily composed questionnaire mailed (often addressed to "occupant") according to a conveniently chosen mailing list, with no provisions for following up on those who do not respond. The result is bad data, riddled with biases that have unknowable effects. Lots of statistical methods can be applied to such data, but the well-known adage applies: garbage in, garbage out. Getting reasonable samples of human populations is a considerable art (and a fairly substantial industry); we will not try to capture the essence of that art in a couple of pages.

There are other valid and useful sampling methods, such as stratified sampling. Some of these are discussed in Chapter 19. Most uses of probability theory in the remaining chapters are based on simple random sampling. The basic ideas remain the same when other sampling methods are used. The formulas just get a little more complicated.

SECTION 6.1 EXERCISES

6.1 Suppose that we want to select a random sample of $n = 10$ entities from a population of 800 individuals. Use Appendix Table 9 to identify the individuals to be sampled. Start in row 5, column 1, and read down.

6.2 City officials sample the opinions of homeowners in a community about the possibility of raising taxes to improve the quality of local schools. A directory of all homes in the city is used; a computer generates random numbers to identify the addresses to be sampled. An interviewer visits each home between the hours of 3:00 P.M. and 6:00 P.M.; if no one is home, the address is eliminated from the sample and replaced by another randomly chosen address. Does this process approximate random sampling?

6.3 A university bookstore manager is mildly concerned about the number of textbooks that were underordered and thus unavailable two days after the beginning of classes. The manager instructs an employee to pick a random number, go to the place where that number book is shelved, and examine the next 50 titles, recording how many titles are unavailable.

a. Technically, this process doesn't yield a random sample of the books in the store. Why not?

b. How could a truly random sample be obtained?

6.4 A professional baseball team has a 20-game ticket plan and a 40-game plan. The sales director wants to assess fan interest in a combination plan by which two separate purchasers of 20-game plans can pool their money and buy a 40-game plan at a modest discount. The target population to be sampled is all current 20-game plan purchasers. An up-to-date list of the 4256 current purchasers is available. Explain how to obtain a random sample of the current purchasers.

6.5 One way to sample the purchasers in Exercise 6.4 is to develop a list of the seat numbers held by purchasers and to take a random sample of seat numbers. Most likely this will not yield a random sample of purchasers. Explain why not.

6.6 A building manager for a 2526-office complex hires a new cleaning contractor. The manager wants to get a rough idea of how satisfactory the contractor's weekend cleaning efforts are. One possible strategy is to select 3 offices on the first three floors and to examine those offices and the 10 offices on either side. Another strategy, requiring roughly equal time, is to select 15 offices completely at random. It is argued that the first strategy is better because it allows for inspection of more offices. Is the argument valid?

6.2 SAMPLE STATISTICS AND SAMPLING DISTRIBUTIONS

Once a sample has been selected and numerical data obtained, the first task is to summarize it. In Chapter 2 we defined several summary measures, such as the sample mean and the sample standard deviation. Each of these is an example of a **sample statistic**.

sample statistic

The numerical value that a sample statistic will have cannot be predicted exactly in advance. Even if we know that a population mean μ is \$216.37 and that the population standard deviation σ is \$32.90—even if we know the complete population distribution—we cannot say that the sample mean $\bar{Y}$ will be exactly \$216.37. A sample statistic is a random variable; it is subject to random variation because it is based on a random sample of measurements selected from the population of interest, And, like any other random variable, a sample statistic has a probability distribution. We call the theoretical probability distribution of a sample statistic the **sampling distribution** of that statistic.

sampling distribution

The actual mathematical derivation of sampling distributions is one of the basic problems of mathematical statistics. Techniques employed include the basic probability methods of Chapter 3, Monte Carlo methods (discussed in Section 6.6), and numerous other mathematical manipulations. We first illustrate how the sampling distribution for $\bar{Y}$ can be obtained for a simplified population. Later in the chapter several general results are presented.

EXAMPLE 6.3 The sample mean $\bar{y}$ is to be calculated from a random sample of size 2 taken from a population consisting of the 5 values (\$2, \$3, \$4, \$5, \$6). Find the sampling distribution of $\bar{Y}$, based on sample of size 2.

Solution One way to find $P_{\bar{Y}}(\bar{y})$ is by counting. There are $\binom{5}{2} = 10$ possible samples of 2 items from the 5 items. These are shown here:

Possible samples of size 2	2, 3	2, 4	2, 5	2, 6	3, 4	3, 5	3, 6	4, 5	4, 6	5, 6
Value of $\bar{y}$	2.5	3	3.5	4	3.5	4	4.5	4.5	5	5.5

Assuming that each sample of size 2 is equally likely, it follows that the sampling distribution for $\bar{Y}$ based on $n = 2$ observations selected from this population is as indicated here:

$\bar{y}$:	2.5	3	3.5	4	4.5	5	5.5
$P_{\bar{Y}}(\bar{y})$:	1/10	1/10	2/10	2/10	2/10	1/10	1/10

This sampling distribution is shown as a graph in Figure 6.1. □

Usually a sample statistic is used as an estimate of a population parameter. For example, a sample mean can be used to estimate the corresponding mean μ of the population from which the sample was drawn. The sampling distribution of a sample statistic is then used to determine how accurate the estimate is likely to be. In Example 6.3 the population mean μ is known to be \$4. Obviously, we don't ever know μ in practice. Still, we can use the sampling distribution of $\bar{Y}$ to determine the probability that, for example, the computed value of the sample mean will be more than \$.50 away from μ. For Example

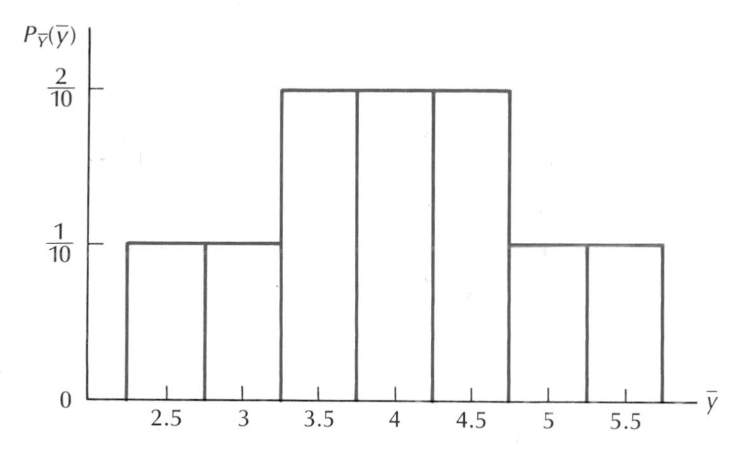

FIGURE 6.1 Sampling Distribution for $\bar{Y}$; Example 6.3

6.3, this probability is

$$P_{\bar{Y}}(2.5) + P_{\bar{Y}}(3) + P_{\bar{Y}}(5) + P_{\bar{Y}}(5.5) = \frac{4}{10}$$

In general, a sample statistic is used to make inferences about a population parameter. The sampling distribution of the statistic is crucial in determining how good the inference is likely to be.

interpretations of a sampling distribution

Sampling distributions can be interpreted in at least two ways. One way uses the long-run relative-frequency approach. Imagine taking repeated samples of a fixed size from a given population and calculating the value of the sample statistic for each sample. In the long run, the relative frequencies for the possible values of the sample statistic approach the corresponding sampling distribution probabilities. For example, if one takes a large number of samples from the population distribution probabilities of Example 6.3 and the sample mean for each sample, computes approximately 20% have $\bar{y} = 3.5$.

The other way to interpret a sampling distribution makes use of the classical interpretation of probability. Imagine listing all possible samples that can be drawn from a given population. The probability that a sample statistic has a particular value (say, that $\bar{y} = 3.5$) is then the proportion of all possible samples that yield that value. In Example 6.3, $P_{\bar{Y}}(3.5) = 2/10$ corresponds to the fact that 2 of the 10 samples have a sample mean equal to 3.5. Both the repeated-sampling and the classical approach to finding probabilities for a sample statistic are legitimate.

In practice, though, a sample is taken only once, and only one value of the sample statistic is calculated. **A sample distribution is not something you can see in practice; it is not an empirically observed distribution. Rather it is a theoretical concept**, a set of probabilities derived from assumptions about the population and about the sampling method.

There's an unfortunate similarity between the phrase *sampling distribution*, meaning the theoretically derived probability distribution of a statistic, and the phrase *sample distribution*, which refers to the histogram of **individual** values actually observed in a particular sample. The two phrases mean very different things. To avoid confusion, we refer

sample histogram to the distribution of sample values as the **sample histogram** rather than as the sample distribution.

EXAMPLE 6.4 Refer to Example 6.3. How could one use a computer to approximate the sampling distribution for $\bar{Y}$ based on $n = 2$ observations?

Solution Suppose that we decide to simulate 10,000 samples of size 2 from the population of Example 6.3. We can use a computer to generate 10,000 random digits (0, 1, . . . , 9) from a uniform distribution. As noted in Example 6.3, there are 10 possible samples. The 2, 3 sample ($\bar{y} = 2.5$) can be assigned to each 0 digit, the 2, 4 sample ($\bar{y} = 3$) to each 1 digit, . . . , and the 5, 6 sample ($\bar{y} = 5.5$) to each 9 digit. Thus each possible sample has the same probability. This process results in 10,000 $\bar{y}$-values. The values and corresponding relative frequencies are shown here:

Value of $\bar{y}$:	2.5	3.0	3.5	4.0	4.5	5.0	5.5
Frequency:	1024	991	2006	2018	1975	1005	981
Relative frequency:	.1024	.0991	.2006	.2018	.1975	.1005	.0981

Note that these simulated relative frequencies are very close to the theoretical probabilities .1, .1, .2, .2, .2, .1, and .1. This method always yields approximate distributions with the approximation getting better (and more expensive) as the number of repetitions increases. □

EXAMPLE 6.5 We found that the sampling distribution of $\bar{Y}$ in Example 6.4 assigned probability 2/10 to $\bar{y} = 4.5$. What is the interpretation of this result?

Solution There are at least two useful interpretations. First, we could suppose that a long series (conceptually, an infinitely long series) of samples of size 2 had been taken from this population. In such a case, the long-run fraction of samples yielding a sample mean of 4.5 would be .2. Alternatively, we could suppose that a list of all possible samples of size 2 was prepared; 2 of every 10 samples would yield a sample mean of 4.5. □

In this book we derive only some relatively simple theoretical (sampling) distributions. More complicated sampling distributions require sophisticated mathematics, and that's not our priority. Instead, we focus on the assumptions underlying the theory—and the consequences of violating these assumptions. For the derivation of the results we suggest consulting a good mathematical statistics book such as Mood, Graybill, and Boes (1974).

Of course, sampling distributions can also be calculated for samples taken with replacement. If, in Example 6.3, the sample is taken with replacement, the sampling distribution of $\bar{Y}$ is as follows:

$\bar{y}$:	2.0	2.5	3.0	3.5	4.0	4.5	5.0	5.5	6.0
$P_{\bar{Y}}(\bar{y})$:	.04	.08	.12	.16	.20	.16	.12	.08	.04

A computer-simulation study could be done assuming sampling with replacement. For example, we could draw 10,000 2-digit random numbers, letting each of the numbers 00, 01, 02, and 03 correspond to $\bar{y} = 2$, each of 04 through 11 to $\bar{y} = 2.5$, and so on. The relative frequencies from such a study would be close to the probabilities shown. □

SECTION 6.2 EXERCISES

6.7 The owner of a chain of laundromats estimates the average between breakdowns of washing machines by finding the time since the previous repair of the 100 most recently repaired machines. What bias may there be in this procedure?

6.8 Can you suggest a better approach to collecting a sample of times to breakdown for 100 machines in Exercise 6.7?

6.9 One way to audit 1% of all transactions passing through a brokerage house is to check all transactions with serial numbers ending in 00. If you were an embezzler working in the back room, what would you think of this approach?

6.10 Suggest an idealized way to sample 1% of all transactions in Exercise 6.9.

6.11 A random sample of size 3 is to be drawn, without replacement, from the population of Example 6.3. The sampling distribution of $\bar{Y}$ can be proved to be

$\bar{y}$:	3.000	3.333	3.667	4.000	4.333	4.667	5.000
$P_{\bar{Y}}(\bar{y})$:	.10	.10	.20	.20	.20	.10	.10

a. Plot histograms of this sampling distribution and the sampling distribution of Example 6.3.
b. Find the probability that $\bar{Y}$ is no more than \$.50 away from the population mean \$4.00, assuming $n = 3$.

6.12 Compute the expected values and variances for the sampling distributions of Example 6.3 and Exercise 6.11. How does the difference in sample size affect the expected values and variances?

6.13 A random sample of size 8 is to be taken with replacement from a population with the following probability distribution:

Value:	4	8	12	16
Probability:	.50	.30	.15	.05

The sampling distributions of $\bar{Y}$, the sample mean, can be shown to be (to 4 decimal places):

$\bar{y}$:	4.0	4.5	5.0	5.5	6.0	6.5	7.0	7.5	8.0
$P_{\bar{Y}}(\bar{y})$:	.0039	.0188	.0488	.0898	.1293	.1535	.1550	.1359	.1048
$\bar{y}$:	8.5	9.0	9.5	10.0	10.5	11.0	11.5	12.0	12.5
$P_{\bar{Y}}(\bar{y})$:	.0718	.0439	.0242	.0119	.0053	.0021	.0008	.0002	.0001
$\bar{y}$:	13.0	13.5	14.0	14.5	15.0	15.5	16.0		
$P_{\bar{Y}}(\bar{y})$:	.0000	.0000	.0000	.0000	.0000	.0000	.0000		

a. Plot histograms of the population distribution and the sampling distribution of $\bar{Y}$.
b. Verify that the population mean is 7.0.
c. Find the expected value and variance of $\bar{Y}$.
d. Find the probability that $\bar{Y}$ is no more than 1.5 units away from the population mean μ.

6.14 Refer to the histograms plotted in Exercise 6.13, part (a).
a. Which histogram shows a smaller variance (determine this without arithmetic)?
b. Which histogram shows less skewness?

6.15 A computer simulation of the sampling process of Exercise 6.13 yields the following:

Mean:	4.0	4.5	5.0	5.5	6.0	6.5	7.0	7.5	
Frequency:	3	25	45	91	122	138	165	137	
Mean:	8.0	8.5	9.0	9.5	10.0	10.5	11.0	11.5	12.0
Frequency:	115	71	38	23	12	9	3	2	1

a. Plot the simulation frequencies in a histogram.
b. Compare this histogram to the theoretical probabilities shown in Exercise 6.13.

6.3 EXPECTED VALUES AND STANDARD ERRORS OF SAMPLE SUMS AND SAMPLE MEANS

The sampling distribution of a sample statistic is a probability distribution. Its exact form depends on the population distribution being sampled. Fortunately, the basic properties of the most important sampling distributions, those of sample sums and sample means, can be derived from minimal assumptions about the population. In this section we find expected values and variances for these distributions. In the next section we show that the normal distribution is often a good approximation to the exact shapes of these sampling distributions.

The mathematical results shown in Appendix 4A of Chapter 4 extend immediately to give us the desired results for the sampling distribution of a sum. Suppose that we take a random sample of size n from a population of sufficient size N that n is small relative to N. Let Y_i denote the ith observation in the sample and let T denote the sample sum $Y_1 + Y_2 + \ldots + Y_n$. Then it can be shown that the expected value and variance of T are as shown below:

Expected Value and Variance of a Sample Sum

If T is the sum of n values drawn at random from a population with mean μ and variance σ^2, then the expected value, variance, and standard deviation for T are

$$E(T) = \mu + \mu + \ldots + \mu = n\mu$$
$$\text{Var}(T) = \sigma^2 + \sigma^2 + \ldots + \sigma^2 = n\sigma^2$$
$$\sigma_T = \sqrt{\text{Var}(T)} = \sqrt{n}\sigma$$

EXAMPLE 6.6 Suppose that the long-run average of the number of Medicare claims submitted per week to a regional office is 62,000, and that the standard deviation is 7000. If it is assumed that the weekly claims submissions during a 4-week period constitute a random sample of size 4, what are the expected value and standard deviation of the total number of claims in the period?

Solution We are given that $\mu = 62{,}000$, $\sigma = 7000$, and $n = 4$. It follows that

$$E(T) = (4)(62{,}000) = 248{,}000$$
$$\sigma_T = (\sqrt{4})(7000) = 14{,}000$$

The critical part of the assumption of random sampling is the assumption of independence of the numbers of claims submitted from one week to another. If independence doesn't hold, the expected value of T is still correct but, according to Appendix 4A (Chapter 4), the standard deviation is wrong. □

At this point, it is very handy to introduce a new name for an existing concept. We specify many sample statistics and many sampling distributions in the next several chapters; it turns out that there are many different formulas for the standard deviations of the sampling distributions of these statistics. Most of the formulas (like the one for the standard deviation of T) involve the population standard deviation, and it becomes difficult to distinguish between the different standard deviations. From here on, then, we use the term **standard error** to denote the theoretically derived standard deviation of the sampling distribution of a statistic. The standard error of the sample sum T is the standard deviation of its sampling distribution $\sigma_T = \sqrt{n}\sigma$.

standard error

The sampling distribution of a sample mean is even more important than the sampling distribution of a sum; the sample average is the most widely used of all statistics. The expected value, variance, and standard error of this sampling distribution can be found quite easily using the results about a sample sum and the principles introduced in Appendix 4A. If $Y_1, \ldots, Y_n$ represent the n individual values from a random sample, the sample mean is $\bar{Y} = (Y_1 + \ldots + Y_n)/n = T/n$, the sample total divided by the sample size. From the Appendix 4A, it follows that we can find the expected value and standard error of the sample mean by dividing those for the corresponding sample sum T by n.

Expected Value and Standard error of $\bar{Y}$

If a random sample of size n is drawn from a population, the expected value and standard error of $\bar{Y}$ are

$$E(\bar{Y}) = \frac{n\mu}{n} = \mu$$

$$\sigma_{\bar{Y}} = \frac{\sqrt{n}\sigma}{n} = \frac{\sigma}{\sqrt{n}}$$

EXAMPLE 6.7 In the situation of Example 6.6, find the expected value and standard error of the average weekly number of claims over a 4-week period.

Solution In Example 6.6, $\mu = 62{,}000$, $\sigma = 7000$, and $n = 4$. If $\bar{Y}$ is the 4-week average of the weekly number of claims, then

$$E(\bar{Y}) = \mu = 62{,}000$$

$$\sigma_{\bar{Y}} = \frac{\sigma}{\sqrt{n}} = 7000/\sqrt{4} = 3500$$

□

The fact that $E(\bar{Y}) = \mu$ means that the sample mean estimates the population mean correctly on average. In one particular sample, the sample mean may overestimate the population mean. In another, the sample mean may underestimate. But in the long run there is no **systematic** tendency for a sample mean to overestimate or underestimate the population mean. This is true regardless of the sample size.

The standard deviation of the sampling distribution of the mean (the standard error of the sample mean) is crucial in determining the probable amount of error in an estimate. We just said that there is no systematic tendency to over- or underestimate μ with $\bar{Y}$. This wouldn't be much of a consolation if we knew that half the time we made a huge overestimate, the other half an equally huge underestimate! The standard error of a sample mean $\sigma_{\bar{Y}}$, in conjunction with the Empirical Rule of Chapter 2, can be used to give a good indication of the probable deviation of a particular sample mean from the population mean.

EXAMPLE 6.8 Suppose that a supermarket manager is interested in estimating the mean checkout time for the nonexpress checkout lanes. An assistant manager obtains a random sample of 25 checkout times. If previous data suggest that the population standard deviation is 1.10 minutes, describe the probable deviation of $\bar{Y}$ from the unknown population mean μ.

Solution The Empirical Rule indicates that approximately 95% of the time $\bar{Y}$ is within two standard errors ($2\sigma_{\bar{Y}}$) of the population mean μ. For $n = 25$,

$$2\sigma_{\bar{Y}} = \frac{2\sigma}{\sqrt{n}} = \frac{2(1.10)}{5} = .44$$

The probable error for $\bar{Y}$ is no more than .44 minute. □

The probable accuracy of a sample mean, as measured by its standard error, is affected by the sample size. Because the standard error of the sample mean is the population standard deviation divided by the square root of the sample size, the standard error decreases as the sample size increases. For example, if the sample size had been either 50 or 100 instead of 25 in Example 6.8, the probable errors ($2\sigma_{\bar{Y}}$) would have been, respectively, .31 or .22.

As the sample size increases to infinity, the standard error of the sample mean decreases toward zero. For a large sample size, the standard error of the mean is very small, and the sample mean based on a huge sample is very close to the true population mean, with very high probability.

SECTION 6.3 EXERCISES

6.16 Refer to the sampling distribution of Exercise 6.13. Show that the expected value and variance found in Exercise 6.13 agree with the theoretical results of this section.

6.17 An automobile insurer has found that repair claims have an average of \$927 and a standard deviation of \$871. Suppose that the next 50 claims can be regarded as a random sample from the long-run claims process.

a. Find the expected value and standard error of the total of the next 50 claims.

b. Find the expected value and standard error of the average of the next 50 claims.

6.18 A computer simulation can itself be regarded as a sampling process. Suppose that a study is done concerning the time required to complete a research and development project. There is considerable variability in the times required to complete the various pieces of the project, so the overall completion time has considerable variablity. Assume that the time to completion has a mean of 28.2 months and a standard deviation of 6.9 months.

a. If the simulation involves 1000 independent trials of the project, find the expected value and standard error of the simulation (sample) mean.

b. Find the expected value and standard error if 4000 trials are simulated.

6.4 SAMPLING DISTRIBUTIONS FOR MEANS AND SUMS

In the last section we stated the appropriate expected values and standard errors for two sample statistics: the sample sum T and the sample mean $\bar{Y}$. In this section we show that in most situations a normal distribution provides a good approximation to the sampling distribution for T or $\bar{Y}$.

According to a theorem of mathematical statistics, if a population distribution is (exactly) normal, then the sampling distributions for the sample sum T and the sample mean $\bar{Y}$ are also (exactly) normal. The relevant expected values and standard errors are those given in the previous section.

EXAMPLE 6.9 A timber company is planning to harvest 400 trees from a very large 50-year-old stand. The yield of lumber from each tree is largely determined by its diameter. Assume that the distribution of diameters in the stand is normal with mean 44 inches and standard deviation 4 inches. Also assume (perhaps unrealistically) that the selection of the 400 trees is effectively random. Find the probability that the average diameter of the harvested trees is between 43.5 and 44.5 inches.

Solution The population distribution (of the diameters of all trees in the stand) is assumed to be normal. It follows from the previous result that the sampling distribution of $\bar{Y}$ is also normal. The appropriate expected value and standard error are

$$\mu_{\bar{Y}} = \mu = 44$$

$$\sigma_{\bar{Y}} = \frac{\sigma}{\sqrt{n}} = \frac{4}{\sqrt{400}} = .20$$

As usual, we calculate normal probabilities by calculating z-scores (see Figure 6.2):

$$P(43.5 \leq \bar{Y} \leq 44.5) = P\left(\frac{43.5 - 44}{.20} \leq Z \leq \frac{44.5 - 44}{.20}\right)$$
$$= P(-2.50 \leq Z \leq 2.50)$$
$$= 2(.4938) = .9876$$

□

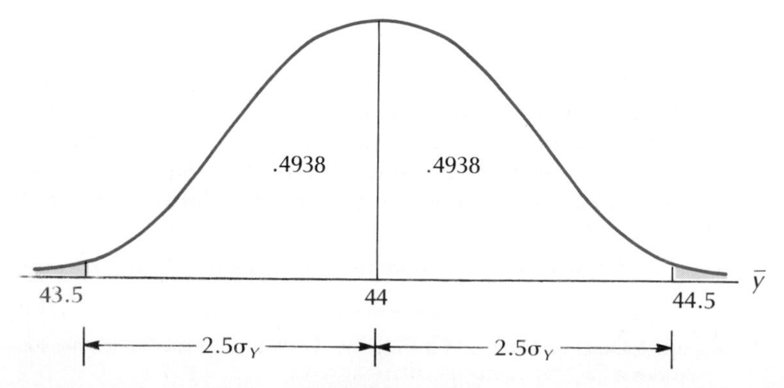

FIGURE 6.2 Probability Calculation for Example 6.9

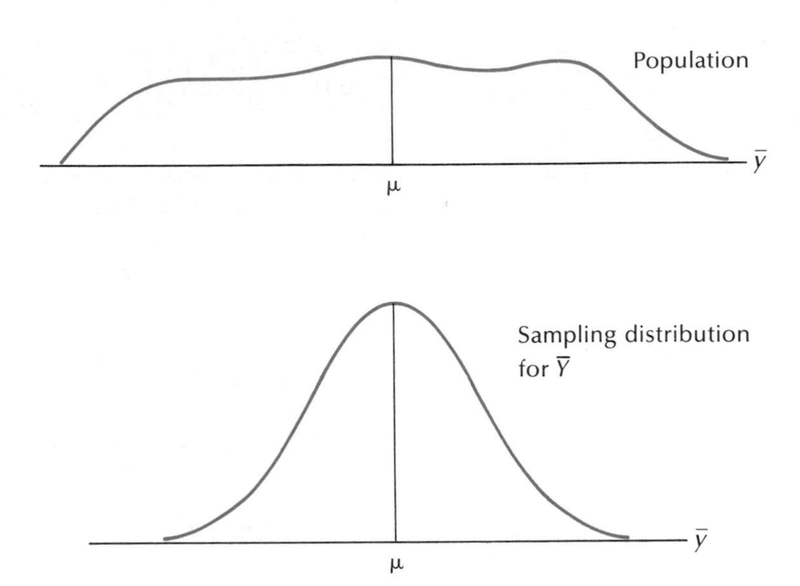

FIGURE 6.3 An Illustration of the Central Limit Theorem

Use of this theorem, as stated, requires the assumption that the population distribution is exactly normal. In practice, no distribution is exactly normal. Another theorem, called a **Central Limit Theorem**, implies that the assumption of a normal population is not crucial.

Central Limit Theorem

Central Limit Theorem for Sums and Means
For *any* population (with finite mean μ and standard deviation σ), the sampling distributions of the sample sum and of the sample mean are approximately normal if the sample size n is sufficiently large.

This is a rather remarkable theorem. **Regardless of the nature of the population distribution—discrete or continuous, symmetric or skewed, unimodal or multimodal—the sampling distributions for T and $\bar{Y}$ are always normal as long as the sample size is large enough.** This is illustrated in Figure 6.3 for the sample mean. The condition that the population mean and standard deviation must be finite is almost always satisfied.*

An obvious question is how large a sample is sufficiently large? The Central Limit Theorem is a mathematical theorem—"n sufficiently large" is translated into "as n goes

* The only exception we know of is the case of so-called stable laws, which are sometimes used as models in finance.

to infinity"—so it does not contain the answer to this question. An enormous number of studies have tried to answer the question, using other mathematical theorems and computer simulations. Many textbooks give a blanket rule: Use the normal approximation anytime n exceeds 30.

This rule is a good basic rule for using the Central Limit Theorem. A better rule would consider the effect of skewness. If the population distribution is very skewed, the actual sampling distribution for $n = 30$ or for $n = 40$ is also somewhat skewed, less so than the population distribution, but enough to make the normal approximation mediocre. If the population distribution is symmetric, the sampling distribution even with $n = 10$ or so is remarkably close to normal. A better rule would be based on a plot of the sample data, and drawing a picture of the data is always a good idea. If a histogram of the sample data shows obvious skewness (and hence suggests skewness for the population distribution), a normal approximation should be used skeptically unless n is up around 100. If the histogram has little skewness, the normal approximation may be used confidently, even with an n of 15 or 20.*

EXAMPLE 6.10 A computer program was used to draw 1000 samples each, with sample sizes 4, 10, 30, and 60, from an exponential population having mean and standard deviation both equal to 1. (A discussion of how such computer simulations are done is in Section 7.6.) Histograms of the sample means are shown in Figure 6.4. As the sample size increases, how does the shape of the theoretical (sampling) distribution of means change? How does the variability of sample means change?

Solution For $n = 4$, the distribution of means is clearly right skewed, although not as skewed as the exponential distribution itself. As the sample size increases, the skewness decreases. For a sample of size 60, the distribution of means appears to be very close to normal. The Central Limit Theorem indicates that the theoretical distribution of sample means should, indeed, approach a normal distribution as the sample size increases.

From the scale at the bottom of each histogram, we can assess the variability of sample means. As n increases, the range of sample means decreases, indicating that variability decreases. The fact that the standard error of the sample mean decreases as n increases indicates that the variability of sample means should decrease as n increases. □

Remember that you cannot plot the sampling distribution itself. That is the theoretical, long-run distribution arising from repeated sampling; in practice you take only one sample. The data plot that we refer to is the sample histogram; it is useful as a rough indicator of the population shape, which is known to have an effect on the quality of the Central Limit Theorem normal approximation.

* The quality of a normal approximation is also slightly affected by how heavy the tails are in the population. Even if a population is nearly symmetric, it may contain many more extremely large and extremely small values than would a near-normal distribution. A heavy-tailed distribution is suggested by the presence in a sample of outliers—a few individual values that fall very far from the bulk of the data. We discuss the treatment of outliers in later chapters.

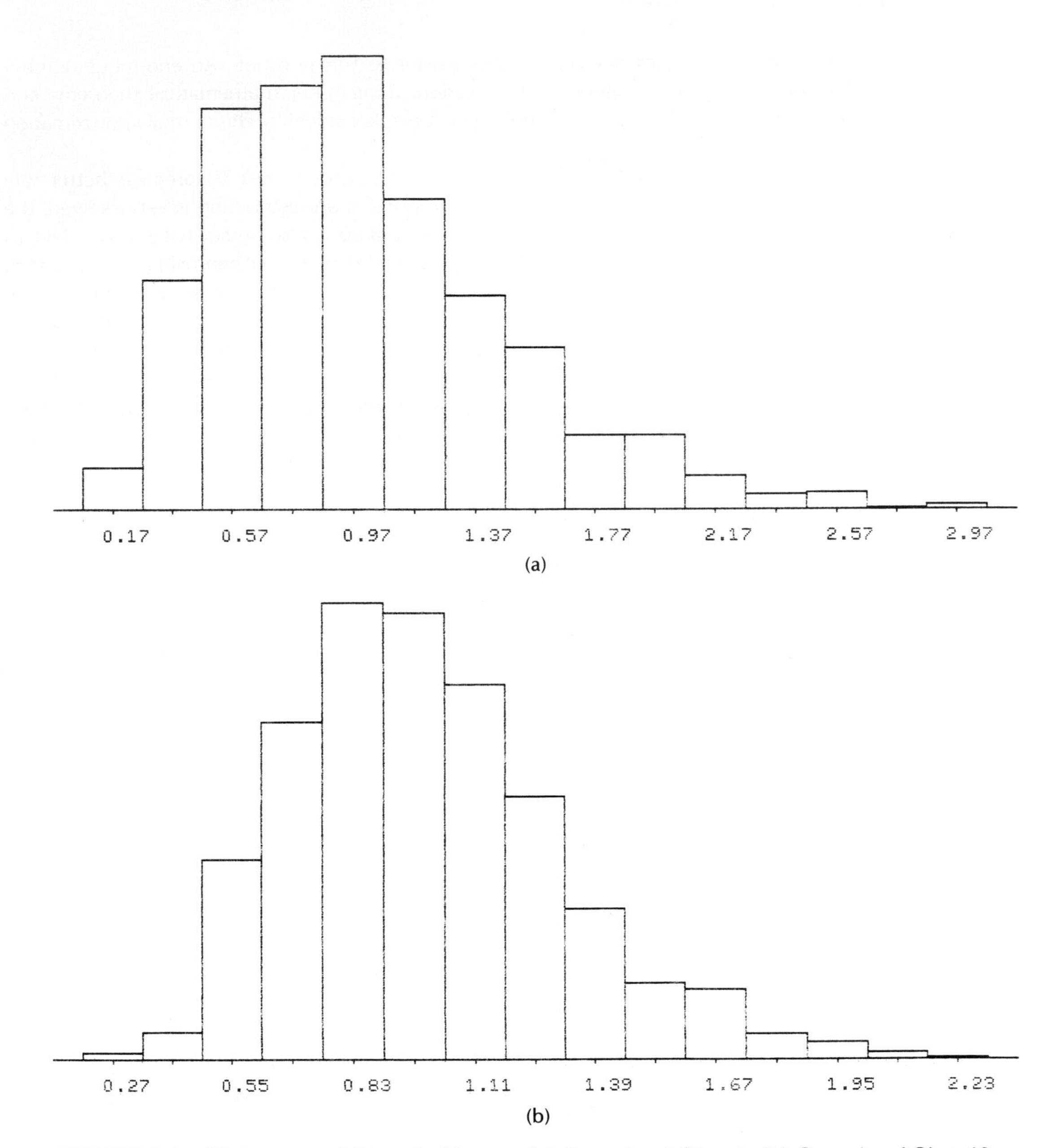

FIGURE 6.4 **Histogram of Sample Means: (a) Sample of Size 4; (b) Sample of Size 10; (c) Sample of Size 30; (d) Sample of Size 60**

EXAMPLE 6.11 In the supermarket checkout time situation of Example 6.8, the following actual times in minutes were observed ($n = 25$): .4, .4, .5, .5, .5, .6, .6, .7, .8, .9, 1.1, 1.2, 1.4, 1.5, 1.8, 2.0, 2.3, 2.6, 2.9, 3.4, 4.2, 5.0, 6.6, 9.2, 16.3 ($\bar{y} = 2.70$). Does it appear that a normal approximation to the sampling distribution of $\bar{Y}$ (for future samples of size $n = 25$, for instance) would be satisfactory?

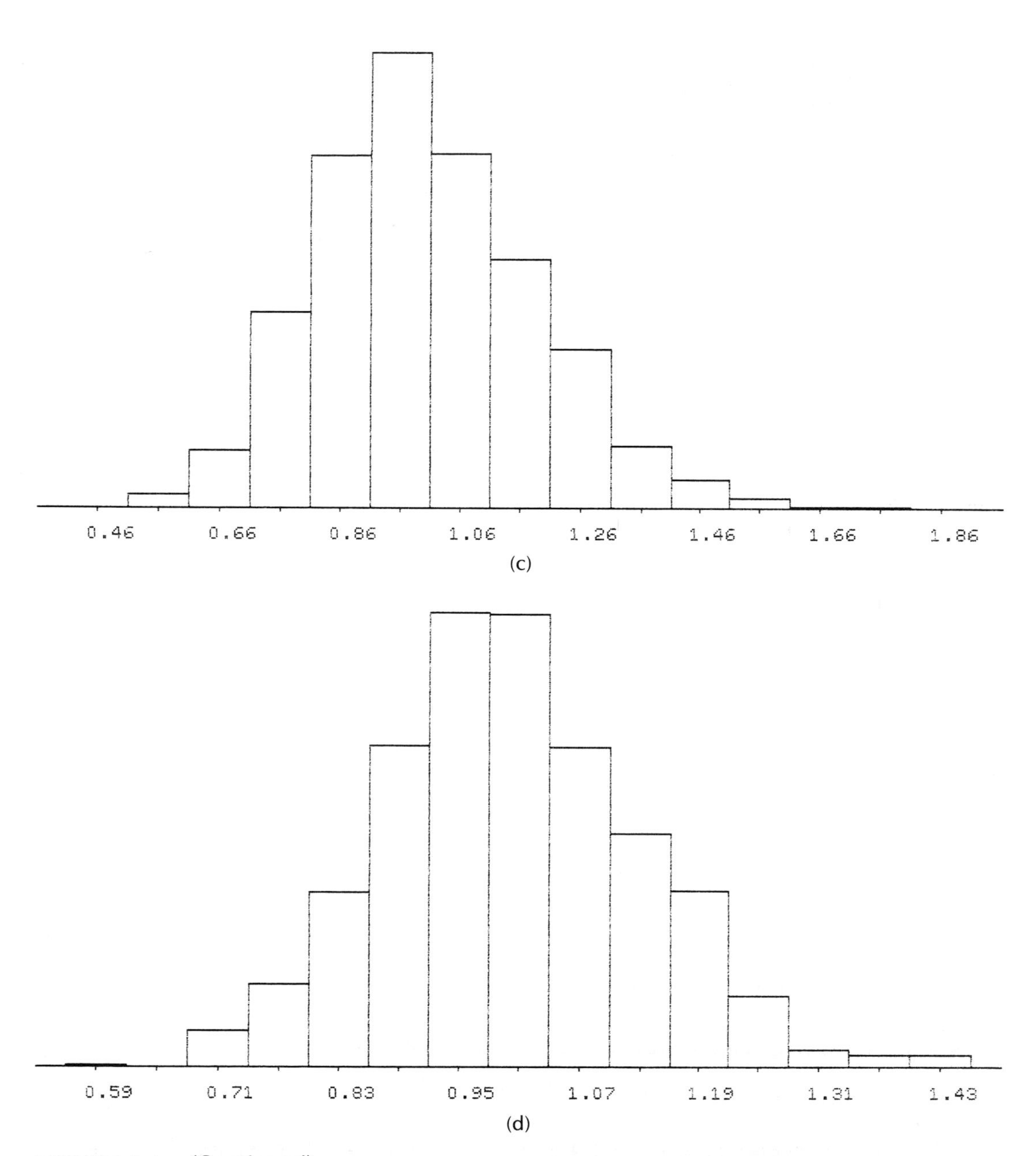

FIGURE 6.4 **(Continued)**

Solution The sample data suggest that the population distribution of checkout times is likely to be highly skewed. See the histogram in the computer output on p. 178. Most times are quite brief, but there are a few people who really slow things up. A sample of 25 is not enough to *deskew* the sampling distribution. Even a sample of 50 isn't really enough in this situation. Therefore the Empirical Rule probabilities (which are based on the normal distribution) in Example 6.8 are most likely inaccurate for $n = 25$ and $n = 50$; for $n = 100$, the probabilities should be fairly close.

```
COLUMN          C1
COUNT           25

      0.4000      0.4000      0.5000      0.5000      0.5000      0.6000
      0.6000      0.7000      0.8000      0.9000      1.1000      1.2000
      1.4000      1.5000      1.8000      2.0000      2.3000      2.6000
      2.9000      3.4000      4.2000      5.0000      6.6000      9.2000
     16.3000

MTB > desc c1
   C1          N =   25          MEAN =   2.6960          ST.DEV. =   3.56
MTB > hist c1

    MIDDLE OF      NUMBER OF
    INTERVAL       OBSERVATIONS
        0.            10    **********
        2.             9    *********
        4.             2    **
        6.             2    **
        8.             0
       10.             1    *
       12.             0
       14.             0
       16.             1    *
```

□

EXAMPLE 6.12 A firm that sells frozen 9-ounce steaks to restaurants is concerned about the fat content of individual steaks. It has claimed that the fat content has mean 8.1% and standard deviation 1.0%. Use a normal approximation to find the probability that the mean fat content in a random sample of 25 steaks exceeds 8.5%. Would you expect the normal approximation to be accurate?

Solution The appropriate expected value and standard error are

$$\mu_{\bar{Y}} = \mu = 8.1\%$$

$$\sigma_{\bar{Y}} = \frac{\sigma}{\sqrt{n}} = \frac{1.0}{\sqrt{25}} = 0.2\%$$

The normal approximation yields

$$P(\bar{Y} > 8.5) = P\left(Z > \frac{8.5 - 8.1}{0.2}\right)$$

$$= P(Z > 2.00) \approx .0228$$

In this situation we would expect the distribution to be fairly symmetric; we would not expect to see fat contents of (say) 15% or more, at least for a firm that stays in business, nor fat contents of practically 0%. (Of course a plot of actual data would be useful in checking our guesses.) If our expectation is correct, the normal approximation should be quite good for $n = 25$. □

SECTION 6.4 EXERCISES

6.19 The number of column-inches of classified advertisements appearing on Mondays in a certain daily newspaper is roughly normally distributed with mean 327 inches and standard deviation 34 inches. Assume that the results for 10 consecutive Mondays can be regarded as a random sample.

a. Find the expected value and standard error of the total number of column-inches of classified advertisement for 10 Mondays.
b. Find the probability that the total is between 3150 and 3390 inches.
c. Find the probability that the average number of column-inches per Monday is between 314 and 339.

6.20 Refer to Exercise 6.19. Find a range of the form $327 - k$ to $327 + k$ such that

$$P(327 - k \leq \bar{Y} \leq 327 + k) \approx .95$$

6.21 Suppose that a certain population has the following distribution:

Value:	200	300	400	500	600
Relative Frequency:	.60	.20	.12	.06	.02

The population mean is 270, and the population standard deviation is 102.470. Exact probability computations show the following:

n	$\sigma_{\bar{Y}}$	$P(\bar{Y} < \mu - 2\sigma_{\bar{Y}})$	$P(\bar{Y} < \mu - \sigma_{\bar{Y}})$	$P(\bar{Y} > \mu + \sigma_{\bar{Y}})$	$P(\bar{Y} > \mu + 2\sigma_{\bar{Y}})$
2	72.46	0	0	.2160	.0336
4	51.23	0	.1296	.1965	.0521
8	36.23	0	.1460	.1594	.0319
16	25.62	.0173	.1876	.1486	.0295
32	18.11	.0127	.1543	.1473	.0340

a. Draw a histogram of the population distribution. What is the obvious feature of this histogram?
b. For each sample size, compute the probability that $\bar{Y}$ falls within two standard errors of μ. How good is the normal approximation for various values of n?
c. Repeat part (b) for $\bar{Y}$ within one standard error of μ.

6.22 In Exercise 6.17 we considered an automobile insurer whose repair claims averaged \$927 over the past with a standard deviation of \$871. A random sample of 50 new claims is taken.

a. Describe the sampling distribution for $\bar{Y}$.
b. Use a normal approximation to calculate $P(\bar{Y} > 1100)$.

6.23 How good would you expect the normal approximation in Exercise 6.22 to be?

6.24 Refer to Exercise 6.22. Suppose $\bar{y} = \$1100$ is observed for the 50 new claims. What do you conclude about repair claims for this year? Would your conclusions change if $\bar{y} = \$1000$?

6.5 USES AND MISUSES OF THE CENTRAL LIMIT THEOREM

The Central Limit Theorem introduced in Section 6.4 was used to justify normal approximations to the sampling distributions of sample sums and means. The same mathematical theorem can be interpreted to indicate situations in which a **population** distribution can be assumed to be approximately normal. Variations on this theorem can be used to show that sampling distributions of other statistics are approximately normal. But the theorem can also be misinterpreted; we now indicate a couple of additional uses for the Central Limit Theorem and a common misconception.

Here is the formal mathematical statement of the Central Limit Theorem: If Y_1, $Y_2, \ldots, Y_n$ are independent random variables with the same probability density functions $f_Y(y)$, then $T = Y_1 + Y_2 + \ldots + Y_n$ and $\bar{Y} = (Y_1 + Y_2 + \ldots + Y_n)/n$ have approximately normal distributions for sufficiently large values of n. When we are talking about sampling distributions, $Y_1, Y_2, \ldots, Y_n$ represent individual values drawn in a sample of size n, and T and $\bar{Y}$ are interpreted as the sample total and mean. There are other practical interpretations of the same mathematical theorem.

One such situation occurs when each **individual** value in a population can be thought of as a sum of n independent terms. Example 6.9 is such a case; the diameter of an individual tree is the sum of a large number of independent terms, each term being the yearly growth of that tree. It's not unreasonable to assume that the year-by-year increases in diameter are independent, with roughly identical probability distributions.* Therefore the individual values of tree diameters, that is, the population, can be expected to have a roughly normal distribution. This interpretation of the Central Limit Theorem gives a reason why some (but by no means all) populations can be expected to have roughly normal distributions.

The normal approximation to the binomial probability distribution that was presented in Chapter 5 is a consequence of the Central Limit Theorem. Suppose that we assign the value 1 to all the successes in the population and the value 0 to all the failures. The population mean is the total number of 1s divided by the population size; this is exactly the population proportion of successes π. The population variance turns out to be $\pi(1 - \pi)$. In a sample of size n, the sample mean $\bar{Y}$ is just the sample proportion $\hat{\pi}$. So

$$Z = \frac{\bar{Y} - \mu}{\frac{\sigma}{\sqrt{n}}} = \frac{\hat{\pi} - \pi}{\frac{\sqrt{\pi(1 - \pi)}}{\sqrt{n}}}$$

has an approximately normal distribution if n is large enough. Furthermore, the sample sum is Y, the number of successes.

$$Z = \frac{Y - n\pi}{\sqrt{n\pi(1 - \pi)}}$$

is an equivalent, approximately normal statistic.

* And fancier versions of the Central Limit Theorem hold even if these assumptions aren't satisfied exactly.

Another important consequence of this Central Limit Theorem is that the Empirical Rule has wide applicability, particularly the 95% portion of the rule. Since the sampling distribution for $\bar{Y}$ is near normal with a single mode and not too much skewness for most reasonable sample sizes, the interval $\mu \pm 2\sigma_{\bar{Y}}$ should contain approximately 95% of the possible values of $\bar{Y}$. This implies that the maximum probable error for estimating μ with $\bar{Y}$ is $2\sigma_{\bar{Y}}$. There is a compensating factor working in our favor. Even though the remaining .05 probability is not evenly split between the two tails, the sum of these tail probabilities is approximately .05, making the Empirical Rule work. Thus the Empirical Rule works well even when the sampling distribution for $\bar{Y}$ is somewhat skewed, leaving one tail probability near zero. The other tail probability is near .05.

Versions of a Central Limit Theorem also apply to sample statistics other than sums and means. Mathematical statistics contain many theorems that conclude that such-and-such statistic has an approximately normal distribution when n is sufficiently large. Sample proportions, sample medians, sample variances, and many other statistics all have approximately normal distributions for large samples. Once expected values and standard errors have been found for these statistics, approximate normal probabilities can be calculated.

But the normal distribution does not always apply. The Central Limit Theorem has been misinterpreted to suggest that every distribution—population, observed data, or whatever—must be normal. In particular, some students believe that *any* large population must have a normal distribution. Central Limit Theorems typically refer to sums or averages of many terms, but unless a sum or average is involved, mere largeness does not imply normality. For example, all individuals living in the United States constitute a large population, but the distribution of wealth among these individuals is extremely skewed. In spite of skewness in the distribution of wealth among these individuals, the Central Limit Theorem guarantees that the distribution of the sample mean income ($\bar{Y}$) is approximately normal for sufficiently large values of n. When we deal with individual data, data plots are the best way to examine normality. The normality of theoretical sampling distributions can be tested by simulation methods, which we consider in the next section.

6.6 COMPUTER SIMULATIONS

Thus far in this chapter, the focus has necessarily been theoretical. The concept of a sampling distribution is inevitably a theoretical one. A sampling distribution is best understood as the distribution of a statistic arising from taking many samples under given conditions. In practice, it's unlikely that one would take multiple samples. But one can use a computer to take multiple samples from a specified population, compute any specified statistic for each sample, and calculate the distribution of the results. In this section we discuss the application of such methods to sampling distributions.

procedure for simulating values

The **procedure for simulating values** is based on the assumed cdf $F_Y(y)$. We begin with a random variable that has a uniform distribution (with all values having equal probability). Many methods are known for drawing what look like uniformly distributed variables.

One way is to multiply a 15-digit number by another 15-digit number and to take the middle ten digits as the random number. Another way is to use prepared tables of random numbers. The trick is to convert these uniformly distributed numbers into simulated values from a random variable having the assumed cdf $F_Y(y)$.

Suppose that $F_Y(y)$ is assumed to be

y	9	10	11	12
$F_Y(y)$	.37	.79	.94	1.00
$P_Y(y)$	.37	.42	.15	.06

Also suppose that there is a source of a uniformly distributed 2-decimal random variable U with probabilities $P_U(u) = .01$, for $u = .01, .02, \ldots, 1.00$. Now proceed as follows:

$$
\begin{aligned}
&\text{if } .01 \le U \le .37, && \text{assign } Y = 9 \\
&\text{if } .38 \le U \le .79, && \text{assign } Y = 10 \\
&\text{if } .80 \le U \le .94, && \text{assign } Y = 11 \\
&\text{if } .95 \le U \le 1.00, && \text{assign } Y = 12
\end{aligned}
$$

Thus, a simulated value, $u = .59$, which falls between .37 and .79, is assigned the value $y = 10$ (see Figure 6.5). All other simulated values are assigned y-values in a similar way. The cutoff values for U, namely .37, .79, .94, and 1.00, obviously were taken from the $F_Y(y)$ values shown in Figure 6.5).

This process does yield random Y-values; we must still show that these values have the correct probabilities.

$$
\begin{aligned}
P(Y = 9) &= P(.01 \le U \le .37) = P(U = .01) + \ldots + P(U = .37) \\
&= .01 + \ldots + .01 = 37(.01) = .37 \\
P(Y = 10) &= P(.38 \le U \le .79) = P(U = .38) + \ldots + P(U = .79) \\
&= .01 + \ldots + .01 = 42(.01) = .42
\end{aligned}
$$

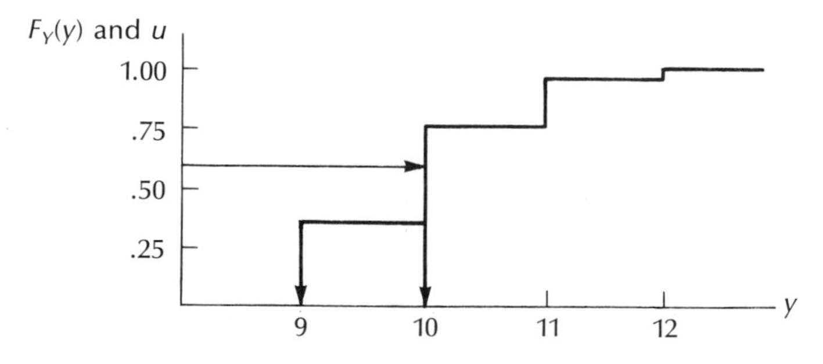

FIGURE 6.5 Simulating Values from a Random Variable with cdf $F_Y(y)$

The probabilities do match the desired ones. The idea can best be seen on a graph such as Figure 6.5.

A uniformly distributed random number is drawn, and then the appropriate value of Y is assigned from the $F_Y(y)$ table.

EXAMPLE 6.13 Simulate six values of Y in the coronary-care case illustration of Section 4.2 (p. 79). Assume that a uniform random-number generator yields the following values: .579, .286, .413, .107, .962, .494. The $F_Y(y)$ table is reproduced below:

y:	0	1	2	3	4	5	6	7	8
$F_Y(y)$:	.001	.003	.006	.011	.024	.061	.139	.224	.336

y:	9	10	11	12	13	14	15	16	17
$F_Y(y)$:	.510	.672	.782	.870	.925	.964	.988	.997	1.000

Solution

$u = .579$ is between .510 and .672; assign $y = 10$
$u = .286$ is between .224 and .336; assign $y = 8$
$u = .413$ is between .336 and .510; assign $y = 9$
$u = .107$ is between .061 and .139; assign $y = 6$
$u = .962$ is between .925 and .964; assign $y = 14$
$u = .494$ is between .336 and .510; assign $y = 9$

If this process had been carried out over a very large number of trials (uniform random numbers), $.672 - .510 = .162$ of the Y assignments would have been $y = 10$, as should happen according to the assumed Y probabilities. □

Monte Carlo approach

This **Monte Carlo approach**—and why that particular gambling casino is so honored, we don't know—is an enormously and widely useful trick. Once appropriate assumptions are made, it is fairly easy to simulate almost any random situation. By varying assumptions, one can test for sensitivity—which aspects are crucial, which are less so. There are many uses of this technique. We can use the computer to calculate the average value of a statistic, averaged over, say, 1000 samples. This average is a good approximation to the expected value of the statistic, which is the theoretical average value over an infinite number of samples. We can use the computer to calculate the standard deviation of the statistic for 1000 (or whatever number) samples. This standard deviation is a good approximation to the standard error of the statistic; remember that the standard error of a statistic is its theoretical standard deviation over infinitely many samples. To check the shape of the theoretical distribution of a statistic, we may compute the skewness or outlier-proneness of the statistic, or draw histograms. Computer simulation is an extremely flexible way to check the validity of any theoretical results.

Suppose that a computer takes 1000 simple random samples, each of size 25, from a population having a normal distribution with mean 100 and standard deviation 15, and computes the mean for each sample. Further, suppose that after computing all these sample means, the computer calculates the average sample mean to be 99.921, the standard deviation of the sample means to be 3.014, and the median of the sample means to be 100.003. Theoretically, the long-run average should be the expected value of the sample mean, namely, the population mean. The simulation average, 99.921, is very close to the theoretical value, 100. Also, the simulation standard deviation of the sample mean, 3.014, is very close to the theoretical standard error of the sample mean, $15.0/\sqrt{25} = 3.000$. Finally, the theoretical (sampling) distribution should have a normal shape; in particular, the skewness should be zero. The simulation "median of means," 100.003, is very close to the simulation "mean of means." 99.921, suggesting that the distribution of means is at least very close to symmetric. A histogram of the means could be constructed to check on symmetry.

EXAMPLE 6.14 A computer program calculates the sample median for 1000 samples of size 25 taken from a normal population having mean 100 and standard deviation 15. The average median is 100.081, the standard deviation of the medians is 3.763, and a stem-and-leaf display of the medians appears nearly normal. What does each result indicate about the sampling distribution of the sample median in this case?

Solution The simulation average should approximate the expected value of the sample median; by symmetry, the expected value should be the same as the population mean, 100. The simulation standard deviation should approximate the standard error of the sample median; it can be shown that theoretically the standard error should be 3.760. The histogram suggests that the sampling distribution of a sample median is also normal, at least when sampling from this population. As mentioned in Section 6.5, for large samples the theoretical sampling distribution of a median is also normal. Apparently, when sampling from a normal distribution, a sample size of 25 is enough to rate as a "large sample." □

We present the results of many computer simulations in this book. We are concerned not only with evidence about the correctness of expected value and standard error formulas, but especially about the correctness of theoretical results about the shape of the sampling distribution. If the sampling distribution of a certain statistic theoretically should be normal, but simulation results indicate that the actual distribution is clearly nonnormal, that indicates that statistical inferences based on the statistic could be seriously wrong.

A **normal probability plot** is an even better way to assess the normality of a distribution than is a histogram. In a normal plot, the actual values are plotted against the values that would be expected assuming a normal distribution. If, in fact, the distribution is normal, the actual values are very close (within random variation) to the expected values, and the normal plot is a straight line. If, however, the actual distribution is not normal, the actual values depart from expected, and the normal plot shows some kind of curve. Figure 6.6 shows normal plots for normally distributed data, right-skewed data, and outlier-prone data. Note that, except for "wiggles and jiggles and bumps," the normal

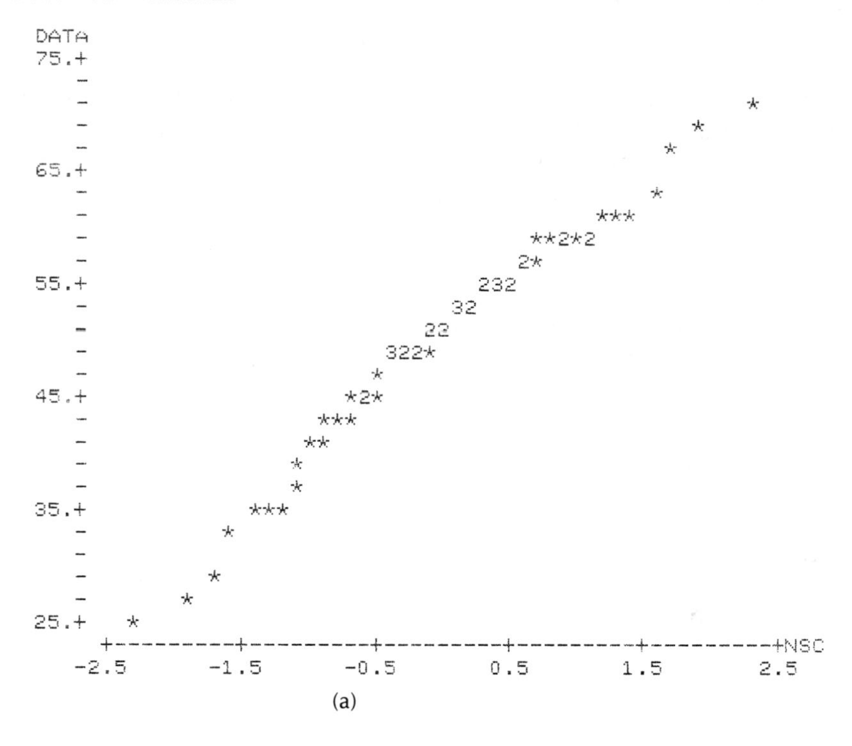

(a)

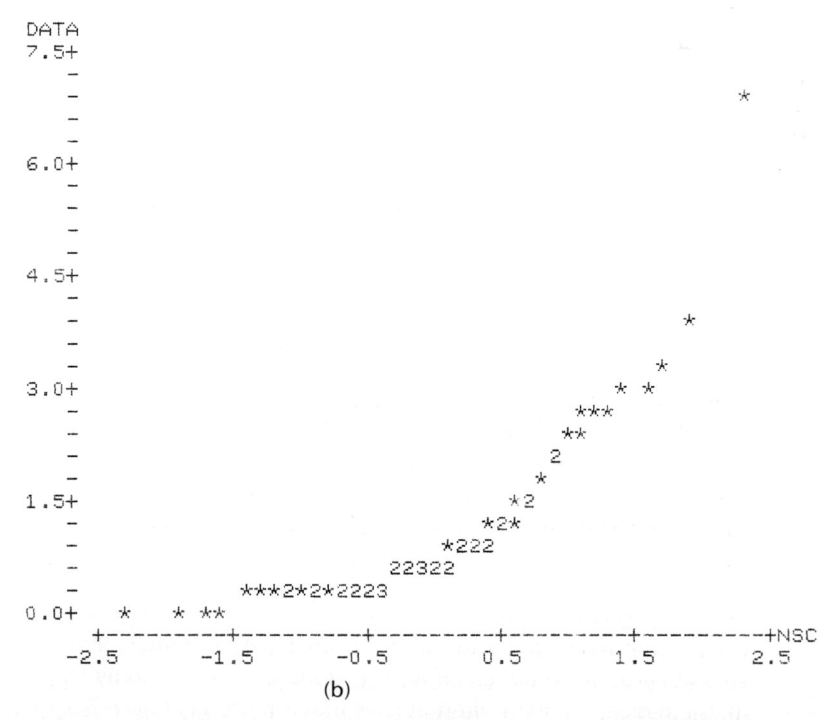

(b)

FIGURE 6.6 Normal Plots for (a) Normal, (b) Skewed, and (c) Outlier-prone Data

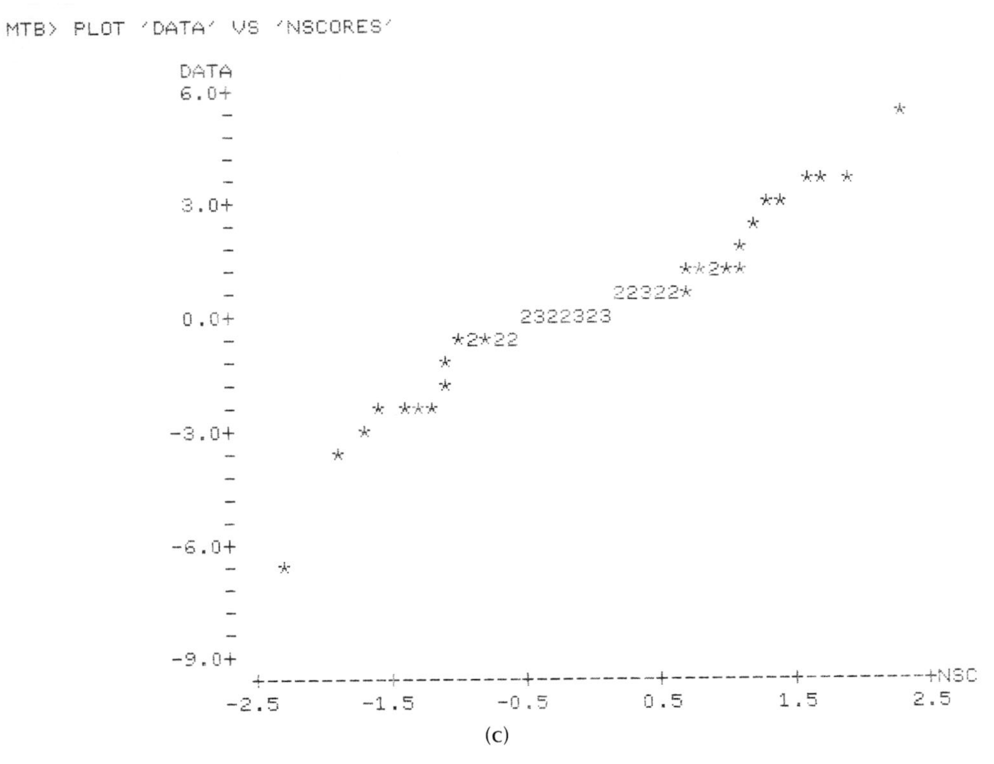

FIGURE 6.6 **(Continued)**

data yield a straight-line plot, the skewed data yield a curved plot, and the outlier-prone data yield an **S**-shaped plot.

EXAMPLE 6.15 A normal plot of 1000 means, each based on a sample of size 10 taken from an exponential-shaped population (see Section 5.7) is shown in Figure 6.7. Does the normal plot indicate that the sampling distribution of means is approximately normal in this situation?

Solution No. There is a clear curve in the plot, indicating that the sampling distribution of the sample means is clearly skewed in this situation. □

Computer simulations are a useful supplement to mathematical derivations of sampling distributions, not a substitute for them. A computer simulation necessarily involves very specific assumptions about the statistic and the underlying population, whereas a mathematical theorem often applies much more generally. But as a supplement to, and illustration of, mathematical results, computer simulations can be extremely valuable.

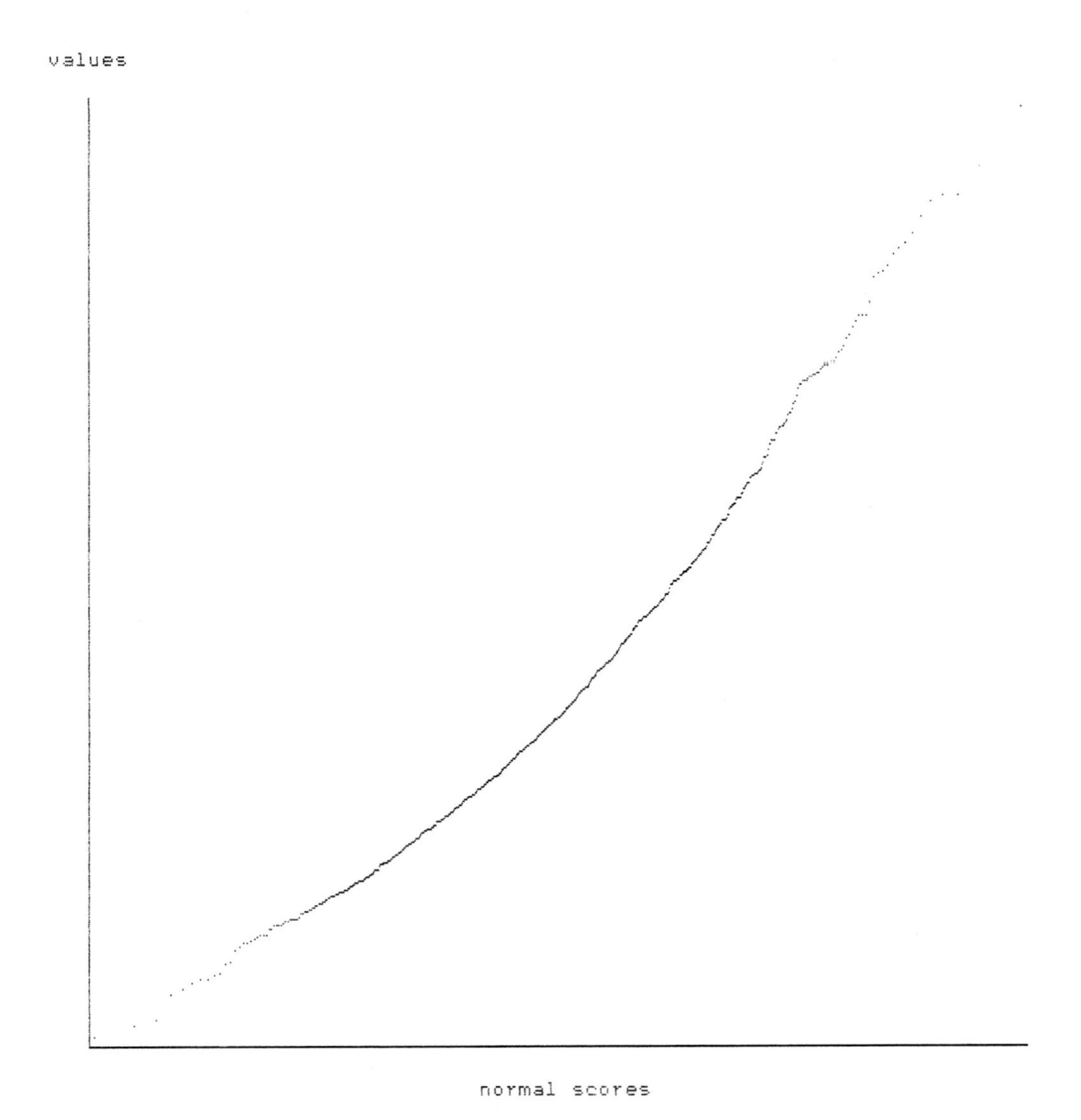

FIGURE 6.7 Normal Plot of Means for Example 6.15

EXAMPLE 6.16 Normal plots of 1000 means, based on samples of size 30 and 60 from the exponential distribution, are shown in Figure 6.8. What is the effect of increasing sample size?

Solution As the sample size increases, the normal plot comes closer to a straight line, indicating that the theoretical (sampling) distribution of the sample mean approaches the normal distribution as the sample size increases. This is precisely what the Central Limit Theorem states. □

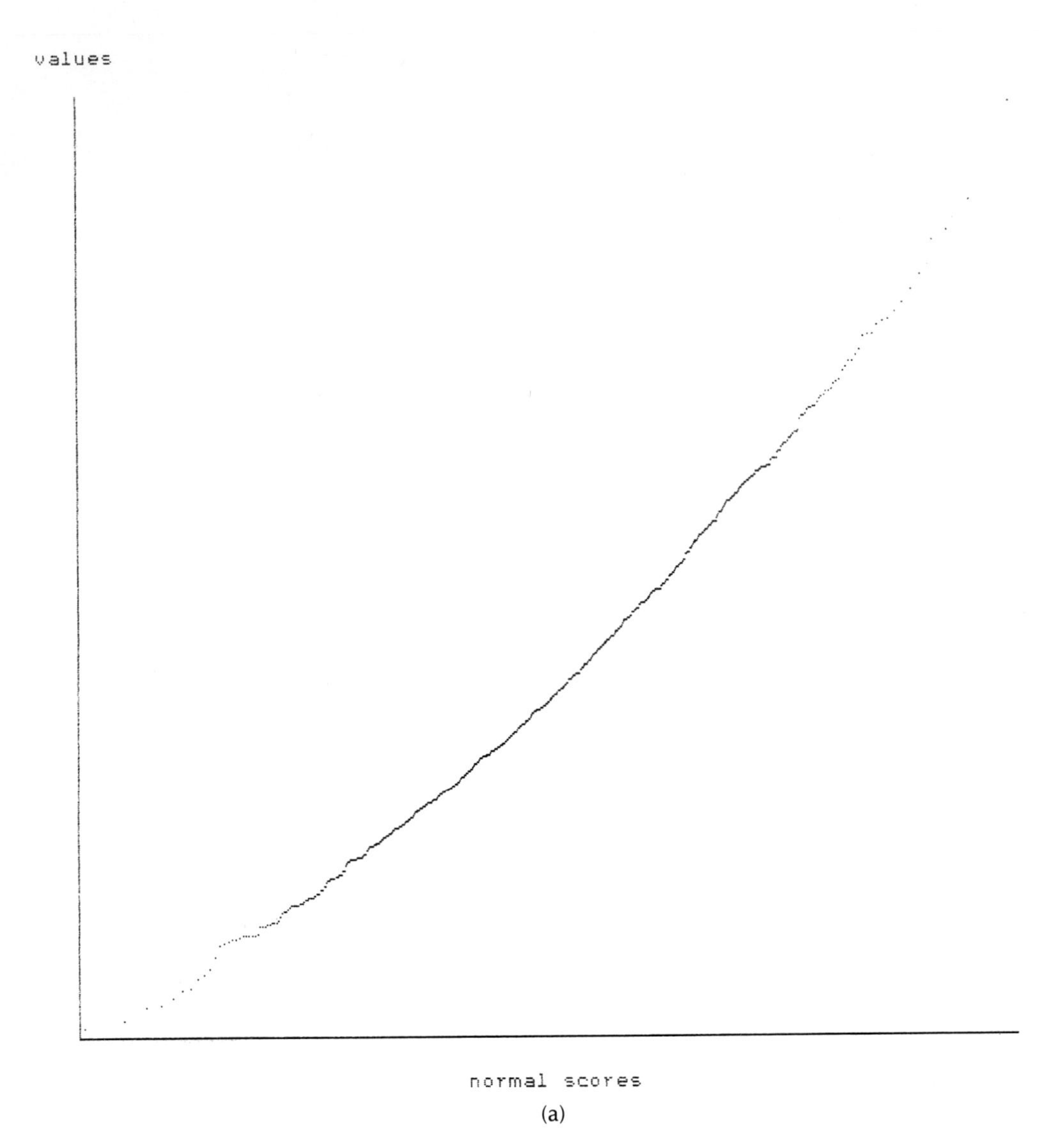

(a)

FIGURE 6.8 **Normal Plot of Means of Example 6.16: (a) Sample of Size 30; (b) Sample of Size 60**

SECTION 6.6 EXERCISES

6.25 Sample means are found for 1000 samples of size 4 taken from a normally distributed population with mean 50 and standard deviation 10. The average of these means is 50.1643 and the standard deviation is 5.0104.

a. What theoretical quantities are being approximated by the average of the means and by the standard deviation of the means?

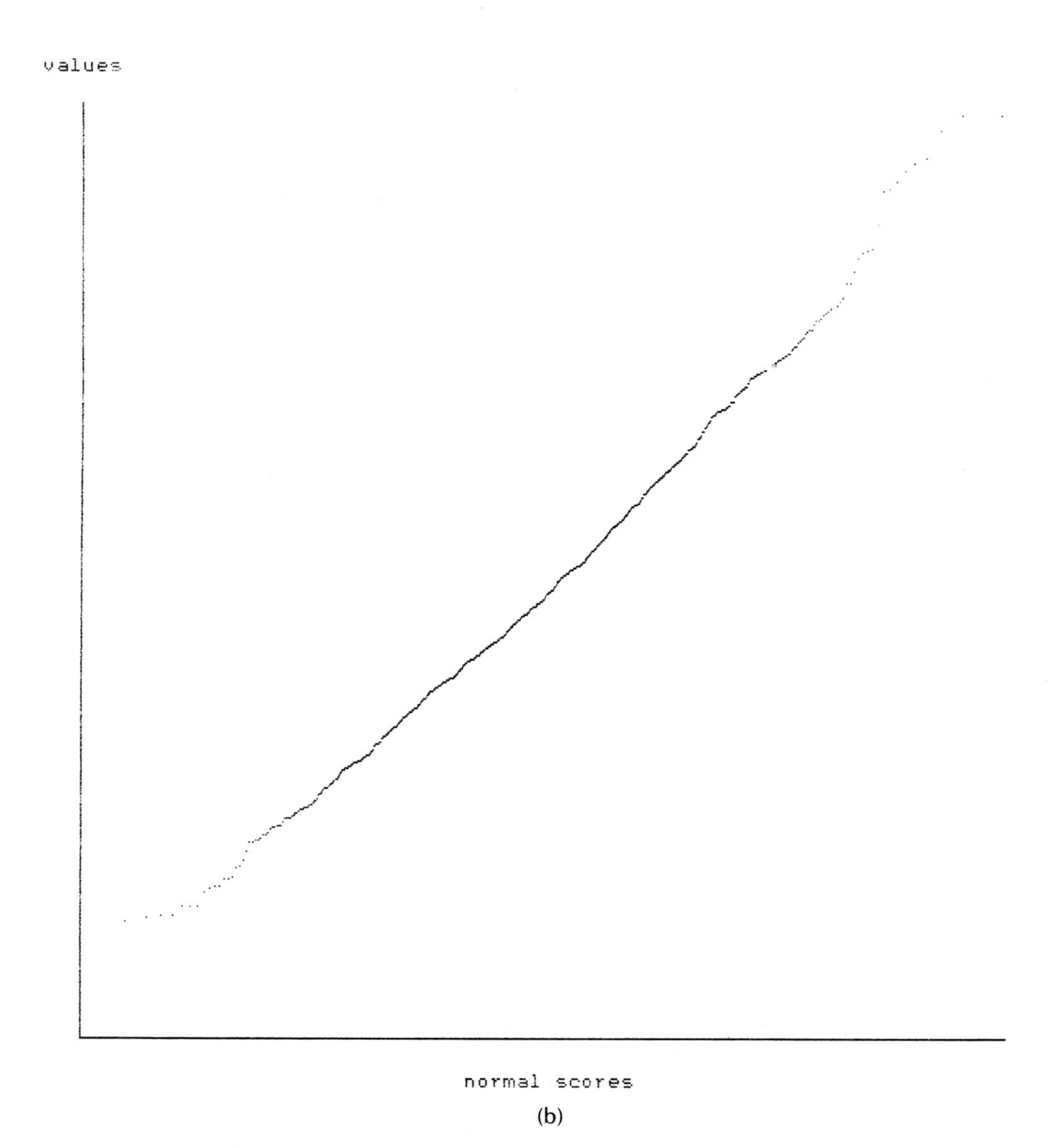

(b)

FIGURE 6.8 **(Continued)**

b. What are the theoretical values of these quantities? Do the simulation results closely approximate these theoretical (infinitely many samples) values?

6.26 A histogram of the means calculated in Exercise 6.25 is shown in Figure 6.9. Does it appear that the sample means are normally distributed? Should they be, given the population from which the samples are drawn?

6.27 Sample medians are found for 1000 samples of size 30 taken from a Laplace population (a symmetric, moderately outlier-prone population). The average of the medians is .0082 and

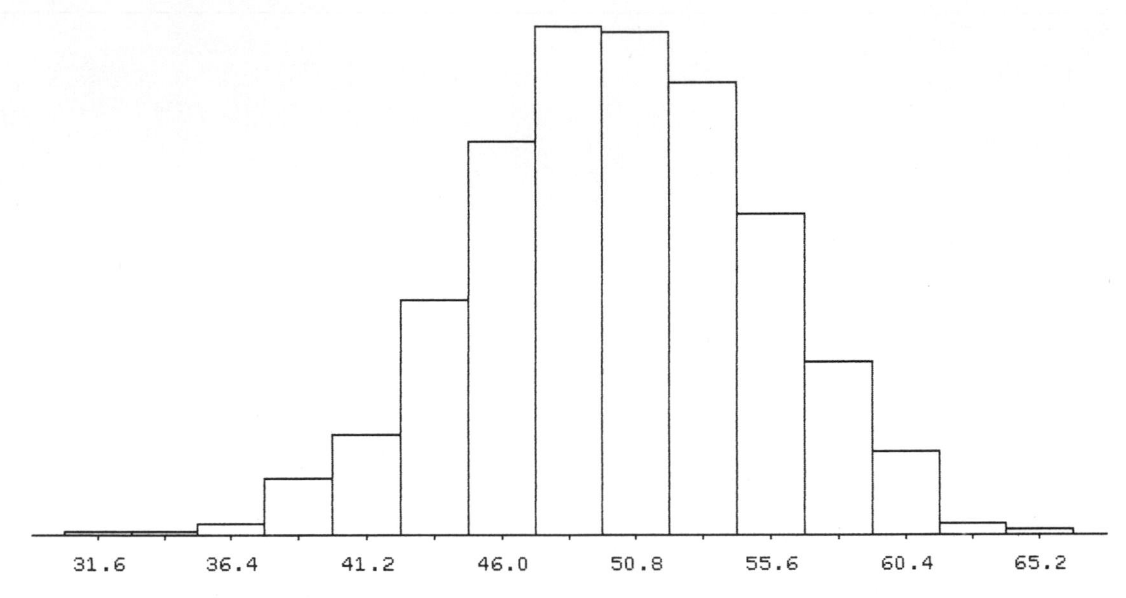

FIGURE 6.9 Histogram of Means from a Normal Population; Exercises 6.25–6.26

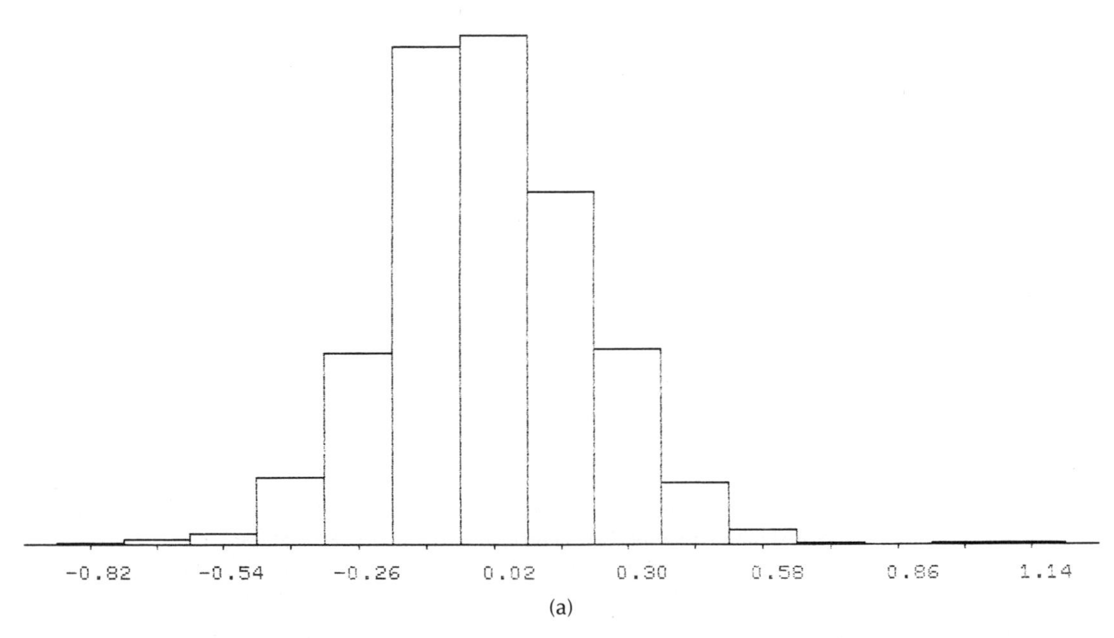

FIGURE 6.10 Histogram (a) and Normal Plot (b) of Medians from a Laplace Population; Exercise 6.28

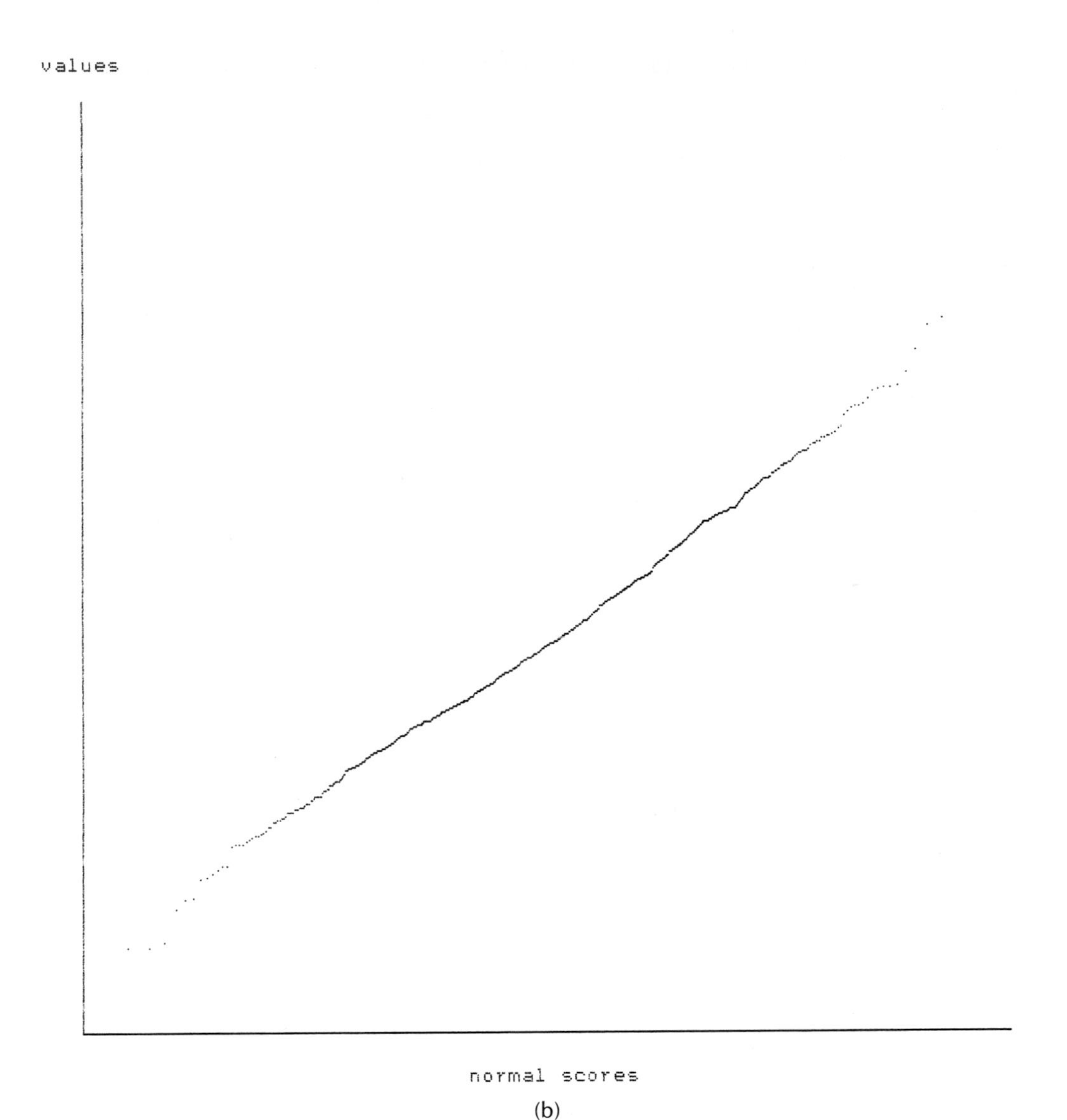

FIGURE 6.10 (Continued)

the standard deviation is .2070. What do these results indicate about the theoretical (sampling) distribution of the median?

6.28 A histogram and normal plot of the medians calculated in Exercise 6.27 are shown in Figure 6.10. Do they indicate that the theoretical distribution of the median is approximately normal in this case?

6.29 Sample means are calculated (for 1000 samples each) from samples of sizes 10 and 30 taken from a discrete population having possible values 1, 2, 3, 4, and 5, with respective probabilities

.1, .2, .4, .2, and .1. The averages and standard deviations of these means are as follows:

Sample Size	Mean	Standard Deviation
10	3.0076	.3563
30	2.9986	.2006

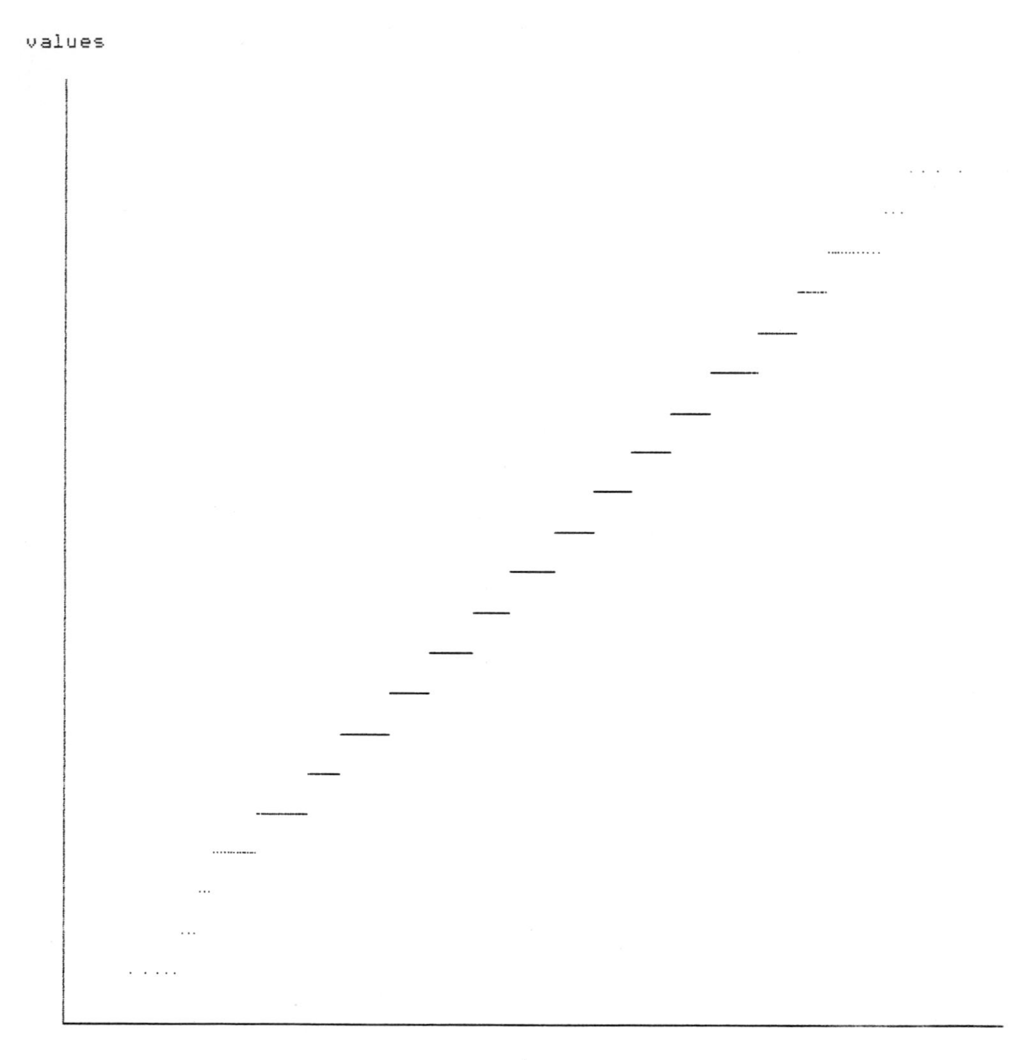

(a)

FIGURE 6.11 Normal Plots of Means [(a) Sample Size 10; (b) Sample Size 30]; Exercise 6.29

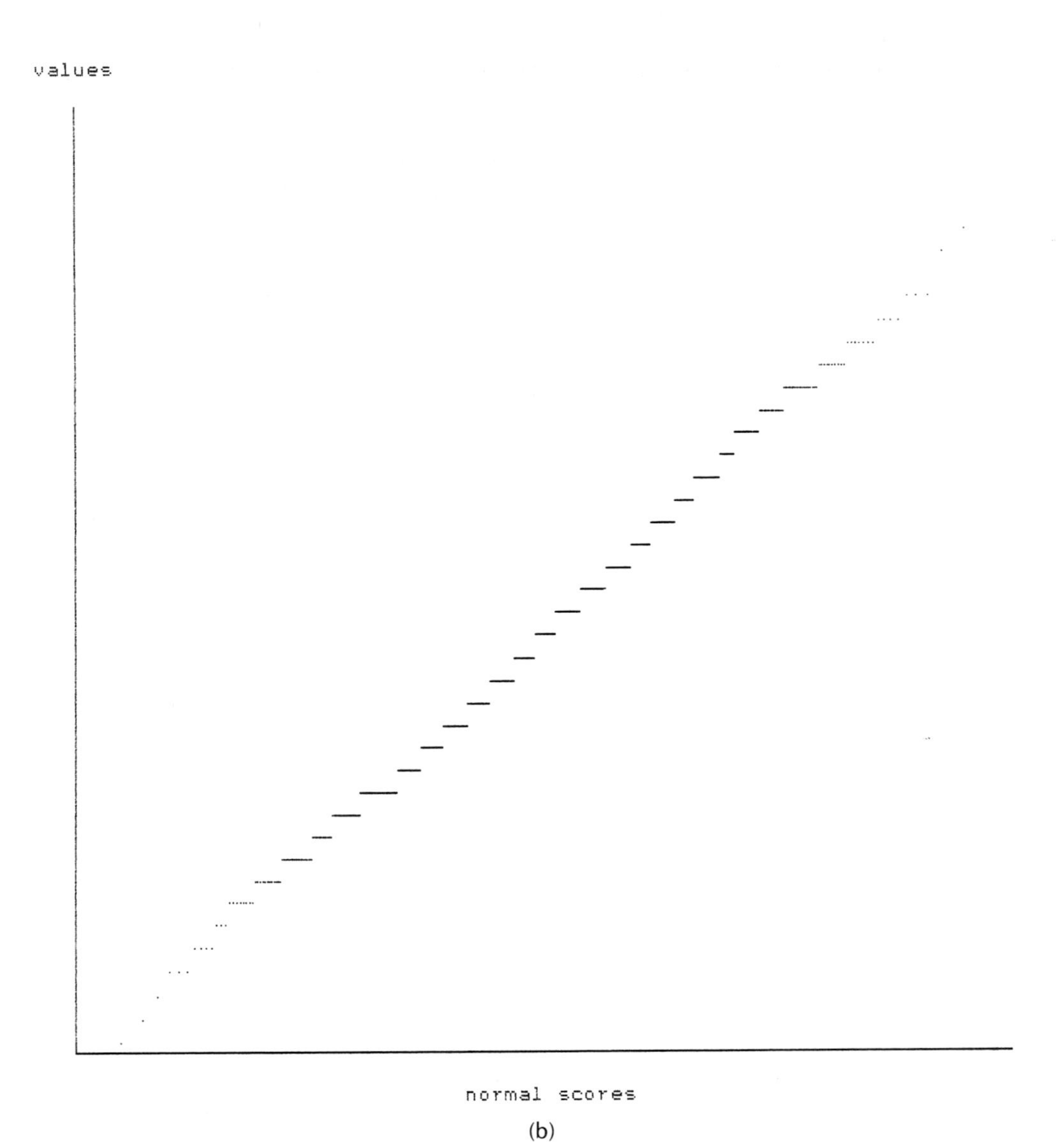

(b)

FIGURE 6.11 **(Continued)**

a. Calculate the population mean and standard deviation.
b. What are the expected values and standard errors of the sample means for each sample size?
c. How closely do the simulation results agree with the theoretical values calculated in part (b)?

6.30 Normal plots of the means calculated in Exercise 6.29 are shown in Figure 6.11.

a. What explains the "staircase" pattern in the plots?
b. Except for this pattern, does it appear that the theoretical distribution of the sample mean is approximately normal for this population and these sample sizes?

6.31 Sample means are calculated for 1000 samples of size 10 taken from a Laplace population. The population mean is zero and the population variance is 2.00. The average value of the means is .0100 and the standard deviation is .4366.

a. Calculate the theoretical expected value and standard error of the sample mean.

b. How closely do the simulation results agree with the theoretical results?

6.32 A histogram and a normal plot of the means of the data from Exercise 6.31 are shown in Figure 6.12. The Laplace population is somewhat outlier-prone but symmetric. Does it appear that the theoretical distribution of the means is close to normal?

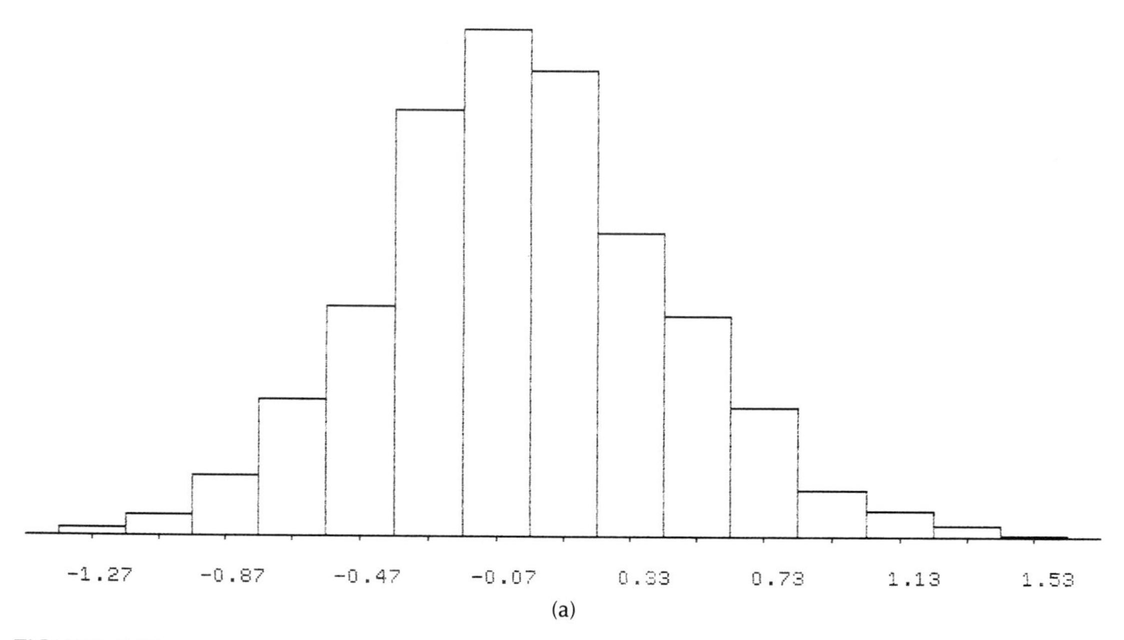

FIGURE 6.12 **Histogram (a) and Normal Plot (b) for Means from a Laplace Population; Exercises 6.31–6.32**

SUMMARY

Chapter 6 presented the very important concept of a sampling distribution for a sample statistic. The sampling distributions for the sample mean $\bar{Y}$ and sample sum T were studied in detail. If repeated random samples of size n are selected from a population with finite population mean μ and standard deviation σ, then expected values and standard errors for $\bar{Y}$ and T are, respectively,

$$E(\bar{Y}) = \mu, \qquad E(T) = n\mu$$

$$\sigma_{\bar{Y}} = \frac{\sigma}{\sqrt{n}}, \qquad \sigma_T = \sqrt{n}\sigma$$

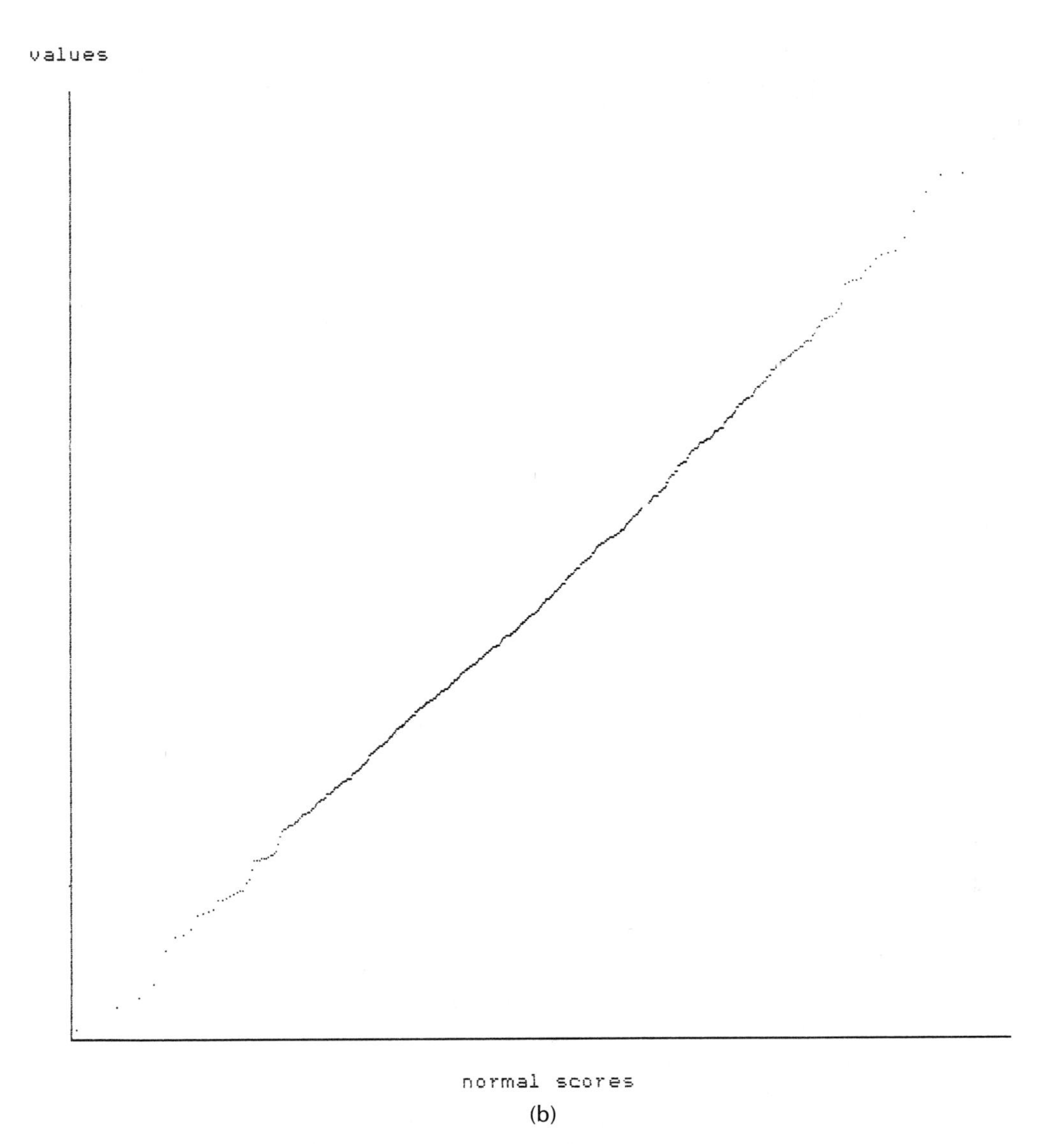

(b)

FIGURE 6.12 **(Continued)**

In addition, a Central Limit Theorem states that when n is large the sampling distributions for $\bar{Y}$ and T are approximately normal. Thus it is possible to determine the probable deviation of $\bar{Y}$ from μ. The widely useful technique of Monte Carlo (computer) simulation allows us to verify mathematical results about the expected value, standard error, and sampling distribution of a statistic. Simulation is also extremely helpful when mathematical results are not known and in checking the effect of violation of assumptions. The results of this chapter enable us to make inferences about an unknown population mean μ based on the sample mean $\bar{Y}$.

KEY TOPICS AND FORMULAS: Random Sampling and Sampling Distributions

1. Statistic: a summary value computed from a random sample
2. Selection bias: any systematic tendency to overrepresent or underrepresent some part of a population when taking a sample
3. Sampling distribution of a statistic: the theoretical probability distribution of a statistic
4. Standard error of a statistic: the theoretical standard deviation of the statistic's sampling distribution
5. Expected value and standard error of a sample sum T:

$$\mu_T = n\mu$$
$$\sigma_T = \sqrt{n}\sigma$$

6. Expected value and standard error of a sample mean $\bar{Y}$:

$$\mu_{\bar{Y}} = \mu$$
$$\sigma_{\bar{Y}} = \frac{\sigma}{\sqrt{n}}$$

7. Central Limit Theorem for sums and means: For any population, the sampling distributions of T and $\bar{Y}$ are approximately normal if n is sufficiently large.

CHAPTER 6 EXERCISES

6.33 A newspaper columnist claims that directors of Fortune 1000 firms averaged only 19.1% of gross income in federal income tax the previous year. From the published column, it can be inferred that the claimed standard deviation is 6.8%. Suppose that you, as an Internal Revenue Service officer, are charged with taking a sample of 200 such executives to test this claim.

a. The names of all Fortune 1000 firm directors are publicly available. How can you use such a list to select a random sample? What problems might you encounter?

b. Can you assume that, since $n = 200$ is a fairly large sample, the distribution of percentage of gross income paid in income tax in the sample is approximately normal? Explain.

6.34 In Exercise 6.19 we assumed that the number of column-inches of classified advertisements appearing on Mondays in a certain daily newspaper has a normal distribution. Do you think that this is a reasonable approximation? Explain.

6.35 In Exercise 5.57 (p. 150) we assumed that the hourly pay rate of workers is normally distributed with a mean of \$5.10 and a standard deviation of \$0.40. Suppose that the union presents data from a committee of 22 unhappy workers showing that these 22 workers have an average pay of \$4.86. Can you think of an explanation?

6.36 In Exercise 5.74 (p. 158) we assumed that the weekly demand for 5-pound sacks of flour at a particular supermarket is normally distributed with mean 72.0 cases and standard deviation 1.6 cases.

a. Do you think that a normal distribution might be a decent model?

b. How could you take a reasonable random sample of size 15 from this population?

6.37 Assume that a random sample of size 15 has been taken in Exercise 6.36.
a. What is $P(\bar{Y} > 73.0)$?
b. Find a 95% range for $\bar{Y}$; that is, find a value k such that $P(72.0 - k \leq \bar{Y} \leq 72.0 + k) = .95$.

6.38 A department store expects the average "inventory shrinkage" (a euphemism for theft by employees and customers) to be 2.2%. The standard deviation from one sale category to another is assumed to be 1.6%. The store has 2571 sales categories, from which a sample of 100 categories is to be selected for detailed inventory checking.
a. How would you select such a random sample?
b. Is a simple random sample desirable here? Granted that you don't know about fancier sampling methods, can you think of other considerations for sampling?

6.39 Suppose that a random sample of 100 categories is chosen in Exercise 6.38.
a. Find the expected value and standard error of the sample average shrinkage.
b. How much difference does it make in part (a) whether you use sampling without replacement or sampling with replacement?
c. Use a normal approximation to calculate $P(\bar{Y} > 2.4\%)$.

6.40 How good do you expect the normal approximation in Exercise 6.39 to be? How would you use the sample data to help indicate how much faith you have in the approximation?

6.41 Suppose that a population has the following distribution:

Values:	10	80	90	100	110	120	190
Relative frequency:	.02	.10	.20	.36	.20	.10	.02

a. Verify that the population mean is 100 and the population standard deviation is 21.07.
b. Draw a histogram of this distribution. What is the obvious feature?

6.42 Refer to Exercise 6.41. The exact sampling distribution of the sample mean (for sample sizes 2, 4, and 8) has the following properties:

Sample Size	Standard Error	$P(\bar{Y} < \mu - 2\sigma_{\bar{Y}})$	$P(\bar{Y} < \mu - \sigma_{\bar{Y}})$	$P(\bar{Y} > \mu + \sigma_{\bar{Y}})$	$P(\bar{Y} > \mu + 2\sigma_{\bar{Y}})$
2	14.90	.0388	.0888	.0888	.0388
4	10.54	.0488	.0899	.0899	.0488
8	7.45	.0333	.1456	.1456	.0333

a. For each sample size, compute the exact probability that $\bar{Y}$ is within one standard error of μ.
b. How good is the normal approximation, for each n?
c. Repeat parts (a) and (b) for $\bar{Y}$ within two standard errors of μ.

APPENDIX: STANDARD ERROR OF A MEAN

Some students want to understand *why* the formula for the standard error of $\bar{Y}$ comes out as it does. For their benefit, this appendix contains a sketchy proof. There are two key ideas; each is stated in terms of variance. First, multiplying a random variable by a constant multiplies its variance by the *square* of the constant. Second, the variance of a sum of *independent* random variables is the sum of the component variances. Therefore,

$$\operatorname{Var}(\bar{Y}) = \operatorname{Var}\left(\frac{\sum Y_i}{n}\right)$$

$$= \frac{1}{n^2}\operatorname{Var}\left(\sum Y_i\right) \quad \text{because dividing by } n \text{ is equivalent to multiplying by } 1/n$$

$$= \frac{1}{n^2}\sum \operatorname{Var}(Y_i) \quad \text{because the individual variables are, by assumption, independent}$$

$$= \frac{1}{n^2}(\sigma^2 + \sigma^2 + \ldots + \sigma^2) \quad \text{because the variables are drawn from a population with variance } \sigma^2$$

$$= \frac{n\sigma^2}{n^2} = \frac{\sigma^2}{n}$$

Taking a square root yields the standard error of $\bar{Y}$:

$$\sigma_{\bar{Y}} = \sqrt{\operatorname{Var}(\bar{Y})}$$

$$= \frac{\sigma}{\sqrt{n}}$$

REVIEW EXERCISES CHAPTERS 4–6

R21 Scores on an aptitude test for assembly workers are roughly normally distributed with a mean of 200 and a standard deviation of 40.

a. Find the probability that a randomly chosen individual scores above 210.

b. Find the probability that the mean of a random sample of 25 individuals is larger than 210.

R22 If the distribution of scores in Exercise R21 is not exactly normal, which answer is a poorer approximation? Why?

R23 Suppose that a random sample of 50 price changes is selected under the conditions of Exercise R3 (p. 70). The number of changes made during the period is so large that it doesn't matter whether the sample is taken with or without replacement.

a. Write an expression for the probability that three or fewer changes are posted incorrectly.

b. Find a numerical value for the probability in part (a).

c. What assumptions were made in answering part (a)? Under what conditions might any of these assumptions be in error?

R24 A certain part is kept in inventory at an automobile dealership. The number in stock at a given time follows the probability distribution

$$f_X(x) = \frac{(x+1)}{66}, \qquad x = 0, 1, \ldots, 10$$

a. Find the mean and standard deviation of the number in stock at a given time.

b. If the dealer requires 3 of this part on a given day, what is the probability that there is enough in stock?

R25 Suppose that the auto dealer in Exercise R24 has 4 separate parts in stock and that the probability distribution for the number of each in stock is the distribution specified in Exercise R24.

a. Find the mean and variance of the total stock of the 4 parts.

b. What additional assumptions, if any, did you make in answering part (a)? For each assumption, is the assumption more critical in determining the mean or in determining the variance?

R26 Now suppose that the dealer in Exercise R24 has 200 separate parts, and that the availability of each part is given by the probability distribution of Exercise R24. Find the approximate probability that the average number in stock (averaged over the 200 parts) is greater than 7. Should the approximation be a good one?

∫ **R27** The daily demand for propane gas from a particular dealer (in appropriate units of measure) is random, with probability density

$$f_Y(y) = .0012y^2(10 - y), \qquad 0 < y < 10$$

a. A student attempts to find the probability that the demand is between 5 and 8 units (both included) by calculating $f_Y(5) + f_Y(6) + f_Y(7) + f_Y(8)$. Explain why this procedure doesn't give the right answer.

b. Calculate the probability sought in part (a).

c. Find the mean and standard deviation of Y.

∫ **R28** Of those days when demand exceeds 5 units in Exercise R27, what fraction have a demand less than 8 units?

R29 Suppose that the manufacturing process for glass wire used in fiber-optic transmission introduces impurities at an average rate of .0002 impurities per foot of wire. The wire is cut into 1000-foot sections; if any impurity is found in a section, that section is recycled. What is the probability that a randomly chosen section contains no impurities?

R30 What additional assumptions beyond the stated ones did you make in answering Exercise R29?

R31 A certain radio show has a catalog from which fans of the show may order records or tapes, as well as souvenir items. Suppose that 40% of the orders involve no records or tapes, 30% involve 1 record or tape, 15% involve 2, 10% involve 3, and 5% involve 4. For each possible number of records or tapes ordered, the percentage of orders of 0, 1, 2, 3, 4, or 5 souvenir items is given in the following table:

	Souvenirs Ordered (%)					
Records/Tapes Ordered	0	1	2	3	4	5
0	0	60	30	5	3	2
1	10	40	25	15	5	5
2	5	30	40	10	8	7
3	3	15	22	30	20	10
4	1	4	15	30	40	10

Calculate the joint probability distribution of X = number of records or tapes ordered and Y = number of souvenirs ordered, in table form. Are the two types of orders independent?

R.32 Find the expected number of souvenirs ordered, assuming the probabilities shown in Exercise R31. Also find the standard deviation of the number of souvenirs ordered.

R33 A manufacturing process that is working properly produces 5% defective items because of impurities in materials or other random factors. Suppose that 20 items are selected from the output of the process and inspected. Assume that the process is working properly.

a. Find the probability that 2 or more of the selected items are defective.

b. What did you assume in answering part (a)?

R34 An alternative inspection method for the process in Exercise R33 is to inspect every item and stop the process whenever 2 defectives have been found within the most recent 10 inspected. Does this inspection method satisfy the assumptions of a binomial random variable?

R35 In the ratings discussed in Exercise R12 (p. 71), suppose that scores of 3, 2, and 1 are assigned to performance ratings of "excellent," "satisfactory," and "unsatisfactory," respectively. Find the mean and variance of the score of a randomly chosen junior executive.

∫ **R36** The availability and actual stocking of generic grocery products by supermarkets is surveyed. X is defined as the fraction of all products stocked by a randomly chosen supermarket that are available to that market as generics, and Y is defined as the fraction of available generic products that are actually stocked; the survey results indicate that the joint probability density of X and Y can be approximated by

$$f(x, y) = 6(1 - x - 2y + 2xy + y^2 - xy^2), \qquad 0 < x < 1, 0 < y < 1$$

a. Find the probability that X is less than .3 and Y less than .5.

b. Find the marginal probability density of Y.

c. Are X and Y assumed to be independent in this joint density? Support your answer.

∫ **R37** Using the density given in Exercise R36, find the mean and standard deviation of the fraction actually stocked.

∫ **R38** Find the expected value and variance of $W = 40X + 20Y$ for the joint density given in Exercise R36.

R39 Suppose that it has been established that the number of program lines per week produced by computer programmers using a commercially available toolbox has a mean of 250 and a standard deviation of 70. Suppose that a random sample of 40 programmers is taken. What is the approximate probability that the average lines produced is greater than 265?

R40 What did you assume in answering Exercise R39? Under what circumstances might the answer to Exercise R39 be a poor approximation?

R41 A realtor believes that, under current conditions, 45% of walk-in customers eventually purchase a home through that realtor. In a random sample of 16 walk-in customers, what is the probability that 3 or fewer eventually purchase a home through the realtor? Provide a numerical answer.

R42 Specify all assumptions you made in answering Exercise R41. Are there any assumptions that appear grossly unreasonable?

R43 A bond broker occasionally calls clients to try to place tax-exempt bonds. Define X = number of calls made to a particular client in a three-month period and Y = number of orders made in that period by the client. Assume that

$$f(x, y) = \frac{(4 - x)(xy + 1)}{30(1 + 2x)}, \qquad x = 1, 2, 3, y = 0, 1, 2, 3, 4$$

a. Find the conditional distribution of Y given X, in either mathematical or tabular form.

b. Are X and Y independent? In context, should they be?

R44 Find the mean and variance of Y in Exercise R43.

R45 A computer software wholesaler occasionally gets special handling orders that must be shipped by air. Such orders are expensive and unprofitable. Records indicate that such orders occur at an average rate of 1.6 per workday. In a week with five workdays, what is the probability that there are 10 or more special handling orders? Provide a numerical answer.

R46 Carefully specify the assumptions you made in answering Exercise R45. Are any of these assumptions obviously wrong?

∫ **R47** When data files are transferred between computers, the files are broken up into packets for transmission. When a packet is received at the destination computer, it goes through a checking program. Define X = time required to transmit a randomly chosen packet and Y = time required to check the packet, both measured in thousandths of a second. Assume that the joint probability density of X and Y is

$$f(x, y) = .00000012y(100 - y)e^{-.02(x-120)}, \qquad x > 120,\ 0 < y < 100$$

Find the probability that X is between 200 and 300 and Y is between 30 and 50.

∫ **R48** Find the conditional density of Y given X for the random variables shown in Exercise R47. What does this joint density indicate about the dependence of X and Y?

∫ **R49** Find the expected values and variances of X, Y, and $T = X + Y$ for the random variables of Exercise R47.

∫ **R50** Suppose that we redefine X and Y in Exercise R47 to express them in seconds rather than thousandths of a second. Thus $X' = .001X$ and $Y' = .001Y$. Find the expected values and variances of X', Y', and $T' = X' + Y'$.

R51 Refer again to Exercise R47. A random sample of 250 packets is chosen and the transmission time for each is recorded.

a. Find the approximate probability that the average transmission time in the sample is larger than .180 seconds.

b. Should the approximate probability calculated in part (a) be a good approximation to the unknown exact probability? Explain why.

7

POINT ESTIMATION

Now we are ready to discuss the basic problems of statistical inference. The objective of statistics is to make inferences about one or more population parameters based on observable sample data. These inferences take several related forms. Conceptually, the simplest inference method is point estimation: the best single guess one can give for the value of the population parameter. Point estimation is the topic of this chapter. Other related inference procedures are interval estimation, in which one uses a point estimate and an allowance for random error to specify a reasonable range for the value of a parameter, and hypothesis testing, in which one isolates a particular possible value for the parameter and asks if this value is plausible given the data. Interval estimation is the topic of Chapter 8 and hypothesis testing is the topic of Chapter 9; both of these chapters depend greatly on the results established in this chapter. Chapters 10 through 16, as well as Chapter 19, all extend the basic principles stated in Chapters 7–9 to a number of commonly occurring situations.

Within this chapter, we begin in Section 7.1 by discussing some criteria for good procedures for estimating a population parameter. Section 7.2 is a discussion of the relative merits of sampling with and without replacement, and of the relative importance of the absolute sample size compared to the fraction of the population that is being sampled; the results are surprising to many people. Section 7.3 introduces a very general method, called maximum likelihood, which usually yields good point estimates.

7.1 POINT ESTIMATORS

The simplest statistical inference is **point estimation**, where we compute a single value (statistic) from the sample data to estimate a population parameter. How do we decide which sample statistic to compute to give a single, numerical estimate for a population parameter? Suppose that we are trying to estimate a population mean and that we are willing to assume that the population distribution is normal. One natural summary statistic that can be used to estimate the population mean is the sample mean. Since the population mean for a normal distribution is also the population median, the sample median is also a plausible estimating statistic. So is an 80% trimmed mean, the average of the middle 80% of the values. Even if the population is symmetric, the sample is almost sure to be somewhat asymmetric, because of random variation. Thus, for any particular sample, the three methods yield somewhat different estimates. The mean is heavily influenced by outliers. A trimmed mean is less influenced by outliers, but it wastes data by ignoring (for instance) 20% of the data. The median can be thought of as an extremely trimmed mean, for which one discards all but the middle one or two data points. Which method should be used?

To begin the discussion, we need a technical definition. We use θ as the generic symbol for a population parameter. We use $\hat{\theta}$ to indicate an estimate of θ based on sample data.

Estimator

An **estimator** $\hat{\theta}$ of a parameter θ is a function of random sample values $Y_1, Y_2, \ldots, Y_n$ that yields a point estimate of θ. An estimator is itself a random variable and therefore has a theoretical (sampling) distribution.

There is a technical distinction between an *estimator* as a function of random variables and an *estimate* as a single number. It is the distinction between a process (the estimator) and the result of that process (the estimate). The important aspect of this definition is that we can only define good processes (estimators), not guarantee good results (estimates). We will show, for example, that when one samples from a normal population, the sample mean is the best estimator; however, we cannot guarantee that the result is always optimal—that is, we cannot guarantee that, in every single sample, the sample mean is always closer to the population mean than, say, the sample median. The best we can do is find estimators that give good results in the long run.

EXAMPLE 7.1 If Y_1, Y_2, and Y_3 are the (random) results of a sample of three individuals from a population, define a sample mean estimator. If, in a particular sample, the values 106.8, 102.0, and 105.0 are obtained, what is the resulting estimate?

Solution The estimator

$$\bar{Y} = \frac{Y_1 + Y_2 + Y_3}{3}$$

can be interpreted as the process "take a sample of three values and average them." In the particular sample, $y_1 = 106.8$, $y_2 = 102.0$, and $y_3 = 105.0$ yield $\bar{y} = 104.6$ as an estimate of the population mean from this particular sample. □

The first property that we want an estimator (and its sampling distribution) to have is that it estimate the population parameter correctly on the average. For example, it seems wrong to use the sample 90th percentile to estimate the median (50th percentile) of a population as opposed to using the sample median. While it is conceivable that, in a particular sample, the 90th percentile is closer to the population median than is the sample median, generally the sample 90th percentile is too large; that is, the 90th percentile of the sample tends to overestimate the median of the population. We want to use an estimating statistic that does not systematically overestimate or underestimate the desired population parameter.

Unbiased Estimator

An estimator $\hat{\theta}$ that is a function of the sample data $Y_1, Y_2, \ldots, Y_n$ is called **unbiased** for the population parameter θ if its expected value equals θ; that is, $\hat{\theta}$ is an unbiased estimator of the parameter θ if $E(\hat{\theta}) = \theta$.

An unbiased estimator is correct on the average. The expected value of $\hat{\theta}$ can be thought of as the average of $\hat{\theta}$ values for all possible samples, or alternatively as the long-run average of $\hat{\theta}$ values for repeated samples. The condition that the estimator $\hat{\theta}$ be unbiased says that the **average** $\hat{\theta}$ value is exactly correct. It does not say that a **particular** $\hat{\theta}$ value is exactly correct (see Figure 7.1). If the estimator is biased, the amount of bias is Bias $(\hat{\theta}) = E(\hat{\theta}) - \theta$.

EXAMPLE 7.2 Suppose that $Y_1, Y_2, \ldots, Y_n$ represent the values obtained by a simple random sample from a population having mean μ and variance σ^2. Verify that $\bar{Y}$, the sample mean, is an unbiased estimator of μ.

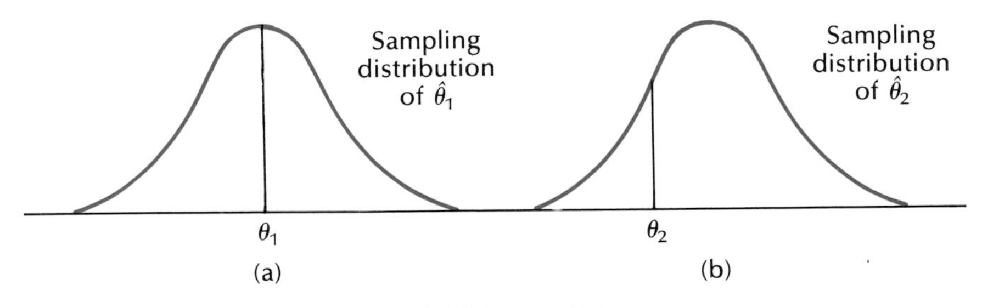

FIGURE 7.1 Illustration of (a) Unbiased and (b) Biased Estimators

Solution In Chapter 4, we showed that $E(\bar{Y}) = \mu$. Thus by definition the sample mean is an unbiased estimator of the population mean. □

The requirement that an estimator be unbiased is not very restrictive and does not rule out many potential estimators. Usually there are many unbiased estimators of any population parameter. For example, when sampling from a normal population, the sample mean, median, and trimmed mean are all unbiased estimators of the population mean μ.

Lack of bias is not the only property that we want an estimator to possess. An estimator that is unbiased but grossly overestimates the parameter of interest half the time and grossly underestimates it the other half isn't a very good estimator. A second property that we require of an estimator is that it have a sampling distribution with most of its probability concentrated near the parameter to be estimated. One measure of this concentration of the sampling distribution of an estimator is given by its standard error: the smaller the standard error, the more concentration of probability there is near the parameter of interest. Figure 7.2 shows the sampling distributions for two hypothetical unbiased estimators of a population parameter θ. It is obvious that $\sigma_{\hat{\theta}_1} > \sigma_{\hat{\theta}_2}$ and hence that $\hat{\theta}_2$ is a more desirable estimator of θ than is $\hat{\theta}_1$.

The standard error of an estimator is also related to the probable degree of error of an estimator: the smaller the standard error, the smaller the probable degree of error. Therefore we would like to find an unbiased estimator with the smallest possible standard error, or equivalently, the smallest probable error.

Efficient Estimator

An estimator is called **most efficient** for a particular problem if it has the smallest standard error of all possible unbiased estimators.

The word *efficient* is used because the estimator makes the best possible use of the sample data in a given situation. A most efficient unbiased estimator is usually preferred to any other, according to standard statistical theory. Given some very specific assumptions, it is possible to find most efficient estimators. For example, if the population from

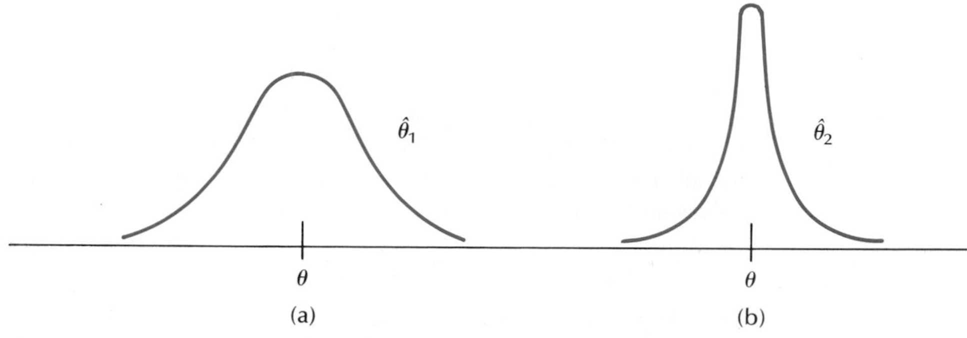

FIGURE 7.2 **Sampling Distributions of $\hat{\theta}_1$ and $\hat{\theta}_2$**

which the sample measurements are drawn is normal, the sample mean has a smaller standard error than the sample median, any sample trimmed mean, or any other unbiased estimator. Therefore, if there is good reason to assume a normal population, the sample mean is the best estimator of the population mean.

EXAMPLE 7.3 A computer program draws 1000 samples, each of size 30, from a normally distributed population having mean 50 and standard deviation 10. For each sample, the mean, median, and trimmed mean (average of the middle 80% of the sample data) are computed. The average value and standard deviation of each set of estimates for the 1000 samples are as follows:

Statistic	Average Value	Standard Deviation
Mean	50.1254	1.8373
Median	50.1696	2.2607
Trimmed mean	50.1196	1.8947

Do the three statistics appear to be unbiased? Which one appears to be most efficient?

Solution The average value of each estimator is (a simulation approximation to) its expected value. The average value of each estimator is very close to the population mean, 50, so all three estimators appear to be unbiased, at least in this situation. The standard deviation of each estimator is (a simulation approximation to) its standard error. The sample mean has the smallest standard error, so it seems to be most efficient in this situation. □

Unfortunately, efficiency claims are heavily dependent on assumptions. The sample mean is not always most efficient when the population distribution is not normal. In particular, when the population distribution has heavy tails, the sample mean is less efficient than a trimmed mean (though it still is unbiased). Heavy-tailed distributions tend to yield lots of extreme, "oddball" values, which influence a mean more than a trimmed mean. A great deal of research is being conducted to find so-called **robust estimators**: statistics that are nearly unbiased and nearly efficient for a wide variety of possible population distributions. There is not yet any general agreement on any most robust estimators, but it's reasonable to assume that such methods will be used increasingly in the near future. We do not spend much space on these methods despite their potential usefulness. The formulas involved in robust estimation are more complicated than those we present, and few such formulas are known, but the basic principles for using the formulas are the same.

robust estimator

EXAMPLE 7.4 A computer is programmed to draw 1000 samples, each of size 30, from an extremely heavy-tailed outlier-prone population having mean zero and standard deviation 9.95. Sample means, medians, and trimmed means are computed for each sample. The average values and standard deviations of the estimates are shown here:

Statistic	Average Value	Standard Deviation
Mean	.0228	1.8757
Median	.0148	.4510
Trimmed mean	.0081	.5667

What do these results indicate about the bias and efficiency of the three estimators when sampling from this population?

Solution All three averages, approximations to the expected values, are close to zero, so all three estimators seem to be unbiased. In this case the standard error of the median appears to be much smaller than the standard error of the mean and somewhat smaller than that of the trimmed mean. Thus, for this outlier-prone population, the sample median appears to be somewhat more efficient than the trimmed mean and much more efficient than the sample mean. □

SECTION 7.1 EXERCISES

7.1 A random sample of 20 vice-presidents of Fortune 500 firms is taken. The amount each vice-president paid in federal income taxes as a percentage of gross income is determined. The data are

16.0	18.1	18.6	20.2	21.7	22.4	22.4	23.1	23.2	23.5
24.1	24.3	24.7	25.2	25.9	26.3	27.9	28.0	30.4	33.7

a. Compute the sample mean and median.
b. Compute the 20% trimmed mean; that is, delete the lowest 10% and highest 10% of the data and find the mean of the remainder.

7.2 Refer to the data of Exercise 7.1.
a. Construct a histogram using about 6 classes.
b. Is there evidence of nonnormality in the data?
c. Which of the sample statistics computed in Exercise 7.1 would you select to estimate the population mean?

7.3 A Monte Carlo study involves 10,000 random samples of size 16 from a normal population with $\mu = 100$ and $\sigma = 20$. For each sample, the mean, the median, and the 20% trimmed mean are calculated, with the following results:

Estimator	Mean	Median	Trimmed Mean
Average	100.23	99.96	99.98
Variance	26.52	40.61	27.49

a. What does the study suggest about the bias of the three estimators in this situation?
b. Which of the three estimators appears most efficient?

7.4 A sample of 30 editions of a weekly newspaper reveals the following numbers of column-inches of classified advertising:

171	185	193	199	204	210	216	218	221	223
225	228	228	230	234	235	237	240	241	243
245	249	251	254	257	262	263	271	280	379

a. Compute the mean and median.
b. Compute the 20% trimmed mean, the average of the middle 80% of the values.

7.5 Refer to the data of Exercise 7.4.
a. Construct a stem-and-leaf display. What is the most conspicuous aspect of the display?
b. Do the data suggest that the mean is the most efficient estimator for this situation?

7.6 Suppose that Y_1, Y_2, Y_3, and Y_4 represent a random sample of four observations from a population with mean μ and standard deviation σ. Two estimators of the mean might be considered:

$$\hat{\mu}_1 = \bar{Y} = \frac{Y_1 + \ldots + Y_4}{4}$$

and

$$\hat{\mu}_2 = .2Y_1 + .3Y_2 + .3Y_3 + .2Y_4$$

The results of Chapter 4 indicate that, for independent random variables,

$$E(\textstyle\sum c_i Y_i) = (\sum c_i)\mu$$
$$\text{Var}(\textstyle\sum c_i Y_i) = (\sum c_i^2)\sigma^2$$

a. Are both estimators unbiased?
b. Which estimator has the smaller variance?

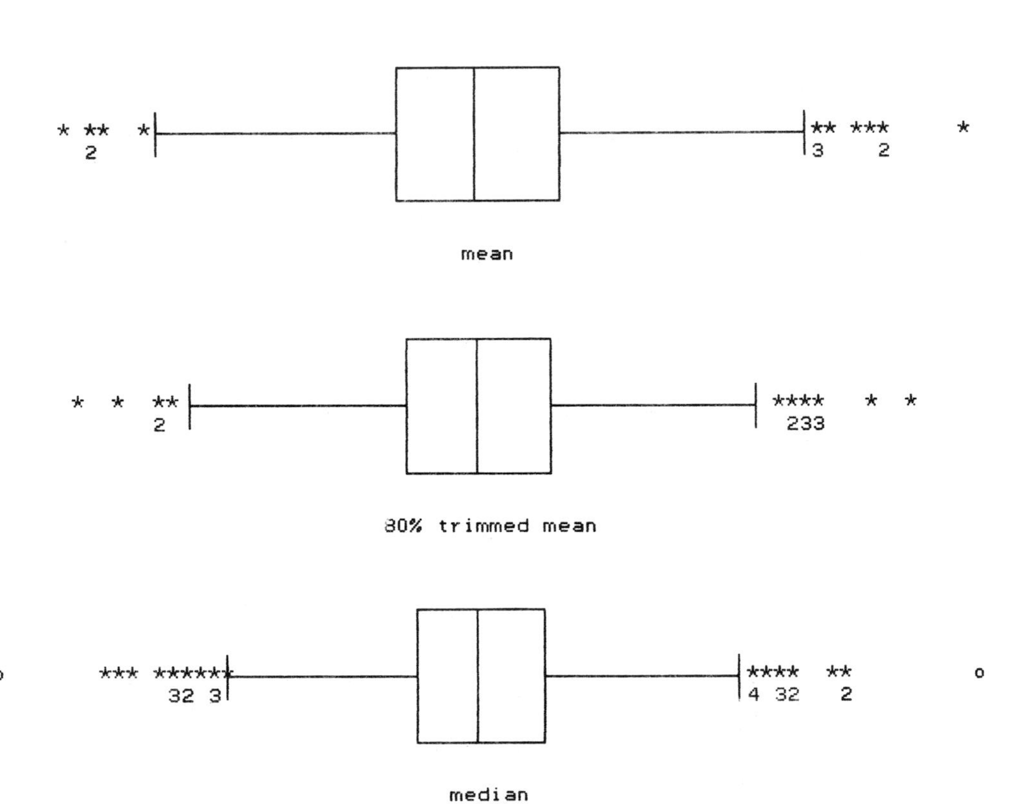

FIGURE 7.3 Box Plots for Three Estimators; Laplace Population, $n = 10$

7.7 Box plot of means, trimmed means (with the top 10% and bottom 10% of the data deleted), and medians for samples of size 10 from a Laplace (mildly outlier-prone) population are shown in Figure 7.3. The mean of this population is zero.

a. Do the three estimators appear to be unbiased?

b. Which estimator appears to be most efficient?

7.8 The averages and standard deviations for the three estimators in Exercise 7.7 are as follows;

Estimator	Average	Standard Deviation
Mean	.0100	.4366
Trimmed mean	.0040	.3899
Median	.0032	.3704

Are these results consistent with your answers to Exercise 7.7?

7.9 Box plots of means, trimmed means, and medians for samples of size 60 from a Laplace population are shown in Figure 7.4.

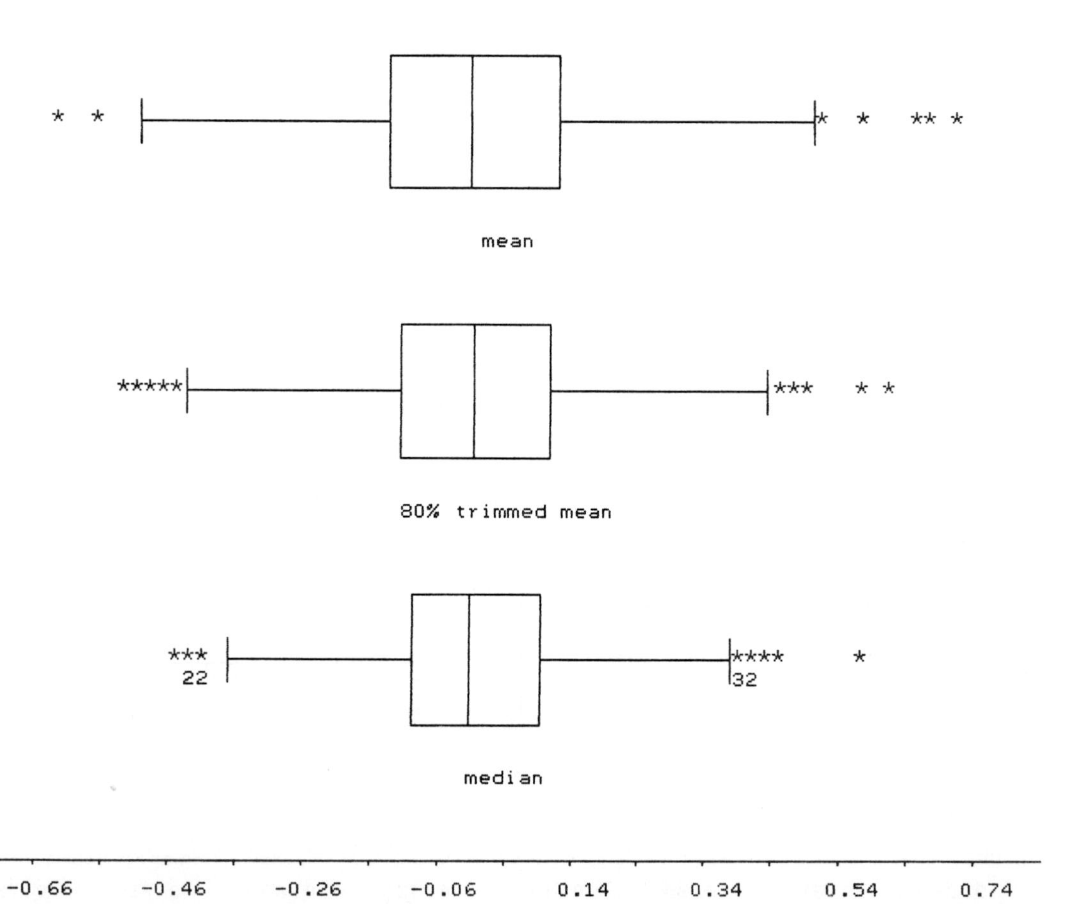

FIGURE 7.4 Box Plots for Three Estimators; Laplace Population, $n = 60$

a. Which estimator appears to be most efficient?
b. Compare your answer to the answers of the preceding exercises. Does the choice of an efficient estimator depend heavily on the sample size?

7.10 The averages and standard deviations for the three estimators in Exercise 7.9 are as follows:

Estimator	Average	Standard Deviation
Mean	−.0025	.1845
Trimmed mean	.0002	.1575
Median	.0009	.1408

Are these results consistent with your answers to Exercise 7.9?

7.11 Samples of size 30 are chosen from a uniform population. The population mean is .500 and the population variance is .08333. The population shape is symmetric and absolutely flat; there are no values less than zero or greater than one, so there is no possibility of outliers. The averages and standard deviations on the top of the next page are obtained:

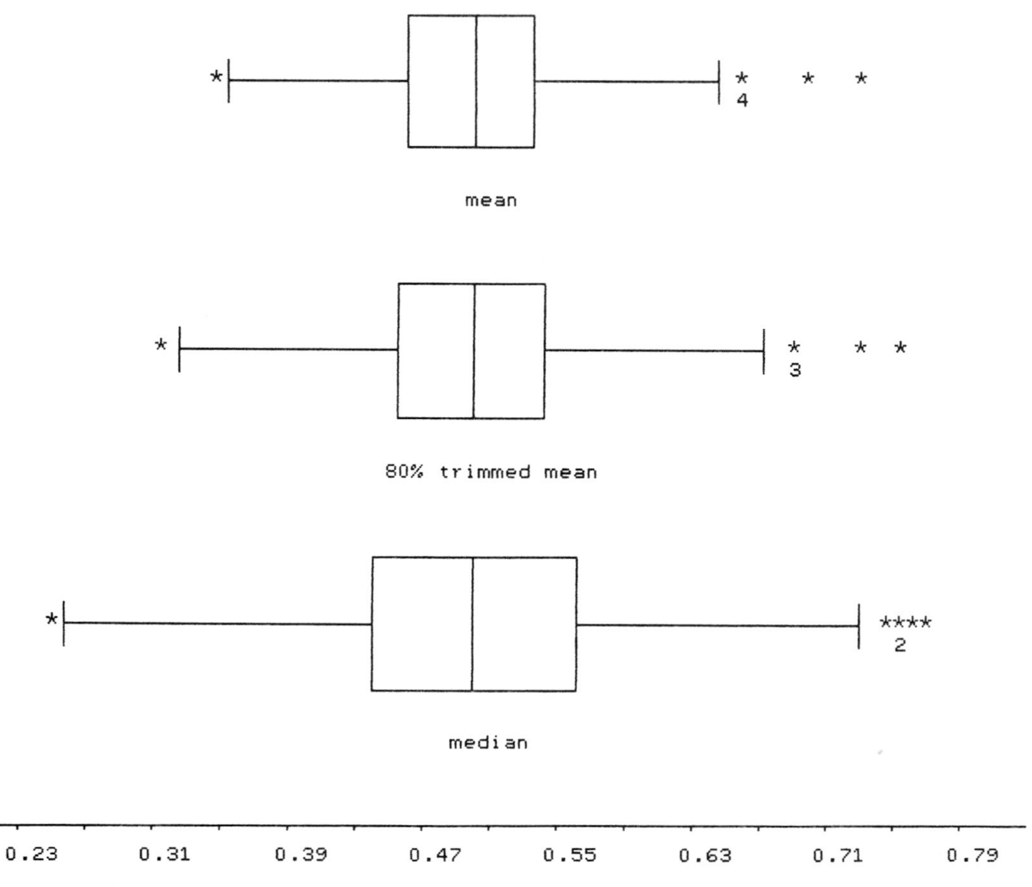

FIGURE 7.5 Box Plots for Three Estimators; Uniform Population, $n = 30$

Estimator	Average	Standard Deviation
Mean	.5015	.0504
Trimmed mean	.5017	.0611
Median	.5043	.0644

a. Should the estimators be unbiased, given the nature of the population? Do they appear to be?

b. Which of the three estimators appears to be most efficient?

7.12 Box plots for the estimators of Exercise 7.11 are shown in Figure 7.5. Do these plots support your answers to Exercise 7.11?

7.2 SAMPLING WITH AND WITHOUT REPLACEMENT

In Chapter 6 we derived the formula for the standard error of the sample mean ($\sigma_{\bar{Y}} = \sigma/\sqrt{n}$) under the assumption that successive values of the sampled random variables $Y_1, \ldots, Y_n$ were independent. The assumption is literally correct, in the random-sampling situation, only when sampling with replacement. Yet sampling with replacement seems to be a poor idea in most contexts. In this section we show that the distinction between sampling with and without replacement is not important. As a side benefit, we can clarify the relative importance of the absolute sample size and the fraction of the population that is sampled.

According to the idea of efficiency, unbiased estimators (in our case sample means taken with and without replacement) may be compared on the basis of variances, or equivalently, of standard errors. It can be proved that the variance of a sample mean (based on a sample size n, a population size N, and a population variance σ^2) is σ^2/n when sampling with replacement, and $(\sigma^2/n)(N - n)/(N - 1)$ when sampling without replacement. A standard error can be found, as always, by taking the square root of the associated variance.

variance of $\bar{Y}$, with and without replacement

EXAMPLE 7.5 Suppose a sample of size 200 is to be taken from a population of size 20,000. Compare the variances and standard errors for the sample mean based on sampling with replacement and sampling without replacement. Which sampling procedure yields the more efficient estimator of μ?

Solution

Sampling Method	With Replacement	Without Replacement
$\text{Var}(\bar{Y})$	$\frac{\sigma^2}{200} = .005\sigma^2$	$\frac{\sigma^2}{200}\left(\frac{20{,}000 - 200}{20{,}000 - 1}\right) = .00495\sigma^2$
$\sigma_{\bar{Y}}$	$\frac{\sigma}{\sqrt{200}} = .0707\sigma$	$\frac{\sigma}{\sqrt{200}}\sqrt{\frac{20{,}000 - 200}{20{,}000 - 1}} = .0704\sigma$

Sampling without replacement gives the smaller standard error, but the difference appears in the third significant digit (the fourth decimal place) of the standard error. It's unlikely that this difference would have practical importance. □

EXAMPLE 7.6 Consider the following two sampling situations where $\bar{Y}$ is used to estimate μ:

a. sampling 100 items (without replacement) from a population of 1000
b. sampling 1000 items (without replacement) from a population of 1,000,000

If everything else is assumed equal, which procedure yields the smaller standard error for $\bar{Y}$?

Solution In situation (a) you would sample a larger **fraction** of the population, while in situation (b) you would have a larger sample **size**. To see which effect is more important, we can compute the standard error of $\bar{Y}$ for the two procedures:

$$\text{(a)} \quad \frac{\sigma}{\sqrt{100}} \sqrt{\frac{1000 - 100}{1000 - 1}} = .0949\sigma$$

$$\text{(b)} \quad \frac{\sigma}{\sqrt{1000}} \sqrt{\frac{1{,}000{,}000 - 1000}{1{,}000{,}000 - 1}} = .0316\sigma$$

Assuming the two populations have equal variances (and therefore equal standard deviations), the standard error of $\bar{Y}$ is much smaller in situation (b). □

Two general conclusions are illustrated by the preceding examples. First, **the distinction between sampling with and without replacement can safely be ignored except in the rare case that the sample size is at least 10% of the population size.** This is the point of Example 7.5.

The second basic conclusion, which is illustrated in Example 7.6, is surprising to most people. **The absolute sample size is much more important in determining probable accuracy than is the fraction of the population that is sampled.** Of course, for a fixed population size, increasing the sample size also increases the sampling fraction. For virtually any realistic situation, the numerical effect of the change in absolute size is more important. The sample size, not the fraction, determines how much information is in the data.

EXAMPLE 7.7 A computer is programmed to draw 1000 samples of size 20 from a population, sampling both with and without replacement. The population distribution is

Individuals:	00–09	10–29	30–69	70–89	90–99
Value:	1	2	3	4	5

(The population mean is 3.) Sample means are calculated for all the samples, and averages and standard deviations are 2.9996 and .249 with replacement and 3.008 and .216 without replacement.

a. Does the sample mean appear to be unbiased?
b. Does sampling with replacement or without appear to be more efficient?

Solution a. The averages both are very close to 3.0, so the sample mean appears to be unbiased in this case.

b. The standard deviations (simulation approximations to the standard errors) are smaller in the case of sampling without replacement, so sampling without replacement appears to be more efficient. □

EXAMPLE 7.8 The simulation of Example 7.7 is repeated, but with the following changes. The population becomes

Individuals:	0–99	100–299	300–699	700–899	900–999
Value:	1	2	3	4	5

(Again, the population mean is 3.) The sample size in this simulation is 50. Note that the sample size in this example is 2.5 times the sample size in Example 7.7, but that we are sampling only 5% of the population as compared to 20% in Example 7.7. The following results are obtained: with replacement, average = 3.002 and standard deviation = .153; without replacement, average = 2.997 and standard deviation = .150.

What does a comparison of this example to Example 7.7 indicate about the relative importance of sample fraction and sample size?

Solution In each case, the simulation approximation to the standard error is smaller in this example than in Example 7.7. Thus the effect of increased sample size is shown to be more important than the effect of sampling a smaller fraction of the population. □

SECTION 7.2 EXERCISES

7.13 A sample is taken of 90 individuals who are responsible for forecasting within large regional banks. There are 650 individuals who might be sampled. Each forecaster in the sample indicates a predicted percentage growth in real disposable income for the next year. Assume that the population standard deviation of forecasts is .4%.

a. In practice, would the sample be chosen with replacement or without?
b. Assuming sampling with replacement, calculate the standard error of the mean.
c. Assuming sampling without replacement, calculate the standard error of the mean.
d. Which standard error is smaller? by how much?

7.14 Viewers of election-night coverage on television have often heard projections of a statewide winner based on a small percentage of the vote. For states such as New York and California, the projection may be based on 1% of the vote, but for states such as Wyoming and Delaware, the projection is based on a much larger percentage. Why?

7.3 MAXIMUM LIKELIHOOD ESTIMATION (∂)

In Section 7.1 we discussed two desirable properties for estimators of parameters; we want unbiased estimators with small standard errors. How does one go about finding an estimator of a parameter θ in a given sampling situation? One general procedure for selecting an estimator is called the **method of maximum likelihood**.

method of maximum likelihood

There are several reasons why one might use the maximum likelihood estimator for a parameter. Although maximum likelihood estimators are not always unbiased and efficient, they are usually about the best that one can find, because of the following properties: As the sample size increases, the bias of the maximum likelihood estimator tends to zero, its standard error approaches the smallest possible standard error, and its sampling distribution approaches normality. It is because of these properties that many statisticians favor the use of maximum likelihood estimators in many sampling situations.

One of the simplest ways to illustrate the concept of maximum likelihood estimation is by showing how to find a maximum likelihood estimator for a given problem. Suppose that we have a binomial experiment with an unknown probability of success π and that we obtain $y = 2$ successes in $n = 5$ trials. We can use Appendix Table 1 to evaluate the probability of two successes in five trials. For $\pi = .05$, the probability is .0214; for $\pi = .10$, the probability is .0729. As we proceed across the table, the probability of two successes in five trials increases to a maximum (of .3456) at $\pi = .40$ and then decreases. By definition, $\hat{\pi} = .40$ is the maximum likelihood estimate (at least among the values shown in Appendix Table 1) of π when the data are two successes in five trials.

EXAMPLE 7.9 Find the maximum likelihood estimate of π in a binomial experiment with $n = 20$ and $y = 16$.

Solution If we start using Appendix Table 1 for $n = 20$ and $y = 16$ at $\pi = .05$, we find that the probability of 16 successes in 20 trials increases as we go from $\pi = .05$ to $\pi = .50$. To continue reading the table, we must look at the right-hand edge for the y value and at the bottom for the π value. The probability of 16 successes in 20 trials increases as we go back across the table until it reaches a maximum (of .2182) at $\pi = .80$. Among the values for π shown in Appendix Table 1, $\hat{\pi} = .80$ is the maximum likelihood estimate when $y = 16$ and $n = 20$. □

To define maximum likelihood estimation more generally, we need some other definitions.

Likelihood Function

For discrete data $y_1, y_2, \ldots, y_n$, the likelihood function L is the probability of observing the data that are in fact observed:

$$L(y_1, y_2, \ldots, y_n, \theta) = P(y_1, y_2, \ldots, y_n)$$

regarded as a function of the unknown population parameter θ. If the data are drawn from a continuous distribution $f_Y(y)$, the probability distribution P is replaced by the probability density f:

$$L(y_1, y_2, \ldots y_n, \theta) = f(y_1, y_2, \ldots, y_n)$$

Assuming that the sample values are drawn independently, the probability P or the density f may be obtained as a product:

$$L(y_1, y_2, \ldots, y_n, \theta) = P(y_1)P(y_2) \ldots P(y_n)$$

or

$$L(y_1, y_2, \ldots, y_n, \theta) = f(y_1)f(y_2) \ldots f(y_n)$$

If, in a binomial experiment with $n = 5$, we obtain $y = 2$, then the likelihood is simply the probability of two successes in five trials taken as a function of the unknown population probability of success π.

EXAMPLE 7.10 Suppose that the number of new orders arriving at a small machine shop on a given day follows a Poisson distribution with unknown mean μ, independently of arrivals on other days. Suppose that one order arrives on the first day of a sample and that four orders arrive on the second (and last) day of the sample. Write down the likelihood function.

Solution Remember that the Poisson distribution is discrete with

$$P(y) = e^{-\mu} \frac{\mu^y}{y!}$$

The observed values are $y_1 = 1$ and $y_2 = 4$. The likelihood is

$$L(1, 4, \mu) = e^{-\mu} \frac{\mu^1}{1!} e^{-\mu} \frac{\mu^4}{4!}$$

which can be simplified to

$$L(1, 4, \mu) = e^{-2\mu} \frac{\mu^5}{1!4!}$$

□

Maximum Likelihood Estimate of θ

For observed sample values $y_1, y_2, \ldots, y_n$, the maximum likelihood estimate of a parameter θ is the value $\hat{\theta}$ that maximizes the likelihood function $L(y_1, y_2, \ldots, y_n)$.

In a binomial experiment with $n = 5$ and $y = 2$, the maximum likelihood estimate of π appears from Appendix Table 1 to be .40.

EXAMPLE 7.11 Refer to the likelihood function found in Example 7.10. Indicate how one should use a table of Poisson probabilities to find the maximum likelihood estimate of μ.

Solution The likelihood found in Example 7.11 was the product of the Poisson probabilities of obtaining $y_1 = 1$ and $y_2 = 4$. Using Appendix Table 2, we can calculate these probabilities and obtain the following:

μ	2.3	2.4	2.5	2.6	2.7
$P_Y(1)$	.2306	.2177	.2052	.1931	.1815
$P_Y(4)$	.1169	.1254	.1336	.1414	.1488
$L(1, 4, \mu)$	.0270	.0273	.0274	.0273	.0270

The value of μ that maximizes the likelihood function appears to be 2.5. Thus the maximum likelihood estimate is $\hat{\mu} = 2.5$. □

In principle, one can always find maximum likelihood estimators by numerical computation of the likelihood function. But often it's easier to use elementary calculus to find them. Recall from calculus that to find the maximum of a function one sets the first derivative of the function equal to zero and solves the resulting equation. One should also use a second-derivative check to make sure that one is obtaining a maximum, not a minimum; by and large, the solution of the first-derivative equation gives a maximum likelihood estimator rather than a minimum.

In likelihood problems, it is often convenient to work with the natural logarithm of the likelihood rather than with the likelihood itself. Because the logarithm is an increasing function, as the likelihood increases to its maximum, so does the log-likelihood. For example, consider again a binomial experiment with $n = 5$ and $y = 2$. The likelihood is

$$L(2, \pi) = \frac{5!}{2!3!}\pi^2(1 - \pi)^3$$

Denoting the log-likelihood by $l(2, \pi)$, we have

$$l(2, \pi) = \log(5!) - \log(2!3!) + 2 \log \pi + 3 \log(1 - \pi)$$

Set the first derivative equal to zero:

$$\frac{2}{\pi} - \frac{3}{1 - \pi} = 0$$

The solution of this equation is $\hat{\pi} = 2/5 = .40$. Thus calculus yields the same maximum likelihood estimate, $\hat{\pi} = .40$, as is found by numerical methods.

EXAMPLE 7.12 Refer to Example 7.10. Find the maximum likelihood estimate of μ using calculus.

Solution Again, it is convenient to use the log-likelihood. We found the likelihood in Example 7.10 to be

$$L(1, 4, \mu) = e^{-2\mu}\frac{\mu^5}{1!4!}$$

so the log-likelihood is

$$l(1, 4, \mu) = -2\mu + 5 \log(\mu) - \log(1!4!)$$

Setting the first derivative equal to zero gives the equation

$$\frac{\partial l(1, 4, \mu)}{\partial \mu} = -2 + \frac{5}{\mu}$$

so the maximum likelihood estimate of μ is $\hat{\mu} = 5/2 = 2.5$. Once again, the result found by calculus agrees with the result found numerically. □

An important problem in statistical theory is estimation of the population mean μ based on a random sample from a normal population. The normal distribution is based on a continuous random variable, with density

$$f(y) = \frac{1}{\sqrt{2\pi}\sigma} e^{-.5(y-\mu)^2/\sigma^2}$$

The log-likelihood function for a random sample from a normal population can be found by routine algebra. It is

$$l(y_1, y_2, \ldots, y_n) = \left(\frac{-n}{2}\right)\log(2\pi) - n\log(\sigma) - .5\sum\frac{(y_i - \mu)^2}{\sigma^2}$$

For any specified value of σ, say, 3.72, the derivative of the log-likelihood with respect to μ is

$$\frac{\partial l(\mu)}{\partial \mu} = 0 + 0 + \sum\frac{(y_i - \mu)}{3.72^2}(-1)$$

The maximum likelihood estimator is found by solving this expression set to zero:

$$-\sum\frac{(y_i - \mu)}{3.72^2} = 0$$

One way to see that $\hat{\mu} = \bar{y}$ is the solution to this equation is to note that $\sum(y_i - \bar{y}) = 0$. Obviously the choice of any particular value for σ is irrelevant to the computation of the maximum likelihood estimator of μ; our arbitrary choice of $\sigma = 3.72$ had no effect on the calculation. In general, if the population shape is normal, the sample mean is the most efficient estimator and also the maximum likelihood estimator. Thus we know that the sample mean is a maximum likelihood estimator of the population mean when the population is normal, and also that the sample proportion is the maximum likelihood estimator of the probability of success in a binomial experiment.

SECTION 7.3 EXERCISES

7.15 The times $Y_1, Y_2, \ldots$ between arrivals of customers to a store can often be assumed to have a negative exponential distribution:

$$f_{Y_i}(y) = \theta e^{-\theta y}, \qquad 0 < y < \infty$$

Using a process similar to the one for discrete distributions, we can find the maximum likelihood estimator here for a sample of n measurements $y_1, y_2, \ldots, y_n$ by maximizing the likelihood

$$f_{Y_1}(y_1)f_{Y_2}(y_2)\ldots f_{Y_n}(y_n)$$

Suppose that a sample of size $n = 4$ values yields

$$y_1 = 2.4, \quad y_2 = .8, \quad y_3 = .2, \quad y_4 = 4.6$$

a. Show that the likelihood when $\theta = .5$ is $(.5)^4 e^{-4} = .00114$.

b. Construct a table of likelihoods for the problem, with

$$\theta = .1, .2, .3, .4, .5, .6, .7, .8, .9, 1.0$$

(You will need a calculator that computes exponentials.)

c. What do you think the maximum likelihood estimate of θ is in this situation?

∂ **7.16** a. Use calculus to prove that the maximum likelihood estimate of θ in Exercise 7.15 is $\hat{\theta} = .5$.

b. Show that for a sample of size n from the negative exponential distribution, the maximum likelihood estimator of θ is

$$\hat{\theta} = \frac{n}{\sum_i Y_i} = \frac{1}{\bar{Y}}$$

7.17 One form of the lognormal probability density is given by the following mathematical function:

$$f_Y(y) = \frac{1}{\sqrt{2\pi}y} e^{-1/2[(\log y) - \theta]^2}, \qquad 0 < y < \infty$$

where $\pi = 3.14159\ldots$ and $\log y$ is the natural logarithm of y. Suppose that a sample of size 2 yields $y_1 = 4.28$ and $y_2 = 4.69$.

a. Use a calculator that computes natural logarithms to verify the following values. Hint: The likelihood for $n = 2$ observations is

$$\left(\frac{1}{2\pi y_1 y_2} e^{-1/2(\log y_1 - \theta)^2} e^{-1/2(\log y_2 - \theta)^2}\right)$$

θ:	1.2	1.3	1.4	1.5	1.6
Likelihood:	.00723	.00760	.00783	.00791	.00783

b. What is the maximum likelihood estimate of θ, according to the values?

∂ **7.18** a. Use calculus to maximize the likelihood for Exercise 7.17.

b. Show that for general n the maximum likelihood estimator of θ in the lognormal density is

$$\hat{\theta} = \frac{\sum_{i=1}^{n} \log y_i}{n}$$

Hint: To maximize the likelihood, minimize

$$\sum_{i=1}^{n} (\log y_i - \theta)^2$$

7.19 The Laplace density

$$f_Y(y) = .5e^{-|y-\theta|}$$

is symmetric around the value of θ but outlier-prone compared to a normal distribution. Suppose that five observations from a Laplace density yield $y_1 = 2.6$, $y_2 = 5.1$, $y_3 = 4.7$, $y_4 = 9.6$, and $y_5 = 5.0$.

a. Find the mean and median for the sample data.

b. Compute the likelihood function when θ equals each of the values found in part (a).

c. Can the sample mean be the maximum likelihood estimator for a Laplace population? Explain your reasoning.

SUMMARY

In this chapter we have presented important concepts related to point estimation of a population mean μ and a population proportion π. The choice of the best point estimator revolves about whether or not some basic requirements are satisfied. It seems reasonable to demand that a point estimator be correct on the average and that the standard deviation of the sampling distribution of the point estimator be small. A point estimator $\hat{\theta}$ of a parameter θ is unbiased if $E(\hat{\theta}) = \theta$. If the standard error $\sigma_{\hat{\theta}}$ of $\hat{\theta}$ is smaller than the standard error for any other point estimator of θ, then $\hat{\theta}$ is said to be a most efficient estimator.

Potential candidates for a point estimator can be developed in a variety of ways. The procedure discussed in this chapter is the method of maximum likelihood. Other methods, not discussed in this text, include the method of moments and Bayes' estimation.

KEY FORMULAS: Point Estimation

1. Unbiased estimator $\hat{\theta}$ of a parameter θ

   $E(\hat{\theta}) = \theta$

2. Standard error of $\bar{Y}$

   sampling with replacement $\sigma_{\bar{y}} = \dfrac{\sigma}{\sqrt{n}}$

   sampling without replacement $\sigma_{\bar{y}} = \dfrac{\sigma}{\sqrt{n}}\sqrt{\dfrac{N-n}{N-1}}$

3. Likelihood function

   $L(y_1, y_2, \ldots, y_n, \theta) = P(y_1)P(y_2) \ldots P(y_n)$

   For continuous random variables, replace the probability $P(y_i)$ by the probability density $f(y_i)$.

CHAPTER 7 EXERCISES

7.20 A Monte Carlo study involves 5000 samples, each of size 30, from a heavy-tailed population having a mean of 300 and a standard deviation of 25. For each sample, the mean, median, and 20% trimmed mean are computed:

Statistic	Mean	Median	Trimmed Mean
Average value	298.91	300.74	299.09
Variance	35.79	41.27	28.47

a. Do the estimators all appear to be unbiased, or nearly so?
b. Which estimator appears to be most efficient?

7.21 Samples of size 30 are taken from an exponential population having mean 1. Recall from Chapter 5 that the exponential distribution is quite severely right-skewed.
a. Should the sample mean, sample trimmed mean, and median all be unbiased estimators of the population mean?
b. The average values (over 1000 samples) of these three estimators are 1.0026 for the mean, .8489 for the trimmed mean, and .7149 for the median. Do these results agree with your answer to part (a)?

7.22 Samples of size 60 are taken from a symmetric but severely outlier-prone population having mean zero and standard deviation 9.95. Box plots of the resulting means, trimmed means, and medians are shown in Figure 7.6.
a. Do the estimators appear to be unbiased?
b. Would the sample mean be the most efficient estimator of the population mean when sampling from this population?

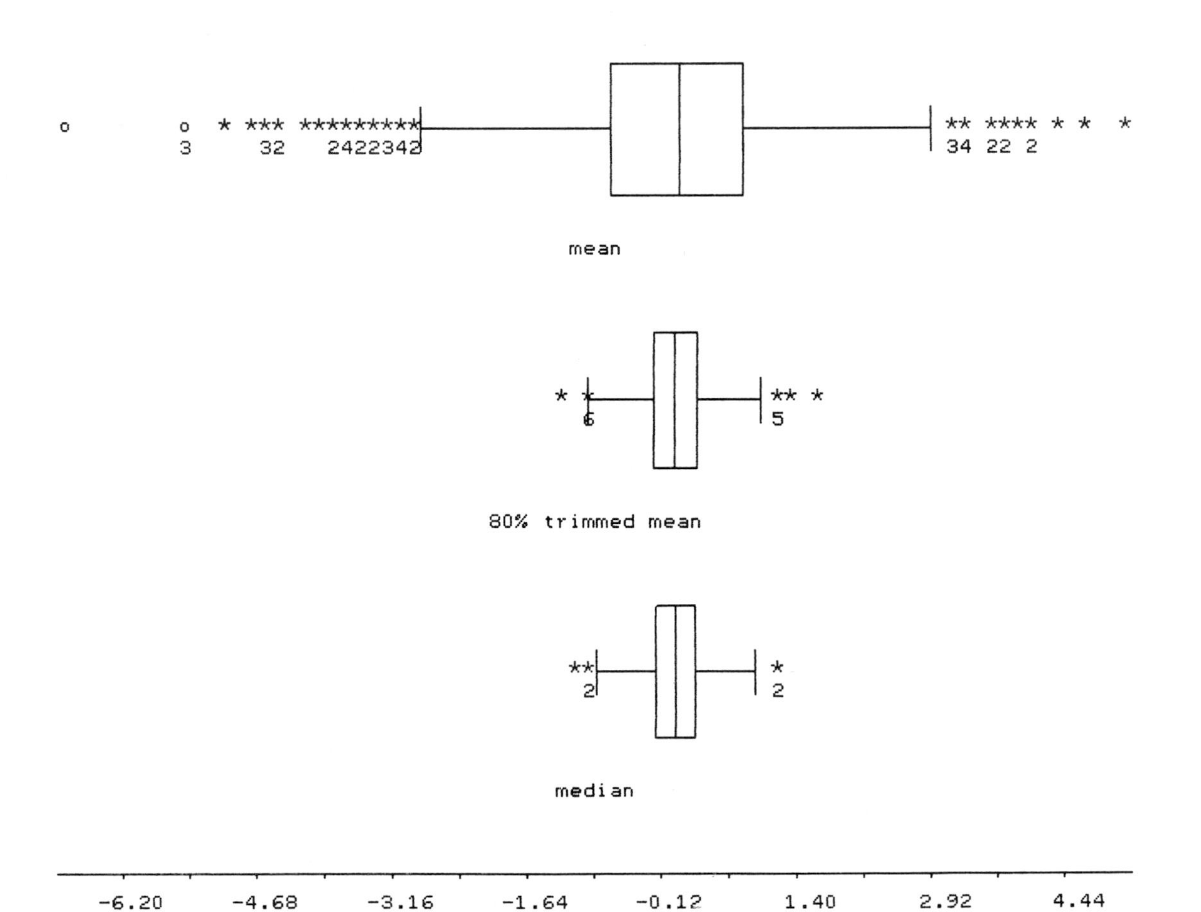

FIGURE 7.6 Box Plots for Three Estimators; Outlier-prone Population, $n = 60$

7.23 The averages and standard deviations for the three estimators in Exercise 7.22 are shown below:

Estimator	Average	Standard Deviation
Mean	−.0285	1.3165
Trimmed mean	−.0014	.3699
Median	.0003	.3262

Are these results consistent with your answers to Exercise 7.22?

7.24 A random sample of 50 observations is to be obtained from a population of interest. Under what data shapes might you prefer to use the sample median rather than the sample mean as an estimate of the "center" of the population? For each case, explain why you would prefer to use the median.

7.25 A sample of 400 individuals is taken from a target group of all individuals in the United States that hold driver's licenses. The purpose of the study is to assess the awareness of a new product. There are many, many millions of people in the target group; a sample of 400 is only a microscopic fraction of the population. It is therefore argued that the sample cannot possibly give an accurate estimate of the proportion of individuals who are aware of the product. Is this a valid argument?

7.26 A study yields two random variables, Y_1 and Y_2. Each has expected value μ, the population mean of interest. The variance of Y_1 is 2 and the variance of Y_2 is 8. Three estimators are proposed:

$$\hat{\theta}_1 = .5Y_1 + .5Y_2$$
$$\hat{\theta}_2 = .2Y_1 + .8Y_2$$
$$\hat{\theta}_3 = .8Y_1 + .2Y_2$$

Use the properties of expected values discussed in Chapter 4 to determine which of these estimators is unbiased.

7.27 Which of the estimators in Exercise 7.26 has the smallest variance? Use the properties of variances discussed in Chapter 4. Which random variable should be given greater weight, the one with the large variance or the one with the small variance?

7.28 A food service company that operates company dining rooms tries out a new procedure for bidding on contracts. Each of the four offices of the company uses the procedure until three successful bids are made. Office 1 requires 7 bids, office 2 requires 9 bids, office 3 requires 6 bids, and office 4 requires 14 bids. It is assumed that the negative binomial distribution (Chapter 5) should apply to Y = required number of bids. Recall that for the negative binomial distribution,

$$P_Y(y) = \frac{(y-1)!}{(k-1)!(y+k)!}\pi^k(1-\pi)^{y-k}$$

where k = number of successes to be obtained and π is the probability of success.

a. What is the value for k in this problem?

b. Calculate the likelihood of the actual results for $\pi = .10, .11, \ldots, .15$.

c. What value appears to be the maximum likelihood estimate of π?

∂ **7.29** Use calculus to find the maximum likelihood estimator of π in Exercise 7.28.

7.30 An insurance firm uses the Pareto distribution

$$f_Y(y) = \theta(y+1)^{-(\theta+1)}, \qquad y > 0$$

as the assumed probability distribution for Y = dollar amount of settlements for personal liability claims (in thousands). For one class of policies, three claims have $y_1 = .82$, $y_2 = .63$, and $y_3 = 7.55$.

a. Write an expression for the likelihood as a function of θ.

b. Calculate the likelihood for $\theta = 1$, 2, and 3. Which of these values of θ appears closest to the maximum likelihood estimate?

∂ **7.31** Use calculus to find the maximum likelihood estimator of θ in Exercise 7.30. How does the answer generalize to any sample of size n from the Pareto distribution?

7.32 A chain of shoe stores sets a budgeted sales figure for each store. Data are collected on the actual sales of a sample of stores as a fraction of the budgeted sales. The following Minitab output is obtained:

```
MTB > print 'sales'
sales
  100.8    98.2    99.9    99.5   100.1   100.1   100.7   103.9    80.5
   99.9   100.0    98.3   142.3   101.3    98.2    98.2    98.7    93.2
  101.8    93.6   101.2    99.7    97.7   100.3   119.4    99.3   101.6
   84.0   100.5   100.3    99.1    99.6   100.1   100.2   102.1   100.2
  100.9    99.5   104.8   102.7   109.3    98.0

MTB > describe 'sales'

                  N     MEAN   MEDIAN   TRMEAN    STDEV   SEMEAN
sales            42   100.71   100.10   100.09     8.61     1.33

                MIN      MAX       Q1       Q3
sales         80.50   142.30    98.60   101.23

MTB > histogram of 'sales'

Histogram of sales   N = 42

Midpoint   Count
      80       1  *
      85       1  *
      90       0
      95       2  **
     100      32  ********************************
     105       3  ***
     110       1  *
     115       0
     120       1  *
     125       0
     130       0
     135       0
     140       1  *
```

a. Locate the sample mean and median.

b. Which of these values should be the better estimate of the population mean?

7.33 A normal plot for the data of Exercise 7.32 is shown in Figure 7.7.

a. What does the shape of the normal plot indicate about the shape of the sample data?

b. Does the normal plot confirm your answer to part (b) of Exercise 7.32?

7.34 A sales representative for a coffee producer measures the fraction of shelf space for all coffees devoted to the producer's brands. The results on page 223 are obtained:

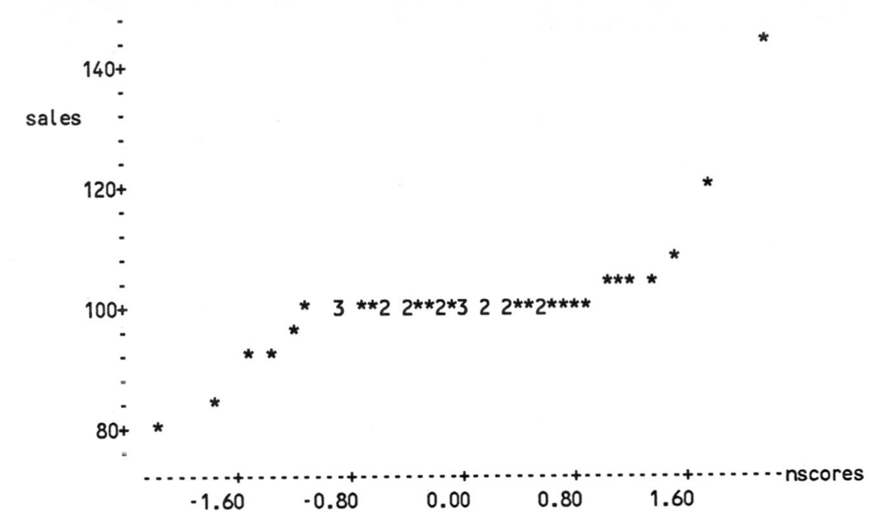

FIGURE 7.7 Normal Plot for Exercises 7.32–7.33

```
MTB > print 'fraction'
fraction
  0.157895   0.160000   0.105263   0.142857   0.250000   0.166667   0.210526
  0.142857   0.200000   0.200000   0.160000   0.210526   0.142857   0.227273
  0.090909   0.125000   0.142857   0.142857   0.210526   0.130435   0.136364
  0.269231   0.136364   0.176471   0.227273   0.120000   0.176471   0.210526
  0.238095   0.210526   0.111111   0.187500   0.142857   0.095238   0.210526
  0.130435   0.192308   0.187500   0.210526   0.117647   0.200000   0.125000
  0.230769   0.190476   0.062500   0.166667   0.083333   0.142857   0.150000
  0.153846

MTB > stem and leaf of 'fraction'

Stem-and-leaf of fraction  N  = 50
Leaf Unit = 0.010

    1    0 6
    4    0 899
    7    1 011
   14    1 2223333
   24    1 4444444555
  (6)    1 666677
   20    1 8899
   16    2 0001111111
    6    2 2233
    2    2 5
    1    2 6

MTB > describe 'fraction'

                N     MEAN   MEDIAN   TRMEAN    STDEV   SEMEAN
fraction       50  0.16623  0.16000  0.16631  0.04644  0.00657

              MIN      MAX       Q1       Q3
fraction  0.06250  0.26923  0.13488  0.21053
```

Minitab for Exercise 7.34

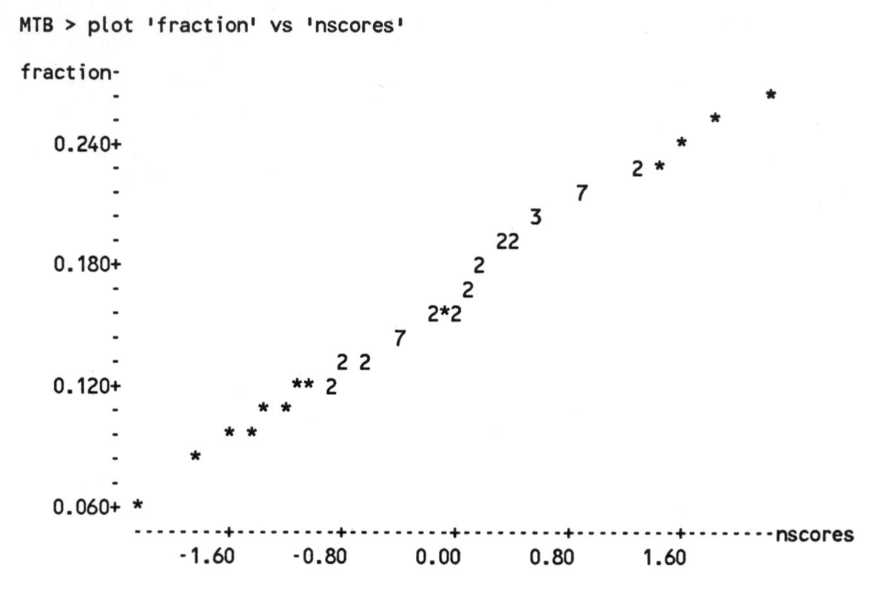

FIGURE 7.8 Normal Plot for Exercises 7.34–7.35

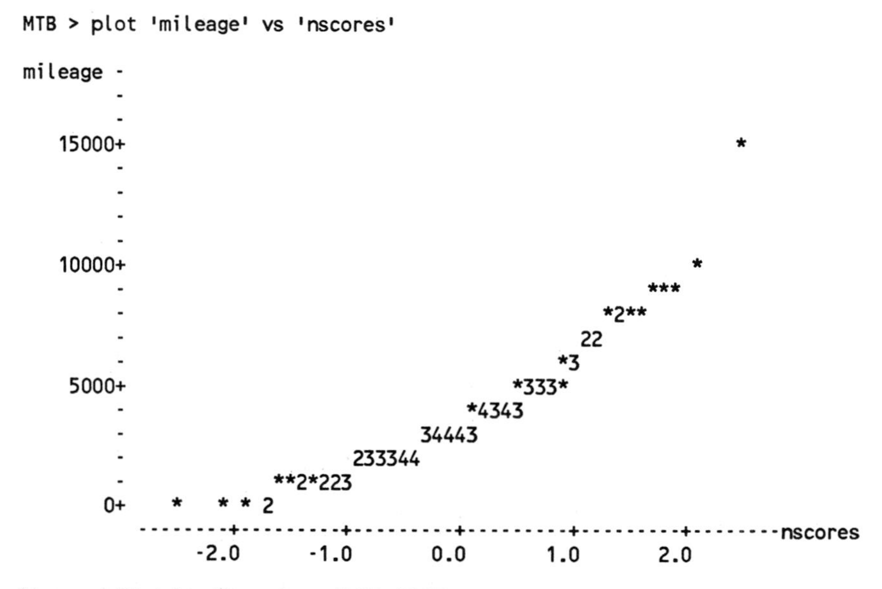

FIGURE 7.9 Normal Plot for Exercises 7.36–7.37

Is there any reason to think that the sample mean is an inefficient estimator of the population mean?

7.35 A normal plot of the data in Exercise 7.34 is shown in Figure 7.8. Does the shape of the normal plot confirm your answer to Exercise 7.34?

7.36 A motorists' club collects data for a sample of its members on the number of miles of pleasure driving in a one-year period. The following results are obtained:

```
MTB > print 'mileage'
mileage
   5153    1340    8245    2596    2069    3897    2487    2264    3243
   1345    6588    3039    2861    6528    1804    9529     355    1558
   4184     355    3269     264    8255    2003    5042    2669     330
   8927    1986    7569    3006    2480    6304    2105    2569    3593
   5055    4995    4342    1107    9399    3201    2760    7956    3755
   8445     956    1976    4087    8638    3708    1806    4233    7063
   3655    4563    3222    3897    5380    3274     877    5874   14725
   3128    4000    4737    1705     334    1147    1939    4031    3056
    968    4268    1803     804     522    6156    2809    4427    3207
   5167    1772    2518    5375    2206    3708    5646    5358     715
   2308    1318    2971    1754    6774    2217    5174    1046

MTB > describe 'mileage'

                N     MEAN   MEDIAN   TRMEAN    STDEV   SEMEAN
mileage        98     3713     3204     3534     2548      257

              MIN      MAX       Q1       Q3
mileage       264    14725     1906     5079

MTB > histogram of 'mileage'

Histogram of mileage   N = 98

Midpoint   Count
       0      11  ***********
    2000      33  *********************************
    4000      28  ****************************
    6000      15  ***************
    8000       8  ********
   10000       2  **
   12000       0
   14000       1  *
```

Is it reasonable to use the sample median as an estimate of the population mean? Why?

7.37 A normal plot of the data of Exercise 7.36 is shown in Figure 7.9. Does the shape of the plot confirm your answer to Exercise 7.36?

8

INTERVAL ESTIMATION

The discussion in Chapter 7 focused on one type of statistical inference—point estimation. There the task was to find the best single-number guess for the value of a population parameter. The methods developed in Chapter 7 didn't include any explicit indication of the probable degree of error. In this chapter we develop methods that allow for reporting not only the best estimate of a parameter value, but also the probable degree of error of that estimate. The methods of this chapter allow one to state not merely the best single guess for a population parameter, but also an interval of reasonable guesses for that parameter. Because these methods yield a whole numerical interval of reasonable possibilities for a parameter, they are called interval-estimation methods.

In this chapter we build on the point-estimation methods of Chapter 7 to develop interval estimates for parameters. Once we specify general methods, we turn to some specific methods for the most important parameters, such as population means and proportions. Section 8.1 introduces the general concept of a confidence interval; the discussion happens to be stated in terms of means. Section 8.2 extends the concept of a confidence interval to the problem of estimating a proportion. In Section 8.3 we turn to a basic problem in planning statistical studies: How large a sample is needed to achieve a desired accuracy? Section 8.4 contains the basic theory of Student's t distribution, a widely used distribution in establishing confidence intervals and in performing other kinds of statistical inference; the consequences of the use of Student's t distribution in confidence interval problems are discussed in Section 8.5. In Section 8.6 we discuss the assumptions underlying confidence-interval methods, the consequences of violating those assumptions, and methods for detecting possible violations. In Section 8.7 we discuss interval estimation for a population median, as contrasted to methods for a mean.

8.1 INTERVAL ESTIMATION OF A POPULATION MEAN WITH KNOWN STANDARD DEVIATION

probable range

interval estimate

The ideas discussed in the previous chapter dealt with point estimation: finding a best guess for a population parameter. Such point estimates are almost inevitably in error to some degree. Specification of a **probable range** for the parameter—a plus-or-minus range for error—is crucial in indicating the reliability of estimates. A statement like "the estimated response rate is 28%" is less useful than one like "the estimated response rate is 28% ± 2%." And 28% ± 2% indicates a much more reliable estimate than 28% ± 15%. In this section we use the idea of a sampling distribution to construct an **interval estimate** for a population mean. We discuss confidence intervals for proportions later in the chapter.

The idea is best introduced by an example. Suppose that a random sample of size 36 is to be taken, and that the sampling distribution of $\bar{Y}$ is normal.* (If the population can be assumed to be symmetric, the Central Limit Theorem should apply.) Somewhat artificially, we assume that the population standard deviation is known to be 18.0. The expected value of $\bar{Y}$ is the population mean μ, the parameter being estimated, and the standard error of $\bar{Y}$ is $\sigma_{\bar{Y}} = \sigma/\sqrt{n} = 18/\sqrt{36} = 3$. From the properties of a normal distribution, there is a 95% chance that $\bar{Y}$ is within 1.96 standard errors of μ (see Figure 8.1):

$$P[\mu - 1.96(3) \leq \bar{Y} \leq \mu + 1.96(3)] = .95$$

Looked at differently, any time the observed sample mean $\bar{y}$ lies in the interval $\mu \pm 1.96(3)$, the interval $\bar{y} \pm 1.96(3)$ encloses μ. This is shown in Figure 8.2. Because there is a 95% chance that $\bar{Y}$ lies in the interval $\mu \pm 1.96(3)$, there is a 95% chance that the interval $\bar{Y} \pm 1.96(3)$ encloses μ. In practice, we take only a single sample from the population of

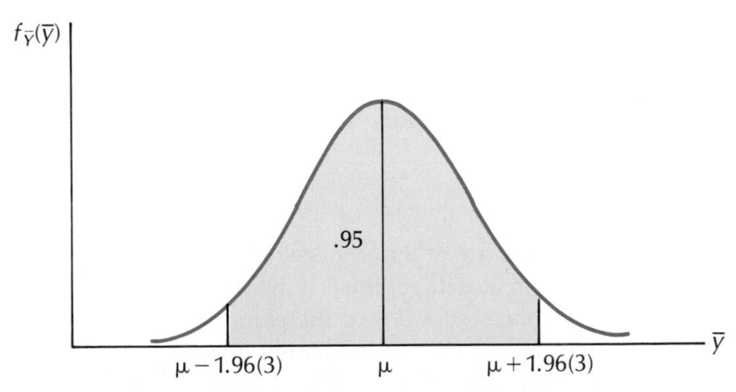

FIGURE 8.1 **Sampling Distribution of $\bar{Y}$**

* Again, we use capital letters for random variables and lowercase letters for the resulting values. Thus, when planning to take a sample, we consider probabilities about $\bar{Y}$. When the sample yields values 90, 96, 100, and 106, $\bar{y} = 98$.

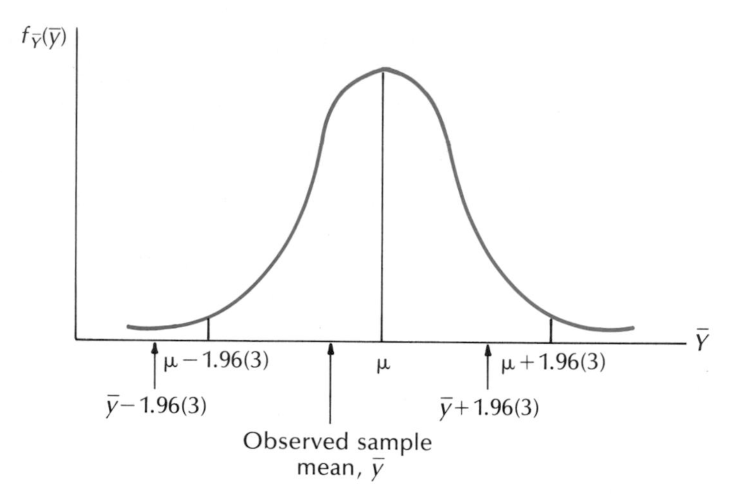

FIGURE 8.2 Sampling Distribution of $\bar{Y}$

95% confidence interval for μ

interest. The interval $\bar{y} \pm 1.96(3)$ that we construct using the observed sample mean is called a **95% confidence interval for μ**.

The general formula for a confidence interval for a population mean is derived in the same way. The formula is exactly correct only when the population distribution is normal and the population standard deviation is known. However, it provides an excellent approximation when the population distribution is symmetric or only modestly skewed, for sample sizes of, say, 30 or more.

100(1 − α)% Confidence Interval for μ, with σ Known
$\bar{y} - z_{\alpha/2}\sigma_{\bar{Y}} \leq \mu \leq \bar{y} + z_{\alpha/2}\sigma_{\bar{Y}}$ where $\sigma_{\bar{Y}} = \sigma/\sqrt{n}$ and $z_{\alpha/2}$ is the tabulated value cutting off a right-tail area of $\alpha/2$ in the standard normal distribution.

EXAMPLE 8.1 An airline needs an estimate of the average number of passengers on a newly scheduled flight. Its experience is that data for the first month of flights are unreliable, but that thereafter the passenger load settles down. Therefore the mean passenger load is calculated for the first 20 weekdays of the second month (regarded as a random sample of 20 days from a hypothetical population of weekdays) after initiation of this particular new flight. If the sample mean is 112.0 and the population standard deviation is assumed to be 25, find a 90% confidence interval for the true, long-run mean number of passengers on this flight.

Solution We assume that the hypothetical population of daily passenger loads for weekdays is not badly skewed. Then the sampling distribution for $\bar{Y}$ is approximately normal and the confidence-interval results are approximately correct, even for a sample size of only 20 weekdays. For this example $\bar{y} = 112.0$, $\sigma = 25$, and $\sigma_{\bar{Y}} = \sigma/\sqrt{20} = 5.59$. Then for a 90% confidence interval, we use $z_{.05} = 1.645$ in the formula to obtain

$$112 \pm 1.645(5.59), \quad \text{or} \quad 102.80 \quad \text{to} \quad 121.20$$

We are 90% confident that the long-run mean μ lies in this interval. □

There is a small logical problem in interpreting the "90%" of a 90% confidence interval. In Example 8.1 it is tempting to write

$$P(102.80 \le \mu \le 121.20) = .90$$

but, as written, there is no random quantity in the expression: μ is an unknown population constant, while 102.80 and 121.20 are simply numbers. Therefore, in strict logic, we can't apply a probability to one particular interval.

The "90%" refers to the **process** of constructing confidence intervals. Each particular confidence interval either does or does not include the true value of the parameter being estimated. In the long run, 90% of the intervals so constructed include the population value. So we say that we have 90% confidence that $102.80 \le \mu \le 121.20$. This is shorthand for "the interval $102.80 \le \mu \le 121.20$ is the result of a process that in the long run has 90% probability of being correct."

interpretation of a confidence interval

EXAMPLE 8.2 A Monte Carlo study considers 5000 samples, each of size 40, from a near-normal population. For each sample, 90% and 95% confidence intervals for the population mean are calculated. A count is made of those samples for which the true mean falls below, within, and above the confidence interval:

	Below	Within	Above
90% interval	236	4513	251
95% interval	129	4753	118

What are the expected frequencies? Compare the theoretical (expected) and the observed frequencies.

Solution The expected frequencies can be found by multiplying the theoretical probabilities by 5000:

	Below	Within	Above
90% interval	250	4500	250
95% interval	125	4750	125

The simulation frequncies are all quite close to the expected frequencies. □

The discussion in this section has included one rather unrealistic assumption—namely, that the population standard deviation is known. In practice, it's difficult to find situations in which the population mean is unknown but the standard deviation is known.

Usually both the mean and the standard deviation must be estimated from the sample. Because σ is estimated by the sample standard deviation s, the actual standard error of the mean $\sigma/\sqrt{n}$ is naturally estimated by $s/\sqrt{n}$. This estimation introduces another source of random error (s varies randomly, from sample to sample, around σ) and, strictly speaking, invalidates our confidence-interval formula. Fortunately, the formula is still a very good approximation for large sample sizes. As a very rough rule, we can use this formula when n is larger than 30*; a better way to handle this issue is described in Section 8.5.

substituting s for σ

EXAMPLE 8.3 Suppose that the airline in Example 8.1, takes a sample of 40 days and finds a sample mean of 112.0 and a sample standard deviation of 25. Find a 95% confidence interval for the true mean.

Solution For $\bar{y} = 112$, $s = 25$, and $n = 40$, $\sigma_{\bar{Y}} \approx 25/\sqrt{40} = 3.95$. Then using $z_{.025} = 1.96$, the 95% confidence interval for μ is

$$112 \pm 1.96(3.95) \quad \text{or} \quad 104.26 \quad \text{to} \quad 119.74$$

□

SECTION 8.1 EXERCISES

8.1 Consider the data of Exercise 7.1 (p. 207). Assume that the population standard deviation is 4.0.
a. Calculate a 95% confidence interval for μ.
b. Calculate a 99% confidence interval for μ.

8.2 Give a careful verbal interpretation of the confidence interval in Exercise 8.1, part (a).

8.3 From the appearance of the data of Exercise 7.1, is it reasonable to assume that the sampling distribution of $\bar{Y}$ is nearly normal? (This assumption underlies the confidence-interval procedure.)

8.4 Refer to Exercise 7.13 (p. 213). Suppose that the sample of 90 forecasts yields an average prediction of a 2.7% growth in real disposable income. Assume that the population standard deviation is .4%. Calculate a 90% confidence interval for the population mean forecast.

8.5 In an audit of inventories, an internal auditor takes a sample of 36 items and determines the "shrinkage" (loss due to shoplifting or employee theft) for each item in percentage terms. The sample mean is 5.8% and the standard deviation is 4.2%. Calculate a 95% confidence interval for the true mean shrinkage.

8.6 Do you believe that the sampling distribution of $\bar{Y}$ in Exercise 8.5 would be approximately normal?

8.2 CONFIDENCE INTERVALS FOR A PROPORTION

The confidence-interval method in Section 8.1 can be adapted quite directly to give a confidence interval for a population proportion. The method is based on a normal approximation to the sampling distribution of a sample proportion. As such, it is an approximation, and some rules are needed for its use.

* This rule happens to coincide with a standard rule for appealing to the Central Limit Theorem. The latter rule is good only for symmetric or modestly skewed population distributions.

expected value and standard error of $\hat{\pi}$

The sample proportion of successes, denoted $\hat{\pi}$, is just the number Y of successes divided by the sample size. As the mean and standard deviation for the binomial random variable Y are $n\pi$ and $\sqrt{n\pi(1-\pi)}$, respectively, it follows (from the properties developed in Chapter 4) that the **expected value and standard error of $\hat{\pi}$** are, respectively,

$$E(\hat{\pi}) = \pi \qquad \text{and} \qquad \sigma_{\hat{\pi}} = \sqrt{\pi(1-\pi)/n}$$

For sufficiently large n, $\hat{\pi}$ has an approximately normal distribution; so, for instance

$$P\left(-1.96 \le \frac{\hat{\pi} - \pi}{\sigma_{\hat{\pi}}} \le 1.96\right) \approx .95$$

Equivalently,

$$P(\hat{\pi} - 1.96\sigma_{\hat{\pi}} \le \pi \le \hat{\pi} + 1.96\sigma_{\hat{\pi}}) \approx .95$$

This looks very much like a confidence-interval formula, but there is the problem that the standard error $\sigma_{\hat{\pi}} = \sqrt{\pi(1-\pi)/n}$ involves the unknown population parameter π. Just as we can replace σ by s in $\sigma_{\bar{Y}}$ when n is large, so can we replace $\pi(1-\pi)$ by $\hat{\pi}(1-\hat{\pi})$ in $\sigma_{\hat{\pi}}$. This yields a usable confidence-interval formula for the population proportion:

$100(1-\alpha)$% Confidence Interval for a Proportion

$$\hat{\pi} - z_{\alpha/2}\sqrt{\frac{\hat{\pi}(1-\hat{\pi})}{n}} \le \pi \le \hat{\pi} + z_{\alpha/2}\sqrt{\frac{\hat{\pi}(1-\hat{\pi})}{n}}$$

This is the same "sample statistic $\pm$ table value times standard error" that occurs in the confidence interval for a mean. The sample mean $\bar{y}$ is replaced by the sample proportion $\hat{\pi}$. Similarly, $\sigma_{\bar{Y}}$ is replaced by $\sigma_{\hat{\pi}}$.

EXAMPLE 8.4 Suppose that in a sample of 2200 households with one or more television sets, 471 watch a particular network's show at a given time. Find a 95% confidence interval for the population proportion of households watching this show.

Solution The sample proportion is $\hat{\pi} = 471/2200 = .214$, and $\sqrt{\hat{\pi}(1-\hat{\pi})/n} = .00874$. The z table value that cuts off a right-tail area of .025 is 1.96. The confidence interval is

$$.214 - 1.96(.00874) \le \pi \le .214 + 1.96(.00874)$$

or

$$.197 \le \pi \le .231$$

You might check the rankings of current television shows to see how much the difference between a 19.7% share and a 23.1% share would make in a show's ranking. □

normal approximation to binomial distributions

This confidence-interval method is based on a **normal approximation to a binomial distribution** that is appropriate for sufficiently large n. The rule is that both $n\pi$ and $n(1-\pi)$ should be at least 5, but, because π is the unknown population proportion, the rule has to be based on $n\hat{\pi}$ and $n(1-\hat{\pi})$ instead. Usually a sample size that violates this rule (or even comes close) yields a confidence interval that is too wide to be informative. For example, if $n = 20$ and $\hat{\pi} = .20$, then $n\hat{\pi} = 4$ and the 95% confidence interval for π is $.025 \leq \pi \leq .375$. This confidence interval is practically useless; we know of few product managers who would consider "your product's market share is between 2.5% and 37.5%" to be very informative. However, even if a sample size satisfies the rule, we are not assured that the interval is informative. The rule judges only the adequacy of the sample size and the accuracy of the confidence interval based on the normal approximation. It is possible to use binomial probabilities to develop exact, if very wide, 90% or 95% confidence intervals.

SECTION 8.2 EXERCISES

8.7 The sales manager for a hardware wholesaler finds that 229 of the previous 500 calls to hardware store owners resulted in new product placements. Assuming that the 500 calls represent a random sample, find a 95% confidence interval for the long-run proportion of new product placements.

8.8 Give a careful verbal interpretation of the confidence interval found in Exercise 8.7.

8.9 As part of a market research study, in a sample of 125, 84 individuals are aware of a certain product. Calculate a 90% confidence interval for the proportion of individuals in the population who are aware of the product.

8.10 Should the normal approximation underlying the confidence interval of Exercise 8.9 be adequate?

8.11 In a sample of 40 middle managers of a large firm, it is found that 8 are actively involved in local civic or charitable organizations. Calculate a 90% confidence interval for the proportion of all middle managers who are so involved.

8.3 HOW LARGE A SAMPLE IS NEEDED?

Information is expensive. Gathering it is costly in terms of salaries, expenses, and time (and profits) lost. Obviously some information is crucial is making management decisions. So the question of how much information (how large a sample) to gather is basic. The confidence interval provides a convenient method for answering this question.

Suppose that an operations officer of a large multibranch bank is concerned about the daily average level of checks left at branches on weeknights. Each day, armored cars take each branch's receipts to a processing center, where checks are recorded and sent to a clearinghouse. The cars must visit some of the branches before the end of banking hours, but a substantial volume of checks can remain uncollected until the next day. The lost interest can be costly. For how many days must the volume of uncollected checks be calculated to get a reasonable idea of the true daily average?

There are two related aspects of the phrase *reasonable idea* to consider in the context of a confidence interval. First, what confidence level should be selected? Second, how wide a confidence interval can be tolerated?

The confidence level is often set at 95% or 90%. In part this is primitive tribal custom, passed on by generations of statistics textbooks. In part it's a decent translation of reasonable certainty. It's fairly easy to understand 90 (or 95) chances in 100, but hard to comprehend 999,999 chances in 1,000,000.

tolerable width

The **tolerable width** depends heavily on the context of the problem. Plus or minus \$80,000 (or a width of \$160,000) is moderately large for the nightly idle-check volume of an entire bank, enormous for the nightly idle volume of one branch, and tiny for the average daily amount cleared by all U.S. banks. The tolerance must be determined by a manager who knows the situation.

When considering a confidence interval for a population mean μ, the plus-or-minus term of a confidence interval is $z_{\alpha/2}\sigma_{\bar{Y}}$, where $\sigma_{\bar{Y}} = \sigma/\sqrt{n}$. Three quantities determine the value of the plus-or-minus term: the desired confidence level (which determines the z table value used), the standard deviation σ, and the sample size (which together with σ determines the standard error $\sigma_{\bar{Y}}$). Usually a guess must be made about the size of the population standard deviation. (Sometimes an initial sample is taken to estimate the standard deviation; this estimate provides a basis for determining the additional sample size that is needed). For a given tolerable width, once the confidence level is specified and an estimate of σ supplied, the required sample size can be calculated by trial and error or by a formula.

The trial-and-error approach can be illustrated with the idle-check example. Suppose that a 95% confidence interval is desired, with a width of no more than \$5000 (a plus-or-minus range no greater than \$2500), and that the long-run standard deviation is assumed to be \$10,000. Suppose that we first try $n = 16$. The confidence interval is $\bar{y} \pm 1.96(10{,}000/\sqrt{16})$ or $\bar{y} \pm 4900$. This interval is about twice as wide as desired; to halve the width of the confidence interval, we must quadruple the sample size because the sample size appears in the standard error formula as $\sqrt{n}$. With $n = 64$, the 95% confidence interval is $\bar{y} \pm 1.96(10{,}000/\sqrt{64})$ or $\bar{y} \pm 2450$, which is about what is wanted. Because the assumption that the standard deviation is \$10,000 is just a guess, there's not much point in arguing over whether n should be 64 or 63 or 65; a "ballpark" value for n serves the purpose.

Calculation of the required sample size can be done by formula. Set $z_{\alpha/2}\sigma/\sqrt{n}$ equal to the specified plus-or-minus tolerance E and solve for n.

Sample Size Required for a $100(1 - \alpha)\%$ Confidence Interval of Given Width for μ

The **sample size** required to obtain a $100(1 - \alpha)\%$ confidence interval for a population mean μ of the form $\bar{y} \pm E$, where $E = z_{\alpha/2}\sigma/\sqrt{n}$, is

$$n = \frac{z_{\alpha/2}^2\sigma^2}{E^2}$$

The width of the confidence interval is $2E$.

EXAMPLE 8.5 Union officials are concerned about reports of inferior wages being paid to employees of a company under its jurisdiction. How large a sample is needed to obtain a 90% confidence interval for the population mean hourly wage μ with width equal to \$1.00? Assume that $\sigma = \$4.00$.

Solution The desired width is $2E = 1.00$ and $\sigma = 4.00$. Substituting into the sample size formula with $z_{\alpha/2} = 1.645$, we obtain

$$n = \frac{(1.645)^2(4^2)}{(.5)^2} \approx 173$$ □

EXAMPLE 8.6 How large a sample is needed to obtain a 95% confidence interval for μ with a width of two-tenths of a (population) standard deviation?

Solution The desired width $2E = .2\sigma$, so $E = .1\sigma$. Therefore

$$n = \frac{(1.96)^2\sigma^2}{(.1\sigma)^2} = \frac{(1.96)^2}{(.1)^2} \approx 384$$ □

Determining sample size for a confidence interval for a proportion is a similar process. The corresponding formula is

$$n = \frac{z_{\alpha/2}^2\hat{\pi}(1 - \hat{\pi})}{E^2}$$

The only problem is that the sample size depends on $\hat{\pi}$. Until the sample size is determined and the sample taken, we do not know $\hat{\pi}$. There are several possible solutions to our problem. We can substitute $\hat{\pi} = .5$ into the sample size formula, which results in a conservative sample size that is usually larger than is actually required. Another possibility is to substitute a value of $\hat{\pi}$ obtained from either a previous study or a pilot study. The sample-size formula for estimating a binomial proportion is shown here:

Sample Size Required for a 100(1 − α)% Confidence Interval of Given Width for π

The sample size required to obtain a 100(1 − α)% confidence interval for π of the form $\hat{\pi} \pm E$, where

$$E = z_{\alpha/2}\sqrt{\frac{\hat{\pi}(1 - \hat{\pi})}{n}}$$

is

$$n = \frac{z_{\alpha/2}^2\hat{\pi}(1 - \hat{\pi})}{E^2}$$

Note: Use $\hat{\pi} = .5$ for a conservative (large) sample size or use the value of $\hat{\pi}$ from a previous (or pilot) study.

EXAMPLE 8.7 A direct-mail sales company must determine its credit policies quite carefully. Suppose that the firm suspects that advertisements in a certain magazine have led to an excessively high rate of write-offs (accounts that are regarded as uncollectible). The firm wants to establish a 90% confidence interval for this magazine's write-off proportion that is accurate to $\pm.02$.

a. How many accounts must be sampled to guarantee this goal?
b. If this many accounts are sampled and 10% of the sampled accounts are determined to be write-offs, what is the resulting 90% confidence interval?

Solution a. The sample size formula is

$$n = \frac{z_{\alpha/2}^2 \hat{\pi}(1 - \hat{\pi})}{E^2}$$

Using the conservative estimate $\hat{\pi} = .5$ and substituting $E = .02$ with $z_{\alpha/2} = 1.645$, the required sample size is

$$n = \frac{(1.645)^2(.5)^2}{(.02)^2} \approx 1691$$

b. If a sample of 1691 accounts shows 169 (essentially 10%) write-offs, the 90% confidence interval for the true write-off proportion is

$$.10 \pm 1.645\sqrt{(.10)(.90)/1691} = .10 \pm .012$$

The conservative nature of the confidence interval that results from a sample size determined by setting $\hat{\pi} = .5$ in the formula is indicated here. The actual confidence interval has $E = .012$, whereas the target was $E = .02$. Had the firm been willing to make an initial guess that $\hat{\pi}$ would be about .10, it could have used a smaller sample size:

$$n = \frac{(1.645)^2(.1)(.9)}{(.02)^2} = 609$$

□

As Example 8.7 indicates, **basing a sample-size determination on the assumption that $\hat{\pi}$ is .5 can be excessively conservative**. Whenever there is information to suggest that the sample proportion differs from .5, the substitution $\hat{\pi} = .5$ results in a large (conservative) sample size. The corresponding confidence interval has a smaller width than the target width.

SECTION 8.3 EXERCISES

8.12 a. Refer to Example 8.5. How large a sample is needed to obtain a 90% confidence interval with width \$0.50? with width \$0.25? with width \$0.125?

b. In general, how much must one increase a sample size to cut the width of a confidence interval in half (using a specified confidence level)?

8.13 Refer to Example 8.6. How large a sample is needed to obtain a width of three-tenths of a standard deviation? four-tenths?

8.14 An automobile insurance firm wants to find the average amount per claim for auto body repairs. Its summary records combine amounts for body repair with all other amounts, so a sample of individual claims must be taken. A 95% confidence interval with a width no greater than \$50 is wanted. A "horseback guess" says that the standard deviation is about \$400. How large a sample is needed?

8.15 Suppose that the guess of the standard deviation in Exercise 8.14 is somewhere between \$300 and \$450.

a. Compute required sample sizes for $\sigma = 300$ and for $\sigma = 450$.

b. What would happen to the width of the confidence interval if the n corresponding to $\sigma = 450$ was used but in fact the standard deviation came out \$300?

8.16 Do you think that the sample size used in Exercise 8.14 would be adequate to assume that $\bar{Y}$ had approximately a normal sampling distribution?

8.17 A manufacturer of boxes of candy is concerned about the proportion of imperfect boxes—those containing cracked, broken, or otherwise unappetizing candies.

a. How large a sample is needed to get a 95% confidence interval for this proportion with a width no greater than .02? Use the conservative substitution.

b. How does the answer to part (a) change if we assume that the proportion of imperfect boxes is at least .005 and no more than .08?

8.4 THE *t* DISTRIBUTION

The estimation procedures for a population mean μ presented in Section 8.1 are based on the assumption that either σ is known or that there is a sufficient number of measurements (e.g., 30 or more) so that the sample standard deviation s can replace σ in the standard error for $\bar{y}$, $\sigma/\sqrt{n}$. Sometimes it is impossible or uneconomical to obtain a large sample when making an inference about a population mean. For example, in a study of rush-hour traffic patterns around a bridge on Friday evenings, it would take more than six months to generate 30 observations on the total Friday evening rush-hour traffic volume. This may be too long before some corrective remedies are proposed.

W. S. Gosset faced a similar problem around the turn of the twentieth century when, as a chemist for Guinness Breweries, he was asked to make judgments on the mean quality of various brews. He was supplied with only small sample sizes to reach his conclusions.

Gosset believed that, for small samples, when he used the z statistic

$$\frac{\bar{Y} - \mu_0}{\sigma/\sqrt{n}}$$

with σ replaced by s, he was incorrectly rejecting the null hypothesis H_0: $\mu = \mu_0$ at a much higher rate than that specified by the predetermined value of α. He became intrigued by the problem and set out to derive the sampling distribution of the quantity

$$\frac{\bar{Y} - \mu_0}{s/\sqrt{n}}$$

particularly for $n < 30$.

substituting s for σ

The **substitution of s for σ** in the z statistic

$$z = \frac{\bar{Y} - \mu_0}{\sigma/\sqrt{n}}$$

adds a second source of variability in addition to $\bar{Y}$. We suggested the substitution of s for σ provided $n \geq 30$ in Section 8.1. Now we give confidence intervals for μ for any value of $n > 1$. The procedures are very similar to the large-sample results presented earlier.

Gosset derived the sampling distribution for the statistic

$$\frac{\bar{Y} - \mu}{s/\sqrt{n}}$$

and he published his results in 1908 under the nom de plume "Student" because it was against company policy to publish his results. The statistic

$$t = \frac{\bar{Y} - \mu}{s/\sqrt{n}}$$

Student's t

is frequently referred to as **Student's t** and its distribution as Student's t distribution. We can summarize the properties of a t distribution by comparing it to a standard normal (z) distribution.

Properties of Student's t Distribution

1. The t distribution, like the z distribution, is symmetric about the mean $\mu = 0$.
2. The t distribution is more variable than the z distribution (see Figure 8.3).
3. There are many different t distributions. We specify a particular one by its degrees of freedom, d.f. If a random sample is taken from a normal population, then the statistic

   $$t = \frac{\bar{Y} - \mu}{s/\sqrt{n}}$$

   has a t distribution with d.f. $= n - 1$.
4. As n increases (or equivalently as the d.f. increase), the distribution of t approaches the distribution of z.

degrees of freedom

A general definition of the term **degrees of freedom** requires n-dimensional geometry and an understanding of linear algebra. We don't go into such detail; rather we try to

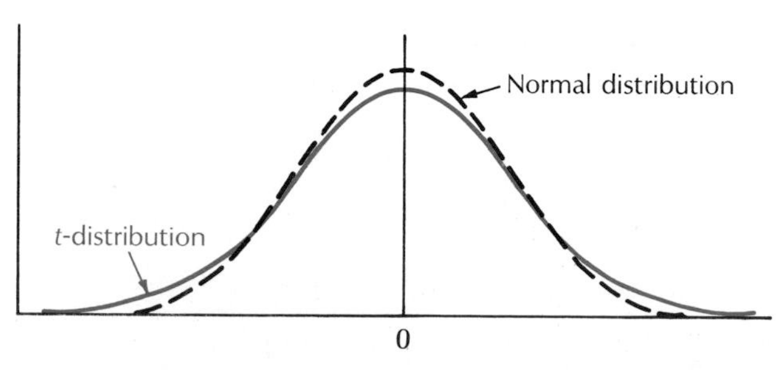

FIGURE 8.3 **A t Distribution with a Normal Distribution Superimposed**

give you an intuitive understanding of the concept. The d.f. refers to a parameter of a t distribution and is used to indicate the number of data points that are free to vary in a sample. For a sample of size n, there are n pieces of information (data points), but when the sample mean is known, only $n - 1$ of these are free to vary. The last (nth) is determined by the preceding $n - 1$. Because of this property, the t statistic is said to have $n - 1$ degrees of freedom.

Although a mathematical formula can be given for the probability density function of the t distribution, it is not important since there are tables to evaluate t probabilities. Because of the symmetry of t, only upper-tail percentage points (probabilities or areas) of the distribution of t have been tabulated. These appear in Appendix Table 4. The degrees of freedom (d.f.) are listed along the left-hand column of the page. An entry in the table specifies a value of t, say, t_a, such that an area a lies to its right (see Figure 8.4). Various values of a appear across the top of the page. Thus, for example, with d.f. = 7, the value of t with an area .05 to its right is 1.895 (found in the a = .05 column and d.f. = 7 row).

EXAMPLE 8.8 If a random sample of size $n = 15$ is taken from a normally distributed population, find

$$P\left(\frac{\bar{Y} - \mu}{s/\sqrt{n}} > 2.145\right)$$

and

$$P\left(-2.145 \leq \frac{\bar{Y} - \mu}{s/\sqrt{n}} \leq 2.145\right)$$

Solution We must use the t table, Appendix Table 4, with $n - 1 = 14$ d.f. The table indicates values cutting off specific right-tail areas. In particular, $P(t_{14\text{ d.f.}} > 2.145)$ is shown to be .025, so

$$P\left(\frac{\bar{Y} - \mu}{s/\sqrt{n}} > 2.145\right) = .025$$

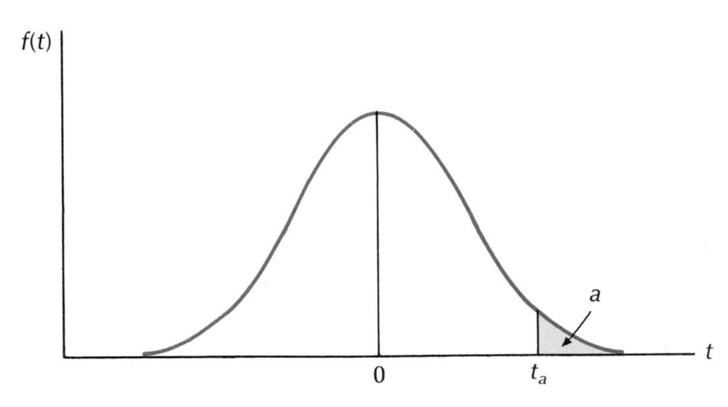

FIGURE 8.4 **Illustration of Area Tabulated in Appendix Table 4 for the *t* Distribution**

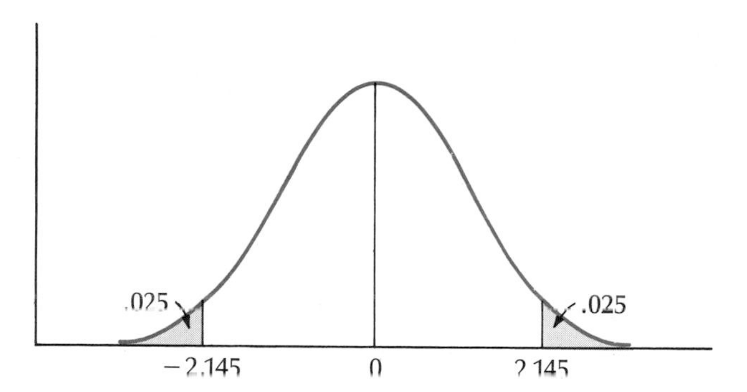

FIGURE 8.5 *t* Distribution with 14 d.f.

(see Figure 8.5). The t distribution is symmetric around zero, so the left-tail area $P(t_{14\text{ d.f.}} < -2.145) = .025$ also. The remaining area after both tails are cut off is .95, so

$$P\left(-2.145 \leq \frac{\bar{Y} - \mu}{s/\sqrt{n}} \leq 2.145\right) = .95$$ □

An examination of the t table indicates the effect of changing from

$$z = \frac{\bar{Y} - \mu}{\sigma/\sqrt{n}}$$

to

$$t = \frac{\bar{Y} - \mu}{s/\sqrt{n}}$$

For very small n, t table values are quite large, and for d.f. = 2, the right-tail area of .025 is cut off at 4.303. As the d.f. increases, the t table values for a given tail area decrease. At the bottom line, infinite degrees of freedom, the t table contains the normal distribution (z) values.

This phenomenon can be explained by considering how the t distribution arises. We get a t statistic by replacing the true standard deviation σ by the sample standard deviation s, thus introducing an additional source of random variation. When n is small, the value of s can vary widely from the value of σ, and the t distribution must have quite a large variance. As n gets larger, there is less random variation of s from σ and the t distribution's variance gets smaller. As n approaches infinity, s approaches σ and the only important source of randomness is $\bar{Y}$; the z distribution accounts for the variation of $\bar{Y}$ around μ.

When we first discussed replacing σ by s, we used a rule that we could use the z tables if n was greater than or equal to 30. The t table values for 30 d.f. are fairly close to normal table values, except for very small tail areas. But there's no need to preserve this rule any longer. We may as well use the t tables routinely for all t statistics. If the actual degrees of freedom are not shown in the table, rough interpolation is accurate enough for most purposes. If the sample size is very large, it is obvious that normal table values are close enough.

EXAMPLE 8.9 For a random sample of size 46 from a normal population, find the cutoff point for a right-tail area of .025; that is, find $t_{.025}$ such that

$$P\left(\frac{\bar{Y}-\mu}{s/\sqrt{n}} > t_{.025}\right) = .025$$

Solution There are $n - 1 = 45$ d.f. The table entries for 40 and 60 d.f. are 2.021 and 2.000, respectively. As the actual 45 d.f. is closer to 40 than to 60, $t_{.025}$ ought to be closer to 2.021 than to 2.000. We can use a simple diagram, such as that shown here, to find the interpolated value of t.

t value	2.021	$t = ?$	2.000
d.f.	40	45	60

Because 45 is 1/4 the distance from d.f. = 40 to d.f. = 60, we want a t value that is 1/4 the distance from 2.021 to 2.000. With $.021/4 = .005$, the desired t value is $2.021 - .005 = 2.016$. □

SECTION 8.4 EXERCISES

8.18 A random sample of size 4 is to be taken from a normal population with mean $\mu = 100$. Let

$$t = \frac{\bar{Y} - 100}{s/\sqrt{4}}$$

Evaluate the following probabilities:

a. $P(t > 1.638)$
b. $P(t > 5.841)$
c. $P(t < -2.353)$
d. $P(-2.353 < t < 2.353)$
e. $P(|t| > 3.182)$
f. $P(|t| > 4.541)$

Draw pictures.

8.19 Suppose that the t statistic in Exercise 8.18 is mistakenly assumed to have a normal (z) distribution. Evaluate $P(t > 1.638)$ and $P(|t| > 1.638)$ under this erroneous assumption. Does this assumption cause an overstatement or an understatement of the probabilities?

8.20 A Monte Carlo study is made by taking 1100 samples, each of size 4, from the normal population of Exercise 8.18. The t statistic is defined in that exercise. The results of the study are summarized as follows:

Event	Frequency
$t < -2.353$	44
$-2.353 < t < -1.638$	59
$-1.638 < t < 1.638$	896
$1.638 < t < 2.353$	47
$t > 2.353$	54

a. What are the theoretical relative frequencies?

b. Is there any evidence of a systematic departure from these theoretical relative frequencies?

8.21 For a t statistic with 72 degrees of freedom, use rough interpolation to find 90th, 95th, and 99th percentiles.

8.5 CONFIDENCE INTERVALS WITH THE t DISTRIBUTION

The mathematical development of Section 8.4 can be used to state inference procedures for a mean, which can be used in the typical case that the population standard deviation is not known. This section is devoted to confidence intervals.

The changes in the confidence-interval procedure when σ is unknown are easy enough. Replace the unknown σ by s to get an estimated standard error $s/\sqrt{n}$; use t tables instead of z tables.

EXAMPLE 8.10 Calculate a 95% confidence interval for the population mean if a sample of size $n = 25$ from a nearly normal population yields a sample mean of 96.2; assume that the population standard deviation is 15.0. Recalculate the interval assuming that the population standard deviation is unknown and that the sample standard deviation is 16.2.

Solution We may use the procedures of Section 8.1 for the first problem:

$$\bar{y} - z_{\alpha/2}\frac{\sigma}{\sqrt{n}} \leq \mu \leq \bar{y} + z_{\alpha/2}\frac{\sigma}{\sqrt{n}}$$

$$96.2 - 1.96\frac{(15.0)}{\sqrt{25}} \leq \mu \leq 96.2 + 1.96\frac{(15.0)}{\sqrt{25}}$$

$$90.32 \leq \mu \leq 102.08$$

In the second problem, we need the t table value for $n - 1 = 24$ d.f. This value must cut off combined left- and right-tail areas of .05 so the desired right-tail area is .025. The t table value is 2.064. Replace $\sigma = 15.0$ by $s = 16.2$ and $z_{.025} = 1.96$ by $t_{.025} = 2.064$.

$$96.2 - 2.064\frac{(16.2)}{\sqrt{25}} \leq \mu \leq 96.2 + 2.064\frac{(16.2)}{\sqrt{25}}$$

$$89.51 \leq \mu \leq 102.89$$

This interval is wider than the previous one because $t_{.025} > z_{.025}$, and also because in this case $s = 16.2$ happens to be greater than the assumed $\sigma = 15.0$. □

The general t confidence interval for μ based on a t distribution with d.f. $= n - 1$ is shown here:

100(1 − α)% Confidence Interval for μ, with σ Unknown

$$\bar{y} - t_{\alpha/2}\frac{s}{\sqrt{n}} \leq \mu \leq \bar{y} + t_{\alpha/2}\frac{s}{\sqrt{n}}$$

where $t_{\alpha/2}$ is the tabulated t value cutting off a right-tail area of $\alpha/2$, with $n - 1$ d.f.

small-sample confidence interval

This formula is often called a **small-sample** confidence interval for the mean, but it is valid for *any* sample size. For a large sample size, the difference between using t tables and z tables is negligible, so the importance of the t versus z distinction is greatest for small sample sizes. The assumption of a normal population is most crucial for small sample sizes, where the Central Limit Theorem has relatively little effect.

EXAMPLE 8.11 An airline has four ticket counter positions at a particular airport. In an attempt to reduce waiting lines for customers, the airline introduces the "snake system." Under this system, all customers enter a single waiting line that winds back and forth in front of the counter. A customer who reaches the front of the line proceeds to the first free position.

Each weekday for three weeks, the airline customer-relations manager finds the waiting time in minutes for the first customer entering after 4.00 P.M. One observation is excluded because of an unusual condition: the airport was fogged in and many flight plans had to be changed. The data are

4.3 5.2 2.1 6.2 5.8 4.7 3.8 9.3 5.0 4.1 6.0 8.7 0.5 4.9

Find a 95% confidence interval for the long-run mean waiting time on weekdays under normal conditions.

Solution First calculate $\bar{y} = 5.043$ and $s = 2.266$. The t table value (13 d.f., one-tail area .025) is 2.160. The interval is

$$5.043 - 2.160\frac{(2.266)}{\sqrt{14}} \leq \mu \leq 5.043 + 2.160\frac{(2.266)}{\sqrt{14}}$$

or

$$3.735 \leq \mu \leq 6.351$$

It would be better to report this, rounded off to roughly the accuracy of the data, as $3.7 \leq \mu \leq 6.4$. □

sample size required for estimating μ

One of the important uses of confidence intervals is in determining the sample size required to yield a desired degree of statistical accuracy. Accuracy is defined by the level of confidence and the width of the interval. Recall that when we assume σ is known and specify the degree of confidence $100(1 - \alpha)\%$ and the desired confidence interval width

$2E$, we find the desired sample size by solving the equation

$$\frac{z_{\alpha/2}\sigma}{\sqrt{n}} = E$$

for n. Now we would like to find n by solving $t_{\alpha/2}s/\sqrt{n} = E$, but there are two difficulties. First, s is not known until the sample is taken, and second, we do not have the d.f. for $t_{\alpha/2}$ until n is specified. The first problem can be handled either by using a rough, ballpark guess for s or by specifying the desired width as some fraction of a standard deviation. (An error in estimating a mean to within .01 standard deviation would be dwarfed by the variation of individual values from the mean, while an error of 1.00 standard deviation would be pretty substantial.) The second problem can be solved by making a preliminary assumption that n is large enough that z can be substituted for t. If the resulting n turns out to be small, trial and error (in the direction of increasing n) usually gets an answer quickly.

EXAMPLE 8.12 Suppose that in Example 8.11 a 95% confidence interval with a plus-or-minus tolerance of half a standard deviation is desired. What sample size is needed?

Solution E is to be $.5s$. For the moment, assume that we can use the z table value 1.96 as an approximation to $t_{.025}$. Solving the equation

$$\frac{1.96s}{\sqrt{n}} = .5s$$

for n, we get

$$n = \frac{(1.96s)^2}{(.5s)^2} = 15.4$$

For $n = 16$ (15 d.f.) we would use $t_{.025} = 2.131$ instead of 1.96 and get an actual value of E equal to

$$\frac{2.131s}{\sqrt{16}} \approx .533(s)$$

which is a bit too large. Try $n = 18$ (17 d.f.); $t_{.025} = 2.110$, and

$$\frac{2.110s}{\sqrt{18}} \approx .497(s)$$

so $n = 18$ will do. □

SECTION 8.5 EXERCISES

8.22 A manufacturer of cookies and crackers does a small survey of the age at sale of one of its brands. A random sample of 23 retail markets in a particular region is chosen. In each store, the number of days since manufacture of the frontmost box of crackers is determined by a date code on the box.

The data (age in days, arranged from lowest to highest) are

27 34 36 36 38 39 39 39 40 40 42 45 47 51 52
57 63 71 75 84 96 110 147

a. Verify that $\bar{y} = 56.87$ and $s = 28.97$.
b. Calculate a 99% confidence interval for the true mean age.

8.23 Suppose that the manufacturer in Exercise 8.22 wants to obtain a 90% confidence interval with a width of no more than six days. Assuming that the sample standard deviation does not change, how large a sample is needed?

8.24 A consumer group wants to estimate the average delivered price of a certain model of refrigerator in the New York metropolitan area. Prices are determined by comparison shoppers at 14 randomly selected stores in the area. The dollar prices (including taxes) are

341 347 319 331 326 298 335 351 316 307 335 320 329 346

Calculate a 95% confidence interval for the true mean.

8.25 A random sample of 20 taste-testers rate the quality of a proposed new product on a 0–100 scale. The ordered scores are

16 20 31 50 50 50 51 53 53 55 57 59 60 60 61 65
67 67 81 92

a. Calculate a 95% confidence interval for the population mean score. Should t tables or z tables be used?
b. Plot the data. Is there any reason to think that the use of a mean-based confidence interval is a poor idea?

8.26 A furniture mover calculates the actual weight as a proportion of estimated weight for a sample of 31 recent jobs. The sample mean is 1.13 and the sample standard deviation is .16.
a. Calculate a 95% confidence interval for the population mean using t tables.
b. Assume that the population standard deviation is .16. Calculate a 95% confidence interval for the population mean using z tables.
c. Are the intervals calculated in parts (a) and (b) of roughly similar size?

8.27 When the data underlying Exercise 8.26 are plotted, the plot shows a strong skewness to the right. Does this indicate that the nominal 95% confidence level may be in error?

8.6 ASSUMPTIONS FOR INTERVAL ESTIMATION

Any statistical method involves assumptions. Some assumptions are general and apply to a wide variety of methods; others are specific to a particular method. We'll have a lot to say about assumptions in future chapters. Because interval estimation of a single parameter (whether it be a mean, a proportion, or a median), is a relatively simple concept; many of the issues of assumptions and assumption violation can be dealt with most clearly in this context.

First, we should emphasize that the methods in this chapter apply only to random samples. The allowance for error inherent in confidence intervals is only an allowance for random error; no allowance is made for any biases in data collection. If the data underlying a confidence interval have been collected in a lazy, convenience sample, the confidence interval is very likely to be wrong simply because of the biases in data collection. There are no known methods to compensate for the biases in badly chosen samples.

Within the context of legitimately random samples, there are some specific assumptions that can be problematic. One key assumption is **independence within samples**. All the methods described in this chapter assume that the observations are independent of each other. Not all random-sampling methods yield independent observations. For example, suppose that a real-estate assessor chooses 22 city blocks of homes to evaluate, from the tax lists of a city, and then assesses the market value of all homes in each block. Assuming that the assessor does, in fact, choose the blocks randomly, there is no systematic bias in favor of low-value homes or high-value homes. But there is a dependence problem. Given the well-established tendency of high-value homes to cluster together (and also of low-value homes to occur in bunches), if one home in the sample has higher than average values, so do adjacent homes. The assessment may involve, say, 300 homes; however, the method does not give 300 separate, independent measurements of home values. In fact, the data arising from the assessor's evaluations would be more appropriately evaluated by the cluster-sampling methods described in Chapter 19.

The most common source of problems with the assumption of independence occurs in **time-series** data, data collected in a well-defined chronological order. Suppose, for example, that we measure the dollar volume of back orders for a particular manufacturer on 20 consecutive Friday afternoons. It's reasonable to suppose that a high back-order volume on one Friday is likely to be followed by high back-order volumes on succeeding Fridays, and the same for low volumes. The standard error formulas that we use in confidence intervals depend very heavily on the assumption of independence of observations. When there is dependence, the standard error formulas may underestimate the actual uncertainty in an estimate. Even for modest dependencies, the degree of underestimation may be serious.

Beyond the assumption of independence, methods for means involve an assumption that the underlying population is normally distributed. **In practice, no population is exactly normal.** When we use t-distribution methods for a mean, we are assuming that the underlying population is normal, and this assumption is guaranteed to be more or less wrong.

There are two types of issues to consider when populations are assumed to be nonnormal. First, what kind of nonnormality is assumed, and second, what possible effects do these specific forms of nonnormality have on the t-distribution procedures? The most important deviations from normality are skewed distributions and heavy-tailed distributions. (Heavy-tailed distributions show up in otherwise roughly symmetric data by the occurrence of outliers.)

In order to evaluate the effect of nonnormality as exhibited by skewness or heavy tails, we consider whether the t-distribution procedures are still approximately correct for these forms of nonnormality and whether there are other more efficient procedures. Even if a confidence interval for μ based on t gives nearly correct results for, say, a heavy-tailed population distribution, there may be a more efficient procedure (which gives a smaller confidence interval width) based on a trimmed mean.

The question of approximate correctness of t procedures has been studied for quite a long time. The general conclusion of these studies is that the probabilities specified by the t procedures, particularly the confidence level, are fairly accurate even when the population distribution is heavy-tailed (or light-tailed). In contrast, skewness, particularly with small sample sizes, can have a nasty effect on these probabilities, particularly in one-tailed procedures. A t distribution is symmetric, of course. When the population distribution is skewed, the actual sampling distribution of a t statistic is skewed. The skewness decreases as the sample size increases, but there is no magic sample size that completely deskews the actual sampling distribution.

sensitivity of one-tailed procedures

The second question, that of the efficiency of t procedures, has only recently been studied seriously. There has been a near-unanimous conclusion from these studies. When the population distribution is symmetric but heavy-tailed, various **robust** procedures are more efficient than the standard t procedures. Virtually all robust procedures eliminate or give low weight to the few largest and smallest observations in the sample. The ordinary sample mean gives equal weight to all observations and is very sensitive to extreme sample values. Therefore when the population distribution is heavy-tailed, robust procedures tend to give more accurate estimates and have smaller standard errors than the ordinary sample mean.

robust methods

Unfortunately, less work has been done on the effectiveness of these robust procedures when the population distribution is skewed. A 20% trimmed mean, which averages the middle 80% of the data values, is unquestionably a biased estimator of the population mean when the population is skewed. Whether this bias is compensated for by a lower standard error is an open mathematical and conceptual question. However, it would be worrisome to have an estimator with a small standard error that always overestimated the population parameter.

So what is a nonexpert manager to do? First of all, *look at the data*. One of the serious dangers of using available statistical software is that statistical analyses may be done, untouched by human minds. A simple histogram of the data, or some other plotting device, reveals any gross skewness or extreme outliers. We can also calculate the coefficient of skewness for the sample data (see Chapter 2). If there's no blatant nonnormality, the nominal t-distribution probabilities should be reasonably correct and the t procedure should be reasonably efficient. If the data values are obviously skewed or heavy-tailed, the t-distribution probabilities and the efficiency of the t procedure are highly suspect. Whenever possible, you should try something else in these situations. For example, we present a confidence interval for a population median in Section 8.7. Since the median is not as affected by outliers as is the mean, this can be used as an alternative to the interval based on the t statistic.

Other robust procedures are mentioned in this text, but we cannot do complete justice to them. We expect that these procedures will be integrated into some of the statistical software systems in the next few years. If such programs are not available, a manager can at least cultivate an alert skepticism about the accuracy of the stated probabilities.

8.7 CONFIDENCE INTERVALS FOR A MEDIAN

As early as Chapter 2 we noted that a median is less sensitive than a mean to gross skewness or outliers in data. Therefore we sometimes want to make inferences based on a median rather than a mean. Inferences about a median turn out to be relatively robust methods when the assumption of population normality is shaky. That is, median inferences are generally more believable than mean inferences when the assumptions underlying mean inference are violated; even when there are large sample sizes, so that mean inferences can be based on the Central Limit Theorem, median inferences may be more precise and efficient. This phenomenon occurs particularly when the data are outlier-prone.

One might expect the confidence interval for a median to follow the conventional "estimate plus or minus table value times standard error" format. It doesn't. Instead, a reformulation of the problem, based on the idea that roughly half the sample data should be above the population median and half below, reduces the problem to a binomial situation.

order statistics

This interval is based on **order statistics**. The kth-order statistic in a sample of n measurements is the kth smallest value in the sample. Thus the smallest measurement is the first-order statistic, the largest measurement is the nth-order statistic, and for an odd number of measurements the sample median is the $(n + 1)/2$th-order statistic. A procedure for using order statistics to get a confidence interval for a median is outlined here.

$100(1 - \alpha)\%$ Confidence Interval for a Median

1. For a given sample size, use binomial tables for $\pi = .50$ and the sample size n to find the value k that satisfies two conditions:

   a. $P(\text{number of successes} \leq k) \leq \alpha/2$;
   b. $P(\text{number of successes} \leq k + 1) > \alpha/2$.

   If the sample size exceeds the values of n displayed in Table 1, the value of k is the greatest integer less than $.5n - .5z_{\alpha/2}\sqrt{n}$.
2. The $100(1 - \alpha)\%$ confidence interval for the population median has lower and upper confidence limits of $(k + 1)$th-order statistic and $(n - k)$th-order statistic, respectively.

EXAMPLE 8.13

An insurance adjuster obtains estimates from two garages for repairs on foreign cars that had been damaged in a collision. The adjuster was most interested in the difference of the estimates. A sample of 15 differences was obtained.

Car:	1	2	3	4	5	6	7	8	9	10	11	12	13	14	15
Difference:	.3	1.1	1.1	−.2	.3	.5	.4	.9	.2	.6	.3	1.1	.8	.9	.9

Find a 95% confidence interval for the population median of differences.

Solution

For $n = 15$ and $\alpha/2 = .025$, we can see from binomial tables that

$$P(y \leq 3 | n = 15, \pi = .50) = .0176$$

and

$$P(y \leq 4 | n = 15, n = .50) = .0592$$

We therefore take $k = 3$. Then the 95% confidence interval for the population median has lower and upper limits of the 4th- and 12th-order statistics, respectively. For these data, the differences, arranged in order, are

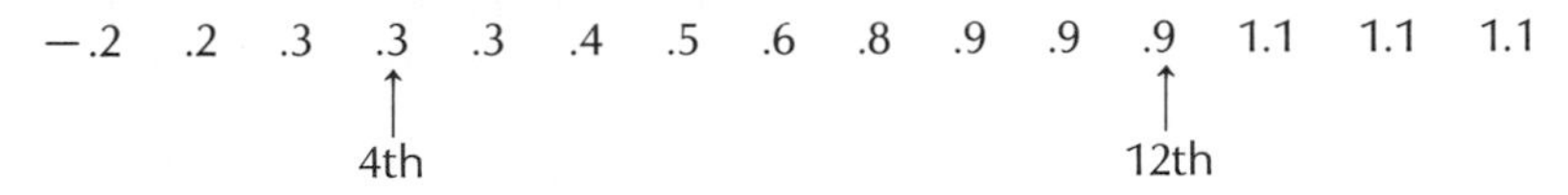

The 95% confidence interval for the population median is thus

$$.3 \leq \text{population median} \leq .9$$ □

EXAMPLE 8.14 Give the order statistics that provide the lower and upper 90% confidence limits for a population median based on $n = 100$ measurements.

Solution We could use Appendix Table 1. Alternatively, the value of k is the greatest integer less than

$$.5n - .5z_{\alpha/2}\sqrt{n}$$

For $n = 100$ and $\alpha/2 = .05$, we have

$$.5(100) - .5(1.645)10 = 41.775$$

so $k = 41$. The appropriate confidence limits are the 42nd- and 59th-order statistics. □

There's no reason to use median-based intervals when sampling from a normal population. In that situation, as we know from Chapter 7, the sample mean is the most efficient (smallest standard error) unbiased estimator; when the population is normal, a t (or z) interval has the narrowest, most precise range. But when the data are severely outlier-prone, mean intervals may well have a larger plus-or-minus range than median intervals. Once again, it's useful to plot the data to see which method is most meaningful and appropriate.

EXAMPLE 8.15 A taste test of a proposed new product involved 50 customers rating taste appeal on a 0.0 (best) to 9.9 (worst) scale. The following ratings were obtained.

0.8	1.0	2.9	3.0	3.4	3.5	3.7	3.7	3.8	3.9	4.1	4.1	4.2
4.2	4.2	4.3	4.3	4.3	4.3	4.3	4.3	4.4	4.4	4.4	4.4	4.4
4.4	4.5	4.5	4.5	4.5	4.5	4.5	4.6	4.6	4.6	4.6	4.7	4.7
4.7	4.8	4.8	4.8	4.9	4.9	4.9	5.1	5.1	6.7	6.8		

The mean rating for the 50 tasters is 4.300; the standard deviation is 0.966.

a. Plot the data. What is the general shape? Are there any outliers?
b. Calculate a 90% confidence interval for the mean.
c. Calculate a 90% confidence interval for the median.
d. According to part (a), which interval should be shorter? Is it?

Solution a. A stem-and-leaf display or a histogram indicates that the data are close to symmetric. By eye, there seem to be several outliers. The "hinges" (25th and 75th percentiles) are 4.2

and 4.7, giving an IQR of 4.7 − 4.2 = 0.5. The inner fences are 4.2 − 1.5(0.5) = 3.45 and 4.7 + 1.5(0.5) = 6.45; 7 of the 50 scores fall beyond these fences, strongly suggesting that we are sampling from an outlier-prone population.

b. There is no t table entry for 50 − 1 = 49 d.f.; conservatively, we use the entry for 40 d.f. and a (1 − .90)/2 = .05, 1.684. The interval is

$$4.300 - 1.684\frac{(0.966)}{\sqrt{50}} \leq \mu \leq 4.300 + 1.684\frac{(0.966)}{\sqrt{50}}$$

or

$$4.07 \leq \mu \leq 4.53$$

c. Adding binomial probabilities with $n = 50$ and $\pi = .50$ from the table, we find $P(Y \leq 18) = .0313$ and $P(Y \leq 19) = .0583$. Thus we take $k = 18$ and the 90% confidence interval ranges from the $k + 1 = 19$th value, 4.3, to the $n - k = 32$nd value, 4.5.

$$4.3 \leq \text{median} \leq 4.5$$

d. We have indicated that using a median is more efficient (gives shorter confidence intervals) when the data are outlier-prone. The confidence interval for the median has width 4.5 − 4.3 = 0.2; the confidence interval for the mean has width 0.46. □

EXAMPLE 8.16 A normal plot for the data of Example 8.15, constructed using Minitab, is shown here. Does the plot confirm the judgment of the shape of the data made in Example 8.15?

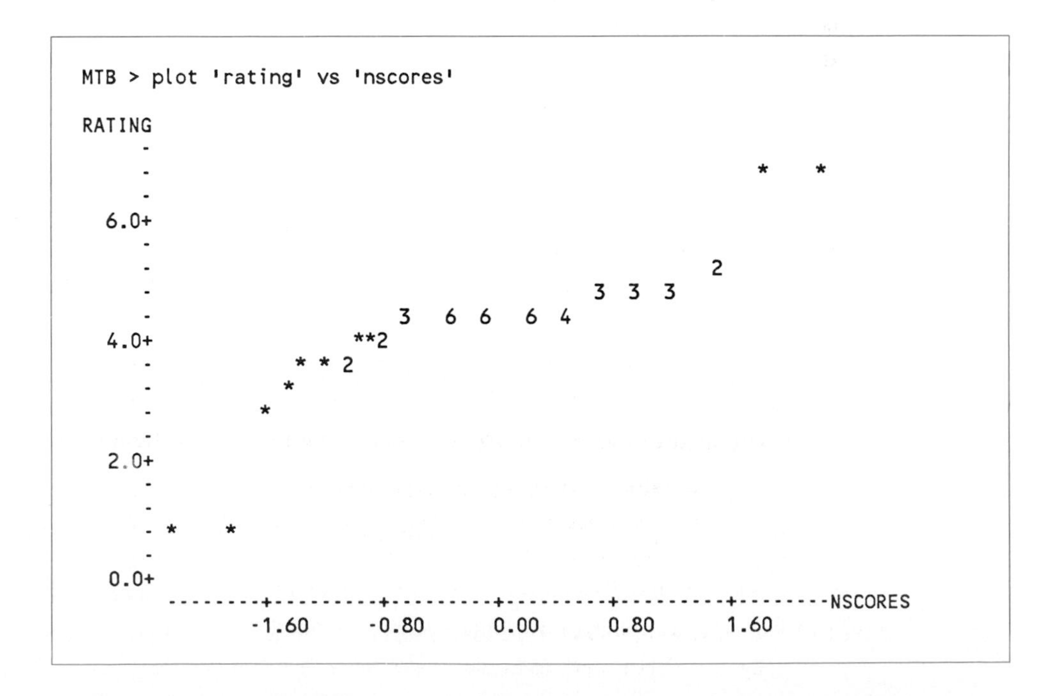

Solution In Exercise 8.15 we concluded that the data were outlier-prone. The normal plot clearly shows the **S**-shape that is characteristic of outlier-prone data. □

SECTION 8.6 AND 8.7 EXERCISES

8.28 Refer to the data of Exercise 8.24.

a. Plot a histogram or stem-and-leaf display of the data.

b. Is there any obvious reason to doubt the approximate correctness of the 95% confidence level?

8.29 Using the data of Exercise 8.24, find a 95% confidence interval for the true median price.

8.30 For the data of Exercise 8.22, calculate a 90% confidence interval for the median age.

8.31 A clothing manufacturer tested a computerized system for cutting bolts of cloth. The main concern was the amount of wasted cloth; a skilled (and expensive) cutter averaged 11.3% waste. A random sample of 18 patterns is chosen and the waste percentage of the computized system is found. The data are as follows.

10.6	11.3	11.6	11.6	11.8	12.0	12.1	12.2	12.4
12.6	12.8	13.0	13.2	13.3	13.6	13.9	14.3	14.9

(mean 12.622, standard deviation 1.113)

a. Calculate a 95% confidence interval for the mean.

b. Calculate a 95% confidence interval for the median.

8.32 Consider the data of Exercise 8.31.

a. Construct a stem-and-leaf display of the data. Is there clear evidence of nonnormality?

b. What does your answer to part (a) indicate about the relative efficiency of the sample mean versus the sample median? Is your answer confirmed by the relative widths of the two confidence intervals found in Exercise 8.31?

8.33 If a computer program is available to you, construct a normal plot of the data of Exercise 8.31. Does the plot show obvious nonnormality?

SUMMARY

In this chapter we have presented important concepts of interval estimation for a population mean, a population proportion, and a population median. Construction of confidence intervals is based on selection of a good point estimator, using the ideas in Chapter 7. In particular, an unbiased, efficient estimator usually leads to a good (narrow, precise) confidence interval. An understanding of the general population shape, which can often be obtained by plotting the sample data, is important in selecting a good confidence interval method. Once an estimator has been chosen, the sampling distribution of the estimator can be used to obtain a confidence interval. If the sampling distribution is approximately normal, a typical confidence interval is of the form

$$\text{estimator} \pm (\text{table value})(\text{standard error})$$

where the table value comes from the z table if the population standard deviation is known or from the t table if it is not. Some intervals, such as that for the median, have a different form. Deciding on the desired width of a confidence interval can be a useful way of determining how large a sample is needed.

KEY FORMULAS: Interval Estimation

1. $100(1-\alpha)\%$ confidence interval for μ, with σ known

$$\bar{y} \pm z_{\alpha/2}\sigma_{\bar{Y}}, \qquad \text{where } \sigma_{Y} = \frac{\sigma}{\sqrt{n}}$$

Note that $z_{\alpha/2}$ is the normal table value cutting off a right-tail area equal to $\alpha/2$.

2. $100(1-\alpha)\%$ confidence interval for π

$$\hat{\pi} \pm z_{\alpha/2}\sqrt{\frac{\hat{\pi}(1-\hat{\pi})}{n}}$$

3. Sample size required to obtain a $100(1-\alpha)\%$ confidence interval with width $2E$ for μ

$$n = \frac{z_{\alpha/2}^2\sigma^2}{E^2}$$

4. Sample size required to obtain a $100(1-\alpha)\%$ confidence interval with width $2E$ for π

$$n = \frac{z_{\alpha/2}^2\hat{\pi}(1-\hat{\pi})}{E^2}$$

where $\hat{\pi}$ may be estimated from previous information or, conservatively, taken as .50

5. $100(1-\alpha)\%$ confidence interval for μ

$$\bar{y} \pm t_{\alpha/2}\frac{s}{\sqrt{n}}$$

where $t_{\alpha/2}$ is based on d.f. $= n - 1$

6. $100(1-\alpha)\%$ confidence interval for median

$(k+1)$th-order statistic $\le$ median $\le$ $(n-k)$th-order statistic

where

$$P(\text{number of successes} \le k) \le \frac{\alpha}{2}$$

and

$$P(\text{number of successes} \le k+1) > \frac{\alpha}{2}$$

in Appendix Table 1, $\pi = .50$. For large n, $k \approx .5n - .5z_{\alpha/2}\sqrt{n}$.

CHAPTER 8 EXERCISES

8.34 A random sample of the year-end statements of 22 small businesses (under $500,000 in yearly sales) in a city shows that the sample mean of gross margin on sales is 5.2% and the standard deviation is 3.3%. Use these results to calculate a 90% confidence interval for the population mean, where the population is (the gross margin of) the several thousand small businesses in the city.

8.35 Refer to Exercise 8.34. Obviously the gross margin of a functioning business can't be negative. The Empirical Rule for two standard deviations would indicate that a substantial fraction of the businesses have negative gross margins.

a. Is it likely that the sample data would appear nearly normal?
b. What does your answer to part (a) indicate about the confidence interval calculated in Exercise 8.34?

8.36 A research project for an insurance company wishes to investigate the mean value of the personal property held by urban apartment renters. A previous study suggested that the population standard deviation should be roughly $10,000. A 95% confidence interval with a width of $1000 (a plus or minus of $500) is desired. How large a sample must be taken to obtain such a confidence interval?

8.37 It could be argued that the data of Exercise 8.36 would be quite skewed, with a few individuals having very large personal-property values. Therefore, the argument goes, the confidence interval would be completely invalid. Is the argument correct?

8.38 Many individuals over the age of 40 develop an intolerance for milk and milk-based products. A dairy has developed a line of lactose-free products that are more tolerable to such individuals. To assess the potential market for these products, the dairy commissions a market research study of individuals over 40 in its sales area. A random sample of 250 individuals shows that 86 of them suffer from milk intolerance. Calculate a 90% confidence interval for the population proportion that suffers milk intolerance based on the sample results.

8.39 A follow-up study to the survey of Exercise 8.38 is planned. A 90% confidence interval is to be constructed. What sample size is needed to estimate the population proportion with an error of no more than .02, under the following conditions?

a. Assume that the sample proportion is approximately the same as that found in Exercise 8.38.
b. Now assume that the population proportion may be anything.

8.40 Shortly before April 15 of the previous year, a team of sociologists conduct a survey to study their theory that tax cheaters tend to allay their guilt by holding certain beliefs. A total of 500 adults are interviewed and asked under what situations they think cheating on an income tax return is justified. The responses include these:

56% agree that "other people don't report all their income."
50% agree that "the government is often careless with tax dollars."
46% agree that "cheating can be overlooked if one is generally law-abiding."

Assuming that the data are a simple random sample of the population of taxpayers (or tax-nonpayers), calculate 95% confidence intervals for the population proportion that agrees with each statement.

8.41 An editorial writer, commenting on the study in Exercise 8.40, claims that the opinion of 500 individuals out of the total number of taxpayers in the United States is virtually worthless; these might be the "cheatingest" 500 people in the entire country. Criticize this editorial stand.

8.42 The caffeine content (in milligrams) of a random sample of 50 cups of black coffee dispensed by a new machine is measured. The mean and standard deviation are 100 milligrams and 7.1 milligrams, respectively. Construct a 98% confidence interval for the true (population) mean caffeine content per cup dispensed by the machine.

8.43 The machine in Exercise 8.42 is capable of dispensing 3000 cups per day. The caffeine content varies because of variation in caffeine content of the ground coffee beans and because of variation in brewing time.

a. Is the study in Exercise 8.43 questionable because such a small fraction of the machine's output is analyzed?

b. The 50 cups sampled are taken consecutively from the machine. Does this make the study questionable?

8.44 A random sample of the year-end financial statements of a sample of 22 small (under $500,000 in sales) retail businesses in a city show that the average net margin on sales is .0210 and the standard deviation is .0114. Find a 90% confidence interval for the mean net margin for all small retail businesses in the city.

8.45 Tax records indicate that there were 9783 small retail businesses in Exercise 8.44. The sample was taken without replacement. Is it crucial to correct the confidence-interval computations for the without-replacement sampling?

8.46 The data in Exercise 8.44 indicate that the distribution of net margins has a peak at about .015, with some businesses having much larger margins but none having lower margins. What does this fact indicate about the claimed 90% confidence interval?

8.47 A random sample of 100 scores was obtained. The data and Minitab output are shown here.

```
40  42  45  47  48  48  49  49  50  51  51  52  53
54  55  55  55  55  56  56  56  56  56  56  57  57
57  57  58  58  58  58  58  59  59  59  59  59  59
60  60  60  60  60  60  60  60  60  61  61  61  61
61  62  62  62  62  63  63  63  64  64  64  65  65
65  65  65  66  66  66  66  66  67  67  67  67  67
67  67  68  68  68  68  69  69  69  69  70  70  72
72  72  73  73  74  76  79  81  81

MTB > describe 'values'

                N     MEAN   MEDIAN   TRMEAN    STDEV   SEMEAN
VALUES        100   61.460   61.000   61.478    7.845    0.784

              MIN      MAX       Q1       Q3
VALUES     40.000   81.000   57.000   67.000

MTB > tinterval with 95% confidence for 'values'

             N      MEAN     STDEV  SE MEAN   95.0 PERCENT C.I.
VALUES     100    61.460     7.845    0.784  ( 59.903,  63.017)
```

a. Draw a stem-and-leaf display or a histogram. What is the general shape of the data?

b. Should the confidence interval for the mean be wider or narrower than a 95% confidence interval for the median?

c. Calculate a 95% confidence interval for the median. Does the result agree with your answer to part (b)?

8.48 A normal plot of the data in Exercise 8.47, done using Minitab, is shown here. Does the plot confirm your judgment made in part (a) of Exercise 8.47?

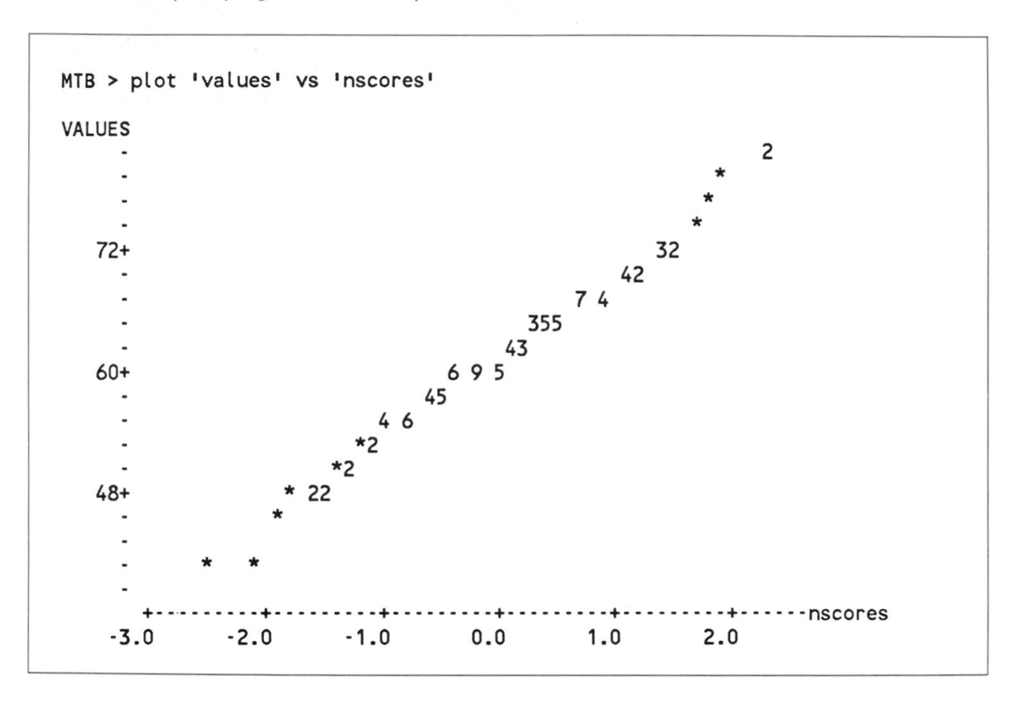

APPENDIX: THE CHI-SQUARE AND *t* DISTRIBUTIONS

In this appendix we present a brief sketch of the mathematics underlying the *t* distribution and the related chi-square (χ^2) distribution. This is the basis for the methods of the preceding chapter and of the next chapters. It also should help in clarifying the concept of degrees of freedom.

The formal mathematical assumption is that there are n random variables $Y_1, Y_2, \ldots, Y_n$, each having a normal distribution with mean μ and standard deviation σ; in particular, the assumptions hold for a random sample from a normal population. The corresponding z scores are $Z_1 = (Y_1 - \mu)/\sigma, \ldots, Z_n = (Y_n - \mu)/\sigma$.

First we define the χ^2 distribution with n degrees of freedom (d.f.).

χ^2 Distribution

The χ^2 distribution with n d.f. is the distribution of

$$W = Z_1^2 + \ldots + Z_n^2$$

$$= \frac{\sum_i (Y_i - \mu)^2}{\sigma^2}$$

Two properties of the χ^2 distribution follow directly from the definition:

properties of χ^2 distribution

1. If $\bar{Y} = \sum_i Y_i/n$, then $n(\bar{Y} - \mu)^2/\sigma^2$ has a χ^2 distribution with 1 d.f.
2. If W_1 and W_2 are independent random variables having χ^2 distribution with n_1 and n_2 d.f., respectively, then $W_1 + W_2$ has a χ^2 distribution with $n_1 + n_2$ d.f.

The first property follows because $(\bar{Y} - \mu)/(\sigma/\sqrt{n})$ is a z statistic, as shown in Chapter 6. Thus $[(\bar{Y} - \mu)/(\sigma/\sqrt{n})]^2 = n(\bar{Y} - \mu)^2/\sigma^2$ is a Z^2 statistic and can be regarded as a sum with only one term. The definition of a χ^2 distribution implies that $n(\bar{Y} - \mu)^2/\sigma^2$ has a χ^2 distribution with 1 d.f.

The second property follows because both W_1 and W_2 are sums of Z^2 terms, so that $W_1 + W_2$, as a sum of sums, is itself a sum of $(n_1 + n_2)$ Z^2 terms. The assumption of independence of W_1 and W_2 can be shown to allow the assumption that all the Z^2 terms are independent, so that $W_1 + W_2$ has the defining properties of a χ^2 distribution with $n_1 + n_2$ d.f.

These results allow us to relate the χ^2 distribution to the sampling distribution of s^2, the sample variance.

> **Sampling Distribution of s^2**
>
> For a random sample of size n from a normal population, the statistic
>
> $$(n - 1)\frac{s^2}{\sigma^2}$$
>
> has a χ^2 sampling distribution with $n - 1$ d.f.

Recall that $s^2 = \sum (Y_i - \bar{Y})^2/(n - 1)$, so that $(n - 1)(s^2/\sigma^2) = \sum (Y_i - \bar{Y})^2/\sigma^2$. The proof of the sampling distribution is based on the ideas already developed in this appendix. While we cannot provide a rigorous proof within the scope of this text, we can indicate the basic idea.

The key to the proof is the identity

$$\frac{\sum (Y_i - \mu)^2}{\sigma^2} = \frac{\sum (Y_i - \bar{Y})^2}{\sigma^2} + \frac{n(\bar{Y} - \mu)^2}{\sigma^2}$$

which can be proved by writing $\sum (Y_i - \mu)^2$ as $\sum [(Y_i - \bar{Y}) + (\bar{Y} - \mu)]^2$, expanding the square, and noting that $\sum_{i=1}^{n} (\bar{Y} - \mu)(Y_i - \bar{Y}) = (\bar{Y} - \mu) \sum_{i=1}^{n} (Y_i - \bar{Y}) = 0$.

The quantity $\sum (Y - \mu)^2/\sigma^2$ has, by definition, a χ^2 distribution with n d.f. As we have noted, the $n(\bar{Y} - \mu)^2/\sigma^2$ term has a χ^2 distribution with 1 d.f. Because independent χ^2 statistics add to form χ^2 statistics with more degrees of freedom, it is plausible that the other term $\sum (Y - \bar{Y})^2/\sigma^2$ also has a χ^2 distribution. Further, since degrees of freedom are additive, this term should have $n - 1$ d.f. The missing step of the argument is the proof that the two terms on the right side of the equation are independent; this fact is proved only in advanced texts.

These preliminary results on χ^2 distributions are used in developing the t distribution. We first define the t distribution with v d.f.:

t Distribution with v d.f.

The t distribution with v d.f. is the distribution of

$$t = \frac{Z}{\sqrt{W/v}}$$

where Z has a standard normal distribution ($\mu = 0$, $\sigma = 1$) and W a χ^2 distribution with v d.f. Z and W are required to be independent.

We take Z to be $(\bar{Y} - \mu)/(\sigma/\sqrt{n})$ and W to be $(n - 1)(s^2/\sigma^2)$, which was just shown to have a χ^2 distribution with $v = n - 1$ d.f. (It can be shown that Z and W are independent.) By definition,

$$t = \frac{\dfrac{(\bar{Y} - \mu)}{\sigma/\sqrt{n}}}{\sqrt{(n - 1)\dfrac{s^2}{\sigma^2}\Big/(n - 1)}} = \frac{\bar{Y} - \mu}{s/\sqrt{n}}$$

which is the t statistic used in this chapter.

9

HYPOTHESIS TESTING

The ideas of point and interval estimation of Chapters 7 and 8 are used to find the best single guess and a reasonable range for the value of a population parameter. In this chapter we introduce hypothesis-testing methods, which can be used to assess whether sample data are reasonably compatible with a specific, assumed value of a parameter. We begin with some important concepts and definitions, then turn to the application of these methods in single-sample situations. Then we discuss the concepts of *p*-value and statistical significance, and how these ideas can be misinterpreted. Finally, we relate hypothesis testing and confidence intervals.

This is a key chapter; many concepts introduced here are used repeatedly in the rest of the book.

9.1 A TEST FOR A BINOMIAL PROPORTION

Preliminary market research for a new product often involves having a sample of consumers compare the product to a competitor or to an older version of the product. Suppose that 100 customers indicate a preference either for the new product or for an older version. What results from the sample would conclusively indicate that the new product was superior?

This problem can most easily be handled using hypothesis testing ideas. There are four key concepts: the null hypothesis, the research hypothesis, the test statistic, and the rejection region.

research hypothesis

The market researchers have a **research hypothesis**—that the new product is superior to the old one. Formally, a research hypothesis, denoted H_a, is a statement about a population parameter. In the example, the relevant parameter is the proportion π of customers in the entire population (not just the sample) who prefer the new product. To say that the new product is superior is to say that more than half of all customers prefer the new product; that is, H_a: $\pi > .50$.

null hypothesis

The **null hypothesis**, denoted H_0, is the denial of the research hypothesis H_a. As the name suggests, the null hypothesis often has a negative quality. In the market research example, if $\pi \leq .50$, the new product is *not* preferred to the older version. We call H_0: $\pi \leq .50$ the null hypothesis because it negates or denies our research hypothesis. Later, we see that the boundary value between H_0 and H_a ($\pi = .50$ in the market research example) is the crucial value of π for the test.

one-sided and two-sided hypotheses

The research hypothesis may be either **one-sided** (directional) or **two-sided** (nondirectional). In the example, we specify a particular direction for H_a relative to H_0. H_a: $\pi > .50$ is a one-sided hypothesis. In contrast, had we specified H_a: $\pi \neq .50$, we would have had a two-sided, nondirectional research hypothesis. The purpose of the study determines the choice of one-sided or two-sided research hypotheses. In comparing a new product to an old one, the intent is to see if the new product is better, so a one-sided research hypothesis is used. If we are comparing two versions of a new product, we want to test whether either version is clearly superior to the other. If π is the population proportion favoring version 1, we want to know if either $\pi > .50$ or $\pi < .50$; thus, we use the two-sided H_a: $\pi \neq .50$.

EXAMPLE 9.1 A very large supermarket chain supplies freshly baked bread. Inventory theory calculations indicate that, to balance the costs of stale bread and of lost sales and customer goodwill, the chain should run out of bread by closing 20% of the time. A random sample of 50 stores is to be chosen and the stockout rate determined in these stores on a particular date. Formulate a "no problem" null hypothesis. Should the research hypothesis be taken as one-sided or two-sided?

Solution The null hypothesis refers to π, the proportion of all stores in the chain (not just the sampled ones) that are out of stock on the date. The desired value for π is .20, so the "no problem" null hypothesis is H_0: $\pi = .20$. In this problem, we should be concerned with stockout rates that are either too low (resulting in too much stale bread) or too high (resulting in lost sales and lost customer goodwill). Therefore, we should take the two-sided H_a: $\pi \neq .20$. □

The basic strategy in hypothesis testing is to attempt to support the research hypothesis by "contradicting" the null hypothesis. The null hypothesis is "contradicted" if the sample data are highly unlikely given H_0 and more likely given H_a. Thus, to support H_a: $\pi > .50$, we would need to find that the sample results are highly improbable assuming that H_0: $\pi \leq .50$ is true.

test statistic

The data must be summarized in a **test statistic** (T.S.). This statistic is calculated to see if it is reasonably compatible with the null hypothesis. When we are testing a proportion, the test statistic is very simple; we count the number of successes in the sample to find T.S.: Y = number of successes. In the example of comparing new and old products, given the assumption that the old product is at least as good, it is very unlikely for Y, the number of customers in the sample who prefer the new product, to be very large. Thus, if Y comes out very large, we reject the null hypothesis and support the research hypothesis that the new product is better. To repeat, the basic logic is as follows:

1. Assume that H_0: $\pi \leq .50$ is true;
2. Calculate the value of T.S.: Y = number of customers in the sample who prefer the new product;
3. If this value is highly unlikely (which, in this case, means very large), reject H_0 and support H_a.

EXAMPLE 9.2 In Example 9.1, π represented the proportion of all stores running out of bread on a particular date. If the null hypothesis is H_0: $\pi = .20$ and the research hypothesis H_a is two-sided, what is an appropriate test statistic? What sort of values of this test statistic would contradict the null hypothesis and therefore support the research hypothesis?

Solution The natural test statistic is T.S.: Y = number of stores in the sample of 50 that are out of bread on that date. Assuming that H_0: $\pi = .20$ is true, it is likely that Y is close to $50(.20) = 10$. Values of Y far below 10 or far above 10 are unlikely given H_0 and tend to contradict H_0. □

It's necessary in hypothesis testing to draw the line between values of the test statistic that are relatively likely given the null hypothesis and values that are relatively unlikely. At what value of the test statistic do we start to say that the data support the research hypothesis? Knowledge of the sampling distribution of the test statistic is used to answer this question. Values of the test statistic that are sufficiently unlikely given the null hypothesis (as determined by the sampling distribution) form a **rejection region** (R.R.) for the statistical test.

rejection region

Specification of a rejection region must recognize the possibility of error. Suppose that, for a sample of 100 customers, we set the rejection region at $y = 59$ or more customers preferring the new product. Even if the null hypothesis H_0: $\pi \leq .50$ is true, there is a small probability of observing $y \geq 59$. If such a situation were to occur, the market researchers would erroneously think that the new product was superior to the old one. This error—rejecting a null hypothesis that is, in fact, true—is called a **Type I error**. In establishing a rejection region, an investigator must specify the maximum tolerable probability of a Type I error; this maximum probability is denoted by α.

Type I error

The α probability in a test of a proportion can be computed simply by adding binomial probabilities from Appendix Table 1, because the test statistic Y satisfies all the assumptions for a binomial random variable. For the example of preference for a new product over an old one, if we have $n = 100$, H_0: $\pi \leq .50$, and R.R.: $y \geq 59$, then the α risk is*

$$\alpha = \max_{\pi} P(Y \geq 59 \mid \pi \leq .50)$$

* Technically, the conditional probability notation $P(Y \geq 59 \mid \pi \leq .50)$ isn't correct, because $\pi \leq .50$ is not a random event in a sample space. The notation is very convenient, though, so we continue to use it. To be technically correct, read the "|" symbol as *assuming* rather than as *given*.

In principle, to find α we must calculate $P(Y \geq 59)$ for every value of $\pi \leq .50$. Looking in Appendix Table 1 with $n = 100$ and $\pi = .50$, we find

$$P(Y \geq 59 | \pi = .50) = .0159 + .0108 + \ldots = .0444$$

Looking in the $\pi = .45$ column, we find

$$P(Y \geq 59 | \pi = .45) = .0016 + .0009 + \ldots = .0034$$

Similarly,

$$P(Y \geq 59 | \pi = .40) = .0001 + .0000 + \ldots = .0001$$

Note that the highest value of $P(Y \geq 59)$ occurs at the boundary value of H_0: $\pi \leq .50$, namely, at $\pi = .50$. Thus, if the rejection region is R.R.: reject H_0 if $Y \geq 59$, then $\alpha = .0444$.

Usually, α, the maximum allowable value for the probability of Type I error, is specified and then a suitable rejection region found. For example, in the product-comparison example, suppose that we specify $\alpha = .10$. We know that the important value for π is .50; adding binomial probabilities with $\pi = .50$, we find that

$$P(Y \geq 56 | \pi = .50) = .1358 \quad \text{and} \quad P(Y \geq 57 | \pi = .50) = .0968.$$

Therefore, a suitable rejection region is $Y \geq 57$.

For most studies, α is specified to be .10, .05, or .01, although the value chosen is somewhat arbitrary. We say more about choosing α in Section 9.11.

EXAMPLE 9.3 Find a rejection region corresponding to $\alpha = .10$ in the bread-inventory problem of Examples 9.1 and 9.2.

Solution We noted in Example 9.2 that the rejection region should include both very large and very small y values. The most natural way to proceed is to locate the two parts of the rejection region at equal distances from the expected value of Y under H_0, namely $50(.20) = 10$. From Appendix Table 1 with $n = 50$ and $\pi = .20$, we find

$$P(Y \geq 16 | \pi = .20) = .0308 \quad \text{and} \quad P(Y \leq 4 | \pi = .20) = .0185$$

and

$$P(Y \geq 15 | \pi = .20) = .0607 \quad \text{and} \quad P(Y \leq 5 | \pi = .20) = .0490$$

We choose the rejection region R.R.: $Y \leq 4$ or $Y \geq 16$, because

$$P(Y \leq 4 \text{ or } Y \geq 16) = .0308 + .0185 = .0493$$

whereas

$$P(Y \leq 5 \text{ or } Y \geq 15) = .0607 + .0490 = .1097$$

which is larger than the allowable α (.10). □

The last step in hypothesis testing is to obtain the data and come to a conclusion. For example, suppose that 66 of the 100 customers sampled in the product-comparison example indicate a preference for the new product. We have established a rejection region (for $\alpha = .10$) as R.R.: $Y \geq 59$. The value $y = 66$ falls well within that region. Thus we have contradicted H_0, and the data support the research hypothesis H_a: $\pi > .50$.

EXAMPLE 9.4 Does finding that 14 of the 50 stores were out of stock in the bread-inventory problem of Examples 9.1–9.3 support the research hypothesis?

Solution We found the R.R.: $Y \leq 4$ or $Y \geq 16$ in Example 9.3. The value $y = 14$ is not in this region. Therefore the data do not support the research hypothesis. □

The process of performing a hypothesis test can be summarized conveniently in a five-step list, which we illustrate for the product-comparison example.

Five Steps of a Statistical Test

1. null hypothesis H_0: $\pi \leq .50$
2. research hypothesis H_a: $\pi > .50$
3. test statistic T.S.: $Y =$ number of customers preferring the new product
4. rejection region R.R.: for $\alpha = .10$, $Y \geq 57$
5. conclusion: Because $y = 66$, reject H_0 and support H_a.

This list incorporates the basic strategy of hypothesis testing. We need to formulate a null hypothesis, choose either a one-sided or a two-sided research hypothesis, and select an appropriate test statistic. Then selection of a tolerable α probability allows us to specify a rejection region—those potential values of the test statistic that we will declare to contradict the null hypothesis. Finally, obtaining the actual data allows us to reach a conclusion.

SECTION 9.1 EXERCISES

9.1 Suppose that the prevailing opinion among stock market analysts is that only 35% of all rumored takeover bids actually result in a takeover. One group of analysts believes that even this figure is too high. To test this belief, the group plans to track the next 20 rumored takeover bids to see how many actually result in takeovers.

a. Define the relevant parameter for a statistical hypothesis test.
b. Formulate "the prevailing opinion is correct" as a null hypothesis.
c. Formulate a research hypothesis for the group that believes that the prevailing opinion is wrong.
d. Suppose that a rejection region is established to reject the null hypothesis if 3 or fewer of the 20 rumored bids actually result in takeovers. What is the corresponding α probability?

9.2 Two of the rumored bids in Exercise 9.1 result in takeovers. Is there sufficient evidence in the data to support the research hypothesis?

9.3 A large chain of realtors offers a guaranteed-purchase plan. Houses that have been listed but not sold for six weeks are bought by the realtor at a predetermined price. Over time, 5% of the houses that the realtor lists are purchased under this plan. Because houses with swimming pools are sometimes hard to sell, it is suspected that a larger fraction of such houses must be bought by the realtor under the plan. The chain has listed 50 houses with pools under the plan, and it can determine how many were bought under the plan.

a. What is the relevant population parameter for this problem?
b. Formulate the appropriate research hypothesis. Should it be one-sided or two-sided?
c. State the null hypothesis.
d. Use binomial tables to determine a rejection region corresponding to a tolerable $\alpha = .05$.

9.4 Suppose that 7 of 50 houses in Exercise 9.3 are purchased under the plan. Does this evidence support the research hypothesis? Can H_0 be rejected at $\alpha = .05$?

9.5 The long-time city manager of a small city had been awarded grants on 50% of the applications for aid submitted to the federal government. A new city manager is appointed and submits 18 applications to the federal government in the first year. The city council wanted to test whether there is a change in the success rate under the new manager.
a. Define the appropriate population parameter for a statistical test.
b. Formulate an appropriate null hypothesis.
c. Should the research hypothesis be one-sided or two-sided?
d. Find a rejection region corresponding to $\alpha = .05$, using the test statistic Y = number of successful applications out of the 18 submitted.

9.6 Suppose that 7 of the 18 applications in Exercise 9.5 result in grants. Can H_0 be rejected at $\alpha = .05$?

9.7 Binomial probabilities were used in Exercise 9.5. Under what conditions might binomial distribution be a poor assumption?

9.8 The manager for research and development of a food company finds that historically only 40% of the potential new products brought to consumer testing are ever marketed. The manager institutes a revised selection method to determine which products should be brought to consumer testing.
a. Formulate the null hypothesis that the revised selection method will have no effect on the proportion of potential new products that are marketed.
b. What arguments can be made in favor of a two-sided research hypothesis?
c. If a one-sided research hypothesis is to be used, what should it be?

9.9 A two-sided research hypothesis is used in the situation of Exercise 9.8. A sample of 20 potential new products are brought to consumer testing. Define Y = number of these products that are eventually marketed.
a. Assuming that the null hypothesis is true, what is the mean (expected value) of Y?
b. Form a rejection region, symmetric around the mean found in part (a), corresponding to a permissible $\alpha = .05$.

9.10 Binomial tables were used in finding the rejection region for Exercise 9.9. It is noted that several of the products brought to consumer testing are competitive versions of one another, so that if one product is marketed, another is very likely not marketed. Does this fact indicate that binomial probabilities may not be applicable? Explain why.

9.2 TYPE II ERROR, β PROBABILITY, AND POWER OF A TEST

Up to this point, we have been concerned about only one kind of error in hypothesis testing—Type I error, rejecting the null hypothesis when it is true. In the product-comparison example, a Type I error would be a claim that the new product was better than the old one when, in fact, it was not. But there is another possible error; the market researchers might claim that the new product is not superior to the old one, when, in fact, it is superior. This error, a **Type II error**, is the failure to reject the null hypothesis when the research hypothesis is true.

Type II error

When the null hypothesis is negative, as it often is, a Type I error can be called a *false positive* error; by coming to the erroneous conclusion that a positive hypothesis H_a is true, we commit a false positive, Type I, error. Similarly, a Type II error can be called a *false negative* error—an erroneous conclusion that a negative hypothesis H_0 is true.

EXAMPLE 9.5 In the bread-inventory problem discussed in Examples 9.1–9.4, what are the consequences of Types I and II errors?

Solution H_0: $\pi = .20$ states that the supermarket chain does not have a problem, and H_a: $\pi \neq .20$ states that there is a problem. A Type I error is therefore the incorrect assertion that there is a problem with bread inventory; in effect, a Type I error is a false alarm. A Type II error is the incorrect assertion that bread inventory is under control; in effect, a Type II error is an erroneous failure to sound an alarm. □

power The probability that a Type II error will be committed, given that the research hypothesis is true, is denoted by β. The quantity $1 - \beta$ is called the **power** of the test; power is the probability that the test will support the research hypothesis, given that it is in fact true. The possible outcomes of a statistical test and the associated probabilities are summarized in Table 9.1.

EXAMPLE 9.6 Refer to Example 9.5. Under certain conditions, the power of the test is .60. What does this mean?

Solution Power refers to the probability that a research hypothesis will be correctly supported. The sentence thus means that, if the research hypothesis is true (i.e., the chain has an inventory problem), there is a 60% chance that the test will discover the existence of that problem. □

In a binomial test for a proportion, β may be calculated by adding binomial probabilities. The conceptual problem is to specify the value to use for π. The research hypothesis in the product-comparison example asserts only that $\pi > .50$. The probability that we will not reject the null hypothesis depends strongly on whether the research hypothesis is "extremely true," such as $\pi = .90$, or "barely true," such as $\pi = .51$. Thus β should be regarded as a *function* of the true value of the population parameter and should be computed for several different values. For example, if we hypothesize that $\pi = .55$, we may find β by adding binomial probabilities for the nonrejection region. (Remember that β is the probability that the null hypothesis is not rejected, given that the research hypothesis is true.) The rejection region for the product-comparison example is $Y \geq 57$; to find β, we must add the probabilities of all values of $y \leq 56$. Reading up the $\pi = .55$

TABLE 9.1 Possible Outcomes and Probabilities for a Hypothesis Test

	Condition	
Conclusion	H_0 is true	H_a is true
Accept H_0	Correct conclusion probability $1 - \alpha$	Type II error probability β
Reject H_0	Type I error probability α	Correct conclusion probability $1 - \beta$ (=power)

column in the $n = 100$ block of Appendix Table 1, we find

$$\beta_{.55} = .0071 + .0108 + .0157 + \ldots = .6172$$

Similar calculations using other columns of Appendix Table 1 determine the values of β shown below:

Value of π in H_a:	.55	.60	.65	.70	.75
β_π:	.6172	.2368	.0389	.0020	.0000

EXAMPLE 9.7 The rejection region corresponding to $\alpha = .10$ in Example 9.3 was found to be $Y \leq 4$ or $Y \geq 16$. Find β and power when $\pi = .30$ and when $\pi = .35$.

Solution To find β, we must add probabilities for all values of y not in the rejection region, namely $5 \leq y \leq 15$. Power is simply $1 - \beta$. Adding probabilities for $y = 5, 6, \ldots, 15$ in the $\pi = .30$ and $\pi = .35$ columns of Appendix Table 1 with $n = 50$, we obtain the following values:

π:	.30	.35
β:	.569	.280
Power:	.431	.720

Note that β decreases (and therefore power increases) as the value of π gets farther from the H_0 value, $\pi = .20$. □

The value of β is influenced by a number of factors.

1. All else equal, if α increases, β decreases. Increasing α makes it easier to reject H_0, thus decreasing the probability that we will *not* reject H_0. Note, however, that it is not true that $\alpha + \beta = 1$. Both α and β are conditional probabilities, defined for different conditions; their sum means little, if anything.
2. All else equal, if n increases, β decreases. With more information, we have a lower risk of error.
3. All else equal, if the hypothetical value of the population parameter moves away from H_0, β decreases. It is easier to detect a large deviation from H_0 than a small one.

SECTION 9.2 EXERCISES

9.11 In Exercise 9.1 we had H_0: $\pi = .35$, H_a: $\pi < .35$, $n = 20$, R.R.: $Y \leq 3$, and an actual α of .0445.
  a. Suppose that, in fact, $\pi = .25$. What is the probability that the null hypothesis will not be rejected? What is the technical name for this probability?
  b. If $\pi = .25$, what is the probability that H_0 will be rejected? What is the technical name for this probability?

9.12 How should the probability found in Exercise 9.11, part (a), change if $\pi = .20$ rather than .25? Base your answer on general principles rather than calculations.

9.13 Assume in Exercise 9.3 that 10% of all houses with swimming pools are, in the long run, bought by the realtor. Find the probability that the null hypothesis will not be rejected. Is this an α or a β probability?

9.14 One kind of error that can be made in the situation of Exercise 9.3 is a claim that houses with swimming pools are more likely to be bought under the plan than are other houses when in fact houses with pools have the same rate as other houses. According to the formulation of Exercise 9.3, is this error a Type I error or a Type II error?

9.15 Suppose that the sample in Exercise 9.3 is enlarged to 100 houses. The value of α remains .05. Assume, as in Exercise 9.13, that 10% of all houses with swimming pools are, in the long run, bought by the realtor.

a. Should the probability that the null hypothesis will not be rejected be larger or smaller than the probability found in Exercise 9.13? Base your answer on general principles rather than calculations.

b. The rejection region with $n = 100$ and $\alpha = .05$ becomes $Y \geq 10$. Calculate the probability that H_0 will not be rejected. Does the result confirm your answer in part (a)?

9.16 Explain what would constitute Type I and Type II errors in Exercise 9.8.

9.17 Find the probability that H_0 in Exercise 9.8 will not be rejected for the following values of π: .45, .50, .55, .65, .75, .80. Sketch a curve of the β probabilities.

9.18 We had H_0: $\pi = .40$, H_a: $\pi \neq .40$, $n = 20$, and R.R.: $Y \leq 3$ or $Y \geq 13$ in Exercise 9.9.

a. On the hypothesis that the actual value of π is .50, find the power of the test.

b. If n is increased to 100 but no other changes are made in the specifications, should the power increase or decrease relative to the power found in part (a)? You shouldn't need to make any calculations to answer this question.

9.3 A TEST FOR A POPULATION MEAN WITH KNOWN STANDARD DEVIATION

We introduced hypothesis-testing concepts in the context of testing a binomial proportion. There are many other population parameters and many other statistical tests. In this section we illustrate basic testing concepts in the context of a statistical test for a population mean.

As usual, we work by example. Suppose that the local bureau of weights and measures is concerned with the actual weight of boxes of a cereal product marked as 16 ounces. There is some variability in weight from one box to another, mostly due to the shape of the cereal pieces. Past experience has shown the standard deviation of box weights to be .1 ounce. While the bureau doesn't require that every box weigh 16 ounces, it does want to assure the public that the average weight of all such cereal boxes is at least 16 ounces. If the bureau suspects the cereal company of short-weighting (underfilling the boxes), how can it go about testing for such short-weighting?

Since the boxes must be opened to test the weight of the contents, the bureau cannot test every box coming off the assembly line. Instead, a random sample of boxes must be chosen and tested. Suppose the bureau's sample data consist of the actual weights for the contents of 25 randomly chosen boxes. How should the bureau proceed?

We can formulate this problem in terms of a statistical test about the population mean weight μ for all cereal boxes produced. The bureau is concerned with the basic problem of short-weighting; in particular, the bureau is interested in supporting the research hypothesis H_a: $\mu < 16$ ounces. For this research hypothesis, the corresponding null hypothesis is H_0: $\mu \geq 16$ ounces. The primary concern is with the boundary value, as we indicated in Section 9.1. We denote the boundary value of the hypothesized mean by μ_0; here $\mu_0 = 16$.

The most plausible test statistic is the sample mean weight $\bar{y}$ of the 25 boxes. Sample means much less than μ_0 are unlikely under H_0 and relatively more likely if H_a: $\mu < 16$ is true. Therefore the rejection region is "reject H_0 if $\bar{y}$ is smaller than could reasonably occur by chance."

To determine the exact rejection region, we need to know the sampling distribution of $\bar{Y}$. Recall from Chapter 6 that if the population distribution of weights is normal with mean μ and standard deviation σ, then the sampling distribution of the sample mean is also normal with expected value equal to the population mean weight ($\mu_Y = \mu$) and with standard error equal to $\sigma_{\bar{Y}} = \sigma/\sqrt{n}$. Even if the population distribution is mildly nonnormal, the Central Limit Theorem helps to make this distribution a good approximation. For the bureau's problem, $\sigma = .1$, $n = 25$, and the crucial value for μ is the boundary null hypothesis value $\mu_0 = 16$. Thus if the null hypothesis is true, the sample mean $\bar{Y}$ is normally distributed with $\mu_{\bar{Y}} = 16$ and $\sigma_{\bar{Y}} = .1/\sqrt{25} = .02$. We can use this information about the sampling distribution of the test statistic $\bar{Y}$ to locate a rejection region.

The entire rejection region for H_0: $\mu = 16$, H_a: $\mu < 16$ is in the lower tail of the distribution of $\bar{Y}$. In particular, from our knowledge of the properties of a normal distribution we know that the boundary of the rejection region is located at a distance of 1.645 standard errors ($1.645\sigma_{\bar{Y}}$) below $\mu = 16$ if α is taken to be .05 (see Figure 9.1).

To determine whether or not to reject the null hypothesis, we can also compute the number of standard errors the observed value of $\bar{y}$ lies below $\mu = 16$. This is done by computing a ***z* statistic** for the observed sample mean $\bar{y}$ using the formula

z statistic

$$z = \frac{\bar{y} - \mu_0}{\sigma/\sqrt{n}} = \frac{\bar{y} - 16}{.02}$$

This suggests two ways to state the rejection region for a statistical test about μ. First, in terms of the test statistic $\bar{y}$, the rejection region is

rejection region using $\bar{y}$

R.R.: For $\alpha = .05$, reject H_0: $\mu \geq 16$ if the observed value of $\bar{y}$ is more than $1.645\sigma_{\bar{Y}}$ below $\mu = 16$ (see Figure 9.1).

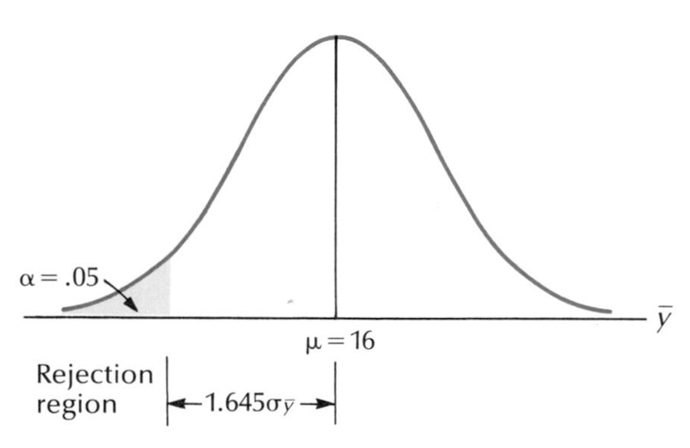

FIGURE 9.1 Rejection Region for the Test Statistic $\bar{y}$ ($\alpha = .05$, one-tailed)

An equivalent way to state the rejection region is in terms of the test statistic $z = (\bar{y} - \mu_0)/\sigma_{\bar{y}}$, also called the z statistic:

rejection region using z

R.R.: For $\alpha = .05$, reject H_0: $\mu = 16$ if the computed value of z is less than or equal to -1.645 (see Figure 9.2).

Because the latter approach is shorter and perhaps simpler, we use it throughout this text.

Finally, suppose that the sample mean weight for a sample of $n = 25$ boxes is 15.83 ounces. What can the bureau conclude concerning the population mean fill? The z statistic

$$z = \frac{15.83 - 16}{.1/\sqrt{25}} = -8.5$$

indicates that the sample mean (15.83) lies 8.5 standard errors below the hypothesized mean $\mu = 16$. Because the computed value of the z statistic (-8.5) lies in the rejection region well beyond the critical value -1.645, the bureau can reject the null hypothesis and claim that the company is short-weighting. A summary list displays the bureau's work.

Summary of One-tailed Test about μ, with σ Known

H_0: $\mu = \mu_0$ ($\mu_0 = 16$ ounces)

H_a: $\mu < \mu_0$

T.S.: $z = \dfrac{\bar{y} - \mu_0}{\sigma_{\bar{y}}}, \qquad \sigma_{\bar{y}} = \dfrac{\sigma}{\sqrt{n}}$

R.R.: For $\alpha = .05$, reject H_0 if $z \leq -1.645$

Conclusion: $z = \dfrac{15.83 - 16}{.1/\sqrt{25}} = -8.5$; reject H_0

Note: For H_0: $\mu = \mu_0$ and H_a: $\mu > \mu_0$, the R.R. for $\alpha = .05$ is $z \geq 1.645$.

For hypothesis tests involving both μ and π, we have noted that the boundary value of the null hypothesis is the important value. In the cereal weight example, suppose that

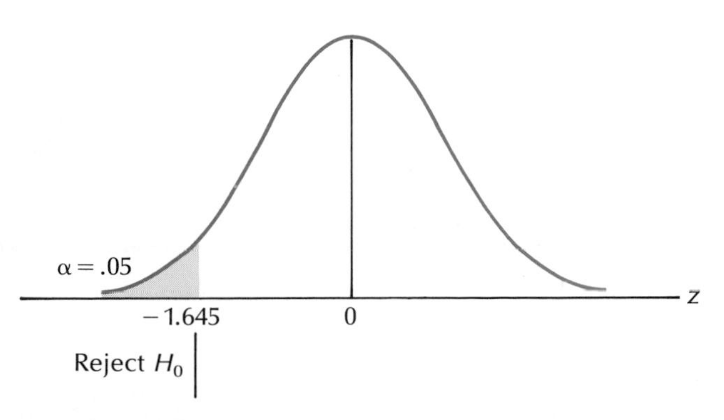

FIGURE 9.2 **Rejection Region for the Test Statistic z ($\alpha = .05$, one-tailed)**

we had taken some other value within H_0 as our mean, such as $\mu = 16.01$. The resulting z statistic would fall even farther into the rejection region:

$$z = \frac{15.83 - 16.01}{.1/\sqrt{25}} = -9.0$$

If the test statistic based on the boundary value leads to rejection of H_0, a test statistic based on any other value in H_0 also leads to rejection of H_0. Up to this point, we have used an inequality sign for H_0 in one-sided tests. Hereafter, we only worry about the crucial boundary value and drop the inequality sign.

EXAMPLE 9.8 A researcher has claimed that the amount of time urban preschool children age 3–5 watch television per week has a mean of 22.6 hours and a standard deviation of 6.1 hours. A market research firm believes that the claimed mean is too low. The television-watching habits of a random sample of 60 urban preschool children are measured, with the parents of each child keeping a daily log of television watching. If the mean weekly amount of time spent watching television is 25.2 hours and if the population standard deviation σ is assumed to be 6.1 hours, should the researcher's claim be rejected at an α value of .01?

Solution The marketing firm's research hypothesis is that 22.6 is too small a value for the population mean. Thus the research hypothesis of interest is H_a: $\mu > 22.6$, and the null hypothesis is H_0: $\mu = 22.6$. We summarize the elements of the statistical test for $\alpha = .01$ as follows:

H_0: $\mu = 22.6$

H_a: $\mu > 22.6$

$$\text{T.S.: } z = \frac{\bar{y} - \mu_0}{\sigma_{\bar{y}}} = \frac{25.2 - 22.6}{6.1/\sqrt{60}} = 3.30$$

R.R.: For $\alpha = .01$, reject H_0 if $z \geq 2.326$

Conclusion: Because $z = 3.30$ is well within the rejection region, we reject H_0: $\mu = 22.6$. □

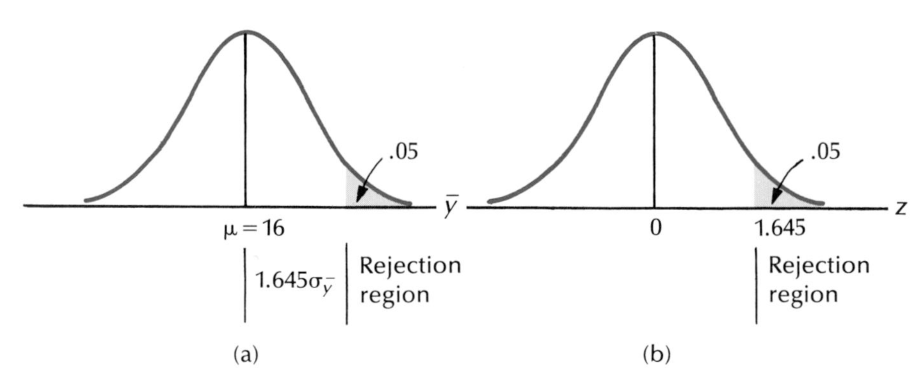

FIGURE 9.3 Rejection Region for H_a: $\mu > 16$, with (a) $\bar{y}$ as Test Statistic; (b) z as Test Statistic

This test procedure for μ can easily be modified to handle other research hypotheses. For example, if the cereal company wants to establish beyond a reasonable doubt that the true mean weight is more than 16 ounces, it could start with the one-sided research hypothesis H_a: $\mu > 16$. Large values of $\bar{y}$ would then indicate rejection of the null hypothesis H_0: $\mu = 16$. In particular for $\alpha = .05$, the rejection region would be values of $\bar{y}$ at least $1.645\sigma_{\bar{Y}}$ *above* $\mu_0 = 16$, or equivalently, values of $z \geq 1.645$ (see Figure 9.3).

A two-tailed test for the research hypothesis H_a: $\mu \neq \mu_0$ follows directly from our discussion of one-tailed tests. For example, the manager of the company who is concerned about possible overfilling or underfilling might well take as a research hypothesis that $\mu \neq 16$. Both large and small values of $\bar{y}$ would indicate rejection of H_0: $\mu = 16$. If we split the rejection region evenly in the tails, the rejection region for $\alpha = .05$ is as shown in Figure 9.4a; the corresponding rejection region based on the z statistic is shown in Figure 9.4b.

The summary chart for the z test can be written to cover all three forms for the research hypothesis. Recall that z_a is the z value that cuts off an area a in the right-hand tail of the z curve; thus $z_{.05} = 1.645$ and $z_{.025} = 1.96$. For a two-tailed test and a given α, the desired cutoff points are $z_{\alpha/2}$ and $-z_{\alpha/2}$. For $\alpha = .05$, we use $z_{.025} = 1.96$ and

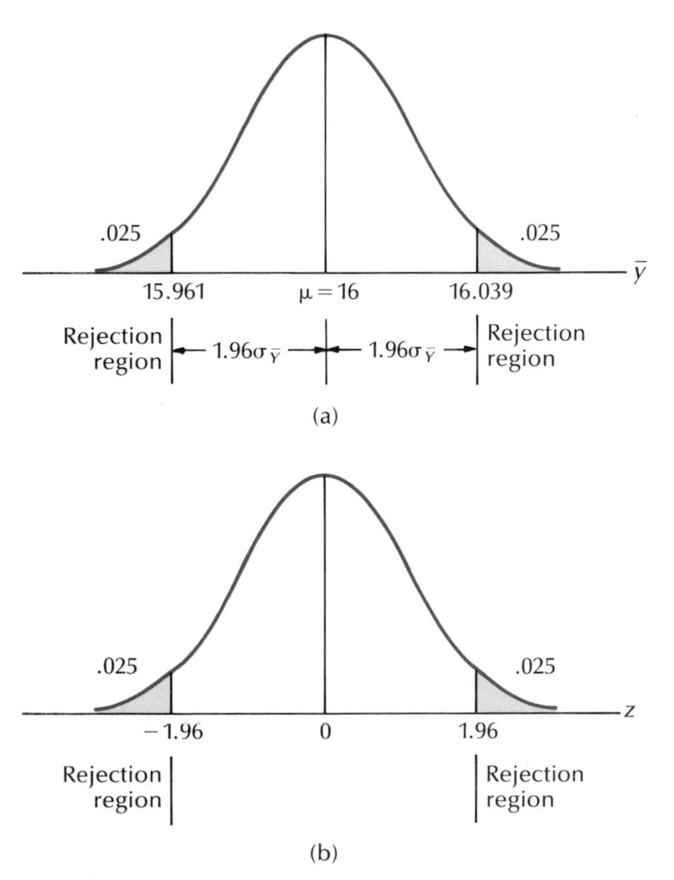

FIGURE 9.4 Rejection Region for $H_0 = 16$, with (a) $\bar{y}$ as Test Statistic; (b) z as Test Statistic

$-z_{.025} = -1.96$. The first four steps of the statistical test for μ (σ known) are shown here. These steps formulate the problem and establish the rejection region; the last step simply involves drawing a conclusion based on the computed value of the z test statistic. If the computed z value falls within the rejection region, we reject the null hypothesis in favor of the research hypothesis. If the z value does not fall within the rejection region, we reserve judgment until β probabilities can be computed for various values of μ.

Summary for z Test, with σ Known

H_0: $\mu = \mu_0$

H_a: 1. $\mu > \mu_0$
2. $\mu < \mu_0$
3. $\mu \neq \mu_0$

T.S.: $z = \dfrac{\bar{y} - \mu_0}{\sigma/\sqrt{n}}$

R.R.: For the probability of a Type I error α, reject H_0 if
1. $z > z_\alpha$
2. $z < -z_\alpha$
3. $z > z_{\alpha/2}$ or $z < -z_{\alpha/2}$

EXAMPLE 9.9 Refer to the television-watching data of Example 9.8. Test the research hypothesis H_a: $\mu \neq 22.6$ using $\alpha = .01$.

Solution The five steps of the solution are summarized here:

H_0: $\mu = 22.6$

H_a: $\mu \neq 22.6$

T.S.: $z = \dfrac{\bar{y} - \mu_0}{\sigma_{\bar{y}}} = 3.30$

R.R.: For $\alpha = .01$, reject H_0 if $z \geq 2.576$ or if $z \leq -2.576$

Conclusion: Because the computed value of z (3.30) falls within the rejection region, we reject H_0: $\mu = 22.6$. Practically speaking, because the sample mean is greater than 22.6 and because we reject H_0: $\mu = 22.6$, we can safely conclude that $\mu > 22.6$. □

The cereal box example is unrealistic in that a population standard deviation is assumed to be known. Population parameters such as the standard deviation usually have unknown values. We managed the problem of unknown standard deviations in Chapter 8 by using the t distribution. Hypothesis-testing methods using the t distribution are discussed in Section 9.6. It is convenient to use z tests as our examples for a bit longer, only to avoid minor complications. Recall that for large samples the difference between t and z tables is negligible. Thus, for samples in the hundreds, it doesn't matter whether we use t or z tables.

Suppose that the bureau of weights and measures takes a sample of $n = 100$ cereal boxes and computes the sample mean to be $\bar{y} = 15.83$ ounces and the **sample** standard deviation to be $s = .12$ ounce. Suppose also that the bureau doesn't want to make any unsupported assumptions about the population standard deviation. A reasonable estimate, based on the data, is that the population standard deviation σ is roughly equal to the sample standard deviation s—that is, roughly equal to .12. It seems reasonable to substitute the value of s, the sample standard deviation, as a best guess for σ in the z statistic. According to a theorem of mathematical statistics, if n is large enough, the resulting z statistic does in fact have a z distribution, at least to a good approximation. The bureau can go ahead and do a z test. The summary follows; this time the bureau is using $\alpha = .01$, just for variety.

Summary for Large-sample Result, with σ Unknown

H_0: $\mu = 16$

H_a: $\mu < 16$

$$\text{T.S.: } z \approx \frac{\bar{y} - \mu_0}{s/\sqrt{n}} = \frac{15.83 - 16}{.12/\sqrt{100}} = -14.167$$

R.R.: For $\alpha = .01$, reject H_0 if $z \leq -2.326$

Conclusion: Because $z = -14.167$ lies within the rejection region, we reject H_0.

You should be able to verify that -2.326 is the appropriate cutoff point corresponding to $\alpha = .01$.

sample-size requirement

The large-sample procedure is easy enough: just substitute the sample standard deviation s for the population standard deviation σ in the z statistic. Further justification for this substitution is given in Section 9.6. For now, assume that this large-sample procedure is a decent approximation if $n \geq 30$ and a good approximation if $n \geq 100$. Because the bureau's test statistic fell so far within the rejection tail, the bureau can reject H_0 without any serious concern.

EXAMPLE 9.10 Suppose that the television-watching sample of Example 9.8 yields a standard deviation of 5.8. Use this value to test the research hypothesis H_a: $\mu > 22.6$, at $\alpha = .01$.

Solution The value of the z statistic with s replacing σ is

$$z = \frac{25.2 - 22.6}{5.8/\sqrt{60}} = 3.47$$

This value falls well within the rejection region $z \geq 2.326$, so H_a is supported. □

9.4 THE β PROBABILITY FOR z TESTS

We introduced the β risk—the probability of not rejecting the null hypothesis when the research hypothesis is true—in the test about a binomial proportion in Section 9.2. The same concepts apply to the z test of this section, but the computation of β is a bit trickier than just adding binomial probabilities. We use the original cereal box example for illustration. Recall that, for H_0: $\mu = 16$, H_a: $\mu < 16$, $\sigma = .1$, $n = 25$, and $\alpha = .05$, the rejection region was $z < -1.645$. To calculate the risk of a Type II error (the probability of incorrectly accepting H_0), we must assume some value for μ under H_a. Once again, the value of β depends on the assumed value of μ in H_a. Specifically, let's assume that $\mu = 15.92$; this corresponds to a short-weighting of .5% of the listed weight. What is the probability that the bureau detects this short-weighting?

The calculation is easier to understand if the rejection region is stated in terms of the sample mean $\bar{y}$ rather than the z statistic. If the rejection region is $z < -1.645$; we reject H_0: $\mu = 16$ for values of $\bar{y}$ at least 1.645 standard errors below $\mu = 16$; that is, we reject H_0 if $\bar{y} \leq 16 - 1.645\,\sigma/\sqrt{n} = 15.9671$. If the true mean is $\mu = 15.92$, the probability β that the sample mean does *not* fall within the rejection region is

$$\beta = P(\bar{Y} > 15.9671 \,|\, \mu = 15.92) = P\left(\frac{\bar{Y} - 15.92}{.1/\sqrt{25}} > \frac{15.9671 - 15.92}{.1/\sqrt{25}}\right)$$

$$= P(z > 2.355) \approx .01$$

The calculation is illustrated in Figure 9.5.

Such calculations can be carried out for any test situation, and they can be summarized in a general formula. If μ_0 is the boundary value of μ under H_0 and μ_a is the selected research hypothesis mean, it can be shown that for a one-tailed test

β for a one-tailed test

$$\beta = P\left(z > -z_\alpha + \frac{|\mu_a - \mu_0|}{\sigma/\sqrt{n}}\right)$$

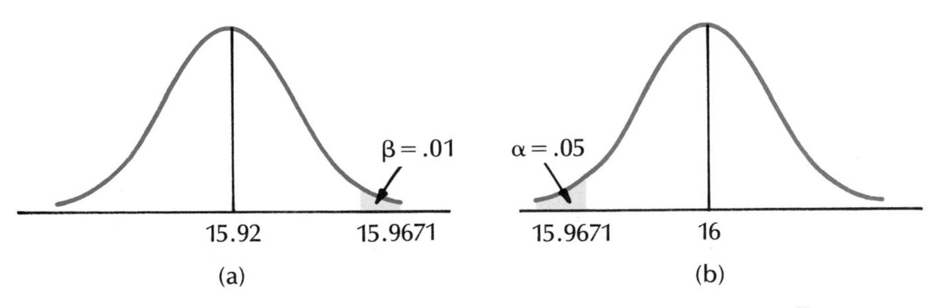

FIGURE 9.5 **Calculation of β for One-tailed z Test: Sample Distribution for $\bar{Y}$ under (a) H_a: $\mu = 15.92$; (b) H_0: $\mu = 16.00$**

In our cereal example, $-z_\alpha = -z_{.05} = -1.645$, while

$$\frac{|\mu_a - \mu_0|}{\sigma/\sqrt{n}} = \frac{|15.92 - 16|}{.1/\sqrt{25}} = \frac{.08}{.02} = 4$$

(Recall that "| |" indicates absolute value, or magnitude of a number without regard to sign.) Hence

$$\beta = P(z > -1.645 + 4) = P(z > 2.355) \approx .01$$

Therefore the bureau has a small probability ($\beta = .01$) of failing to reject H_0 if the degree of short-weighting is .5% of the claimed weight of 16 ounces. In other words, with this test procedure there is a high probability ($1 - \beta = .99$) of detecting a .5% short-weighting if it exists.

A similar calculation can be made for a two-tailed test. The cereal company's production manager had a rejection region (at $\alpha = .05$) of $z \leq -1.96$ or $z \geq 1.96$, or equivalently,

$$\bar{y} \leq \frac{16 - 1.96\,(.1)}{\sqrt{25}} = 15.961 \qquad \text{or} \qquad \bar{y} \geq \frac{16 + 1.96\,(.1)}{\sqrt{25}} = 16.039$$

H_0 is not rejected if $15.961 < \bar{y} < 16.039$. Thus, if the true mean is 15.92, the probability of incorrectly accepting H_0 is

$$\begin{aligned} \beta &= P(15.961 < \bar{Y} < 16.039 \,|\, \mu = 15.92) \\ &= P\left(\frac{15.961 - 15.92}{.1/\sqrt{25}} < z < \frac{16.039 - 15.92}{.1/\sqrt{25}}\right) \\ &= P(2.05 < z < 5.95) \end{aligned}$$

Note that the upper limit 5.95 (which corresponds to the upper limit of $\bar{y} = 16.039$) has practically no effect, since $P(z \geq 5.95)$ is zero to many decimal places. Hence the value of β is $P(2.05 < z < 5.95) \approx P(z > 2.05) = .02$. This calculation is illustrated in Figure 9.6.

Fortunately, a shortcut version of these calculations can be derived. Recall that μ_0 denotes the boundary value of μ under H_0; similarly, let μ_a denote any selected value of

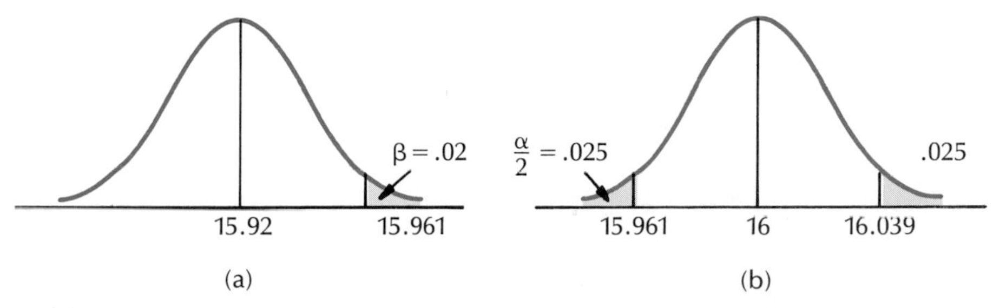

FIGURE 9.6 Calculation of β for Two-tailed z Test: Sample Distribution for $\bar{Y}$ under (a) H_a: $\mu = 15.92$; (b) H_0: $\mu = 16$

μ within the research hypothesis. The shortcut calculation of β is shown in the following box:

Calculation of β for a z Test

One-tailed test:

$$\beta = P\left(z > -z_{\alpha} + \frac{|\mu_a - \mu_0|}{\sigma/\sqrt{n}}\right)$$

Two-tailed test: replace $-z_{\alpha}$ in the one-tailed test by $-z_{\alpha/2}$

EXAMPLE 9.11 Compute β probabilities for the one-tailed tests of Example 9.8. Assume a true mean of 25.0.

Solution Because $\alpha = .01$, $\mu_a = 25.0$, $\mu_0 = 22.6$, $\sigma = 6.1$, and $n = 60$,

$$\beta = P\left(z \geq -2.326 + \frac{|25.0 - 22.6|}{6.1/\sqrt{60}}\right)$$

$$= P(z \geq .72) = .2358$$

□

In Chapter 8 we used confidence intervals to decide how big a sample to take. Alternatively, hypothesis testing can be used to choose a sample size. For example, a chain of inexpensive restaurants featuring steaks makes much of its profit from "add-ons"—additional-cost items ordered by diners. Currently, add-ons average \$4.24 per check, with a standard deviation of \$2.00. An incentive program for waitresses and waiters is being considered, one designed to sell more add-ons (i.e., increase the mean). There is also a concern that the program may backfire and lead to actual reductions. The target is a mean of \$4.50. How many checks should be sampled under a trial run of the incentive scheme?

Because of the concern that the program might backfire, a two-sided hypothesis test should be used. To calculate the required sample size, it is necessary to specify both α and β. Suppose we formulate the null hypothesis as H_0: $\mu = \$4.24$ and specify α to be .05 and β to be .10. We take $\mu_a = \$4.50$ as the target and assume that σ continues to be \$2.00.

We could proceed by trial and error. Suppose that we try $n = 400$. Then

$$\beta = P\left(z > -1.96 + \frac{|4.50 - 4.24|}{2.00/\sqrt{400}}\right)$$

$$= P(z > .64) = .5000 - .2389 = .2611$$

The β probability is higher than desired; a larger sample is needed. We could keep trying other sample sizes and eventually figure out the required n. Fortunately, it's possible to calculate the required n directly.

Sample Size Required for a Specified α and β

One-tailed test:

$$n = \frac{(z_\alpha + z_\beta)^2\sigma^2}{(\mu_a - \mu_0)^2}$$

Two-tailed test: replace z_α in the one-tailed test by $z_{\alpha/2}$

For the restaurant example, we have a two-tailed test with $\alpha = .05$; therefore we use $z_{\alpha/2} = z_{.05/2} = 1.96$ and $z_\beta = z_{.10} = 1.28$:

$$n = \frac{(1.96 + 1.28)^2(2.00)^2}{(4.50 - 4.24)^2} = 621.16$$

To ensure that the β probability is no greater than .10, we must round n *up* to 622.

EXAMPLE 9.12 A light-duty parts manufacturing company sometimes uses temporary workers hired through an agency. These workers have a mean production of 2250 items per day and a standard deviation of 260 items per day. Managers at the company propose to get a sample of temporaries from a different agency, one that claims to use more rigorous standards, to see if productivity increases. They propose to use a one-tailed test with $\alpha = .01$. If the long-run (population) mean productivity of temporaries from the new agency is in fact 2380 (half a standard deviation higher), the managers want the test to have power .95. How many temporaries must be tested?

Solution We assume that the standard deviation for temporaries from the new agency is the same as from the old one, namely $\sigma = 260$. The natural null hypothesis is that the temporaries from the new agency are no better than those from the old agency; H_0: $\mu = \mu_0 = 2250$. We take $\mu_a = 2380$. From Appendix Table 3, $z_{.01} = 2.33$ and $z_{.05} = 1.64$ or 1.65; let's call it 1.645. Thus

$$n = \frac{(2.33 + 1.645)^2(260)^2}{(2380 - 2250)^2} = 63.20$$

Rounding up, we find that the managers need a sample of 64 temporaries. Note that it might be difficult to get a random sample; to get new business, the agency might provide its most productive workers. □

SECTIONS 9.3 AND 9.4 EXERCISES

9.19 The manager of a health maintenance organization has set as a target that the mean waiting time of nonemergency patients not exceed 30 minutes. In spot checks, the manager finds the waiting times for 22 patients; the patients are selected randomly on different days. Assume that the population standard deviation of waiting times is 10 minutes.

a. What is the relevant parameter to be tested?

b. Formulate null and research hypotheses.

c. State the test statistic and the rejection region corresponding to $\alpha = .05$.

9.20 Suppose that the mean waiting time for the 22 patients in Exercise 9.19 is 38.1 minutes. Can H_0 be rejected?

9.21 For the test procedure of Exercise 9.19, find the probability that H_0 will not be rejected, assuming a true mean waiting time of 34 minutes. Do the same for other values of μ, and sketch a β curve.

9.22 It was stated in Exercise 9.19 that the 22 patients were selected on different days. Why would one not want to select 22 patients on one randomly chosen day?

9.23 A radio station wants to control the time allotted to unpaid public-service commercials. If there are too many such commercials, the station loses revenue; if there are too few, the station loses points with the Federal Communications Commission. The target figure is an average of 1.5 commercial minutes per hour. A sample of 18 hours gives the following times (in minutes) allotted to public-service commercials:

.0 .0 .0 .0 .0 .0 .5 .5 .5 1.0 1.5 1.5
1.5 2.0 2.0 2.5 3.0 6.5 (mean = 1.278)

Assume that the population standard deviation is 1.60. State all parts of a z test of H_0: $\mu = 1.5$. Should H_a be one- or two-tailed? Use $\alpha = .05$.

9.24 Refer to Exercise 9.23. Calculate β probabilities for $\mu = 1.0$, 1.2, 1.4, 1.6, 1.8, and 2.0. Sketch a β curve.

9.25 The theory underlying the test in Exercise 9.23 assumes that $\bar{Y}$ has an approximately normal distribution. From the appearance of the data, do you believe that the approximation is a good one for this problem?

9.5 THE *p*-VALUE FOR A HYPOTHESIS TEST

In the hypothesis-testing problems we've considered so far, we always come to a reject–don't reject decision, without regard to the conclusiveness of the decision. In practice, this is often an oversimplification. In the product-comparison example of Section 9.1 we specified H_0: $\pi = .50$, H_a: $\pi > .50$, and $n = 100$, and we chose $\alpha = .05$. The corresponding rejection region was $y \geq 59$. Formally, $y = 59$ leads to exactly the same conclusion as $y = 99$. Evidently, the farther the value of the test statistic extends into the rejection region, the more conclusive is the rejection of the null hypothesis. How can we measure the weight of the sample evidence for rejecting a null hypothesis in favor of a research hypothesis?

p-value

The weight of evidence, or conclusiveness index, for rejecting a null hypothesis is called the ***p*-value** or attained-significance level. The *p*-value is the probability (assuming H_0) of a test statistic value equal to or more extreme than the actually observed value. As the test statistic gets farther into the rejection region, the weight of evidence for rejecting the null hypothesis gets more conclusive and the *p*-value gets smaller. In the product-comparison example, suppose that the observed y is 59. The rejection region is $Y \geq 59$, so formally we would (barely) reject the null hypothesis. Based on the binomial table for $n = 100$ and $\pi = .50$, the probability of the observed y of 59 or larger is

$$P(Y \geq 59 \mid \pi = .50) = .0444$$

But now suppose that the y value is, instead, 84, far out in the rejection region. The *p*-value for that result is

$$P(Y \geq 84) \mid \pi = .50) = .0000$$

to four decimal places. Of course the probability is not exactly zero, so we might better indicate $p = .0001$. The farther within the rejection region the test statistic falls, the smaller the p-value is, and the stronger evidence we have to reject the null hypothesis.

Very small p-values indicate strong evidence for rejecting the null hypothesis. The reason is that a small p-value indicates that the observed value of the test statistic is very unlikely if the null hypothesis is true. In the product-comparison example, an observed y of 84 ($p < .0001$) would be much more conclusive than an observed y of 59 ($p = .0444$). A small p-value indicates that the null hypothesis may be rejected quite conclusively.

Although no null hypothesis can ever be absolutely disproven, a very small p-value leads to its rejection and to support of the research hypothesis beyond a reasonable doubt.

EXAMPLE 9.13 Find the p-value if 18 stockouts are observed in a sample of 50 stores in Example 9.1.

Solution The null and research hypotheses are H_0: $\pi \leq .20$ and H_a: $\pi > .20$. Using binomial tables with $n = 50$ and probability of success (stockout), $\pi = .20$, $P(Y \geq 18) = .0062$. The p-value is even less than .01, indicating that H_0 can be rejected rather conclusively. □

two-tailed p-value

When the rejection region of the test is two-tailed, the p-value computation must be modified slightly. A more extreme value than the observed test statistic could be in the same tail or in the opposite tail. If the sampling distribution of the test statistic is symmetric (which is the case for most two-tailed tests), the **two-tailed p-value** can be computed by doubling the one-tailed value. In the product-comparison example, the one-tailed p-value corresponding to $y = 59$ is .0444; if H_a had been taken as two-sided, H_a: $\pi \neq .50$, the p-value would be .0888.

The computation of p-values is also simple for a z test. In the cereal box example for testing mean fill weight, the bureau of weights and measures used a one-tailed test of H_0: $\mu = 16$ and H_a: $\mu < 16$. The z statistic actually observed was -8.5. Since the last entry in Appendix Table 3, 3.09, corresponds to a tail area of .001, the p-value, $P(z \leq -8.5 | \mu = 16)$, is smaller than .001 for this one-tailed test.

For a two-tailed test, such as that for H_0: $\mu = 16$, H_a: $\mu \neq 16$ in the cereal box example, the p-value, $P(z \leq -8.5 \text{ or } z \geq 8.5 | \mu = 16)$, is less than .002. More extensive z tables indicate that the one-tailed p-value is less than .0000000001, one chance in ten billion. Based on the sample data, the bureau can reject H_0 with remarkable confidence. The computation of p-values based on the z statistic proceeds as follows:

p-Values for z Test

1. If H_a: $\mu > \mu_0$, p-value $= P(z > z_{\text{actual}})$
2. If H_a: $\mu < \mu_0$, p-value $= P(z < z_{\text{actual}})$
3. If H_a: $\mu \neq \mu_0$, p-value $= 2P(z > |z_{\text{actual}}|)$

Similarly, p-values for a test about π using the z statistic are computed for one- and two-tailed tests as shown here with π replacing μ.

The computed value of the z statistic is denoted by z_{actual}.

Most computer programs automatically compute p-values. A very small p-value indicates that the null hypothesis may be rejected at any plausible α value; a large p-value, such as .4 or .6, indicates that the null hypothesis should not be rejected at plausible α values. A very general principle relates p-values to α; the principle is so general that it deserves to be called the Universal Rejection Region.

Universal Rejection Region

If α has been specified, reject the null hypothesis if and only if the p-value is less than the specified α.

EXAMPLE 9.14 Many computer software programs for statistical analyses routinely compute p-values, usually in two-tailed form.

a. For the following output, find the appropriate one-tailed p-value.
b. Verify the p-value computation using z tables.
c. Can the null hypothesis be rejected at $\alpha = .05$?

```
      MEANTEST
ENTER NULL HYPOTHESIS MEAN
□:
      150:
  DO YOU WANT A Z TEST OR A T TEST? ENTER Z OR T
Z
ENTER ASSUMED POPULATION SIGMA.
□:
      20
ONE-TAILED OR TWO-TAILED TEST? ENTER 1 OR 2
□:
      2
ENTER DATA
□:
      112  119  124  133  137  138  146  148  150  151  152  154  156
□:
      156  158  161  164  167  171  173  175  182  189  197  199
SAMPLE MEAN IS 156.48
Z STATISTIC EQUALS 1.62
P-VALUE IS 0.1052322439
```

Solution a. The p-value is shown (to an excessive number of digits) as .105. The one-tailed p-value is half that or about .053.

b. The z statistic is shown as 1.62. From the normal tables, the area to the right of 1.62 is $.5000 - .4474 = .0526$.

c. Because the p-value is not less than .05, we cannot reject H_0 at $\alpha = .05$, although we can come close. Note also that the p-value, .0526, is not less than .05, although it is close. □

As stated previously, the p-value is called the attained-significance level of a statistical test. The results of a statistical test are often summarized by stating that the result is statistically significant at the specified p-value. For example, in the product-comparison example, a y of 59 is statistically significant at $p = .0444$, using a one-tailed test. In the cereal weight example, $z = -8.5$ is statistically significant at $p = .0000000001$. The smaller the p-value, the more conclusive the rejection of the null hypothesis.

The phrase *statistically significant* is unfortunate. The word *significant* suggests "important," "interesting," and "large." Statistical significance does not necessarily imply importance, relevance, or practical significance. Statistical significance only implies that a null hypothesis can be rejected with a specified low risk of error. A better phrase would be *statistically detectable*. To say that a difference is statistically significant or statistically detectable is to say that the observed result cannot reasonably be attributed to random variation alone: The cereal weight problem is a case in point. With a z value of -8.5, the test is statistically significant at the $p \leq .0000000001$ level. This allows the bureau to conclude with great confidence that the company is guilty of short-weighting, although the sample mean weight of 15.83 ounces is only about 1% less than the nominal package weight. It's a small but conclusively underfilling of the packages.

One should recognize that rarely is any null hypothesis exactly true. For this reason, with a large enough sample size, almost any null hypothesis can be rejected. What does this mean? If the null hypothesis is rejected, it means that a difference has been established fairly conclusively, but no judgment has been made as to the importance or practical significance of the declared difference.

Conversely, a sample result with associated p-value $> .05$ (and considered by some to indicate "not statistically significant") could but may not have been the result of random fluctuation; that is, even though the p-value is greater than .05, there still may be an underlying effect. The problem is that we have not established it beyond a reasonable doubt. All in all, you should be careful not to read too much into statistical significance. The p-value gives the weight of the sample evidence for rejection of the null hypothesis. The experimenter must still judge the practical significance of observed results that are declared statistically significant.

SECTION 9.5 EXERCISES

9.26 Find the p-value for the test of Exercise 9.20.

9.27 Find the p-value for the two-tailed test of Exercise 9.23.

9.28 A finance company finds that 15% of its customers fall behind in their payments. A revised loan policy is tried on a random sample of 50 customers. If 4 of the sample fall behind in payments, give a p-value for a statistical test of H_0: $\pi = .15$ versus H_a: $\pi < .15$.

9.29 Use normal tables to find the approximate p-value for Exercise 9.28.

9.30 A sales manager believes that a firm's sales representatives should spend about 40% of their working days traveling. If they are on the road for much less, new orders decline and the service and news-gathering functions of the representatives are not adequately met. If they travel much more than 40% of the time, expense accounts eat up any incremental profit. A study of the previous five months (110 working days) shows the following data (number of traveling days by each representative):

32 36 41 45 48 48 51 54 57 64
($\bar{y} = 47.6$, $s = 9.65$, $n = 10$)

A computer output for these data is shown (based on an assumed population standard deviation of 10.0).

```
        MEANTEST
ENTER NULL HYPOTHESIS MEAN
□:
        44
  DO YOU WANT A Z TEST OR A T TEST? ENTER Z OR T
Z
ENTER ASSUMED POPULATION SIGMA.
□:
        10.0
ONE-TAILED OR TWO-TAILED TEST? ENTER 1 OR 2
□:
        2
ENTER DATA
□:
        32   36   41   45   48   48   51   54   57   64
SAMPLE MEAN IS 47.6
Z STATISTIC EQUALS 1.138
P-VALUE IS 0.2549
```

a. Identify the value of the z statistic.
b. Identify the p-value.
c. Is a one-tailed or a two-tailed p-value more appropriate for this problem?

9.31 The sales manager of Exercise 9.30 concludes that the discrepancy between the observed average of 47.6 and the desired average of 44.0 is not statistically significant and therefore that the study proves that the travel-days situation is under control.

a. Do you agree that the result is not statistically significant (at the usual α levels)?
b. Do you agree that the study proves that the travel-days situation is under control?

9.32 The battery pack of a hand calculator is supposed to perform 20,000 calculations before needing recharging. A test of 114 battery packs gives an average of 19,695 calculations and a standard deviation of 1103.

a. Formulate null and research hypotheses.
b. Calculate the appropriate test statistic and p-value.

9.33 Is the result in Exercise 9.32 statistically significant at the usual α levels? Would you call the result practically significant?

9.6 HYPOTHESIS TESTING WITH THE t DISTRIBUTION

The modifications of normal (z) procedures to get t-distribution confidence intervals also apply to hypothesis tests. Once again, we replace σ by s and use t tables instead of z tables. In this section we summarize the procedure and take care of some other small differences in mechanics.

The basic procedure for any hypothesis-testing method requires formulating null and research hypotheses (H_0 and H_a), choosing a test statistic (T.S.), defining a rejection region (R.R.), calculating the T.S. value, and finally stating a conclusion. Here we are concerned with testing hypotheses about a population mean; we are still making the formal mathematical assumption that the population distribution is exactly normal.

Small-sample Test of Hypotheses about μ

H_0: $\mu = \mu_0$

H_a: 1. $\mu > \mu_0$

2. $\mu < \mu_0$

3. $\mu \neq \mu_0$

T.S.: $t = \dfrac{\bar{y} - \mu_0}{s/\sqrt{n}}$

R.R.: For a given probability α of a Type I error, reject H_0 if

1. $t > t_\alpha$
2. $t < -t_\alpha$
3. $|t| \geq t_{\alpha/2}$

where t_α cuts off a right-tail area of a in a t distribution with $n - 1$ d.f.

EXAMPLE 9.15 An airline institutes a "snake system" waiting line at its counters to try to reduce the average waiting time. The mean waiting time under specific conditions with the previous system was 6.1 minutes. A sample of 14 waiting times is taken; the times are measured at widely separated times to eliminate the possibility of dependent observations. The resulting sample mean is 5.043 and the standard deviation is 2.266. Test the null hypothesis of no change against an appropriate research hypothesis, using $\alpha = .10$. Assume that the population of waiting times is approximately normal.

Solution The population parameter of interest is μ, the long-run mean waiting time under normal conditions using the snake system. The research hypothesis is that the mean is lower than the previous mean, 6.1, so H_a: $\mu < 6.1$. We may take the null hypothesis to be H_0: $\mu = 6.1$ (no change). As usual, we need worry only about the boundary value of the null hypothesis.

$$H_0: \mu = 6.1$$
$$H_a: \mu < 6.1$$
$$\text{T.S.: } t = \frac{5.043 - 6.1}{2.266/\sqrt{14}} = -1.75$$

R.R.: For $\alpha = .10$ and d.f. $= 13$, reject H_0 for $t \leq -1.350$

Because the observed value of t, -1.75, is less than -1.350, we reject H_0 and conclude that the apparent reduction in mean waiting time (from 6.1 to about 5 minutes) is not merely a statistical fluke. □

p-value for a t test

Earlier in this chapter we introduced the *p*-value as an index of the degree of support for a research hypothesis from a given data set. There we were able to use z tables to compute *p*-values. Now we must use t tables, which are much less extensive; for given degrees of freedom, a t table gives only a few values. With these tables we can get only approximate *p*-values (although most statistical software systems give precise *p*-values). The key to the approximation is the fact that the *p*-value is the smallest α value that allows rejection of the null hypothesis. If the null hypothesis can be rejected at a particular α level, the *p*-value must be less than that α. If the null hypothesis cannot be rejected at a particular α level, the *p*-value must be greater than that α. Therefore we can often bracket the *p*-value between two numbers. All that is needed is to locate the actually observed t statistic between two t table values. The bounds on the *p*-value can be read directly.

EXAMPLE 9.16 Find bounds on the *p*-value in Example 9.15.

Solution In Example 9.15 we found that we could reject H_0 at $\alpha = .10$ because $t = -1.75$ was below $-t_{.10,13\text{ d.f.}} = -1.350$. Therefore $p < .10$. When we try $\alpha = .05$, we find that we cannot quite reject H_0; the tabulated t value is $-t_{.05,13\text{ d.f.}} = -1.771$. Therefore $p > .05$. We can summarize the approximate *p*-value as $.05 < p < .10$. □

EXAMPLE 9.17 An insurance adjuster in a small city uses two different garages to handle repairs to foreign cars damaged in collisions. To test whether the garages are competitive in cost, the adjuster obtains estimates from both garages for repair cost on each of 15 such cars. The data are shown in the table below. Test the null hypothesis that the mean difference is zero against an appropriate research hypothesis. What can be said about a *p*-value?

	Repair Estimates (in Hundreds of Dollars)						
Car	1	2	3	4	5	6	7
Garage 1	7.6	10.2	9.5	1.3	3.0	6.3	5.3
Garage 2	7.3	9.1	8.4	1.5	2.7	5.8	4.9
Difference, d	.3	1.1	1.1	−.2	.3	.5	.4

	Repair Estimates (in Hundreds of Dollars)							
Car	8	9	10	11	12	13	14	15
Garage 1	6.2	2.2	4.8	11.3	12.1	6.9	7.6	8.4
Garage 2	5.3	2.0	4.2	11.0	11.0	6.1	6.7	7.5
Difference, d	.9	.2	.6	.3	1.1	.8	.9	.9

Solution The null hypothesis is that the true mean difference $\mu_d = 0$. As no particular direction has been specified for the research hypothesis, take H_a: $\mu_d \neq 0$. We base the test on the differences (which are designated by d rather than y here). The test statistic is

$$t = \frac{\bar{d} - 0}{s_d/\sqrt{n}}$$

and is based on $n - 1 = 14$ d.f. Routine calculations give $\bar{d} = .613$ and $s_d = .394$, so

$$t = \frac{.613}{.394/\sqrt{15}} = 6.03$$

The largest tabled t-value for 14 d.f. is 2.977, corresponding to a one-tail area of .005. Thus even for a (two-tailed) α of .01, H_0 could easily be rejected. The p-value must be less than .01; in fact, we suspect that the p-value is much smaller than .01. Formally, we conclude that the two garages have different average estimates. Practically, it is clear that garage 1 has higher average estimates than garage 2. □

Evaluation of β and power is more difficult for t tests than for z tests. The method for calculating β stated in Section 9.4 is strictly valid only for z tests, but it can be used as an approximation for t tests. Because a t statistic is more variable than a z statistic, the formula tends to underestimate β and therefore to overestimate power. The easiest way to use the method is to specify a value for

$$\frac{\mu_a - \mu_0}{\sigma}$$

and a value for α. For example, suppose that a t test is run using $n = 25$ and $\alpha = .05$ (two-tailed) and that we hypothesize that the true population mean is .8 standard deviations above the null hypothesis mean;

$$\frac{\mu_a - \mu_0}{\sigma} = .8$$

Then, approximately,

$$\beta = P\left(z > -z_{\alpha/2} + \frac{|\mu_a - \mu_0|}{\sigma/\sqrt{n}}\right)$$

$$= P\left(z > -1.96 + \frac{.8}{1/\sqrt{25}}\right)$$

$$= P(z > 2.04) = .0207$$

It follows that power is approximately $1 - .0207 = .9793$ under these conditions. As we indicated, the calculation underestimates β and overestimates power. Thus the power is not quite as good as the calculation indicates.

EXAMPLE 9.18 In a computer simulation, 1000 samples of size 30 are drawn from a normal population having mean 55 and standard deviation 10. The null hypothesis that the population mean is 50 is tested, based on each sample. The following results are obtained:

Mu	Sigma	n
55.000	10.0000	30

number of times H_0: "mean is 50" is rejected in favor of

alpha	"mean > 50"	"mean < 50"	total (alpha doubled)
0.100	919	0	919
0.050	856	0	856
0.025	752	0	752
0.010	624	0	624
0.005	541	0	541

The indicated α values are for one-tailed tests. The total shown in the output corresponds to a two-tailed test; as shown, the α value should be doubled. What probability is being approximated by the fraction 919/1000? How close is this approximation to the theoretical probability calculated by formula?

Solution In this simulation the null hypothesis is false; μ is 55, not 50. The fraction 919/1000 approximates the probability that the test will reject the null hypothesis when it is false; by definition, that probability is $1 - \beta$, the power of the test. We can calculate the theoretical β by formula. For a one-tailed test with $\alpha = .10$, the required table value is $z_{.10} = 1.28$; $\mu_0 = 50$, $\mu_a = 55$, $\alpha = 10$, and $n = 30$. Therefore

$$\beta = P\left(z > -1.28 + \frac{|55 - 50|}{10/\sqrt{30}}\right) = P(z > 1.46)$$

$$= .0721$$

So power = 1 − .0721 = .9279. The simulation value, .919, is quite close to the calculated power. □

SECTION 9.6 EXERCISES

9.34 A dealer in recycled paper places empty trailers at various sites; these are gradually filled by individuals who bring in old newspapers and the like. The trailers are picked up (and replaced by empties) on several schedules. One such schedule involves pickup every second week. This schedule is desirable if the average amount of recycled paper is more than 1600 cubic feet per two-week period. The dealer's records for 18 two-week periods show the following volumes (in cubic feet) at a particular site:

1660	1820	1590	1440	1730	1680	1750	1720	1900
1570	1700	1900	1800	1770	2010	1580	1620	1690

($\bar{y} = 1718.3$, $s = 137.8$)

Assume that these figures represent the results of a random sample. Do they support the research hypothesis that $\mu > 1600$, using $\alpha = .10$? Write out all parts of the hypothesis-testing procedure.

9.35 Place an upper bound on the p-value of Exercise 9.34. Would you say that $\mu > 1600$ is strongly supported?

9.36 A federal regulatory agency is investigating an advertised claim that a certain device can increase the gasoline mileage of cars. Seven such devices are purchased and installed in seven cars belonging to the agency. Gasoline mileage for each of the cars under standard conditions is recorded both before and after installation.

	Car						
	1	2	3	4	5	6	7
Mpg before	19.1	19.9	17.6	20.2	23.5	26.8	21.7
Mpg after	20.0	23.7	18.7	22.3	23.8	19.2	24.6
Change	.9	3.8	1.1	2.1	.3	−7.6	2.9

The mean change is .50 miles per gallon and the standard deviation is 3.77.

a. Formulate appropriate null and research hypotheses.

b. Is the advertised claim supported at $\alpha = .05$? Carry out the steps of a hypothesis test.

9.37 Use the data of Exercise 9.36 to construct a 90% confidence interval for the mean change. On the basis of this interval, can one reject the hypothesis of no mean change? (Note that the two-sided 90% confidence interval corresponds to a one-tailed $\alpha = .05$ test.)

9.38 Would you say that the agency of Exercises 9.36 and 9.37 has conclusively established that the device has no effect on the average mileage of cars? What does the width of the interval in Exercise 9.37 have to do with your answer?

9.39 A small manufacturer has a choice between shipping via the postal service and via a private shipper. As a test, 10 destinations are chosen and packages shipped to each by both routes.

The delivery times, in days, are as follows:

	Destination									
	1	2	3	4	5	6	7	8	9	10
Postal service	3	4	5	4	8	9	7	10	9	9
Private shipper	2	2	3	5	4	6	9	6	7	6
Difference	1	2	2	−1	4	3	−2	4	2	3

a. Calculate the mean and standard deviation of the differences.
b. Test the null hypothesis of no mean difference in delivery times against the research hypothesis that the private shipper has a shorter average delivery time. Use $\alpha = .01$.

9.40 A standard computer package (BMDP) is used to analyze the data of Exercise 9.39. The relevant portion of the output follows:

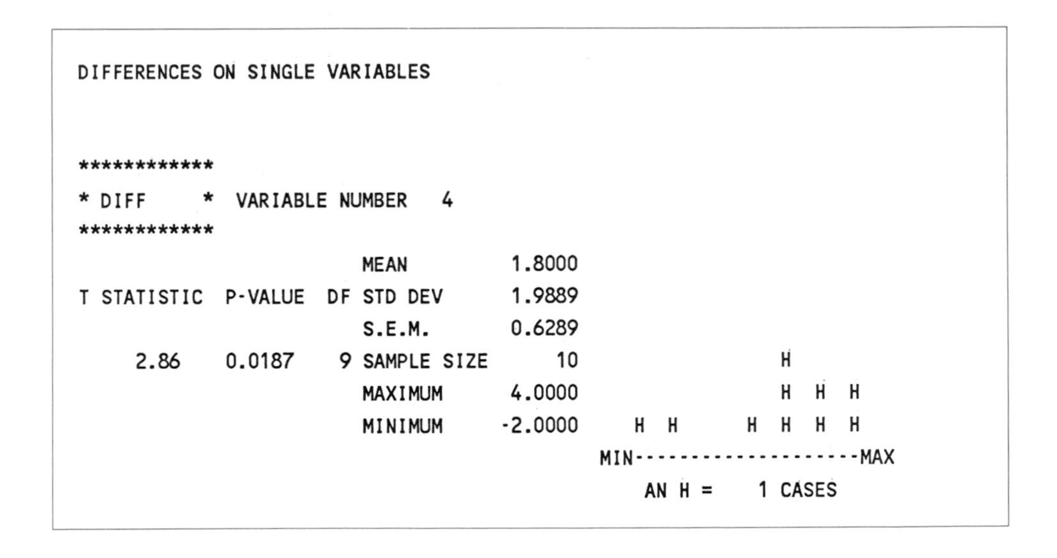

```
DIFFERENCES ON SINGLE VARIABLES


************
* DIFF     *  VARIABLE NUMBER   4
************
                                MEAN          1.8000
T STATISTIC  P-VALUE  DF  STD DEV       1.9889
                                S.E.M.        0.6289
     2.86     0.0187    9  SAMPLE SIZE       10              H
                                MAXIMUM       4.0000              H  H  H
                                MINIMUM      -2.0000    H  H     H  H  H  H
                                                       MIN------------------MAX
                                                          AN H =    1 CASES
```

a. Locate the calculated value of the t statistic.
b. Locate the p-value. Is it based on a one-tailed or two-tailed test? If necessary, convert the p-value to the appropriate number of tails.

9.7 THE EFFECT OF POPULATION NONNORMALITY

We discussed the effect of population nonnormality on t confidence intervals in Section 8.6. Exactly the same conclusions apply to t tests. The nominal α value and p-value are reasonably accurate if the population is symmetric but heavy- (or light-) tailed relative to the normal distribution. In this case, a t test may be inefficient; inefficiency in hypothesis testing terms means that some other test—such as the sign (median) test discussed in the next section—has better power at the same α level. We illustrate these effects of nonnormality with several simulation studies.

EXAMPLE 9.19 A simulation study takes 1000 samples of size 30 from a Laplace population, a symmetric, moderately outlier-prone population. The following results are obtained:

Checking Alpha

Simulation of One Sample t Test (1000 samples)
Population shape is moderately outlier prone

Mu	Sigma	n
50.000	10.0000	30

one-tailed:

number of times H_0: "mean is 50" is rejected in favor of

alpha	"mean > 50"	"mean < 50"	total (alpha doubled)
0.100	104	95	199
0.050	51	51	102
0.025	28	24	52
0.010	7	6	13
0.005	4	3	7

average t is 0.0077 with variance of 1.086943

Which hypothesis is true in the simulation? Does the outlier-proneness of the Laplace population have a serious effect?

Solution The output indicates that H_0 is $\mu = 50$, and indeed the population mean is 50. Therefore, fractions such as 104/1000 are approximating α, the probability of Type I error; the fractions are approximations because they are based on 1000 samples, not on an infinite number. Notice that all the fractions are very close to the nominal α values. For example, with a one-tailed α of .025, the observed fractions are .028 and .024. □

EXAMPLE 9.20 Another simulation study involves samples of size 30 from a Laplace population. In this study, the mean is 55, so H_0: $\mu = 50$ is false. The results of a t test and also of a sign test (a test for the median, which is also 55 by the symmetry of the Laplace population) are shown here:

Results for t Test

Mu	Sigma	n
55.000	10.0000	30

number of times H_0: "mean is 50" is rejected in favor of

alpha	"mean > 50"	"mean < 50"	total (alpha doubled)
0.100	913	0	913
0.050	831	0	831
0.025	745	0	745
0.010	629	0	629
0.005	537	0	537

Simulation of Sign Test (1000 samples)

Simulation results using the normal approximation
number of times H_0: "median is 50" is rejected in favor of

alpha	"median > 50"	"median < 50"	total (alpha doubled)
0.100	956	0	956
0.050	905	0	905
0.025	816	0	816
0.010	686	0	686
0.005	519	0	519

Which test appears to have better power, in general?

Solution Recall that power is the probability that the null hypothesis will be rejected, assuming that it is false. We note that for every α except .005 (one-tailed), the sign test rejects the hypothesis more frequently than does the t test. Therefore the sign test appears generally more powerful for this moderately outlier-prone population. □

EXAMPLE 9.21 A simulation study chooses samples of various sizes from a squared-exponential population. This population is strongly right-skewed. The following results are obtained for a t test:

Mu	Sigma	n
50.000	10.0000	10

one-tail:

number of times H_0: "mean is 50" is rejected in favor of

alpha	"mean > 50"	"mean < 50"	total (alpha doubled)
0.100	27	308	335
0.050	6	249	255
0.025	2	210	212
0.010	0	176	176
0.005	0	155	155

Mu	Sigma	n
50.000	10.0000	30

one-tail:

number of times H_0: "mean is 50" is rejected in favor of

alpha	"mean > 50"	"mean < 50"	total (alpha doubled)
0.100	40	237	277
0.050	6	189	195
0.025	0	156	156
0.010	0	120	120
0.005	0	99	99

Mu	Sigma	n
50.000	10.0000	60

one-tail:

number of times H_0 "mean is 50" is rejected in favor of

alpha	"mean > 50"	"mean < 50"	total (alpha doubled)
0.100	53	190	243
0.050	9	145	154
0.025	2	114	116
0.010	1	70	71
0.005	0	59	59

What effect does the skewness of this population have? What happens as the sample size increases?

Solution The skewness causes the nominal α probabilities to be seriously wrong, especially for one-tailed tests. The effect is worst for the smallest n, 10, but it is still severe when n is 60.

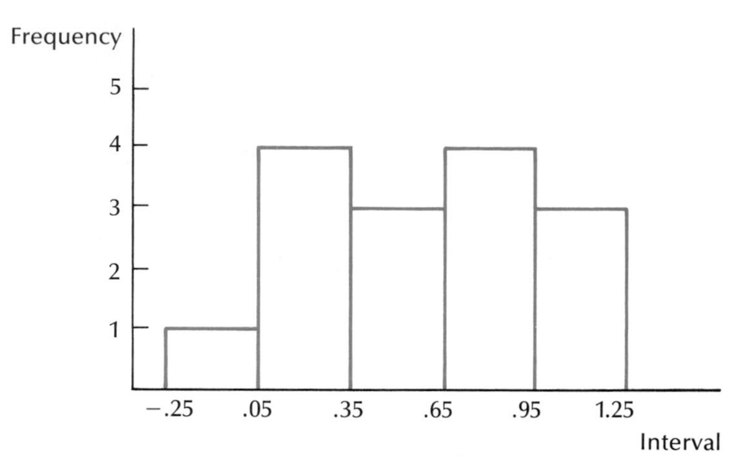

FIGURE 9.7 **Histogram for Example 9.22**

EXAMPLE 9.22 Plot a histogram of the data of Example 9.17. Does the histogram indicate that the conclusion of Example 9.17 is suspect?

Solution We group the data into five intervals and obtain the following:

Interval:	−.2 to 0	.1 to .3	.4 to .6	.7 to .9	1.0 to 1.2
Frequency:	1	4	3	4	3

The histogram is pretty much flat (Figure 9.7), with no gross outliers and no blatant skewness. Therefore the t probabilities should be decent approximations and the procedure tolerably efficient. The conclusion of Example 9.17 was based on T.S.: $t = 6.03$. This falls well within the rejection region for any reasonable α. Thus we can safely reject the null hypothesis. □

9.8 TESTS ABOUT A POPULATION MEDIAN

The median is more robust than the mean. The specific values attained by the few largest and smallest observations affect the mean considerably but the median not at all. In some situations, a population median is a more useful description of the center of a distribution than is the population mean.

As with confidence intervals about a median, hypothesis tests about a median do *not* follow the standard format of inferences about a mean. Instead, a reformulation reduces the problem to a binomial situation. To illustrate, suppose that the null hypothesis is that the median score on a programming aptitude test is 60, with a research (alternative) hypothesis that the median is smaller. The idea is that every observation in a random sample is compared to the hypothesized median of 60. Then by letting every observation above 60 denote a success and every observation below 60 denote a failure*, the problem involving a population median is translated into one related to a binomial random variable. By definition of a median, the null hypothesis yields a .5 probability of success. If the research hypothesis is true, the probability of a success is something less than .5. Therefore the binomial test described earlier in this chapter applies.

stating median test in terms of a binomial distribution

EXAMPLE 9.23 Perform a median test for the data of Example 9.17.

Solution We take H_0: *median* difference is zero against a two-sided H_a. Define a success as a difference that is greater than zero. Since no difference equals zero, $n = 15$ and the total number of successes Y in the 15 trials is $y = 14$. From a binomial table, with $n = 15$, $\pi = .5$, $P_Y(y \geq 14) \approx .0005$. For a two-tailed test, we add the two probabilities $P_Y(y \leq 1)$ and $P_Y(y \geq 14)$ to obtain a p-value of .0010. Again H_0 can safely be rejected. □

EXAMPLE 9.24 Refer to Example 9.20. What does the simulation study indicate about the desirability of using a median test rather than a t test for the outlier-prone population?

Solution As we observed in Example 9.20, the power of the median (sign) test is generally better for outlier-prone data. □

SECTION 9.8 EXERCISES

9.41 Refer to the data of Exercise 8.24 (p. 244).

a. Plot a histogram or stem-and-leaf display of the data.

b. Is there any obvious reason to doubt the approximate correctness of the 95% confidence level?

9.42 Refer again to the data of Exercise 8.24.

a. Test the null hypothesis that the mean price is 315 against a two-sided H_a. Find bounds on the p-value.

b. Test the null hypothesis that the median price is 315 against a two-sided H_a. Find bounds on the p-value.

c. Is there a serious difference in the conclusions of parts (a) and (b)?

9.43 Find a 95% confidence interval for the true median price, using the data of Exercise 8.24.

9.44 Refer to the data of Exercise 8.22 (p. 243).

a. Perform a t test of H_0: $\mu \leq 45$ versus H_a: $\mu > 45$. Use $\alpha = .05$.

b. Plot the data. Is there reason to be skeptical of your conclusion in part (a)?

* If there is one observation exactly equal to 60, it is discarded and the sample size reduced by one. With two values equal to 60, count one success and one failure; with three, count one success, one failure, and one discard; and so on.

9.45 Again refer to Exercise 8.22.

a. Test the null hypothesis that the true median is less than or equal to 45. Use $\alpha = .05$.

b. Is there a major difference between your conclusions here and in Exercise 9.44? If there is, what explains the difference?

9.46 Random samples of size 30 are drawn from a normal population having mean 55 and standard deviation 10. Both t and median (sign) tests are performed. The following results are obtained:

Simulation of One Sample t Test (1000 samples)

Mu	Sigma	n
55.000	10.0000	30

number of times H_0: "mean is 50" is rejected in favor of

alpha	"mean > 50"	"mean < 50"	total (alpha doubled)
0.100	919	0	919
0.050	856	0	856
0.025	752	0	752
0.010	624	0	624
0.005	541	0	541

Simulation of Sign Test (1000 samples)

Mu	Sigma	n
55.000	10.0000	30

Simulation results using binomial probabilities

number of times H_0: "median is 50" is rejected in favor of

alpha	"median > 50"	"median < 50"	total (alpha doubled)
0.100	698	0	698
0.050	698	0	698
0.025	564	0	564
0.010	409	0	409
0.005	256	0	256

a. What probabilities are being approximated in this simulation?

b. What do the results indicate about the relative desirability of the t and sign tests when sampling from a normal population?

9.47 In a computer simulation study, samples of size 10, 30, and 60 are drawn from a skewed population. The following results are obtained for a t test:

Mu	Sigma	n
50.000	10.0000	10

one-tail:

number of times H_0: "mean is 50" is rejected in favor of

alpha	"mean > 50"	"mean < 50"	total (alpha doubled)
0.100	68	118	186
0.050	44	78	122
0.025	20	48	68
0.010	6	21	27
0.005	5	16	21

Mu	Sigma	n
50.000	10.0000	30

one-tail:

number of times H_0: "mean is 50" is rejected in favor of

alpha	"mean > 50"	"mean < 50"	total (alpha doubled)
0.100	86	103	189
0.050	36	61	97
0.025	15	33	48
0.010	7	19	26
0.005	3	13	16

Mu	Sigma	n
50.000	10.0000	60

one-tail:

number of times H_0: "mean is 50" is rejected in favor of

alpha	"mean > 50"	"mean < 50"	total (alpha doubled)
0.100	103	103	206
0.050	42	47	89
0.025	16	26	42
0.010	3	15	18
0.005	1	8	9

a. Show that the simulation study is evaluating α.
b. How accurate are the nominal one-tailed α probabilities? Does the answer change as the sample size changes?
c. How accurate are the nominal two-tailed α probabilities?

9.48 A simulation study involves drawing samples of size 10, 30, and 60 from a symmetric, discrete distribution, taking only 5 possible values. For each sample, a *t* test is performed. The following results are obtained:

Mu	Sigma	n
50.000	10.0000	10

one-tail:

number of times H_0: "mean is 50" is rejected in favor of

alpha	"mean > 50"	"mean < 50"	total (alpha doubled)
0.100	100	107	207
0.050	53	56	109
0.025	26	31	57
0.010	9	10	19
0.005	5	8	13

Mu	Sigma	n
50.000	10.0000	30

one-tail:

number of times H_0: "mean is 50" is rejected in favor of

alpha	"mean > 50"	"mean < 50"	total (alpha doubled)
0.100	110	100	210
0.050	59	44	103
0.025	28	26	54
0.010	9	13	22
0.005	6	5	11

Mu	Sigma	n
50.000	10.0000	60

one-tail:

number of times H_0: "mean is 50" is rejected in favor of

alpha	"mean > 50"	"mean < 50"	total (alpha doubled)
0.100	90	104	194
0.050	36	52	88
0.025	18	27	45
0.010	6	10	16
0.005	3	5	8

a. Show that this simulation is evaluating α for a t test in this population.
b. Are the nominal values of α seriously in error? Does your answer change as the sample size changes?

9.9 TESTING A POPULATION PROPORTION USING A NORMAL APPROXIMATION

In the first section of this chapter we did a test of a population proportion using binomial tables. The practical problem is that complete binomial tables are not always available, and if available they are cumbersome and necessarily limited. For example, how would you test the null hypothesis that $\pi = .373$ with $n = 277$? Even with a good computer program, binomial probabilities are relatively slow and expensive to compute. The normal-distribution approximation to the binomial allows approximate tests to be run; the method is much like that of Section 9.3 for a statistical test related to μ.

The product-comparison example of Section 9.1 illustrates the method. The null hypothesis is H_0: $\pi = .50$ and the research hypothesis is H_a: $\pi > .50$. Hence we want a one-tailed test. In Chapter 5 we showed that if n is large and π is not too close to 0 or 1, the z statistic for the binomial random variable Y,

$$z = \frac{Y - n\pi}{\sqrt{n\pi(1 - \pi)}}$$

is approximately standard (tabled) normal. This z can be used as the test statistic instead of y; the relevant value for π is the (boundary) null hypothesis value, $\pi_0 = .50$. As with a statistical test for μ, the one-tailed rejection region for $\alpha = .05$ is $z > 1.645$. A little algebra shows that $z > 1.645$ is equivalent to $y > n\pi_0 + 1.645\sqrt{n\pi_0(1 - \pi_0)}$, or, in the product-comparison problem, $y \geq 58.225$. Since y can assume only integer values, this rejection region is equivalent to $y \geq 59$, which is the rejection region we got in Section 9.1 for $\alpha = .05$. Similarly, for $\alpha = .10$, the rejection region is $z \geq 1.282$, or $y \geq 56.41$. Since y can assume only integer values, this is equivalent to $y \geq 57$, which again is the rejection region we got in Section 9.1 for $\alpha = .10$. The observed y value of 68 corresponds to a z score of 3.6:

$$z = \frac{y - n\pi_0}{\sqrt{n\pi_0(1 - \pi_0)}} = \frac{68 - 50}{\sqrt{100(.5)(.5)}} = 3.6$$

Hence we reject H_0: $\pi = .50$ for $\alpha = .05$ (and for $\alpha = .01$). The approximate procedure for testing a population proportion using a z statistic is summarized below.

Summary for Test of a Population Proportion Using the Normal Approximation

H_0: $\pi = \pi_0$
H_a: 1. $\pi > \pi_0$
2. $\pi < \pi_0$
3. $\pi \neq \pi_0$

$$\text{T.S.: } z = \frac{y - n\pi_0}{\sqrt{n\pi_0(1 - \pi_0)}}$$

R.R.: For the probability of a Type I error α, reject H_0 if
1. $z > z_\alpha$
2. $z < -z_\alpha$
3. $z > z_{\alpha/2}$ or $z < -z_{\alpha/2}$

Note: π_0 is the (boundary) null-hypothesis value of the population proportion π.

There's another way to write the test statistic z. If $\hat{\pi}$ is the sample proportion (so $\hat{\pi} = y/n$), then z can be written

$$z = \frac{\hat{\pi} - \pi_0}{\sqrt{\pi_0(1 - \pi_0)/n}}$$

For the product-comparison example, $\hat{\pi} = 68/100 = .68$ and again $z = 3.6$. The two forms of z are algebraically equal, so they always give the same answer.

We said that the z test for π is approximate and works best if n is large and π_0 is not too near 0 or 1. A natural next question is, When can we use it? There are several rules to answer the question; none of them should be considered sacred. Our sense of the many studies that have been done is this: if either $n\pi_0$ or $n(1 - \pi_0)$ is less than about 2, treat the results of a z test very skeptically. If $n\pi_0$ and $n(1 - \pi_0)$ are at least 5, the z test should be reasonably accurate. For the same sample size, tests based on extreme values of π_0 (e.g., .001) are less accurate than tests for values of π_0 such as .05 or .10. For example, a test of H_0: $\pi = .0001$ with $n\pi_0 = 1.2$ is much more suspect than one for H_0: $\pi = .10$ with $n\pi_0 = 50$. If the issue becomes crucial, it's best to interpret the result skeptically.

sample-size requirement

9.10 THE RELATION BETWEEN HYPOTHESIS TESTS AND CONFIDENCE INTERVALS

We now have two forms of inference: confidence intervals and hypothesis tests. Both can be performed on the same data. How are they related? For the cereal box example, a 95% confidence interval for the true mean weight is $\mu \pm 1.96\sigma_{\bar{y}}$. Substituting $\bar{y} = 15.83$,

$n = 25$, and $\sigma = .1$, we have

$$15.83 - 1.96 \frac{(.1)}{\sqrt{25}} < \mu < 15.83 + 1.96 \frac{(.1)}{\sqrt{25}}$$

or

$$15.791 < \mu < 15.869$$

In our statistical test of H_0: $\mu \geq 16$, the boundary value $\mu_0 = 16$ does not fall within the 95% confidence interval, so it seems plausible to reject H_0. What is the probability of a Type I error for a statistical test based on a 95% confidence interval? In general, a particular null-hypothesis value, say, θ_0, of any population parameter θ may be rejected with the probability of a Type I error α if and only if θ_0 does not fall in a $(1 - \alpha) \times 100\%$ confidence interval for θ.

For example, because the 95% confidence interval $15.791 < \mu < 15.869$ does not include $\mu_0 = 16$, we can reject H_0: $\mu = 16$, based on $\alpha = .05$. In fact, this is a general method for constructing confidence intervals; a 95% confidence interval can be defined as the set of nonrejectible ($\alpha = .05$) null-hypothesis values. For instance, in a two-tailed test a particular μ value is not rejected using $\alpha = .05$ if the z statistic lies in the interval

$$-1.96 < \frac{\bar{y} - \mu}{\sigma/\sqrt{n}} < 1.96$$

A little algebra shows that this is equivalent to

$$\bar{y} - 1.96 \frac{\sigma}{\sqrt{n}} < \mu < \bar{y} + 1.96 \frac{\sigma}{\sqrt{n}}$$

which is the 95% confidence interval for μ. (There's a slight problem with using $<$ or $\leq$. However, the probability that z is exactly equal to 1.96 is so small—theoretically it's zero—that we don't worry about it.) In this sense, confidence intervals and hypothesis tests give equivalent results.

The usual confidence interval is two-sided. As above, such confidence intervals correspond to two-tailed tests. There is such a thing as a one-sided confidence interval. The nonrejection region for a left-tailed test at $\alpha = .05$ is

$$\frac{\bar{y} - \mu}{\sigma/\sqrt{n}} > -1.645$$

When solved for μ, this is

$$\mu < \bar{y} + 1.645 \frac{\sigma}{\sqrt{n}}$$

one-sided confidence interval

which is a **one-sided confidence interval**. In the cereal box example, this becomes $\mu < 15.863$; because the boundary value $\mu_0 = 16$ does not fall within this interval, H_0: $\mu = 16$ may be rejected using $\alpha = .05$, one-tailed. For the remainder of this text, we use two-sided confidence intervals, which can be used to test two-sided research hypotheses.

EXAMPLE 9.25 For the television-watching data of Example 9.8, use a 99% confidence interval to test H_0: $\mu = 22.6$ versus H_a: $\mu \neq 22.6$.

Solution The two-sided research hypothesis implies that a two-sided confidence interval may be used to test the null hypothesis. In Example 9.8, $\bar{y} = 25.2$, σ is assumed to be 6.1, and $n = 50$. The 99% confidence interval is

$$25.2 - 2.576 \frac{6.1}{\sqrt{60}} < \mu < 25.2 + 2.576 \frac{6.1}{\sqrt{60}}$$

or

$$23.2 < \mu < 27.2$$

Because the value of μ under H_0, 22.6, does not fall within the interval, we reject H_0 using $\alpha = .01$. Of course, the same conclusion was obtained in Example 9.8. □

confidence interval and β

When the null hypothesis is not rejected, confidence intervals are useful in giving a crude measure of the risk of a Type II error. Roughly speaking, a wide 95% confidence interval indicates a high degree of uncertainty and therefore a high probability β of Type II error. (Of course 95% confidence fixes α at .05.) For example, if a seller of high-intensity lights for portable television cameras claims a mean life of 40 hours and a sample of 10 lights yields a 95% confidence interval of $28.0 < \mu < 44.0$, the seller's claim cannot be rejected using an $\alpha = .05$ level. Note, too, that the interval is very wide; the lower limit of 28.0 is 30% below the claimed value. If the difference between a mean life of 40 hours and, say, a mean life of 30 hours was crucial in deciding whether to buy, the buyer would not be comfortable in accepting $\mu = 40$. The probability of a Type II error corresponding to $\mu = 30$ would undoubtedly be quite large.

SECTION 9.10 EXERCISES

9.49 Calculate a 95% confidence interval for the true population mean based on the data of Exercise 9.23. Show that this confidence interval is consistent with the conclusion of Exercise 9.23.

9.50 Calculate a 90% confidence interval for the mean, based on the data of Exercise 9.30, assuming $\sigma = 10$. Can the null hypothesis be rejected at $\alpha = .10$?

9.51 Calculate a 99% confidence interval for the mean, based on the data of Exercise 9.32. Use it to test H_0: $\mu = 20{,}000$ against a two-sided research hypothesis.

9.11 HYPOTHESIS TESTING AS A DECISION METHOD

In this chapter, hypothesis testing has been presented as a method for supporting or not supporting research hypotheses. Such hypotheses need not be directly related to a decision. However, there are many situations, such as the television-camera light example in the previous section, where a decision such as buy or don't buy is directly involved. In such situations, the research hypothesis is usually called the alternative hypothesis*: the problem is more one of deciding between two hypotheses than of supporting or not supporting a research hypothesis. The two hypotheses usually correspond directly to two possible actions.

The television-camera light problem illuminates how to use a statistical test for decision making. The null hypothesis can be taken as $\mu \geq 40$. Let's assume that the television station would be happy to buy the lights if this null hypothesis is true. The relevant alternative is H_a: $\mu < 40$; if the true mean life is substantially less than 40 hours, the station presumably wouldn't want to buy. Suppose that the buyer plans to observe a sample of 8 lights before reaching a decision. In addition, assume that $\sigma = 10$ hours. (Obviously, in practice we would not know σ and hence would need a larger n to substitute s for σ in the z statistic.) If the station sets $\alpha = .10$, the rejection region for a statistical test of H_0: $\mu \geq 40$ and H_a: $\mu < 40$ is $z \leq -1.282$. Thus, in terms of a decision procedure, the station will buy the light bulbs if $z > -1.282$ and will not buy if $z \leq -1.282$. In this situation, a Type I error is failing to buy if the claim (H_0: $\mu \geq 40$) is correct.

Suppose that a true mean life of 30 hours would make the lights uneconomical. A Type II error in this situation is buying lights that are uneconomical. The television station would be concerned with the β risk. According to our formula in Section 9.4,

$$\beta_{30} = P\left(z > -1.282 + \frac{|30 - 40|}{10/\sqrt{8}}\right)$$

$$= P(z > 1.55) \approx .06$$

Thus, even a sample of size 8 is enough to yield reasonably low error risks ($\alpha = .10$, $\beta_{30} = .06$), because $\mu_a = 30$ and $\mu_0 = 40$ are rather far apart. The problem of deciding whether or not to buy, based on results from the sample, can be interpreted as a hypothesis-testing problem.

The practical difficulty with this approach is the choice of desirable α and β risks. Of course we'd like to have both risks very low. Usually, though, the only way to do this is to take a very large sample, which can be expensive or impossible. As we've noted previously, for a fixed sample size a decrease in α can be obtained only by an increase in β. It's often hard to see what the right choice is. The choice depends, not only on the relative costs of the two errors, but also on the relative plausibility of the hypotheses. If the cost of bad lights is high and the loss in passing up good lights is low, a Type II error is more costly than a Type I error. So the β risk should be chosen much smaller than the α risk. If the seller is known to be very reliable in its claims, so that the null hypothesis is very plausible, the television station may be relatively more willing to consider a low sample mean as a fluke and reject the null hypothesis only for very extreme z statistic values. A rejection region that rejects the null hypothesis only for very extreme values of a test statistic has a very low α value. The proper choice of α and β depends, not only on the relative costs of the two types of error, but also on the **prior probabilities** of the two hypotheses.

prior probability

* That's why we use the symbol H_a instead of, perhaps, H_r.

This issue, which is part of decision theory, is discussed in Chapters 17 and 18. Until we get there, we'll have to settle for rather rough and arbitrary choices of α (and β) when we use hypothesis tests for decision making.

SECTION 9.11 EXERCISES

9.52 A flour miller's contract with a grain seller requires that the average protein content of winter wheat sold to the miller be at least 13.5%. The miller has several samples from each grain shipment analyzed. If the protein content is below 13.5% (to a statistically significant extent), the miller deducts a penalty from the payment to the seller.

a. Formulate null and alternative hypotheses for this problem.
b. What actions follow from rejecting and from not rejecting the null hypothesis?
c. What would be the consequence of setting α at a very small value?

9.53 Assume the following conditions in Exercise 9.52:

i. Invoking a penalty can lead to legal actions; if the penalty is claimed and the shipment is subsequently found to meet the protein standard, substantial damages must be paid by the miller.
ii. If the full shipment is deficient in protein, the miller incurs a modest additional cost to supplement the protein content.
iii. The shipment comes from an established seller and is drawn from grain of an excellent harvest.

What do these considerations imply about the choice appropriate α and β values?

9.54 A former commissioner of the Food and Drug Administration commented that Congress complains violently whenever the FDA mistakenly allows an unsafe or ineffective drug onto the market but never says a word if a safe, effective drug is not allowed onto the market. If we take the research hypothesis as "drug is safe and effective," what does the ex-commissioner's comment imply about the relative costs of Type I and Type II errors?

SUMMARY

In this chapter we have presented the fundamental concepts of a statistical test of a hypothesis. We began by illustrating the mechanics of a test of a binomial proportion. After specifying the research and null hypotheses and selecting a test statistic, we saw how specification of α, the probability of a Type I error, fixes the location of the rejection region. If the computed value of the test statistic for the sample data falls within the rejection region, we reject the null hypothesis in favor of the research hypothesis. If not, we reserve judgment until values of β (the probability of a Type II error) can be computed for various values of the parameter under test.

After discussing the specifics of a binomial test, the hypothesis-testing procedure was formalized using a z statistic as the test statistic for a large sample test of a population mean μ and a binomial proportion π. The summary used in this chapter for H_0, H_a, T.S., and R.R. is a standard format for most statistical tests in this text.

The important concept of the attained-significance level of a statistical test was then presented. Many professional journals summarize the results of a statistical test by stating the p-value. The p-value, or attained-significance level, is an index of the weight of the sample evidence for contradicting the null hypothesis. The smaller the p-value, the greater the weight of evidence for rejecting the null hypothesis. But keep in mind that statistical significance does not address the question of the practical significance of the sample results. Hence a small p-value does not imply that the findings have practical significance.

KEY FORMULAS: Statistical Tests

1. Test for μ, σ known
   Null hypothesis: $\mu = \mu_0$

$$\text{Test statistic:} \qquad z = \frac{\bar{y} - \mu_0}{\sigma/\sqrt{n}}$$

2. Test for μ, σ unknown, $n \geq 30$: the procedure is the same as for σ known, except that s replaces σ in the formula for the test statistic.
3. Calculation of β for a test on μ
   a. For a two-tailed test,

$$\beta \approx P\left(z > -z_{\alpha/2} + \frac{|\mu_a - \mu_0|}{\sigma/\sqrt{n}}\right)$$

   b. For a one-tailed test,

$$\beta = P\left(z > -z_{\alpha} + \frac{|\mu_a - \mu_0|}{\sigma/\sqrt{n}}\right)$$

4. Sample size required for a specified α and β
   One-tailed test:

$$n = \frac{(z_\alpha + z_\beta)^2\sigma^2}{(\mu_a - \mu_0)^2}$$

   Two-tailed test: Replace z_α by $z_{\alpha/2}$
5. Test for median: Choose as test statistic $Y =$ number of sample values exceeding null hypothesis value of median. Test H_0: $\pi = .50$.
6. Test for π, normal approximation
   Null hypothesis: $\pi = \pi_0$

$$\text{Test statistic:} \qquad z = \frac{y - n\pi_0}{\sqrt{n\pi_0(1 - \pi_0)}}$$

7. p-values for z tests
   a. H_a: $\mu > \mu_0$ (or $\pi > \pi_0$), $\quad p = P(z > z_{\text{actual}})$
   b. H_a: $\mu < \mu_0$ (or $\pi < \pi_0$), $\quad p = P(z < z_{\text{actual}})$
   c. H_a: $\mu \neq \mu_0$ (or $\pi \neq \pi_0$), $\quad p = 2P(z > |z_{\text{actual}}|)$

CHAPTER 9 EXERCISES

9.55 A manufacturer of yogurt products stamps a sell date on every container. Yogurt not sold by this date is discarded. As a check on this dating system, 50 containers are kept for 8 days beyond the sell date, the maximum length of time the yogurt should be kept in a home refrigerator. Under such severe conditions, the manufacturer is willing to concede that 10% of the containers will have deteriorated quality. A higher percentage indicates a need for a change in sell-date policy. Assume that 9 of the 50 containers show deteriorated quality. Carry out a statistical test using binomial tables and $\alpha = .05$.

9.56 Find the p-value in Exercise 9.55.

9.57 EPA miles-per-gallon ratings are obtained for all models of cars sold in the United States. One of these figures purports to represent mileage in combined city-country driving. Suppose a consumer group test-drives 8 cars of a model with an EPA rating of 28.2 miles per gallon. If H_0 is $\mu = 28.2$, what argument would lead to a one-sided research hypothesis? What would lead to a two-sided research hypothesis?

9.58 Assume that the population standard deviation is 2.1 and that the mean gas mileage for the 8 cars is 26.7 in Exercise 9.57. Can the two-sided research hypothesis be supported at $\alpha = .01$?

9.59 Find the p-value in Exercise 9.58.

9.60 An official of the consumer group interprets the result of Exercise 9.58 as being not statistically significant. The official concludes that therefore it can reliably be assumed that the true mean is 28.2. Do you agree?

9.61 Compute a 99% confidence interval for the true mean mileage in Exercise 9.58. Use this interval to confirm the result of that exercise. What can we "reliably assume" about the true mean mileage?

9.62 In a nationwide opinion poll based on a random sample of 2417 people, one question is "How do you rate the ethics of business executives of large companies?" A rating of 3 means "no better or worse than most people," a rating of 1 is "much better than most poeple," and 5 is "much worse than most people." The mean rating is 3.05 and the standard deviation is 0.62.

a. Calculate a 95% confidence interval for the population mean rating.

b. Can H_0: $\mu = 3.00$ be rejected (against a two-sided alternative) at $\alpha = .05$?

9.63 A newspaper reporting on the poll of Exercise 9.62 reports that "respondents rated the ethics of big business significantly worse than average."

a. Is this statement true in the statistical sense?

b. Do you think it might mislead the general public?

9.64 What can be said about the p-value of Exercise 9.62?

REVIEW EXERCISES CHAPTERS 7–9

R52 A large computer software firm installs a new editor for use by a random sample of its programmers. After the programmers have learned to use the editor comfortably, the firm measures the number of lines of debugged code produced by each programmer. (The programming tasks are of comparable difficulty.) The data are

178	183	199	201	204	210	218	218	219	220	225	227	231
232	232	233	233	235	238	239	241	243	244	246	247	249
250	251	264	266	270	271	271	273	275	276	277	279	283
284	285	286	289	289	298	303	306	315	315	345		

For these data, the sample size is 50, the sample mean is 253.32, and the sample standard deviation is 36.1.

a. The population standard deviation using the previous editor was 35.4. Assume that this population standard deviation applies to the new editor as well. Calculate a 99% confidence interval for the population mean using the new editor.
b. Is there clear evidence in the data that the sample mean is likely to be an inefficient estimator of the population mean?

R53 Refer to the confidence interval calculated in part (a) of Exercise R52. The population mean using the old editor was 230.2. Can we reject the null hypothesis that the mean for the new editor is 230.2, using $\alpha = .01$, based on the confidence interval?

R54 Carry out a formal hypothesis test of the null hypothesis that the population mean remains 230.2 against the research hypothesis that it is not equal to 230.2 for the data of Exercise R52. Use $\alpha = .01$ and assume that the population standard deviation is 35.4.

R55 For the hypothesis test performed in Exercise R54, state a p value. Should it be one-sided or two-sided?

R56 Redo Exercises R52–R55 without making the assumption that the population standard deviation is 35.4. Do any of your conclusions change substantially?

R57 Using the data of Exercise R52, calculate a 99% confidence interval for the population median. On the basis of this interval, can one reject the null hypothesis that the population median is 230, using $\alpha = .01$?

R58 Which 99% confidence interval is wider, the interval calculated in Exercise R56 or the one in R57? What does your answer suggest about the efficiency of the sample mean as compared to the sample median in this particular case?

R59 It is claimed that 45% of all walk-in customers at a particular real estate office eventually buy a home through that office. To test the claim, it is planned to regard the next 100 customers as a random sample. Set up a formal test of the research hypothesis that the population proportion is less than .45, using $\alpha = .05$.

R60 How does the rejection region in Exercise R59 change if the maximum allowable α is set at .01?

R61 Suppose that the head of the real estate office in Exercise R59 determines that 32 of the 100 walk-in customers eventually buy homes through the office. Does this fact support H_a if α is set at .05?

R62 Does the result indicated in Exercise R61 lead to rejection of H_0 if α is set at .01? What does your answer indicate about the p-value for the data?

R63 A forester needs to test a new method for growing pine trees for lumber, one designed to minimize the loss to browsing deer. A five-year trial is needed. A sample of 25 stands is to be planted and tended using the new method. The current yield has a mean of 272.6 and a standard deviation of 67.3, in appropriate units of measurement.

a. Formulate reasonable null and research hypotheses. In particular, should the research hypothesis be one-sided or two-sided?
b. Set up the first four parts of a formal test of the null hypothesis. Assume that the population standard deviation remains unchanged. The desired α value is .05.

R64 Suppose that the population mean yield under the new growing method in Exercise R63 is 305. Calculate the probability that the test in Exercise R63 will not reject the null hypothesis. What is the technical name for this probability?

R65 Assume that the population mean yield is 295 rather than the 305 assumed in Exercise R64. Should the probability that the null hypothesis will not be rejected using the test in Exercise R63 be larger or smaller than the probability calculated in Exercise R64?

R66 Assume that the experiment of Exercise R64 results in the following yield data:

135 185 231 247 262 285 300 304 310 312 313 319 322
322 324 328 335 362 366 368 370 384 384 385 401
($\bar{y} = 314.16$, $s = 64.0$)

What is the conclusion of the test specified in Exercise R64?

R67 Refer to Exercises R64 and R66. Find the probability that a sample mean is equal to or larger than 314.16, given a population mean of 272.6. What is the technical name for this probability?

R68 Plot the data of Example R66. Is there any reason to believe that the sample mean is not the best estimator of the population mean?

R69 A market research firm is trying to estimate the proportion of shoppers at a suburban mall that are regular readers of two particular newspapers in order to judge the relative merits of the two papers as advertising vehicles. Employees of the firm are instructed to sample shoppers randomly until they find a shopper who is a regular reader of both newspapers; once a regular reader is found, a value of $X =$ number of other shoppers interviewed is recorded. Under reasonable assumptions, the probability distribution of X is geometric:

$$f(x) = \pi(1 - \pi)^x \qquad \text{for } x = 0, 1, 2, \ldots,$$

Suppose that six values of X are recorded: $x_1 = 5$, $x_2 = 8$, $x_3 = 2$, $x_4 = 9$, $x_5 = 0$, $x_6 = 6$. Calculate the likelihood of these values for values of π ranging from .2 through .5. From the numerical results, what seems to be the best estimate of π?

R70 a. For the data of Exercise R69, use calculus to find the maximum likelihood estimate.
b. Generalize the derivation in part (a) to arbitrary values of $x_1, \ldots, x_n$.

R71 It can be proved that the long-run average (over many samples) of the estimator found in Exercise R70 is larger than π, although the long-run average decreases rapidly toward π as the sample size increases. What desirable property of an estimator is (slightly) violated by this particular estimator?

R72 A computer center that serves, among other clients, small savings and loan associations needs to know the proportion of jobs from these businesses that require intervention by the computer operator. In a random sample of 133 such jobs, 22 require operator intervention. Calculate a 95% confidence interval for the population proportion of jobs requiring intervention.

R73 Suppose that the confidence interval found in Exercise R72 is felt to be too wide. Calculate the sample size required to obtain a 95% confidence interval with a width of .06 (a plus or minus of .03) under each of two assumptions:
a. Assume that the sample proportion continues to equal 22/133.
b. Assume that the sample proportion may take any value.

R74 A normal approximation was used in answering Exercises R72 and R73. Can we be confident that the approximation is a good one?

R75 An auditor wishes to verify transaction records of a firm. The transactions are placed in random order. An auditor trainee keeps a cumulative total of the dollar amounts of the transactions; every time the total moves over a $100,000 increment (that is, when the total passes $100,000, $200,000, $300,000, etc.), the transaction is set aside for verification. Show that this process does not yield a random sample of the transactions.

R76 The process of Exercise R75 yields 241 transactions. The mean size of the transactions is $5381 and the standard deviation is $2271. The amounts, when plotted, show substantial right skewness.
a. Calculate a supposed 95% confidence interval for the population mean transaction size.
b. This interval is in fact quite likely not to include the actual mean transaction size of the population. Explain why.

R77 A chemical manufacturer doing a pilot study of yields obtains a sample of 26 small batches. The yields, expressed as percentages of the theoretical maximum, are

67.6	68.5	74.7	77.6	78.4	79.3	79.5	80.3	80.3	80.7	80.8	80.8	80.9
81.2	81.4	81.4	81.5	82.5	82.5	82.9	82.9	83.8	84.4	84.4	85.4	86.0

($\bar{y} = 80.37$, $s = 4.37$)

Calculate a 90% confidence interval for the population mean yield.

R78 Use the data in Exercise R77 to calculate a 90% confidence interval for the population median.

R79 Compare the widths of the intervals in Exercises R77 and R78. What do the widths indicate about the relative efficiency of using the sample mean versus the sample median in this situation? Does a plot of the data indicate the same thing?

R80 Use the data in Exercise R77 to test the null hypothesis that the population mean is 82.0 against a two-sided research hypothesis. State bounds on the *p*-value.

R81 The complete pilot study for Exercise R77 eventually involves a sample of 150 batches. Assuming that the population standard deviation is about 4.4, and that the population mean is 80.4, find the probability that H_0: population mean = 82.0 will be rejected. Assume an α of .05.

R82 The probability calculated in Exercise R81 is *not* a *p*-value. Explain why not.

∂ **R83** The Pareto density is sometimes used as a model in insurance situations that can result in many small claims and a few enormous ones. One form of Pareto density is

$$f(y) = \frac{1}{\theta}(y+1)^{-1-1/\theta} \qquad \text{for } y > 0$$

Given a sample with $y_1 = 2.730$, $y_2 = 5.124$, $y_3 = .798$, and $y_4 = 36.215$, find a good estimate of θ.

∂ **R84** Find a good estimator of the parameter θ in Exercise R83 for arbitrary values $y_1, \ldots, y_n$.

R85 The estimator of Exercise R84 has the following properties: Its long-run average value is θ, and among all estimators with average value θ it has the smallest standard error. What do these properties indicate about this estimator?

10

COMPARING TWO SAMPLES

The statistical inference procedures we have discussed so far deal with inferences about a single parameter, based on data from a single random sample. These confidence-interval and hypothesis-testing methods treat population parameters one by one. In this chapter we develop statistical inference procedures for comparing two quantities.

The idea of comparing two sets of data is a fundamental one. Whether it's a problem of comparing consumer acceptance of two possible formulations of a new product, comparing productivity from two piece-rate pay schemes, or comparing the yields from two states' tax-auditing procedures, most managers face many such situations.

The basic principles for analyzing such problems have been developed in previous chapters. We present several specific procedures in this chapter and discuss which methods are appropriate in which situations.

10.1 COMPARING THE MEANS OF TWO POPULATIONS WITH KNOWN STANDARD DEVIATIONS

The basic principles underlying the statistical methods for two-sample inferences are the same as those developed in the preceding chapters for single-sample inferences. The formulas are more complicated, but the essential ideas are familiar by now. To make the transition to two-sample problems easier, we initially make the artificial assumption that both population standard deviations are known. As in the one-sample problem, this assumption is not crucial for large sample sizes. As you might expect, when we drop this assumption we use sample variances and t tables; the more realistic situation of unknown standard deviations is the topic of the next section.

The procedures of this section are based on the following formal mathematical assumptions:

Formal Assumptions for Two-sample Inferences Concerning Means

1. A random sample is selected from each of the two populations. The samples are independent. The notation is

Group 1		Group 2
μ_1	population mean	μ_2
σ_1^2	population variance	σ_2^2
n_1	sample size	n_2
$\bar{y}_1$	sample mean	$\bar{y}_2$
s_1^2	sample variance	s_2^2

2. Each population distribution is normal.
3. We are interested in inferences concerning $\mu_1 - \mu_2$, the difference of the two population means.

We discuss the effect of violations of these assumptions at the end of Section 10.2.

The obvious way to form a point estimator of the difference in population means $\mu_1 - \mu_2$ is to use the difference in sample means $\bar{Y}_1 - \bar{Y}_2$. Because each sample mean is an unbiased estimator of the corresponding population mean, $\bar{Y}_1 - \bar{Y}_2$ is an **unbiased estimator of $\mu_1 - \mu_2$**.

unbiased estimator of $\mu_1 - \mu_2$

$$E(\bar{Y}_1 - \bar{Y}_2) = E(\bar{Y}_1) - E(\bar{Y}_2) = \mu_1 - \mu_2$$

The key to inference problems related to $\mu_1 - \mu_2$ is that the variances of $\bar{Y}_1$ and $\bar{Y}_2$ can be added, since the random variables $\bar{Y}_1$ and $\bar{Y}_2$ are independent.

Variance of $\bar{Y}_1 - \bar{Y}_2$; Independent Samples

$$\text{Var}(\bar{Y}_1 - \bar{Y}_2) = \text{Var}(\bar{Y}_1) + \text{Var}(\bar{Y}_2) = \frac{\sigma_1^2}{n_1} + \frac{\sigma_2^2}{n_2}$$

standard error of $\bar{Y}_1 - \bar{Y}_2$

The **standard error of $\bar{Y}_1 - \bar{Y}_2$** is

$$\sigma_{\bar{Y}_1 - \bar{Y}_2} = \sqrt{\frac{\sigma_1^2}{n_1} + \frac{\sigma_2^2}{n_2}}$$

EXAMPLE 10.1 Suppose that samples of respective sizes 50 and 32 are drawn from populations with the following characteristics:

	Group 1	Group 2
Population mean	85	75
Population standard deviation	10	16

Find the expected value and standard error of the difference in sample means, $\bar{Y}_1 - \bar{Y}_2$.

Solution

$$E(\bar{Y}_1 - \bar{Y}_2) = 85 - 75 = 10$$

$$\sigma_{\bar{Y}_1 - \bar{Y}_2} = \sqrt{\frac{\sigma_1^2}{n_1} + \frac{\sigma_2^2}{n_2}}$$

$$= \sqrt{\frac{(10)^2}{50} + \frac{(16)^2}{32}} = \sqrt{10.00} = 3.16$$

Note that the standard error of the difference in means is not the sum of the standard errors (and certainly not the difference of them). □

The crucial condition for the use of this formula for the standard error is that the two samples are independent. The formula is not appropriate when there is any matching up or pairing entities in the two samples, as in "before" and "after" measurements on the same individuals. Methods for **paired samples** are discussed in Section 10.4.

paired samples

Given this standard error formula, confidence-interval and hypothesis-testing procedures are direct extensions of the one-sample methods. We first state these two-sample procedures under the artificial assumption that both population variances are known. As in the one-sample case, these procedures can be used with sample variances replacing population variances, provided that both sample sizes are large, say, 30 or more. Procedures for smaller sample sizes are discussed in the next section.

We presented one-sample estimation procedures in Chapters 7 and 8. These results can be summarized as follows: If θ represents the unknown parameter, $\hat{\theta}$ is an unbiased estimator of θ, and $\sigma_{\hat{\theta}}$ is the standard error of the estimator $\hat{\theta}$, then the large sample $100(1 - \alpha)\%$ confidence interval for θ is $\hat{\theta} \pm z_{\alpha/2}\sigma_{\hat{\theta}}$. This same format holds for two-sample problems with $\theta = \mu_1 - \mu_2$, $\hat{\theta} = \bar{Y}_1 - \bar{Y}_2$, and $\sigma_{\hat{\theta}} = \sigma_{\bar{Y}_1 - \bar{Y}_2}$.

$100(1 - \alpha)\%$ Confidence Interval for $\mu_1 - \mu_2$; Large, Independent Samples

$$\bar{y}_1 - \bar{y}_2 - z_{\alpha/2}\sigma_{\bar{Y}_1 - \bar{Y}_2} \le \mu_1 - \mu_2 \le \bar{y}_1 - \bar{y}_2 + z_{\alpha/2}\sigma_{\bar{Y}_1 - \bar{Y}_2}$$

where

$$\sigma_{\bar{Y}_1 - \bar{Y}_2} = \sqrt{\frac{\sigma_1^2}{n_1} + \frac{\sigma_2^2}{n_2}}$$

If both n_1 and n_2 exceed 30, unknown population variances (σ_1^2 and σ_2^2) may be replaced by the respective sample variances s_1^2 and s_2^2.

EXAMPLE 10.2 A health insurer gathers data on the length of hospitalization (in days) of patients operated on for appendicitis. Random samples from two different hospitals yielded the following results:

	Hospital 1	Hospital 2
Sample mean	8.2	9.4
Sample standard deviation	3.6	2.9
Sample size	56	38

Find a 90% confidence interval for $\mu_1 - \mu_2$, the difference in long-run average lengths of hospitalization.

Solution The difference in sample means is $\bar{y}_1 - \bar{y}_2 = 8.2 - 9.4 = -1.2$ and the appropriate z value for a 90% confidence interval is $z_{.05} = 1.645$. Since both sample sizes exceed 30, we can replace the unknown variances (σ_1^2 and σ_2^2) by the corresponding sample variances to obtain

$$\sigma_{\bar{Y}_1 - \bar{Y}_2} \approx \sqrt{\frac{(3.6)^2}{56} + \frac{(2.9)^2}{38}} = .673$$

Substituting into the formula, we obtain the 90% confidence interval for $\mu_1 - \mu_2$:

$$-1.2 - 1.645(.673) \le \mu_1 - \mu_2 \le -1.2 + 1.645(.673)$$

or

$$-2.3 \le \mu_1 - \mu_2 \le -.1$$

We are 90% confident that the long-run length of hospitalization is from .1 to 2.3 days shorter for hospital 1. □

Hypothesis testing for two samples is also a direct extension of the one-sample procedure. If $\hat{\theta}$ is an unbiased estimator of the unknown parameter of interest, and $\sigma_{\hat{\theta}}$ represents the standard error of $\hat{\theta}$, then the large-sample statistical test for θ takes this form:

H_0: $\theta = \theta_0$ (θ_0 is specified)
H_a: 1. $\theta > \theta_0$
2. $\theta < \theta_0$
3. $\theta \neq \theta_0$

T.S.: $z = \dfrac{\hat{\theta} - \theta_0}{\sigma_{\hat{\theta}}}$

R.R.: For Type I error rate α
1. Reject H_0 if $z > z_\alpha$
2. Reject H_0 if $z < -z_\alpha$
3. Reject H_0 if $|z| > z_{\alpha/2}$

This same format applies to a two-sample test with $\theta = \mu_1 - \mu_2$, $\hat{\theta} = \bar{y}_1 - \bar{y}_2$, and $\sigma_{\hat{\theta}} = \sigma_{\bar{Y}_1 - \bar{Y}_2}$.

Hypothesis Test for $\mu_1 - \mu_2$; Large, Independent Samples

H_0: $\mu_1 - \mu_2 = D_0$ (D_0 is specified; often $D_0 = 0$)
H_a: 1. $\mu_1 - \mu_2 > D_0$
2. $\mu_1 - \mu_2 < D_0$
3. $\mu_1 - \mu_2 \neq D_0$

$$\text{T.S.: } z = \frac{(\bar{y}_1 - \bar{y}_2) - D_0}{\sqrt{\dfrac{\sigma_1^2}{n_1} + \dfrac{\sigma_2^2}{n_2}}}$$

R.R.: For a Type I error rate α
1. Reject H_0 if $z > z_\alpha$
2. Reject H_0 if $z < -z_\alpha$
3. Reject H_0 if $|z| > z_{\alpha/2}$

If n_1 and n_2 both exceed 30, σ_1^2 and σ_2^2 may be replaced by s_1^2 and s_2^2, respectively.

EXAMPLE 10.3 Refer to Example 10.2. Test the null hypothesis of equal long-run average stay against a general (two-sided) alternative; use $\alpha = .10$.

Solution

H_0: $\mu_1 - \mu_2 = 0$
H_a: $\mu_1 - \mu_2 \neq 0$

$$\text{T.S.: } z = \frac{(\bar{y}_1 - \bar{y}_2) - 0}{\sqrt{\dfrac{s_1^2}{n_1} + \dfrac{s_2^2}{n_2}}} = \frac{(8.2 - 9.4)}{\sqrt{\dfrac{(3.6)^2}{56} + \dfrac{(2.9)^2}{38}}} = -1.78$$

R.R.: Reject H_0 if $|z| > z_{.05} = 1.645$

Conclusion: $|z| = 1.78 > 1.645$; reject H_0. We conclude that the mean lengths of stay differ. (Practically, it is clear that the mean stay for hospital 1 is shorter.)

This conclusion also follows from the fact that the 90% confidence interval for $\mu_1 - \mu_2$ in Example 10.2 does not include the value zero. □

SECTION 10.1 EXERCISES

10.1 A manufacturer of puffed cereal tries two different preventive-maintenance approaches on two of the "guns" used in processing. The number of hours of operation between required shutdowns is recorded:

	Machine 1	Machine 2
Mean	62.4	55.8
Standard deviation	37.1	42.2
n	126	155

a. Calculate a 95% confidence interval for the difference in true means.
b. Can the hypothesis of equal mean operation times be rejected at $\alpha = .05$?

10.2 Calculate the p-value in Exercise 10.1.

10.3 A wholesaler specializing in drugstore items incurs considerable labor costs in "picking" every order. In an effort to reduce the time per order, a minicomputer is programmed to list the items in an efficient order. Two programs, based on different efficiency principles, are tested. One hundred orders are run through each program, and the total labor time per order is recorded. The data yield the following:

	Program 1	Program 2
Mean (hours)	1.64	1.89
Standard deviation	1.02	0.94

a. Calculate a 90% confidence interval for the difference in long-run mean times.
b. Test the null hypothesis of equal means against a two-sided alternative. Use $\alpha = .10$.

10.4 Find the p-value of the test in Exercise 10.3, part (b).

10.5 What would you guess the skewness of the data underlying Exercise 10.3 might be? If you feel that there may be substantial skewness, does this invalidate your conclusions in Exercises 10.3 and 10.4?

10.6 Two agencies supply temporary workers for light manufacturing. Both agencies give their workers a test of dexterity. The two populations at a particular time have the following means and standard deviations.

	Agency K	Agency R
Mean	78.28	75.79
Standard deviation	9.63	11.25

Samples of 8 workers from agency K and 12 from agency R are sent to a manufacturer on one particular day. Each worker's dexterity score is recorded.

a. What is the expected value of the difference of mean dexterity scores (agency K − agency R)?
b. Find the standard error of this difference.

10.7 Suppose that the sample in Exercise 10.6 had been 12 from agency K and 8 from agency R. How, if at all, would your answers to Exercise 10.6 change?

10.8 A trucking firm specializes in LTL (less-than-truckload) shipments for mail order firms. The firm uses two priority classes and charges fees accordingly. The mean delivery time for class 1 is 7.20 working days and the standard deviation is 4.21 working days. For class 2, the mean is 11.41 and the standard deviation is 6.33. In checking to see whether these times have changed, the firm routinely takes separate random samples of 25 shipments from each of the two priority classes and finds the sample mean delivery times.

a. What is the expected difference in the sample means?

b. What is the variance of the difference of the means?

10.9 Suppose that the samples in Exercise 10.8 are not taken completely randomly. Instead, all shipments are placed into 25 distance categories (with category 1 being nearest and category 25 being farthest). Then one class 1 shipment and one class 2 shipment are chosen randomly from each category. Do your answers to Exercise 10.8 remain valid?

10.2 COMPARING THE MEANS OF TWO POPULATIONS WITH UNKNOWN STANDARD DEVIATIONS

The confidence interval for a population mean μ is of the form $\bar{y} \pm z_{\alpha/2}\sigma/\sqrt{n}$. In Chapter 8 we learned that when the unknown σ is replaced by s, $t_{\alpha/2}$ values (with $n - 1$ d.f.) are used instead of $z_{\alpha/2}$ values in confidence intervals and the statistical test. It would perhaps seem natural that, in the two-sample case with small samples sizes when σ_1 and σ_2 are replaced by the sample standard deviations s_1 and s_2, we would use $t_{\alpha/2}$ values rather than $z_{\alpha/2}$ values for the corresponding confidence intervals and test. However, this is *not* the case.

assumption of equal variances

When the standard deviations are unknown, inferences about $\mu_1 - \mu_2$ based on independent random samples require an additional assumption besides independence and population normality. We also assume that the two unknown population variances are equal: $\sigma_1^2 = \sigma_2^2$. The common unknown variance is designated σ^2.

To help motivate the results, note that if $\sigma_1^2 = \sigma_2^2 = \sigma^2$, the large-sample confidence interval for $\mu_1 - \mu_2$ can be written

$$\bar{y}_1 - \bar{y}_2 - z_{\alpha/2}\sigma\sqrt{\frac{1}{n_1} + \frac{1}{n_2}} < \mu_1 - \mu_2 < \bar{y}_1 - \bar{y}_2 + z_{\alpha/2}\sigma\sqrt{\frac{1}{n_1} + \frac{1}{n_2}}$$

The corresponding confidence interval for $\mu_1 - \mu_2$ is shown here.

100(1 − α)% Confidence Interval for $\mu_1 - \mu_2$, with σ's Unknown and Independent Samples

$$\bar{y}_1 - \bar{y}_2 - t_{\alpha/2}s_p\sqrt{\frac{1}{n_1} + \frac{1}{n_2}} < \mu_1 - \mu_2 < \bar{y}_1 - \bar{y}_2 + t_{\alpha/2}s_p\sqrt{\frac{1}{n_1} + \frac{1}{n_2}}$$

where

$$s_p = \sqrt{\frac{(n_1 - 1)s_1^2 + (n_2 - 1)s_2^2}{n_1 + n_2 - 2}}$$

and

$$\text{d.f.} = n_1 + n_2 - 2$$

Note: This procedure can be used for all sample sizes.

The quantity s_p in the confidence interval for $\mu_1 - \mu_2$ is an estimate of the common population standard deviation σ and is formed by combining information from the two independent samples. The two estimates s_1^2 and s_2^2 are weighted by their respective degrees of freedom to form the **pooled variance** s_p^2. For the special case in which the sample sizes are the same ($n_1 = n_2$), the formula for s_p^2 reduces to $s_p^2 = (s_1^2 + s_2^2)/2$, the average of the two sample variances. The degrees of freedom for s_p^2 combine the degrees of freedom for s_1^2 and s_2^2; d.f. $= (n_1 - 1) + (n_2 - 1) = n_1 + n_2 - 2$.

pooled variance

EXAMPLE 10.4 A taxicab company wants to test two programs for improving the gasoline mileage of its drivers. Under program A, drivers are assigned a target mileage and receive modest bonuses for better performance. Under program B, drivers are allowed a maximum monthly quota of gasoline; if it runs out, a driver has to pay for extra gasoline out of pocket. All taxis used are standard models and are given standard maintenance. After three months, each driver's mileage per gallon is calculated. The data are as follows:

A:	15.9	17.5	19.1	16.9	18.3	17.3	17.0	16.2	16.8	17.1
B:	16.1	15.8	15.3	16.5	14.9	15.5	16.4	16.0	16.7	17.2

Find a 95% confidence interval for the difference in mean gasoline mileage.

Solution

Program	Mean	Variance	Standard Deviation	Sample Size
A	17.21	.8788	.9374	10
B	16.04	.4804	.6931	10

The required table value for a 95% confidence interval with 10 + 10 − 2 = 18 d.f. is $t_{.025}$; from Appendix Table 4, $t_{.025} = 2.101$. The pooled sample variance is

$$s_p^2 = \frac{9(.8788) + 9(.4804)}{18} = .6796$$

so

$$s_p = \sqrt{.6796} = .8244$$

The confidence interval is

$$(17.21 - 16.04) - 2.101(.8244)\sqrt{\frac{1}{10} + \frac{1}{10}} < \mu_A - \mu_B < (17.21 - 16.04) + 2.101(.8244)\sqrt{\frac{1}{10} + \frac{1}{10}}$$

or

$$.40 < \mu_A - \mu_B < 1.94$$

□

The corresponding t test for comparing μ_1 and μ_2 based on independent samples with the standard deviations unknown is summarized here:

Hypothesis Test for $\mu_1 - \mu_2$; Independent Samples and σ's Unknown

H_0: $\mu_1 - \mu_2 = D_0$ (D_0 is specified; often $D_0 = 0$)

H_a: 1. $\mu_1 - \mu_2 > D_0$
2. $\mu_1 - \mu_2 < D_0$
3. $\mu_1 - \mu_2 \neq D_0$

$$\text{T.S.: } t = \frac{(\bar{y}_1 - \bar{y}_2) - D_0}{s_p\sqrt{\dfrac{1}{n_1} + \dfrac{1}{n_2}}}$$

R.R.: 1. $t > t_\alpha$
2. $t < -t_\alpha$
3. $|t| > t_{\alpha/2}$

where t_a cuts off a right-tail area a for the t distribution with $n_1 + n_2 - 2$ d.f.
Note: This method can be used for all sample sizes.

EXAMPLE 10.5 Refer to Example 10.4. Test the research hypothesis that program A yields a higher mean mileage than program B. Use $\alpha = .10$.

Solution

H_0: $\mu_A - \mu_B = 0$

H_a: $\mu_A - \mu_B > 0$

$$\text{T.S.: } t = \frac{(\bar{y}_A - \bar{y}_B) - 0}{s_p\sqrt{\dfrac{1}{n_1} + \dfrac{1}{n_2}}} = \frac{(17.21 - 16.04)}{.8244\sqrt{\dfrac{1}{10} + \dfrac{1}{10}}} = 3.17$$

R.R.: Reject H_0 if $t > t_{.10,\, 18\text{ d.f.}} = 1.330$

Conclusion: Because $t = 3.17 > 1.33$, the mean mileage under A is significantly greater than under B based on $\alpha = .10$. In fact, since H_0 can be rejected at $\alpha = .01$ ($t_{.01,18} = 2.552$), the p-value is less than .01. □

The two-sample t test and confidence interval are based on several mathematical assumptions. Once again, these assumptions are not exactly satisfied in practice. The most crucial assumption is the independence of the two samples. If this assumption is not valid, the procedures can be grossly erroneous. If the samples are taken from different populations, and if there is no connection between the elements of one sample and those of the other, the **independence assumption** should be valid. But if the two measurements are taken on the same elements at different times or if there is any connection between elements of the samples, the two-sample t test is not appropriate and other methods of analysis must be used. For example, if one sample represents measurements on product awareness for individuals before an advertising campaign and the second sample represents measurements of product awareness on these same individuals after advertising exposure, the two-sample t-test is not appropriate. The paired-sample procedures of Section 10.4 should be used to compare the difference in mean awareness before and after.

independence assumption

normality assumption

The assumption that both populations are normally distributed is less crucial because of the Central Limit Theorem. Even if the populations are not normal, the sampling distributions of $\bar{Y}_1$ and $\bar{Y}_2$ are approximately normal for modestly large sample sizes. In fact, we believe that the quality of the Central Limit Theorem normal approximation is determined largely by the total sample size $n_1 + n_2$. Even moderate population skewness is not a serious problem. If both populations are skewed in the same direction, the fact that we are dealing with a difference in means tends to make the sampling distribution of $\bar{Y}_1 - \bar{Y}_2$ more symmetric. Generally, if $n_1 + n_2$ is at least 30 or so, we are confident of the t probabilities; if the data are reasonably symmetric within each sample, even a total sample of 15 or so should serve. Of course, with small samples, confidence intervals are wide and β probabilities high. The point is that the t probabilities are reasonably accurate. A nonparametric alternative to the two-sample t test (called Wilcoxon's rank sum test) that does not require normality of the two populations is presented in Section 10.3.

assumption of equal variances

The new assumption is that of equal population variances; even though two population variances are equal, the sample variances differ because of random variation. Many studies have been made about the effect of unequal population variances. The universal conclusion is that for equal sample sizes, even substantial differences in variances (such as $\sigma_1^2 = 3\sigma_2^2$) have remarkably little effect. The most dangerous situation is one in which the larger population variance is associated with the smaller sample size. If n_1 is only half the size of n_2 but σ_1^2 is, say, twice σ_2^2, the nominal t probabilities may be seriously in error. The best cure for this problem is to take equal sample sizes.

When the sample variances (s_1^2 and s_2^2) suggest that there may be a problem in assuming that the two population variances are equal, we can modify the usual t statistic to obtain an approximate t test. Welch (1938) showed that the distribution of the statistic

$$t' = \frac{\bar{y}_1 - \bar{y}_2}{\sqrt{\dfrac{s_1^2}{n_1} + \dfrac{s_2^2}{n_2}}}$$

can be approximated by a t distribution using Welch's approximation:

Welch's Approximation for the t' Statistic; Independent Samples

1. H_0: $\mu_1 - \mu_2 = 0$
2. The test statistic is

$$t' = \frac{\bar{y}_1 - \bar{y}_2}{\sqrt{(s_1^2/n_1) + (s_2^2/n_2)}}$$

3. The rejection region for t' can be obtained from Appendix Table 4 for

$$\text{d.f.} = \frac{(n_1 - 1)(n_2 - 1)}{(n_2 - 1)c^2 + (1 - c)^2(n_1 - 1)}$$

where

$$c = \frac{s_1^2/n_1}{s_1^2/n_1 + s_2^2/n_2}$$

Note: If d.f. is not an integer, round *down* to the nearest integer.

The test based on the t' statistic is sometimes referred to as the **separate variance *t* test** since the test statistic is identical to that (presented in Section 10.1) for large samples. The statistic has each population variance (σ_1^2 and σ_2^2) replaced by the separate sample variances s_1^2 and s_2^2.

EXAMPLE 10.6 A firm has a generous but rather complicated policy concerning end-of-year bonuses for its lower-level managerial personnel. The policy's key factor is a subjective judgment of "contribution to corporate goals." A personnel officer takes samples of 24 female and 36 male managers to see if there are any difference in bonuses, expressed as a percentage of yearly salary. The data are listed here:

Gender	Bonus Percentage								
F	9.2	7.7	11.9	6.2	9.0	8.4	6.9	7.6	7.4
	8.0	9.9	6.7	8.4	9.3	9.1	8.7	9.2	9.1
	8.4	9.6	7.7	9.0	9.0	8.4			
M	10.4	8.9	11.7	12.0	8.7	9.4	9.8	9.0	9.2
	9.7	9.1	8.8	7.9	9.9	10.0	10.1	9.0	11.4
	8.7	9.6	9.2	9.7	8.9	9.2	9.4	9.7	8.9
	9.3	10.4	11.9	9.0	12.0	9.6	9.2	9.9	9.0

A computer program yields the following output:

```
                                         TTEST PROCEDURE

VARIABLE: PERCENT

SEX      N          MEAN        STD DEV      STD ERROR        MINIMUM          MAXIMUM    VARIANCES        T

F       24    8.53333333     1.18895887     0.24269521     6.20000000      11.90000000    UNEQUAL    -3.9013
M       36    9.68333333     1.00384973     0.16730829     7.90000000      12.00000000    EQUAL      -4.0367

FOR HO: VARIANCES ARE EQUAL, F'=     1.40 WITH 23 AND 35 DF     PROB > F'= 0.3584

                                    WILCOXON SCORES (RANK SUMS)

                               SUM OF   EXPECTED    STD DEV        MEAN
          LEVEL           N    SCORES   UNDER HO   UNDER HO       SCORE

          F              24    481.00     732.00      66.15       20.04
          M              36   1349.00    1098.00      66.15       37.47
```

a. Identify the value of the pooled-variance t statistic.
b. Identify the value of the t' statistic.
c. Use both statistics to test the research hypothesis of unequal means at $\alpha = .05$ and at $\alpha = .01$. Does the conclusion depend on which statistic is used?

Solution

a. The pooled-variance statistic is $t = -4.0367$.
b. $t' = -3.9013$.

c. The t statistic based on the pooled variance has d.f. $= 24 + 36 - 2 = 58$. For a two-sided H_a, we reject H_0 at $\alpha = .05$ if $|t| > t_{.025} \approx 2.00$; with $\alpha = .01$, reject if $|t| > t_{.005} \approx 2.66$. Because $|t| = 4.037$, we can easily reject H_0 even at $\alpha = .01$. For the t' statistic based on separate variances, the degrees of freedom can be computed using the formula

$$\text{d.f.} = \frac{(n_1 - 1)(n_2 - 1)}{(n_2 - 1)c^2 + (1 - c)^2(n_1 - 1)}, \qquad \text{with } c = \frac{s_1^2/n_1}{s_1^2/n_1 + s_2^2/n_2}$$

For these data,

$$c = \frac{1.4137/24}{1.4137/24 + 1.0076/36} = \frac{.0589}{.0589 + .0280} = .6778$$

and

$$c^2 = .4594, \qquad (1 - c)^2 = .1038$$

Then

$$\text{d.f.} = \frac{23(35)}{35(.4594) + .1038(23)} = \frac{805}{16.0790 + 2.3876} = \frac{805}{18.4671} = 43.59$$

Rounding down to the nearest integer, d.f. $= 43$. For a two-sided research hypothesis and $\alpha = .05$, we reject H_0 if $|t'| > t_{.025,43} \approx 2.02$. For $\alpha = .01$, we reject H_0 if $|t'| > t_{.005,43} = 2.70$. Since $|t'| = 3.90$, we can easily reject H_0 even at $\alpha = .01$. The conclusions from the t and t' tests are essentially the same and the research hypothesis is quite conclusively supported. □

We have presented several different approaches in the last two sections. In this section we developed pooled-variance t methods based on an assumption of equal population variances. In addition, we introduced the t' statistic for an approximate t when the variances are not equal. In Section 10.1 we used separate variances in a z statistic and appealed to large-sample theory. Confidence intervals and hypothesis tests based on these different procedures (t, t', or z) need not give identical results. Standard computer packages often report the results of both the pooled-variance and separate-variance t tests. Which should a manager believe?

The choice depends on the evidence about underlying assumptions. If plots of each sample appear roughly normal, and if the sample variances are roughly equal, the pooled-variance t test should be valid and most efficient. If plots of each sample are normal but the sample variances are clearly different (especially if the sample sizes differ), the separate-variance t' test is more believable. If the sample sizes are equal, the pooled-variance and separate-variance t tests will usually give the same results; in fact, the test statistics are algebraically equal for equal sample sizes. But if the data in one or both samples are obviously non-normal, the rank sum approach discussed in the next section is preferred. As usual, a little thought and some careful looks at the data will let a manager make a reasonable choice.

EXAMPLE 10.7 A simulation study involves samples from independent, normal populations. The following results are obtained:

```
              Checking Alpha (different sample sizes and sigmas)

Simulation of Two Sample t-test (1000 samples)

Popn         Mu          Sigma          n
  1        50.000       14.1421        10
  2        50.000       10.0000        20

using the pooled variance t test

one-tail:

       number of times H0:  "mu1-mu2 is   0" is rejected in favor of
alpha   "mu1-mu2 >   0"    "mu1-mu2 <   0"    total (alpha doubled)
0.100         137                127                 264
0.050          70                 71                 141
0.025          36                 34                  70
0.010          14                 16                  30
0.005          11                 10                  21

using separate variances (t')

one-tail:

       number of times H0:  "mu1-mu2 is   0" is rejected in favor of
alpha   "mu1-mu2 >   0"    "mu1-mu2 <   0"    total (alpha doubled)
0.100         104                 99                 203
0.050          44                 41                  85
0.025          20                 21                  41
0.010          10                 10                  20
0.005           4                  4                   8


average pooled t is 0.0047 with variance of    1.354355
average t' is 0.0071 with variance of    1.122882
```

What do these results indicate about the choice of pooled-variance t versus t'?

Solution One of the assumptions underlying the pooled-variance t test has been violated; the population variances aren't equal. The null hypothesis is true, because the population means are equal. The pooled-variance t test rejects the null hypothesis more frequently than the nominal α value would indicate. For example, for a nominal α of .05 (one-tailed test), we would expect 50 and 50 rejections, but we get 70 and 71 rejections. The t' test rejects the null hypothesis just about as often as α indicates. □

EXAMPLE 10.8 Another simulation study is done with samples from independent normal populations, with the following results.

```
                 Checking Alpha (different sample sizes and sigmas)

Simulation of Two Sample t-test (1000 samples)

Popn          Mu          Sigma          n
  1         50.000       10.0000        10
  2         50.000       14.1421        20

using the pooled variance t test

one-tail:

             number of times H0:  "mu1-mu2 is    0" is rejected in favor of
alpha      "mu1-mu2 >    0"     "mu1-mu2 <    0"      total (alpha doubled)
0.100              71                   84                  155
0.050              32                   36                   68
0.025              13                   16                   29
0.010               4                    7                   11
0.005               1                    3                    4

using separate variances (t')

one-tail:

             number of times H0:  "mu1-mu2 is    0" is rejected in favor of
alpha      "mu1-mu2 >    0"     "mu1-mu2 <    0"      total (alpha doubled)
0.100             103                  104                  207
0.050              43                   48                   91
0.025              21                   20                   41
0.010               7                    9                   16
0.005               1                    5                    6

average pooled t is -0.0196 with variance of    0.873156
average t' is -0.0226 with variance of    1.088475
```

What do these results indicate about t versus t'?

Solution Again, the assumption of equal variances is violated. This time, however, the smaller variance is associated with the smaller sample size; in Example 10.7 the smaller variance was associated with the larger sample size. In this example, we find that the number of false rejections of the null hypothesis is consistently *smaller* than would be indicated by the nominal α. Again, the t' test rejects the null hypothesis just about as often as α indicates. □

SECTION 10.2 EXERCISES

10.10 A processor of recycled aluminum cans is concerned about the levels of impurities (principally other metals) contained in lots from two sources. Laboratory analysis of sample lots yields the following data (kilograms of impurities per hundred kilograms of product):

Source I: 3.8 3.5 4.1 2.5 3.6 4.3 2.1 2.9 3.2 3.7 2.8 2.7
mean = 3.267, standard deviation = .676)

Source II: 1.8 2.2 1.3 5.1 4.0 4.7 3.3 4.3 4.2 2.5 5.4 4.6
mean = 3.617, standard deviation = 1.365)

a. Calculate the pooled variance and standard deviation.
b. Calculate a 95% confidence interval for the difference in mean impurity levels.
c. Can the processor conclude, using $\alpha = .05$, that there is a nonzero difference in means?

10.11 Calculate the p-value in Exercise 10.10, part (c).

10.12 A simulation study involves samples from independent normal populations. The sample sizes are unequal, as are the population variances.
a. Should the pooled-variance t test or the t' test be better in this situation? What does "better" mean?
b. Does the output shown below confirm your judgment in part (a)?

```
              Checking Alpha (different sample sizes and sigmas)

Popn        Mu          Sigma         n
  1       50.000       20.0000        5
  2       50.000       10.0000       25

using the pooled variance t test

one-tail:

        number of times H0:  "mu1-mu2 is   0" is rejected in favor of
alpha    "mu1-mu2 >   0"    "mu1-mu2 <   0"    total (alpha doubled)
0.100          211                209                420
0.050          152                142                294
0.025          101                102                203
0.010           65                 63                128
0.005           44                 43                 87

using separate variances (t′)

one-tail:

        number of times H0:  "mu1-mu2 is   0" is rejected in favor of
alpha    "mu1-mu2 >   0"    "mu1-mu2 <   0"    total (alpha doubled)
0.100          106                 97                203
0.050           53                 46                 99
0.025           23                 25                 48
0.010           13                  9                 22
0.005           12                  6                 18

average pooled t is 0.0166 with variance of    2.711859
average t′ is 0.0336 with variance of    1.843635
average approx. df is 4.5040 with variance of    3.407391
```

10.13 Construct separate plots of the impurities data from each source in Exercise 10.10. Which of the assumptions (if any) of the t test seem suspect? Do you think that there is serious reason to doubt the conclusion of Exercise 10.10, part (c)?

10.14 Company officials are concerned about the length of time a particular drug retains its potency. A random sample (sample 1) of 10 bottles of the product is drawn from current production and analyzed for potency. A second sample (sample 2) is obtained, stored for one year, and then analyzed. The readings obtained are

Sample 1:	10.2	0.5	10.3	10.8	9.8	10.6	10.7	10.2	10.0	10.6
Sample 2:	9.8	9.6	10.1	10.2	10.1	9.7	9.5	9.6	9.8	9.9

The data are analyzed by a standard program package (SAS). The relevant output is shown on the top of the next page.

```
VARIABLE: POTENCY

SAMPLE    N          MEAN      STD DEV     STD ERROR      MINIMUM      MAXIMUM   VARIANCES       T     DF  PROB > |T|

  1      10   10.37000000   0.32335052    0.10225241   9.80000000  10.80000000   UNEQUAL    4.2368   16.6     0.0006
  2      10    9.83000000   0.24060110    0.07608475   9.50000000  10.20000000   EQUAL      4.2368   18.0     0.0005

FOR HO: VARIANCES ARE EQUAL, F'=    1.81 WITH 9 AND 9 DF     PROB > F'= 0.3917
```

a. Identify the sample means and standard deviations.

b. Locate the value of the t statistic. Is the pooled-variance t statistic identified as "equal variance" or "unequal variance"?

c. Locate the value of the t' statistic.

d. Why are these two statistics equal in this case?

10.15 a. Plot the data of Exercise 10.14. Use separate plots for the two samples.

b. Does it seem that there are serious violations of the assumptions underlying the pooled-variance test?

10.16 a. Locate the p-value for the pooled-variance t test in the output of Exercise 10.14. Is it one-tailed or two-tailed?

b. What conclusion would you reach concerning the possibility of a decrease in mean potency over one year?

10.17 To compare the performance of microcomputer spreadsheet programs, teams of three students each choose whatever spreadsheet program they wish. Each team is given the same set of standard accounting and finance problems to solve. The time (minutes) required for each team to solve the set of problems is recorded. The following data are obtained for the two most widely used programs:

Program	Time										$\bar{y}$	s	n
A	39	57	42	53	41	44	71	56	49	63	51.50	10.46	10
B	43	38	35	45	40	28	50	54	37	29			
	36	27	52	33	31	30					38.00	8.67	16

a. Calculate the pooled variance.

b. Use this variance to find a 99% confidence interval for the difference of population means.

c. According to this interval, can the null hypothesis of equal means be rejected at $\alpha = .01$?

10.18 Redo parts (b) and (c) of Exercise 10.17 using a separate-variance (t') method. Which method is more appropriate in this case? How critical is it which method is used?

10.19 A manufacturer of modems uses microcomputer chips from two difference sources. As part of quality-control testing, the manufacturer obtains data on the rate of defective chips per thousand for each lot of chips. The following results are obtained; note that the rate is not necessarily an integer number, because the lot sizes are not exactly 1000 chips per lot.

Source	Number of defectives/1000									
I	9.8	9.9	10.2	10.5	10.7	10.8	11.7	13.9	19.2	27.6
II	10.6	11.0	11.5	11.8	11.9	12.7	14.2	16.8	21.7	29.9

a. Calculate means and standard deviations for both sources.

b. Calculate the pooled-variance t statistic for testing the null hypothesis of equal means.

c. Calculate the t' statistic for testing the same hypothesis.
d. Explain why t and t' are equal for these data.

10.20 a. Can the null hypothesis in Exercise 10.19 be rejected at $\alpha = .05$ in favor of a two-sided research hypothesis? Use the pooled-variance t statistic.
b. State bounds on the p-value.
c. Calculate the approximate d.f. for the t' statistic.
d. Redo parts (a) and (b) using the t' statistic.

10.21 Another simulation study comparing the pooled-variance t and t' tests involves independent samples from normal populations. The population variances are unequal, but the sample sizes are equal. The following results are obtained:

```
                 Checking Alpha (same sample sizes; different sigmas)

 Simulation of Two Sample t-test (1000 samples)

 Popn          Mu          Sigma          n
   1        50.000       20.0000         10
   2        50.000       10.0000         10

 using the pooled variance t test

 one-tail:

         number of times H0:  "mu1-mu2 is   0" is rejected in favor of
 alpha     "mu1-mu2 >    0"   "mu1-mu2 <   0"     total (alpha doubled)
 0.100          112                101                 213
 0.050           44                 47                  91
 0.025           20                 24                  44
 0.010            7                 10                  17
 0.005            3                  5                   8

 using separate variances (t')

 one-tail:

         number of times H0:  "mu1-mu2 is   0" is rejected in favor of
 alpha     "mu1-mu2 >    0"   "mu1-mu2 <   0"     total (alpha doubled)
 0.100          103                 97                 200
 0.050           41                 45                  86
 0.025           15                 21                  36
 0.010            6                  9                  15
 0.005            3                  3                   6

 where both methods agree

 one-tail:

         number of times H0:  "mu1-mu2 is   0" is rejected in favor of
 alpha     "mu1-mu2 >    0"   "mu1-mu2 <   0"     total (alpha doubled)
 0.100          103                 97                 200
 0.050           41                 45                  86
 0.025           15                 21                  36
 0.010            6                  9                  15
 0.005            3                  3                   6


 average pooled t is 0.0165 with variance of    1.118298
 average t' is 0.0165 with variance of    1.118298
 average approx. df is 13.1030 with variance of    4.681072
```

a. Which test is better here? How important is the choice?
b. What would the d.f. be for the pooled-variance t test? How does that number compare to the average d.f. for the t' test?

10.3 A NONPARAMETRIC ALTERNATIVE: THE WILCOXON RANK SUM TEST

The two-sample t test described in the previous section is based on several mathematical assumptions. In particular, we assume that both populations have normal distributions with equal variances. When the assumptions are not satisfied, the t test may still be valid, in the sense that the nominal probabilities are approximately correct, particularly if the sample sizes are large and equal. Even so, there is another hypothesis-testing method that requires weaker mathematical assumptions, is almost as powerful when the t assumptions are satisfied, and is more powerful in other situations. We describe this test, called the *Wilcoxon rank sum test*, in this section.

The mathematical assumption for this test is that independent random samples are taken from two populations; the null hypothesis is that the two population distributions are identical (but not necessarily normal). The Wilcoxon rank sum test probabilities are exactly correct for any two populations with identical continuous distributions and are generally conservative for two populations with identical discrete distributions.

ranking sample data

The test is based on the ranks of the sample data values. The rank of an individual observation is its position in the combined sample: rank 1 indicates the smallest value, rank 2 indicates the next smallest value, and so on. As the phrase *rank sum test* indicates, the Wilcoxon rank sum test is based on the sum of the ranks in either sample. Under the null hypothesis of identical population distributions, the sum of the ranks in one sample is proportional to the sample size. If one population is shifted to the right of another, that is, if the first population tends to yield larger observations, the rank sum for the first sample tends to be large. Of course a small rank sum for the first sample indicates that the first population is shifted to the left of the second. Define T to be the sum of the ranks in the first sample. Under the null hypothesis, the expected value and variance of T have been determined:

$$\mu_T = \frac{n_1(n_1 + n_2 + 1)}{2}$$

$$\sigma_T^2 = \frac{n_1 n_2}{12}(n_1 + n_2 + 1)$$

If both n_1 and n_2 are 10 or larger, the sampling distribution of T is approximately normal. This allows use of a z statistic in testing the hypothesis of equal distributions.

Wilcoxon Rank Sum Test

H_0: The two populations are identical

H_a: 1. Population 1 is shifted to the right of population 2
2. Population 1 is shifted to the left of population 2
3. Population 1 is shifted to the right or left of population 2

T.S.: $z = \dfrac{T - \mu_T}{\sigma_T}$

where T denotes the rank sum for sample 1

R.R.: 1. $z > z_\alpha$
2. $z < -z_\alpha$
3. $|z| > z_{\alpha/2}$

Note: The normal approximation is reasonably accurate if $n_1 \geq 10$ and $n_2 \geq 10$. Special tables are available for smaller values of n_1 and n_2 (e.g., Hollander and Wolfe, 1973).

EXAMPLE 10.9 Perform a rank sum test for Example 10.5.

Solution The first step is to rank the observations. It helps in doing the ranking to order the values in each sample from lowest to highest.

	Program A									
Value:	15.9	16.2	16.8	16.9	17.0	17.1	17.3	17.5	18.3	19.1
Rank:	5	8	12	13	14	15	17	18	19	20

	Program B									
Value:	14.9	15.3	15.5	15.8	16.0	16.1	16.4	16.5	16.7	17.2
Rank:	1	2	3	4	6	7	9	10	11	16

The sum of the ranks in the A sample is

$$T = 5 + 8 + \ldots + 20 = 141$$

Under the null hypothesis

$$\mu_T = \frac{10(10 + 10 + 1)}{2} = 105$$

$$\sigma_T^2 = \frac{(10)(10)}{12}(10 + 10 + 1) = 175$$

$$\sigma_T = \sqrt{175} = 13.23$$

So

$$z = \frac{T - \mu_T}{\sigma_T} = \frac{141 - 105}{13.23} = 2.72$$ □

In Example 10.5, the research hypothesis H_a was that the mean for program A was larger than the mean for program B. The corresponding research hypothesis for the rank sum test is that the program A distribution is shifted to the right of the program B distribution. This research hypothesis is supported and the null hypothesis is rejected if T (and therefore z) is too large to be attributed to chance. The one-tailed p-value for $z = 2.72$ is .0033, so the null hypothesis is rejected for $\alpha = .10, .05, .01$, or even .005. As $n_1 = n_2 = 10$, we are barely within the adequacy range of the normal approximation. It is conceivable that the real p-value is a bit larger than .0033, but still the null hypothesis should be rejected at any conventional α level.

EXAMPLE 10.10 Refer to the computer output of Example 10.6.

a. Identify the value of the rank sum statistic.
b. Find the test statistic for the null hypothesis of equal distribution of bonuses by gender.
c. State an approximate two-tailed p-value for the test in part (b).
d. How does the conclusion of this test compare with that found in Example 10.6?

Solution

a. The sum of the ranks in sample 1 is shown as 481.0.
b. Because $n_1 = 24$ and $n_2 = 36$, a normal approximation should be quite good.

$$\mu_T = \frac{24(24 + 36 + 1)}{2} = 732.0$$

$$\sigma_T^2 = \frac{(24)(36)(24 + 36 + 1)}{12} = 4392.0$$

$$z = \frac{481.0 - 732.0}{\sqrt{4392.0}} = -3.79$$

c. The z value is beyond the range of our z tables. The largest table entry, 3.09, corresponds to a one-tailed area of .001. Therefore a z value of -3.79 must correspond to a two-tailed area less than 2(.001) = .002.
d. As in Example 10.6, we have conclusive support for the research hypothesis. □

treatment of ties

The theory behind the rank sum test assumes that the population distributions are continuous, so there is zero probability that two observations are exactly equal. In practice there are often ties—two or more equal observations. Each observation in a set of tied values is assigned the average of the ranks for the set. If two observations are tied for ranks 2 and 3, each is given rank 2.5; the next larger value gets rank 4, and so on. There is a correction to the variance formula for the case of tied ranks (see Ott, 1984). The variance formula given above is generally conservative and usually very close, unless there are many, many ties.

The Wilcoxon rank sum test is a direct competitor of the two-sample t test. Both tests are sensitive to differences in location (mean or median) as opposed to dispersion or spread. The rank sum test requires fewer assumptions than the t test (in particular it does not assume population normality), but it uses less information from the data; only ordering information is relevant to the rank sum test.* When the assumptions underlying the t test are close to correct, the t test is better. Both theoretical results and simulations clearly indicate that a t test (using the pooled variance if the sample variances are equal or if the sample sizes are equal, but separate variances if both variances and sample sizes aren't equal) will have correct α values and optimal power for normal populations. For obviously non-normal data, the rank sum test has a more believable α value (especially for small samples) and usually has better power.

EXAMPLE 10.11 To investigate the effect of skewness on the pooled-variance t test as well as the rank sum test, 1000 samples are drawn from a squared-exponential population; this population is extremely right-skewed. The following results are obtained.

```
                 Checking Alpha (different sample sizes; same sigmas)

 Simulation of Two Sample t-test (1000 samples)

 Popn          Mu          Sigma             n
   1         50.000       10.0000            5
   2         50.000       10.0000           25

 using the pooled variance t test

 one-tail:

        number of times H0:  "mu1-mu2 is    0" is rejected in favor of
 alpha    "mu1-mu2 >    0"   "mu1-mu2 <    0"     total (alpha doubled)
 0.100          146                  35               181
 0.050           95                   3                98
 0.025           51                   0                51
 0.010           25                   0                25
 0.005           15                   0                15

 Results of Wilcoxon Rank Sum Test using Z as test statistic

       number of times H0:  "two populations are identical" rejected in fa
 alpha    Popn1 rt of Popn2   Popn1 left of Popn2    total (alpha doubled)
 0.100          102                  93                   195
 0.050           37                  51                    88
 0.025           15                  30                    45
 0.010            5                  14                    19
 0.005            4                   3                     7
```

What do the results indicate about the effect of skewness on the two tests?

Solution The null hypothesis is true in this simulation; both means are 50. The actual number of rejections of the null hypothesis by the t test is far from what is indicated by the nominal α value for one-tailed probabilities. The rank sum test, which doesn't assume normal populations, appears to be rejecting the null hypothesis the correct number of times. □

* Therefore the rank sum test can be used when the observations are qualitative and ordinal, as when 1 = strongly opposed, 2 = opposed, 3 = neutral, and so on.

EXAMPLE 10.12 A simulation study investigating the effect of outliers on the *t* and rank sum tests involves independent samples from Laplace (mildly outlier-prone) populations. One part of the study has both population means equal; a second part involves different means. The following results are obtained:

```
                   Checking Alpha (same sample sizes and sigmas)

Simulation of Two Sample t-test (1000 samples)

Popn         Mu         Sigma         n
  1       50.000       10.0000       30
  2       50.000       10.0000       30

using the pooled variance t test

one-tail:

      number of times H0:  "mu1-mu2 is   0" is rejected in favor of
alpha   "mu1-mu2 >   0"   "mu1-mu2 <   0"    total (alpha doubled)
0.100          106                98               204
0.050           50                47                97
0.025           22                22                44
0.010            7                10                17
0.005            3                 5                 8
Results of Wilcoxon Rank Sum Test using Z as test statistic

     number of times H0:  "two populations are identical" rejected in fa
alpha   Popn1 rt of Popn2   Popn1 left of Popn2   total (alpha doubled)
0.100            111                98                 209
0.050             53                47                 100
0.025             21                21                  42
0.010              7                12                  19
0.005              5                 4                   9

                   Checking Power (same sample sizes and sigmas)

Simulation of Two Sample t-test (1000 samples)

Popn         Mu         Sigma         n
  1       50.000       10.0000       30
  2       60.000       10.0000       30

using the pooled variance t test

one-tail:

      number of times H0:  "mu1-mu2 is   0" is rejected in favor of
alpha   "mu1-mu2 >   0"   "mu1-mu2 <   0"    total (alpha doubled)
0.100            0               996               996
0.050            0               986               986
0.025            0               962               962
0.010            0               912               912
0.005            0               864               864
Results of Wilcoxon Rank Sum Test using Z as test statistic

     number of times H0:  "two populations are identical" rejected in fa
alpha   Popn1 rt of Popn2   Popn1 left of Popn2   total (alpha doubled)
0.100              0               999                 999
0.050              0               998                 998
0.025              0               993                 993
0.010              0               966                 966
0.005              0               937                 937
```

What do these results indicate about the choice of the t or rank sum test when the populations are outlier-prone?

Solution The results for both tests when the null hypothesis is true indicate that the nominal α is (very close to) correct; the simulation obtained just about the expected number of false rejections. When the research hypothesis is true, as in the second part of the study, we want to reject the null hypothesis. The rank sum test consistently yields more rejections than does the t test. The rank sum test is more powerful in this situation. □

SECTION 10.3 EXERCISES

10.22 Refer to Exercise 10.10.

a. Calculate the sum of the ranks in each sample.

b. Test the null hypothesis of equal locations, using a two-tailed test with $\alpha = .05$.

10.23 a. Compare the conclusion of Exercise 10.10, part (c), with that of Exercise 10.22, part (b).

b. The data of Exercise 10.10 were plotted in Exercise 10.13. On the basis of these plots, would you tend to believe the conclusion of the rank test or of the t test?

10.24 Find the p-value for the test in Exercise 10.22.

10.25 The computer package used in Exercise 10.14 also calculated rank sums. The relevant output is shown here:

```
                         DRUG POTENCY DATA

 ANALYSIS FOR VARIABLE POTENCY CLASSIFIED BY VARIABLE  SAMPLE

                 AVERAGE SCORES WERE USED FOR TIES


                    WILCOXON SCORES (RANK SUMS)


                        SUM OF    EXPECTED    STD DEV        MEAN
LEVEL          N        SCORES    UNDER H0    UNDER H0       SCORE

1             10        146.00      105.00       13.17       14.60
2             10         64.00      105.00       13.17        6.40

          WILCOXON 2-SAMPLE TEST (NORMAL APPROXIMATION)
          (WITH CONTINUITY CORRECTION OF .5)
          S=  146.00       Z= 3.0743       PROB >|Z|=0.0021

          T-TEST APPROX. SIGNIFICANCE=0.0062

          KRUSKAL-WALLIS TEST (CHI-SQUARE APPROXIMATION)
          CHISQ=   9.69     DF=  1     PROB > CHISQ=0.0019
```

a. Identify the rank sums.

b. Locate the value of the z statistic.

c. Formulate appropriate null and research hypotheses.
d. Is H_a supported at $\alpha = .01$?

10.26 a. Locate the p-value in the output of Exercise 10.25. Is it one-tailed or two-tailed? What p-value should be reported in Exercise 10.25?
b. What conclusion would be reached in using the rank sum test? How does it compare to the conclusion of the t test? Does it matter much which test is used?

10.27 The data for Exercise 10.17 are reproduced here:

Program	Time										$\bar{y}$	s	n
A	39	57	42	53	41	44	71	56	49	63	51.50	10.46	10
B	43	38	35	45	40	28	50	54	37	29			
	36	27	52	33	31	30					38.00	8.67	16

a. Find the ranks of the combined data. It's much easier if you sort the data in each sample first.
b. Find the rank sums.
c. Is there a statistically significant difference ($\alpha = .01$) between programs according to the rank sum test?

10.28 Do the data of Exercise 10.17 (and 10.27) indicate that the rank sum test is preferable to a t test? Explain, preferably with pictures.

10.29 The data of Exercise 10.19 are as follows:

Source	Number of Defectives/1000									
I	9.8	9.9	10.2	10.5	10.7	10.8	11.7	13.9	19.2	27.6
II	10.6	11.0	11.5	11.8	11.9	12.7	14.2	16.8	21.7	29.9

a. Use a rank test to test the null hypothesis that both sources have the same distribution of defectives per 1000. Use $\alpha = .05$ and a two-sided research hypothesis.
b. Find the two-tailed p-value.

10.30 Is there reason to think that a rank test is more appropriate than a t test for the data of Exercise 10.19 (and 10.29)?

10.31 A simulation study tests the effect of severe skewness on the relative usefulness of the pooled-variance t test, the t' test, and the rank sum test. Independent samples are taken from the severely skewed squared-exponential distribution. The results are as follows:

```
                    Checking Alpha (same sample sizes and sigmas)

Popn        Mu          Sigma         n
  1       50.000      10.0000        10
  2       50.000      10.0000        10

using the pooled variance t test

one-tail:

      number of times H0:  "mu1-mu2 is   0" is rejected in favor of
alpha   "mu1-mu2 >   0"   "mu1-mu2 <   0"     total (alpha doubled)
0.100          99                 108                207
0.050          32                  37                 69
0.025          11                  12                 23
0.010           3                   3                  6
0.005           0                   0                  0

using separate variances (t')

one-tail:

      number of times H0:  "mu1-mu2 is   0" is rejected in favor of
alpha   "mu1-mu2 >   0"   "mu1-mu2 <   0"     total (alpha doubled)
0.100          94                  98                192
0.050          27                  27                 54
0.025           6                  10                 16
0.010           1                   1                  2
0.005           0                   0                  0

Results of Wilcoxon Rank Sum Test using Z as test statistic

     number of times H0:  "two populations are identical" rejected in fa
alpha   Popn1 rt of Popn2   Popn1 left of Popn2   total (alpha doubled)
0.100             105                110                  215
0.050              44                 50                   94
0.025              24                 29                   53
0.010               6                 12                   18
0.005               1                  3                    4
```

For any of the tests, are the nominal α values grossly wrong?

10.32 Another simulation study compares the pooled-variance t, t', and rank sum tests when the research hypothesis of unequal means is true. The populations are both the severely skewed squared-exponential shape. The results are as follows:

```
                  Checking Power (same sample sizes and sigmas)

Popn         Mu          Sigma         n
  1       50.000       10.0000        10
  2       60.000       10.0000        10

using the pooled variance t test

one-tail:

      number of times H0:  "mu1-mu2 is    0" is rejected in favor of
alpha   "mu1-mu2 >    0"   "mu1-mu2 <    0"     total (alpha doubled)
0.100             0                 858                  858
0.050             0                 789                  789
0.025             0                 728                  728
0.010             0                 655                  655
0.005             0                 586                  586

using separate variances (t')

one-tail:

      number of times H0:  "mu1-mu2 is    0" is rejected in favor of
alpha   "mu1-mu2 >    0"   "mu1-mu2 <    0"     total (alpha doubled)
0.100             0                 850                  850
0.050             0                 779                  779
0.025             0                 716                  716
0.010             0                 625                  625
0.005             0                 550                  550

Results of Wilcoxon Rank Sum Test using Z as test statistic

     number of times H0:  "two populations are identical" rejected in fa
alpha   Popn1 rt of Popn2   Popn1 left of Popn2   total (alpha doubled)
0.100              0                  984                   984
0.050              0                  971                   971
0.025              0                  940                   940
0.010              0                  879                   879
0.005              0                  760                   760
```

a. Why is the "Checking Power" title justified?
b. Which of the three tests appears to have the best power in this situation?

10.4 PAIRED-SAMPLE METHODS

The methods of the preceding two sections are appropriate for the analysis of two independent samples. We have emphasized that those methods are not appropriate for situations such as before–after measurements in which each measurement in one sample is matched or paired with a corresponding measurement in the other. In this section we discuss methods for paired-sample data.

control of variability

The advantage of pairing observations is the **control of variability** that would otherwise obscure a real difference in means. For example, suppose that an office manager wants to test two new typewriter models to find which one yields greater average speed. One test procedure would be to assign 10 secretaries randomly to one model and another 10 secretaries to the other. This procedure would yield two independent samples. Another procedure would be to have 10 randomly chosen secretaries type on both models; the 10 typing speeds on each model would constitute paired or matched samples. Of course, there are large differences in typing speed among secretaries. These differences would cause large variability in the independent-samples experiment and would tend to conceal any real differences between the two models. In the paired-sample experiment, the manager can calculate the difference in the two models' speeds for the same secretaries; individual variability in speed cancels out of the difference. The individual-variability factor does not cause random variability in the paired-sample experiment.

As indicated in the secretary example, statistical methods for working with paired samples are all based on the same idea. Calculate all differences of matched scores and apply single-sample methods to the resulting sample of differences. In particular, the t-distribution methods for confidence intervals and hypothesis tests described in Chapters 8 and 9 may be used.

EXAMPLE 10.13 Insurance adjusters investigate the relative automobile repair costs at two garages. Each of 15 cars recently involved in accidents is taken to both garages 1 and 2 for separate estimates of repair costs. The resulting data are analyzed incorrectly as coming from two independent samples, and correctly as coming from paired samples. Use the following computer printouts to compare the resulting t statistics. What accounts for the difference in these statistics? (Costs are entered in hundreds of dollars.)

```
      TWOSAMPLE
ENTER VALUES IN SAMPLE 1.
□
      7.6  10.2  9.5  1.3  3.0  6.3  5.3  6.2  2.2  4.8  11.3  12.1  6.9  7.6  8.4
ENTER VALUES IN SAMPLE 2.
□
      7.3   9.1  8.4  1.5  2.7  5.8  4.9  5.3  2.0  4.2  11.0  11.0  6.1  6.7  7.5

SAMPLE                    1            2

MEAN                 6.8467       6.2333

ST. DEV.             3.2040       2.9413

SAMPLE SIZE              15           15

SUM OF RANKS          249.0        216.0

POOLED-VARIANCE T STATISTIC = 0.546

SEPARATE-VARIANCE T STATISTIC = 0.546

PAIRED SAMPLE

DIFFERENCES    0.300    1.100    1.100    -0.200    0.300    0.500    0.400
       0.900   0.200    0.600    0.300     1.100    0.800    0.900    0.900

MEAN DIFF =    0.613      ST. DEV. OF DIFFS =      0.394

ENTER HYPOTHESIZED MEAN DIFFERENCE.
□

      0

VALUE OF T STATISTIC = 6.023

SIGNED RANK STATISTIC (ZERO DIFFERENCES EXCLUDED)

EFFECTIVE N =    15      T PLUS =    118.5      T MINUS =    -1.5
```

Solution The pooled-sample and separate-sample (t') statistics shown under TWOSAMPLE are equal because $n_1 = n_2$; the value is only .546. For 28 d.f., this value does not approach statistical significance at any reasonable α level. The PAIREDSAMPLE t statistic equals 6.023, which is "off the tables"—significant at all reasonable α levels. The reason is that there is huge variability in the severity of damage to the 15 cars. This source of variability makes the TWOSAMPLE standard error very large and therefore the TWOSAMPLE t quite small. Because the PAIREDSAMPLE t is based on differences between the two garages' estimates on the same cars, it is not affected by the variability among cars. □

EXAMPLE 10.14 A tasting panel of 15 people is asked to rate two new kinds of tea on a scale ranging from 0 to 100; 25 means "I would try to finish it only to be polite," 50 means "I would drink it but not buy it," 75 means "it's about as good as any tea I know," and 100 means "it's superb; I would drink nothing else." (What 0 means is left to your imagination.) The ratings are as follows:

	Person							
	1	2	3	4	5	6	7	8
Tea S	85	40	75	81	42	50	60	15
Tea J	65	50	43	65	20	65	35	38
Difference	+20	−10	+32	+16	+22	−15	+25	−23

	Person						
	9	10	11	12	13	14	15
Tea S	65	40	60	40	65	75	80
Tea J	60	47	60	43	53	61	63
Difference	+5	−7	0	−3	+12	+14	+17

a. Calculate a 95% confidence interval for the population difference in mean ratings.
b. Test the null hypothesis of no difference against a two-sided alternative using $\alpha = .05$.
c. What advantage does matching have in this situation?

Solution a. If we call the differences d_i, then $\bar{d} = 7.00$ and $s_d = 16.08$. Of course

$$\bar{d} = \frac{\sum d_i}{15} \quad \text{and} \quad s_d^2 = \frac{\sum (d_i - \bar{d})^2}{14}$$

The population mean of the differences is the same as the difference in means, so $\mu_d = \mu_S - \mu_J$. Because our calculations are based on 15 differences, there are 14 d.f., and the required t table value is 2.145. The confidence interval is

$$7.00 - 2.145\frac{16.08}{\sqrt{15}} < \mu_S - \mu_J < 7.00 + 2.145\frac{16.08}{\sqrt{15}}$$

or

$$-1.9 \leq \mu_S - \mu_J \leq 15.9$$

b. Because the value 0 is included in this 95% confidence interval, it follows that $H_0: \mu_S - \mu_J = 0$ cannot be rejected at $\alpha = .05$ using a two-tailed test. The t statistic is

$$t = \frac{7.00 - 0}{16.08/\sqrt{15}} = 1.69$$

which has a two-tailed p-value a little larger than .10.

c. The matching is somewhat useful in accounting for individual differences in taste. There is some tendency for those who give high scores to S to also give high scores to J, and for those who give low S scores to also give low J scores. Had we erroneously used the two-sample formula, we would have had a standard error

$$s_p\sqrt{\frac{1}{15} + \frac{1}{15}} = 6.21$$

rather than the correct standard error

$$s_d/\sqrt{15} = 4.15$$

□

The formal statement of these matched-pairs procedures merely requires replacing y's by d's in the one sample t-distribution procedures. These are summarized here:

$100(1 - \alpha)\%$ Confidence Interval for μ_d Based on Matched Samples

$$\bar{d} - t_{\alpha/2}s_d/\sqrt{n} < \mu_d < \bar{d} + t_{\alpha/2}s_d/\sqrt{n}$$

where n is the number of pairs of observations (and therefore the number of differences) and $t_{\alpha/2}$ cuts off a right-tail area of $\alpha/2$ for the t distribution with $n - 1$ d.f.

Hypothesis Test for Matched Samples

H_0: $\mu_d = D_0$ (D_0 is specified; often $D_0 = 0$)

H_a: 1. $\mu_d > D_0$
2. $\mu_d < D_0$
3. $\mu_d \neq D_0$

T.S.: $t = \dfrac{\bar{d} - D_0}{s_d/\sqrt{n}}$

R.R.: 1. $t > t_\alpha$
2. $t < -t_\alpha$
3. $|t| > t_{\alpha/2}$

In fact, Example 9.16 was an illustration of the paired-sample t test.

There are some competitors to the paired-sample t test. One could test that the median difference is zero using the test described in Section 9.8. That was done in Example 9.22. This test is often called the **sign test**. There is also a rank test alternative called the *Wilcoxon signed-rank test*.

sign test

signed-rank test

The formal null hypothesis for the **Wilcoxon signed-rank test** is that the true distribution of differences is symmetric around a specified number D_0; almost always D_0 is taken to be zero. The test is primarily sensitive to the distribution being shifted to the right or left of D_0; one- or two-sided research hypotheses may be tested. Again the test works with differences (if D_0 is not zero, D_0 is subtracted from each difference). Discard all differences that are exactly zero and reduce n accordingly. Then the differences are ranked in order of absolute value, smallest to largest. The appropriate sign is attached to each rank. Define

T_+ and T_-

T_+ = the sum of the positive ranks; if there are no positive ranks, $T_+ = 0$

T_- = the sum of the negative ranks; if there are no negative ranks, $T_- = 0$

and

n = the number of nonzero differences

The Wilcoxon signed-rank test is presented next.

Wilcoxon Signed-rank Test

H_0: The distribution of differences is symmetric around D_0 (D_0 is specified; usually D_0 is zero)

H_a: 1. the differences tend to be larger than D_0
2. the differences tend to be smaller than D_0
3. the differences tend to be shifted ways from D_0

T.S.: 1. $T = |T_-|$
2. $T = T_+$
3. $T =$ smaller of $|T_-|, T_+$

R.R.: ($n \leq 50$): For a specified value of α (one-tailed .05, .025, .01, or .005; two-tailed .10, .05, .02, .01) and fixed number of nonzero differences n, reject H_0 if the value of T is less than or equal to the appropriate entry in Appendix Table 7. ($n > 50$): Compute the test statistic

$$z = \frac{T - \frac{n(n+1)}{4}}{\sqrt{\frac{n(n+1)(2n+1)}{24}}}$$

For cases 1 and 2, reject H_0 if $z < -z_\alpha$; for case 3 reject H_0 if $z < -z_{\alpha/2}$.

EXAMPLE 10.15 Refer to the data of Example 10.14. Use the signed-rank test to test the null hypothesis of symmetry around $D_0 = 0$ against a two-sided alternative. Use $\alpha = .05$.

Solution The differences and their signed ranks are presented below:

	Person									
	1	2	3	4	5	6	7	8	9	10
Difference	+20	−10	+32	+16	+22	−15	+25	−23	+5	−7
Signed rank	+10	−4	+14	+8	+11	−7	+13	−12	+2	−3

	Person				
	11	12	13	14	15
Difference	0	−3	+12	+14	+17
Signed rank	X	−1	+5	+6	+9

The 0 difference (person 11) is discarded, so $n = 14$.

$$T_+ = 10 + 14 + 8 + 11 + 13 + 2 + 5 + 6 + 9 = 78$$
$$T_- = -4 - 7 - 12 - 3 - 1 = -27$$

(A good check is that $T_+ - T_-$ must always equal $n(n+1)/2$, which equals 105 here.) For a two-sided research hypothesis, $T = \{\text{smaller of } |-27| \text{ and } 78\} = 27$. As $n = 14 < 50$, we find the $\alpha = .05$ (two-sided) entry in Appendix Table 7; it is 21. Because $T = 27 > 21$, we cannot reject H_0 at $\alpha = .05$. We cannot reject at $\alpha = .10$ either; the table value is 26. While we do not need the large-sample ($n > 50$) approximation in this problem, it can be computed:

$$z = \frac{27 - (14)(15)/4}{\sqrt{\dfrac{(14)(15)(29)}{24}}} = -1.60$$

For $\alpha = .10$ (two-tailed), we reject if $z < -z_{.05} = -1.645$; if we had used the z approximation we would not have rejected H_0 at $\alpha = .10$. □

EXAMPLE 10.16 Use the signed-rank information in the computer printout of Example 10.13 to test the research hypothesis that garage 1's estimates tend to be higher than garage 2's. How does the result of the signed-rank test compare with the result of the t test of Example 10.13?

Solution For this one-sided research hypothesis, the test statistic $T = |T_-|$; from the printout, $T_- = -1.5$, so $T = 1.5$. According to Appendix Table 7 of critical values for the signed rank test, H_0 is rejected (at $\alpha = .005$ with $n = 15$) if $T \leq 16$. Since 1.5 is much less than 16, we can reject H_0 conclusively. While $n = 15$ is too small to give us much faith in a normal approximation,

$$z = \frac{1.5 - (15)(16)/4}{\sqrt{\dfrac{(15)(16)(31)}{24}}} = -3.32$$

This z value is "off the table." Even if the normal approximation is poor, this strongly indicates that the null hypothesis can be rejected even at extremely low α values. The same conclusion applies to the t value of 6.023 found in Example 10.13. □

The choice of the appropriate paired-sample test from this section follows the guidelines of Chapter 9. If the assumptions of the t test are satisfied—in particular, if the distribution of differences is roughly normal—the t test is more powerful. If the distribution of differences is grossly skewed, the nominal t probabilities may be misleading. If the distribution is roughly symmetric but has heavy tails (as indicated by the presence of outliers), the signed-rank test may be more powerful. Often, as in Examples 10.14 and 10.15, the tests yield essentially the same conclusion.

choice of method

Even with this discussion you may still be confused as to which statistical test (or confidence interval) to apply in a situation in which there is a choice of two or more methods. When in doubt, do several different tests; computing costs are usually minimal, especially with the availability of many different statistical software packages such as Minitab, SAS, and SPSS. If the results from the different analyses yield different results, you should be concerned about identifying the peculiarities of the data set to understand why the results differ. If the results agree, and if there are no blatant violations of assumptions, you should be very confident in your conclusions.

SECTION 10.4 EXERCISES

10.33 A manufacturer of an air compressor and tire pump wants to test two possible point-of-purchase displays. The product is sold through independent auto parts stores, which vary greatly in sales volume. A total of thirty stores agree to feature the display for one month. The stores are matched on the basis of annual sales volume. One of the two largest stores is randomly chosen to receive display A, while the other received B. The same thing is done for the third and fourth largest stores, and so on down to the two smallest. Sales for the one-month period are recorded:

	Pairing														
Display	1	2	3	4	5	6	7	8	9	10	11	12	13	14	15
A	46	39	40	37	32	26	21	23	20	17	13	15	11	8	9
B	37	42	37	38	27	19	20	17	20	12	12	9	7	2	6
Difference	+9	−3	+3	−1	+5	+7	+1	+6	0	+5	+1	+6	+4	+6	+3

The mean of the differences is 3.47 and the standard deviation is 3.31.

a. Use a paired sample t test to test the research hypothesis of unequal means. Use $\alpha = .10$.

b. Calculate a 90% confidence interval for the true mean difference.

10.34 a. Carry out the signed-rank test for the data of Exercise 10.33. Use a two-tailed test with $\alpha = .10$.

b. How does the conclusion of this test compare with that of the t test in Exercise 10.33?

10.35 Place bounds on the p-values for the tests in Exercises 10.33 and 10.34.

10.36 Consider the situation of Exercise 10.33. An alternative approach would be to assign display A to 15 randomly chosen stores and display B to the rest.

a. Suppose that the data of Exercise 10.33 were obtained in this way. Carry out the appropriate *t* test.

b. Does there seem to be any advantage to the pairing process actually used in Exercise 10.33?

10.37 Again refer to Exercise 10.33. Carry out a binomial (sign) test of the null hypothesis that the proportion of positive differences equals the proportion of negative differences. What should you do about a zero difference?

10.38 A simulation study of the *t* and signed-rank tests involved taking samples from the mildly outlier-prone Laplace population. A simulation with H_0 being true indicated that both tests had α values close to the nominal value. The following additional results were obtained.

```
                              Checking Power

Simulation of One Sample t-test (1000 samples)

Data in file Laplace.SAM
Population shape is moderately outlier prone.

           Mu          Sigma          n
        55.000       10.0000         30

t test

one-tail:

        number of times H0:  "mean is  50" is rejected in favor of
alpha     "mean >  50"     "mean <  50"          total (alpha doubled)
0.100         913                 0                  913
0.050         831                 0                  831
0.025         745                 0                  745
0.010         629                 0                  629
0.005         537                 0                  537

average t is 2.8985 with variance of    1.412205

Simulation of Wilcoxon Signed Rank Test (1000 samples)
Simulation results using the tabled critical values

one-tail:

        number of times H0: "symmetric around  50" rejected in favor of
alpha     "tend to be >  50"    "tend to be <  50"     total (alpha doubled)
0.100            954                     0                  954
0.050            907                     0                  907
0.025            839                     0                  839
0.010            723                     0                  723
0.005            630                     0                  630
```

What do these results indicate about the relative desirability of the two tests for this mildly outlier-prone population?

10.39 A manufacturer can use either the postal service or a private shipper to deliver its small shipments. To help in making a choice, the manufacturer selects 10 destinations and ships parcels to each destination by each carrier. The delivery times (in days) are

	Destination									
	1	2	3	4	5	6	7	8	9	10
Postal	3	4	5	4	8	9	7	10	9	9
Private	2	2	3	5	4	6	9	6	7	6

Computer output (BMD) for this study is shown below:

```
************
* DIFF     *  VARIABLE NUMBER   4
************
                            MEAN          1.8000
T STATISTIC  P-VALUE   DF  STD DEV       1.9889
                            S.E.M.        0.6289
     2.86     0.0187     9  SAMPLE SIZE      10                H
                            MAXIMUM       4.0000                H  H  H
                            MINIMUM      -2.0000    H  H     H  H  H  H
                                                  MIN--------------------MAX
                                                      AN H =    1 CASES

 WILCOXON SIGNED RANKS TEST RESULTS


 NUMBER OF NON-ZERO DIFFERENCES


               POSTDAYS   PRIVDAYS

                    2          3
POSTDAYS   2        0
PRIVDAYS   3       10          0


 SMALLER SUM OF LIKE-SIGNED RANKS


               POSTDAYS   PRIVDAYS

                    2          3
POSTDAYS   2     0.00
PRIVDAYS   3     6.00       0.00


 LEVEL OF SIGNIFICANCE OF WILCOXON SIGNED RANKS
TEST USING NORMAL APPROXIMATION (TWO-TAIL)


               POSTDAYS   PRIVDAYS

                    2          3
POSTDAYS   2   1.0000
PRIVDAYS   3   0.0284     1.0000
```

a. Locate the value of the t statistic.
b. Locate the value of the signed-rank statistic.
c. Use each statistic separately to test the hypothesis of equal mean delivery times. Use a two-sided H_a and $\alpha = .05$.

10.40 Find p-values for both test statistics in Exercise 10.39. How conclusive would you say the apparent difference in means is?

10.41 Why is the experiment in Exercise 10.39 more effective than randomly assigning 10 packages to the postal service and another 10 to the private shipper? How would you select the 10 destinations?

10.42 A direct mail company tests two different versions of a special-offer catalog. A sample of ZIP codes is chosen. Each version of the catalog is sent to half the people on the firm's mailing list within each of the selected ZIP codes. The response per thousand catalogs is recorded for each ZIP code. The data are as follows:

ZIP code:	1	2	3	4	5	6	7	8	9	10	11	12
Version A:	10.8	13.4	8.9	10.6	17.0	14.1	11.2	13.4	9.9	10.7	11.3	14.2
Version B:	11.3	15.0	9.9	10.0	17.7	12.6	11.8	13.7	10.4	9.9	12.8	14.9

a. Why should this experiment be regarded as a paired-sample study?
b. Calculate the mean and standard deviation of the differences.
c. Can the research hypothesis that version B yields a higher average response rate be supported using a t test at $\alpha = .05$?
d. State bounds on the p-value for testing the research hypothesis in part (c).

10.43 a. Perform a signed-rank test for the data of Exercise 10.42. Use $\alpha = .05$ and a one-sided H_a.
b. Put bounds on the p-value.
c. Do the signed-rank and t tests come to the same conclusion?

10.44 An analyst of the data in Exercise 10.42 interprets the results as proving that version B is no better than version A. Do you agree?

10.45 An organization keeps an extensive file of volunteers coded according to skills and desired activities. Two data-base management programs are tested. A sample of 10 combinations of skills and activities is specified; then both programs are used to find all volunteers with the specified combination. The time required (seconds) is recorded for every search. The data are as follows:

Combination:	1	2	3	4	5	6	7	8	9	10	Mean	s
Program 1:	136	298	187	192	100	170	240	200	102	155	178.0	60.756
Program 2:	158	391	206	128	120	194	272	221	119	184	199.3	83.225

a. Explain why this is a paired-sample experiment.
b. Calculate a 95% confidence interval for the population mean of differences.
c. According to this interval, can the null hypothesis that the means for programs 1 and 2 are equal be rejected in favor of a two-sided research hypothesis? What is the effective α?

10.46 Perform a signed-rank test, using the data of Exercise 10.45, of the null hypothesis that the differences are symmetric around zero. Can the hypothesis be rejected at $\alpha = .05$ using a two-tailed test?

10.47 What plot(s) of the data should be drawn to decide whether a t test or a signed-rank test will be more effective for the data of Exercise 10.45? Which test would you choose? Does the choice of test affect the conclusion?

10.48 a. Using the data of Exercise 10.45 and a pooled-variance t method, calculate a 95% confidence interval for the difference of population means.
b. Compare the width of this interval to the width of the interval found in Exercise 10.45. What does the comparison indicate about the desirability of pairing in this study?

10.5 TWO-SAMPLE PROCEDURES FOR PROPORTIONS

So far in this chapter we have concentrated entirely on tests and confidence intervals for means. In this section we consider procedures for proportions. The methods are based on the normal approximation to the binomial distribution, so some consideration of required sample size is necessary.

We assume that two **independent** random samples of sizes n_1 and n_2 are taken. The respective sample proportions are denoted as $\hat{\pi}_1$ and $\hat{\pi}_2$, and the (unknown) population proportions are called π_1 and π_2. Our goal is to make inferences about the difference, if any, in the population proportions. The natural estimator is the difference in sample proportions $\hat{\pi}_1 - \hat{\pi}_2$. It is unbiased: $E(\hat{\pi}_1 - \hat{\pi}_2) = \pi_1 - \pi_2$. To calculate confidence intervals and perform hypothesis tests, we need a standard error formula. Recall from Chapter 6 that the variance of a sample proportion $\hat{\pi}$ is $\pi(1 - \pi)/n$. By the assumed independence of the two samples, we may add the variances of $\hat{\pi}_1$ and $\hat{\pi}_2$ to obtain

standard error of $\hat{\pi}_1 - \hat{\pi}_2$

$$\text{Var}(\hat{\pi}_1 - \hat{\pi}_2) = \frac{\pi_1(1 - \pi_1)}{n_1} + \frac{\pi_2(1 - \pi_2)}{n_2}$$

$$\sigma_{\hat{\pi}_1 - \hat{\pi}_2} = \sqrt{\frac{\pi_1(1 - \pi_1)}{n_1} + \frac{\pi_2(1 - \pi_2)}{n_2}}$$

The confidence interval for $\hat{\pi}_1 - \hat{\pi}_2$ follows the familiar form $\hat{\theta} \pm z_{\alpha/2}\sigma_{\hat{\theta}}$, where $\hat{\theta}$ is now $\hat{\pi}_1 - \hat{\pi}_2$ and $\sigma_{\hat{\theta}}$ is $\sigma_{\hat{\pi}_1 - \hat{\pi}_2}$. Since the standard error ($\sigma_{\hat{\pi}_1 - \hat{\pi}_2}$) depends on the unknown population proportions π_1 and π_2, in practice we must substitute the sample proportions $\hat{\pi}_1$ and $\hat{\pi}_2$ into the standard-error formula. If both sample sizes are sufficiently large (say, at least 30), this substitution can be made without affecting the normal approximation.

100(1 − α)% Confidence Interval for $\pi_1 - \pi_2$

$$(\hat{\pi}_1 - \hat{\pi}_2) - z_{\alpha/2}\sigma_{\hat{\pi}_1 - \hat{\pi}_2} < \pi_1 - \pi_2 < (\hat{\pi}_1 - \hat{\pi}_2) + z_{\alpha/2}\sigma_{\hat{\pi}_1 - \hat{\pi}_2}$$

where

$$\sigma_{\hat{\pi}_1 - \hat{\pi}_2} \approx \sqrt{\frac{\hat{\pi}_1(1 - \hat{\pi}_1)}{n_1} + \frac{\hat{\pi}_2(1 - \hat{\pi}_2)}{n_2}}$$

EXAMPLE 10.17 A new product is test-marketed in the Grand Rapids, Michigan, and Wichita, Kansas, metropolitan areas. Advertising in the Grand Rapids area is based almost entirely on television commercials. In Wichita, a roughly equal dollar amount is spent on a balanced mix of television, radio, newspaper, and magazine ads. Two months after the ad campaign begins, surveys are taken to determine consumer awareness of the product.

	Grand Rapids	Wichita
Number interviewed	608	527
Number aware	392	413

Calculate a 95% confidence interval for the regional difference in the proportion of all consumers who are aware of the product.

Solution The sample awareness proportion is higher in Wichita, so let's take Wichita as region 1.

$$\hat{\pi}_1 = 413/527 = .784; \qquad \hat{\pi}_2 = 392/608 = .645$$

The estimated standard error is

$$\sqrt{\frac{(.784)(.216)}{527} + \frac{(.645)(.355)}{608}} = .0264$$

Therefore the 95% confidence interval is

$$(.784 - .645) - 1.96(.0264) < \pi_1 - \pi_2 < (.784 - .645) + 1.96(.0264)$$

or

$$.087 < \pi_1 - \pi_2 < .191$$

indicating that somewhere between 8.7% and 19.1% more Wichita consumers than Grand Rapids consumers are aware of the product. □

rule for sample sizes

This confidence-interval method is based on the normal approximation to the binomial distribution. In Chapter 6 we indicated as a general rule that $n\hat{\pi}$ and $n(1 - \hat{\pi})$ should both at least 5 to use this normal approximation. For this confidence interval to be used, the rule should hold for each sample. In practice, sample sizes that come even close to violating this rule aren't very useful, because they lead to excessively wide confidence intervals. For instance, even though $n\hat{\pi}$ and $n(1 - \hat{\pi})$ are greater than 5 for both samples when $n_1 = 30$, $\hat{\pi}_1 = .20$ and $n_2 = 60$, $\hat{\pi}_2 = .10$, the 95% confidence interval is $-.06 < \pi_1 - \pi_2 < .26$; π_1 could be anything from 6 percentage points lower than π_2 to 26 percentage points higher.

Hypothesis testing about the difference between two population proportions is based on the z statistic from a normal approximation. The typical null hypothesis is that there is no difference between the population proportions, though any specified value for $\pi_1 - \pi_2$ may be hypothesized. The procedure is summarized below.

Hypothesis Test for $\pi_1 - \pi_2$

H_0: $\pi_1 - \pi_2 = D_0$ (D_0 is specified; often $D_0 = 0$)

H_a: 1. $\pi_1 - \pi_2 > D_0$
2. $\pi_1 - \pi_2 < D_0$
3. $\pi_1 - \pi_2 \neq D_0$

$$\text{T.S.: } z = \frac{(\hat{\pi}_1 - \hat{\pi}_2) - D_0}{\sqrt{\dfrac{\hat{\pi}_1(1 - \hat{\pi}_1)}{n_1} + \dfrac{\hat{\pi}_2(1 - \hat{\pi}_2)}{n_2}}}$$

R.R.: 1. $z > z_\alpha$
2. $z < -z_\alpha$
3. $|z| > z_{\alpha/2}$

Note: This test should be used only if $n_1\hat{\pi}_1$, $n_1(1 - \hat{\pi}_1)$, $n_2\hat{\pi}_2$, and $n_2(1 - \hat{\pi}_2)$ are all at least 5.

EXAMPLE 10.18 Refer to Example 10.17. Test the hypothesis of equal population-awareness proportions against a two-sided alternative. State a p-value.

Solution $\hat{\pi}_1 = .784$, $n_1 = 527$, $\hat{\pi}_2 = .645$, and $n_2 = 608$. The general rule for using a z test is amply met; the smallest of the four indicators is $n_1(1 - \hat{\pi}_1) = 114$.

$$z = \frac{(.784 - .645) - 0}{\sqrt{\dfrac{(.784)(.216)}{527} + \dfrac{(.645)(.355)}{608}}}$$

$$= \frac{.139}{.0264} = 5.26$$

A z value of 5.26 is far beyond the range of our z table. The p-value is some very small number. □

There is a slight variation on this test. Under the null hypothesis H_0: $\pi_1 - \pi_2 = 0$, the two populations have an equal proportion, call it π, of successes. The natural estimator of this common proportion is $\bar{\pi}$, the total number of successes divided by the total sample size. This estimator is, by easy algebra, a weighted average of the two sample proportions $\hat{\pi}_1$ and $\hat{\pi}_2$:

$$\bar{\pi} = \frac{n_1\hat{\pi}_1 + n_2\hat{\pi}_2}{n_1 + n_2}$$

The modification to the z test uses a standard error:

$$\sqrt{\frac{\bar{\pi}(1 - \bar{\pi})}{n_1} + \frac{\bar{\pi}(1 - \bar{\pi})}{n_2}}$$

that is, the unknown population proportions π_1 and π_2 are replaced by $\bar{\pi}$, rather than by $\hat{\pi}_1$ and $\hat{\pi}_2$, in the standard-error formula. In practice, the numerical effect of this modification is usually negligible.

EXAMPLE 10.19 Perform the modified z test for Example 10.17. How does the modification affect your conclusion?

Solution The total sample size is $608 + 527 = 1135$; the total number of successes is $392 + 413 = 805$. Note that $413 = n_1\hat{\pi}_1 = 527(.784)$ and $392 = n_2\hat{\pi}_2 = 608(.645)$. Therefore,

$$\bar{\pi} = \frac{805}{1135} = \frac{527(.784) + 608(.645)}{527 + 608}$$

$$= .709$$

The modified standard error is

$$\sqrt{\frac{\bar{\pi}(1 - \bar{\pi})}{n_1} + \frac{\bar{\pi}(1 - \bar{\pi})}{n_2}} = \sqrt{\frac{(.709)(.291)}{527} + \frac{(.709)(.291)}{608}}$$

$$= .0270$$

compared to the previous standard error of .0264. The modified z statistic is

$$z = \frac{(.784 - .645)}{.0270} = 5.15$$

compared to the previous z value of 5.26. The conclusion remains the same: The null hypothesis is emphatically rejected. The p-value is so small as to go beyond our normal table. □

SECTION 10.5 EXERCISES

10.49 A consumer finance company considers its bad-debt experience for married and unmarried couples. A sample of 3200 loans yields the following data:

Status	Number of Loans	Bad Debts
Married	2128	102
Unmarried	1072	31

Calculate a 90% confidence interval for the true difference in proportions of bad debts.

10.50 Refer to Exercise 10.49.

a. Test the null hypothesis of equal proportions. Let H_a be two-sided. Use $\alpha = .10$.

b. In performing this test, how important is the modification of the standard error (using the pooled proportion $\bar{\pi}$)? Does this modification affect the conclusion?

10.51 Find p-values for the unmodified and modified z statistics in Exercise 10.50.

10.52 In a survey, it is found that 1697 of 2961 urban-area residents regularly watch a network television news program, whereas 674 of 983 rural or small-town residents are regular watchers.

a. Calculate a 95% confidence interval for the difference in proportions.

b. Test the research hypothesis that a higher percentage of rural (small-town) residents are regular watchers. Use $\alpha = .05$.

10.53 Find the p-value in Exercise 10.52. How conclusive is the evidence favoring the research hypothesis?

10.54 Recalculate the test statistic and p-value for the data of Exercise 10.52 using the modified ($\bar{\pi}$) standard error. How important is the modification?

10.55 A retail computer dealer is trying to decide between two methods for servicing customers' equipment. The first method emphasizes preventive maintenance; the second emphasizes quick response to problems. Samples of customers are served by each of the two methods; of course, each customer is served by only one method. After six months, it is found that 171 of 200 customers served by the first method are very satisfied with the service, as compared to 153 of 200 customers served by the second method.

a. Test the research hypothesis that the population proportions are different. Use $\alpha = .10$. State your conclusion carefully.

b. State a p-value for the test in part (a).

10.56 Redo Exercise 10.55 using $\bar{\pi}$, the pooled proportion, in the standard error. How much difference does it make which method is used?

10.57 The media-selection manager for an advertising agency inserts the same advertisement for a client bank in two magazines. The ads are similarly placed in each magazine. One month later, a market research study finds that 226 of 473 readers of the first magazine are aware of the banking services offered in the ad, as are 165 of 439 readers of the second magazine (readers of both magazines are excluded).

a. Calculate a 95% confidence interval for the difference of proportions of readers who are aware of the advertised services.
b. Are the sample sizes adequate to use the normal approximation?
c. Does the confidence interval indicate that there is a statistically significant difference using $\alpha = .05$?

10.58 Using the data of Exercise 10.57, perform a formal test of the null hypothesis of equal populations. Use $\alpha = .05$. How important is it whether or not the pooled proportion is used in the standard error?

10.59 Samples of 30 electric motors for dot matrix printers are subjected to severe testing for reliability. Of the motors from supplier 1, 22 pass the test; of the motors from supplier 2, only 16 pass.
a. Show that the difference is not statistically significant at $\alpha = .05$ (two-tailed).
b. Can we claim to have shown that the two suppliers provide equally reliable motors?

10.60 Use the data of Exercise 10.59 to calculate a 95% confidence interval for the difference of proportions. Interpret the result, carefully, in terms of the relative reliability of the two suppliers.

SUMMARY

In this chapter we have considered inferences for comparing two populations. The large-sample estimation procedures and statistical test for inferences about the difference between two population means $\mu_1 - \mu_2$ or two population proportions $\pi_1 - \pi_2$ follow directly from the large-sample procedures for μ (and π) presented in Chapters 8 and 9.

Not all examples involve large samples. The so-called small-sample procedures are very similar to the large-sample procedures, usually with the z value (or statistic) replaced by a t. In order to develop a t statistic or confidence interval for $\mu_1 - \mu_2$ based on the t distribution, we made an additional assumption that the two population variances σ_1^2 and σ_2^2 are equal. Then the standard error for $\bar{Y}_1 - \bar{Y}_2$ is estimated by $s_p\sqrt{1/n_1 + 1/n_2}$.

The large- and small-sample statistical tests and estimation procedures for $\mu_1 - \mu_2$ in Sections 10.1–10.3 and the procedures for $\pi_1 - \pi_2$ in Section 10.5 dealt with independent samples. Inferences about $\mu_1 - \mu_2$ based on paired data were discussed in Section 10.4.

KEY FORMULAS: Comparing Two Samples

1. Confidence interval for $\mu_1 - \mu_2$ (large, independent samples)

$$(\bar{y}_1 - \bar{y}_2) \pm z_{\alpha/2}\sqrt{\frac{\sigma_1^2}{n_1} + \frac{\sigma_2^2}{n_2}}$$

If $n_1 \geq 30$ and $n_2 \geq 30$, s_1^2 and s_2^2 can replace σ_1^2 and σ_2^2, respectively.

2. Statistical test for $\mu_1 - \mu_2$ (large, independent samples)

Null hypothesis: $\mu_1 - \mu_2 = D_0$ (D_0 is specified)

Test statistic: $$z = \frac{(\bar{y}_1 - \bar{y}_2) - D_0}{\sqrt{\dfrac{\sigma_1^2}{n_1} + \dfrac{\sigma_2^2}{n_2}}}$$

3. Confidence interval for $\mu_1 - \mu_2$ (independent samples, standard deviations unknown)

$$(\bar{y}_1 - \bar{y}_2) \pm t_{\alpha,2} s_p \sqrt{\frac{1}{n_n} + \frac{1}{n_2}}$$

where

$$s_p = \sqrt{\frac{(n_1 - 1)s_1^2 + (n_2 - 1)s_2^2}{n_1 + n_2 - 2}}$$

and

d.f. $= n_1 + n_2 - 2$

4. Statistical test for $\mu_1 - \mu_2$ (independent samples, standard deviations unknown)

Null hypothesis: $\mu_1 - \mu_2 = D_0$

Test statistic: $t = \dfrac{(\bar{y}_1 - \bar{y}_2) - D_0}{s_p \sqrt{\dfrac{1}{n_1} + \dfrac{1}{n_2}}}$

5. Wilcoxon rank sum test (independent samples, $n_1, n_2 \geq 10$)

Null hypothesis: The two populations are identical

Test statistic: $z = \dfrac{T - \mu_T}{\sigma_T}$

where T = rank sum for sample 1

$$\mu_T = \frac{n_1(n_1 + n_2 + 1)}{2}$$

$$\sigma_T = \sqrt{\frac{n_1 n_2}{12}(n_1 + n_2 + 1)}$$

6. Confidence interval for μ_d (paired samples)

$$\bar{d} \pm t_{\alpha/2} \frac{S_d}{\sqrt{n}}$$

7. Statistical test for μ_d (paired data)

Null hypothesis: $\mu_d = D_0$

Test statistic: $t = \dfrac{\bar{d} - D_0}{s_d/\sqrt{n}}$

8. Wilcoxon signed-rank test (paired data)

Null hypothesis: The distribution of d values is symmetric about D_0

Test statistic: a. For $n \leq 50$, T (see Appendix Table 7 for critical values)

b. For $n > 50$, $z = \dfrac{T - \mu_T}{\sigma_T}$

where

$$\mu_T = \frac{n(n+1)}{4}$$

$$\sigma_T = \sqrt{\frac{n(n+1)(2n+1)}{24}}$$

9. Confidence interval for $\pi_1 - \pi_2$

$$(\hat{\pi}_1 - \hat{\pi}_2) \pm z_{\alpha/2}\sigma_{\hat{\pi}_1 - \hat{\pi}_2}$$

where

$$\sigma_{\hat{\pi}_1 - \hat{\pi}_2} \approx \sqrt{\frac{\hat{\pi}_1(1-\hat{\pi}_1)}{n_1} + \frac{\hat{\pi}_2(1-\hat{\pi}_2)}{n_2}}$$

10. Statistical test for $\pi_1 - \pi_2$

Null hypothesis: $\pi_1 - \pi_2 = D_0$

Test statistic: $z = \dfrac{(\hat{\pi}_1 - \hat{\pi}_2) - D_0}{\sigma_{\hat{\pi}_1 - \hat{\pi}_2}}$

CHAPTER 10 EXERCISES

10.61 An auditor for a national bank credit card samples the accounts of two local banks that process cardholders' accounts. The results are

Bank	Accounts Audited	Accounts in Error	Mean Error	Standard Deviation
A	475	41	\$41.27	\$19.42
B	384	39	\$60.38	\$31.68

The mean and standard deviations are based on only those accounts that are in error.

a. Calculate a 95% confidence interval for the difference in true error proportions.

b. Can the research hypothesis of unequal proportions be supported using $\alpha = .05$?

10.62 Find the p-value for the test in Exercise 10.61.

10.63 Refer to the data of Exercise 10.61.

a. Calculate a 90% confidence interval for the difference in means.

b. Give a careful interpretation of this confidence interval. To what population(s) does it apply?

c. Test the research hypothesis of unequal means, using the large-sample test of Section 10.1. Use $\alpha = .05$.

10.64 Find the p-value for the test of Exercise 10.63.

10.65 a. Use the pooled-variance t test of the hypothesis of Exercise 10.63.

b. Use the t' statistic for this hypothesis. How does t' relate to the large-sample z test of Section 10.1?

10.66 A simulation study compares the power of the paired-sample t test and the signed-rank test when the population of differences is spread uniformly over a certain range. This population has no values that are even close to being outliers. The study first verifies that both tests are satisfactory as far as α probabilities are concerned. Then the study obtains the following output:

Checking Power
Simulation of One Sample t Test (1000 samples)

Mu	Sigma	n
5.000	10.0000	30

one-tail:

number of times H_0: "mean is 0" is rejected in favor of

alpha	"mean >0"	"mean <0"	total (alpha doubled)
0.100	940	0	940
0.050	857	0	857
0.025	762	0	762
0.010	609	0	609
0.005	503	0	503

Simulation of Wilcoxon Signed Rank Test (1000 samples)

Simulation results using the tabled critical values

one-tail:

number of times H_0: "symmetric around 0" rejected in favor of

alpha	"tend to be >0"	"tend to be <0"	total (alpha doubled)
0.100	894	0	894
0.050	787	0	787
0.025	686	0	686
0.010	535	0	535
0.005	414	0	414

Given that the population is not at all outlier-prone, which test should have better power? Does it, in this study?

10.67 A fruit grower plants 12 stands of each of two varieties of apple tree. At maturity, the following yields are observed (in bushels per 100 trees):

Variety R:	64.2	71.1	59.8	74.6	37.1	58.7	61.6	54.0	47.3	53.2	68.0	61.1
Variety K:	59.9	72.0	62.1	66.7	32.4	49.0	57.4	50.8	49.0	48.6	61.9	60.0

Assume that the 24 stands are randomly selected from the grower's available acreage and that the yields are listed in an arbitrary order.

a. Use the t test to test the research hypothesis that the mean yield of variety R exceeds that of variety K. Use $\alpha = .10$.

b. Use an appropriate rank test for the same hypothesis. Again use $\alpha = .10$.

10.68 Plot the data of Exercise 10.67. Which of the tests in that exercise seems more appropriate? Does it matter (to the conclusion) which test is used?

10.69 Find p-values for the tests of Exercise 10.67.

10.70 Refer to the data of Exercise 10.67. Now assume that the grower plants the two varieties side by side on 12 plots, and that the data are presented by plot number.

a. Use a t test for the research hypothesis that the mean yield of variety R exceeds that of variety K. Use $\alpha = .10$.

b. Use a rank test for this hypothesis, again using $\alpha = .10$.

10.71 Plot the relevant data for Exercise 10.70. Which of the tests in that exercise seems more appropriate? Does it matter (to the conclusion) which test is used?

10.72 Find p-values for the tests of Exercise 10.70.

10.73 Exercises 10.67 and 10.70 indicate two alternative experimental designs. What is the advantage of the design in Exercise 10.70? If this design is adopted, how would you select the plots to ensure that the yield ratings were reasonably valid for the grower's entire farm?

10.74 Two possible methods for retrofitting jet engines to reduce noise are being considered. Identical planes are fitted with two systems. Noise recording devices are installed directly under the flight path of a major airport. Each time one of the planes lands at the airport, a noise level is recorded. The data are analyzed by a computer package (SAS). The relevant output is:

VARIABLE: DBREAD

SYSTEM	N	MEAN	STD DEV	STD ERROR	MINIMUM	MAXIMUM	VARIANCES	T	DF	PROB > |T|
H	42	100.90476190	2.99438111	0.46204304	95.00000000	110.00000000	UNEQUAL	4.4491	21.5	0.0002
R	20	92.50000000	8.19178022	1.83173774	79.00000000	111.00000000	EQUAL	5.9126	60.0	0.0001

a. Locate the t statistic.

b. Locate the t' statistic.

c. Can the research hypothesis of unequal means be supported using $\alpha = .01$? Does it matter which statistic is used?

10.75 Based on the output of Exercise 10.74, which of the test statistics in that exercise seems more reliable? How crucial is the choice?

10.76 The data of Exercise 10.74 are also used in a rank sum test. The relevant output is shown below:

WILCOXON SCORES (RANK SUMS)

LEVEL	N	SUM OF SCORES	EXPECTED UNDER H0	STD DEV UNDER H0	MEAN SCORE
H	42	1608.00	1323.00	66.22	38.29
R	20	345.00	630.00	66.22	17.25

WILCOXON 2-SAMPLE TEST (NORMAL APPROXIMATION)
(WITH CONTINUITY CORRECTION OF .5)
S= 345.00 Z=-4.2963 PROB >|Z|=0.0000

T-TEST APPROX. SIGNIFICANCE=0.0001

a. Locate the rank sum statistic.

b. Locate the z statistic and p-value.

c. How does the conclusion of the rank sum test compare to the conclusion reached in Exercise 10.74?

APPENDIX: THE MATHEMATICS OF POOLED-VARIANCE *t* METHODS

In developing the pooled-variance *t* statistic, we made the seemingly arbitrary assumption that the population variances were equal. The χ^2 and *t*-distribution results of the appendix to Chapter 8 can be used to indicate why this assumption was made.

The definition of a *t* statistic is

$$t = \frac{Z}{\sqrt{W/\nu}}$$

where W has a χ^2 distribution with ν d.f. As the true standard error of $\bar{Y}_1 - \bar{Y}_2$ is $\sqrt{\sigma_1^2/n_1 + \sigma_2^2/n_2}$, the desired Z statistic is

$$Z = \frac{(\bar{Y}_1 - \bar{Y}_2) - (\mu_1 - \mu_2)}{\sqrt{\frac{\sigma_1^2}{n_1} + \frac{\sigma_2^2}{n_2}}}$$

Because $(n_1 - 1)s_1^2/\sigma_1^2$ and $(n_2 - 1)s_2^2/\sigma_2^2$ are (independently) χ^2 distributed, as argued in the appendix to Chapter 8, the appropriate χ^2 statistic is the sum

$$W = (n_1 - 1)\frac{s_1^2}{\sigma_1^2} + (n_2 - 1)\frac{s_2^2}{\sigma_2^2}$$

with $(n_1 - 1) + (n_2 - 1) = n_1 + n_2 - 2$ d.f.

Substituting these expressions for Z and W into the definition of a *t* statistic yields the following result:

$$t = \frac{\dfrac{(\bar{Y}_1 - \bar{Y}_2) - (\mu_1 - \mu_2)}{\sqrt{\frac{\sigma_1^2}{n_1} + \frac{\sigma_2^2}{n_2}}}}{\sqrt{\left((n_1 - 1)\frac{s_1^2}{\sigma_1^2} + (n_2 - 1)\frac{s_2^2}{\sigma_2^2}\right)\Big/(n_1 + n_2 - 2)}}$$

Without further assumptions, this can't be simplified, and the *t* statistic depends on the unknown population variances. To solve this problem, we assumed $\sigma_1^2 = \sigma_2^2$. Then the common variance can be factored out of the square roots. With a little algebra, the *t* definition becomes

$$t = \frac{(\bar{Y}_1 - \bar{Y}_2) - (\mu_1 - \mu_2)}{\sqrt{\frac{(n_1 - 1)s_1^2 + (n_2 - 1)s_2^2}{n_1 + n_2 - 2}}\sqrt{\frac{1}{n_1} + \frac{1}{n_2}}}$$

$$= \frac{(\bar{Y}_1 - \bar{Y}_2) - (\mu_1 - \mu_2)}{s_p\sqrt{\frac{1}{n_1} + \frac{1}{n_2}}}$$

which is the form used in Chapter 10. In summary, the assumption that $\sigma_1^2 = \sigma_2^2$ is made so that the defining condition for *t* yields a *t* statistic that does not depend on unknown population variances.

11

CHI-SQUARE AND *F* METHODS

The inference methods presented so far have focused largely on means, occasionally on proportions and medians. In this chapter we develop additional methods; some deal with inferences about variances, others with inferences about several proportions.

11.1 CHI-SQUARE METHODS FOR A VARIANCE

Many managerial statistical problems are concerned with variability of a response rather than with the average response. If a cereal-packaging line producing nominal 20-ounce packages actually produces 50% 10-ounce fills and 50% 30-ounce fills, the average is 20, but those consumers who buy 10-ounce boxes are apt to be rather noisily unhappy. Or consider a city bus route where the schedule calls for a bus arrival every 15 minutes; a system with an average gap of 15 minutes and a standard deviation of 1 minute is a lot more reliable than one with an average gap of 15 minutes and a standard deviation of 5 minutes. We assess variability by calculating variances or standard deviations. In this section we develop methods for making inferences about a population variance and standard deviation using data from a single sample.

The methods of this section are based on the chi-square (χ^2) distribution. In the Appendix to Chapter 8 we showed that under certain assumptions the quantity $(n - 1)(s^2/\sigma^2)$ has a χ^2 distribution with $n - 1$ degrees of freedom. The inference methods of this section require use of χ^2 tables. In contrast to the z and t distributions, the χ^2 distribution is not symmetric, and the use of χ^2 tables is a little more complicated.

Appendix Table 5 gives percentage points—values that cut off a specified right-tail area α. For instance, the value that cuts off a right-tail area of $\alpha = .05$ for a χ^2 distribution with 5 d.f. is 11.07 (see Figure 11.1a). To find a *left*-tail point we must look up the complementary area. For example, the value that cuts off a left-tail area of .05 corresponds to a right-tail area of .95. The tabled value for $\alpha = .95$ and d.f. = 5 is 1.15. A quick sketch like Figure 11.1b clarifies the situation. For large d.f., an approximate value is

$$\chi^2_a \left(\frac{z_a}{2} - \sqrt{\text{d.f.} - .5}\right)^2.$$

The tabled χ^2 values can be used to find a confidence interval for a population variance, and the corresponding confidence interval for a population standard deviation can be found by taking square roots of the upper and lower confidence limits.

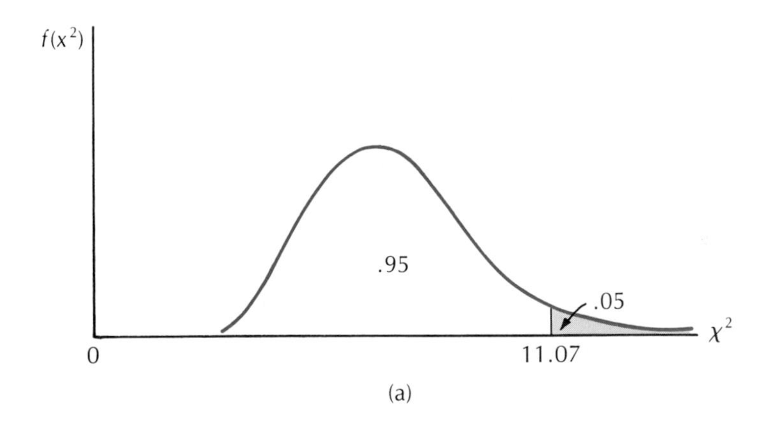

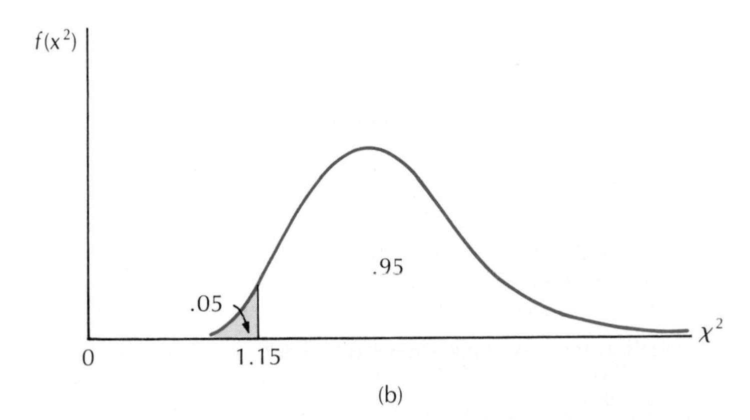

FIGURE 11.1 χ^2 Distribution with d.f. = 5: (a) Upper .05 Value; (b) Lower .05 Value

100(1 − α)% Confidence Intervals for σ^2 and σ

For σ^2: $(n-1)\dfrac{s^2}{\chi^2_{\alpha/2}} < \sigma^2 < (n-1)\dfrac{s^2}{\chi^2_{1-\alpha/2}}$

For σ: $\sqrt{(n-1)\dfrac{s^2}{\chi^2_{\alpha/2}}} < \sigma < \sqrt{(n-1)\dfrac{s^2}{\chi^2_{1-\alpha/2}}}$

where χ^2_a cuts off a right-tail area a in the χ^2 distribution with $n - 1$ d.f.

EXAMPLE 11.1 A pharmaceutical company manufactures a certain kind of antihistamine tablet. It is important to limit the variability of potency from one tablet to another. The quality-control department routinely tests random samples of tablets from each batch. The nominal potency of each tablet is 25 milligrams and the measured potencies of 30 tablets in one sample are

24.1	27.2	26.7	23.6	26.4	25.2
25.8	27.3	23.2	26.9	27.1	26.7
22.7	26.9	24.8	24.0	23.4	25.0
24.5	26.1	25.9	25.4	22.9	24.9
26.4	25.4	23.3	23.0	24.3	23.8

Construct a 95% confidence interval for the standard deviation of tablet potency.

Solution The usual calculations show that $\bar{y} = 25.097$ and $s^2 = 2.1583$. Because $n = 30$, there are 29 d.f. From Appendix Table 5, the $\chi^2_{1-\alpha/2}$ and $\chi^2_{\alpha/2}$ values for a 95% confidence interval are $\chi^2_{.975} = 16.05$ and $\chi^2_{.025} = 45.72$. Therefore the 95% confidence interval for σ^2 is

$$(29)\frac{2.1583}{45.72} < \sigma^2 < (29)\frac{2.1583}{16.05}$$

or

$$1.369 < \sigma^2 < 3.900$$

Taking square roots, the 95% confidence interval for σ is

$$\sqrt{1.369} < \sigma < \sqrt{3.900}$$

or

$$1.17 < \sigma < 1.97$$

□

The justification for this confidence interval is that the event

$$(n-1)\frac{s^2}{\chi^2_{\alpha/2}} < \sigma^2 < (n-1)\frac{s^2}{\chi^2_{1-\alpha/2}}$$

is algebraically identical to the event

$$\chi^2_{1-\alpha/2} < \frac{(n-1)s^2}{\sigma^2} < \chi^2_{\alpha/2}$$

We know that $(n-1)s^2/\sigma^2$ has a χ^2 distribution with $n-1$ d.f. and that $\chi^2_{1-\alpha/2}$ and $\chi^2_{\alpha/2}$ cut off left- and right-tail areas* of $\alpha/2$, respectively. The probability of the event (the confidence level) is therefore the desired value $1-\alpha$.

The χ^2 distribution can also be used to test hypotheses about a variance or standard deviation. We state the test in terms of variances; tests for a standard deviation follow by taking appropriate square roots.

Hypothesis Test for σ^2

H_0: $\sigma^2 = \sigma_0^2$

H_a: 1. $\sigma^2 > \sigma_0^2$

2. $\sigma^2 < \sigma_0^2$

3. $\sigma^2 \neq \sigma_0^2$

T.S.: $\chi^2 = (n-1)s^2/\sigma^2$

R.R.: 1. Reject H_0 if $\chi^2 > \chi^2_\alpha$

2. Reject H_0 if $\chi^2 < \chi^2_{1-\alpha}$

3. Reject H_0 if $\chi^2 > \chi^2_{\alpha/2}$ or $\chi^2 < \chi^2_{1-\alpha/2}$

where χ^2_a cuts off right-tail area a in the χ^2 distribution with $n-1$ d.f.

EXAMPLE 11.2 Suppose that the pharmaceutical company in Example 11.1 wants the standard deviation of potencies in any lot of tablets to be no more than 1.40. Can this H_0 be rejected for $\alpha = .10$?

Solution The elements of the statistical test for σ^2 are

H_0: $\sigma^2 = (1.4)^2 = 1.96$

H_a: $\sigma^2 > 1.96$

$$\text{T.S.: } \chi^2 = (n-1)\frac{s^2}{\sigma_0^2} = \frac{29(2.1583)}{1.96} = 31.934$$

R.R.: The χ^2_α tabulated value for $\alpha = .10$ and d.f. $= 29$ is 39.09, so there is insufficient evidence to reject H_0.

In this problem, a good case can be made for the argument that $\sigma^2 = (1.40)^2$ is a **maximum** allowable variance and that the research hypothesis should be H_a: $\sigma^2 < 1.96$. A batch of tablets would not be released unless quality control could support $\sigma^2 < 1.96$. This is a tougher standard; for the data of Example 11.1, $\sigma^2 < 1.96$ is not supported at all. □

* We use equal tail areas even though the distribution isn't symmetric. By juggling tail areas, it is possible to narrow the confidence interval a little bit, but the equal-tails procedure is simpler and nearly as good.

The χ^2 methods for inferences about σ^2 (and σ) are based on the assumption that the population distribution is normal. This assumption is much more important for inferences about variances than it is for inferences about means. The Central Limit Theorem helps greatly in normalizing the sampling distribution of a mean, but there is no comparable theorem for variances. Population nonnormality, in the form of skewness or heavy tails, can have serious effects on the nominal significance and confidence probabilities for a variance (or standard deviation). If a plot of the sample data shows substantial skewness or outliers, the nominal probabilities given by the χ^2 distribution are suspect. There are some computationally elaborate inference procedures about a variance (such as the so-called jackknife method) that are less sensitive to the normality assumption. These may well replace the χ^2-based methods as computation costs decrease and computer programs become more widely available.

EXAMPLE 11.3 A simulation study involves 1000 samples of size 51 each from a moderately outlier-prone population. The population variance is 64.8. A χ^2 test of the variance is performed for each sample. The results are as follows:

one-tail:

	number of times H_0: "variance = 64.8 is rejected in favor of		
alpha	"variance < 64.8"	"variance > 64.8"	total (alpha doubled)
0.100	205	289	494
0.050	162	221	383
0.025	127	171	298
0.010	106	111	217
0.005	87	86	173

What do the results indicate about the test?

Solution The nominal α probabilities are much, much smaller than the actual fractions. This is evidence that the claimed α value of the χ^2 test of a variance is quite sensitive to nonnormality. □

SECTION 11.1 EXERCISES

11.1 Suppose that Y has a χ^2 distribution with 27 d.f.

a. Find $P(Y > 46.96)$.

b. Find $P(Y > 18.11)$.

c. Find $P(Y < 12.88)$.

d. What is $P(12.8786 < Y < 46.9630)$?

11.2 For a χ^2 distribution with 11 d.f.,

a. find $\chi^2_{.025}$;

b. find $\chi^2_{.975}$.

11.3 Suppose that Y has a χ^2 distribution with 277 d.f. Find approximate values for $\chi^2_{.025}$ and $\chi^2_{.975}$.

11.4 A sample of 25 observations is drawn from a normal population with unknown mean μ and variance σ^2. Define

$$\chi^2 = \frac{(n-1)s^2}{\sigma^2}$$

Find the following probabilities:

a. $P(\chi^2 > 12.4)$
b. $P(\chi^2 < 36.4)$
c. $P(9.89 < \chi^2 < 45.56)$

11.5 A packaging line fills nominal 32-ounce tomato juice jars with an actual mean of 32.30 ounces. The process should have a standard deviation smaller than .15 ounce per jar (a larger standard deviation leads to too many underweight and overfilled jars). Samples of 61 jars are regularly taken to test the process. One such sample yields a sample mean of 32.28 ounces and a standard deviation of .132 ounce. Does this indicate (using $\alpha = .05$) that $\sigma < .15$? Carry out a formal hypothesis test.

11.6 Suppose that the research hypothesis in Exercise 11.5 is formulated as $\sigma > .15$. Does this reformulation tend to be more or less generous in terms of what sample results cause the packaging line to be shut down for adjustment?

11.7 A certain part for a small assembly should have a diameter of 4.000 millimeters, and a maximum standard deviation of .011 millimeter is allowed by specifications. A random sample of 26 parts shows the following diameters:

3.952	3.978	3.979	3.984	3.987	3.991	3.995	3.997	3.999	3.999	3.999
4.000	4.000	4.000	4.001	4.001	4.002	4.002	4.003	4.004	4.006	4.009
4.010	4.012	4.023	4.041							

a. Calculate the sample mean and standard deviation.
b. Can the research hypothesis that $\sigma > .011$ be supported (at $\alpha = .05$) by these data? State all parts of a statistical hypothesis test.

11.8 Calculate 90% confidence intervals for the true variance and for the true standard deviation for the data of Exercise 11.7.

11.9 Plot the data of Exercise 11.7. Does the plot suggest any violation of the assumptions underlying your answers to Exercises 11.7 and 11.8? Would such a violation have a serious effect on the validity of your answers?

11.10 Baseballs vary somewhat in their rebounding coefficient. A "dead ball" has a relatively low rebound, while a "rabbit ball" has a high rebound. A standard test has been developed. A purchaser of large quantities of baseballs requires that the mean value be 85 and the standard deviation be less than 2 units. A sample of 81 baseballs is tested. The mean value is 84.91 and the standard deviation is 1.80. Can the research hypothesis that $\sigma < 2$ be supported using $\alpha = .05$? Carry out the steps of a formal hypothesis test.

11.11 Place bounds on the p-value in Exercise 11.10.

11.2 COMPARING VARIANCES OR STANDARD DEVIATIONS FOR TWO INDEPENDENT SAMPLES

In this section we consider tests and confidence intervals for comparing two variances (or standard deviations) based on data from two independent samples. These procedures are based on another theoretical distribution, called the *F distribution*. Formally, an ***F* statistic**

***F* statistic**

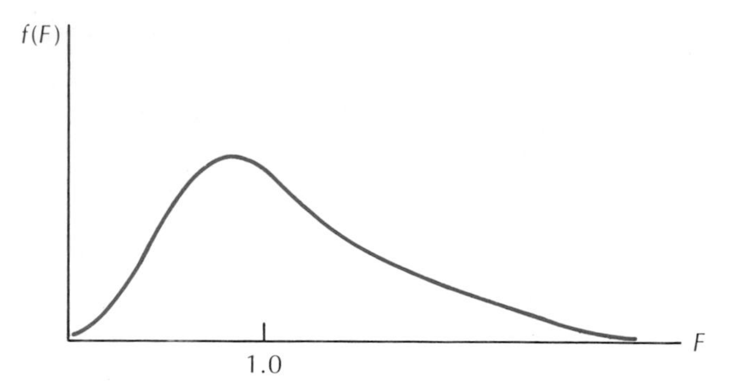

FIGURE 11.2 **Typical *F* Distribution**

is defined as a ratio:

$$F = \frac{\chi_1^2/\text{d.f.}_{\cdot 1}}{\chi_2^2/\text{d.f.}_{\cdot 2}}$$

where χ_1^2 and χ_2^2 are independent statistics having χ^2 distributions with respective degrees of freedom $\text{d.f.}_{\cdot 1}$ and $\text{d.f.}_{\cdot 2}$. The F distribution has degrees of freedom for both numerator and denominator. An F statistic cannot take on negative values; the F distribution is typically skewed to the right, as in Figure 11.2. The mean of this distribution is close to 1; the mode and median are somewhat less.

***F* distribution**

Because the ***F* distribution** depends on two d.f. numbers, tables are either bulky or sparse. Appendix Table 6 gives upper-tail areas of .05, .025, and .01. The numerator degrees of freedom ($\text{d.f.}_{\cdot 1}$) indexes columns, and the denominator degrees of freedom ($\text{d.f.}_{\cdot 2}$) indexes rows. For example, for $\text{d.f.}_{\cdot 1} = 3$ and $\text{d.f.}_{\cdot 2} = 9$, $P(F > 3.86) = .05$ and $P(F > 6.99) = .01$.

A test of the null hypothesis of equal variances for two populations follows from the definition of an F statistic and the fact that (under certain assumptions) the statistic $(n-1)s^2/\sigma^2$ has a χ^2 distribution with $n-1$ d.f. Suppose that we have two independent samples of respective sizes n_1 and n_2. The population variances are σ_1^2 and σ_2^2, and the sample variances are s_1^2 and s_2^2. Then

$$F = \frac{\chi_1^2/\text{d.f.}_{\cdot 1}}{\chi_2^2/\text{d.f.}_{\cdot 2}} = \frac{(n_1-1)\dfrac{s_1^2}{\sigma_1^2}\Big/(n_1-1)}{(n_2-1)\dfrac{s_2^2}{\sigma_2^2}\Big/(n_2-1)}$$

$$= \frac{s_1^2/\sigma_1^2}{s_2^2/\sigma_2^2}$$

Under the null hypothesis H_0: $\sigma_1^2 = \sigma_2^2$, the test statistic becomes $F = s_1^2/s_2^2$. The test procedure is outlined next:

F Test for Equality of Variances

H_0: $\sigma_1^2 = \sigma_2^2$

H_a: 1. $\sigma_1^2 > \sigma_2^2$
2. $\sigma_1^2 < \sigma_2^2$
3. $\sigma_1^2 \neq \sigma_2^2$

T.S.: $F = \dfrac{s_1^2}{s_2^2}$

R.R.: 1. Reject H_0 if $F > F_{\alpha, n_1-1, n_2-1}$, the tabled value cutting off a right-tail area of the F distribution with d.f.$_1 = n_1 - 1$ and d.f.$_2 = n_2 - 1$.
2. Reject H_0 if $F < 1/F_{\alpha, n_2-1, n_1-1}$. Note that F_{α, n_2-1, n_1-1} is the tabled F value with **numerator** d.f. $= n_2 - 1$ and **denominator** d.f. $= n_1 - 1$.
3. Reject H_0 if $F > F_{\alpha/2, n_1-1, n_2-1}$ or if $F < 1/F_{\alpha/2, n_2-1, n_1-1}$.

The reason for the transposition of degrees of freedom is that the choice of which sample variance to put in the numerator is arbitrary.

EXAMPLE 11.4 A study compares the variabilities in strengths of 1-inch-square sections of a synthetic fiber produced under two different procedures. Samples of 9 squares from each procedure are obtained, yielding the following strengths (in pounds per square inch):

Procedure 1:	74	90	103	86	75	102	97	85	69
Procedure 2:	59	66	73	68	70	71	82	69	74

Test the research hypothesis of unequal variability using $\alpha = .10$.

Solution Standard computations yield

	$\bar{y}$	s^2	n
Procedure 1	86.778	153.944	9
Procedure 2	70.222	38.944	9

The test can be summarized as follows:

H_0: $\sigma_1^2 = \sigma_2^2$

H_a: $\sigma_1^2 \neq \sigma_2^2$

T.S.: $F = \dfrac{s_1^2}{s_2^2} = \dfrac{153.944}{38.944} = 3.95$

R.R.: Because d.f.$_1$ = d.f.$_2 = 8$ and $F_{.05,8,8} = 3.44$, we reject H_0 if $s_1^2/s_2^2 > 3.44$ or if $s_1^2/s_2^2 < 1/3.44 = .291$

Conclusion: Because $F = 3.95 > 3.44$, we reject H_0 and conclude that the variability for procedure 1 is greater than that for procedure 2. □

It is convenient to use the ratio of two sample variances, rather than their difference, to construct a confidence interval for comparing two population variances.

100(1 − α)% Confidence Interval for σ_1^2/σ_2^2 or σ_1/σ_2

$$\frac{s_1^2}{s_2^2}\frac{1}{F_{\alpha/2,n_1-1,n_2-1}} < \frac{\sigma_1^2}{\sigma_2^2} < \frac{s_1^2}{s_2^2}F_{\alpha/2,n_2-1,n_1-1}$$

For comparing standard deviations, take square roots throughout.

EXAMPLE 11.5 Calculate a 90% confidence interval for the ratio of procedure standard deviations based on the data of Example 11.4.

Solution In Example 11.4 we found that the ratio of sample variances was $s_1^2/s_2^2 = 3.95$ and $F_{.05,8,8} = 3.44$. The 90% confidence interval for the ratio of the population variances is

$$3.95\left(\frac{1}{3.44}\right) < \frac{\sigma_1^2}{\sigma_2^2} < 3.95(3.44)$$

or

$$1.148 < \frac{\sigma_1^2}{\sigma_2^2} < 13.588$$

Taking square roots,

$$1.072 < \frac{\sigma_1}{\sigma_2} < 3.686$$

Note that a ratio of 1.00 is not included in the 90% confidence interval. This corresponds to the rejection of H_0: $\sigma_1^2 = \sigma_2^2$ at $\alpha = .10$ in Example 11.4. □

normality assumption

The critical assumption for these procedures is the assumption of normal population distributions. If either or both of the populations are moderately skewed or heavy-tailed, the nominal *F* probabilities can be alarmingly erroneous. As for the one-sample case, it is quite possible that alternative procedures for comparing variances are less sensitive to the normal assumption, and hence more useful. We do not know of widely available computer programs for such alternative methods.

EXAMPLE 11.6 Do the data values in Example 11.4 indicate that the *F* procedures of Examples 11.4 and 11.5 may be misleading?

Solution There isn't much data available to assess the distribution. Plots of what little data we have seem tolerably close to normal. The procedure 2 data in particular appear close to normal, while the procedure 1 data suggest some slight skewness. If these were the only values available, we would regard the nominal .10 probability as a half-decent approximation. However, we guess that in the long run there will be a modest number of exceptionally weak squares, yielding left-skewed population distributions. Therefore a method that is less sensitive (more robust) to nonnormality is preferable. □

SECTION 11.2 EXERCISES

11.12 Refer to the data of Exercise 10.10 (p. 319). Test the hypothesis of equal variances against a two-sided alternative. Use $\alpha = .05$.

11.13 Refer to Exercise 11.12. How would you use F tables to place bounds on the p-value?

11.14 Plot the data of Exercise 10.10. Is there reason to be suspicious of the conclusion of Exercise 11.12?

11.15 The total sales for antidepressants throughout the United States during a given year represents a sizable amount of money, and competition among drug companies is keen. Suppose that one company markets drug A and a second company markets drug B. Both of these prescription drugs are for the relief of depression. One factor in the treatment of depressed patients is the patient's perceived benefit from the therapy; the time to first perception of relief (onset of relief) can therefore be an important variable to examine. The table below summarizes the perceived time to onset of relief (in numbers of days) for a random sample of 20 patients treated with drug A, and for a second random sample of 20 patients treated with drug B.

	n	$\bar{y}$	s
Drug A	20	8.3	3.1
Drug B	20	10.2	2.5

a. Would you expect the data to be normal?

b. Assuming that the data are approximately normal, run a two-sided test of H_0: $\sigma_1^2 = \sigma_2^2$ using $\alpha = .05$.

11.16 Refer to the data of Exercise 11.15. Construct a 95% confidence interval for the ratio σ_1^2/σ_2^2.

11.17 A simulation study involves 1000 independent samples of size 20 from each of two populations. Both populations have the same variance. Both are moderately outlier-prone. The following results are obtained for the variance ratio (F) test.

one-tail:

number of times H_0: "variances equal" is rejected in favor of

alpha	"var1 > var2"	"var1 < var2"	total (alpha doubled)
0.100	194	196	390
0.050	139	134	273
0.025	97	93	190
0.010	63	65	128
0.005	41	43	84

Does the presence of outliers have a serious effect on the accuracy of the claimed α for the variance ratio test?

11.3 TESTS FOR SEVERAL PROPORTIONS

There is an entirely different set of hypothesis-testing procedures that also happen to use χ^2 tables. These are methods for testing several proportions. Two variants of this test are discussed in this section and the next.

goodness-of-fit test

The χ^2 **goodness-of-fit** test is used to test the hypothesis that several proportions have specified numerical values. For instance, suppose that a life insurance company has a mix of 40% whole life policies, 25% level term policies, 15% decreasing term policies, and 20% other types. A change in this mix could signal a need to change commission, reserve, or investment practices, but the company does not want to react to random short-term fluctuations. If the company does a study of the last 1200 policies issued (regarded as a random sample), the χ^2 goodness-of-fit test can be used to test the statistical significance of deviations from the historical percentages.

In Chapter 9 we developed a test for a single binomial proportion, often using a z approximation. We could run a separate test on the proportion for each policy type. The trouble with such a procedure is that although the Type I error is controlled at some level α for each test, the overall probability of error (incorrectly rejecting the null hypothesis in at least one test) may be much larger than α. If four separate tests are run at $\alpha = .05$, what is the probability of at least one Type I error? The tests are not independent. If one sample proportion is too high, the others tend to be too low, because the sample proportions must add to 1. But there is no way to get an accurate assessment of the overall probability of error from combining several z tests. The goodness-of-fit method yields a combined test of all proportions with a specified overall α level.

multinomial sampling

The χ^2 procedure assumes **multinomial sampling**. This is the extension of binomial sampling to more than two categories. Now we have k categories, and in each of n independent trials, the probability of observing a member of category i is π_i. This probability is assumed to be constant over trials.

expected number

The procedure works by comparing the observed number in each category to the **expected number** in that category. If there are n items in a sample and the probability of any items falling in category i is π_i, then, by the binomial distribution, the expected number in category i is

$$E_i = n\pi_i$$

The test procedure is summarized next:

Goodness-of-fit Test for Several Proportions

H_0: $\pi_i = \pi_{i,0}$ for categories $i = 1, \ldots, k$; $\pi_{i,0}$ are specified probabilities or proportions

H_a: H_0 is not true

T.S.: $\chi^2 = \sum_i \frac{(n_i - E_i)^2}{E_i}$ where n_i is the observed number in category i and $E_i = n\pi_{i,0}$ is the expected number under H_0.

R.R.: Reject H_0 if $\chi^2 > \chi^2_\alpha$, the right-tail α percentage point of a χ^2 distribution with d.f. $= k - 1$. Note that the d.f. depends on k, the number of categories, not on n, the number of observations.

EXAMPLE 11.7 Suppose that, in the insurance company illustration at the beginning of this section, the previous 1200 policies issued consist of 439 whole life policies, 323 level term policies,

197 decreasing term policies, and 241 others. Assess the statistical significance (at $\alpha = .10$) of any shift from the historical policy mix. What actual conditions might lead to violations of the multinomial assumptions?

Solution The following table summarizes the calculations:

Category, i	Whole Life	Level Term	Decreasing Term	Other	Total
Historical proportion, $\pi_{i,0}$	.40	.25	.15	.20	1.00
Observed number, n_i	439	323	197	241	1200
Expected number, $E_i = 1200\pi_{i,0}$	480	300	180	240	1200
$n_i - E_i$	−41	+23	+17	+1	0
$\frac{(n_i - E_i)^2}{E_i}$	3.502	1.763	1.606	.004	6.875

The number of categories is $k = 4$, so there are $k - 1 = 3$ d.f.; $\chi^2_{.10} = 6.25$ for 3 d.f. Because $\chi^2 = 6.875 > 6.25$, H_0 is rejected, and the shift from the historical mix is statistically significant at $\alpha = .10$. It appears that term policies are becoming more popular and whole life policies less so. (Note: The $\alpha = .05$ percentage point is 7.81. H_0 could not be rejected at $\alpha = .05$, therefore the p-value is somewhere between .05 and .10.) □

multinomial assumptions

As with the binomial distribution, the key **multinomial assumptions** are independence of trials and constant probabilities over trials. Independence would be violated if several of the policies were sold to the same person or within the same family. Constant probability would be violated if the 1200 policies were sold over a long enough period that a time trend could matter. The most serious possible violation would occur if the policies were sold by relatively few agents and if there were major differences in product mix among agents.

The statistic for the χ^2 goodness-of-fit test is the sum of k terms, which is why d.f. depends on k, not n. There are $k - 1$ d.f. instead of k because the sum of the $n_i - E_i$ terms must always equal $n - n = 0$; $k - 1$ of these terms are free to vary but the kth is determined by this requirement.

The mathematics underlying this test is based on an approximation that is a lineal descendant of the normal approximation to the binomial distribution. The quality of this approximation has been extensively studied. A fairly conservative general rule is that the approximation is adequate if all E_i are at least 5.0.

The goodness-of-fit test has been used extensively to test the adequacy of various scientific theories. One problem of such applications is that the hypothesis of interest is formulated as the null hypothesis, not the research hypothesis. If a scientist has a pet theory and wants to show that it gives a good fit to the data, the scientist wants to accept the null hypothesis. But the potential error in accepting it is Type II, and β probabilities of Type II errors are hard to calculate. In general, the null hypothesis tends to be accepted (the β probability is high) if n is small, or if there are many categories. Even if the general rule that all E_i are at least 5.0 is satisfied, the β risk can be large. **A "good fit" conclusion is always suspect.** Perhaps the best procedure for a manager is to look at deviations of sample proportions from theoretical proportions and to use the χ^2 test results as an indicator of the degree of potential random variation.

EXAMPLE 11.8 Suppose that in a test of a random walk theory of stock price changes, 125 security analysts are each asked to select four stocks listed on the New York Stock Exchange that are expected to outperform the Standard and Poor's Index over a 90-day period. According to one random walk theory of stock price changes, the analysts should do no better than coin flipping; the number of correct guesses by any particular analyst should follow a binomial distribution with $n = 4$ and probability of success .50. The data are

Number correct:	0	1	2	3	4
Frequency:	3	23	51	39	9

Test the random walk hypothesis at $\alpha = .05$ and at $\alpha = .10$.

Solution

Number correct	0	1	2	3	4
Theoretical proportion (based on binomial probability)	.0625	.2500	.3750	.2500	.0625
Observed proportion	.0240	.1840	.4080	.3120	.0720
n_i	3	23	51	39	9
E_i	7.8125	31.250	46.875	31.250	7.8125
$\frac{(n_i - E_i)^2}{E_i}$	2.9645	2.1780	0.3630	1.922	0.1805

$$\chi^2 = \sum \frac{(n_i - E_i)^2}{E_i} = 7.608$$

The tabled .05 and .10 points for χ^2 with $5 - 1 = 4$ d.f. are 9.49 and 7.78, respectively. Thus H_0 cannot be rejected at $\alpha = .05$ or even at $\alpha = .10$, and the data might be declared "a good fit" to the random walk theory. But the actual proportions of analysts who are correct for two, three, or four stocks are larger than those predicted by the theory. The result of the χ^2 test indicates that this fact could conceivably have been a collective lucky break for the analysts. We would say that such data would suggest, though by no means conclusively prove, that analysts can do somewhat better than the random walk theory would indicate. □

SECTION 11.3 EXERCISES

11.18 The director of a data-processing center wants to test the Poisson distribution model for arrivals of jobs to a central computer. The mean arrival rate during the relevant period is 3.8 jobs per minute. Records are kept on the number of arrivals in each of 2000 1-minute periods. The results are

Arrivals:	0	1	2	3	4	5	6	7	8	9+
Frequency:	38	155	328	392	415	399	170	61	27	15

a. Use Poisson tables ($\lambda = 3.8$) to calculate probabilities for each category of number of arrivals.
b. Calculate expected frequencies.
c. Is the Poisson model a good fit to the data? Use $\alpha = .01$.

11.19 Can you detect any systematic discrepancy between the observed frequencies and the expected (Poisson model) frequencies in the data of Exercise 11.18?

11.20 A gift shop owner believes that 30% of the customers who enter the shop buy no items, 45% buy 1 item, and 25% buy 2 or more items. Observation of 25 customers yields the following data:

Number of purchases:	0	1	2+
Frequency:	10	6	9

a. Calculate expected frequencies, assuming that the owner's hypothesis is valid.
b. Test the owner's hypothesis using $\alpha = .05$. Can the owner claim a good fit based on these data?

11.21 Suppose that the data in Exercise 11.20 are based on 250 customers, with respective frequencies 100, 60, and 90.
a. Test the owner's hypothesis at $\alpha = .05$.
b. Explain the discrepancy in conclusion between this exercise and Exercise 11.20.

11.4 CHI-SQUARE TESTS OF INDEPENDENCE

dependence

In Chapter 3 we introduced the idea of independence. In particular we discussed the idea that **dependence** of variables means that one variable has some value for predicting the other. With sample data there usually appears to be some degree of dependence. In this section we develop a χ^2 test that assesses whether the perceived dependence in sample data may be a fluke—the result of random variability rather than real dependence.

First, the frequency data are to be arranged in a cross-tabulation with r rows and c columns. The possible values of one variable determine the rows of the table and the possible values of the other determine the columns. We denote the population proportion (or probability) falling in row i, column j as π_{ij}. The total proportion for row i is $\pi_{i.}$, and the total proportion for column j is $\pi_{.j}$. If the row and column proportions (probabilities) are independent, then $\pi_{ij} = \pi_{i.}\pi_{.j}$. For instance, suppose that a personnel manager for a large firm wants to assess the popularity of three alternative flexible time-scheduling (flextime) plans among clerical workers in four different offices. The following indicates a set of proportions (π_{ij}) exhibiting independence. The proportion of all clerical workers who favor plan 2 and work in office 1 is $\pi_{21} = .03$; the proportion of all workers favoring plan 2 is $\pi_{2.} = .30$ and the proportion working in office 1 is $\pi_{.1} = .10$. Independence holds for that cell because $\pi_{21} = .03 = (\pi_{2.})(\pi_{.1}) = (.30)(.10)$. Independence also holds for all other cells.

	Office				
Favored Plan	1	2	3	4	Total
1	.05	.20	.15	.10	.50
2	.03	.12	.09	.06	.30
3	.02	.08	.06	.04	.20
Total	.10	.40	.30	.20	

The null hypothesis for this χ^2 test is independence. The research hypothesis specifies only that there is some form of dependence, that is, that it is not true that $\pi_{ij} = \pi_{i.}\pi_{.j}$ in every cell of the table. The test statistic is once again the sum over all cells of

$$(\text{Observed value} - \text{expected values})^2/\text{expected value}$$

The computation of expected values E_{ij} under the null hypothesis is different for the independence test than for the goodness-of-fit test. The null hypothesis of independence does not specify numerical values for the row probabilities $\pi_{i.}$ and column probabilities $\pi_{.j}$, so these probabilities must be estimated by the row and column relative frequencies. If $n_{i.}$ is the actual frequency in row i, estimate $\pi_{i.}$ by $\hat{\pi}_{i.} = n_{i.}/n$; similarly $\hat{\pi}_{.j} = n_{.j}/n$. Assuming the null hypothesis of independence is true, it follows that $\hat{\pi}_{ij} = \hat{\pi}_{i.}\hat{\pi}_{.j} = (n_{i.}/n)(n_{.j}/n)$.

Estimated Expected Values $\hat{E}_{ij}$

Under the hypothesis of independence, the estimated expected value in row i, column j is

$$\hat{E}_{ij} = n\hat{\pi}_{ij} = n\frac{(n_{i.})}{n}\frac{(n_{.j})}{n} = \frac{(n_{i.})(n_{.j})}{n}$$

the row total multiplied by the column total divided by the grand total.

EXAMPLE 11.9 Suppose that in the flexible time-scheduling illustration a random sample of 216 workers yields the following frequencies:

	Office				
Favored Plan	1	2	3	4	Total
1	15	32	18	5	70
2	8	29	23	18	78
3	1	20	25	22	68
Total	24	81	66	45	216

Calculate a table of $\hat{E}_{ij}$ values.

Solution For row 1, column 1 the estimated expected number is

$$\hat{E}_{11} = \frac{(\text{row 1 total})(\text{column 1 total})}{\text{grand total}} = \frac{(70)(24)}{216} = 7.78$$

Similar calculations for all cells yield the table below.

	Office				
Plan	1	2	3	4	Total
1	7.78	26.25	21.39	14.58	70.00
2	8.67	29.25	23.83	16.25	78.00
3	7.56	25.50	20.78	14.17	68.01
Total	24.01	81.00	66.00	45.00	216.01

Note that the row and column totals in the $\hat{E}_{ij}$ table equal (except for round-off error) the corresponding totals in the observed (n_{ij}) table. □

The χ^2 test of independence is summarized in the following:

χ^2 Test of Independence

H_0: The row and column variables are independent

H_a: The row and column variables are dependent (associated)

T.S.: $\chi^2 = \sum_{i,j} (n_{ij} - \hat{E}_{ij})^2/\hat{E}_{ij}$

R.R.: Reject H_0 if $\chi^2 > \chi^2_\alpha$, where χ^2_α cuts off area α in a χ^2 distribution with $(r-1)(c-1)$ d.f.; r = number of rows, c = number of columns.

EXAMPLE 11.10 Carry out the χ^2 test of independence for the data of Example 11.9. First use $\alpha = .05$, then obtain a bound for the p-value.

Solution The term for cell (1, 1) is $(n_{11} - \hat{E}_{11})^2/\hat{E}_{11} = (15 - 7.78)^2/7.78 = 6.70$. Similar calculations are made for each cell. Substituting into the test statistic, we find $\chi^2 = 6.70 + \ldots + 4.33 = 27.12$. For $(3-1)(4-1) = 6$ d.f., the tabled χ^2 value (6 d.f., $\alpha = .05$) is 12.59. The observed χ^2 value of 27.12 far exceeds 12.59, so H_0 is rejected at $\alpha = .05$. In fact, 27.12 exceeds the tabled value even for $\alpha = .005$ (the smallest α in the table), namely 18.55. Therefore H_0 is rejected even for $\alpha = .005$ and p-value $< .005$. □

The degrees of freedom for the χ^2 test of independence relate to the number of cells in the two-way table that are free to vary while the marginal totals remain fixed. For example, in a 2×2 table (2 rows, 2 columns) only one cell entry is free to vary. Once that entry is fixed, the remaining cell entries are determined by subtracting from the corresponding row or column total (see Figure 11.3a). Similarly, with a 2×3 table (2 rows, 3 columns), two of the cell entries are free to vary. Once these entries are set, the remaining cell entries are determined by subtracting from the appropriate row or column total (see Figure 11.3b). In general, for a table with r rows and c columns, $(r-1)(c-1)$ of the cell

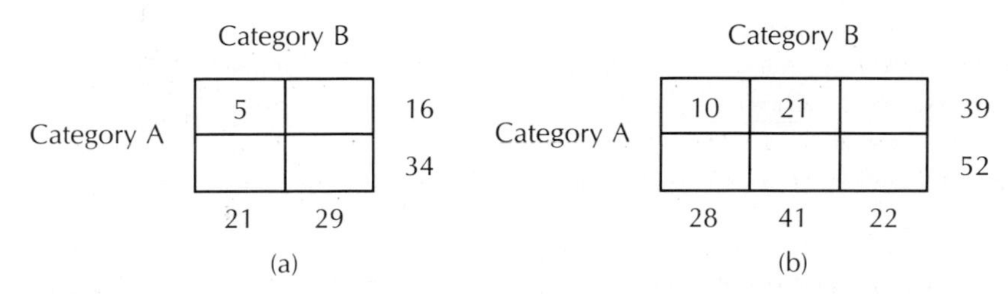

FIGURE 11.3 (a) One Degree of Freedom for a 2×2 Table; (b) Two Degrees of Freedom for a 2×3 Table

entries are free to vary. This number represents the degrees of freedom for the χ^2 test of independence.

This χ^2 test of independence is also based on an approximation. A conservative rule is that each $\hat{E}_{ij}$ must be at least 5 to use the approximation comfortably. Standard practice if some $\hat{E}_{ij}$'s are too small is to lump together those rows (or columns) with small totals until the rule is satisfied.

The only function of this χ^2 test is to determine whether apparent dependence in sample data may be a fluke, plausibly a result of random variation. Rejection of the null hypothesis indicates only that the apparent association is not reasonably attributable to chance. It does not indicate anything about the **strength or** type of **association**. The percentage analysis and the λ measure defined in Section 11.5 are used to indicate how strong the dependence is.

strength of association

The same χ^2 test statistic applies to a slightly different sampling procedure. An implicit assumption of our discussion surrounding the χ^2 test of independence has been that the data result from a single random sample from the whole population. Often, separate random samples are taken from the subpopulations defined by the column (or row) variable. In the flextime Example 11.9, the data might well have resulted from separate samples (of respective sizes 24, 81, 66, and 45) from the four offices rather than from a single random sample of 216 workers. The null hypothesis of independence is then stated (in an equivalent form) as H_0: the conditional probability of row i given column j is the same for all columns j. The test is called a **test of homogeneity** of distributions (that is, that the probabilities or proportions by column are equal). The mechanics and conclusions of the test are identical, so the distinction is minor.

test of homogeneity

EXAMPLE 11.11

A poll of attitudes toward five possible energy policies is taken. Random samples of 200 individuals from major oil- and natural-gas-producing states, 200 from coal states, and 400 from other states are drawn. Each respondent indicates a most preferred alternative from among the following:

1. primarily emphasize conservation
2. primarily emphasize domestic oil and gas exploration
3. primarily emphasize investment in solar-related energy
4. primarily emphasize nuclear-energy development and safety
5. primarily reduce environmental restrictions and emphasize coal-burning activities

The results are as follows:

Policy Choice	Oil/gas States	Coal States	Other States	Total
1	50	59	161	270
2	88	20	40	148
3	56	52	188	296
4	4	3	5	12
5	2	66	6	74
Total	200	200	400	800

Conduct a χ^2 test of homogeneity of distributions for the three groups of states. Give the p-value for this test.

Solution A test that the corresponding population distributions are different makes use of the following table of expected values:

Policy Choice	Oil/gas States	Coal States	Other States
1	67.5	67.5	135
2	37	37	74
3	74	74	148
4	3	3	6
5	18.5	18.5	37

The table violates our general rule that all $\hat{E}_{ij}$'s be at least 5. There is no obvious choice for combining policy 4 with some other. Therefore we leave the table as is, but we realize that the nominal χ^2 probabilities are slightly suspect. The test procedure is outlined here:

H_0: The column distributions are homogeneous

H_a: The column distributions are not homogeneous

$$\text{T.S.: } \chi^2 = \sum (n_{ij} - \hat{E}_{ij})^2/\hat{E}_{ij}$$
$$= (50 - 67.5)^2/67.5 + (88 - 37)^2/37 + \ldots + (6 - 37)^2/37$$
$$= 289.22$$

R.R. and Conclusion: Because the tabled value of χ^2 for d.f. $= 8$ and $\alpha = .005$ is 21.96, the p-value is $< .005$. □

Even recognizing the limited accuracy of the χ^2 approximation, we can reject the hypothesis of homogeneity at some very small p-value. Percentage analysis, particularly of state type for a given policy choice, shows dramatic differences; for instance, 1% of those living in oil/gas states favor policy 5, compared to 33% of those living in coal states and favoring policy 5.

The χ^2 test described in this section has a limited, but important, purpose. This test only assesses whether the data indicate that there's a statistically detectable (significant) relation among various categories. It doesn't measure how strong the apparent relation might be. A weak relation in a large data set may be detectable (significant); a strong relation in a small data set may be nonsignificant. Methods for assessing the apparent strength of relation in a data set are discussed in Section 11.5.

SECTION 11.4 EXERCISES

11.22 A personnel director for a large, research-oriented firm categorizes colleges and universities as most desirable, good, adequate, and undesirable for purposes of hiring their graduates. Data are collected on 156 recent graduates, and each is rated by a supervisor.

School	Rating		
	Outstanding	Average	Poor
Most desirable	21	25	2
Good	20	36	10
Adequate	4	14	7
Undesirable	3	8	6

Output from a standard computer package (SPSS) is shown below:

```
                   RATING
          COUNT   |
                  |OUTSTAND AVERAGE  POOR          ROW
                  |ING                             TOTAL
                  |   1.00|    2.00|    3.00|
SCHOOL     --------+--------+--------+--------+
            1.00  |   21   |   25   |    2   |     48
  MOST DESIRABLE  |        |        |        |   30.8
                  +--------+--------+--------+
            2.00  |   20   |   36   |   10   |     66
  GOOD            |        |        |        |   42.3
                  +--------+--------+--------+
            3.00  |    4   |   14   |    7   |     25
  ADEQUATE        |        |        |        |   16.0
                  +--------+--------+--------+
            4.00  |    3   |    8   |    6   |     17
  UNDESIRABLE     |        |        |        |   10.9
                  +--------+--------+--------+
          COLUMN      48       83       25        156
           TOTAL    30.8     53.2     16.0      100.0

 CHI-SQUARE    D.F.      SIGNIFICANCE      MIN E.F.     CELLS WITH E.F.< 5
 ----------    ----      ------------      --------     -----------------

   15.96704      6          0.0139           2.724      2 OF    12 ( 16.7%)

                                                WITH SCHOOL        WITH RATING
        STATISTIC                 SYMMETRIC     DEPENDENT          DEPENDENT
        ---------                 ---------     ------------       ------------

LAMBDA                             0.00613        0.01111            0.00000
UNCERTAINTY COEFFICIENT            0.04716        0.04211            0.05357
SOMERS' D                          0.26573        0.28627            0.24794
ETA                                               0.31150            0.30639
```

a. Locate the value of the χ^2 statistic.
b. Locate the *p*-value.
c. Can the director safely conclude that there is a relation between school type and rating?
d. Is there any problem in using the χ^2 approximation?

11.23 Calculate rating percentages for each school. Do the percentages reflect the existence of the relation found in Exercise 11.22?

11.24 A study of potential age discrimination considers promotions among middle managers in a large company. The data are

	Age				
	Under 30	30–39	40–49	50 or Over	Total
Promoted	9	29	32	10	80
Not promoted	41	41	48	40	170
Total	50	70	80	50	

a. Find the expected numbers under the hypothesis of independence.
b. Calculate the degrees of freedom.
c. Is there a statistically significiant relation between age and promotions, using $\alpha = .05$?

11.25 Place bounds on the p-value in Exercise 11.24.

11.26 The data of Exercise 11.24 are combined as follows:

	Age		
	Up to 39	40 or Over	Total
Promoted	38	42	80
Not promoted	82	88	170
Total	120	130	

a. Can the hypothesis of independence be rejected using a reasonable α?
b. What is the effect of combining age categories? Compare the answers to Exercises 11.24 and 11.26.

11.5 MEASURING STRENGTH OF RELATION

The χ^2 test discussed in Section 11.4 has a built-in limitation. By design, the test only answers the question, Is there a statistically detectable (significant) relation among the categories? It cannot answer the question of whether the relation is strong, interesting, or relevant. This is not a criticism of the test; no hypothesis test can answer these questions. In this section we discuss methods for assessing the strength of relation shown in cross-tabulated data.

The simplest (and often the best) method for assessing the strength of a relation is simple percentage analysis. If there is no relation (that is, if complete independence holds), then percentages by row or by column show no relation. For example, suppose that a direct mail company tests two different offers to see if the response rates differ. Their results are as shown here:

	Response		
Offer	Yes	No	Total
A	40	160	200
B	80	320	400
Total	120	480	600

To check the relation, if any, we calculate percentages of response for each offer. We see that (40/200) = .20 (that is, 20%) respond to offer A and (80/400) = .20 respond to offer B. Because the percentages are exactly the same, there is no indication of a relation. Alternatively, we note that one-third of the "yes" respondents and one-third of the "no" respondents were given offer A. Because these fractions are exactly the same, there is no indication of a statistical relation.

Of course it is rare to have data that show absolutely no relation in the sample. More commonly the percentages by row or by column differ, suggesting some relation. For example, a firm planning to market a cleaning product commissions a market research study of the leading current product. The variables of interest are the frequency of use and the rating of the leading product. The data are shown below:

	Rating			
Use	Fair	Good	Excellent	Total
Rare	64	123	137	324
Occasional	131	256	129	516
Frequent	209	171	45	425
Total	404	550	311	1265

One natural analysis of the data takes the frequencies of use as givens and looks at the ratings as functions of use. The analysis essentially looks at conditional probabilities of the rating factor, given the use factor, but it recognizes that the data are only a sample, not the population. When use is rare, the best estimate is that 64/324 = .1975 (or 19.75%) will rate the product as fair, that 123/324 = .3796 will rate it good, and that 137/324 = .4228 will rate it excellent. The corresponding proportions for occasional users are 131/516 = .2539, 256/516 = .4961, and 129/516 = .2500. For frequent users the proportions are .4918, .4024, and .1059. The proportions (or percentages, if one multiplies by 100) are quite different for the three use categories, indicating that rating is related to use. Alternatively, we may calculate the use categories as percentages of the rating categories. In either case, there appears to be a relation. Because the proportions of ratings differ quite a bit as one varies use (or the proportions of use differ quite a bit as one varies rating), there is a suggestion that there is a fairly strong relation between use and rating.

Another way to analyze relations in data is to consider predictability. The stronger the relation exhibited in data, the better one can predict the value of one variable from the value of the other. We could imagine a situation where every rare user rated the product as excellent, every occasional user rated the product as good, and every frequent user rated the product as fair. In such a case, there would be a perfect statistical relation; in terms of predictability, given the use, one could predict the rating exactly. Of course in practice prediction and relation are not perfect; we need a measure of the strength of relation defined as degree of predictability.

dependent and independent variables

We need to distinguish between a **dependent variable**—the variable one is trying to predict—and an **independent variable**—the variable one is using to make the prediction. If one is trying to predict rating given use, use serves as the independent variable and rating as the dependent variable. No cause-and-effect connotations are intended; the choice of independent and dependent variables is entirely up to the person analyzing the data.

The simplest prediction rule is to predict the most common value (the mode) of the dependent variable; this rule is the basis of the λ (lambda) predictability measure. In our use–rating example, when use is rare, the most common rating is excellent, with 137 responses; if one predicts excellent for every rare case, one makes $187 = 64 + 123 = 324 - 137$ prediction errors. Similarly for occasional use, a prediction of rating as good gives $260 = 131 + 129 = 516 - 25$ errors; given frequent use, a prediction of rating as fair gives $216 = 171 + 45 = 425 - 209$ errors. The total number of errors is $187 + 260 + 216 = 663$. By comparison, if use is not known, we would have to predict the most common rating, namely good, we would make $715 = 404 + 311 = 1265 - 550$ errors. Reasonably enough, given information about use, we do better in predicting rating than when we have no information. The λ measure indicates how much better we predict. When use is the independent variable and rating is the dependent variable, the difference in prediction errors is $715 - 663 = 52$; we take this difference as a fraction of the errors made not knowing the independent variable, namely 715. If use is the independent variable,

$$\lambda = \frac{\text{errors with unknown independent variable} - \text{errors with known independent variable}}{\text{errors with unknown independent variable}}$$

$$= \frac{715 - 663}{715} = .115$$

The value of λ ranges between 0 and 1; $\lambda = 1$ indicates that, at least in the sample, the dependent variable is predicted perfectly given the independent variable. A value of $\lambda = 0$ occurs if there is independence in the data, or if the same value of the dependent variable is always predicted. To interpret other values of λ, note that it is a proportionate reduction in error (PRE) measure. The value $\lambda = .115$ that we found means that we make 11.5% fewer errors predicting rating given use than we would predicting rating without information about use. Values of λ above about .30 are rare in real data; thus $\lambda = .115$ indicates a modest relation between rating and use. Note that the value of λ depends on which variable is taken as the dependent variable. For rating predicting use, $\lambda = .073$.

Calculation of λ

For every level of the independent variable (every row or every column), find the modal value of the dependent variable. If there are two or more modal values with equal frequency, choose one arbitrarily. Add the frequencies in all nonmodal cells to find K = number of prediction errors with known independent variable value.

Refer to the marginal (total) frequencies of the dependent variable. Add the frequencies in all nonmodal categories to find U = number of prediction errors with unknown independent variable values.

$$\lambda = (U - K)/U$$

EXAMPLE 11.12 An internal survey of samples of clerical workers, supervisory personnel, and junior managers is taken to obtain opinions on a proposed flextime schedule. The following output (SAS) is obtained:

TABLE OF OPINION BY LEVEL

OPINION / LEVEL FREQUENCY EXPECTED COL PCT	clerical	superv	junior manager	TOTAL
strongly oppose	5 9.8 12.50	8 7.3 26.67	9 4.9 45.00	22
oppose	8 10.2 20.00	10 7.7 33.33	5 5.1 25.00	23
favor	14 11.1 35.00	7 8.3 23.33	4 5.6 20.00	25
strongly favor	13 8.9 32.50	5 6.7 16.67	2 4.4 10.00	20
TOTAL	40	30	20	90

STATISTICS FOR TABLE OF OPINION BY LEVEL

STATISTIC	DF	VALUE	PROB
CHI-SQUARE	6	12.110	0.060

STATISTIC	VALUE	ASE
GAMMA	-0.428	0.114
LAMBDA ASYMMETRIC C|R	0.120	0.106
LAMBDA ASYMMETRIC R|C	0.123	0.079
LAMBDA SYMMETRIC	0.122	0.083

SAMPLE SIZE = 90
ASE IS THE ASYMPTOTIC STANDARD ERROR.
R|C MEANS ROW VARIABLE DEPENDENT ON COLUMN VARIABLE.

a. What does the χ^2 test indicate about the relation shown in the data?
b. What can be seen by examining the column percentages?
c. Locate the value for λ with opinion taken as the dependent variable. Interpret the number.
d. Verify the calculation of λ.

Solution a. The χ^2 statistic is shown as CHI-SQUARE = 12.110 with 6 d.f. The p-value is shown as PROB = .060. We cannot reject the null hypothesis of independence at $\alpha = .05$, but we can reject it at $\alpha = .10$. There is only limited evidence that the apparent relation is non-random; therefore we should not place too much reliance on any apparent dependence.

b. The clerical staff seems to be largely in favor (35%) or strongly in favor (32.5%). The junior managers are largely strongly opposed (45%) or opposed (25%) and the supervisors are more or less evenly spread out over various opinions. Again, the result of the χ^2 test indicates that we can't rely too heavily on this apparent relation; it could conceivably have arisen by sheer random variation.

c. Here the row variable is the dependent variable. The statement that R|C MEANS ROW VARIABLE DEPENDENT ON COLUMN VARIABLE indicates that we want LAMBDA ASYMMETRIC R|C, which is .123.

d. The predicted cells, with the column listed first, are (CLERICAL, FAVOR), (SUPERV, OPPOSE), (JUNIOR MANAGER, STRONGLY OPPOSE). Adding the frequencies in all other cells, we find $K = 5 + 8 + 13 + 8 + 7 + 5 + 5 + 4 + 2 = 57$. In the TOTAL column for OPINION, we find that the most frequent opinion is FAVOR. Adding the frequencies for the other categories, we find $U = 22 + 23 + 20 = 65$. Therefore

$$\lambda = \frac{65 - 57}{65} = .123$$ □

Percentage analyses and values of λ play a fundamentally different role than does the χ^2 test. The point of a χ^2 test is to see how much evidence there is that there *is* a relation, whatever the size may be. The point of percentage analyses and λ is to see *how strong* the relation appears to be, taking the data at face value. The two types of analyses are complementary.

SUMMARY

Several important techniques were presented for making inferences about a single population variance (or standard deviation) and the ratio of two population variances (or standard deviations). The one-sample confidence interval and test procedure utilizes a χ^2 distribution based on $n - 1$ degrees of freedom. The corresponding distribution for making inferences based on two-sample variances is the F distribution. The important point to remember when applying these techniques is that nonnormality of the distribution(s) can play havoc with the nominal confidence coefficient or p-value implied by the inference. Again, you should protect yourself by plotting the sample data before proceeding with the analysis.

The χ^2 distribution plays a very important role in statistical inference. Three applications involving χ^2 were presented: a goodness-of-fit test, a test of independence, and a test of homogeneity of distributions. Although the tests apply to different situations, the test statistics are similar. Care should be taken to ensure that the expected cell values are large enough to use the stated test. If they are not, judicious combining of cells can sometimes alleviate the problem.

For frequency data, measures of the strength of relation complement the χ^2 test. Percentage calculations often reveal the structure of a relation. The λ measure, an proportionate reduction in error (PRE) measure, indicates how well a dependent variable may be predicted, given knowledge of the value of an independent variable.

KEY FORMULAS: Chi-Square and *F* Methods

1. $100(1-\alpha)\%$ confidence interval for σ^2 and σ

$$\frac{(n-1)s^2}{\chi^2_{\alpha/2}} < \sigma^2 < \frac{(n-1)s^2}{\chi^2_{1-\alpha/2}}$$

$$\sqrt{\frac{(n-1)s^2}{\chi^2_{\alpha/2}}} < \sigma < \sqrt{\frac{(n-1)s^2}{\chi^2_{1-\alpha/2}}}$$

2. Statistical test for σ^2

H_0: $\sigma^2 = \sigma_0^2$

T.S.: $\chi^2 = \dfrac{(n-1)s^2}{\sigma_0^2}$

where d.f. $= n-1$

3. *F* test for equality of variances

H_0: $\sigma_1^2 = \sigma_2^2$

T.S.: $F = s_1^2/s_2^2$ with (n_1-1) numerator and (n_2-1) denominator degrees of freedom

4. Confidence interval for σ_1^2/σ_2^2

$$\frac{s_1^2}{s_2^2}\frac{1}{F_{\alpha/2,n_1-1,n_2-1}} < \frac{\sigma_1^2}{\sigma_2^2} < \frac{s_1^2}{s_2^2}F_{\alpha/2,n_2-1,n_1-1}$$

5. χ^2 goodness-of-fit test

H_0: $\pi_i = \pi_{i0}, \quad i = 1, \ldots, k$

T.S.: $\chi^2 = \sum_i \dfrac{(n_i - E_i)^2}{E_i}$

where $E_i = n\pi_{i0}$ and d.f. $= k-1$

6. χ^2 test of independence

H_0: The row and column variables are independent

$$\text{T.S.: } \chi^2 = \sum_{i,j} \frac{(n_{ij} - \hat{E}_{ij})^2}{\hat{E}_{ij}}$$

where

$$\hat{E}_{ij} = \frac{(n_{i.})(n_{.j})}{n}$$

and

$$\text{d.f.} = (r - 1)(c - 1)$$

7. χ^2 test of the homogeneity of the column (row) distributions

H_0: The column (row) distributions are homogeneous

$$\text{T.S.: } \chi^2 = \sum_{i,j} \frac{(n_{ij} - \hat{E}_{ij})^2}{\hat{E}_{ij}}$$

where

$$\hat{E}_{ij} = \frac{(n_{i.})(n_{.j})}{n}$$

and

$$\text{d.f.} = (r - 1)(c - 1)$$

CHAPTER 11 EXERCISES

11.27 Refer to the computer output of Exercise 10.74 (p. 350).

a. Why should a regulatory agency be concerned with the variability in noise levels as well as their mean?

b. Test the null hypothesis of equal standard deviations against a two-sided alternative. Use $\alpha = .01$.

c. Find the p-value for this test.

11.28 Plot the data of Exercise 10.74 separately for each of the two samples. Is there any reason to doubt the conclusion of Exercise 11.27?

11.29 Use the data of Exercise 10.74 to test the research hypothesis that the standard deviation for the first engine is greater than 2.0.

11.30 A state consumer-protection agency tests two processes that could be used to determine the water content of canned hams. Each process is used to estimate the water content of 14 chunks of ham. The 28 chunks are randomly chosen from a uniform source. The data and a printout from a standard computer package (BMDP) follow:

Process	Water Content (%)													
1	9.81	9.90	9.93	9.99	10.21	10.03	10.16	9.87	10.00	10.36	9.96	9.89	10.22	10.18
2	9.71	10.35	10.16	9.88	9.99	10.13	10.27	9.84	9.89	10.01	10.22	9.86	9.93	9.80

```
DIFFERENCES ON SINGLE VARIABLES

************
* WATER    *  VARIABLE NUMBER    2      GROUP     1 *1.00000  2 *2.00000       1 *1.00000(N=  14)              2 *2.00000(N=  14)
************                            MEAN         10.0364     10.0029
        STATISTICS    P-VALUE    DF     STD DEV       0.1627      0.1936
                                        S.E.M.        0.0435      0.0517
T (SEPARATE)     0.50 0.6237   25.3     SAMPLE SIZE       14          14
T (POOLED)       0.50 0.6235   26       MAXIMUM      10.3600     10.3500          H  H      H                          X
                                        MINIMUM       9.8100      9.7100        HHHHHHH   HHH   H               X XXXXX XX  XX X X X
                                                                           MIN--------------------MAX     MIN--------------------MAX
                                                                              AN H =      1 CASES            AN X =      1 CASES
```

a. Calculate the F statistic.

b. Can one conclude that process 2 has greater variability? Use $\alpha = .05$.

11.31 Refer to Exercise 11.30. Test the research hypothesis that the population standard deviation for process 1 exceeds .08. Use $\alpha = .05$.

11.32 Plot the data of Exercise 11.30. Is there any reason to be suspicious of the conclusion of that exercise?

11.33 A speaker who advises managers on how to avoid being unionized claims that only 25% of industrial workers favor union membership, 40% are indifferent, and 35% are opposed. In addition, the advisor claims that these opinions are independent of actual union membership. A random sample of 600 industrial workers yields the following data:

	Favor	Indifferent	Opposed	Total
Members	140	42	18	200
Nonmembers	70	198	132	400
Total	210	240	150	600

a. What part of the data is relevant to the 25%, 40%, 35% claim?

b. Test this hypothesis using $\alpha = .01$.

11.34 What can be said about the p-value in Exercise 11.33?

11.35 Test the hypothesis of independence in the data of Exercise 11.33. How conclusively is it rejected?

11.36 Calculate (for the data of Exercise 11.33) percentages of workers in favor of unionization, indifferent, and opposed; separately for members and for nonmembers. Do the percentages suggest that there is a strong relation between membership and opinion?

11.37 Calculate the λ value for predicting opinion given membership, using the data of Exercise 11.33. Does the value of λ indicate that there is a strong relation between membership and opinion?

11.38 Three different television commercials have been prepared to advertise an established product. The commercials are shown separately to theater panels of consumers; each consumer views only one of the possible commercials and then states an opinion of the product. Opinions range from 1 (very favorable) to 5 (very unfavorable). The data are

	Opinion					
Commercial	1	2	3	4	5	Total
A	32	87	91	46	44	300
B	53	141	76	20	10	300
C	41	93	67	36	63	300
Total	126	321	234	102	117	900

a. Calculate expected frequencies under the null hypothesis of independence.
b. How many degrees of freedom are available for testing this hypothesis?
c. Is there evidence that the opinion distributions are different for the various commercials? Use $\alpha = .01$.

11.39 State bounds on the p-value for Exercise 11.38.

11.40 In your judgment, is there a strong relation between type of commercial and opinion in the data of Exercise 11.38? Support your judgment with computation of percentages and a λ value.

12

INTRODUCTION TO THE ANALYSIS OF VARIANCE

Methods for comparing two population means, based on independent random samples, were presented in Chapter 10. Very often the two-sample problem is a simplification of what we encounter in practical situations. For example, suppose we wished to compare the mean hourly wage for nonunion farm laborers from three different ethnic groups (black, white, and Spanish-American) employed by a large produce company. Independent random samples of farm laborers would be selected from each of the three ethnic groups (populations). Then using the information from the three sample means, we would try to make an inference about the corresponding population mean hourly wages. Most likely, the sample means would differ, but this does not necessarily imply a difference among the population means for the three ethnic groups. How do we decide whether the differences among the sample means are large enough to imply that the corresponding population means are different? We answer this question using a statistical testing procedure called an **analysis of variance**(ANOVA).

ANOVA

12.1 TESTING THE EQUALITY OF SEVERAL POPULATION MEANS

Suppose that a manufacturing firm is reconsidering its basic policy of selecting supervisors for work areas. Three possible policies are A, to promote from within the work force and use an in-house training program; B, to promote from within and require that supervisors study for an associate degree at a community college; and C, to hire only experienced supervisors from outside. The firm currently has supervisors who have been hired under each of the three conditions. Random samples of size 7 from each of the three categories are chosen, and effectiveness ratings are determined for each supervisor. The data are

								Mean	Variance
Policy A	39	51	58	61	65	72	86	61.71	225.24
Policy B	22	38	43	47	49	54	72	46.43	232.95
Policy C	18	31	41	43	44	54	65	42.29	229.24

It appears that the A group has substantially higher ratings on the average, but there is a good deal of variability within each group and the sample sizes are small. Is it possible that the apparent average superiority of the A group is merely the result of random variation?

One way to try to answer this question is to run t tests on all pairs of means. The data are almost ideal for t testing; each sample yields a nearly symmetric distribution, the sample sizes are equal, and the sample variances are very nearly equal. The result of two-tailed t tests are shown below:

Comparison	t Statistic	Conclusion at $\alpha = .05$(12 d.f.)
H_0: $\mu_A - \mu_B = 0$	1.89	Retain H_0 ($.05 < p < .10$)
H_0: $\mu_A - \mu_C = 0$	2.41	Reject H_0 ($p < .05$)
H_0: $\mu_B - \mu_C = 0$	.51	Retain H_0 ($p > .20$)

We conclude that the A versus B difference may conceivably be a fluke. The A versus C difference probably isn't, and there is no apparent difference between policies B and C.

There is a fundamental objection to this procedure. When running many t-tests, it's impossible to know what the **overall risk** of Type I error is; the more tests one runs on a given set of data, the larger the risk of committing a Type I error for at least one of the comparisons. If we had 15 groups instead of 3, we could run $\binom{15}{2} = 105$ t tests. If the tests were independent and each one was run using $\alpha = .05$, we would expect $105(.05) = 5.25$ such errors even if the null hypothesis was true for each test. The ANOVA method of this section yields a single test statistic for comparing all the means so that the overall risk of Type I error is controlled. For this reason, it is preferred to the multiple t test method.

overall α risk

The analysis of variance is based on sums of squares (SS). For sample data displayed as in Table 12.1, we use the following notation.

y_{ij} = jth sample measurement in group i

$$i = 1, 2, \ldots, I; \quad j = 1, 2, \ldots, n_i$$

n_i = sample size for group i

$\bar{y}_i$ = sample mean for group i

$\bar{y}$ = average of all the sample measurements

n = the total sample size. For the above data $n = n_1 + n_2 + \ldots + n_I$

With this notation it is possible to express the variability of the n sample measurements about $\bar{y}$ as

SS(Total)

$$\text{SS(Total)} = \sum_{i,j} (y_{ij} - \bar{y})^2$$

TABLE 12.1 **Notation for Sample Data in an ANOVA**

Group	Sample Data	Sample Mean	Unknown Group Mean
1	$y_{11}\ y_{12} \cdots y_{1n_1}$	$\bar{y}_1$	μ_1
2	$y_{21}\ y_{22} \cdots y_{2n_2}$	$\bar{y}_2$	μ_2
$\vdots$	$\vdots \quad \vdots \quad\quad \vdots$	$\vdots$	$\vdots$
I	$y_{I1}\ y_{I2} \quad\quad y_{In_I}$	$\bar{y}_I$	μ_I

and to partition this quantity into two components, SS(Between) and SS(Within).

$$\sum_{i,j} (y_{ij} - \bar{y})^2 = \sum_i n_i(\bar{y}_i - \bar{y})^2 + \sum_{i,j} (y_{ij} - \bar{y}_i)^2$$

$$\text{SS(Total)} \quad = \text{SS(Between)} + \text{SS(Within)}$$

SS(Within) Because **SS(Within)** is based on deviations from specific group means, it is not affected by possible deviations of these sample means from each other; SS(Within) reflects only random variation within the samples. In contrast, SS(Between) is strongly affected by discrepancies among group means. If the true group means μ_i are equal, the sample means $\bar{y}_i$ are close to each other and therefore close to the grand mean $\bar{y}$, and SS(Between) tends to be small. But if the true group means are different, the sample means tend to be far apart and **SS(Between)** tends to be large. If SS(Between) is large relative to SS(Within), the null hypothesis that the true group means are equal should be rejected.

SS(Between)

To make a precise statement of the analysis of variance (ANOVA) test, we need to do a little mathematics. We make the following formal mathematical assumptions:

Assumptions for an ANOVA

1. The sample measurements $y_{i1}, y_{i2}, \ldots, y_{in_i}$ are selected from a normal population ($i = 1, 2, \ldots, I$).
2. The samples are independent.
3. The unknown population mean and variance for the measurements from sample i are μ_i and σ^2, respectively.

The null hypothesis for analysis of variance is H_0: $\mu_1 = \mu_2 = \ldots = \mu_I$. By assumption, the true variances are also equal. Therefore, under the null hypothesis, all the sample measurements y_{ij} may be regarded as large sample from a normal population. It follows that

$$\frac{\sum_{i,j} (y_{ij} - \bar{y})^2}{\sigma^2} = \frac{\text{SS(Total)}}{\sigma^2}$$

has a χ^2 distribution with $n - 1$ degrees of freedom, just as in Chapters 8 and 11. In the Appendix to this chapter we show that

$$\frac{\text{SS(Total)}}{\sigma^2} = \frac{\text{SS(Between)}}{\sigma^2} + \frac{\text{SS(Within)}}{\sigma^2}$$

According to a very useful mathematical fact, Cochran's Theorem, these components are independent and have χ^2 distributions.

d.f. for SS(Within) and SS(Between)

The degrees of freedom for SS(Within) can be found by realizing that there are $n_i - 1$ d.f. for squared deviations within group i and that d.f.'s can be added across the groups. Therefore, there are $(n_1 - 1) + (n_2 - 1) + \ldots + (n_I - 1) = n - I$ d.f. for SS(Within). The quantity SS(Between) $= \sum_i n_i(\bar{y}_i - \bar{y})^2$ has I terms, but since the constraint $\sum_i n_i(\bar{y}_i - \bar{y}) = 0$ determines one term (once the remaining $I - 1$ have been found), there are $I - 1$ d.f. for SS(Between).

We can use these results to develop a test statistic for the null hypothesis H_0: $\mu_1 = \mu_2 = \ldots = \mu_I$. In Chapter 11 the F statistic was defined to be the ratio of two independent χ^2 variables each divided by its degrees of freedom:

$$F = \frac{\chi_1^2/\text{d.f.}_1}{\chi_2^2/\text{d.f.}_2}$$

Therefore

$$F = \frac{\dfrac{\text{SS(Between)}}{\sigma^2}\Big/(I-1)}{\dfrac{\text{SS(Within)}}{\sigma^2}\Big/(n-I)} = \frac{\text{SS(Between)}/(I-1)}{\text{SS(Within)}/(n-I)}$$

mean square (MS)

is an F statistic under H_0: $\mu_1 = \mu_2 = \ldots = \mu_I$. The term **mean square (MS)** has been given to any sum of squares divided by its degrees of freedom. With this terminology, the F statistic for our analysis of variance is

$$F = \frac{\text{MS(Between)}}{\text{MS(Within)}}$$

Large values of MS(Between) relative to MS(Within) indicate differences among the population means and lead to the rejection of the null hypothesis.

The ANOVA test of the null hypothesis is summarized next:

ANOVA for Testing the Equality of I Group Means

H_0: $\mu_1 = \mu_2 = \ldots = \mu_I$
H_a: Not all μ_i are equal

$$\text{T.S.: } F = \frac{\text{MS(Between)}}{\text{MS(Within)}} = \frac{\sum_i n_i(\bar{y}_i - \bar{y})^2/(I-1)}{\sum_{ij} (y_{ij} - \bar{y}_i)^2/(n-I)}$$

R.R.: For specified α, reject H_0 if $F > F_\alpha$, where F_α cuts off a right-tail area of α in the F distribution with $I - 1$ numerator and $n - I$ denominator d.f.

We calculated the necessary sums of squares for the supervisor-effectiveness data given near the beginning of this section. The results are usually organized in a so-called ANOVA table, as shown here:

Source	SS	d.f.	MS	F
Between	1466.00	2	733.00	3.20
Within	4124.57	18	229.14	
Total	5590.57	20		

For a test at $\alpha = .05$, we compare the F value of 3.20 to the tabled .05 point (2 and 18 d.f.), namely 3.55. Because $3.20 < 3.55$, we would retain the null hypothesis; notice that this result contradicts the conclusion of the inappropriate multiple t tests.

The ANOVA arithmetic is often done by a computer program, but without such help, the pain of hand computation can be eased by shortcut formulas.

Shortcut Formulas for ANOVA Sums of Squares

$$\text{SS(Within)} = \sum_{i,j} y_{ij}^2 - \frac{\sum_i \left(\sum_j y_{ij}\right)^2}{n_i}$$

$$\text{SS(Between)} = \frac{\sum_i \left(\sum_j y_{ij}\right)^2}{n_i} - \frac{\left(\sum_{i,j} y_{ij}\right)^2}{n}$$

$$\text{SS(Total)} = \sum_{i,j} y_{ij}^2 - \frac{\left(\sum_{i,j} y_{ij}\right)^2}{n}$$

$$= \text{SS (Within)} + \text{SS (Between)}$$

Alternatively, we can compute the mean $\bar{y}_i$ and variance s_i^2 for each group, indexed by i.

Alternative Method for Computation of Sums of Squares

$$\text{SS(Within)} = \sum_i (n_i - 1)s_i^2$$

$$\text{SS(Between)} = \sum_i n_i(\bar{y}_i - \bar{y})^2$$

$$= \sum_i (n_i\bar{y}_i^2) - n\bar{y}^2$$

where

$$\bar{y} = \frac{\sum_i n_i\bar{y}_i}{n},$$

and $n = \sum n_i$

EXAMPLE 12.1 A panel of potential cereal-product consumers is asked to rate one of four potential new products. Each member of the panel rates only one of the products, comparing it to a standard, existing cereal on a 100-point scale. The scores are

Product I: 16 31 57 62 67 71 73 75
Product II: 30 35 52 60 64 65 65 67 70 71 75 82
Product III: 43 51 53 54 56 58 61 64 64 67 68 70 71 75 79
Product IV: 29 39 46 50 59 61 62

Test the hypothesis of equal mean product scores by an ANOVA. First use $\alpha = .05$, then find a p-value.

Solution A worksheet is shown below:

Product	$\sum_j y_{ij}$	$\left(\sum_j y_{ij}\right)^2$	n_i	$\left(\sum_j y_{ij}\right)^2 / n_i$	$\sum_j y_{ij}^2$
I	452	204,304	8	25,538.000	28,794
II	736	541,696	12	45,141.333	47,754
III	934	872,356	15	58,157.067	59,508
IV	346	119,716	7	17,102.286	18,024
Total	2,468		42	145,938.686	154,080

$$\text{SS(Within)} = 154{,}080 - 145{,}938.686 = 8141.314$$
$$\text{SS(Between)} = 145{,}938.686 - (2468)^2/42 = 914.305$$

We can also use a calculator to find the following results:

Product	$\bar{y}_i$	s_i^2	n_i
I	56.50000	465.1429	8
II	61.33333	237.5152	12
III	62.26667	96.4952	15
IV	49.42857	153.6190	7

The grand mean is

$$\bar{y} = [8(56.50000) + 12(61.33333) + 15(62.26667) + 7(49.42857)]/42$$
$$= 58.76191$$
$$\text{SS(Within)} = 7(465.1429) + 11(237.5152) + 14(96.4952) + 6(153.6190)$$
$$= 8141.314$$
$$\text{SS(Between)} = 8(56.50000)^2 + 12(61.33333)^2 + 15(62.26667)^2 + 7(49.42857)^2 - 42(58.76191)^2$$
$$= 914.305$$

The ANOVA table is

Source	SS	d.f.	MS	F
Between	914.305	3	304.768	1.422
Within	8141.314	38	214.245	
Total	9055.619	41		

Because $F = 1.422$ is less than 2.84, the $F_{.05}$ value based on 3 and 38 d.f., H_0 is retained. The p-value is larger than .05, most likely quite a bit larger. The observed deviations of sample means are well within the range of random variation. □

EXAMPLE 12.2 Suppose that a drug company, in testing two new products against a currently used drug, is concerned about the possible side effect of increased blood pressure. In a clinical trial of the drugs, 12 patients are tested under each drug, at standard dosages. Blood pressures are recorded initially and again one hour after a single dose of the assigned drug product. Blood pressure changes are as follows:

Drug	Blood Pressure Changes											
A	−10	−10	−8	−5	0	3	5	7	8	10	12	15
B	0	5	8	10	12	15	16	17	20	22	25	25
C	−5	−1	0	2	5	8	8	10	14	16	20	20

An ANOVA is performed on the data, using the SAS software, and the output is shown below. Identify the following: the sample means, the mean squares MS(Between) and MS(Within), the value of the F statistic, and the p-value for H_0: $\mu_A = \mu_B = \mu_C$.

ONE FACTOR ANOVA OF BLOOD PRESSURE INCREASES

ANALYSIS OF VARIANCE PROCEDURE

DEPENDENT VARIABLE: CHGBLPR

SOURCE	DF	SUM OF SQUARES	MEAN SQUARE	F VALUE	PR > F	R-SQUARE	C.V.
MODEL	2	913.55555556	456.77777778	6.61	0.0039	0.286055	100.0806
ERROR	33	2280.08333333	69.09343434		ROOT MSE		CHGBLPR MEAN
CORRECTED TOTAL	35	3193.63888889			8.31224605		8.30555556

MEANS

DRUG	N	CHGBLPR
A	12	2.2500000
B	12	14.5833333
C	12	8.0833333

Solution The sample means are shown as MEANS and are, respectively, 2.2500000, 14.5833333, and 8.0833333. The mean square for MODEL in this case is MS (Between), namely 456.77777778. MS (Within) is the ERROR mean square, 69.09343434. F is 6.61, and the p-value is PR > F, 0.0039. Thus the evidence for differences in mean blood-pressure change with different drugs is quite conclusive. □

Like all statistical inference procedures, this F test is based on certain assumptions. The three basic assumptions are population normality, equal group variances, and independence of observations. The **normality assumption** is perhaps the least crucial. The ANOVA test is a test on means (despite its name); the Central Limit Theorem has its effect.

normality assumption

If the populations are badly skewed and if the sample sizes are small (such as 10 each), the Central Limit Theorem effect doesn't take over and the nominal F probabilities may be in error. If the population distributions are roughly symmetric but have heavy tails, as indicated by outliers in the sample data, the F probabilities are reasonably accurate, but alternative procedures may be more efficient (make better use of the data).

assumption of equal variance
balanced design

The assumption of equal true variances for each group is important if the sample sizes are substantially different. Many studies have indicated that when all n_i's are equal, in a so-called **balanced design**, the effect of even grossly unequal variances is minimal. However, if the n_i's are substantially different—say, if the largest n_i is at least twice the smallest—then unequal variances can cause major distortions in the nominal F probabilities. The worst case is when large variances occur in groups with small sample sizes. The best way to avoid problems is to strive for equal n_i's.

independence assumption

Violation of the **assumption of independence** can cause a great deal of trouble. When the data arise from a cross-sectional random sample taken at a specific time, independence holds virtually by definition. If the data arise by measurements taken repeatedly over time, there is a possibility of dependence from one measurement to the next. In such situations, nominal F probabilities are very suspect.

EXAMPLE 12.3 Refer to the data of Example 12.1. Is there any indication of trouble because of violation of assumptions?

Solution The data are cross-sectional, so independence is not a problem. But the data appear to be left-skewed, particularly in groups I and IV, where the sample sizes are small. The variances appear substantially different, and the big variances go with the small n's. The conclusions of the F test in Example 12.1 are very shaky. □

12.2 COMPARING SEVERAL DISTRIBUTIONS BY A RANK TEST

In Section 13.1, we noted that one assumption underlying the ANOVA F test was that all populations were normal. If that assumption is incorrect, as evidenced by substantial skewness or outliers in the samples, the nominal probabilities given by the F table may be in error. In addition, the F test may not be efficient. In this section, we introduce the Kruskal-Wallis rank sum test, which doesn't require the normal-population assumption.

The Kruskal-Wallis test is an extension of the Wilcoxon rank sum test described in Section 10.3. The formal null hypothesis is that all the populations have the same distribution, not necessarily a normal distribution. The formal research hypothesis is that the populations differ in any way; the test is largely sensitive to differences of location (means or medians), rather than differences of variances.

To carry out the test, all the data values are combined and ranked from lowest to highest. In case of ties, average ranks are assigned, as in the Wilcoxon rank sum test. The Kruskal-Wallis statistic may be thought of as

$$H = \frac{12}{n(n+1)} \text{SS(Between, ranks)}$$

where SS(Between, ranks) is the SS(Between) obtained for the rankings rather than the original data, and n is the combined sample size. Thus if the average rank in every sample is equal, H will be 0, but if the average ranks differ greatly among samples (indicating a clear difference in locations), H will be large. Conventionally, the Kruskal-Wallis statistic is stated in terms of T_i, the *sum* of the ranks in group i, rather than in terms of the average rank; of course, the average is simply the sum divided by the relevant sample size, so the test could be stated either way.

Kruskal-Wallis Test

H_0: The distributions are identical (effectively, the populations have equal means)
H_a: The distributions differ in location

$$\text{T.S.: } H = \left\{ \frac{12}{n(n+1)} \sum_i \frac{T_i^2}{n_i} \right\} - 3(n+1)$$

where n_i = sample size in sample i, ($i = 1, 2, \ldots, I$), n = total sample size, and T_i = sum of combined-sample ranks for measurements in sample i

R.R.: For specified α, reject H_0 if $H > \chi^2_\alpha$ where χ^2_α cuts off a right-tail area α for the χ^2 distribution with $I - 1$ d.f.

EXAMPLE 12.4 Perform a Kruskal-Wallis test for the data of Example 12.1. First use $\alpha = .05$, then find a p-value.

Solution The data, ranks, and T_i values are shown below:

I		II		III		IV	
Score	Rank	Score	Rank	Score	Rank	Score	Rank
16	1	30	3	43	7	29	2
31	4	35	5	51	10	39	6
57	15	52	11	53	12	46	8
62	21.5	60	18	54	13	50	9
67	29	64	24	56	14	59	17
71	35	65	26.5	58	16	61	19.5
73	37	65	26.5	61	19.5	62	21.5
75	39	67	29	64	24	$T_1 =$	83.0
$T_1 =$	181.5	70	32.5	64	24		
		71	35	67	29		
		75	39	68	31		
		82	42	70	32.5		
		$T_1 =$	291.5	71	35		
				75	39		
				79	41		
				$T_1 =$	347.0		

Ties are handled by averaging ranks. For instance, the three 64 scores are tied for ranks 23, 24, and 25 and are all assigned rank 24. Here $n = 42$ (and, as a check, the highest rank is 42).

$$H = \frac{12}{(42)(43)}\left\{\frac{(181.5)^2}{8} + \frac{(291.5)^2}{12} + \frac{(347)^2}{15} + \frac{(83)^2}{7}\right\} - 3(43)$$

$$= \frac{12}{(42)(43)}(20{,}210.212) - 3(43) = 5.287$$

The right-tail .05 point for χ^2 with $4 - 1 = 3$ d.f. is 7.81; because $H = 5.287 < 7.81$, H_0 is retained. At $\alpha = .10$, the table value is 6.25, so the p-value $> .10$. The result agrees with the F test result of Example 12.1. Had there been a conflict, we would have believed the Kruskal-Wallis result, because of the major violations of F test assumptions discussed in Example 12.3. □

SECTIONS 12.1 AND 12.2 EXERCISES

12.1 A test is made of five different incentive-pay schemes for piecerate workers. Eight workers are assigned randomly to each plan. The total number of items produced by each worker over a 20-day period is recorded.

	Plan				
	A	B	C	D	E
Production	1,106	1,214	1,010	1,054	1,210
	1,203	1,186	1,069	1,101	1,193
	1,064	1,165	1,047	1,029	1,169
	1,119	1,177	1,120	1,066	1,223
	1,087	1,146	1,084	1,082	1,161
	1,106	1,099	1,062	1,067	1,200
	1,101	1,161	1,051	1,109	1,189
	1,049	1,153	1,029	1,083	1,197
Mean	1,104.38	1,162.62	1,059.00	1,073.88	1,192.75
Variance	2,136.55	1,116.84	1,137.71	662.41	409.93
$\sum y_{ij}$	8,835	9,301	8,472	8,591	9,542
$\sum y_{ij}^2$	9,772,109	10,821,393	8,979,812	9,230,297	11,384,090

a. Calculate the grand mean.

b. Calculate SS(Between) from the definition (p. 382).

c. Use the fact that SS(Within) $= \sum_i (n_i - 1)s_i^2$ to calculate SS(Within).

d. What are the appropriate degrees of freedom for these sums of squares?

12.2 Refer to the data of Exercise 12.1. Use the shortcut formula (p. 384) to calculate SS(Between) and SS(Within).

12.3 Perform the ANOVA F test for the data of Exercise 12.1. State bounds on the p-value. What would you conclude?

12.4 Plot the data of Exercise 12.1 by plan. Do there appear to be any blatant violations of assumptions?

12.5 Perform a Kruskal-Wallis test on the data of Exercise 12.1. Place bounds on the p-value.

12.6 How do the results of the F test and Kruskal-Wallis test on the data of Exercise 12.1 compare? Does it matter much which test is used?

12.7 A data-processing firm administers an aptitude test to all applicants for programming jobs. The results of these tests are analyzed by the source (general advertising, trade journal advertising, employment agency, personal recommendation, or off the street).

Source	Score														
Genadv	36	47	38	51	62	78	60	47	49	53	26	38	61	39	43
Tradej	58	64	62	47	71	90	65	82	61	59					
Agency	47	59	48	81	66	50	42	53							
Recomm	67	61	82	97	65	72	54	69	58						
Street	38	47	80	41	38	66	50								

An SAS analysis of this data is given below:

```
                                  ANALYSIS OF VARIANCE PROCEDURE

DEPENDENT VARIABLE: SCORE

SOURCE            DF    SUM OF SQUARES       MEAN SQUARE      F VALUE         PR > F      R-SQUARE          C.V.

MODEL              4     3478.17505669      869.54376417         4.92         0.0023      0.308890       23.1247

ERROR             44     7782.06984127      176.86522367                    ROOT MSE                  SCORE MEAN

CORRECTED TOTAL   48    11260.24489796                                   13.29906853                 57.51020408


                                          MEANS

                                SOURCE         N          SCORE

                                AGENCY         8     55.7500000
                                GENADV        15     48.5333333
                                RECOMM         9     69.4444444
                                STREET         7     51.4285714
                                TRADEJ        10     65.9000000


                                 PROGRAMMER APTITUDE SCORE

                   ANALYSIS FOR VARIABLE SCORE CLASSIFIED BY VARIABLE  SOURCE

                               AVERAGE SCORES WERE USED FOR TIES


                                WILCOXON SCORES (RANK SUMS)

                                  SUM OF    EXPECTED     STD DEV       MEAN
                 LEVEL       N    SCORES    UNDER H0    UNDER H0      SCORE

                 GENADV     15    247.00      375.00       46.05      16.47
                 TRADEJ     10    331.50      250.00       40.27      33.15
                 AGENCY      8    187.00      200.00       36.93      23.37
                 RECOMM      9    329.50      225.00       38.69      36.61
                 STREET      7    130.00      175.00       34.96      18.57

                    KRUSKAL-WALLIS TEST (CHI-SQUARE APPROXIMATION)
                    CHISQ=  16.10      DF=  4     PROB > CHISQ=0.0029
```

a. Find the sums of squares.

b. Locate the value of the F statistic.

c. Test the null hypothesis of equal mean aptitude, using $\alpha = .01$.

12.8 Plot the data of Exercise 12.7 by source. Are there any obvious violations of the ANOVA assumptions?

12.9 a. Locate the value of the Kruskal-Wallis statistic for the output of Exercise 12.7.
b. Can the hypothesis of equal locations be rejected at $\alpha = .05$?
c. Compare the conclusions of the F test and the Kruskal-Wallis test. Does it matter much which test is used?

12.3 SPECIFIC COMPARISONS AMONG MEANS

The ANOVA F test and the Kruskal-Wallis test developed in Section 12.1 test for an overall pattern of discrepancies among the group means. Rejection of the hypothesis that all population (group) means are equal does not indicate specifically which means are not equal. In this section we outline one of many possible methods for assessing differences among specified means. Several other methods are discussed in Ott (1984), and there are other minor variations available. We state this method in terms of confidence intervals. As usual, we can use the resulting intervals to perform hypothesis tests.

One approach would be to use the t methods of Chapter 10 to construct t-type confidence intervals for all possible pairs of means. The objection to this is essentially the same as our objection to multiple t tests in the previous section. While the confidence level for each interval separately may be 95%, there is no way to measure the overall confidence that all the intervals are correct. The Tukey and Scheffé methods of this section are used to yield a desired overall confidence level.

Formally, we again assume that the data constitute independent random samples from I groups or subpopulations. The groups are labeled by the index i, where $i = 1, 2, \ldots I$, and the sample size for group i is n_i. We assume that the population distribution in group i is normal with mean μ_i and variance σ^2. There is no subscript on σ^2 because we assume that the population variances for all groups are identical. The estimate $\hat{\sigma}^2$ is MS (Within), which is a generalization of the pooled variance defined in Chapter 10.

Tukey method

The **Tukey method** is designed specifically to compare any two means, say, $\bar{y}_i$ and $\bar{y}_{i*}$. The underlying mathematical theory assumes that all sample sizes are equal: $n_1 = n_2 = \ldots = n_I$. The common sample size is usually denoted n. Note that here n means the individual sample size, *not* the overall sample size.

100(1 − α)% Confidence Intervals for All Pairs of Means, Using Tukey's Method

$$(\bar{y}_i - \bar{y}_{i*}) - q_\alpha\,(I, \text{d.f.}_2)\sqrt{\text{MS(Within)}/n} < \mu_i - \mu_{i*} < (\bar{y}_i - \bar{y}_{i*}) + q_\alpha\,(I, \text{d.f.}_2)\sqrt{\text{MS(Within)}/n}$$

where n is the sample size for each sample, I is the number of samples, d.f._2 is the degrees of freedom for MS(Within) and $q_\alpha\,(I, \text{d.f.}_2)$ is found in Appendix Table 8. There is $100(1 - \alpha)$% confidence that **all possible** intervals comparing two means are correct.

EXAMPLE 12.5 Construct (overall) 95% confidence intervals for the differences in mean blood-pressure increase for the data of Example 12.2.

Solution The relevant summary statistics are

Drug i	A	B	C
$\bar{y}_i$	2.25	14.58	8.08
n_i	12	12	12
MS(Within) = 69.09			

The desired q_α $(I, \text{d.f.}_2)$ value has $\alpha = .05$, $I = 3$, and $\text{d.f.}_2 = 33$. There is no $\text{d.f.}_2 = 33$ entry in Appendix Table 8, so we use $\text{d.f.}_2 = 30$, $q_{.05}(3,30) = 3.49$. It's clear from the table that $q_{.05}(3,30)$ is just a little larger than the desired $q_{.05}(3,33)$, so we err slightly on the conservative (wider interval) side. The desired 95% confidence intervals are

$$(14.58 - 2.25) - 3.49\sqrt{69.09/12} < \mu_B - \mu_A < (14.58 - 2.25) + 3.49\sqrt{69.09/12}$$

$$(8.08 - 14.58) - 3.49\sqrt{69.09/12} < \mu_C - \mu_B < (8.08 - 14.58) + 3.49\sqrt{69.09/12}$$

$$(8.08 - 2.25) - 3.49\sqrt{69.09/12} < \mu_C - \mu_A < (8.08 - 2.25) + 3.49\sqrt{69.09/12}$$

or

$$3.96 < \mu_B - \mu_A < 20.70$$

$$-14.87 > \mu_C - \mu_B < 1.87$$

$$-2.54 < \mu_C - \mu_A < 14.20$$

Recall that confidence intervals may be used to conduct (two-sided) hypothesis tests. The natural null hypothesis, H_0: $\mu_i - \mu_{i*} = 0$, is rejected if zero does not fall in the interval. In Example 12.5, the 95% confidence intervals indicate that at $\alpha = 0.05$ we can reject $\mu_B - \mu_A = 0$ but must retain $\mu_C - \mu_B = 0$ and $\mu_C - \mu_A = 0$. We conclude that the mean blood-pressure increase under drug B is higher than under drug A. Drug C is in the middle, and the difference of $\bar{y}_C$ from either $\bar{y}_A$ or $\bar{y}_B$ could be the result of random variation. □

The formal assumption in the Tukey method that $n_1 = n_2 = \ldots = n_I$ may be relaxed somewhat. As long as the n_i's are not drastically different, say, if the largest n_i is no more than twice the smallest, the Tukey method may be used with

$$n = \frac{1}{\dfrac{1}{n_1} + \dfrac{1}{n_2} + \ldots + \dfrac{1}{n_I}}$$

harmonic mean

the so-called **harmonic mean** of the n_i's.

If the n_i's are drastically different, or if a more complicated comparison such as the average of means 1 and 2 versus the average of means 3, 4, and 5 is desired, a more general but more conservative approach, the **Scheffé method**, must be used.

Scheffé method

A confidence interval for comparing two population means can be constructed using the Scheffé method as shown here:

$100(1-\alpha)\%$ Confidence Interval for $\mu_i - \mu_{i^*}$, Using the Scheffé Method

$$(\bar{y}_i - \bar{y}_{i^*}) - \sqrt{(I-1)F_\alpha}\sqrt{\text{MS(Within)}\left(\frac{1}{n_i}+\frac{1}{n_{i^*}}\right)} < \mu_i - \mu_{i^*} < (\bar{y}_i - \bar{y}_{i^*})$$
$$+ \sqrt{(I-1)F_\alpha}\sqrt{\text{MS(Within)}\left(\frac{1}{n_i}+\frac{1}{n_{i^*}}\right)}$$

where F_α cuts off a right-tail area α for the F distribution with $I-1$ and $n_1 + \ldots + n_I - I = n - I$ d.f.

EXAMPLE 12.6 Construct (overall) 95% confidence intervals for the differences in mean blood-pressure increase for the data of Example 12.2.

Solution The relevant summary statistics are

	Drug A	Drug B	Drug C
Sample mean, $\bar{y}_i$	2.25	14.58	8.08
Sample size, n_i	12	12	12
	MS(Within) = 69.09		

The desired F_α is based on 2 and 33 d.f. The d.f. combination is not shown in our F table, so we use 2 and 30 d.f., and $F_\alpha = 3.32$. It's clear from that table that $F_{2,30}$ is a little larger than $F_{2,33}$, so we err on the conservative (wider interval) side.

The desired intervals are

$$(14.58 - 2.25) - \sqrt{(3-1)3.32}\sqrt{69.09\left(\frac{1}{12}+\frac{1}{12}\right)} < \mu_B - \mu_A$$
$$< (14.58 - 2.25) + \sqrt{(3-1)3.32}\sqrt{69.09\left(\frac{1}{12}+\frac{1}{12}\right)}$$

$$(8.08 - 14.58) - \sqrt{(3-1)3.32}\sqrt{69.09\left(\frac{1}{12}+\frac{1}{12}\right)} < \mu_C - \mu_B$$
$$< (8.08 - 14.58) + \sqrt{(3-1)3.32}\sqrt{69.09\left(\frac{1}{12}+\frac{1}{12}\right)}$$

$$(8.08 - 2.25) - \sqrt{(3-1)3.32}\sqrt{69.09\left(\frac{1}{12}+\frac{1}{12}\right)} < \mu_C - \mu_A$$
$$< (8.08 - 2.25) + \sqrt{(3-1)3.32}\sqrt{69.09\left(\frac{1}{12}+\frac{1}{12}\right)}$$

or

$$3.59 < \mu_B - \mu_A < 21.07$$
$$-15.24 < \mu_C - \mu_B < 2.24$$
$$-2.91 < \mu_C - \mu_A < 14.57$$

Using this procedure, there is 95% confidence that all three intervals are correct. □

The Scheffé method is quite conservative, even when compared to the Tukey method. This can be seen from the results of the confidence-interval calculations in Examples 12.5 and 12.6.

The Scheffé method can also be used to construct a confidence interval for a linear combination of means of the form $\sum_i c_i\mu_i$, where the c_i's are constants with $\sum_i c_i = 0$. This general comparison is a weighted sum (not a weighted average), where the weights may be chosen arbitrarily, subject to the condition that their sum is zero. The estimate of $\sum_i c_i\mu_i$ is

$$\hat{\ell} = \sum_i c_i\bar{y}_i$$

Thus, for example,

$$\hat{\ell}_1 = \left(\frac{1}{2}\bar{y}_1 + \frac{1}{2}\bar{y}_2\right) - \left(\frac{1}{3}\bar{y}_3 + \frac{1}{3}\bar{y}_4 + \frac{1}{3}\bar{y}_5\right)$$

$$\hat{\ell}_2 = \left(\frac{10}{11}\bar{y}_1 + \frac{1}{11}\bar{y}_2\right) - \left(\frac{6}{13}\bar{y}_3 + \frac{7}{13}\bar{y}_5\right)$$

and

$$\hat{\ell}_3 = \bar{y}_2 - \bar{y}_4$$

are all comparisons in this sense. The respective choices of c weights are

$$\hat{\ell}_1\text{: } c_1 = \frac{1}{2} \quad c_2 = \frac{1}{2} \quad c_3 = -\frac{1}{3} \quad c_4 = -\frac{1}{3} \quad c_5 = -\frac{1}{3}$$

$$\hat{\ell}_2\text{: } c_1 = \frac{10}{11} \quad c_2 = \frac{1}{11} \quad c_3 = -\frac{6}{13} \quad c_4 = 0 \quad c_5 = -\frac{7}{13}$$

$$\hat{\ell}_3\text{: } c_1 = 0 \quad c_2 = 1 \quad c_3 = 0 \quad c_4 = -1 \quad c_5 = 0$$

The most common use of the Scheffé method in practice involves the comparison of an average of one group of means with another, such as $\hat{\ell}_1$. In theory, a comparison of weighted sums like $\hat{\ell}_2$ is also possible. The Scheffé method can also be used for pairwise comparisons such as $\hat{\ell}_3$, but it is more conservative (gives wider confidence intervals) than the Tukey procedure when the latter can also be used.

Confidence Intervals for Comparison of Several Means: Scheffé Method

Comparison: $\hat{\ell} = \sum_i c_i\bar{y}_i$, with $\sum_i c_i = 0$

$$\hat{\ell} - \sqrt{(I-1)F_\alpha}\sqrt{\text{MS(Within)}\sum_i (c_i^2/n_i)} < \sum_i c_i\mu_i$$

$$< \hat{\ell} + \sqrt{(I-1)F_\alpha}\sqrt{\text{MS(Within)}\sum_i (c_i^2/n_i)}$$

where I is the number of sample means that can be compared (whether or not all means have nonzero weight in $\hat{\ell}$) and F cuts off a right-tail area α for the F distribution with $I - 1$ and $n - I$ d.f. There is $100(1 - \alpha)\%$ confidence that all possible intervals of this form are correct.

EXAMPLE 12.7 Calculate a 95% confidence interval for $\hat{\ell} = \bar{y}_A - (1/2)\bar{y}_B - (1/2)\bar{y}_C$ for the data of Example 12.5. Can we conclude that the true A mean differs significantly from the average of the true B and C means?

Solution $\hat{\ell} = 2.25 - (1/2)14.58 - (1/2)8.08 = -9.08$. The desired F_α is based on $\alpha = .05$ and 2 and 33 d.f. That d.f. combination is not shown in our F table, so we use 2 and 30 d.f. and $F_{.05} = 3.32$. It's clear from the F table that $F_{2,30}$ is a bit larger than $F_{2,33}$, so we are erring just a little on the conservative (wider interval) side.

$$\sqrt{(I-1)F_{.05}}\sqrt{\text{MS(Within)}\sum_i c_i^2/n_i} = \sqrt{2(3.32)}\sqrt{69.09\left(\frac{(1)^2}{12} + \frac{(-1/2)^2}{12} + \frac{(-1/2)^2}{12}\right)}$$

$$= 7.57$$

The interval is

$$-9.08 - 7.57 < \sum_i c_i\mu_i < -9.08 + 7.57$$

or

$$-16.65 < \mu_A - \left(\frac{1}{2}\mu_B + \frac{1}{2}\mu_C\right) < -1.51$$

This interval does not include zero, so at $\alpha = .05$ we conclude that μ_A is statistically significantly different from (in fact, less than) the average of μ_B and μ_C. □

The Tukey method and especially the Scheffé method are conservative. It is possible to have an F test indicate significant differences among means and yet have all pairwise difference confidence intervals include zero. (It can be proved that there is always some comparison, perhaps irrelevant or plain ridiculous, that yields a significant difference in such a case.) The convervatism of these methods does, however, give ample protection against even fairly serious violations of assumptions. Unless there is gross skewness or inequality of variances in the data, the nominal confidence level is still valid.

SECTION 12.3 EXERCISES

12.10 Refer to the data of Exercise 12.1.

a. Calculate overall 95% confidence intervals for all possible differences of means.

b. Which differences can be declared significant at $\alpha = .05$?

12.11 a. Instead of the Scheffé method of Exercise 12.10, calculate 10 separate 99.5% confidence intervals for all 10 possible differences in means. The required t table value is 2.977.

b. If each interval has probability .005 of being wrong, do you think that the overall probability of error exceeds .05? Justify your answer.

12.12 Compare the widths of the intervals in Exercises 12.10 and 12.11.

12.13 Refer to the data of Exercise 12.7. Which pairs of means can be declared significantly different using $\alpha = .01$?

12.4 TWO-FACTOR ANOVA: BASIC IDEAS

ANOVA methods can be extended to deal with more complicated experimental situations; such experiments are often conducted in market research and can be conducted usefully in many other managerial applications. For example, suppose a researcher is interested in the average annual overhead per loan (excluding advertising expenses) for four different nationwide finance companies (A, B, C, and D). We can take a random sample of n_i loans in a given year from each company and compute the overhead per loan. The methods in Section 12.1 illustrate how to test H_0: $\mu_1 = \mu_2 = \mu_3 = \mu_4$ using an F test.

two-factor experiment

This same problem can be extended to a **two-factor experiment**. Suppose we wish to classify loans into one of several categories (home improvement, auto, consolidation, and all other). Then it should be possible to examine the average annual overhead for each loan category and company. If we let μ_{ij} denote the population mean for observations taken from the ith level of factor 1 and the jth level of factor 2, the population means can be displayed as shown here for this two-factor experiment.

	Factor 2 (Loan Category)			
Factor 1 (Company)	1	2	3	4
A	μ_{11}	μ_{12}	μ_{13}	μ_{14}
B	μ_{21}	μ_{22}	μ_{23}	μ_{24}
C	μ_{31}	μ_{32}	μ_{33}	μ_{34}
D	μ_{41}	μ_{42}	μ_{43}	μ_{44}

As with the single-factor experiments of Section 12.1, we are interested in inferences about the unknown population means using an ANOVA.

As a second example, suppose that the drug company of Example 12.2 analyzes mean blood-pressure changes under drugs A, B, and C, considering female (F) and male (M) patients separately. The variable in this two-factor experiment is blood pressure change. One factor is type of drug and the other is gender of patient. There are six factor-level combinations (AF, AM, BF, BM, CF, and CM). For each combination, there is, in principle, a whole population of blood pressure changes from which a sample is taken. We are primarily concerned with inferences about these population means.

For these problems we again partition the total sum of squares SS(Total) into component parts. The one-factor ANOVA leads us to consider SS(Total) = SS(Between) + SS(Within), where the group variable is the single factor of interest. Now we are dealing with a second factor: SS(Total) becomes SS(Between, factor 1) + SS(Between, factor 2) + SS(Remainder). Before we develop formulas for these separate sums of squares, we reexpress the population means in terms of a model for a two-factor experiment.

two-factor model

We first assume that the unknown population mean for the observations from the ith level of factor 1 and jth level of factor 2 can be written as

$$\mu_{ij} = \mu + \alpha_i + \beta_j$$

where

μ = an overall mean, which is an unknown constant
α_i = an effect due to the ith level of factor 1; α_i is an unknown constant
β_j = an effect due to the jth level of factor 2; β_j is an unknown constant

The population means for the two-factor experiment concerning blood pressure changes are shown below:

	Factor 2 (Gender)	
Factor 1 (Drug)	Female	Male
A	$\mu_{11} = \mu + \alpha_1 + \beta_1$	$\mu_{12} = \mu + \alpha_1 + \beta_2$
B	$\mu_{21} = \mu + \alpha_2 + \beta_1$	$\mu_{22} = \mu + \alpha_2 + \beta_2$
C	$\mu_{31} = \mu + \alpha_3 + \beta_1$	$\mu_{32} = \mu + \alpha_3 + \beta_2$

Several comments should be made concerning these means. First, for any given drug product, the difference in population means for females and males is $\beta_1 - \beta_2$. For example, using drug B, the difference in population means for females and males is

$$\mu_{21} - \mu_{22} = (\mu + \alpha_2 + \beta_1) - (\mu + \alpha_2 + \beta_2) = \beta_1 - \beta_2$$

Second, for a given gender, the difference in population means for two drug products i and i^* is $\alpha_i - \alpha_{i^*}$. For example, the difference between the population means for drugs A and C in males is

$$\mu_{12} - \mu_{32} = (\mu + \alpha_1 + \beta_2) - (\mu + \alpha_3 + \beta_2) = \alpha_1 - \alpha_3$$

Not everyone would be willing to assume the model $\mu_{ij} = \mu + \alpha_i + \beta_j$ in a two-factor experiment. For example, we indicated that **this model assumes that the differences in population means for males and females are the same no matter which of the three drugs we consider**. Similarly, we might not be willing to assume that, given two drugs, the difference between their population means is the same for both genders. If either of these conditions is violated, the model $\mu_{ij} = \mu + \alpha_i + \beta_j$ does not fit the experimental situation. A model that is more suited employs the notion of an **interaction**. Two factors are said to interact if, for example, the difference in population means for two levels of factor 1 depends on which level of factor 2 is considered. For example, if the difference

interaction

in mean blood-pressure change for drugs A and B is different for males and females, then the factors drug product and gender are said to interact and the model $\mu_{ij} = \mu + \alpha_i + \beta_j$ is inappropriate.

two-factor model with interaction

A model that allows for the possibility of interaction in a two-factor experiment expresses the population mean for the ith level of factor 1 and jth level of factor 2 as

$$\mu_{ij} = \mu + \alpha_i + \beta_j + \delta_{ij}$$

The terms μ, α_i, and β_j are defined as they were previously, and δ_{ij} is an effect due to the combination of the ith level of factor 1 and the jth level of factor 2. The population means for the blood pressure experiment could be written as follows:

	Factor 2 (Gender)	
Factor 1 (Drug)	Female	Male
A	$\mu + \alpha_1 + \beta_1 + \delta_{11}$	$\mu + \alpha_1 + \beta_2 + \delta_{12}$
B	$\mu + \alpha_2 + \beta_1 + \delta_{21}$	$\mu + \alpha_2 + \beta_2 + \delta_{22}$
C	$\mu + \alpha_3 + \beta_1 + \delta_{31}$	$\mu + \alpha_3 + \beta_2 + \delta_{32}$

Note that, with the interaction model, the difference between drugs A and B depends on gender.

For females:
$$\mu_{11} - \mu_{21} = (\mu + \alpha_1 + \beta_1 + \delta_{11}) - (\mu + \alpha_2 + \beta_1 + \delta_{21})$$
$$= (\alpha_1 - \alpha_2) + (\delta_{11} - \delta_{21})$$

For males:
$$\mu_{12} - \mu_{22} = (\mu + \alpha_1 + \beta_2 + \delta_{12}) - (\mu + \alpha_2 + \beta_2 + \delta_{22})$$
$$= (\alpha_1 - \alpha_2) + (\delta_{12} - \delta_{22})$$

The concept of interaction is crucial for statistical thinking. For instance, in studying the effect of two factors, residential density (high or low) and parking lot size (large or small), on the sales volume in dollars per day, we would expect that the difference in mean sales volumes for stores with large and small parking lots would be lower for stores located near high-density residential areas than for those located near low-density areas. For this situation, a model that allows for interaction would be appropriate.

profile plot

A convenient graphic check for interaction is a **profile plot**. Label the horizontal axis by levels of one factor, and label the vertical axis for means. Plot each mean as a point and connect all the points corresponding to levels of the other factor. Two possible profile plots of the sample data for the two-factor experiment on blood pressure changes are shown in Figure 12.1 If there is no interaction, the population profile plot lines are exactly parallel, as in Figure 12.1a, because the differences in means are constant. In Figure 12.1b the lines are not parallel, suggesting the presence of interaction between the two factors. Since the lines are fairly close to parallel in this case, the interaction may not be numerically large.

There are a number of different hypotheses one might wish to test concerning the population means for a two-factor experiment. The ANOVA methods of the next section spell these out in detail.

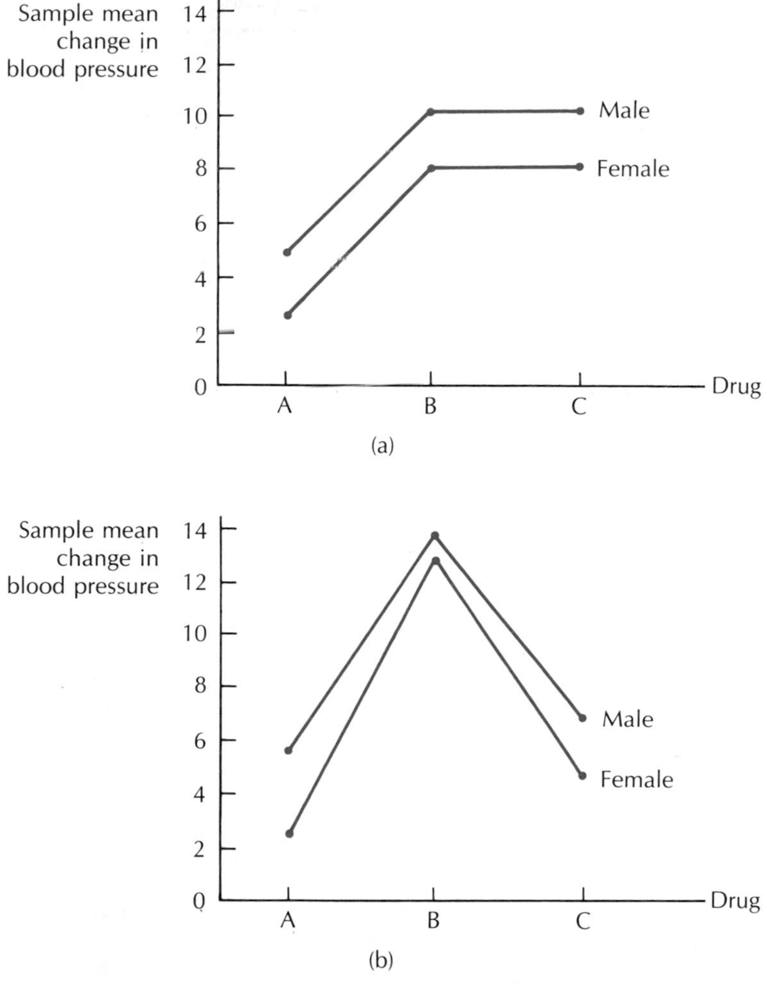

FIGURE 12.1 **Profile Plots for Blood Pressure Means: (a) No Interaction; (b) Interaction**

12.5 TWO-FACTOR ANOVA: METHODS

The previous section contains some basic concepts of two-factor ANOVA, stated in terms of population means. In this section we present inference methods based on sample data selected from these populations. The formal mathematical assumptions are the usual ones: normality of population distributions, equal population variances, and statistical independence of observations. Additionally, we assume a **balanced design**—an equal number of observations for each cell or combination of factor levels. Two-factor ANOVA with unbalanced designs is tricky, but the effort required to get equal *n*'s is amply repaid by the relative ease of analysis and interpretation.

balanced design

To illustrate, consider an experiment on the heat loss of four different types of commercial thermal pane. Five different levels of exterior temperature setting are used with three panes of each type tested at each exterior temperature setting. The data are shown below:

Exterior Temperature i	Pane Type j			
	A	B	C	D
60	3.2	7.5	10.3	12.4
	3.9	8.3	11.5	13.6
	4.3	8.0	10.9	12.9
50	5.7	10.6	12.9	15.0
	6.6	10.1	13.4	14.8
	6.1	9.7	13.0	15.5
40	7.5	11.0	13.4	15.5
	7.0	11.6	14.6	16.1
	6.5	10.7	14.0	16.2
30	8.0	12.0	14.6	17.5
	7.7	12.4	15.0	17.1
	8.2	11.7	15.3	16.4
20	10.1	13.6	17.1	18.8
	9.9	14.0	17.0	19.2
	10.0	14.5	16.8	18.8

The model for this two-factor experiment can be written as

$$\mu_{ij} = \mu + \alpha_i + \beta_j + \delta_{ij}$$

where α_i is the effect due to the ith temperature setting ($i = 1, 2, \ldots 5$), β_j is the effect due to the jth type of thermal pane ($j = 1, 2, 3, 4$), and δ_{ij} is the interaction effect corresponding to the ith temperature and jth thermal pane type.

EXAMPLE 12.8 Refer to the heat-loss experiment just described. Give the μ_{ij}'s for each combination of i and j.

Solution The μ_{ij} model with interaction terms is used to compute the following table.

Exterior Temperature i	Pane Type j		
	A	...	D
60	$\mu_{11} = \mu + \alpha_1 + \beta_1 + \delta_{11}$	...	$M_{14} = \mu + \alpha_1 + \beta_4 + \delta_{14}$
50			
40	$\vdots$		$\vdots$
30			
20	$M_{51} = \mu + \alpha_5 + \beta_1 + \delta_{51}$	...	$M_{54} = \mu + \alpha_5 + \beta_4 + \delta_{54}$

□

One of the first hypotheses of interest in the heat-loss example relates to the interaction between the two factors. If the two factors do not interact, then the difference in mean heat loss between two panes is the same for any of the five exterior settings. Similarly, with no temperature-by-pane interaction, the difference in mean heat loss at two temperatures is the same for any given pane type. In other words, with no interaction, the δ_{ij}'s of the model $\mu_{ij} = \mu + \alpha_i + \beta_j + \delta_{ij}$ are all equal and for practical purposes can be assumed to be zero. A test of whether or not interaction is present is then a test of whether or not the δ_{ij}'s all equal zero.

If interaction is not present, the model can be written

$$\mu_{ij} = \mu + \alpha_i + \beta_j$$

main effects Separate tests for the effects, called **main effects**, of the two factors, temperature setting and thermal pane type, are most easily interpreted when no interaction is present. A test of the thermal pane main effect is a test of the null hypothesis that there are no differences in mean heat loss for any of the four thermal panes. One could also test the rather implausible hypothesis that there are no differences in mean heat losses for any of the five exterior temperatures.

All of these hypotheses, the one related to interaction and the separate ones for main effects of the two factors, can be tested using an ANOVA. The two-factor ANOVA works with the following partition of the total sum of squares:

partitioning SS(Total)

$$\text{SS(Total)} = \text{SS(Between, factor 1)} + \text{SS(Between, factor 2)} + \text{SS(Interaction)} + \text{SS(Within)}$$

The formulas for these sums of squares use the following notation:

y_{ijk}: kth observation at the ith level (row) of factor 1 and the jth level (column) of factor 2

$i = 1, 2, \ldots, I$

$j = 1, 2, \ldots, J$

$k = 1, 2, \ldots, n$

Note that n is the number of observations per cell, not the overall sample size.

$\bar{y}_{ij.}$: sample mean response for the ith level of factor 1 and the jth level of factor 2

$$\bar{y}_{ij.} = \sum_k \frac{y_{ijk}}{n}$$

$\bar{y}_{i..}$: sample mean response for the ith level of factor 1

$$\bar{y}_{i..} = \sum_{j,k} \frac{y_{ijk}}{nJ}$$

$\bar{y}_{.j.}$: sample mean response for the jth level of factor 2

$$\bar{y}_{.j.} = \sum_{i,k} \frac{y_{ijk}}{nI}$$

$\bar{y}_{...}$: grand mean for all sample observations

$$\bar{y}_{...} = \sum_{i,j,k} \frac{y_{ijk}}{nIJ}$$

two-factor sums of squares With this notation, the sums of squares for a two-factor ANOVA are defined as follows:

$$\text{Factor 1: SS(Rows)} = nJ \sum_i (\bar{y}_{i..} - \bar{y}_{...})^2$$

$$\text{Factor 2: SS(Columns)} = nI \sum_j (\bar{y}_{.j.} - \bar{y}_{...})^2$$

$$\text{Interaction: SS(Interaction)} = n \sum_{i,j} (y_{ij.} - \bar{y}_{i..} - \bar{y}_{.j.} + \bar{y}_{...})^2$$

$$\text{SS(Within)} = \sum_{i,j,k} (y_{i,j,k} - \bar{y}_{ij.})^2$$

EXAMPLE 12.9 Compute the sums of squares for an ANOVA of the heat-loss data (p. 400).

Solution The following table of means can be used to compute these sums of squares. Entries in the body of the table are the $\bar{y}_{ij.}$'s. Note that $n = 3$, $I = 5$, and $J = 4$.

Exterior Temperature i	Pane Type j				
	A	B	C	D	$\bar{y}_{i..}$
60	3.800	7.933	10.900	12.967	8.900
50	6.133	10.133	13.100	15.100	11.117
40	7.000	11.100	14.000	15.933	12.008
30	7.967	12.033	14.967	17.000	12.992
20	10.000	14.033	16.967	18.933	14.983
$\bar{y}_{.j.}$	6.980	11.047	13.987	15.987	$\bar{y}_{...} = 12.000$

Substituting (unrounded) means from this table we find

$$\begin{aligned}\text{SS(Rows)} &= (3)(4)[(-3.100)^2 + (-.883)^2 + (.008)^2 + (.992)^2 + (2.983)^2] \\ &= 243.288\end{aligned}$$

$$\begin{aligned}\text{SS(Columns)} &= (3)(5)[(-5.020)^2 + (-.953)^2 + (1.987)^2 + (3.987)^2] \\ &= 689.244\end{aligned}$$

$$\text{SS(Interaction)} = 3[(-.080)^2 + (-.014)^2 + \ldots + (-.037)^2] = .074$$

The within-cell sum of squares must be calculated from the original data. For this example,

$$\begin{aligned}\text{SS(Within)} &= (3.2 - 3.800)^2 + (7.5 - 7.933)^2 + \ldots + (3.9 - 3.800)^2 + (8.3 - 7.933)^2 \\ &\quad + (4.3 - 3.800)^2 + (8.0 - 7.933)^2 + \ldots + (18.8 - 18.933)^2 \\ &= 7.353\end{aligned}$$

□

By now, an alert manager will have detected the desirability of delegating the arithmetic to a computer. But if it must be done by hand, the following computing formulas make the work less painful:

Shortcut Formulas for Computing Sums of Squares

Let

$$T = \sum_{i,j,k} y_{ijk}, \quad \text{where there are } nIJ \text{ observations in } T$$

$$A_i = \sum_{j,k} y_{ijk}, \quad \text{where there are } nJ \text{ observations in } A_i$$

$$B_j = \sum_{i,k} y_{ijk}, \quad \text{where there are } nI \text{ observations in } B_j$$

$$C_{ij} = \sum_{k} y_{ijk}, \quad \text{where there are } n \text{ observations in } C_{ij}$$

Then

$$\text{SS(Rows)} = \frac{\sum_i A_i^2}{nJ} - \frac{T^2}{nIJ}$$

$$\text{SS(Columns)} = \frac{\sum_j B_j^2}{nI} - \frac{T^2}{nIJ}$$

$$\text{SS(Within)} = \sum_{i,j,k} y_{ijk}^2 - \frac{\sum_{i,j} C_{ij}^2}{n}$$

$$\text{SS(Total)} = \sum_{i,j,k} y_{ijk}^2 - \frac{T^2}{nIJ}$$

$$\text{SS(Interaction)} = \text{SS(Total)} - \text{SS(Rows)} - \text{SS(Columns)} - \text{SS(Within)}$$

EXAMPLE 12.10 A state environmental agency tests two different methods of burning bituminous coal to generate electricity, in connection with four different "scrubbers" to reduce the resulting air pollution. The primary concern is the emission of particulate matter. Four trials are run with each scrubber combined with each burning method. Particulate emission is measured for each trial. The data are

	Scrubber			
Method	1	2	3	4
A	18.9	8.8	23.7	23.6
	16.1	16.5	15.9	22.1
	14.7	11.7	16.2	16.7
	16.9	13.0	18.0	18.9
B	24.3	24.0	9.3	18.4
	21.1	27.1	12.1	8.6
	18.0	22.6	15.6	15.1
	16.2	23.1	12.4	9.9

Computer output from an ANOVA program is shown below:

DEPENDENT VARIABLE: EMISSION

SOURCE	DF	ANOVA SS	F VALUE	PR > F
METHOD	1	1.16281250	0.12	0.7366
SCRUBBER	3	48.03093750	1.59	0.2168
METHOD*SCRUBBER	3	475.47343750	15.78	0.0001

MEANS

METHOD	N	EMISSION
A	16	16.9812500
B	16	17.3625000

SCRUBBER	N	EMISSION
1	8	18.2750000
2	8	18.3500000
3	8	15.4000000
4	8	16.6625000

METHOD	SCRUBBER	N	EMISSION
A	1	4	16.6500000
A	2	4	12.5000000
A	3	4	18.4500000
A	4	4	20.3250000
B	1	4	19.9000000
B	2	4	24.2000000
B	3	4	12.3500000
B	4	4	13.0000000

a. Use the indicated means to construct a profile plot.
b. Verify the computation of the various sums of squares.

Solution a.

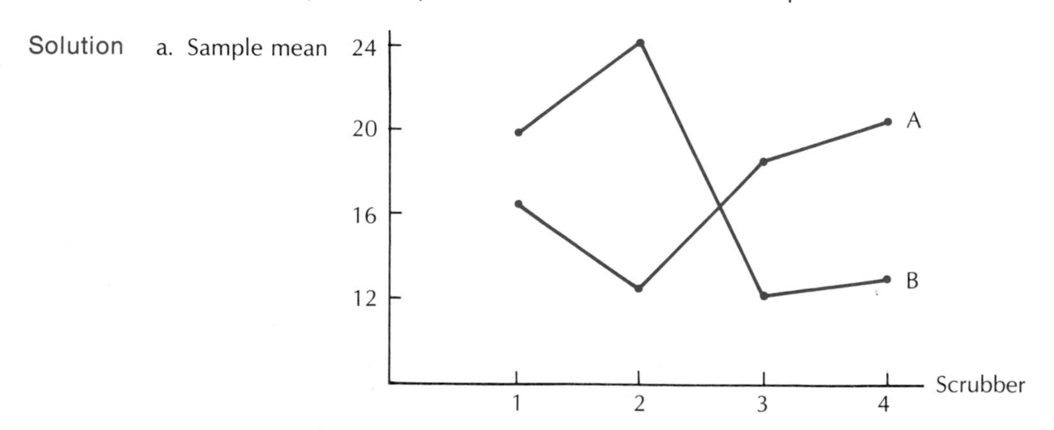

FIGURE 12.2 Solution to Example 12.10, Part (a)

b. Let the factor method be indexed by i. Then $A_1 = 271.7$, $A_2 = 277.8$, and $T = 549.5$.

$$SS(\text{Method}) = \frac{(271.7)^2 + (277.8)^2}{(4)(4)} - \frac{(549.5)^2}{(4)(2)(4)} = 1.162813$$

SS(Scrubber) is computed similarly, as 48.030937.

The C_{ij}'s are

	1	2	3	4
A	66.6	50.0	73.8	81.3
B	79.6	96.8	49.4	52.0

By direct computation, $\sum_{i,j,k} y_{ijk}^2 = 10{,}201.59$.

$$SS(\text{Within}) = 10{,}201.59 - \frac{(66.6)^2 + (50.0)^2 + \ldots + (52.0)^2}{4} = 240.97750$$

$$SS(\text{Total}) = 10{,}201.59 - \frac{(549.5)^2}{4(2)(4)} = 765.64469$$

The interaction sum of squares, SS(METHOD*SCRUBBER) in the printout, is computed by subtraction. □

These sums of squares are the basis for F tests of the various effects. The theory behind these tests is **Cochran's Theorem**. It can be proved with some tedious algebra that

Cochran's Theorem

$$\frac{SS(\text{Total})}{\sigma^2} = \frac{SS(\text{Rows})}{\sigma^2} + \frac{SS(\text{Columns})}{\sigma^2} + \frac{SS(\text{Interaction})}{\sigma^2} + \frac{SS(\text{Within})}{\sigma^2}$$

SS(Total) is $\sum_{i,j,k} (y_{ijk} - \bar{y}_{...})^2$. If there are no true effects (all α's, β's, and δ's = 0), the overall data constitute one large random sample from a normal population. Therefore, SS(Total)/σ^2 has a χ^2 distribution with $nIJ - 1$ degrees of freedom. Also by Cochran's Theorem, each separate SS/σ^2 has a χ^2 distribution independent of all other components. Once we have the appropriate degrees of freedom for sum of squares, we can construct F ratios for all the tests.

The degrees of freedom for I rows is $I - 1$. Similarly, the degrees of freedom for J rows is $J - 1$. The degrees of freedom for the interaction sum of squares is $(I - 1)(J - 1)$. This formula is very similar to the degrees of freedom in the χ^2 independence test of Chapter 11, and the theory behind it is also similar. Finally, the degrees of freedom for SS(Within) is $\sum_{i,j} (n - 1) = IJ(n - 1)$. All these results can be summarized in an ANOVA table (Table 12.2).

TABLE 12.2 **ANOVA Table, Two-factor Experiment**

Source	SS	df	MS
Rows	SS(Rows)	$I - 1$	SS(Rows)/$(I - 1)$
Columns	SS(Columns)	$J - 1$	SS(Columns)/$(J - 1)$
Interaction	SS(Interaction)	$(I - 1)(J - 1)$	SS(Interaction)/$[(I - 1)(J - 1)]$
Within	SS(Within)	$IJ(n - 1)$	SS(Within)/$[IJ(n - 1)]$
Total	SS(Total)	$nIJ - 1$	

The F tests for the null hypotheses of no effects follow directly from previous work. **Note that the test for interaction should usually precede the tests for the row and column main effects.**

Two-factor ANOVA F Tests

H_0: 1. All $\delta_{ij} = 0$ (no interaction effect)
2. All $\alpha_i = 0$ (no row effect)
3. All $\beta_j = 0$ (no column effect)

H_a: H_0 is not true

T.S.: 1. $F = \text{MS(Interaction)}/\text{MS(Within)}$
2. $F = \text{MS(Rows)}/\text{MS(Within)}$
3. $F = \text{MS(Columns)}/\text{MS(Within)}$

R.R.: For a fixed value of α, reject H_0 if $F > F_\alpha$ where F_α cuts off an area α in the upper tail of an F distribution with d.f.$_1$ = d.f. for numerator MS and d.f.$_2$ = d.f. for MS(Within)

EXAMPLE 12.11 Refer to the computer output of Example 12.10.

a. Summarize the conclusions of F tests (each at $\alpha = .05$) of the hypotheses of no interaction, no method effect, and no scrubber effect.
b. Identify p-values for each of the tests of part (a).

Solution

Null Hypothesis	F	Conclusion	p-value
No interaction	15.78	$F > 3.01$, reject H_0	.0001
No method (row) effect	.12	$F < 4.26$, retain H_0	.7366
No scrubber (column) effect	1.59	$F < 3.01$, retain H_0	.2168

□

significant interaction

The results of the F tests for row and column effects must be interpreted carefully when a **significant interaction** is present. The first thing one should do with these (and any other) interactions is construct a profile plot of the sample means $\bar{y}_{ij}$. If, for example, a significant interaction is detected in the ANOVA and the data appear as shown in Figure 12.3a, then a comparison of the level of factor 1 (rows) may be meaningful since the row 1 mean is always larger than either rows 2 or 3 even though the magnitude of the difference varies for levels of factor 2 (columns). Similarly, the F test of the main effect for factor 2 may also be of interest even though the difference between levels depends on the level of factor 1.

Not all interactions are so orderly. When interactions are both statistically significant and as large as those shown in Figure 12.3b, tests of main effects may or may not be relevant. The null hypothesis of no main effect of a factor states that the factor means (averaged with equal weights, over the categories of the other factor) are equal. The presence of substantial interaction means that the effects of one factor depend substantially

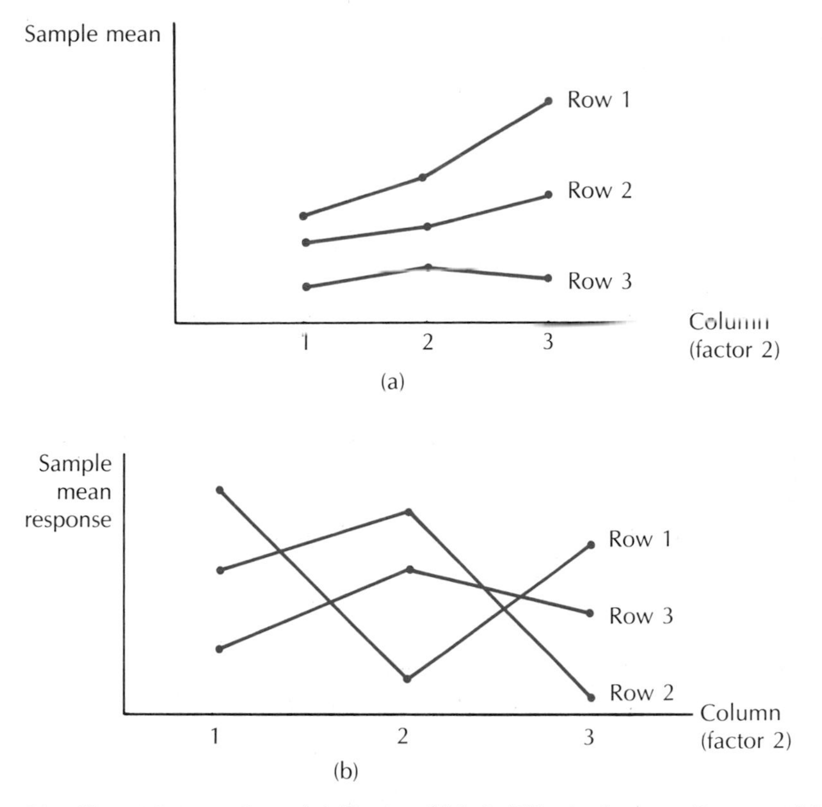

FIGURE 12.3 **Significant Interaction: (a) Tests of Main Effects Appear Reasonable; (b) Tests on Main Effects Are Not Reasonable**

on the level of the other factor. In such a case, judgment must be used as to whether the equally weighted row or column means are of much interest.

Even if interaction is present, it is still possible to use the Tukey or Scheffé methods to make comparisons of cell means. A general comparison of cell means is now written as $\hat{\ell} = \sum_{i,j} c_{ij}\bar{y}_{ij.}$ with $\sum_{i,j} c_{ij} = 0$. For a pairwise comparison, take one c_{ij} equal to $+1$, another equal to -1, and the rest equal to zero.

100% $(1 - \alpha)$% Confidence Intervals for $\mu_{ij} - \mu_{i*j*}$ in a Two-factor ANOVA Using the Tukey Method

$$(\bar{y}_{ij} - \bar{y}_{i*j*}) - q_\alpha(IJ, \text{d.f.}_2)\sqrt{\text{MS(Within)}/n} < \mu_{ij} - \mu_{i*j*}$$
$$< (\bar{y}_{ij} - \bar{y}_{i*j*}) + q_\alpha(IJ, \text{d.f.}_2)\sqrt{\text{MS(Within)}/n}$$

where IJ is the number of cells, $\text{d.f.}_2 = IJ(n - 1)$ is the d.f. for MS(Within), $q_\alpha(IJ, \text{d.f.}_2)$ is given by Appendix Table 8, and n = sample size per cell

EXAMPLE 12.12 Because there is considerable interaction in the situation of Example 12.10 (page 404), we would like to compare the performance of each method when mated with the best scrubber. Construct a 95% Tukey confidence interval for the difference in best-scrubber performance for each method.

Solution The best scrubber for Method A is scrubber 2, with $\bar{y}_{A2} = 12.500$. The best scrubber for Method B is scrubber 3, with $\bar{y}_{B3} = 12.350$. For a 95% confidence interval with $IJ = 8$, d.f.$_2 = 24$, the value of $q_{.05}$ (8, 24) is 4.68. Substituting into the formula with $n = 4$ and MS(Within) = 10.0407, we have

$$(12.500 - 12.350) - 4.68\sqrt{10.0407/4} < \mu_{A2} - \mu_{B3} < (12.500 - 12.350) + 4.68\sqrt{10.0407/4}$$

or

$$-7.265 < \mu_{A2} - \mu_{B3} < 7.565$$ □

When interaction is not substantial or significant or when a manager is willing to ignore it and compare equally weighted means, the Tukey and Scheffé methods still apply. A general comparison of row means, say, is $\hat{\ell} = \sum_i c_i \bar{y}_{i..}$.

$100(1-\alpha)$% Confidence Intervals for Comparisons of Row Means

Tukey method for pairwise comparison

$$(\bar{y}_{i..} - \bar{y}_{i^*..}) - q_\alpha(I, \text{d.f.}_2)\sqrt{\text{MS(Within)}/nJ} < \mu_{i..} - \mu_{i^*.} < (\bar{y}_{i..} - \bar{y}_{i^*.}) + q_\alpha(I, \text{df}_2)\sqrt{\text{MS(Within)}/nJ}$$

Scheffé method for general comparison of row means

$$\hat{\ell} - \sqrt{(I-1)F_\alpha(I-1, \text{d.f.}_2)}\sqrt{\text{MS(Within)}\sum_i c_i^2/nJ} < \sum_i c_i\mu_{i..} < \hat{\ell} + \sqrt{(I-1)F_\alpha(I-1, \text{d.f.}_2)}\sqrt{\text{MS(Within)}\sum_i c_i^2/nJ}$$

Note that MS(Within) is divided by nJ, the effective sample size for each row mean. For column means replace J by I.

Scheffé method for general comparison of cell means

$$\hat{\ell} - \sqrt{\text{d.f.}_1 F_\alpha(\text{d.f.}_1, \text{d.f.}_2)}\sqrt{\text{MS(Within)}\sum_{i,j} c_{ij}^2/n} < \sum_{i,j} c_{ij}\mu_{ij} < \hat{\ell} + \sqrt{\text{d.f.}_1 F_\alpha(\text{d.f.}_1, \text{d.f.}_2)}\sqrt{\text{MS(Within)}\sum_{i,j} c_{ij}^2/n}$$

For general comparisons, d.f.$_1 = IJ - 1$ and d.f.$_2 = IJ(n-1)$. F_α(d.f.$_1$, d.f.$_2$) is found in Appendix Table 6.

EXAMPLE 12.13 In Examples 12.10 and 12.11 there is substantial interaction. Compare the mean performances of each method when mated to its optimal scrubber. Construct a 95% Scheffé confidence interval for the difference in true means. Then compare your result to those of Example 12.11.

Solution The desired means are $\bar{y}_{A2} = 12.500$ and $\bar{y}_{B3} = 12.350$. We use $IJ - 1 = 7$ numerator d.f. and the 24 denominator error d.f. For a 95% interval, we need $F_{.05,7,24}$. From the tables, this equals 2.42. MS(Within), or equivalently MS(Error), is 10.0407. Each $n = 4$. The interval is

$$(12.500 - 12.350) - \sqrt{7(2.42)}\sqrt{(10.0407)(1/4 + 1/4)} < \mu_{A2} - \mu_{B3}$$
$$< (12.500 - 12.350) + \sqrt{7(2.42)}\sqrt{(10.0407)(1/4 + 1/4)}$$

or

$$-9.072 < \mu_{A2} - \mu_{B3} < 9.372$$

Notice that the interval is extremely wide. □

SECTIONS 12.4–12.5 EXERCISES

12.14 Three products are tested in each of five geographic regions. A mean approval rating is found in each case. The sample sizes are large enough that the means can be regarded essentially as population means.

	Product		
Region	A	B	C
1	68	80	77
2	55	64	58
3	62	72	67
4	67	76	70
5	58	68	63

a. Calculate the grand mean. Interpret this value.
b. Calculate region effects and product effects.
c. What does the difference in effects of products A and B represent?

12.15 Refer to the means of Exercise 12.14. Calculate a table of interaction effects.

12.16 Construct a profile plot for the means of Exercise 12.14. (Arbitrarily, put products along the horizontal axis.)

12.17 a. Is there interaction in the means? Refer to Exercise 12.14.
b. Is there a consistent superiority of any one product over all regions?

12.18 An experiment on workfare programs is undertaken by a state welfare agency. Three plans are tested. Plan A requires that all physically able welfare recipients work at assigned jobs for minimum wages. Plan B allows recipients to select or reject jobs, and welfare benefits are reduced by $1 for every $2 earned. Plan C is a "control": welfare officials visit plan C recipients as often as subjects in plans A and B and encourage job searching, but they offer no special job-taking incentives. Each plan is used with 20 families in each of four geographic areas of the state. For each family, the one-year increase in income is found. The following means (in thousands of dollars) are observed:

	Area			
Plan	1	2	3	4
A	2.1	1.7	1.3	1.7
B	1.9	1.8	1.7	1.8
C	.8	.7	.6	.7

An ANOVA table follows:

Source	SS	d.f.	MS
Plan	59.20	2	29.60
Area	4.80	3	1.60
Interaction	2.40	6	.40
Error	273.60	228	1.20
	340.00	239	

a. Find the grand mean.
b. Find the row and column effects.
c. Verify the computation of SS(Plan).

12.19 Refer to Exercise 12.18. Calculate F tests for the effects of (a) plan, (b) area, and (c) interaction. Use $\alpha = .05$ in each case. State a conclusion for each test.

12.20 a. Construct a profile plot for the data of Exercise 12.18. Arrange plans along the horizontal axis.
b. Does this plot suggest the possibility that interaction is present?
c. Does the F test of Exercise 12.19 indicate that interaction is present? Is there a conflict between the plot and the test? If so, what is a possible explanation?

12.21 Refer to the data of Exercise 12.18. Use a Tukey test with $\alpha = .05$ to see if any of the plan mean differences can be declared statistically significant.

12.22 An experiment tests four types of tanks for commercial growing of shrimp. One of each type of tank is placed at each of eight locations. The shrimp yields (in pounds) are recorded:

Tank Type	Location 1	2	3	4	5	6	7	8
1	150	138	159	180	97	180	84	103
2	130	135	171	170	89	160	85	89
3	125	116	149	129	78	129	77	94
4	116	128	139	151	73	147	75	84

An SPSS output is shown below.

SOURCE OF VARIATION	SUM OF SQUARES	DF	MEAN SQUARE	F	SIGNIF OF F
MAIN EFFECTS	33236.250	10	3323.625	40.19067	0.000
TANKTYPE	3259.375	3	1086.458	13.13791	0.000
LOCATION	29976.875	7	4282.411	51.78471	0.000
EXPLAINED	33236.250	10	3323.625	40.19067	0.000
RESIDUAL	1736.625	21	82.696		
TOTAL	34972.875	31	1128.157		

a. Locate the sums of squares for location and tank type.
b. Locate the values of F statistics for location and tank type.
c. Can you conclude that there are significant ($\alpha = .01$) differences among the tank types?

12.23 Refer to the computer output of Exercise 12.22. Locate the p-value for the test of tank-type effect.

12.24 Refer again to the data of Exercise 12.22. Suppose that the data had been incorrectly analyzed as a one-factor ANOVA, ignoring the location effect. Would the conclusion have been different? Carry out the analysis.

12.25 a. Calculate the mean yields by tank type for the data of Exercise 12.22.
b. Which pairs of means can be declared significantly different ($\alpha = .01$) using the Scheffé method?

12.6 RANDOMIZED BLOCK DESIGNS

randomized block design

The ANOVA methods discussed in this chapter extend the basic methods of comparing two means to the more general problem of comparing several means. We have applied the procedure in two experimental settings—one-factor ANOVA as described in Section 12.1 and two-factor ANOVA as described in Sections 12.4 and 12.5. In this section we describe an experimental design that is a hybrid of these settings, the **randomized block design**.

A chain of personal-computer stores wants to test three possible point-of-purchase displays of a new type of holder for computer disks. The chain managers propose to test-market the displays in three stores in each of three sales districts. One approach is to assign displays completely randomly to stores. The only factor is the type of display; the one-factor ANOVA methods of Section 12.1 can be used to analyze the resulting sales-volume data. One problem with this approach is that there is a chance of obtaining an assignment such as the following:

	Products Displayed		
District 1:	A	A	A
District 2:	B	B	B
District 3:	C	C	C

This would be a disastrous assignment. Because all the displays of a given type are in one district, there is no way to tell whether differences in sales volume are due to differences in displays or to differences in districts; that is, with this assignment, effect of display and effect of district are **confounded**. But even if the assignment of displays to stores does not turn out to be confounded, there is another objection to a one-factor experiment. By not controlling for district, this design implies that any district-to-district differences are part of random error. If there are, in fact, large district-to-district differences, MS(Error) is large; by failing to control for district, this design leads to very wide confidence intervals and to tests with very poor power.

The one-factor experimental design can be modified to control for differences among districts. The chain's managers can restrict the random assignment of displays to stores by requiring that each display be used once in each district. One such randomization is shown below. Note that each display is placed in one (randomly chosen) store in each district.

	Products Displayed		
District 1:	A	C	B
District 2:	B	A	C
District 3:	A	B	C

This experimental setting is referred to as a randomized block design. Characteristically, there is one factor, called the **treatment factor**, that is of primary interest; in this example, the treatment factor is type of display. In addition, there is another factor, called a **block factor**, that is of less interest in itself but should be controlled to avoid confounding and to reduce random error. In this example, districts would be considered block factors. Randomized block methods can be thought of as extensions of the paired-sample methods of Chapter 10; had the chain been considering only displays A and B, the sales-volume data could be regarded as paired by district. One reason for blocking is the same as the reason for pairing—to reduce the amount of random variation.

treatment factor

block factor

A randomized block is a two-factor (treatments and blocks) experiment with a one-factor focus; that is, the primary aim of the design is to compare means for the treatment factor. The block factor tends to be regarded as a "nuisance" factor; it is controlled, not so much to find out what its effects are as to avoid having those effects contaminate the analysis of the treatment factor.

There is another difference between a randomized block design and the two-factor designs discussed in Sections 12.4 and 12.5. There is typically only $n = 1$ observation per cell in a randomized block design. In the ANOVA table for a two-factor design, the d.f. for error is $IJ(n - 1)$; if we used two-factor methods for a randomized block design, we would have no d.f. for error, no MS(Error) number, and no way to do tests or confidence intervals. To avoid this problem, we assume that there is no interaction between treatments and blocks; the degrees of freedom that would have been used up in estimating interaction become available, given the assumption of no interaction, for estimating error.

The model for a randomized block design is

$$Y_{ij} = \mu + \alpha_i + \beta_j + \epsilon_{ij}$$

where Y_{ij} is the response for treatment i in block j, μ is the population grand mean, α_i is the true (population) effect of treatment i, β_j is the true effect of block j, and ϵ_{ij} is random error. Note that, by assumption, there is no interaction term in the model.

The ANOVA computations for a randomized block design are similar to those of a two-factor ANOVA, with the exception that there is, by assumption, no SS(Interaction).

Computation of Sums of Squares for a Randomized Block Design

$$\text{SS(Treatment)} = J \sum (\bar{y}_{i.} - \bar{y}_{..})^2, \quad \text{with d.f.} = I - 1$$
$$\text{SS(Block)} = I \sum (\bar{y}_{.j} - \bar{y}_{..})^2, \quad \text{with d.f.} = J - 1$$
$$\text{SS(Error)} = \sum (y_{ij} - \bar{y}_{i.} - \bar{y}_{.j} + \bar{y}_{..})^2, \quad \text{with d.f.} = (I - 1)(J - 1)$$
$$\text{SS(Total)} = \sum (y_{ij} - \bar{y}_{..})^2$$
$$= \text{SS(Treatment)} + \text{SS(Block)} + \text{SS(Error)}, \quad \text{with d.f.} = IJ - 1$$

In these formulas, $\bar{y}_{i.}$ is the mean for treatment i, $\bar{y}_{.j}$ is the mean for block j, and $\bar{y}_{..}$ is the grand mean; I is the number of treatments and J is the number of blocks. Usually, the easy way is to compute SS(Treatment), SS(Block), and SS(Total), and then find SS(Error) by subtraction.

Suppose that the sales-volume data for the three different displays in three different districts is as follows:

	District			
Display	1	2	3	Average
A	86	97	96	93.0
B	55	82	79	72.0
C	60	88	77	75.0
Average	67.0	89.0	84.0	80.0

Then

$$\text{SS(Treatment)} = \text{SS(Display)}$$
$$= 3[(93.0 - 80.0)^2 + (72.0 - 80.0)^2 + (75.0 - 80.0)^2]$$
$$= 774.0$$

$$\text{SS(Block)} = \text{SS(District)}$$
$$= 3[(67.0 - 80.0)^2 + (89.0 - 80.0)^2 + (84.0 - 80.0)^2]$$
$$= 798.0$$

$$\text{SS(Total)} = (86 - 80.0)^2 + (97 - 80.0)^2 + \ldots + (77 - 80.0)^2$$
$$= 1684.0$$

$$\text{SS(Error)} = \text{SS(Total)} - \text{SS(Treatment)} - \text{SS(Block)}$$
$$= 1684.0 - 774.0 - 798.0$$
$$= 112.0$$

The ANOVA table and F tests are completed just as in previous sections. The results for the sales-volume data are as follows:

Source	SS	d.f.	MS	F
Treatment (display)	774.0	2	387.0	13.82
Block (district)	798.0	2	399.0	14.25
Error	112.0	4	28.0	
Total	1684.0	8		

The F statistics are obtained, as usual, by dividing the indicated MS by MS(Error); the degrees of freedom for the F statistic are those indicated by the ANOVA table. For the sales-volume data, both F statistics have d.f.$_1 = 2$ and d.f.$_2 = 4$; for these d.f., $F_{.025} = 10.65$ and $F_{.01} = 18.00$. For the display factor, the F statistic falls between 10.65 and 18.00; the p-value is therefore between .025 and .01. Thus the chain's managers have fairly conclusive evidence that there are real, more than random, differences in mean sales volume by type of display. The F test for the district factor also shows that there are differences in mean volume by district; the managers knew that very well already. Note that the block factor does indeed account for a large fraction of SS(Total); had the managers not adopted a randomized block design, the SS(Error) would have been much larger, possibly concealing the effect of the display factor.

Most computer packages can compute the ANOVA table for a randomized block design. Alternatively, the definitions or the following shortcut formulas may be used:

Shortcut Formulas for a Randomized Block Design

$$\text{SS(Treatment)} = \frac{\sum A_i^2}{J} - \frac{T^2}{IJ}$$

$$\text{SS(Block)} = \frac{\sum B_j^2}{I} - \frac{T^2}{IJ}$$

$$\text{SS(Total)} = \sum y_{ij}^2 - \frac{T^2}{IJ}$$

$$\text{SS(Error)} = \text{SS(Total)} - \text{SS(Treatment)} - \text{SS(Block)}$$

where

$$A_i = \sum_j y_{ij}$$

$$B_j = \sum_i y_{ij}$$

$$T = \sum_{i,j}$$

SECTION 12.6 EXERCISES

12.26 An information-systems manager is testing four data-base management systems for possible use. A key variable is speed of execution of programs. The manager chooses six representative tasks and writes programs within each management system. The following times are recorded:

	Task					
System	1	2	3	4	5	6
I	58	324	206	94	39	418
II	47	331	163	75	30	397
III	73	355	224	106	59	449
IV	38	297	188	72	25	366

a. Compute means for each system and each task. Find the grand mean.
b. Use the definitions to find SS(System), SS(Task), and SS(Total). Find SS(Error).
c. Is there a statistically detectable (significant) difference among system means? State bounds on the p-value.

12.27 Recompute the sums of squares in Exercise 12.26 using the shortcut formulas.

12.28 Test all pairs of means for significant differences in Exercise 12.26 using $\alpha = .05$.

12.29 An experiment compares four different mixtures of the components of a rocket propellant; the mixtures contain differing proportions of oxidizer, fuel, and binder. To compare the mixtures, five different samples of propellant are prepared for each mixture. Each of five investigators is randomly assigned one sample of each of the four mixtures and is asked to measure the propellant thrust. The data are shown here:

	Investigator				
Mixture	1	2	3	4	5
1	2340	2355	2362	2350	2348
2	2658	2650	2665	2640	2653
3	2449	2458	2432	2437	2445
4	2403	2410	2418	2397	2405

a. Identify the blocks and treatments for this design.
b. Why would one want to use this design, as opposed to assigning mixtures completely randomly to investigators?

12.30 a. Refer to Exercise 12.29. Use the computer output shown below to conduct an ANOVA. Use $\alpha = .05$.
b. Which mixture appears to have the best (highest) mean? Is its mean significantly ($\alpha = .05$) higher than the average of the other three means?

```
                          ANOVA FOR ROCKET PROPELLANT DATA

                          ANALYSIS OF VARIANCE PROCEDURE

DEPENDENT VARIABLE: THRUST

SOURCE            DF   SUM OF SQUARES     MEAN SQUARE    F VALUE    PR > F    R-SQUARE        C.V.

MODEL              7  261713.45000000  37387.63571429     542.96    0.0001    0.996853      0.3368

ERROR             12     826.30000000     68.85833333              ROOT MSE            THRUST MEAN

CORRECTED TOTAL   19  262539.75000000                            8.29809215          2463.75000000

SOURCE            DF         ANOVA SS    F VALUE    PR > F

MIXT               3  261260.95000000    1264.73    0.0001
INVEST             4     452.50000000       1.64    0.2273

                                   MEANS

                      MIXT          N        THRUST

                      1             5    2351.00000
                      2             5    2653.20000
                      3             5    2444.20000
                      4             5    2406.60000
```

12.31 Is there evidence of a significant investigator effect in the computer output of Exercise 12.30? What would such an effect indicate about the accuracy of the investigators?

12.32 As one part of a taste-testing experiment, three different formulations of a new frozen dinner product are tested. Samples of each formulation are given, in random order, to each of 12 testers. Each tester gives a score to each formulation. A Minitab analysis of the data is shown here:

```
MTB > TWOWAY ANOVA OF 'RATING' BY 'FORM' AND 'PERSON'

ANALYSIS OF VARIANCE  RATING

SOURCE        DF        SS        MS
FORM           2     767.4     383.7
PERSON        11    5301.2     481.9
ERROR         22    1532.6      69.7
TOTAL         35    7601.2
MTB > TABLE BY 'FORM';
SUBC> MEANS OF 'RATING'.

 ROWS: FORM

      RATING
        MEAN

  1   57.667
  2   46.917
  3   49.250
 ALL  51.278
```

a. Perform an F test of the null hypothesis of equal score (RATING) means by formulation (FORM). State bounds on the p-value.

b. Are there statistically significant ($\alpha = .01$) differences among taste-tester (PERSON) means? What does the result of the test indicate about the people in the experiment?

12.33 a. Calculate 95% confidence intervals for all pairwise differences of score means for the three formulations, using the results in Exercise 12.32. Note that each mean is an average of 12 scores.

b. According to these intervals, which pairwise differences, if any, are statistically significant at $\alpha = .05$?

12.34 Use the means shown in Exercise 12.32 to verify SS(FORM). Note that, because of rounding off, your answer may differ from that shown in the output.

12.7 MORE COMPLEX EXPERIMENTS

Carefully designed experiments have become a more important tool for managers in the past few years. Experiments have always been relevant in the context of the research and development laboratory; recently, they have also become important in market research and in quality control. The most important concepts of experimental design have already been introduced in this chapter; in this section, we sketch some elaborations of the basic ideas.

Perhaps the most widely known role of designed experiments is in quality control. The quality of a manufactured product is the result of a number of different factors, including the quality of the parts supplied, the design of the product, and the skill of the workers who assemble the product. One of the key ideas from the modern quality-control literature is that finding a quality problem by varying one factor at a time is not an effective approach. Instead, there should be carefully planned experiments that control all critical factors. For example, suppose that the quality of a video monitor is measured by an index that reflects the brightness of display, the size of flaws on the screen, and the appearance of the case. After some consideration, three key factors are isolated: which of three suppliers of the internal electronic "gun" is used, which of two types of glass is used for the screen, and which of four assembly teams puts together the product.

It would not be wise to have teams A and B always use glass 1 and teams C and D always use glass 2, with team A using guns from supplier 1, team B guns from supplier 2, and teams C and D guns from supplier 3. In this situation, one can't separate the effect of the supplier from the effect of the gun, and neither can be separated from the effect of the assembly team. Or suppose that only the source of the gun varies across assembly teams. If teams A and B consistently put out inferior products in such a situation, it can't be determined whether their work is poor or whether glass 1 is inferior. Instead, all teams should use all gun suppliers and all glass types. The most convenient and effective way to vary team, supplier, and glass type is in a balanced design.

Measuring the quality index as a function of team, gun supplier, and glass type in a balanced design is a **factorial experiment** with three factors. The data can be analyzed in terms of main effects, two-way interactions (team by supplier, team by glass type, and supplier by glass type), and a three-way interaction. If a quality problem arises, the analysis will suggest the appropriate remedy. For example, if the mean for supplier 1 is consistently low, that supplier should bring up quality standards, but if there is a strong supplier-by-glass-type interaction, the technical compatibility of the guns and glass type should be investigated. The computations required for a multifactor experiment are similar to those for a two-factor experiment, but they are tedious enough to be left to a computer. Analyses by F tests and Tukey and Scheffé methods are obvious extensions of these methods for two-factor studies. In particular, one should check for interactions before considering overall averages (main effects).

For example, suppose that samples of eight monitors each are taken from the production of each combination of team, supplier, and glass type, and that the following computer output is obtained.

```
                                THREE-WAY ANOVA FOR QUALITY CONTROL DATA

                                     ANALYSIS OF VARIANCE PROCEDURE

DEPENDENT VARIABLE: QUALITY

SOURCE               DF     SUM OF SQUARES       MEAN SQUARE    F VALUE       PR > F     R-SQUARE           C.V.

MODEL                23     24557.95312500    1067.73709239       10.94       0.0001     0.599609         4.3901

ERROR               168     16398.62500000      97.61086310                 ROOT MSE                QUALITY MEAN

CORRECTED TOTAL     191     40956.57812500                                 9.87982101               225.04687500

SOURCE               DF          ANOVA SS      F VALUE     PR > F

TEAM                  3     1957.89062500         6.69     0.0003
SUPPLIER              2    20509.96875000       105.06     0.0001
TEAM*SUPPLIER         6      203.15625000         0.35     0.9109
TYPE                  1      435.00520833         4.46     0.0362
TEAM*TYPE             3      303.05729167         1.03     0.3786
SUPPLIER*TYPE         2      460.01041667         2.36     0.0979
TEAM*SUPPLIER*TYPE    6      688.86458333         1.18     0.3213


                                         MEANS

                                  TEAM        N        QUALITY

                                  1          48     229.645833
                                  2          48     220.625000
                                  3          48     225.145833
                                  4          48     224.770833

                               SUPPLIER        N        QUALITY

                               1              64     231.953125
                               2              64     210.437500
                               3              64     232.750000


                                  TYPE        N        QUALITY

                                  1          96     223.541667
                                  2          96     226.552083
```

First, we look for interactions. SAS uses an asterisk to denote interactions; we see that TEAM*SUPPLIER, TEAM*TYPE, SUPPLIER*TYPE, and TEAM*SUPPLIER*TYPE all have p-values (shown as $PR > F$) that are relatively large. The only interaction that is even close to statistically significant is SUPPLIER*TYPE (p-value 0.0979).

Then, we look for main effects. There is a huge SUPPLIER effect; the SS for SUPPLIER is extremely large, the F value is enormous at 105.06, and the p-value ($PR > F$) is tiny at 0.0001. The computed MEANS clearly show that SUPPLIER 2 has a very low quality average. There also is a clear evidence of a significant effect of TEAM ($PR > F = 0.0003$); the TEAM 1 mean is high and the TEAM 2 mean is low. There is a marginally significant ($PR > F = 0.0362$) TYPE effect; TYPE 2 has a slightly higher mean than TYPE 1. Tukey-method tests could be performed for each main effect to verify that the apparent differences of means are statistically detectable.

Careful experimentation isn't easy. Everyone who is involved in the study must understand the importance of balanced design and of randomization. The effort required has a real payoff. A well-designed, well-executed experiment can reveal the important variables much more clearly than any other data-gathering method.

SECTION 12.7 EXERCISES

12.35 An experiment is designed to determine factors affecting electricity usage in single-family homes. In a new suburban development, there are two basic home layouts. equal numbers of houses were built with two levels of insulation (*M* being the suggested minimum and *H* a larger amount). Another factor is the size of family living in the home (0, 1, 2 or more children). The electricity usage for each house is determined. An SAS analysis of the data is shown below:

a. Locate the *F* statistics.

b. Which factors and interactions are significant at $\alpha = .01$?

ELECTRICITY USAGE STUDY

ANALYSIS OF VARIANCE PROCEDURE

DEPENDENT VARIABLE: ELECUSE

SOURCE	DF	SUM OF SQUARES	MEAN SQUARE	F VALUE	PR > F	R-SQ
MODEL	11	2529.16666667	229.92424242	17.74	0.0001	0.84
ERROR	36	466.50000000	12.95833333		ROOT MSE	
CORRECTED TOTAL	47	2995.66666667			3.59976851	

SOURCE	DF	ANOVA SS	F VALUE	PR > F
INSUQUAL	1	432.00000000	33.34	0.0001
LAYOUT	1	320.33333333	24.72	0.0001
INSUQUAL*LAYOUT	1	0.33333333	0.03	0.8735
FAMTYPE	2	1147.79166667	44.29	0.0001
INSUQUAL*FAMTYPE	2	0.37500000	0.01	0.9856
LAYOUT*FAMTYPE	2	627.54166667	24.21	0.0001
INSUQ*LAYOU*FAMTY	2	0.79166667	0.03	0.9699

EXERCISE 12.35 (continued on next page)

MEANS

INSUQUAL	N	ELECUSE
H	24	51.4166667
M	24	57.4166667

LAYOUT	N	ELECUSE
A	24	51.8333333
B	24	57.0000000

INSUQUAL	LAYOUT	N	ELECUSE
H	A	12	48.9166667
H	B	12	53.9166667
M	A	12	54.7500000
M	B	12	60.0833333

FAMTYPE	N	ELECUSE
0	16	49.4375000
1	16	52.7500000
2	16	61.0625000

INSUQUAL	FAMTYPE	N	ELECUSE
H	0	8	46.3750000
H	1	8	49.8750000
H	2	8	58.0000000
M	0	8	52.5000000
M	1	8	55.6250000
M	2	8	64.1250000

LAYOUT	FAMTYPE	N	ELECUSE
A	0	8	42.8750000
A	1	8	49.3750000
A	2	8	63.2500000
B	0	8	56.0000000
B	1	8	56.1250000
B	2	8	58.8750000

MEANS

INSUQUAL	LAYOUT	FAMTYPE	N	ELECUSE
H	A	0	4	39.7500000
H	A	1	4	46.7500000
H	A	2	4	60.2500000
H	B	0	4	53.0000000
H	B	1	4	53.0000000
H	B	2	4	55.7500000
M	A	0	4	46.0000000
M	A	1	4	52.0000000
M	A	2	4	66.2500000
M	B	0	4	59.0000000
M	B	1	4	59.2500000
M	B	2	4	62.0000000

12.36 Refer to the output of Exercise 12.35.

a. Draw the interaction profiles for insulation quality by layout.

b. Draw the interaction profiles for layout by family type.

12.37 Refer to the output of Exercise 12.35.

a. Use the Scheffé method to calculate a 95% confidence interval for the difference in mean electricity usage for the two insulation levels.

b. Do you think it makes sense to compare the layout means ignoring family type?

c. Treat the means by layout and family type as a set of six means. Which pairs differ significantly ($\alpha = .05$)?

SUMMARY

One-factor analysis of variance (ANOVA) methods for comparing the population means from different groups provide an extension to the two-sample t test for comparing two population means (Chapter 10). One important reason for using any ANOVA F test for H_0: $\mu_1 = \mu_2 = \ldots = \mu_I$, instead of doing a series of t tests for all pairwise comparisons of the population means, is to control the overall probability of a Type I error. If an F test is statistically significant, a multiple-comparison procedure, such as the Tukey or Scheffé methods, can be used to investigate which means differ.

The two-factor ANOVA also deals with population means μ_{ij}. A model for the μ_{ij}'s that allows for interaction between the two factors is

$$\mu_{ij} = \mu + \alpha_i + \beta_j + \delta_{ij}$$

A test for no interaction is a test of H_0: all $\delta_{ij} = 0$. Comparisons of the μ_{ij}'s can be made after finding a significant interaction using an extension of the Scheffé method of Section 12.2. Tests for the row and column main effects make use of the null hypotheses H_0: all $\alpha_i = 0$ and H_0: all $\beta_j = 0$, respectively.

The Scheffé and Tukey methods can be used to make specific comparisons of the row (or column) means.

A useful design is the randomized block. This experimental design controls for both a fundamental "treatment" factor and a nuisance "blocking" factor. Typically, in a randomized block design, there is only one observation per treatment per block, so that any treatment-block interaction must be ignored; however, controlling block effects can often reduce the error variance substantially, allowing for more precise inferences.

Not all problems fall neatly into one- or two-factor experiments. Many times managers are faced with experiments in which there are many factors affecting the response of interest. Judicious use of a statistical consultant could help in designing an experiment to investigate the effects of certain important factors on the response. Extensions of ANOVA techniques presented here would likely be used to draw conclusions.

KEY FORMULAS: Introduction to the Analysis of Variance

1. Testing the equality of I group means

   H_0: $\mu_1 = \mu_2 = \ldots = \mu_I$

   $$\text{T.S.: } F = \frac{\text{MS(Between)}}{\text{MS(Within)}}$$

   where

   $$\text{MS(Between)} = \sum_i n_i(\bar{y}_i - \bar{y})^2/(I - 1)$$

   $$\text{MS(Within)} = \sum_{i,j} (y_{ij} - \bar{y}_i)^2/(n - I)$$

2. Shortcut formulas for ANOVA sums of squares

   $$\text{SS(Total)} = \sum_{i,j} y_{ij}^2 - \frac{\left(\sum_{i,j} y_{ij}\right)^2}{n}$$

   $$\text{SS(Between)} = \sum_i \left(\sum_j y_{ij}\right)^2 \Big/ n_i - \frac{\left(\sum_{i,j} y_{ij}\right)^2}{n}$$

   $$\text{SS(Within)} = \text{SS(Total)} - \text{SS(Between)}$$

3. $\text{SS(Between)} = \sum n_i(\bar{y}_i - \bar{y})^2$

   $\text{SS(Within)} = \sum (n_i - 1)s_i^2$

   $\text{SS(Total)} = \text{SS(Between)} + \text{SS(Within)}$

   where

   n_i is the number of observations in level i
   $\bar{y}_i$ is the sample mean for level i
   s_i^2 is the sample variance for level i

4. Kruskal-Wallis test (a nonparametric alternative to the ANOVA for testing I group means)

   H_0: The distributions are identical (effectively, the populations have equal means)

   $$\text{T.S.: } H = \frac{12}{n(n + 1)} \sum_i \frac{T_i^2}{n_i} - 3(n + 1)$$

   where T_i is the sum of the ranks for the n_i measurements in sample i and d.f. $= I - 1$

5. Tukey's $100(1-\alpha)\%$ confidence interval for $\mu_i - \mu_{i*}$

$$(\bar{y}_i - \bar{y}_{i*}) \pm q_\alpha(T, \text{d.f.}_2)\sqrt{\text{MS(Within)}/n}$$

6. Scheffé's $100(1-\alpha)\%$ confidence interval for $\mu_i - \mu_{i*}$

$$(\bar{y}_i - \bar{y}_{i*}) \pm \sqrt{(I-1)F_\alpha}\sqrt{\text{MS(Within)}\left(\frac{1}{n_i} + \frac{1}{n_{i*}}\right)}$$

7. Scheffé's confidence interval for comparing several means, $\sum c_i\mu_i$

$$\sum c_i\bar{y}_i \pm \sqrt{(I-1)F_\alpha}\sqrt{\text{MS(Within)}\sum c_i^2/n_i}$$

8. Shortcut formulas for sums of squares in two-factor ANOVA

$$\text{SS(Rows)} = \sum_i \frac{A_i^2}{nJ} - \frac{T^2}{nIJ}$$

$$\text{SS(Columns)} = \sum_j \frac{B_j^2}{nI} - \frac{T^2}{nIJ}$$

$$\text{SS(Within)} = \sum_{i,j,k} y_{ijk}^2 - \sum_{i,j} \frac{C_{ij}^2}{n}$$

$$\text{SS(Total)} = \sum_{i,j,k} y_{ijk}^2 - \frac{T^2}{nIJ}$$

$$\text{SS(Interaction)} = \text{SS(Total)} - \text{SS(Rows)} - \text{SS(Columns)} - \text{SS(Within)}$$

9. Two-factor ANOVA

H_0: 1. all $\delta_{ij} = 0$ (no interaction effect)
2. all $\alpha_i = 0$ (no row effect)
3. all $\beta_j = 0$ (no column effect)

T.S.: 1. $F = \text{MS(Interaction)}/\text{MS(Within)}$
2. $F = \text{MS(Rows)}/\text{MS(Within)}$
3. $F = \text{MS(Columns)}/\text{MS(Within)}$

10. Tukey $100(1-\alpha)\%$ confidence interval for $\mu_{ij} - \mu_{i*j*}$

$$(\bar{y}_{ij} - \bar{y}_{i*j*}) \pm q_\alpha(IJ, \text{d.f.}_2)\sqrt{\text{MS(Within)}/n}$$

11. Randomized block design

$$SS(\text{Treatment}) = J\sum(\bar{y}_{i.} - \bar{y}_{..})^2 = \frac{(\sum A_i^2)}{J} - \frac{T^2}{IJ}$$

where $A_i = \sum_j y_{ij}$, $T = \sum_i \sum_j y_{ij}$

$$SS(\text{Block}) = I\sum(\bar{y}_{.j} - \bar{y}_{..})^2 = \frac{(\sum B_j^2)}{I} - \frac{T^2}{IJ}$$

where $B_j = \sum_i y_{ij}$, $T = \sum_i \sum_j y_{ij}$

$$SS(\text{Total}) = \sum(y_{ij} - \bar{y}_{..})^2 = \sum(y_{ij}^2) - \frac{T^2}{IJ}$$

$$SS(\text{Error}) = \sum(y_{ij} - \bar{y}_{.j} - \bar{y}_{i.} + \bar{y}_{..})^2$$
$$= SS(\text{Total}) - SS(\text{Treatment}) - SS(\text{Block})$$

12. F tests for randomized block design

For H_0: No treatment effect,

$$F = \frac{SS(\text{Treatment})/(I-1)}{SS(\text{Error})/(I-1)(J-1)}$$

For H_0: No block effect

$$F = \frac{SS(\text{Block})/(J-1)}{SS(\text{Error})/(I-1)(J-1)}$$

CHAPTER 12 EXERCISES

12.38 Three different designs of video recording equipment are subjected to accelerated use testing, and the times to failure (in hours) of each unit are recorded:

	Time									
Design A:	226	400	462	489	510	541	547	563	581	603
Design B:	329	366	409	451	465	490	517	546	577	615
Design C:	421	484	506	566	589	605	619	634	651	660

a. Calculate means and variances for each design.

b. Use these results to calculate SS (Between) and SS (Within).

c. Use the shortcut formulas to calculate the sums of squares.

12.39 Refer to Exercise 12.38. Is the research hypothesis of unequal means supported? First use $\alpha = .05$, then place bounds on the p-value.

12.40 Perform the Kruskal-Wallis test for the hypothesis of Exercise 12.39. Is the conclusion substantially different from that of Exercise 12.39?

12.41 Plot the data of Exercise 12.38. Do there appear to be any major violations of ANOVA assumptions?

12.42 Use the Scheffé procedure to calculate overall 95% confidence intervals for the pairwise differences among the means of Exercise 12.38.

12.43 In a study of the effects of television commercials on 7-year-old children, the attention span of children watching commercials for clothing, food products, and toys is measured. To reduce the effects of outliers, only the median attention span for each commercial is used.

Commercial	Median Attention Span (Seconds)											
Clothes	21	30	23	37	21	18	30	42	36			
Food	32	51	46	30	25	41	38	50	45	53	57	41
Toys	48	59	51	47	58	56	49	55	52	49	60	

SAS output is shown below:

```
DEPENDENT VARIABLE: ATTSPAN     ATTENTION SPAN

SOURCE              DF    SUM OF SQUARES      MEAN SQUARE     F VALUE      PR > F     R-SQUARE        C.V.

MODEL                2     2953.64299242    1476.82149621       23.10      0.0001     0.614386     18.9378

ERROR               29     1853.82575758      63.92502612                 ROOT MSE              ATTSPAN MEAN

CORRECTED TOTAL     31     4807.46875000                                7.99531276               42.21875000


                                    MEANS

                        COMMTYPE          N        ATTSPAN

                        DUDS              9     28.6666667
                        FOODS            12     42.4166667
                        TOYS             11     53.0909091


                         WILCOXON SCORES (RANK SUMS)

                                  SUM OF    EXPECTED    STD DEV       MEAN
              LEVEL          N    SCORES    UNDER H0    UNDER H0     SCORE

              TOYS          11    274.50      181.50       25.19     24.95
              FOODS         12    193.50      198.00       25.67     16.12
              DUDS           9     60.00      148.50       23.84      6.67

                  KRUSKAL-WALLIS TEST (CHI-SQUARE APPROXIMATION)
                  CHISQ=  18.87     DF=  2     PROB > CHISQ=0.0001
```

a. Locate the value of the F statistic for the null hypothesis of equal means.
b. Can this hypothesis be rejected using $\alpha = .01$?
c. Locate the p-value for this test.

12.44 a. Calculate the value of the Kruskal-Wallis statistic in Exercise 12.43.
b. Can the hypothesis of equal means (or more properly "locations") be rejected using $\alpha = .01$?

12.45 a. Plot the data of Exercise 12.43.
b. Do there appear to be serious violations of ANOVA assumptions?
c. Does it matter much whether an F test or a Kruskal-Wallis test is used?

12.46 Calculate 99% Scheffé confidence intervals for differences in means for the data of Exercise 12.43. Which differences, if any, are significant at $\alpha = .01$?

12.47 A township manager had four appraisers estimate the fair market value of 12 houses. The estimates, in thousands of dollars, are

	Home											
Appraiser	1	2	3	4	5	6	7	8	9	10	11	12
A	86	76	93	110	73	55	96	74	96	140	88	72
B	81	75	95	105	70	53	91	75	95	120	75	68
C	89	80	97	110	80	61	89	80	100	130	90	75
D	90	76	96	108	78	63	99	77	99	135	94	75

Computer output from the SAS package is shown below:

```
                         HOME VALUATIONS BY APPRAISERS

                         ANALYSIS OF VARIANCE PROCEDURE

DEPENDENT VARIABLE: VALUE      FAIR MARKET VALUE OF HOUSE

SOURCE              DF      SUM OF SQUARES        MEAN SQUARE     F VALUE        PR > F      R-SQUARE          C.V.

MODEL               14      16467.62500000      1176.25892857       98.60        0.0001      0.976651        3.9166

ERROR               33        393.68750000        11.92992424                  ROOT MSE                  VALUE MEAN

CORRECTED TOTAL     47      16861.31250000                                   3.45397224                 88.18750000

SOURCE              DF           ANOVA SS       F VALUE      PR > F

APPRAISR             3       381.56250000         10.66      0.0001
HOME                11     16086.06250000        122.58      0.0001

                         ANALYSIS OF VARIANCE PROCEDURE

                                    MEANS

                        APPRAISR         N          VALUE

                        1               12     88.2500000
                        2               12     83.5833333
                        3               12     90.0833333
                        4               12     90.8333333

                          HOME           N          VALUE

                          1              4      86.500000
                          2              4      76.750000
                          3              4      95.250000
                          4              4     108.250000
                          5              4      75.250000
                          6              4      58.000000
                          7              4      93.750000
                          8              4      76.500000
                          9              4      97.500000
                          10             4     131.250000
                          11             4      86.750000
                          12             4      72.500000
```

a. How relevant is the null hypothesis of equality of house prices? Is this hypothesis rejected?
b. Locate the value of the F statistic for testing the null hypothesis of the equality of appraiser effects.
c. Can the null hypothesis be rejected at typical α levels? What is the indicated p-value?

12.48 Use the Scheffé method to test ($\alpha = .05$) the significance of pairwise differences in appraiser effects for the data of Exercise 12.47.

12.49 The data of Exercise 12.47 were also analyzed incorrectly as a one-way ANOVA experiment, with the following output:

```
                              HOME VALUATION IGNORING HOME FACTOR
                               ANALYSIS OF VARIANCE PROCEDURE
DEPENDENT VARIABLE: VALUE      FAIR MARKET VALUE OF HOUSE
SOURCE              DF     SUM OF SQUARES     MEAN SQUARE    F VALUE      PR > F     R-SQUARE        C.V.
MODEL                3       381.56250000    127.18750000       0.34      0.7968     0.022629     21.9453
ERROR               44     16479.75000000    374.53977273                ROOT MSE               VALUE MEAN
CORRECTED TOTAL     47     16861.31250000                             19.35303007              88.18750000
```

a. Would the null hypothesis of equal appraiser means be rejected at $\alpha = .05$?
b. How important is it to control for the effect of house differences in Exercise 12.47.

12.50 As part of an environmental impact study, an offshore oil driller tests three different drilling-rig designs at two different depths. The dependent variable of interest is the fish catch in a small net alongside the rig. The data are

	Design		
Depth	A	B	C
1	1025	896	925
	988	914	984
	1104	953	963
2	721	630	665
	785	687	741
	655	652	706

SAS output is given below:

```
                              FISH CATCH AROUND OIL RIGS

                            ANALYSIS OF VARIANCE PROCEDURE

DEPENDENT VARIABLE: CATCH

SOURCE              DF    SUM OF SQUARES      MEAN SQUARE     F VALUE      PR > F    R-SQUARE        C.V.

MODEL                5   378554.00000000   75710.80000000       38.65      0.0001    0.941536      5.3132

ERROR               12    23506.00000000    1958.83333333                ROOT MSE              CATCH MEAN

CORRECTED TOTAL     17   402060.00000000                              44.25870912            833.00000000

SOURCE              DF          ANOVA SS     F VALUE    PR > F

DEPTH                1   350005.55555556      178.68    0.0001
DESIGN               2    24892.00000000        6.35    0.0131
DEPTH*DESIGN         2     3656.44444444        0.93    0.4200

                                        MEANS

                              DEPTH   DESIGN        N        CATCH

                              1       A             3   1039.00000
                              1       B             3    921.00000
                              1       C             3    957.33333
                              2       A             3    720.33333
                              2       B             3    656.33333
                              2       C             3    704.00000
```

a. Plot an interaction profile.
b. Test the null hypothesis of no interaction at $\alpha = .05$.
c. Does it appear that there is substantial interaction?

12.51 Refer to the output of Exercise 12.50.
a. Are there significant effects of design? State the p-value.
b. Are there significant depth effects at $\alpha = .05$?

12.52 Refer to the output of Exercise 12.50. Calculate overall 95% confidence intervals for differences among design effects.

12.53 A food manufacturer tests several formulations of an orange juice product. Three different sweetness levels are tested, in combination with two acidity levels and two color levels. Six panels rate each combination on a 1–99 scale. The ratings are shown below:

Sweetness	Acidity	Color	Rating					
1	1	1	40	35	45	42	48	45
1	1	0	62	56	60	54	60	65
1	2	1	38	32	50	40	46	35
1	2	0	60	50	48	61	55	53
2	1	1	45	56	50	45	55	48
2	1	0	72	56	63	75	67	68
2	2	1	47	53	46	52	56	47
2	2	0	60	66	56	64	72	70
3	1	1	35	50	35	40	43	35
3	1	0	56	48	52	45	59	52
3	2	1	25	35	30	24	34	34
3	2	0	40	36	32	31	34	38

The BMDP analysis is shown below:

SOURCE	SUM OF SQUARES	DEGREES OF FREEDOM	MEAN SQUARE	F	TAIL PROB.
MEAN	171307.55556	1	171307.55556	6234.40	0.0000
SWEET	4149.52778	2	2074.76389	75.51	0.0000
ACIDITY	624.22222	1	624.22222	22.72	0.0000
COLOR	3200.00000	1	3200.00000	116.46	0.0000
SA	488.52778	2	244.26389	8.89	0.0004
SC	203.08333	2	101.54167	3.70	0.0307
AC	80.22222	1	80.22222	2.92	0.0927
SAC	24.19444	2	12.09722	0.44	0.6459
ERROR	1648.66667	60	27.47778		

CELL MEANS FOR 1-ST DEPENDENT VARIABLE

SWEET =	LOW	LOW	LOW	LOW	MEDIUM	MEDIUM	MEDIUM	MEDIUM	HIGH	HIGH
ACIDITY =	*1.00000	*1.00000	*2.00000	*2.00000	*1.00000	*1.00000	*2.00000	*2.00000	*1.00000	*1.00000
COLOR =	NATURAL	ENHANCED	NATURAL	ENHANCED	NATURAL	ENHANCED	NATURAL	ENHANCED	NATURAL	ENHANCED
RATING	59.50000	42.50000	54.50000	40.16667	66.83333	49.83333	64.66667	50.16667	52.00000	39.66667

a. Does there appear to be a significant three-factor interaction?

b. Plot the acidity—color interaction profile. Does it appear that this interaction is substantial?

12.54 Summarize the results of the *F* tests shown in the output of Exercise 12.53.

12.55 Refer to Exercise 12.53.

a. Calculate 95% confidence intervals for the mean difference in rating by acidity level. Which differences are statistically significant at $\alpha = .05$?

b. Which differences among the 12 cell means are statistically significant ($\alpha = .05$) according to the Scheffé method?

APPENDIX: SUMS OF SQUARES

In this appendix we sketch a proof that SS(Total) = SS(Between) + SS(Within) for a one-factor design. Similar proofs apply to more complicated ANOVA problems, as long as the design is balanced. The quantity SS(Total) can be written as follows

$$\begin{aligned}
\text{SS(Total)} &= \sum_{i,j} (y_{ij} - \bar{y})^2 \\
&= \sum_{i,j} (y_{ij} - \bar{y}_i + \bar{y}_i - \bar{y})^2 \\
&= \sum_{i,j} (y_{ij} - \bar{y}_j)^2 + 2 \sum_i \left[\sum_j (y_{ij} - \bar{y}_i) \right] (\bar{y}_i - \bar{y}) + \sum_{i,j} (\bar{y}_i - \bar{y})^2
\end{aligned}$$

by expanding $(a + b)^2$ as $a^2 + 2ab + b^2$. For a fixed group i, the quantity in square brackets of the second term is a sum of deviations from a mean, which is always zero. Therefore the middle term is zero. The first term is SS(Within) by definition. The last term involves

$\sum_j (\bar{y}_i - \bar{y})^2$; the quantity being summed is a constant (with respect to j) summed n_i times. Therefore the last term is SS(Between), and the desired result holds.

REVIEW EXERCISES CHAPTERS 10–12

R86 A sample of 40 testers rates the thirst-quenching property of an old formulation of a soft drink and a new formulation. The drinks are scored on a 0–100 scale. The results are

Formulation	Mean	Standard Deviation
Old	41.15	14.7
New	45.55	17.8

a. Calculate a 95% confidence interval for the difference of the true (population) means using a pooled-variance procedure.
b. Based on this interval, can the null hypothesis of equal means be rejected, using $\alpha = .05$?
c. In fact, it's a very bad idea to use a pooled-variance procedure for these data. Why?

R87 The differences between ratings of the new formulation and the old one in Exercise R86 are shown here:

4	3	−1	9	−4	7	−6	4	−2	7	2	4	14	1
2	12	−7	11	12	9	4	10	−5	−7	−2	−3	10	
−1	23	2	7	5	15	4	0	1	5	−4	16	15	

The mean of the differences is 4.40 and the standard deviation is 7.05.
a. Calculate a 95% confidence interval for the population mean difference.
b. Based on this interval, can the null hypothesis that the mean difference is zero be rejected using $\alpha = .05$?
c. Is there any evidence of a serious violation of assumptions for this procedure?

R88 Perform an appropriate rank test for the difference data of Exercise R87, using $\alpha = .05$ (two-sided). Does this test lead to the same conclusion as the t test of that exercise?

R89 Place bounds on the p-values in Exercises R87 and R88. Are the p-values similar in the two exercises?

R90 Compare the widths of the confidence intervals found in Exercises R86 and R87. What do the relative sizes of the intervals indicate about the effect of having the same tasters rate both formulations (as opposed to having different tasters rating each formulation)?

R91 A large corporation has a pool of individuals responsible for most word-processing jobs. As a means of providing a pleasant and productive atmosphere, the company plays taped music during the workday. Some individuals complain that the music occasionally becomes distracting. As an experiment, the company provides varying degrees of control over the music's volume (ranging from 1 = no control to 4 = complete control) to samples of size 16 each of word-processing operators. An efficiency score is obtained for each person.

Efficiency	Degree of Control 1	2	3	4
	42	55	63	66
	57	50	57	63
	37	65	24	64
	52	22	55	57
	58	65	64	64
	58	56	56	60
	56	63	61	62
	57	58	60	62
	41	65	63	58
	49	57	64	54
	53	52	67	65
	55	61	66	60
	53	64	66	63
	42	57	52	64
	48	65	47	49
	48	66	65	61
$\bar{y}$	50.38	57.56	58.13	60.75
s	6.80	10.76	10.71	4.45

a. Test the null hypothesis that the four population means are equal, using $\alpha = .01$. State the conclusion carefully.

b. Place bounds on the p-value.

c. Is there any indication that any of the formal mathematical assumptions have been violated? If so, do the violations make your answers to parts (a) and (b) seriously wrong?

R92 a. Using the data of Exercise R91, calculate simultaneous 99% confidence intervals for the differences of all pairs of means.

b. Which pairs of means, if any, are significantly different at $\alpha = .05$?

R93 a. In Exercise R91, define a comparison between the mean for the no-control group and the average of the means of the other groups.

b. Test the hypothesis that the value of this comparison is zero, using $\alpha = .05$.

R94 For the data of Exercise R91, the value of the Kruskal-Wallis statistic is 17.18.

a. Can the null hypothesis that the four groups' distributions are equal be rejected at $\alpha = .01$?

b. Place bounds on the p-value of the statistic.

c. Does this test yield essentially the same conclusion as the test in Exercise R91?

R95 In the study of Exercise R91, the participants who have some degree of control also rate that control as not useful, somewhat useful, or very useful. The frequencies are

Rating	Degree of Control 1	2	3	Total
Not useful	7	3	2	12
Somewhat useful	4	7	5	16
Very useful	5	6	9	20
Total	16	16	16	48

a. Is there a statistically significant relation between rating and degree of control, using $\alpha = .05$?
b. Place bounds on the p-value for the test statistic used in part (a).
c. Is there any reason to be skeptical of the (approximate) p-value bounds?

R96 Do the frequencies in Exercise R95 indicate a moderately strong relation, whether or not the relation is statistically significant?

R97 One person interprets the results of Exercise R95 as proving that there is no relation between rating and degree of control. Is this interpretation appropriate?

R98 A real-estate firm in the headquarters city of a large corporation contracts the corporation to find housing and mortgages for the corporation's newly arriving managers. One concern is the length of mortgage contracts. Samples are taken of managers arriving by transfer from other corporation offices and of newly hired managers. The data are taken over a time when mortgage availability and terms are stable. The following lengths (in months) of mortgage contracts are obtained:

Transfers: 180 240 300 360 240 180 144 300 240 240 360 180 180
300 240 ($n = 15$, $\bar{y} = 245.6$, $s = 66.9$)

New hires: 360 360 360 240 270 300 360 360 300 360 360 300 300
240 300 360 360 360 360 360 300 300 360 240 360 360
360 360 300 360 360 300 ($n = 32$, $\bar{y} = 329.1$, $s = 41.4$)

Use pooled-variance methods to calculate a 95% confidence interval for the difference of population means. Based on this interval, can you conclude that there is a statistically significant difference? What α are you using?

R99 The use of pooled-variance methods in Exercise R98 is a poor idea for (at least) two reasons. Discuss the reasons.

R100 Recalculate the confidence interval of Exercise R98 using a more appropriate method. Does the new interval lead to the same conclusion about the significance of the difference?

R101 When the data of Exercise R98 are ranked, the sum of the ranks in the transfers sample is 198.0. Test the null hypothesis that the distribution of mortgage lengths is the same in the transfers and new-hires populations against a general research hypothesis. State a bound on the p-value.

R102 Test the null hypothesis of equal variances, using the data of Exercise R98. Use a two-sided research hypothesis.
a. Can the null hypothesis be rejected at $\alpha = .01$?
b. Is there any reason to suspect that the nominal α of .01 might be in error? Why?

R103 An income-tax preparation service tests three microcomputer programs designed to help its staff prepare state and federal tax returns. Random samples of 10 experienced and 10 inexperienced preparers are assigned to each of three programs. The time required to prepare a standard return is obtained for each of the 60 preparers. The following means and (standard deviations) are obtained:

Program	Experienced	Inexperienced
A	36.60 (5.190)	40.50 (3.342)
B	28.30 (6.201)	41.70 (3.164)
C	34.00 (6.600)	40.60 (4.971)

a. Verify that SS(Program) = 129.7 and that SS(Error) = 1393.5.
b. Is there a statistically significant difference among the program means, using $\alpha = .05$?

R104 Which pairs of program means in Exercise R103, if any, are significantly different, using $\alpha = .05$?

R105 Refer to the means of Exercise R103. Construct a profile plot. Is there any reason to think that the overall program means might not be a good indication of the relative merits of the programs? It might be relevant to know that the service has relatively low turnover among its preparers; only about 15% of the preparers are inexperienced.

R106 Is there evidence of a nonconstant variance problem in Exercise R103?

R107 An analyst of microcomputer software firms keeps track of the number of test cycles of new programs before release for public sale. The following frequencies are obtained:

Number of cycles:	1	2	3	4	5+	
Frequency:	39	16	5	3	12	(total = 75)

One theory indicates that the probability that there are exactly x cycles is $.4(.6)^{x-1}$, for $x = 1, 2, \ldots$. Are the actual frequencies consistent with this theory if one allows a .05 risk of Type 1 error?

R108 One possible objection to the test used in Exercise R107 is that there are fewer than five observations in the $x = 4$ category. Is this a valid objection?

R109 In a study of a robot welding arm, data are collected on the height of the center of the weld. The desired height is 36 inches; the desired standard deviation is no more than .100 inch. In a sample of 29 welds, the mean height is 36.021 inches and the standard deviation is .112 inch. Calculate a 90% confidence interval for the population standard deviation.

R110 Using the data of Exercise R109, test the null hypothesis that the population standard deviation is less than or equal to .100. State bounds on the resulting p-value.

R111 It is claimed that the result of Exercise R110 proves that the robot welding arm is operating within tolerance, at least as far as variability is concerned. Is this claim justified?

R112 Examination of the data of Exercise R109 indicates that a very large fraction of the welds are within .05 inch of the target, but also that a few are .2 inch or more away from the target. Does this fact indicate that there is a problem with using the method of Exercises R109 and R110?

R113 Data are collected on the price to earnings ratio (P/E) of common stocks of companies in two industries: electric utilities and computer services. The data are

Utilities:	5.8	6.6	6.1	5.7	5.4	6.0	6.0	5.5	5.9	5.6	6.0	6.0	5.8
	5.8	5.9	6.2	6.2	6.0	6.0	6.1	($n = 20$, $\bar{y} = 5.92$, $s = .291$)					
Computer services:	6.2	6.7	5.9	6.8	6.7	6.2	6.5	6.5	6.8	6.6	7.5	6.2	7.9
	6.4	6.9	4.7	6.9	6.5	6.5	6.4	($n = 20$, $\bar{y} = 6.54$, $s = .624$)					

Basic financial concepts suggest that the mean P/E should be higher for the computer service industry than for utilities. Do the data support this hypothesis? Use the appropriate t method. State a bound on the p-value.

R114 Test the hypothesis in Exercise R113 using rank methods. Again, state a bound on the p-value.

R115 From the appearance of the data in Exercise R113, is a t test or a rank test more appropriate? How much difference does the choice of test make in the conclusion?

R116 The P/E ratios should be more variable in the computer services stocks than in the utilities in Exercise R113. Is this hypothesis supported by the data, using $\alpha = .01$?

R117 What are the critical assumptions underlying your method in Exercise R116? Do these assumptions appear reasonable?

R118 A package-delivery company tests its current dispatching rule against a computerized rule. One of the methods is selected randomly for use on a given day; a key customer records the service as excellent, good, fair, or poor for each day. The following frequencies are obtained:

	Excellent	Good	Fair	Poor	Total
Current rule	36	39	15	10	100
Computerized rule	48	42	8	2	100

a. Calculate a 95% confidence interval for the difference of proportions of ratings of excellent between the current and computerized rules.
b. Use this interval to test the null hypothesis of equal proportions. What conclusion can be reached?

R119 Perform a formal hypothesis test for the null hypothesis of equal proportions of ratings of excellent using the data of Exercise R118. Assume that $\alpha = .05$.

R120 In gathering the data of Exercise R118, it is noted that there may be carryover effects from one day to the next such that one poor day may tend to be followed by another poor day. If in fact there are carryover effects, does this violate any of the assumptions underlying the method of Exercises R118 and R119?

R121 Using the data of Exercise R118, test the null hypothesis that the distribution of opinion is the same for the current rule as for the computerized rule. State bounds on the p-value.

R122 Other than the potential problem cited in Exercise R120, are there any violations of assumptions for the test in Exercise R121?

R123 A law firm tests two models of printers for use in its office. A random sample of 20 documents is chosen; each document is printed out by each printer. The time required (in seconds) is recorded. The data are

	Document													
	1	2	3	4	5	6	7	8	9	10	11	12	13	14
Printer A	24	40	16	28	28	43	18	25	19	17	17	21	37	25
Printer B	22	36	29	21	20	36	16	27	15	13	11	13	30	20
A − B	2	4	−13	7	8	7	2	−2	4	4	6	8	7	5

	15	16	17	18	19	20	$\bar{y}$	s
Printer A	43	22	38	30	32	41	28.20	9.32
Printer B	36	23	29	24	25	30	23.80	7.83
A − B	7	−1	9	6	7	11	4.40	5.21

a. Perform an appropriate t test for the null hypothesis of equal means against a two-sided research hypothesis. Use $\alpha = .05$.
b. State a bound on the p-value for the test in part (a). Can one safely come to a conclusion about the relative speed of the two printers?

R124 Use a rank test for the data of Exercise R123 to test the null hypothesis that the mean difference is zero. Does this test give the same conclusion as the t test of Exercise R123?

R125 Compare the standard deviation of the differences to the A and B standard deviations in Exercise R123. What does this comparison indicate about the desirability of printing out the same 20 documents on both printers rather than using one set of 20 documents for one printer and a different set for the other printer?

R126 a. Use the data of Exercise R123 to calculate a 95% confidence interval for the standard deviation of the differences.

b. Is there any indication from Exercise R123 that there is an important violation of an assumption of the procedure in part (a)?

R127 As part of a performance review, junior managers are rated on a 50-point scale of managerial potential. The following results are obtained:

Undergraduate Major	n	$\bar{y}$	s
Business	12	21.17	8.26
Engineering	9	15.44	7.32
Liberal arts	18	27.39	8.15

Source	SS	d.f.	MS
Between	897.7	2	448.9
Within	2306.2	36	64.1
Total	3203.9	38	

a. Verify the computation of the sums of squares.

b. Is there a statistically significant difference among the means, using $\alpha = .05$?

R128 The rank sums for the data of Exercise R127 are business, 217.0; engineering, 101.0; and liberal arts, 462.0. Perform the appropriate rank test of the null hypothesis that the distribution of managerial potential is the same in the three groups.

R129 Examination of the data in Exercise R127 shows a definite right-skewness in all three samples. What does this fact indicate about the relative appropriateness of the tests in Exercises R127 and R128?

R130 a. What is the appropriate method for making pairwise comparisons of the means in Exercise R127?

b. Are any pairs of means significantly different, using $\alpha = .05$?

R131 The scale underlying the data of Exercise R127 is interpreted such that a score under 20 indicates little potential, a score between 20 and 29 indicates some potential, and a score above 30 indicates high potential. Examination of the data give the following frequencies:

	Potential			
	Little	Some	High	Total
Business	7	3	2	12
Engineering	8	0	1	9
Liberal arts	2	9	7	18
Total	17	12	10	39

Test the null hypothesis that rated managerial potential is unrelated to type of education. Use $\alpha = .05$.

R132 Is there evidence that the nominal α probability in Exercise R131 is a poor approximation? How critical is the quality of the approximation to the general conclusion?

13

LINEAR REGRESSION AND CORRELATION METHODS

One of the most important uses of statistics for managers is prediction. A manager may want to forecast the cost of a particular contracting job given the size of that job, to forecast the sales of a particular product given the current rate of growth of gross national product, or to forecast the number of parts that will be produced given a certain size work force. The statistical method most widely used in making predictions is **regression analysis**.

regression analysis

In the regression approach, past data on the relevant variables are used to develop and evaluate a prediction equation. The variable that is being predicted by this equation is the **dependent variable**; a variable that is used to make the prediction is an **independent variable**. In this chapter we discuss regression methods involving a single independent variable. In Chapter 14 we extend these methods to multiple regression, which is the case of more than one independent variable.

dependent and independent variables

There are a number of tasks which can be accomplished in a regression study.

1. The data can be used to obtain a prediction equation.
2. The data can be used to estimate the amount of variability or uncertainty around the equation.
3. Because the data are only a sample, inferences can be made about the true population values for the regression quantities.
4. The prediction equation can be used to predict a reasonable range of values for future values of the dependent variable.
5. The data can be used to estimate the degree of correlation between dependent and independent variables, which indicates how strong the relation is.

In this chapter these tasks are carried out for the case of one independent variable.

Like any statistical method, regression analysis is based on assumptions. We begin in Section 13.1 by describing these assumptions. Methods for estimating the prediction equation and estimating the variability around it are given in Section 13.2. Basic inference methods for regression are discussed in Section 13.3. In Section 13.4 we deal with prediction of future values of the dependent variable. Section 13.5 contains methods for assessing correlation. Finally in Section 13.6 we examine correlation based on ranks, an approach that is sometimes useful when assumptions underlying the ordinary correlation coefficient aren't met.

13.1 ASSUMPTIONS IN LINEAR REGRESSION PROBLEMS

In this chapter we consider **linear** regression analysis, in which the equation for predicting a dependent y given an independent variable x is the equation of a straight line. Suppose, for example, that the director of a county highway department wants to predict the cost of a resurfacing contract that is up for bids. Cost could be predicted as a function of the miles of road to be resurfaced, and specifically as a linear function. Let y = total cost of a project, x = number of miles to be resurfaced, and $\hat{y}$ = predicted cost of a project. To say that the predicted y is a linear function of the x value is to say $\hat{y} = \hat{\beta}_0 + \hat{\beta}_1 x$, where $\hat{\beta}_0$ is the **intercept** term and $\hat{\beta}_1$ is the **slope** (see Figure 13.1). The quantity $\hat{\beta}_1$ is the predicted change in y per unit increase in x, and $\hat{\beta}_0$ is the predicted value of y when $x = 0$. In accounting terms, $\hat{\beta}_0$ is the fixed cost of the resurfacing project and $\hat{\beta}_1$ is the variable cost per mile. To obtain a prediction of the cost of resurfacing projects, the highway department director needs to use sample data to determine the slope $\hat{\beta}_1$ and intercept $\hat{\beta}_0$.

intercept
slope

The basic idea of linear regression analysis is to fit a prediction line relating a dependent variable Y to an independent variable X. The first and most crucial assumption is that the underlying relation between Y and X is, in fact, linear. Suppose, for example, that we are predicting Y, the cost of a road resurfacing project, from X, the mileage involved. The

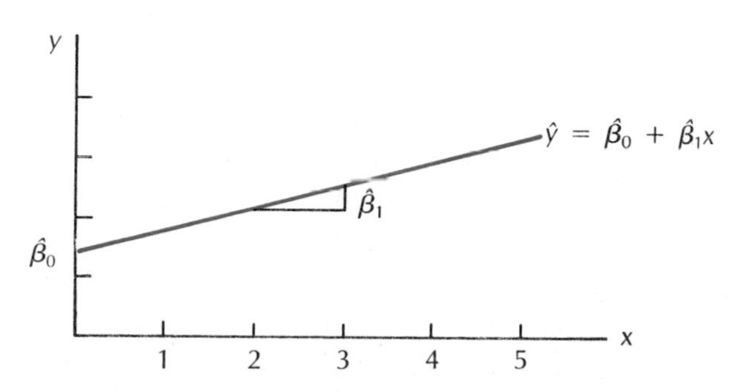

FIGURE 13.1 Linear Prediction Function

slope of this regression equation can be interpreted as the variable cost per mile, while the intercept is the fixed cost. According to the **assumption of linearity**, the slope of the equation does not change as mileage changes.* In economic terms, we are assuming that there are no economies or diseconomies from projects of longer mileage. There is little point in doing a regression study unless this assumption makes sense, at least roughly, both in the economic or practical situation and on examination of a plot of the data.

linearity assumption

For example, one of the most important factors in determining the mileage per gallon of gasoline yielded by a car is the car's weight. A prediction equation with weight as the independent variable and miles per gallon as the dependent variable might be adequately approximated as linear. It is also possible that taking gallons per mile as the dependent variable would yield a more nearly linear relationship.

Assuming linearity, we would like to write $y = \beta_0 + \beta_1 x$. However, according to such an equation y is an exact linear function of x; no room is left for the inevitable errors (deviations of actual y values from their predicted values). Therefore, corresponding to each y we introduce a **random error term** ϵ_i and assume the model

random error term

$$Y_i = \beta_0 + \beta_1 x_i + \epsilon_i$$

The random variable Y is assumed to be made up of a predictable part (a linear function of x) and an unpredictable part (the random error ϵ_i). The coefficients β_0 and β_1 are interpreted as the true underlying intercept and slope. The error term ϵ_i includes the effects of all other factors, known or unknown. In the road resurfacing project, unpredictable factors such as strikes, weather conditions, and equipment breakdowns would contribute to ϵ_i; so would factors, such as hilliness or pre-repair condition of the road, that might have been used in prediction but were not. The combined effects of unpredictable and ignored factors yield the random error terms ϵ_i.

For example, one way to measure the mileage per gallon of gas attained by various new cars would be to assign each car to a different driver, say, for a one-month period. What unpredictable and ignored factors might contribute to prediction error? Unpredictable (random) factors in this study would include the driving habits and skills of the drivers, the type of driving done (city vs. highway), and the number of stoplights encountered. Factors that would be ignored in a regression analysis of mileage and weight would include engine size and type of transmission (manual vs. automatic).

In regression studies, the values of the independent variable (the x_i values) are usually taken as predetermined constants, so the only source of randomness is the ϵ_i terms. While most economic and business applications have fixed x_i values, this is not always the case. For example, suppose that x_i is the score of an applicant on an aptitude test and Y_i is the productivity of the applicant. If the data are based on a random sample of applicants, X_i (as well as Y_i) is a random variable. The question of fixed versus random in regard to X is not crucial for regression studies. If the X_i's are random, we can simply regard all probabilit', statements as conditional on the observed x_i's.

the assumption of fixed independent variables

* In terms of calculus, the first derivative of the equation (of total cost with respect to mileage) is assumed to be constant.

When we assume that the x_i's are constants, the only random portion of the model for Y_i is the random error term ϵ_i. The following formal assumptions are made:

Formal Assumptions of Regression Analysis

1. The errors all have expected value zero; $E(\epsilon_i) = 0$ for all i.
2. The errors all have the same variance; $\text{Var}(\epsilon_i) = \sigma_\epsilon^2$ for all i.
3. The errors are independent of each other.
4. The errors are all normally distributed; ϵ_i is normally distributed for all i.

These are formal assumptions, made in order to derive the significance tests and prediction methods that follow. In practice, any or all of them may not hold exactly. We discuss detection and effects of deviations from these assumptions in Chapter 15.

The assumptions on ϵ_i may be translated into probability statements about the dependent variable $Y_i = \beta_0 + \beta_1 x_i + \epsilon_i$. Since x_i is a constant, so is $\beta_0 + \beta_1 x_i$. Adding a constant $(\beta_0 + \beta_1 x_i)$ to a normally distributed random variable (ϵ_i) yields another normally distributed random variable (Y_i). From properties of expectations, it follows that $E(Y_i) = E(\beta_0 + \beta_1 x_i + \epsilon_i) = \beta_0 + \beta_1 x_i$ and $V(Y_i) = V(\beta_0 + \beta_1 x_i + \epsilon_i) = \sigma_\epsilon^2$. By assumption then, **$Y_i$ is normally distributed with expected value $\beta_0 + \beta_1 x_i$ and variance σ_ϵ^2 and is independent of other Y values.** For example, suppose $\beta_0 = 1.5$, $\beta_1 = 2.5$, and $\sigma_\epsilon^2 = 6.1$. Then for $x_i = 2.0$, Y_i is normal with mean $1.5 + 2.5(2.0) = 6.5$ and variance 6.1. For $x_i = 3.0$, Y_i is normal with mean $1.5 + 2.5(3.0) = 9.0$ and the same variance, 6.1. Figure 13.2 shows that there is a theoretical normal distribution of Y's at each x value. The expected values for these theoretical distributions are assumed to be along the true regression line $y = \beta_0 + \beta_1 x$; the variances are all the same.

In practice, only sample data are available. The population parameters β_0, β_1, and σ_ϵ^2 all have to be estimated from limited sample data. The formal assumptions made in this section allow us to make inferences about the true parameter values from the sample data.

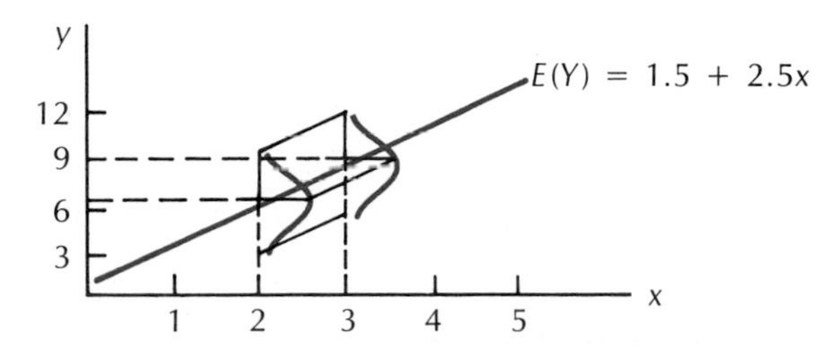

FIGURE 13.2 Theoretical Distribution of Y in Regression

13.2 ESTIMATING MODEL PARAMETERS

The quantities β_0 and β_1 in the regression model

$$Y = \beta_0 + \beta_1 x + \epsilon$$

are population quantities. We must estimate these values from sample data. The error variance σ_ϵ^2 is another population parameter that must be estimated. The first regression problem is to obtain estimates of the slope, intercept, and variance; we discuss how to do so in this section.

The road resurfacing example of Section 13.1 is a convenient illustration. Suppose the following data for similar resurfacing projects in the recent past are available:

Cost y_i (in thousands of dollars):	6.0	14.0	10.0	14.0	26.0
Mileage x_i (in miles):	1.0	3.0	4.0	5.0	7.0

scatter plot

A first step in examining the relation between y and x is to plot the data as a **scatter plot**, where each point represents the (x, y) coordinates of one data entry, as in Figure 13.3. The plot makes it clear that there is an imperfect but generally increasing relation between x and y. A straight-line relation appears plausible.

The regression analysis problem is to find the best straight-line prediction. The criterion for "best" is based on squared prediction error. We find the equation of the prediction line, that is, the slope $\hat{\beta}_1$ and intercept $\hat{\beta}_0$ that minimize the total squared prediction error. The method that accomplishes this goal is called the **least-squares** method, because it chooses $\hat{\beta}_0$ and $\hat{\beta}_1$ to minimize the quantity

least-squares

$$\sum_i (y_i - \hat{y}_i)^2 = \sum_i [y_i - (\hat{\beta}_0 + \hat{\beta}_1 x_i)]^2$$

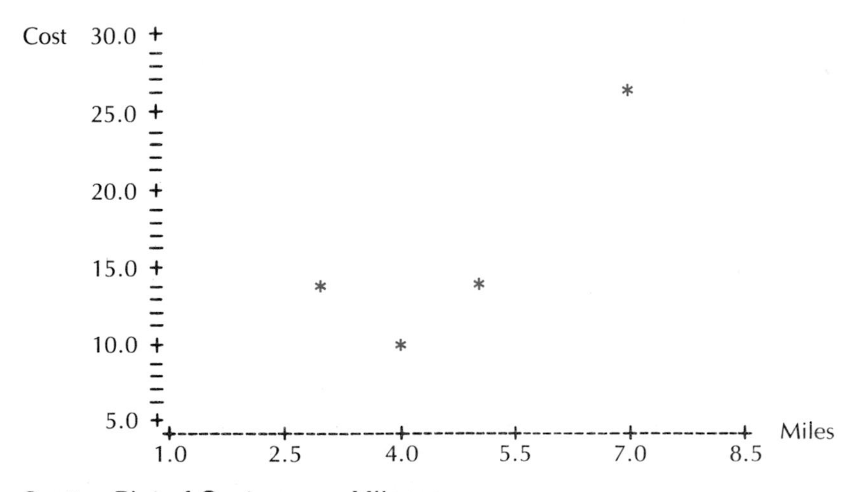

FIGURE 13.3 **Scatter Plot of Cost versus Mileage**

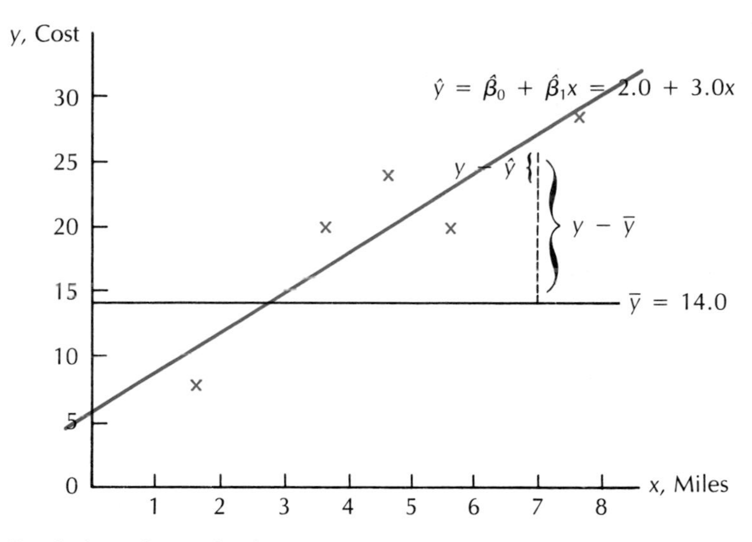

FIGURE 13.4 **Deviations from the Least-squares Line and from the Mean**

The prediction errors are shown on the plot of Figure 13.4 as vertical deviations from the line. The deviations are taken as vertical distances because we're trying to predict y values and errors should be taken in the y direction. For these data the least-squares line can be shown to be $\hat{y} = 2.0 + 3.0x$; one of the deviations from it is indicated by the smaller brace. For comparison, the mean $\bar{y} = 14.0$ is also shown; deviation from the mean is indicated by the larger brace. The least-squares estimates are obtained as follows:

$$\hat{\beta}_1 = \frac{S_{xy}}{S_{xx}} \quad \text{and} \quad \hat{\beta}_0 = \bar{y} - \hat{\beta}_1 \bar{x}$$

where

$$S_{xy} = \sum_i x_i y_i - \frac{\sum_i x_i \sum_i y_i}{n} \quad \text{and} \quad SS_{xx} = \sum_i x_i^2 - \frac{\left(\sum_i x_i\right)^2}{n}$$

For the road resurfacing data, $n = 5$ and

$$\sum x_i = 1.0 + \ldots + 7.0 = 20.0$$

Similarly,

$$\sum y_i = 70.0$$

Also

$$\sum x_i^2 = (1.0)^2 + \ldots + (7.0)^2 = 100.0$$

and

$$\sum x_i y_i = (1.0)(6.0) + \ldots (7.0)(26.0) = 340.0$$

Therefore,

$$S_{xy} = 340.0 - \frac{(20.0)(70.0)}{5} = 60.0$$

and

$$S_{xx} = 100.0 - \frac{(20.0)^2}{5} = 20.0$$

So

$$\hat{\beta}_1 = \frac{60.0}{20.0} = 3.0$$

and

$$\hat{\beta}_0 = \frac{70.0}{5} - (3.0)\frac{20.0}{5} = 2.0$$

The equation $\hat{y} = 2.0 + 3.0x$ is also shown in Figure 13.4.

EXAMPLE 13.1 Data from a sample of 10 pharmacies are used to examine the relation between prescription sales volume and the percentage of prescription ingredients purchased directly from the supplier. The sample data are shown here:

Pharmacy	Sales Volume, y (in \$1000)	% of Ingredients Purchased Directly, x
1	25	10
2	55	18
3	50	25
4	75	40
5	110	50
6	138	63
7	90	42
8	60	30
9	10	5
10	100	55

a. Find the least-squares estimates for the regression line $\hat{y} = \hat{\beta}_0 + \hat{\beta}_1 x$.
b. Predict sales volume for a pharmacy that purchases 15% of its prescription ingredients directly from the supplier.
c. Plot the x, y data and the prediction equation, $\hat{y} = \hat{\beta}_0 + \hat{\beta}_1 x$.

Solution a. The least-squares estimates can be obtained from the calculations performed in the following table:

	y	x	xy	x^2
	25	10	250	100
	55	18	990	324
	50	25	1,250	625
	75	40	3,000	1,600
	110	50	5,500	2,500
	138	63	8,694	3,969
	90	42	3,780	1,764
	60	30	1,800	900
	10	5	50	25
	100	55	5,500	3,025
Totals	713	338	30,814	14,832

$$S_{xx} = \sum x^2 - \frac{(\sum x)^2}{n} = 14832 - \frac{(338)^2}{10} = 3407.6$$

$$S_{xy} = \sum xy = \sum xy - \frac{(\sum x)(\sum y)}{n} = 30814 - \frac{(338)(713)}{10} = 6714.6$$

Substituting into the formulas for $\hat{\beta}_0$ and $\hat{\beta}_1$,

$$\hat{\beta}_1 = \frac{S_{xy}}{S_{xx}} = \frac{6714.6}{3407.6} = 1.9704778, \text{ rounded to } 1.97$$

$$\hat{\beta}_0 = \bar{y} - \hat{\beta}_1\bar{x} = 71.3 - 1.9704778(33.8) = 4.6978519, \text{ rounded to } 4.70$$

b. When $x = 15\%$, the predicted sales volume is $\hat{y} = 4.70 + 1.97(15) = 34.25$ (i.e., \$34,250). The prediction equation is shown in Figure 13.5. □

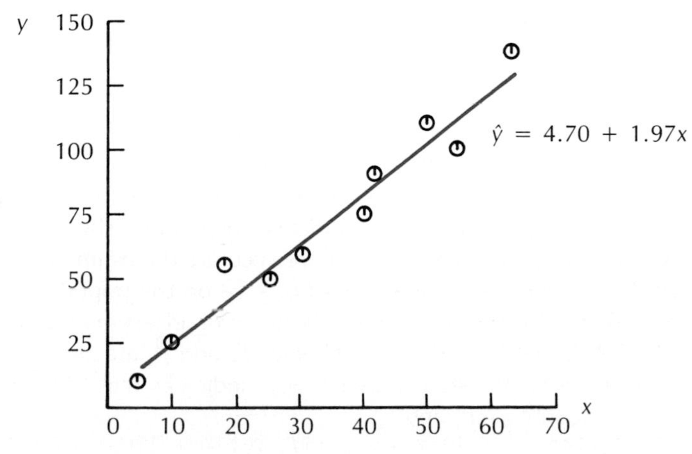

FIGURE 13.5 **Sample Data and Least-squares Prediction Equation for Example 13.1**

EXAMPLE 13.2 Use the SAS output shown below to identify the least-squares estimates for the road resurfacing data.

DEPENDENT VARIABLE: TOTCOST

SOURCE	DF	SUM OF SQUARES	MEAN SQUARE	F VALUE	PR > F	R-SQUARE	C.V.
MODEL	1	180.00000000	180.00000000	12.27	0.0394	0.803571	27.3551
ERROR	3	44.00000000	14.66666667		ROOT MSE		TOTCOST MEAN
CORRECTED TOTAL	4	224.00000000			3.82970843		14.00000000

SOURCE	DF	TYPE I SS	F VALUE	PR > F	DF	TYPE III SS	F VALUE	PR > F
MILES	1	180.00000000	12.27	0.0394	1	180.00000000	12.27	0.0394

PARAMETER	ESTIMATE	T FOR H0: PARAMETER=0	PR > |T|	STD ERROR OF ESTIMATE
INTERCEPT	2.00000000	0.52	0.6376	3.82970843
MILES	3.00000000	3.50	0.0394	0.85634884

Solution The intercept is shown as $\hat{\beta}_0 = 2.000$. The slope (coefficient of $x =$ miles is $\hat{\beta}_1 = 3.000$. □

It is possible to prove that both estimators $\hat{\beta}_0$ and $\hat{\beta}_1$ are unbiased,

$$E(\hat{\beta}_0) = \beta_0, \qquad E(\hat{\beta}_1) = \beta_1$$

standard errors of $\hat{\beta}_0, \hat{\beta}_1$

and that in linear regression, the variances and standard errors for $\hat{\beta}_0$ and $\hat{\beta}_1$ are, respectively

$$\text{Var}(\hat{\beta}_0) = \sigma_\epsilon^2 \left[\frac{1}{n} + \frac{\bar{x}^2}{S_{xx}}\right], \qquad \sigma_{\hat{\beta}_0} = \sigma_\epsilon \sqrt{\frac{1}{n} + \frac{\bar{x}^2}{S_{xx}}}$$

and

$$\text{Var}(\hat{\beta}_1) = \sigma_\epsilon^2 \left[\frac{1}{S_{xx}}\right], \qquad \sigma_{\hat{\beta}_1} = \sigma_\epsilon \sqrt{\frac{1}{S_{xx}}}$$

Finally, it can be proved that the sampling distributions of $\hat{\beta}_0$ and $\hat{\beta}_1$ are both normal. These facts are most readily proved using matrix concepts (see Johnston, 1977).

The variances for $\hat{\beta}_0$ and $\hat{\beta}_1$ indicate their probable accuracy as estimates of the true parameter values. Both of these variances depend on the theoretical error variance σ_ϵ^2. The smaller the error variance, the more accurate the estimates. For the extreme case where all the errors ϵ_i are zero, all values of Y_i lie on the population regression line $Y_i = \beta_0 + \beta_1 x_i$ and $\sigma_\epsilon^2 = 0$. Obviously there is no error in estimating the values β_0 and β_1 from sample data, and the error variances for $\hat{\beta}_0$ and $\hat{\beta}_1$ are zero.

The formula for the standard error of $\hat{\beta}_1$ indicates that the variability of $\hat{\beta}_1$ is influenced by two quantities, σ_ϵ^2 and S_{xx}.

$$\sigma_{\hat{\beta}_1} = \frac{\sigma_\epsilon}{\sqrt{S_{xx}}}$$

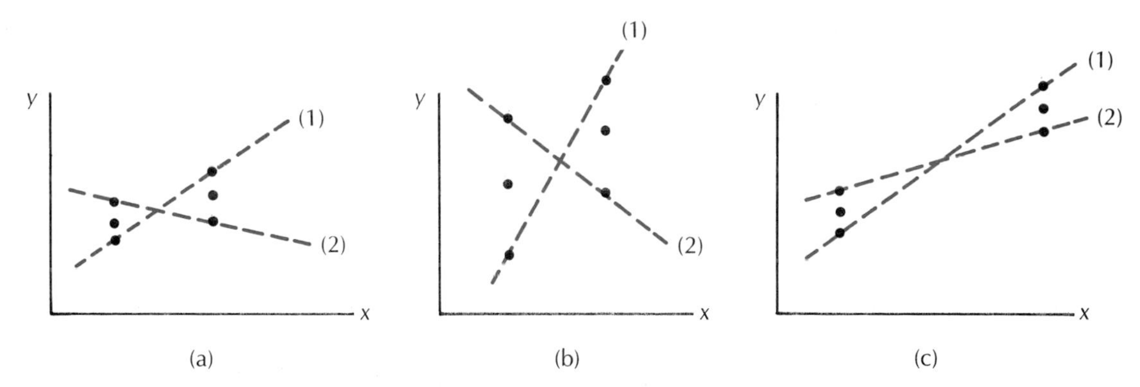

FIGURE 13.6 **The Effects of σ_ϵ and S_{xx} on $\sigma_{\hat{\beta}_1}$: (a) Small σ_ϵ, Small S_{xx}; (b) Large σ_ϵ, Small S_{xx}; (c) Small σ_ϵ, Large S_{xx}**

The greater the variability σ_ϵ of the y value for a given value of x, the larger $\sigma_{\hat{\beta}_1}$ is. The greater the spread in the x value (as measured by S_{xx}), the smaller $\sigma_{\hat{\beta}_1}$ is. These results are illustrated in Figure 13.6.

In each part of Figure 13.6, the two dashed lines represent the extremes for fitted lines "through" the sample observations. For example, by comparing dashed lines in Figure 13.6a and 13.6b, we see that the greater the variability in the y value for a given value of x, the greater the variability in the estimated slope. Similarly, by comparing the dashed lines in Figure 13.6a and 13.6c, we see that the greater the spread of the x values and hence the larger the value of S_{xx}, the smaller the variability in the estimated slope.

The variance of the estimated intercept $\hat{\beta}_0$ is influenced by n, naturally, and also by the size of $\bar{x}^2$ relative to S_{xx}. The predicted y value when $x = 0$ is represented by $\hat{\beta}_0$; if all the x_i are, for instance, large positive numbers, predicting Y at $x = 0$ is a huge extrapolation from the actual data. Such extrapolation magnifies small errors, and $\mathrm{Var}(\hat{\beta}_0)$ is large. The ideal situation for estimating $\hat{\beta}_0$ is when $\bar{x} = 0$.

EXAMPLE 13.3 Suppose that in a particular regression study $\sigma_\epsilon^2 = 6.1$. (In practice, the true error variance would not be known, so this is a hypothetical exercise.) The x values for the regression study are

18.1 20.0 20.8 21.5 22.0 22.4 22.9 24.0 25.4 27.3

Find the variances and standard errors for $\hat{\beta}_0$ and $\hat{\beta}_1$.

Solution For these data, $\sum x_i = 224.4$ and $\sum x_i^2 = 5099.12$. It follows that

$$\bar{x} = \frac{224.4}{10} = 22.44$$

and

$$S_{xx} = 5099.12 - \frac{(224.4)^2}{10} = 63.584$$

Substituting $\sigma_\epsilon^2 = 6.1$, $\bar{x} = 22.44$, and $S_{xx} = 63.584$, we have

$$\text{Var}(\hat{\beta}_0) = 6.1\left[\frac{1}{10} + \frac{(22.44)^2}{63.584}\right] = 48.92$$

and

$$\text{Var}(\hat{\beta}_1) = \frac{6.1}{63.584} = .0959, \qquad \sigma_{\hat{\beta}_1} = \sqrt{.0959} = .31$$

□

To this point, we have considered only the estimates of intercept and slope. We also have to estimate the true error variance σ_ϵ^2. This quantity can be thought of as "variance around the line," or as the mean squared prediction error. The estimate of σ_ϵ^2 is based on the **residuals** $y_i - \hat{y}_i$, which are the prediction errors in the sample. The estimate of σ_ϵ^2 based on the sample data is

residuals

$$s_\epsilon^2 = \frac{\sum_i (y_i - \hat{y}_i)^2}{n-2} = \frac{\text{SS(Residual)}}{n-2}$$

where SS again stands for sum of squares. The quantity SS(Residual) is sometimes called SS(Error). In the computer output for Example 13.2, SS(Error) is shown to be 44.0.

Just as we divide by $n - 1$ rather than n in the ordinary sample variance s^2 (in Chapter 2), we divide by $n - 2$ in s_ϵ^2, the estimated variance around the line. To see why, suppose our sample size is $n = 2$. No matter how large or small σ_ϵ^2 may be, the estimated regression line goes exactly through the two points and the residuals are automatically zero. Thus for $n = 2$ we simply don't have enough information to estimate σ_ϵ^2 at all. In our definition, s_ϵ^2 is undefined for $n = 2$, as it should be. Another argument for dividing by $n - 2$ is that

$$E(s_\epsilon^2) = \sigma_\epsilon^2$$

Dividing by $n - 2$ makes s_ϵ^2 an unbiased estimator of σ_ϵ^2. In the computer output of Example 13.2, $n - 2$ is shown as DF (degrees of freedom) for ERROR, and s_ϵ^2 is shown as MEAN SQUARE for ERROR. The formula for hand calculation of s_ϵ^2 is

$$s_\epsilon^2 = \frac{\text{SS(Residual)}}{n-2} = \frac{\sum y_i^2 - \hat{\beta}_0 \sum y_i - \hat{\beta}_1 \sum x_i y_i}{n-2}$$

As indicated previously, s_ϵ^2 is an unbiased estimator of σ_ϵ^2. The square root s_ϵ of the sample variance is called the **sample standard deviation around the regression line** or the **residual standard deviation.** Since s_ϵ estimates σ_ϵ, the standard deviation of Y_i, s_ϵ estimates the standard deviation of the population of y values associated with a given value of the independent variable x. The SAS output labels s_ϵ as STD. DEV.; for our Example 13.2, s_ϵ is 3.830.

The estimates $\hat{\beta}_0$, $\hat{\beta}_1$, and s_ϵ are basic in regression analysis. They specify the regression line and the probable degree of error associated with y values for a given value of x. The next step is to use these sample estimates to make inferences about the true parameters.

EXAMPLE 13.4 For the following data,

a. construct a scatter plot;
b. compute the least-squares estimates for β_0 and β_1 in the model $Y = \beta_0 + \beta_1 x + \epsilon$;
c. predict y for $x = 19.5$;
d. compute s_ϵ, the sample standard deviation about the regression line.

y:	31.5	33.1	27.4	24.5	27.0	27.8	23.3	24.7	16.9	18.1
x:	18.1	20.0	20.8	21.5	22.0	22.4	22.9	24.0	25.4	27.3

Solution a. A scatter plot is shown in Figure 13.7; the indicated line will be shown to be the least-squares line in part (b).

b. First we must calculate $\hat{\beta}_0$ and $\hat{\beta}_1$. Direct calculations yield $\sum x_i = 224.4$, $\sum x_i^2 = 5099.12$, $\sum y_i = 254.3$, $\sum y_i^2 = 6706.91$, and $\sum x_i y_i = 5595.30$. Substituting these values,

$$\hat{\beta}_1 = \frac{S_{xy}}{S_{xx}} = \frac{\sum x_i y_i - (\sum x_i)(\sum y_i)/n}{\sum x_i^2 - (\sum x_i)^2/n}$$

$$= \frac{5595.30 - (224.4)(254.3)/10}{5099.12 - (224.4)^2/10} = -1.7487$$

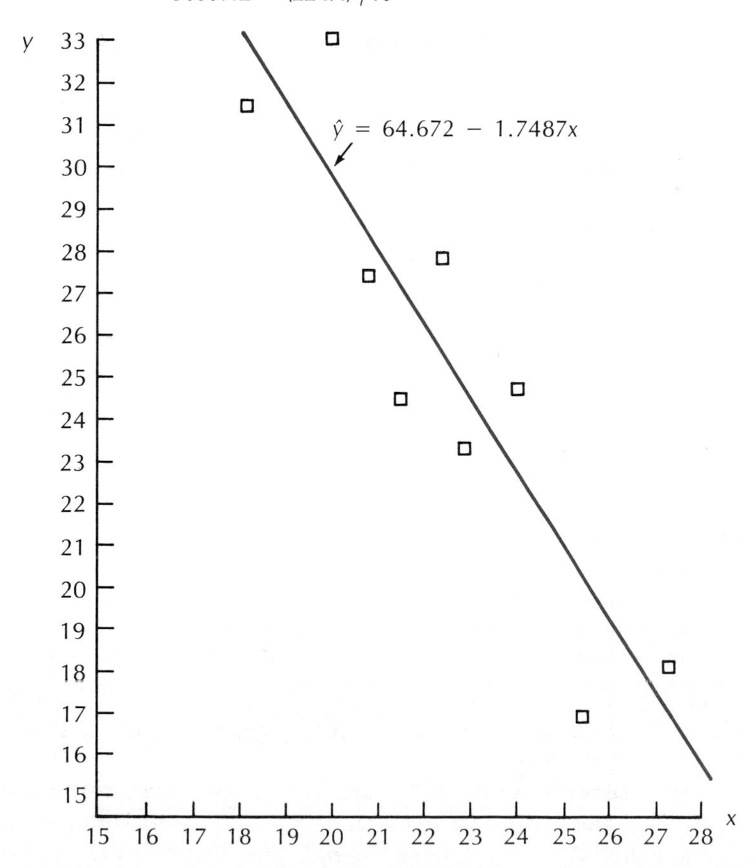

FIGURE 13.7 Scatter Plot for the Data of Example 13.4

and

$$\hat{\beta}_0 = \bar{y} - \hat{\beta}_1\bar{x}$$
$$= 25.43 - (-1.7487)(22.44) = 64.672$$

c. The least-squares prediction equation is

$$\hat{y} = 64.672 - 1.7487x$$

Substituting $x = 19.5$, the predicted value of y is $\hat{y} = 30.57$.

d. The hand-calculation formula for s_ϵ^2 yields

$$s_\epsilon^2 = \frac{\sum y_i^2 - \hat{\beta}_0 \sum y_i - \hat{\beta}_1 \sum x_i y_i}{n-2}$$

$$= \frac{6706.91 - (64.672)(254.3) - (-1.7487)(5595.30)}{10-2}$$

$$= 5.6652$$

Therefore, $s_\epsilon = \sqrt{s_\epsilon^2} = \sqrt{5.6652} = 2.380$.

Note that the prediction equation is shown in Figure 13.7. □

SECTIONS 13.1 AND 13.2 EXERCISES

13.1 The following data are obtained:

x:	1	1	1	3	3	3
$x' = \log_{10} x$:	.000	.000	.000	.477	.477	.477
y:	13.5	15.4	16.1	18.3	19.9	20.9
x:	5	5	5	7	7	7
$x' = \log_{10} x$:	.699	.699	.699	845	.845	.845
y:	20.8	23.1	22.1	22.8	24.9	24.5

a. Plot y versus x. Draw an approximate regression line through the plot.
b. Separately, plot y versus x'. Draw an approximate regression line through the plot.
c. Which plot appears more nearly linear to you?

13.2 Refer to the data of Exercise 13.1.
a. Calculate the least-squares equation $\hat{y} = \hat{\beta}_0 + \hat{\beta}_1 x$.
b. Calculate the residual standard deviation.

13.3 Refer to the data of Exercise 13.1.
a. Calculate the least-squares equation $\hat{y} = \hat{\beta}_0 + \hat{\beta}_1 x'$.
b. Calculate the residual standard deviation.

13.4 Compare the residual standard deviations s_ϵ in Exercises 13.2 and 13.3. Which is smaller? Does this confirm your opinion about the choice of model based on the plots of Exercise 13.1?

13.5 As one part of a study of commercial bank branches, data are obtained on the number of independent businesses (x) located in sample ZIP code areas and the number of bank branches (y) located in these areas. The commercial centers of cities are excluded.

x:	92	116	124	210	216	267	306	378	415	502	615	703
y:	3	2	3	5	4	5	5	6	7	7	9	9

$\sum x_i = 3{,}944$, $\sum y_i = 65$, $\sum x_i y_i = 26{,}208$

$\sum x_i^2 = 1{,}732{,}524$, $\sum y_i^2 = 409$, $n = 12$

a. Plot the data. Does a linear equation relating y to x appear plausible?
b. Calculate the regression equation (with y as the dependent variable).
c. Calculate the sample residual standard deviation s_ϵ.

13.6 Does it appear that variability of y increases with x in the data plot of Exercise 13.5? (This would violate the assumption of constant variance.)

13.3 INFERENCES ABOUT PARAMFTERS

The t distribution can be used to make significance tests and confidence intervals for the true slope and intercept. One natural null hypothesis is that the true slope $\beta_1 = 0$. If this H_0 is true, a change in x yields no predicted change in y, and it follows that x has no value in predicting y. We know from the previous section that the sample slope $\hat{\beta}_1$ has expected value β_1 and standard error

$$\sigma_{\hat{\beta}_1} = \sigma_\epsilon \sqrt{\frac{1}{S_{xx}}}$$

***t* test for β_1**

In practice, σ_ϵ is not known and must be estimated by s_ϵ in this formula. Combining these facts, we get the t statistic

$$t = \frac{\hat{\beta}_1 - \beta_1}{\text{estimated standard error } (\hat{\beta}_1)} = \frac{\hat{\beta}_1 - \beta_1}{s_\epsilon \sqrt{\frac{1}{S_{xx}}}}$$

The use of this statistic in testing H_0: $\beta_1 = 0$ is indicated in the following summary.

Summary for t Test of H_0: $\beta_1 = 0$

H_0: $\beta_1 = 0$
H_a: 1. $\beta_1 > 0$
2. $\beta_1 < 0$
3. $\beta_1 \neq 0$

T.S.: $t = \dfrac{\hat{\beta}_1 - 0}{s_\epsilon \sqrt{\frac{1}{S_{xx}}}}$

R.R.: For d.f. $= n - 2$ and Type I error α,
1. reject H_0 if $t > t_\alpha$
2. reject H_0 if $t < -t_\alpha$
3. reject H_0 if $|t| > t_{\alpha/2}$

EXAMPLE 13.5 Use the computer output of Example 13.2 to locate the value of the t statistic for testing H_0: $\beta_1 = 0$ in the road resurfacing example. Give the observed level of significance for the test.

Solution It is clear from the output that the value of the test statistic in the column labeled T FOR H0: PARAMETER = 0 is $t = 3.50$. The p-value for the two-tailed alternative H_a: $\beta_1 \neq 0$, labeled as PR > |T|, is .0394. Because this value is small, we can reject the hypothesis that mileage has no effect in predicting cost. □

EXAMPLE 13.6 The following data show mean ages of executives of 15 firms in the food industry and the previous year's percentage increase in earnings per share of the firms. Test the hypothesis that executive age has no predictive value. Should a one-sided or two-sided alternative be used? (Davis [1979] reports on this topic, but the data are hypothetical.)

Mean age, x:	38.2	40.0	42.5	43.4	44.6	44.9	45.0	45.4
Change, earnings per share, y:	8.9	13.0	4.7	−2.4	12.5	18.4	6.6	13.5
x:	46.0	47.3	47.3	48.0	49.1	50.5	51.6	
y:	8.5	15.3	18.9	6.0	10.4	15.9	17.1	

Solution In the model $Y = \beta_0 + \beta_1 x + \epsilon$, the null hypothesis is H_0: $\beta_1 = 0$. As Davis points out, the myth in American business is that younger managers tend to be more aggressive and harder driving, but it is also possible that the greater experience of the older executives leads to better decisions. Therefore there is good reason to choose a two-sided research hypothesis, H_a: $\beta_1 \neq 0$.

For the t test, we need $\hat{\beta}_1$, s_ϵ, and S_{xx}. The necessary summary figures are

$$\sum x_i = 683.8, \quad \sum y_i = 167.3, \quad \sum x_i y_i = 7741.74$$
$$\sum x_i^2 = 31358.58, \quad \sum y_i^2 = 2349.61, \quad n = 15$$

$$\hat{\beta}_1 = \frac{\sum x_i y_i - (\sum x_i)(\sum y_i)/n}{\sum x_i^2 - (\sum x_i)^2/n}$$

$$= \frac{115.09067}{186.41733} = .6173818$$

Note that

$$S_{xx} = \sum x_i^2 - (\sum x_i)^2/n = 186.41733$$

and that

$$\hat{\beta}_0 = \bar{y} - \hat{\beta}_1 \bar{x} = -16.991045$$

The interpretation of $\hat{\beta}_0$ is rather interesting in this example; younger is not necessarily better.

Now

$$s_\epsilon^2 = \frac{\sum y_i^2 - \hat{\beta}_0 \sum y_i - \hat{\beta}_1 \sum x_i y_i}{n - 2}$$

$$= \frac{412.60245}{13} = 31.73865$$

and $s_\epsilon = \sqrt{31.73865} = 5.6337066$.

Therefore

$$t = \frac{\hat{\beta}_1}{s_\epsilon\sqrt{\frac{1}{S_{xx}}}} = \frac{.6173818}{5.6337066\sqrt{\frac{1}{186.41733}}} = 1.4962451$$

or about 1.50. The tabulated t value (13 d.f., two-sided $\alpha = .10$) is 1.771. Therefore we cannot reject H_0 even at $\alpha = .10$. □

It is also possible to calculate a confidence interval for the true slope:

100(1 − α)% Confidence Interval for Slope β_1

$$\hat{\beta}_1 - t_{\alpha/2}s_\epsilon\sqrt{\frac{1}{S_{xx}}} < \beta_1 < \hat{\beta}_1 < t_{\alpha/2}s_\epsilon\sqrt{\frac{1}{S_{xx}}}$$

EXAMPLE 13.7 Compute a 95% confidence interval for the slope β_1 using the output from Example 13.2.

Solution In the SAS output the estimated standard error of $\hat{\beta}_1$ is $s_\epsilon\sqrt{1/S_{xx}} = .856$, rounded off. The corresponding confidence interval β_1 is then

$$3.00 \pm 3.182(.856), \quad \text{or} \quad .276 \text{ to } 5.724$$

The predicted cost per additional mile of resurfacing could be anywhere from $276 to $5724. The large width of this interval results largely from the t value associated with the small sample size. □

There is an alternative test, an F test, for the null hypothesis of no predictive value. This test gives the same result as a two-sided t test of H_0: $\beta_1 = 0$ in simple linear regression. The F test is summarized next.

F Test for H_0: $\beta_1 = 0$

H_0: $\beta_1 = 0$
H_a: $\beta_1 \neq 0$

$$\text{T.S.: } F = \frac{\text{SS(Regression)}/1}{\text{SS(Residual)}/(n-2)} = \frac{\text{MS(Regression)}}{\text{MS(Residual)}}$$

R.R.: with $\text{d.f.}_1 = 1$ and $\text{d.f.}_2 = n - 2$, reject H_0 if $F > F_\alpha$

Note: $\text{SS(Regression)} = \hat{\beta}_0 \sum y_i + \hat{\beta}_1 \sum x_i y_i - (\sum y_i)^2/n$ and $\text{SS(Residual)} = \sum y_i^2 - \hat{\beta}_0 \sum y_i - \hat{\beta}_1 \sum x_i y_i$

Virtually all computer packages calculate this F statistic. In the road resurfacing example, the SAS output shows $F = 12.27$ with a p-value of .0394. Again the hypothesis of no predictive value can be rejected. It is always true for simple linear regression problems that $F = t^2$; in the example, $12.27 \approx (3.50)^2$. Therefore the F and two-sided t tests are equivalent here.

EXAMPLE 13.8 For the data of Example 13.4, use the F test for H_0: $\beta_1 = 0$. Show that $t^2 = F$.

Solution The required sums of squares are

$$\begin{aligned}\text{SS(Regression)} &= \hat{\beta}_0 \sum y_i + \hat{\beta}_1 \sum x_i y_i - (\textstyle\sum y_i)^2/n \\ &= (64.672)(254.3) + (-1.7487)(5595.30) - (254.3)^2/10 \\ &= 194.74 \\ \text{SS(Residual)} &= \textstyle\sum y_i^2 - \hat{\beta}_0 \sum y_i - \hat{\beta}_1 \sum x_i y_i \\ &= 6706.91 - (64.672)(254.3) - (-1.7487)(5595.30) \\ &= 45.32.\end{aligned}$$

Therefore,

$$F = \frac{\text{SS(Regression)}/1}{\text{SS(Residual)}/(n-2)} = \frac{194.74/1}{45.32/8}$$

$$= 34.37$$

This value of the F statistic lies beyond all tabled values with d.f.$_1 = 1$ and d.f.$_2 = 8$. Hence we may reject H_0. For the t statistic, we have

$$t = \frac{\hat{\beta}_1}{s_\epsilon \sqrt{1/S_{xx}}} = \frac{-1.7487}{2.380\sqrt{1/63.58}} = -5.86$$

Note that $F = 34.37 = t^2 = (-5.86)^2$. □

You should be able to work out comparable hypothesis-testing and confidence-interval formulas for the intercept β_0 using the estimated standard error of $\hat{\beta}_0$ as $s_\epsilon\sqrt{1/n + \bar{x}^2/S_{xx}}$. Often, in practice, this parameter is of less interest than the slope. In particular, there often is no reason to hypothesize that the true intercept is zero (or any other particular value). Computer packages almost always test H_0: $\beta_1 = 0$, but some don't bother with a test on the intercept term.

It is also possible to calculate a confidence interval for the true error variance σ_ϵ^2. The only change from the χ^2 variance-inference methods of Chapter 11 is in degrees of freedom. Now s_ϵ^2 has $n - 2$ d.f.

$100(1-\alpha)\%$ Confidence Interval for σ_ϵ^2

$$s_\epsilon^2 \frac{n-2}{\chi^2_{\alpha/2}} < \sigma_\epsilon^2 < s_\epsilon^2 \frac{n-2}{\chi^2_{1-\alpha/2}}$$

where χ_a^2 cuts off area a in the right tail of the χ^2 distribution with $n - 2$ d.f. Note: $(n-2)s_\epsilon^2 = \text{SS(Residual)}$.

EXAMPLE 13.9 For the data of Example 13.6, calculate 95% confidence intervals for σ_ϵ^2 and for σ_ϵ.

Solution The tabled $\chi^2_{.025}$ and $\chi^2_{.975}$ values for $n - 2 = 13$ d.f. are, respectively, 24.74 and 5.01. Since s_ϵ^2 is 31.73865 in Example 13.6, the 95% confidence interval for σ_ϵ^2 is

$$(31.73865)\left(\frac{13}{24.74}\right) < \sigma_\epsilon^2 < (31.73865)\left(\frac{13}{5.01}\right)$$

or

$$16.677544 < \sigma_\epsilon^2 < 82.355778$$

The confidence interval for σ_ϵ may be found by taking square roots:

$$4.0838149 < \sigma_\epsilon < 9.0750085$$

□

SECTION 13.3 EXERCISES

13.7 Refer to the data of Exercise 13.5.

a. Calculate a 90% confidence interval for β_1.
b. What is the interpretation of H_0: $\beta_1 = 0$ in Exercise 13.5?
c. What is the natural research hypothesis H_a for that problem?
d. Do the data support H_a at $\alpha = .05$?

13.8 Find the p-value of the test of H_0: $\beta_1 = 0$ for Exercise 13.8.

13.9 Calculate a 95% confidence interval for σ_ϵ using the data of Exercise 13.5.

13.10 A firm that prints automobile bumper stickers investigates the relation between the total direct cost of a lot of stickers and the number produced in the printing run. The data are analyzed by a standard computer package (BMDP). The relevant output is shown below:

```
MULTIPLE R              0.9982            STD. ERROR OF EST.          12.2068
MULTIPLE R-SQUARE       0.9964

ANALYSIS OF VARIANCE
                     SUM OF SQUARES     DF     MEAN SQUARE          F RATIO    P(TAIL)
          REGRESSION  1167746.5000       1    1167746.5000         7836.946     0.0000
          RESIDUAL       4172.1484      28        149.0053

                                                  STD. REG
   VARIABLE          COEFFICIENT   STD. ERROR       COEFF          T    P(2 TAIL) TOLERANCE

INTERCEPT              99.77704
RUNSIZE       1        51.91785      0.58647        0.998      88.527     0.0000     1.00000
```

a. Plot the data; do you detect any difficulties with using a linear regression model? Can you see any blatant violations of assumptions? The raw data are

Runsize:	2.6	5.0	10.0	2.0	.8	4.0	2.5	.6	.8	1.0	2.0	
Total cost:	230	341	629	187	159	327	206	124	155	147	209	
Runsize:	3.0	.4	.5	5.0	20.0	5.0	2.0	1.0	1.5	.5	1.0	1.0
Total cost:	247	135	125	366	1146	339	208	150	179	128	155	143
Runsize:	.6	2.0	1.5	3.0	6.5	2.2	1.0					
Total cost:	131	219	171	258	415	226	159					

b. Write the estimated regression equation indicated in the output. Find the residual standard deviation.

c. Calculate a 95% confidence interval for the true slope. What are the interpretations of the intercept and slope in this problem?

13.11 Refer to the computer output of Exercise 13.10.

a. Locate the value of the t statistic for testing H_0: $\beta_1 = 0$.

b. Locate the p-value for this test. Is the p-value one-tailed or two-tailed? If necessary, calculate the p-value for the appropriate number of tails.

13.12 Refer to the computer output of Exercise 13.10.

a. Locate the value of the F statistic and the associated p-value.

b. How do the p-values for this F test and the t test of Exercise 13.11 compare? Why should this relation hold?

13.13 Calculate a 90% confidence interval for σ_ϵ using the information in Exercise 13.10.

13.4 PREDICTING NEW y VALUES USING REGRESSION

In all the regression analyses done so far, we have been summarizing and making inferences about relations in data that have already been observed. Thus we've been predicting the past. One of the most important uses of regression is in trying to forecast the future. In the road resurfacing example, the county highway director wants to predict the cost of a new contract that is up for bids. In this section we discuss how to make such regression forecasts and how to determine the plus-or-minus probable error factor.

There are two possible interpretations of a y prediction based on a given x. Suppose that the highway director substitutes $x = 6$ miles in the regression equation $\hat{y} = 2.0 + 3.0x$ and gets $\hat{y} = 20$. This can be interpreted as either

"the average cost $E(Y)$ of *all* resurfacing contracts for 6 miles of road will be \$20,000,"

or

"the cost y of *this specific* resurfacing contract for 6 miles of road will be \$20,000."

The best-guess prediction in either case is 20, but the plus-or-minus factor differs. It's easier to predict an average value $E(Y)$ than an individual Y value. We discuss the plus-or-minus range for predicting an average first, with the understanding that this is an intermediate step toward solving the specific-value problem.

In the mean-value forecasting problem, suppose that the value of the predictor x is known. Since the previous values of x have been designated $x_1, \ldots, x_n$, call the new value x_{n+1}. Then $\hat{y}_{n+1} = \hat{\beta}_0 + \hat{\beta}_1 x_{n+1}$ is used to predict $E(Y_{n+1})$. As $\hat{\beta}_0$ and $\hat{\beta}_1$ are unbiased, $E(\hat{y}_{n+1}) = \beta_0 + \beta_1 x_{n+1} = E(Y_{n+1})$; thus, $\hat{y}_{n+1}$ is an unbiased predictor of $E(Y_{n+1})$. The standard error of $\hat{y}_{n+1}$ can be shown to be

$$\sigma_\epsilon \sqrt{\frac{1}{n} + \frac{(x_{n+1} - \bar{x})^2}{S_{xx}}}$$

where S_{xx} is the sum of squared deviations of the original n values of x_i. Again σ_ϵ must be estimated by s_ϵ, and t tables must be used. The usual approach to forming a confidence interval, namely, estimate $\pm t$ (standard error of estimate), yields a confidence inverval for $E(Y_{n+1})$ as shown here:

$100(1 - \alpha)\%$ Confidence Interval for $E(Y_{n+1})$

$$\hat{y}_{n+1} - t_{\alpha/2} s_\epsilon \sqrt{\frac{1}{n} + \frac{(x_{n+1} - \bar{x})^2}{S_{xx}}} < E(Y_{n+1}) < \hat{y}_{n+1} + t_{\alpha/2} s_\epsilon \sqrt{\frac{1}{n} + \frac{(x_{n+1} - \bar{x})^2}{S_{xx}}}$$

where t_a cuts off area a in the right tail of the t distribution with $n - 2$ d.f.

For the resurfacing example, the following computer output shows the estimated *value of* $E(Y)$ to be 20 when $x = 6$. The corresponding 95% confidence interval on $E(Y)$ is 12.29 to 27.21.

PARAMETER	ESTIMATE	T FOR H0: PARAMETER=0	PR > |T|	STD ERROR OF ESTIMATE
INTERCEPT	2.00000000	0.52	0.6376	3.82970843
MILES	3.00000000	3.50	0.0394	0.85634884

OBSERVATION	OBSERVED VALUE	PREDICTED VALUE	RESIDUAL	LOWER 95% CL FOR MEAN	UPPER 95% CL FOR MEAN
1	6.00000000	5.00000000	1.00000000	-4.82631770	14.82631770
2	14.00000000	11.00000000	3.00000000	4.90597646	17.09402354
3	10.00000000	14.00000000	-4.00000000	8.54933964	19.45066036
4	14.00000000	17.00000000	-3.00000000	10.90597646	23.09402354
5	26.00000000	23.00000000	3.00000000	13.17368230	32.82631770
6 *	.	20.00000000	.	12.29160220	27.70839780

The forecasting plus-or-minus term in the confidence interval for $E(y_{n+1})$ depends on the sample size n and the standard deviation around the regression line, as one might expect. It also depends on the squared distance of x_{n+1} from $\bar{x}$ (the mean of the previous x_i values) relative to S_{xx}. As x_{n+1} gets farther from $\bar{x}$, the term $(x_{n+1} - \bar{x})^2/S_{xx}$ gets larger. When x_{n+1} is far away from the other x values, so that this term is large, the prediction is a considerable extrapolation from the data. Small errors in estimating the regression line are magnified by the extrapolation, as in Figure 13.8. The term $(x_{n+1} - \bar{x})^2/S_{xx}$ could be called an **extrapolation penalty** since it increases with the degree of extrapolation. This penalty term actually understates the risk of extrapolation. It is based on the assumption of a linear relation, and that assumption gets very shaky for large extrapolations.

extrapolation penalty

EXAMPLE 13.10 For the data of Example 13.4, calculate a 95% confidence interval for $E(Y_{n+1})$ based on an assumed x_{n+1} of 22.4. Compare the width of this interval to one based on an assumed x_{n+1} of 30.4.

* Observation 6 designates the predicted value $\hat{y}_{n+1} = 20$ and the 95% confidence limits for $E(Y_{n+1})$.

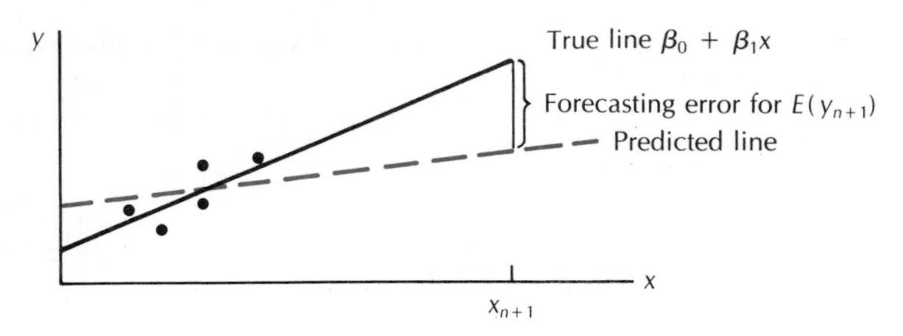

FIGURE 13.8 **Effect of Extrapolation on Forecast Error**

Solution For $x_{n+1} = 22.4$,

$$\hat{y}_{n+1} = 64.672 + (-1.7487)(22.4) = 25.5$$

The estimated standard error for estimating $E(Y_{n+1})$ is

$$s_\epsilon \sqrt{\frac{1}{n} + \frac{(x_{n+1} - \bar{x})^2}{S_{xx}}} = 2.380 \sqrt{\frac{1}{10} + \frac{(22.4 - 22.44)^2}{5099.12 - (224.4)^2/10}} = .753$$

The tabled t value ($n - 2 = 8$ d.f., $\alpha = .025$) is 2.306. The 95% confidence interval is

$$25.5 - (2.306)(.753) < E(Y_{n+1}) < 25.5 + (2.306)(.753) \quad \text{or} \quad 23.764 < E(Y_{n+1}) < 27.236$$

For $x_{n+1} = 30.4$,

$$\hat{y}_{n+1} = 64.672 + (-1.7487)(30.4) = 11.5$$

with standard error

$$2.380\sqrt{1/10 + (30.4 - 22.44)^2/[5099.12 - (224.4)^2/10]} = 2.492$$

The 95% confidence interval is

$$11.5 - (2.306)(2.492) < E(Y_{n+1}) < 11.5 + (2.306)(2.492), \quad \text{or} \quad 5.753 < E(Y_{n+1}) < 17.247$$

The interval for $E(Y_{n+1})$ when $x_{n+1} = 22.4$ is much shorter than that for $E(Y_{n+1})$ when $x_{n+1} = 30.4$. An x value of 30.4 is well outside the range of x values in the data, so the extrapolation penalty is severe. □

Usually, the more relevant forecasting problem is that of predicting an individual Y_{n+1} value rather than $E(Y_{n+1})$. The same best-guess $\hat{y}_{n+1}$ is used but the forecasting plus-or-minus term is larger when predicting Y_{n+1} rather than $E(Y_{n+1})$. In fact, it can be shown that the plus-or-minus forecasting error using $\hat{y}_{n+1}$ to predict Y_{n+1} is

$$\sigma_\epsilon \sqrt{1 + \frac{1}{n} + \frac{(x_{n+1} - \bar{x})^2}{S_{xx}}}$$

If the population standard deviation σ_ϵ is estimated by s_ϵ, the prediction interval for Y_{n+1} is as follows:

100(1 − α)% Prediction Interval for Y_{n+1}

$$\hat{y}_{n+1} - t_{\alpha/2}s_\epsilon\sqrt{1 + \frac{1}{n} + \frac{(x_{n+1} - \bar{x})^2}{S_{xx}}} < Y_{n+1} < \hat{y}_{n+1} + t_{\alpha/2}s_\epsilon\sqrt{1 + \frac{1}{n} + \frac{(x_{n+1} - \bar{x})^2}{S_{xx}}}$$

where t_a cuts off area a in the right tail of the t distribution with $n - 2$ d.f.

In the road resurfacing example, the corresponding 95% prediction limits for Y_{n+1} when $x = 6$ are 5.58 to 34.42, as shown in the output below. The 95% intervals for $E(Y_{n+1})$ and for Y_{n+1} are shown in Figure 13.9; the inner curves are for $E(Y_{n+1})$ the outer ones for Y_{n+1}.

The only difference between prediction of a mean $E(Y_{n+1})$ and prediction of an individual value Y_{n+1} is the term +1 in the standard error formula. The presence of this

OBSERVATION	OBSERVED VALUE	PREDICTED VALUE	RESIDUAL	LOWER 95% CL INDIVIDUAL	UPPER 95% CL INDIVIDUAL
1	6.00000000	5.00000000	1.00000000	-10.65582994	20.65582994
2	14.00000000	11.00000000	3.00000000	-2.62665089	24.62665089
3	10.00000000	14.00000000	-4.00000000	0.64866336	27.35133664
4	14.00000000	17.00000000	-3.00000000	3.37334911	30.62665089
5	26.00000000	23.00000000	3.00000000	7.34417006	38.65582994
6 *	.	20.00000000	.	5.57890821	34.42109179

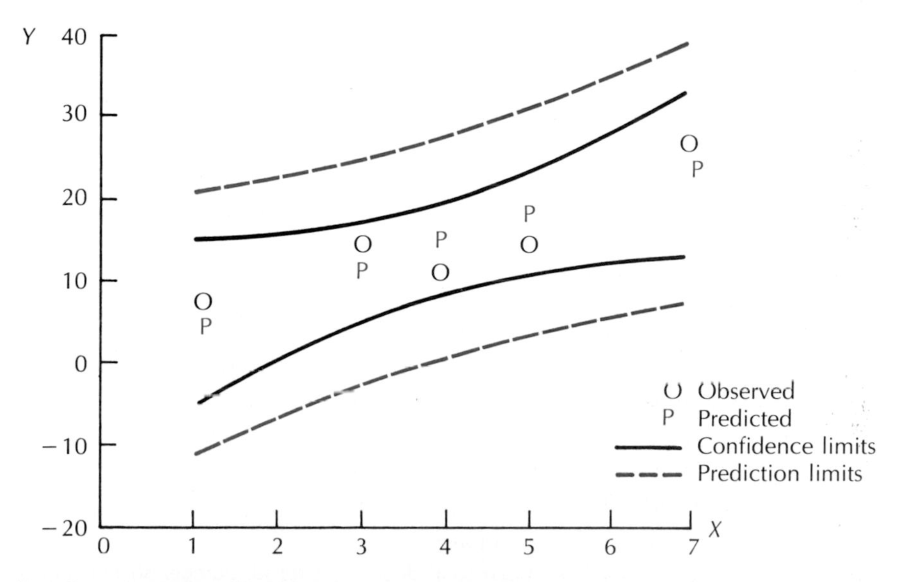

FIGURE 13.9 **Predicted versus Observed Values with 95% limits; Example 13.2**

extra term indicates that predictions of individual values are less accurate than predictions of means. The extrapolation penalty term still applies, as does the warning that it understates the risk of extrapolation. If n is large and the extrapolation term is small, the $+1$ term dominates the square root factor in the prediction interval. In such cases the interval becomes, approximately, $\hat{y}_{n+1} - t_{\alpha/2}s_\epsilon < Y_{n+1} < \hat{y}_{n+1} + t_{\alpha/2}s_\epsilon$. Thus, for large n, roughly 68% of the residuals (forecast errors) are less than $\pm 1s_\epsilon$ and 95% less than $\pm 2s_\epsilon$. There isn't much point in devising rules for when to ignore the other terms in the square root factor. They are easy to calculate and it does no harm to include them.

EXAMPLE 13.11 For the data of Example 13.4, find a 95% prediction interval for Y_{n+1} with $x_{n+1} = 22.4$, and find the interval with $x_{n+1} = 30.4$. Compare these to widths estimated by the $\pm 2s_\epsilon$ rules above.

Solution As in Example 13.10, $\hat{y}_{n+1} = 25.5$ if $x_{n+1} = 22.4$. The estimated standard error for predicting Y_{n+1} is

$$s_\epsilon\sqrt{1 + \frac{1}{n} + \frac{(x_{n+1} - \bar{x})^2}{S_{xx}}} = 2.380\sqrt{1 + \frac{1}{10} + \frac{(22.4 - 22.44)^2}{5099.12 - (224.4)^2/10}}$$
$$= 2.496$$

The tabled t value ($n - 2 = 8$ d.f., $a = .025$) is 2.306. The prediction interval is

$$25.5 - (2.306)(2.496) < Y_{n+1} < 25.5 + (2.306)(2.496), \quad \text{or} \quad 19.744 < Y_{n+1} < 31.256$$

The $\pm 2s_\epsilon$ range is

$$25.5 - (2)(2.380) < Y_{n+1} < 25.5 + (2)(2.380), \quad \text{or} \quad 20.74 < Y_{n+1} < 30.26$$

The latter is too narrow, mostly because the tabled t value with only 8 d.f. is larger than 2.

For $x_{n+1} = 30.4$, $\hat{y}_{n+1} = 11.5$ as in Example 13.10. The estimated standard error for predicting Y_{n+1} is

$$2.380\sqrt{1 + \frac{1}{10} + \frac{(30.4 - 22.44)^2}{5099.12 - (224.4)^2/10}} = 3.446$$

The 95% prediction interval is

$$11.5 - (2.306)(3.446) < Y_{n+1} < 11.5 + (2.306)(3.446), \quad \text{or} \quad 3.554 < Y_{n+1} < 19.446$$

The $\pm 2s_\epsilon$ range is

$$11.5 - (2)(2.380) < Y_{n+1} < 11.5 + (2)(2.380), \quad \text{or} \quad 6.74 < Y_{n+1} < 16.26$$

The latter is much too narrow. Not only is the tabled t value larger than 2, but also the large extrapolation penalty is not reflected. □

SECTION 13.4 EXERCISES

13.14 Refer to the data of Exercise 13.10.

a. Predict the mean total direct cost for all bumper sticker orders with a printing run of 2000 stickers (that is, with RUNSIZE = 2.0).

b. Calculate a 95% confidence interval for this mean.

13.15 Does the prediction in Exercise 13.14 represent a major extrapolation?

13.16 Refer to Exercise 13.14.

a. Predict the total direct cost for a particular bumper sticker order with a printing run of 2000 stickers. Calculate a 95% prediction interval.

b. Would an actual total direct cost of $250 be surprising for this order?

13.17 Refer to the data of Exercise 13.5.

a. Predict the number of bank branches located in a ZIP code area having 1267 independent businesses. Give a 90% prediction interval.

b. Do you think that this is a reasonable prediction, given the data?

13.18 Refer to the data of Exercise 13.1.

a. If the model with Y a linear function of x is adopted (as in Exercise 13.2), what is the predicted Y when $x = 20$?

b. If the model with Y a linear function of $x' = \log_{10} x$ is adopted (as in Exercise 13.3), what is the predicted value of Y when $x = 20$?

c. Which of the two predictions seems more reasonable (or perhaps less unreasonable)?

13.19 Give a 95% prediction interval for the prediction you selected in Exercise 13.18.

13.5 CORRELATION

Once we've found the prediction line, we need to measure how well it predicts actual values. In the road resurfacing example, if we were given the mileage values x_i, we could use the prediction equation $\hat{y}_i = 2.0 + 3.0x_i$ to predict costs. The deviations of actual values from predicted values, called **residuals**, measure prediction errors. These errors are summarized by the sum of squared residuals SS(Residual) $= \sum (y_i - \hat{y}_i)^2$, which is 44 for these data. For comparison, if we were not given the x_i values, the best squared error predictor of y would be the mean value $\bar{y} = 14$, and the sum of squared prediction errors would, in this case, be $\sum_i (y_i - \bar{y})^2 =$ SS(Total) $= 224$. The proportionate reduction in error would be

residuals

$$\frac{\text{SS(Total)} - \text{SS(Residual)}}{\text{SS(Total)}} = \frac{224 - 44}{224} = .804$$

indicating a fairly strong relation between the mileage to be resurfaced and the cost of resurfacing.

correlation coefficient

This proportionate reduction in error is closely related to the **correlation coefficient** of x and y. **The general idea of a correlation is that it measures the strength of the linear relation between x and y.** The stronger the correlation, the better x predicts y. The mathematical definition of the correlation coefficient, denoted r_{yx}, is

$$r_{yx} = \frac{S_{xy}}{\sqrt{S_{xx}S_{yy}}}$$

where S_{xy} and S_{xx} are defined as before and

$$S_{yy} = \sum_i y_i^2 - \frac{\left(\sum_i y_i\right)^2}{n}$$

In the example, $r_{yx} = 60/\sqrt{(20)(224)} = .896$.

Generally, the correlation r_{yx} is a positive number if y tends to increase as x increases, r_{yx} is negative if y tends to decrease as x increases, and r_{yx} is zero if either there is no relation between changes in x and changes in y or there is a nonlinear relation such that patterns of y increase and y decrease (as x increases) cancel each other.

EXAMPLE 13.12 Consider the following data:

y:	25	41	47	59	54	56	49	43	30
x:	10	20	20	30	30	30	40	40	50

a. Should the correlation be positive or negative?
b. Calculate the correlation.

Solution a. Notice that as x increases from 10 to 50, y first increases, then decreases. Therefore the correlation should be small. The y values don't decrease quite back to where they started, so the correlation should be positive.

b. Calculations show $\sum x = 270$, $\sum x^2 = 9300$, $\sum y = 404$, $\sum y^2 = 19{,}198$, $\sum xy = 12{,}260$, and $n = 9$. Therefore

$$S_{xx} = 9300 - \frac{(270)^2}{9} = 1200$$

$$S_{yy} = 19{,}198 - \frac{(404)^2}{9} = 1062.8889$$

$$S_{xy} = 12{,}260 - \frac{(270)(404)}{9} = 140$$

$$r_{yx} = \frac{140}{\sqrt{(1200)(1062.8889)}} = .1240$$

The correlation is indeed a small positive number. □

Correlation and regression predictability are closely related. The proportionate reduction in error for regression defined earlier is called the **coefficient of determination**. The coefficient of determination is simply the square of the correlation coefficient,

coefficient of determination

$$r_{yx}^2 = \frac{\text{SS(Total)} - \text{SS(Residual)}}{\text{SS(Total)}}$$

which is the proportionate reduction in error. In the resurfacing example, $r_{yx} = .896$ and $r_{yx}^2 = .804$.

A correlation of zero indicates no predictive value using the equation $y = \hat{\beta}_0 + \hat{\beta}_1 x$; that is, one can predict y as well without knowing x as one can knowing x. A correlation of 1 or -1 indicates perfect predictability—a 100% reduction in error attributable to knowledge of x. A correlation coefficient should routinely be interpreted in terms of its squared value, the coefficient of determination. Thus, correlation of $-.3$, say, indicates only a 9% reduction in squared prediction error. Many books and most computer programs use the equation

$$\text{SS(Total)} = \text{SS(Residual)} + \text{SS(Regression)}$$

where

$$\text{SS(Regression)} = \sum_i (\hat{y}_i - \bar{y})^2$$

Since the equation can be expressed as SS(Residual) $= (1 - r_{yx}^2)$ SS(Total), it follows that SS(Regression) $= r_{yx}^2$ SS(Total), which again says that regression on x explains a proportion r^2 of the total squared error of y.

EXAMPLE 13.13 Find SS(Total), SS(Regression), and SS(Residual) for the data of Example 13.12.

Solution SS(Total) $= S_{yy}$, which we computed to be 1062.8889 in Example 13.12. We also found that $r_{yx} = .1240$, so $r_{yx}^2 = (.1240)^2 = .0154$. Using the fact that SS(Regression) $= r_{yx}^2$ SS(Total), we have SS(Regression) $= (.0154)(1062.8889) = 16.3685$. Because SS(Residual) = SS(Total) − SS(Regression), SS(Residual) $= 1062.8889 - 16.3685 = 1046.5204$.

Note that SS(Regression) and r_{yx}^2 are very small. This suggests that x is not a good predictor of y. The reality, though, is that the relation between x and y is extremely nonlinear. A *linear* equation in x does not predict y very well, but a nonlinear equation would do far better. □

The sample correlation r_{yx} is the basis for estimation and significance testing of the population correlation ρ_{yx}. Statistical inferences are always based on formal mathematical assumptions. The formal assumptions of regression analysis—linear relation between x and y, and constant variance around the regression line, in particular—are also assumed in correlation inference. In regression analysis, the x values are regarded as predetermined constants. In correlation analysis, we regard the x values as randomly selected (and the regression inferences are conditional on the sampled x values). If the x's are not drawn randomly, it is possible that the correlation estimates are biased. In some texts, the additional assumption is made that the x values are drawn from a normal population. (This assumption, together with the regression assumptions, implies that the x, y values are drawn from a bivariate normal population.) The inferences we make do not depend crucially on this normality assumption.

assumptions for correlation inference

bias in estimating ρ

The most basic inference problem is potential bias in estimation of ρ_{yx}. Under the assumptions we've given, the sample correlation is not quite an unbiased estimator. However, the degree of bias is small, and not worth correcting. A more serious problem arises when the x values are predetermined, as often happens in regression analysis. The choice of x values can systematically increase or decrease the sample correlation. In general, a wide range of x values tends to increase the magnitude of the correlation coefficient and a small range to decrease it. This effect is shown in Figure 13.10. If all the points in this scatter plot are included, there is an obvious, strong correlation between x and y. But suppose we consider only x values in the range between the dashed vertical lines. By eliminating the outside parts of the scatter diagram, the sample correlation coefficient is still positive, but much weaker. It's clear then that the correlation coefficient (and the coefficient of determination) can be affected by systematic choices of x values. Thus it's

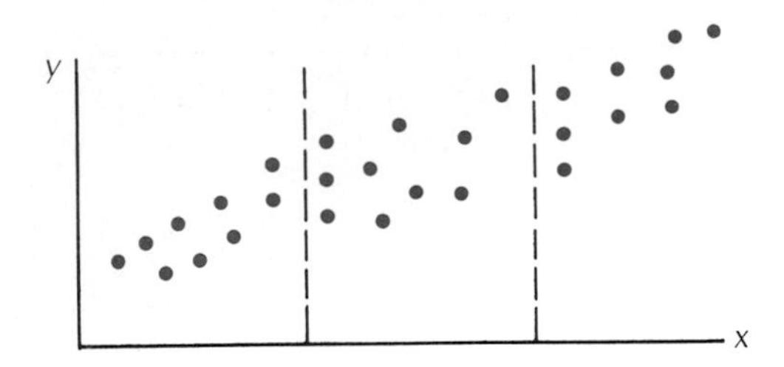

FIGURE 13.10 **Effect of Limited x Range on Sample Correlation Coefficient**

a good idea to consider the residual standard deviation s_ϵ and the magnitude of the slope in deciding how well a linear regression line predicts y.

EXAMPLE 13.14 Suppose that a company has the following data on productivity y and aptitude test score x for 12 keypunch operators:

y:	41	39	47	51	43	40	57	46	50	59	61	52
x:	24	30	33	35	36	36	37	37	38	40	43	49

Find the correlation coefficient and the residual standard deviation based on all 12 observations. Compare these to the correlation and residual standard deviation based on the 6 observations with the highest x scores.

Solution For all 12 observations,

$$\sum x_i = 438, \quad \sum x_i^2 = 16{,}414, \quad \sum y_i = 586$$
$$\sum y_i^2 = 29{,}232, \quad \sum x_iy_i = 21{,}720$$

Therefore

$$r_{yx} = \frac{\sum x_iy_i - (\sum x_i)(\sum y_i)/n}{\sqrt{\sum x_i^2 - (\sum x_i)^2/n}\sqrt{\sum y_i^2 - (\sum y_i)^2/n}}$$

$$= \frac{21{,}720 - (438)(586)/12}{\sqrt{16{,}414 - (438)^2/12}\sqrt{29{,}232 - (586)^2/12}}$$

$$= .646$$

Also, $\hat{\beta}_1 = .775$, $\hat{\beta}_0 = 20.54$, and $s_\epsilon = 6.021$. For the six highest x scores

$$\sum x_i = 244, \quad \sum x_i^2 = 10{,}032, \quad \sum y_i = 325$$
$$\sum y_i^2 = 17{,}771, \quad \sum x_iy_i = 13{,}242$$

Therefore

$$r_{yx} = \frac{13{,}242 - (244)(325)/6}{\sqrt{10{,}032 - (244)^2/6}\sqrt{17{,}771 - (325)^2/6}} = .188$$

Also, $\hat{\beta}_1 = .232$, $\hat{\beta}_0 = 44.74$, and $s_\epsilon = 6.292$.

In going from all 12 observations to the six observations with the highest x values, the correlation has decreased drastically but the residual standard deviation has hardly changed at all. □

Just as it is possible to test the null hypothesis that a true slope is zero, we can also test H_0: $\rho_{yx} = 0$. The procedure is summarized in the following:

Summary for a Test of H_0: $\rho_{yx} = 0$

H_0: $\rho_{yx} = 0$

H_a: 1. $\rho_{yx} > 0$

2. $\rho_{yx} < 0$

3. $\rho_{yx} \neq 0$

T.S.: $t = r_{yx}\sqrt{n-2}/\sqrt{1-r_{yx}^2}$

R.R.: With $n - 2$ d.f. and Type I error probability α,

1. $t > t_\alpha$

2. $t < -t_\alpha$

3. $|t| > t_{\alpha/2}$

We tested the hypothesis that the true slope is zero (in predicting resurfacing cost from mileage) in Example 13.5; the resulting t statistic was 3.50. For that data, r_{yx} can be calculated as .896421 and r_{yx}^2 as .803571; hence the correlation t statistic is $.896\sqrt{3}/\sqrt{1 - .803571} = 3.50$. In general, the t tests for a slope and for a correlation give identical results; it doesn't matter which form is used. It follows that the t test is valid for any choice of x values. The bias mentioned previously does not affect the sign of the correlation.

EXAMPLE 13.15 Perform t tests for the null hypotheses of zero correlation and zero slope for the data of Example 13.14 (all observations). Use an appropriate one-sided alternative.

Solution First, the appropriate H_a ought to be $\rho_{yx} > 0$ (and therefore $\beta_1 > 0$). It would be nice if an aptitude test had a positive correlation with the productivity score it was predicting! In Example 13.15, $n = 12$, $r_{yx} = .646$, and

$$t = \frac{.646\sqrt{12-2}}{\sqrt{1-(.646)^2}} = 2.68$$

Because this value falls between the tabled t values for d.f. = 10, $\alpha = .025$ (2.228) and for d.f. = 10, $\alpha = .01$ (2.764), the p-value lies between .010 and .025. Hence H_0 may be rejected.

The t statistic for testing the slope β_1 is

$$t = \frac{\hat{\beta}_1}{s_\epsilon\sqrt{\dfrac{1}{S_{xx}}}} = \frac{.775}{6.021\sqrt{\dfrac{1}{16{,}414 - (438)^2/12}}}$$

$$= 2.66$$

which equals (to within round-off error) the correlation t statistic, 2.68. □

The test for a correlation provides a neat illustration of the difference between statistical significance and statistical importance. Suppose that a psychologist has devised a skills test for production line workers and tests it on a huge sample of 40,000. If the sample correlation between test score and actual productivity is .02, then

$$t = \frac{.02\sqrt{39{,}998}}{\sqrt{1-(.02)^2}} = 4.0$$

We would reject the null hypothesis at any reasonable α level, so the correlation is "statistically significant." However, the test accounts for only $(.02)^2 = .0004$ of the squared error in skill scores, so it is almost worthless as a predictor. Remember, the rejection of the null hypothesis in a statistical test is the conclusion that the sample result cannot plausibly have occurred by chance if the null hypothesis is true. The test itself does not address the practical significance of the result. Clearly, for a sample size of 40,000, even a trivial sample correlation like .02 is not likely to arise by mere luck of the draw. There is no practically meaningful relationship between these test scores and productivity scores.

EXAMPLE 13.16 Suppose that a study is made of forecast profits and actually realized profits for 40 new products introduced by chemical industry firms within the previous four years. Forecast five-year profit (per $1000 invested) is the independent variable x and actual first-year profit (per $1000 invested) is the dependent variable y. Summary results are

$$\sum x_i = 2480, \quad \sum y_i = 1860, \quad \sum x_iy_i = 128{,}860$$
$$\sum x_i^2 = 187{,}660, \quad \sum y_i^2 = 124{,}530, \quad n = 40$$

a. Find the regression equation and residual standard deviation.
b. Find the sample correlation coefficient and coefficient of determination. What does this indicate about the apparent success of current forecasting methods?
c. Test (at $\alpha = .05$) the hypothesis that x has no value in a (linear) prediction of y.
d. Give a 95% prediction interval for the actual first-year profit per thousand dollars invested for a project with a forecast profit of $50.00 (a study like this is reported in Beardsley and Mansfield [1978]).

Solution a.

$$\hat{\beta}_1 = \frac{S_{xy}}{S_{xx}} = \frac{\sum x_iy_i - (\sum x_i)(\sum y_i)/n}{\sum x_i^2 - (\sum x_i)^2/n} = \frac{13{,}540}{33{,}900} = .39941003$$

$$\hat{\beta}_0 = \bar{y} - \hat{\beta}_1\bar{x} = 21.736578$$

So $\hat{y} = 21.74 + .40x$, rounding off to reasonable numbers,

$$s_\epsilon^2 = \frac{\sum y_i^2 - \hat{\beta}_0\sum y_i - \hat{\beta}_i\sum x_iy_i}{n-2} = \frac{32{,}631.988}{38}$$

$$= 858.73654$$

So $s_\epsilon = \sqrt{858.73654} = 29.304207$, or about 29.30.

b. $$r_{yx} = \frac{S_{xy}}{\sqrt{S_{xx}}\sqrt{S_{yy}}} = \frac{13540}{\sqrt{33{,}900}\sqrt{38{,}040}} = .37704967$$

$$r_{yx}^2 = (.37704967)^2 = .14216645$$

This indicates that knowledge of the forecast profits reduces squared error in predicting actual profits by only about 14.2%. This poor predictability was the major finding of the Beardsley and Mansfield article.

c. Because we just calculated the sample correlation, it's convenient to use the correlation t statistic. We take H_0: $\rho_{yx} = 0$. If there is any correlation between forecast and actual, we would hope it is positive, so we take H_a: $\rho_{yx} > 0$.

$$\text{The test statistic } t = \frac{r_{yx}\sqrt{n-2}}{\sqrt{1-r_{yx}^2}} = \frac{.37704967\sqrt{38}}{\sqrt{1-.14216645}}$$

$$= 2.5095095, \quad \text{or about } 2.51$$

The tabled t value for 38 d.f., $\alpha = .05$ (one-tailed) lies somewhere between the values for 30 d.f. and 40 d.f., namely, 1.697 and 1.684; let's assume it is 1.69. The observed t statistic falls beyond either value, so we may reject H_0 at $\alpha = .05$ and conclude that the modest observed correlation is not merely the result of chance. (In fact, the t tables indicate that we could also reject H_0 at $\alpha = .01$ but not at $\alpha = .005$. Therefore the one-tailed p-value would be somewhere between .005 and .01.)

d. For $x_{n+1} = 50.00$, the predicted profit is

$$\hat{y}_{n+1} = 21.736578 + .39941003(50) = 41.70708$$

or about 41.71. By rough approximation, take the two-tailed tabled t value to be 2.02. The 95% prediction interval is

$$41.71 - 2.02(29.30)\sqrt{1 + \frac{1}{40} + \frac{\left(50.00 - \frac{2480}{40}\right)^2}{33{,}900}} < Y_{n+1}$$

$$< 41.71 + 2.02(29.30)\sqrt{1 + \frac{1}{40} + \frac{\left(50.00 - \frac{2480}{40}\right)^2}{33{,}900}}$$

or

$$-18.34 < Y_{n+1} < 101.76$$

This very wide interval confirms the finding in part (b) that current forecasting methods are quite poor. □

13.6 RANK CORRELATION

The ordinary correlation coefficient r_{yx} assesses the degree to which x and y are linearly related. It can happen that y generally increases as x increases, but not necessarily in a linear fashion. In such a case, the ordinary correlation does not completely reflect the extent of the relation between y and x. There are two reasonable approaches for measuring the association between x and y in such a situation: use a nonlinear prediction rule, or find

rank correlation

a general measure of the tendency for y to increase (linearly or not) with x. Nonlinear prediction can be handled by methods discussed in the next two chapters. The **rank correlation coefficient** is another way to measure the extent to which y increases with x.

The calculation of a rank correlation is quite simple. Rank all the x scores from lowest to highest; rank the y scores likewise. Calculate the correlation based on the ranks; this is the rank correlation r_s sometimes referred to as Spearman's rank order correlation coefficient.

EXAMPLE 13.17 Calculate the rank correlation coefficient r_s for the data of Example 13.4.

Solution First, rank the y scores and the x scores separately, from lowest to highest.

y_i:	31.5	33.1	27.4	24.5	27.0	27.8	23.3	24.7	16.9	18.1
rank y_i:	9	10	7	4	6	8	3	5	1	2
x_i:	18.1	20.0	20.8	21.5	22.0	22.4	22.9	24.0	25.4	27.3
rank x_i:	1	2	3	4	5	6	7	8	9	10

Then $\sum \text{rank } x_i = 55$, $\sum (\text{rank } x_i)^2 = 385$, $\sum \text{rank } y_i = 55$, $\sum (\text{rank } y_i)^2 = 385$, and $\sum (\text{rank } x_i)(\text{rank } y_i) = 234$. The rank correlation is

$$\frac{\sum (\text{rank } x_i)(\text{rank } y_i) - (\sum \text{rank } x_i)(\sum \text{rank } y_i)/n}{\sqrt{\sum (\text{rank } x_i)^2 - (\sum \text{rank } x_i)^2/n}\sqrt{\sum (\text{rank } y_i)^2 - (\sum \text{rank } y_i)^2/n}}$$

$$= \frac{234 - (55)(55)/10}{\sqrt{385 - (55)^2/10}\sqrt{385 - (55)^2/10}}$$

$$= -.830$$

□

When there are no ties in rankings, the calculation is even simpler. In this case, it can be proved that

$$r_s = 1 - \frac{6 \sum d_i^2}{n(n^2 - 1)}$$

where d_i is the difference between the y rank and the x rank of observation i.

EXAMPLE 13.18 Use the simplified formula above to recalculate the rank correlation coefficient for the data of Example 13.4.

Solution The ranks and the differences in ranks are

rank y_i:	9	10	7	4	6	8	3	5	1	2
rank x_i:	1	2	3	4	5	6	7	8	9	10
d_i:	8	8	4	0	1	2	−4	−3	−8	−8

$\sum d_i^2 = 302$, so

$$r_s = 1 - \frac{(6)(302)}{10(100 - 1)} = -.830$$

□

The rank correlation may be tested for statistical significance. The null hypothesis is that there is no general tendency for y to either increase or decrease with x. This is true if x and y are independent. The same t statistic and degrees of freedom are used for rank correlation as for ordinary correlation. In the case of rank correlation, the test is approximate, because the formal normality assumptions underlying the t test don't hold for ranks. However, the approximation is accurate enough for $n \geq 10$ or so.

EXAMPLE 13.19 Test the significance of the rank correlation for the data of Example 13.4.

Solution

$$t = \frac{r_s\sqrt{n-2}}{\sqrt{1-r_s^2}}$$

$$= \frac{-.830\sqrt{10-2}}{\sqrt{1-(-.830)^2}} = -4.21$$

This value falls beyond tabulated t values with $10 - 2 = 8$ d.f., so the null hypothesis may be rejected conclusively. □

interpretation of r_s

One problem with rank correlation is that it's not easy to interpret. Ordinary correlation can be interpreted by way of regression and r^2. But what does it mean to predict y rank from x rank, and what does "predicted y rank $= 16.37$" mean? Since we don't know the answers to these questions, we tend to use rank correlation mostly as a significance-testing device, and we don't put much emphasis on its numerical value.

SECTIONS 13.5 AND 13.6 EXERCISES

13.20 Refer to the data of Exercise 13.5. Calculate the correlation coefficient r_{yx}.

13.21 Refer to the data of Exercise 13.5.

a. Test the hypothesis of no true correlation between x and y. Use a one-sided H_a and $\alpha = .05$.

b. Compare the result of this test to that of Exercise 13.7, part (d).

13.22 Refer to the computer output of Exercise 13.10.

a. Locate r_{yx}^2.

b. The estimated slope $\hat{\beta}_1$ is positive; what must be the sign of the sample correlation coefficient?

13.23 Suppose that the study in Exercise 13.10 had been restricted to RUNSIZE values less than 1.8. Would you anticipate a larger or a smaller r_{yx} value?

13.24 Suppose that an advertising campaign for a new product is conducted in 10 test cities. The intensity of advertising X is varied across cities; the awareness percentage Y is found by survey after the ad campaign.

x:	4.0	4.5	5.0	5.5	6.0	6.5	7.0	7.5	8.0	8.5
y:	10.1	10.3	10.4	21.7	36.7	51.5	67.0	68.5	68.2	69.3

$\sum x_i = 62.5$, $\sum y_i = 413.7$, $\sum x_iy_i = 2930.45$

$\sum x_i^2 = 411.25$, $\sum y_i^2 = 23{,}421.27$, $n = 10$

a. Calculate the correlation coefficient r_{yx}.

b. Calculate the rank correlation coefficient r_s.

c. Plot the data. Does the relation appear linear to you? Does it appear to be generally increasing?

SUMMARY

In preceding sections, we considered various statistical inferences (estimation and testing problems) related to linear regression and correlation. The underlying assumptions for a linear regression model $Y = \beta_0 + \beta_1 x + \epsilon$ (namely, fixed x values, random error, normal with mean zero and variance σ_ϵ^2, independence) allow us to state the sampling distributions for the least-squares estimators $\hat{\beta}_0$ and $\hat{\beta}_1$. Based on these results, we are able to specify the form of a statistical test of H_0: $\beta_1 = 0$ and a corresponding confidence interval for the unknown slope.

The prediction of future values of the dependent variable based on specific values of the independent variable x is a problem of particular interest in regression. Prediction of future values using the linear regression prediction equation was discussed and an appropriate prediction interval developed. Finally, we considered inferences related to the population correlation coefficient ρ_{yx}. The rank correlation coefficient provides an alternative to r_{yx} when two variables appear related but are not necessarily linearly related.

KEY FORMULAS: Linear Regression and Correlation Methods

1. Linear regression model

$$Y_i = \beta_0 + \beta_1 x_i + \epsilon_i$$

2. Least-squares estimates of β_0 and β_1

$$\hat{\beta}_1 = \frac{S_{xy}}{S_{xx}},$$

where

$$S_{xy} = \sum xy - \frac{(\sum x)(\sum y)}{n} \quad \text{and} \quad S_{xx} = \sum x^2 - \frac{(\sum x)^2}{n}$$

$$\hat{\beta}_0 = \bar{y} - \hat{\beta}_1 \bar{x}$$

3. $\text{SS(Residual)} = \sum y^2 - \hat{\beta}_0 \sum y - \hat{\beta}_1 \sum xy$
   $\text{SS(Regression)} = \hat{\beta}_0 \sum y + \hat{\beta}_1 \sum xy - (\sum y)^2/n$
4. Estimated standard errors for $\hat{\beta}_0$ and $\hat{\beta}_1$

$$s_{\hat{\beta}_0} = s_\epsilon \sqrt{\frac{1}{n} + \frac{\bar{x}^2}{S_{xx}}}$$

$$s_{\hat{\beta}_1} = s_\epsilon \sqrt{\frac{1}{S_{xx}}}$$

where $s_\epsilon^2 = \dfrac{\text{SS(Residual)}}{n-2}$

5. Inferences about β_1
   a. t test

$$H_0: \beta_1 = 0$$

$$\text{T.S.: } t = \frac{\hat{\beta}_1}{s_\epsilon \sqrt{\dfrac{1}{S_{xx}}}}, \quad \text{where d.f.} = n - 2$$

   b. F test

$$H_0: \beta_1 = 0$$

$$\text{T.S.: } F = \frac{\text{MS(Regression)}}{\text{MS(Residual)}}, \quad \text{where d.f.}_1 = 1 \text{ and d.f.}_2 = n - 2$$

   c. $100(1 - \alpha)\%$ confidence interval

$$\hat{\beta}_1 \pm t_{\alpha/2} s_\epsilon \sqrt{1/S_{xx}}$$

6. $100(1 - \alpha)\%$ confidence interval for σ_ϵ^2

$$s_\epsilon^2 \frac{(n-2)}{\chi^2_{\alpha/2}} < \sigma_\epsilon^2 < s_\epsilon^2 \frac{(n-2)}{\chi^2_{1-\alpha/2}}, \quad \text{where d.f.} = n - 2$$

7. $100(1 - \alpha)\%$ confidence interval for $E(Y_{n+1})$

$$\hat{y}_{n+1} \pm t_{\alpha/2} s_\epsilon \sqrt{\frac{1}{n} + \frac{(x_{n+1} - \bar{x})^2}{S_{xx}}}$$

8. $100(1 - \alpha)\%$ prediction interval for Y_{n+1}

$$\hat{y}_{n+1} \pm t_{\alpha/2} s_\epsilon \sqrt{1 + \frac{1}{n} + \frac{(x_{n+1} - \bar{x})^2}{S_{xx}}}$$

9. Sample correlation coefficient

$$r_{yx} = \frac{S_{xy}}{\sqrt{S_{xx}}\sqrt{S_{yy}}}$$

10. $\text{SS(Total)} = \sum (y_i - \bar{y})^2$
    $= \sum y_i^2 - (\sum y_i)^2/n$

$$\text{Proportionate reduction in error} = \frac{\text{SS(Total)} - \text{SS(Residual)}}{\text{SS(Total)}} = r_{yx}^2$$

11. t test concerning ρ_{yx}

$$H_0: \rho_{yx} = 0$$

$$\text{T.S.: } t = \frac{r_{yx}\sqrt{n-2}}{\sqrt{1 - r_{yx}^2}}$$

12. $r_s = 1 - 6 \sum d_i^2/n(n^2 - 1)$, where d_i is the difference between the y rank and the x rank of observation i (valid only if there are no ties)

CHAPTER 13 EXERCISES

13.25 Consider the data shown here:

x:	10	12	14	15	18	19	23
y:	25	30	36	37	42	50	55

a. Plot the data.
b. Using the data, find the least-squares estimates for the model $Y_i = \beta_0 + \beta_1 x_i + \epsilon_i$.
c. Predict Y when $x = 21$.

13.26 Refer to Exercise 13.25.

a. Calculate s_ϵ, the residual standard deviation.
b. Compute the residuals for these data. Do most lie within $\pm 2s_\epsilon$ of zero?

13.27 A government agency responsible for awarding contracts for much of its research work is under careful scrutiny by a number of private companies. One company examines the relationship between the amount of the contract ($\times$ \$10,000) and the length of time between submission of the contract proposal and contract approval:

Length (in months), y:	3	4	6	8	11	14	20
Size ($\times$ \$10,000), x:	1	5	10	50	100	500	1000

a. Plot y versus x.
b. Fit the line $Y = \beta_0 + \beta_1 x + \epsilon$.
c. Conduct a test of the null hypothesis H_0: $\beta_1 = 0$. Give the p-value for your test for H_a: $\beta_1 > 0$.

13.28 Refer to the data of Exercise 13.27.

a. Plot y versus $\log x$.
b. Fit a linear regression line using $\log x$ as the independent variable.
c. Conduct a test of H_0: $\beta_1 = 0$ and give the level of significance for a one-sided alternative, H_a: $\beta_1 > 0$.

13.29 Use the results of Exercises 13.27 and 13.28 to determine which linear regression line provides the better fit. Give reasons for your choice.

13.30 Refer to the data of Exercise 13.25.

a. Give a 90% confidence interval for β_1, the slope of the linear regression line.
b. Construct a 95% confidence interval for σ_ϵ^2.

13.31 A pharmaceutical company interested in screening compounds developed for use in the treatment of hypertension (high blood pressure) examines a potential candidate using the experiment described here. Six rats selected from a colony of rats bred to be hypertensive are randomly assigned to each of four groups. Group 1 is designated as the control group. These rats are injected with a solution that contains no drug. Groups 2–4 are designated as drug groups. Group 2 rats are all injected with a dose of .1 mg/kg, those in group 3 received .2 mg/kg, and those in group 4 received .4 mg/kg. The response of interest is y, the decrease in blood pressure at two hours post-dosing compared to the corresponding pre-dose blood pressure. The data are

	Blood Pressure Decreases (mm)					
Group 1 (control)	2	5	6	4	3	1
Group 2 (.1 mg/kg)	10	12	15	16	13	11
Group 3 (.2 mg/kg)	25	22	26	19	18	24
Group 4 (.4 mg/kg)	30	32	35	27	26	29

a. Use a computer software package to fit the model

$$Y = \beta_0 + \beta_1 \sqrt{x} + \epsilon$$

b. Conduct a two-sided test of H_0: $\beta_1 = 0$.

13.32 Use the data of Exercise 13.27 to predict the length of time in months before approval of a \$750,000 contract. Give a 95% prediction interval.

13.33 An airline studying fuel usage by a certain type of aircraft obtains data on 100 flights. The air mileage x in hundreds of miles and the actual fuel use y in gallons are recorded. Summary values are

$$\bar{x} = 8.00, \qquad \bar{y} = 550.0$$

$$S_{xx} = 1621.0, \qquad S_{yy} = 4,947,000, \qquad S_{xy} = 81,242$$

a. Calculate the regression equation.

b. Calculate the sample correlation coefficient and coefficient of determination.

13.34 a. Use the fact that $\sum (y_i - \hat{y}_i)^2 = (1 - r_{yx}^2) \sum (y_i - \bar{y})^2$ to calculate SS(Residual) and the residual standard deviation in Exercise 13.33.

b. Calculate the estimated standard error of $\hat{\beta}_1$. Calculate a 95% confidence interval for β_1.

c. Is there any point in testing H_0: $\beta_1 = 0$?

13.35 Refer to the data and calculations of Exercises 13.33 and 13.34.

a. Predict the mean fuel usage of all 1000-mile flights. Give a 95% confidence interval.

b. Predict the fuel usage of a particular 1000-mile flight. Would a usage of 570 gallons be considered exceptionally low? (A 95% prediction interval might help in answering the question.)

13.36 What is the interpretation of $\hat{\beta}_1$ in the situation of Exercise 13.33? Is there a sensible interpretation of $\hat{\beta}_0$?

13.37 A group of investors interested in the acquisition of a corporation examine the quarterly sales volumes and the corresponding advertising volumes for the preceding 10 quarters.

a. Use the data shown here to compute the sample correlation coefficient.

b. Conduct a one-sided test of H_0: $\rho_{yx} = 0$. Give the p-value for your test.

Quarter	Sales (\$10,000) y	Advertising Expense (\$1000) x
1	50	10
2	70	12
3	80	14
4	90	15
5	62	12
6	68	13
7	92	14
8	106	16
9	65	14
10	76	15
11	85	17
12	110	19

13.38 Refer to Exercise 13.37.

a. What conclusion can be drawn from the sample data?

b. What other (uncontrolled) factors might affect sales in addition to the amount of advertising? How are these factors represented in the regression model?

c. In what sense, if any, do these data represent a random sample from the population of interest?

13.39 Refer to Exercise 13.37. Construct a 99% prediction interval for quarterly sales, given a quarterly advertising expense of $18,000.

13.40 The high interest rates being charged for mortgage money in previous years had an impact on the number of housing starts throughout the United States. The data below summarize the prevailing interest rates by quarter and the corresponding number of housing starts for that quarter in a given locale.

Month	Interest Rate (%) x	Number of Housing Starts y
1	11.5	260
2	11.4	250
3	11.6	241
4	12.4	256
5	12.8	270
6	13.2	220
7	13.5	190
8	13.0	195
9	12.7	200
10	12.9	210
11	12.5	230
12	12.0	245

a. Plot the data.

b. Fit a linear regression line to these data. Is β_1 significantly different from zero?

c. Predict the number of housing starts for an interest rate of 11.0%. Use a 95% prediction interval.

13.41 Refer to Exercise 13.40.

a. Calculate the rank correlation coefficient.

b. Run a test of the hypothesis that the number of housing starts y decreases with the interest rate x.

c. Draw conclusions from parts (a) and (b).

13.42 Refer to Exercise 13.31.

a. Use a computer software package to run a one-way analysis of variance.

b. What percentage of the total variability in the y values, SS(Total), is accounted for by the four groups?

13.43 Refer to Exercise 13.37. Additional factors such as interest rates and inflation rate (as measured by the CPI) could help to explain the sales data. One factor that reflects some of these rate changes is the variable "year." For convenience, let the coded variable year x_2 take on the following values.

x_2:	1	1	1	1	2	2	2	2	3	3	3	3
Quarter:	1	2	3	4	5	6	7	8	9	10	11	12

a. Fit the model $Y = \beta_0 + \beta_2 x_2 + \epsilon$.

b. Conduct a test of H_0: $\beta_2 = 0$. Give the level of significance for a two-sided test.

13.44 One use for regression analysis is to find a trend line. The sales y in thousands of units of a certain model of microwave oven are tabulated by month x. A regression line is calculated by a standard computer package (SAS) and is shown below:

VARIABLE	MEAN	VARIANCE	STANDARD DEVIATION
MONTH	10.50000000	35.00000000	5.91607978
SALES	80.00000000	205.68421053	14.34169483

DEP VARIABLE: SALES

ANALYSIS OF VARIANCE

SOURCE	DF	SUM OF SQUARES	MEAN SQUARE	F VALUE	PROB>F
MODEL	1	3552.435	3552.435	179.837	0.0001
ERROR	18	355.5654	19.75363		
C TOTAL	19	3908			

ROOT MSE	4.444506	R-SQUARE	0.9090
DEP MEAN	80	ADJ R-SQ	0.9040
C.V.	5.555633		

PARAMETER ESTIMATES

VARIABLE	DF	PARAMETER ESTIMATE	STANDARD ERROR	T FOR H0: PARAMETER=0	PROB > |T|
INTERCEP	1	55.73158	2.064613	26.994	0.0001
MONTH	1	2.311278	0.1723506	13.410	0.0001

DURBIN-WATSON D 0.348

a. Write the trend equation (regression line).

b. Calculate the residual standard deviation.

13.45 Use the data of Exercise 13.44 to calculate sales forecasts for months 21 and 40. How does the plus-or-minus term increase as one forecasts farther into the future? Note: $S_{xx} = (n - 1)s_x^2$.

13.46 Calculate a 95% confidence interval for β_1 in Exercise 13.44. What does β_1 mean in this problem?

13.47 The management science staff of a grocery products manufacturer is developing a linear programming model for the production and distribution of its cereal products. The model requires transportation costs for a monstrous number of origins and destinations. It is impractical to do the detailed tariff analysis for every possible combination, so sample of 50 routes is selected. For each route, the mileage x and shipping rate y (in dollars per 100 pounds) are found.

A regression analysis is performed, with the following output:

```
Multiple R            .99291       Analysis of Variance
R Square              .98587                        DF      Sum of Squares      Mean Square
Adjusted R Square     .98556       Regression        1         15558.63206      15558.63206
Standard Error       2.20210       Residual         46           223.06460          4.84923

                                   F =     3208.47441       Signif F =  .0000

------------------ Variables in the Equation -----------------

Variable              B       SE B          Beta          T  Sig T

MILEAGE         .050115  8.8475E-04     .992908     56.643  .0000
(Constant)     9.770917     .474035                 20.612  .0000
```

The data are

Mileage:	50	60	80	80	90	90	100	100	100	110	110	110
Rate:	12.7	13.0	13.7	14.1	14.6	14.1	15.6	14.9	14.5	15.3	15.5	15.9
Mileage:	120	120	120	120	130	130	140	150	170	190	200	230
Rate:	16.4	11.1	16.0	15.8	16.0	16.7	17.2	17.5	18.6	19.3	20.4	21.8
Mileage:	260	300	330	340	370	400	440	480	470	510	540	600
Rate:	24.7	24.7	18.0	27.1	28.2	30.6	31.8	32.4	34.5	35.0	36.3	41.4
Mileage:	650	700	720	760	800	810	850	920	960	1050	1200	1650
Rate:	46.4	45.8	46.6	48.0	51.7	50.2	53.6	57.9	56.1	58.7	75.8	89.0

a. Write the regression equation and the residual standard deviation.

b. Calculate a 90% confidence interval for β_1.

13.48 Plot the data of Exercise 13.47. Do you see any problems with the data?

13.49 Predict the shipping rate for a 340-mile route. Calculate a 95% prediction interval. How serious is the extrapolation problem in this exercise?

APPENDIX: THE MATHEMATICS OF LEAST SQUARES (∂)

In this Appendix we use calculus to show that the formulas given in Chapter 13 do in fact give the least-squares intercept $\hat{\beta}_0$ and slope $\hat{\beta}_1$. Then we give an algebraic justification for the interpretation of r^2_{yx}.

First, the least-squares estimate of

$$E(Y_i) = \beta_0 + \beta_1 x_i$$

requires that we choose $\hat{\beta}_0$ and $\hat{\beta}_1$ to minimize

$$\sum_i (y_i - \hat{y}_i)^2 = \sum_i (y_i - \hat{\beta}_0 - \hat{\beta}_1 x_i)^2$$

Standard calculus methods indicate that we should take partial derivatives with respect to $\hat{\beta}_0$ and to $\hat{\beta}_1$ and equate them to zero. (While it is logically possible that this procedure yield a maximum or a saddle point, the solution is a minimum; a second derivative test verifies this.) The resulting equations are

$$\sum 2(y_i - \hat{\beta}_0 - \hat{\beta}_1 x_i)(-1) = 0$$

and

$$\sum 2(y_i - \hat{\beta}_0 - \hat{\beta}_1 x_i)(-x_i) = 0$$

Dividing by 2 and collecting terms yields

$$\sum y_i = n\hat{\beta}_0 + (\sum x_i)\hat{\beta}_1$$

and

$$\sum x_i y_i = (\sum x_i)\hat{\beta}_0 + (\sum x_i^2)\hat{\beta}_1$$

To solve for $\hat{\beta}_1$, multiply the first equation by $(\sum x_i)/n$ and subtract the result from the second to get

$$\frac{\sum x_i y_i - (\sum x_i)(\sum y_i)}{n} = \left[\frac{\sum x_i^2 - (\sum x_i)^2}{n}\right]\hat{\beta}_1$$

or

$$\hat{\beta}_1 = \frac{\sum x_i y_i - (\sum x_i)(\sum y_i)/n}{\sum x_i^2 - (\sum x_i)^2/n}$$

Then $n\hat{\beta}_0 = \sum y_i - \hat{\beta}_1 \sum x_i$ from the first equation, so

$$\hat{\beta}_0 = \bar{y} - \hat{\beta}_1 \bar{x}$$

Thus we have shown that the least-squares principle yields the indicated slope and intercept estimates.

Our next task is to establish that the proportion of the squared error of y values explained by regression is r_{yx}^2. To do so we need to establish a relation between the correlation r_{yx} and the slope $\hat{\beta}_1$. The defining equations are

$$r_{yx} = \frac{S_{xy}}{\sqrt{S_{xx}}\sqrt{S_{yy}}}$$

and

$$\hat{\beta}_1 = \frac{S_{xy}}{S_{xx}}$$

Therefore

$$\hat{\beta}_1 = r_{yx}\frac{\sqrt{S_{yy}}}{\sqrt{S_{xx}}}$$

SS(Regression) is defined as $\sum (\hat{y} - \bar{y})^2$. Substituting for $\bar{y}$, we have

$$\begin{aligned}
\text{SS(Regression)} &= \sum (\hat{\beta}_0 + \hat{\beta}_1 x_i - \bar{y})^2 \\
&= \sum [(\bar{y} - \hat{\beta}_1 \bar{x}) + \hat{\beta}_1 x_i - \bar{y}]^2 \\
&= (\hat{\beta}_1)^2 \sum (x_i - \bar{x})^2 \\
&= r_{yx}^2 \frac{S_{yy}}{S_{xx}} \sum (x_i - \bar{x})^2
\end{aligned}$$

Because $S_{xx} = \sum x^2 - (\sum x)^2/n$ is the shortcut formula for $\sum (x - \bar{x})^2$ and since S_{yy} is the shortcut formula for $\sum (y - \bar{y})^2 = \text{SS(Total)}$, it follows that

$$\begin{aligned}
\text{SS(Regression)} &= r_{yx}^2 \frac{S_{yy}}{S_{xx}} \sum (x - \bar{x})^2 \\
&= r_{yx}^2 \, \text{SS(Total)}
\end{aligned}$$

14

MULTIPLE REGRESSION METHODS

Multiple regression, the use of many independent variables to predict a dependent variable, is probably the statistical technique used most often by managers. In this chapter we set out the foundations of multiple regression, define the statistical model, develop formulas for least-squares estimates of model parameters, and describe a number of tests of significance. In Chapter 15 we look at some of the practical problems that arise in using these multiple regression models.

14.1 THE MULTIPLE REGRESSION MODEL

A multiple regression forecasting equation can take many forms. It may involve only two independent variables, such as $\hat{y} = 28.27 + 1.36x_1 - 2.43x_2$, or it may involve many. For example,

$$\hat{y} = 49.23 + 11.70x_1 + 2.92x_2 - 0.21x_1^2 - 0.32x_1x_2 + 2.27 \log x_2$$

is a perfectly legitimate multiple regression forecasting equation. Our first task is to define the general multiple regression model.

The basic distinction between linear regression and multiple regression is that multiple regression involves several predictor (independent) variables, whereas with linear regression we deal with a single independent variable using the model $Y = \beta_0 + \beta_1 x + \epsilon$. The formal assumptions of linear regression carry over unchanged to multiple regression.

The multiple regression model is a direct extension of the simple linear regression model, $Y_i = \beta_0 + \beta_1 x_{i1} + \epsilon_i$. It is written

$$Y_i = \beta_0 + \beta_1 x_{i1} + \beta_2 x_{i2} + \ldots + \beta_k x_{ik} + \epsilon_i$$

where x_{ij} is the ith observation for the predictor variable x_j. Note that any independent variable may in fact be a function of other ones, such as $x_3 = x_1^2$ or $x_4 = x_1 x_2$. The only restriction is that no independent variable should be a linear function of other x's.

The β parameters in the multiple regression model are most easily interpreted when there are no cross-product or squared terms. The parameter β_0 is the intercept (or constant) term—the expected value of Y when all x's are zero. As we've indicated before, in many situations values of zero for the x's may be meaningless or absurd. In such cases, the intercept term β_0 shouldn't be interpreted too literally. The **partial slopes** $\beta_1, \beta_2, \ldots, \beta_k$ are more important for interpretation. In linear regression, β_1 is the expected change in Y per unit change in x_1. In multiple regression, β_1 is the expected change in Y per unit change in x_1, provided that all other x's stay constant. (The partial slope β_1 is the partial derivative of expected Y with respect to X_1, $\partial E(Y)/\partial x_1$.) Suppose $Y_i = 20 + 4.2x_{i1} + 3.3x_{i2} + \epsilon_i$. Then the expected value of Y, $E(Y)$, when $x_1 = 10$ and $x_2 = 2$ is $20 + 4.2(10) + 3.3(2) = 68.6$. The expected value of Y when x_1 is increased one unit to 11 and x_2 remains constant at 2 is $20 + 4.2(11) + 3.3(2) = 72.8$, an increase in $E(Y)$ of 4.2. Of course, if x_2 is some function of x_1, such as $x_{i2} = (x_{i1})^2$, it wouldn't make sense to speak of changing x_1 but leaving x_2 constant. In such situations, the interpretation of multiple regression coefficients is harder. When powers and cross-products of the predictor variables are included in the model, we do not attempt to attach meaning to the individual coefficients (βs); rather, we interpret the equation as a model for the experimental situation. The goal is to obtain a good prediction of Y using the sample data.

The error term ϵ_i plays exactly the same role in multiple regression as in simple linear regression. It includes all the effects of unpredictable and ignored factors. (The basic hope of multiple regression is to make the ϵ_i values small by including all or most of the relevant predictive factors.) The formal assumptions on the ϵ_i's, which are identical to those stated in Chapter 13, are listed here:

Formal Assumptions for ϵ_i in Multiple Regression

1. $E(\epsilon_i) = 0$ for all i
2. $V(\epsilon_i) = \sigma_\epsilon^2$ for all i
3. The ϵ_i's are independent
4. ϵ_i is normally distributed

$i = 1, 2, \ldots, n$

The effect of violations of these formal assumptions is discussed in Chapter 15.

EXAMPLE 14.1 Find the error terms ϵ_i for the data below under each of the following models:

a. $Y_i = 30 + 4x_{i1} - .6x_{i2} + \epsilon_i$
b. $Y_i = 78 - 36x_{i1} + .4x_{i2} + 6x_{i1}^2 + \epsilon_i$

Y_i	x_{i1}	x_{i2}
35.2	2	10
31.8	3	20
27.7	3	10
36.7	4	20

Solution a. For the first model we have

Y_i	$30 + 4x_{i1} - .6x_{i2}$	ϵ_i
35.2	32.0	3.2
31.8	30.0	1.8
27.7	36.0	−8.3
36.7	34.0	2.7

b. For the second model,

Y_i	$78 - 36x_{i1} + .4x_{i2} + 6x_{i1}^2$	ϵ_i
35.2	34.0	1.2
31.8	32.0	−.2
27.7	28.0	−.3
36.7	38.0	−1.3

Note that the ϵ terms for the second model are much smaller (in absolute value) than those for the first model. □

general linear model

The equation, $Y_i = \beta_0 + \beta_1 x_{i1} + \beta_2 x_{i2} + \ldots + \beta_k x_{ik} + \epsilon_i$, together with the indicated assumptions on the x_{ij} and ϵ_i terms, is often called the **general linear model.** The word *linear* is used in a slightly peculiar way. Since some X predictors may be nonlinear functions of other ones, the expression does not necessarily mean that $E(Y)$ is a linear function of the independent variables. For example, the general linear model $Y_i = \beta_0 + \beta_1 x_{i1} + \beta_2 x_{i1}^2 + \beta_3 x_{i2} + \beta_4 x_{i1} x_{i2} + \epsilon_i$ is certainly not linear in the independent variables x_{i1}, x_{i2}. But if the general linear model is viewed as a function of the β parameters, with the x's regarded as given numbers, it is linear. Any predictive equation that is linear in the β parameters satisfies the general linear model. Thus

$$Y_i = 20 + 5x_i + 2v_i + \epsilon_i$$

and

$$Y_i = 20 + 5x_i + 2x_i^2 + \epsilon_i$$

are both linear models in this sense.

first- and second-order models

When we wish to describe a model that is linear in the x's, we call it a **first-order** model. A model that contains squared or product terms (and possibly linear terms) is a **second-order model.**

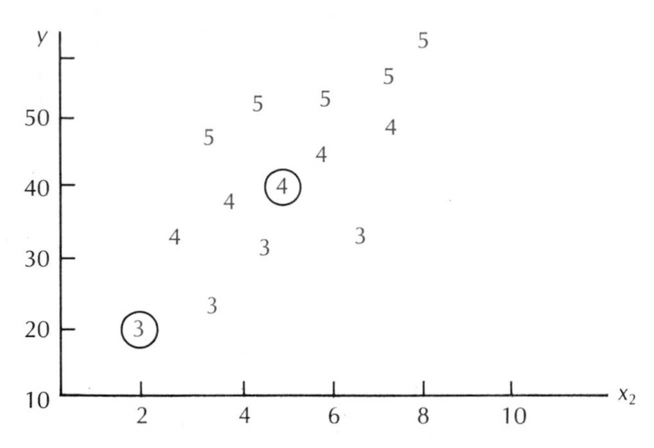

FIGURE 14.1 **Scatter Plot for a First-order Model**

Consider the model $Y_i = \beta_0 + \beta_1 x_{i1} + \beta_2 x_{i2} + \epsilon_i$. The implications of this model, which is linear (first-order) in the two independent variables, are shown in Figure 14.1 using a scatter plot of y versus x_2, with values of x_1 indicated by the numbers 3, 4, and 5.

The 3 in the lower left corner of Figure 14.1 displays the observation $y = 10$ for $x_1 = 3$ and $x_2 = 2$. Similarly, the middle 4 in the figure locates the observation $y = 30$ for $x_1 = 4$ and $x_2 = 5$. Note that when $x_1 = 3$, the y values are roughly linearly related to x_2. Similarly for $x_1 = 4$ and for $x_1 = 5$, y is roughly linearly related to x_2. Furthermore, the lines describing the relation between y and x_2 can be taken to be parallel and equally spaced.

The simplest multiple regression models assume that Y can be predicted by a linear function of the x's. This is an assumption with two parts. First, linearity implies that the effect on Y of a one-unit change in x_j is constant over all values of that x_j, holding all other x's constant. Second, linearity assumes that the effect on Y of a one-unit change in x_j does not depend on the values of the other x's. In the language of Chapter 12, we are assuming that the effects are **additive**; there is no interaction among the independent variables.

additive effects

Consider the scatter plot shown in Figure 14.2. While straight-line relations between y and x_2 seem to hold for $x_1 = 3$, 4, and 5, the lines are clearly not parallel. Instead the slope of the line relating y to x_2 increases as the x_1 value increases. The nonparallelism of these lines indicates that x_1 and x_2 interact and therefore that the model $Y_i = \beta_0 + \beta_1 x_{i1} + \beta_2 x_{i2} + \epsilon_i$ is inappropriate.

When there are many independent variables, it can be very difficult to detect interaction using graphs. Careful consideration of the problem setting can be helpful. Ask what the effect on the dependent variable of changing the independent variable should be. If the answer is, or might be, that it depends on the values of other variables, then interaction should be suspected.

Suppose that a regression model is used to predict yearly sales (Y) of a high-technology product, based on that year's advertising and promotion expenditures (x_1) and the previous

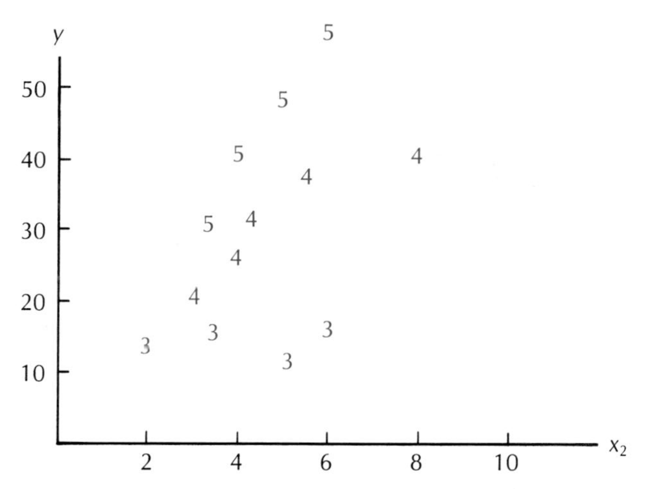

FIGURE 14.2 **Scatter Plot Showing Interaction**

year's research and development expenditures (x_2). A regression model of the form $Y = \beta_0 + \beta_1 x_1 + \beta_2 x_2 + \epsilon$ assumes that the first dollar of advertising and promotion expense (x_1) has the same effect on sales as the millionth dollar. It also implicitly assumes that this effect is the same, regardless of the value of x_2. We are skeptical of both implicit assumptions. First, we expect diminishing returns to apply to advertising and promotion. We also expect that the effect of an increase in advertising and promotion depends on x_2, the previous year's research and development expenditure. If x_2 is small, it seems likely that the products being promoted will be out of date and that the effect of promotion (x_1) will be small. If x_2 is large, it seems likely that the products will be up to date and that promotion will be more effective. Therefore we expect an **x_1x_2 interaction**.

cross-product terms

Multiple regression models that allow possible interaction between the x variables may have **cross-product terms** of the form x_1x_2. For example, using the previous sales situation involving x_1 (advertising and promotion expenditure) and x_2 (research and development expenditure for the previous year), the model $Y = \beta_0 + \beta_1 x_1 + \beta_2 x_2 + \beta_3 x_1 x_2 + \epsilon$ is a valid example of a multiple regression model that allows interaction between the variables x_1 and x_2. Similarly, if there is a third independent variable x_3 (number of salespeople), then the multiple regression model, $Y = \beta_0 + \beta_1 x_1 + \beta_2 x_2 + \beta_3 x_3 + \epsilon$ does not allow interactions among the independent variables, but any of the following models does.

$$Y = \beta_0 + \beta_1 x_1 + \beta_2 x_2 + \beta_3 x_3 + \beta_4 x_1 x_2 + \epsilon$$
$$Y = \beta_0 + \beta_1 x_1 + \beta_2 x_2 + \beta_3 x_3 + \beta_4 x_1 x_2 + \beta_5 x_1 x_3 + \epsilon$$
$$Y = \beta_0 + \beta_1 x_1 + \beta_2 x_2 + \beta_3 x_3 + \beta_4 x_1 x_2 + \beta_5 x_1 x_3 + \beta_6 x_2 x_3 + \epsilon$$

Interactions between independent variables in multiple regression are discussed further in Section 15.3.

EXAMPLE 14.2 Hypothetical sample sales data (Y) are given below for different expenditures on promotion and advertising (x_1) and different research and development expenditures (x_2) for the

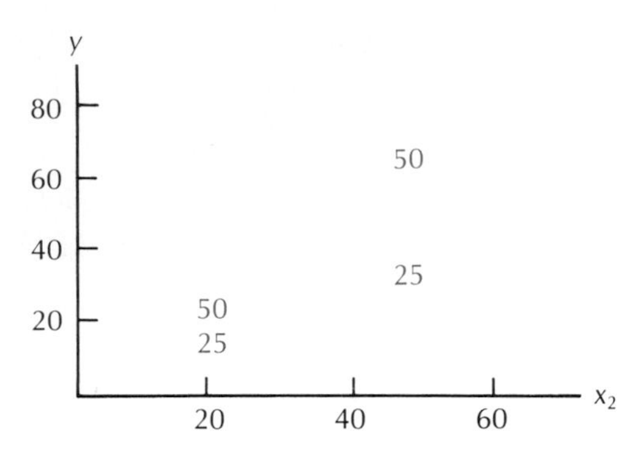

FIGURE 14.3 **Scatter Plot of y versus x_2 for $x_1 = 25$ and 50**

previous year. Do the data suggest an interaction between the variables x_1 and x_2? If so, indicate a possible multiple regression model.

Sales y (in \$1,000,000)	Promotion/Advertising x_1 (in \$1,000)	Research/Development x_2 (in \$1,000)
20	25	20
35	25	45
24	50	20
70	50	45

Solution A scatter plot of the data is shown in Figure 14.3. At the low level of x_2 (20), a doubling of the advertising and promotion from \$25,000 to \$50,000 effects an increase in sales of only \$4,000,000. However, when x_2 is at the higher level, a doubling of advertising and promotion raises sales by \$35,000,000. Clearly there is an interaction between variables x_1 and x_2. The multiple regression model $Y = \beta_0 + \beta_1 x_1 + \beta_2 x_2 + \epsilon$ does not provide a good model for sales forecasting. The model $Y = \beta_0 + \beta_1 x_1 + \beta_2 x_2 + \beta_3 x_1 x_2 + \epsilon$ provides a much better model for forecasting sales in the presence of an $x_1 x_2$ interaction. □

EXAMPLE 14.3 A beverage manufacturer tests possible formulations of a new lime-flavored soft drink. Three possible concentrations of lime flavoring are combined with three different degrees of sweetening. Taste-test panels assign preference scores to each combination on a 0–100 scale. What are the implications of a regression model that predicts the panel score (Y) as a linear function of lime concentration (x_1) and sweetness (x_2)?

Solution Taken to extremes, the model would be absurd. Assume, for example, that the coefficients of x_1 and x_2 were both positive. Then the linearity assumption would imply that increases in lime concentration or sweetness would always yield improved preference, even when x_1 and x_2 were in the tooth-rotting range. Furthermore, the additivity assumption would imply that the effect of changing sweetness would be the same for low levels of lime concentration as for high levels. Perhaps within a narrow range of variation of x_1 and x_2 these would be decent approximations, but surely they are shaky assumptions over the whole range of possible values of x_1 and x_2. □

In this section we have discussed the form of a multiple regression prediction equation. Obviously, in any given situation the choice of the particular model is of utmost importance and requires thoughtful consideration by the manager. The ultimate test is whether or not it provides good forecasts.

We turn our attention now to the important problem of determining estimates for the coefficients (the intercept β_0 and the partial slopes $\beta_1, \ldots, \beta_k$) in the multiple regression model in order to obtain the multiple regression forecasting equation.

14.2 ESTIMATING MULTIPLE REGRESSION COEFFICIENTS

The first task of a multiple regression analysis is to estimate the intercept β_0, the partial slopes $\beta_1, \beta_2, \ldots, \beta_k$, and the error variance σ^2_ϵ. The basic principle is least-squares estimation, just as in linear regression. The arithmetic involved in multiple regression gets heavy very quickly, so the work is usually done by a computer. The only examples we work out by hand here involve a small number of independent variables and unrealistically few observations.

normal equations

In linear regression, the least-squares principle yields two equations, the **normal equations**, that must be solved for the estimated intercept $\hat{\beta}_0$ and the estimated slope $\hat{\beta}_1$ (see the Appendix to Chapter 13). In multiple regression, there are $k + 1$ equations to be solved for the estimated intercept $\hat{\beta}_0$ and the k estimated partial slopes $\hat{\beta}_1, \hat{\beta}_2, \ldots, \hat{\beta}_k$. The $k + 1$ equations are indicated in Table 14.1; the use of the variable names on the top and to the left of the equations is explained below. All sums are taken over all observations ($i = 1, 2, \ldots, n$). There is a pattern to these equations that makes them a bit less formidable than they seem. Each entry in the normal equations can be obtained by multiplying the variables for the indicated row and column and summing. For example, in the last equation, the value $\sum x_{ik}y_i$ on the left side is obtained by multiplying the indicated row entry x_{ik} by the indicated column entry y_i and summing. This pattern even works for the coefficient of $\hat{\beta}_0$ in the first equation: multiplying 1 by $\hat{\beta}_0$ and adding n times yields the term $n\hat{\beta}_0$.

TABLE 14.1 **Normal Equations for Multiple Regression Coefficients**

	y_i		$\hat{\beta}_0$		$x_{i1}\hat{\beta}_1$		$x_{i2}\hat{\beta}_2$		$\ldots$		$x_{ik}\hat{\beta}_k$
1	$\sum y_i$	=	$n\hat{\beta}_0$	+	$\sum x_{i1}\hat{\beta}_1$	+	$\sum x_{i2}\hat{\beta}_2$	+	$\ldots$	+	$\sum x_{ik}\hat{\beta}_k$
x_{i1}	$\sum x_{i1}y_i$	=	$\sum x_{i1}\hat{\beta}_0$	+	$\sum x_{i1}^2\hat{\beta}_1$	+	$\sum x_{i1}x_{i2}\hat{\beta}_2$	+	$\ldots$	+	$\sum x_{i1}x_{ik}\hat{\beta}_k$
x_{i2}	$\sum x_{i2}y_i$	=	$\sum x_{i2}\hat{\beta}_0$	+	$\sum x_{i1}x_{i2}\hat{\beta}_1$	+	$\sum x_{i2}^2\hat{\beta}_2$	+	$\ldots$	+	$\sum x_{i2}x_{ik}\hat{\beta}_k$
$\vdots$	$\vdots$										
x_{ik}	$\sum x_{ik}y_i$	=	$\sum x_{ik}\hat{\beta}_0$	+	$\sum x_{i1}x_{ik}\hat{\beta}_1$	+	$\sum x_{i2}x_{ik}\hat{\beta}_2$	+	$\ldots$	+	$\sum x_{ik}^2\hat{\beta}_k$

EXAMPLE 14.4 Use the following data to estimate the coefficients of the multiple regression model $Y = \beta_0 + \beta_1 x_1 + \beta_2 x_2 + \epsilon$.

y	x_1	x_2
3	−2	−1
10	−1	0
12	0	0
12	1	0
13	2	1

Solution There are three equations to solve:

	y_i		$\hat{\beta}_0$		$x_{i1}\hat{\beta}_1$		$x_{i2}\hat{\beta}_2$
1	$\sum y_i$	=	$n\hat{\beta}_0$	+	$\sum x_{i1}\hat{\beta}_1$	+	$\sum x_{i2}\hat{\beta}_2$
x_{i1}	$\sum x_{i1}y_i$	=	$\sum x_{i1}\hat{\beta}_0$	+	$\sum x_{i1}^2\hat{\beta}_1$	+	$\sum x_{i1}x_{i2}\hat{\beta}_2$
x_{i2}	$\sum x_{i2}y_i$	=	$\sum x_{i2}\hat{\beta}_0$	+	$\sum x_{i1}x_{i2}\hat{\beta}_1$	+	$\sum x_{i2}^2\hat{\beta}_2$

The required sums for these equations are $\sum y_i = 50$, $\sum x_{i1} = 0$, $\sum x_{i2} = 0$, $\sum x_{i1}y_i = 22$, $\sum x_{i1}^2 = 10$, $\sum x_{i1}x_{i2} = 4$, $\sum x_{i2}y_i = 10$, and $\sum x_{i2}^2 = 2$; $n = 5$. Substituting these quantities into the normal equations, we have

$$50 = 5\hat{\beta}_0 + 0\hat{\beta}_1 + 0\hat{\beta}_2$$
$$22 = 0\hat{\beta}_0 + 10\hat{\beta}_1 + 4\hat{\beta}_2$$
$$10 = 0\hat{\beta}_0 + 4\hat{\beta}_1 + 2\hat{\beta}_2$$

The solution for $\hat{\beta}_0$, $\hat{\beta}_1$, and $\hat{\beta}_2$ is outlined here:

1. The first equation yields $\hat{\beta}_0 = 50/5 = 10$.
2. Multiplying the third equation by 2 and subtracting it from the second equation yields $2 = 2\hat{\beta}_1$, so $\hat{\beta}_1 = 1$.
3. Now we can substitute $\hat{\beta}_1 = 1$ in, say, the third equation to get $10 = 4(1) + 2\hat{\beta}_2$, or $\hat{\beta}_2 = 3$. The multiple regression forecasting equation is thus $\hat{y} = 10 + 1x_1 + 3x_2$. □

Literally thousands of computer programs have been written to perform the calculations needed in multiple regression. The output of such programs typically has a list of variable names, together with the estimated partial slopes, labeled COEFFICIENTS (or ESTIMATES or PARAMETERS). The intercept term $\hat{\beta}_0$ is usually called INTERCEPT (or CONSTANT); sometimes it is shown along with the slopes, but with no variable name.

EXAMPLE 14.5 The data underlying Example 14.3 (shown next) are analyzed with the Minitab software package. Identify the estimates of the partial slopes and the intercept.

y:	25	34	28	40	36	42	44	53	49
x_1:	−10	−10	−10	0	0	0	10	10	10
x_2:	−5	0	5	−5	0	5	−5	0	5

```
MTB > regress c1 on 2 variables in c2 and c3

The regression equation is
Y = 39.0 + 0.983 X1 + 0.333 X2

Predictor        Coef          Stdev      t-ratio
Constant       39.000          1.256        31.05
X1             0.9833         0.1538         6.39
X2             0.3333         0.3076         1.08

s = 3.768        R-sq = 87.5%      R-sq(adj) = 83.3%

Analysis of Variance

SOURCE        DF          SS          MS
Regression     2      596.83      298.42
Error          6       85.17       14.19
Total          8      682.00

SOURCE        DF      SEQ SS
X1             1      580.17
X2             1       16.67
```

Solution This output gives the values twice, once by printing THE REGRESSION EQUATION IS Y = $39.0 + 0.983x_1 + 0.333x_2$ and once in the column headed COEF. Note that the intercept value 39.0000 does not have a variable name; the estimated partial slopes .9833 and .3333 are associated with x_1 and x_2, respectively. □

The coefficient of an independent variable x_j in a multiple regression equation does not, in general, equal the coefficient that would apply to that variable in a simple linear regression. In multiple regression, the coefficient refers to the effect of changing that x_j variable while other independent variables stay constant. In simple linear regression, all other potential independent variables are ignored. If other independent variables are correlated with x_j (and therefore don't tend to stay constant while x_j changes), simple linear regression with only x_j as an independent variable captures not only the direct effect of changing x_j but also the indirect effect of the associated changes in other x's. Multiple regression is a method for isolating only the direct effect of changing x_j alone.

EXAMPLE 14.6

a. Fit the simple linear regression model $Y_i = \beta_0 + \beta_1 x_{i1} + \epsilon_i$ using the data of Example 14.4.
b. Compare the coefficient of x_1 for this model and the one obtained for the multiple regression model in Example 14.4.

Solution a. Using the formulas $\hat{\beta}_1 = S_{x_1y}/S_{x_1x_1}$ and $\hat{\beta}_0 = \bar{y} - \hat{\beta}_1\bar{x}_1$ from Chapter 13, it can be shown that the simple linear regression equation is $\hat{y} = 10 + 2.2x_1$.

b. The coefficients for the simple linear regression model are different (10 and 2.2) because the coefficient of x_1 in the simple linear regression represents the expected change in y for a unit change in x_1 (allowing x_2 to vary), while in the multiple regression equation the coefficient of x_1 is the expected change in y when x_2 is held constant. □

In addition to estimating the intercept and partial slopes, it is important to estimate the residual standard deviation s_ϵ. The residuals are defined as before, the difference between the observed value and the predicted value of Y:

$$y_i - \hat{y}_i = y_i - (\hat{\beta}_0 + \hat{\beta}_1 x_{i1} + \hat{\beta}_2 x_{i2} + \ldots + \hat{\beta}_k x_{ik})$$

The sum of squared residuals SS(Residual), also called SS(Error), is defined exactly as it sounds:

$$\begin{aligned} \text{SS(Residual)} &= \sum (y_i - \hat{y}_i)^2 \\ &= \sum [y_i - (\hat{\beta}_0 + \hat{\beta}_1 x_{i1} + \hat{\beta}_2 x_{i2} + \ldots + \hat{\beta}_k x_{ik})]^2 \end{aligned}$$

A shortcut computing formula is

$$\text{SS(Residual)} = \sum y_i^2 - \hat{\beta}_0 \sum y_i - \hat{\beta}_1 \sum x_{i1}y_i - \hat{\beta}_2 \sum x_{i2}y_i - \ldots - \hat{\beta}_k \sum x_{ik}y_i$$

The d.f. for this sum of squares is $n - (k + 1)$. One d.f. is subtracted for the intercept and one d.f. is subtracted for each of the k partial slopes. The mean square residual MS(Residual), also called MS(Error), is the residual sum of squares divided by $n - (k + 1)$. Finally, the residual standard deviation s_ϵ is the square root of MS(Residual).

$$\begin{aligned} s_\epsilon &= \sqrt{\text{MS(Residual)}} \\ &= \sqrt{\frac{\text{SS(Residual)}}{n - (k + 1)}} \end{aligned}$$

EXAMPLE 14.7

a. Calculate the SS(Residual) both from the definition and by the shortcut formula for the data of Example 14.4.
b. Calculate the residual standard deviation.

Solution a. To get the residuals, we need the predicted values $\hat{y}_i$. These values can be found by plugging the appropriate x_{i1} and x_{i2} values into the equation $\hat{y}_i = 10 + 1x_1 + 3x_2$.

y_i	$\hat{y}_i$	$y_i - \hat{y}_i$	$(y_i - \hat{y}_i)^2$
3	5	−2	4
10	9	1	1
12	10	2	4
12	11	1	1
13	15	−2	4
		0	14

As shown, SS(Residual) = 14. Also

$$\begin{aligned}\text{SS(Residual)} &= \sum y_i^2 - \hat{\beta}_0 \sum y_i - \hat{\beta}_1 \sum x_{i1}y_i - \hat{\beta}_2 \sum x_{i2}y_i \\ &= 566 - 10(50) - 1(22) - 3(10) \\ &= 14\end{aligned}$$

b. The d.f. is 5 − (2 + 1) = 2, so the residual standard deviation is

$$s_\epsilon = \sqrt{\frac{14}{2}} = 2.646$$

□

EXAMPLE 14.8 Identify SS(Residual) and s_ϵ in the output of Example 14.5.

Solution In the section of the output labeled ANALYSIS OF VARIANCE, SS(Residual) is shown as 85.17, with 6 d.f. The residual standard deviation is indicated by $s = 3.768$. □

The residual standard deviation is crucial in determining the probable error of a prediction using the regression equation. The precise standard error to be used in forecasting an individual Y value is stated in Section 14.6. As in simple linear regression, this standard error depends on the sample size and the degree of extrapolation involved in the forecast. The standard error must be multiplied by an appropriate t table value to give the probable error. A rough approximation, ignoring extrapolation and d.f. effects, is that the probable error is $\pm 2s_\epsilon$. This approximation can be used as a rough indicator of the forecasting quality of a regression model.

EXAMPLE 14.9 The admissions office of a business school develops a regression model that uses aptitude test scores and class rank to predict the grade average (4.00 = straight A; 2.00 = C average, the minimum graduation average; 0.00 = straight F). The residual standard deviation is $s_\epsilon = .46$. Does this value suggest highly accurate prediction?

Solution A measure of the probable error of prediction is $2s_\epsilon = .92$. For example. if a predicted average is 2.80, then an individual's grade is roughly between 2.80 − .92 = 1.88 (not good enough to graduate) and 2.80 + .92 = 3.72 (good enough to graduate magna cum laude). This is not an accurate forecast. □

coefficient of determination

The final topic to be discussed in this section is the **coefficient of determination**, R^2. This quantity is defined and interpreted very much like the r^2 value in Chapter 13. (The customary notation is R^2 for multiple regression and r^2 for linear regression.) As in Chapter 13, the coefficient of determination is defined as the proportional reduction in the squared error of Y obtained by knowledge of the values of $x_1, \ldots, x_k$. For example, if we have the multiple regression model $Y = \beta_0 + \beta_1 x_1 + \beta_2 x_2 + \beta_3 x_3 + \epsilon$ and $R^2_{Y \cdot x_1 x_2 x_3} = .736$, then 73.6% of the variability of the Y values is accounted for by variability in x_1, x_2, and x_3. Formally

$$R^2_{Y \cdot x_1 \ldots x_k} = \frac{\text{SS(Total)} - \text{SS(Residual)}}{\text{SS(Total)}}$$

where

$$\text{SS(Total)} = \sum y_i^2 - \frac{(\sum y_i)^2}{n}$$

EXAMPLE 14.10 Calculate $R^2_{y \cdot x_1 x_2}$ for the data of Example 14.4.

Solution In Example 14.7, SS(Residual) was calculated to be 14. SS(Total) can be calculated as

$$\sum y_i^2 - \frac{(\sum y_i)^2}{n} = 566 - \frac{(50)^2}{5} = 66$$

Therefore

$$R^2_{y \cdot x_1 x_2} = \frac{66 - 14}{66} = .788$$

□

EXAMPLE 14.11 Locate the value of $R^2_{y \cdot x_1 x_2}$ in the computer output of Example 14.5.

Solution We want R-SQ = 87.5%, not the one that is ADJ. Alternatively, SS(Total) = 682.0 and SS(Residual) = 85.2 are shown on the output and we can compute $R^2_{y \cdot x_1 x_2} = (682.0 - 85.17)/682.0 = .875$. □

There is no general relation between the multiple R^2 from a multiple regression equation and the individual coefficients of determination $r^2_{yx_1}, r^2_{yx_2}, \ldots, r^2_{yx_k}$ other than $R^2 \geq \max_j(r^2_{yx_j})$. If all the independent variables are themselves uncorrelated, then coefficients of determination can be added. If the x's are correlated, it is difficult to take apart the overall predictive value of $x_1, x_2, \ldots, x_k$ as measured by $R^2_{y \cdot x_1 \ldots x_k}$, into separate pieces that can be attributable to x_1 alone, to x_2 alone, . . . , to x_k alone. When the independent variables are themselves correlated, **collinearity** (sometimes called **multicollinearity**) is present. Collinearity is usually present to some degree in a multiple regression study, being only a slight problem for slightly correlated x's but a more severe one for highly correlated x's. Thus, if collinearity occurs in a regression study, and it usually does to some degree, it is almost impossible to take apart the overall $R^2_{y \cdot x_1 x_2 \ldots x_k}$ into separate components associated with each x variable. The correlated x's account for overlapping pieces of the variability in y, so that often, but not inevitably,

collinearity

$$R^2_{y \cdot x_1 x_2 \ldots x_k} < r^2_{yx_1} + r^2_{yx_2} + \ldots + r^2_{yx_k}$$

EXAMPLE 14.12

a. Calculate r_{yx_1}, r_{yx_2}, and $r_{x_1 x_2}$ for the data of Example 14.4.
b. Is collinearity present?
c. Does $R^2_{y \cdot x_1 x_2} = r^2_{yx_1} + r^2_{yx_2}$?

Solution a. The formula for r_{yx} in Chapter 13 becomes

$$r_{yx_1} = \frac{S_{x_1 y}}{\sqrt{S_{x_1 x_1}}\sqrt{S_{yy}}}$$

$$= \frac{22 - 0(50)/5}{\sqrt{10 - (0)^2/5}\sqrt{566 - (50)^2/5}}$$

$$= .8563$$

Similarly,

$$r_{yx_2} = \frac{S_{x_2y}}{\sqrt{S_{x_2x_2}}\sqrt{S_{yy}}}$$

$$= \frac{10 - (0)(50)/5}{\sqrt{2 - (0)^2/5}\sqrt{566 - (50)^2/5}}$$

$$= .8704$$

and

$$r_{x_1x_2} = \frac{S_{x_1x_2}}{\sqrt{S_{x_1x_1}}\sqrt{S_{x_2x_2}}}$$

$$= \frac{4 - (0)(0)/5}{\sqrt{10 - (0)^2/5}\sqrt{2 - (0)^2/5}}$$

$$= .8944$$

b. Because $r_{x_1x_2} \neq 0$, collinearity is present, and indeed it is fairly serious.
c. Because

$$r^2_{yx_1} = (.8563)^2 = .733$$
$$r^2_{yx_2} = (.8704)^2 = .758$$

and

$$R^2_{y \cdot x_1x_2} = .788$$

$R^2_{y \cdot x_1x_2}$ is less than the sum of $r^2_{yx_1}$ and $r^2_{yx_2}$. □

SECTIONS 14.1 AND 14.2 EXERCISES

14.1 A manufacturer of industrial chemicals investigates the effect on its sales of promotion activities (primarily direct contact and trade show), direct development expenditures, and short-range research effort. Data are assembled for 24 quarters (6 years) and analyzed by a standard multiple regression program (SAS) as shown below (in $100,000 per quarter).

DEPENDENT VARIABLE: SALES

SOURCE	DF	SUM OF SQUARES	MEAN SQUARE	F VALUE	PR > F	R-SQUARE	C.V.
MODEL	3	43901.76502960	14633.92167653	22.28	0.0001	0.769693	8.1489
ERROR	20	13136.23497040	656.81174852		ROOT MSE		SALES MEAN
CORRECTED TOTAL	23	57038.00000000			25.62833878		314.50000000

PARAMETER	ESTIMATE	T FOR H0: PARAMETER=0	PR > |T|	STD ERROR OF ESTIMATE
INTERCEPT	326.38940741	1.35	0.1918	241.61283337
PROMO	136.09828861	4.84	0.0001	28.10758630
DEVEL	-61.17527129	-1.20	0.2438	50.94101324
RESEARCH	-43.69507838	-0.90	0.3766	48.32295140

a. Write the estimated regression equation.
b. Calculate the residual standard deviation.
c. Locate SS(Residual).

14.2 State the interpretation of $\hat{\beta}_1$, the estimated coefficient of promotion expenses, of Exercise 14.1.

14.3 The following artificial data are designed to illustrate the effect of correlated and uncorrelated independent variables:

y:	17	21	26	22	27	25	28	34	29	37	38	38
x:	1	1	1	1	2	2	2	2	3	3	3	3
w:	1	2	3	4	1	2	3	4	1	2	3	4
v:	1	1	2	2	3	3	4	4	5	5	6	6

a. Plot x versus w, x versus v, and w versus v.
b. Which of these plots indicate zero correlations?
c. Show that the simple linear regression equation for predicting y from x is

$$\hat{y} = 14.5 + 7.0x$$

The following sums are needed:

$$\sum x_i = 24, \quad \sum y_i = 342, \quad \sum x_i y_i = 740$$
$$\sum x_i^2 = 56, \quad \sum y_i^2 = 10{,}282, \quad n = 12$$

d. Calculate the residual standard deviation.
e. Calculate r_{yx}^2.

14.4 Refer to the data of Exercise 14.3.

a. Suppose that the multiple regression equation $\hat{y} = \hat{\beta}_0 + \hat{\beta}_1 x + \hat{\beta}_2 w + \epsilon$ is used. Do you think that the coefficient of x will change from the value shown in Exercise 14.3? Will the intercept change? Why?
b. Verify that the coefficient $\hat{\beta}_0 = 8.5$, $\hat{\beta}_1 = 7.0$, and $\hat{\beta}_2 = 2.4$ satisfy the normal equations. The following sums are needed, in addition to those given in Exercise 14.3:

$$\sum w_i = 30, \quad \sum w_i y_i = 891$$
$$\sum w_i^2 = 90, \quad \sum w_i x_i = 60$$

c. Calculate the residual standard deviation.
d. Calculate $R_{y \cdot xw}^2$.

14.5 Refer to Exercise 14.3.

a. Suppose that the multiple regression equation $\hat{y} = \hat{\beta}_0 + \hat{\beta}_1 x + \hat{\beta}_2 v$ is used to predict Y given x and v. Do you think that the coefficient of X will change from the value shown in Exercise 14.3? Will the intercept change? Why?
b. Verify that the coefficients $\hat{\beta}_0 = 17.0$, $\hat{\beta}_1 = -3.0$, and $\hat{\beta}_2 = 5.0$ satisfy the least-squares equations. The following sums are needed, in addition to those given in Exercise 14.3.

$$\sum v_i = 42, \quad \sum v_i y_i = 1324$$
$$\sum v_i^2 = 182, \quad \sum v_i x_i = 100$$

c. Calculate the residual standard deviation s_ϵ.
d. Calculate $R_{y \cdot xy}^2$.

14.6 The Minitab computer package yields the following output for the data of Exercise 14.3.

```
MTB > correlations of c1-c4

              Y         X         W
X         0.856
W         0.402     0.000
V         0.928     0.956     0.262

MTB > regress c1 on 3 variables in c2 c3 c4

The regression equation is
Y = 10.0 + 5.00 X + 2.00 W + 1.00 V

Predictor        Coef           Stdev      t-ratio
Constant       10.000           5.766         1.73
X               5.000           6.895         0.73
W               2.000           1.528         1.31
V               1.000           3.416         0.29

s = 2.646        R-sq = 89.5%      R-sq(adj) = 85.6%

Analysis of Variance

SOURCE        DF          SS             MS
Regression     3      479.00         159.67
Error          8       56.00           7.00
Total         11      535.00
```

a. Write the least-squares prediction equation. Locate the residual standard deviation s_ϵ.

b. Show that the multiple R^2 is larger than the R^2 of Exercise 14.5. Is the residual standard deviation smaller?

14.7 A chemical firm tests the yield resulting from the presence of varying amounts of two catalysts. Yields are measured for five different amounts of catalyst 1 paired with four different amounts of catalyst 2. A second-order model is fit, to approximate the anticipated nonlinear relation. The variables are y = yield, x_i = amount of catalyst 1, x_2 = amount of catalyst 2, $x_3 = x_1^2$, $x_4 = x_1x_2$, and $x_5 = x_2^2$. The data are analyzed by a standard computer package (SAS). Selected output is shown below:

```
DEPENDENT VARIABLE: YIELD

SOURCE            DF   SUM OF SQUARES   MEAN SQUARE   F VALUE   PR > F       R-SQUARE        C.V.

MODEL              5     448.19324829   89.63864966     17.55   0.0001       0.862437      3.7611

ERROR             14      71.48900671    5.10635762             ROOT MSE                 YIELD MEAN

CORRECTED TOTAL   19     519.68225500                           2.25972512              60.08150000


                                       T FOR HO:     PR > |T|      STD ERROR OF
      PARAMETER           ESTIMATE     PARAMETER=0                 ESTIMATE

      INTERCEPT        50.01950000         11.39      0.0001       4.39050111
      CAT1              6.64357143          3.30      0.0052       2.01211633
      CAT2              7.31450000          2.67      0.0183       2.73977110
      CAT1SQ           -1.23142857         -4.08      0.0011       0.30196847
      CAT1CAT2         -0.77240000         -2.42      0.0299       0.31957339
      CAT2SQ           -1.17550000         -2.33      0.0355       0.50528990
```

a. Write the estimated regression equation.
b. Locate SS(Residual) and the residual standard deviation.

14.8 Refer to Exercise 14.7. Calculate $R^2_{y \cdot x_1 x_2 x_3 x_4 x_5}$.

14.3 INFERENCES IN MULTIPLE REGRESSION

The ideas of the preceding section involve point (best-guess) estimation of the regression coefficients, the standard deviation s_ϵ, and the coefficients of determination R^2. In this section we discuss inferences about the partial slope parameters in the multiple regression model.

First we present a test of an overall null hypothesis about the partial slopes ($\beta_1, \beta_2, \ldots, \beta_k$) in the multiple regression model. According to this hypothesis, H_0: $\beta_1 = \beta_2 = \ldots = \beta_k = 0$, none of the variables included in the multiple regression has any predictive value at all. The research hypothesis is a very general one, namely, H_a: at least one $\beta_j \neq 0$. The test statistic is an F statistic very similar to the F statistics of Chapter 12. To state the test, we must first define the sum of squares attributable to the regression of Y on the variables $x_1, x_2, \ldots, x_k$. We designate this sum of squares SS(Regression).

SS(Regression)

$$\text{SS(Regression)} = \sum (\hat{y}_i - \bar{y})^2$$
$$\text{SS(Total)} = \sum (y_i - \bar{y})^2$$
$$= \text{SS(Regression)} + \text{SS(Residual)}$$

Unlike SS(Total) and SS(Residual), SS(Regression) is not interpreted in terms of prediction error. Rather, it measures the extent to which the predictions $\hat{y}_i$ vary as the x's vary. If SS(Regression) = 0, the predicted y values ($\hat{y}$) are all the same. In such a case, information about the x's is useless in predicting y. If SS(Regression) is large relative to SS(Residual), the indication is that there is real predictive value in the independent variables $x_1, x_2, \ldots, x_k$. The test statistic is stated in terms of mean squares rather than sums of squares. As always, a mean square is a sum of squares divided by the appropriate d.f.

***F* Test of H_0: $\beta_1 = \beta_2 = \ldots = \beta_k = 0$**

H_0: $\beta_1 = \beta_2 = \ldots = \beta_k = 0$

H_a: At least one $\beta_j \neq 0$

$$\text{T.S.: } F = \frac{\text{SS(Regression)}/k}{\text{SS(Residual)}/[n - (k + 1)]} = \frac{\text{MS(Regression)}}{\text{MS(Residual)}}$$

R.R.: With $\text{d.f.}_1 = k$ and $\text{d.f.}_2 = n - (k + 1)$, reject H_0 if $F > F_\alpha$

EXAMPLE 14.13 Carry out the F test of H_0: $\beta_1 = \beta_2 = 0$ for the data of Example 14.4. Use $\alpha = .05$.

Solution From previous work in Example 14.10 we found that SS(Total) = 66, and in Example 14.7 we computed SS(Residual) = 14. It follows that

$$\begin{aligned} \text{SS(Regression)} &= \text{SS(Total)} - \text{SS(Residual)} \\ &= 66 - 14 = 52 \end{aligned}$$

The test procedure is then

H_0: $\beta_1 = \beta_2 = 0$

H_a: At least one of the β's $\neq 0$

$$\text{T.S.: } F = \frac{\text{MS(Regression)}}{\text{MS(Residual)}} = \frac{52/2}{14/2} = 3.714$$

with $n = 5$ and $k = 2$

R.R.: For d.f.$_1$ = d.f.$_2$ = 2 and $\alpha = .05$, the rejection region is $F > 19.00$.

Conclusion: Because the computed value of F is less than 19.00, we cannot reject H_0. Of course, since n is so small, the risk of a Type II error in accepting H_0 could be quite high. □

EXAMPLE 14.14

a. Locate SS(Regression) in the computer output of Example 14.5.
b. Calculate the F statistic.
c. Can we safely conclude that the independent variables x_1 and x_2 together have predictive power?

Solution a. SS(Regression) is shown in the ANALYSIS OF VARIANCE section of the output as 596.83.

b. The MS(Regression) and MS(Residual) values are also shown there:

$$F = \frac{\text{MS(Regression)}}{\text{MS(Residual)}} = \frac{298.42}{14.19} = 21.0$$

c. For d.f.$_1$ = 2, d.f.$_2$ = 6, and $\alpha = .01$, the tabled F value is 10.92. Therefore we have strong evidence (p-value well below .01) to reject the null hypothesis and to conclude that the x's collectively have at least some predictive value. □

Rejection of the null hypothesis of this F test is not an overwhelmingly impressive conclusion. **This rejection merely indicates that there is some degree of predictive value somewhere among the independent variables**, without giving any indication of which individual independent variables are useful. The next task, therefore, is to make inferences about the individual partial slopes.

To make these inferences, we need the estimated standard error of each partial slope. These standard errors are computed and shown by most regression computer programs. The formula that we present for the standard error is useful in considering the effect of collinearity (correlated independent variables), but it is not a good way to do the computation.

Estimated Standard Error of $\hat{\beta}_j$ in a Multiple Regression

$$s_{\hat{\beta}_j} = s_\epsilon \sqrt{\frac{1}{\sum (x_{ij} - \bar{x}_j)^2 (1 - R^2_{x_j \cdot x_1 \ldots x_{j-1} x_{j+1} \ldots x_k})}}$$

where $R^2_{x_j \cdot x_1 \ldots x_{j-1} x_{j+1} \ldots x_k}$ is the R^2 value obtained by letting x_j be the **dependent** variable in a multiple regression with all other x's independent variables. Note that s_ϵ is the residual standard deviation for the multiple regression of y on $x_1, x_2, \ldots, x_k$.

EXAMPLE 14.15 Compute the estimated standard error of $\hat{\beta}_1$ for the data of Example 14.4.

Solution The standard deviation, from Example 14.7, is $s_\epsilon = 2.646$, and $\sum (x_{i1} - \bar{x}_1)^2 = \sum x_{i1}^2 - (\sum x_{i1})^2/n = 10$. The R^2 value obtained by treating x_1 as a dependent variable and x_2 (the only other x variable) as independent is simply $r^2_{x_1 x_2}$. From Example 14.12, this is $(.8944)^2 = .800$. Therefore the estimated standard error of $\hat{\beta}_1$ is

$$s_{\beta_1} = 2.646 \sqrt{\frac{1}{10(1 - .800)}} = 1.871$$

□

effect of collinearity

The most important use of the formula for estimated standard error is to illustrate the **effect of collinearity**. If the independent variable x_j is highly collinear with one or more other independent variables, $R^2_{x_j \cdot x_1 \ldots x_{j-1} x_{j+1} \ldots x_k}$ is by definition very large and $1 - R^2_{x_j \cdot x_1 \ldots x_{j-1} x_{j+1} \ldots x_k}$ is near zero. Division by a near-zero number yields a very large standard error. Thus the most important effect of severe collinearity is that it results in very large standard errors of partial slopes.

A large standard error for any estimated partial slope indicates a large probable error for the estimate. The partial slope $\hat{\beta}_j$ of x_j estimates the effect of increasing x_j by one unit while all other x's remain constant. If x_j is highly collinear with other x's, when x_j increases the other x's also vary rather than staying constant. Therefore, it is difficult to estimate β_j, and its probable error is large, when x_j is severely collinear with other independent variables.

The standard error of each estimated partial slope $\hat{\beta}_j$ is used in a confidence interval and statistical test for β_j. The confidence interval follows the familiar format of estimate $\pm$ (table value)(estimated standard error).

100(1 − α)% Confidence Interval for β_j

$$\hat{\beta}_j - t_{\alpha/2} s_{\hat{\beta}_j} < \beta_j < \hat{\beta}_j + t_{\alpha/2} s_{\hat{\beta}_j}$$

where $t_{\alpha/2}$ cuts off area $\alpha/2$ in the tail of a t distribution with d.f. $= n - (k + 1)$.

EXAMPLE 14.16 Calculate a 95% confidence interval for β_1 for the data of Example 14.4.

Solution $\hat{\beta}_1$ was found to be 1.00 in Example 14.4, and $s_{\hat{\beta}_1}$ was 1.871. The t value that cuts off an area of .025 in a t distribution with d.f. $= n - (k + 1) = 5 - (2 + 1) = 2$ is 4.303. The confidence interval is $1.00 - 4.303(1.871) < \beta_1 < 1.00 + 4.303(1.871)$, or $-7.051 < \beta_1 < 9.051$. □

EXAMPLE 14.17 Locate the estimated partial slope for x_2 and its standard error in the output of Example 14.5. Calculate a 90% confidence interval for β_2.

Solution $\hat{\beta}_2$ is .3333 with standard error (labeled STDEV .3076. The tabled t value is 1.943 [tail area .05, 9 − (2 + 1) = 6 d.f.]. The desired interval is $.3333 - 1.943(.3076) < \beta_2 < .3333 + 1.943(.3076)$, or $-.2644 < \beta_2 < .9310$. □

interpretation of H_0: $\beta_j = 0$

The usual null hypothesis for inference about β_j is H_0: $\beta_j = 0$. This hypothesis does not assert that x_j has no predictive value by itself. It asserts that it has no additional predictive value over and above that contributed by the other independent variables; that is, if all other x's had already been used in a regression model and then x_j was added last, no improvement in prediction would result. We interpret H_0: $\beta_j = 0$ to mean that x_j has no additional predictive value as the "**last predictor** in." The t test of this H_0 is summarized below.

last predictor

Summary for Testing H_0: $\beta_j = 0$

H_0: $\beta_j = 0$

H_a: 1. $\beta_j > 0$

2. $\beta_j < 0$

3. $\beta_j \neq 0$

T.S.: $t = \hat{\beta}_j / s_{\hat{\beta}_j}$

R.R.: 1. $t > t_\alpha$

2. $t < -t_\alpha$

3. $|t| > t_{\alpha/2}$

where t_a cuts off a right-tail area a in the t distribution with d.f. $= n - (k + 1)$

EXAMPLE 14.18

a. Use the information given in Example 14.16 to test H_0: $\beta_1 = 0$ at $\alpha = .05$. Use a two-sided alternative.

b. Is the conclusion of the test compatible with the confidence interval?

Solution a. The test statistic for H_0: $\beta_1 = 0$ versus H_a: $\beta_1 \neq 0$ is $t = \hat{\beta}_1 / s_{\hat{\beta}_1} = 1.00/1.871 = .534$. Because the .025 point for the t distribution with $5 - (2 + 1) = 2$ d.f. is 4.303, H_0 must be retained; x_1 does not appear to have any additional predictive power in the presence of the other independent variable x_2.

b. The 95% confidence interval includes zero, which also indicates that H_0: $\beta_1 = 0$ must be retained at $\alpha = .05$, two-tailed. □

EXAMPLE 14.19 Locate the t statistic for testing H_0: $\beta_2 = 0$ in the output of Example 14.5. Can H_a: $\beta_2 > 0$ be supported at any of the usual α levels?

Solution The t statistics are shown under the heading T−RATIO. For x_2 the t statistic is 1.08. The t table value for 6 d.f. and $\alpha = .10$ is 1.440, so H_0 cannot be rejected even at $\alpha = .10$. □

The multiple regression F and t tests that are discussed in this chapter test different null hypotheses. It sometimes happens that the F test results in the rejection of

H_0: $\beta_1 = \beta_2 = \cdots = \beta_k = 0$, while no t test of H_0: $\beta_j = 0$ is significant. In such a case, we can conclude that there is predictive value in the equation as a whole, but we cannot identify the specific variables having predictive value. Remember that each t test is testing "last predictor in" value. When the predictor variables are highly correlated among themselves, it often happens that no x_j can be shown to have significant "last in" predictive value, even though the x's together have been shown to be useful.

SECTION 14.3 EXERCISES

14.9 Refer to the computer output of Exercise 14.1.

a. Locate the F statistic.
b. Can the hypothesis of no overall predictive value be rejected at $\alpha = .01$?
c. Locate the t statistic for the coefficient of promotion $\hat{\beta}_1$.
d. Test the research hypothesis that $\beta_1 \neq 0$. Use $\alpha = .05$.
e. State the conclusion of the test in part (d).

14.10 Locate the p-value for the test of Exercise 14.9, part (d). It is one-tailed or two-tailed?

14.11 Summarize the results of the t tests in Exercise 14.9. What null hypotheses are being tested?

14.12 Refer to the computer output of Exercise 14.6.

a. Locate MS(Regression) and MS(Residual).
b. What is the value of the F statistic?
c. Determine the p-value for the F test.
d. What conclusion can be established from the F test?

14.13 A metalworking firm conducts an energy study using multiple regression methods. The dependent variable is Y = energy consumption cost per day (in thousands of dollars), and the independent variables are x_1 = tons of metal processed in the day, x_2 = average external temperature − 60°F (a union contract requires cooling of the plant whenever outside temperatures reach 60°, x_3 = rated wattage for machinery in use, and $x_4 = x_1x_2$. The data are analyzed by a standard program (SAS). Selected output is shown here.

DEPENDENT VARIABLE: ENERGY

SOURCE	DF	SUM OF SQUARES	MEAN SQUARE	F VALUE	PR > F	R-SQUARE	C.V.
MODEL	4	257.04891106	64.26222777	9.86	0.0001	0.663592	4.9894
ERROR	20	130.31108894	6.51555445		ROOT MSE		ENERGY MEAN
CORRECTED TOTAL	24	387.36000000			2.55255841		51.16000000

PARAMETER	ESTIMATE	T FOR H0: PARAMETER=0	PR > |T|	STD ERROR OF ESTIMATE
INTERCEPT	7.20439073	0.41	0.6855	17.53227919
METAL	1.36291836	1.47	0.1559	0.92438439
TEMP	0.30588336	0.19	0.8522	1.62104565
WATTS	0.01024173	2.16	0.0427	0.00473218
METXTEMP	-0.00277728	-0.04	0.9717	0.07722576

a. Write the estimated model.
b. Summarize the results of the various t tests.

14.4 INFERENCES BASED ON THE COEFFICIENT OF DETERMINATION

The methods of the previous section yield tests for the null hypothesis of no overall predictive value of all variables (F test) and of no incremental predictive value of one variable (t test). Other null hypotheses, such as that of no incremental predictive value of a set of variables, may be tested by way of the coefficient of determination R^2.

Recall that $R^2_{y \cdot x_1 \ldots x_k}$ measures the reduction in squared error for y attributed to knowledge of all the x predictors. The F test of Section 14.3 can be stated in terms of $R^2_{y \cdot x_1 \ldots x_k}$. Since the regression of y on the x's accounts for a proportion $R^2_{y \cdot x_1 \ldots x_k}$ of the total squared error in y,

$$\text{SS(Regression)} = R^2_{y \cdot x_1 \ldots x_k} \text{SS(Total)}$$

The remaining fraction, $1 - R^2$, is incorporated in the residual squared error

$$\text{SS(Residual)} = (1 - R^2_{y \cdot x_1 \ldots x_k})\ \text{SS(Total)}$$

***F* and *R*²** The F test statistic of Section 14.3 can be rewritten as

$$F = \frac{\text{MS(Regression)}}{\text{MS(Residual)}} = \frac{R^2_{y \cdot x_1 \ldots x_k}/k}{(1 - R^2_{y \cdot x_1 \ldots x_k})/[n - (k + 1)]}$$

This statistic is to be compared with tabulated F values for d.f.$_1 = k$ and d.f.$_2 = n - (k + 1)$.

EXAMPLE 14.20 A large city bank studies the relation of average account size in each of its branches to per capita income in the corresponding ZIP code area, number of business accounts, and number of competitive bank branches. The data are analyzed by a multiple regression program (BMDP), shown below:

```
 DEPENDENT VARIABLE. . . . . . . . . . . . . . .          1 ACCTSIZE

MULTIPLE R              0.8699          STD. ERROR OF EST.          2.9108
MULTIPLE R-SQUARE       0.7567

ANALYSIS OF VARIANCE
                      SUM OF SQUARES      DF      MEAN SQUARE        F RATIO     P(TAIL)
         REGRESSION        447.9653        3         149.3218         17.624      0.0000
         RESIDUAL          144.0339       17           8.4726

                                                     STD. REG
   VARIABLE          COEFFICIENT    STD. ERROR        COEFF           T    P(2 TAIL) TOLERANCE

INTERCEPT              1.15862
INCOME       2         2.66675       1.35716          1.017       1.965     0.0660     0.05343
BUSIN        3        -0.13453       0.22016         -0.193      -0.611     0.5493     0.14313
COMPET       4         0.03470       1.56340          0.007       0.022     0.9825     0.13187
```

a. Identify the multiple regression prediction equation.
b. Use the R^2 value shown to test $H_0: \beta_1 = \beta_2 = \beta_3 = 0$. (Note: $n = 21$.)

Solution a. From the output, the multiple regression forecasting equation is

$$\hat{y} = 1.15862 + 2.66675x_1 - 0.13453x_2 + 0.03470x_3$$

b. The test procedure based on R^2 is

$H_0: \beta_1 = \beta_2 = \beta_3 = 0$
H_a: At least one β_j differs from zero

$$\text{T.S.: } F = \frac{R^2_{y \cdot x_1x_2x_3}/3}{(1 - R^2_{y \cdot x_1x_2x_3})/(21 - 4)}$$

$$= \frac{.7567/3}{.2433/17} = 17.624$$

R.R.: For $\text{d.f.}_1 = 3$ and $\text{d.f.}_2 = 17$, the critical .05 value of F is 3.20.

Since the computed F statistic, 17.624, is greater than 3.20, we reject H_0 and conclude that one or more of the x values has some predictive power. Note that the F value we compute is the same as that shown in the output. □

***F* test for several β_j's**

There is another F test for the hypothesis that several ($<k$) β_j's are zero. For example, if we have the multiple regression model

$$y = \beta_0 + \beta_1x_1 + \beta_2x_2 + \beta_3x_3 + \beta_4x_4 + \beta_5x_5 + \epsilon$$

with $k = 5$ independent variables, we can test the null hypothesis

$$H_0: \beta_4 = \beta_5 = 0$$

According to this null hypothesis, the independent variables x_4 and x_5 together have no predictive value once x_1, x_2, and x_3 are included as predictors. The t test of Section 14.3 tests a single coefficient on a "last predictor in" basis. Now we are testing x_4 and x_5 on a "last 2 predictors in" basis.

The idea is to compare the R^2 values when x_4 and x_5 are excluded and when they are included in the prediction equation. When they are included, the R^2 is automatically at least as large as the R^2 when they are excluded, because we can predict at least as well with more information as with less. The F test for this null hypothesis tests whether the gain in R^2 is more than could be expected by chance alone. In general, let k be the total number of predictors, and let g be the number of predictors with coefficients not hypothesized to be zero ($g < k$). Then $k - g$ represents the number of predictors with coefficients that are hypothesized to be zero. The idea is to find R^2 values using all predictors (the **complete model**) and using only the g predictors that do not appear in the null hypothesis (the **reduced model**). Once these have been computed, the test proceeds as outlined below. The notation is easier if we assume that the reduced model contains $\beta_1, \beta_2, \ldots, \beta_g$.

complete and reduced models

An F Test of H_0: $\beta_{g+1} = \beta_{g+2} = \ldots = \beta_k = 0$

H_0: $\beta_{g+1} = \beta_{g+2} = \ldots = \beta_k = 0$

H_a: H_0 is not true

$$\text{T.S.: } F = \frac{(R^2_{\text{complete}} - R^2_{\text{reduced}})/(k-g)}{(1 - R^2_{\text{complete}})/[n-(k+1)]}$$

R.R.: $F > F_\alpha$, where F cuts off a right-tail of area α of the F distribution with $\text{d.f.}_1 = (k-g)$ and $\text{d.f.}_2 = [n-(k+1)]$

EXAMPLE 14.21 A state fisheries commission wants to estimate the number of bass caught in a given lake during a season in order to restock the lake with the appropriate number of young fish. The commission could get a fairly accurate assessment of the seasonal catch by extensive "netting sweeps" of the lake before and after a season, but this technique is much too expensive to be done routinely. Therefore the commission samples a number of lakes and records y, the seasonal catch (thousands of bass per square mile of lake area); x_1, the number of lakeshore residences per square mile of lake area; x_2, the size of the lake in square miles; $x_3 = 1$ if the lake has public access, 0 if not; and x_4, a structure index. (*Structures* are weed beds, sunken trees, dropoffs, and other living places for bass.) The data are

y	x_1	x_2	x_3	x_4
3.6	92.2	.21	0	81
.8	86.7	.30	0	26
2.5	80.2	.31	0	52
2.9	87.2	.40	0	64
1.4	64.9	.44	0	40
.9	90.1	.56	0	22
3.2	60.7	.78	0	80
2.7	50.9	1.21	0	60
2.2	86.1	.34	1	30
5.9	90.0	.40	1	90
3.3	80.4	.52	1	74
2.9	75.0	.66	1	50
3.6	70.0	.78	1	61
2.4	64.6	.91	1	40
1.9	50.0	1.10	1	22
2.0	50.0	1.24	1	50
1.9	51.2	1.47	1	37
3.1	40.1	2.21	1	61
2.6	45.0	2.46	1	39
3.4	50.0	2.80	1	53

The commission is convinced that x_1 and x_2 are important variables in predicting y because they both reflect how intensively the lake has been fished. There is some question as to whether x_3 and x_4 are useful as additional predictor variables. Therefore regression models (with all x's entering linearly) are run with and without x_3 and x_4. Relevant portions of the Minitab output are shown on the top of the next page.

```
MTB > regress c1 on 4 variables in c2-c5

The regression equation is
Y = - 1.94 + 0.0193 X1 + 0.332 X2 + 0.836 X3 + 0.0477 X4

Predictor        Coef         Stdev      t-ratio
Constant      -1.9378        0.9081        -2.13
X1            0.01929       0.01018         1.90
X2             0.3323        0.2458         1.35
X3             0.8355        0.2250         3.71
X4           0.047714      0.005056         9.44

s = 0.4336       R-sq = 88.2%       R-sq(adj) = 85.0%

Analysis of Variance

SOURCE        DF          SS          MS
Regression     4     21.0474      5.2619
Error         15      2.8206      0.1880
Total         19     23.8680

MTB > regress c1 on 2 variables in c2 and c3

The regression equation is
Y = - 0.11 + 0.0310 X1 + 0.679 X2

Predictor        Coef         Stdev      t-ratio
Constant       -0.107         2.336        -0.05
X1            0.03102       0.02650         1.17
X2             0.6794        0.6178         1.10

s = 1.138        R-sq = 7.7%        R-sq(adj) = 0.0%

Analysis of Variance

SOURCE        DF          SS          MS
Regression     2       1.845       0.922
Error         17      22.023       1.295
Total         19      23.868
```

a. Write the complete and reduced models.
b. Write the null hypothesis for testing that the omitted variables have no (incremental) predictive value.
c. Perform an F test for this null hypothesis.

Solution a. The complete and reduced models are, respectively,

$$Y_i = \beta_0 + \beta_1 x_{i1} + \beta_2 x_{i2} + \beta_3 x_{i3} + \beta_4 x_{i4} + \epsilon_i$$

and

$$Y_i = \beta_0 + \beta_1 x_{i1} + \beta_2 x_{i2} + \epsilon_i$$

The corresponding multiple regression forecasting equations based on the sample data are

$$\text{Complete: } \hat{y} = -1.94 + .0193x_1 + .332x_2 + .836x_3 + .0477x_4$$

$$\text{Reduced: } \hat{y} = -.11 + .0310x_1 + .679x_2$$

b. The appropriate null hypothesis of no predictive power for x_3 and x_4 is
H_0: $\beta_3 = \beta_4 = 0$

c. The test statistic for the H_0 of part (b) makes use of $R^2_{\text{complete}} = .882$, $R^2_{\text{reduced}} = .077$, $k = 4$, $g = 2$, and $n = 20$:

$$\text{T.S.: } F = \frac{(R^2_{\text{complete}} - R^2_{\text{reduced}})/(4-2)}{(1 - R^2_{\text{complete}})/(20-5)}$$

$$= \frac{(.882 - .077)/2}{(1 - .882)/15} = 51.165$$

R.R.: The .01 F value for d.f.$_1 = 2$ and d.f.$_2 = 15$ is 6.36, so we can reject H_0 convincingly. The variables x_3 and x_4 are useful predictors in the multiple regression equation. □

SECTION 14.4 EXERCISES

14.14 Refer to the output shown in Exercise 14.1.

a. Locate the R^2 value.
b. Calculate the F statistic (based on R^2) for the null hypothesis that none of the independent variables has predictive value.
c. Compare the value of this F statistic to that shown in the output.

14.15 Another regression analysis of the data of Exercise 14.1 is done, using only developmental expenditures as an independent variable. SAS output is shown below:

SOURCE	DF	SUM OF SQUARES	MEAN SQUARE	F VALUE	PR > F	R-SQUARE	C.V.
MODEL	1	20094.34500769	20094.34500769	11.97	0.0022	0.352298	13.0298
ERROR	22	36943.65499231	1679.25704511		ROOT MSE		SALES MEAN
CORRECTED TOTAL	23	57038.00000000			40.97873894		314.50000000

a. Locate R^2 for the reduced model.
b. Use this R^2 and that given in Exercises 14.1 to calculate an F statistic.
c. What is the mathematical null hypothesis of this test? Give a careful interpretation of this hypothesis in terms of predictive value.
d. What is the conclusion of this F test using $\alpha = .01$?

14.16 Another regression equation is calculated using BMDP based on the data of Example 14.20 (page 497). The only independent variable used is INCOME. The output is shown below:

```
  DEPENDENT VARIABLE. . . . . . . . . . . . . . .      1 ACCTSIZE

 MULTIPLE R           0.8628          STD. ERROR OF EST.          2.8219
 MULTIPLE R-SQUARE    0.7444

 ANALYSIS OF VARIANCE
                     SUM OF SQUARES     DF    MEAN SQUARE        F RATIO    P(TAIL)
          REGRESSION      440.7036       1       440.7036         55.344     0.0000
          RESIDUAL        151.2956      19         7.9629
```

a. Locate R^2 for the reduced model.
b. Locate R^2 for the complete model in the output of Example 14.20.
c. Calculate the F statistic based on the incremental R^2. What null hypothesis is being tested?

14.17 For the chemical data of Exercise 14.7, a regression model is fit using only the first-order variables x_1 and x_2. Selected output is shown below:

DEPENDENT VARIABLE: YIELD

SOURCE	DF	SUM OF SQUARES	MEAN SQUARE	F VALUE	PR > F	R-SQUARE	C.V.
MODEL	2	305.80784100	152.90392050	12.15	0.0005	0.588452	5.9036
ERROR	17	213.87441400	12.58084788		ROOT MSE		YIELD MEAN
CORRECTED TOTAL	19	519.68225500			3.54694909		60.08150000

PARAMETER	ESTIMATE	T FOR H0: PARAMETER=0	PR > |T|	STD ERROR OF ESTIMATE
INTERCEPT	70.31000000	27.36	0.0001	2.57000878
CAT1	-2.67600000	-4.77	0.0002	0.56082189
CAT2	-0.88020000	-1.24	0.2315	0.70938982

a. Write the estimated complete model from Exercise 14.7.
b. Write the estimated reduced model.
c. Locate R^2 for the complete model in Exercise 14.8 and R^2 for the reduced model here.
d. Is there convincing evidence that the addition of the second-order terms x_3, x_4, and x_5 has improved the predictive value of the model?

14.5 FORECASTING USING MULTIPLE REGRESSION

One of the major uses of multiple regression models is in forecasting a y value given certain values of the independent x variables. The best-guess forecast is easy; just substitute the assumed x values into the estimated regression equation. In this section we discuss the relevant standard errors for such forecasts.

As in Chapter 13, the problem of forecasting y given values of the x's can be interpreted in two ways. We can think of the resulting $\hat{y}$ value as a best guess of $E(Y)$, the long-run average y value that results from averaging infinitely many observations of y when the x's have the designated values. Alternatively, and usually more interesting, we can think of $\hat{y}$ as a forecast of the actual y value that occurs given the x's. Of course it's harder to forecast an individual value than a mean, so the standard error for individual y forecasts is larger than the one for $E(Y)$. Since the relevant formulas for standard error require an understanding of matrix algebra that is not required for this text, we present crude approximations to these standard errors that ignore the major problem of extrapolation.

Extrapolation is not a problem if each x_j value is equal to the mean of the sample x_j values used in fitting the multiple regression model. In this case (and only in this case) the standard errors are quite simple. The estimated standard errors in forecasting $E(Y)$ and individual y values when there is no extrapolation are, respectively,

$$s_\epsilon\sqrt{\frac{1}{n}} \quad \text{and} \quad s_\epsilon\sqrt{1+\frac{1}{n}}$$

In fact, these no-extrapolation standard errors would be obtained by setting the extrapolation penalty term to zero in the Chapter 13 formulas.

However, the relevant values for $x_1, x_2, \ldots, x_k$ used in a forecast are hardly ever precisely equal to the respective sample means, so some degree of extrapolation is required. Although it would be helpful to use matrices to compute approximate standard errors, we rely on computer output to get numerical values for these standard errors. Computer programs typically give a standard error for an individual y forecast. While this information can also be used to find a standard error for estimating $E(Y)$, the individual y forecast is usually more relevant. The appropriate plus-or-minus term for forecasting can be found by multiplying the standard error by a tabled t value with d.f. $= n - (k + 1)$. In fact, many computer programs give the plus-or-minus term directly.

EXAMPLE 14.22 An advertising manager for a manufacturer of prepared cereals wants to develop an equation to predict sales (s) based on advertising expenditures for children's television (c), daytime television (d), and newspapers (n). Data are collected monthly for the previous 30 months (and divided by a price index to control for inflation). A multiple linear regression is fit, yielding the following computer output (SAS):

PARAMETER	ESTIMATE	T FOR H0: PARAMETER=0	PR > |T|	STD ERROR OF ESTIMATE
INTERCEPT	0.34954867	0.65	0.5214	0.53776104
C	0.06488372	3.80	0.0008	0.01706376
D	0.00628634	0.26	0.7949	0.02393904
N	0.05785077	2.74	0.0110	0.02113293

OBSERVATION	OBSERVED VALUE	PREDICTED VALUE	RESIDUAL	LOWER 95% CL INDIVIDUAL	UPPER 95% CL INDIVIDUAL
27	2.34000000	2.42060156	-0.08060156	2.02320918	2.81799394
28	1.99000000	1.93924981	0.05075019	1.53145179	2.34704784
29	2.15000000	2.07381038	0.07618962	1.67063254	2.47698822
30	2.35000000	2.49656474	-0.14656474	2.11575518	2.87737430
31 *	.	3.08658507	.	2.61561332	3.55755682

DEPENDENT VARIABLE: S

OBSERVATION	OBSERVED VALUE	PREDICTED VALUE	RESIDUAL	LOWER 95% CL FOR MEAN	UPPER 95% CL FOR MEAN
27	2.34000000	2.42060156	-0.08060156	2.28427805	2.55692507
28	1.99000000	1.93924981	0.05075019	1.77504681	2.10345282
29	2.15000000	2.07381038	0.07618962	1.92144316	2.22617759
30	2.35000000	2.49656474	-0.14656474	2.42120349	2.57192599
31 *	.	3.08658507	.	2.79939561	3.37377453

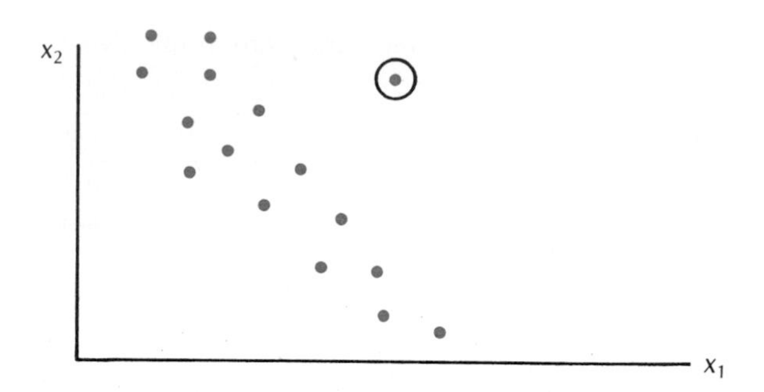

FIGURE 14.4 **Extrapolation in Multiple Regression**

a. Write the regression equation.
b. Locate the predicted y value ($\hat{y}$) when $c = 31$, $d = 5$, and $n = 12$. Locate the lower and upper limits for a 95% confidence interval for $E(Y)$ and the upper and lower 95% prediction limits for an individual Y value.

Solution a. The column labeled ESTIMATE yields the equation

$$\hat{y} = .34955 + .06488c + .00629d + 0.5755n$$

b. The predicted y value is shown as OBSERVATION 31. Note that no OBSERVED VALUE is shown for this observation. As can be verified by substituting $c = 31$, $d = 5$, and $n = 12$ into the equation, the predicted y is 3.0866. The 95% confidence limits for the mean $E(Y)$ are shown as 2.7994 to 3.3738, while the wider prediction limits for an individual y value are 2.6156 to 3.5576. □

extrapolation in multiple regression

The notion of extrapolation is more subtle in multiple regression than in simple linear regression. Briefly, the extrapolation error depends not only on the range of each separate x_j predictor used in developing the regression equation, but also on the correlations among the x_j values. In Figure 14.4, the circled point has an x_1 value well within the range of previous x_1 values, as it does for the x_2 value. Yet the point itself lies far outside the range of previous points, because of the very strong (negative, in this case) correlation of previous x_1 and x_2 values. To avoid gross extrapolation, a manager must select x values that are within (or at least close to) the range of previous values not only variable by variable, but also in combination. Those who try to forecast without regard to the data base on which the regression equation is built display a touching, childlike, and possibly expensive faith in statistical magic.

SECTION 14.5 EXERCISES

14.18 Refer to the chemical firm data of Exercise 14.7. Predicted yields for $x_1 = 3.5$ and $x_2 = 3.5$ (observation 21) and also for $x_1 = 3.5$ and $x_2 = 2.5$ (observation 22) are calculated based on the model $\hat{y} = 50.0195 + 6.6436x_1 + 7.3145x_2 - 1.2314x_1^2 - .7724x_1x_2 - 1.1755x_2^2$ of Exercise 14.7 and on the model $\hat{y} = 70.3100 - 2.6760x_1 - .8802x_2$ of Exercise 14.17. Selected output is shown below:

OBSERVATION	OBSERVED VALUE	PREDICTED VALUE	RESIDUAL	LOWER 95% CL INDIVIDUAL	UPPER 95% CL INDIVIDUAL
1	58.01000000	60.79824286	-2.78824286	54.78096039	66.81552532
2	65.89000000	62.97512857	2.91487143	57.49639919	68.45385795
3	62.70000000	62.68915714	0.01084286	57.24575991	68.13255438
4	58.22000000	59.94032857	-1.72032857	54.46159919	65.41905795
5	55.83000000	54.72864286	1.10135714	48.71136039	60.74592532
6	66.69000000	63.81384286	2.87615714	58.28481389	69.34287182
7	63.93000000	65.21832857	-1.28832857	60.00318599	70.43347115
8	62.83000000	64.15995714	-1.32995714	58.89199760	69.42791669
9	63.52000000	60.63872857	2.88127143	55.42358599	65.85387115
10	52.96000000	54.65464286	-1.69464286	49.12561389	60.18367182
11	62.53000000	64.47844286	-1.94844286	58.94941389	70.00747182
12	65.27000000	65.11052857	0.15947143	59.89538599	70.32567115
13	65.71000000	63.27975714	2.43024286	58.01179760	68.54771669
14	58.04000000	58.98612857	-0.94612857	53.77098599	64.20127115
15	51.09000000	52.22964286	-1.13964286	46.70061389	57.75867182
16	63.74000000	62.79204286	0.94795714	56.77476039	68.80932532
17	63.43000000	62.65172857	0.77827143	57.17299919	68.13045795
18	56.72000000	60.04855714	-3.32855714	54.60515991	65.49195438
19	55.16000000	54.98252857	0.17747143	49.50379919	60.46125795
20	49.36000000	47.45364286	1.90635714	41.43636039	53.47092532
21 *	.	59.92597500	.	54.70811858	65.14383142
22 *	.	62.36787500	.	57.08292311	67.65282689

OBSERVATION	OBSERVED VALUE	PREDICTED VALUE	RESIDUAL	LOWER 95% CL INDIVIDUAL	UPPER 95% CL INDIVIDUAL
1	58.01000000	66.75380000	-8.74380000	58.42067813	75.08692187
2	65.89000000	64.07780000	1.81220000	56.00061828	72.15498172
3	62.70000000	61.40180000	1.29820000	53.41175332	69.39184668
4	58.22000000	58.72580000	-0.50580000	50.64861828	66.80298172
5	55.83000000	56.04980000	-0.21980000	47.71667813	64.38292187
6	66.69000000	65.87360000	0.81640000	57.81376993	73.93343007
7	63.93000000	63.19760000	0.73240000	55.40267825	70.99252175
8	62.83000000	60.52160000	2.30840000	52.81700484	68.22619516
9	63.52000000	57.84560000	5.67440000	50.05067825	65.64052175
10	52.96000000	55.16960000	-2.20960000	47.10976993	63.22943007
11	62.53000000	64.99340000	-2.46340000	56.93356993	73.05323007
12	65.27000000	62.31740000	2.95260000	54.52247825	70.11232175
13	65.71000000	59.64140000	6.06860000	51.93680484	67.34599516
14	58.04000000	56.96540000	1.07460000	49.17047825	64.76032175
15	51.09000000	54.28940000	-3.19940000	46.22956993	62.34923007
16	63.74000000	64.11320000	-0.37320000	55.78007813	72.44632187
17	63.43000000	61.43720000	1.99280000	53.36001828	69.51438172
18	56.72000000	58.76120000	-2.04120000	50.77115332	66.75124668
19	55.16000000	56.08520000	-0.92520000	48.00801828	64.16238172
20	49.36000000	53.40920000	-4.04920000	45.07607813	61.74232187
21 *	.	57.86330000	.	50.02807100	65.69852900
22 *	.	58.74350000	.	51.05254533	66.43445467

* OBSERVATION WAS NOT USED IN THIS ANALYSIS

a. Locate the 95% confidence limits for individual prediction in the model $\hat{y} = 50.0195 + 6.6436x_1 + 7.3145x_2 - 1.2314x_1^2 - .7724x_1x_2 - 1.1755x_2^2$.

b. Locate the 95% confidence limits for individual prediction in the model $\hat{y} = 70.3100 - 2.6760x_1 - .8802x_2$.

c. Are the confidence limits for the model of part (a) much tighter than those for the model of part (b)?

14.6 SOME MULTIPLE REGRESSION THEORY

In this section matrix notation is used to sketch some of the mathematics underlying multiple regression. The ideas in this section indicate how multiple regression calculations are actually done, whether by hand or by computer. We do not prove most of the results; proofs are available in many specialized texts, such as Draper and Smith (1981).

The starting point for the use of matrix notation is the multiple regression model itself. Recall that a model relating a response Y to a set of independent variables of the form

$$Y = \beta_0 + \beta_1 x_1 + \beta_2 x_2 + \ldots + \beta_k x_k + \epsilon$$

is called the general linear model. The least-squares estimates $\hat{\beta}_0, \hat{\beta}_1, \ldots, \hat{\beta}_k$ of the intercept and partial slopes in the general linear model can be obtained using matrices.

Let the $n \times 1$ matrix $\mathbf{Y}$

$$\mathbf{Y} = \begin{bmatrix} y_1 \\ y_2 \\ \vdots \\ y_n \end{bmatrix}$$

be the matrix of observations, and let the $n \times (k + 1)$ matrix $\mathbf{X}$

$$\mathbf{X} = \begin{bmatrix} 1 & x_{11} & \cdots & x_{1k} \\ 1 & x_{21} & \cdots & x_{2k} \\ \vdots & \vdots & & \vdots \\ 1 & x_{n1} & \cdots & x_{nk} \end{bmatrix}$$

be a matrix of settings for the independent variables augmented with a column of 1's. The first row of $\mathbf{X}$ contains a 1 and the settings for the k independent variables for the first observation. Row 2 contains a 1 and corresponding settings on the independent variables for y_2. Similarly, the other rows contain settings for the remaining observations.

Next we turn to the least-squares estimates $\hat{\beta}_0, \hat{\beta}_1, \ldots, \hat{\beta}_k$ of the intercept and partial slopes in the multiple regression model. Recall that the least-squares principle involves choosing the estimates to minimize the sum of squared residuals. Those familiar with the calculus will see that the solution can be found by differentiating SS(Residual) with respect to $\hat{\beta}_j$ ($j = 0, \ldots, k$) and setting the result to zero. The resulting normal equations, in matrix notation, are

$$(\mathbf{X}'\mathbf{X})\,\hat{\boldsymbol{\beta}} = \mathbf{X}'\mathbf{Y}$$

where

$$\hat{\boldsymbol{\beta}} = \begin{bmatrix} \hat{\beta}_0 \\ \hat{\beta}_1 \\ \vdots \\ \hat{\beta}_k \end{bmatrix}$$

is the desired vector of estimated coefficients. (This set of equations expressed in matrix notation is identical to the normal equations given in Section 14.2.) Provided that the matrix **X′X** has an inverse (it does as long as no x_j is perfectly collinear with other x's), the solution is

$$\hat{\boldsymbol{\beta}} = (\mathbf{X}'\mathbf{X})^{-1}\,\mathbf{X}'\mathbf{Y}$$

EXAMPLE 14.23 Suppose that in a given experimental situation,

$$\mathbf{Y} = \begin{bmatrix} 25 \\ 19 \\ 33 \\ 23 \end{bmatrix} \quad \text{and} \quad \mathbf{X} = \begin{bmatrix} 1 & -2 & 5 \\ 1 & -2 & -5 \\ 1 & 2 & 5 \\ 1 & 2 & -5 \end{bmatrix}$$

Obtain the least-squares estimates for the prediction equation

$$\hat{y} = \hat{\beta}_0 + \hat{\beta}_1 x_1 + \hat{\beta}_2 x_2$$

Solution For these data,

$$\mathbf{X}'\mathbf{X} = \begin{bmatrix} 4 & 0 & 0 \\ 0 & 16 & 0 \\ 0 & 0 & 100 \end{bmatrix}$$

$$\mathbf{X}'\mathbf{Y} = \begin{bmatrix} 100 \\ 24 \\ 80 \end{bmatrix}$$

The **X′X** matrix is a diagonal one, so inverting the matrix is easy. The solution is

$$\hat{\boldsymbol{\beta}} = (\mathbf{X}'\mathbf{X})^{-1}\,\mathbf{X}'\mathbf{Y}$$

$$= \begin{bmatrix} .25 & 0 & 0 \\ 0 & .0625 & 0 \\ 0 & 0 & .01 \end{bmatrix} \begin{bmatrix} 100 \\ 24 \\ 80 \end{bmatrix} = \begin{bmatrix} 25 \\ 1.5 \\ 0.8 \end{bmatrix}$$

and the prediction equation is

$$\hat{y} = 25 + 1.5x_1 + .8x_2.$$

□

The hard part of the arithmetic in multiple regression is computing the inverse of **X′X**. For most realistic multiple regression problems, this task takes hours by hand and fractions of a second by computer. This is the major reason that most multiple regression problems are done with computer software.

Once the inverse of the $\mathbf{X'X}$ matrix is found and the $\hat{\boldsymbol{\beta}}$ vector is calculated, the next task is to compute the residual standard deviation. The hard work is to compute SS(Residual) $= \sum (y_i - \hat{y}_i)^2$. The shortcut formula given in Section 14.2 can be written as SS(Residual) $= \mathbf{Y'Y} - \hat{\boldsymbol{\beta}}' (\mathbf{X'Y})$.

EXAMPLE 14.24 Compute SS(Residual) for the data of Example 14.23.

Solution $\hat{\boldsymbol{\beta}}$ and $\mathbf{X'Y}$ were calculated to be

$$\begin{bmatrix} 25 \\ 1.5 \\ 0.8 \end{bmatrix} \quad \text{and} \quad \begin{bmatrix} 100 \\ 24 \\ 80 \end{bmatrix}$$

respectively, and

$$\mathbf{Y'Y} = [25 \quad 19 \quad 33 \quad 23] \begin{bmatrix} 25 \\ 19 \\ 33 \\ 23 \end{bmatrix} = 2604$$

The shortcut formula yields

$$\text{SS(Residual)} = 2604 - [25 \quad 1.5 \quad 0.8] \begin{bmatrix} 100 \\ 24 \\ 80 \end{bmatrix} = 4$$

□

Similar calculations yield SS(Total) and SS(Regression). While the formulas for these sums can be expressed artificially in pure matrix notation, the easy way is to mix matrix and algebraic notation:

$$\text{SS(Regression)} = \hat{\boldsymbol{\beta}}'(\mathbf{X'Y}) - \frac{(\sum y_i)^2}{n}$$

$$\text{SS(Total)} = \mathbf{Y'Y} - \frac{(\sum y_i)^2}{n}$$

EXAMPLE 14.25 Calculate SS(Regression) and SS(Total) for the data of Example 14.23.

Solution $\sum y_i = 100$ and $n = 4$. The relevant matrix calculations were performed in the previous example.

$$\text{SS(Regression)} = 2600 - \frac{(100)^2}{4} = 100$$

$$\text{SS(Total)} = 2604 - \frac{(100)^2}{4} = 104$$

Note that SS(Total) $= 104 = 100 + 4 =$ SS(Regression) + SS(Residual). □

These sum-of-squares calculations are needed for making inferences based on R^2 using F tests. For inferences about individual coefficients using t tests, the estimated standard errors of the coefficients are needed. In Section 14.3 we presented a conceptually useful but computationally cumbersome formula for these estimated standard errors. There is a much easier way to compute them that involves only the standard deviation s_ϵ and the main diagonal elements of the $(\mathbf{X}'\mathbf{X})^{-1}$ matrix.

Estimated Standard Error of $\hat{\beta}_j$

$$s_{\hat{\beta}_j} = s_\epsilon \sqrt{v_{jj}}$$

where s_ϵ is the standard deviation from the regression equation and v_{jj} is the entry in row $j + 1$, column $j + 1$ of $(\mathbf{X}'\mathbf{X})^{-1}$.

$$(\mathbf{X}'\mathbf{X})^{-1} = \begin{bmatrix} v_{00} & & & \\ & v_{11} & & \\ & & \ddots & \\ & & & v_{kk} \end{bmatrix}$$

Because the $(\mathbf{X}'\mathbf{X})^{-1}$ matrix must be computed in obtaining the $\hat{\beta}_j$'s, it is easy to get the estimated standard errors.

EXAMPLE 14.26 Calculate the residual standard deviation s_ϵ and the estimated standard errors of $\hat{\beta}_0$, $\hat{\beta}_1$, and $\hat{\beta}_2$ for the data of Example 14.23.

Solution From Example 14.24, we know that SS(Residual) $= 4$ with d.f. $= n - (k + 1) = 4 - (2 + 1) = 1$, so MS(Residual) $= 4/[4 - (2 + 1)] = 4$ and $s_\epsilon = \sqrt{\text{MS(Residual)}} = 2.0$. The $(\mathbf{X}'\mathbf{X})^{-1}$ matrix is

$$\begin{bmatrix} .25 & 0 & 0 \\ 0 & .0625 & 0 \\ 0 & 0 & .01 \end{bmatrix}$$

so the standard errors for $\hat{\beta}_0$, $\hat{\beta}_1$, and $\hat{\beta}_2$ are, respectively,

$$s_{\hat{\beta}_0} = 2.0\sqrt{.25} = 1.0$$
$$s_{\hat{\beta}_1} = 2.0\sqrt{.0625} = .50$$
$$s_{\hat{\beta}_2} = 2.0\sqrt{.01} = 0.20$$

□

One other use of matrix notation occurs when a multiple regression model is used to estimate the expected value of Y at new settings of the x's or to predict an individual value of Y at new settings of the x's. In Chapter 13, standard errors were shown for the linear regression case. These standard errors involve the residual standard deviation, the sample size, and an extrapolation penalty. The matrix notation formulas for these standard errors once again involve the $(\mathbf{X}'\mathbf{X})^{-1}$ matrix. We designate the future value of Y and $E(Y)$ as Y_{n+1} and $E(Y_{n+1})$, respectively.

Standard Error for Estimating $E(Y_{n+1})$

$$s_{\hat{Y}_{n+1}} = s_\epsilon \sqrt{\mathbf{X}_{n+1}(\mathbf{X}'\mathbf{X})^{-1}\mathbf{X}_{n+1}}$$

where

$$\mathbf{X}_{n+1} = \begin{bmatrix} 1 \\ x_{n+1,1} \\ x_{n+1,2} \\ \vdots \\ x_{n+1,k} \end{bmatrix}$$

is the vector of settings of the independent variables at which $E(Y_{n+1})$ is to be predicted.

Standard Error for Predicting Y_{n+1}

$$s_{\hat{Y}_{n+1}} = s_\epsilon \sqrt{1 + \mathbf{X}'_{n+1}(\mathbf{X}'\mathbf{X})^{-1}\mathbf{X}_{n+1}}$$

When each x_j is equal to its sample mean $\bar{x}_j$, the term $\mathbf{X}'_{n+1}(\mathbf{X}'\mathbf{X})^{-1}\mathbf{X}_{n+1}$ reduces to $1/n$. Otherwise there is an extrapolation penalty.

EXAMPLE 14.27 Find standard errors for $E(Y_{n+1})$ and Y_{n+1} for the data of Example 14.23.

a. Assume that x_1 and x_2 are both zero (that is, when $x_{n+1,1} = \bar{x}_1$ and $x_{n+1,2} = \bar{x}_2$).
b. Examine the extrapolation penalty when $x_{n+1,1} = 1$ and $x_{n+1,2} = 6$.

Solution a. The residual standard deviation has been found to be $s_\epsilon = 2.0$. The desired $\mathbf{X}_{n+1}$ vector is

$$\mathbf{X}_{n+1} = \begin{bmatrix} 1 \\ x_{n+1,1} \\ x_{n+1,2} \end{bmatrix} = \begin{bmatrix} 1 \\ 0 \\ 0 \end{bmatrix}$$

Thus

$$\mathbf{X}'_{n+1}(\mathbf{X}'\mathbf{X})^{-1}\mathbf{X}_{n+1} = [1 \quad 0 \quad 0] \begin{bmatrix} .25 & 0 & 0 \\ 0 & .0625 & 0 \\ 0 & 0 & .01 \end{bmatrix} \begin{bmatrix} 1 \\ 0 \\ 0 \end{bmatrix} = .25 = \frac{1}{n}$$

The standard errors for estimating $E(Y_{n+1})$ and predicting Y_{n+1} are, respectively,

$$2.0\sqrt{.25} = 1.0$$

and

$$2.0\sqrt{1 + .25} = 2.236$$

b. For $x_{n+1,1} = 1$ and $x_{n+1,2} = 6$,

$$\mathbf{X}_{n+1} = \begin{bmatrix} 1 \\ 1 \\ 6 \end{bmatrix}$$

and

$$\mathbf{X}'_{n+1}(\mathbf{X}'\mathbf{X})^{-1}\mathbf{X}_{n+1} = [1 \quad 1 \quad 6]\begin{bmatrix} .25 & 0 & 0 \\ 0 & .0625 & 0 \\ 0 & 0 & .01 \end{bmatrix}\begin{bmatrix} 1 \\ 1 \\ 6 \end{bmatrix} = .6725$$

Thus the estimated standard errors for estimating $E(Y_{n+1})$ and predicting Y_{n+1} when $x_{n+1,1}$ 1 and $x_{n+1,2} = 6$ are, respectively,

$$2.0\sqrt{.6725} = 1.6401$$

and

$$2.0\sqrt{1 + .6725} = 2.5865$$

Note that these errors are larger than their counterparts when there is no extrapolation. □

SECTION 14.6 EXERCISES

14.19 The effects of temperature settings and lighting levels on office productivity are studied. The variables are y, a measure of work production; x_1, office temperature $-68°$; and x_2, lighting level -60. The data are

y:	43	42	49	39	47	57	46	48	53
x_1:	−4	−4	−4	−2	−2	−2	0	0	0
x_2:	−5	0	5	−5	0	5	−5	0	5

y:	49	51	63	56	54	64
x_1:	2	2	2	4	4	4
x_2:	−5	0	5	−5	0	5

a. Write the **X** matrix and **Y** vector for these data.

b. Verify that

$$\mathbf{X}'\mathbf{X} = \begin{bmatrix} 15 & 0 & 0 \\ 0 & 120 & 0 \\ 0 & 0 & 250 \end{bmatrix} \quad \text{and} \quad \mathbf{X}'\mathbf{Y} = \begin{bmatrix} 761 \\ 200 \\ 265 \end{bmatrix}$$

c. Calculate the $\hat{\boldsymbol{\beta}}$ vector.

14.20 Refer to the data of Exercise 14.19.

a. Calculate $\mathbf{Y}'\mathbf{Y}$.

b. Calculate SS(Residual) $= \mathbf{Y}'\mathbf{Y} - \hat{\boldsymbol{\beta}}'(\mathbf{X}'\mathbf{Y})$.

c. Calculate s_ϵ. The d.f. for SS(Residual) is $n - 3 = 12$.

14.21 Refer to the data of Exercise 14.19.

a. Calculate the value of the F statistic.

b. Can we reject the null hypothesis that neither x_1 nor x_2 has predictive value? Use $\alpha = .01$.

14.22 Perform t tests for the partial slope coefficients $\hat{\beta}_1$ and $\hat{\beta}_2$ in Exercise 14.19. What conclusion follows from each test?

14.23 In a small regression study, the following data are observed:

y:	15	18	12	13	19	17	16	25	20	21	26	21	22	20	29	26
x_1:	−3	−2	−2	−1	−1	−1	0	0	0	0	1	1	1	2	2	3
x_2:	−5	−3	−3	−2	−2	−1	−1	0	0	1	1	2	2	3	3	5

a. Verify that

$$\mathbf{X}'\mathbf{X} = \begin{bmatrix} 16 & 0 & 0 \\ 0 & 40 & 64 \\ 0 & 64 & 106 \end{bmatrix} \quad \text{and} \quad \mathbf{X}'\mathbf{Y} = \begin{bmatrix} 320 \\ 91 \\ 148 \end{bmatrix}$$

b. Calculate the least-squares coefficient vector $\hat{\boldsymbol{\beta}}$.
c. Calculate the residual standard deviation.

14.24 Refer to the data of Exercise 14.23.
a. Calculate the estimated standard errors of $\hat{\beta}_1$ and $\hat{\beta}_2$.
b. Calculate 90% confidence intervals for β_1 and β_2.
c. Test the hypothesis that $\beta_1 > 0$. Use $\alpha = .05$.

14.25 Refer to Exercise 14.23.
a. Calculate SS(Regression), SS(Residual), and SS(Total).
b. Calculate $R^2_{y \cdot x_1 x_2}$.
c. Test H_0: $\beta_1 = \beta_2 = 0$ at $\alpha = .01$.

14.26 Refer to Exercise 14.23.
a. Predict Y when $x_{n+1,1} = 2$ and $x_{n+1,2} = 3$.
b. Give a 90% prediction interval for the prediction in part (a).
c. Predict Y when $x_{n+1,1} = 2$ and $x_{n+1,2} = -3$.
d. Give a 90% prediction interval for the prediction in part (c).

SUMMARY

In this chapter we introduced the multiple regression model

$$Y = \beta_0 + \beta_1 x_1 + \beta_2 x_2 + \ldots + \beta_k x_k + \epsilon$$

where β_0 is the intercept and $\beta_1, \ldots, \beta_k$ are the partial slopes.

Many inferences are possible using sample data and a multiple regression equation. We can make an overall F test of the null hypothesis

$$H_0: \beta_1 = \beta_2 = \ldots = \beta_k = 0$$

Or, in other words, we can test the null hypothesis that none of the independent variables contributes to the prediction of y. We can also be more specific. The sample data can be used to test H_0: $\beta_j = 0$ or to place a confidence interval about the parameter β_j. These last two inferences relate to a single β and focus on the contribution of an individual predictor variable when entered last.

Not all inferences related to a multiple regression equation can be phrased in terms of all the x's or a single x. Sometimes we must consider the contribution of a subset of predictor variables. We discussed an F test of the null hypothesis that a subset of $(k - g)$ β's are identically zero $(g < k)$.

Finally, in this chapter we expanded on the interpretation of the coefficient of determination and we presented the important concepts of forecasting (with a corresponding prediction error) using a multiple regression. The underlying mathematics of multiple regression was discussed briefly using matrix notation.

KEY FORMULAS: Multiple Regression

1. Multiple regression model

$$Y_i = \beta_0 + \beta_1 x_{i1} + \beta_2 x_{i2} + \ldots + \beta_k x_{ik} + \epsilon_i$$

2. Normal equations (see Table 14.1, p. 483)

3. $\text{SS(Residual)} = \sum y_i^2 - \hat{\beta}_0 \sum y_i - \hat{\beta}_1 \sum x_{i1}y_i - \hat{\beta}_2 \sum x_{i2}y_i - \ldots - \hat{\beta}_k \sum x_{ik}y_i$
4. $s_\epsilon = \sqrt{\text{MS(Residual)}} = \sqrt{\text{SS(Residual)}/[n - (k + 1)]}$
5. Coefficient of determination

$$R^2_{Y \cdot x_1 x_2 \ldots x_k} = \frac{\text{SS(Total)} - \text{SS(Residual)}}{\text{SS(Total)}}$$

6. F test for $\beta_1, \ldots, \beta_k$

$$H_0: \beta_1 = \beta_2 = \ldots = \beta_k = 0$$

$$\text{T.S.: } F = \frac{\text{MS(Regression)}}{\text{MS(Residual)}}$$

where d.f.$_1 = k$ and d.f.$_2 = n - (k + 1)$

7. Estimated standard error of $\hat{\beta}_j$

$$s_{\hat{\beta}_j} = s_\epsilon \sqrt{\frac{1}{\sum (x_{ij} - \bar{x}_j)^2 (1 - R^2_{x_j \cdot x_1 \ldots x_{j-1} x_{j+1} \ldots x_k})}}$$

8. $100(1 - \alpha)\%$ confidence interval for β_j

$$\hat{\beta}_j \pm t_{\alpha/2} s_{\hat{\beta}_j}$$

9. Statistical test for β_j

$$H_0: \beta_j = 0$$

$$\text{T.S.: } t = \frac{\hat{\beta}_j}{s_{\hat{\beta}_j}}, \text{ where d.f.} = n - (k + 1)$$

10. F test for $\beta_1, \ldots, \beta_k$ based on R^2

$$H_0: \beta_1 = \beta_2 = \ldots = \beta_k = 0$$

$$\text{T.S.: } F = \frac{\text{MS(Regression)}}{\text{MS(Residual)}} = \frac{R^2_{y \cdot x_1 \ldots x_k}/k}{(1 - R^2_{y \cdot x_1 \ldots x_k})/[n - (k + 1)]}$$

11. F test for a set of β's

$$H_0: \beta_{g+1} = \beta_{g+2} = \ldots = \beta_k = 0$$

$$\text{T.S.: } F = \frac{(R^2_{\text{complete}} - R^2_{\text{reduced}})/(k - g)}{(1 - R^2_{\text{complete}})/[n - (k + 1)]}$$

12. Formulas using matrices

$\hat{\boldsymbol{\beta}} = (\mathbf{X}'\mathbf{X})^{-1}\mathbf{X}'\mathbf{Y}$

$\text{SS(Residual)} = \mathbf{Y}'\mathbf{Y} - \hat{\boldsymbol{\beta}}'\mathbf{X}'\mathbf{Y}$

$s_{\hat{\beta}_j} = s_\epsilon \sqrt{v_{jj}}$

$s_{\hat{Y}_{n+1}} = s_\epsilon \sqrt{\mathbf{X}'_{n+1}(\mathbf{X}'\mathbf{X})^{-1}\mathbf{X}_{n+1}}$ for estimating $E(Y_{n+1})$

$s_{\hat{Y}_{n+1}} = s_\epsilon \sqrt{1 + \mathbf{X}'_{n+1}(\mathbf{X}'\mathbf{X})^{-1}\mathbf{X}_{n+1}}$ for predicting Y_{n+1}

CHAPTER 14 EXERCISES

14.27 A study of demand for imported subcompact cars consists of data from 12 metropolitan areas. The variables are

DEMAND:	imported subcompact car sales as a percentage of total sales
EDUC:	average number of years of schooling completed by adults
INCOME:	per capita income
POPN:	area population
FAMSIZE:	average size of intact families

BMDP output is shown below:

```
MULTIPLE R              0.9811           STD. ERROR OF EST.          2.6863
MULTIPLE R-SQUARE       0.9625

ANALYSIS OF VARIANCE
                     SUM OF SQUARES    DF    MEAN SQUARE        F RATIO    P(TAIL)
        REGRESSION      1295.7041       4      323.9260          44.890     0.0000
        RESIDUAL          50.5125       7        7.2161

                                                STD. REG
   VARIABLE        COEFFICIENT   STD. ERROR       COEFF          T   P(2 TAIL) TOLERANCE

INTERCEPT            -1.31949
EDUC          2       5.54983      2.70228        0.446      2.054     0.0791     0.11363
INCOME        3       0.88514      1.30846        0.099      0.676     0.5205     0.25194
POPN          4       1.92483      1.37081        0.131      1.404     0.2031     0.61124
FAMSIZE       5     -11.38928      6.66931       -0.438     -1.708     0.1314     0.08156
```

a. Write the regression equation. Place the standard error of each coefficient below the coefficient, perhaps in parentheses.

b. Locate R^2 and the residual standard deviation.

14.28 Summarize the conclusions of the F test and the various t tests in the output of Exercise 14.27.

14.29 Another analysis of the data of Exercise 14.27 uses only EDUC and FAMSIZE to predict DEMAND. The output is shown below:

```
MULTIPLE R              0.9707           STD. ERROR OF EST.          2.9389
MULTIPLE R-SQUARE       0.9423

ANALYSIS OF VARIANCE
                     SUM OF SQUARES    DF    MEAN SQUARE        F RATIO    P(TAIL)
        REGRESSION      1268.4828       2      634.2414          73.432     0.0000
        RESIDUAL          77.7338       9        8.6371

                                                STD. REG
   VARIABLE        COEFFICIENT   STD. ERROR       COEFF          T   P(2 TAIL) TOLERANCE

INTERCEPT           -19.16498
EDUC          2       7.79256      2.49031        0.626      3.129     0.0121     0.16014
FAMSIZE       5      -9.46411      5.20709       -0.364     -1.818     0.1025     0.16014
```

a. Locate the R^2 value for this reduced model.

b. Test the null hypothesis that the true coefficients of INCOME and POPN are zero. Use $\alpha = .05$. What is the conclusion?

14.30 The manager of documentation for a computer software firm wants to forecast the time required to document moderate-size computer programs. Records are available for 26 programs. The variables are y = number of writer-days needed, x_1 = number of subprograms, x_2 = average number of lines per subprogram, $x_3 = x_1x_2$, $x_4 = x_2^2$, and $x_5 = x_1x_2^2$. A portion of the output from a regression analysis (SAS) of the data is shown below:

DEPENDENT VARIABLE: Y

SOURCE	DF	SUM OF SQUARES	MEAN SQUARE	F VALUE	PR > F	R-SQUARE	C.V.
MODEL	5	2546.02735209	509.20547042	44.31	0.0001	0.917195	11.9597
ERROR	20	229.85726330	11.49286316		ROOT MSE		Y MEAN
CORRECTED TOTAL	25	2775.88461538			3.39011256		28.34615385

PARAMETER	ESTIMATE	T FOR H0: PARAMETER=0	PR > |T|	STD ERROR OF ESTIMATE
INTERCEPT	-16.81979712	-1.45	0.1636	11.63104920
X1	1.47018752	4.02	0.0007	0.36594367
X2	0.99477822	1.63	0.1194	0.61144114
X1X2	-0.02400705	-1.01	0.3243	0.02375645
X2SQ	-0.01031004	-1.40	0.1774	0.00737400
X1X2SQ	0.00024957	0.71	0.4862	0.00035178

a. Write the multiple regression model and locate the residual standard deviation.

b. What does the variable x_3 represent in terms of the problem?

c. Does x_3 have a statistically significant predictive value as "last predictor in"?

14.31 The model $Y = \beta_0 + \beta_1x_1 + \beta_2x_2 + \epsilon$ is fit to the data of Exercise 14.30. Selected output is shown here:

DEPENDENT VARIABLE: Y

SOURCE	DF	SUM OF SQUARES	MEAN SQUARE	F VALUE	PR > F	R-SQUARE	C.V.
MODEL	2	2516.12160091	1258.06080045	111.39	0.0001	0.906422	11.8558
ERROR	23	259.76301448	11.29404411		ROOT MSE		Y MEAN
CORRECTED TOTAL	25	2775.88461538			3.36066126		28.34615385

PARAMETER	ESTIMATE	T FOR H0: PARAMETER=0	PR > |T|	STD ERROR OF ESTIMATE
INTERCEPT	0.84008527	0.24	0.8089	3.43374955
X1	1.01583472	12.81	0.0001	0.07929252
X2	0.05582624	1.08	0.2897	0.05150660

a. Write the complete and reduced-form estimated models.

b. Is the improvement in R^2 obtained by adding x_3, x_4, and x_5 statistically significant at $\alpha = .05$? What is the p-value for this test?

14.32 A study of the effects of air pollution in a metropolitan area involve data on a daily pollution index and respiratory illness admissions to area hospitals. Relevant parts of the printout (SAS) are shown below. POLINDEX is that day's pollution index, POLLLAG1 is the index for the previous day, and POLLLAG2 is the index for the day before that.

PARAMETER	ESTIMATE	T FOR H0: PARAMETER=0	PR > |T|	STD ERROR OF ESTIMATE
INTERCEPT	9.85081550	8.67	0.0001	1.13636639
POLINDEX	2.49902738	9.54	0.0001	0.26204465

DEPENDENT VARIABLE: ADMITS

SOURCE	DF	SUM OF SQUARES	MEAN SQUARE	F VALUE	PR > F	R-SQUARE	C.V.
MODEL	1	1159.34045339	1159.34045339	90.95	0.0001	0.727886	18.7092
ERROR	34	433.40954661	12.74733961		ROOT MSE		ADMITS MEAN
CORRECTED TOTAL	35	1592.75000000			3.57034167		19.08333333

PARAMETER	ESTIMATE	T FOR H0: PARAMETER=0	PR > |T|	STD ERROR OF ESTIMATE
INTERCEPT	-2.67027671	-2.61	0.0135	1.02333140
POLLLAG1	0.72545042	3.10	0.0040	0.23424056

DEPENDENT VARIABLE: ADMITS

SOURCE	DF	SUM OF SQUARES	MEAN SQUARE	F VALUE	PR > F	R-SQUARE	C.V.
MODEL	3	1273.64879574	424.54959858	46.46	0.0001	0.822896	15.6192
ERROR	30	274.11591014	9.13719700		ROOT MSE		ADMITS MEAN
CORRECTED TOTAL	33	1547.76470588			3.02277968		19.35294118

PARAMETER	ESTIMATE	T FOR H0: PARAMETER=0	PR > |T|	STD ERROR OF ESTIMATE
INTERCEPT	7.81896016	6.51	0.0001	1.20152608
POLINDEX	1.99503279	7.32	0.0001	0.27271180
POLLLAG1	0.82254859	2.56	0.0157	0.32111307
POLLLAG2	0.26587278	0.99	0.3324	0.26985082

X'X INVERSE MATRIX

DEPENDENT VARIABLE : ADMITS

	INTERCEPT	POLINDEX	POLLLAG1	POLLLAG2
INTERCEPT	0.15799866	-0.01571841	-0.00352192	-0.01527707
POLINDEX	-0.01571841	0.00813945	-0.00528721	0.00140857
POLLLAG1	-0.00352192	-0.00528721	0.01128504	-0.00518463
POLLLAG2	-0.01527707	0.00140857	-0.00518463	0.00796956

a. Write down the regression equation predicting admissions from the same-day pollution index.
b. Locate the residual standard deviation and R^2 value for this equation.
c. Write down the regression equation predicting admissions from the previous-day pollution index.

14.33 Refer to the printout of Exercise 14.32.
a. Write down the regression equation predicting admission using the same-day index and the values for the two previous days.
b. What do the t statistics indicate about the predictive value of the various lagged pollution index values?

14.34 Refer to Exercise 14.32.
a. Locate the $(\mathbf{X}'\mathbf{X})^{-1}$ matrix for the model that regresses admissions at day t on pollution at day t, pollution at day $t - 1$, and pollution at day $t - 2$.
b. Locate the standard deviation for this model.
c. Predict the number of admissions if today's index is 5, yesterday's was 4, and the day before yesterday's was 3.
d. Calculate a 95% interval for the predicted value in part (c).

14.35 Again refer to Exercise 14.32.
a. Predict today's admissions if today's pollution index is 9, yesterday's was 1, and the day before yesterday's was 9.
b. Calculate a 95% interval for this prediction.

14.36 Compare the widths of the prediction intervals found in Exercises 14.34 and 14.35. Why is the interval in 14.35 wider?

14.37 A chain of small convenience food stores performs a regression analysis to explain variation in sales volume among 16 stores. The variables in the study are

SALES:	Average daily sales volume of a store, in thousands of dollars
SIZE:	Floor space in thousands of square feet
PARKING:	Number of free parking spaces adjacent to the store
INCOME:	Estimated per household income of the ZIP code area of the store

Output from a regression program (BMDP) follows:

```
MULTIPLE R              0.8895          STD. ERROR OF EST.          0.7724
MULTIPLE R-SQUARE       0.7912

ANALYSIS OF VARIANCE
                      SUM OF SQUARES    DF    MEAN SQUARE        F RATIO    P(TAIL)
         REGRESSION         27.1296      3         9.0432         15.158     0.0002
         RESIDUAL            7.1592     12         0.5966

                                                   STD. REG
   VARIABLE          COEFFICIENT   STD. ERROR        COEFF          T    P(2 TAIL) TOLERANCE

INTERCEPT             0.87272
SIZE            2     2.54794        1.20083         0.385      2.122      0.0554     0.52842
PARKING         3     0.22028        0.15539         0.249      1.418      0.1817     0.56477
INCOME          4     0.58932        0.17806         0.479      3.310      0.0062     0.83117
```

a. Write down the regression equation. Indicate the standard errors of the coefficients.
b. Carefully interpret each coefficient.
c. Locate R^2 and the residual standard deviation.

14.38 Summarize the results of the F and t tests for the output of Exercise 14.37.

14.39 A producer of various feed additives for cattle conducts a study of the number of days of feedlot time required to bring beef cattle to market weight. Eighteen steers of essentially identical age and weight are purchased and brought to a feedlot. Each steer is fed a diet with a specific combination of protein content, antibiotic concentration, and percentage of feed supplement. The data are

STEER:	1	2	3	4	5	6	7	8	9
PROTEIN:	10	10	10	10	10	10	15	15	15
ANTIBIO:	1	1	1	2	2	2	1	1	1
SUPPLEM:	3	5	7	3	5	7	3	5	7
TIME:	88	82	81	82	83	75	80	80	75

STEER:	10	11	12	13	14	15	16	17	18
PROTEIN:	15	15	15	20	20	20	20	20	20
ANTIBIO:	2	2	2	1	1	1	2	2	2
SUPPLEM:	3	5	7	3	5	7	3	5	7
TIME:	77	76	72	79	74	75	74	70	69

Computer output from a regression analysis follows:

```
MULTIPLE R             0.9490          STD. ERROR OF EST.          1.7096
MULTIPLE R-SQUARE      0.9007


ANALYSIS OF VARIANCE
                    SUM OF SQUARES     DF     MEAN SQUARE         F RATIO     P(TAIL)
        REGRESSION       371.0832       3        123.6944          42.323      0.0000
        RESIDUAL          40.9166      14          2.9226


                                                  STD. REG
   VARIABLE        COEFFICIENT   STD. ERROR        COEFF          T   P(2 TAIL) TOLERANCE

INTERCEPT            102.70834
PROTEIN      2        -0.83333      0.09870       -0.711     -8.443     0.0000     1.00000
ANTIBIO      3        -4.00000      0.80590       -0.418     -4.963     0.0002     1.00000
SUPPLEM      4        -1.37500      0.24675       -0.469     -5.572     0.0001     1.00000
```

a. Write down the regression equation.
b. Find the standard deviation.
c. Find the R^2 value.

14.40 Refer to Exercise 14.39.

a. Predict the feedlot time required for a steer fed 15% protein, 1.5% antibiotic concentration, and 5% supplement.
b. Do these values of the independent variables represent a major extrapolation from the data?
c. Give a 95% confidence interval for the mean time predicted in part (a).

14.41 The data of Exercise 14.39 are also analyzed by a regression model using only protein content as an independent variable, with the following output:

```
MULTIPLE R             0.7111          STD. ERROR OF EST.          3.5678
MULTIPLE R-SQUARE      0.5057

ANALYSIS OF VARIANCE
                  SUM OF SQUARES      DF     MEAN SQUARE         F RATIO    P(TAIL)
       REGRESSION      208.3332        1        208.3332          16.367     0.0009
       RESIDUAL        203.6667       16         12.7292

                                                STD. REG
  VARIABLE        COEFFICIENT   STD. ERROR       COEFF           T   P(2 TAIL) TOLERANCE

INTERCEPT           89.83334
PROTEIN      2      -0.83333      0.20599        -0.711     -4.046     0.0009     1.00000
```

a. Write the regression equation.

b. Find the R^2 value.

c. Test the null hypothesis that the coefficients of ANTIBIO and SUPPLEM are zero at $\alpha = .05$.

14.42 A sex discrimination suit alleges that a small college discriminated in salaries against women. A regression study considers the following variables:

SALARY:	base salary per year (thousands of dollars)
SENIOR:	seniority at the college (years)
SEX:	1 for men, 0 for women
RANKD1:	1 for full professors, 0 for others
RANKD2:	1 for associate professors, 0 for others
RANKD3:	1 for assistant professors, 0 for others
DOCT:	1 for holders of doctorate, 0 for others

Note that lecturers and instructors have value 0 for all 3 RANKD variables. Computer output from the study is shown below:

```
DEP VARIABLE: SALARY
                                          ANALYSIS OF VARIANCE

                                     SUM OF           MEAN
                     SOURCE     DF   SQUARES          SQUARE         F VALUE        PROB>F

                     MODEL       6   2119.347         353.2245       64.646         0.0001
                     ERROR      23   125.6717         5.463985
                     C TOTAL    29   2245.019

                        ROOT MSE     2.337517        R-SQUARE        0.9440
                        DEP MEAN     31.92667        ADJ R-SQ        0.9294
                        C.V.         7.321519

                                          PARAMETER ESTIMATES

                                  PARAMETER          STANDARD        T FOR H0:
                 VARIABLE  DF      ESTIMATE             ERROR      PARAMETER=0       PROB > |T|

                 INTERCEP   1      18.67841          1.378787           13.547           0.0001
                 SENIOR     1     0.5420351        0.07615423            7.118           0.0001
                 SEX        1      1.207423          1.064854            1.134           0.2685
                 RANKD1     1      8.777947           1.93803            4.529           0.0001
                 RANKD2     1      4.421096          1.779685            2.484           0.0207
                 RANKD3     1      2.716527          1.423856            1.908           0.0690
                 DOCT       1     0.9224809          1.258893            0.733           0.4711
```

a. Write down the regression equation.
b. What is the interpretation of the coefficient of SEX?
c. What is the interpretation of the coefficient of RANKD1?

14.43 Refer to Exercise 14.42.
a. Test the hypothesis that the coefficient of SEX is positive. Use $\alpha = .05$.
b. What does the conclusion of this test indicate about allegations of discrimination?

14.44 a. Locate the value of the F statistic in Exercise 14.42.
b. What null hypothesis is being tested by this statistic?
c. Is this null hypothesis rejected at $\alpha = .01$? How plausible is this null hypothesis?

14.45 Another regression model of the data of Exercise 14.42 omits SEX and DOCT from the list of independent variables. The output is shown below:

SOURCE	DF	SUM OF SQUARES	MEAN SQUARE	F VALUE	PROB>F
MODEL	4	2110.925	527.7313	98.389	0.0001
ERROR	25	134.0934	5.363737		
C TOTAL	29	2245.019			

ROOT MSE	2.315974	R-SQUARE	0.9403
DEP MEAN	31.92667	ADJ R-SQ	0.9307
C.V.	7.254044		

PARAMETER ESTIMATES

VARIABLE	DF	PARAMETER ESTIMATE	STANDARD ERROR	T FOR H0: PARAMETER=0	PROB > |T|
INTERCEP	1	19.71134	1.077636	18.291	0.0001
SENIOR	1	0.5571643	0.07439458	7.489	0.0001
RANKD1	1	9.241436	1.821404	5.074	0.0001
RANKD2	1	5.105024	1.587501	3.216	0.0036
RANKD3	1	3.224291	1.32044	2.442	0.0220

a. Locate R^2 for this reduced model.
b. Test the null hypothesis that the true coefficients of SEX and DOCT are zero. Use $\alpha = .01$.

15

CONSTRUCTING A MULTIPLE REGRESSION MODEL

This chapter presents some suggestions for actually creating a useful multiple regression model. It is, we hope, a practical chapter, one that builds on and extends the material of Chapter 14.

four steps in a multiple regression study

A typical multiple regression study passes through at least four steps. First, potentially useful predictor (independent) variables are selected. Qualitative variables can be incorporated by the "dummy variable" device discussed in Section 15.1. Additional variables may be created (in time-series data) by lagging independent variables; this process is discussed in Section 15.2. The second step is tentative selection of plausible forms for the multiple regression model. This may require transformation or combination of either the dependent variable or the independent variables, as discussed in Section 15.3. The third step, which usually requires fitting several candidate models, is the selection of a particular model that is appropriate and useful for the given situation. This step may be done simply by comparing the results of several models or, more elaborately, by the process of stepwise regression discussed in Section 15.4. The fourth step is checking the selected model for potential violations of assumptions. At this step, the residuals (actual y values minus predicted y values) are examined for evidence of severe nonnormality or nonconstant variance, as discussed in Section 15.5. If, in addition, the multiple regression model involves time-series data, it is also necessary to check the residuals for nonindependence (autocorrelation), as discussed in Section 15.6. With luck, a satisfactory multiple regression model can be found by one pass through this four-step process. Usually, several passes are needed.

A fifth step should ideally be made before the model is put into use. This step is checking the results of the model on new data, as discussed in Section 15.7. This is to confirm that the selected model has the indicated predictive value and is not merely an artifact of too much "data massaging."

Obviously, a great deal of calculating is required in the process of selecting a model, and access to statistical software packages is a practical necessity.

15.1 SELECTING CANDIDATE INDEPENDENT VARIABLES (step 1)

Perhaps the most critical decision in constructing a multiple regression model is the initial selection of independent variables. In later sections of this chapter we consider many methods for refining a multiple regression analysis, but, first of all, a decision must be made about which independent (x) variables to consider for inclusion and hence which data to gather. While initially it may appear that an optimum strategy might be to construct a monstrous multiple regression model with very many variables, these models are difficult to interpret and are much more costly from a data-gathering and computer-usage standpoint. How can a manager make a reasonable selection of initial variables to include in a regression analysis?

selection of the independent variables

Knowledge of the problem area is critically important in initial selection of data. First, identify the dependent variable to be studied. Individuals, who have had experience with this variable by observing it, trying to predict it, and trying to explain changes in it often have remarkably good insight as to what factors (independent variables) affect the variable. As a consequence, the first step involves consulting those who have the most experience with the dependent variable of interest. For example, suppose that the problem is to forecast the quarterly sales volume of a certain brand of portable electric typewriter. The quarterly sales volume of this brand is the dependent variable being studied. Certain variables such as price and advertising budget are obvious candidates for independent variables to be included in a multiple regression forecasting equation. A good district sales manager would very likely be able to suggest several other predictors.

collinearity

A more technical aspect of initial selection of variables has to do with **collinearity** the issue of correlated independent variables discussed in Chapter 14. If several very similar independent variables can be used, and they vary closely together, it makes little difference which is chosen. For example, if a measure of national economic activity is to be used to predict typewriter sales, it shouldn't matter much whether gross national product, net national product, or national disposable income is used, since these three variables are nearly perfectly correlated. It could be argued that one might as well include all three variables initially and let the regression analysis select which one is to be used in the final model. This is fair enough, but gathering data on many variables and entering them into a computer is time-consuming. The effort would be better spent on variables that might make a real difference in the quality of prediction.

An ideal set of independent variables shows low correlations among themselves but higher correlations between each such variable and the dependent variable. One way to approach this ideal is to try to find separate "dimensions" of variability for the dependent variable. In predicting portable typewriter sales, it might be useful to guess the market segments that would buy such typewriters. Suppose that three such segments are identified: small offices, home use, and gifts. The natures of these segments suggest different potential independent variables. For instance, the office segment is likely to be affected by the price of word-processing equipment, the home-use segment is likely to be affected by per capita disposable income, and the gift segment is likely to be affected by the number of graduating high school seniors. There is reason to believe that each of these independent variables will correlate with typewriter sales, and there is no obvious reason why they should be heavily correlated with each other.

EXAMPLE 15.1 A firm that sells and services minicomputers is concerned about the volume of service calls. The firm maintains several district service branches within each sales region, and computer owners requiring service call the nearest branch. The branches are staffed by technicians trained at the main office. The key problem is whether technicians should be assigned to main-office duty or to service branches; assignment decisions have to be made monthly. The required number of service-branch technicians grows in almost exact proportion to the number of service calls. Discussion with the service manager indicates that the key variables in determining the volume of service calls seem to be the number of computers in use, the number of new installations, whether or not a model change has been introduced recently, and the average temperature. (High temperatures, or possibly the associated high humidity, lead to more frequent computer troubles, especially in imperfectly air-conditioned offices.) Which of these variables can be expected to be correlated with each other?

Solution It is hard to imagine why temperature should be correlated with any of the other variables. There should be some correlation between number of computers in use and number of new installations, if only because every new installation is a computer in use. Unless the firm has been growing at an increasing rate, we wouldn't expect a severe correlation (we would, however, like to see the data). The correlation of model change to number in use and new installations isn't at all obvious; surely data should be collected and correlations analyzed. □

dummy variables

One special type of independent variable that could be included in the multiple regression model is the **dummy,** or **indicator, variable.** Such variables are used to represent qualitative (categorical) variables such as geographic region, type of incentive plan, or "protected-class" membership in a discrimination suit. The simplest dummy-variable situation occurs when the qualitative variable has only two categories, such as female–male or protected–not protected; then the dummy variable is defined by assigning one category of the qualitative variable the value 1, and the other the value 0. For example, suppose that y is the total production cost (in dollars) of a printing run, x_1 is the number of items printed (in thousands of items), and x_2 is a dummy variable that equals 1 when the run is on a rush basis and 0 when it is on a regular basis. Assume that the multiple regression prediction equation is

$$\hat{y} = 86.2 + 5.1x_1 + 20.5x_2$$

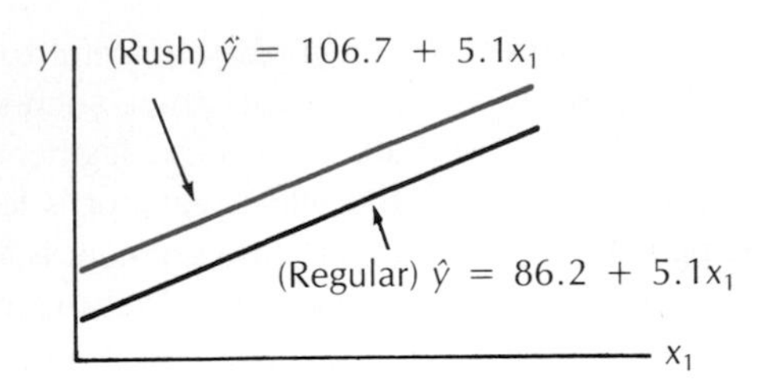

FIGURE 15.1 **Effect of a Two-value Dummy Variable**

The coefficient $\hat{\beta}_2 = 20.5$ of x_2 may be interpreted as the estimated difference in cost between a rush job ($x_2 = 1$) and a regular job ($x_2 = 0$) for any specified run size (x_1).

By substituting $x_2 = 1$ into the multiple regression prediction equation, we get an equation relating y to x_1 for rush jobs. The corresponding equation relating y to x_1 for regular jobs is obtained by substituting $x_2 = 0$.

Rush jobs: $\hat{y} = 86.2 + 5.1x_1 + 20.5(1) = 106.7 + 5.1x_1$

Regular jobs: $\hat{y} = 86.2 + 5.1x_1 + 20.5(0) = 86.2 + 5.1x_1$

Note that the two prediction equations (corresponding to $x_2 = 0$ or $x_2 = 1$) are parallel lines with different intercepts. These equations are shown in Figure 15.1.

If a qualitative variable can take on more than two levels, definition of the dummy variable is a bit more complicated. We do not want to code residence as 0 = urban, 1 = suburban, and 2 = rural and then use the resulting x in a regression. A one-unit increase in this x could mean either a change from urban to suburban or a change from suburban to rural. There is no reason to assume that the two possible changes would predict the same change in any y; the coefficient of such a variable wouldn't mean much. Instead, we could use two dummy variables to define residence: define $x_1 = 1$ if residence = suburban and $x_1 = 0$ otherwise, and define $x_2 = 1$ if residence = rural and $x_2 = 0$ otherwise. If both x_1 and x_2 are zero, it follows by elimination that residence = urban. In general, with this scheme, a qualitative variable with k categories can be coded using $k - 1$ dummy variables. If all $k - 1$ dummies are zero for an observation, the observation must fall in the kth category.

interpreting coefficients

The interpretation of regression coefficients in the prediction equation requires some thought. Suppose that x_1 and x_2 are the dummy variables for the qualitative variable residence, x_3 is education in years, and y is income in thousands of dollars. An appropriate multiple regression model is

$$Y = \beta_0 + \beta_1 x_1 + \beta_2 x_2 + \beta_3 x_3 + \epsilon$$

where

$$x_1 = \begin{cases} 1 \text{ if suburban} \\ 0 \text{ otherwise} \end{cases} \qquad x_2 = \begin{cases} 1 \text{ if rural} \\ 0 \text{ otherwise} \end{cases}$$

The interpretations assigned to the β's for the dummy variables x_1 and x_2 can be seen by examining the corresponding expectations $E(Y)$.

For urban residence ($x_1 = x_2 = 0$): $E(Y) = \beta_0 + \beta_3 x_3$
For suburban residence ($x_1 = 1, x_2 = 0$): $E(Y) = \beta_0 + \beta_1 + \beta_3 x_3$
For rural residence ($x_1 = 0, x_2 = 1$): $E(Y) = \beta_0 + \beta_2 + \beta_3 x_3$

It follows that β_1 is the difference in expected income between suburban and urban residents for a fixed number of years of education x_3. Similarly, β_2 is the difference in expected income for rural and urban residents for a fixed number of years of education x_3. It follows, too, that $\beta_2 - \beta_1$ is the difference in expected income for rural and suburban residents at a fixed education level.

Knowing the interpretations of the β's, one can make sense of a least-squares prediction equation. Suppose that the least-squares prediction equation based on sample data is

$$\hat{y} = 3.05 + 5.91x_1 - 1.84x_2 + .12x_3$$

The coefficient of x_1 is the estimated difference in incomes (in thousands of dollars) for suburban residents when compared to urban residents with the same years of education. Similarly, the coefficient of x_2 is the predicted difference in average income for rural and urban residents with the same education. With years of education x_3 fixed, the regression equation tells us that suburban residents average \$5910 higher income than urban residents, while rural residents average \$1840 lower than urban residents. The "undummied" category (urban residents in the example) serves as the comparison standard. A comparison of rural and suburban residents can be made by comparing each to an urban resident. For education held constant, a suburban resident is predicted to earn $5.91 - (-1.84) = 7.75$ thousand dollars more than a rural resident.

EXAMPLE 15.2 The following model relates a person's salary increase y to one's seniority in years x_1, gender x_2, and location of residence (urban, suburban, or rural).

$$Y = \beta_0 + \beta_1 x_1 + \beta_2 x_2 + \beta_3 x_3 + \beta_4 x_4 + \epsilon$$

$$x_2 = \begin{cases} 1 & \text{if female} \\ 0 & \text{otherwise} \end{cases} \qquad x_3 = \begin{cases} 1 & \text{if suburban} \\ 0 & \text{otherwise} \end{cases} \qquad x_4 = \begin{cases} 1 & \text{if rural} \\ 0 & \text{otherwise} \end{cases}$$

Interpret the coefficients (β's) in the general linear model.

Solution Because there are several dummy variables and one quantitative independent variable, the easiest way to interpret the coefficients is to obtain expected values $E(Y)$ for each combination of settings of the dummy variables. These expected values are listed below.

	Urban ($x_3 = 0, x_4 = 0$)	Suburban ($x_3 = 1, x_4 = 0$)	Rural ($x_3 = 0, x_4 = 1$)
Male ($x_2 = 0$)	$E(Y) = \beta_0 + \beta_1 x_1$	$E(Y) = \beta_0 + \beta_1 x_1 + \beta_3$	$E(Y) = \beta_0 + \beta_1 x_1 + \beta_4$
Female ($x_2 = 1$)	$E(Y) = \beta_0 + \beta_1 x_1 + \beta_2$	$E(Y) = \beta_0 + \beta_1 x_1 + \beta_2 + \beta_3$	$E(Y) = \beta_0 + \beta_1 x_1 + \beta_2 + \beta_4$

From this table, it is clear that β_2 is the difference in mean salary increase $E(Y)$ for females and males, for a given level of seniority (x_1) and location of residence. For example, at a given level of x_1 in an urban district, the difference in mean salary increases for females and males is

$$(\beta_0 + \beta_1 x_1 + \beta_2) - (\beta_0 + \beta_1 x_1) = \beta_2$$

This result also applies for any other fixed geographic location. Similarly, β_3 is the difference in mean salary increase between suburban and urban residences for females (or males) at a given seniority level, and β_4 is the difference in mean salary increase in the rural and urban districts for females (or males) at a given seniority level. □

EXAMPLE 15.3 The minicomputer firm of Example 15.1 is concerned with recent model changes as a source of service calls. It is felt that most of the effects of model changes occur in the first two months. After that time, minor manufacturing changes eliminate any blatant problems. Discuss how dummy variables could be used in a regression model (based on monthly data) to treat the issue of model change.

Solution There are three possibilities in any given month: model has changed that month, model has changed the previous month, or neither of these. Two dummies are needed, and one way to define them is

$$x_1 = \begin{cases} 1 \text{ if model has changed this month} \\ 0 \text{ if not} \end{cases}$$

$$x_2 = \begin{cases} 1 \text{ if model has changed the previous month} \\ 0 \text{ if not} \end{cases}$$

A problem with this approach is that it does not reflect the *number* of new installations in the model-change months. □

15.2 LAGGED PREDICTOR VARIABLES (step 1)

lagged variables

In many time-series regression problems, the independent variables in the equation should be **lagged**. A regional sales manager may try to forecast monthly sales (s) given the number of initial calls (c) and the number of sales presentations (p) by salespeople. For many types of products, the calls and presentations will result in sales some months later, rather than immediately. The sales manager does not want an equation of the form

$$\hat{s}_t = \hat{\beta}_0 + \hat{\beta}_1 c_t + \hat{\beta}_2 p_t$$

but rather an equation something like

$$\hat{s}_t = \hat{\beta}_0 + \hat{\beta}_1 c_{t-2} + \hat{\beta}_2 p_{t-1}$$

In this example, initial calls are lagged two months behind sales, and presentations one month behind. The implicit idea is that initial calls in one month tend to generate presentations the next month, which in turn tend to generate sales a month later.

Lagging variables is often necessary when regression is to be used for forecasting. A sales forecasting equation like

$$\hat{s}_t = \hat{\beta}_0 + \hat{\beta}_1 c_t + \hat{\beta}_2 p_t$$

is most likely useless, because the values of c_t and p_t are not known when the forecast has to be made. To make regression useful in forecasting, the predictor variables must be lagged at least far enough so that their values can be known at forecast time.

The major problem with lagging variables is deciding the number of periods to lag. Should the sales manager lag the presentation variable by one month, two months, or what? It's tempting to include many lags in a single regression equation, such as

$$\begin{aligned}\hat{s}_t = \hat{\beta}_0 &+ \hat{\beta}_1 c_{t-1} + \hat{\beta}_2 c_{t-2} + \hat{\beta}_3 c_{t-3} + \hat{\beta}_4 c_{t-4} + \hat{\beta}_5 c_{t-5} \\ &+ \hat{\beta}_6 p_{t-1} + \hat{\beta}_7 p_{t-2} + \hat{\beta}_8 p_{t-3} + \hat{\beta}_9 p_{t-4} + \hat{\beta}_{10} p_{t-5}\end{aligned}$$

There are two problems with this strategy. First, the number of independent variables is large. Since one must have more observations than variables in a regression (the d.f. for error is number of observations − number of predictors − 1), this strategy requires a lot of data. Second, the lagged variables are likely to be severely correlated (for further discussions of problems related to correlated predictor variables, see Section 14.2).

Econometricians have proposed several other strategies for deciding on lags (see Johnston, 1977). These strategies typically involve rather severe assumptions that are difficult to verify. If one has no inkling of plausible lag structures, these methods are worth a try.

Often it's possible to find reasonable lags by thoughtful trial and error. The sales manager may well try lagging presentations by one month in one regression equation, and by both one and two in a second. If the second equation doesn't forecast much better than the first, and if most sales result within a month or so of presentations, there is no reason to try further lags. Knowledge of the basic process involved is almost always useful in choosing lags.

EXAMPLE 15.4 An automobile dealer offers a 12-month or 12,000-mile warranty on new cars and a 3-month or 3000-mile warranty on used cars. As a consequence, the dealer's repair department handles a certain number of warranty-covered repair jobs each month. Data are collected on y_t, the number of warranty repair jobs in month t; x_{t1}, the number of new car sales in month t; and x_{t2}, the number of used car sales in month t. The data are analyzed by two regression models:

$$Y_t = \beta_0 + \beta_1 x_{t1} + \beta_2 x_{t2} + \epsilon_t$$
$$Y_t = \beta_0 + \beta_1 x_{t1} + \beta_2 x_{t-1,1} + \beta_3 x_{t-2,1} + \beta_4 x_{t2} + \beta_5 x_{t-1,2} + \beta_6 x_{t-2,2} + \epsilon_t$$

Computer output is shown below: variable 7 is y_t, variable 1 is x_{t1}, variable 4 is x_{t2}; variables 2 and 3 are $x_{t-1,1}$ and $x_{t-2,1}$; variables 5 and 6 are $x_{t-1,2}$ and $x_{t-2,2}$.

```
MTB > regress c7 on 2 variables in c1 and c4

The regression equation is
Y = 0.33 + 0.312 X1 + 0.171 X2

Predictor        Coef          Stdev     t-ratio
Constant        0.326          6.535        0.05
X1             0.3115         0.1412        2.21
X2             0.1715         0.1857        0.92

s = 5.174        R-sq = 38.3%      R-sq(adj) = 34.3%

Analysis of Variance

SOURCE        DF         SS          MS
Regression     2     514.50      257.25
Error         31     829.74       26.77
Total         33    1344.24

MTB > regress c7 on 6 variables in c1-c6

The regression equation is
Y = - 25.1 + 0.176 X1 + 0.141 X1lag1 + 0.123 X1lag2 + 0.264 X2 - 0.003 X2lag1
           + 0.259 X2lag2

Predictor        Coef          Stdev     t-ratio
Constant      -25.052          5.776       -4.34
X1            0.17643        0.08728        2.02
X1lag1        0.14055        0.09187        1.53
X1lag2        0.12297        0.08970        1.37
X2             0.2644         0.1082        2.44
X2lag1        -0.0030         0.1060       -0.03
X2lag2         0.2590         0.1241        2.09

s = 2.903        R-sq = 83.1%      R-sq(adj) = 79.3%

Analysis of Variance

SOURCE        DF         SS          MS
Regression     6    1116.72      186.12
Error         27     227.51        8.43
Total         33    1344.24
```

a. Write the regression equation for the nonlagged model, together with the residual standard deviation and R^2 values.
b. Do the same for the lagged model.
c. Is there reason to believe that the lagged variables are useful in prediction?

Solution a. For this model, the regression equation is

$$\hat{y}_t = .33 + .312x_{i1} + .171x_{i2}$$

The corresponding values of s_ϵ and R^2 are

$$s_\epsilon = 5.174 \qquad \text{and} \qquad R^2 = .383$$

b. The regression equation for the second model is

$$\hat{y}_t = -25.1 + .176x_{t1} + .141x_{t-1,1} + .123x_{t-2,1} + .264x_{t2} - .003x_{t-1,2} + .259x_{t-2,2}$$

The values of s_ϵ and R^2 are

$$s_\epsilon = 2.903 \qquad \text{and} \qquad R^2 = .831$$

c. The inclusion of lagged variables has helped in the prediction of y. The value of R^2 has been increased considerably, and s_ϵ is much smaller. □

EXAMPLE 15.5 In Example 15.3 it was indicated that new-model introduction affected serivce calls for two months. What lags should be used in a (monthly) data base?

Solution New installations reflect only the current month. To capture the two-month effect, the new-installations variable should also be lagged by a month. As a check on the belief that the new-model effect lasts only two months, it wouldn't be totally inappropriate to lag new installations by another month as well. □

SECTIONS 15.1 AND 15.2 EXERCISES

15.1 A city probation office tries to use various reported crime rates to forecast the volume of new cases. The total number of nonviolent crimes reported in the city is determined for 12 quarters, as is the volume of new cases. The data are

Quarter, t:	1	2	3	4	5	6	7	8	9	10	11	12
Crimes (thousands), x_t:	6.4	5.6	5.8	6.6	7.0	6.7	6.5	7.1	7.2	6.9	6.8	6.7
New cases (hundreds), y_t:	13.1	13.5	12.7	12.9	14.3	14.8	14.4	13.9	15.0	15.5	15.1	14.6

a. Calculate the correlation between x_t and y_t.
b. Calculate the "lag 1" correlation—the correlation between x_{t-1} and y_t. Note that the effective sample size is 11, not 12.
c. Which of the correlations is stronger? Does this result seem sensible?

15.2 Refer to the data of Exercise 15.1.
a. Calculate the regression equation $\hat{y}_t = \hat{\beta}_0 + \hat{\beta}_1 x_{t-1}$.
b. Calculate the residual standard deviation.
c. Give a 95% prediction interval for the volume of new cases in quarter 13.

15.3 Refer to the data of Exercise 15.1 What would happen if you tried to calculate lag 10 and lag 11 correlations?

15.4 A textbook publisher begins a sales forecasting system. The chief editors for each of three divisions prepares forecast first-year sales for all new books published in a certain year. These forecasts are later compared to actual sales. Computer output from a regression study (SAS) follows:

DEPENDENT VARIABLE: ACTUAL

SOURCE	DF	SUM OF SQUARES	MEAN SQUARE	F VALUE	PR > F	R-SQUARE	C.V.
MODEL	3	894.74508876	298.24836292	109.47	0.0001	0.924030	13.9768
ERROR	27	73.56200801	2.72451882		ROOT MSE		ACTUAL MEAN
CORRECTED TOTAL	30	968.30709677			1.65061165		11.80967742

PARAMETER	ESTIMATE	T FOR H0: PARAMETER=0	PR > |T|	STD ERROR OF ESTIMATE
INTERCEPT	-1.86308649	-1.86	0.0738	1.00171014
FORECAST	1.35417285	16.33	0.0001	0.08292934
DIV2	-4.00656296	-5.67	0.0001	0.70677705
DIV3	0.91579423	1.25	0.2205	0.73021786

a. Identify the dummy variables.

b. What is the interpretation of the coefficients of these dummy variables?

c. Do the *t* statistics suggest that the dummy variables separately have some predictive value?

15.5 For the data of Exercise 15.4, what additional computer output would you need to test the null hypothesis that the dummy variables collectively have no predictive value once the forecast value has been included in the regression equation?

15.6 Refer to Exercise 15.4.

a. What null hypothesis is being tested by the *F* statistic? What does this hypothesis say about the forecasting system?

b. Can this null hypothesis be rejected at $\alpha = .01$? Locate the *p*-value.

15.7 Refer to Exercise 15.4.

a. If the forecast sales had been exactly correct on the average within each division, what would the regression equation be? In particular, what would the coefficient of forecast equal?

b. Calculate a 95% confidence interval for the forecast coefficient. Does it include the value you specified in part (a)?

15.3 NONLINEAR REGRESSION MODELS (step 2)

In Section 14.1, we pointed out that use of a first-order regression model (one containing only linear terms in the *x*'s) assumes linearity in each variable separately and also assumes that there are no interactions among the independent variables. In this section we discuss how to modify a first-order model to handle some kinds of nonlinearities.

In some situations it is evident that a model that is linear in all the independent variables is inadequate. In Example 14.3 we pointed out that a regression model that predicts y, a taste-test preference score for a lime drink, as a linear function of x_1, lime concentration, and x_2, sweetness, is very dubious. In other cases, scatter plots of the data reveal nonlinearities. In linear regression, the ordinary plot of y versus x is adequate. In multiple regression, the effect of variations in other x's can obscure the nonlinearity. For this reason, a standard strategy is to fit a first-order model and then plot the residuals from this model against each independent variable. If a more appropriate model contains, for instance, a term in x_1^2, then the plot of residuals against x_1 shows a nonlinear pattern. Because the use of a first-order model removes the linear effect of the other independent variables, nonlinearities often show up more clearly in these residual plots.

residual plots

Interactions among variables are harder to detect in scatter plots. If x_1 and x_2 interact in determining y, three variables are involved; unfortunately, three-dimensional plots are hard to draw. Common sense and the use of the stepwise regression procedures of Section 15.4 may be the best approach to determining whether interactions are present. For the special case in which one of the independent variables is a qualitative variable represented by one or more dummy variables, interaction may be detected by plotting residuals (from a first-order model) against the other independent variables. Separate plots should be made for observations in each category of the qualitative variable. The first-order, no-interaction model implies that these plots should be parallel. If the separate residual plots are not close to parallel, the possibility of interaction should be considered.

EXAMPLE 15.6 In Example 15.3 we noted that the new-model dummy variables did not reflect the number of new installations in new-model months. Data for 29 months are collected on the following variables.

Variable	Description
7	total number of service calls in the month
1	number of minicomputers in use, beginning of month
2	number of new installations, this month
3	number of new installations, previous month
4	1 if model change in this month, 0 if not
5	1 if model change in previous month, 0 if not
6	average temperature, this month

A first-order model is fit to the data. Plots of residuals against variables 2 and 3 are shown in Figure 15.2. Is there evidence that interaction terms should be used?

To see if the effect of this month's installations depends on whether or not there was a model change this month, we plot residuals against variable 2 and circle the cases for variable 4 = 1. Similarly, to see if the effect of installations in the previous month depends on whether or not there was a model change last month, we plot residuals against variable 3 and circle the cases corresponding to variable 5 = 1.

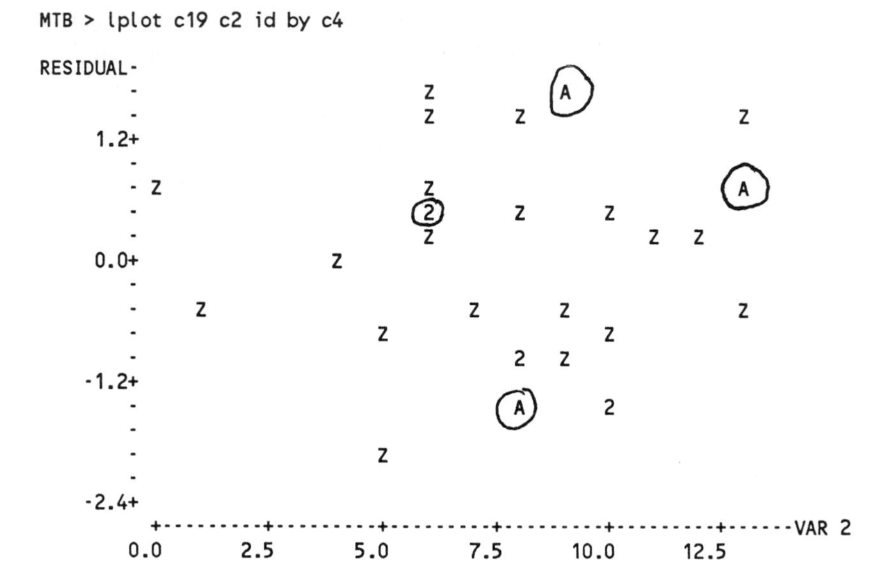

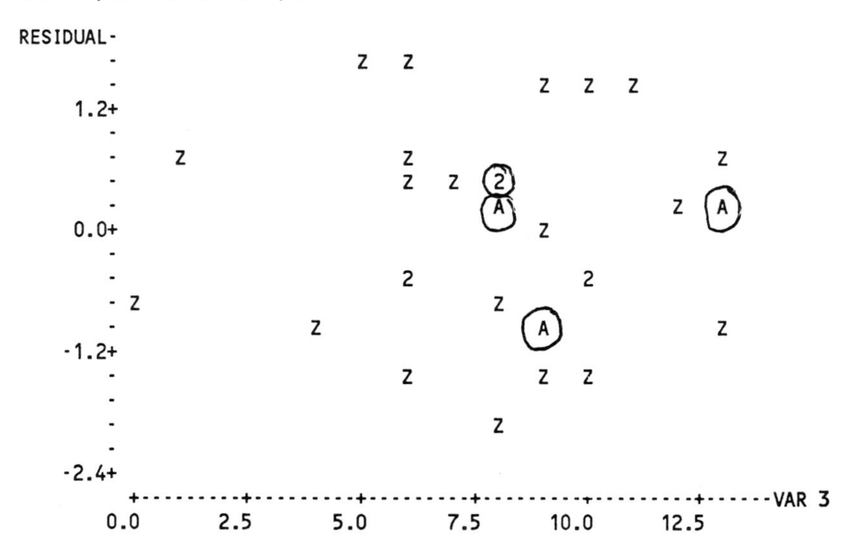

FIGURE 15.2 Residual Plot for Example 15.6

Solution Certainly, a strong pattern doesn't exactly leap off the page at you. There is no obvious positive or negative slope in either plot. There is no visually obvious reason to include the interaction terms, but the theoretical reasoning behind the idea is strong enough that we may want to test it anyway. □

Once potential nonlinearity has been discovered, it may be dealt with either by transforming existing variables or by adding additional, higher-order terms in the x's. Often both of these strategies are used. Transformations may be suggested either by the nature of the problem or by the look of the data.

With economic variables and time-series data, growth often occurs at a roughly constant percentage rather than at an absolute rate. For instance, total sales of a large company may grow at the rate of 8% per year, as opposed to $8 million per year. Of course, if initial sales are $100 million, and 8% growth in the first year is the same as $8 million growth. But in later years, constant percentage growth and constant additive growth differ. The table below shows the total sales for a company over a five year period, based on each of the two types of growth.

	Year					
	0	1	2	3	4	5
8% growth	100.0	108.0	116.6	126.0	136.0	146.9
$8 million growth	100.0	108.0	116.0	124.0	132.0	140.0

Most basic finance books show that if a quantity y grows at a rate r per unit time (continuously compounded), the value of y at time t is

$$y_t = y_0 e^{rt}$$

where y_0 is the initial value. This relation may be converted into a linear relation between Y_t and t by a **logarithmic transformation**

logarithmic transformation

$$\log y_t = (\log y_0) + rt$$

The linear regression method of Chapter 13 can be used to fit data for this regression model with $\beta_0 = \log y_0$ and $\beta_1 = r$. When y is an economic variable such as total sales, the logarithmic transformation is often used in a multiple regression model:

$$\log Y_i = \beta_0 + \beta_1 x_{i1} + \beta_2 x_{i2} + \ldots + \beta_k x_{ik} + \epsilon_i$$

The Cobb-Douglas production function is another standard example of a nonlinear model that can be transformed into a regression equation:

$$y = c\, l^{\alpha}\, k^{\beta}$$

where y is production, l is labor input, k is capital input, and α and β are unknown constants. Again, to transform the dependent variable we take logarithms to obtain

$$\begin{aligned} \log y &= (\log c) + \alpha(\log l) + \beta(\log k) \\ &= \beta_0 + \beta_1(\log l) + \beta_2(\log k) \end{aligned}$$

This suggests that a regression of log production on log labor and log capital is linear.

EXAMPLE 15.7 An important economic concept is the *price elasticity of demand*, defined as the negative of the percentage change in quantity demanded per percentage change in price. It can be shown that a price elasticity of 1 means that a (small) price change yields no change in total revenue. An inelastic demand (elasticity less than 1) means that a small price increase yields an increase in revenue; elastic demand is the opposite.

Data are obtained on y, daily demand for lettuce (in heads sold per hundred customers) for varying levels of price (dollars per head). As much as possible, other conditions that might affect demand are held constant; all participating stores are located in middle-class suburbs, no competitors are running sales on lettuce, and so on. The data are

x:	.79	.79	.84	.84	.89	.89	.94	.94	.99	.99
y:	40.2	37.1	37.4	34.9	32.8	35.5	30.6	34.2	31.2	29.8
xy:	31.758	29.309	31.416	29.316	21.192	31.595	28.764	32.148	30.888	29.502

a. What economic quantity does xy represent?
b. Does there appear to be any trend in xy values as x increases?
c. If xy is constant, what is true of $\log x + \log y$?
d. If a product has price elasticity equal to 1, what does the regression equation of $\log y$ versus $\log x$ look like?

Solution a. The term xy is price per head times heads per hundred customers. Therefore it represents revenue per hundred customers.

b. No trend is apparent in a plot of the data. Revenue, xy, appears constant.

c. Since $\log xy = \log \text{constant} = \log x + \log y$, $\log x + \log y$ should be constant.

d. A price elasticity of 1 means that $\log y = \text{constant} - \log x$. The regression equation with $\log y$ as dependent variable and $\log x$ independent should have a slope nearly equal to -1 (plus-or-minus random error). Thus a regression model in $\log y$ and $\log x$ is useful in elasticity studies. □

A fairly extensive discussion of possible transformations is found in Tukey (1977).

Sometimes nonlinearity can be handled by adding extra terms in the independent variables. Quadratic terms, such as x_1^2, are particularly useful when it is thought that y increases to a maximum and then decreases (or decreases to a minimum and then increases) as x_1 increases. If a plot of residuals (from a first-order model) versus an independent variable shows a parabolic pattern, a quadratic term should be added.

EXAMPLE 15.8 For the service-call situation of Example 15.6, the effect of temperature may not be linear. A regression model is calculated using the independent variables and interaction terms indicated in Example 15.6. Residuals are plotted against average temperature, as shown in Figure 15.3. Does this plot suggest that a quadratic term would be a useful predictor?

Solution Again, there is certainly nothing blatant here. If not for the point at the lower right, there would not be any pattern at all. We would not base any decision on a couple of odd points. □

Interaction terms may also be added to the multiple regression model to cure some forms of nonlinearity. Unfortunately, multiple regression was not designed to capture interaction effects. The experimental design, ANOVA approach (where every possible value of

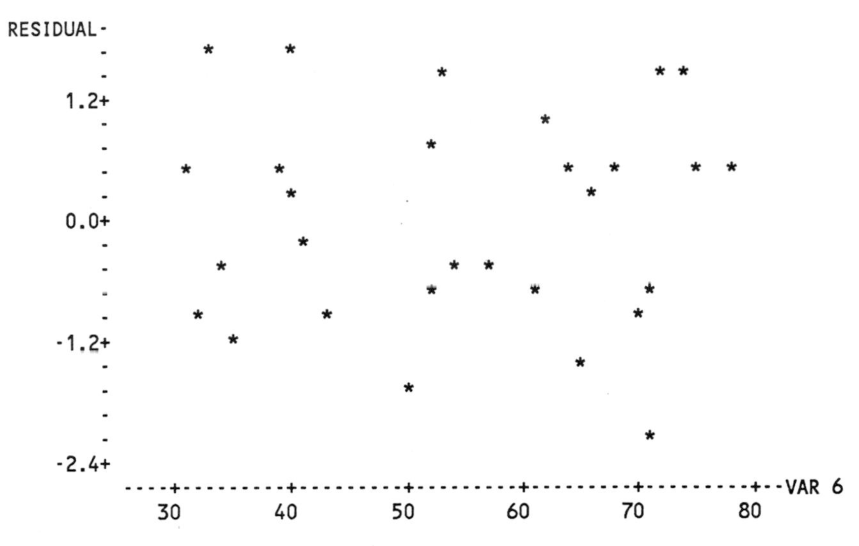

FIGURE 15.3 Residual Plot for Example 15.8

one predictor is combined with each possible value of every other predictor) yields a much more clearly intelligible treatment of this form of nonlinearity. In practice, however, it may not be possible to plan the neatly balanced data-gathering schemes of the ANOVA approach. In such situations, the regression approach may be modified to yield some information (however imperfect) about interaction effects.

cross-product terms

One basic but mechanical device is to include **cross-product terms**, as we indicated in Chapter 14. A sales forecasting equation such as

$$\hat{y} = .40 + .04c + .01d + .005cd$$

allows for a certain type of interaction between d and c. If $c = 200$, the predicted change in y per unit change in d is $.01 + .005(200)$. If $c = 500$, the predicted change in sales y per unit change in d is $.01 + .005(500)$. In general, if

$$\hat{y} = \hat{\beta}_0 + \hat{\beta}_1 c + \hat{\beta}_2 d + \hat{\beta}_3 cd$$

the predicted change in y per unit change in d is $\hat{\beta}_2 + \hat{\beta}_3 c$ for a given value of c. Thus the effect of a change in d depends in a certain way on the level of c. This is one form of interaction, though by no means the only one. A cross-product term such as cd may be treated as just another predictor variable. Its coefficient may be tested for statistical significance by a t test, and its predictive value may be assessed by the increase in squared correlation. Therefore this approach to the problem of interaction can be done routinely within the ordinary regression structure.

EXAMPLE 15.9 Salary data for 427 teachers are examined for a seniority and unionization study. Product terms are created: $x_6 = x_1x_2$ and $x_7 = x_1x_3$. A summary computer output is shown for the model $Y = \beta_0 + \beta_1 x_1 + \beta_2 x_2 + \ldots + \beta_7 x_7 + \epsilon$. Test the hypotheses that x_6 and x_7 have no incremental predictive value.

VARIABLE	COEFF	ST.ERROR	T-STATISTIC
INTERCEPT	10.243	------	------
X1	2.070	7.912	0.262
X2	2.963	0.031	16.065
X3	1.475	0.203	7.249
X4	1.932	0.396	5.590
X5	0.808	0.301	2.686
X6	0.177	0.032	5.554
X7	0.093	0.011	8.183

Solution For these data, $n = 427$. The d.f. for the t test is $427 - (7 + 1) = 419$, so the relevant t table is effectively the normal table. The t statistics for x_6 and x_7 fall far beyond normal table values. Therefore we may conclude that some degree of interaction between x_1 and x_2 and between x_1 and x_3 has been shown. □

Insertion of cross-product terms into the regression equation is sometimes thought to be the only way to handle interaction in regression. Certainly this approach does not handle all possible kinds of interaction, but it does provide a useful approximation to solving the problem of interaction within regression. A manager who believes that interaction effects are crucial in predicting a dependent variable may well spend the extra money to gather data in the neatly balanced form of ANOVA methods.

Finally, thoughtful consideration of underlying economic relations may suggest other combinations of variables to address questions of nonlinearity. For example, suppose that a regression study is made of the total yearly expenditures of cities (y) on water supply systems. Natural independent variables are x_1, the population size; x_2, the total water consumption; and x_3, the number of miles of water lines (of course, there are other possibilities). A regression analysis based on these variables would be bedeviled by collinearity; every other variable would be strongly correlated with city size. A better analysis would take the dependent variable as y/x_1, the per capita expenditure. Natural independent variables would be x_2/x_1, x_3/x_1, and perhaps x_1 itself.

EXAMPLE 15.10 The data of Example 15.6 indicated (to no one's surprise) that variable 7, the number of service calls, increases as variable 1, the number of minicomputers in use, increases. What would be the interpretation of a new variable, defined as variable 7 divided by variable 1? Why might the new variable be an appropriate dependent variable?

Solution The new variable represents the number of service calls per computer. For a growing business such as the minicomputer firm, defining the variables as fractions of the number of computers in use might well reduce collinearity. □

EXAMPLE 15.11 A manufacturer of feed for chickens faces a great deal of month-to-month variability in sales. A regression study attempts to forecast monthly sales volumes. The feed is used largely for chickens age 20–50 days, so the number of chicken starts in the previous month is expected to be a critical predictor variable. Monthly data on starts are available. In addition, feed sales are expected to be quite sensitive to price. The prices of the manufacturer's feed and the primary competitor's feed, as well as the wholesale price of chickens, are very plausible predictor variables.

The working group charged with performing the regression study argues over the form of the regression equation. The following suggestions are made:

1. a linear (first-order) model in starts, price, competitor's price, and chicken price
2. a first-order model in starts, difference in price (between manufacturer and competitor), and chicken price
3. a first-order model in starts and the price difference as a fraction of chicken price

Write the three suggested models. Is there a simple relation between one model and any other?

Solution Let

y = monthly sales of the manufacturer's feed
x_1 = starts in the previous month
x_2 = current price of manufacturer's feed
x_3 = current price of competitor's feed
x_4 = current price of chickens

Model 1 is a first-order model in these variables:

$$Y = \beta_0 + \beta_1 x_1 + \beta_2 x_2 + \beta_3 x_3 + \beta_4 x_4 + \epsilon$$

Model 2 involves x_2 and x_3 only through their difference:

$$Y = \beta_0^* + \beta_1^* x_1 + \beta_2^*(x_2 - x_3) + \beta_3^* x_4 + \epsilon$$

Model 3 involves the difference as a fraction of chicken price x_4:

$$Y = \beta_0^{**} + \beta_1^{**} x_1 + \beta_2^{**}(x_2 - x_3)/x_4$$

Model 2 is equal to model 1 if $\beta_2^* = \beta_2$ and $-\beta_2^* = \beta_3$.
Model 3 is not a first-order model in $x_1, \ldots, x_4$, so there is no simple relation between it and the others. □

SECTION 15.3 EXERCISES

15.8 A consultant who specializes in corporate gifts to charities, schools, cultural institutions, and the like is often asked to suggest an appropriate dollar amount. The consultant undertakes a regression study to try to predict the amount contributed by corporations to colleges and universities and is able to obtain information on the contributions of 38 companies. Financial information about these companies is available from their annual reports. Other information is obtained from such sources as business magazines. From experience, the consultant believes that the level of contributions is affected by the profitability of a firm, the size of the firm, whether the firm is in a high-education industry (such as data processing, electronics, or chemicals), the educational level of the firm's executives, and whether the firm matches the contributions of employees. Profitability can be measured by pre-tax or post-tax income, size by number of employees or gross sales, and educational level by average number of years of education or by percentage of executives holding advanced degrees.

a. Would you expect pre-tax and post-tax income to be highly correlated? How about number of employees and gross sales?
b. Discuss how to define profitability, size, and educational level so that the correlations among these variables are not automatically huge.

15.9 The consultant of Exercise 15.8 proposes to define an industry-type variable as follows:

$$\text{INDUSTRY} = \begin{cases} 3 \text{ if firm is primarily in the electronics industry} \\ 2 \text{ if firm is primarily in the data-processing industry} \\ 1 \text{ if firm is primarily in the chemical industry} \\ 0 \text{ otherwise} \end{cases}$$

a. Explain why this is not a good idea.
b. Suggest an alternative approach for indicating these industries.
c. How could the factor of whether or not the firm matches employee contributions be incorporated into a regression model?

15.10 The consultant of Exercise 15.8 collects data on the following variables:

CONTRIB:	millions of dollars contributed
INCOME:	pre-tax income, in millions of dollars
SIZE:	number of employees, in thousands
DPDUMMY:	1 if firm is primarily in the data-processing industry 0 if not
ELDUMMY:	1 if firm is primarily in the electronics industry 0 if not
CHDUMMY:	1 if firm is primarily in the chemical industry 0 if not
EDLEVEL:	proportion of executives holding advanced degrees
MATCHING:	1 if firm matches employee contributions 0 if not

a. Does it seem like a good idea to take CONTRIB as the dependent variable, with all other variables as independent variables?
b. What does the variable CONTRIB/INCOME represent?

15.11 Refer to Exercise 15.10. The consultant suspects that the effect of SIZE on CONTRIB/INCOME differs greatly among firms in the data-processing, electronics, chemical, and other industries. How can the regression model be modified to test this suspicion?

15.12 Refer to Exercise 15.10. The consultant suspects that the effect of increasing EDLEVEL is itself increasing; that is, all else being equal, there is little difference in CONTRIB/INCOME for firms with EDLEVEL = .2 versus .3, more for firms with EDLEVEL = .4 versus .5, and still more for firms with EDLEVEL = .6 versus .7.

a. How can a regression model be formulated to test this suspicion?
b. If the consultant's suspicion is correct, and if the residuals from a first-order regression model are scatter-plotted against EDLEVEL, what pattern of residuals would you expect to see?

15.13 A company that has developed a plastic film for use in wrapping food (such as crackers and cookies) has a problem with film stiffness. To be useful with modern packaging machines, stiffness (as given by an accepted measure) must be high. Stiffness is thought to be the result of certain variables of the production process. A regression study attempts to predict film stiffness for various combinations of these variables. A total of 32 pilot plant runs are made. Data are recorded on the following variables:

STIFF:	stiffness	REPEL:	percentage of recycled pelletized material used
MELT:	melt temperature (°F)	SPEED:	line production speed (feet per minute)
CHILL:	chill temperature (°F)	KNIFE:	setting of vacuum knife

There is considerable uncertainty among the firm's chemical engineers as to the mathematical form of the relation among these variables. The following output is obtained for a first-order model:

```
PEARSON CORRELATION COEFFICIENTS / PROB > |R| UNDER HO:RHO=0 / N = 32

            STIFF      MELT     CHILL     REPEL     SPEED     KNIFE

STIFF     1.00000   0.05933   0.13753  -0.88640   0.02966  -0.30752
           0.0000    0.7470    0.4529    0.0001    0.8720    0.0869

MELT      0.05933   1.00000   0.00000   0.00000   0.00000   0.00000
           0.7470    0.0000    1.0000    1.0000    1.0000    1.0000

CHILL     0.13753   0.00000   1.00000   0.00000   0.00000   0.00000
           0.4529    1.0000    0.0000    1.0000    1.0000    1.0000

REPEL    -0.88640   0.00000   0.00000   1.00000   0.00000   0.00000
           0.0001    1.0000    1.0000    0.0000    1.0000    1.0000

SPEED     0.02966   0.00000   0.00000   0.00000   1.00000   0.00000
           0.8720    1.0000    1.0000    1.0000    0.0000    1.0000

KNIFE    -0.30752   0.00000   0.00000   0.00000   0.00000   1.00000
           0.0869    1.0000    1.0000    1.0000    1.0000    0.0000
```

a. How much collinearity is present in these data?
b. The 32 observations involved one measurement of each combination of MELT = 510, 530, 550, 570 with CHILL = 70, 80, 90, 100, and REPEL = 20, 30. How much correlation should there be between MELT and CHILL and between MELT and REPEL?

15.14 A first-order model is fit to the data of Exercise 15.13. The following (SAS) output is obtained and plotted in Figure 15.4. Is there any evidence, by eye, of nonlinearity? RESSTIFF is the name of the residuals.

DEPENDENT VARIABLE: STIFF

SOURCE	DF	SUM OF SQUARES	MEAN SQUARE	F VALUE	PR > F	R-SQUARE	C.V.
MODEL	5	3106.40000000	621.28000000	48.73	0.0001	0.903581	2.8780
ERROR	26	331.47500000	12.74903846		ROOT MSE		STIFF MEAN
CORRECTED TOTAL	31	3437.87500000			3.57057957		124.06250000

EXERCISE 15:14 (continued on facing page)

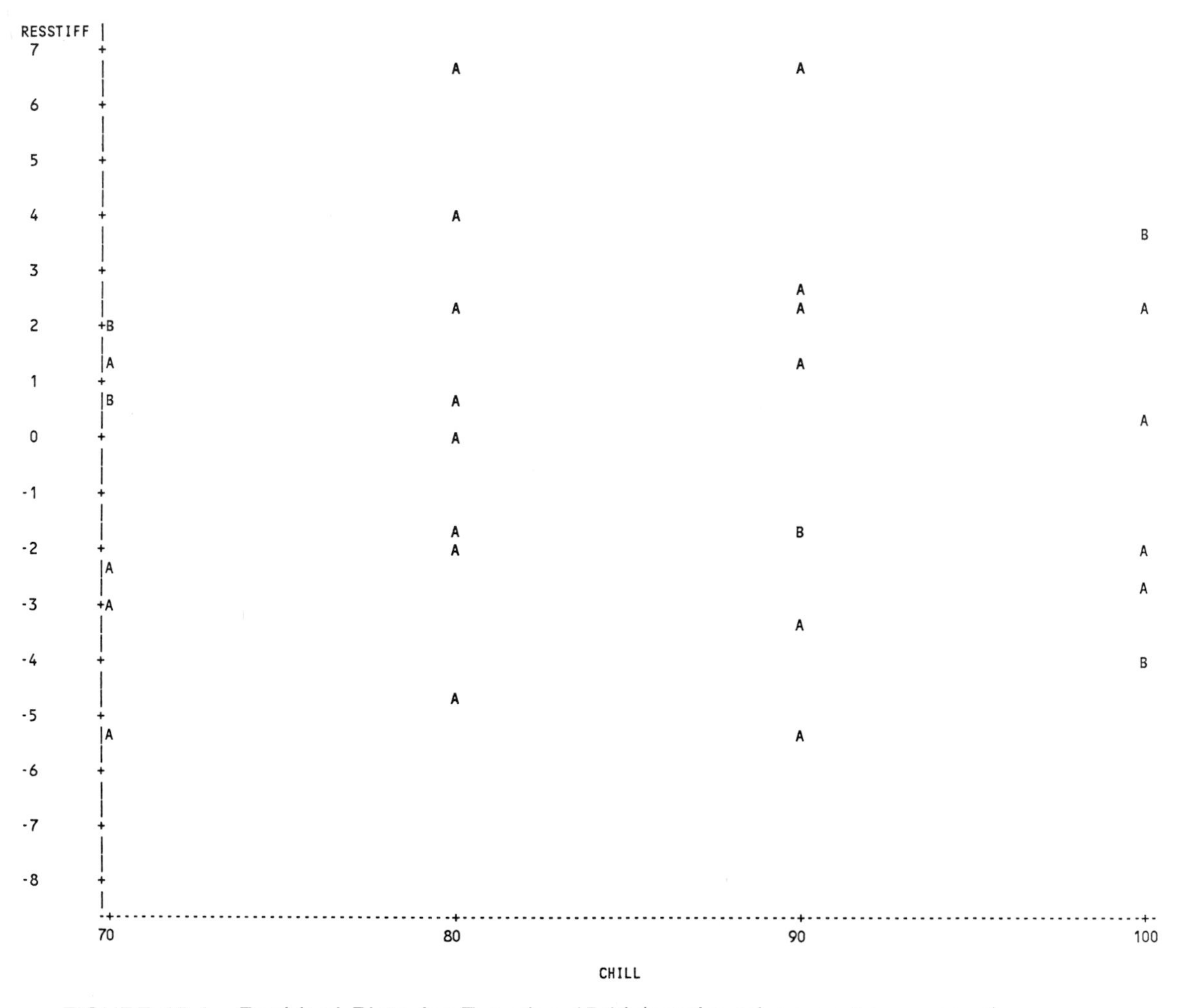

FIGURE 15.4 Residual Plots for Exercise 15.14 (continued on next two pages)

PARAMETER	ESTIMATE	T FOR H0: PARAMETER=0	PR > |T|	STD ERROR OF ESTIMATE
INTERCEPT	170.96250000	8.34	0.0001	20.50947556
MELT	0.02750000	0.97	0.3389	0.02822791
CHILL	0.12750000	2.26	0.0325	0.05645582
REPEL	-1.83750000	-14.56	0.0001	0.12623905
SPEED	0.00687500	0.49	0.6303	0.01411396
KNIFE	-0.31875000	-5.05	0.0001	0.06311953

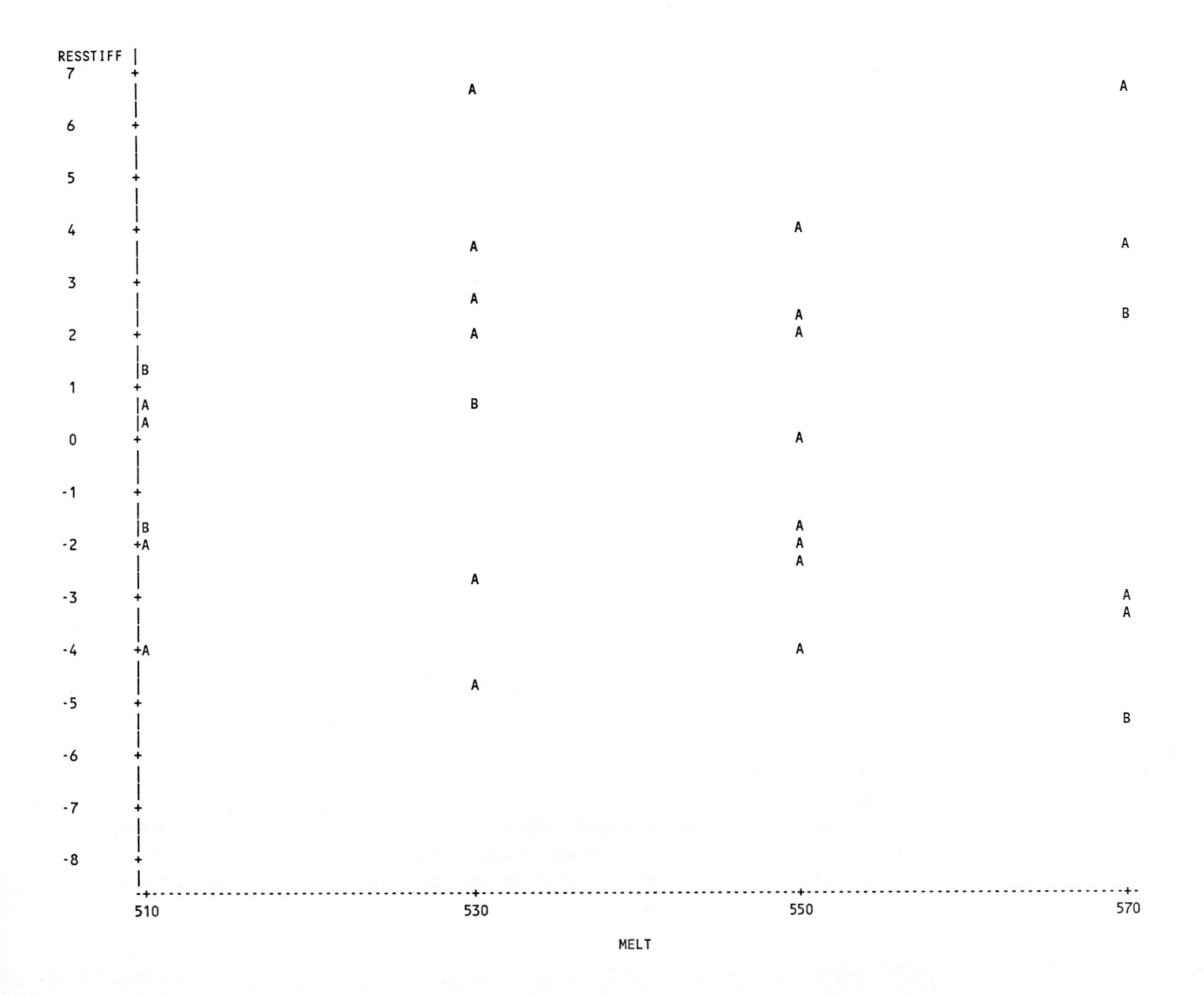

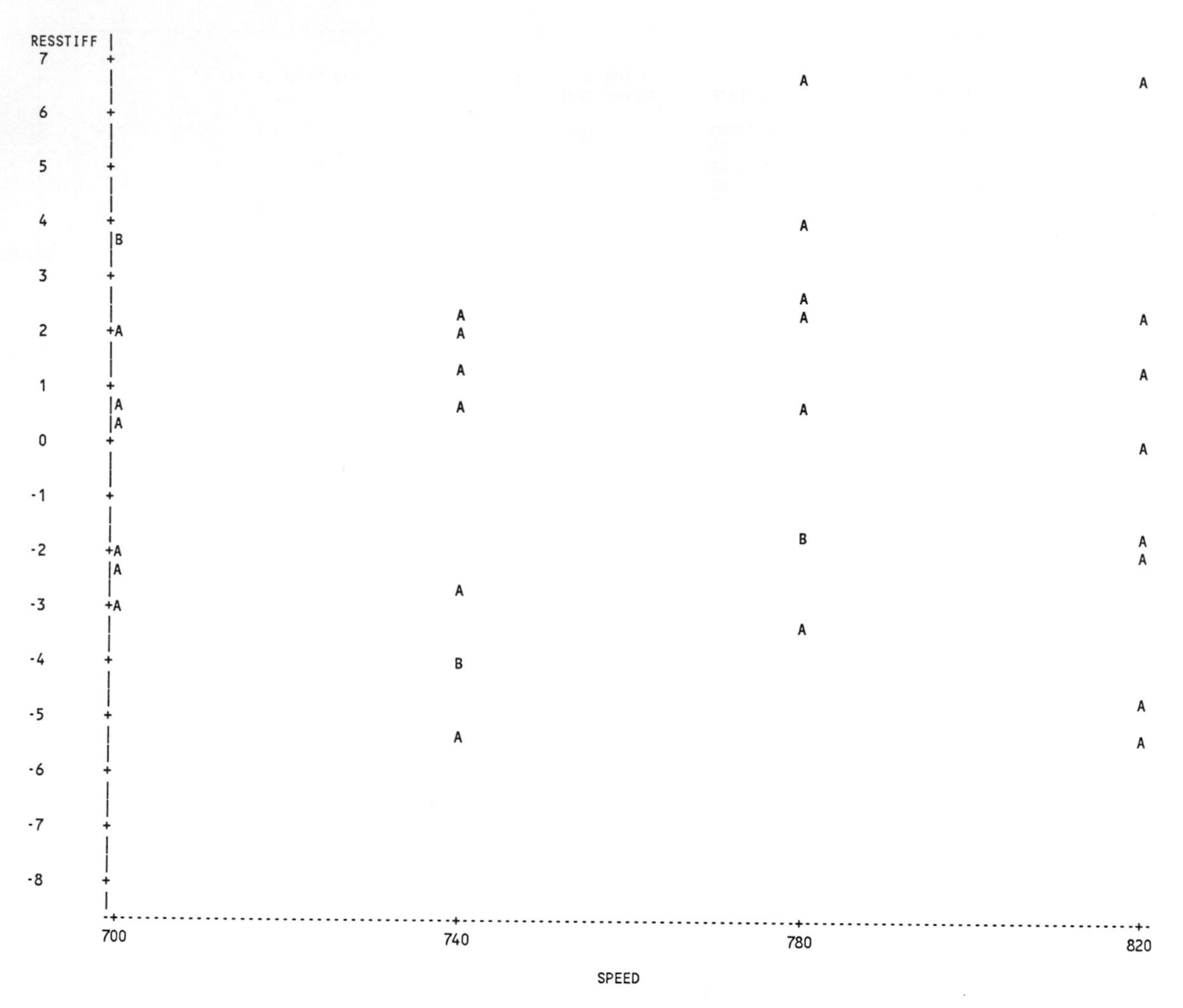

FIGURE 15.4 (continued)

15.15 In an attempt to detect nonlinearity in the data of Exercise 15.13, a second-order model (containing squared MELT and CHILL terms) is run, and the output on the top of the next page is obtained.

a. How much larger is the R^2 for this model than the R^2 for the first-order model of Exercise 15.14?

b. Use the F test for complete and reduced models to test the null hypothesis that the addition of the squared terms yields no additional predictive value. Use $\alpha = .05$.

c. Do the t statistics indicate that either squared term is a statistically significant ($\alpha = .05$) predictor as "last predictor in"?

DEPENDENT VARIABLE: STIFF

SOURCE	DF	SUM OF SQUARES	MEAN SQUARE	F VALUE	PR > F	R-SQUARE	C.V.
MODEL	7	3141.62500000	448.80357143	36.36	0.0001	0.913828	2.8319
ERROR	24	296.25000000	12.34375000		ROOT MSE		STIFF MEAN
CORRECTED TOTAL	31	3437.87500000			3.51336733		124.06250000

PARAMETER	ESTIMATE	T FOR H0: PARAMETER=0	PR > |T|	STD ERROR OF ESTIMATE
INTERCEPT	-308.00000000	-0.67	0.5086	459.03091353
MELT	1.37750000	0.82	0.4195	1.67714997
CHILL	3.63375000	1.54	0.1370	2.36158010
REPEL	-1.83750000	-14.79	0.0001	0.12421629
SPEED	-0.03437500	-1.11	0.2793	0.03105407
KNIFE	-0.31875000	-5.13	0.0001	0.06210815
MELTSQ	-0.00125000	-0.81	0.4287	0.00155270
CHILLSQ	-0.02062500	-1.49	0.1505	0.01388780

15.4 CHOOSING AMONG REGRESSION MODELS (step 3)

In the previous sections of this chapter we have suggested many reasons to include variables in a regression study. The variables may be defined directly or they may be defined by using dummy variables, lagged variables, transformed variables, interaction cross-product variables, or some other combination. Most of these suggestions involve adding more independent variables to the study. A manager who begins a regression study may well suffer from the "kitchen-sink syndrome" and try to throw every possible variable into a regression model. Some sensible guidelines are needed for selecting the independent variables, from all candidates, to be used in the final regression model.

Stepwise regression is a device for computerized selection of the "best" independent variables. There are several types of stepwise regression procedure; we concentrate on the simplest one, **forward selection**.

forward selection

Forward-selection stepwise regression selects independent variables for inclusion in a regression model one at a time. The first variable included is the one that has the highest r^2 value for predicting y; assume that this variable is called x_1. The second variable included is the one that, when combined with x_1, yields the highest R^2 value; call the second variable x_2. If there is any degree of collinearity among the x's, x_2 may not have the second-largest r^2_{yx} value. In fact $R^2_{y \cdot x_1 x_2}$ may not even have the highest R^2 value of any two-x regression model.

The third variable included by foward selection yields the highest R^2 value when combined with x_1 and x_2. The process continues in this same manner. Obviously, a considerable amount of computation is needed. That is why we would never attempt stepwise regression without access to a computer.

When should the selection process stop? There are several criteria that may be used, depending in part on which stepwise computer program is used. A common one uses the t test for statistical significance of a single regression coefficient. When a variable is selected for possible inclusion, a t test is performed on the coefficient. If the null hypothesis, H_0: $\beta_j = 0$, can be rejected at an α level specified by the user, the variable x_j is included and stepwise selection continues. If the null hypothesis is retained, the variable is not included and selection stops. To avoid stopping too early, a relatively large α such as .10 or .20 is typically used. Of course it is also possible to force the procedure to include all variables, one at a time, if the α value is too large.

EXAMPLE 15.12 A forward-selection stepwise regression analysis is performed for the data of Example 15.6 (first-order model). The variables are as indicated in that example, and the row labeled R-SQ gives the value at each step of R^2. Thus variable 1 explains .6419 of the variation in variable 7, variable 2 an additional .2513 (for a total of .8932), and so on.

STEP	1	2	3	4	5	6
CONSTANT	72.39	-83.73	-101.94	-98.54	-97.08	-97.47
var1	1.137	1.527	1.484	1.467	1.450	1.451
T-RATIO	6.96	14.72	16.51	16.37	16.19	15.39
var2		9.0	8.3	8.0	7.5	7.5
T-RATIO		7.82	8.25	7.82	7.06	5.91
var6			0.61	0.63	0.70	0.70
T-RATIO			3.23	3.39	3.65	3.25
var5				10.5	13.2	13.1
T-RATIO				1.31	1.60	1.54
var4					10.7	10.8
T-RATIO					1.26	1.22
var3						0.1
T-RATIO						0.07
S	30.8	17.1	14.7	14.5	14.3	14.6
R-SQ	64.19	89.32	92.46	92.97	93.42	93.42

a. Which variables are not statistically significant at $\alpha = .10$ as "last predictor in"?
b. How much higher an R^2 value does the six-variable model have than the model with only variables 1, 2, and 6?

Solution a. The tabled t value with $29 - (6 + 1) = 22$ d.f. and one-tailed area .05 (presumably a two-tailed test should be used) is 1.717. Only variables 1, 2, and 6 have t statistics larger than that in the STEP 6 column.

b. The six-variable R^2 is .9342 as compared to the R^2 of .9246 given by variables 1, 2, and 6 alone. The difference is .0094. This modest gain in R^2 is not close to being statistically significant. The F statistic for incremental R^2 (discussed in Section 14.4) is 1.05, with a p-value greater than .25. □

backward elimination

There are many variations in stepwise procedures. Variables may be successively removed from the model in **backward elimination**. This process begins with all variables included in the model. Then, one at a time, variables that offer very little predictive value are deleted. Additionally, foward or backward selection can be modified to use various check-up procedures for retesting variables already included in, or excluded from, the model. For example, in forward selection a variable included at one step can be found at a later step to be of little use; a check-up procedure allows later elimination of the variable. There are sophisticated stepwise procedures that incorporate these and other checks.

Stepwise regression can be of great value in suggesting reasonable models. Like any other statistical method, it requires thinking and judgment for proper use.

stepwise bias

One technical problem is that some biases are introduced by stepwise regression. Because stepwise regression selects variables to yield a large R^2 value, it is quite likely that the resulting R^2 is an overestimation of the actual predictive value of the variables in the model. The magnitudes of the resulting coefficients also tend to be too large.

Additionally, stepwise regression involves decisions that are based on differences of R^2 values. These differences contain an element of random error, which can be quite large for small differences in R^2 values. For example, suppose that in a forward selection based on 28 y values, the correlation between y and the first selected variable x_1 is $r_{yx_1} = .6$ (hence $r^2_{yx_1} = .36$). A 95% confidence interval on $r^2_{yx_1}$, calculated by a method not shown in this text, ranges from .09 to .64. Because of this large sampling variation, one variable could be selected over another for inclusion in the regression model, even though the selected variable has less actual predictive value than others not selected.

Stepwise methods usually involve some form of hypothesis test to select among models. An alternative approach to model selection is based on Mallows' (1973) C_p statistic. If a model contains p coefficients—typically one intercept and $p - 1$ slopes corresponding to $p - 1$ independent variables—then

$$C_p = \frac{\text{SS(Residual, } p \text{ coefficients)}}{\text{MS(Residual, all coefficients)}} - (n - 2p)$$

$$= \frac{(n - p)\ \text{MS(Residual, } p \text{ coefficients)}}{\text{MS(Residual, all coefficients)}} - (n - 2p)$$

where the last step follows because the d.f. for MS (Residual, p coefficients) is $n - (k + 1) = n - (p - 1 + 1) = n - p$. If the p-coefficient model contains all the useful predictors, MS(Residual, p coefficients) is essentially the same as MS(Residual, all coefficients). In this case, C_p roughly equals $(n - p) - (n - 2p) = p$. But if the p-coefficient model is inadequate, C_p is substantially larger than p. One plausible model-selection strategy is to select the regression model with the fewest independent variables having C_p approximately equal to p.

EXAMPLE 15.13 Assume that data are collected for 20 independent pharmacies in an attempt to predict prescription volume (sales per month). The independent variables are total floor space, percentage of floor space allocated to the prescription department, number of available parking spaces, whether or not the pharmacy is in a shopping center, and per capita income for the surrounding community. The data and SAS output are shown below. What does the C_p statistic suggest as the most reasonable model?

OBS	VOLUME	FLOOR_SP	PRESC_RX	PARKING	SHOPCNTR	INCOME
1	22	4900	9	40	1	18
2	19	5800	10	50	1	20
3	24	5000	11	55	1	17
4	28	4400	12	30	0	19
5	18	3850	13	42	0	10
6	21	5300	15	20	1	22
7	29	4100	20	25	0	8
8	15	4700	22	60	1	15
9	12	5600	24	45	1	16
10	14	4900	27	82	1	14
11	18	3700	28	56	0	12
12	19	3800	31	38	0	8
13	15	2400	36	35	0	6
14	22	1800	37	28	0	4
15	13	3100	40	43	0	6
16	16	2300	41	20	0	5
17	8	4400	42	46	1	7
18	6	3300	42	15	0	4
19	7	2900	45	30	1	9
20	17	2400	46	16	0	3

ALL SUBSETS PREDICTION OF DRUGSTORE VOLUME

N=20 REGRESSION MODELS FOR DEPENDENT VARIABLE: VOLUME MODEL: MODEL1

NUMBER IN MODEL	R-SQUARE	C(P)	VARIABLES IN MODEL
1	0.00480421	30.4539	PARKING
1	0.03353172	29.1129	FLOOR_SP
1	0.04105340	28.7618	SHOPCNTR
1	0.14798995	23.7702	INCOME
1	0.43933184	10.1709	PRESC_RX
2	0.04210776	30.7126	PARKING SHOPCNTR
2	0.06855667	29.478	FLOOR_SP PARKING
2	0.20543099	23.089	PARKING INCOME
2	0.23487329	21.7147	FLOOR_SP INCOME
2	0.25653635	20.7035	FLOOR_SP SHOPCNTR
2	0.49576794	9.53661	SHOPCNTR INCOME
2	0.53142435	7.87224	PRESC_RX PARKING
2	0.54748785	7.12242	PRESC_RX INCOME
2	0.64706473	2.47436	PRESC_RX SHOPCNTR
2	0.66566267	1.60624	FLOOR_SP PRESC_RX
3	0.25569607	22.7427	FLOOR_SP PARKING INCOME
3	0.26507110	22.3051	FLOOR_SP PARKING SHOPCNTR
3	0.49828073	11.4193	PARKING SHOPCNTR INCOME
3	0.50012580	11.3332	FLOOR_SP SHOPCNTR INCOME
3	0.60243233	6.55772	PRESC_RX PARKING INCOME
3	0.64711563	4.47198	PRESC_RX SHOPCNTR INCOME
3	0.66259120	3.74961	PRESC_RX PARKING SHOPCNTR
3	0.66641145	3.57129	FLOOR_SP PRESC_RX INCOME
3	0.67943313	2.96346	FLOOR_SP PRESC_RX PARKING
3	0.69072432	2.43641	FLOOR_SP PRESC_RX SHOPCNTR
4	0.50128901	13.2789	FLOOR_SP PARKING SHOPCNTR INCOME
4	0.66300855	5.73013	PRESC_RX PARKING SHOPCNTR INCOME
4	0.68058567	4.90966	FLOOR_SP PRESC_RX PARKING INCOME
4	0.69326657	4.31774	FLOOR_SP PRESC_RX SHOPCNTR INCOME
4	0.69873952	4.06228	FLOOR_SP PRESC_RX PARKING SHOPCNTR
5	0.70007369	6	FLOOR_SP PRESC_RX PARKING SHOPCNTR INCOME

Solution Note that for k variables in the model, $p = k + 1$; there are k slopes and one intercept. For the one-variable models, no C_p is close to $p = 2$. The two-variable model using FLOOR SP and PRESC RX has $C_p = 1.606$, actually below $p = 3$. On the C_p criterion, this model appears to be a good one. Note also that the R^2 value for this model is almost as large as the R^2 value for the "kitchen-sink" model involving all variables. □

Mallows (1973) points out the C_p statistic is as susceptible to random variation as any other statistic; it is not an infallible guide. Neither is any other statistical method. In selecting a regression model, a manager should use experience and judgment as well as statistical results. If one model involves highly reasonable relations and variables, yet does somewhat less well than another, less plausible model on a purely statistical basis, a manager might well choose the first model anyway.

SECTION 15.4 EXERCISES

15.16 A forward-selection stepwise regression is run using a first-order model for the data of Exercise 15.13. The following SAS output is obtained:

```
                         STEPWISE REGRESSION OF FILM STIFFNESS DATA

                   FORWARD SELECTION PROCEDURE FOR DEPENDENT VARIABLE STIFF

STEP 1    VARIABLE REPEL ENTERED      R SQUARE = 0.78569611       C(P) =     29.78867185

                                  B VALUE          STD ERROR          TYPE II SS              F        PROB>F

                  INTERCEPT    170.00000000
                  REPEL         -1.83750000       0.17520821       2701.12500000         109.99       0.0001

-------------------------------------------------------------------------------------------------------------

STEP 2    VARIABLE KNIFE ENTERED      R SQUARE = 0.88026761       C(P) =      6.28674862

                                  B VALUE          STD ERROR          TYPE II SS              F        PROB>F

                  INTERCEPT    201.87500000
                  REPEL         -1.83750000       0.13320081       2701.12500000         190.30       0.0001
                  KNIFE         -0.31875000       0.06660041        325.12500000          22.91       0.0001

-------------------------------------------------------------------------------------------------------------

STEP 3    VARIABLE CHILL ENTERED      R SQUARE = 0.89918191       C(P) =      3.18636398

                                  B VALUE          STD ERROR          TYPE II SS              F        PROB>F

                  INTERCEPT    191.03750000
                  CHILL          0.12750000       0.05562951         65.02500000           5.25       0.0296
                  REPEL         -1.83750000       0.12439138       2701.12500000         218.21       0.0001
                  KNIFE         -0.31875000       0.06219569        325.12500000          26.27       0.0001

-------------------------------------------------------------------------------------------------------------

STEP 4    VARIABLE MELT ENTERED       R SQUARE = 0.90270152       C(P) =      4.23727280

                                  B VALUE          STD ERROR          TYPE II SS              F        PROB>F

                  INTERCEPT    176.18750000
                  MELT           0.02750000       0.02782635         12.10000000           0.98       0.3318
                  CHILL          0.12750000       0.05565269         65.02500000           5.25       0.0300
                  REPEL         -1.83750000       0.12444320       2701.12500000         218.03       0.0001
                  KNIFE         -0.31875000       0.06222160        325.12500000          26.24       0.0001

-------------------------------------------------------------------------------------------------------------

NO OTHER VARIABLES MET THE 0.5000 SIGNIFICANCE LEVEL FOR ENTRY INTO THE MODEL.
```

a. List the order in which the independent variables enter the regression model.
b. List the independent variables from largest (in absolute value) to smallest correlation with STIFF. Correlations are shown in Exercise 15.13.
c. Compare the ordering of the variables given by the two lists.

15.17 Refer to Exercise 15.16. Use the F test for complete and reduced models described in Section 14.4 to test the hypothesis that the last two variables entered in the stepwise regression have no predictive value.

15.18 The consultant of Exercise 15.10 runs a regression model with CONTRIB/INCOME as the dependent variable. The key to the variables in the following output is as follows:

1 = INCOME
2 = SIZE
3 = DPDUMMY
4 = ELDUMMY
5 = CHDUMMY
6 = EDLEVEL
7 = MATCHING
8 = CONTRIB
9 = CONTRIB/INCOME

```
Predictor        Coef          Stdev     t-ratio
Constant     0.024569       0.004707        5.22
INCOME     -0.0000711      0.0001946       -0.37
SIZE         0.001146       0.002087        0.55
DPDUMMY      0.006555       0.006736        0.97
ELDUMMY       0.01557        0.01126        1.38
CHDUMMY      0.007371       0.005187        1.42
EDLEVEL      -0.03033        0.02558       -1.19
MATCHING     0.000284       0.002013        0.14

s = 0.005533     R-sq = 19.2%      R-sq(adj) = 0.4%

Analysis of Variance

SOURCE       DF          SS           MS
Regression    7  0.00021843   0.00003120
Error        30  0.00091855   0.00003062
Total        37  0.00113698
```

a. Can the hypothesis that none of the independent variables has predictive value be rejected (using reasonable α values)?
b. Which variables have been shown to have statistically significant (say, $\alpha = .10$) predictive value as "last predictor in"?

15.19 A simpler regression model than that of Exercise 15.18 is obtained by regressing the dependent variable (variable 9) on the independent variables 3, 4, 6, and 7. The following output is obtained.

```
Predictor        Coef        Stdev     t-ratio
Constant     0.022158     0.002713        8.17
DPDUMMY     -0.004729     0.004531       -1.04
ELDUMMY     -0.002143     0.007182       -0.30
EDLEVEL       0.01237      0.01678        0.74
MATCHING    -0.000254     0.001948       -0.13

s = 0.005692     R-sq = 6.0%      R-sq(adj) = 0.0%

Analysis of Variance

SOURCE        DF           SS            MS
Regression     4   0.00006774    0.00001693
Error         33   0.00106924    0.00003240
Total         37   0.00113698
```

a. What is the increment to R^2 for the model of Exercise 15.18, as opposed to the model considered here?
b. Is this increment statistically significant at $\alpha = .05$?
c. Which model do you think is more sensible, given the information you have?

15.5 RESIDUALS ANALYSIS: NONNORMALITY AND NONCONSTANT VARIANCE (step 4)

Once independent variables, including any polynomial or cross-product terms, have been defined and a tentative model selected, the next step in a careful regression analysis is to check for any gross violations of assumptions. The basic method for this check is analysis of the residuals from the model.

We have already discussed residuals analysis in Section 15.3. There we suggested plotting the residuals from a first-order (linear terms only) model against each independent variable, looking for evidence of nonlinearity. In this section we discuss the use of residuals analysis for detecting nonnormality (including outliers) and nonconstant variance.

skewness

A simple histogram of the residuals reveals severe **skewness** or wild outliers. Skewness is not a terribly serious problem for sample sizes of 30 or more. The Central Limit Theorem allows us to use normal-distribution methods to make inferences about means, even if the population distribution is not normal. A more complicated version of the theorem allows us to make inferences about the coefficients and correlations if the distribution of errors isn't normal. In particular, the t and F tests of Chapter 14 are valid to a good approximation for even modestly large sample sizes ($n \geq 30$.) The guidelines given in discussing inferences about means also apply in regression.

Nonnormality arising from skewness may have some effect when predicting individual y values. Because the prediction is about one particular y value, there is no averaging involved, and the Central Limit Theorem doesn't apply. If serious skewness is detected in a histogram of residuals, the "95%" of a 95% prediction interval must be taken with a grain of salt.

outlier

An **outlier** is a data point that falls far away from the rest of the data. Sometimes it is possible to isolate the reason for the outlier, other times it is not. For example, such a point may arise because of an error in recording the data or in entering it into a computer, it may arise because the observation is obtained under conditions quite different from those under which the other observations are obtained. No matter what the source or reason for outliers, if they go undetected they can cause serious distortions in a regression equation.

EXAMPLE 15.14 Suppose the data for a regression study are as shown below.

x:	10	13	16	18	20	22	24	27	30
y:	31	35	42	45	51	53	59	31	70

Draw a scatter plot of the data, identify the outlier, and fit the model $Y = \beta_0 + \beta_1 x + \epsilon$ with and without the outlier point.

Solution A scatter plot of the data (Figure 15.5) shows that any line with slope about 2 and intercept about 10 fits all the data points fairly well, except for the $x = 27$, $y = 31$ point. If that point is included, the least-squares equation is

$$\hat{y} = 19.94 + 1.32x$$

If it is excluded, the equation is

$$\hat{y} = 9.93 + 2.00x$$

The scatter plot shows clearly that the observation (27,31) is an outlier and that the regression equation is distorted by inclusion of this point. □

Outliers cause particularly serious distortions because regression is based on minimizing total squared error. Rather than fitting a line with many small errors and one or two large ones (which yield huge squared errors), the least-squares method accepts numerous moderate errors to avoid large ones. The effect is to twist the line in the direction of the outlier.

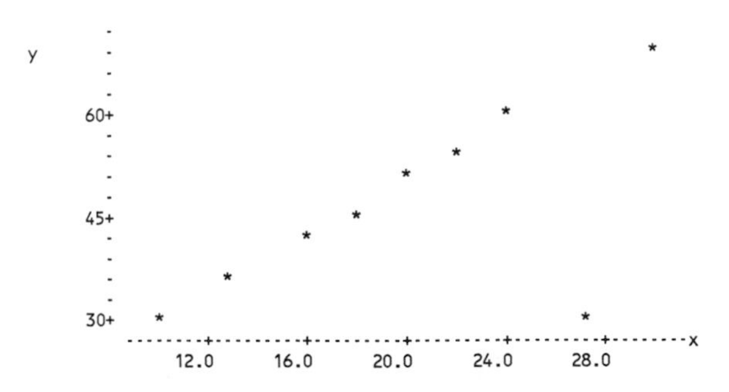

FIGURE 15.5 **Effect of an Outlier; Example 15.14**

The first problem with outliers is detection. In simple linear regression, a scatter plot of y versus x reveals outliers very clearly. In multiple regression, since it is not possible to plot all variables at once, scatter plots of y against each x separately are sometimes helpful. However, examination of the residuals (prediction errors) often provides more information regarding outliers than does separate scatter plots. A large residual suggests that the data point may be an outlier, but the fact that the regression line is twisted toward outliers (as indicated in the sample linear regression situation) sometimes causes an outlier to have a modest residual while residuals for perfectly legitimate data points are larger.

EXAMPLE 15.15 Use the computer output shown below and the residual $(y - \hat{y})$ for each observation to identify potential outliers for the data of Example 15.14.

VARIABLE	DF	PARAMETER ESTIMATE	STANDARD ERROR	T FOR H0: PARAMETER=0	PROB > |T|
INTERCEP	1	19.9428	12.29563	1.622	0.1488
X	1	1.319527	0.5878062	2.245	0.0597

OBS	ACTUAL	PREDICT VALUE	RESIDUAL
1	31	33.1381	-2.1381
2	35	37.0966	-2.0966
3	42	41.0552	0.944773
4	45	43.6943	1.30572
5	51	46.3333	4.66667
6	53	48.9724	4.02761
7	59	51.6114	7.38856
8	31	55.57	-24.57
9	70	59.5286	10.4714

Solution From the output the residual for the data point $x = 27$, $y = 31$ is $31 - 55.57 = 24.57$. The next largest residual is 10.47 for the data point $x = 30$, $y = 70$. Most other residuals are quite a bit smaller. Note that the outlier data point $x = 27$, $y = 31$ "pulls" the least-squares line down, making larger residuals for the data points near $x = 27$. □

jackknife method

Another approach to detecting outliers is the **"jackknife" method**. This involves calculating a series of regression equations, each time excluding one data point. When an outlier is excluded, coefficients in the regression equation change substantially. In principle, one could try excluding two or three points at a time, but the number of equations to calculate and examine would become prohibitive. The one-at-a-time jackknife method may not always catch multiple outliers, but often it does.

In practice, it may be necessary to consider a combination of techniques for examining the sample data for outliers. First, simple x, y scatter plots may suggest that certain observations are outliers. An examination of the residuals may (or may not) confirm this suspicion. If neither the scatter plots nor residuals suggest the existence of one or more outliers, one can probably end the search. However, identification of possible outliers could require additional work with jackknife techniques to isolate specific outliers.

If outliers are detected, what should be done with them? Of course recording or transcribing errors should simply be corrected. Sometimes an outlier obviously comes from a different population than the other data points. For example, a Fortune 500 conglomerate firm doesn't belong in a study of small manufacturers. In such situations, the outliers can reasonably be omitted from the data. Unless a compelling reason can be found, throwing out a data point seems like cheating. There are "**robust regression**" methods that retain possible outliers and try to minimize distortions caused by them. One such method minimizes the sum of absolute (rather than squared) deviations. These methods should be used if there appear to be outliers that cannot be justifiably excluded from the data.

robust regression

EXAMPLE 15.16 Apply a jackknife procedure, eliminating one data point at a time, to the data of Example 15.15. Examine the estimated slopes and intercepts to locate possible outliers.

Solution The estimated slopes and intercepts are listed below. Note that the last two data points appear to be outliers.

Data Point Excluded	Slope	Intercept
10,31	1.21286	22.47672
13,35	1.26116	21.42333
16,42	1.33281	19.55234
18,45	1.32834	19.60120
20,51	1.31953	19.35947
22,53	1.29235	19.97601
24,59	1.21563	21.04531
27,31	2.00354	9.93239
30,70	.79712	28.42905

Thus, while the scatter plot of Figure 15.5 identified one potential outlier (the point 27, 31), an examination of the residuals as well as the jackknife procedure detects a second potential outlier (the point 30, 70). An examination of residuals from the regression omitting the point 27, 31 indicates that the point 30, 70 is not in fact an outlier. □

Another formal assumption of regression analysis is that the (true, population) error variance σ_ϵ^2 is constant, regardless of the values of the x predictors. This assumption may

constant variance

also be violated in practice. In particular, it often occurs that combinations of x values leading to large predicted values of y also lead to relatively large variance around the predicted value. We here consider the consequences, detection, and possible cure of the problem of nonconstant error variance.

When the variance around the prediction equation is not constant, there are two basic consequences. Ordinary least-squares regression does not give the most accurate possible estimate of the regression equation, and the plus-or-minus error of prediction given in Chapters 13 and 14 may be seriously in error.

The estimation problem is less serious. If the error variance is not constant, the usual least-squares estimates are still valid in the sense of being unbiased. Furthermore, various studies have indicated that the F and t statistics still give about the same conclusions. The issue here is one of "opportunity cost"; if **heteroscedasticity** (nonconstant variance) is recognized, it is possible to improve the estimation of the regression equation and the various related statistics. The technique of **weighted least squares**, as described in Johnston (1977), for example, yields somewhat more accurate estimates of the regression coefficients than does ordinary least-squares regression (more accurate estimates have smaller standard errors). The same technique makes the F and t statistics more powerful for testing the appropriate null hypotheses. Weighted least squares, in the presence of heteroscedasticity around the equation, makes more efficient use of the data.

heteroscedasticity

weighted least squares

The more serious problem arises in making forecasts. The best-guess forecast based on ordinary least-squares regression is still unbiased, but (given nonconstant error variance) the usual plus-or-minus formulas can be badly wrong. If the forecast y value falls in a high-variance zone, the theoretical plus-or-minus term may be much too small.

Probably the best way to detect heteroscedasticity is by eye and by data plot. The most useful plot is predicted y versus actual y, or predicted y versus residual $y - \hat{y}$. Most standard statistical computer programs can calculate predicted, actual, and residual values. Some have commands that produce the desired plots. In such plots, look for evidence that the variability of actual y values (or of residuals) increases as predicted y increases. There are several statistical significance tests for the research hypothesis of nonconstant variance. They generally tend to confirm the evidence of the "eyeball test," and the theory behind these tests leans heavily (uncomfortably so) on the normal-distribution assumption. We tend to prefer the eyeball method for detecting heteroscedasticity.

EXAMPLE 15.17 A very crude model for predicting the price of common stocks might take price per share (y) as a linear function of previous year's earnings per share (x_1), change in earnings per share (x_2), and asset value per share (x_3). A scatter plot of residuals versus predicted y values for a regression study of 26 stocks is shown in Figure 15.6. Is there evidence of a problem of heteroscedasticity?

Solution In the plot of residuals versus predicted values, there is a general tendency for the magnitude of the residuals to increase as $\hat{y}$ increases. All residuals on the left are small. Therefore there seems to be a problem of nonconstant variance. □

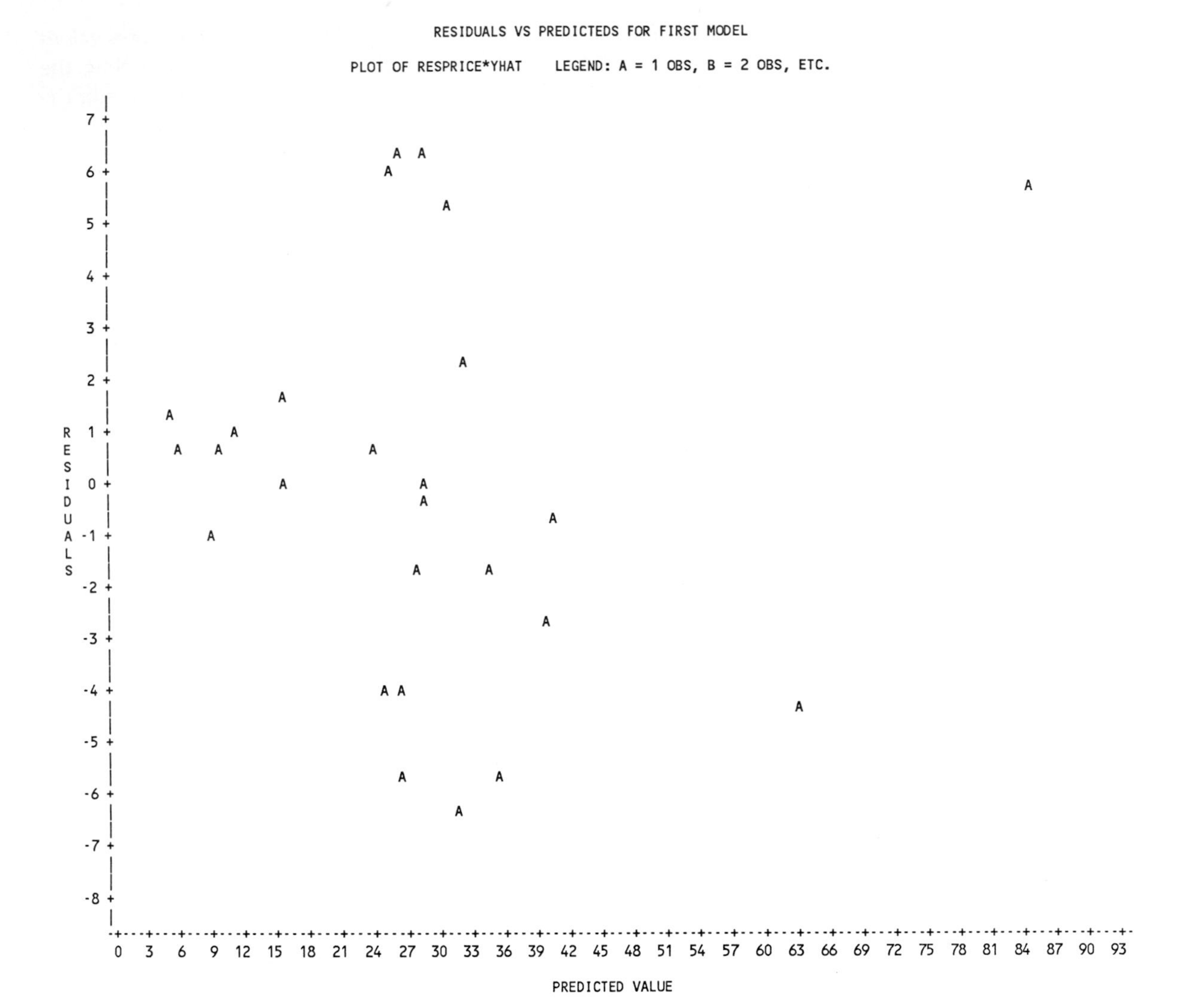

FIGURE 15.6 **Residuals versus Predicted Values; Example 15.17**

Once the problem of nonconstant variance is detected, there are two basic cures. One is weighted least squares. The other, which often turns out to be equivalent to weighted least squares is appropriate reexpression of the dependent variable. For example, one may try to predict the number of airline tickets sold for a particular airport from the population of the relevant metropolitan area, the average disposable income in the area, the number of Fortune 500 companies in the area, and so on. Almost certainly, there will be larger variance in number of tickets sold at larger airports. If the dependent variable is redefined as number of tickets sold per capita, the problem of heteroscedasticity may well disappear. A little thought in defining the regression equation often goes a long way.

EXAMPLE 15.18 The dependent variable in Example 15.17 is redefined to be price per share divided by earnings per share (the P/E ratio). The SAS output is shown on the top of the next page, and the revised model is plotted in Figure 15.7.

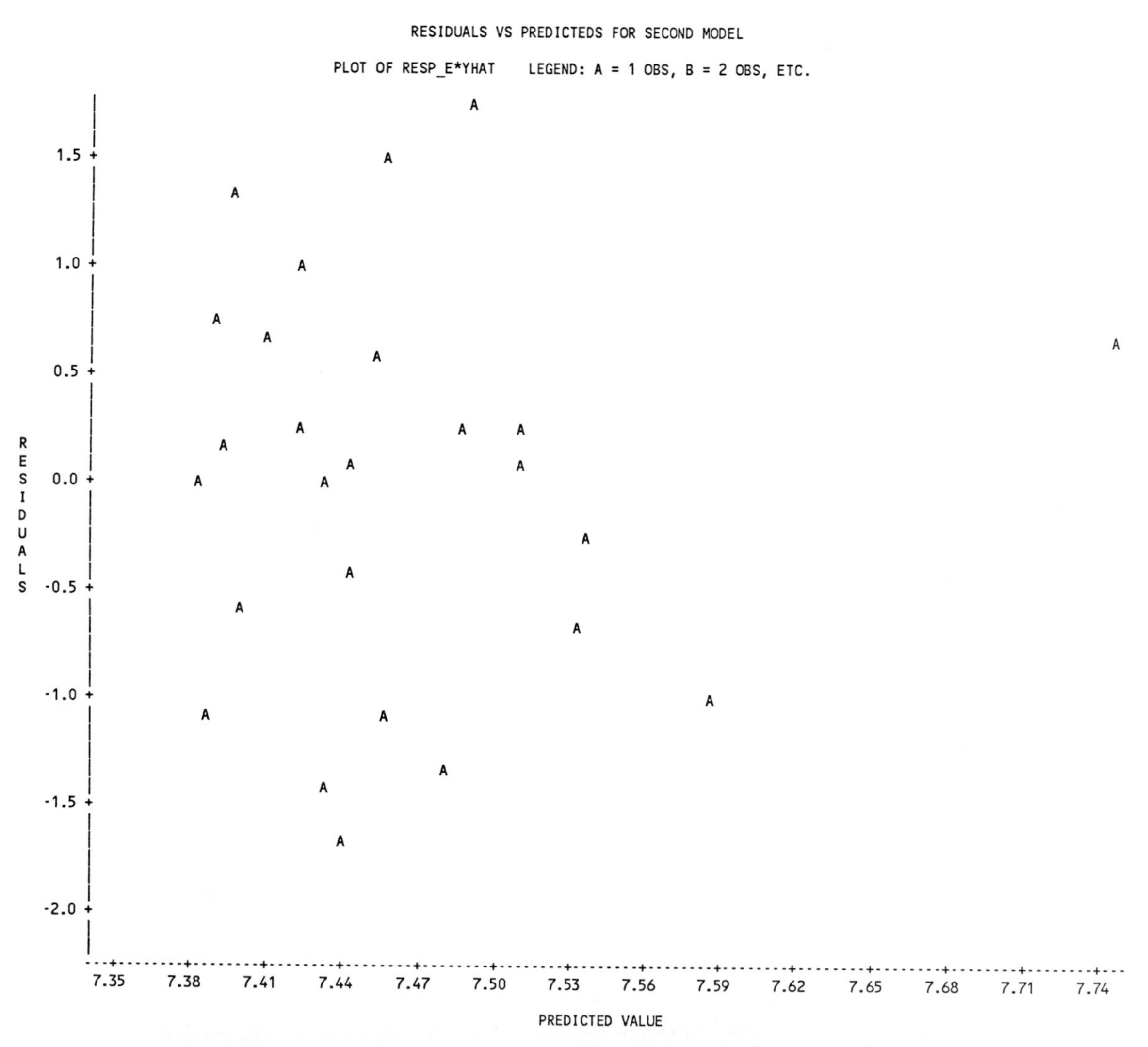

FIGURE 15.7 **Residuals versus Predicted Values; Example 15.18**

```
                                SECOND MODEL

DEP VARIABLE: P_E
                            ANALYSIS OF VARIANCE

                              SUM OF          MEAN
SOURCE       DF             SQUARES         SQUARE      F VALUE      PROB>F

MODEL         2           0.1520641     0.07603204        0.083      0.9209
ERROR        23            21.15736      0.9198854
C TOTAL      25            21.30943

      ROOT MSE            0.9591066       R-SQUARE       0.0071
      DEP MEAN              7.46351       ADJ R-SQ      -0.0792
      C.V.                 12.85061

                            PARAMETER ESTIMATES

                      PARAMETER       STANDARD      T FOR H0:
VARIABLE  DF           ESTIMATE          ERROR    PARAMETER=0    PROB > |T|

INTERCEP   1           7.372642      0.3967074         18.585        0.0001
DEARN      1          -0.459169        1.85963         -0.247        0.8072
ASSET      1         0.00891629     0.02395633          0.372        0.7132
```

a. Write the model for Example 15.17.
b. Divide this model equation by the earnings variable to obtain the revised model.
c. Identify the estimated coefficients in the revised model.
d. Does there appear to be a problem of nonconstant variance with the revised model (see Figure 15.7)?
e. Use the revised model to predict the price of a stock with earnings 3.00, growth .27, and assets 14.25.

Solution a. $Y = \beta_0 + \beta_1 x_1 + \beta_2 x_2 + \beta_3 x_3 + \epsilon$.

b. $Y/x_1 = \beta_1 + \beta_2(x_2/x_1) + \beta_3(x_3/x_1) + \epsilon$, where we omit the term β_0/x_1.

c. $\hat{\beta}_1 = 7.373$, $\hat{\beta}_2 = -.459$, and $\hat{\beta}_3 = .0089$.

d. The residual plot is much better for the revised model.

e. Substituting $x_1 = 3.00$, $x_2 = .27$, and $x_3 = 14.25$ into the ratio model, we have $\hat{y} = 7.373 - .459(.27/3) + .0089(14.25/3) = 7.374$. □

EXAMPLE 15.19 A first-order model is developed for predicting feed sales using the independent variables defined in Example 15.11. A plot of residuals versus starts is shown in Figure 15.8. Is there evidence of nonconstant variance?

Solution There appears to be a definite increase in variability as starts increase. This seems reasonable, and we would not be surprised to see more variability in high-activity months. □

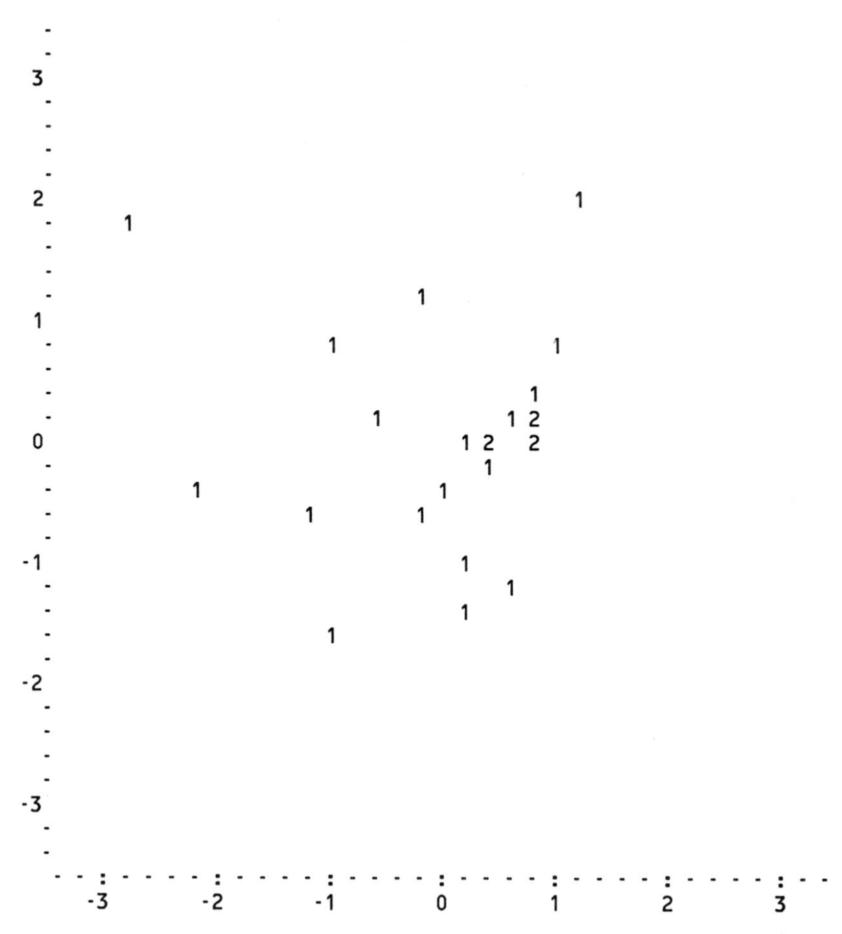

FIGURE 15.8 Scatter Plot of 24 Standardized Values of Residuals versus Starts; Example 15.19

15.6 RESIDUALS ANALYSIS: AUTOCORRELATION (step 4)

One of the crucial assumptions of regression analysis is that the error terms ϵ_i, which might be called the "true residuals," are independent. Much of the statistical theory of regression depends heavily on this assumption. In practice the assumption may be wrong. Times-series data, where the data points are measured at successive times often shown more-or-less cyclic behavior. If such behavior is shown by a dependent variable y, and if no x variable matches the apparent cycles, the sample residuals show evidence of dependence. This problem, which is largely restricted to time-series data, is called **autocorrelation**.

autocorrelation

To see why autocorrelation (dependence of the error terms) is a problem, assume that we have the linear regression model

$$Y_i = \beta_0 + \beta_1 x_i + \epsilon_i$$

where x_i represents the time when observation y_i is obtained and the least-squares prediction equation is

$$\hat{y}_i = \hat{\beta}_0 + \hat{\beta}_1 x_i$$

Now, for illustration purposes, suppose that the values of the parameters β_0 and β_1 are known, so that we can draw the line $E(Y_i) = \beta_0 + \beta_1 x_i$. Two such regression lines are shown by solid lines, one in Figure 15.9a and a second in Figure 15.9b. The observations in these figures are designated by data points, and the least-squares prediction equations are shown by dashed lines.

In each figure the y values (and hence the error terms ϵ) show a definite nonrandom pattern about the true regression line $E(Y) = \beta_0 + \beta_1 x$. In Figure 15.9a, a series of several y values above the true regression line (positive errors) is followed by a series of y values below the true regression line (negative errors). In Figure 15.9b the negative errors occur first, followed by the positive errors. If the errors (ϵ) are independent, the pattern of positive

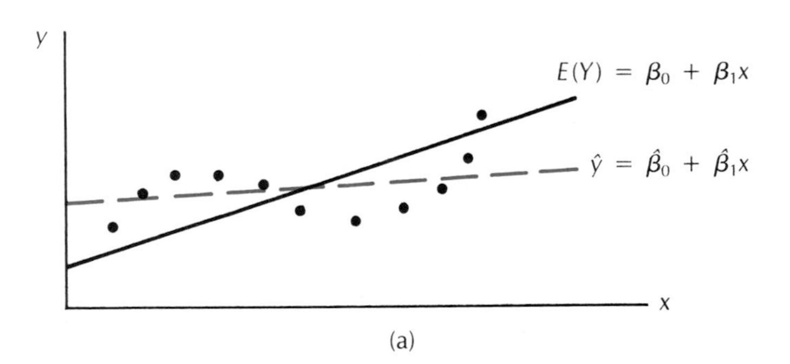

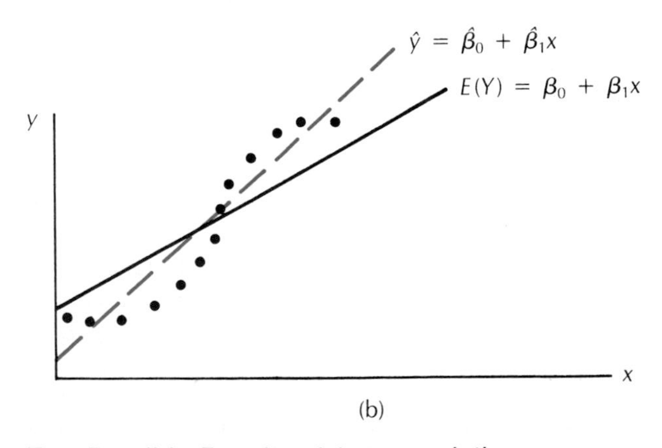

FIGURE 15.9 Two Possible Results of Autocorrelation

and negative errors should be random. The cyclic pattern of the y values (and hence the errors) clearly indicates that autocorrelation is present.

effect of autocorrelation

What are the **effects of autocorrelation** on the usual types of inferences that we make in regression? Figure 15.9 illustrates some of the difficulties. In both situations (where autocorrelation exists), the least-squares prediction equation provides too good a fit to the sample data; that is, the least-squares line is much closer to the y values than is the true regression line. Because of this, the residuals (the observed errors, $y - \hat{y}$) are smaller than the true errors (the ϵ's) and the residual standard deviation s_ϵ provides an underestimate of the population standard deviation σ_ϵ. The net effect is that all formulas for standard error involving s_ϵ underestimate the actual standard errors. Further, if the residual standard deviation is too small in the presence of autocorrelation, the coefficient of determination is too large.

In practice, the detection of autocorrelation is based on the residuals, since the true errors (the ϵ's) are unknown. If a plot of the residuals versus time shows a cyclic, nonrandom pattern, it is likely that the true errors are dependent, and hence that autocorrelation is present.

EXAMPLE 15.20

Suppose that the data for a simple regression study are as follows:

y:	6.1	6.0	6.1	6.3	6.8	6.8	7.0	7.1	7.0	6.7	6.8	7.0	7.2	7.4	7.5
x:	1	2	3	4	5	6	7	8	9	10	11	12	13	14	15

Calculate and graph the least-squares regression line on a scatter plot of the data. Does there seem to be an autocorrelation problem?

Solution

The regression line $\hat{y} = 5.98 + .10x$. It is shown on a scatter plot of the data in Figure 15.10. The cyclic pattern (several negative residuals followed by several positive residuals, etc.) clearly indicates that autocorrelation is present. □

Durbin-Watson statistic

A formal test for autocorrelation uses the **Durbin-Watson statistic**. This statistic is based on the idea that, given (positive) autocorrelation, any one residual tends to be close to the following residual; a large positive residual tends to be followed by another large positive one, and so on. Therefore the squared differences of successive residuals tend to

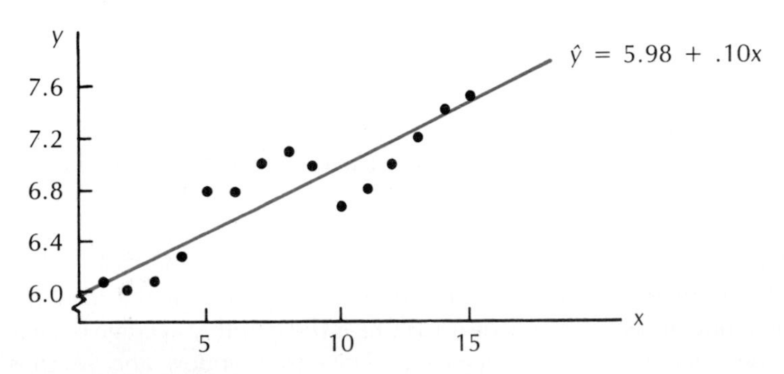

FIGURE 15.10 **Correlated Residuals; Example 15.20**

be smaller under positive autocorrelation than they are when independent. The Durbin-Watson statistic is

$$d = \frac{\sum_{t=1}^{n-1} (\hat{\epsilon}_{t+1} - \hat{\epsilon}_t)^2}{\sum_{t=1}^{n} \hat{\epsilon}_t^2}$$

where $\hat{\epsilon}_t$ is the residual at time t. If the true errors are in fact independent, the expected value of d is about 2.0. Positive autocorrelation tends to make $\hat{\epsilon}_{t+1}$ close to $\hat{\epsilon}_t$ and therefore to make d less than 2.0. Tables for a formal hypothesis test based on d are available (see Johnston, 1977). In practice, though, we would hope to accept the null hypothesis of zero autocorrelation. Since accepting any null hypothesis leads to the nasty question of Type II errors, we prefer to use the Durbin-Watson statistic as an index rather than as a formal test. Any value of d less than 1.5 or 1.6 leads us to suspect autocorrelation.

EXAMPLE 15.21 Calculate the Durbin-Watson statistic for the data of Example 15.20. Does it indicate an autocorrelation problem?

Solution

x_t	y_t	$\hat{y}_t = 5.98 + .10x_t$	$\hat{\epsilon}_t = y_t - \hat{y}_t$	$\hat{\epsilon}_{t+1} - \hat{\epsilon}_t$	$(\hat{\epsilon}_{t+1} - \hat{\epsilon}_t)^2$	$\hat{\epsilon}^2$
1	6.1	6.08	+.02	−.20	.04	.0004
2	6.0	6.18	−.18	.00	.00	.0324
3	6.1	6.28	−.18	+.10	.01	.0324
4	6.3	6.38	−.08	+.40	.16	.0064
5	6.8	6.48	+.32	−.10	.01	.1024
6	6.8	6.58	+.22	+.10	.01	.0484
7	7.0	6.68	+.32	.00	.00	.1024
8	7.1	6.78	+.32	−.20	.04	.1024
9	7.0	6.88	+.12	−.40	.16	.0144
10	6.7	6.98	−.28	.00	.00	.0784
11	6.8	7.08	−.28	+.10	.01	.0784
12	7.0	7.18	−.18	+.10	.01	.0324
13	7.2	7.28	−.08	+.10	.01	.0064
14	7.4	7.38	+.02	.00	.00	.0004
15	7.5	7.48	+.02	—	—	.0004
					.46	.638

The Durbin-Watson statistic $d = .46/.638 = .721$. This value is far below the ideal value of 2.0 and the cutoff of 1.5. Autocorrelation is clearly a problem. □

EXAMPLE 15.22 Data for 24 months for the feed manufacturer of Example 15.11 are shown below. STARTS is the number of starts in the previous month, RELPRI is the manufacturer's feed price in the month (relative to an index), CHICKP is the monthly average price of chickens, COMPPR is the chief competitor's price (also relative to an index), and SALES is monthly feed sales by the manufacturer.

ROW	STARTS	RELPRI	CHICKP	COMPPR	SALES
* 1 *	6.46000	16.21000	0.49300	15.99000	241.00000
* 2 *	7.20000	16.19000	0.51700	16.31000	264.00000
* 3 *	6.68000	16.06000	0.46200	16.26000	259.00000
* 4 *	7.01000	15.97000	0.49000	16.12000	258.00000
* 5 *	7.47000	16.31000	0.53600	16.41000	265.00000
* 6 *	7.68000	16.58000	0.59400	16.49000	255.00000
* 7 *	7.65000	16.97000	0.57000	17.00000	267.00000
* 8 *	7.49000	17.21000	0.53800	17.01000	243.00000
* 9 *	7.38000	17.08000	0.49900	16.96000	251.00000
* 10 *	7.46000	17.00000	0.48600	17.21000	268.00000
* 11 *	7.58000	17.15000	0.52500	17.47000	277.00000
* 12 *	7.56000	17.31000	0.49000	17.22000	260.00000
* 13 *	7.60000	17.08000	0.47300	17.11000	269.00000
* 14 *	7.31000	17.11000	0.43100	17.01000	252.00000
* 15 *	7.04000	16.97000	0.45600	16.99000	248.00000
* 16 *	7.03000	16.90000	0.46400	17.16000	278.00000
* 17 *	7.36000	16.84000	0.47700	17.24000	295.00000
* 18 *	7.53000	17.17000	0.50900	17.38000	277.00000
* 19 *	7.68000	17.52000	0.49200	17.46000	264.00000
* 20 *	7.73000	17.67000	0.47400	17.81000	284.00000
* 21 *	7.51000	17.65000	0.51000	17.70000	267.00000
* 22 *	7.84000	17.34000	0.49500	17.47000	291.00000
* 23 *	7.67000	17.59000	0.50100	17.50000	263.00000
* 24 *	7.70000	17.52000	0.42300	17.63000	279.00000

A first-order model is run with independent variables STARTS, RELPRI, CHICKP, and COMPPR (using the IDA package). The following selected output is obtained. The plot of residuals is shown in Figure 15.11. Is there evidence of autocorrelation?

```
> coef

VARIABLE  B(STD.V)      B        STD.ERROR(B)    T

STARTS      0.6048  2.9846E+01  5.5488E+00    5.379
RELPRI     -2.2814 -7.5555E+01  6.5053E+00  -11.614
CHICKP     -0.1547 -6.7000E+01  3.4754E+01   -1.928
COMPPR      2.3187  7.5660E+01  6.8608E+00   11.028
CONSTANT    0       7.0308E+01  4.5134E+01    1.558

> summ

             MULTIPLE R  R-SQUARE
UNADJUSTED     0.9651      0.9315
 ADJUSTED      0.9576      0.9171

STD. DEV. OF RESIDUALS = 4.8684E+00
N =   24


> durb

DURBIN-WATSON STAT. =       1.5574
```

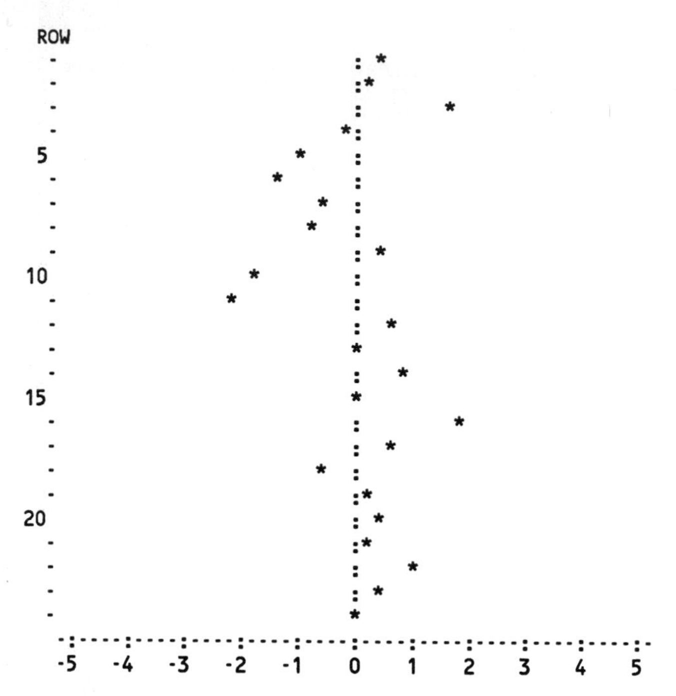

FIGURE 15.11 Sequence Plot of Standardized Values of Residuals; Example 15.22

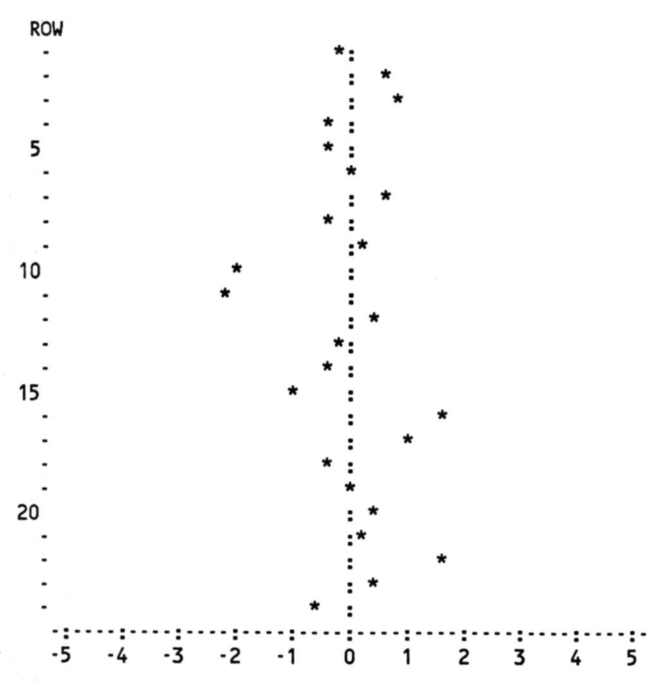

FIGURE 15.12 Sequence Plot of Standardized Values of Residuals; Example 15.23

Solution There is some suggestion of autocorrelation. The Durbin-Watson statistic is 1.5574 and there is perhaps a hint of a pattern in the sequence plot of residuals; that is, there seems to be some roughly cyclic behavior in the sequence of residuals. □

EXAMPLE 15.23 In an attempt to solve the problem of nonconstant variance detected in Example 15.19, a revised model is run. The dependent variable is SHARE = SALES/STARTS. The independent variables are RELPRI and RELDIF = (RELPRI − COMPPR)/CHICKP. Selected output is shown below, and the residuals are plotted in Figure 15.12 on the preceding page. Is there evidence of an autocorrelation problem?

```
> coef

VARIABLE  B(STD.V)      B       STD.ERROR(B)    T

RELPRI    -0.0672 -2.5317E-01  3.2385E-01  -0.782
RELDIF    -0.9103 -5.3179E+00  5.0192E-01 -10.595
CONSTANT    0       3.9351E+01  5.5075E+00   7.145

> durb

DURBIN-WATSON STAT. =       1.4008

> plts
WANT EXPLANATION ? n
* GIVE NAME OR COL NUMBER FOR VARIABLE TO BE
  PLOTTED : residu
PLOT ALL ROWS ? y
```

Solution The Durbin-Watson statistic is 1.4 and the sequence plot of residuals clearly suggests autocorrelation. □

If autocorrelation is suspected, either because of a plot of residuals versus time or because of a Durbin-Watson statistic less than about 1.5 or 1.6, what should be done about the regression model? Ideally, an autocorrelation model for the error terms should be adopted. One simple error model is the **first-order autoregressive model**

first-order autoregressive model

$$\epsilon_t = u_t + \rho u_{t-1}$$

where the u_t values are independent and ρ is a model parameter; $\rho > 0$ yields positive autocorrelation.* If this model is correct and if (miraculously) ρ is known, then it can be proven that

$$Y_t - \rho Y_{t-1} = \beta_0(1-\rho) + \beta_1(x_{t1} - \rho x_{t-1,1}) + \ldots + \beta_k(x_{tk} - \rho x_{t-1,k}) + (\epsilon_t - \rho\epsilon_{t-1})$$

is a model satisfying the assumption of independent error. In practice, the problem is to estimate the unknown error parameter ρ.

A quick approach to the problem is to assume $\rho = 1$. This leads to a regression of the differences $y_t - y_{t-1}$ on the differences $x_{t1} - x_{t-11}, \ldots, x_{tk} - x_{t-1,k}$. Often, using differences eliminates any autocorrelation problems. This method also tends to reduce collinearity, which is often a major problem with time-series data.

use of differences

* This is the simplest of the Box-Jenkins models discussed in Chapter 16.

Cochran-Orcutt method

Regression of differences is a crude approach, because it assumes that the autocorrelation parameter $\rho = 1$. A more sophisticated approach is the Cochran-Orcutt method. In this method, we begin with a raw-data regression, estimate ρ, re-regress on $y_t - \rho y_{t-1}$, reestimate ρ, and so on to convergence. Alternatively, it is possible to search for the least-squares value of ρ. For details, consult a time-series specialist.

Alternatively, the presence of cyclic residuals (autocorrelation) sometimes suggests new independent variables that should be included in the regression model. It may well be that inclusion of another variable, whose "highs" match the positive residuals and whose "lows" match the negative ones, will improve the model. There should be an explanation, that is, a theoretical reason, why the newly included variable relates to the dependent variable. Of course it's always risky to throw new variables into a model blindly.

EXAMPLE 15.24 First differences of the variables of Example 15.23 are calculated and a regression run. Selected IDA output is shown on the top of the next page. Note that although there were 24 observations initially, we now have 23 first differences. The variables are DSHARE, DRELPR, and DRELDF, where the initial D in the variable names indicates that the quantities are differences. Is there evidence of an autocorrelation problem of a problem of nonconstant variance (see Figure 15.13)?

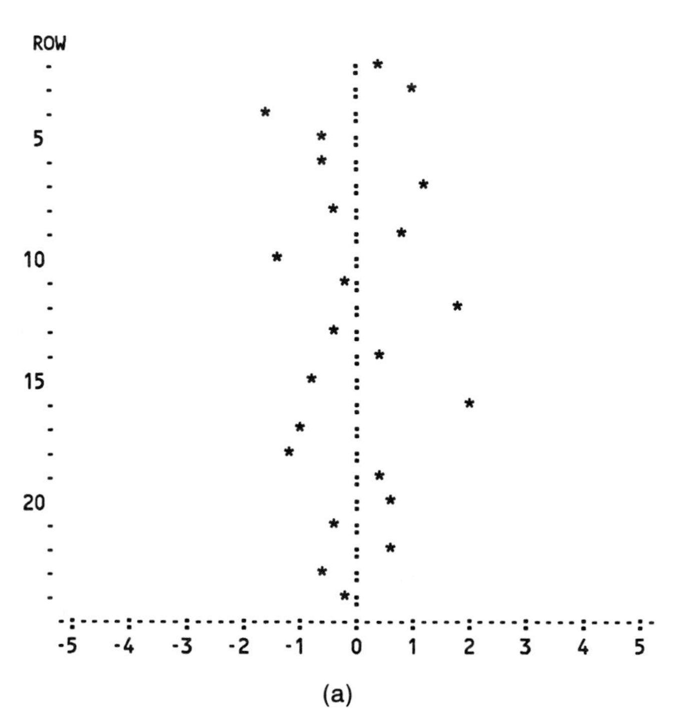

(a)

FIGURE 15.13 Example 15.24: (a) Sequence Plot of Standardized Values of Residuals; (b) Scatter Plot of 23 Standardized Values of Residuals versus Starts

```
> coef

VARIABLE  B(STD.V)      B        STD.ERROR(B)    T

DRELPR    -0.0811  -8.7320E-01   1.1592E+00   -0.753
DRELDF    -0.8684  -4.7008E+00   5.8276E-01   -8.066
CONSTANT    0       4.0462E-02   2.0865E-01    0.194

> durb

DURBIN-WATSON STAT. =       2.6979

> plts
```

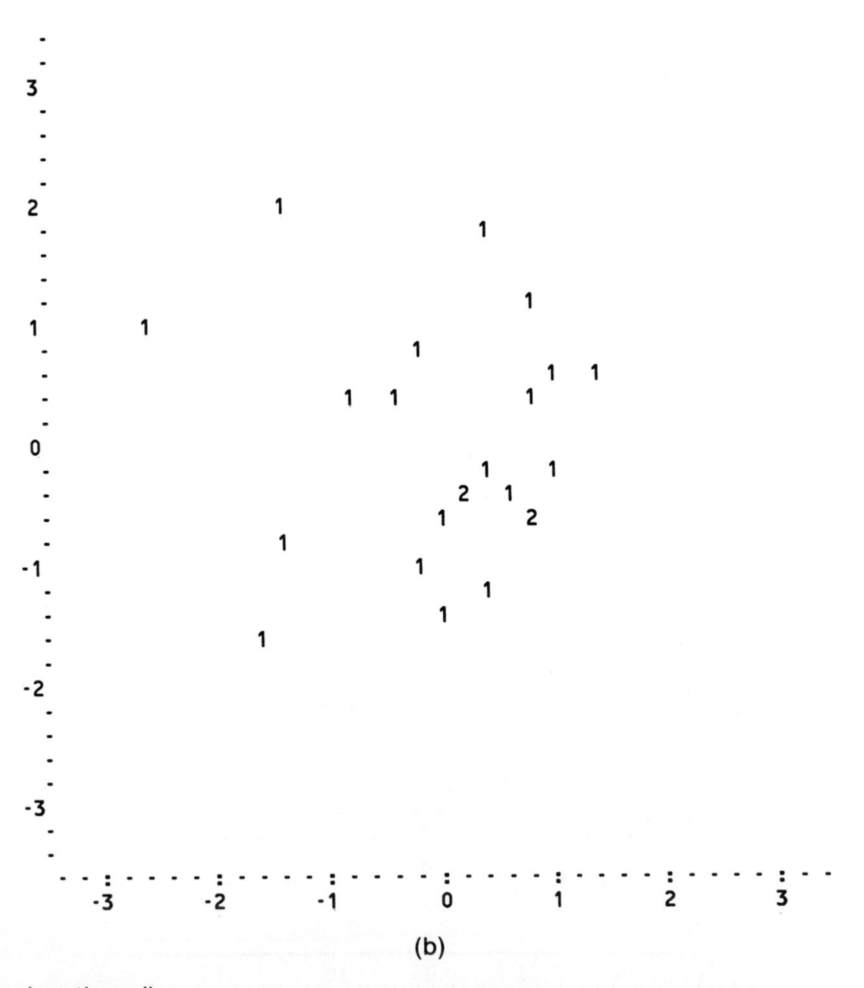

(b)

FIGURE 15.13 **(continued)**

Solution The Durbin-Watson statistic is just about 2.7, and the sequence plot of residuals (Figure 15.13a) shows a pattern of alternating positive and negative signs. If anything, the differencing method has produced overkill; positive autocorrelation has been converted to apparent negative autocorrelation. Negative autocorrelation at least yields conservative results. A Cochran-Orcutt or search method might make more efficient use of the data.

The plot of residuals versus starts (Figure 15.13b) shows no problem of nonconstant variance. A plot of residual versus predicted values might be worthwhile, to see if variability increases with increasing values of $\hat{y}$. □

SECTIONS 15.5 AND 15.6 EXERCISES

15.20 Residual plots for the regression model of Exercise 15.19 are shown in Figure 15.14 (a few of the marks on the plot represent two observations rather than one).

a. Is there any strong suggestion of nonlinearity?

b. Is there any strong suggestion of nonconstant variance?

15.21 The district sales office for a particular automobile is interested in predicting the sales of the "top of the line" luxury car in the district. It is obvious that sales are affected by the rated gasoline mileage of the car and by the car loan interest rate charged by the company's financing agency. It also seems plausible that sales are affected by gasoline prices and by the price of the car. Data are collected for 48 months; the last 6 months are reserved for model validation (to be discussed in Section 15.7). The variables are

MILAGE:	rated gas mileage of the car
GASPRI:	average price per gallon (in cents) in the district
PREGAS:	average gas price in the previous month
INTRAT:	interest rate (percent per year)
CARPRI:	sticker price divided by the consumer price index

A first-order model is fit, using the IDA package. Selected output is shown on the top of the next page. Figure 15.15 shows the sequence plot of residuals.

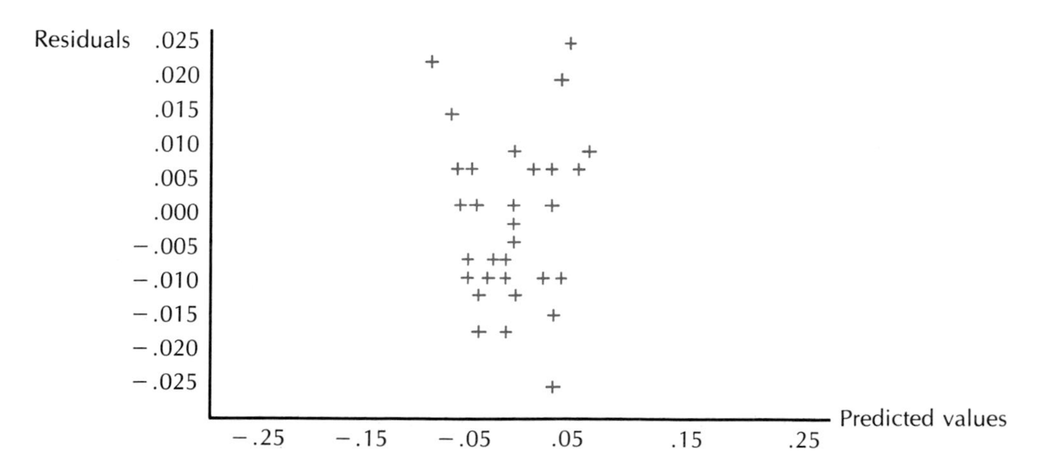

FIGURE 15.14 **Residual Plot; Exercise 15.20**

VARIABLE	B(STD.V)	B	STD.ERROR(B)	T
MILAGE	0.5277	2.3079E+02	5.3907E+01	4.281
GASPRI	-0.4004	-8.4934E+00	1.7006E+00	-4.994
PREGAS	-0.6884	-1.4538E+01	1.6374E+00	-8.878
INTRAT	-0.5310	-2.5170E+02	9.4153E+01	-2.673
CARPRI	-0.3382	-5.6437E+01	4.2268E+01	-1.335
CONSTANT	0	5.7597E+03	2.0978E+03	2.746

MEAN = 0.0000E+00
STD. DEV. = 4.7543E+01
SAMPLE SIZE = 42
DURBIN-WATSON STAT. = 0.7999

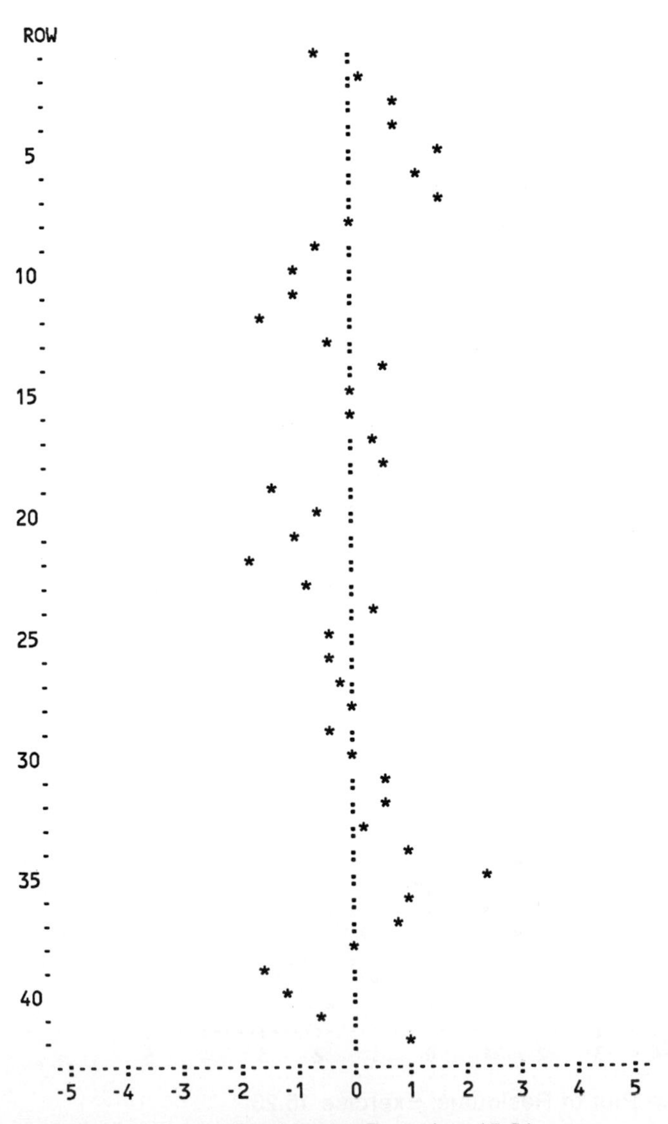

FIGURE 15.15 Sequence Plot of Residuals; Exercise 15.21

a. Write the estimated regression model.
b. Locate the standard deviation.

15.22 Refer to the output for the model for Exercise 15.21 and to Figure 15.15.

a. Does the Durbin-Watson statistic indicate that there is a problem of (positive) autocorrelation?
b. Does the sequence plot of residuals indicate that autocorrelation is a problem?

15.23 The auto-sales forecasting model of Exercise 15.21 is modified in two ways. First, the gas price variables are divided by the mileage figure to yield rated gas price per mile driven. The current

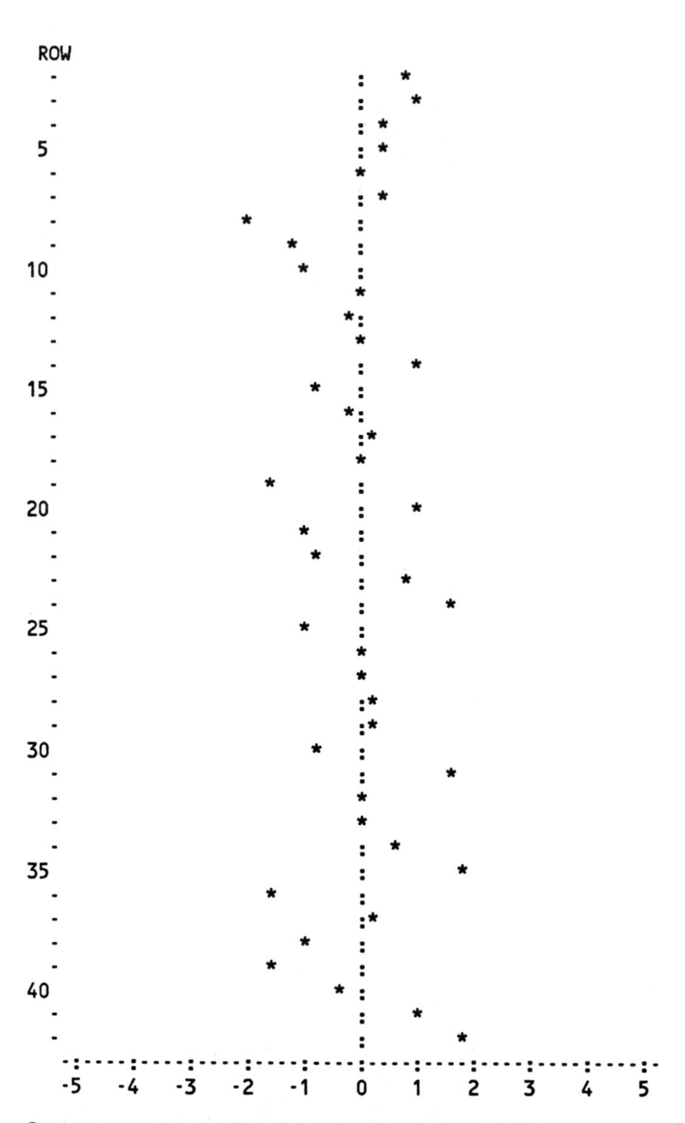

FIGURE 15.16 Sequence Plot of Residuals; Exercise 15.23

month's price per mile is called PRIMIL, and the previous month's is LAGPRM. Second, differences of the variables SALES, PRIMIL, LAGPRM, INTRAT, and CARPRI are calculated; the names are prefaced by a C. A regression model based on these differences is fit, with the following results (IDA). The sequence plot of residuals is shown in Figure 15.16, on the preceding page.

VARIABLE	B(STD.V)	B	STD.ERROR(B)	T
CPRIMI	-0.3537	-8.9662E+01	1.6951E+01	-5.289
CLAGPR	-0.8760	-2.2195E+02	1.5782E+01	-14.064
CINTRA	-0.0865	-1.0950E+02	8.5126E+01	-1.286
CCARPR	0.0095	5.8950E+00	3.9183E+01	0.150
CONSTANT	0	-6.6169E-01	6.6030E+00	-0.100

	MULTIPLE R	R-SQUARE
UNADJUSTED	0.9276	0.8605
ADJUSTED	0.9192	0.8450

STD. DEV. OF RESIDUALS = 3.7339E+01
N = 41

DURBIN-WATSON STAT. = 1.8867

a. Write the regression equation.
b. Locate the residual standard deviation.
c. Is the residual standard deviation larger for the difference model than for the original? Why would this be a common result?

15.7 MODEL VALIDATION

In this computer era, data can be "massaged" repeatedly at little cost. Regressions can be run with every conceivable combination of variables, transformations, and lags. It isn't unusual to have a hundred different regression equations run to try to predict a variable y. The statistical theory of regression analysis implicitly assumes that the choice of predictors x and dependent variable y has been made, once and for all, and that a single regression equation is found. The gap between theory and practice is wide.

The practice of calculating many regression equations and selecting the best one often leads to overoptimism. The apparent predictability decreases when the equation is used with new data—for two reasons. First, any regression equation is based on predicting the past; the equation is chosen to yield the best fit to available data. Second, when one selects the best of many equations, one runs a risk of "capitalizing on chance." Even purely random phenomena can be explained after the fact by some combination of variables, if one searches long enough.

A good practical approach to selecting an appropriate prediction equation is to validate the chosen regression equation with new data. In a time-series study, the most recent data can be withheld from the original regression study. Then the chosen regression equation can be used to predict recent values. The resulting standard deviation gives a good indication of the future predictive value of the equation. The same procedure can be used with cross-section data. Select some subset of observations (perhaps 10 or 20%) at random, withhold them from the original study, and use them for validation. The chosen regression typically does not perform quite as well in the validation study. If it still works reasonably well, that's good grounds for believing that it will be useful in practice.

EXAMPLE 15.25 An additional 12 months' data are collected for the feed manufacturer of Example 15.11. The difference model of Example 15.24 is used with these data. The results are shown below.

	Actual DSHARE	Predicatd DSHARE	Error
	.6243	1.4735	−.8492
	2.7931	1.7694	1.0237
	−1.4616	−1.6785	.2169
	−2.1039	−.4668	−1.6371
	1.0487	.2520	.7967
	3.4225	1.3979	2.0246
	.8994	1.9018	−1.0024
	−1.1583	−.7206	−.4377
	1.2767	.4526	.8241
	4.0023	4.2146	−.2123
	−2.9046	−3.9520	1.0474
	−.8726	−.5465	−.3261
Mean	.4638	.3414	.1224
Standard deviation	2.2153	2.0665	1.0520

a. Is there any flagrant bias in the predictions?
b. The residual standard deviation for the model of Example 15.24 is .9293. Is the error standard deviation grossly different?

Solution a. There's no obvious, systematic error. The mean error is small (.1224) relative to the size of the actual values and to the error standard deviation. Exactly half of the errors are positive.

b. The error standard deviation (1.0520) is only slightly larger than .9293. □

SECTION 15.7 EXERCISES

15.24 The auto-sales difference model of Exercise 15.23 is tested against data for months 43–48, which have been reserved for validation purposes. Selected output is shown here.

ROW	Y (OBSERVED)	YPRED (PREDICTED)	ERROR (Y-PRED)	SE (STD.ERROR)	ERROR/SE SQUARED
43	1.1000E+02	1.4230E+02	-3.2302E+01	3.9539E+01	6.6743E-01
44	-5.0000E+00	6.0607E+01	-6.5607E+01	3.8738E+01	2.8683E+00
45	2.7000E+01	2.8318E+01	-1.3177E+00	3.8475E+01	1.1729E-03
46	-4.5000E+01	-1.2990E+01	-3.2010E+01	4.3213E+01	5.4873E-01
47	8.1000E+01	-2.0403E+01	1.0140E+02	3.7939E+01	7.1439E+00
48	1.1000E+01	-6.8245E+00	1.7825E+01	3.8408E+01	2.1537E-01
MEAN	2.9833E+01		-2.0016E+00		1.9075E+00
S.D.	5.6958E+01		5.8224E+01		2.7649E+00

a. Is there strong evidence that the model systematically over- or underestimates the new y values?

b. Is the residual standard deviation about the same size as that of Exercise 15.23?

15.25 The consultant of Exercise 15.8 obtains data on eight more firms for purposes of model validation. These firms have the same general characteristics as the firms in the original study; no outrageous extrapolation is needed. The output results are obtained for the full model of Exercise 15.18 and the simpler model of Exercise 15.19.

ROW	Y (OBSERVED)	YPRED (PREDICTED)	ERROR (Y-PRED)	SE (STD.ERROR)	ERROR/SE SQUARED
39	2.1689E-02	2.2258E-02	-5.6888E-04	7.6159E-03	5.5796E-03
40	1.7102E-02	3.4139E-02	-1.7037E-02	9.9448E-03	2.9350E+00
41	1.4938E-02	2.1126E-02	-6.1886E-03	6.1308E-03	1.0190E+00
42	3.3333E-02	3.3550E-02	-2.1669E-04	1.5649E-02	1.9172E-04
43	1.5500E-02	1.7715E-02	-2.2154E-03	7.0702E-03	9.8182E-02
44	2.4333E-02	2.4698E-02	-3.6496E-04	1.7031E-02	4.5922E-04
45	6.4453E-03	2.5253E-02	-1.8808E-02	7.9858E-03	5.5467E+00
46	1.4333E-02	1.9342E-02	-5.0086E-03	6.6630E-03	5.6506E-01
MEAN	1.8459E-02		-6.3010E-03		1.2713E+00
S.D.	8.0226E-03		7.5135E-03		1.9935E+00

ROW	Y (OBSERVED)	YPRED (PREDICTED)	ERROR (Y-PRED)	SE (STD.ERROR)	ERROR/SE SQUARED
39	2.1689E-02	2.2246E-02	-5.5692E-04	6.3994E-03	7.5735E-03
40	1.7102E-02	2.2736E-02	-5.6335E-03	8.1305E-03	4.8010E-01
41	1.4938E-02	2.1881E-02	-6.9436E-03	6.2911E-03	1.2182E+00
42	3.3333E-02	2.5490E-02	7.8428E-03	6.3051E-03	1.5472E+00
43	1.5500E-02	2.5243E-02	-9.7431E-03	6.1848E-03	2.4817E+00
44	2.4333E-02	2.3854E-02	4.7956E-04	7.0329E-03	4.6495E-03
45	6.4453E-03	2.4755E-02	-1.8309E-02	6.0808E-03	9.0664E+00
46	1.4333E-02	2.4260E-02	-9.9268E-03	5.9659E-03	2.7687E+00
MEAN	1.8459E-02		-5.3489E-03		2.1968E+00
S.D.	8.0226E-03		7.9447E-03		2.9647E+00

a. Is there any systematic bias obvious from either prediction?
b. The standard deviation is estimated to be .006 in both models. Is there any evidence that the standard deviation has increased in the validation?
c. Is there any strong reason to abandon the simpler model?

SUMMARY

In this chapter we have discussed some of the "tricks of the trade" for constructing satisfactory multiple regression models. To summarize, we present a series of steps to take in attempting a regression analysis.

Step 1 **a.** Try to identify and collect data on the most relevant independent variables. Ask those who are familiar with the problem what might be important predictors. Make sure the variables are well defined, but don't worry much about delicate distinctions. If two variables differ only in fine detail, either one should be about as satisfactory as the other.

b. Consider what dummy variables or lagged variables should be added to the model.

c. Think about whether a first-order model makes sense. Are additional nonlinear terms required? What transformations of the dependent variable might be sensible?

Step 2 Calculate a first-order model involving all the variables and plot the residuals against each variable. Look for nonlinear patterns of the residuals. (This is a good spot to look for outliers as well.) Based on an interpretation of these plots, add appropriate quadratic terms, cross products, or other transformed variables to the model.

Step 3 Either by trial and error or by formal stepwise regression methods, add or delete variables. Consider both the technical issues (values of t statistics, C_p, and changes in R^2, in particular) and also whether the model is reasonable in context.

Step 4 **a.** Once a tentative model has been selected, fit the model and look at the residuals from that model. Look again for outliers or gross skewness. Plot residuals against predicted values to look for nonconstant variance. If this problem appears serious, consider possible transformations of the dependent variable and redo the analysis using the transformed Y.

b. With time-series data, plot the data against time, looking for autocorrelation. If this plot and the Durbin-Watson statistic indicate that autocorrelation is a problem, reanalyze the data using differences or the Cochran-Orcutt method (if an appropriate computer program is available).

Step 5 Once a model has been chosen, validate it using new data. Expect to find some modest increase in the residual standard deviation. If the model consistently underpredicts or overpredicts or if the errors become much larger in magnitude, try whatever model you identified as second best.

This agenda may seem impossibly long and time-consuming. In fact, most of the time and effort are required at the first step—deciding which variables to use, collecting the sample data, entering the data, and constructing the necessary data files prior to analysis. The actual machine-run time may only be a small portion of this time. The extra effort involved in developing a thoughtful, nonmechanical regression analysis is not that great, and the payoff can be large.

CHAPTER 15 EXERCISES

15.26 The residuals from the regression study of Example 14.20 (p. 497) are plotted against the predicted values, as shown in Figure 15.17.

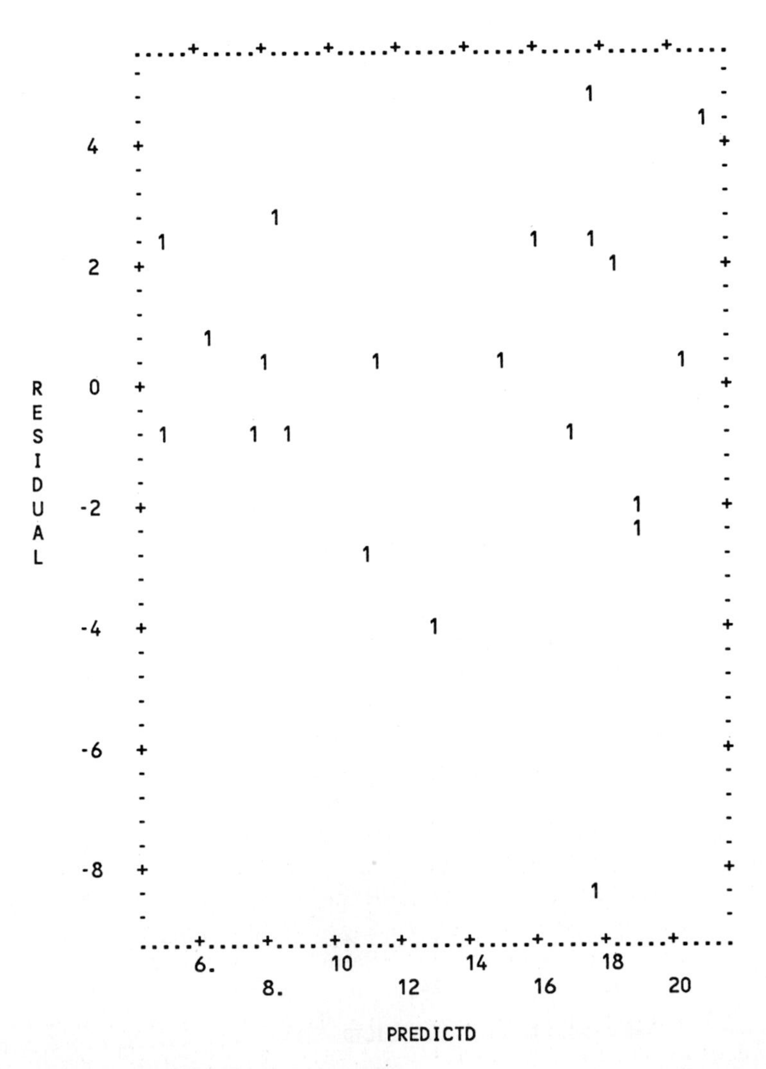

FIGURE 15.17 **Scatter Plot for Exercise 15.26**

a. Is there evidence of any problem with this regression study? If so, what?
b. Recall that the dependent variable in Example 14.20 is average account size per branch. Suppose that the dependent variable is redefined as average account size as a fraction of per capita income. What effect would you expect this redefinition to have on the problem of part (a)?

15.27 Refer to the feedlot study of Exercise 14.39 (p. 518).
a. Is there a problem of collinearity?
b. Would you expect to find a problem of autocorrelation?
c. What additional information would you need to assess possible heteroscedasticity? Might this be a problem?

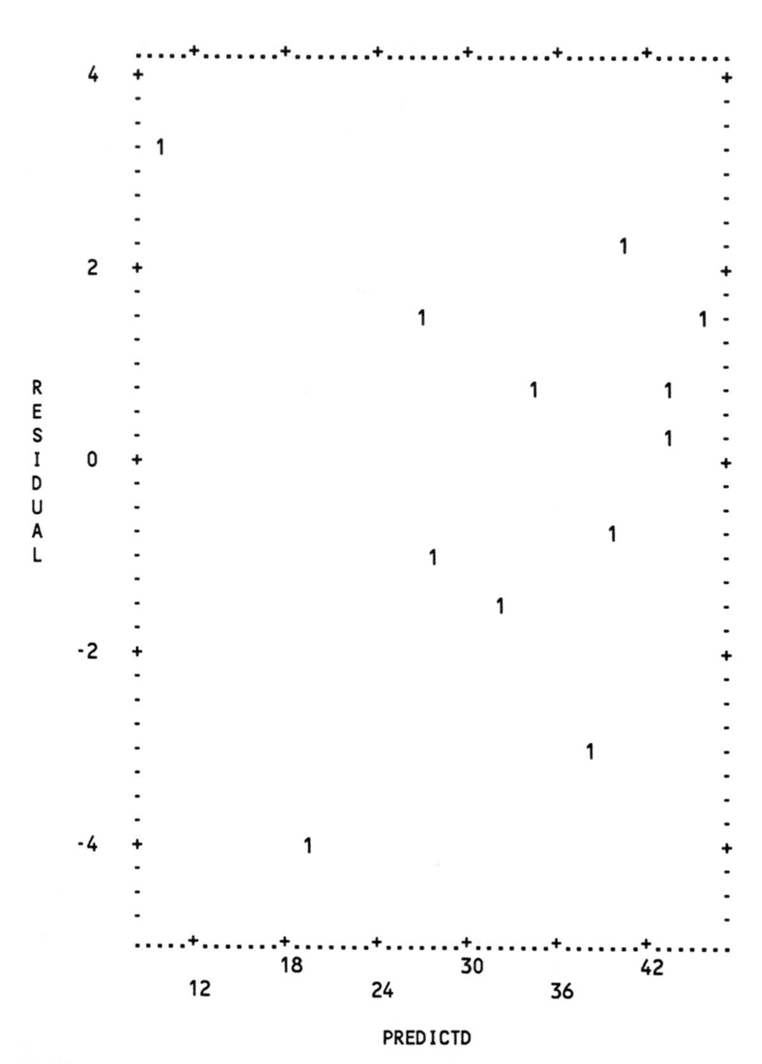

FIGURE 15.18 Scatter Plot for Exercise 15.28

15.28 Residuals from the regression of Exercise 14.27 are plotted in Figure 15.18 on the preceding page.

a. What potential problems could be detected in such a plot?

b. Do you find any evidence of such problems?

CASE STUDY (EXERCISES 15.29–15.37)

15.29 A firm that manufactures and sells moderately sophisticated word-processing equipment budgets a certain dollar amount for post-sale support activites. This support is provided by field representatives who train users initially and by home-office representatives who answer questions by telephone. Individuals shuttle between home and field-representative positions frequently, so it is not very meaningful to separate the positions in the budget. A regression study is made in an attempt to forecast the required support budget.

Analysis of the support requirements suggests that the training aspect of post-sales support typically involves users who have installed the equipment in the current and preceding months. For the next four months, there is substantial call-back support as users forget aspects of training or add new equipment operators. After that period, support activity is mostly trouble-shooting and new application work. Therefore the installation data to be used as independent variables are broken down as

1. number of installations in budget month (known in previous month)
2. number of installations in previous month
3. number of installations in preceding four months

These data are collected separately for the two levels of sophistication (A and B) of equipment (the A class requires more sophisticated training). Thus six independent variables are entered.

There is some question as to whether support costs increase in proportion to the number of installations. One opinion holds that the cost per installation decreases as installations increase because of improved user's manuals and more efficient training methods.

a. Identify any lagged variables.

b. Does the description of the situation indicate that any severe interaction can be expected?

c. Is there any indication of possible need for nonlinear terms in a regression model?

15.30 The firm of Exercise 15.29 collects data on the budgeted number of representatives (y) and the six independent variables described in that exercise. Data are available for 36 months, and the most recent 6 months' data are reserved for validation. A first-order regression model is fit to the remaining 30 months' data. A plot of the residuals against time is shown in Figure 15.19.

a. Is there any indication of possible outliers?

b. Is there any indication of autocorrelation?

15.31 Investigation of the data of Exercise 15.30 shows that the support budget in months 16 and 17 had been cut back drastically in a spasm of cost cutting. This action led to many user complaints, so the attempt was abandoned. The budget numbers of those months are changed in the data base to the figures that had been planned before the cost-cutting attempt. A first-order model yields the plots of residuals versus predicted values shown in Figure 15.20. Is there any indication of possible heteroscedasticity (nonconstant variance)?

15.32 The square of the current month's installations (both A and B types) are added as independent variables. A stepwise (forward selection) regression analysis yields the SPSS output on the top of page 577.

```
         -3.                -                3.
Case #   O:.............:  :.............:O   SUPP     *PRED      *RESID
     1   .                 ..*               .    75    74.6499       .3501
     2   .                *..                .    91    92.2644     -1.2644
     3   .                *..                .    78    78.8807     -0.8807
     4   .                 ..*               .    78    77.7459       .2541
     5   .               * ..                .   102   103.9882     -1.9882
     6   .                 ..*               .    90    88.7810      1.2190
     7   .                 ..  *             .    74    70.5757      3.4243
     8   .                 ..*               .    83    81.6864      1.3136
     9   .                 ..  *             .    97    94.0584      2.9416
    10   .                *..                .    98    98.6061     -0.6061
    11   .               * ..                .    88    89.7106     -1.7106
    12   .                 .. *              .    90    87.5255      2.4745
    13   .               * ..                .   104   105.7701     -1.7701
    14   .                *..                .   118   118.9869     -0.9869
    15   .                *..                .   107   107.9193     -0.9193
    16   *                 ..                .    78    99.4699    -21.4699
    17   .  *              ..                .    80    98.4080    -18.4080
    18   .                 ..*               .   102   101.8631       .1369
    19   .                 ..  *             .   102    97.8875      4.1125
    20   .                 ..   *            .   119   113.5058      5.4942
    21   .                 .. *              .   126   123.6658      2.3342
    22   .                 ..    *           .   119   112.4479      6.5521
    23   .                 ..    *           .   103    97.0311      5.9689
    24   .                 ..    *           .   101    95.1445      5.8555
    25   .                 ..  *             .   117   113.5252      3.4748
    26   .                 ..       *        .   130   119.3081     10.6919
    27   .             *   ..                .   141   144.4656     -3.4656
    28   .               * ..                .   128   130.5836     -2.5836
    29   .                *..                .   142   142.9883     -0.9883
    30   .                 ..*               .   154   153.5564       .4436
Case #   O:.............:  :.............:O   SUPP     *PRED      *RESID
         -3.                -                3.
```

FIGURE 15.19 Residuals versus Time; Exercise 15.30

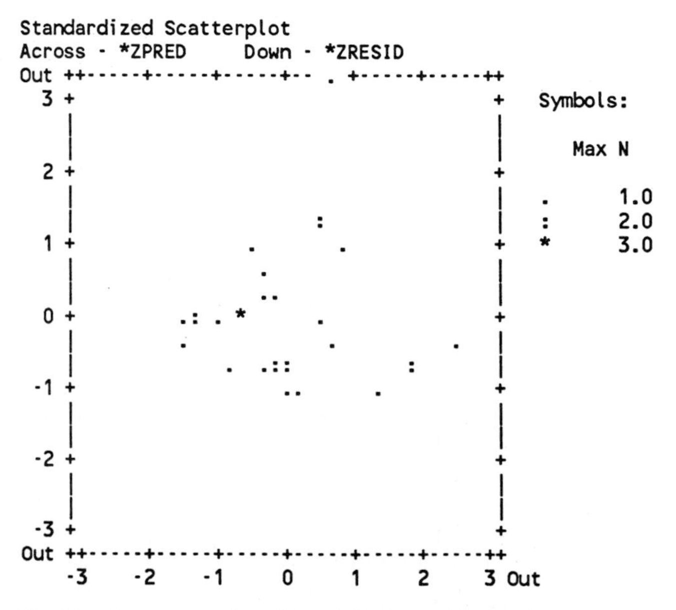

FIGURE 15.20 Residuals versus Predicted Values; Exercise 15.31

Step	MultR	Rsq	AdjRsq	F(Eqn)	SigF	RsqCh	FCh	SigCh		Variable	BetaIn	Correl	
1	.8375	.7015	.6908	65.792	.000	.7015	65.792	.000	In:	BCURRSQ	.8375	.8375	
2	.9118	.8314	.8189	66.572	.000	.1299	20.808	.000	In:	ACURRSQ	.4532	.7944	
3	.9581	.9180	.9085	96.964	.000	.0866	27.427	.000	In:	BPREV	.3546	.7245	PREVIOUS MONTH B CLASS SALES
4	.9723	.9454	.9367	108.230	.000	.0275	12.571	.002	In:	APREV	.2086	.6749	PREVIOUS MONTH A CLASS SALES
5	.9840	.9682	.9616	146.075	.000	.0228	17.185	.000	In:	BPREC4	.2141	.7762	PREV 4 MONTHS B CLASS SALES
6	.9888	.9777	.9719	168.171	.000	.0095	9.833	.005	In:	BCURR	.8028	.8338	CURRENT MONTH B CLASS SALES
7	.9903	.9807	.9746	159.923	.000	.0030	3.439	.077	In:	ACURR	.3074	.7704	CURRENT MONTH A CLASS SALES
8	.9918	.9837	.9775	158.815	.000	.0030	3.892	.062	In:	APREC4	.0871	.6130	PREV 4 MONTHS A CLASS SALES

Variable(s) Entered on Step Number 8.. APREC4 PREV 4 MONTHS A CLASS SALES

Multiple R	.99184
R Square	.98374
Adjusted R Square	.97755
Standard Error	3.08822

Analysis of Variance

	DF	Sum of Squares	Mean Square
Regression	8	12117.08731	1514.63591
Residual	21	200.27936	9.53711

F = 158.81494 Signif F = .0000

------------------ Variables in the Equation ------------------

Variable	B	SE B	Beta	T	Sig T
BCURRSQ	-0.020725	.017516	-0.290969	-1.183	.2500
ACURRSQ	.005785	.024706	.038526	.234	.8171
BPREV	.573122	.149819	.156053	3.825	.0010
APREV	.986435	.153308	.226510	6.434	.0000
BPREC4	.205807	.062177	.167540	3.310	.0033
BCURR	2.264419	.912878	.597190	2.481	.0217
ACURR	1.372116	.661190	.323884	2.075	.0504
APREC4	.148455	.075248	.087054	1.973	.0618
(Constant)	-12.629123	12.608849		-1.002	.3279

a. List the sequence in which the variables are added.

b. How much of an increment to R^2 is obtained by inclusion of the last four variables?

15.33 Additional output from the regression analysis of Exercise 15.32, for the model with all variables included, yields scatter plots shown in Figure 15.21.

a. Is there evidence of serious autocorrelation?

b. Is there evidence of nonconstant variance?

15.34 Differences for all variables in the data of Exercise 15.33 are calculated, and a stepwise (forward selection) regression is run. The output includes the following:

Variable(s) Entered on Step Number 8.. DBPREC4

Multiple R	.97020
R Square	.94129
Adjusted R Square	.91781
Standard Error	3.35551

Analysis of Variance

	DF	Sum of Squares	Mean Square
Regression	8	3610.60417	451.32552
Residual	20	225.18894	11.25945

F = 40.08416 Signif F = .0000

------------------ Variables in the Equation ------------------

Variable	B	SE B	Beta	T	Sig T
DACURR	1.334738	.540328	.623942	2.470	.0226
DAPREV	1.063336	.152473	.497357	6.974	.0000
DBCURR	2.025487	.685888	1.004195	2.953	.0079
DBPREV	.396343	.172939	.173751	2.292	.0329
DBCURRSQ	-0.020037	.012464	-0.542414	-1.608	.1236
DAPREC4	.119333	.111170	.065704	1.073	.2959
DACURRSQ	.009662	.019943	.121371	.484	.6333
DBPREC4	.058331	.140928	.027911	.414	.6833
(Constant)	.520634	.762649		.683	.5026

Is the sequence in which the variables are added similar to that found in Exercise 15.32?

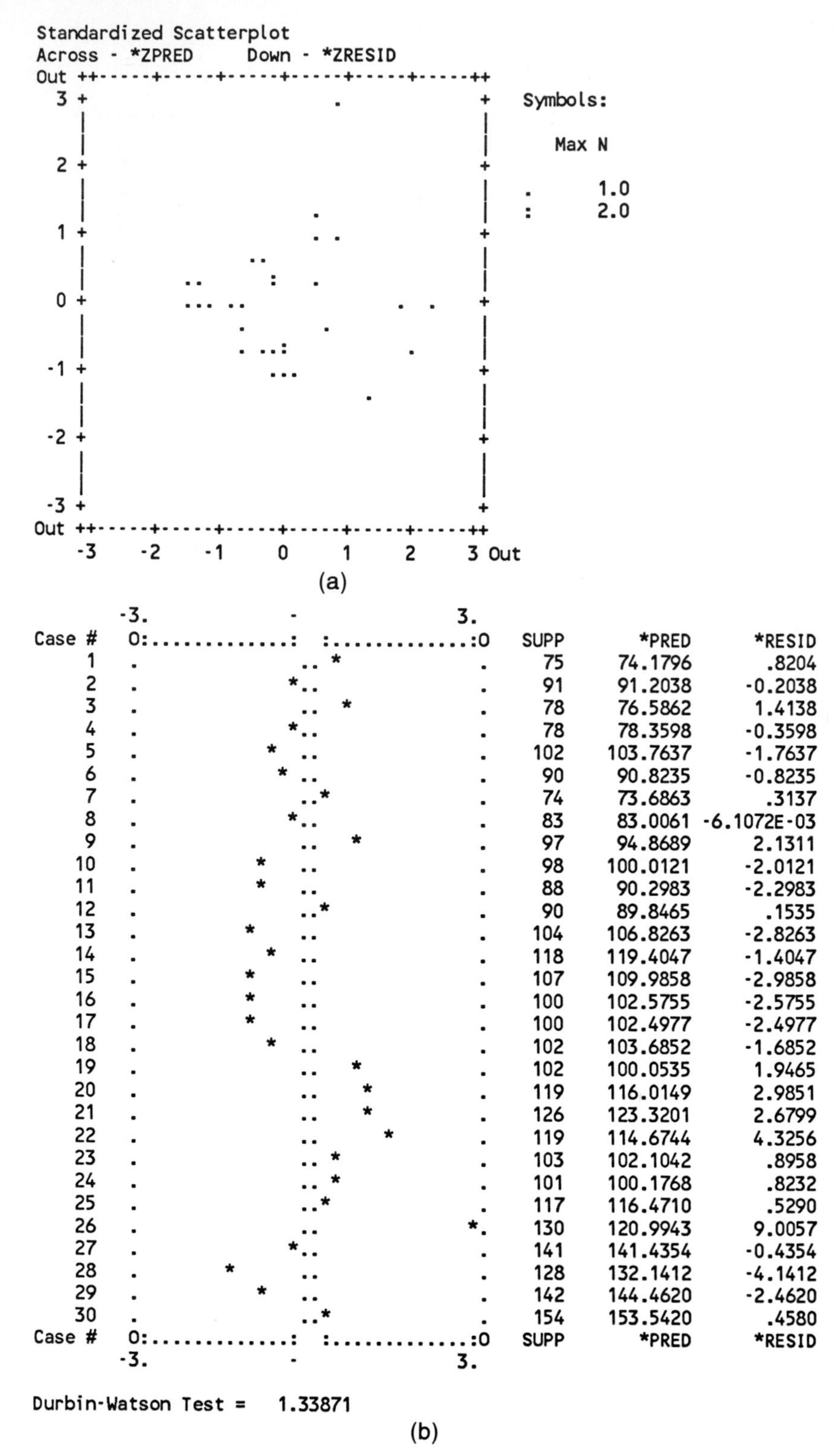

Case #	SUPP	*PRED	*RESID
1	75	74.1796	.8204
2	91	91.2038	-0.2038
3	78	76.5862	1.4138
4	78	78.3598	-0.3598
5	102	103.7637	-1.7637
6	90	90.8235	-0.8235
7	74	73.6863	.3137
8	83	83.0061	-6.1072E-03
9	97	94.8689	2.1311
10	98	100.0121	-2.0121
11	88	90.2983	-2.2983
12	90	89.8465	.1535
13	104	106.8263	-2.8263
14	118	119.4047	-1.4047
15	107	109.9858	-2.9858
16	100	102.5755	-2.5755
17	100	102.4977	-2.4977
18	102	103.6852	-1.6852
19	102	100.0535	1.9465
20	119	116.0149	2.9851
21	126	123.3201	2.6799
22	119	114.6744	4.3256
23	103	102.1042	.8958
24	101	100.1768	.8232
25	117	116.4710	.5290
26	130	120.9943	9.0057
27	141	141.4354	-0.4354
28	128	132.1412	-4.1412
29	142	144.4620	-2.4620
30	154	153.5420	.4580

Durbin-Watson Test = 1.33871

(b)

FIGURE 15.21 Residual Plots for Exercise 15.33: (a) Residuals and Predicted Values; (b) Residuals and Time

15.35 Additional output from the all-variables model of Exercise 15.34 is shown in Figure 15.22.

a. Is there an autocorrelation problem with the differenced data?

b. Is there evidence of heteroscedasticity?

15.36 Consider the output from the regression study of Exercise 15.34.

a. Write the regression model.

b. Can the null hypothesis that all partial slopes are zero be rejected at reasonable α levels?

c. Which variables have coefficients that differ significantly (at $\alpha = .05$) from zero as "last predictor in"?

d. Identify and interpret the R^2 value.

e. Identify the residual standard deviation.

15.37 In Exercise 15.30 it was noted that the most recent six months' data were reserved for validation. These values, in difference form, are shown below, along with the predicted values given by the model of Exercise 15.34.

ACURR	APREV	APREC4	BCURR	BPREV	BPREC4	ACURRSQ	BCURRSQ	Actual Support	Predicted Support
−2	−3	9	5	3	1	−80	−135	2	5.78
5	−2	3	−6	5	1	215	−80	−3	−.97
−4	5	−1	4	−6	3	−176	215	5	2.99
−4	−4	−7	−6	4	8	−144	−176	−5	−9.94
14	−4	−4	7	−6	6	644	−144	17	15.29
−2	14	−5	4	7	−3	−116	644	11	12.78

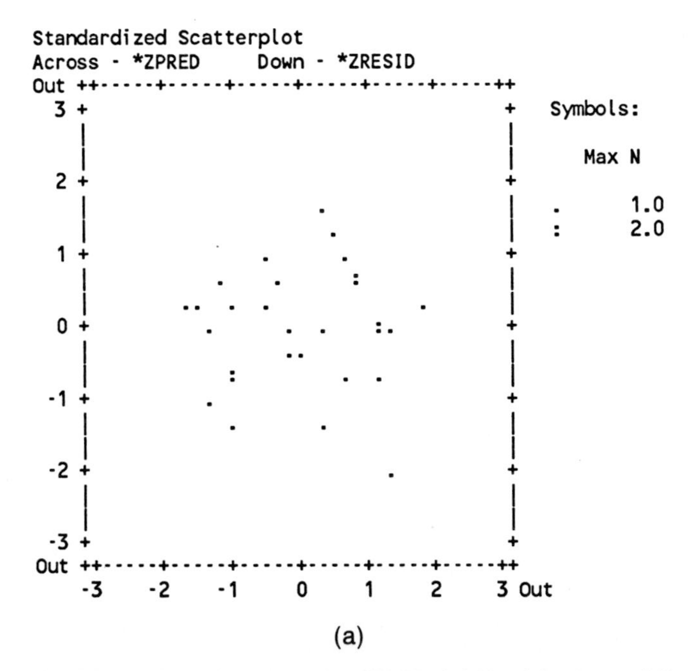

FIGURE 15.22 Residual plots for Exercise 15.35: (a) Residuals and Predicted Values; (b) Residuals and Time (on p. 580)

```
Casewise Plot of Standardized Residual

        -3.          -            3.
Case #  O:...........: :..........:O      DSUPP      *PRED     *RESID
     2  .            ..*           .      16.00    15.9080      .0920
     3  .            .. *          .     -13.00   -14.2197     1.2197
     4  .           *..            .        .00      .3697    -0.3697
     5  .            .. *          .      24.00    22.9906     1.0094
     6  .           *..            .     -12.00   -11.5631    -0.4369
     7  .            .. *          .     -16.00   -16.8264      .8264
     8  .          * ..            .       9.00    10.8579    -1.8579
     9  .            ..    *       .      14.00    11.0276     2.9724
    10  .      *     ..            .       1.00     5.9603    -4.9603
    11  .         *  ..            .     -10.00    -7.8641    -2.1359
    12  .            ..   *        .       2.00    -0.2706     2.2706
    13  .         *  ..            .      14.00    16.0303    -2.0303
    14  .            ..   *        .      14.00    11.4172     2.5828
    15  .          * ..            .     -11.00    -9.0065    -1.9935
    16  .            .. *          .      -7.00    -8.1911     1.1911
    17  .          * ..            .        .00      .7565    -0.7565
    18  .          * ..            .       2.00     3.0275    -1.0275
    19  .            ..    *       .        .00    -3.1818     3.1818
    20  .           *..            .      17.00    17.0217    -0.0217
    21  .           *..            .       7.00     7.3773    -0.3773
    22  .            ..    *       .      -7.00    -9.7461     2.7461
    23  .        *   ..            .     -16.00   -13.1824    -2.8176
    24  .            .. *          .      -2.00    -2.8116      .8116
    25  .           *..            .      16.00    16.0501    -0.0501
    26  .            ..       *    .      13.00     6.8943     6.1057
    27  .   *        ..            .      11.00    18.1027    -7.1027
    28  .      *     ..            .     -13.00    -8.1258    -4.8742
    29  .            ..  *         .      14.00    12.2196     1.7804
    30  .            ..    *       .      12.00     7.9780     4.0220
Case #  O:...........: :..........:O      DSUPP      *PRED     *RESID
        -3.          -            3.
```

(b)

FIGURE 15.22 (continued)

a. Calculate the residuals and their standard deviation.
b. Is this residual standard deviation much larger than the one found in Exercise 15.34?
c. Previous forecasts of the budget were often in error by as much as ± 20 units. Does it appear that the regression model will be useful in predicting the budget figure?

CASE STUDY (EXERCISES 15.38–15.43)

15.38 A wholesale hardware company fills orders at a large warehouse. The volume of orders fluctuates substantially from day to day, and the company has a policy of filling each order on the day of receipt. Orders are phoned or mailed, and received by 10 A.M. The warehouse supervisor estimates the total time required for each order and assigns the required number of workers to order filling. Excess workers, if any, are assigned to other tasks and cannot be recalled until the next day. The supervisor's estimates of time per order are often quite erroneous, so that sometimes workers are idle and other times substantial overtime must be paid. A regression study is attempted to improve the prediction of required time per order. The minicomputer used for order processing and inventory control can easily calculate predicated times once a model is determined.

Items in the order are classified as frequently, moderately, or rarely ordered. The rarely ordered items are stored in the most distant parts of the warehouse and therefore require more time to obtain. Both the number of items ordered in each category and the average size per item are thought to influence the order-filling time. Most items ordered are in carton units, and less-than-carton items (which are mostly rarely ordered) are filled as "loose-box" items. Certain orders require special packing to protect fragile items. The supervisor assigns each such order to either packing station A or B. All items are assembled at a central station and placed on skids. A forklift is used to move the skids from the assembly station to delivery trucks.

A sample of 50 orders is selected; care is taken to include relatively extreme types of order (small and large, mostly loose-box and mostly carton, mostly frequent and mostly rare). Values for each of the following variables are recorded:

TIME:	time (in worker-minutes) needed to fill order
NUMFREQ:	number of frequently ordered items in the order
NUMMOD:	number of moderately ordered items in the order
NUMRARE:	number of rarely ordered items in the order
NUMLOOSE:	number of loose-box items
AVSZCAR:	average size of carton items (cartons/item)
AVSZLB:	average size of loose-box items (pieces/item)
SPECIALS:	0 if no special packing need
	1 if special packing done at station A
	2 if special packing done at station B
SKIDS:	number of skids needed

The supervisor thinks that the most important predictor variables are the NUM ones, because much of the order-filling time is taken up by workers moving to the appropriate locations. However, this travel time is not likely to be directly proportional to the number of items, because a worker can combine items found in the same general area of the warehouse. The order size of each item is also relevant, because some time is used in taking each item from the shelf. This time is expected to be in proportion to AVSZ. The time spent moving skids is expected to be in proportion to SKIDS. The only other major time factor is assembly-station time, which is expected to depend on a combination of NUM and AVSZ variables.

a. Explain why the SPECIALS variable should be recoded as two variables:

   SPECIALA = 1 if special packing done at station A, 0 if not

   SPECIALB = 1 if special packing done at station B, 0 if not

b. Explain the interpretation of the coefficients of these two variables.
c. According to the information given by the supervisor, which independent variables might be transformed?
d. Is there any indication that interaction terms might be useful?

MULTIPLE R	0.9652	STD. ERROR OF EST.	4.3409
MULTIPLE R-SQUARE	0.9316		

ANALYSIS OF VARIANCE

	SUM OF SQUARES	DF	MEAN SQUARE	F RATIO	P(TAIL)
REGRESSION	10273.3750	9	1141.4861	60.577	0.0000
RESIDUAL	753.7462	40	18.8437		

VARIABLE		COEFFICIENT	STD. ERROR	STD. REG COEFF	T	P(2 TAIL)	TOLERANCE
INTERCEPT		12.58383					
NUMFREQ	2	0.13947	0.05319	0.207	2.622	0.0123	0.27536
NUMMOD	3	0.23438	0.10860	0.147	2.158	0.0370	0.36939
NUMRARE	4	0.54236	0.15506	0.183	3.498	0.0012	0.62762
NUMLOOSE	5	1.17128	0.25393	0.241	4.613	0.0000	0.62756
AVSZCAR	6	-0.21619	0.62200	-0.019	-0.348	0.7300	0.59143
AVSZLB	7	0.49755	0.31877	0.075	1.561	0.1264	0.73237
SPECIALA	8	0.70119	1.78378	0.017	0.393	0.6963	0.88128
SPECIALB	9	1.81956	2.30071	0.040	0.791	0.4337	0.67423
SKIDS	10	1.35125	0.23116	0.476	5.846	0.0000	0.25816

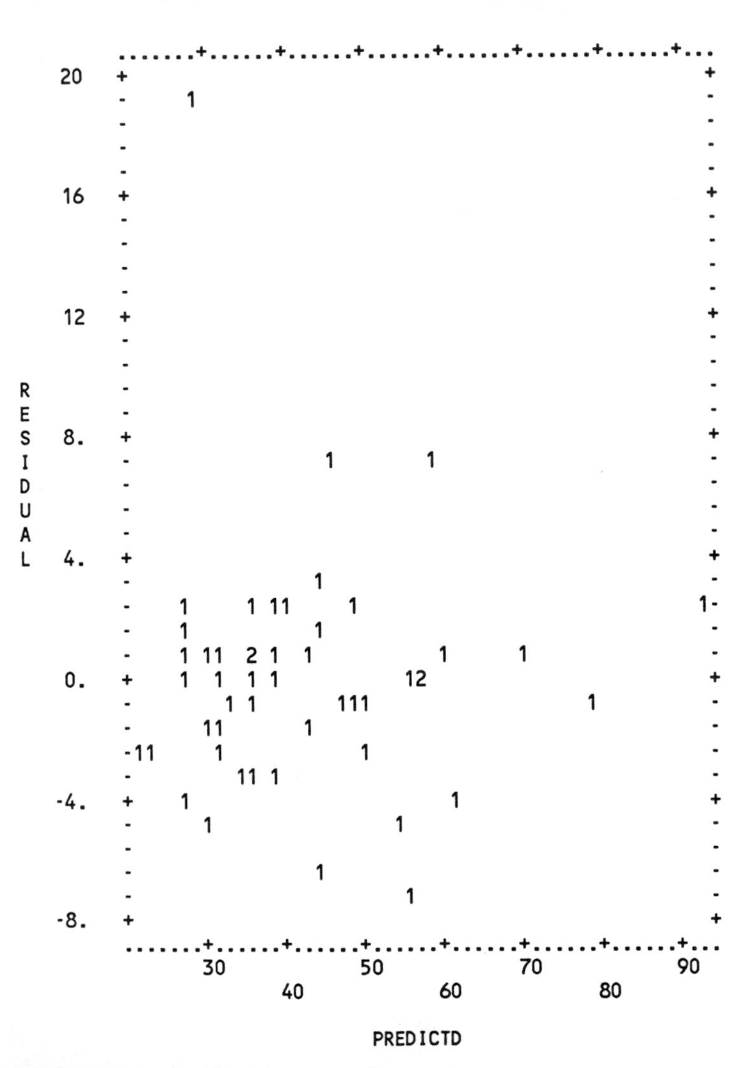

FIGURE 15.23 Scatter Plot for Exercise 15.39

15.39 The data for the study of Exercise 15.38 are shown below. A first-order model is fit, and the BMDP output and scatter plot shown in Figure 15.23 (p. 582) are obtained.

CASE NO. LABEL	1 TIME	2 NUMFREQ	3 NUMMOD	4 NUMRARE	5 NUMLOOSE	6 AVSZCAR	7 AVSZLB	8 SPECIALA	9 SPECIALB	10 SKIDS
1	27	16	6	3	4	1.600	3.500	1	0	2
2	57	70	25	2	2	3.200	2.700	0	0	18
3	39	25	6	8	4	2.400	3.300	0	0	8
4	20	10	2	1	1	6.200	4	0	0	4
5	55	50	30	6	7	2.600	3	0	0	12
6	95	85	43	16	10	4.700	6.200	0	1	25
7	19	12	3	1	1	2.400	3	0	0	3
8	35	10	12	13	3	5	6	0	0	5
9	61	87	12	4	6	3	4	1	0	16
10	44	50	22	8	2	4	1.500	0	0	9
11	23	12	8	8	2	1.400	3.500	0	0	2
12	47	27	6	12	14	1.600	7.200	0	0	4
13	53	64	8	1	2	3	2.500	0	0	14
14	70	71	34	7	10	2.100	4.600	0	0	16
15	32	25	1	2	2	2.800	5	0	0	10
16	39	16	10	3	8	2	4.700	1	0	6
17	32	25	12	4	2	1.800	6	0	0	4
18	30	40	6	1	1	2.600	3	0	0	7
19	38	72	21	8	3	2	2.700	0	0	6
20	78	64	37	21	6	3	4.300	0	1	20
21	30	23	6	2	1	4.200	2	0	0	6
22	38	35	13	7	1	1.400	8	0	0	5
23	25	16	4	2	4	1.600	9.300	0	0	3
24	47	19	17	10	6	3.200	2.700	1	0	8
25	42	31	12	16	3	1.400	2.300	0	1	4
26	46	46	8	11	2	4	2.500	0	0	13
27	48	12	6	1	5	2	6.200	0	0	3
28	28	16	4	4	2	6.100	1.500	0	0	5
29	28	37	8	2	1	2.200	2	1	0	3
30	32	21	9	6	4	1.800	6.500	0	0	3
31	51	58	4	3	6	4	8.300	0	0	11
32	48	24	15	1	8	3	9.100	0	1	10
33	49	36	12	9	2	4	6	0	0	15
34	42	51	18	3	5	1.800	2.800	0	0	8
35	58	77	30	15	12	1.200	1.200	1	0	6
36	37	24	6	2	2	4	6	0	0	10
37	37	16	8	5	1	6	10	0	0	9
38	46	36	12	4	6	2.600	5.100	0	1	8
39	58	74	15	16	4	1.800	1.500	0	0	13
40	49	51	10	3	8	3.700	4.500	0	0	16
41	41	24	6	8	3	4.100	3.300	1	0	9
42	31	36	8	1	1	2.700	5	0	0	6
43	31	21	4	5	6	1.600	1.300	0	0	3
44	49	42	10	2	2	4	8.500	0	1	18
45	29	16	3	4	2	2.700	3.500	0	0	7
46	29	28	15	2	1	3	6	0	0	5
47	66	73	18	15	5	1.900	4.200	0	0	12
48	31	15	8	8	6	1.200	2.300	0	0	4
49	36	23	4	10	4	1.500	2.800	0	0	6
50	36	44	9	2	1	4	5	1	0	10

a. Is there any indication of outliers?

b. Is there any indication of nonconstant variance?

c. Is there any indication of autocorrelation? Would autocorrelation be expected in this study?

15.40 It was discovered that observation 27 of the data in Exercise 15.39 was taken during a brief strike. The order was filled by the president and vice-president of the firm, who couldn't find most of the items. This observation is deleted from the data set. Regression runs are made using logarithms of the NUM variables and using the square roots of these variables. The resulting residual standard deviations and R^2 values are

	Logarithms	Square Roots
s_ϵ	3.4804	3.0561
R^2	.9570	.9669

Which transformation appears more effective?

15.41 A forward-selection stepwise regression is run on the data resulting from Exercise 15.40. The square-root transformation is used. The following BMDP output is obtained:

```
STEP NO.    1
--------------
VARIABLE ENTERED    10 SKIDS

MULTIPLE R              0.8478
MULTIPLE R-SQUARE       0.7188
ADJUSTED R-SQUARE       0.7128
STD. ERROR OF EST.      8.1101


                  VARIABLES IN EQUATION FOR TIME

                                   STD. ERROR  STD REG
    VARIABLE      COEFFICIENT      OF COEFF    COEFF      TOLERANCE
(Y-INTERCEPT        20.91346 )
SKIDS      10        2.43382        0.2220     0.848      1.00000

STEP NO.    2
--------------
VARIABLE ENTERED     5 NUMLOOSE

MULTIPLE R              0.9317
MULTIPLE R-SQUARE       0.8680
ADJUSTED R-SQUARE       0.8623
STD. ERROR OF EST.      5.6156


                  VARIABLES IN EQUATION FOR TIME

                                   STD. ERROR  STD REG
    VARIABLE      COEFFICIENT      OF COEFF    COEFF      TOLERANCE
(Y-INTERCEPT        15.38700 )
NUMLOOSE    5        1.93220        0.2679     0.397      0.94548
SKIDS      10        2.16751        0.1581     0.755      0.94548

STEP NO.    3
--------------
VARIABLE ENTERED    12 SQTNUMM

MULTIPLE R              0.9615
MULTIPLE R-SQUARE       0.9244
ADJUSTED R-SQUARE       0.9193
STD. ERROR OF EST.      4.2979
```

(continued)

VARIABLES IN EQUATION FOR TIME

VARIABLE	COEFFICIENT	STD. ERROR OF COEFF	STD REG COEFF	TOLERANCE
(Y-INTERCEPT	8.69073)			
NUMLOOSE 5	1.44632	0.2215	0.297	0.80983
SKIDS 10	1.74510	0.1413	0.608	0.69350
SQTNUMM 12	3.76762	0.6506	0.308	0.59406

STEP NO. 4

VARIABLE ENTERED 13 SQTNUMR

MULTIPLE R	0.9740
MULTIPLE R-SQUARE	0.9488
ADJUSTED R-SQUARE	0.9441
STD. ERROR OF EST.	3.5777

VARIABLES IN EQUATION FOR TIME

VARIABLE	COEFFICIENT	STD. ERROR OF COEFF	STD REG COEFF	TOLERANCE
(Y-INTERCEPT	5.81847)			
NUMLOOSE 5	1.22869	0.1904	0.253	0.75932
SKIDS 10	1.79848	0.1182	0.626	0.68674
SQTNUMM 12	2.74049	0.5863	0.224	0.50699
SQTNUMR 13	2.88554	0.6306	0.187	0.69785

STEP NO. 5

VARIABLE ENTERED 11 SQTNUMF

MULTIPLE R	0.9804
MULTIPLE R-SQUARE	0.9612
ADJUSTED R-SQUARE	0.9567
STD. ERROR OF EST.	3.1494

VARIABLES IN EQUATION FOR TIME

VARIABLE	COEFFICIENT	STD. ERROR OF COEFF	STD REG COEFF	TOLERANCE
(Y-INTERCEPT	1.80256)			
NUMLOOSE 5	1.19112	0.1679	0.245	0.75656
SKIDS 10	1.55219	0.1234	0.541	0.48826
SQTNUMF 11	1.52926	0.4119	0.179	0.38643
SQTNUMM 12	1.84518	0.5697	0.151	0.41612
SQTNUMR 13	2.99877	0.5559	0.194	0.69575

a. How much does inclusion of the last variable increase R^2?
b. Test the null hypothesis that the last two variables have no incremental predictive value.

15.42 The independent variable in Exercise 15.41 is redefined to be the time *per item*, that is

$$\frac{\text{TIME}}{\text{NUMFREQ} + \text{NUMMOD} + \text{NUMRARE}}$$

All independent variables (including the dummy variables) are divided by (NUMFREQ + NUMMOD + NUMRARE). The following output results from a regression run on the transformed data.

```
MULTIPLE R            0.9854          STD. ERROR OF EST.          0.0492
MULTIPLE R-SQUARE     0.9710

ANALYSIS OF VARIANCE
                       SUM OF SQUARES    DF    MEAN SQUARE       F RATIO    P(TAIL)
          REGRESSION           3.1638     9         0.3515       145.132     0.0000
          RESIDUAL             0.0945    39         0.0024

                                                   STD. REG
   VARIABLE          COEFFICIENT    STD. ERROR       COEFF          T    P(2 TAIL) TOLERANCE

INTERCEPT               0.02436
SQTNFPER     12         1.80133      0.32271         0.256      5.582     0.0000     0.35369
SQTNMPER     13         1.63110      0.55029         0.138      2.964     0.0052     0.34265
SQTNRPER     14         2.73480      0.40116         0.259      6.817     0.0000     0.51381
NUMLPER      15         1.14912      0.13448         0.291      8.545     0.0000     0.64017
ASZCRPER     16         0.36002      0.16331         0.105      2.205     0.0335     0.32741
ASZLBPER     17         0.21039      0.12790         0.070      1.645     0.1080     0.40985
SPECAPER     18         1.87525      0.85495         0.068      2.193     0.0343     0.78185
SPECBPER     19         2.92493      1.41125         0.064      2.073     0.0449     0.78115
SKIDSPER     20         1.28820      0.12342         0.385     10.437     0.0000     0.54602
```

a. Write the regression model and the residual standard deviation.
b. Is any violation of assumptions shown by the residual plots on pp. 587–594?

15.43 A validation study of the model obtained in Exercise 15.42 is based on an additional 10 orders. The results are shown below.

Actual Time	Time per Item	Regression Forecast	Superintendent's Forecast
36	.6316	.6691	.7895
24	1.0000	.9057	.8333
26	.7647	.8482	.7692
42	.7925	.6558	.9434
34	.5667	.5608	.8333
31	.9688	.9021	.7813
27	.8182	.9247	.9091
32	.6531	.7567	.8163
34	.8293	.9272	.7317
38	.7451	.7479	.7843

(*problem continued on p. 594*)

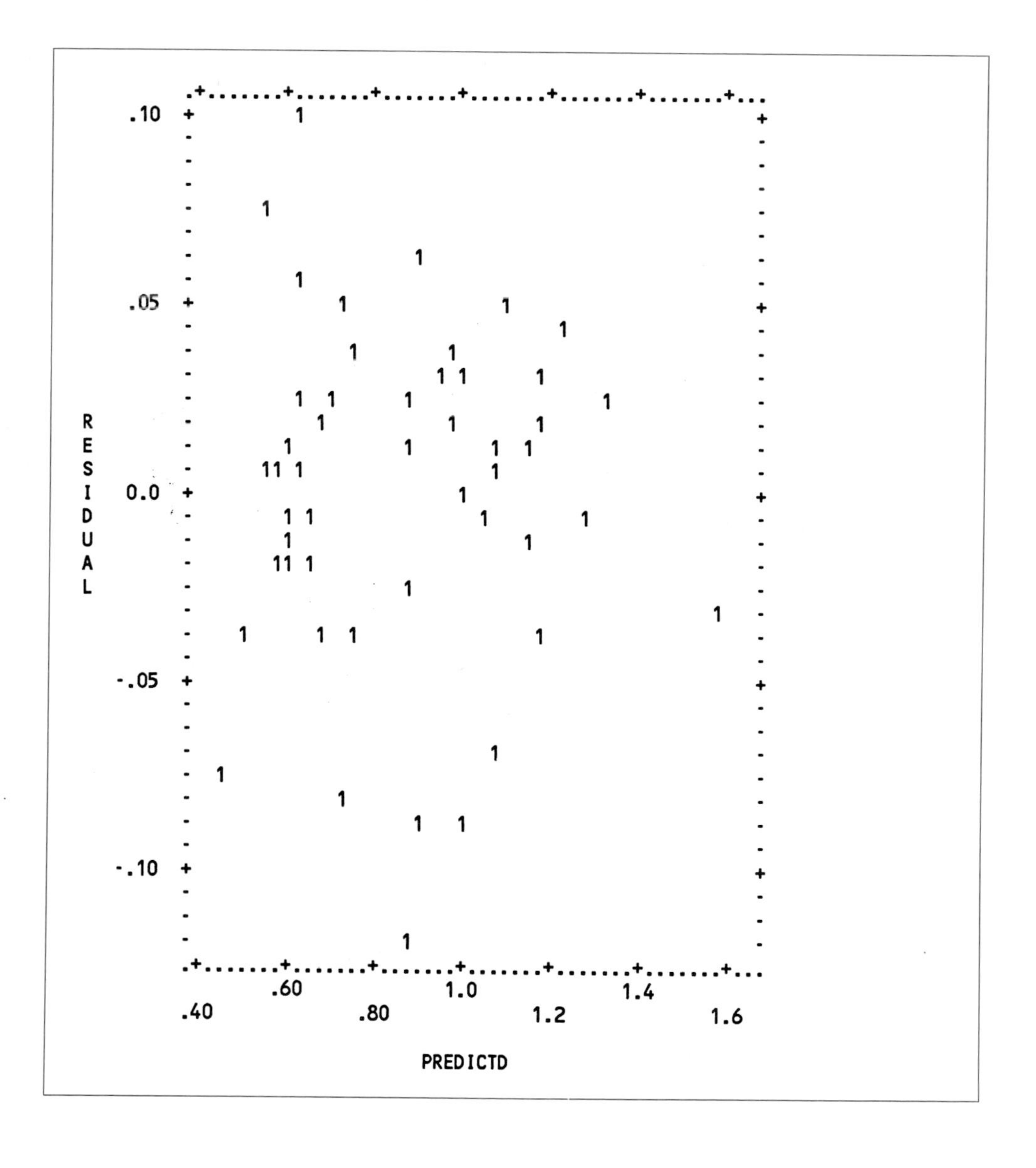
.10
.05
0.0
-.05
-.10
RESIDUAL
.40
.60
.80
1.0
1.2
1.4
1.6
PREDICTD

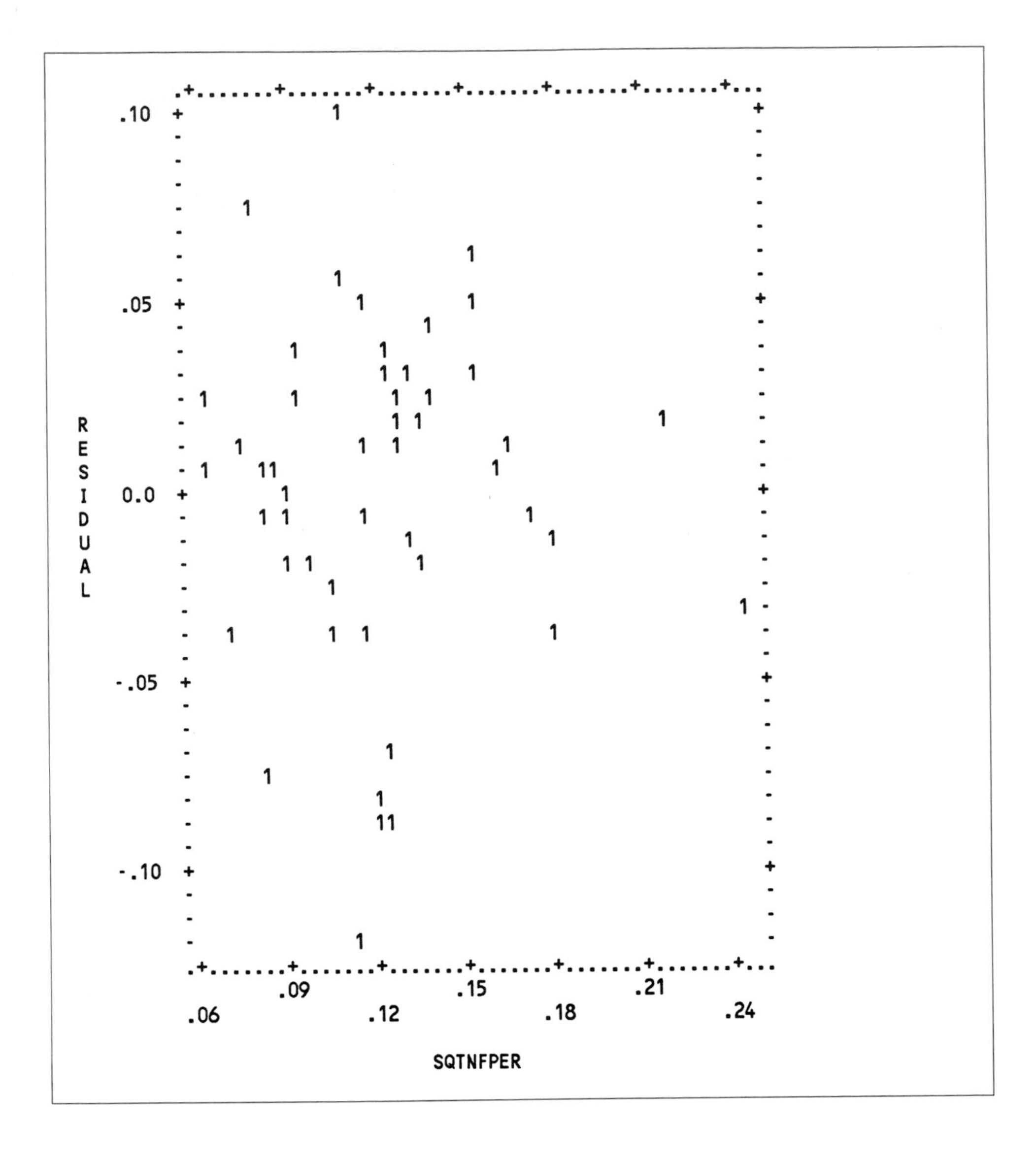
RESIDUAL
.10
.05
0.0
-.05
-.10
.06
.09
.12
.15
.18
.21
.24
SQTNFPER

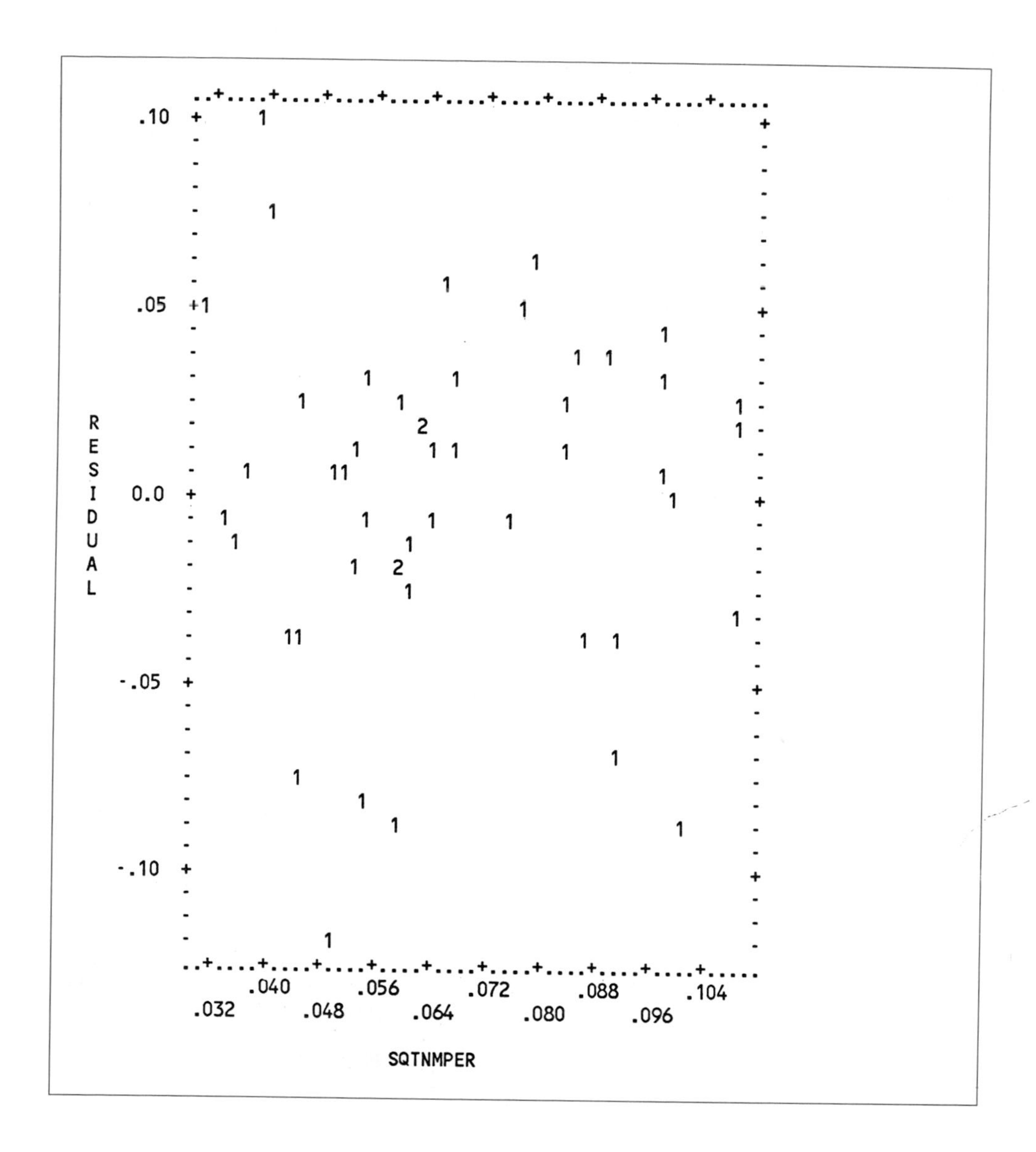
.10
.05
0.0
-.05
-.10
RESIDUAL
.032
.040
.048
.056
.064
.072
.080
.088
.096
.104
SQTNMPER

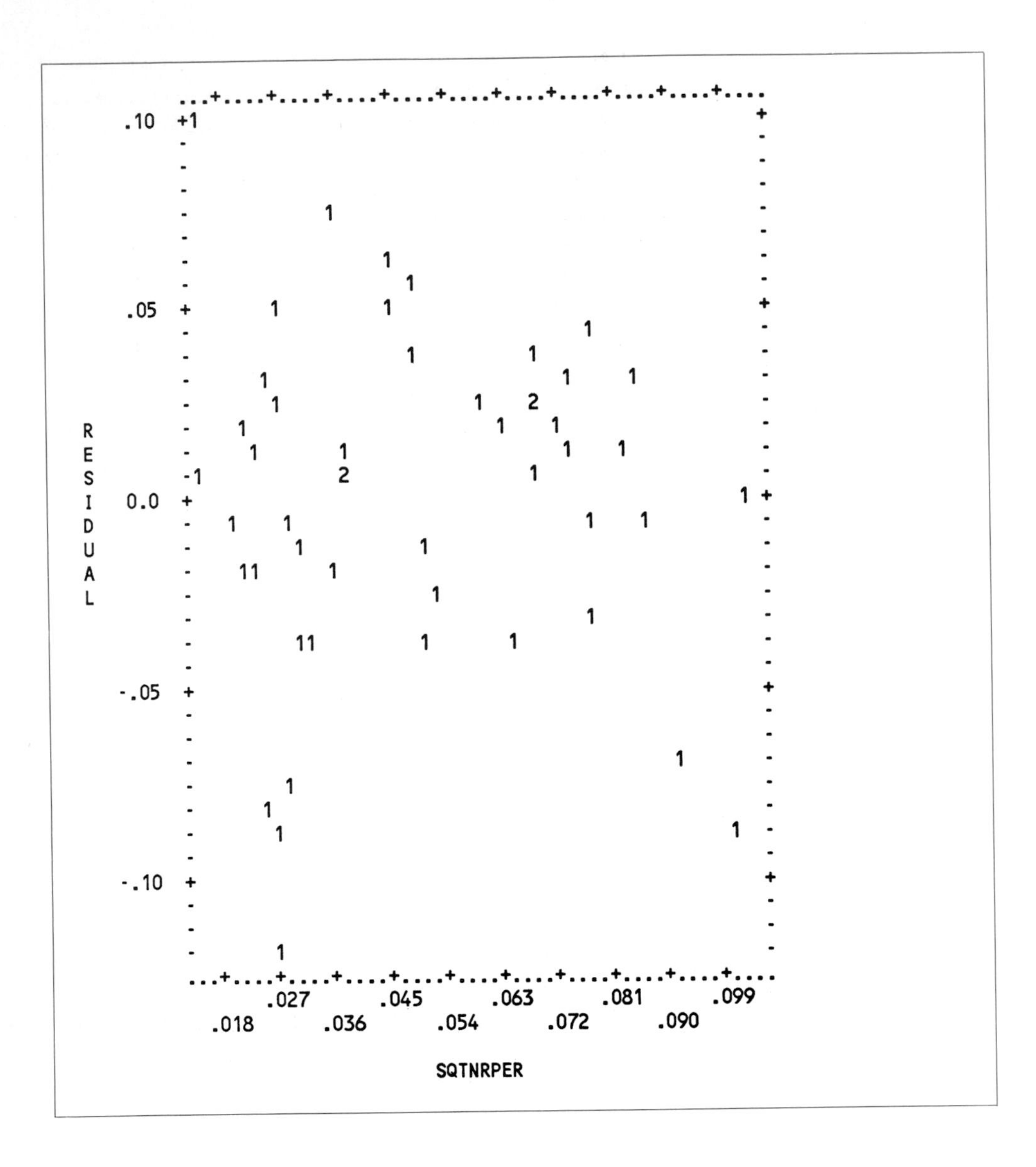
.10
.05
0.0
-.05
-.10
RESIDUAL
.018
.027
.036
.045
.054
.063
.072
.081
.090
.099
SQTNRPER

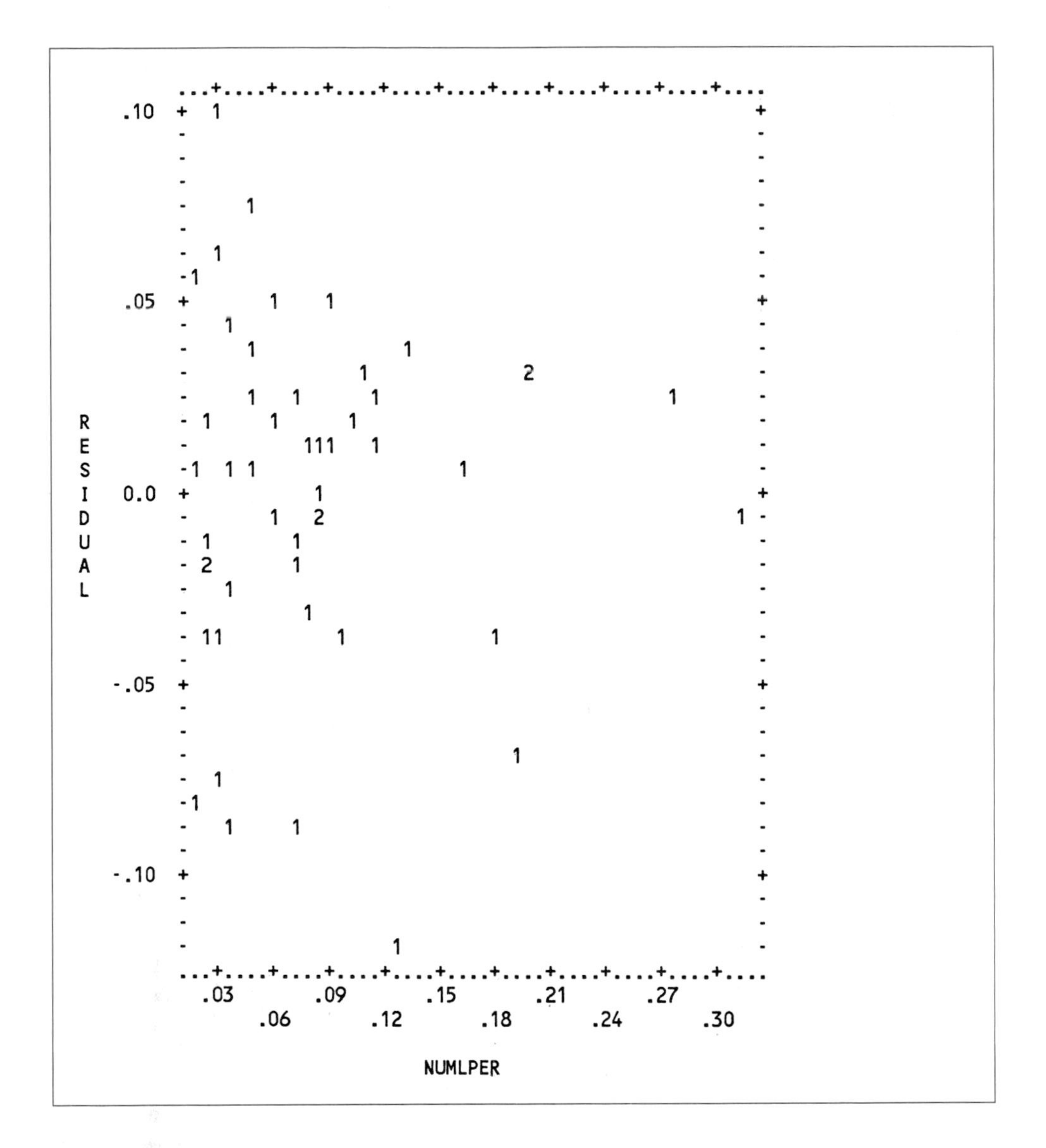

.10
.05
0.0
-.05
-.10
RESIDUAL
.03
.06
.09
.12
.15
.18
.21
.24
.27
.30
NUMLPER

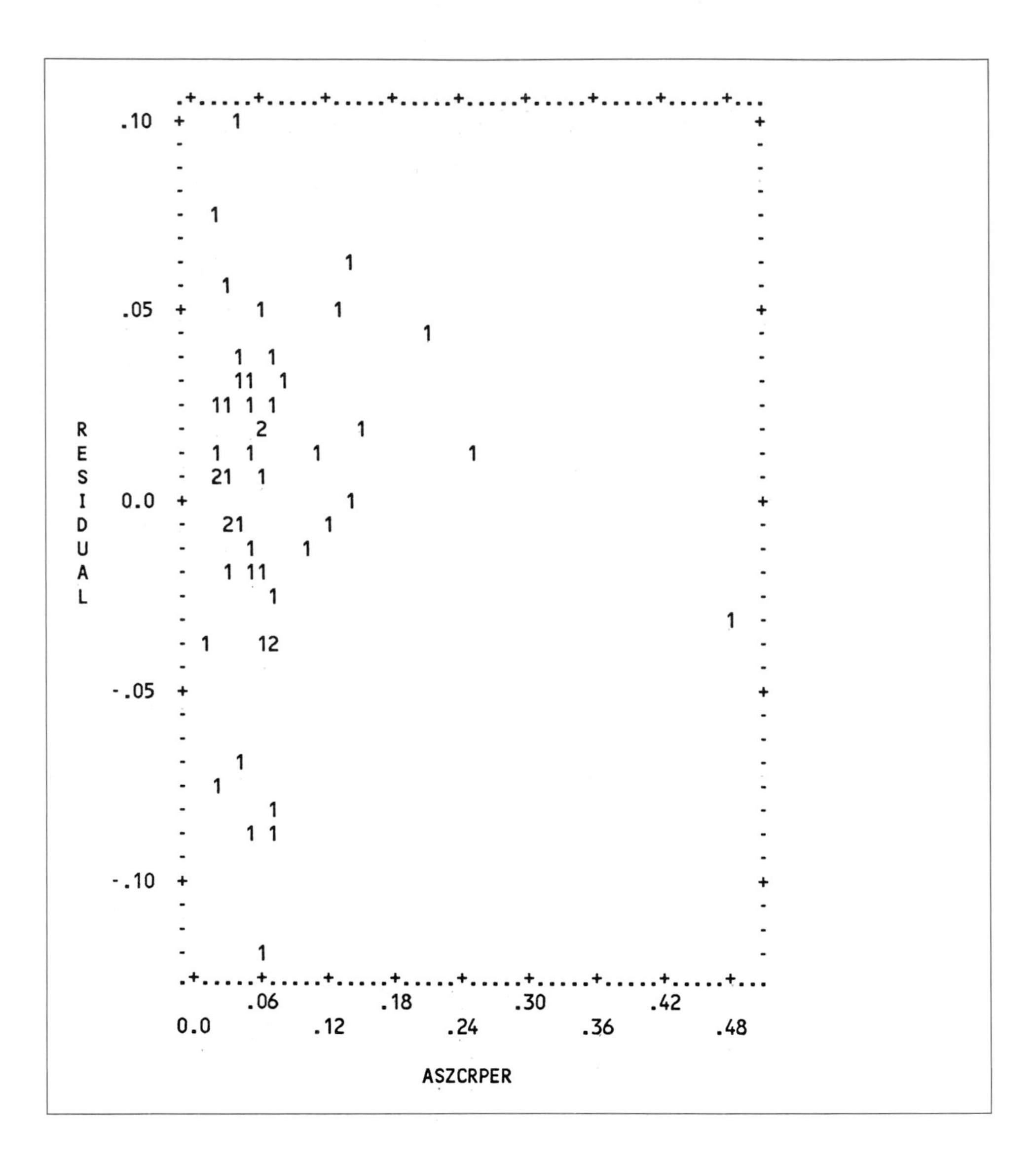
.10
.05
0.0
-.05
-.10
RESIDUAL
.06
.18
.30
.42
0.0
.12
.24
.36
.48
ASZCRPER

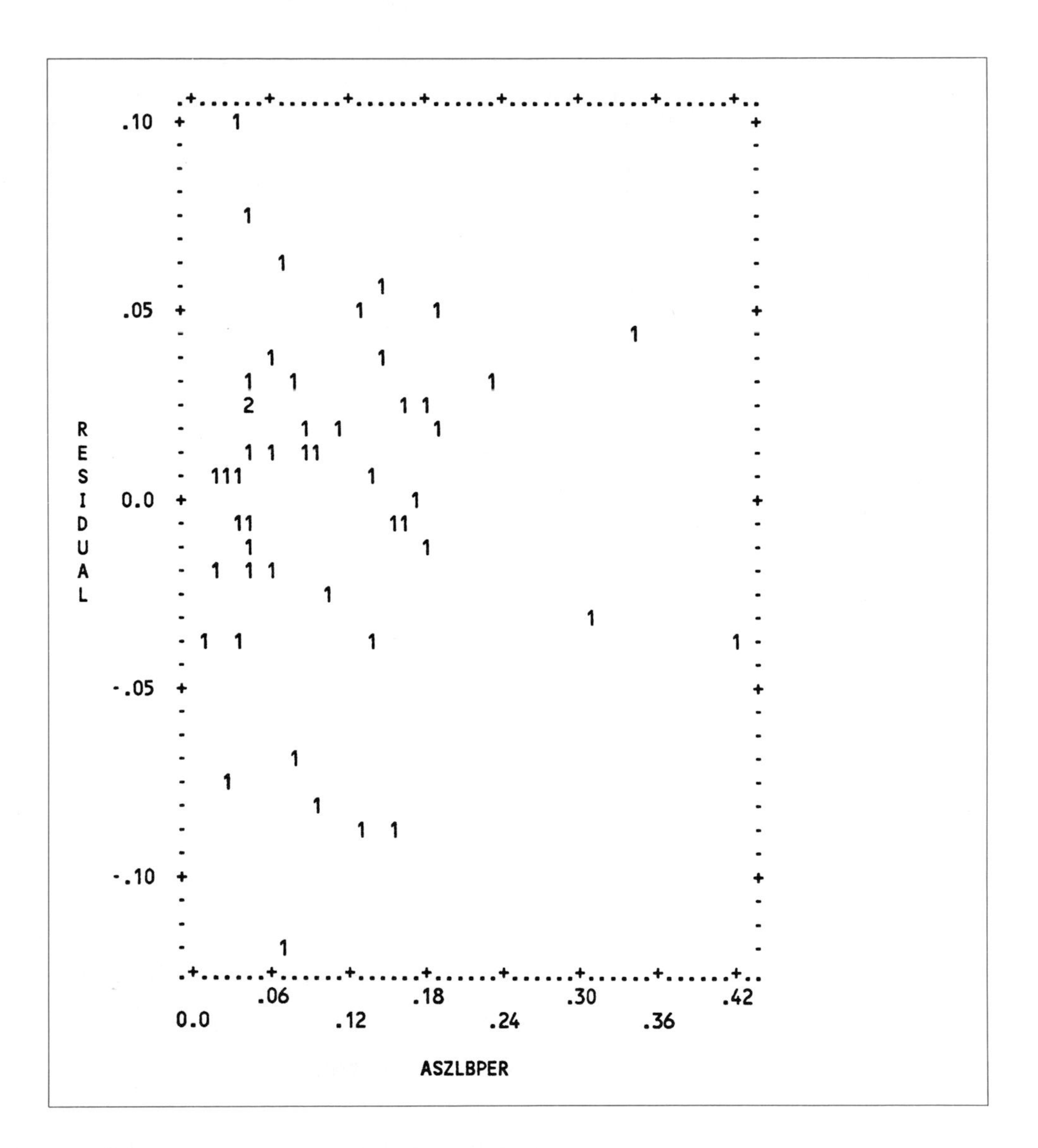
.10
.05
0.0
-.05
-.10
RESIDUAL
.06
.18
.30
.42
0.0
.12
.24
.36
ASZLBPER

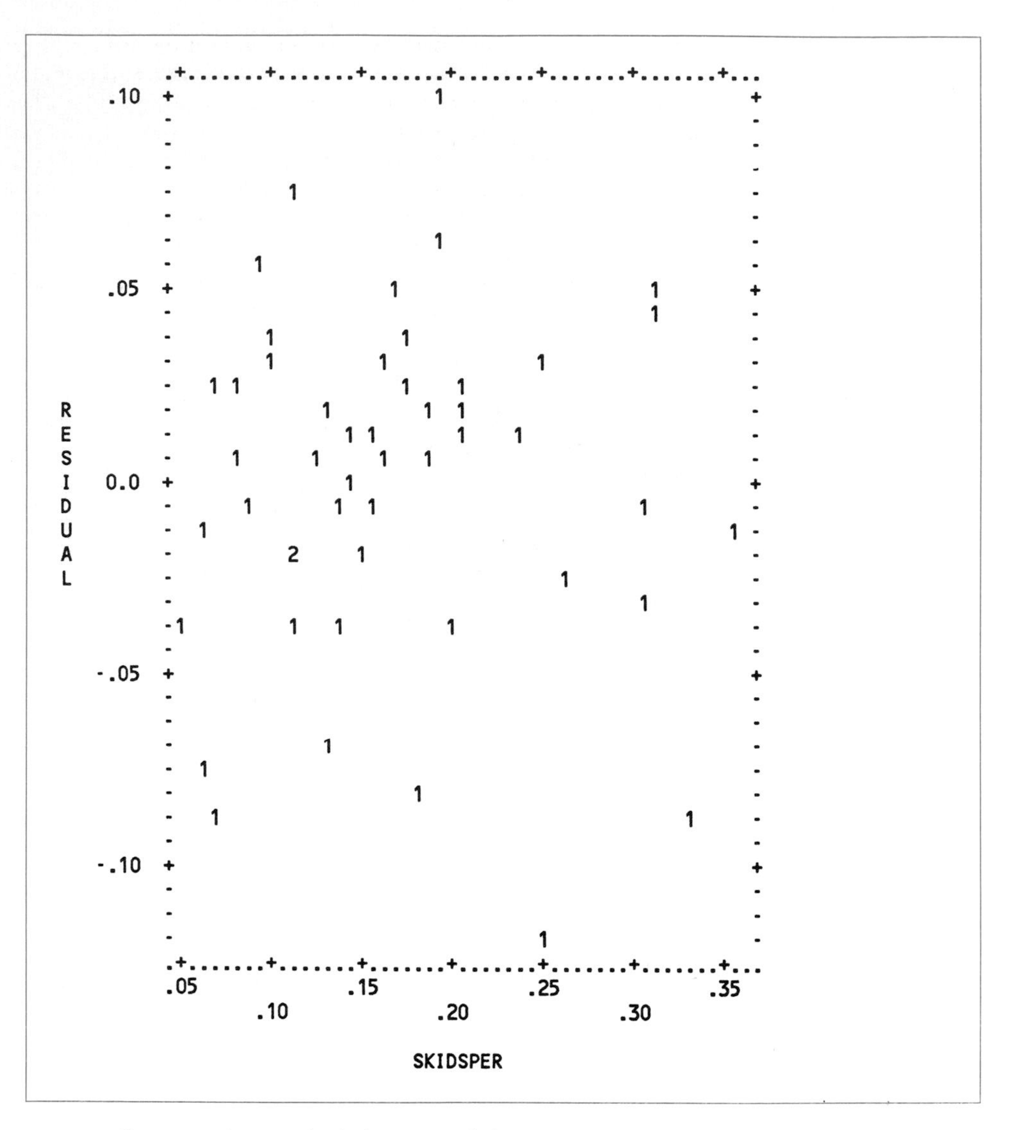

a. Compute the standard deviation of the regression prediction errors and the standard deviation of the superintendent's prediction errors.

b. Has the regression standard deviation increased from the standard deviation shown in Exercise 15.42?

c. Does the result of this study suggest that the regression model will yield better forecasts than the superintendent's forecasts?

CASE STUDY

15.44 The marketing managers of an office products company have some difficulty in evaluating the field sales representatives' performance. The representatives travel among the outlets for the company's products, creating displays, trying to increase volume, introducing new products, and discovering any problems that the outlets are having with the company's products. The job involves a great deal of travel time. The marketing managers believe that one important factor in the representatives' performance is the motivation to spend a great deal of time on the road. Other variables also have an effect. Some sales districts have more potential than others, either because of differences in population or differences in the number of retail outlets. Large districts are difficult because of the extra travel time.

One important variable is compensation. Some of the representatives are paid a salary plus a commission on sales; others work entirely for a larger commission on sales. The marketing managers suspect that there is a difference between the two groups in their effectiveness, although some of them argue that the important factor is the combination of commission status and number of outlets. In particular, they suspect that commission-only representatives with many outlets to cover are highly productive.

Data are collected on 51 representatives. The data include DISTRICT number, PROFIT, (net profit margin for all orders placed through the representative the dependent variable of interest), AREA (of the district in thousands of square miles), POPN (millions of people in the district), OUTLETS (number of outlets in the district) and COMMIS, which is 1 for full-commission representatives and 0 for partially salaried representatives.

Use the tabulated data to perform a multiple regression analysis. Find out if the variables suspected by the managers as having an effect on PROFIT actually do have an effect; in particular, try to discover if there is a combination effect of COMMIS and OUTLETS. Omit variables that show little predictive value. Locate and, if possible, correct any serious violations of assumptions. Write a brief report to the marketing managers explaining your findings; the managers are not familiar with the technical language of statistics, although they do have a fair idea of what a standard deviation is.

DIST	PROFIT	AREA	POPN	OUTLETS	COMMIS
1	1011	16.96	3.881	213	1
2	1318	7.31	3.141	158	1
3	1556	7.81	3.766	203	1
4	1521	7.31	4.587	170	1
5	979	19.84	3.648	142	1
6	1290	12.37	3.456	159	1
7	1596	6.15	3.695	178	1
8	1155	14.21	3.609	182	1
9	1412	7.45	3.801	181	1
10	1194	14.43	3.322	148	1
11	1054	6.12	5.124	227	0
12	1157	11.71	4.158	139	1
13	1001	9.36	3.887	179	0
14	831	19.14	2.230	124	1
15	857	11.75	4.468	205	0
16	188	40.34	.297	85	1
17	1030	7.16	4.224	211	0
18	1331	9.37	3.427	145	1
19	643	7.62	4.031	205	1
20	992	27.54	2.370	166	1
21	795	15.97	3.903	149	1
22	1340	12.97	3.423	186	1
23	689	17.36	2.390	141	0
24	1726	6.24	4.947	223	1
25	1056	11.20	4.166	176	0
26	989	18.09	4.063	187	1

(continued)

DIST	PROFIT	AREA	POPN	OUTLETS	COMMIS
27	895	13.32	3.105	131	1
28	1028	14.97	4.116	170	0
29	771	21.92	1.510	144	1
30	484	34.91	.741	126	1
31	917	8.46	5.260	234	0
32	1786	7.52	5.744	210	0
33	1063	14.43	2.703	141	1
34	1001	15.37	3.583	158	0
35	1052	11.20	4.469	167	1
36	1610	7.20	4.951	174	1
37	1486	13.49	3.474	211	1
38	1576	6.56	4.637	172	1
39	1665	9.35	3.900	185	1
40	878	11.12	3.766	166	0
41	849	10.58	3.876	189	0
42	775	17.82	2.753	164	0
43	1012	10.03	4.449	193	0
44	1436	10.01	4.680	157	1
45	798	10.70	4.806	200	0
46	519	24.38	2.367	142	0
47	1701	6.57	5.563	199	0
48	1387	6.64	4.357	166	1
49	1717	9.24	4.670	221	1
50	1032	11.62	3.993	180	0
51	973	12.85	3.923	193	0

REVIEW EXERCISES CHAPTERS 13–15

R133 A contractor bids on many small jobs. The current process of preparing bids is expensive and time-consuming. An attempt is made to predict y, the total direct cost of a job, based on x, the direct labor hours required. Data are collected on 26 jobs:

x:	214	228	235	239	247	248	278	289	291
y:	7444	7223	10,509	8931	9674	8084	11,784	10,067	11,344
x:	298	306	314	319	333	353	364	464	495
y:	7355	14,946	15,088	7409	15,475	11,524	13,209	16,012	22,570
x:	505	607	625	651	738	771	796	840	
y:	26,285	18,427	22,892	20,689	33,636	28,465	22,018	29,744	

The following sums are obtained:

$\sum x = 11{,}048 \quad \sum y = 410{,}804$

$\sum x^2 = 5{,}705{,}438 \quad \sum y^2 = 7{,}991{,}850{,}900$

$\sum xy = 209{,}738{,}350$

a. Calculate the least-squares regression equation.
b. What is the economic interpretation of the slope coefficient?

c. What is the economic interpretation of the intercept coefficient?
d. Calculate the residual standard deviation. Interpret its numerical value.

R134 Find the correlation between x and y for the data of Exercise R133. Interpret the resulting number.

R135 Refer again to Exercise R133.
a. Calculate the estimated standard error of the slope.
b. Find a 95% confidence interval for the true value of the slope.

R136 In Exercise R133, is the null hypothesis that the slope is zero economically plausible? Can this hypothesis be rejected conclusively by the data?

R137 a. The contractor in Exercise R133 has a new job with $x = 890$ hours. Calculate the predicted y value and a 90% prediction interval for the actual Y value.
b. The contractor has another new job with $x = 436$ hours. Calculate a 90% prediction interval for the actual Y value.
c. Which of the intervals calculated in parts (a) and (b) is wider? Why?

R138 A plot of the residuals from the regression analysis of the data in Exercise R133 against the x values shows that residuals corresponding to small x values are small positive or negative numbers, but that several of the residuals corresponding to large x values are relatively large in magnitude.
a. What regression assumption is called into question by this finding?
b. Which of your answers to Exercises R133–R137 is most questionable because of this finding?

R139 A bank that offers charge cards to customers studies the yearly purchase amount on the card as related to the age, income, and years of education of the cardholder, and whether or not the cardholder owns or rents a home. The following Minitab output is obtained; the variables are self-explanatory, except for OWNER, which equals 1 if the cardholder owns a home and 0 if the cardholder rents a home.

```
MTB > CORRELATIONS OF C1-C5

            PURCH        AGE    INCOME     OWNER
AGE         0.932
INCOME      0.928      0.837
OWNER       0.462      0.212     0.686
EDUCN       0.222      0.057     0.310     0.476

MTB > REGRESS C1 ON 4 VARIABLES C2-C5, RESIDS TO C19, PREDS TO C20

The regression equation is
PURCH = - 0.744 + 0.0329 AGE + 0.00900 INCOME + 0.115 OWNER + 0.00818 ED

Predictor        Coef          Stdev      t-ratio
Constant     -0.74439        0.06978       -10.67
AGE          0.032896       0.003809         8.64
INCOME       0.008999       0.005075         1.77
OWNER         0.11502        0.04982         2.31
EDUCN        0.008176       0.003611         2.26

s = 0.09263      R-sq = 94.6%      R-sq(adj) = 94.4%

Analysis of Variance

SOURCE        DF          SS          MS
Regression     4     23.1295      5.7824
Error        155      1.3301      0.0086
Total        159     24.4596
```

a. Locate the least-squares regression equation.
b. Explain what each slope coefficient means.
c. How meaningful is the intercept term?

R140 Refer to the output of Exercise R139.
a. What would the hypothesis that all slopes are zero mean about predictability in this context?
b. Show that this null hypothesis may be rejected emphatically.
c. What do the various t statistics indicate about the incremental predictive value of the variables?

R141 Is there evidence of collinearity in the data of Exercise R139?

R142 Additional predictor variables for the problem of Exercise R139 are created by multiplying OWNER by each of the other independent variables. Minitab output is shown below.

```
MTB > REGRESS C1 ON 7 VARS IN C2-C8 RESIDS TO C19 PREDS TO C20

The regression equation is
PURCH = - 0.916 + 0.0272 AGE + 0.0204 INCOME + 0.247 OWNER + 0.00536 EDU
           + 0.00043 X1X3 - 0.0066 X2X3 + 0.00510 X3X4

Predictor        Coef          Stdev      t-ratio
Constant      -0.9156         0.1332        -6.87
AGE          0.027249       0.005535         4.92
INCOME       0.020373       0.008714         2.34
OWNER          0.2468         0.2083         1.18
EDUCN        0.005361       0.005250         1.02
X1X3         0.000432       0.009902         0.04
X2X3         -0.00656        0.01312        -0.50
X3X4         0.005105       0.007254         0.70

s = 0.09217      R-sq = 94.7%      R-sq(adj) = 94.5%

Analysis of Variance

SOURCE         DF          SS          MS
Regression      7     23.1682      3.3097
Error         152      1.2914      0.0085
Total         159     24.4596

SOURCE         DF      SEQ SS
AGE             1     21.2586
INCOME          1      1.7745
OWNER           1      0.0524
EDUCN           1      0.0440
X1X3            1      0.0327
X2X3            1      0.0018
X3X4            1      0.0042
```

a. What is the reason for introducing the product terms X1X3, X2X3, and X3X4?
b. Test the null hypothesis that the coefficients of all the product terms are zero.

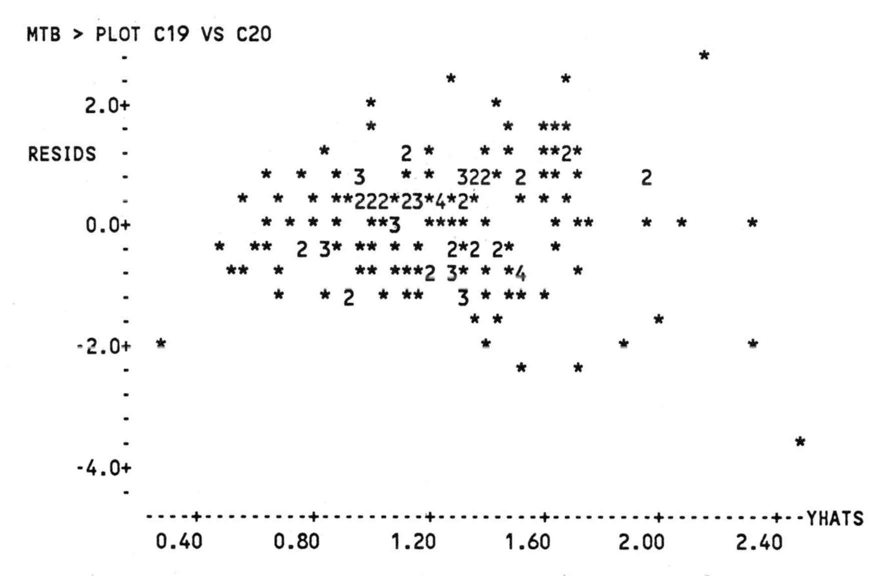

FIGURE R.1 Residuals versus Predicted Values; Exercises R142–R143

R143 Residuals for the model in Exercise R142 are plotted against predicted values, as shown in Figure R.1. Are there any obvious violations of regression assumptions?

R144 A revised regression model for Exercise R142 is attempted. The original variables in the study are each divided by INCOME, then a regression model is calculated for the transformed variables. A plot of the residuals against the predicted values is shown in Figure R.2. What problem is evident in the plot?

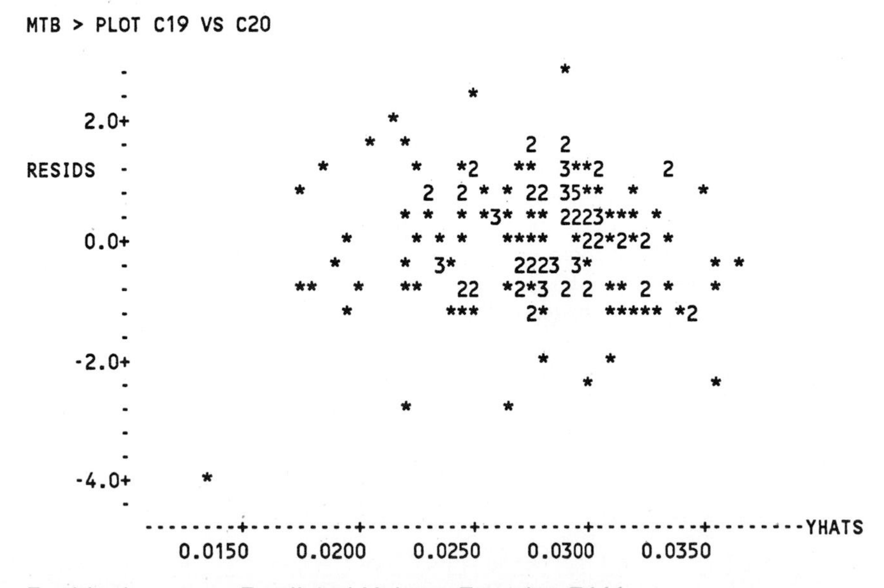

FIGURE R.2 Residuals versus Predicted Values; Exercise R144

R145 Another regression model for the data of Exercise R.142 is attempted using the natural logarithm of INCOME, rather than INCOME, as an independent variable. The output from Minitab is shown here:

```
MTB > REGRESS C1 ON 4 VARS C2 C6 C4 C5 RESIDS TO C19 YHATS TO C20

The regression equation is
PURCH = - 3.38 + 0.0247 AGE + 0.913 LOGINC + 0.0118 OWNER + 0.00784 EDUCN

Predictor        Coef          Stdev      t-ratio
Constant      -3.3834         0.6754        -5.01
AGE          0.024719       0.003765         6.57
LOGINC         0.9134         0.2273         4.02
OWNER         0.01183        0.04916         0.24
EDUCN        0.007844       0.003462         2.27

s = 0.08904      R-sq = 95.0%      R-sq(adj) = 94.8%

Analysis of Variance

SOURCE        DF          SS           MS
Regression     4     23.2306       5.8076
Error        155      1.2290       0.0079
Total        159     24.4596

SOURCE        DF      SEQ SS
AGE            1     21.2586
LOGINC         1      1.9296
OWNER          1      0.0016
EDUCN          1      0.0407
```

a. As compared to the original model, has predictive value improved?

b. As compared to the original model, is the incremental predictive value of the LOGINC variable greater?

c. Is the LOGINC variable more statistically significant than the INCOME variable in the original model?

R146 A regression model is constructed to predict the spread between interest rates of government bonds (risk-free) and corporate bonds rated BAA (somewhat risky). The independent variables are a privately constructed leading economic indicator (LEADING), a measure of the relative supply of corporate and governmental bonds in a given month (SUPPLY), the actual rate of government bonds in that month (RATE), and the month number (MONTH). SAS output is shown on the top of the next page.

CORRELATIONS OF ORIGINAL VARIABLES

PEARSON CORRELATION COEFFICIENTS / PROB > |R| UNDER H0:RHO=0 / N = 24

	SPREAD	LEADING	SUPPLY	RATE	MONTH
SPREAD	1.00000	-0.85042	0.67591	0.87240	-0.86171
	0.0000	0.0001	0.0003	0.0001	0.0001
LEADING	-0.85042	1.00000	-0.22698	-0.98189	0.99688
	0.0001	0.0000	0.2862	0.0001	0.0001
SUPPLY	0.67591	-0.22698	1.00000	0.25868	-0.24415
	0.0003	0.2862	0.0000	0.2223	0.2502
RATE	0.87240	-0.98189	0.25868	1.00000	-0.98638
	0.0001	0.0001	0.2223	0.0000	0.0001
MONTH	-0.86171	0.99688	-0.24415	-0.98638	1.00000
	0.0001	0.0001	0.2502	0.0001	0.0000

Is there evidence of collinearity in the data?

R147 A regression model is constructed using the data of Exercise R146, omitting MONTH. SAS output is shown below.

REGRESSION FOR ORIGINAL VARIABLES

DEP VARIABLE: SPREAD

ANALYSIS OF VARIANCE

SOURCE	DF	SUM OF SQUARES	MEAN SQUARE	F VALUE	PROB>F
MODEL	3	1.245096	0.4150321	321.310	0.0001
ERROR	20	0.02583373	0.001291686		
C TOTAL	23	1.27093			

ROOT MSE	0.03594004	R-SQUARE	0.9797
DEP MEAN	3.181728	ADJ R-SQ	0.9766
C.V.	1.129576		

PARAMETER ESTIMATES

VARIABLE	DF	PARAMETER ESTIMATE	STANDARD ERROR	T FOR H0: PARAMETER=0	PROB > |T|
INTERCEP	1	-0.0141918	2.156388	-0.007	0.9948
LEADING	1	-0.00826552	0.007199793	-1.148	0.2645
SUPPLY	1	3.75408	0.2566032	14.630	0.0001
RATE	1	0.2269971	0.07023813	3.232	0.0042

DURBIN-WATSON D	0.807
(FOR NUMBER OF OBS.)	24
1ST ORDER AUTOCORRELATION	0.518

a. Show that the null hypothesis that all slopes are zero can be rejected at any reasonable level of significance.

b. Do all the *t* tests of individual slopes lead to rejection of the null hypothesis?

R148 Locate the residual standard deviation in Exercise R147. What does it indicate about the predictive value of the equation? (The SPREAD variable has a standard deviation of about .2.)

R149 The residuals for the model in Exercise R147 are plotted against time in Figure R.3.

a. Does there appear to be a violation of regression assumptions?

b. Show that the output of Exercise R147 indicates a violation.

c. What are the consequences of this violation?

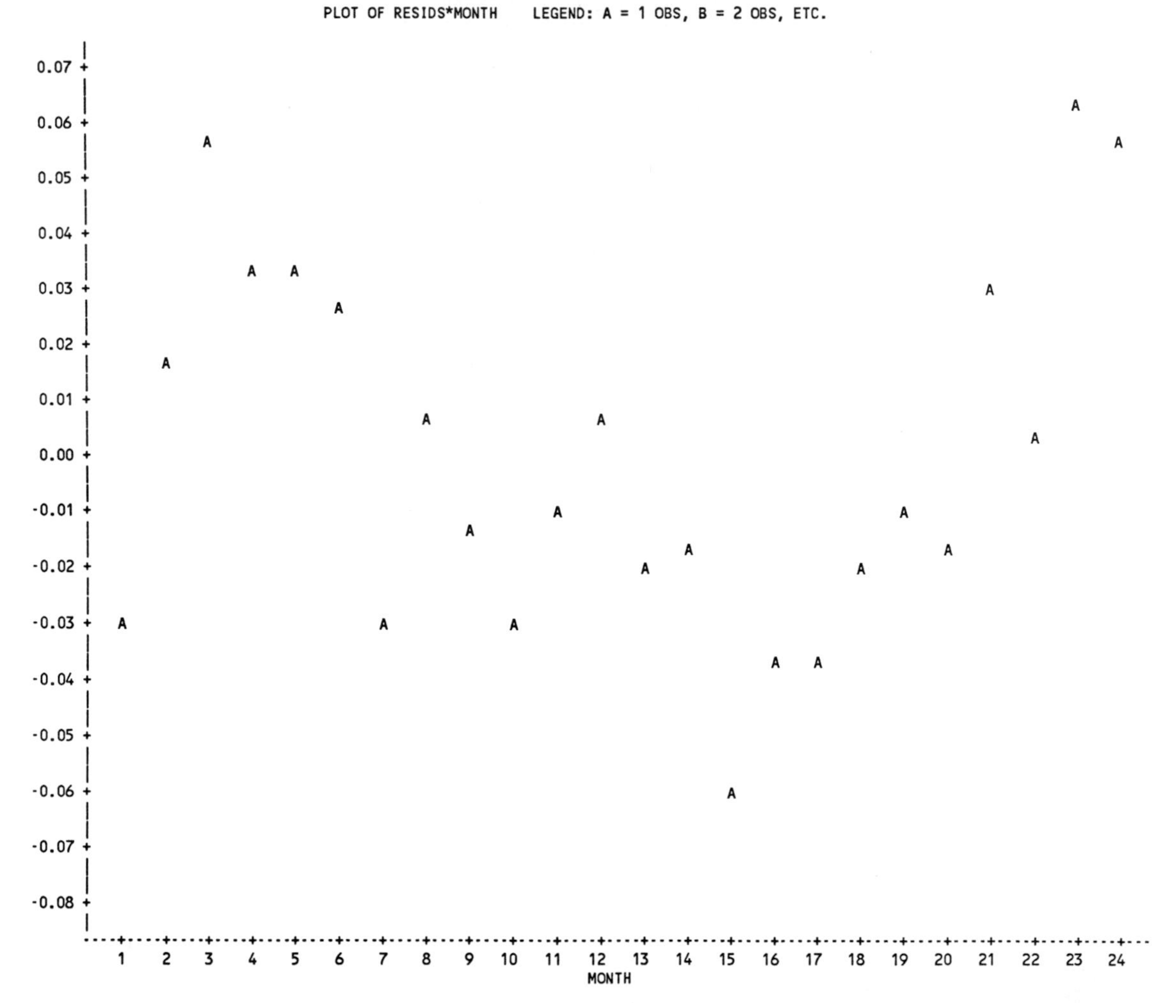

FIGURE R.3 **Residuals versus Time; Exercises R147–R149**

R150 The data of Exercise R147 are converted to differences and the model is recalculated. SAS output includes the following:

```
                              REGRESSION FOR DIFFERENCED VARIABLES

DEP VARIABLE: DSPREAD
                                    ANALYSIS OF VARIANCE

                                 SUM OF         MEAN
              SOURCE     DF     SQUARES        SQUARE        F VALUE       PROB>F

              MODEL       3   0.5252236     0.1750745        260.674       0.0001
              ERROR      19   0.01276081 0.0006716217
              C TOTAL    22   0.5379844

                  ROOT MSE    0.02591566       R-SQUARE       0.9763
                  DEP MEAN    -0.0260657       ADJ R-SQ       0.9725
                  C.V.          -99.4246

                                    PARAMETER ESTIMATES

                              PARAMETER        STANDARD      T FOR H0:
          VARIABLE  DF         ESTIMATE           ERROR    PARAMETER=0     PROB > |T|

          INTERCEP   1      0.004044871     0.009158125          0.442         0.6637
          DLEADING   1       -0.0106223     0.008161656         -1.301         0.2087
          DSUPPLY    1         3.351632       0.1313994         25.507         0.0001
          DRATE      1        0.2139063      0.03623192          5.904         0.0001

DURBIN-WATSON D             2.003
(FOR NUMBER OF OBS.)           23
1ST ORDER AUTOCORRELATION  -0.072
```

a. How much have the coefficients changed as compared to those in Exercise R147?

b. Have the results of *F* and *t* tests changed, as compared to the results in Exercise R147? If so, which set of tests are more believable?

c. Does it appear that working with differences has cured the violation of assumptions found in Exercise R149?

R151 The residuals from the model of Exercise R150 are plotted against month number in Figure R.4. Is there a clear pattern in this plot? Should there be, given the results in Exercise R150?

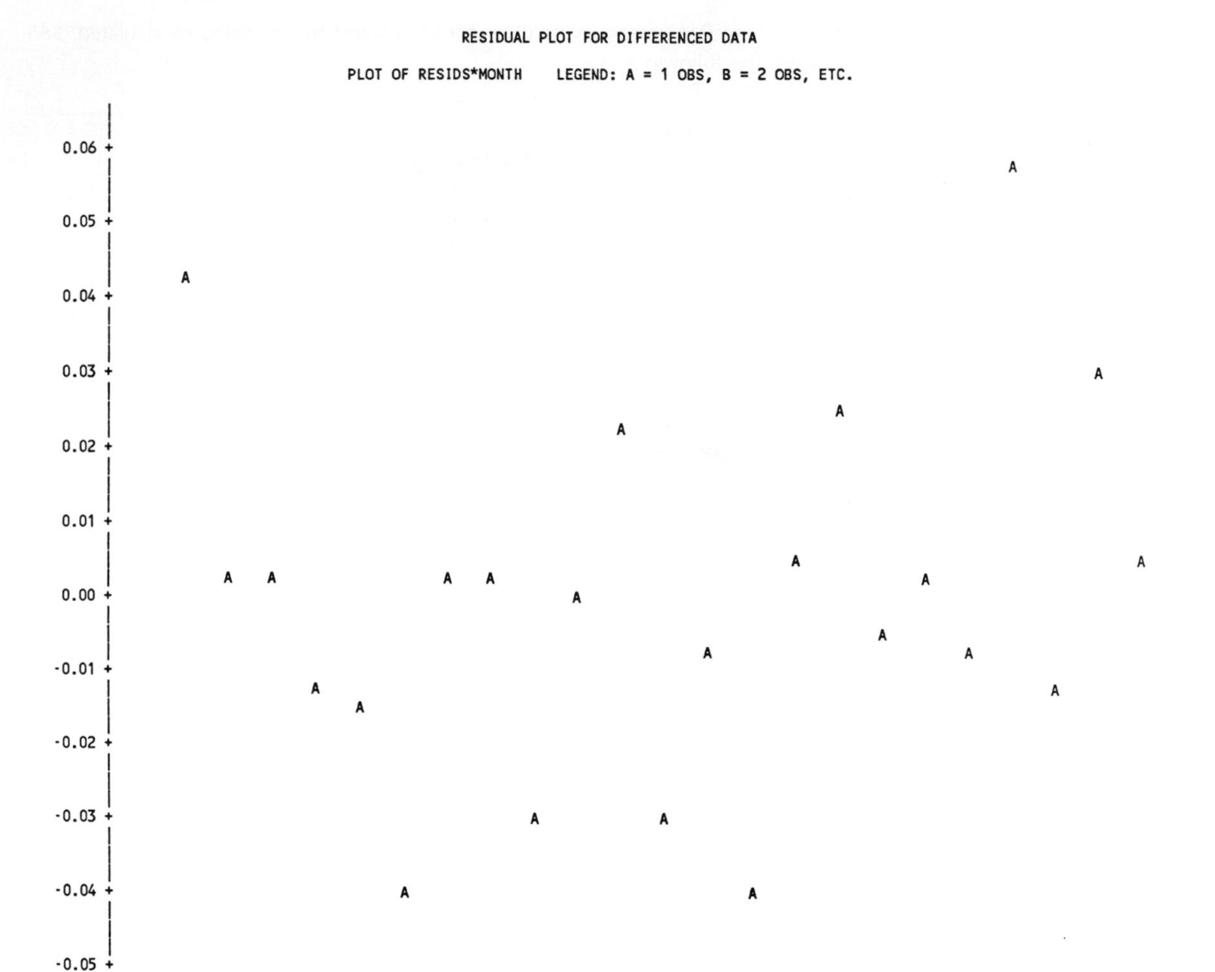

FIGURE R.4 **Residuals versus Time; Exercises R150–R151**

R152 Correlations for the difference data are shown below. Has differencing decreased collinearity?

```
                    CORRELATIONS OF DIFFERENCED VARIABLES

PEARSON CORRELATION COEFFICIENTS / PROB > |R| UNDER HO:RHO=0 / NUMBER OF OBSERVATIONS

                     DSPREAD DLEADING  DSUPPLY    DRATE

         DSPREAD     1.00000  0.25674  0.96542  0.34370
                      0.0000   0.2370   0.0001   0.1083
                          23       23       23       23

         DLEADING    0.25674  1.00000  0.29280  0.12873
                      0.2370   0.0000   0.1752   0.5583
                          23       23       23       23

         DSUPPLY     0.96542  0.29280  1.00000  0.14566
                      0.0001   0.1752   0.0000   0.5072
                          23       23       23       23

         DRATE       0.34370  0.12873  0.14566  1.00000
                      0.1083   0.5583   0.5072   0.0000
                          23       23       23       23
```

R153 An oil company does a small study of the sale of kerosene at its service stations. The dependent variable is the monthly sales (thousands of gallons) of kerosene. The independent variables are monthly sales of gasoline (thousands of gallons), average income in the census tract where the station is located, and a rough index of the amount of traffic past the station. The sample is of 21 company stations in a particular month. SPSS output is shown on the next page.

```
Correlation:

             KERO      GAS    AVGINC   TRAFFIC

KERO        1.000     .756     .374      .560
GAS          .756    1.000     .532      .566
AVGINC       .374     .532    1.000      .377
TRAFFIC      .560     .566     .377     1.000
```

```
Variable(s) Entered on Step Number  1..    GAS        GASOLINE SALES

Multiple R              .75594          Analysis of Variance

R Square                .57145                           DF      Sum of Squares      Mean Square

Adjusted R Square       .54890          Regression        1           457.42848        457.42848

Standard Error         4.24908          Residual         19           343.03818         18.05464

Variable(s) Entered on Step Number  2..    TRAFFIC    TRAFFIC COUNT

Multiple R              .77279          Analysis of Variance

R Square                .59720                           DF      Sum of Squares      Mean Square

Adjusted R Square       .55245          Regression        2           478.04254        239.02127

Standard Error         4.23231          Residual         18           322.42412         17.91245

                                        F =       13.34386       Signif F =  .0003

Variable(s) Entered on Step Number  3..    AVGINC     AVERAGE INCOME IN TERRITORY

Multiple R              .77451          Analysis of Variance

R Square                .59987                           DF      Sum of Squares      Mean Square

Adjusted R Square       .52925          Regression        3           480.17195        160.05732

Standard Error         4.34061          Residual         17           320.29472         18.84087

                                        F =        8.49522       Signif F =  .0011

---------------- Variables in the Equation -----------------

Variable              B          SE B         Beta          T  Sig T

GAS             .079375       .024098      .674541      3.294  .0043
TRAFFIC        7.861368      7.304205      .201466      1.076  .2968
AVGINC        -0.131951       .392495    -0.061280     -0.336  .7408
(Constant)    -3.332787      8.681572                  -0.384  .7058
```

a. Locate the least-squares regression equation. Interpret each of the slopes.

b. Negative sales are impossible; should we be concerned about a negative intercept?

R154 Refer to the output of Exercise R153.

a. Locate SS(Regression).

b. Find the sequential contribution of each independent variable to SS(Regression).

c. What do these results suggest about deleting one or more independent variables from the model?

d. Do the results of *t* tests also suggest deleting one or more independent variables from the model?

R155 a. Does the output in Exercise R153 indicate a serious collinearity problem?

b. Does the output indicate a serious autocorrelation problem? Would one expect autocorrelation in this study?

R156 A plot of residuals versus predicted values is shown in Figure R.5. Is there evidence of nonconstant variance? Are there any potentially serious outliers?

R157 It is discovered that station number 9 in the sample of Exercise R153 has a rather large contract to supply kerosene to a group of stores. No other station known to the company has a similar contract. Therefore station 9 is deleted from the sample and the regression is recalculated. The variables are given an N- (no contracts) prefix. Output is shown on the next page.

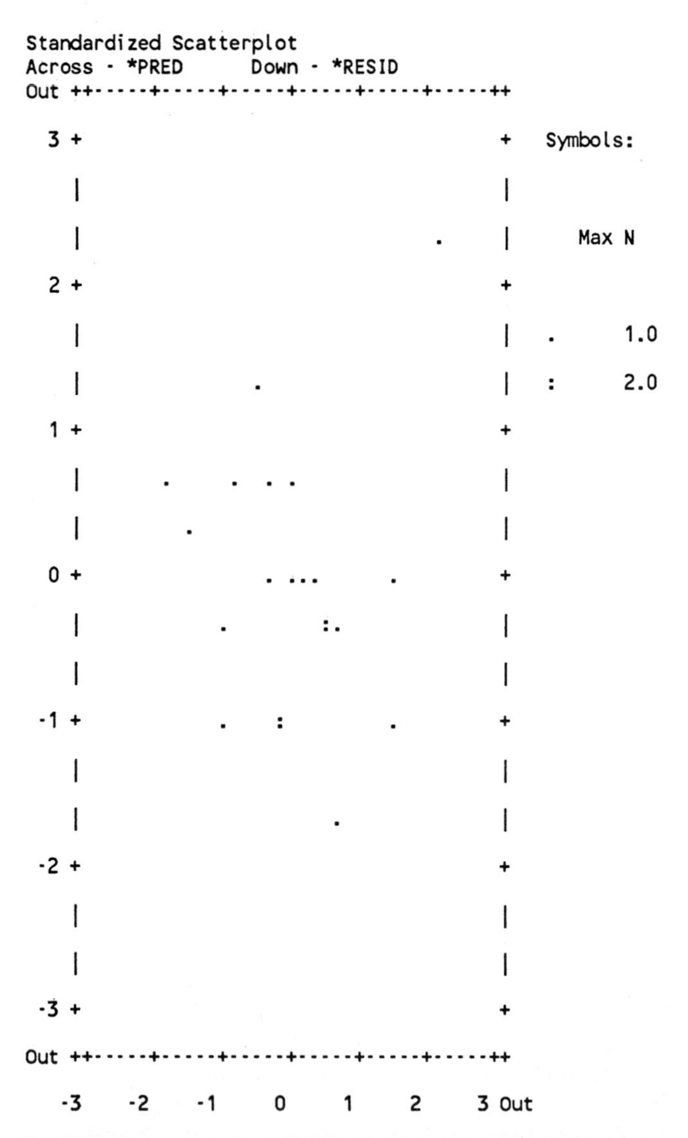

FIGURE R.5 Residuals versus Predicted Values; Exercises R153–R156

```
Correlation:

              NKERO      NGAS   NAVGINC  NTRAFFIC

NKERO         1.000      .693      .163      .383
NGAS           .693     1.000      .440      .452
NAVGINC        .163      .440     1.000      .258
NTRAFFIC       .383      .452      .258     1.000

Variable(s) Entered on Step Number  1..    NGAS        GASOLINE SALES

Multiple R             .69313        Analysis of Variance

R Square               .48044                          DF     Sum of Squares     Mean Square

Adjusted R Square      .45157        Regression         1          167.72082       167.72082

Standard Error        3.17439        Residual          18          181.38118        10.07673

                                     F =       16.64437       Signif F =  .0007

Variable(s) Entered on Step Number  2..    NAVGINC     AVERAGE INCOME IN TERRITORY

Multiple R             .71084        Analysis of Variance

R Square               .50529                          DF     Sum of Squares     Mean Square

Adjusted R Square      .44709        Regression         2          176.39843        88.19922

Standard Error        3.18732        Residual          17          172.70357        10.15903

                                     F =        8.68185       Signif F =  .0025

Variable(s) Entered on Step Number  3..    NTRAFFIC    TRAFFIC COUNT

Multiple R             .71649        Analysis of Variance

R Square               .51335                          DF     Sum of Squares     Mean Square

Adjusted R Square      .42211        Regression         3          179.21275        59.73758

Standard Error        3.25854        Residual          16          169.88925        10.61808

                                     F =        5.62603       Signif F =  .0079

----------------- Variables in the Equation -----------------

Variable              B          SE B         Beta          T    Sig T
NGAS            .064006       .018546      .727894      3.451    .0033
NAVGINC       -0.279260       .297238    -0.182919     -0.940    .3614
NTRAFFIC       2.903315      5.639369      .100943       .515    .6137
(Constant)     7.817130      7.159078                   1.092    .2910
```

a. Have the regression slopes changed because of the omission of station 9?
b. How has the residual standard deviation changed?
c. How has the coefficient of determination changed?

R158 Refer to the output in Exercise R157. Test the null hypothesis that the coefficients of NAVGINC and NTRAFF are both zero. Can this hypothesis be rejected at the usual α values?

R159 Data are collected on the yield of a chemical under various combinations of temperature and pressure. The data were as follows:

TEMP	PRES	YIELD	TEMP	PRES	YIELD	TEMP	PRES	YIELD
2200	3.8	75.50	2250	3.8	76.80	2300	3.8	78.50
2200	4.2	77.90	2250	4.2	79.20	2300	4.2	80.20
2200	3.8	75.90	2250	3.8	76.00	2300	3.8	78.80
2200	4.2	77.90	2250	4.2	78.90	2300	4.2	80.20

a. Plot TEMP versus PRES.
b. What must the value of $r_{\text{TEMP, PRES}}$ be?
c. How severe is the collinearity problem for these data?

R160 A regression model is fit to the data of Exercise R159. The output is shown here.

```
MTB > REGRESS C3 ON 2 VARS IN C1 C2 PUT RESIDS IN C19 YHATS IN C20

The regression equation is
YIELD = - 2.41 + 0.0262 TEMP + 5.33 PRES

Predictor        Coef          Stdev     t-ratio
Constant       -2.412          6.915       -0.35
TEMP         0.026250       0.002889        9.09
PRES           5.3333         0.5897        9.04

s = 0.4085        R-sq = 94.8%      R-sq(adj) = 93.7%

Analysis of Variance

SOURCE        DF          SS          MS
Regression     2      27.435      13.717
Error          9       1.502       0.167
Total         11      28.937

SOURCE        DF      SEQ SS
TEMP           1      13.781
PRES           1      13.653
```

a. Locate the prediction equation.
b. Roughly, what is the width of a 95% prediction interval for a value of YIELD? Assume that any extrapolation penalty can be neglected.

R161 Perform F and t tests for Exercise R160. What do the results indicate about the predictive value of TEMP and PRES?

R162 A residual plot against values of TEMP for the model of Exercise R160 is shown in Figure R.6. Is there any evidence of a problem shown by this plot?

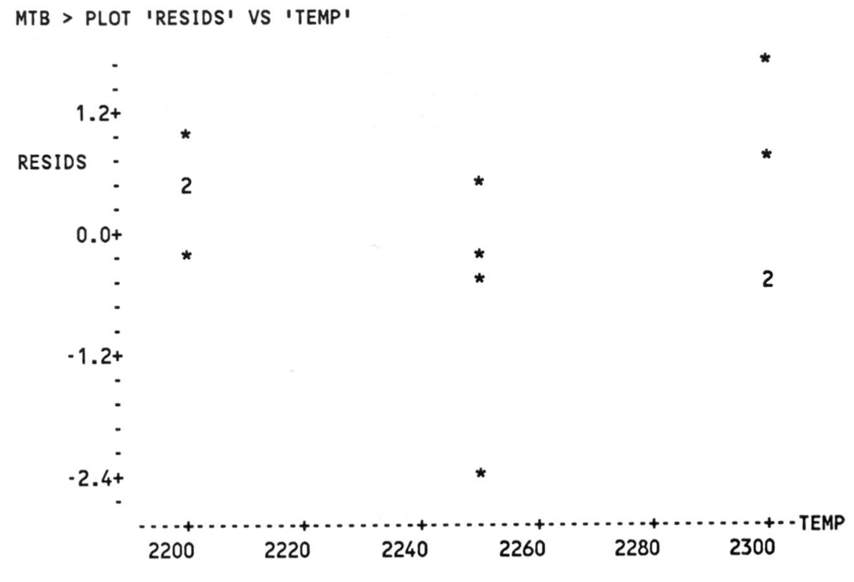

FIGURE R.6 Residuals versus Temperature; Exercises R159–R162

R163 Based on the plot of Exercise R162, a new variable, TEMPSQ = $(\text{TEMP} - 2250)^2$, is created. A new regression analysis is performed. The output is shown here:

```
MTB > REGRESS C3 ON 3 VARS C1 C2 C4 RESIDS TO C19 YHATS TO C20

The regression equation is
YIELD = - 2.67 + 0.0262 TEMP + 5.33 PRES +0.000155 TEMPSQ

Predictor        Coef          Stdev     t-ratio
Constant       -2.671          6.283       -0.43
TEMP         0.026250       0.002624       10.00
PRES           5.3333         0.5356        9.96
TEMPSQ     0.00015500     0.00009090        1.71

s = 0.3711       R-sq = 96.2%      R-sq(adj) = 94.8%

Analysis of Variance

SOURCE        DF          SS           MS
Regression     3     27.8350       9.2783
Error          8      1.1017       0.1377
Total         11     28.9366

SOURCE        DF      SEQ SS
TEMP           1     13.7812
PRES           1     13.6533
TEMPSQ         1      0.4004
```

a. Have the coefficients of TEMP and PRESS changed much?

b. Does adding TEMPSQ greatly improve the prediction of YIELD? Indicate the parts of the output that support your judgment.

R164 A store manager for a supermarket chain does a regression study of the weekly sales of the store and the volume of promotional activity (advertising and coupons) the previous week. Minitab output is shown below.

```
MTB > REGRESS C1 ON 1 VAR IN C2 RESIDS C19 YHATS IN C20;
SUBC> DW.

The regression equation is
SALES = - 7.8 + 5.31 PROMO

Predictor        Coef          Stdev     t-ratio
Constant        -7.79          24.95       -0.31
PROMO          5.3129         0.6382        8.33

s = 12.96       R-sq = 75.9%      R-sq(adj) = 74.8%

Analysis of Variance

SOURCE        DF         SS          MS
Regression     1      11647       11647
Error         22       3697         168
Total         23      15345

Durbin-Watson statistic = 0.43
```

a. Is there evidence of a statistically significant predictive value of PROMO for predicting SALES?

b. How much of the variability (squared error) of SALES is accounted for by PROMO?

R165 The output of Exercise R164 indicates that there is a serious violation of at least one assumption. What is it, how do you know that this violation has occurred, and what are the consequences of the violation on your answers in Exercise R164?

R166 Differences are calculated for the data of Exercise R164 and a new regression equation is found. The output is shown below.

```
MTB > REGRESS C11 ON 1 VAR IN C12 PUT RESIDS IN C19 YHATS IN C20;
SUBC> DW.

The regression equation is
C11 = - 1.70 + 5.02 C12

23 cases used 1 cases contain missing values

Predictor        Coef          Stdev      t-ratio
Constant       -1.696          1.737        -0.98
C12            5.0187         0.2764        18.16

s = 8.328        R-sq = 94.0%     R-sq(adj) = 93.7%

Analysis of Variance

SOURCE        DF          SS          MS
Regression     1       22870       22870
Error         21        1457          69
Total         22       24327

Durbin-Watson statistic = 1.43
```

a. Is there statistically significant predictive value of DPROMO (C12) for predicting DSALES?
b. How much of the variability of DSALES is accounted for by variation in DPROMO (C11)?
c. How do your answers here compare to the answers of Exercise R164?

R167 Did the use of differences eliminate the violation of assumptions in Exercise R165?

R168 In a paper mill, a liquid slurry of wood fibers is forced through a screen. The yield of fibers is known to increase as the difference in pressure between the two sides of the screen increases. Data are collected and a regression equation found. The output is shown here.

```
                              REGRESSION WITH ORIGINAL DATA

DEP VARIABLE: YIELD
                                    ANALYSIS OF VARIANCE

                              SUM OF         MEAN
           SOURCE     DF     SQUARES       SQUARE      F VALUE      PROB>F

           MODEL       1    2455.976     2455.976      414.234      0.0001
           ERROR      28    166.0107     5.928955
           C TOTAL    29    2621.987

              ROOT MSE      2.434945     R-SQUARE       0.9367
              DEP MEAN         78.11     ADJ R-SQ       0.9344
              C.V.          3.117328

                                    PARAMETER ESTIMATES

                        PARAMETER       STANDARD      T FOR H0:
      VARIABLE  DF       ESTIMATE          ERROR    PARAMETER=0     PROB > |T|

      INTERCEP   1       60.78444      0.9603551         63.294         0.0001
      CHGPRES    1      0.6300202     0.03095505         20.353         0.0001
```

a. What would the null hypothesis that the true slope is zero mean in this situation?
b. Show that this hypothesis can be conclusively rejected.

R169 Locate and interpret the residual standard deviation in the output of Exercise R168.

R170 A plot of the residuals from Exercise R168 versus values of CHGPRES is shown in Figure R.7.
a. Is there an indication that a nonlinear equation may give a better fit to the data?
b. Are there any severe outliers?

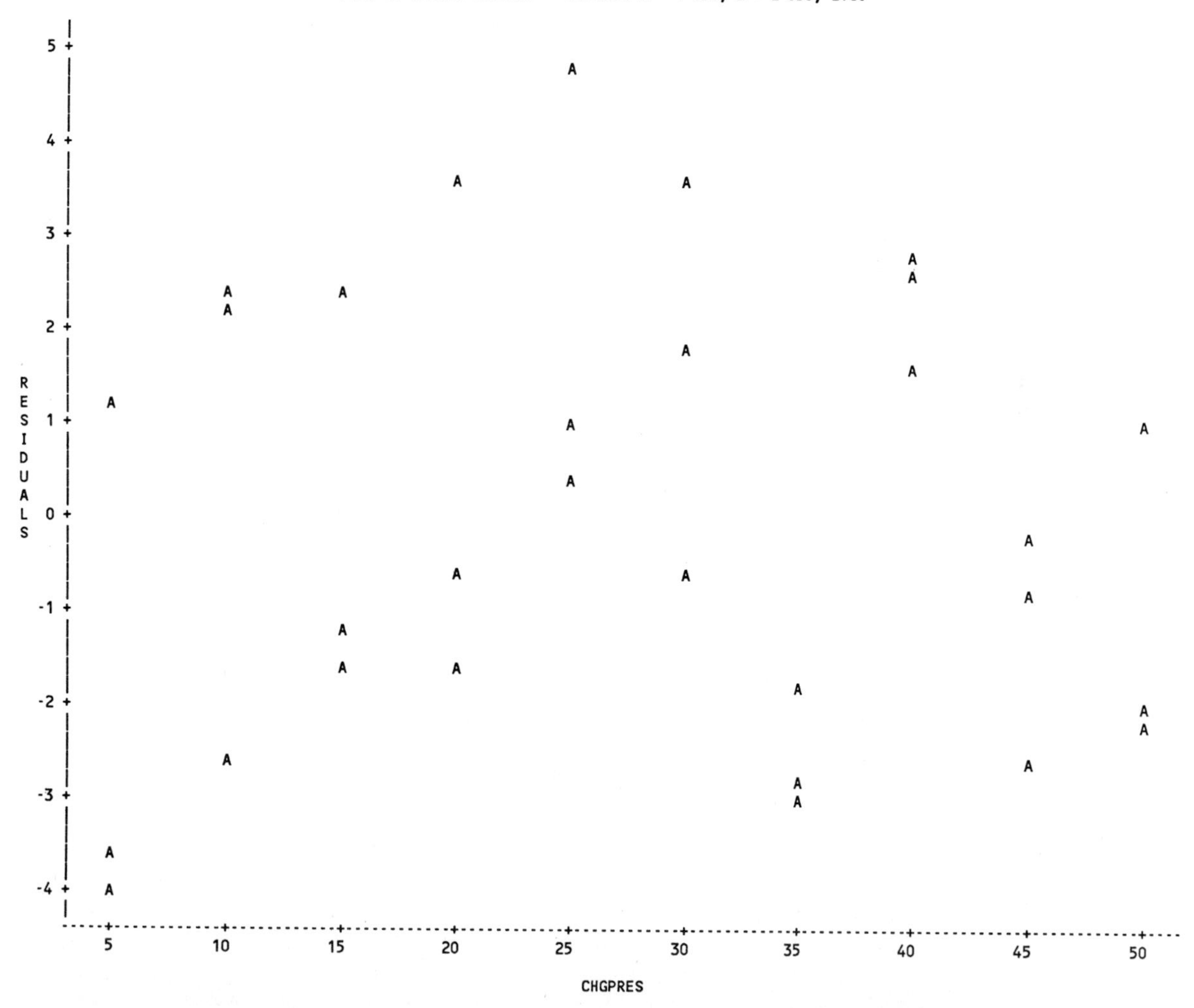

FIGURE R.7 Residuals versus Change in Pressure; Exercises R168–R170

R171 The square root of the pressure change is calculated for the data of Exercise R168 and a new regression is calculated. The output is as shown here.

```
                              REGRESSION WITH TRANSFORMED DATA

DEP VARIABLE: YIELD
                                  ANALYSIS OF VARIANCE

                              SUM OF         MEAN
          SOURCE     DF      SQUARES        SQUARE       F VALUE       PROB>F

          MODEL       1       2474.3        2474.3       469.102       0.0001
          ERROR      28     147.6871      5.274541
          C TOTAL    29     2621.987

             ROOT MSE       2.296637      R-SQUARE        0.9437
             DEP MEAN          78.11      ADJ R-SQ        0.9417
             C.V.            2.94026

                                  PARAMETER ESTIMATES

                         PARAMETER          STANDARD      T FOR H0:
      VARIABLE  DF        ESTIMATE             ERROR    PARAMETER=0      PROB > |T|

      INTERCEP   1        47.75158          1.463042         32.639          0.0001
      SQRTPRES   1        6.042607         0.2789912         21.659          0.0001
```

Has the square-root transformation improved the fit of the model?

16

TIME-SERIES ANALYSIS

One part of statistical analysis of great relevance to managers is analysis of **time series** data collected on the same variable for many time periods. From macroeconomic data such as disposable income to microeconomic data such as weekly sales of one particular product at one particular store, time series are basic data for managers. In Chapter 15 we discussed how multiple regression methods can be applied to time-series data. The emphasis there was on the special problems that such data presented, particularly the problem of autocorrelated errors.

If time-series data present problems, they also present opportunities. One of the very best predictors of the future behavior of a variable is its past behavior. Intelligent analysis of time-series data can often yield insights that help in understanding and predicting the future values of the series. In this chapter we discuss some useful ways of looking at time-series data.

The key aspect of this chapter is that the behavior of a time series is considered pretty much in isolation. We don't look at prediction of one time series by other series, as we did in regression. Instead, we use past values of a time series to predict future values of the same series. One major virtue of this approach is that the data requirements are minimal. As long as the series have been defined consistently over time (as, unfortunately, some series aren't), there's no need to bring together many different sets of data. Therefore the methods in this chapter are very useful in situations such as inventory control, where very many forecasts must be made and the value of a little extra accuracy is not huge.

One part of time-series analysis is the construction of index numbers to reflect changes in prices over time. Methods for constructing and using such indices are discussed in Section 16.1. In Section 16.2 we turn to the classical breakdown of time series into trend, seasonal, cyclic, and irregular components. In Section 16.3 we discuss some smoothing methods that can be used as simple forecasting methods and also as methods for clearing up the effect of irregular, random movements of the series. The Box-Jenkins methods of Section 16.4 are potentially very valuable, both in the context of a single series and as a way to combine concepts of regression and time series.

16.1 INDEX NUMBERS

Index numbers are probably the best-known results of statistical reasoning. The value of the Dow-Jones Industrial Index is reported daily by newspapers and broadcasts. News radio stations have taken to interrupting their regular broadcasts to give the monthly Consumer Price Index as soon as it is released. Many cost-of-living adjustments are based on such indices, as are Social Security payments. In this section we discuss some basic ideas of index-number construction, with primary emphasis on price indices.

There are essentially two types of indices—price indices and quantity indices. A price index measures the change in the prices of a group of items over time; a quantity index measures the change in the amount of a group of goods or services produced over time. Stock market indices such as the Dow Jones Industrial Average or the Standard and Poor's 500 are price indices; production indices for automobiles, steel, and the like are quantity indices.

A price index is useful in its own right as an indicator of the general level of prices. Also, measurements of general economic activity such as the gross national product are divided by a price index to separate the effect of price changes from actual changes in activity. If all prices change by a constant percentage, any price index serves these functions; the interpretation of a price index gets more delicate when prices vary differently.

Once again we work by example. Suppose that we want to construct an index for the price of entertainment. The following price data are available:

	Price		
Item	1984	1985	1986
Color television set 19″ console	$500	$550	$525
Movie ticket suburban theater	$4.00	$4.40	$4.50
Baseball ticket reserved seat	$4.00	$4.40	$5.00
Theater admission season subscription	$40.00	$44.00	$60.00

There is no problem when comparing 1984 and 1985 entertainment prices. Every price shown increased by 10% from 1984 to 1985. No matter how the entertainment index is calculated, if 1984 is taken as the base year with an index value of 100, the 1985 index value must be 110, a 10% increase.

There is a problem with comparing 1986 to 1984 prices, because the price changes are not consistent. Color televisions are up 5% (1986 over 1984), movie tickets 12.5%, baseball tickets 25%, and theater admissions 50%. One solution, not a very good one, is to take the ratio of total prices in 1986 to total prices in 1984. Conventionally, this ratio is multiplied by 100 so that the index is stated in percentage terms. This index value is

$$\frac{525 + 4.50 + 5.00 + 60.00}{500 + 4.00 + 4.00 + 40.00} \times 100 = 108.5$$

indicating an 8.5% rise in prices from 1984 to 1986. This is an example of a simple aggregate index. A formal definition follows:

Simple Aggregate Price Index

A simple aggregate price index for year k, denoted I_k, is the ratio of a sum of prices in year k to the sum of prices in the base year (year 0), expressed as a percentage:

$$I_k = \frac{\sum p_{ki}}{\sum p_{0i}} \times 100$$

where p_{ki} and p_{0i} are the prices of item i in year k and year 0 respectively.

The Dow-Jones Industrial Index began as a simple aggregate index of the prices of 30 blue chip stocks. So many modifications have been made over the years that it is no longer recognizable as such an index.

EXAMPLE 16.1 Closing 1984 and 1985 stock prices for a group of nonprescription drug wholesalers are shown here. Construct a simple aggregate price index for 1985 using 1984 as the base year.

	Closing Stock Price	
Wholesaler	1984	1985
Begley	$15\frac{1}{2}$	$16\frac{1}{4}$
Bindley	$9\frac{7}{8}$	19
Durr-Fillauer	$10\frac{1}{4}$	$15\frac{1}{4}$
Ketchum	$15\frac{1}{4}$	$20\frac{1}{4}$
Med. Shoppe	19	24

Solution

$$I_{1985} = \frac{16\frac{1}{4} + 19 + 15\frac{1}{4} + 20\frac{1}{4} + 24}{15\frac{1}{2} + 9\frac{7}{8} + 10\frac{1}{4} + 15\frac{1}{4} + 19} = 1.356$$

A stockholder owning one share of each wholesaler would have seen a 35.6% increase in the price of the portfolio. □

The obvious objection to a simple aggregate index is that it does not reflect the amounts of various goods that are typically purchased. Our entertainment price index implicitly assumes that the relevant consumer buys as many color televisions as movie tickets. For most purposes, it is more reasonable to weight each price by an appropriate quantity to form a weighted aggregate index.

Weighted Aggregate Price Index

A weighted aggregate price index for year k is the ratio of a weighted sum of prices in year k to a weighted sum of prices in the base year 0 expressed as a percentage. The weights q_i are appropriately chosen quantities of each item:

$$I_k = \frac{\sum p_{ki}q_i}{\sum p_{0i}q_i} \times 100$$

EXAMPLE 16.2 In constructing an entertainment index using the data in this section, assume that a representative family buys a new color television set every eight years, ten movie tickets and two baseball tickets per year, and a subscription theater admission once every five years. Calculate the weighted aggregate price index values for 1985 and 1986.

Solution The appropriate quantities can all be translated into amounts per year; for television sets the quantity is $1/8 = .125$ per year, for movie and baseball tickets ten and two per year, and for theater admissions $1/5 = .2$ per year. The price index value for 1985 is

$$I_{1985} = \frac{550(.125) + 4.40(10) + 4.40(2) + 44.00(.2)}{500(.125) + 4.00(10) + 4.00(2) + 40.00(.2)} \times 100 = 110.0$$

The price index value for 1986 is

$$I_{1986} = \frac{525(.125) + 4.50(10) + 5.00(2) + 60(.2)}{500(.125) + 4.00(10) + 4.00(2) + 40(.2)} = 111.9$$

The 1985 index value reflects the uniform 10% price increase. The 1986 index value (111.9) is substantially higher than the simple aggregate value (108.5) found previously. The basic reason is that the price of television sets increased relatively little. This price receives much less weight in the weighted aggregate index than it does in the simple aggregate index, while the larger increases receive more weight. □

The major difficulty in defining a weighted aggregate index is, of course, the specification of appropriate items and the weights attached to the items. The Bureau of Labor Statistics, in preparing the Consumer Price Index, selected about 300 carefully defined items. The quantities q_i are obtained by a sample of wage earners and their families. The selection of items and quantities inevitably becomes less appropriate as time goes by. Thus almost any index must be revised periodically.

choice of weights

In a sense, the whole exercise of fixed-quantity weighted price indices is a denial of elementary economics. When a particular commodity becomes more expensive, we buy less of it. When a new product of superior quality appears on the market, we change our purchasing behavior. On the other hand, if product quality declines, we may receive less utility for a given expenditure. Ideally, a price index would not be so closely tied to specific products but rather would reflect the cost of obtaining certain goals. Instead of basing a food price index on the price of so much steak, so much beans, and so many apples, we would prefer a food price index based on the total cost of a diet meeting specified nutrition and taste standards. The difficulties involved in pinning down such elusive goals, however, are formidable. At least the current fixed-quantity approach to price indices provides an objective, if somewhat arbitrary, standard. As we noted earlier in this section, the choice of weights matters only to the extent that price changes differ among products. All in all, we suggest that price indices should be treated like any other statistical quantities—as estimates that are subject to error.

SECTION 16.1 EXERCISES

16.1 Data for a price index based on eight commodities are collected over a three-year period, yielding the following results:

	Commodity							
	1	2	3	4	5	6	7	8
Price								
year 1	6.00	7.25	6.60	10.50	4.60	12.50	25.00	300.00
year 2	6.58	7.80	7.25	11.58	5.05	13.75	27.48	329.50
year 3	7.25	8.59	7.99	12.75	5.50	15.10	30.21	362.45
Quantity								
per year	6.25	4.00	2.50	1.00	.80	.25	.10	.01

a. Compute simple aggregate price indices for years 2 and 3, using year 1 as the base period.
b. Compute weighted aggregate price indices for years 2 and 3, using year 1 as the base period.

16.2 An analyst of the computer-calculator industry gathers data on the cost and quantity sold of calculating and computing devices for a four-year period. The data are collected in six major categories, as follows:

	Category					
	A	B	C	D	E	F
Year 1						
price	20.00	50.68	989	35,416	195,626	651,928
quantity	2,060,000	121,200	86,104	82,147	21,047	1,306
Year 2						
price	18.64	48.21	1,021	37,215	206,114	721,200
quantity	2,547,000	142,900	89,216	81,021	21,926	1,339
Year 3						
price	16.93	47.03	1,096	40,462	215,963	790,087
quantity	2,997,000	163,800	95,114	76,050	20,875	1,575
Year 4						
price	16.61	46.89	1,129	41,943	229,120	864,326
quantity	3,451,000	177,500	94,397	71,194	19,975	1,498

a. Compute simple aggregate price indices for years 2, 3, and 4, using year 1 as the base period.
b. Compute weighted aggregate price indices for these years, using year 1 quantities as weights. (Base-year weights are used in Laspeyres indices.)
c. Compute weighted aggregate price indices for these years, using the quantities for each year as weights for that year. (Current-year weights as used in Paasche indices.)

16.3 Refer to the price indices computed in Exercise 16.2.

a. Is there a serious difference between the unweighted, simple aggregate index and the index that is weighted by yearly quantities? Why?
b. Is there a serious difference between the year 1 weighted index and the year 4 weighted index? Why?

16.4 A manufacturer of small electric appliances uses mostly stainless steel sheets, copper wire, a plastic substance, and glass. As a part of its pricing policy, the company routinely calculates a raw-material cost index. Price data (P) for the previous five years and utilization data (Q) are as follows. The quantities are measured in special units used only by the manufacturer.

	Steel		Copper		Plastic		Glass	
Year	P	Q	P	Q	P	Q	P	Q
1	280	220	450	30	24	40	48	21
2	306	244	430	34	38	43	51	24
3	327	256	492	36	36	45	53	25
4	360	280	582	38	39	49	54	30
5	396	308	573	44	44	52	56	35

a. Compute simple aggregate price index values for years 2–5, using year 1 as the base period.
b. Compute weighted aggregate price indices (using year 1 quantity weights) for these years.
c. Explain any major discrepancies you find between the two sets of values.

16.2 THE CLASSICAL TREND, CYCLIC, AND SEASONAL APPROACH

One way to examine a time series is to break it into components. A standard approach is to find components corresponding to a long-term trend, any cyclic behavior, seasonal behavior, and a residual, irregular part. For example, data have been collected on the ridership of the PATCO public transit line running between southern New Jersey and Philadelphia. The data (supplied by Bruce Allen of the Wharton School) are recorded in 13 four-week blocks per year:

	Year						
Period	'78	'79	'80	'81	'82	'83	'84
1	835,938	811,652	866,418	833,187	822,984	882,712	820,304
2	869,525	833,518	862,057	891,476	860,086	854,016	806,592
3	886,450	872,598	880,454	914,372	873,017	861,124	807,453
4	898,116	854,777	891,191	893,255	853,588	832,343	804,755
5	867,477	866,441	888,444	897,121	881,955	824,050	800,477
6	816,262	808,512	832,906	844,012	800,938	800,650	754,132
7	798,688	856,082	836,486	841,386	846,146	763,825	750,522
8	796,355	831,873	835,888	847,642	805,389	778,466	
9	751,609	793,173	806,469	832,761	795,165	747,695	
10	860,696	916,420	927,439	858,855	832,338	844,658	
11	867,974	899,469	973,009	897,555	875,602	870,317	
12	833,632	865,423	872,488	849,891	838,034	814,071	
13	844,517	868,392	860,692	862,413	838,066	797,018	

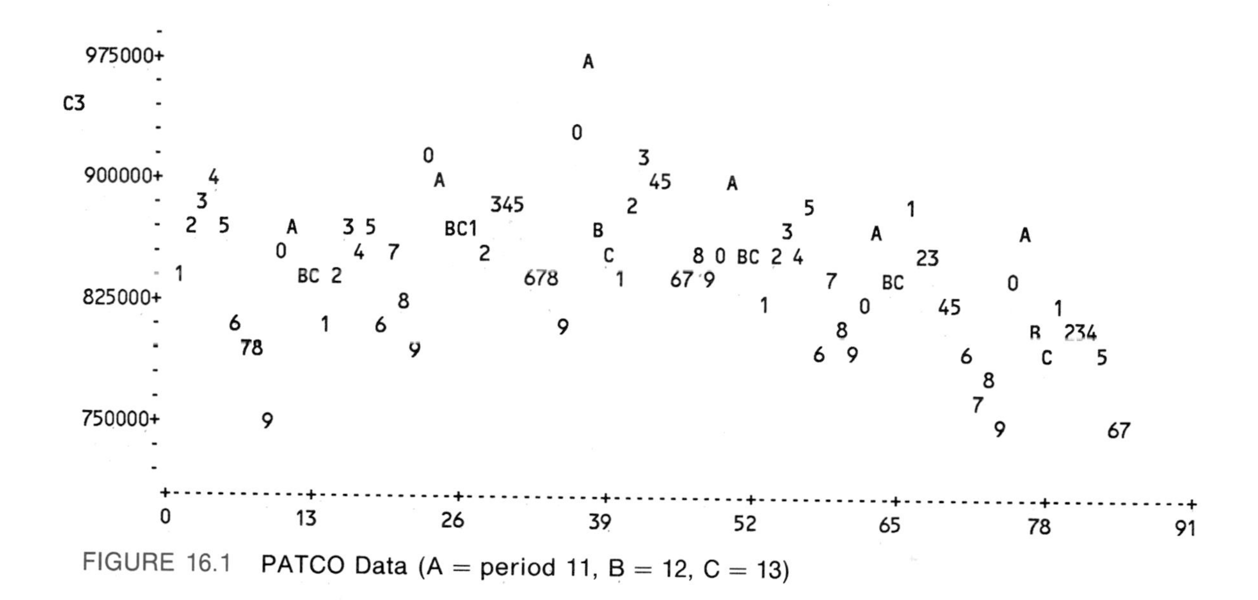

FIGURE 16.1 PATCO Data (A = period 11, B = 12, C = 13)

What can be seen from the plot of these data shown in Figure 16.1? First, there appears to be a modest downward **trend**. Second, there is a **seasonal** pattern to the data. The values in periods 6 through 9 of each year tend to be somewhat below the trend; these periods fall during the summer, when potential riders are on vacation. There may be a longer-term **cyclic** effect. Ridership may be somewhat above the trend in the middle years, and below the trend in the latest, most recent years. Finally, there is a **random**, irregular aspect; even if the trend, seasonal, and cyclic behavior were known exactly, it wouldn't be possible to predict ridership exactly.

EXAMPLE 16.3 Monthly data on the number of paid admissions (in thousands) to an indoor sports and entertainment arena for 72 months are given below and plotted in Figure 16.2. Identify any apparent trend, seasonal, and cyclic effects.

J	F	M	A	M	J	J	A	S	O	N	D
89	101	116	111	94	59	44	44	73	78	99	93
96	110	118	116	107	69	52	54	85	92	113	109
120	136	155	155	129	98	69	78	116	143	154	166
183	199	227	219	198	148	94	108	160	201	215	246
242	264	359	308	265	193	150	146	243	260	332	293
323	377	470	422	345	239	176	182	288	342	380	379

Solution The upward trend in admissions is obvious in the figure. The trend may not be linear, however. Rather, it may be S-shaped, involving initial slow growth, a period of more rapid growth, and then a tapering-off. (One consideration suggests that the trend cannot increase

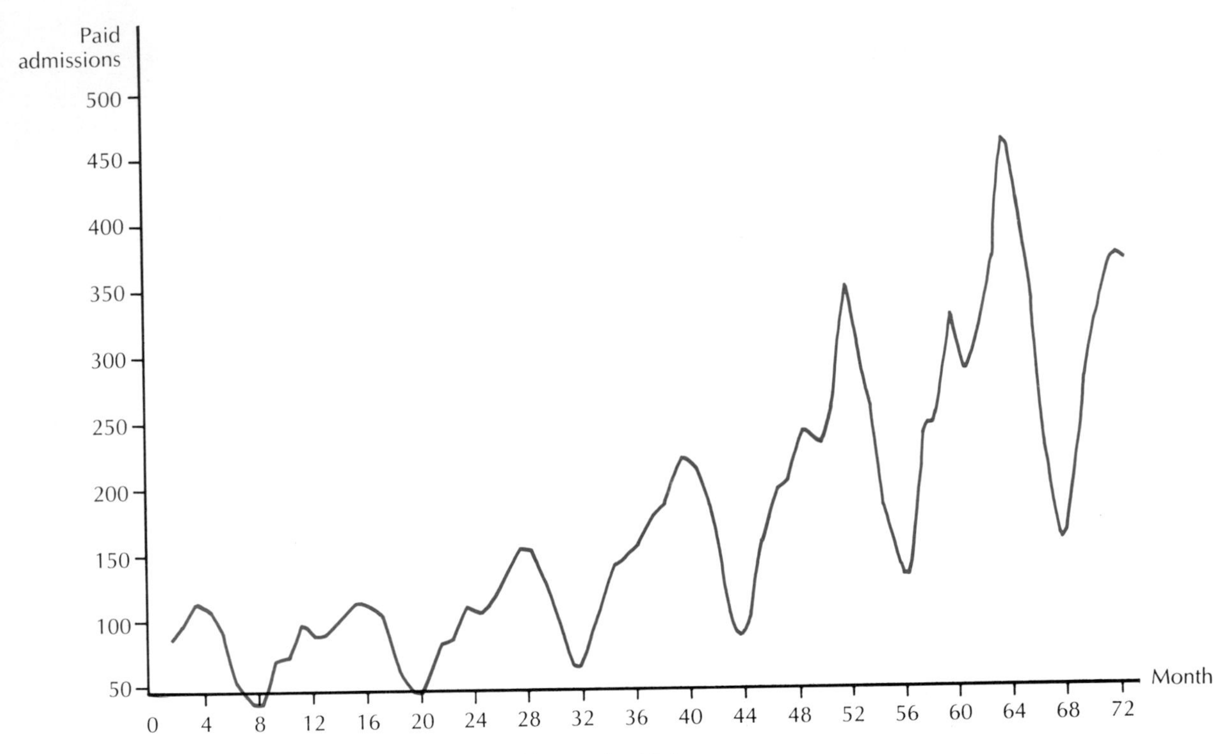

FIGURE 16.2 **Admissions to Arena; Example 16.3**

forever. Paid admissions to the arena are limited by the available number of dates and seats.) There is a strong seasonal effect; admissions drop dramatically in the summer months. If there's any cyclic componeı ıt, it's not apparent in the figure. □

One natural approach to the analysis of a time series is to remove the trend first, then the seasonal effects. The hope is that this procedure will bring out any longer-term cycles more clearly. The key decision that must be made in removing trend is the specification of the mathematical form of a trend equation. Denote the values of the time-series variable as y_t, where t runs from time 1 to some time T. Several mathematical forms for trends are widely used:

Linear: $Y_t = \beta_0 + \beta_1 t + \epsilon_t$

Exponential: $Y_t = \beta_0 e^{\beta_1 t} \epsilon_t$

Logistic: $Y_t = \dfrac{\beta_0}{1 + \beta_1 e^{\beta_2 t}} \epsilon_t$

Gompertz: $Y_t = \beta_0 e^{\beta_1 e^{\beta_2 t}} \epsilon_t$

linear trend

The **linear trend** equation is the simplest. The coefficient β_1 represents the average increase in y per unit time. This increase is measured in absolute terms (e.g., dollars) rather than in percentage terms. A linear trend equation should be used when there is no evident curvature in the plot of y_t against t. A good way to estimate β_0 and β_1 is by least squares, as in simple linear regression, with t as the independent variable.

EXAMPLE 16.4 The PATCO ridership data shown previously in this section are used in a linear regression, with period number as the independent variable. The output shown here indicates the equation and the trend values.

```
The regression equation is
C3 = 868007 - 558 C4

867450  866892  866334  865776  865219  864661  864103  863545  862987  862430  861872  861314  86075
860198  859641  859083  858525  857967  857409  856852  856294  855736  855178  854621  854063  85350
852947  852389  851832  851274  850716  850158  849601  849043  848485  847927  847369  846812  84625
845696  845138  844580  844023  843465  842907  842349  841792  841234  840676  840118  839560  83900
838445  837887  837329  836771  836214  835656  835098  834540  833983  833425  832867  832309  83175
831194  830636  830078  829520  828962  828405  827847  827289  826731  826174  825616  825058  82450
823942  823385  822827  822269  821711  821153  820596
```

Write the estimated trend equation. Interpret the slope.

Solution The trend equation is ridership = 868,007 − 558t, where t is the time period. The slope indicates that, over time, the transit line is losing about 558 riders per four-week period. □

exponential trend

The **exponential trend** equation is used when the trend reflects a roughly constant percentage growth; e^{β_1} is the percentage growth rate per time, while β_0 is the value of the trend at time $t = 0$. For instance, if at time 0 the trend value is 250, and the trend reflects an 8% growth rate per year, the trend equation is

$$\hat{y}_t = 250(1.08)^t$$

Since $e^{.077} = 1.08$, or equivalently $\log_e 1.08 = .077$, the trend equation can be rewritten as

$$\hat{y}_t = 250\, e^{.077t}$$

The exponential trend equation should be used if a plot of the natural logarithm of y_t against t appears roughly linear. If

$$Y_t = \beta_0 e^{\beta_1 t} \epsilon_t$$

then

$$\log Y_t = (\log \beta_0) + \beta_1 t + \log \epsilon_t$$

The coefficients ($\log \beta_0$) and β_1 can be estimated by regressing $\log y_t$ against $x_t = t$.

EXAMPLE 16.5 An exponential trend is fit to the PATCO ridership data. The output shown below results from a least-squares fit of $\log Y_t$ to $t(t = 1, \ldots, 85)$. Trend values are shown by period.

```
The regression equation is
C5 = 13.7 -0.000671 C4

867095  866513  865932  865350  864770  864190  863610  863030  862451  861872  861294  860717  86013
859562  858985  858408  857832  857257  856682  856107  855532  854958  854384  853812  853239  85266
852094  851522  850950  850379  849809  849239  848669  848099  847530  846961  846394  845826  84525
844691  844124  843557  842992  842426  841861  841295  840731  840167  839604  839040  838477  83791
837352  836790  836229  835668  835107  834546  833986  833427  832868  832309  831750  831192  83063
830077  829520  828964  828407  827851  827296  826740  826186  825632  825078  824524  823971  82341
822865  822313  821761  821210  820659  820108  819558
```

a. Identify the least-squares estimates of ($\log \beta_0$) and β_1.
b. Write the estimated trend equation and identify the monthly percentage increase in sales.
c. Does a linear trend or an exponential trend appear more appropriate?

Solution a. The intercept term is 13.7, while the slope is $-.000671$.

b. The prediction equation is obtained by taking $e^{13.7} = 890{,}900$ for the initial constant. The equation is

$$\hat{y}_t = -.000671t$$

The estimated change in ridership is $e^{-.000671}$, or .99933, a .067 percent per period decrease.

c. The choice is not blatantly obvious. The plot of the data and the trend equations suggest that a linear trend fits the data as well as anything. □

It is risky to assume that exponential growth will continue indefinitely. The sales history of a product may reveal a consistent 15% growth rate in unit sales per year, but sooner or later the sales must approach some saturation level and the percentage growth must slow down. The **logistic** and **Gompertz trend** equations yield very similar S-shaped trends. Figure 16.3 shows typical logistic and Gompertz curves.

logistic and Gompertz trends

Fitting either logistic or Gompertz trend equations is a bit harder than fitting linear or exponential trends. It is not possible to transform either S-shaped curve into a regression model. There are more complicated numerical methods for fitting such equations.

The trend, cyclic, seasonal, and irregular aspects of a time series can be combined in either of two ways. The **multiplicative model** is

multiplicative model

$$y_t = T_t C_t S_t I_t$$

where T_t, C_t, S_t, and I_t are the trend, cyclic, seasonal, and irregular components at time t. This model can be understood in terms of percentage changes. For instance, a seasonal index for a particular month equal to 1.08 means that the series tends to be 8% above

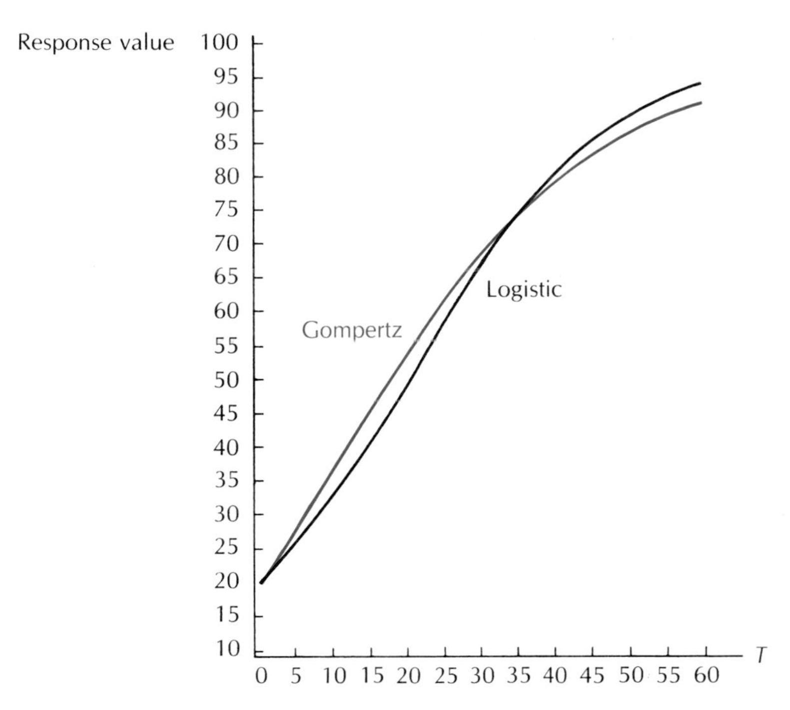

FIGURE 16.3 **Typical Logistic and Gompertz Curves**

the value predicted by trend and cycle alone. (By definition, the irregular component is unpredictable.) The **additive model** is

additive model

$$y_t = T_t + C_t + S_t + I_t$$

This model can be understood in absolute units as opposed to percentages. For instance, a seasonal index for daily water consumption in a city might have a value of 8 million gallons per day for August. That means that the water-consumption series for August would tend to be 8 million gallons above what was predicted by trend and cycle. A multiplicative model may be converted to an additive one by taking logarithms (to any base):

$$\log y_t = \log T_t + \log C_t + \log S_t + \log I_t$$

An additive model is very convenient when the trend equation is linear; a multiplicative model is convenient for an exponential trend. (Recall that we estimated an exponential trend by regressing the logarithm of y_t on t.) For most economic series, percentage changes tend to be more stable, and the multiplicative model is more commonly used. Therefore most of our examples use the multiplicative model.

Once a trend equation has been estimated, the influence of trend can be removed from the data, yielding a **"detrended" time series**. For the multiplicative model, divide the

detrended series

actual y_t by the trend value:

$$\frac{y_t}{T_t} = \frac{T_t C_t S_t I_t}{T_t} = C_t S_t I_t$$

contains no trend. If $y_t/T_t = 1.052$, for example, then the actual y_t is 5.2% above trend. For the additive model, subtract the trend value from y_t:

$$y_t - T_t = T_t + C_t + S_t + I_t - T_t = C_t + S_t + I_t$$

If $y_t - T_t = 620$, then y_t is 620 units above trend.

seasonal index

Next, the detrended values can be used to construct a **seasonal index**. There are several methods that can be used. The simplest method applies to an additive model. To construct a seasonal index value for, say, October, simply average all the available detrended October values. For instance, if 60 months of data are available and the five October detrended values are .982, .965, .961, .976, and .966, the seasonal index is the mean, .970. The same method could perhaps be used in a multiplicative model. Alternatively, since the multiplicative model is additive in logarithms, we can average the logarithms of detrended values, then take the antilogarithm to obtain the index. The logarithms (base 10) of the five October detrended values are $-.0078885$, $-.0154727$, $-.0172766$, $-.0105502$, and $-.0150229$, which have a mean of $-.0132422$. Take the antilogarithm by calculating $10^{-.0132422} = .970$. The two methods do not yield identical answers (in this case the difference is in the fifth decimal place), but if the detrended values are not too variable, as is the case here, the difference is small. The interpretations of the additive and multiplicative seasonal indices are very different. The additive index of .970 means that October values tend to be .970 units above the trend; the multiplicative index of .970 means that October values tend to be 97.0% of the trend.

For the PATCO ridership data shown at the beginning of this section, the detrended values are

-31512	2633	20116	32340	2259	-48399	-65415	-67190	-111378	-1734	6102	-27682	-16239
-48546	-26123	13515	-3748	8474	-48898	-770	-24421	-62563	61242	44848	11360	14887
13471	9668	28622	39917	37728	-17252	-13114	-13155	-42016	79512	125640	25676	14438
-12509	46338	69792	49232	53656	1105	-963	5851	-8473	18179	57437	10331	23410
-15461	22199	35688	16817	45741	-34718	11048	-29151	-38818	-1087	42735	5725	6315
51518	23380	31046	2823	-4912	-27755	-64022	-48823	-79036	18485	44701	-10987	-27482
-3638	-16793	-15374	-17514	-21234	-67021	-70074						

To compute a seasonal index, we simply average the values for each period, obtaining

Period:	1	2	3	4	5	6	7
Index:	-6626	8798	26,239	17,160	17,413	$-35,102$	$-29,015$
	8	9	10	11	12	13	
	$-29,440$	$-57,008$	29,137	53,612	2437	2585	

Thus ridership in period 1 tends to be 6626 passengers below the trend. Notice that in periods 6 through 9, the summer periods, ridership is substantially lower.

deseasonalized data

Once seasonal indices have been calculated, detrended data can be **deseasonalized**. Simply divide by the index value (in the multiplicative model) or subtract the index value (in the additive model).

cyclic patterns

Detrended, seasonally adjusted values are useful in attempting to identify **cyclic patterns**. Only cyclic and irregular (random) components remain in such values. Even so, identification of cyclic patterns is a tricky business. The basic problem is identifying the peak-to-peak or trough-to-trough length (called the *period*) of a supposed cycle. Seasonal patterns don't present any such problem. By definition, seasonal cycles have a one-year period. But identifying longer-term cyclic behavior is difficult. What do you do with data that appear to encompass $3\frac{1}{2}$ cycles, with peak-to-peak distances of $2\frac{1}{12}$ years, $3\frac{7}{12}$ years, and $4\frac{10}{12}$ years? One simpleminded idea is to take the mean distance ($3\frac{1}{2}$ years) as the apparent period of the cycle. But there is no guarantee that this is a very accurate estimate.

spectral analysis

There is a much more sophisticated version of time-series analysis, **spectral analysis**, which is related to the idea of cyclic behavior. The sophisticated mathematics of spectral analysis, called *Fourier analysis*, is borrowed from physics and electrical engineering. The idea can be stated, without too much distortion, in terms of regression. A sine curve is a cyclic pattern, as shown in Figure 16.4; any period may be specified for a sine curve.

If several sine curves with different periods are specified and weights are assigned to each curve, very complicated cyclic patterns can be reconstructed as weighted sums of sine curves. Spectral analysis can be regarded as the result of regressing a (detrended or seasonally adjusted) time series on all possible sine curves. The **spectrum** of the time series indicates how much of the variation in y_t is explained by the sine curve having each possible period. If y_t shows a very strong cyclic pattern with a given period, spectral analysis shows that a sine curve with that period explains much of the variation in y_t, and the spectrum has a sharp "peak" corresponding to that period. Estimating the spectrum of a time series is not a job for amateurs. For our purposes, it is enough to say that spectral analysis is an available method for analyzing cyclic data.

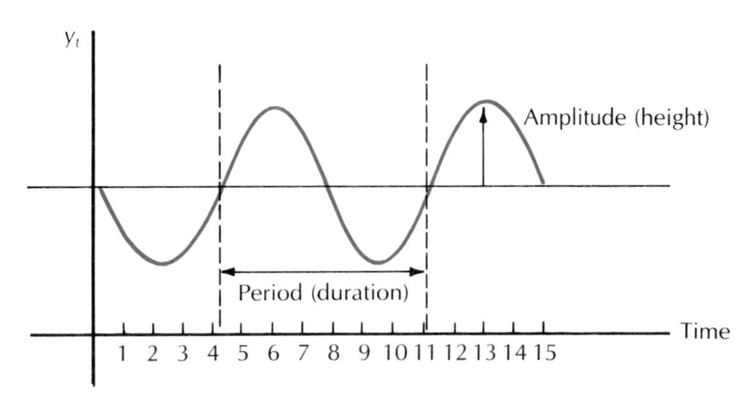

FIGURE 16.4 **Sine Wave**

SECTION 16.2 EXERCISES

16.5 A supplier of high-quality audio equipment for automobiles accumulates monthly sales data on speakers and receiver-amplifier units for five years. The data (in thousands of units per month) are shown here:

Year	J	F	M	A	M	J	J	A	S	O	N	D
1	101.9	93.0	93.5	93.9	104.9	94.6	105.9	116.7	128.4	118.2	107.3	108.6
2	109.0	98.4	99.1	110.7	100.2	112.1	123.8	135.8	124.8	114.1	114.9	112.9
3	115.5	104.5	105.1	105.4	117.5	106.4	118.6	130.9	143.7	132.2	120.8	121.3
4	122.0	110.4	110.8	111.2	124.4	112.4	124.9	138.0	151.5	139.5	127.7	128.0
5	128.1	115.8	116.0	117.2	130.7	117.5	131.8	145.5	159.3	146.5	134.0	134.2

Plot the sales data. Do you see any overall trend in the data? Do there seem to be any cyclic or seasonal effects?

16.6 Fit a linear trend equation to the data of Exercise 16.5. The following summary figures will be useful; y_t represents sales in month t, $t = 1, \ldots, 60$.

$$\sum y_t = 7122.0, \quad \sum t = 1830, \quad \sum y_t^2 = 858{,}693.02$$
$$\sum t^2 = 73{,}810, \quad \sum ty_t = 227{,}951.30$$

16.7 The data of Exercise 16.5 are detrended by dividing by the trend equation values of Exercise 16.6, with the following results:

Year	J	F	M	A	M	J	J	A	S	O	N	D
1	1.014	.920	.919	.918	1.019	.913	1.017	1.114	1.218	1.115	1.007	1.013
2	1.011	.908	.909	.909	1.004	.904	1.006	1.104	1.205	1.101	1.002	1.003
3	1.003	.903	.903	.901	.999	.900	.998	1.096	1.198	1.096	.997	.996
4	.996	.897	.896	.895	.996	.896	.991	1.089	1.199	1.091	.994	.991
5	987	.888	.886	.891	.989	.885	.988	1.086	1.183	1.083	.986	.984

Plot the detrended values against time. Do they show a seasonal effect?

16.8 A simple seasonal index is constructed by averaging the detrended values of Exercise 16.7. The index values are

J	F	M	A	M	J	J	A	S	O	N	D
1.002	.903	.903	.903	1.001	.900	1.000	1.098	1.199	1.097	.997	.997

a. Construct detrended, deseasonalized data for year 5.
b. Construct forecast values for year 6 by calculating the trend values and then multiplying by the seasonal factor.

16.9 Plot the detrended, seasonally adjusted data from Exercise 16.5. Is there any visual evidence of a cyclic pattern?

16.10 A machine-tool firm that produces a variety of products for manufacturers has quarterly records of total activity for the previous eight years. The data reflect activity rather than price, so inflation is irrelevant. The data are

	Quarter			
Year	1	2	3	4
1	97.2	100.2	102.8	102.6
2	106.1	107.8	110.5	110.6
3	116.5	117.3	119.9	119.3
4	126.1	125.7	128.3	132.1
5	133.2	133.8	141.1	142.1
6	144.2	146.1	151.6	154.0
7	155.8	158.6	165.8	167.0
8	171.1	172.6	176.5	179.7

a. Plot the data against time (quarters 1–32).
b. Does there appear to be a clear trend? If so, what form of trend equation would you suggest?
c. Can you detect cyclic or seasonal features?

16.11 Fit an exponential trend to the data of Exercise 16.10. The following summary figures are relevant:

$$\sum \log y_t = 156.41, \quad \sum t = 528, \quad \sum (\log y_t)^2 = 765.59$$
$$\sum t^2 = 11{,}440, \quad \sum t \log y_t = 2634.48$$

16.12 Detrended values for the data of Exercise 16.10 are calculated by dividing by trend values, using the trend equation of Exercise 16.11. The detrended values are

	Quarter			
Year	1	2	3	4
1	.9939	1.0046	1.0106	.9890
2	1.0028	.9990	1.0041	.9854
3	1.0177	1.0048	1.0070	.9825
4	1.0182	.9952	.9960	1.0055
5	.9941	.9792	1.0125	.9998
6	.9948	.9882	1.0055	1.0015
7	.9935	.9916	1.0165	1.0038
8	1.0084	.9974	1.0001	.9984

a. Plot the detrended values against time (quarters 1–32).
b. Can you detect possible seasonal or cyclic effects?

16.13 Use the trend equation of Exercise 16.11 to forecast activity for quarters 33–36.

16.3 SMOOTHING METHODS

The classical approach described in the previous section does not lead directly to forecasts, though forecasts can be derived as a by-product. An alternative approach to time-series analysis is **smoothing**, which attempts to get rid of the irregular, random component of the series but does not concern itself with details of trends, seasons, and cycles.

Most smoothing methods yield forecasts that are, in one sense or another, averages of past values. If the data show a pronounced trend, these forecasts therefore tend to lag behind the trend. In addition, these methods usually ignore seasonal factors. We discuss smoothing methods for no-trend, no-seasonal data.

moving averages

One of the most widely used smoothing methods is **moving averages**: averaging the M most recent values to forecast the next value.

EXAMPLE 16.6 One of the key elements of the budget of a small-town television station is the monthly advertising-time sales. In the previous 36 months, the monthly sales (measured in minutes per day) of paid, local advertising have been the following:

Month:	1	2	3	4	5	6	7	8	9	10	11	12
Sales:	86	80	85	93	96	102	97	89	96	87	82	81
Month:	13	14	15	16	17	18	19	20	21	22	23	24
Sales:	87	89	97	88	95	87	81	79	82	85	96	93
Month:	25	26	27	28	29	30	31	32	33	34	35	36
Sales:	99	103	96	85	78	83	90	96	85	82	89	96

Construct 3-month and 5-month moving-average forecasts. Plot forecast values against actual values by month.

Solution The first 3-month moving average is $(86 + 80 + 85)/3 = 83.6667$, shown as 83.7 in the output. The first 5-month moving average is $(86 + 80 + 85 + 93 + 96)/5 = 88.0$, as shown.

3-Month Moving Average:

—	—	83.7	86.0	91.3	97.0	98.3	96.0	94.0	90.7	88.3	93.3
83.3	85.7	91.0	91.3	93.3	90.0	87.7	82.3	80.7	82.0	87.7	91.3
96.0	98.3	99.3	94.7	86.3	82.0	83.7	89.7	90.3	87.7	85.3	89.0

5-Month Moving Average:

—	—	—	—	88.0	91.2	94.6	95.4	96.0	94.2	90.2	87.0
86.6	85.2	87.2	88.4	91.2	91.2	89.6	86.0	84.8	82.8	84.6	87.0
91.0	95.2	97.4	95.2	92.2	89.0	86.4	86.4	86.4	87.2	88.4	89.6

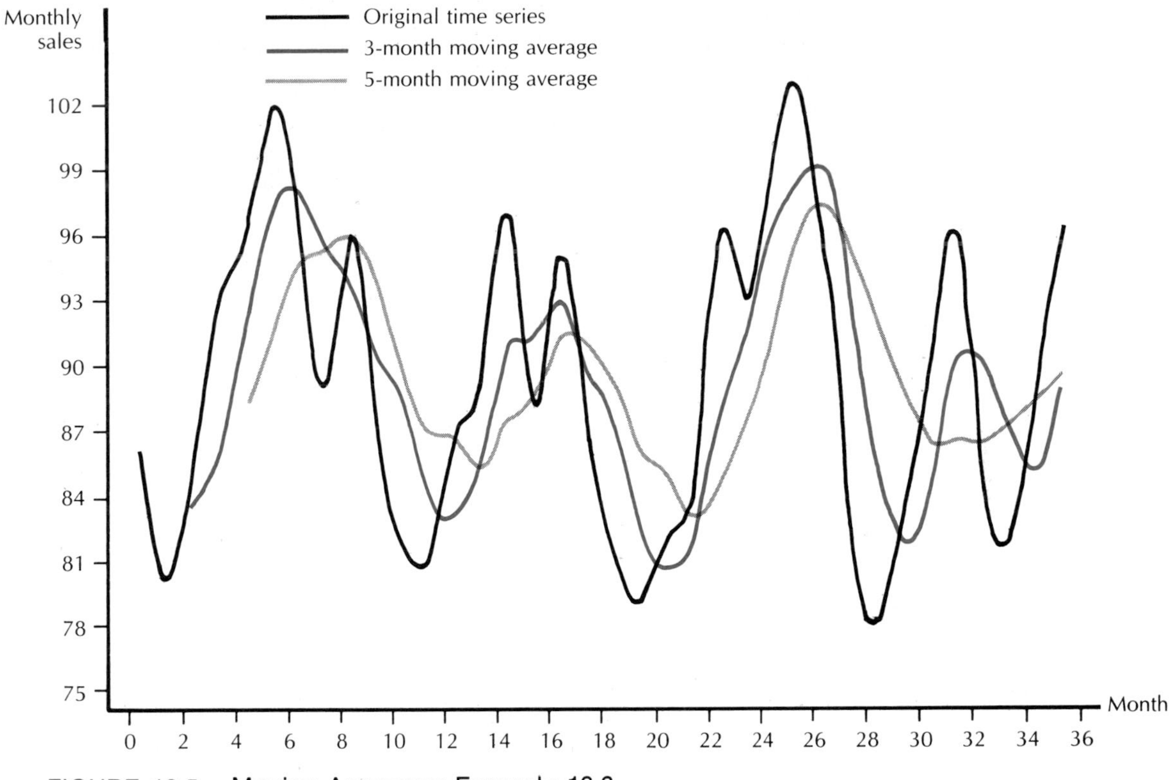

FIGURE 16.5 **Moving Averages; Example 16.6**

The 3-month and 5-month moving averages are plotted, with the original time series, in Figure 16.5. □

running medians

A variation on the idea of moving-average smoothing is the method of moving (or running) medians. The **running-median** smoothing method involves calculating the median, instead of the mean, of the *M* most recent values of the time series. Because a median is not affected by extreme values, as a mean is, running-median forecasts are unaffected by the occasional fluke value, whether that value is extremely large or extremely small.

EXAMPLE 16.7 Calculate 3-month and 5-month running medians for the data of Example 16.6. How do these forecasts compare with the moving-average forecasts of that example?

Solution The first 3-month running median is median (86, 80, 85) = 85. The first 5-month running median is median (86, 80, 85, 93, 96) = 86. The series of running medians is shown here.

3-Month Running Median:

—	—	85	85	93	96	97	97	96	89	87	82	82
87	89	89	95	88	87	81	81	81	81	82	85	93
96	99	99	96	85	83	83	90	90	85	85	89	

5-Month Running Median:

—	—	—	—	86	93	96	96	96	96	89	87
87	87	87	88	89	89	88	87	82	82	82	85
93	96	96	96	96	85	85	85	85	85	89	89

The 3-month and 5-month running medians are plotted, with the original time series, in Figure 16.6. □

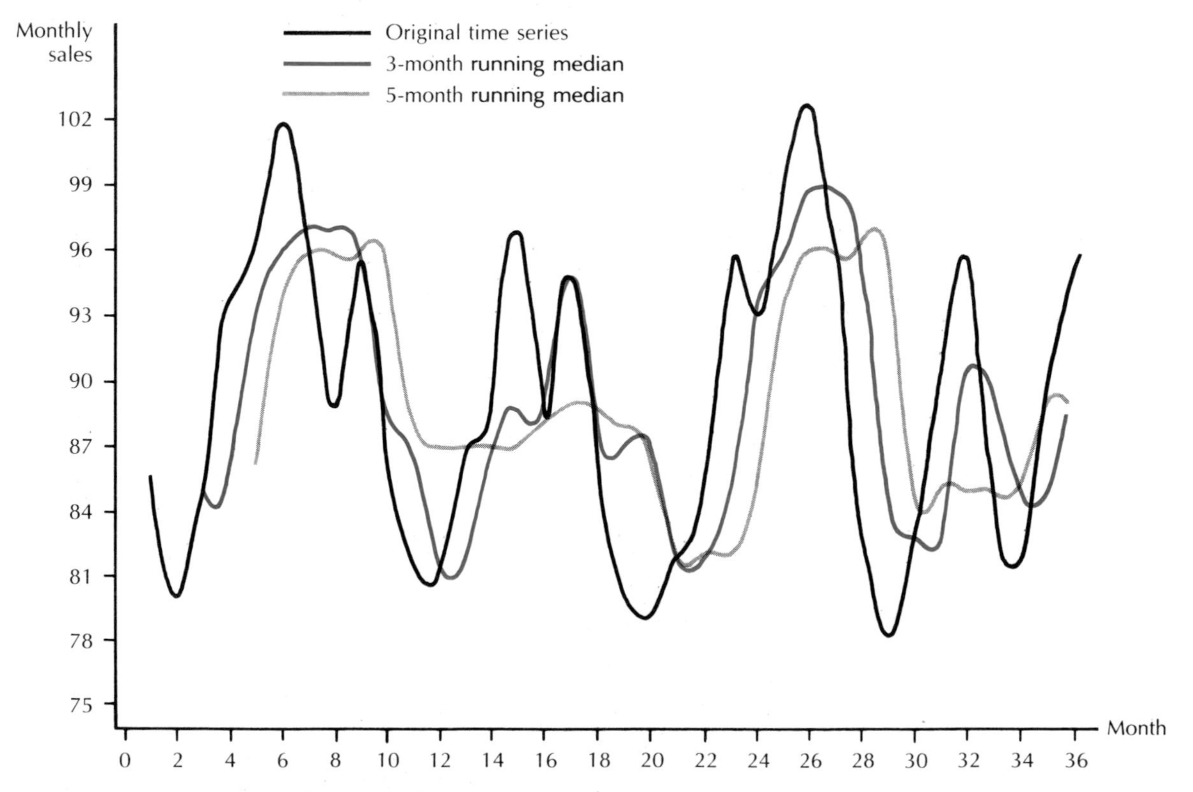

FIGURE 16.6 **Running Medians; Example 16.7**

choice of time periods

How many time periods should be used in a moving average or running median? An extremely large value of M (the number of periods) causes the forecast to change very slowly, and such forecasts are not effective in picking up short-term variations. An extremely small value for M causes the forecasts to be very "jittery." In the extreme case of $M = 1$, the forecast for the next period is simply the current value. This forecast incorporates all the random, irregular features of the time series. If the series moves in fairly smooth waves with modest randomness, a small number of periods (such as 3) is desirable. If the series exhibits a great deal of randomness around a vaguely cyclic pattern, greater averaging (such as 5 or 7 periods) is preferred. Comparison of the past forecasting performance of several possible values of M often clarifies the choice.

The moving-average and running-median smoothing methods are open to criticism on the grounds that they give the same weight to relatively old values as to the most recent values. The forecasts tend to be somewhat slow in responding to short-run trends or cycles. An alternative method, **exponential smoothing**, gives higher weight to the most recent values and often can be more effective in reacting to shifts or cycles in the series.

exponential smoothing

The exponential-smoothing forecast $\hat{y}_{t+1}$ of the actual value y_{t+1} is defined as

$$\hat{y}_{t+1} = \alpha y_t + \alpha(1-\alpha)y_{t-1} + \alpha(1-\alpha)^2 y_{t-2} + \ldots$$

smoothing constant

The number α (no relation to α, the probability of Type I error) lies between 0 and 1 and is called the **smoothing constant**; an appropriate choice of α is discussed shortly. For example, suppose that α is chosen to be .6. Then

$$\hat{y}_{t+1} = .6y_t + .24y_{t-1} + .096y_{t-2} + .0384y_{t-3} + \ldots$$

The most recent value of the time series is given the heaviest weight, and the weights given to previous values drop off rapidly. In fact, the weights α, $\alpha(1-\alpha)$, $\alpha(1-\alpha)^2$, $\alpha(1-\alpha)^3, \ldots$ decrease exponentially to zero. Hence the name *exponential smoothing*.

There is an alternative way of finding exponentially smoothed forecasts that is easier computationally and also helps to indicate the effect of various choices of the smoothing constant α:

$$\hat{y}_{t+1} = \alpha y_t + (1-\alpha)\hat{y}_t$$

The next forecast value is weighted average of the current actual value and the current forecast value. To use this formula in computation, an initial (time 1) forecast must be chosen; the effect of the initial forecast dies out quite rapidly. One good choice is $\hat{y}_1 = y_1$. Then

$$\begin{aligned}
\hat{y}_2 &= \alpha y_1 + (1-\alpha)\hat{y}_1 = y_1 \\
\hat{y}_3 &= \alpha y_2 + (1-\alpha)\hat{y}_2 = \alpha y_2 + (1-\alpha)y_1 \\
\hat{y}_4 &= \alpha y_3 + (1-\alpha)\hat{y}_3 = \alpha y_3 + \alpha(1-\alpha)y_2 + (1-\alpha)^2 y_1 \\
\hat{y}_5 &= \alpha y_4 + (1-\alpha)\hat{y}_4 = \alpha y_4 + \alpha(1-\alpha)y_3 + \alpha(1-\alpha)^2 y_2 + (1-\alpha)^3 y_1
\end{aligned}$$

and so on.

With this choice of $\hat{y}_1$, the forecasts rapidly approach the infinite series value

$$\hat{y}_{t+1} = \alpha y_t + \alpha(1-\alpha)y_{t-1} + \alpha(1-\alpha)^2 y_{t-2} + \ldots$$

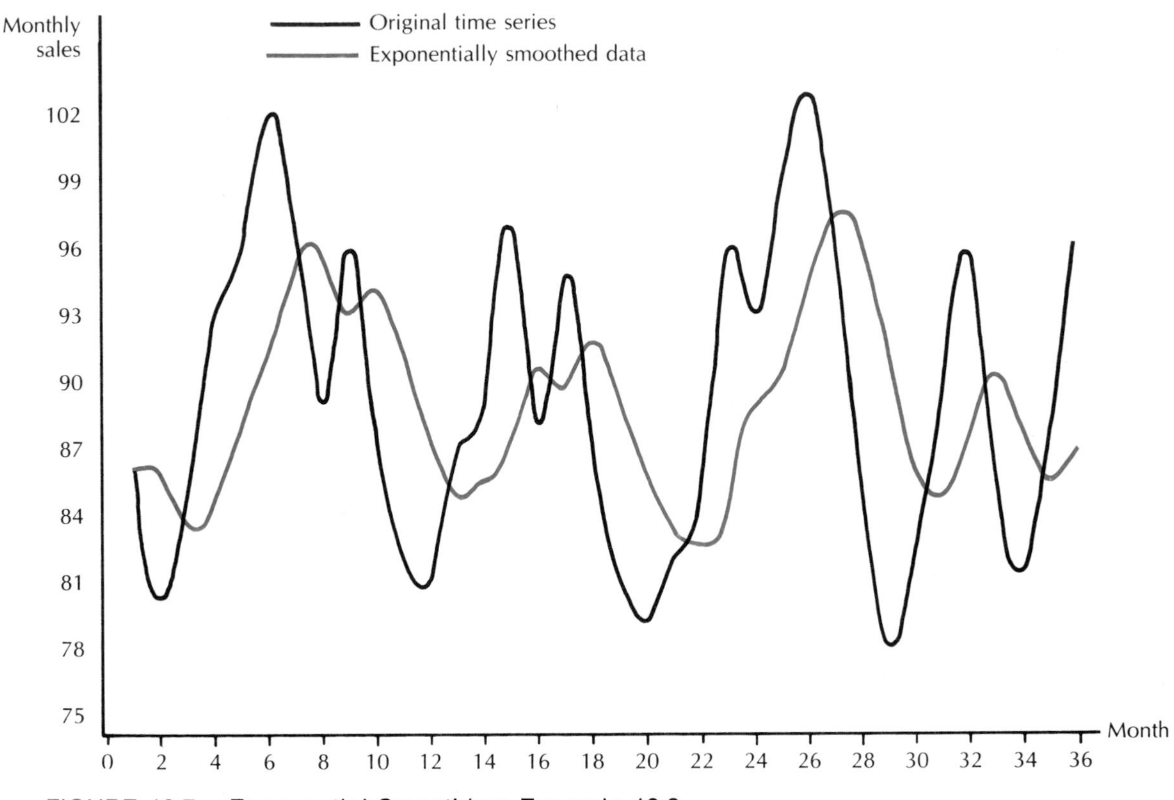

FIGURE 16.7 **Exponential Smoothing; Example 16.8**

EXAMPLE 16.8 Compute the exponentially smoothed forecast values for the data of Example 16.6, using $\alpha = .4$. Plot the observed time series and the smoothed time series.

Solution The exponentially smoothed values, calculated by a computer, are plotted with the original time series in Figure 16.7. □

choice of α

The **choice of α** is important in exponential smoothing. A large value of α, such as .7, makes the next forecast value very sensitive to the current value. Therefore the forecast picks up shifts or short-run trends in the series quite quickly, at the cost of being very sensitive to random fluctuations in the series. A small value of α, such as .2, makes the forecast less affected by random fluctuations, but also less effective in picking up shifts or trends. One fairly simple way to select α is to try various α's on historical data; that is, select the α that minimizes average absolute or average squared error in the data.

SECTION 16.3 EXERCISES

16.14 As part of an inventory-management method, the monthly demands for various products are recorded and used for future forecasting. The demands for the previous 24 months of one particular product are

Month:	1	2	3	4	5	6	7	8	9	10	11	12
Demand:	89	97	101	168	120	107	100	96	89	97	143	105
Month:	13	14	15	16	17	18	19	20	21	22	23	24
Demand:	96	84	93	110	125	110	93	95	89	93	105	110

a. Calculate 3-month and 5-month moving averages.

b. Plot the original time series, together with the 3-month and 5-month moving averages.

c. Calculate the average squared error for the 3-month moving average used as a forecast of the next month's demand. Do the same for the 5-month moving average.

16.15 Refer to Exercise 16.14.

a. Calculate 3-month and 5-month running medians.

b. Plot the original, 3-month, and 5-month smoothed data.

c. Calculate the average squared error for the 3-month running median used as a forecast of the next month's demand. Do the same for the 5-month moving average.

16.16 a. Calculate exponentially smoothed forecasts of the demands in Exercise 16.14. Use a smoothing constant of .8.

b. Compute the average squared error for this forecast.

16.17 Compare the average squared error for the forecasts generated in Exercises 16.14–16.16. On the basis of this criterion, which forecast appears to be preferred?

16.18 A large supermarket tries to forecast volume (total number of transactions) to assist in its short-term staffing decisions. Of particular interest are the weekday evening (Monday–Thursday, 5 P.M.–11 P.M.) volumes, which fluctuate because of competitors' actions, weather, and many unknown factors. The average weekday evening volume is collected for 40 weeks. These volumes and 3- and 5-week moving averages are shown below:

Data:

40.2	43.1	44.2	43.6	45.1	47.3	45.9	46.3	45.7	43.9
44.1	43.8	43.1	41.5	42.1	41.6	40.8	42.4	44.5	46.9
49.4	50.0	51.6	52.3	51.8	53.1	51.9	48.7	49.2	47.1
45.2	41.1	43.8	46.0	41.7	39.4	41.2	43.1	42.5	44.9

3-week moving average:

—	—	42.5	43.6	44.3	45.3	46.1	46.5	46.0	45.3
44.6	43.9	43.7	42.8	42.2	41.7	41.5	41.6	42.6	44.6
46.9	48.8	50.3	51.3	51.9	52.4	52.3	51.2	49.9	48.3
47.2	44.5	43.4	43.6	43.8	42.4	40.8	41.2	42.3	43.5

5-week moving average:

—	—	—	—	43.2	44.7	45.2	45.6	46.1	45.8
45.2	44.8	44.1	43.3	42.9	42.4	41.8	41.7	42.3	43.2
44.8	46.6	48.5	50.0	51.0	51.8	52.1	51.6	50.9	50.0
48.4	46.3	45.3	44.6	43.6	42.4	42.4	42.3	41.6	42.2

a. Plot the actual data and the moving averages on the same graph.

b. Which moving average appears to track the actual values better?

16.19 The data of Exercise 16.18 leads to the following 3- and 5-week running medians:

3-week running median:

—	—	43.1	43.6	44.2	45.1	45.9	46.3	45.9	45.7
44.1	43.9	43.8	43.1	42.1	41.6	41.6	41.6	42.4	44.5
46.9	49.4	50.0	51.6	51.8	52.3	51.9	51.9	49.2	48.7
47.1	45.2	43.8	43.8	43.8	41.7	41.2	41.2	42.5	43.1

5-week running median:

—	—	—	—	43.6	44.2	45.1	45.9	45.9	45.9
45.7	44.1	43.9	43.8	43.1	42.1	41.6	41.6	41.6	42.1
42.4	44.5	46.9	49.4	50.0	51.6	51.8	51.9	51.9	51.8
44.2	48.7	47.1	45.2	45.2	43.8	41.7	41.7	41.7	42.5

a. Plot the actual data and running medians on the same graph.
b. Which running median seems to be giving better forecasts?

16.20 The average squared error and average absolute error for the forecasts of Exercises 16.18 and 16.19 are shown below.

	Mean Squared Error	Mean Absolute Error
3-week moving average	.626	.604
5-week moving average	2.183	1.166
3-week running median	.608	.438
5-week running median	2.420	1.154

Does the same forecast method appear to be best on each criterion?

16.21 An evening newspaper records the number of pages of display (unclassified) advertisements over a 70-month period. Monthly averages are computed. The data, plus exponentially smoothed forecasts, are shown below:

Sales:

69.2	67.6	68.0	70.5	73.4	65.2	68.7	72.5	74.1	76.9
70.0	73.2	78.0	75.6	74.9	72.2	70.9	74.3	69.1	63.5
66.7	68.0	64.7	68.9	69.0	69.7	71.3	74.5	70.8	75.6
74.8	73.5	72.8	69.9	71.5	70.4	68.5	67.1	63.5	66.8
67.9	69.0	68.3	71.1	73.2	72.9	74.6	76.1	75.3	77.9
78.3	75.4	72.9	73.6	71.5	69.9	67.3	69.3	67.2	67.4
63.8	64.7	65.8	68.0	67.2	68.7	70.2	72.5	73.6	71.7

Exponential smoothing, $\alpha = .2$:

—	69.20	68.88	68.70	69.06	69.93	68.98	68.93	69.64	70.53
71.81	71.45	71.80	73.04	73.55	73.82	73.50	72.98	73.24	72.41
70.63	69.84	69.48	68.52	68.60	68.68	68.88	69.37	70.39	70.47
71.50	72.16	72.43	72.50	71.98	71.89	71.59	70.97	70.20	68.86
68.45	68.34	68.47	68.44	68.97	69.81	70.43	71.27	72.23	72.85
73.86	74.75	74.88	74.48	74.30	73.74	72.98	71.84	71.33	70.51
69.88	68.67	67.87	67.46	67.57	67.49	67.74	68.23	69.08	69.99

Exponential smoothing, $\alpha = .8$:

—	69.20	67.92	67.98	70.00	72.72	66.70	68.30	71.66	73.61
76.24	71.25	72.81	76.96	75.87	75.09	72.78	71.28	73.70	70.02
64.80	66.32	67.66	65.29	68.18	68.84	69.53	70.95	73.79	71.40
74.76	74.79	73.76	72.99	70.52	71.30	70.58	68.92	67.46	64.29
66.30	67.58	68.72	68.38	70.56	72.67	72.85	74.25	75.73	75.39
77.40	78.12	75.94	73.51	73.58	71.92	70.30	67.90	69.02	67.56
67.43	64.53	64.67	65.57	67.51	67.26	68.41	69.84	71.97	73.27

a. Plot the actual data. Is there a pronounced trend?
b. Plot the exponentially smoothed forecasts for $\alpha = .2$ and $\alpha = .8$.
c. Which forecast method seems to overrespond to random variation?

16.22 The mean squared error and mean absolute error for the most recent two years of data in Exercise 16.21 are calculated for each exponentially smoothed forecast.

	Mean Squared Error	Mean Absolute Error
$\alpha = .2$	11.735	2.852
$\alpha = .8$	7.224	2.221

Which α value seems better?

16.4 THE BOX-JENKINS APPROACH

Both the classical analysis, which takes into account the trend, seasonal, and cyclic components of time-series data, and the smoothing methods of the previous section have the problem that they don't reflect any theoretical structure. Instead, they are rough-and-ready methods for cleaning up time-series data, without much regard to the process that generated the data in the first place. In this section we sketch some of the basic ideas of what is known as the Box-Jenkins approach to the analysis of time-series data.

autoregressive model

The simplest Box-Jenkins model is the **autoregressive model**. The idea of an autoregressive model is to use the past values of a time series as independent variables in predicting future values. The simplest autoregressive model is a first-order model, designated AR(1):

$$Y_t = \phi_0 + \phi_1 Y_{t-1} + \epsilon_t$$

The *order* of a Box-Jenkins model refers to the maximum time lag used, not to the maximum power of a variable as in regression analysis. In this model only the most recent y value is assumed to be a useful predictor. The ϵ_t term, which reflects only pure random error, is assumed to have mean zero, constant variance, no autocorrelation, and a normal distribution, just as in regression. The ϵ_t process is sometimes called **white noise**.

white noise

stationary series

As stated, the autoregressive model does not contain a trend. Box-Jenkins models were designed for **stationary** time series—ones with no trend, constant variability, and stable correlations over time. In particular, the series should be detrended. If the time series shows only a linear trend, taking (first) differences $y_t - y_{t-1}$ often yields an approximately stationary series. For an exponential trend, differences in logarithms of y values will serve.

Autoregressive models often yield data with a distinct cyclic pattern, despite the lack of any cyclic function (such as a sine curve) in the model.

EXAMPLE 16.9 The following 25 values of a time series are available:

129.60	139.79	147.89	152.69	157.22	163.91	173.49	182.40	189.16
195.74	202.14	207.51	212.02	215.44	221.54	229.93	237.05	244.31
250.46	257.50	260.77	265.78	268.97	272.17	278.49		

a. Is there evidence of a trend?
b. Convert the data to approximately stationary form.

Solution a. A plot (not shown) of the data indicates a roughly linear trend.
b. The first differences of the series are

10.19	8.10	4.80	4.53	6.69	9.58	8.92	6.76	6.58	6.39	5.37	4.51
3.42	6.10	8.40	7.11	7.26	6.15	7.03	3.27	5.01	3.19	3.20	6.32

□

EXAMPLE 16.10 Plot the difference data of Example 16.9 against time. Does it appear cyclic?

Solution A plot of the differences is shown in Figure 16.8. The dashed line indicates the mean value. There appears to be a definite cyclic pattern. □

Earlier in this chapter we noted that trying to determine the period of a particular type of cycle might not be fruitful, and that an alternative strategy is to seek models that tend to produce roughly cyclic behavior. Box-Jenkins models provide one very rich class of possible models.

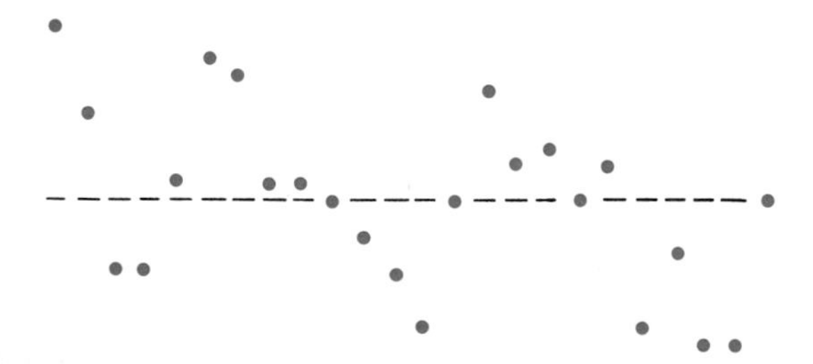

FIGURE 16.8 First Differences; Example 16.10

correlogram
autocorrelation coefficients

A key idea in Box-Jenkins methods is the **correlogram**. This is a plot (against time) of the **autocorrelation coefficients** of a time series y_t. Autocorrelation coefficients are almost, but not quite, the same as ordinary correlations of the time series with the same series lagged by a number (j) of periods. For example, we might take the series shown below and lag it one period, as indicated:

Time, t:	1	2	3	4	5	6	7	8	9
Series, y_t:	86	81	82	85	90	93	89	86	—
Lagged series, y_{t-1}:	—	86	81	82	85	90	93	89	86

Omitting times 1 and 9 (because one series or the other does not have a value there), we can compute the correlation of y_t and y_{t-j} using the sample-correlation formula of Chapter 13. The autocorrelation coefficient r_j of y_t and a variable y_{t-j} is only slightly different than the sample correlation between y_t and y_{t-j}. The formula is

$$r_j = \frac{\sum_{t=j+1}^{T} (y_t - \bar{y})(y_{t-j} - \bar{y})}{\sum_{t=1}^{T} (y_t - \bar{y})^2}$$

where T is the number of periods of data available and

$$\bar{y} = \frac{\sum_{t=1}^{T} y_t}{T}$$

You should be able to verify that $\bar{y} = 86.5$ for the data shown in this paragraph and that the lag 1 autocorrelation is

$$r_1 = \frac{(81 - 86.5)(86 - 86.5) + \ldots + (86 - 86.5)(89 - 86.5)}{(86 - 86.5)^2 + \ldots + (86 - 86.5)^2} = .5855$$

(The ordinary correlation is .5867.) While shortcut formulas can be used, the calculation of autocorrelations is almost always done by computer.

The correlogram is simply a scatter plot of lags versus autocorrelations r_j.

EXAMPLE 16.11 The autocorrelations [up to a maximum lag of 10 for the (difference)] data of Example 16.9 are

Lag j:	1	2	3	4	5	6	7	8	9	10
r_j:	.455	−.008	−.189	−.121	−.047	−.144	−.077	.137	.333	.206

Construct the correlogram.

Solution The correlogram is shown in Figure 16.9. □

The sample correlogram in Figure 16.9 should be compared to a theoretical correlogram. For any particular Box-Jenkins model, theoretical autocorrelation coefficients ρ_j can

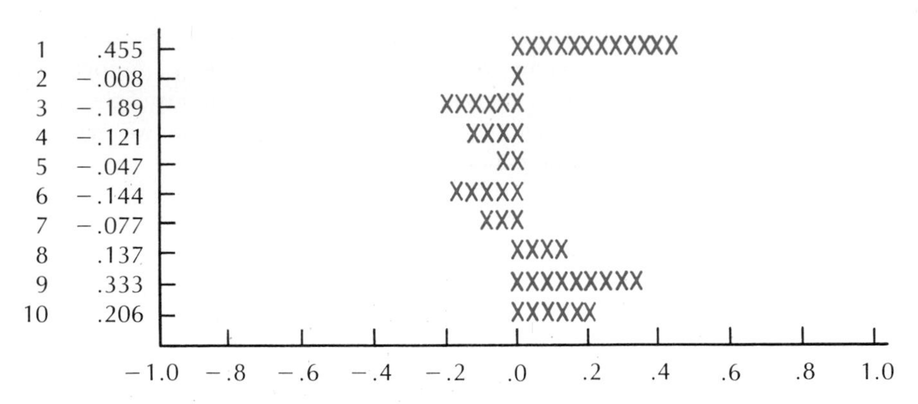

FIGURE 16.9 **Sample Autocorrelations; Example 16.11**

be derived (derivations are given in Nelson, 1977, and Box and Jenkins, 1970). For the first-order autoregressive model $Y_t = \phi_0 + \phi_1 Y_{t-1} + \epsilon_t$, it can be proved that these theoretical autocorrelation coefficients are given by the geometric series

$$\rho_j = \phi_1^j, \qquad j = 1, 2, \ldots$$

Other Box-Jenkins models yield different theoretical autocorrelations.

EXAMPLE 16.12 Calculate the theoretical autocorrelations, for lags up to 10, for an AR(1) model with $\phi_1 = .80$. Do these autocorrelations resemble those found in Example 16.11?

Solution The theoretical autocorrelations are just successive powers of .80.

lag j:	1	2	3	4	5	6	7	8	9	10
ρ_j:	.8000	.6400	.5120	.4096	.3277	.2621	.2097	.1678	.1342	.1074

These values of ρ_j are plotted in Figure 16.10. Note how the theoretical autocorrelations for an AR(1) model with $\phi_1 > 0$ gradually decay toward zero. Although we never

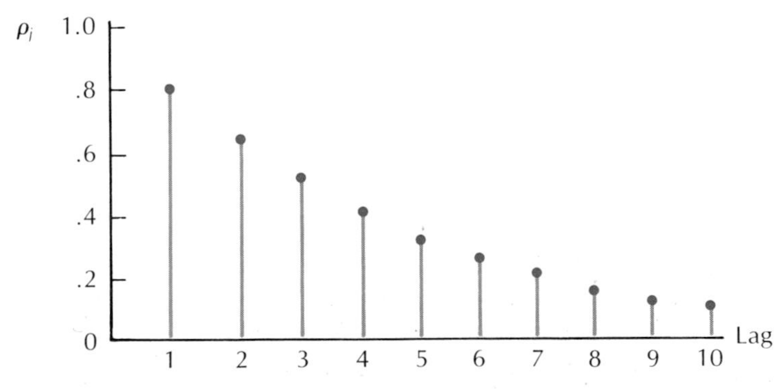

FIGURE 16.10 **Correlogram of the Theoretical Autocorrelations for an AR(1) Model with $\phi_1 = .80$; Example 16.12**

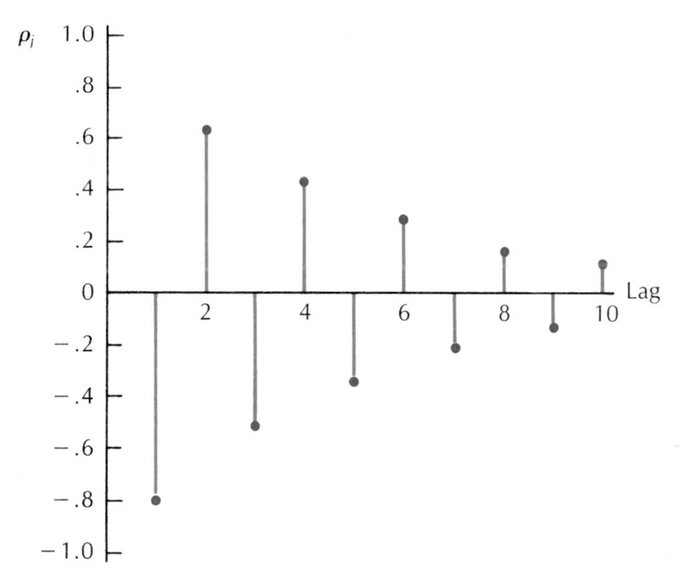

FIGURE 16.11 **Correlogram of the Theoretical Autocorrelations for an AR(1) model with $\phi_1 = -.80$**

know the actual value of ϕ_1 (or equivalently the ρ_j's), the sample autocorrelations for an AR(1) model should possess a pattern similar to that for the ρ_j's. The autocorrelations in Example 16.11 do not possess such a pattern, so the AR(1) model is not reasonable. □

The theoretical autocorrelations for an AR(1) model with $\phi_1 < 0$ also decay toward zero, but the sign of ρ_j alternates from positive to negative (see Figure 16.11).

higher-order autoregressive models

There are many other Box-Jenkins models. The autoregressive idea can be extended to higher-order (longer lags) models. The pth-order autoregressive model AR(p) is

$$Y_t = \phi_0 + \phi_1 Y_{t-1} + \ldots + \phi_p Y_{t-p} + \epsilon_t$$

where the errors ϵ_t are assumed to satisfy the "white noise" assumptions.

moving average model

There is also a different type of model, called a **moving-average model**.* The first-order moving-average model MA(1) is

$$Y_t = \theta_0 + \epsilon_t - \theta_1 \epsilon_{t-1}$$

where the ϵ's are assumed to be white noise.†

The difference between autoregressive and moving-average models can be seen by considering the effect of past ϵ values on current y values. Think of the ϵ values as random "shocks" input to an economic system, and y values as the output of the system. By

* The phrase *moving average* is used in a different sense in Box-Jenkins models than in smoothing models.

† The minus sign on θ_1 is conventional, allowing for a convenient "backshift" notation in Box and Jenkins (1970).

appropriate substitutions (for Y_{t-1}, then Y_{t-2}, etc.) in a first-order autoregressive model, the weight given the lag j shock is seen to be ϕ_1^j. Assuming that $|\phi_j| < 1$, it follows that the weights of long-ago shocks decline steadily. In contrast, the first-order moving-average model gives a weight of $-\theta_1$ to ϵ_{t-1} and no weight at all to shocks with longer lags. In a pure moving-average model, the effect of a shock persists for a specified number of time periods, then disappears suddenly. In a pure autoregressive model, the effect of a shock declines gradually. A moving-average model may be appropriate for a machine shop where ϵ_t represents a sudden burst of new orders and Y_t represents production. If all orders are filled in either the current month or the next month, the effect of two-month-old orders on production is zero. An autoregressive model may be appropriate for a model of sales of a certain kind of front-wheel-drive car. A positive ϵ shock may represent an upsurge in interest in such cars. Such a shock may reasonably be assumed to signal a longer-term increase in sales, but the effect should gradually decline because of such factors as new competitors and new technology.

The theoretical correlogram of a moving-average model reflects the difference between moving-average and autoregressive models. The general qth-order moving-average model, MA(q), is

$$Y_t = \theta_0 + \epsilon_t - \theta_1\epsilon_{t-1} - \ldots - \theta_{t-q}\epsilon_q$$

The effect of a shock ϵ is assumed to persist up to lag q, then suddenly drop to zero. The theoretical autocorrelation reflects the drop-to-zero pattern. For the first-order moving-average model, $y_t = \theta_0 + \epsilon_t - \theta_1\epsilon_{t-1}$, the theoretical autocorrelations are

$$\rho_j = \begin{cases} 1, & \text{for } j = 0 \\ \dfrac{-\theta_1}{1+\theta_1^2}, & \text{for } j = 1 \\ 0, & \text{for } j \geq 2 \end{cases}$$

Thus the correlogram for a MA(1) model has a single spike; all autocorrelations beyond $j = 1$ are zero, as shown in Figure 16.12. Note the difference between the correlograms

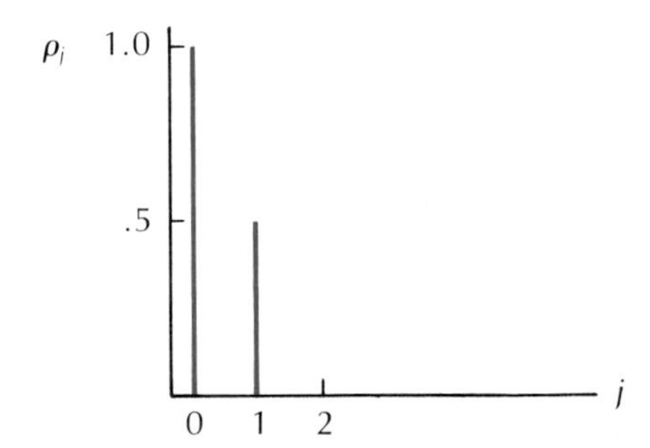

FIGURE 16.12 **Correlogram for a MA(1) Model with $\theta_1 = -.80$**

for AR(1) and MA(1) models. The theoretical correlogram for a pure moving-average process MA(q) tapers off gradually for $j < q$, then suddenly drops to zero for $j > q$. The sample correlogram of a pure moving-average process behaves similarly, but with a degree of random error. For any sample autocorrelations, $1/\sqrt{T}$ provides a rough approximation for the standard error of the sample autocorrelation (assuming the true autocorrelation is zero), where T is the number of observations. A sample correlogram that suddenly drops to within $\pm 1/\sqrt{T}$ of zero is a good candidate for a pure moving-average model.

EXAMPLE 16.13 The following data represent weekly sales of prime grade steak (in hundreds of pounds) from a butcher shop. The shop does not run special sales. It often receives orders to be filled the next week, but it rarely defers orders for two or more weeks.

1	6994.34990	16	6791.99290	31	7083.94930	46	6875.10270
2	6819.13960	17	6638.27280	32	6905.46870	47	6672.32360
3	6627.78400	18	6582.60120	33	6795.83390	48	6795.10270
4	6903.73050	19	6632.21910	34	7343.88200	49	6789.81020
5	6803.14080	20	7052.63850	35	7038.02370	50	6702.24150
6	6824.55020	21	7150.15790	36	7209.44100	51	7009.76710
7	6673.45040	22	6505.48590	37	7077.89760	52	6986.90840
8	7016.24510	23	6771.52660	38	7180.69140	53	6765.66510
9	7102.40200	24	6952.83790	39	7102.35700	54	6916.00870
10	6703.55110	25	6791.50430	40	6895.33970	55	7028.07120
11	7006.72310	26	6549.27090	41	6792.29880	56	6678.83590
12	7212.95430	27	6895.44530	42	6848.69020	57	6898.03080
13	7162.08620	28	7355.78130	43	6797.43900	58	6999.29990
14	7023.15720	29	7201.25080	44	6485.33360	59	6850.31260
15	6553.87670	30	7028.80190	45	6514.19600	60	6675.29850

The theoretical autocorrelations (based on an estimated $\theta_1 = .600$) are

ORDER	AUTOCORR
1,	.4412
2 and higher	.0000

a. Plot theoretical (T) and sample (S) correlograms on the same plot.
b. Does the "sudden-zero" property of moving-average models show up in the data?

Solution a. The sample correlations are plotted in Figure 16.13. The first-order theoretical autocorrelation of .4412 is slightly larger than the sample value of .353, while all other theoretical autocorrelations are zero.

Order	Autocorr
1	.353
2	.043
3	.204
4	.171
5	−.087
6	−.183
7	.061
8	.059
9	−.120
10	−.129
11	−.172
12	−.104

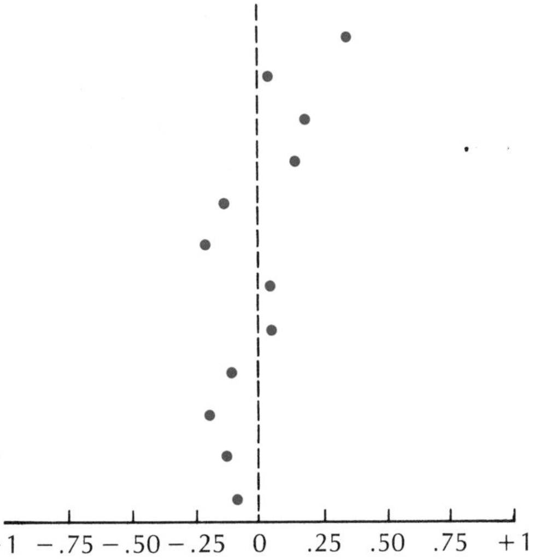

FIGURE 16.13 **Sample Autocorrelations for Data of Example 16.13**

b. The second-order sample autocorrelation is very near zero. The higher-order autocorrelations wander around with no obvious pattern. □

EXAMPLE 16.14 Refer to the correlogram of Example 16.11 shown in Figure 16.9. Identify an appropriate candidate for a Box-Jenkins model.

Solution For $T = 24$, $1/\sqrt{24} = .20$ represents an approximate standard error for any autocorrelation coefficient r_j. Since all autocorrelations are within $2(.20) = .40$ of zero except $r_1 = .455$ and only the long lag (9 and 10) autocorrelations are greater than .20, an MA(1) is a likely candidate model. □

general model Autoregressive and moving-average models may be combined. The general autoregressive–moving-average model ARMA (p, q) is written

$$Y_t = \phi_0 + \phi_1 Y_{t-1} + \phi_2 Y_{t-2} + \ldots + \phi_p Y_{t-p} + \epsilon_t - \theta_1 \epsilon_{t-1} - \theta_2 \epsilon_{t-2} - \ldots - \theta_q \epsilon_{t-q}$$

If the lags p and q are large, very complex models can result. The usual strategy is to try to find a simple model (with very small p and q lags) that fits the data adequately and approximates the autocorrelations of the series. There are several strategies for identifying such models. One is based on an examination of the estimated correlogram. As we have seen, the theoretical correlogram for a pure autoregressive process tapers gradually to zero as the lag increases, while the theoretical correlogram of a pure moving-average process drops suddenly to zero. A combination autoregressive and moving-average process exhibits a combination of these phenomena in its correlogram: the correlogram tapers off, then drops sharply, then tapers off again. The lag just before the drop-off is a good candidate for the maximum lag q of the moving-average part. The maximum lag for the autoregressive component is harder to identify in the correlogram. Other Box-Jenkins methods, such as partial autocorrelation, are needed for precise identification of reasonable Box-Jenkins models. Box and Jenkins (1970) and Nelson (1977) contain good tips on the difficult task of model identification.

least-squares fit

Once the maximum lags for an ARMA model have been specified, the next task is to estimate the ϕ or θ model parameters. The basic idea is to search for the **least-squares fit**. The process of finding the least-squares estimates is complex, especially if there is a moving-average component in the model. Any of the standard Box-Jenkins computer programs estimates the model parameters.

EXAMPLE 16.15 A time series contains the following data for a 40-month period:

134.59	97.60	74.03	55.75	84.24	126.61	141.67	128.72	146.22	157.66
158.57	157.85	154.02	113.37	109.29	160.06	177.27	161.21	123.35	124.43
170.74	175.41	201.49	227.60	233.43	239.80	184.09	133.22	119.65	123.52
139.92	175.82	166.81	163.33	122.41	111.96	151.67	160.21	180.46	188.86

The following autocorrelations are obtained:

Lag:	1	2	3	4	5	6	7	8	9	10
Correlation:	.767	.382	.093	−.040	−.014	.022	.035	.035	−.019	−.050

Plot the correlogram. What ARMA models are indicated?

Solution See Figure 16.14. Only the first two (possibly three) autocorrelations differ from zero by any substantial amount. Given the steep drop to zero, one might try a moving average of two or three terms. Perhaps a first-order (or at most a second-order) autoregressive

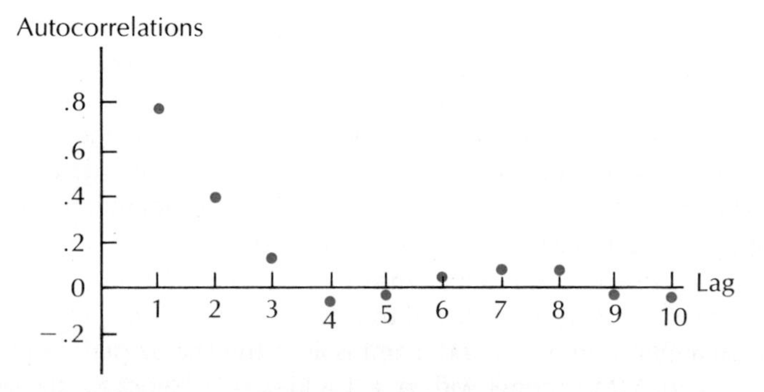

FIGURE 16.14 **Correlogram for Example 16.15**

term might be tried, given that there is some suggestion of a decay toward zero in the correlogram. □

EXAMPLE 16.16 ARMA models with $q = 1$ were estimated for $p = 1$ and 2 for the data of Example 16.15. The respective average squared errors were 432.95 and 424.65. Does the more complex model with $p = 2$ do much better?

Solution The average squared error for the more complicated model is only very slightly (about 2%) smaller, which hardly seems to be very useful. □

Box-Jenkins seasonal models

In our discussion of Box-Jenkins models, we have indicated how trends can be removed by differencing (of logs, if necessary) and how such models can capture apparent cyclic effects. We haven't said anything about seasonal effects; Box-Jenkins models can be extended to capture seasonal effects, too. In fact, one of the most popular deseasonalizing methods used by U.S. government agencies, the X-11 procedure of the census bureau, is closely related to Box-Jenkins methods (see Pierce, 1980). We do not go into detail here, except to note that the basic idea is to include autoregressive and moving-average terms with lag 4 for quarterly data, lag 12 for monthly data, and so on. The Nelson (1977) and Box and Jenkins (1970) texts contain full discussions.

Box-Jenkins forecasting

Once a model has been adopted and the coefficients estimated, a Box-Jenkins model can be used for forecasting. For the first-order autoregressive model, the one-period-ahead forecast is $\hat{y}_{t+1} = \hat{\phi}_0 + \hat{\phi}_1 y_t$; the two-period-ahead forecast is

$$\hat{y}_{t+2} = \hat{\phi}_0 + \hat{\phi}_1 y_{t+1}$$

but y_{t+1} is not yet known and must be replaced by $\hat{y}_{t+1}$ (at the cost of additional random error). This process can be continued for forecasts of any period ahead; the obvious extension works for any higher-order autoregressive model. Moving-average models are trickier to use for forecasting. In the first-order moving-average model,

$$y_{t+1} = \hat{\theta}_0 - \hat{\theta}_1 \epsilon_t$$

but ϵ_t is the unknown true error at time t and must itself be estimated. Estimation of error terms in moving-average (or general ARMA) models is done by a rather elaborate computerized method involving (among other things) backward forecasting. Any good Box-Jenkins program can handle the technical problems. One good approach to checking Box-Jenkins models uses the concept of validation of Chapter 15: fit the model based on, say, the first 80% of the data, then generate forecasts of the remaining data. While there are more sophisticated model-selection methods available, this approach should detect any gross deviations from a model.

EXAMPLE 16.17 An ARMA model with $p = 1$ and $q = 1$ is fit to the first 32 periods of the data of Example 16.15, as is an ARMA model with $p = 1$ and $q = 2$. Forecasts are generated for periods 33–40, with the following results:

Actual	ARMA(1, 1)	Error	ARMA(1, 2)	Error
166.81	188.55	−21.74	191.71	−24.90
163.33	175.40	−12.07	180.67	−17.34
122.41	166.94	−44.53	165.43	−43.02
111.96	161.51	−49.55	154.83	−42.87
151.67	158.02	−6.35	149.84	1.83
160.21	155.78	4.43	148.59	11.62
180.46	154.33	26.13	149.00	31.46
188.86	153.41	35.45	149.79	39.07
Mean		−8.53		−5.52
Standard deviation		30.41		31.67

Is there compelling evidence in favor of the more complicated ARMA(2, 1) model?

Solution Neither model seems to grossly over- or underestimate the data in the validation sample. Both commit some rather large errors. While the average error for the ARMA(1, 2) model is slightly better than that for the ARMA(1, 1) model, the standard deviation is higher. We don't feel compelled to use the more complicated model. □

The Box-Jenkins approach to time-series data is in many ways attractive. The general ARMA structure is rich and allows for a wide variety of autocorrelation patterns. Many kinds of cyclic and seasonal patterns can be captured by Box-Jenkins models. Yet it is important for a manager to realize that these models aren't magic. A typical Box-Jenkins model used in forecasting a time series is based entirely on the past history of the series itself, and it doesn't allow for any predictive value of other variables. The virtue of such an approach is that it requires only readily available data—only the past history of the series. The weakness of the approach is that it doesn't take advantage of the predictive value of other, related series. The Box-Jenkins approach is a useful technique, and it can be applied within a regression context. But it is not a cure-all.

SECTION 16.4 EXERCISES

16.23 Examination of the sales records of a firm for 60 months yields the following data on monthly sales of a standard product:

90.8	94.8	100.1	104.8	112.2	121.9	124.6	137.5	156.0	165.3	173.7	182.4
178.7	174.0	175.4	173.3	175.7	176.0	179.7	178.2	186.4	195.7	204.9	214.3
222.5	227.3	222.6	219.8	220.5	220.4	228.3	227.2	228.3	235.7	241.2	248.8
257.3	267.4	276.9	288.8	299.6	315.8	322.8	332.2	347.0	357.7	363.2	366.7
371.9	373.7	371.8	371.4	359.7	355.4	348.4	342.6	337.7	329.8	327.7	324.5

a. Plot the data against time. Is there an evident trend?
b. The month-to-month changes in sales are

3.9	5.3	4.7	7.4	9.7	2.7	12.9	18.5	9.3	8.4	8.7	−3.7
−4.7	1.4	−2.1	2.4	.3	3.7	−1.5	8.2	9.2	9.1	9.4	8.2
4.8	−4.7	−2.8	.7	.1	7.9	−1.1	1.1	7.4	5.5	7.6	8.5
10.1	9.5	11.9	10.8	16.2	7.0	9.4	14.8	10.7	5.5	3.5	5.2
1.8	−1.9	−.4	−11.7	−4.3	−7.0	−5.8	−4.9	−7.9	−2.1	−3.2	

Plot change against time. Is there an evident linear trend in the changes? Does there appear to be some cyclic behavior?

16.24 Autocorrelations are calculated for the change data of Exercise 16.23. The results are

Lag:	1	2	3	4	5	6
Correlation:	.693	.581	.490	.285	.154	.049
Lag:	7	8	9	10	11	12
Correlation:	−.042	−.208	−.232	−.250	−.284	−.231

a. Verify the calculation of the lag 1 correlation.
b. Plot the correlations against lag numbers. Is there any lag at which the lag correlations drop off suddenly?

16.25 A first-order autoregressive model is fit to the change data of Exercise 16.23. The estimated autoregressive model is

$\hat{y}_t = 1.093 + .709\, y_{t-1}$

The predicted values are

3.8	3.9	4.9	4.4	6.3	7.9	3.1	10.2	14.2	7.7	7.0	7.3
−1.6	−2.2	2.1	−.4	2.8	1.3	3.7	.0	6.9	7.7	7.6	7.8
6.9	4.5	−2.2	−.9	1.6	1.1	6.6	.3	1.9	6.3	5.0	6.5
7.1	8.2	7.9	9.5	8.8	12.6	6.0	7.7	11.6	8.7	5.0	3.6
4.8	2.4	−.3	.8	−7.2	−1.9	−3.9	−3.0	−2.4	−4.5	−.4	

with a mean squared error of 21.67. Plot predicted and actual change values over time. Does it appear that the model yields useful forecasts of changes in sales?

16.26 The estimated lag correlations for the model of Exercise 16.25 are

Lag:	1	2	3	4	5	6
Correlation:	.690	.556	.472	.260	.121	.023
Lag:	7	8	9	10	11	12
Correlation:	−.094	−.229	−.258	−.265	−.289	−.240

How do the estimated lag correlations based on the model compare to the computed lag correlations of Exercise 16.24?

16.27 A manufacturer of novelty items produces (among other items) pennants for sports teams. Weekly production figures include both long-term runs, which are items that sell fairly steadily with slow fluctuation, and short-term runs, which are items that sell in sudden one- or two-week bursts. Data are collected on weekly production for the preceding 156 weeks. There is virtually no trend in the data. Lag correlations are as follows:

Lag:	1	2	3	4	5	6	7	8	9	10
Correlation:	.826	.556	.397	.258	.166	.117	.094	.131	.183	.222

a. Plot these correlations against lag number. Is there any lag at which correlations drop off suddenly?
b. What autoregressive and moving-average terms would you suggest for inclusion in a Box-Jenkins model for these data?

16.28 The data of Exercise 16.27 are fitted to three possible Box-Jenkins models: ARMA(1, 1), ARMA(1, 2), and ARMA(2, 1) (see Figure 16.15). Do the more complicated models—ARMA(1, 2) and ARMA(2, 1)—yield obviously better correlograms than the ARMA (1, 1) model?

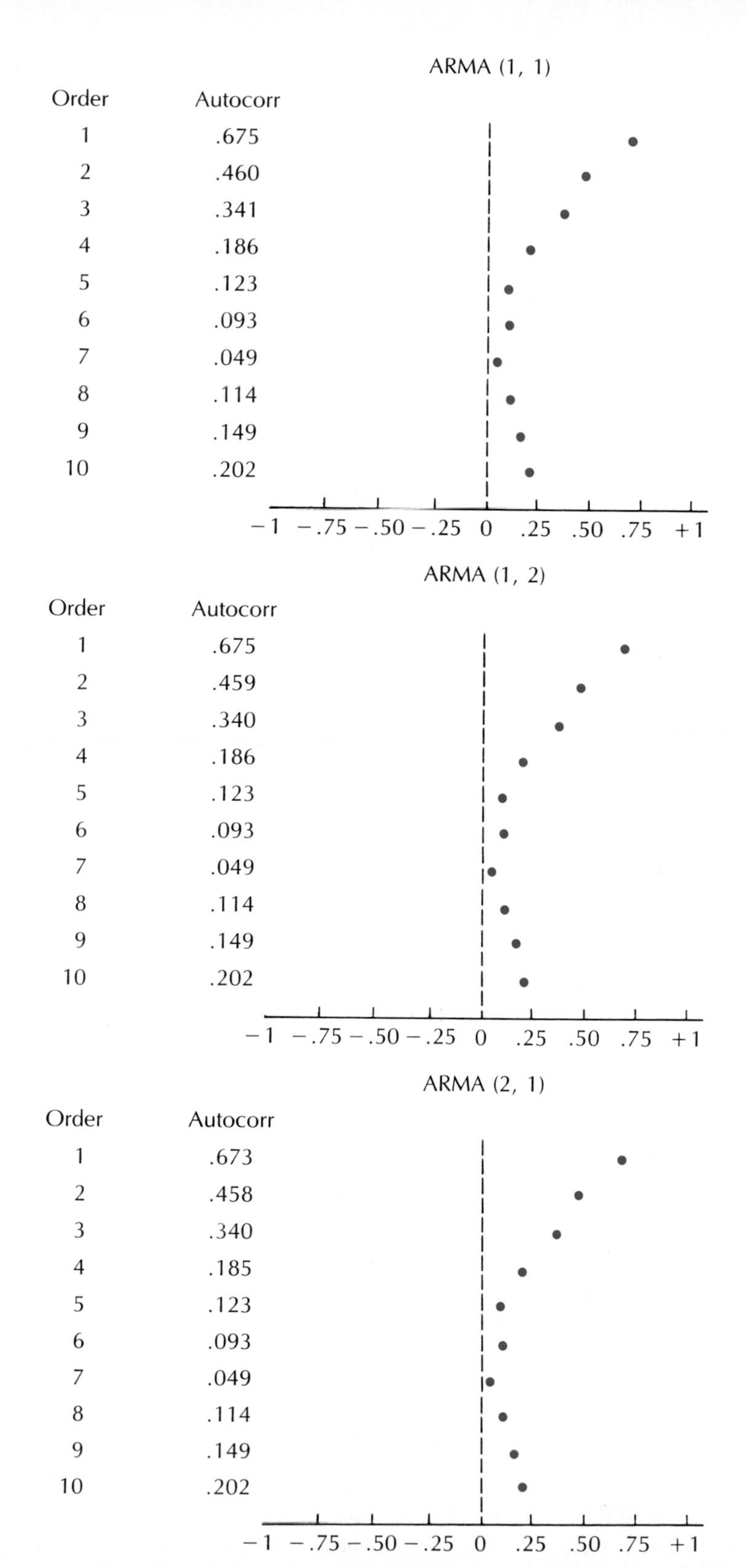

FIGURE 16.15 Box-Jenkins Models; Exercise 16.28

16.29 The sum of squared errors for the three models of Exercise 16.28 are as follows:

Model:	ARMA (1, 1)	ARMA(1, 2)	ARMA(2, 1)
SSE:	15,279.47	15,279.64	15,279.41

Is there any evidence to indicate that the more complicated models fit much better?

SUMMARY

This chapter deals with time-series data: data collected on one or more variables over time. We begin our consideration of such data with a discussion of index numbers; these numbers are the result of attempts to summarize complicated patterns of price or quantity movement.

The special price-level problem of economic time series can be partially remedied by the use of price indices. The construction of price indices involves a choice of quantity weights. Because such choices are not logically inevitable, they are another source of variability and uncertainty in time-series analysis.

The analysis of time-series data requires sophisticated approaches, and a number of methods have been proposed. The classical trend-cycle-season-irregularity approach is useful in isolating trends and correcting for seasonal factors, but it is less successful in dealing with cyclic behaviors. Various smoothing techniques, including moving-average, running-median, and exponential-smoothing methods, are useful in generating forecasts once the time series has been purged of trend and seasonal factors.

The Box-Jenkins approach is an alternative forecasting scheme that involves explicit models for the evolution of a time series. This style of time-series analysis raises an interesting question about cycles. Though no explicit cyclic feature appears in any Box-Jenkins model, data that apparently follow a cyclic pattern can fit such models very well.

KEY FORMULAS: Time-Series Analysis

1. Simple aggregate price index in year k, with base year 0

$$I_k = \frac{\sum p_{ki}}{\sum p_{0i}} \times 100$$

2. Weighted aggregate price index

$$I_k = \frac{\sum p_{ki}q_i}{\sum p_{0i}q_i} \times 100$$

3. Multiplicative model involving trend, cyclic, seasonal, and irregular components

$$y_t = T_t C_t S_t I_t$$

4. Additive model

$$y_t = T_t + C_t + S_t + I_t$$

5. Exponential smoothing

$$\begin{aligned}\hat{y}_{t+1} &= \alpha y_t + \alpha(1-\alpha)y_{t-1} + \alpha(1-\alpha)^2 y_{t-2} + \ldots \\ &= \alpha y_t + (1-\alpha)\hat{y}_t\end{aligned}$$

6. First-order autoregressive model, AR(1)

$$Y_t = \phi_0 + \phi_1 Y_{t-1} + \epsilon_t$$

7. jth-order sample autocorrelation coefficient

$$r_j = \frac{\sum_{t=j+1}^{T} (y_t - \bar{y})(y_{t-j} - \bar{y})}{\sum_{t=1}^{T} (y_t - \bar{y})^2}$$

8. Theoretical autocorrelation coefficients for an AR(1) model

$$\rho_j = \phi_1^j, \qquad j = 0, 1, 2, \ldots$$

9. pth-order autoregressive model, AR(p)

$$Y_t = \phi_0 + \phi_1 Y_{t-1} + \ldots + \phi_p Y_{t-p} + \epsilon_t$$

10. First-order moving-average model, MA(1)

$$Y_t = \theta_0 + \epsilon_t - \theta_1 \epsilon_{t-1}$$

11. Theoretical autocorrelation coefficients for a MA(1)

$$\rho_j = \begin{cases} 1, & \text{for } j = 0 \\ \dfrac{-\theta_1}{1+\theta_1^2}, & \text{for } j = 1 \\ 0, & \text{for } j \geq 2 \end{cases}$$

12. qth-order moving-average model, MA(q)

$$Y_t = \theta_0 + \epsilon_t - \theta_1 \epsilon_{t-1} - \ldots - \theta_q \epsilon_{t-q}$$

13. Mixed autoregressive, moving-average model, ARMA (p, q)

$$Y_t = \phi_0 + \phi_1 Y_{t-1} + \ldots + \phi_p Y_{t-p} + \epsilon_t - \theta_1 \epsilon_{t-1} - \ldots - \theta_q \epsilon_{t-q}$$

14. Forecasting using Box-Jenkins models

AR(1): $\hat{y}_{t+1} = \hat{\phi}_0 + \hat{\phi}_1 y_t$, $\hat{y}_{t+2} = \hat{\phi}_0 + \hat{\phi}_1 \hat{y}_{t+1}, \ldots$

MA(1): $\hat{\theta}_0 - \hat{\theta}_1 \hat{\epsilon}_t$, where $\hat{\epsilon}_t$ is estimated using backward forecasting

CHAPTER 16 EXERCISES

16.30 A manufacturer of snack foods compiles dollar-sales volume monthly for four years. Because prices have changed substantially, the firm uses a price index to deflate the dollar figures. The results are shown below (sales in thousands of dollars per month):

Year 1	J	F	M	A	M	J	J	A	S	O	N	D
Sales	421.9	447.0	439.3	479.3	478.5	481.2	506.8	477.5	480.5	502.5	527.9	521.8
Deflated sales	413.7	440.4	429.1	465.4	454.9	466.3	486.9	457.9	455.1	477.3	498.1	487.7
Index	102.0	101.5	102.4	103.0	105.2	103.2	104.1	104.3	105.6	105.3	106.0	107.0

Year 2												
Sales	526.3	527.7	543.3	519.5	579.3	570.1	570.7	603.1	609.8	600.7	637.1	655.8
Deflated sales	492.4	491.8	503.6	481.1	525.7	520.2	515.6	538.5	545.0	530.7	555.5	568.3
Index	106.9	107.3	107.9	108.0	110.2	109.6	110.7	112.0	111.9	113.2	114.7	115.4

Year 3												
Sales	632.6	631.5	661.8	670.5	677.2	701.4	711.1	700.3	712.2	750.9	729.7	764.2
Deflated sales	548.7	544.4	560.9	563.5	565.8	583.6	592.1	582.2	586.7	609.5	588.0	608.0
Index	115.3	116.0	118.0	119.0	119.7	120.2	120.1	120.3	121.4	123.2	124.1	125.7

Year 4												
Sales	738.8	784.8	780.8	795.9	832.9	828.4	845.0	874.8	885.6	884.5	884.5	895.4
Deflated sales	586.4	620.4	615.3	612.3	634.9	630.5	637.8	658.3	662.9	660.6	653.3	652.2
Index	126.0	126.5	126.9	130.0	131.2	131.4	132.5	132.9	133.6	133.9	135.4	137.3

a. Plot the dollar-volume sales figures over time (months 1–48). What sort of trend equation seems appropriate?

b. Plot the deflated sales figures over time. What sort of trend equation seems appropriate for these deflated values?

c. Can you explain the discrepancy in the trends of parts (a) and (b)?

16.31 a. Calculate a linear trend equation for the deflated sales values of Exercise 16.30. The following summary figures have been obtained:

$\sum y_t = 26{,}359.5, \quad \sum ty_t = 690{,}868.6, \quad \sum t = 1176$

$\sum t^2 = 38{,}024, \quad n = 48$

b. Plot the trend line on the same plot as the deflated sales values. Does there appear to be any seasonal or cyclic fluctuation around the trend?

16.32 The deflated sales figures of Exercise 16.30 are divided by the trend values obtained in Exercise 16.31 to obtain the following detrended values:

Year 1:	.95	1.00	.97	1.04	1.00	1.02	1.05	.98	.96	1.00	1.03	1.00
Year 2:	1.00	.99	1.00	.95	1.03	1.01	.99	1.02	1.02	.99	1.03	1.04
Year 3:	.99	.98	1.00	1.00	.99	1.01	1.02	.99	.99	1.02	.98	1.00
Year 4:	.96	1.01	.99	.98	1.01	.99	1.00	1.02	1.02	1.01	.99	.98

a. Calculate a seasonal index by averaging the values for each month separately.

b. Calculate detrended, deseasonalized values for the last six months of year 4.

16.33 An intercity transportation company maintains quarterly records on the number of riders on its "milk run" (short distance, many stops) bus routes. The data (in thousands of riders per quarter) are

	Quarter			
Year	1	2	3	4
1	3.83	4.61	5.68	3.90
2	4.17	4.92	6.29	4.18
3	4.63	5.45	6.79	4.55
4	4.95	5.92	7.32	4.98
5	5.41	6.55	7.94	5.41
6	6.16	6.98	8.71	6.06
7	6.45	7.63	9.55	6.43
8	6.75	8.38	10.67	7.04

a. Plot the data (ridership vs. quarters 1–32).

b. What sort of trend equation seems appropriate?

16.34 Fit an exponential trend curve to the data of Exercise 16.33. The following summary figures have been calculated:

$\sum \log y_t = 57.313,\quad \sum t = 528,\quad n = 32$

$\sum t \log y_t = 1004.30,\quad \sum t^2 = 11{,}440$

16.35 The following detrended values are obtained by dividing the data of Exercise 16.33 by trend values obtained using the trend equation obtained in Exercise 16.34.

	Quarter			
Year	1	2	3	4
1	.891	1.050	1.266	.851
2	.891	1.028	1.287	.837
3	.907	1.045	1.275	.836
4	.890	1.042	1.261	.840
5	.893	1.058	1.255	.837
6	.933	1.034	1.263	.860
7	.896	1.038	1.271	.838
8	.861	1.044	1.303	.840

a. Does there appear to be a strong seasonal component in the data?

b. Calculate seasonal indices for each quarter by simple averaging.

16.36 a. Calculate exponentially smoothed forecasts of the detrended values of Exercise 16.35. Use $\alpha = .5$.

b. Do these forecasts work well for quarter 3 in various years? Can you think of an explanation?

16.37 a. Calculate 5-quarter moving averages for the data of Exercise 16.35.

b. Calculate 5-quarter running medians for these data.

c. Would these figures, used as forecasts, do well for quarter 3?

d. What further step would you suggest for improving moving-average or running-median forecasting for this situation?

16.38 A food-service company provides vending machines for a growing number of companies in a metropolitan area. Data have been collected for the past seven years on the quarterly prices of the major lines of business—sandwiches, hot meals, beverages, and desserts. In addition, total dollar-sales data are available. The data on prices are as follows:

	Quarter			
Year 1	1	2	3	4
Sandwiches	.62	.62	.65	.66
Hot meals	.80	.83	.86	.88
Beverages	.20	.20	.20	.20
Desserts	.35	.35	.36	.38
Year 2				
Sandwiches	.65	.68	.69	.71
Hot meals	.88	.91	.93	.99
Beverages	.20	.20	.21	.23
Desserts	.38	.38	.39	.41
Year 3				
Sandwiches	.73	.75	.78	.80
Hot meals	1.01	1.04	1.07	1.11
Beverages	.23	.25	.25	.26
Desserts	.41	.42	.45	.45
Year 4				
Sandwiches	.82	.85	.85	.87
Hot meals	1.14	1.19	1.20	1.22
Beverages	.30	.30	.30	.32
Desserts	.45	.47	.48	.50
Year 5				
Sandwiches	.90	.94	.97	1.00
Hot meals	1.26	1.32	1.35	1.40
Beverages	.32	.35	.35	.37
Desserts	.50	.51	.53	.56
Year 6				
Sandwiches	1.02	1.05	1.08	1.10
Hot meals	1.43	1.47	1.53	1.55
Beverages	.37	.38	.39	.39
Desserts	.57	.58	.61	.63
Year 7				
Sandwiches	1.12	1.15	1.20	1.23
Hot meals	1.56	1.61	1.70	1.74
Beverages	.40	.40	.41	.41
Desserts	.65	.66	.69	.71

The data on total sales (in thousands of dollars per quarter) are

	Quarter			
Year	1	2	3	4
1	138.93	143.60	150.47	155.77
2	158.00	164.89	171.77	184.48
3	189.45	198.38	209.69	220.37
4	231.40	244.36	249.67	260.99
5	271.25	289.20	300.49	316.63
6	327.25	340.49	357.53	367.32
7	376.85	389.49	411.34	422.54

a. Assume that the relative quantity weights are sandwiches, .30; hot meals, .28; beverages, .26; and desserts, .16. Calculate price indices for the fourth quarter in years 1 to 7.
b. Recalculate the price indices using the relative quantity weights sandwiches, .25; hot meals, .35; beverages, .28; and desserts, .12.
c. How much difference does it make which set of weights is used? Why?

16.39 The total-sales data of Exercise 16.38 are deflated by a price index based on the quantity weights shown in part (a). The deflated values are

	Quarter			
Year	1	2	3	4
1	138.93	141.48	143.17	144.76
2	147.52	149.63	152.41	155.42
3	156.82	158.96	162.05	165.69
4	168.29	171.36	173.98	176.94
5	179.40	182.70	185.03	187.46
6	190.38	192.91	195.27	197.38
7	199.29	201.08	202.73	203.63

a. Plot the values. What form of trend equation seems appropriate?
b. Calculate a linear trend equation.

16.40 Differences of successive quarterly values for the data of Exercise 16.39 are calculated. The following lag correlations are computed for these differences:

Lag:	0	1	2	3	4	5	6
Correlation:	1.000	.318	.152	.170	.127	.050	.148
Lag:	7	8	9	10	11	12	
Correlation:	−.058	−.198	−.160	−.140	−.123	−.307	

a. Plot the correlations against lag number.
b. Is there any visually obvious place where the lag correlations fall off sharply?
c. What Box-Jenkins model would you fit to the difference data?

16.41 Autoregressive models of order $p = 1$ and $p = 2$ are fit to the data of Exercise 16.40. The following results are obtained:

	AR(1) Model	
Parameter	Estimate	Standard Error
Autoregressive lag 1	.409	.211
Constant	1.401	—
Residual standard deviation	.602	(25 d.f.)
Sum of squared residuals	9.076	

	AR(2) Model	
Parameter	Estimate	Standard Error
Autoregressive lag 1	.378	.222
Autoregressive lag 2	.108	.228
Constant	1.205	—
Residual standard deviation	.612	(24 d.f.)
Sum of squared residuals	9.000	

a. Write the respective models.
b. Does the more complicated AR(2) model have a dramatically smaller average squared error?
c. Which of the two models would you use?

17

SOME BASIC IDEAS OF DECISION THEORY

Most management decisions must be made in the face of uncertainty. Prices and models for new automobiles must be selected on the basis of shaky forecasts of consumer preference, national economic trends, and competitive actions. The size and allocation of a hospital nursing staff must be decided with limited information on patient load. The inventory level of a product must be set in the face of uncertainty about demand.

Probability is the language of uncertainty. The concepts of probability theory developed earlier in this book can be used to make decisions under uncertainty. In this chapter we discuss some basic concepts of statistical decision theory. Section 17.1 deals with definitions and concepts. In Section 17.2 we discuss decision making based solely on expected values, ignoring risk. Decision making in the face of risk is measured in Section 17.3 by way of expected values and variances. Finally, in Section 17.4 we introduce some of the basic ideas of utility theory.

17.1 THE COMPONENTS OF DECISION THEORY

Our first task is to provide a reasonable framework for analyzing decisions in situations involving risk and uncertainty. In this section we define and illustrate the basic language of statistical decision theory.

The first essential ingredient of decision theory is the decision to be made. To use decision theory, it is necessary to specify the possible actions (decisions) that may be

action space taken. The set of possible actions is called the **action space** A; individual actions are labeled a_1, a_2, and so on:

$$A = \{a_1, a_2, \ldots, a_n\}$$

It is assumed that the actions are defined in such a way that exactly one action will be chosen.

EXAMPLE 17.1 A grocery manager must decide how much bread to stock each day. Available space accommodates up to 84 loaves. Define an action space.

Solution Since the manager must select the number of loaves to stock, the possible actions are the integers between 0 and 84, inclusive.

$$A = \{0, 1, 2, \ldots, 83, 84\}$$

□

EXAMPLE 17.2 A product manager has two new frozen dinner products that might be marketed. For each product, the manager can take on of the following actions: introduce it, scrap it, or send it back to the laboratory kitchens for further development. Define an action space that encompasses the two new products.

Solution The requirement that one and only one action is taken makes it convenient to combine the various possible decisions for each frozen dinner product into overall actions. Since there are three possible decisions for each product, there are nine possible actions. These are listed in the action space.

$A = \{$(introduce 1, introduce 2), (introduce 1, scrap 2), (introduce 1, redevelop 2),
(scrap 1, introduce 2), (scrap 1, scrap 2), (scrap 1, redevelop 2),
(redevelop 1, introduce 2), (redevelop 1, scrap 2), (redevelop 1, redevelop 2)$\}$

□

Decision theory has to do with decisions made under uncertainty. To use it, there must be uncertainty about something. The something may be relatively concrete, such as tomorrow's demand for bread, or relatively amorphous, such as next year's consumer confidence level. The idea is that the eventual degree of success of any selected action depends on which one of several future outcomes* actually occurs. **The second essential element of decision theory is the list of possible outcomes.** This set is essentially the sample space introduced in Chapter 3. It is customary to denote the possible **outcomes** as θ_1, θ_2, ... The set of possible outcomes is $\Theta = \{\theta_1, \theta_2, \ldots\}$.

EXAMPLE 17.3 In Example 17.1, the store's profit from stocking a certain number of loaves of bread depends on the actual demand of customers for bread. Define an appropriate sample space.

* These outcomes are sometimes called *states of nature*, but there is no requirement for them to refer to natural phenomena.

Solution One good way to define the set of possible outcomes is to specify the possible numbers of loaves demanded.

$$\Theta = \{0, 1, 2, \ldots, 83, 84 \text{ or more}\}$$

□

EXAMPLE 17.4 In Example 17.2, the success of the possible actions depends on consumer preference for the new products compared to current competitors. It also depends on consumer preferences for potential redeveloped products, but for simplicity we ignore that. Assume that preference can adequately be specified as very high, high, medium, low, or very low for each of the two products. Define an appropriate set of possible outcomes.

Solution Again, we want to define the set so that exactly one outcome occurs. One possibility is that both products receive very high preference; call this outcome (VH_1, VH_2). Using this code, we can write the possible outcomes as

$$\Theta = \{(VH_1, VH_2), (VH_1, H_2), (VH_1, M_2), (VH_1, L_2), (VH_1, VL_2), (H_1, VH_2), (H_1, H_2), \ldots, (VL_1, L_2), (VL_1, VL_2)\}$$

□

The set comprises 25 different outcomes.

One of the virtues of decision theory is that it forces a manager to be explicit about the possible actions that may be taken. It is worthwhile to take some time to consider whether the action space A really includes all the possible actions that may plausibly be taken. Having specified the possible actions, a manager must consider outcomes associated with these actions. Fortunately, it is usually not so crucial to list all the possible outcomes; often even a considerable simplification of the outcome set does not seriously distort the relative desirability of the various actions. In Example 17.4 we could have defined 50 acceptance levels for each product rather than five. It is unlikely that the increased complexity would lead to any substantial change in the relative merits of the actions.

payoff function

The third component of statistical decision theory is the payoff function. For each possible action and each resulting outcome, one must specify a numerical value

$$v(a_i, \theta_j)$$

that represents the profit (or desirability) of action a_i given that outcome θ_j actually occurs. Often the payoffs are arranged in a matrix and sometimes they are specified by a mathematical function.

EXAMPLE 17.5 Suppose that the store in the bread inventory problem of Examples 17.1 and 17.3 earns a profit of 7 cents for each loaf that is sold and loses 3 cents per loaf of unsold bread (which must be discounted on the day-old shelf). For simplicity, limit the action space for loaves ordered to $A = \{0, 1, 2\}$ and the outcome space for loaves demanded to $\Theta = \{0, 1, 2\}$. Compute the values for the payoff function $v(a_i, \theta_j)$.

Solution If $a_i = 0$, no bread is ordered and there is no profit (or loss) no matter what the outcome. Hence

$$v(0, \theta_j) = 0, \qquad \text{for all } \theta_j$$

If $a_i = 1$ and $\theta_j = 0$, the one loaf ordered is not sold, and

$$v(1, 0) = -3$$

If $a_i = 1$ and $\theta_j = 1$, the one loaf ordered is sold, so

$$v(1, 1) = 7$$

Note also that

$$v(1, 2) = 7$$

If $a_i = 2$, then it is easy to see that

$$v(2, 0) = -6, \quad v(2, 1) = 7 - 3 = 4, \quad \text{and} \quad v(2, 2) = 14$$

In general, the profit depends on whether the demand is greater than or less than the number of loaves ordered. If $\theta_j \geq a_i$, all a_i loaves are sold, for a total profit of $7a_i$. If $\theta_j < a_i$, only θ_j loaves are sold and $a_i - \theta_j$ are left unsold, for a profit of $7\theta_j - 3(a_i - \theta_j)$. It follows that

$$v(a_i, \theta_j) = \begin{cases} 7a_i, & \text{if } \theta_j \geq a_i \\ 7\theta_j - 3(a_i - \theta_i), & \text{if } \theta_j < a_i \end{cases}$$

A portion of the payoff matrix for Examples 17.1 and 17.3 is shown below.

	Outcome θ_j					
Action a_i	0	1	2	3	4	...
0	0	0	0	0	0	
1	−3	7	7	7	7	
2	−6	4	14	14	14	...
3	−9	1	11	21	21	
4	−12	−2	8	18	28	
⋮			⋮			

□

numerical payoffs

The requirement of a **numerical payoff** function is a limitation on the applicability of statistical decision theory. Most managers have a very hard time assigning numerical, quantitative values (the v's) to intangible, qualitative results. While an insurer can quantify the value of the loss of a worker's life as the stated policy coverage, we could hope that the managers of the worker's plant have a more complicated, less numerical loss. There are two extreme positions that can be (and are) taken concerning the applicability of decision theory in light of the requirement of a numerical payoff function.

1. Decision theory should restrict itself to the readily quantified aspects of a decision. The manager can weigh these results against the qualitative aspects of the problem.
2. Decisions involving intangibles are made every day. These decisions implicitly put a quantitative value $v(a_i, \theta_j)$ on each qualitative, intangible result. If decision theorists are clever enough, they can obtain these values and incorporate them in a numerical analysis.

We suspect that most decision theorists hold something like the second position; it leads to a wide scope of application and to many interesting research activities. But a skeptical manager might do well to stick with something like the first position, even while testing out methods for evaluating intangible results. It may not be wise to make hard decisions based on soft numbers.

We believe there is a third position. It usually is not necessary to specify each value $v(a_i, \theta_j)$ to the nearest penny. By a **sensitivity analysis**, which we discuss in the next section, it is often possible to find an action that looks good over a fairly wide range of possible values of $v(a_i, \theta_j)$ assigned to the intangible results (a_i, θ_j). Such analyses are particularly useful when payoffs can be specified only within limits.

sensitivity analysis

The final element in the development of the basic concepts of decision theory is probability. Since statistical decision theory is useful when there is an element of uncertainty, we need to incorporate probability. A probability $P(\theta_j)$ must be specified for each possible outcome θ_j in Θ. These probabilities may be determined subjectively, by statistical methods that may or may not use forecasting, or by a combination of approaches. In practice it is rare that a manager can assess probabilities with complete accuracy. Indeed, two managers may well have different subjective probability assessments. Sensitivity analysis can also be useful in determining how important such differences are in selecting a desirable action.

probability for uncertainty

With development of the basic elements of decision theory (action space, outcome space, payoff function, and probability) we can illustrate how to use these elements in making decisions.

SECTION 17.1 EXERCISES

17.1 An independent automobile rental firm must decide how many cars to purchase three months in advance of a new model year. The company currently uses 80 cars, which all must be replaced. A maximum of 100 cars can be stored and serviced, and a minimum of 40 cars must be on hand to provide competitive service. The desired number of cars depends heavily on the level of business travel when the cars become available. The level of travel may be very light, light, moderate, heavy, or very heavy.

a. Identify an action space.

b. Identify the set of possible outcomes (states of nature).

17.2 The car rental firm of Exercise 17.1 analyzes the possible results for ordering 50, 60, 70, 80, 90, and 100 cars. The following approximate payoff matrix (in thousands of dollars profit per month) is obtained:

	Outcome, Travel Level				
Action, Cars Ordered	VL	L	M	H	VH
50	30	40	40	40	40
60	35	50	55	55	55
70	20	40	60	70	70
80	25	30	55	80	85
90	−10	20	50	80	110
100	−25	10	45	80	105

a. Show that ordering 50 cars is an unreasonable action. (Hint: Compare the payoffs to 50 and 60 cars for each possible travel level.) In the language of decision theory, ordering 50 cars is an inadmissible action.
b. Are there any other actions that are clearly unreasonable?

17.3 The concession holder at a major league baseball stadium must decide how many refreshment stands to open for each game. One stand is adequate for about 1000 fans. Every stand that is open, with enough fans in attendance, yields a profit of $800. Each excess stand results in a net loss of $200. So, if there are 10 stands open but only 9000 fans, 9 stands make a total of $7200 in profits but the tenth stand loses $200, for a net profit of $7000. Assume that the maximum possible crowd is 30,000, that there are 30 stands that can be opened, and that the number of fans is an exact multiple of 1000.
a. Specify an action set.
b. Specify a set of possible outcomes.
c. Construct the payoff matrix.

17.4 Are there any obviously unreasonable actions in the action set of Exercise 17.3; that is, can any action be ruled out as always being inferior to another one for all possible outcomes?

17.5 An investor has the opportunity to buy all or part of an apartment complex that can possibly be converted into condominiums; the investor can buy a 100%, 75%, 50%, 25%, or 0% share. The return on investment depends on two factors—whether or not a condominium-conversion law is passed in the city and whether rental levels rise or stay the same. If the law is not passed, the complex will be converted, and the profit for a 100% share will be $100,000 (present value). If the law is passed and the rental level rises, the profit for a 100% share will be $40,000. If the law is passed and the rental level stays the same, there will be a loss of $20,000 for a 100% share. Profits or losses for partial shares are proportional to the size of a share; thus, for instance, the profit to a 75% share of a converted complex is $75,000.
a. Specify the action space.
b. Specify the set of possible outcomes.
c. Construct the payoff matrix.

17.6 A firm is facing a lawsuit charging discrimination against older employees. Attorneys for the firm believe that an out-of-court settlement can be reached at a cost of $80,000. Alternatively, the firm can contest the suit. If it is found not guilty, legal fees will cost $20,000. If the suit is tried by Judge A and the firm is found guilty, the cost is expected to be $150,000. If it is tried by Judge B and the firm is found guilty, the cost is expected to be $180,000.
a. Describe an action space and a set of possible outcomes.
b. Construct a payoff matrix. Express costs as negative profits.

17.2 DECISION MAKING USING EXPECTED VALUES

expected payoff

In this section we combine the elements defined in Section 17.1 to assess the merits of various possible actions. The basic selection criterion we use is the action a_i that results in the largest **expected payoff**. This criterion combines the payoff and probability aspects of a decision problem to give a numerical measure of the expected return to each possible action.

Recall from Chapter 4 that the expected value of any (discrete) random variable Y with probability distribution $P_Y(y)$ is

$$E(Y) = \sum yP_Y(y)$$

that is, $E(Y)$ is the probability-weighted average of the possible Y values. In decision theory, the random variable is the payoff $v(a_i, \theta_j)$ associated with a particular action a_i and outcome θ_j. The possible outcomes are assigned probabilities $P(\theta_j)$. The expected payoff, or **expected return**, to action a_i is often denoted $R(a_i)$.

expected return

Expected Return to Action a_i

$$R(a_i) = \sum_j v(a_i, \theta_j)P(\theta_j)$$

where $v(a_i, \theta_j)$ is the payoff for action a_i if outcome θ_j occurs.

EXAMPLE 17.6 The manager of a clothing store must decide how many extra salespoeple to schedule for the store's annual clearance sale. If the number of additional customers is small, the existing sales force can handle the load. If the crowds turn out to be heavier, additional salespeople are needed to avoid the loss of sales and goodwill. The manager's estimated payoffs (in hundreds of dollars per day) are listed in the following table for the action space $A = \{0, 1, 2, 3\}$ and outcome space $\Theta = \{\text{small, moderate, large}\}$. The manager's estimates for the outcome probabilities $P(\theta_j)$ are also listed.

Actions, Additional Salespeople	Outcome, Additional Customers		
	Small	Moderate	Large
0	1	1	1
1	−1	2	3
2	−3	3	6
3	−5	3	10
$P(\theta_j)$	.2	.6	.2

Compute the expected return for each action and determine the action with the highest expected return.

Solution The expected payoffs can be obtained for this manager's problem using the definition

$$R(a_i) = \sum_j v(a_i, \theta_j)P(\theta_j)$$

Taking the actions, payoffs, and outcome probabilities, we obtain the following expected payoffs.

Action, Additional Salespeople	Expected Payoff $R(a_i)$
0	$1(.2) + 1(.6) + 1(.2) = 1.0$
1	$-1(.2) + 2(.6) + 3(.2) = 1.6$
2	$-3(.2) + 3(.6) + 6(.2) = 2.4$
3	$-5(.2) + 3(.6) + 10(.2) = 2.8$

The action with the highest expected payoff is to schedule three extra salespeople. The selection of the action with the highest expected return can also be done using an **expected-value tree**. For these same data, the expected-value tree would be as shown here:

expected-value tree

Action a_i	Outcome θ_j	Payoff $v(a_i, \theta_j)$	Probabilities $P(\theta_j)$	Expected Return $R(a_i)$
Add 0	small	1	.2	
	moderate	1	.6	1.0
	large	1	.2	
Add 1	small	−1	.2	
	moderate	2	.6	1.6
	large	3	.2	
Add 2	small	−3	.2	
	moderate	3	.6	2.4
	large	6	.2	
Add 3	small	−5	.2	
	moderate	3	.6	2.8
	large	10	.2	

□

As we noted previously, the expected-value criterion is to select the action that has the highest expected return. This criterion is appealing in several respects. The required calculations are very easy. More important, this criterion is desirable in the long run. A manager who takes action based on the expected-value criterion will have higher average payoffs in the long run than one who takes action on some other basis. This argument has particular force for decisions that must be made repeatedly under identical (or at least similar) conditions. In such cases, the long-run average return per action approximates the expected return, so of course selecting the highest expected return gives the best long-run result.

risk

The hitch in this argument is that applying a strategy that maximizes the expected payoff ignores short-run **risk**. A long-run argument isn't much consolation to the owners of a small firm that goes bankrupt taking a short-run loss. We discuss the assessment of risk in the next section.

We have seen that the expected return to a particular action is determined by the payoffs to various actions given various outcomes and by the probabilities of those outcomes. These quantities can be estimated in many situations, but it is rare that they are known exactly. Instead, these quantities are more or less rough estimates. The potential effect of misestimation on the supposedly optimal action (and its near-optimal competitors) can be found by **sensitivity analysis**.

sensitivity analysis

The idea of sensitivity analysis is to vary the basic quantities of the analysis to see if the best action remains optimal, or nearly so. The underlying quantity estimates can be varied one at a time or in combination. There are usually many ways to vary the estimates. Ideally, the best action remains optimal (or nearly optimal) when the estimates are altered in ways that are unfavorable to that action. In such a case a manager can be confident that the selected action is a good one.

EXAMPLE 17.7 In Example 17.6, the highest expected return was obtained by scheduling three additional salespeople. Suppose that the store manager is uncertain about the payoffs when there is a large number of additional customers, and also about the respective probabilities of small, moderate, and large numbers of additional customers. The payoffs (in hundreds of dollars per day) for the action space $A = \{\text{schedule 0, 1, 2, or 3 additional salespeople}\}$ and the outcome (large) can be as low as 1, 2.5, 6, and 9 or as high as 1, 4, 8, and 12. The probabilities $P(\theta_j)$ for small, moderate, and large numbers can be as pessimistic as .3, .6, and .1 or as optimistic as .1, .5, and .4. Conduct a sensitivity analysis by computing the expected payoffs under these extreme situations when there are large numbers of additional customers.

Solution There are four combinations of changes in payoffs and outcome probabilities that can be made:

Payoffs, $v(a_i, \text{large})$		Outcome Probabilities, $P(\theta_j)$
(1, 2.5, 6, 9)	with	(.3, .6, .1)
(1, 2.5, 6, 9)	with	(.1, .5, .4)
(1, 4, 8, 12)	with	(.3, .6, .1)
(1, 4, 8, 12)	with	(.1, .5, .4)

We can (and you should) calculate the expected returns for each combination to see if a_4 (schedule 3 additional salespeople) remains optimal. In this example, the critical combination is clear. The reason that a_4 was originally declared optimal is that there was a high payoff for a large number of additional customers, and P(large) was fairly high. We should calculate expected returns assuming the lower (1, 2.5, 6, 9) payoffs and the pessimistic (.3, .6, .1) probabilities, as indicated in the following revised payoff table:

Action, Additional Salespeople	Outcome, Additional Customers		
	Small	Moderate	Large
0	1	1	1
1	−1	2	2.5
2	−3	3	9
3	−5	3	9
$P(\theta_j)$	.3	.6	.1

The expected payoffs using this revised table are shown below:

Action, Additional Salespeople	Expected Payoff $R(a_i)$
0	$1(.3) + 1(.6) + 1(.1) = 1.0$
1	$-1(.3) + 2(.6) + 2.5(.1) = 1.15$
2	$-3(.3) + 3(.6) + 6(.1) = 1.5$
3	$-5(.3) + 3(.6) + 9(.1) = 1.2$

By changing payoffs and probabilities in an unfavorable direction, we observe that the action a_4 (schedule 3 additional salespeople) becomes the second-best decision to a_3 (schedule 2 additional salespeople). The difference in payoffs between a_3 and a_4 is \$30 per day. Since the advantage of a_3 over a_4 is small even when we loaded the deck in its favor, however, we would conclude that a_4 is a near-optimal solution that is reasonably insensitive to changes in the underlying payoffs and probabilities. □

SECTION 17.2 EXERCISES

17.7 A drug manufacturer has applied to the Food and Drug Administration for approval of a new prescription drug. A ruling to be made in 18 months will either allow the firm to market the drug immediately, require additional information, or completely reject the drug. The firm can choose to begin full production now, so that on approval the drug can be marketed immediately, go into limited production, or await the final decision. Payoffs to the various actions are as follows (in hundreds of thousands of dollars):

	Approved	Additional Information	Disapproved
Full production	9.7	−2.9	−3.9
Limited production	3.6	−0.7	−1.7
Await decision	0	0	0

The probability of approval is assessed at .7, of additional information at .2, and of disapproval at .1. Based on the expected-value criterion which is the best action?

17.8 Refer to Exercise 17.7. An executive of the firm claims that the payoffs to full production are closer to 6.0 (approved), −3.5 (additional information), and −4.5 (disapproved). Another executive asserts that the respective probabilities are .5, .2, and .3.

a. Which is the best action (in expected value), assuming the revised payoffs and the original probabilities?
b. Which is the best action (in expected value), assuming the original payoffs and the revised probabilities?
c. Which is the best action (in expected value), assuming the revised payoffs and the revised probabilities?
d. Summarize the results of this sensitivity analysis.

17.9 A firm is considering three price levels for a new product. The payoffs depend on the time required for another company to market its competitive product. The relevant assumed values are

	Profits (Millions of Dollars, Present Value)		
Price	at 1 Year	at 1.5 Year	at 2 Years
Low	3.2	3.4	3.5
Moderate	2.9	4.5	4.8
High	2.0	4.0	5.5
Probabilities	.25	.50	.25

What price level yields the highest expected profit?

17.10 Refer to Exercise 17.9. The firm's marketing staff believes that the profits for moderate prices are close to correct but that the profits for low and high prices could be in error by as much as half a million dollars in either direction for each amount of time required.

a. In which direction should the profits for low and high prices be changed to perform a sensitivity analysis?

b. Does the originally optimal action remain so with the modified profit figures?

17.11 Refer to Exercises 17.9 and 17.10.

a. What action has the highest expected payoff if the probabilities for 1 year, 1.5 years, and 2 years are .4, .3, and .3, respectively, and if the payoffs of Exercise 7.9 are correct?

b. Which action has the highest expected payoff if these modified probabilities hold and if the profit figures for low and high prices are each raised by half a million dollars?

c. Summarize the results of this sensitivity analysis.

17.12 Refer to Exercise 17.6. Assume that the case will be assigned to Judge A with probability .7 and to Judge B with probability .3. Whichever judge handles the case, the probability that the firm will be found guilty is .4. Which action (settlement or going to trial) has the lower expected cost?

17.13 Refer to Exercise 17.12. Does the optimal action change if the probability that the firm will be found guilty is increased to .6? decreased to .3?

17.3 THE ELEMENT OF RISK

In the last section we suggested that expected return is a good, basic, long-run criterion for selecting a desirable action. The difficulty with expected return as a criterion is that it ignores the element of risk. According to the expected-payoff criterion, an investment strategy with an action that has payoffs of \$500 or \$1500 with equal probabilities (1/2) is no more or less desirable than an action that has a \$50,000 loss or \$52,000 gain with equal probabilities. Both have an expected payoff of \$1000. However, when one considers the severe risk associated with the win \$52,000–lose \$50,000 situation, it is not difficult to select the less risky investment. In this section we discuss one way to evaluate risk. In the next section we discuss some methods that have been proposed for combining the aspects of expected return and risk.

Generally speaking, for a given expected return, risk increases as the probabilities of both extremely good and extremely bad payoffs increase; the more spread out the payoff probabilities, the riskier the decision. Beginning in Chapter 2, we have used variance (or standard deviation) as a measure of the spread in sample data. In the same spirit, variance can be considered as a measure of risk; the greater the variance (variability) for a given action, the larger the risk associated with that action.

variance and risk

The variance for a particular action a_i is defined below in the notation of decision theory:

Variance of an Action Var(a_i)

$$\text{Var}(a_i) = \sum_j [v(a_i, \theta_j) - R(a_i)]^2 P(\theta_j)$$

A shortcut formula is

$$\text{Var}(a_i) = \sum_j [v(a_i, \theta_j)]^2 P(\theta_j) - [R(a_i)]^2$$

$$\text{Standard deviation } (a_i) = \sqrt{\text{Var}(a_i)}$$

EXAMPLE 17.8 Calculate Var(a_i) for each of the actions a_i for the scheduling problem discussed in Example 17.6.

Solution The payoffs, outcome probabilities, and expected payoffs for the action space A = {schedule 0, 1, 2, or 3 additional salespeople} are summarized here:

Action, Additional Salespeople	Outcome, Additional Customers			Expected Payoff $R(a_i)$
	Small	Moderate	Large	
0	1	1	1	1.0
1	−1	2	2	1.6
2	−3	3	6	2.4
3	−5	3	10	2.8
$P(\theta_j)$	.2	.6	.2	

Using the shortcut formula for Var(a_i), we can compute the risk for each action as follows:

Action, Additional Salespeople	Risk Var(a_i)
0	$1(.2) + 1(.6) + 1(.2) - (1)^2 = 0$
1	$1(.2) + 4(.6) + 9(.2) - (1.6)^2 = 1.84$
2	$9(.2) + 9(.6) + 36(.2) - (2.4)^2 = 8.64$
3	$25(.2) + 9(.6) + 100(.2) - (2.8)^2 = 22.56$

Although action a_4 (schedule 3 additional salespeople) has the highest expected payoff, it also has the highest risk. You, as a manager, may be willing to take an action with a lower expected payoff and correspondingly lower risk. □

Variance is a particularly useful measure for assessing the combined risk of a portfolio. In this context we need the concept of covariance introduced in Chapter 4. Recall that the covariance of two random variables X and Y, Cov(X, Y), is a measure of the strength of the (linear) relation between them. In the Appendix of Chapter 4 we proved that, for any random variables X and Y,

$$\text{Var}(X + Y) = \text{Var}(X) + \text{Var}(Y) + 2\,\text{Cov}(X, Y)$$

Thus, if the returns to two investments tend to be either both large or both small, the covariance is positive, inflating the variance of the total return. But if one of the investments tends to give a large return when the other gives a small return, the covariance is negative, and the variance of the total return is smaller.

We need some definitions. Suppose that a manager has d dollars that can be distributed among k different investments. Let

Y_i = return per dollar assigned to investment i ($i = 1, 2, \ldots, k$) a random variable with mean μ_i and variance σ_i^2

σ_{ij} = covariance between the return on investments Y_i and Y_j

d_i = dollar amount assigned to investment i ($i = 1, 2, \ldots, k$)

$d = \sum_i d_i$

In this situation, an action is a choice of a portfolio of investments $(d_1, d_2, \ldots, d_k)$. An outcome is an entire set of returns to the investments $(Y_1, Y_2, \ldots, Y_k)$. Therefore the action set and outcome set are

$$A = \{(d_1, d_2, \ldots, d_k)\}$$
$$\Theta = \{(Y_1, Y_2, \ldots, Y_k)\}$$

The payoff (return to the portfolio) for any given action $(d_1, d_2, \ldots, d_k)$ and outcome $(Y_1, Y_2, \ldots, Y_k)$ is

$$d_1Y_1 + d_2Y_2 + \ldots + d_kY_k = \sum d_iY_i$$

where k is the number of distinct investments.

EXAMPLE 17.9 Suppose an investor has \$50,000 to invest and would like to invest in some combination of stock A (1), stock B (2), bond (3), and treasury bill (4). Assume that the expected dollar returns for these investments over a one-year period are

Investment:	Stock A (1)	Stock B (2)	Bond (3)	Bill (4)
Expected return/dollar, μ_i:	1.130	1.120	1.095	1.088

Also assume that the variances and covariances for the returns per dollar invested are as shown next:

	Investment			
	1	2	3	4
Investment				
1	.0597	.0369	.0051	.0000
2		.0421	.0013	.0000
3			.0049	.0000
4				.0000

Identify μ_1, μ_3, σ_2^2, σ_4^2, σ_{13}, and σ_{23}.

Solution The expected returns per dollar invested are given in the first table above: $\mu_1 = 1.130$ and $\mu_3 = 1.095$. Variances and covariances are presented in the second table. Note that variances appear along the diagonal and covariances off the diagonal.

$$\sigma_2^2 = .0421, \qquad \sigma_4^2 = .0000$$

and

$$\sigma_{13} = .0051, \qquad \sigma_{23} = .0013$$

□

According to the basic properties of expected value developed in Chapter 4, the expected return to the portfolio is just the sum of dollars invested times expected earnings per dollar:

$$E\left(\sum_i d_i Y_i\right) = \sum_i d_i E(Y_i) = \sum d_i \mu_i$$

Variance (or risk) computations are more complicated, because both variances and covariances are involved. For a portfolio with $k = 2$ investments, it can be shown that the variance (risk) of the portfolio is

portfolio risk

$$\begin{aligned} \text{Var}(d_1Y_1 + d_2Y_2) &= d_1^2 \text{ Var}(Y_1) + d_2^2 \text{ Var}(Y_2) + 2d_1d_2 \text{ Cov}(Y_1, Y_2) \\ &= d_1^2\sigma_1^2 + d_2^2\sigma_2^2 + 2d_1d_2\sigma_{12} \end{aligned}$$

In general, the variance of the portfolio payoff is

$$\begin{aligned} \text{Var}\left(\sum_i d_iY_i\right) &= \sum_i d_i^2 \text{ Var}(Y_i) + 2\sum_{i<j} d_id_j \text{ Cov}(Y_i, Y_j) \\ &= \sum_i d_i^2\sigma_i^2 + 2\sum_{i<j} d_id_j\sigma_{ij} \end{aligned}$$

where the notation $\sum_{i<j}$ means sum over all possible pairs where i is less than j.

EXAMPLE 17.10 Refer to the investment problem of Example 17.9. Suppose that the manager of the account wants to place \$20,000 in stock A (1), \$10,000 in stock B (2), \$15,000 in bond (3), and \$5000 in treasury bill (4).

a. Identify d.
b. Identify the particular action (d_1, d_2, d_3, d_4).
c. Compute the expected return for this investment action.
d. Compute the variance (and standard deviation) of the investment return for this action.

Solution a. The total number of dollars available for investment is $d = \$50{,}000$.

b. The investment action chosen by the manager is $(d_1 = \$20{,}000,\ d_2 = \$10{,}000,\ d_3 = \$15{,}000,\ d_4 = \$5000)$.

c. The expected return for this action is

$$E\left(\sum d_i Y_i\right) = \sum d_i \mu_i$$

Substituting for the d_i and μ_i, we obtain

$$E\left(\sum d_i Y_i\right) = 20{,}000(1.130) + 10{,}000(1.120) + 15{,}000(1.095) + 5000(1.088) = \$55{,}665$$

d. We can use the expression for $\text{Var}(\sum d_i Y_i)$ to compute the variance for the investment return associated with the action $(d_1 = \$20{,}000,\ d_2 = \$10{,}000,\ d_3 = \$15{,}000,\ d_4 = \$5000)$.

$$\text{Var}\left(\sum d_i Y_i\right) = \sum d_i^2 \text{ Var}(Y_i) + 2\sum_{i<j} d_i d_j \text{ Cov}(Y_i, Y_j)$$

The first term in this formula is

$$\begin{aligned}\sum d_i^2 \text{ Var}(Y_i) &= (20{,}000)^2(.0597) + (10{,}000)^2(.0421) \\ &\quad + (15{,}000)^2(.0049) + (5000)^2(.0000) \\ &= 29{,}192{,}500\end{aligned}$$

The second term is

$$\begin{aligned}2\sum_{i<j} d_i d_j \text{ Cov}(Y_i, Y_j) &= 2[20{,}000(10{,}000)(.0369) + 20{,}000(15{,}000)(.0051) \\ &\quad + 20{,}000(5000)(.0000) + 10{,}000(15{,}000)(.0013) \\ &\quad + 10{,}000(5000)(.0000) + (15{,}000)(5000)(.0000)] \\ &= 18{,}210{,}000\end{aligned}$$

Hence $\text{Var}(\sum d_i Y_i) = 29{,}192{,}500 + 18{,}210{,}000 = 47{,}402{,}500$ and the standard deviation of investment return is $\sqrt{47{,}402{,}500} = \$6884.95$.

In summary, the expected return for the investment action $(d_1 = \$20{,}000,\ d_2 = \$10{,}000,\ d_3 = \$15{,}000,\ d_4 = \$5000)$ is \$55,665 and the standard deviation is \$6884.95. □

risk and return

There is typically a positive relation between risk (variance) and expected return; individual investments with high returns also carry high risks, and low-risk investments yield low returns. But by judicious combination of investments, it may be possible to obtain a fairly high expected return with a fairly low variance.

EXAMPLE 17.11 Refer to Examples 17.9 and 17.10. Compute the expected return, variance, and standard deviation for an investment action that places all \$50,000 in treasury bills (4).

Solution The particular action chosen is ($d_1 = 0$, $d_2 = 0$, $d_3 = 0$, $d_4 = \$50{,}000$). The expected payoff and variance for this action are easy to compute, since $d_1 = d_2 = d_3 = 0$.

$$E\left(\sum d_i Y_i\right) = \sum d_i \mu_i = \$50{,}000(1.088) = \$54{,}400$$

$$\text{Var}\left(\sum d_i Y_i\right) = d_4^2 \text{Var}(Y_4) = (50{,}000)^2(.0000) = .0000$$

Compared to the action ($d_1 = \$20{,}000$, $d_2 = \$10{,}000$, $d_3 = \$15{,}000$, $d_4 = \$5000$), the action ($d_1 = d_2 = d_3 = 0$, $d_4 = \$50{,}000$) has a lower expected return, but the net gain of \$4400 (versus \$5665) is without risk. □

Variance (or standard deviation) has been widely used as the basic measure of risk in financial portfolio analysis. As a risk measure, it is fairly easy to calculate. One very convenient property is that the portfolio variance is calculated from individual variances and from covariances of pairs of investment outcomes Y_i, Y_j. A limitation of variance as a risk measure is that it treats above-expected and below-expected returns symmetrically. For example, suppose the payoff $\sum d_i Y_i$ for a particular action ($d_1, d_2, \ldots, d_k$) has a probability distribution as indicated in Figure 17.1a. The expected payoff and variance for the investment action are $\sum d_i \mu_i$ and $\text{Var}(\sum d_i Y_i)$, respectively. Consider a second investment action ($d_1', d_2', \ldots, d_k'$), which has the same expected payoff and variance but a different probability distribution (Figure 17.1b). Should we consider the two investment actions to be equally risky? The second investment action clearly has a higher "down-side risk" than the first investment action, even though the risks (as measured by the variances) are identical. When the probability distribution of payoffs is highly skewed, variance may be a

skewness problem

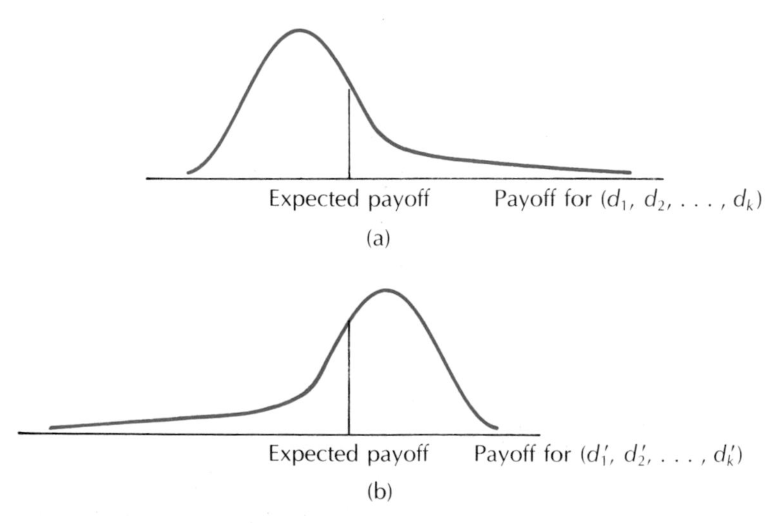

FIGURE 17.1 **Possible Payoff Distribution**

somewhat deceptive measure of risk. Fortunately, the return to portfolio of many investments is likely to have a roughly symmetric, normal distribution, thanks to the Central Limit Theorem. For these symmetric distributions, it seems eminently reasonable to equate higher variance with higher risk. Thus, for portfolio analysis in particular, variance appears to be a very suitable and convenient risk measure.

SECTION 17.3 EXERCISES

17.14 A fuel oil dealer must decide how much oil to purchase at the beginning of the heating season. The basic options are full capacity (enough oil to fill all storage tanks), 90% capacity, 80% capacity, or 70% capacity. The crucial unpredictable variable is the weather during the heating season. In a severe winter, the dealer can sell the full-capacity amount at a good profit, but with a warm winter, excess oil must be carried over to the following summer at a serious interest penalty. Also, if too little oil is ordered initially, a refill of more expensive oil must be purchased. The dealer classifies the possible outcomes as severe, cold, normal, mild, or warm winter. The payoff matrix (profits in thousands of dollars) and assessed probabilities of the possible outcomes are

	Winter				
Order	Severe	Cold	Normal	Mild	Warm
Full	120	90	60	30	0
90%	80	90	80	70	60
80%	50	70	75	70	65
70%	30	40	50	60	70
Probabilities	.10	.20	.40	.20	.10

a. Which order quantity yields the highest expected profit?
b. Which order quantity yields the lowest standard deviation (risk)?
c. Is there any conflict between the goals of maximizing expected return and minimizing risk?

17.15 Refer to Exercise 17.14. Suppose that, because of forecasts of a relatively cold winter, the assessed outcome probabilities are revised to .20, .30, .30, .15, and .05 for severe, cold, normal, mild, and warm, respectively.

a. Recalculate expected profits and standard deviations for the various actions.
b. Is there any reason to modify the choice of optimal action found in Exercise 17.14?

17.16 A revised forecast of interest rates and replacement fuel oil prices leads to the following payoff matrix for Exercise 17.14:

	Winter				
Order	Severe	Cold	Normal	Mild	Warm
Full	120	80	40	0	−40
90%	75	90	75	60	45
80%	40	65	75	65	55
70%	15	30	45	60	70

Use the probabilities shown in Exercise 17.14.

a. Which order quantity yields the highest expected return?

b. Which quantity has the lowest risk (standard deviation)?

c. Is the preferred action sensitive to this modification of the payoff matrix?

17.17 The random variables Y_1 and Y_2 represent one-year net returns to $1000 investments in project 1 and project 2. Probability distribution $P_{Y_1}(y_1)(y_1, y_2)$ is as follows:

		y_2							
		70	80	90	100	110	120	130	$Py_1(y_1)$
	80	.01	.01	.03	.09	.06	.07	.03	.30
y_1	100	.01	.02	.11	.12	.11	.02	.01	.40
	120	.03	.07	.06	.09	.03	.01	.01	.30
	$P_{Y_2}(y_2)$	.05	.10	.20	.30	.20	.10	.05	

a. Calculate the covariance of Y_1 and Y_2.

b. Calculate the correlation of Y_1 and Y_2.

17.18 a. Refer to Exercise 17.17. Calculate the expected return of a portfolio consisting of $3000 invested in project 1 and $2000 invested in project 2.

b. Calculate the variance of this portfolio.

c. Compare the return and risk of this portfolio to that of an investment of $2000 in project 1 and $3000 in project 2.

17.19 The treasurer of a certain firm has a short-term surplus of $1 million in cash to invest. The money will be needed in one year, but not before. Three possible investments for the cash have been suggested. The expected returns per dollar, variances, and covariances are shown below:

	Investment		
	A	B	C
Expected return per dollar, μ_i	1.088	1.091	1.095
Var(A) = .00	Cov(A, B) = .00		Cov(A, C) = .00
	Var(B) = .022		Cov(B, C) = .020
			Var(C) = .068

a. Find the expected return and variance for a portfolio with equal amounts in each investment.

b. Compute the expected return and risk for (1) .5 million dollars invested in each of A and C; (2) 1 million dollars invested in B. Compare your results.

17.20 Refer to Exercise 17.2. Assume that the five travel levels have equal probabilities.

a. Calculate expected returns for each number of cars ordered.

b. Find the standard deviations of returns for each number of cars ordered.

c. Is there an order quantity that has both the highest expected return and the lowest risk?

17.21 We concluded in Exercise 17.2 that ordering 50 cars was an unreasonable action because ordering 60 cars always yielded a higher profit. Compare expected returns and standard deviations for these two actions. Do the results show the inferiority of ordering 50 cars?

17.22 An investor has a choice of three stocks to purchase. Expected returns, variances, and covariances (per dollar invested) are

Stock:	1	2	3
Expected return:	1.090	1.088	1.091

Variance-Covariance Matrix

	1	2	3
1	.095	.020	.080
2	.020	.104	.030
3	.080	.030	.095

Thus, for example, the variance per dollar invested in stock 2 is .104 and the covariance for stocks 1 and 2 is .020.

a. What is the expected return to a portfolio with $1000 invested in each of the three stocks?
b. What is the variance of this portfolio?

17.23 In Exercise 17.22, stock 2 has both the lowest expected return and the highest variance. Even so, show that an investor who desires low risk might prefer the portfolio of Exercise 17.22 to a portfolio with $1500 invested in each of stocks 1 and 3.

17.4 THE BASIC IDEAS OF EXPECTED UTILITY THEORY

In the two previous sections, expected payoff and variance were interpreted as measures of return and risk, respectively. Nothing was said about the balancing—the trade-off—of these two aspects of an investment. The idea of expected utility in decision theory can be regarded as an attempt to reflect managers' desires about this trade-off in an analyzable way and to help managers be logically consistent in making their decisions. This kind of utility theory originated with von Neumann and Morganstern (1967).

The decision-theory framework of this chapter has forced us to consider a set of possible actions a_i, outcomes θ_j, payoffs $v(a_i, \theta_j)$, and the associated outcome probabilities $P(\theta_j)$. With this information we calculated two numbers, the expected value and variance, for each action.

expected utility

In utility theory, a single number, called **expected utility**, is assigned to each of several actions. The expected utility for an action reflects both the risk and the expected payoff. The key to expected utility is the specification of a **utility function**. This function assigns a numerical value to each possible payoff. The expected utility of any action is the expected value of the assigned utilities. Once a utility function has been specified, it's easy to determine a manager's most preferred action; calculate the expected utilities and choose the action with the highest expected utility.

utility function

EXAMPLE 17.12

Suppose that a manager of a multinational firm has accounts receivable (six months from now) of several million French francs. Three possible "hedging" actions can be taken against the possibility of changes in currency exchange rates, by purchasing futures contracts: the entire balance due can be hedged, or half can be, or none. The payoffs (including commission charges) to each action and the relevant outcome probabilities can be summarized as follows:

	Outcome θ_j		
Action a_i	Franc Loses Value	No Change	Franc Gains Value
Full hedge	−20	−20	−20
Half hedge	−50	−10	30
No hedge	−80	0	80
$P(\theta_j)$	.30	.40	.30

Assume that the following utility function applies:

Payoff:	−80	−50	−20	−10	0	30	80
Utility:	0	40	65	74	80	90	100

a. Find the expected payoff for each action. Which action has the highest expected payoff?
b. Find the expected utility for each action. Which action is preferred?

Solution

a.

Action	Expected Payoff
Full hedge	−20(.3) − 20(.4) − 20(.3) = −20
Half hedge	−50(.3) − 10(.4) + 30(.3) = −10
No hedge	−80(.3) + 0(.4) + 80(.3) = 0 (highest)

The expected-payoff criterion selects the no-hedge action.

b. Replace each payoff by its utility and again compute expected values.

Action	Expected Utility
Full hedge	65(.3) + 65(.4) + 65(.3) = 65.0
Half hedge	40(.3) + 74(.4) + 90(.3) = 68.6 (highest)
No hedge	0(.3) + 80(.4) + 100(.3) = 62.0

The half-hedge action is preferred according to the expected-utility criterion, apparently as a balancing of small expected loss and modest risk. □

A utility function reflects, rather than determines, a manager's willingness to accept risks in pursuit of profits. The idea is not that a manager should possess this utility function as opposed to that one. Rather, the idea is that a utility function corresponds to a manager's specified risk willingness. The technical details of utility-function construction are presented in the Appendix to this chapter.

The utility functions of Figure 17.2 curve downward, substantially for U_1, slightly for U_2. Technically, they are called concave functions. It's not hard to show that use of a

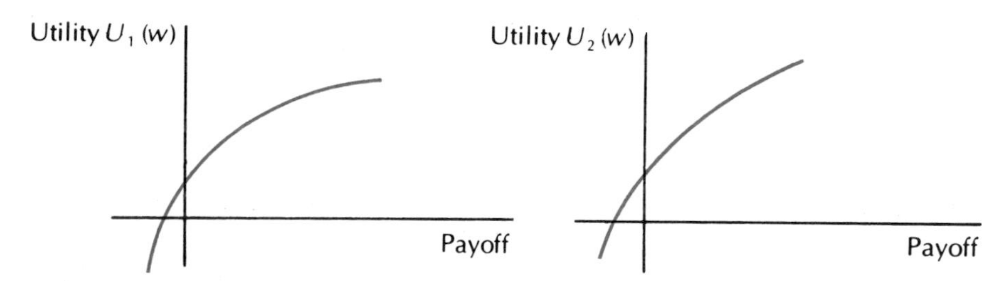

FIGURE 17.2 **Typical Utility Functions**

risk and utility linear utility function (which has no curvature) is equivalent to use of the expected-payoff criterion. In this situation, the element of risk is ignored. A concave utility function does take risk into account.

EXAMPLE 17.13 For the problem of Example 17.12, the following three utility functions are considered:

Payoff $v(a_i, \theta_j)$	−80	−50	−20	−10	0	30	80
Utility function U_1	0	40	65	74	80	90	100
Utility function U_2	0	30	55	65	70	83	100
Utility function U_3	0	18.75	37.50	43.75	50	68.75	100

a. Draw smooth curves representing each utility function. Compare the degrees of curvature.

b. Compute expected utilities for each action under each utility function.

Solution a. The utility functions are shown in Figure 17.3. Clearly U_1 has the most severe curvature, U_2 less severe curvature, and U_3 none at all.

b. Expected utilities for U_1 were computed in Example 17.12. Similar computations give the following table of expected utilities:

Action	U_1	U_2	U_3
Full hedge	65.0	55.0	37.50
Half hedge	68.6	59.9	43.75
No hedge	62.0	58.0	50.00

□

risk aversion The degree of curvature of the utility function reflects the decision maker's attitude toward risk. A downward curving (concave) utility function implies **risk aversion**. A risk-averse manager having a choice between two actions with equal expected payoffs prefers the action with less variability. A severely curved utility function indicates a strongly risk-averse decision maker; a linear function indicates no risk aversion at all.*

* It is possible to have an upward-curving (convex) utility function—for a manager who actually prefers risk.

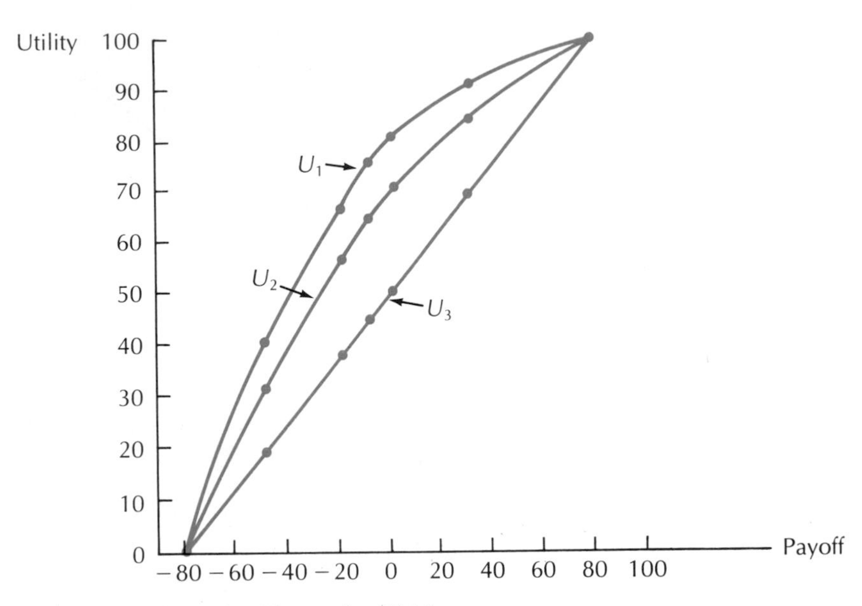

FIGURE 17.3 **Utility Functions for Example 17.13**

EXAMPLE 17.14 Which utility function in Example 17.13 indicates strongest risk aversion? which least?

Solution Function U_1 is most risk-averse because it has the strongest curvature. Function U_3, the linear utility function, shows no curvature and no risk aversion at all. Function U_2 is in between. Note that the no-hedge action has the highest return and highest risk. It is the least preferred action under U_1, second under U_2, and most preferred under U_3. □

The main contribution of the expected-utility approach to decision making is that it forces managers to make consistent, explicit, quantitative judgments about risk. If decisions are made in an organization on an ad hoc basis, some of them are likely to be extremely conservative while others are wildly risky. The effort involved in selecting a utility function is repaid by the consistency of resulting decisions. Of course utility theory is not a magical cure-all. It's not usually possible to assign exact utilities to all possible payoffs, particularly when intangible, nonquantitative outcomes are involved. Utility calculations must be regarded as approximations and must be used thoughtfully.

SECTION 17.4 EXERCISES

17.24 A company with a current net worth of $2 million is considering an expansion program. The success of this program depends largely on future demand for the company's product line. The payoff matrix and assessment of outcome probabilities are given next, where payoffs are expressed as net worth in millions of dollars after three years.

Action	Demand		
	Steady	Small Increase	Large Increase
Expansion	1.50	2.40	3.80
No expansion	2.10	2.40	2.70
Probabilities	.40	.30	.30

Assume that the firm's utility function can be taken as $U(x) = 1 - e^{-0.5x}$, where x is net worth after three years.

a. Show that expansion has a higher expected payoff but also a higher standard deviation.

b. Which action has higher expected utility?

17.25 Refer to Exercise 17.24. Are the conclusions of that exercise affected if the demand probabilities are changed to .50, .30, and .20 for steady, small increase, and large increase, respectively?

17.26 Refer to Exercise 17.24. Are the conclusions of that exercise affected if the utility function is changed to $U(x) = 1 - e^{-0.6x}$?

17.27 An investor has a current wealth of \$400,000 and is considering five possible investment strategies. The wealth position after one year depends on the chosen strategy and on the change in the economic climate over the year, as follows (in thousands of dollars):

Strategy	Economic Climate Change			
	Decline	Constant	Small Increase	Strong Increase
1	270	370	470	570
2	310	385	460	535
3	350	400	450	500
4	390	415	440	465
5	430	430	430	430

The assumed prior probabilities of decline, constant, small increase, and strong increase in economic climate are .10, .20, .40, and .30, respectively.

a. Show that the expected return is highest for strategy 1 and then declines to lowest for strategy 5.

b. Show that the variance is also highest for strategy 1 and declines to lowest for strategy 5.

17.28 Refer to Exercise 17.27. If the investor's utility as a function of wealth position is expressed as $U(x) = \log_{10} x$, which investment strategy is preferred?

SUMMARY

In this chapter we have considered decision making in the face of uncertainty. The basic elements of any decision analysis are

1. set of possible actions A
2. set of possible outcomes Θ
3. payoffs assigned to each $v(a_i, \theta_j)$
4. outcome probabilities $P(\theta_j)$

With this groundwork we developed the notion of expected payoff and discussed decisions based on maximizing the expected payoff.

Decisions based on the expected payoff ignore the element of risk. One widely accepted measure of the risk associated with any action a_i is $\text{Var}(a_i)$, or equivalently, the standard deviation of a_i. Utility theory attempts to blend the notions of expected payoff and risk by replacing an expected payoff for a given action with a manager's utility for that action. Thus, in the face of increased risk, a manager may give an action with higher expected payoff and lower utility.

KEY FORMULAS: Some Basic Ideas of Decision Theory

1. Expected return to action a_i, $R(a_i)$

$$R(a_i) = \sum_j v(a_i, \theta_j)P(\theta_j)$$

2. Variance of action a_i

$$\text{Var}(a_i) = \sum_j [v(a_i, \theta_j)]^2P(\theta_j) - [R(a_i)]^2$$

3. $$\text{Cov}(X, Y) = \sum_{x,y} (x - \mu_x)(y - \mu_y)P_{XY}(x, y) = \left(\sum_{x,y} xyP_{XY}(x, y)\right) - \mu_X\mu_Y$$

4. Portfolio analysis
   a. Expected return for action $(d_1, d_2, \ldots, d_k)$

$$E\left(\sum_i d_iY_i\right) = \sum_i d_i\mu_i$$

   b. Variance of action $(d_1, d_2, \ldots, d_k)$

$$\text{Var}\left(\sum_i d_iY_i\right) = \sum_i d_i^2\,\text{Var}(Y_i) + 2\sum_{i<j} d_id_j\,\text{Cov}(Y_i, Y_j)$$

APPENDIX: MORE ON EXPECTED-UTILITY THEORY

In any decision-theory problem we begin with a set of mutually exclusive actions A, a set of possible outcomes Θ, payoffs $v(a_i, \theta_j)$ associated with each action a_i and outcome θ_j, and outcome probabilities $P(\Theta_j)$. It is then possible to compute the expected payoff and variance for each action. Utility theory is used to bridge the gap between a decision rule using expected payoffs (which ignores risk) and one that uses a risk criterion, the variance of payoffs, but ignores the expected payoff. Utility theory offers a compromise between these two outcomes by incorporating risk into a payoff decision rule. This is done by converting the expected payoff for a given action into an expected utility.

gambles

Utility theory is based on **gambles**. A gamble is an action defined by a set of possible payoffs and a set of probabilities associated with the payoffs. The payoffs in this book are numerical results (e.g., dollars). In principle, the payoffs need not be numerical, but it appears to be somewhat difficult to evaluate nonnumerical gambles in terms of numerical utilities.

utility

The **utility** of a gamble to a particular manager can be inferred from the manager's preference for a gamble relative to another action that has a fixed, certain payoff. For example, suppose as an investment manager you are faced with the problem of choosing between the following two actions with corresponding payoffs:

a_1: receive \$10 for certain
a_2: receive \$120 with probability .5 or lose \$80 with probability .5

The expected payoffs for the two actions are

$$R(a_1) = \$10 \qquad \text{and} \qquad R(a_2) = 120(.5) - 80(.5) = \$20$$

Since a_2 has the higher expected payoff, you may well choose action a_2 if you can sustain an \$80 loss. For this situation, action a_2 has a higher utility for you than action a_1.

Now suppose that you are faced with a choice between two similar types of actions—but with stakes (payoffs and risks) that are quite a bit higher:

a_1: receive a certain, fixed payoff of \$100,000
a_2: receive of payoff of \$1,000,000 with probability .5 or a loss of \$600,000 with probability .5

For this situation, the expected payoffs are

$$R(a_1) = \$100{,}000$$

and

$$R(a_2) = \$1{,}000{,}000(.5) - \$600{,}000(.5) = \$200{,}000$$

Even though the expected payoff for action a_2 is \$100,000 more than the expected payoff for action a_1, you may consider action a_1 to be more desirable since your firm may not be able to sustain a loss of \$600,000. Under these circumstances, action a_1 has a higher utility for you even though it has a lower expected payoff.

The selection of an action in the last two examples depends on the evaluation of risk relative to the payoff. That's what utility theory is all about. By forcing a manager to make choices between various gambles and other actions with certain payoffs, we can examine how a manager evaluates risk in light of the expected payoffs for various actions.

We can illustrate the assignment of numerical utilities to actions with an example. Suppose that we want to find a manager's utilities for possible payoffs for five different actions. These result in possible changes in the firm's net worth after one year ranging from -2 to $+3$ (millions of dollars). We can assign arbitrary utility values to any two selected payoffs. For example, the outcome -2 (a decrease of 2 million dollars net worth) is least desirable, and the outcome $+3$ is most desirable. A utility value of 0 for outcome -2, $U(-2) = 0$, and a utility value of 1 for outcome $+3$, $U(3) = 1$, may be assigned arbitrarily.* Based on these two utilities and on the manager's preference, we can define certain gambles and determine the utilities for other actions. A gamble that has .5 probabilities for -2 and 3 has expected utility $.5U(-2) + .5U(3) = .5$. Suppose that a manager

* This is true because neither adding a constant to all utilities nor multiplying them by a positive constant changes the relative ranking of expected utilities.

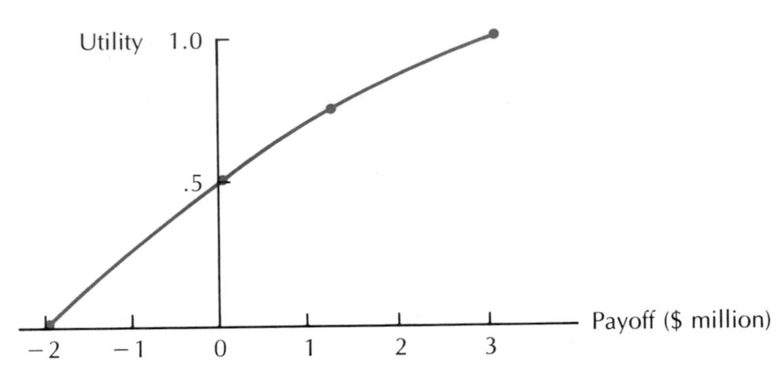

FIGURE 17.4 **Hypothetical Utility Curve**

feels that this gamble is no better and no worse than an action that guarantees 0 change. Because the manager is indifferent about a guaranteed 0 change and the 50–50 gamble between -2 and $+3$, the utility of a guaranteed 0 payoff is $U(0) = .5U(-2) + .5U(3) = .5$.

Now suppose that the manager is indifferent to a 50–50 gamble between 0 and 3 and a guaranteed payoff of 1.2; then the utility of the 1.2 payoff is $U(1.2) = .5U(0) + .5U(3) = .5(.5) + .5(1) = .75$. So far, we have

Payoff:	−2	0	1.2	3
U(Outcome):	.00	.50	.75	1.00

By obtaining the manager's preferences among many such gambles, we can eventually obtain a whole curve of utilities, which might look something like the curve in Figure 17.4.

EXAMPLE 17.15 Suppose that a manager can equate the following 50–50 gambles to actions that would result in the corresponding fixed, certain payoffs.

Payoffs for 50–50 Gambles	Guaranteed Payoff Equivalent
Gamble 1: loss −10, gain 30	0
Gamble 2: lose 0, gain 30	10
Gamble 3: gain 10, gain 30	18
Gamble 4: loss −10, gain 0	−6

Using these specified gambles and the alternate actions with fixed, guaranteed payoffs, develop appropriate utilities and sketch a utility function.

Solution We may arbitrarily set the utility for the lowest payoff $U(-10) = 0$ and the utility for the largest payoff $U(30) = 1$. For the first gamble, $U(0) = .5U(-10) + .5U(30) = .5$. From the second gamble, $U(10) = .5U(0) + .5U(30) = .75$. From the third gamble, $U(18) = .5U(10) + .5U(30) = .875$. Finally, $U(-6) = .5U(-10) + .5U(0) = .25$. The utility function might appear as shown in Figure 17.5. □

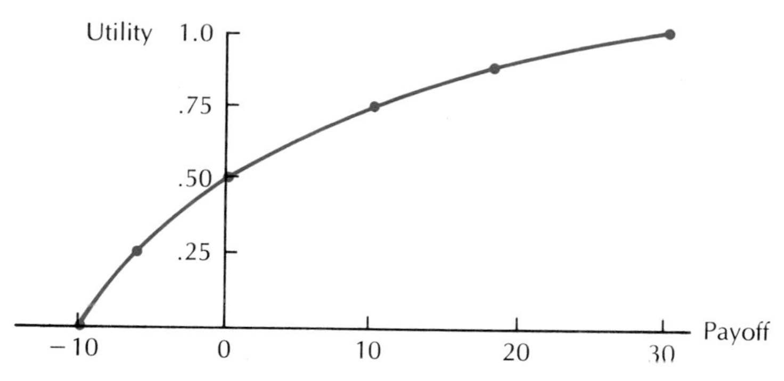

FIGURE 17.5 **Approximate Utility Function for Example 17.15**

CHAPTER 17 EXERCISES

17.29 A consumer-products manufacturer is considering two possible promotion campaigns for its line of skin care products. The aim of each campaign is to increase market share. Success depends heavily on whether the company's two main competitors also have campaigns; both, either one, or neither competitor may do so.

a. Specify an action space.

b. Specify a set of states of nature (outcomes).

17.30 The market research staff of the company of Exercise 17.29 estimates the following market share payoffs:

	Competing Promotions			
	Both	A only	B only	Neither
Promotion 1	28	35	37	44
Promotion 2	34	37	36	40

The company's executives believe that there is a 60% chance that either competitor will have a promotional campaign, independently of the other competitor (and of their own company's decision). The costs of the two promotions are virtually equal.

a. Calculate the expected market share for each of the two promotion campaigns.

b. Calculate the standard deviation of market share for each campaign. Which campaign has lower risk?

17.31 Refer to Exercise 17.30. Assume a utility function for market share x of the form

$$U(x) = x - .02x^2$$

a. Sketch $U(x)$. Does this function reflect risk aversion?

b. Which promotion has a higher expected utility?

c. Does the preferred promotion have higher expected payoff? lower risk? both?

17.32 Refer to Exercises 17.30 and 17.31. Do any of the conclusions change if the probability that either competing firm stages a campaign is decreased to .5 or increased to .7?

17.33 A soft-drink bottler is considering the possibility of either replacing or upgrading its facilities for making nonreturnable bottles. A key element in the decision is whether the state passes a bottle bill requiring returnable, deposit soft-drink bottles. Both 5-cent-deposit and 10-cent-deposit bills have been proposed. The following payoffs have been estimated:

	Bottle Bill		
Action	None	5-Cent	10-Cent
No change	25	20	15
Replace	50	10	−30
Upgrade	35	20	5

The firm believes that there is a 60% chance of no bottle bill being passed, a 30% chance of a 5-cent bill, and a 10% chance of a 10-cent bill.

a. Which action has the highest expected payoff?
b. Which action has the lowest variance?

17.34 Refer to Exercise 17.33. Assume that the bottler's utility, as a function of payoff x, is

$$U(x) = \log_{10}(x + 200)$$

Which action has the highest expected utility?

17.35 Refer to Exercises 17.33 and 17.34. The payoffs for the upgrade action may all be in error by as much as ± 5 units. The probabilities of no bill and the 5-cent bill may be .7 and .2 in one direction or .5 and .4 in the other. Conduct a sensitivity analysis of the bottler's decision problem. Four extreme combinations of payoffs and probabilities may be obtained by varying the upgrade payoffs and the probabilities in both directions.

17.36 The joint probability distribution $P_{XY}(x, y)$ of x = return of $1 invested in stock A and y = return to $1 invested in stock B is as follows:

		y				
		.08	.09	.10	.11	.12
	.08	.00	.00	.02	.04	.09
	.09	.00	.01	.09	.09	.01
x	.10	.00	.06	.18	.06	.00
	.11	.01	.09	.09	.01	.00
	.12	.09	.04	.02	.00	.00

a. Calculate Cov(x, y) and $P_{XY}(x, y)$.
b. Explain why the covariance is negative.

17.37 Refer to the joint distribution of Exercise 17.36.

a. Calculate the expected return and risk for a portfolio with $500 invested in each of the two stocks.
b. Calculate the expected return and risk for an investment of $1000 in stock A.
c. Why is the risk for the portfolio in part (a) so much lower than the risk for the investment in part (b)?

17.38 A growing company plans to issue additional stock in two months, and it must decide whether to offer all common, all preferred, or half of each. The company's treasurer predicts the yields of each choice, depending on the state of the stock market at the time of issue, as follows (all figures in millions of dollars):

Offer	State of Market				
	Down 8%	Down 4%	Steady	Up 4%	Up 8%
Common	92.0	96.0	100.0	104.0	108.0
Preferred	97.0	98.0	99.0	100.0	101.0
Half/half	94.5	97.0	99.5	102.0	104.5

The treasurer assesses the probabilities of the five market states (from lowest to highest) as .25, .30, .20, .15, and .10.

a. Which action has the highest expected yield?

b. Which action has the lowest variance?

17.39 The president of the company of Exercise 17.38 assesses the market probabilities as .10, .30, .40, .15, and .05. Do the answers to that Exercise change if these revised probabilities are used?

17.40 Refer to Exercises 17.38 and 17.39. Assume the utility function $U(x) = 1 - e^{-.01x}$, where $x =$ yield in millions of dollars.

a. Which action has the highest expected utility based on the treasurer's probabilities of Exercise 17.38?

b. Which action has the highest expected utility based on the president's probabilities of Exercise 17.39?

17.41 A trust officer for a bank assesses the expected returns, variances, and covariances for four investment opportunities. The results (expressed as return per dollar) are

Investment:	1	2	3	4
Expected return, μ_i:	1.089	1.090	1.091	1.086

	Variance-Covariance Matrix			
	1	2	3	4
1	.048	.049	.050	−.038
2	.049	.052	.052	−.041
3	.050	.052	.058	−.043
4	−.038	−.041	−.043	.060

(Variances are along the diagonal of this matrix; all other entries are covariances.)

a. Calculate the expected return and standard deviation for a portfolio consisting of $4000 invested in each of investments 1, 2, and 3.

b. Calculate the expected return and standard deviation for a portfolio consisting of $3000 invested in each of investments 1, 2, 3, and 4.

c. Is either portfolio clearly preferred to the other (assuming a somewhat risk-averse investor)?

17.42 Assume that the investor's expected utility function in Exercise 17.41 is equal to

expected return − .001 variance

Which portfolio has the higher expected utility?

18

USING SAMPLE INFORMATION IN MAKING DECISIONS

The elements of decision theory were presented in Chapter 17. We considered action spaces, outcome spaces and the accompanying assessed probability distribution $P(\theta_j)$, and payoff functions. With these essential elements we discussed applications of decision theory that maximize the expected payoff and those that acknowledge the presence of risk as measured by the variance (or standard deviation) of a given action. Utility theory provided a bridge between decisions based entirely on a long-run (expected value) strategy and one based on minimizing risk.

There was one very important factor left out of the discussion of decision theory in Chapter 17. That factor is the role of additional (sample) information in the decision-making process. Bayes' Theorem, which is discussed in Section 18.1, plays an important role in developing this method.

18.1 JOINT PROBABILITIES AND BAYES' THEOREM

In this section we develop Bayes' Theorem, which is a major probability concept used in statistical decision theory. The idea behind Bayes' Theorem is implicit in Section 3.5 on probability trees. You may want to review that section before continuing with this one.

Again we begin with an example, and we use a probability tree in the solution. Suppose that .1% (i.e., .001) of a certain population has tuberculosis (TB). In a fairly reliable screening test, 95% of those who have TB will show positive results and only 2% of those

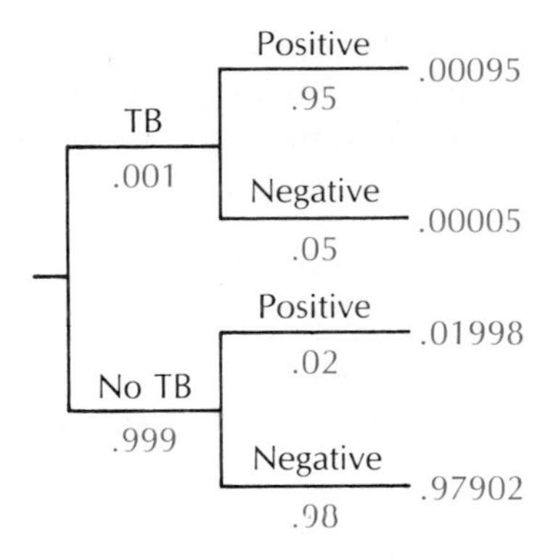

FIGURE 18.1 **Probability Tree for Tuberculosis Screening Example**

who don't have TB will show positive results. We want to answer the following questions:

a. What is the probability that a randomly selected person has both TB and a positive test result?
b. What is the probability of a positive test result?
c. If the result is positive, what is the probability that the person has TB, given a positive test result?

A probability tree for this problem is shown in Figure 18.1. The probability of having both TB and a positive result is shown as .00095. The probability of a positive result is .00095 + .01998 = .02093. The probability of having TB, given a positive test result, is

$$\frac{P(\text{TB and positive})}{P(\text{positive})} = \frac{.00095}{.02093} = .045$$

a surprisingly small value.

prior probability **likelihood** **posterior probability**

All the essential ideas of Bayes' Theorem are reflected in this sample. First, the **prior probability** of TB is .001. The **likelihood** of a positive result is .95 for those having TB and .02 for those not having TB. The **posterior probability** of TB, given a positive test result, is .045. A formula for the posterior probability calculation is

$$\begin{aligned} P(\text{TB}|\text{positive}) &= \frac{P(\text{TB and positive})}{P(\text{positive})} \\ &= \frac{P(\text{positive}|\text{TB})P(\text{TB})}{P(\text{positive}|\text{TB})P(\text{TB}) + P(\text{positive}|\text{no TB})P(\text{no TB})} \\ &= \frac{(.95)(.001)}{(.95)(.001) + (.02)(.999)} = .045 \end{aligned}$$

EXAMPLE 18.1 A book club classifies members as heavy, medium, or light purchasers, and separate mailings are prepared for each of these groups. Overall, 20% of the members are heavy purchasers, 30% medium, and 50% light. A member is not classified into a group until

18 months after joining the club, but a test is made of the feasibility of using the first 3 months' purchases to classify members. The following percentages are obtained from existing records of individuals classified as heavy, medium, or light purchasers.

First 3 Months' Purchases	Group		
	Heavy	Medium	Light
0	5%	15%	60%
1	10%	30%	20%
2	30%	40%	15%
3+	55%	15%	5%

If a member purchases no books in the first 3 months, what is the probability that the member is a light purchaser?

Solution Part of a probability tree is shown in Figure 18.2. Only the zero purchase branches are needed.

$$P(\text{Light} \mid 0) = \frac{P(0 \mid \text{Light})P(\text{Light})}{P(0 \mid \text{Light})P(\text{Light}) + P(0 \mid \text{Medium})P(\text{Medium}) + P(0 \mid \text{Heavy})P(\text{Heavy})}$$

$$= \frac{(.60)(.50)}{(.60)(.50) + (.15)(.30) + (.05)(.20)}$$

$$= .845$$

or more simply, using the probability tree

$$P(\text{Light} \mid 0) = \frac{P(\text{Light and } 0)}{P(0)}$$

$$= \frac{.300}{.300 + .045 + .010} = .845$$

□

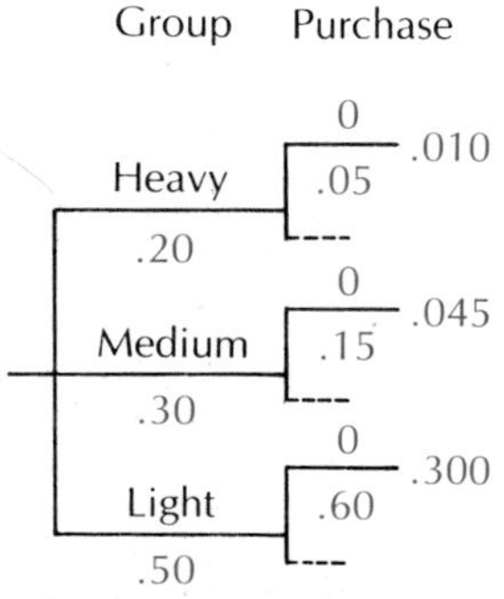

Group Purchase

Heavy .20 — 0 .05 — .010

Medium .30 — 0 .15 — .045

Light .50 — 0 .60 — .300

FIGURE 18.2 **Probability Tree for Example 18.1**

These examples indicate the basic idea of Bayes' Theorem. There is some number k of possible, mutually exclusive, underlying events $A_1, \ldots, A_k$, which are sometimes called the possible **states of nature**. Unconditional probabilities $P(A_1), \ldots, P(A_k)$, often called **prior probabilities**, are specified. There are m possible, mutually exclusive, **observable events** $B_1, \ldots, B_m$. The conditional probabilities of each observable event given each state of nature, $P(B_j|A_i)$, are also specified and these probabilities are called **likelihoods**. The problem is to find the **posterior probabilities** $P(A_i|B_j)$. Prior and posterior refer to probabilities before and after observing an event B_j.

states of nature

observable events

Bayes' Theorem

If $A_1, \ldots, A_k$ are mutually exclusive states of nature and if $B_1, \ldots, B_m$ are m possible mutually exclusive observable events, then

$$P(A_i|B_j) = \frac{P(B_j|A_i)P(A_i)}{P(B_j|A_1)P(A_1) + P(B_j|A_2)P(A_2) + \ldots + P(B_j|A_k)P(A_k)}$$

$$= \frac{P(B_j|A_i)P(A_i)}{\sum_i P(B_j|A_i)P(A_i)}$$

EXAMPLE 18.2

a. Specify the states of nature and observable events for Example 18.1.
b. Specify the prior probabilities and likelihoods.
c. Which posterior probability was calculated in Example 18.1?

Solution

a. The states of nature are the three possible groups: A_1 = heavy, A_2 = medium, and A_3 = light. The observable events are the possible 3-month purchases: $B_1 = 0$, $B_2 = 1$, $B_3 = 2$, and $B_4 = 3$ or more.

b. The prior probabilities are $P(A_1) = .20$, $P(A_2) = .30$, and $P(A_3) = .50$. The likelihoods $P(B_j|A_i)$ are shown in the accompanying table.

	Group A_i		
Purchase B_j	Heavy, A_1	Medium, A_2	Light, A_3
0, B_1	.05	.15	.60
1, B_2	.10	.30	.20
2, B_3	.30	.40	.15
3+, B_4	.55	.15	.05

c. The calculated posterior probability was $P(A_3|B_1)$, or equivalently, $P(\text{light}|0)$. The formula given in the solution is exactly Bayes' Theorem. □

EXAMPLE 18.3 Of all electronic systems produced in a plant, 80% are nondefective, 15% have defect D_1, and 5% have defect D_2. None of the systems has both defects. Defects can be detected with certainty only by destructive testing, but a fairly reliable nondestructive test has been devised. The test has four possible outcomes. The respective likelihoods are

	Defect A_i		
Test Outcome B_j	None, A_1	D_1, A_2	D_2, A_3
1, B_1	.90	.06	.02
2, B_2	.05	.40	.06
3, B_3	.03	.45	.52
4, B_4	.02	.09	.40

If a particular system yields test result 3, what are the posterior probabilities of no defect, defect D_1, and defect D_2?

Solution These three posterior probabilities can be computed using Bayes' Theorem. For example, the probability of no defect given test result 3 is

$$P(\text{None}|3) = P(A_1|B_3) = \frac{P(B_3|A_1)P(A_1)}{P(B_3|A_1)P(A_1) + P(B_3|A_2)P(A_2) + P(B_3|A_3)P(A_3)}$$

$$= \frac{.03(.80)}{.03(.80) + .45(.15) + .52(.05)} = .204$$

Similarly,

$$P(D_1|3) = P(A_2|B_3) = \frac{P(B_3|A_2)P(A_2)}{P(B_3|A_1)P(A_1) + \ldots + P(B_3|A_3)P(A_3)}$$

$$= \frac{.45(.15)}{.03(.80) + \ldots + .52(.05)} = .574$$

and

$$P(D_2|3) = P(A_3|B_3) = \frac{P(B_3|A_3)P(A_3)}{P(B_3|A_1)P(A_1) + \ldots + P(B_3|A_3)P(A_3)}$$

$$= \frac{.52(.05)}{.03(.80) + \ldots + .52(.05)} = .221$$

□

Bayes' Theorem can also be stated in random-variable notation. The states of nature are usually denoted θ and the observable states y.

Bayes' Theorem with Random Variables

For discrete random variables Θ and Y,

$$P_{\Theta|Y}(\theta|y) = \frac{P_\Theta(\theta)P_{Y|\Theta}(y|\theta)}{\sum_\theta P_\Theta(\theta)P_{Y|\Theta}(y|\theta)}$$

For continuous random variables Θ and Y,

$$f_{\Theta|Y}(\theta|y) = \frac{f_\Theta(\theta)f_{Y|\Theta}(y|\theta)}{\int_{-\infty}^{\infty} f_\Theta(\theta)f_{Y|\Theta}(y|\theta)\,d\theta}$$

This is merely a notation shift. No new ideas are involved.

SECTION 18.1 EXERCISES

18.1 One percent of a finance company's loans are defaulted (not completely repaid). The company routinely runs credit checks on all loan applicants. It finds that 30% of defaulted loans went to poor risks, 40% to fair risks, and 30% to good risks. Of the nondefaulted loans, 10% went to poor risks, 40% to fair risks, and 50% to good risks.

a. Use a probability tree to calculate the probability that a poor-risk loan is defaulted.

b. Use Bayes' Theorem to calculate the same probability.

18.2 Refer to Exercise 18.1. Show that the posterior probability of default, given a fair risk, equals the prior probability of default. Explain why this is a reasonable result.

18.3 A manufacturing firm has three machine operators who produce a certain component. Operator A has a 5% defective rate, B has a 3% defective rate, and C has a 2% defective rate. The three operators produce equal numbers of components. Suppose that a randomly selected component is found to be defective. Calculate the posterior probability that the part was produced by A. Compare the result to the prior probability of 1/3.

18.4 Refer to Exercise 18.3. Suppose that a sample of 20 components is taken from a lot produced by one operator and assume that binomial probabilities apply. If no defective components are found in the sample, find the posterior probability that the lot was produced by operator A.

18.5 An underwriter of home insurance policies studies the problem of home fires resulting from wood-burning furnaces. Of all homes having such furnaces, 30% own a Type 1 furnace, 25% a Type 2 furnace, 15% a Type 3, and 30% other types. Five percent of Type 1 furnaces, 3% of Type 2, 2% of Type 3, and 4% of other types have resulted in fires over three years of operation. If a fire occurs in a particular home, what is the probability that a Type 1 furnace is in the home?

18.6 A reviewer of textbooks has a curious "track record." An editor estimates the following rating percentages for highly successful, moderately successful, and unsuccessful books.

	Reviewer's Rating		
Book	Good	Fair	Poor
Highly successful	5%	20%	75%
Moderately successful	15%	40%	45%
Unsuccessful	50%	30%	20%

About 10% of all books are highly successful, 50% are moderately successful, and 40% are unsuccessful.

If this reviewer rates a book as good, calculate the posterior probability that the book is unsuccessful. Compare the result to the prior probability, .40.

18.7 Conditional probabilities can be useful in diagnosing disease. Suppose that three different, closely related diseases (A_1, A_2, and A_3) occur in 25%, 15%, and 12% of the population. In addition, suppose that any one of three mutually exclusive symptom states B_1, B_2, and B_3 may be associated with each of these diseases. Experience shows that the likelihood $P(B_j|A_i)$ of having a given symptom state when the disease is present is as shown in this table:

Symptom State B_j	Disease State A_i		
	A_1	A_2	A_3
B_1	.08	.17	.10
B_2	.18	.12	.14
B_3	.06	.07	.08
B_4 (no symptoms)	.68	.64	.68

Find the probability of disease A_2 given symptom B_1, B_2, B_3, and B_4, respectively.

18.2 BUYING AND USING INFORMATION

The discussion of decision theory in Chapter 17 ignores one important managerial option: a manager can often buy more information. Information can be used to improve forecasts of future outcomes. Since these outcomes determine the relative success or failure of the actions, better information should lead to better decisions. The value of information may be measured by how much influence it has on decision making.

Information is expensive. Gathering and using data almost always cost money. Just as important, the information process takes time. It's easy to forget that while time goes by, profits and gains are being lost and costs are piling up. The relevant question for a manager is, will this information have enough value in decision making to justify the money and time expense to gather it? In this section we introduce some decision-theory ideas about how information can be used to modify decisions and about how to weigh the benefits of information against the money and time costs of gathering it.

money and time costs

Suppose that an oil company has the option of purchasing the rights to all natural gas and oil that may be found under a certain tract of land in southern Wyoming. The company has identified three possible actions: purchase all rights and bear all expenses, purchase a half share of all rights and bear half the expenses, or purchase none of the rights. The payoff to these actions obviously depends on the extent of underground oil and gas deposits actually present. Suppose that the company breaks down the possible outcomes as follows: no gas or oil, gas only, small oil deposit only, large oil deposit only, both oil and gas. Further, suppose that the net payoffs (in millions of dollars) and (prior) probabilities are as follows:

Action a_i	Outcome θ_j No Gas or Oil	Gas Only	Small Oil	Large Oil	Both
Full rights	−10	8	18	40	58
Half rights	−5	4	9	20	29
No rights	0	0	0	0	0
$P(\theta_j)$	.70	.10	.10	.05	.05

We assume that the issue of risk can be neglected, so the expected-value criterion applies. The respective expected returns for the three possible actions are

$E(\text{Return} \mid \text{Full rights})$:
$$(-10)(.70) + 8(.10) + 18(.10) + 40(.05) + 58(.05) = .5$$
$E(\text{Return} \mid \text{Half rights})$:
$$(-5)(.70) + 4(.10) + 9(.10) + 20(.05) + 29(.05) = .25$$
$E(\text{Return} \mid \text{No rights})$:
$$(0)(.70) + 0(.10) + 0(.10) + 0(.05) + 0(.05) = 0$$

The highest expected value corresponds to the full-rights action.

Before embarking on this action, however, it may be desirable for the company to seek additional information. If a "dry hole" (no gas or oil) could be ruled out, the expected profit to a full share would rise dramatically. On the other hand, if it could be known in advance that there was no gas or oil, a considerable loss could be spared. In practice, the company wouldn't be able to know the outcome with certainty unless it bought the rights and drilled. But it's a useful fiction to imagine a source of **perfect information**—a magical worm that can report unerringly about "what's down there." How much would such a beast be worth?

perfect information

According to the prior probabilities, there is a .7 chance that the perfect-information service reports no oil or gas; if so, the proper action is to claim no rights. If the source reports gas only, small oil, large oil, or both gas and oil, the proper action is to buy full rights. Such reports occur with respective probabilities .10, .10, .05, and .05. We can compute the expected return given perfect information using the abbreviated table below.

Action a_i	Outcome θ_j No Gas or Oil	Gas Only	Small Oil	Large Oil	Both
Full rights	—	8	18	40	58
Half rights	—	—	—	—	—
No rights	0	—	—	—	—
$P(\theta_j)$	.70	.10	.10	.05	.05

The expected return for perfect information is

(Payoff to no rights given no oil or gas) P(No oil or gas)
+ (Payoff to full rights given gas only) P(Gas only)
+ (Payoff to full rights given small oil) P(Small oil)
+ (Payoff to full rights given large oil) P(Large oil)
+ (Payoff to full rights given both gas and oil) P(Both gas and oil)
$= 0(.70) + 8(.10) + 18(.10) + 40(.05) + 58(.05)$
$= 7.5$

Expected Value of Perfect Information (EVPI)

The **expected value of perfect information** is the difference between the expected return under perfect information and the maximum expected return based on prior probabilities only.

The expected return under perfect information is

$$\sum_j \text{(maximum payoff given outcome } \theta_j)P(\theta_j)$$

In our illustration, EVPI $= 7.5 - 0.5 = 7.0$.

EXAMPLE 18.4 A manufacturing firm develops a security system for the door locks of hotel and motel rooms. The (present value of) future returns on sales of this system depend heavily on whether the system can be patented. An inventor has developed a somewhat similar system and holds a patent on it. The firm's possible actions are to buy patent rights from the inventor, to apply for its own patent, or to sell the rights to its system to a competitor. The payoffs, in millions of dollars (present value), are estimated as follows:

	Outcome θ_j	
Action a_i	System Patentable	System not Patentable
Buy rights	18.6	16.4
Apply	20.8	8.2
Sell	15.4	15.4
$P(\theta_j)$	.80	.20

The firm's patent attorney estimates that there is a .80 probability that the firm's system is patentable, and therefore a .20 probability that it is not.

a. Under current information conditions, what is the best action to take?
b. What is the EVPI?

Solution a. The respective expected returns on the possible actions are

$$\begin{aligned} E(\text{Return on Buy}): &\quad 18.6(.8) + 16.4(.2) = 18.16 \\ E(\text{Return on Apply}): &\quad 20.8(.8) + \ \ 8.2(.2) = 18.28 \\ E(\text{Return on Sell}): &\quad 15.4(.8) + 15.4(.2) = 15.4 \end{aligned}$$

The highest expected return is for Apply. Neglecting risk, it is the best action.

b. If it were known that the system was patentable, the firm would apply and earn 20.8. If it were known that the system was not patentable, the firm would buy rights and earn 16.4. The expected return under perfect information is $20.8(.8) + 16.4(.2) = 19.92$. Hence $\text{EVPI} = 19.92 - 18.28 = 1.64$. □

The EVPI puts an upper limit on the value of information. If the cost of gathering information, whether perfect or imperfect, exceeds the EVPI, there's no point in paying for information. In the oil and gas example, if the cost of a seismic study of the tract is \$7.7 million, it would be foolish to do the study. Even perfect information is worth only \$7.0 million.

EXAMPLE 18.5 Refer to Example 18.4. Suppose that an opinion from a patent specialist can be obtained for a total cost of \$.4 million. Is the cost justified?

Solution Possibly. The EVPI is \$1.64 million; if the specialist's opinion is reliable enough, it could be worth the \$.4 million cost. □

In practice, information is not perfect. Forecasts are subject to error. Therefore imperfect information is worth less than the EVPI. To assess the value and effect of sample (imperfect) information, we must consider how such information can affect decisions.

The idea is fairly simple: Given any particular result from sample information, revise the probabilities $P(\theta_j)$ of all outcomes according to Bayes' Theorem (Section 18.1). Use these revised (posterior) probabilities $P(\theta_j \mid \text{Sample result } I)$ to select the best decision according to the expected-value criterion. To carry out this idea, we must specify the likelihoods of various sample results conditional on the eventual outcomes.

For instance, in the oil and gas illustration, suppose that a seismic test can be conducted that yields three possible outcomes: poor promise of success, modest promise of success, or good promise of success. Assuming that the test has been used in other explorations, it is possible to assess its reliability by computing the chance (or likelihood) of each test result given a possible outcome. Such an assessment of previous success yields the likelihoods shown next.

Likelihood (Sample Result $I \mid \theta_j$)

	Outcome θ_j				
Sample Result I	No Oil or Gas	Gas Only	Small Oil	Large Oil	Both
Poor	.80	.50	.40	.20	.10
Modest	.15	.30	.40	.30	.30
Good	.05	.20	.20	.50	.60

Bayes' Theorem can be used to calculate the (posterior) probabilities of each outcome given any particular test result. For example,

$$P(\text{No oil or gas} \mid \text{Poor}) = \frac{P(\text{Poor} \mid \text{No oil or gas})P(\text{No oil or gas})}{P(\text{Poor})}$$

where

$$P(\text{Poor}) = P(\text{Poor} \mid \text{No oil or gas})P(\text{No oil or gas}) + P(\text{Poor} \mid \text{Gas})P(\text{Gas}) + \ldots + P(\text{Poor} \mid \text{Both})P(\text{Both})$$

$$P(\text{No oil or gas} \mid \text{Poor}) = \frac{(.80)(.70)}{(.80)(.70) + (.50)(.10) + (.40)(.10) + (.20)(.05) + (.10)(.05)}$$

$$= \frac{.5600}{.6650} = .8421$$

Similar calculations or a probability tree can be made to yield each entry in the following table.

$P(\text{Outcome } \theta_j \mid \text{Sample Result } I)$			
	Sample Result I		
Outcome θ_j	Poor	Modest	Good
No oil or gas	.8421	.5122	.2692
Gas only	.0752	.1463	.1538
Small oil	.0602	.1951	.1538
Large oil	.0150	.0732	.1923
Both	.0075	.0732	.2308
$P(\text{Sample result } I)$	.6650	.2050	.1300

These probabilities are used in the standard expected-value calculations to assess the best action given a particular test result. For example, if the full-rights action is taken in the face of a poor test result, the expected return (using payoffs from p. 693) is

$$-10(.8421) + 8(.0752) + 18(.0602) + 40(.0150) + 58(.0075) = -5.70$$

The remaining expected returns are computed in a similar way.

$E(\text{Return on } a_i \mid \text{Sample Result } I)$			
	Sample Result I		
Action a_i	Poor	Modest	Good
Full rights	−5.70	6.73	22.39
Half rights	−2.85	3.37	11.19
No rights	.00	.00	.00

Based on the expected-return criterion, the best action is to choose No rights given a poor test result and Full rights given either a modest or a good test result.

EXAMPLE 18.6 The manufacturing firm of Example 18.4 has the option of consulting a patent specialist, who will state that the patentability is likely, uncertain, or unlikely. The specialist is something less than infallible. From past history, it appears that the likelihoods of the specialist's opinion, given the eventual outcome, are

Likelihood (Sample Result $I\|\theta_j$)		
	Outcome θ_j	
Sample Result I	Patentable	Not Patentable
Likely	.50	.25
Uncertain	.30	.30
Unlikely	.20	.45

a. Calculate the posterior probabilities $P(\theta_j|\text{Opinion})$.
b. Given each possible opinion of the specialist, compute the expected return for each action based on the posterior probabilities of part (a).

Solution a. For example, the posterior probability

$$P(\text{Patentable}|\text{Likely}) = \frac{P(\text{Likely}|\text{Patentable})P(\text{Patentable})}{P(\text{Likely})}$$

where

$$\begin{aligned} P(\text{Likely}) &= P(\text{Likely}|\text{Patentable})P(\text{Patentable}) \\ &\quad + P(\text{Likely}|\text{Not patentable})P(\text{Not patentable}) \\ &= .50(.80) + .25(.20) = .45 \end{aligned}$$

Hence

$$P(\text{Patentable}|\text{Likely}) = \frac{.50(.80)}{.45} = .889$$

Similar calculations yield the table of the posterior probabilities shown here:

$P(\text{Outcome } \theta_j\|\text{Sample result } I)$			
	Sample Result I		
Outcome θ_j	Likely	Uncertain	Unlikely
Patentable	.889	.800	.640
Not patentable	.111	.200	.360
$P(\text{Sample result } I)$	.45	.30	.25

b. If the opinion is likely and the payoffs are as indicated on p. 694, the expected return to the buy-rights action is

$$E(\text{Return on Buy}|\text{Likely}) = 18.6(.889) + 16.4(.111) = 18.36$$

Similar calculations give the following table.

Expected (Return on a_i | Sample Result I)

Action a_i	Sample Result I Likely	Uncertain	Unlikely
Buy rights	18.36	18.16	17.81
Apply	19.40	18.28	16.26
Sell	15.40	15.40	15.40
P(Sample result I)	.45	.30	.25

The expected value of sample information (EVSI) is defined similarly to the expected value of perfect information (EVPI).

Expected Value of Sample Information (EVSI)

1. Calculate the posterior outcome probabilities P(Outcome θ_j | Sample result I) given each sample (information) result, and also the probabilities of each sample result.
2. Calculate the expected return to each action E(Action a_i | Sample result I) using each set of posterior probabilities.
3. EVSI = the maximum expected return using sample information less the maximum expected return using no sample information.
   = $\sum$ (Maximum expected return | Sample result) P(Sample result) − Maximum expected return using no sample information.

In the oil and gas illustration, we can use the table on p. 696. The best action given a poor result is the choice of no rights, with expected return 0. The best action given a modest result is the choice of full rights (expected return 6.73), and the best result given a good result is the choice of full rights (expected return 22.39). The probabilities of poor, modest, and good results were calculated as .6650, .2050, and .1300, respectively. The maximum expected return using no information was calculated as .50 (see p. 693). Therefore

$$\text{EVSI} = (0)(.6650) + (6.73)(.2050) + (22.39)(.1300) - .50$$
$$= 3.790$$

If the test costs \$4.0 million, it is not worth the cost; if it costs \$3.0 million, it is worth the cost.

EXAMPLE 18.7 Refer to Examples 18.4 and 18.6. Suppose the patent specialist's fee is \$.40 million. Is the information worth the cost?

Solution Calculate the EVSI. The best action for each opinion, the expected return for the best action, and the probability of a given opinion are displayed here. These were obtained from the solution to Example 18.6, part (b).

Opinion	Best Action | Opinion	Expected Return	P(Opinion)
Likely	Apply	19.40	.45
Uncertain	Apply	18.28	.30
Unlikely	Buy	17.81	.25

The maximum expected return with no additional information was computed to be 18.28 in Example 18.4. Hence

$$\begin{aligned} \text{EVSI} &= 19.40(.45) + 18.28(.30) + 17.81(.25) - 18.28 \\ &= .387 \end{aligned}$$

Since the EVSI is less than the cost of the information, the information is not worth the cost. □

SECTION 18.2 EXERCISES

18.8 An entrepreneur is considering building a system of fish farms near major cities to supply freshwater fish (primarily walleyed pike and sauger) to the hotel and restaurant trade. The profitable scale of such an operation depends on the yield of edible fish after a 4-year growth period, and anticipated net profits are as follows (in millions of dollars per year):

	Yield		
Scale	Good	Fair	Poor
Large	10.5	2.2	−5.8
Medium	6.1	2.4	−1.2
Small	3.2	1.9	−0.6
None	.0	.0	.0

After discussion with several specialists, the entrepreneur assesses the probabilities of good, fair, and poor yields as .2, .5, and .3, respectively.

a. What scale of operations has the highest expected return?

b. Imagine that a pilot project could, at the cost of $1.0 million, identify yield exactly. Would such a project be desirable?

18.9 Refer to Exercise 18.8. It is possible to establish a pilot project that will give a fairly accurate indication of yield. The likelihoods (conditional probabilities) of various pilot project results given yields are assumed to be as follows:

	Yield		
Results	Good	Fair	Poor
Favorable	.9	.2	.1
Mediocre	.1	.7	.2
Unfavorable	.0	.1	.7

a. Find the revised (posterior) probabilities of good, fair, and poor yields given a favorable pilot project result.

b. Find the scale that has the greatest expected profit, given a favorable result.

c. What is the probability of a favorable result?

d. Answer parts (a), (b), and (c) for a mediocre result, and then for a poor result.

18.10 The pilot project of Exercise 18.9 costs $.6 million. Is the information provided worth the cost?

18.11 A supermarket chain must decide how much shelf space and what form of display to use for generic food products. These products are offered by some of the chain's major competitors. The crucial unknown factor is the degree to which generic products "cannibalize" sales of national and house brands. A staff report indicates the following display options, three-month profit figures (in hundreds of thousands of dollars), and probabilities:

Display	Degree of Cannibalization 25%	50%	75%	100%
30 ft., consolidated	2.4	1.2	.0	1.2
30 ft., scattered	3.6	2.0	−1.1	−1.8
15 ft., consolidated	1.5	.8	−.4	−.8
15 ft., scattered	2.0	1.2	−.6	−1.0
None	−.3	−.5	−.7	−.9
Probabilities	.50	.25	.20	.05

a. Is any display option always worse than some other option?
b. Which option has the highest expected profit?
c. Suppose an extremely extensive survey costing $240,000 can indicate the degree of cannibalization perfectly. Is such a survey worth its cost?

18.12 Refer to Exercise 18.11. Suppose a limited survey of shoppers at the chain can be undertaken at a cost of $40,000. Because the survey is based on a modest sample and tends to indicate higher substitution of generic products than actually occurs, the following conditional probabilities apply:

Survey Result (How Much Substitution?)	Degree of Cannibalization 25%	50%	75%	100%
Little	.1	.1	.0	.0
Some	.3	.2	.1	.0
Half	.3	.3	.1	.0
Most	.2	.3	.5	.1
All	.1	.1	.3	.9

a. What is the probability that the survey result is Little?
b. If the result is Little, what are the posterior probabilities of each possible degree of cannibalization?
c. If the result is Little, what is the optimal display option?
d. Answer questions (a), (b), and (c) for each of the other possible survey results.

18.13 Is the survey of Exercise 18.12 worth its cost?

18.3 DECISION TREES

The calculations involved in assessing the value of information are not terribly hard, but there are many of them. Any device for keeping the calculations straight is useful. In this section we introduce such a device—decision trees. The concept of a decision tree is very similar to that of a probability tree (Chapter 3) and to that of an expected-value tree (Chapter 17).

Decision trees involve two steps, those leading to decision making and those leading to receiving information. At each decision step, the decision with the greatest expected value (or expected utility) is selected. At each information step, no choice is made. Instead, probabilities are calculated. A very common pattern is to base the first branch on the decision to buy information, the second on the receipt of that sample information, the third on the action decision, and the final branch on the actual future outcomes.

An incomplete decision tree for the security system problem of Examples 18.4, 18.6, and 18.7 is shown in Figure 18.3. The relevant probabilities have been inserted, but the choices at the decision steps have not been indicated.

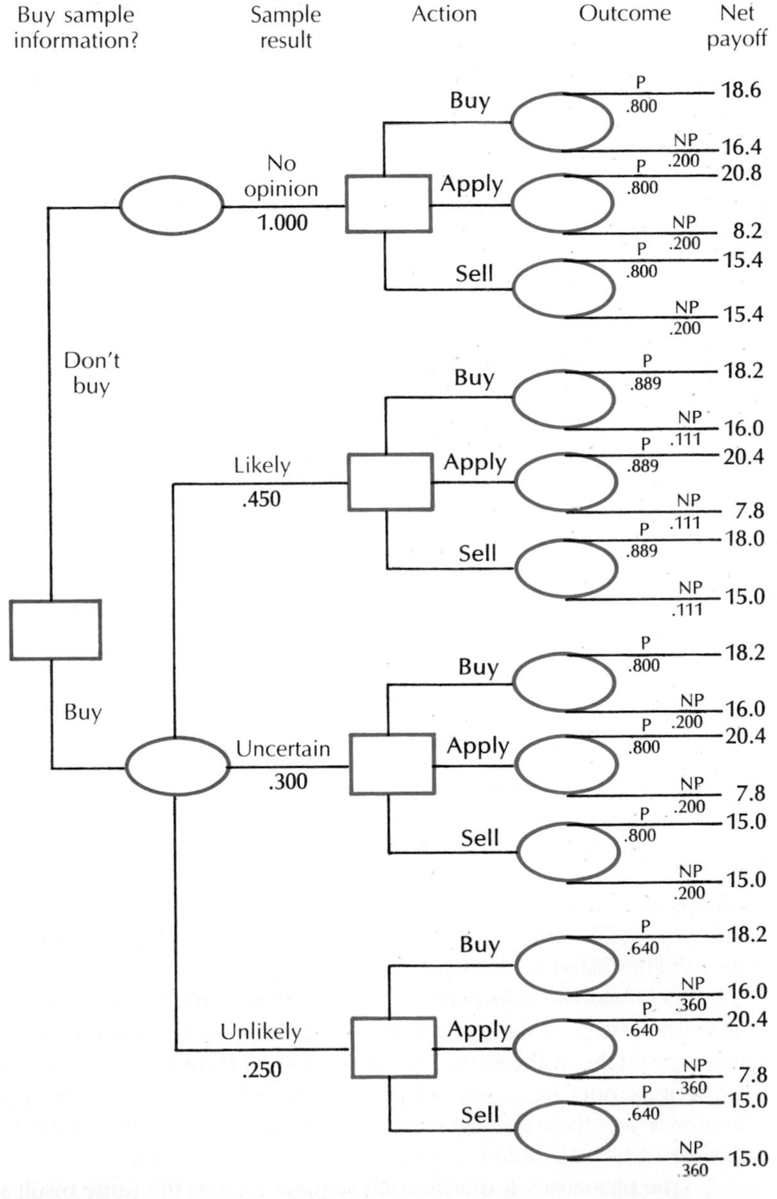

FIGURE 18.3 Incomplete Decision Tree for Examples 18.4, 18.6, and 18.7

tree construction

In order to construct a decision tree such as the one shown in Figure 18.3, we develop the branches of the tree by beginning at the left with the decision about whether or not to purchase information. For each possible decision there are various possible sample outcomes; these possible sample results form the next branches of the decision tree. The next stage of the tree lists the possible actions from the action set. The branches stemming from an action denote the possible outcomes.

At the end of each branch of the decision tree there is a net payoff. This net payoff corresponds to a given decision about buying sample information and a given sample result, action, and outcome. The net payoff for a given action is the actual payoff minus the cost of additional information. For example, all branches of the tree resulting from a decision not to buy sample information have no cost. Hence the net payoffs for the first six action–outcome branches of the tree are the perspective payoffs listed in Example 18.4. All other branches of the decision tree result from the decision to buy additional information. Since the cost of this additional information is .4, the remaining action–outcome branches of the decision tree have net payoffs equal to the original payoff (of Example 18.4) minus .4.

completion of tree

Having filled in the net payoffs, we now work from right to left to complete the decision tree. At the outcome stage of the tree, the probabilities have been inserted. For the decision not to buy additional sample information, the probabilities listed for each outcome (patentable or not) are the outcome probabilities $P(\theta_j)$ given in Example 18.4. For each branch associated with a decision to buy additional sample information, the posterior probability of an outcome given a particular sample result $P(\theta_j | \text{Sample result } I)$ is listed at the outcome stage. For example, the probability of the outcome Patentable given the sample result Likely is .889.

Each ellipse that connects the branches between the outcome and action stages of the decision tree contains the expected net payoff for a particular Buy decision, sample result, and action. For example, at the very top of the decision tree of Figure 18.4, the expected net payoff of the action Buy when no additional sample information is obtained is

$$E(\text{Buy} | \text{No additional information}) = 18.6(.800) + 16.4(.200) = 18.16$$

Similarly, the expected net payoff for the action Apply given a decision to buy sample information and a sample result of Unlikely is

$$E(\text{Apply} | \text{Unlikely}) = 20.4(.640) + 7.8(.360) = 15.86$$

All the expected net payoffs shown in the ellipses of Figure 18.4 between the action and outcome stages of the decision tree are computed using the appropriate net payoffs and probabilities listed to the right.

Refer now to the rectangles between the sample result and action stages of Figure 18.4. The entry for each rectangle is the maximum expected net payoff from the connecting ellipses to the right. The value of 18.28 shown in the first rectangle of Figure 18.4 is the maximum expected net payoff resulting from the action Apply; the Buy and Sell branches are thereafter ignored. The remaining rectangles of Figure 18.4 between the sample result and action stages are filled in the same way.

The probability listed for each sample result at the same result stage was computed

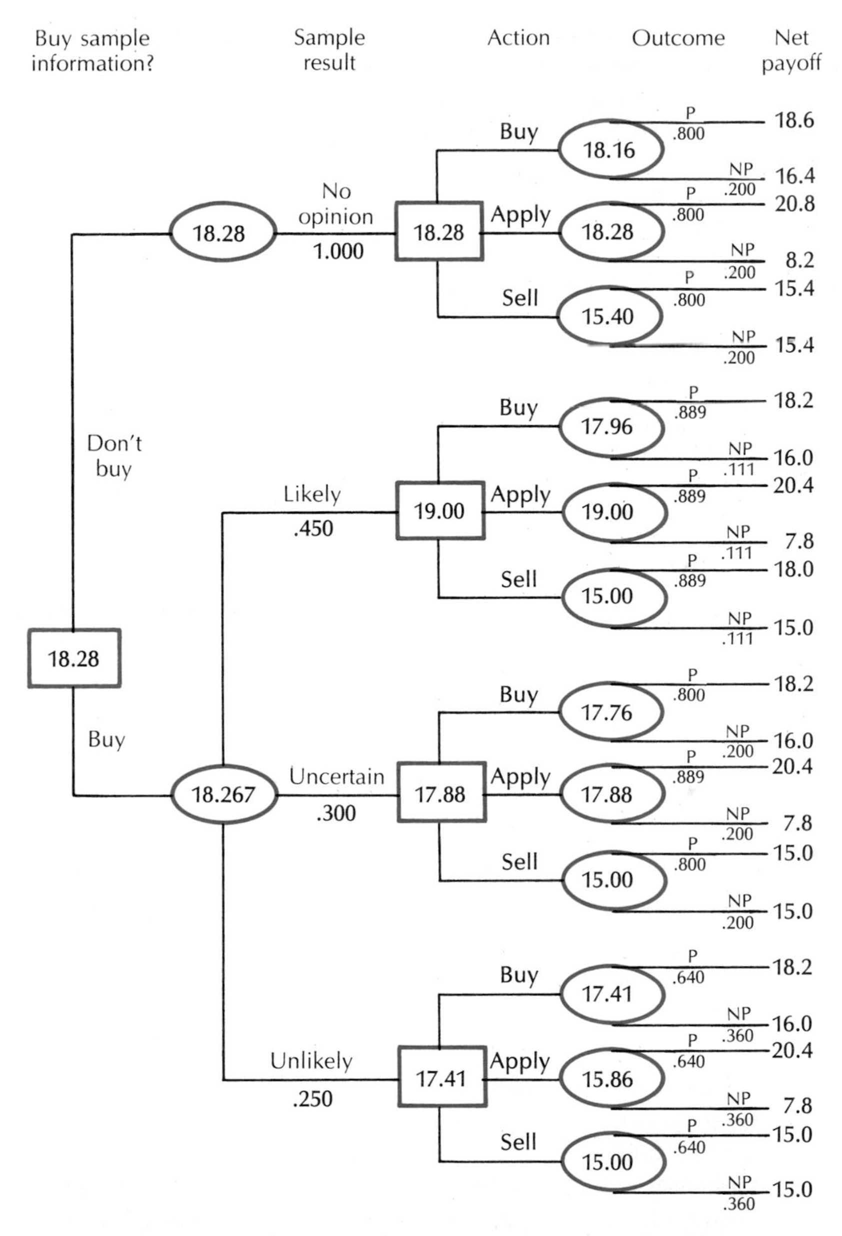

FIGURE 18.4 **Complete Decision Tree for Examples 18.4, 18.6, and 18.7**

in Example 18.6 as part of the calculations for the posterior probabilities $P(\theta_j|\text{Sample result } I)$. For example,

$$\begin{aligned} P(\text{Likely}) &= P(\text{Likely}|\text{Patentable})P(\text{Patentable}) \\ &\quad + P(\text{Likely}|\text{Not Patentable})P(\text{Not patentable}) \\ &= .50(.80) + .25(.20) = .45 \end{aligned}$$

Obviously, the probability of the sample result No opinion is 1.00 when the decision is made *not* to buy additional sample information.

computing EVSI

The ellipses between the Buy Information and sample result stages of the decision tree are used to compute the EVSI. For example, the 18.28 in the first ellipse is 18.28(1.000), and

$$18.267 = 19.00(.450) + 17.88(.300) + 17.41(.250)$$

The difference 18.267 − 18.280 is equal to the EVSI minus .4, the cost of the additional information (see Example 18.7).

The final ellipse of the decision tree contains the maximum expected net payoff for the decision not to buy additional sample information. Since (EVSI − .4) is negative, clearly the best decision about purchasing sample information is not to buy.

To summarize, we use the decision tree by proceeding from right to left. At each information stage we calculate expected values and write these values in the ellipses. At each decision step we choose the maximum expected values and write the values in the rectangles. Poorer actions are ignored.

Constructing a Decision Tree

1. Identify the series of decision and information stages. Usually this series proceeds in a natural chronological order.
2. Construct all branches, with rectangles indicating decisions and ellipses indicating information stages.
3. On the far right, indicate the net payoff for each branch.
4. At each information stage, insert the appropriate result probabilities. Bayes' Theorem calculations or a separate probability tree may be needed.
5. Fill in ellipses and rectangles from right to left. Each ellipse is filled by an expected-value calculation, each rectangle by a choice of the maximum expected value. Nonoptimal decisions are ignored.

EXAMPLE 18.8 A publisher of newspapers serving various suburban communities is considering adding a paper (either a fully separate paper or a spin-off edition of an existing one) to serve a growing area. The success of such a paper depends heavily on the future growth of businesses in the area. The publisher's payoffs and prior probabilities are shown below.

	Outcome θ_j		
Action a_i	Rapid	Moderate	Little
Full paper	5.2	−1.2	−3.6
Spin-off	2.6	1.2	−.4
No paper	.0	.0	.0
$P(\theta_j)$	.50	.30	.20

The publisher can hire a specialist to survey the suburb at a cost of $100,000 (i.e., $.10 million). The specialist will give a positive or negative recommendation. The likelihoods are assumed to be

	Outcome θ_j		
Sample Result I	Rapid	Moderate	Little
Positive	.70	.50	.20
Negative	.30	.50	.80
$P(\theta_j)$	.50	.30	.20

Construct a decision tree for the publisher's problem.

Solution The publisher must decide whether to hire the specialist, receive the opinion (if any), then choose the desired action, and finally observe the actual growth outcome. The tree paths occur in that order. The net payoffs depend on the growth outcome and the specialist's fee. The relevant probabilities are those for the specialist's opinion and for outcomes given the opinion. Given no opinion, the prior probabilities .50, .30, and .20 apply. A probability tree for the specialist's opinions and prior probabilities is shown in Figure 18.5. From the probability tree it follows that

$$P(\text{Positive}) = .35 + .15 + .04 = .54$$
$$P(\text{Negative}) = .15 + .15 + .16 = .46$$

$P(\theta_j)$	P(sample result, $I \mid \theta_j$)	$P(\theta_j$ and sample result, $I)$
Rapid .50	Positive .70	.35
	Negative .30	.15
Moderate .30	Positive .50	.15
	Negative .50	.15
Little .20	Positive .20	.04
	Negative .80	.16

FIGURE 18.5 **Probability Tree for Example 18.8**

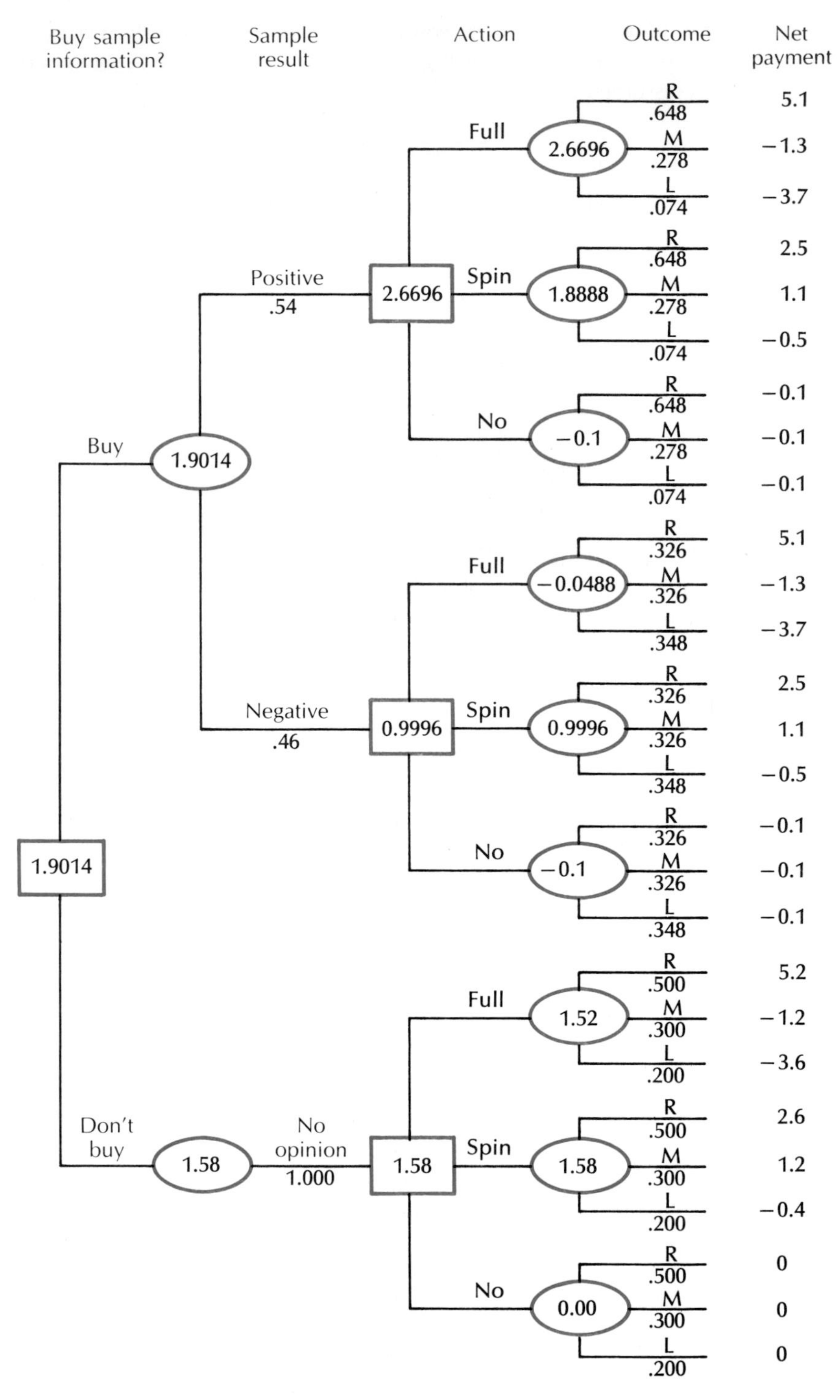

FIGURE 18.6 **Decision Tree for Example 18.8**

Posterior probabilities $P(\theta_j|\text{Sample result } I)$ are then

Outcome θ_j	Sample Result I	
	Positive	Negative
Rapid	$\frac{.35}{.54} = .648$	$\frac{.15}{.46} = .326$
Moderate	$\frac{.15}{.54} = .278$	$\frac{.15}{.46} = .326$
Little	$\frac{.04}{.54} = .074$	$\frac{.16}{.14} = .348$

The resulting decision tree is shown in Figure 18.6.

The optimal action sequence is to buy the opinion—if it is positive, to bring out the full paper, and if it is negative, to use the spin-off edition. □

The virtue of a decision tree is that it forces step-by-step logic on the part of a manager facing a complex decision. Of course, trees can become very large when there are many stages, decisions, and outcomes. It is precisely in these conditions that the step-by-step logic of decision theory is most useful.

SECTION 18.3 EXERCISES

18.14 Refer to Exercise 17.24 (p. 678). The company can undertake a survey of the customers of its product line at a cost of $50,000 to find out the anticipated future demand. Assume the following likelihoods of possible survey results:

Actual Future Demand	Survey Results		
	No Change	Up 10%	Up 20%
Steady	.70	.20	.10
Small increase	.20	.60	.20
Large increase	.10	.20	.70

a. Construct a probability tree to calculate the probability of each survey result and the posterior probability of each actual future demand given each survey result.

b. Construct a decision tree to find the optimal (in the expected-return sense) action given each survey result.

c. Should the survey be undertaken?

18.15 Construct a decision tree to answer Exercises 18.9 and 18.10.

18.16 Suppose that the returns to a large-scale operation, shown in Exercise 18.8, are modified to be 9.9, 2.0, and −6.2 under good, fair, and poor yields, respectively. Modify the decision tree of Exercise 18.15 appropriately. Are any optimal decisions affected?

18.17 Construct a decision tree to answer Exercises 18.12 and 18.13.

18.18 Suppose that the prior probabilities given in Exercise 18.11 are changed to .40, .30, .20, and .10 for 25%, 50%, 75%, and 100%, respectively. Construct a probability tree to recalculate the probabilities for the decision tree of Exercise 18.17. Using the recalculated probabilities, find new optimal decisions. Are there any changes in optimal decisions?

18.4 DECISIONS BASED ON A MEAN OR A PROPORTION ($\int$)

In the examples of first two sections of this chapter, the number of actions and outcomes has been quite small. In some managerial situations, the actions and outcomes may cover an entire numerical interval. The market share of a new product can range anywhere between 0 and 100%. The mean useful lifetime of commercial washing machines can be anywhere from nearly zero to some very large time. The introductory advertising expense for a new product can be almost anything. The possible replacement times for washing machines can range widely. In this section we introduce some decision theory for the case in which either the action space or the outcome space covers a continuous range of possibilities.

The basic elements of continuous-variable decision theory are the same as always; we must consider actions, outcomes, payoffs, and (when we evaluate information) likelihoods. Either the action set A or the outcome set (sample space) Θ may be continuous.

loss

The custom is to talk about losses rather than payoffs. We define the **loss** (sometimes called the **regret**) of a given action under an outcome as the difference between the maximum payoff under an outcome and the payoff to the specified action under the outcome.

EXAMPLE 18.9 Suppose that the following payoff $v(a_i, \theta_j)$ table applies.

Action a_i	Outcome θ_j 1	2	3	4	5	6	Expected Payoff
A	12	16	12	8	4	0	10.0
B	10	15	20	15	10	5	14.5
C	4	12	18	24	18	12	16.0
D	0	7	14	21	28	21	14.7
Prior probability $P(\theta_j)$	.1	.2	.3	.2	.1	.1	

a. Calculate a loss table.
b. Show that action C minimizes expected loss.

Solution a. The maximum payoffs for each outcome are

Outcome:	1	2	3	4	5	6
Best payoff:	12	16	20	24	28	21

The loss function is found by subtracting all payoffs in a column from the maximum payoff in that column.

Action a_i	Outcome θ_j					
	1	2	3	4	5	6
A	0	0	8	16	24	21
B	2	1	0	9	18	16
C	8	4	2	0	10	9
D	12	9	6	3	0	0
Prior probability $P(\theta_j)$	.1	.2	.3	.2	.1	.1

b. The expected losses are A, 10.1; B, 5.6; C, 4.1; D, 5.4. Therefore C is the best action. It minimizes the expected loss, or correspondingly, it maximizes the expected payoff. In fact, the difference in expected losses for different actions equals the difference in expected payoffs. Action C is 6.0 units better than action A, 1.5 better than action B, and 1.3 better than action D. □

It can be proved that the same action that maximizes the expected payoff also minimizes the expected loss, so it really doesn't matter whether we speak of payoffs or of losses.

loss functions

In principle, any loss table or function may be specified by a manager. In keeping with standard decision-theory notation, we label actions as a, outcomes as θ, and loss functions as $L(a, \theta)$. To keep the mathematics relatively simple, we limit our consideration to three types of loss functions for continuous outcomes and action spaces. These loss functions are

$$\text{Squared error:} \quad L(a, \theta) = c(a - \theta)^2$$

$$\text{Absolute error:} \quad L(a, \theta) = c|a - \theta|$$

$$\text{Angle-shaped error:} \quad L(a, \theta) = \begin{cases} c_1(a - \theta), & \text{if } a \geq \theta \\ c_2(\theta - a), & \text{if } a < \theta \end{cases}$$

These loss functions are shown in Figure 18.7. Here the c's represent appropriate constants. The manager should choose an action a to minimize expected value of the selected loss function $L(a, \theta)$.

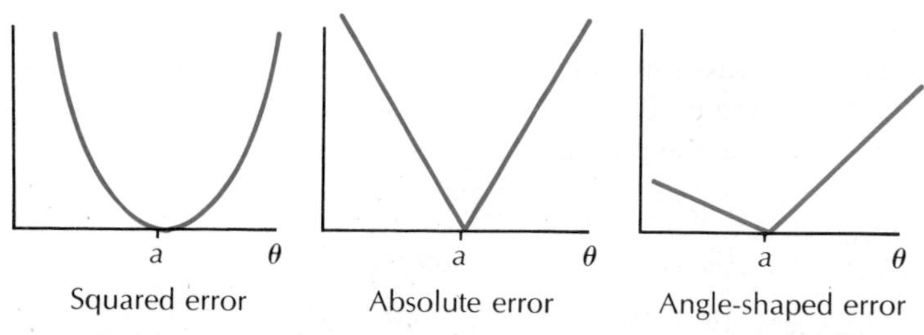

FIGURE 18.7 **Three Possible Loss Functions**

If an action can be selected anywhere in a numerical range, the optimal action is specified by some useful theorems. We assume that the outcome θ ranges over a continuous interval and has a prior probability density $f(\theta)$. Then the optimal action a^* for each of the three types of loss functions is given below.

Squared error loss: $a^* = E(\theta) = \int_{-\infty}^{\infty} \theta f(\theta)\, d\theta$

Absolute error loss: $a^* = \text{median of } f(\theta)$

Angle-shaped loss: $a^* = \left(\frac{c_2}{c_1 + c_2} \times 100\right)\text{th percentile of } f(\theta)$

EXAMPLE 18.10 Suppose that the daily demand θ for beef for a large supermarket chain is assumed to have a normal density with expected value 2000 pounds and standard deviation 150 pounds. Note that both the action space and the outcome space can be considered to be continuous. The action a is the amount of beef supplied. Find the optimal action under each of the following loss functions.

a. $L_1(a, \theta) = .005(a - \theta)^2$
b. $L_2(a, \theta) = .5|a - \theta|$
c. $L_3(a, \theta) = \begin{cases} 2(a - \theta), & \text{if } a \geq \theta \\ 8(\theta - a), & \text{if } a < \theta \end{cases}$

Solution a. For loss function 1, $a^* = E(\theta) = 2000$.

b. For loss function 2, $a^* =$ the median of $f(\theta)$. For a normal density the median equals the expected value, 2000.

c. For loss function 3, $c_1 = 2$ and $c_2 = 8$, so we want the $8/(2 + 8) \times 100 = 80$th percentile of the θ distribution. According to the table of normal probabilities, the 80th percentile is at $z = .84$ standard deviation above the mean, so $a^* = 2000 + .84(150) = 2126$. □

If θ is continuous but there are only a finite number of possible actions, we must compute the expected loss for each value of a and select the action with the smallest expected loss, $EL(a, \theta) = \int_{\text{all } \theta} L(a, \theta) f(\theta)\, d\theta$.

∫ EXAMPLE 18.11 A computer firm develops a new high-speed printer for use with very small minicomputers. The decision whether or not to market the product depends heavily on its potential market share θ. The break-even point is $\theta = .25$. For each .01 below .25, the loss by taking action a_1, marketing the product, is \$20,000. For each .01 above .25, the loss by taking action a_2, not marketing the product, is \$80,000. Thus

$$L(a_1, \theta) = \begin{cases} 20{,}000(.25 - \theta)100, & \text{if } \theta < .25 \\ 0, & \text{if } \theta \geq .25 \end{cases}$$

$$L(a_2, \theta) = \begin{cases} 80{,}000(\theta - .25)100, & \text{if } \theta > .25 \\ 0, & \text{if } \theta \leq .25 \end{cases}$$

The assumed prior probabilities for the market share θ are summarized in the continuous density function $f(\theta) = 12\theta - 24\theta^2 + 12\theta^3$, $0 < \theta < 1$. Which action has the lower expected loss?

Solution

$$EL(a_1, \theta) = \int_0^1 L(a_1, \theta)f(\theta)\,d\theta$$

$$= \int_0^{.25} 20{,}000(.25 - \theta)100(12\theta - 24\theta^2 + 12\theta^3)\,d\theta$$

$$+ \int_{.25}^1 0(12\theta - 24\theta^2 + 12\theta^3)\,d\theta$$

$$= 48{,}047$$

$$EL(a_2, \theta) = \int_0^1 L(a_2, \theta)f(\theta)\,d\theta$$

$$= \int_{.25}^1 80{,}000(\theta - .25)100(12\theta - 24\theta^2 + 12\theta^3)\,d\theta$$

$$+ \int_0^{.25} 0(12\theta - 24\theta^2 + 12\theta^3)\,d\theta$$

$$= 1{,}392{,}187.5$$

Action a_1 (market the product) has a much smaller expected loss. □

The selection of an action, so far in this section, has been based entirely on prior probabilities using an expected-value criterion. We have said nothing about the potential value of information or about how to use it. One concept of value of information, the expected value of perfect information (EVPI), translates very conveniently into terms of loss.

A decision maker who can forecast the final outcome perfectly will pick the action that gives the minimum loss. The minimum loss for any given outcome is zero, by definition of the loss function. Therefore the expected loss under perfect information must be zero. Under current information (using only prior probabilities), the optimal expected loss is $EL(a^*, \theta)$. Therefore

$$\text{EVPI} = EL(a^*, \theta) - 0 = EL(a^*, \theta)$$

The value and use of sample, error-prone information isn't so easy. In principle, the basic idea isn't hard: Use sample information to calculate posterior probabilities using Bayes' Theorem, then select the optimal action using these posterior probabilities. Unfortunately, the mathematics required is rather extensive. We state only two useful special cases. A general treatment is given in DeGroot (1970).

normal prior distribution

One important special case occurs when the outcome θ is a population mean and is assumed to have **normal prior distribution**. Assume that the information consists of a random sample of n observations $Y_1, \ldots, Y_n$, which have expected value θ and known variance* σ^2. Then it can be proven that the posterior density of θ, given the sample

* The more realistic case of unknown variance gives generally similar results, but the math isn't easy. See DeGroot (1970).

information, is also normal with the expected value

$$\frac{\dfrac{\mu_0}{\sigma_0^2} + \bar{y}\left(\dfrac{n}{\sigma^2}\right)}{\dfrac{1}{\sigma_0^2} + \dfrac{n}{\sigma^2}}$$

and variance

$$\frac{1}{\dfrac{1}{\sigma_0^2} + \dfrac{n}{\sigma^2}}$$

where

μ_0 = prior expected value
σ_0^2 = prior variance
$\bar{y}$ = sample mean
σ^2 = variance of each observation (population variance)

The posterior mean (which for a normal distribution is also the median) is a weighted average of the prior mean and the sample mean. The weights depend on the sample size n, the prior variance σ_0^2, and the population variance σ^2.

EXAMPLE 18.12 A new type of bit for digging wells has been developed. The crucial property for such bits is the average useful life—the number of hours the bit can be used before wearing out. Based on general engineering principles, the manufacturer's best guess is that the mean life is 120 hours. Until a field test is conducted, there is uncertainty about the actual mean life. A normal prior distribution for μ with expected value 120 hours and standard deviation 20 hours fairly represents the uncertainty (and a great deal of uncertainty it is; there is a subjective probability of about .05 that the mean is less than 80 hours or greater than 160 hours). A set of 15 bits is tested and the life of each bit is normally distributed with the unknown mean and an assumed standard deviation of 10 hours.

a. Assume that the sample mean lifetime is 135 hours. Find the posterior distribution of the true mean lifetime.
b. An analysis of the potential losses and profits indicates that the bit should be marketed only if the probability that the mean life exceeds 125 hours is at least .9. Should the bit be marketed?

Solution a. The following parameters are specified: the prior mean, $\mu_0 = 120$, the prior standard deviation, $\sigma_0 = 20$, the sample size, $n = 15$, and the population standard deviation, $\sigma = 10$. The posterior distribution of μ is normal, with expected value

$$\frac{\dfrac{\mu_0}{\sigma_0^2} + \bar{y}\left(\dfrac{n}{\sigma^2}\right)}{\dfrac{1}{\sigma_0^2} + \dfrac{n}{\sigma^2}} = \frac{\dfrac{120}{(20)^2} + 135\left[\dfrac{15}{(10)^2}\right]}{\dfrac{1}{(20)^2} + \dfrac{15}{(10)^2}}$$

$$= 134.75$$

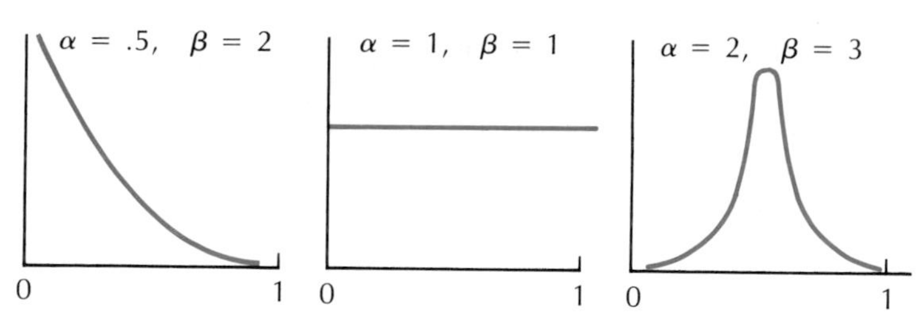

FIGURE 18.8 **Beta Densities**

and variance

$$\frac{1}{\frac{1}{\sigma_0^2} + \frac{n}{\sigma^2}} = 6.557$$

Note that the posterior expected value is heavily weighted toward the sample mean, because the prior uncertainty about the mean is very high.

b. By standard normal calculations

$$P(\theta > 125) = P\left(Z > \frac{125 - 134.75}{\sqrt{6.557}}\right)$$

$$= P(Z > -3.81) \approx 1.00$$

Therefore the company should market the bit. □

Another special case occurs when θ is a population proportion of individuals having some property with assumed prior density

$$f(\theta) = \frac{(\alpha + \beta - 1)!}{(\alpha - 1)!(\beta - 1)!} \theta^{\alpha-1}(1 - \theta)^{\beta-1}, \qquad 0 \leq \theta \leq 1$$

beta prior density

This prior density is called a **beta density**.* A wide variety of shapes for $f(\theta)$ may be generated by taking various values for α and β (see Figure 18.8).

In this situation, assume that the sample information is Y, the number of individuals having the specified property. Variable Y is assumed to have a binomial distribution. Then it can be shown that the posterior density of θ is also a beta density with parameters α' and β', where

$$\alpha' = \alpha + y$$
$$\beta' = \beta + n - y$$

The expected value of this density is $\alpha'/(\alpha' + \beta')$. For large values of α' and β' (say, both α' and β' at least 10), the density is approximately normal with variance $\alpha'\beta'/[(\alpha' + \beta')^2(\alpha' + \beta' + 1)]$.

* α and β may take on noninteger values. In such a case, the factorials must be interpreted by the gamma function.

EXAMPLE 18.13 Refer to Example 18.11. Suppose that the computer firm test-markets the printer to a random sample of 40 potential buyers, and that 8 actually buy the printer while the other 32 choose competitive products.

a. Show that the prior density for θ, the true proportion of buyers, given by

$$f(\theta) = 12\theta - 24\theta^2 + 12\theta^3, \qquad 0 \leq \theta \leq 1$$

is a beta density.

b. Find the posterior distribution of θ.

c. What is the approximate probability that $\theta > .25$, the break-even value?

Solution a. $f(\theta)$ can be rewritten as

$$f(\theta) = 12\theta^{2-1}(1 - \theta)^{3-1}$$

a beta density with $\alpha = 2$ and $\beta = 3$. The expected value for this prior distribution is $\alpha/(\alpha + \beta) = .4$.

b. We have $y = 8$ and $n - y = 32$. The posterior density is again a beta density with

$$\alpha' = 2 + 8 = 10 \qquad \text{and} \qquad \beta' = 3 + 32 = 35$$

The posterior expected value of the true proportion is $\alpha'/(\alpha' + \beta') = .222$. This is somewhat above the sample proportion .200 but well below the prior expected value $2/(2 + 3) = .400$.

c. For these values of α' and β', a normal approximation is decent. The variance is $\alpha'\beta'/[(\alpha' + \beta')^2(\alpha' + \beta' + 1)] = .00376$, so the standard deviation is $\sqrt{.00376} = .0613$.

$$P(\theta > .25) = P\left(Z > \frac{.25 - .222}{.0613}\right)$$

$$= P(Z > .46)$$

$$= .32$$

□

These cases assume some very special probability distributions. It can be proved that these assumptions are not crucial. If the relevant sample size is reasonably large, the posterior distribution is approximately normal. The same general guidelines as those for application of the **Central Limit Theorem** should apply here. Almost always, the posterior expected value falls between the prior expected value and the sample mean. The relative weights given to the prior mean and the sample mean depend on the prior uncertainty and the standard error of the sample mean. As the sample size gets larger, it often occurs that the posterior expected value approximates the sample mean (or more generally, the maximum likelihood estimate of θ). The value of incorporating prior information about θ is greatest when the sample size is small or when there is accurate information about θ.

Central Limit Theorem

SECTION 18.4 EXERCISES

18.19 Refer to Exercise 17.14 (p. 673).

a. Construct a loss table corresponding to the given payoff matrix.

b. Verify that the order quantity that minimizes the expected loss is the same quantity found to be optimal in Exercise 17.14.

18.20 The benefits department of a large corporation is considering modifying the health insurance coverage for its employees. The critical unknown factor in the decision is the proportion of employees who will actually select option A. The corporation must estimate this proportion when selecting the health insurance policy. If the actual proportion turns out to be less than or equal to the estimate, the cost per year to the corporation is $100,000 times the estimated proportion. If the actual proportion exceeds the estimate, the cost per year is $200,000 times the actual proportion minus $100,000 times the estimate.

a. Construct a payoff matrix for actual proportions $\theta = .10, .20, \ldots, .90$ and estimates $a = .10, .20, \ldots, .90$. Note: payoff = negative of cost.

b. Calculate a table of losses (opportunity costs) based on this payoff matrix.

c. Show that the loss function equals the absolute error loss with $c = 100{,}000$.

18.21 Refer to Exercise 18.20. The personnel department is quite uncertain about the actual proportion θ of employees who will select option A. The assumed prior density of θ is

$$f(\theta) = 6(\theta - \theta^2), \qquad \text{for } 0 \le \theta \le 1$$

a. Show that this density is a beta density.

b. What are the mean and median for this density?

c. What is the optimal estimate for the corporation?

18.22 Refer to Exercises 18.20 and 18.21. Suppose that a random sample of 100 employees shows that 40 will select option A. For practical purposes, the sampling distribution may be assumed to be binomial.

a. Calculate the posterior density of θ.

b. Find the mean and median of this density, using a normal approximation.

c. What is the new optimal estimate?

d. Explain why the estimate is above .40 but only slightly so.

18.23 A chain of hardware stores is planning a "do-it-yourself" sale based on heavy local advertising. The crucial decision is how much initial inventory to stock in each store. If demand per store is greater than or equal to stock, the chain's profit per store is $1.00 for each dollar stocked plus $.70 for each dollar of excess demand. If demand is less than stock, the profit is $1.00 for each dollar of demand less $.10 for each dollar of excess stock.

a. Construct a payoff matrix depending on stock a and demand θ. Both stock and demand should range from 5 to 50 (in thousands of dollars per store).

b. Calculate a loss table from this payoff matrix.

c. Show that the loss function is angle-shaped with $c_1 = .1$ and $c_2 = .3$.

18.24 Assume that the prior distribution of demand θ in Exercise 18.23 is normal with mean 30 and standard deviation 8 (in units of thousands of dollars per store). What is the optimal amount to stock?

18.25 Refer to Exercises 18.23 and 18.24. The chain tests the sales campaign in a random sample of 12 stores. The mean demand per store in the sample is $34,000. Assume that the true standard deviation of sales is $12,000.

a. What is the posterior distribution of θ, the demand per store for all stores in the chain?

b. What is the optimal value of a, the amount stocked per store?

18.26 A steel producer is considering installing heat-recovery equipment to reduce energy costs. A critical unknown parameter in the decision is the amount of recoverable energy per ton of steel. This parameter determines the desired size of the equipment. Careful analysis of cost factors indicates that the loss (opportunity cost) function is

$$L(a, \theta) = 1000(a - \theta)^2 \text{ dollars}$$

where a is size and θ is recoverable energy per ton. Engineering estimates suggest that the recoverable energy is normally distributed with mean 2000 units per ton. There is considerable uncertainty in this estimate, however, as indicated by a standard deviation of 150 units per ton.

a. Based on the engineering estimates, what is the optimal size of the equipment?

b. What is the expected loss? Hint: $E(\theta - a)^2 = \text{Var}(\theta) + (E(\theta) - a)^2$.

18.27 Refer to Exercise 18.26. Ten lots of steel are produced. The mean recoverable energy per ton in the sample is 1890 units and the standard deviation (assumed to be the true one) is 120 units.

a. Find the posterior distribution of the heat loss per ton.

b. What is the optimal size of the equipment?

c. What is the expected loss?

18.28 Suppose that the standard deviation in Exercise 18.27 had been 140. How would the answers to that exercise change?

18.5 BAYESIAN DECISION THEORY AND HYPOTHESIS TESTING

Many managerial problems involving statistical information can be stated in terms of either decision theory or hypothesis testing. A natural question is, Which way is better? The question has been debated for a long time to no final conclusion. In this section we make a few suggestions.

The first thing to point out is that often it doesn't matter which approach is used. In many cases, probably the majority, the two approaches yield very similar answers. If the statistical evidence favoring one action over another is strong, either approach so indicates.

advantage of decision theory

Classical significance-testing procedures were not really developed for decision problems. Rather, they were developed for inference, to answer the question, Can this apparent effect be attributable merely to random variation? Some care is needed to adapt hypothesis-testing procedures to a decision situation. As we indicated in Chapter 9, in such situations both the costs of the two kinds of error and the prior probabilities of the hypothesis should be considered in selecting α and β probabilities. Decision theory has the advantage that it incorporates these factors explicitly.

disadvantage of decision theory

One disadvantage of decision-theory methods is that they require more judgment and more computation. In classical methods, once the sample size, test statistic, and α level are specified, the procedure is routine and can be computerized. To use decision theory, a manager must specify prior probabilities, likelihoods, and payoffs or losses. It isn't always easy to make these specifications in any precise, numerical form. While sensitivity analysis can be used to test the effect of changes in specification, the requirements for the use of Bayesian methods are definitely more demanding than those for the use of hypothesis testing. This is particularly true for more complex problems, such as those of multiple regression.

At minimum, the ideas of decision theory can be used in a rough, approximate way to illuminate hypothesis-testing methods. If the cost of falsely rejecting a null hypothesis is high and the prior probability of that null hypothesis being true is substantial, a prudent manager should demand extremely strong evidence (a very small p-value) before rejecting it. If the cost of a Type II error or the prior plausibility of a research hypothesis is high, a manager might well act as if the research hypothesis is true even if the hypothesis test yields very weak evidence favoring it. It may be difficult to put precise numerical values on these high versus low judgments, but ignoring them altogether is not the answer.

Finally, all these methods are intended to help people think clearly in conditions of uncertainty. A manager who understands what the methods are (and are not) intended for, what factors are (and aren't) being considered, and what assumptions are being made should profit from their use.

SUMMARY

The material in this chapter extends the discussion of decision theory begun in Chapter 17 to include additional sample information as part of the decision process. Information is expensive, and it takes time to accumulate it. The utility of the material in this chapter is that it provides the necessary tools for determining whether or not sample information sufficiently improves the decision-making process to justify the time and money necessary to gather and process it.

Applications of decision theory require a great deal of thought and can be time-consuming. It is important that a manager become acquainted with the details of decision theory and understand its limitations. With this familiarity, decision theory can be a valuable analytic aid.

KEY FORMULAS: Using Sample Information in Making Decisions

1. Summary table: Payoff table, $v(a_i, \theta_j)$

	Outcome θ_j		
Action a_i	θ_1	θ_2	...
a_1			
a_2		$v(a_i, \theta_j)$	
⋮			
$P(\theta_j)$	$P(\theta_1)$	$P(\theta_2)$	...

2. Expected value of perfect information (EVPI)

$$\text{EVPI} = \sum_j (\text{maximum payoff given outcome } \theta_j)P(\theta_j)$$

$$- \text{ maximum expected return when no additional information is available}$$

3. Summary table: Likelihood (Sample result $I \mid$ Outcome θ_j)

	Outcome θ_j		
Sample Result I	θ_1	θ_2	...
1			
2		Likelihood (Sample result $I \mid \theta_j$)	
⋮			

4. Bayes' Theorem

$$P(A_i|B_j) = \frac{P(B_j|A_i)P(A_i)}{\sum_i P(B_j|A_i)P(A_i)}$$

5. Posterior probability, $P(\text{Outcome } \theta_j|\text{Sample result } I)$

$$P(\theta_j|\text{Sample result } I) = \frac{P(\text{Sample result } I|\theta_j)P(\theta_j)}{\sum_j P(\text{Sample result } I|\theta_j)P(\theta_j)}$$

	Sample Result I			
Outcome θ_j	1	2	3	. . .
θ_1				
θ_2		$P(\theta_j\|\text{Sample result } I)$		
⋮				
$P(\text{Sample result } I)$				

6. Expected return on action a_i given sample result I

$$E(\text{Return on } a_i|\text{Sample result } I) = \sum_j v(a_i, \theta_j)P(\text{Outcome } \theta_j|\text{Sample result } I)$$

	Sample Result I			
Action a_i	1	2	3	. . .
a_1				
a_2		$E(\text{Return on } a_i\|\text{Sample result } I)$		
⋮				

7. Expected value of sample information (EVSI)

$$\sum_I (\text{Maximum expected return}|\text{Sample result } I)P(\text{Sample result } I)$$

$$- \text{ Maximum expected return using no sample information}$$

8. Loss functions
   a. Squared error: $L(a, \theta) = c(a - \theta)^2$
   b. Absolute error: $L(a, \theta) = c|a - \theta|$
   c. Angle-shaped error: $L(a, \theta) = \begin{cases} c_1(a - \theta), & \text{if } a \geq \theta \\ c_2(\theta - a), & \text{if } a < \theta \end{cases}$

CHAPTER 18 EXERCISES

18.29 The production engineering staff of a manufacturing company has developed two possible redesigns of its production process. Method 1 requires substantial additional training of workers and a relatively modest investment. Method 2 requires less training than even the current method but a heavy investment in automated equipment. The net profit attributable to each method depends on the rate of turnover of workers as shown in the following payoff matrix (in thousands of dollars).

Process	Turnover Rate		
	Down 10%	No Change	Up 10%
Method 1	+80	+40	−100
Method 2	−40	+20	+50
Current	30	0	−40

The labor relations manager believes that there is a 30% chance that, over the relevant time period, the turnover rate will go down by 10%, a 60% chance of no change, and a 10% chance that it will go up by 10%.

a. Which process has the highest expected return?

b. How valuable would perfect information about turnover rate be?

18.30 The company of Exercise 18.29 can hire a consultant to survey a sample of workers and provide an opinion about likely future turnover. The following likelihoods are assumed for the consultant's opinion:

Consultant's Opinion	Actual Turnover Rate		
	Down 10%	No Change	Up 10%
Down sharply	.30	.05	.00
Down slightly	.40	.20	.10
Constant	.20	.50	.20
Up slightly	.10	.20	.40
Up sharply	.00	.05	.30

a. Calculate the probabilities of each possible opinion and the posterior probabilities of actual rates given each opinion. You may want to use a probability tree.

b. Calculate the action that has the highest expected value given each opinion. You may want to use a decision tree.

c. The consultant's fee is $4000. Should the company hire the consultant?

18.31 The likelihoods assumed in Exercise 18.30 are not known with complete accuracy. As part of a sensitivity analysis, the following alternative likelihoods are assumed:

Consultant's Opinion	Actual Turnover Rate		
	Down 10%	No Change	Up 10%
Down sharply	.40	.00	.00
Down slightly	.50	.15	.00
Constant	.10	.70	.10
Up slightly	.00	.15	.50
Up sharply	.00	.00	.40

a. Do these alternative likelihoods indicate greater or lesser value to the consultant's opinion as compared to the likelihoods of Exercise 18.30?

b. Rework Exercise 18.30 assuming the alternative likelihoods. Which decisions, if any, should be changed?

∫ **18.32** The director of media selection for an advertising agency must decide on placement for an ad campaign promoting a "full-flavor" beer. The basic choice is between television commercials on broadcasts of professional football only and commercials on both baseball and football telecasts. If the football audience also constitutes the baseball audience, there is little gain to advertising in both places. After analysis, it is determined that the both-place strategy is preferable only if the proportion θ of baseball watchers who are also football watchers is less than .60. Complete accurate data on this proportion is unavailable. A reasonable prior density for θ is

$$f(\theta) = 20(\theta^3 - \theta^4), \qquad \text{for} \quad 0 \le \theta \le 1$$

a. Verify that this density is a beta density.
b. Sketch the density.
c. What is the prior probability that the proportion θ is less than .6?

18.33 Refer to Exercise 18.32. A random sample of television viewers shows that, of 128 baseball watchers, 70 are also football watchers.

a. Find the posterior distribution of the proportion θ.
b. What is the expected value of θ?
c. Use a normal approximation to calculate the probability that the proportion is less than .60.

18.34 Refer to Exercise 18.33. Do the hypothesis-testing methods of Chapter 9 lead to the rejection of H_0: $\theta \ge .60$ and support of H_a: $\theta < .60$ with an α of .05? Compare the conclusion of such a test to the answer to Exercise 18.33, part (c).

18.35 A trucking firm regularly sends loads from Chicago to Baltimore. The standard route for these loads passes over many toll roads, so the toll cost is substantial. An alternate route has become feasible with the opening of a new interstate highway. The new route will be cheaper if the mean additional travel time is less than 2 hours. Calculation of relative mileage and terrain leads to a normal prior distribution of this mean. The prior expected value is 1.40 hours with a prior standard deviation of .40 hours. Find the prior probability that the true mean is less than 2 hours.

18.36 The trucking firm of Exercise 18.35 tests the new route by sending nine loads. The times required may be regarded as the results of a random sample. The sample mean additional time is 1.33 hours and the standard deviation (which may be assumed to be the true one) is 1.80 hours.

a. What is the posterior distribution of the true mean additional time?
b. What is the probability that the true mean is less than 2 hours?

18.37 a. Use the hypothesis-testing methods of Chapter 9 on the random sample in Exercise 18.36. Show that H_a: $\mu < 2$ is not supported at $\alpha = .05$ or $\alpha = .10$.
b. The answer to part (b) of Exercise 18.36 gave a high probability that the mean is less than 2. Why do the two approaches yield apparently conflicting answers?

18.38 A chain of auto supply stores is preparing a promotion for its line of car radios and cassette decks. A debate arises over whether to aim the promotion primarily at new car buyers or at current owners. The profitability of the two actions depends on new car sales for the last three months of the year. The payoff matrix in thousands of dollars is estimated to be

	Sales				
Aim	Down 10%	Down 5%	Steady	Up 5%	Up 10%
New car buyers	25	75	125	175	225
Current owners	100	80	60	40	20

As of May, when the planning stage for the promotion is in process, the prior probabilities for last-quarter sales are .10, .20, .30, .30, and .10 for down 10%, down 5%, steady, up 5%, and up 10%, respectively.

a. Which action maximizes the expected payoff?

b. Calculate a loss table. Which action minimizes the expected loss?

18.39 Refer to Exercise 18.38. Rather than making an immediate decision, the company can defer its choice until new car-sales data are available for the summer months (June, July, and August). These data may lead to changes in the probabilities for last-quarter sales. Historical data on the relation between summer sales and last-quarter sales yield the following likelihoods:

Summer	Last-Quarter Sales				
Sales	Down 10%	Down 5%	Steady	Up 5%	Up 10%
Down 10%	.30	.20	.10	.15	.05
Down 5%	.30	.25	.20	.20	.15
Steady	.20	.20	.40	.20	.20
Up 5%	.15	.20	.20	.25	.30
Up 10%	.05	.15	.10	.20	.30

a. Calculate posterior probability distributions for last-quarter sales, given each possible value of summer sales.

b. What is the optimal action given each possible value of summer sales?

18.40 The cost to the company in Exercise 18.39 of delaying its choice is $4000, primarily in the cost of preparing duplicate materials. Should the company wait?

18.41 A chain of appliance dealers must place an order for two new models of dishwashers. The total order size is fixed; the manufacturer will supply only 10,000 units. The chain can order any proportion of deluxe and standard units. An analysis of the profitability of sales of the units produces the following table of profits (in thousands of dollars):

Proportion of Deluxe Units	Proportion of Deluxe Units Demanded, θ					
Ordered, a	.05	.10	.15	.20	.25	.30
.05	110	110	110	110	110	110
.10	95	120	120	120	120	120
.15	80	105	130	130	130	130
.20	65	90	115	140	140	140
.25	50	75	100	125	150	150
.30	35	60	85	110	135	160

This table can be extended to other proportions.

a. Calculate a loss table.

b. Show that the loss function is angle shaped, with $c_1 = 300$ and $c_2 = 200$.

c. What is the optimal value of a for a specified prior density $f(\theta)$?

19

SOME ALTERNATIVE SAMPLING METHODS

simple random sample

sample survey design

The basic idea of random sampling was discussed in Chapter 6. There we defined a **simple random sample**: a set of n observations chosen from a population of size N in such a way that all possible samples of size n have an equal chance of being selected. The formulas (particularly those for standard errors) of later chapters were based on an assumption of simple random sampling. There are other ways of choosing random samples, and in this chapter we discuss a few of them. The chapter serves as an introduction to a broad area of statistical theory called **sample survey design**.

stratified sampling

cluster sampling

Why should a manager bother with fancy sample survey designs? The purpose of selecting a sample is to obtain the most accurate information possible for a given expenditure (of both time and money). Accuracy is measured by amount of bias, which ideally is zero, and by probable sampling variability as summarized by a standard error. In some situations it may be possible to break the population down into subpopulations, called **strata**, that have relatively small within-strata variability. Then **stratified sampling** can yield better accuracy for a given sample size. Stratified sampling is the subject of Section 19.2. There are other situations in which simple random sampling is too expensive; for example, if we tried to interview a simple random sample of 1000 adult Americans, the travel costs for interviewers would be horrendous. An alternative sampling procedure is to select 50 ZIP code areas randomly and then interview 20 randomly selected individuals in each area. This procedure is an example of two-stage **cluster sampling**; cluster sampling is discussed in Sections 19.3 and 19.4. The advantage of cluster sampling is that the cost per observation can be much lower than the cost of simple random sampling. These sampling methods (and some other slightly more complex ones discussed in Section 19.5) can be useful in getting good information economically.

One obvious area of application of sample survey design is in market research. Another, rapidly growing, area of application is in accounting, particularly auditing. The use of these methods can be expected to grow even more in the future.

19.1 TAKING A SIMPLE RANDOM SAMPLE

Selecting a good simple random sample takes planning and effort. A sloppy sample can reflect several potential biases. Because most good approaches to sampling involve simple random sampling (from the whole population, or from strata, or of clusters), it is worthwhile to consider some pitfalls to avoid in sampling.

target population
sampling frame

One kind of potential bias arises in selecting the entities (whether people or accounts or whatever) to be sampled. Survey statisticians distinguish the **target population** (the set of entities that ideally should be sampled) from the **sampling frame** (the set of entities that actually can be sampled). In a study of consumer satisfaction with the service departments of new-car dealers, the target population may be all private purchasers of new cars within the past two years. The sampling frame may well be the set of all purchasers for whom dealers have current addresses. Purchasers who have moved (without forwarding addresses) are part of the target population, but since they cannot be sampled, they are not part of the sampling frame. Naturally, the best possible situation is that the target population and sampling frame be identical. This is often possible in auditing situations, but it is very difficult to achieve in sampling human populations. When the target population and sampling frame differ, a manager must usually assume that those entities that cannot be sampled have the same characteristics as those that can. If the assumption is false, one kind of **selection bias** occurs.

selection bias

EXAMPLE 19.1 A chain of paint and wallpaper stores in a mid-size city is considering strategies for pricing its house brand of interior paints. A critical element of this strategy is consumer perception of the quality of its paints. A survey is planned of current homeowners in the city. A list of those who paid real-estate taxes the previous year is used to select a sample of 200 homeowners. Identify the target population and the sampling frame. Are there any major selection biases?

Solution The target population presumably is all current homeowners in the city, but ideally it is the persons in each home who decide on paint purchases. The sampling frame is all homeowners who paid real-estate taxes the previous year. A minor selection bias is the omission of homeowners who paid no taxes, since it seems plausible that their homes would be abandoned and poor candidates for painting. A more serious bias is the omission of those who purchased homes in the current year. These people are likely to be major paint purchasers and quite possibly less familiar with local (as opposed to national) paint brands. □

A subtle form of selection bias can occur if members of the sampling frame have unequal probabilities of being sampled. We would not be very impressed by a study of consumer satisfaction with new-car dealers' service departments that proceeded by taking a random sample of service orders and interviewing the car owners. People who were happy with the service would bring their cars back several times and therefore would have a high probability of being sampled. Conversely, people who didn't like the service would have a low selection probability. This kind of bias is called a **size bias** because it often reflects the size or number of transactions of an individual in the sampling frame.

size bias

EXAMPLE 19.2 A brokerage firm anticipates an increase in the margin requirements for purchasers of common stock and wishes to estimate what fraction of its customers would have been affected by such an increase during the past year. A clerk randomly samples 1000 purchases of common stocks during the year and determines the fraction of these purchases that would have been affected. Identify a size bias.

Solution The target population is customers, not purchases. The sampling procedure is biased toward those customers who purchased stock frequently during the past year. □

non-response bias

There is a different kind of bias that can arise even though no selection biases are present. Once selected, individuals in a sample may not give accurate responses. At the extreme, many individuals may not respond at all. As many as 80% or even more of individuals contacted by mail surveys do not respond; a 30% non-response rate in telephone surveys is considered better than average. In any competent survey, a follow-up mechanism is provided to give some indication as to whether (first-round) non-respondents differ from respondents, but this device has only limited value. The problem of non-response is perhaps the most serious limitation of the usefulness of surveys of human populations. A great deal of ingenuity has been expended to minimize non-response rates, since this is a crucial part of any survey design.

sensitive-question bias

In addition, those who do respond may not respond accurately. Questions can be ambiguous or "loaded" to favor a certain answer. In some situations, certain answers may be more socially acceptable or polite than others. Pre-testing usually reveals any serious defects in questions. The randomized-response technique discussed in Section 19.5 can be used to deal with sensitive questions.

This litany of potential problems doesn't mean that surveys are impossible, but it does mean that they must be done carefully.

Assuming that a survey design has removed potential biases, the remaining source of error is "luck of the draw": pure random variability. Here is where the standard-error formulas of statistical theory are useful. In the next several sections we set out standard errors for various types of survey design. The aim of a survey typically is estimation (rather than hypothesis testing). The desired population parameter may be the population mean μ, the population proportion π, or the total value in the population, which we denote by τ. Since the population mean is the total divided by the population size N (that is, $\mu = \tau/N$), it follows that $\tau = N\mu$. If the sample mean $\bar{y}$ is used to estimate μ, a natural estimate of the population total τ is $N\bar{y}$.

We developed confidence-interval methods for a mean in Chapter 8. We found that t tables should be used for small samples. In most sample surveys, sample sizes are large enough that normal tables may be used. Remember that the normal table "magic number" for 95% confidence intervals is 1.96.

Estimation of μ and τ Using Simple Random Sampling

Point estimate

of μ: $\bar{y}$
of τ: $N\bar{y}$

95% confidence interval

for μ: $\bar{y} \pm 1.96\hat{\sigma}_{\bar{Y}}$
for τ: $N\bar{y} \pm 1.96N\hat{\sigma}_{\bar{Y}}$

where

$$\hat{\sigma}_{\bar{Y}} = \sqrt{\frac{s^2}{n}\left(\frac{N-n}{N-1}\right)}$$

The quantity $(N-n)/(N-1)$, called a **finite population correction factor**, accounts for the fact that there is slightly more information in a sample of 50 from a population of 100 than in a sample of 50 from a population of 10,000.

EXAMPLE 19.3 An industrial firm is concerned about the time per week spent by employees on certain trivial tasks. The time-log sheets of a simple random sample of 50 employees in one week show that the average amount of time spent on these tasks is 10.31 hours with a sample variance of 2.25. If the corporation employs 750 researchers, estimate the total number of hours lost per week on these trivial tasks. Give a 95% confidence interval for τ.

Solution We are told that the population consists of $N = 750$ employees, from which a random sample of $n = 50$ time-log sheets is obtained. The mean amount of work lost for the sampled employees is 10.31. An estimate of the total amount of time lost per week is

$$N\bar{y} = 750(10.31) = 7732.5 \text{ hours}$$

The 95% confidence interval for τ is

$$N\bar{y} \pm 1.96N\hat{\sigma}_{\bar{Y}}$$

where

$$\hat{\sigma}_{\bar{Y}} = \sqrt{\frac{s^2}{n}\left(\frac{N-n}{N-1}\right)}$$

Substituting into the formula for $\hat{\sigma}_{\bar{Y}}$, we obtain

$$\hat{\sigma}_{\bar{Y}} = \sqrt{\frac{2.25}{50}\left(\frac{750-50}{750-1}\right)} = .205$$

The corresponding 95% confidence interval for τ, the total number of hours lost per week on these trivial tasks for the 750 employees, is

$$7732.5 \pm 1.96(750)(.205)$$

or

$$7732.5 \pm 301.4$$

Thus we are 95% confident that the time lost per week is between 7431.1 and 8033.9 hours. This is a staggering amount—approximately 26% of the scheduled work hours. □

Rather than estimating a population mean or total, sample surveys are frequently conducted to estimate the proportion of experimental units in a population that possess a specified characteristic. For example, suppose we are interested in determining a television program's rating by estimating the proportion of families in a given district who watch the program during a given week. Let $y_i = 0$ if the ith family in a simple random sample of n does not possess the characteristic of interest (did not watch the show) and $y_i = 1$ if the ith family does possess the characteristic of interest. Then the proportion of elements in the sample possessing the specified characteristic also represents the sample mean of the y_i's. With this convention, $\hat{\pi}$ is actually $\bar{y}$, and π can be thought of as the mean for the entire population of 0's and 1's. Thus formulas developed for estimating μ can also be used for estimating π. Fortunately, these formulas simplify considerably, as shown below:

Estimation of π Using Simple Random Sampling

Point estimate of π: $\hat{\pi} = \dfrac{y}{n}$

95% confidence interval for π: $\hat{\pi} \pm 1.96\,\hat{\sigma}_{\hat{\pi}}$

where

$$\hat{\sigma}_{\hat{\pi}} = \sqrt{\frac{\hat{\pi}(1-\hat{\pi})}{n}\left(\frac{N-n}{N-1}\right)}$$

EXAMPLE 19.4 A business manager in charge of a large-volume product is concerned about the proportion of customers with delinquent accounts and the proportion of customers with a balance due exceeding \$500. An internal audit is made of a random sample of 120 accounts from a total of 2280 accounts. From this sample, 17 are found to be delinquent and 64 have balances due in excess of \$500. Use these data to give a 95% confidence interval for

a. the proportion of delinquent accounts;
b. the proportion of accounts with balances in excess of \$500.

Solution a. The point estimate of the proportion of customers with delinquent accounts is $17/120 = .14$. The 95% confidence interval for π is

$$.14 \pm 1.96 \sqrt{\frac{.14(.86)}{120}\left(\frac{2280-120}{2280-1}\right)}$$

or

$$.14 \pm .060$$

We are 95% confident that the actual proportion of delinquent accounts is between .080 and .200.

b. In a similar way, the point estimate of the proportion of accounts with balances in excess of \$500 is $64/120 = .53$. The corresponding approximate 95% confidence interval for π is

$$.53 \pm 1.96 \sqrt{\frac{.53(.47)}{120}\left(\frac{2280-120}{2280-1}\right)}$$

or

$$.53 \pm .087$$

We are 95% confident that the actual proportion of accounts with balances in excess of \$500 is between .443 and .617. □

The ideas in this section are essentially a review of ideas from Chapters 6–8. The procedures of the next sections extend these ideas to more complicated sample survey designs. While the formulas become more complicated, the essential idea of estimate plus or minus 1.96 standard errors for 95% confidence still holds.

SECTION 19.1 EXERCISES

19.1 Refer to Example 19.1. Suppose that the paint store chain conducts a telephone survey based on random selection of telephone numbers. (Business phones can be excluded.) Identify the sampling frame and any major selection biases.

19.2 A state commission investigates problems of wheelchair-bound people in obtaining access to public transit and to state offices. A survey of 150 wheelchair-bound people in which each such person keeps a diary recording all instances of inaccessibility is planned. The 150 people are selected from the membership list of a statewide association for handicapped persons. Identify the target population, the sampling frame, and possible selection bias.

19.3 Assume that the survey of the previous exercise can be regarded as a simple random sample. Suppose that 66 of the 150 persons report encountering an inaccessible public building. Give a 95% confidence interval for the population proportion.

19.4 An organization that specializes in arranging rentals and trades of vacation homes wishes to survey the condominium apartments on a certain large island off the Florida coast. Interviewers are instructed to call on randomly chosen apartments during the month of March between 10 A.M. and noon. Follow-up calls are made between 3 P.M. and 5 P.M. the next day, and then again between 7 P.M. and 9 P.M. the following day. Apartment owners are interviewed to determine their interest in renting out their apartments during specified times. What biases could be present in this study?

19.5 Assume that the survey conducted in Exercise 19.4 can be regarded as a simple random sample. Suppose that, of 221 apartment owners who respond, 83 are willing to rent their apartments. The mean availability of the 83 apartments is 4.2 weeks and the standard deviation is 2.8 weeks.
a. Find a 95% confidence interval for the population proportion of rentable apartments.
b. Find a 95% confidence interval for the true mean availability of rentable apartments.

19.6 In an audit of sales for a certain store, an accounting clerk is told to keep a running total of sales accounts. Every sale that takes the total into the next \$10,000 interval is selected for examination. For example, a sale that took the total from \$29,972 to \$30,041 would be selected. The sales are assumed to be in random order, for all practical purposes. Does this procedure yield a practically unbiased simple random sample of all sales?

19.7 Suppose that the sample of Exercise 19.6 yields 267 sales with a mean amount of \$112.24 and a standard deviation of \$91.49. Calculate a 95% confidence interval for the true mean sale amount, assuming that the sample constitutes a simple random sample.

19.8 Did you identify any bias in Exercise 19.6? What effect do you think it has on the confidence interval of Exercise 19.7?

19.2 STRATIFIED RANDOM SAMPLING

Stratification offers an alternative to simple random sampling and in many instances increases the accuracy of information available for estimating μ or τ.

Stratified Random Sample

A stratified random sample is a sample obtained by dividing the population of experimental units into non-overlapping groups, called *strata*. A simple random sample is selected from each stratum.

The first step in selecting a stratified random sample is to clearly specify the strata, making certain that each experimental unit can be classified into only one stratum. For example, in a local election survey we may wish to stratify registered voters according to one of five voting precincts. If precinct boundary lines are clearly defined and the voter registration lists are up to date, there should be no problem in placing registered voters into the appropriate precincts (strata).

After the experimental units are divided into strata, we select a simple random sample of unit from each stratum.

Before presenting the formulas for the estimates of μ (or π) and τ, we need the following notation:

I: number of strata
N_i: number of elements in stratum i ($i = 1, 2, \ldots, I$)
n_i: sample size in stratum i
N: total population size; $N = \sum_i N_i$
n: total sample size; $n = \sum_i n_i$
$\bar{y}_i$: sample mean in stratum i
s_i^2: sample variance in stratum i

The point estimation procedures for μ and τ are given next.

Estimation of μ and τ Using Stratified Random Sampling

Point estimate

of μ: $\bar{y}_{ST} = \frac{\sum_i N_i \bar{y}_i}{N}$

of τ: $N\bar{y}_{ST}$

95% confidence interval

for μ: $\bar{y}_{ST} \pm 1.96\hat{\sigma}_{\bar{y}_{ST}}$

for τ: $N\bar{y}_{ST} \pm 1.96N\hat{\sigma}_{\bar{y}_{ST}}$

where

$$\hat{\sigma}_{\bar{y}_{ST}} = \frac{1}{N}\sqrt{\sum N_i^2 \left(\frac{N_i - n_i}{N_i - 1}\right)\frac{s_i^2}{n_i}}$$

EXAMPLE 19.5 A wholesale food distributor in a large metropolitan area would like to know if demand is great enough to justify adding a new product to the stock. To aid in making the decision, it is planned to add this product to a sample of the stores serviced to estimate average monthly sales. Since only four large chains are serviced in the metropolitan area, it is administratively convenient to stratify the stores, with each chain serving as a stratum. There are 24 stores in stratum 1, 36 in stratum 2, 30 in stratum 3, and 30 in stratum 4. The wholesaler decides that there is enough money to obtain data in a total of 20 retail stores. If we allocate the total sample size among the strata with the stratum sample sizes proportional to the stratum sizes, we obtain sample sizes of 4, 6, 5, and 5 for strata 1–4, respectively. Thus the product is introduced into 4 stores chosen at random from chain 1, 6 stores from chain 2, and 5 stores each from chains 3 and 4. The sales (in hundreds of dollars, after a one-month trial period) are tabulated below. Estimate the average sales for the month and give a 95% confidence interval for μ.

	Stratum (Chain)			
	1	2	3	4
	89	91	108	102
	80	99	96	120
	92	93	100	104
	100	105	93	101
		111	93	123
		101		
Sample mean	89	100	98	110
Sample variance	78.67	55.60	39.50	112.50

Solution The point estimate of μ, the average monthly sales for all stores across the four chains, is

$$\bar{y}_{ST} = \frac{\sum_i N_i \bar{y}_i}{N} = \frac{24(89) + 36(100) + 30(98) + 30(110)}{120} = 99.8$$

The 95% confidence interval for μ is

$$\bar{y}_{ST} \pm 1.96\hat{\sigma}_{\bar{y}_{ST}}$$

where

$$\hat{\sigma}_{\bar{y}_{ST}} = \frac{1}{N}\sqrt{\sum_i N_i^2\left(\frac{N_i - n_i}{N_i - 1}\right)\left(\frac{s_i^2}{n_i}\right)}$$

$$= \frac{1}{120}\left[(24)^2\left(\frac{24-4}{24-1}\right)\frac{78.67}{4} + (36)^2\left(\frac{36-6}{36-1}\right)\frac{55.60}{6}\right.$$

$$\left. + (30)^2\left(\frac{30-5}{30-1}\right)\frac{39.50}{5} + (30)^2\left(\frac{30-5}{30-1}\right)\frac{112.5}{5}\right]^{1/2}$$

$$= \frac{1}{120}\sqrt{43731.0} = 1.74$$

The corresponding confidence interval is

$$99.8 \pm 1.96(1.74), \qquad \text{or} \qquad 99.8 \pm 3.41$$

We are 95% confident that the average monthly sales for these 120 stores are in the interval 96.39 to 103.21 (thousand dollars). □

variability within and between strata

The standard error of $\bar{y}_{ST}$ depends on the estimated variances s_i^2 in the various strata, while the standard error of $\bar{y}$ based on simple random sampling depends on s^2, which estimates the overall population variance. If there is relatively little variability in each strata as compared to the overall population variability, stratification yields a smaller standard error, and hence more accurate estimation, than does simple random sampling. In the language of Chapter 12, it is desirable to have relatively low **variability within strata** as compared to the **variability between strata**. In other words, it is desirable to divide the population into relatively homogeneous (low-variability) strata.

EXAMPLE 19.6 Refer to Example 19.5 and assume that the tabulated data were obtained from a simple random sample of stores. Compute a 95% confidence interval for μ and compare it to the one obtained in Example 19.5 using stratified random sampling.

Solution One can easily verify that the sample mean and variance for the $n = 20$ observations are $\bar{y} = 99.8$ and $s^2 = 111.85$ with $N = 120$. The 95% confidence interval for μ is

$$99.8 \pm 1.96\hat{\sigma}_{\bar{y}}$$

where

$$\hat{\sigma}_{\bar{Y}} = \sqrt{\frac{s^2}{n}\left(\frac{N-n}{N-1}\right)} = \sqrt{\frac{111.85}{20}\left(\frac{120-20}{119}\right)} = 2.17$$

so the 95% confidence interval for μ using simple random sampling is

$$99.8 \pm 1.96(2.17), \qquad \text{or} \qquad 99.8 \pm 4.25$$

Note that this interval is wider than that for stratified random sampling. This will not always be so; but when stratification produces smaller, homogeneous groups, stratified random sampling is an improvement over simple random sampling. □

The mathematics behind the formulas for stratified random sampling follows from the formulas for linear combinations of independent random variables (see p. 115). The point estimate $\bar{y}_{ST}$ is a weighted average of the strata means $\bar{y}_i$; the weights are the strata sizes N_i. Thus

$$\sigma^2_{\bar{Y}_{ST}} = \text{Var}\left(\sum_i \frac{N_i}{N}\bar{y}_i\right) = \sum_i \frac{N_i^2}{N^2}\sigma^2_{\bar{Y}_i}$$

Since a simple random sample is taken within each stratum,

$$\sigma^2_{\bar{Y}_i} = \frac{\sigma_i^2}{n_i}\left(\frac{N_i - n_i}{N_i - 1}\right)$$

The standard-error formula for $\bar{y}_{ST}$ follows by estimating each true stratum variance σ_i^2 by the sample variance s_i^2.

Stratified sampling can also be used to estimate a population proportion π. For example, suppose that the personnel department of a large corporation is interested in estimating the proportion of all vested employees who participate in a company-run stock savings program. If the divisions of the corporation represent strata and a simple random sample of n_i employee records is obtained from stratum i ($i = 1, \ldots, I$), then a point estimate and approximate 95% confidence interval can be obtained using the following formulas:

Estimation of π Using Stratified Random Sampling

Point estimate: $\hat{\pi}_{ST} = \frac{1}{N}\sum N_i\hat{\pi}_i$

95% confidence interval: $\hat{\pi}_{ST} \pm 1.96\hat{\sigma}_{\hat{\pi}_{ST}}$

where

$$\hat{\sigma}_{\hat{\pi}_{ST}} = \frac{1}{N}\sqrt{\sum_i N_i^2\left(\frac{N_i - n_i}{N_i - 1}\right)\frac{\hat{\pi}_i(1-\hat{\pi}_i)}{n_i}}$$

EXAMPLE 19.7 The sample data for the survey of vested employees is shown for each of the three divisions of the corporation. Use these data to estimate π, the proportion of all vested employees who participate in the company-run stock program.

Stratum (Division)	Number N_i of Vested Employees	Sample Size n_i	Number Who Participate
1	450	45	12
2	300	30	15
3	760	76	30

Solution The point estimate of π is

$$\hat{\pi}_{ST} = \frac{1}{N}\sum_i N_i\hat{\pi}_i$$

$$= \frac{1}{1510}[450(.27) + 300(.50) + 760(.39)] = .38$$

The 95% confidence interval for π can be found after we compute $\hat{\sigma}_{\hat{\pi}_{ST}}$.

$$\hat{\sigma}_{\hat{\pi}_{ST}} = \frac{1}{N}\sqrt{\sum_i N_i^2\left(\frac{N_i - n_i}{N_i - 1}\right)\frac{\hat{\pi}_i(1-\hat{\pi}_i)}{n_i}}$$

$$= \frac{1}{1510}\left[(450)^2\left(\frac{450-45}{450-1}\right)\frac{(.27)(.73)}{45} + (300)^2\left(\frac{300-30}{300-1}\right)\frac{(.5)(.5)}{30}\right.$$

$$\left. + (760)^2\left(\frac{760-76}{760-1}\right)\frac{(.39)(.61)}{76}\right]^{1/2}$$

$$= \frac{1}{1510}\sqrt{3106.67} = .037$$

The corresponding 95% confidence interval for π is

$$.38 \pm 1.96(.037), \qquad \text{or} \qquad .38 \pm .073$$

The actual proportion of vested employees in the corporation who participate in the stock program is estimated to be in the interval .307–.453. □

advantages of stratified random sampling

There are several reasons that stratified random sampling often results in an increase in information for a given cost. First, the data often are more homogeneous within each stratum than in the population as a whole. Taking advantage of the reduced variability within each stratum, we obtain estimates that have smaller confidence intervals than comparable estimates from a simple random sample of the same size. Second, the cost of conducting a stratified random sample tends to be less than that for a simple random sample. The elements in each stratum are usually located within a smaller geographic area, and separate teams of interviewers can be sent to the strata for collection of the sample data. Third, separate estimates of population parameters for each stratum can be obtained without additional sampling.

SECTION 19.2 EXERCISES

19.9 A group of college students conducts a survey to determine the average number of college hours (credits) an undergraduate student must earn for various majors to obtain a bachelor's degree from a large university. To do this, the departments of the university are stratified by colleges. Use the sample data shown below to estimate μ, the average number of credit hours required for a bachelor's degree. Construct a 95% confidence interval.

College	Number of Departments	Sample Data
Architecture and fine arts	7	192, 199, 188, 191
Arts and sciences	40	186, 195, 186, 189, 186, 192, 193, 195, 200, 183, 187, 192
Business administration	6	193, 186, 180, 182
Education	8	197, 198, 188, 196
Engineering	20	202, 203, 213, 202, 204, 206, 210, 206

19.10 Refer to Example 19.5. Use the sales data from all 20 stores as a single random sample. Estimate the mean sales with a 95% confidence interval. Compare your results to those of Example 19.5. Does it appear that stratification has helped?

19.11 A study of television-viewing habits stratifies people in a metropolitan area by gender and age. Random samples chosen within strata, and the sampled individuals record one week's prime-time viewing in diaries. The results are as follows:

Gender	Age	Number in Population (Thousands)	Number in Sample	Mean	Standard Deviation
F	18–34	312	127	10.4	4.2
M	18–34	317	129	11.2	3.8
F	35–49	285	116	9.8	3.9
M	35–49	279	114	10.3	4.1
F	50–64	248	101	10.6	4.5
M	50–64	221	90	11.0	4.8
F	65+	189	77	10.1	4.4
M	65+	114	46	10.9	3.9
		1965	800		

a. Calculate a 95% confidence interval for the overall mean viewing time.
b. Calculate a 95% confidence interval for the mean viewing times of all people age 65 and over. Note that only the last two strata are relevant.

19.12 Recalculate a 95% confidence interval for the overall mean viewing time in Exercise 19.11, assuming that the eight strata represent equal proportions of the population. Is there a substantial shift in the interval? Explain why.

19.13 In a study of the 2000 largest nonfinancial corporations, a sample of 250 firms is chosen and the number of firms that value inventory on the LIFO principle is determined. The sample is stratified into five major industries. The results are as follows:

Industry Type	Number of Firms	Number Sampled	Number Using LIFO
A	425	66	16
B	489	61	10
C	443	55	9
D	379	47	11
E	165	21	4

Calculate a 95% confidence interval for the overall proportion of firms using LIFO.

19.14 Refer to the data of Exercise 19.13. Calculate a 95% confidence interval assuming (incorrectly) that the data resulted from a simple random sample. Do the point estimate and confidence interval differ substantially from those found in Exercise 19.13? Can you explain why?

19.3 CLUSTER SAMPLING

A third type of sampling procedure, cluster sampling, sometimes gives more information per unit cost than either simple or stratified random sampling. The population of experimental units is divided into groups, called *clusters*.

Cluster Sample

A cluster sample is a sample obtained by taking a simple random sample of clusters, with an observation obtained from each unit in the sampled clusters.

Note that cluster sampling is similar to stratified random sampling in that we first divide the population of experimental units into groups. However, rather than obtaining a simple random sample of units from each group, we take a simple random sample of groups and sample all units in the chosen groups.

Cluster sampling can be less costly than either simple random sampling or stratified random sampling if the cost of obtaining observations increases as the distance separating experimental units increases or if the cost of obtaining a list of experimental units is high. For example, suppose we wish to estimate the average income per household in a large city. How should we choose the sample? If we use simple random sampling, we need a list of all households in the city, and this may be costly or even impossible to obtain. Even with stratified random sampling, we need a list of all households within each stratum. However, by dividing the city into regions (perhaps blocks), we can obtain a simple random sample of blocks and then interview each household in the sampled blocks. Even if a list of households does exist, it still might be better to use cluster sampling, especially if sampling scattered households increases the travel costs and interviewing time.

Before presenting methods for estimating a population mean or total, we need the following notation:

N: number of clusters
n: number of clusters selected in a simple random sample
m_i: number of elements in cluster i ($i = 1, 2, \ldots, N$)
$\bar{m}$: average cluster size for the sampled clusters; $\bar{m} = \sum_i m_i/n$
M: number of elements in the population; $M = \sum_i m_i$
$\bar{M}$: average cluster size for the population; $\bar{M} = M/N$; if $\bar{M}$ is unknown, it may be estimated by $\bar{m}$
T_i: total for all observations in the ith cluster

The estimation procedures for μ and τ are presented next.

Estimation of μ and τ Using Cluster Sampling

Point estimate

of μ: $\bar{y}_c = \dfrac{\sum T_i}{\sum m_i}$

of τ: $M\bar{y}_c$

95% confidence interval

for μ: $\bar{y}_c \pm 1.96\hat{\sigma}_{\bar{y}_c}$

for τ: $M\bar{y}_c \pm 1.96M\hat{\sigma}_{\bar{y}_c}$

where

$$\hat{\sigma}_{\bar{y}_c} = \sqrt{\left(\frac{N-n}{nN\bar{M}^2}\right)\frac{\sum_i (T_i - \bar{y}_c m_i)^2}{n-1}}$$

EXAMPLE 19.8 Interviews are conducted in each of 25 blocks sampled from a set of 415 blocks in a city. The data on income for adult males are presented below. Use the data to estimate the average income per adult male in the city using a 95% confidence interval.

Cluster i	Number of Adult Males m_i	Total Income per Cluster T_i	Cluster i	Number of Adult Males m_i	Total Income per Cluster T_i
1	8	$96,000	14	10	$49,000
2	12	121,000	15	9	53,000
3	4	42,000	16	3	50,000
4	5	65,000	17	6	32,000
5	6	52,000	18	5	22,000
6	6	40,000	19	5	45,000
7	7	75,000	20	4	37,000
8	5	65,000	21	6	51,000
9	8	45,000	22	8	30,000
10	3	50,000	23	7	39,000
11	2	85,000	24	3	47,000
12	6	43,000	25	8	41,000
13	5	54,000			

$\sum_i m_i = 151 \qquad \sum_i T_i = \$1,329,000$

Solution The best estimate of the population mean μ is

$$\bar{y}_c = \frac{\sum_i T_i}{\sum_i m_i} = \frac{\$1,329,000}{151} = \$8801$$

To calculate the confidence interval, we must compute

$$\sum_i (T_i - \bar{y}_c m_i)^2 = \sum_i T_i^2 - 2\bar{y}_c \sum_i T_i m_i + \bar{y}_c^2 \sum_i m_i^2$$

Thus we have

$$\sum_i T_i^2 = T_1^2 + T_2^2 + \ldots + T_{25}^2$$

$$= (96,000)^2 + (121,000)^2 + \ldots + (41,000)^2 = 82,039,000,000$$

$$\sum_i m_i^2 = m_1^2 + m_2^2 + \ldots + m_{25}^2 = (8)^2 + (12)^2 + \ldots + (8)^2 = 1047$$

$$\sum_i T_i m_i = T_1 m_1 + T_2 m_2 + \ldots + T_{25} m_{25}$$

$$= (96,000)(8) + (121,000)(12) + \ldots + (41,000)(8) = 8,403,000$$

and hence

$$\sum_i (T_i - \bar{y}_c m_i)^2 = 82,039,000,000 - 2(8801)(8,403,000) + (8801)^2(1047)$$

$$= 15,227,502,247$$

Since M is not known, the $\bar{M}$ appearing in the formula for standard error must be estimated by $\bar{m}$, where

$$\bar{m} = \frac{\sum_i m_i}{n} = \frac{151}{25} = 6.04$$

Then

$$\left(\frac{N-n}{Nn\bar{M}^2}\right)\left(\frac{\sum_i (T_i - \bar{y}_c m_i)^2}{n-1}\right) = \left(\frac{415-25}{(415)(25)(6.04)^2}\right)\left(\frac{15{,}227{,}502{,}247}{24}\right)$$
$$= 653{,}785$$

and the approximate 95% confidence interval is

$$8801 \pm 1.96\sqrt{653{,}785}, \quad \text{or} \quad 8801 \pm 1584.8$$

We are 95% confident that the mean income for adult males in the city lies between \$7216.20 and \$10,385.80. Although the width of this confidence interval is rather large, it can be reduced by sampling more clusters, thereby increasing the sample size. □

A population proportion can also be estimated using cluster sampling. For example, in a survey of rank-and-file workers, the leaders of a labor union may be interested in the proportion of members who favor a proposed new benefits package. If the locals of the labor union represent the clusters and a simple random sample of clusters is selected, we can estimate the population proportion π in the following way. Let a_i denote the number of laborers in cluster i who favor the new benefits package ($i = 1, 2, \ldots, n$) and let m_i denote the number of members of the local in the ith cluster. Then an estimate of π is given by

$$\hat{\pi}_c = \frac{\sum_i a_i}{\sum_i m_i}$$

The details are given below:

Estimation of π Using Cluster Sampling

Point estimate: $\hat{\pi}_c = \dfrac{\sum_i a_i}{\sum_i m_i}$

95% confidence interval: $\hat{\pi}_c \pm 1.96\hat{\sigma}_{\hat{\pi}_c}$

where

$$\hat{\sigma}_{\hat{\pi}_c} = \sqrt{\left(\frac{N-n}{nN\bar{M}^2}\right)\frac{\sum (a_i - \hat{\pi}_c m_i)^2}{n-1}}$$

Note: n should be 20 or more unless all the cluster sizes are approximately the same.

EXAMPLE 19.9 Suppose that a random sample of 25 local unions is selected from the total of 520 possible clusters. The sample data from the 25 clusters are shown in the table below. Use these data to construct an approximate 95% confidence interval for π.

Cluster	Number of Members m_i	Number Favoring a_i	Cluster	Number of Members m_i	Number Favoring a_i
1	65	30	14	115	60
2	78	35	15	150	92
3	80	49	16	43	31
4	40	16	17	67	17
5	50	32	18	39	24
6	100	51	19	26	14
7	120	75	20	98	56
8	75	40	21	106	52
9	80	39	22	112	76
10	85	52	23	59	33
11	90	55	24	71	47
12	73	19	25	82	55
13	61	34			

Solution For these data, you should verify that

$$\sum m_i = 1965, \quad \sum m_i^2 = 174{,}499, \quad \bar{m} = 78.60$$
$$\sum a_i = 1084, \quad \sum a_i^2 = 56{,}424, \quad \sum a_i m_i = 97{,}669$$

The estimate of π, the proportion of all union members favoring the new benefits package, is

$$\hat{\pi}_c = \frac{\sum a_i}{\sum m_i} = \frac{1084}{1965} = .55$$

To calculate $\hat{\sigma}_{\hat{\pi}_c}$ we need

$$\begin{aligned}\sum (a_i - \hat{\pi}_c m_i)^2 &= \sum a_i^2 - 2\hat{\pi}_c \sum a_i m_i + \hat{\pi}_c^2 \sum m_i^2 \\ &= 56{,}424 - 2(.55)(97{,}669) + (.55)^2(174{,}499) \\ &= 1774.05\end{aligned}$$

Because we do not know the exact number of union members M, we can use $\bar{m}$ for $\bar{M}$ in the formula for $\hat{\sigma}_{\hat{\pi}_c}$:

$$\begin{aligned}\hat{\sigma}_{\hat{\pi}_c} &= \sqrt{\left(\frac{N-n}{nN\bar{M}^2}\right)\frac{\sum_i (a_i - \hat{\pi}_c m_i)^2}{n-1}} \\ &= \sqrt{\left(\frac{520-25}{25(520)(78.60)^2}\right)\frac{1774.05}{24}} = .21\end{aligned}$$

Hence the approximate 95% confidence interval for π is

$$.55 \pm 1.96(.021), \quad \text{or} \quad .55 \pm .041$$

We are 95% confident that the actual proportion of union members favoring the new benefits package is between .509 and .591. It appears that a majority favor the package. □

Cluster sampling is to be preferred to both simple random sampling and stratified random sampling when a complete list of all elements either is unavailable or is costly to obtain and when the elements of the population are spread out over a large area. In these situations, cluster sampling requires a lower cost to provide the same amount of information.

SECTION 19.3 EXERCISES

19.15 A utility that supplies natural gas to a suburban area needs to estimate the average R value of insulation in the attic area of homes in its service area. A random sample of 18 out of 19,790 blocks is selected from the area, and the R value is found for each house in the selected blocks. The data are as follows:

Block:	1	2	3	4	5	6	7	8	9
Mean:	10.1	12.5	14.0	13.5	9.3	8.0	7.0	9.0	11.2
Number of houses:	20	16	10	8	14	16	22	15	12
Block:	10	11	12	13	14	15	16	17	18
Mean:	12.0	15.0	10.2	16.0	12.0	8.2	10.0	14.2	9.0
Number of houses:	10	7	14	6	12	22	16	10	21

Relevant summary figures are

$$\sum T_i = 2610.8, \quad \sum m_i = 251, \quad \sum T_i m_i = 38{,}556.6$$
$$\sum T_i^2 = 395{,}206.4, \quad \sum m_i^2 = 3931$$

a. Calculate a 95% confidence interval for the true mean R level.
b. Why might this form of sampling be adopted in preference to a simple random sample of houses?

19.16 Suppose that the data of Exercise 19.15 are wrongly assumed to have arisen from a simple random sample.

a. Calculate a 95% confidence interval for the population mean. The sample mean and variance of the R values for the 251 houses included are 10.40 and 29.97, respectively.
b. How does the width of this interval compare to that of the interval in Exercise 19.15?

19.17 A lumber company must precisely estimate the number of board-feet of lumber available in harvest-age trees. The company owns 1200 ten-acre stands of trees, from which 60 stands are selected at random. Inspectors assess the usable board-feet per tree in each stand. Summary figures are shown below.

$$\sum T_i = 4{,}735{,}000, \quad \sum m_i = 47{,}200, \quad \sum T_i m_i = 3{,}806{,}930{,}000$$
$$\sum T_i^2 = 380{,}989{,}000{,}000, \quad \sum m_i^2 = 38{,}074{,}000$$

a. Calculate $\bar{y}_c$.
b. Calculate a 95% confidence interval for the mean number of board-feet per tree.

19.18 A chain of retail stores is considering discontinuing advertising mailings to its inactive accounts (those with less than $25 purchased in the past six months). There are 97 stores in the chain with an average of 1827 accounts per store. It is not possible to process the records of all stores, so a random sample of 20 stores is chosen. Analysis of the records of these stores yields the following data:

	Store									
	1	2	3	4	5	6	7	8	9	10
Number of accounts	2020	1659	1854	1371	2530	1614	1901	2301	1745	1299
Total purchases ($000)	186	153	162	114	259	158	224	261	189	117
Number of inactive accounts	429	371	403	327	580	312	365	417	357	226

	Store									
	11	12	13	14	15	16	17	18	19	20
Number of accounts	1884	1901	1624	1346	1403	2119	1946	1784	1974	1838
Total purchases ($000)	219	198	146	132	265	243	216	168	199	174
Number of inactive accounts	394	373	327	301	415	401	373	361	411	340

a. Calculate $\bar{y}_c$ for purchase amounts.
b. Calculate a 95% confidence interval for the true mean sales per account in the past six months.
c. Calculate a 95% confidence interval for the proportion of inactive accounts.

19.4 SELECTING THE SAMPLE SIZE

So far, we have assumed that the sample size of a survey is known. Of course the selection of a sample size is one of the most important parts of a sample survey design. The issue of sample-size determination for simple random sampling was discussed in Section 8.3. Now we extend the discussion to cover stratified and cluster sampling.

The width of a 95% confidence interval is a useful measure of the probable accuracy of a sample estimator. Because sample information is costly, the aim is to find the smallest sample size that yields a 95% (or whatever level is desired) confidence interval of a specified width, using a particular sample survey design.* In general, the width of a confidence

* If several designs, with possibly differing costs per observation, are available, the required sample sizes and costs can be computed and the least expensive design selected.

interval depends, not only on the design and sample size, but also on one or more unknown variances. These variances may be estimated either in a preliminary study or by a manager's "horseback guess." Then the desired sample size may be found by trial and error or by formula.

half-width

Formulas for the sample size are usually stated in terms of the desired **half-width** E of a confidence interval. An interval of the form point estimate $\pm E$ has width $2E$ and therefore half-width E. Sample sizes needed to yield a desired half-width are shown below. In the case of stratified sampling, it is assumed that the overall sample size is allocated among strata in proportion to strata sizes, so that n_i, the sample size for stratum i, equals nN_i/N. If some other allocation is chosen, the formula is a first approximation. Trial and error can be used to find a more exact value of n.

Sample Sizes for Estimating μ

Simple random sampling:

$$n = \frac{Ns^2}{(N-1)\dfrac{E^2}{4} + s^2}$$

where s^2 is an estimate of the overall population variance.

Stratified random sampling:

$$n = \frac{\sum_i N_i s_i^2}{(N-1)\dfrac{E^2}{4} + \dfrac{1}{N}\sum_i N_i s_i^2}$$

where s_i^2 is an estimate of the variance in stratum i.

Cluster sampling:

$$n = \frac{Ns_c^2}{(N-1)\dfrac{E^2\bar{M}^2}{4} + s_c^2}$$

where s_c^2 is an estimate of the variance of cluster totals. If the estimate is based on a preliminary sample of n' clusters,

$$s_c^2 = \frac{\sum (T_i - \bar{y}_c m_i)^2}{n' - 1}$$

Note: If a confidence level other than 95% is desired, replace 4 by the appropriate z^2 value. For instance, for 90% replace 4 by $(1.645)^2$.

EXAMPLE 19.10 Use the data of Example 19.8 as preliminary information in a pilot study to calculate the sample size required to obtain a 95% confidence interval of width \$500 with cluster sampling.

Solution From Example 19.8 we have $\sum (T_i - \bar{y}_c m_i)^2 = 15{,}227{,}502{,}247$; $N = 415$ and $\bar{m} = 6.04$. Therefore

$$s_c^2 = \frac{\sum (T_i - \bar{y}_c m_i)^2}{n' - 1} = \frac{15{,}227{,}502{,}247}{24}$$

$$= 634{,}479{,}260.3$$

If the desired width is 500, $E = 250$. Substituting into the formula for n with $\bar{M}$ approximated by $\bar{m}$, we obtain

$$n = \frac{415(634{,}479{,}260.3)}{\dfrac{414(250)^2(6.04)^2}{4} + 634{,}479{,}260.3} = 302$$

A cluster sample of size 302 is needed to obtain a 95% confidence interval for μ with width \$500 (i.e., $\pm$\$250). □

In the same way, we can calculate approximate sample sizes for estimating either τ or π for each of the three sample survey designs, as shown in the following.

Approximate Sample Size for Estimating τ

Simple random sampling:

$$n = \frac{Ns^2}{\dfrac{(N-1)E^2}{4N^2} + s^2}$$

Stratified random sampling:

$$n = \frac{\sum_i N_i s_i^2}{\dfrac{(N-1)E^2}{4N^2} + \dfrac{1}{N}\sum N_i s_i^2}$$

Cluster sampling:

$$n = \frac{Ns_c^2}{\dfrac{(N-1)E^2}{4N^2} + s_c^2}$$

where $s_c^2 = [\sum_i (y_i - \bar{y}_c m_i)^2]/(n' - 1)$ from a preliminary sample of n' clusters.

Approximate Sample Size for Estimating π

Simple random sampling:

$$n = \frac{N\hat{\pi}(1 - \hat{\pi})}{\frac{(N - 1)E^2}{4} + \hat{\pi}(1 - \hat{\pi})}$$

where $\hat{\pi}$ is the estimated proportion from a preliminary sample or is a guessed value.

Stratified random sampling:

$$n = \frac{\sum_i N_i\hat{\pi}_i(1 - \hat{\pi}_i)}{\frac{(N - 1)E^2}{4} + \frac{1}{N}\sum_i N_i\hat{\pi}_i(1 - \hat{\pi}_i)}$$

where $\hat{\pi}_i$ is the estimated proportion obtained from a preliminary sample from the ith stratum.

Cluster sampling:

$$n = \frac{Ns_c^2}{\frac{(N - 1)E^2\bar{M}^2}{4} + s_c^2}$$

where $s_c^2 = [\sum_i (a_i - \hat{\pi}_c m_i)^2]/(n' - 1)$ is from a preliminary sample of n' clusters.

Note: If no preliminary information is available and it is difficult to guess π_i, substitute $\hat{\pi}_i = .5$ to obtain a conservative sample size (one that is likely to be larger than needed).

EXAMPLE 19.11 The manager for a chain of department stores wants to conduct an in-house survey to estimate the proportion of accounts that have been delinquent by one month or more at least once in the previous calendar year. The chain consists of five stores. To reduce the cost of sampling, it is decided to use a stratified random sample with the stores serving as strata. Use the information shown here to determine the sample size (and allocation) necessary to achieve a 95% confidence interval for π with a width of .02.

Stratum	Stratum Size N_i	Estimate of π_i from Previous Year
1	1000	.22
2	2500	.35
3	3200	.24
4	1700	.30
5	4100	.15

Solution The formula for n is

$$n = \frac{\sum_i N_i\hat{\pi}_i(1 - \hat{\pi}_i)}{\frac{(N-1)E^2}{4} + \frac{1}{N}\sum_i N_i\hat{\pi}_i(1 - \hat{\pi}_i)}$$

We'll use the previous year's estimates for the $\hat{\pi}_i$'s in the formula and set the interval half-width $E = .01$. Then

$$\sum_i N_i\hat{\pi}_i(1 - \hat{\pi}_i) = 1000(.22)(.78) + 2500(.35)(.65) + \ldots + 4100(.15)(.85)$$

$$= 2203.78$$

The required sample size is

$$n = \frac{2203.78}{12{,}499\frac{(.01)^2}{4} + \frac{2203.78}{12{,}500}} = \frac{2203.78}{.4888} \approx 4509$$

The required total sample size is approximately 4509. The allocation of this sample size to the strata utilizes the formula $n_i = n(N_i/N)$. Thus

$$n_1 = 4509\left(\frac{1000}{12{,}500}\right) \approx 361$$

$$n_2 = 4509\left(\frac{2500}{12{,}500}\right) \approx 902$$

$$n_3 = 4509\left(\frac{3200}{12{,}500}\right) \approx 1154$$

$$n_4 = 4509\left(\frac{1700}{12{,}500}\right) \approx 613$$

$$n_5 = 4509\left(\frac{4100}{12{,}500}\right) = 1479$$

□

SECTION 19.4 EXERCISES

19.19 In planning a stratified sample, a manager guesses that all strata variances equal approximately 40. Each of the 10 strata is made up of 1000 individuals, and equal sample sizes are to be taken in all strata. How large a sample is required to estimate the population mean to within $\pm .5$ unit with 95% confidence?

19.20 The variance for the overall population of Exercise 19.19 is guessed as approximately 80.

a. If a simple random sample is taken, how large a sample is required to estimate the mean to within $\pm .5$ unit with 95% confidence?

b. Is there any major advantage to stratification in this situation? Why?

19.21 A population is divided into 3000 clusters, with an average cluster size of 40. The variance of total cluster scores s_c^2 is estimated to be roughly 20,000. How many clusters must be sampled to yield a 95% confidence interval for the population mean with a half-width of 4?

19.22 Refer to Exercise 19.15. How many blocks have to be sampled to yield a 90% confidence interval with a half-width of 2.0?

19.23 The utility company of Exercise 19.15 also wants to estimate the proportion of homes with insulation that meet minimum government standards. A (very) preliminary sample of five blocks is taken. If the number of houses per block is 20, 15, 16, 18, and 22 and the number of houses per block with at least the minimum insulation is 3, 6, 4, 3, and 5, respectively, how many additional blocks must be sampled to estimate this proportion to within $\pm .2$ with 95% confidence?

19.5 OTHER SAMPLING TECHNIQUES

The sample survey designs that we have discussed in this chapter are among the most widely used designs in business surveys. There are, however, many extensions to these designs and other sampling techniques. We discuss a few of these briefly.

systematic sampling

Systematic sampling provides a useful alternative to simple random sampling that is easier to use and hence less subject to interviewer errors. To select a systematic sample, we imagine the elements of the population numbered from 1 to N. A random selection of one element is made from the first k elements of the population. Every kth element of the population is selected thereafter. This is called a 1-in-k systematic sample. For example, suppose a manager desires to sample $n = 200$ of a total of $N = 1000$ invoices to determine the proportion of invoices with one or more errors. A 1-in-5 systematic sample gives the desired sample size and is easy to obtain. Imagine that the invoices are numbered from 1 to 1000, as shown in Figure 19.1.

To obtain a 1-in-5 systematic sample, we make a random selection of one invoice from the first five. Suppose number 2 is selected. Then we take every fifth invoice from there on. The 200 invoices to be included in the sample are invoices numbered 2, 7, 12,

Invoice #	Invoice sampled
1	
2	2
3	
4	
5	
6	
7	7
8	
9	
10	
⋮	⋮
996	
997	997
998	
999	
1000	

FIGURE 19.1 **A 1-in-5 Systematic Sample**

17, . . . , 997. The formulas for point estimates of μ, τ, and π and the corresponding 95% confidence intervals are the same as those for simple random sampling.

If systematic sampling is easier to use and the formulas are the same as for simple random sampling, why would we ever use simple random sampling? Sometimes the elements of the population, when ordered, have inherent cycles. For example, sales volumes for grocery stores tend to have weekly cycles with greater sales volumes toward the end of the week. Similarly, retail sales of over-the-counter cough syrups and other cold preparations have cyclic sales patterns over the year, as do prices on agricultural commodities. If a systematic sample is used and the sampling pattern corresponds to the inherent cycle (see Figure 19.2), a bias is introduced; the population parameter of interest is either consistently overestimated or underestimated. Systematic sampling should also be avoided if the manager has no information about the population size N, since it is then impossible to determine the sampling rate (value of k) to achieve the desired sample size. For further details, see Scheaffer, Mendenhall, and Ott (1986).

You will recall that, with cluster sampling, the elements of the population are arranged in naturally occurring groups (clusters) and a simple random sample of clusters is selected. Each element in the selected clusters is surveyed. Sometimes, however, the cluster sizes are too large to make it feasible to sample all elements in the selected clusters. In these situations, we employ **two-stage cluster sampling**. First we obtain a simple random sample of clusters, then we select a simple random sample of elements from the selected clusters. Pollsters in national surveys often use two-stage cluster sampling. For example, a national public opinion survey on the mood of the nation could be done using geographic areas (such as states or counties) as clusters and then taking a simple random sample of elements within the sampled clusters.

two-stage cluster sampling

These are obvious extensions to two-stage clustering. In the previous example, we could use multistage cluster sampling by first obtaining a simple random sample of states, then a simple random sample of counties within the sampled states, then a simple random sample of voting districts within the selected counties, and finally a simple random sample of people within the selected voting districts. By subdividing the population a number of times, it is possible to use well-defined sampling units at each stage, and the final stage (e.g., voting district) has a readily available list of elements (frame) from which to draw the simple random sample of opinions. For further details on multistage cluster sampling, see Cochran (1977), Kish (1965), and Scheaffer, Mendenhall, and Ott (1986).

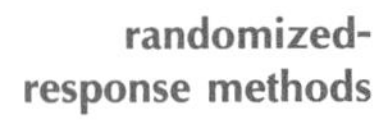

The **randomized-response technique** was developed by Warner (1965) to improve the response rate of individuals surveyed about sensitive or embarrassing questions. Direct

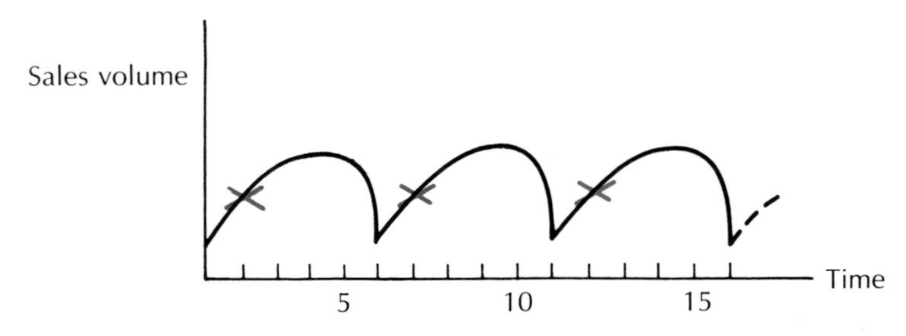

FIGURE 19.2 **Cyclic Pattern in Sales: Systematic Sample Marked in X**

questions about involvement with such activities as shoplifting, abortion, tax evasion, drug usage, and sexual harassment, for example, are often difficult to respond to truthfully.

We illustrate the randomized-response technique for the situation in which a "yes" or "no" answer is required. Extensions to this original work are discussed in Greenberg et al. (1971). The randomized-response technique involves pairing the sensitive question with an innocuous question. For example, in a survey of factory supervisors, those surveyed could be asked to answer "yes" or "no" to one of the two questions:

(sensitive) A: Have you been involved in sexual harassment of employees of the opposite sex?

(nonsensitive) B: Did you ever have chicken pox?

The particular question given an interviewee is unknown to the interviewer. Typically, a box containing a certain number of red and white balls is presented to the interviewee. The interviewee selects one of the balls without revealing the color to the interviewer. If the ball selected is red, the interviewee answers "yes" or "no" to the A question. Otherwise the interviewee answers "yes" or "no" to the B question.

The interesting result is that, without knowing which question individuals answer but knowing the proportion of red balls in the box and the proportion of "yes" responses (to either question) in the sample, it is possible to estimate π, the proportion of persons who have been involved with sexual harassment. The randomized response technique is described in greater detail in Warner (1965) and Greenberg et al. (1971).

SUMMARY

The first and simplest type of sampling procedure discussed was simple random sampling. We obtain a simple random sample of n elements if each sample of size n from the N elements in the population has the same probability of being selected. For estimating a population mean μ or total τ, we use the sample mean $\bar{y}$ or sample total $N\bar{y}$, respectively. The procedure for estimating a population proportion can be thought of as a variation on estimating a population mean. By assigning a 1 to each element possessing the characteristic of interest and a 0 to all others, the population and sample proportions π and $\hat{\pi}$ are also the population and sample mean μ and $\bar{y}$, respectively.

The second sampling procedure we presented was stratified random sampling. We obtain a stratified random sample by separating the population of elements into groups (strata) such that each element belongs to one and only one stratum. A simple random sample is then selected from each of the strata. Stratified random sampling has three major advantages over simple random sampling. First, the variance of estimation procedures for μ and τ are usually reduced because the variance of observations within each stratum is usually smaller than the overall population variance. Second, the cost of collecting observations is often reduced when the population of elements is separated into smaller groups. And finally, separate estimates of parameters in each stratum can also be computed from the same sample data.

The next sampling procedure we presented was cluster sampling. In this procedure the elements of the population are separated into groups called *clusters*, and the experimenter selects a simple random sample of clusters. An observation is obtained from each element in the sampled clusters. Cluster sampling may provide more information per unit cost than simple random sampling or stratified random sampling either when a list of the elements in the population is not available or when the cost of obtaining observations increases as the distance between elements increases.

Sample-size considerations for estimating μ, τ, and π under simple random sampling, stratified random sampling, and cluster sampling were discussed in Section 19.4. For each design, a formula was presented for computing the sample size necessary to achieve a 95% confidence interval of a specified width. These formulas require information from a preliminary sample or guessed values for sample variances.

Finally two extensions to the three sample survey designs and one other sampling technique were presented.

KEY FORMULAS: Some Alternative Sampling Methods

1. Estimation of μ or τ under simple random sampling

   Point estimate

   of μ: $\bar{y}$

   of τ: $N\bar{y}$

   95% confidence interval

   for μ: $\bar{y} \pm 1.96\hat{\sigma}_{\bar{y}}$

   for τ: $N\bar{y} \pm 1.96N\hat{\sigma}_{\bar{y}}$

   where

   $$\hat{\sigma}_{\bar{y}} = \sqrt{\frac{s^2}{n}\left(\frac{N-n}{N-1}\right)}$$

2. Estimation of π using simple random sampling

   Point estimate $\hat{\pi} = \dfrac{y}{n}$

   95% confidence interval $\hat{\pi} \pm 1.96\hat{\sigma}_{\hat{\pi}}$

   where

   $$\hat{\sigma}_{\hat{\pi}} = \sqrt{\frac{\hat{\pi}(1-\hat{\pi})}{n}\left(\frac{N-n}{N-1}\right)}$$

3. Estimation of μ and τ using stratified random sampling

   Point estimate

   $$\text{of } \mu\text{: } \bar{y}_{ST} = \frac{\sum_i N_i\bar{y}_i}{N}$$

   of τ: $N_{\bar{y}_{ST}}$

   95% confidence interval

   for μ: $\bar{y}_{ST} \pm 1.96\hat{\sigma}_{\bar{y}_{ST}}$

   for τ: $N\bar{y}_{ST} \pm 1.96N\hat{\sigma}_{\bar{y}_{ST}}$

where

$$\hat{\sigma}_{\bar{Y}_{ST}} = \frac{1}{N}\sqrt{\sum_i N_i^2\left(\frac{N_i - n_i}{N_i - 1}\right)\frac{s_i^2}{n_i}}$$

4. Estimation of π using stratified random sampling

Point estimate $\hat{\pi}_{ST} = \frac{1}{N}\sum_i N_i\hat{\pi}_i$

95% confidence interval $\hat{\pi}_{ST} \pm 1.96\hat{\sigma}_{\hat{\pi}_{ST}}$

where

$$\hat{\sigma}_{\hat{\pi}_{ST}} = \frac{1}{N}\sqrt{\sum_i N_i^2\left(\frac{N_i - n_i}{N_i - 1}\right)\frac{\hat{\pi}_i(1 - \hat{\pi}_i)}{n_i}}$$

5. Estimation of μ and τ using cluster sampling

Point estimate

of μ: $\bar{y}_c = \frac{\sum_i T_i}{\sum_i m_i}$

of π: $M\bar{y}_c$

95% confidence interval
for μ: $\bar{y}_c \pm 1.96\hat{\sigma}_{\bar{Y}_c}$
for τ: $M\bar{y}_c \pm 1.96M\hat{\sigma}_{\bar{Y}_c}$

where

$$\hat{\sigma}_{\bar{Y}_c} = \sqrt{\left(\frac{N - n}{nN\bar{M}^2}\right)\frac{\sum_i (T_i - \bar{y}_c m_i)^2}{n - 1}}$$

6. Estimation of π using cluster sampling

Point estimate $\hat{\pi}_c = \frac{\sum_i a_i}{\sum_i m_i}$

95% confidence interval $\hat{\pi}_c \pm 1.96\hat{\sigma}_{\hat{\pi}_c}$

where

$$\hat{\sigma}_{\hat{\pi}_c} = \sqrt{\left(\frac{N - n}{nN\bar{M}^2}\right)\frac{\sum_i (a_i - \hat{\pi}_c m_i)^2}{n - 1}}$$

7. Sample sizes for estimating μ

Sample random sampling

$$n = \frac{Ns^2}{(N - 1)\frac{E^2}{4} + s^2}$$

Stratified random sampling

$$n = \frac{\sum_i N_i s_i^2}{(N-1)\frac{E^2}{4} + \frac{1}{N}\sum_i N_i s_i^2}$$

Cluster sampling

$$n = \frac{N s_c^2}{(N-1)\frac{E^2 \bar{M}^2}{4} + s_c^2}$$

8. Sample sizes for estimating τ

Simple random sampling

$$n = \frac{N s^2}{\frac{(N-1)E^2}{4N^2} + s^2}$$

Stratified random sampling

$$n = \frac{\sum_i N_i s_i^2}{\frac{(N-1)E^2}{4N^2} + \frac{1}{N}\sum_i N_i s_i^2}$$

Cluster sampling

$$n = \frac{N s_c^2}{\frac{(N-1)E^2}{4N^2} + s_c^2}$$

9. Sample sizes for estimating π

Simple random sampling

$$n = \frac{N\hat{\pi}(1-\hat{\pi})}{\frac{(N-1)E^2}{4} + \hat{\pi}(1-\hat{\pi})}$$

Stratified random sampling

$$n = \frac{\sum_i N_i \hat{\pi}_i(1-\hat{\pi}_i)}{\frac{(N-1)E^2}{4} + \frac{1}{N}\sum_i N_i \hat{\pi}_i(1-\hat{\pi}_i)}$$

Cluster sampling

$$n = \frac{N s_c^2}{\frac{(N-1)E^2 \bar{M}^2}{4} + s_c^2}$$

CHAPTER 19 EXERCISES

19.24 A publisher of college dictionaries surveys bookstores in the neighborhoods of various colleges and universities to determine the average number of dictionaries displayed on shelves. To do this, the publisher selects 12 colleges and universities at random from a list of all accredited institutions of higher education. Every bookstore on or near these 12 campuses is visited and the number of dictionaries on display is counted.

a. Identify the sample survey method used.

b. Why might this method be preferred to a simple random sample of bookstores?

19.25 The survey of Exercise 19.24 yields the following data:

School	Number of Stores	Number of Dictionaries
1	2	4, 6
2	6	8, 3, 4, 5, 2, 5
3	3	5, 7, 2
4	1	6
5	2	4, 8
6	4	3, 2, 5, 2
7	3	4, 7, 3
8	1	5
9	2	6, 3
10	4	4, 4, 2, 6
11	3	3, 2, 3
12	2	5, 5

a. Calculate a 95% confidence interval for the true mean number of displayed dictionaries.

b. What is the 95% confidence interval if the data are assumed (incorrectly) to be the result of a simple random sample?

c. How large a sample of universities is required to estimate the mean within $\pm .1$ with 95% confidence?

19.26 An auto parts firm wants to estimate the average time required to fill its orders. The orders are classified into four types:

A: single item, off the shelf

B: single item, production required

C: multiple item, off the shelf

D: multiple item, production required

Random samples of 30 orders of each of the four types are selected, and times are determined.

a. Identify the type of sample survey design.

b. Why might this kind of design be preferable to a simple random sample of 120 orders?

19.27 Suppose that the orders of the previous exercise are numbered (without regard to type) from 0001 to 8260, and suppose that you have a table of 4-digit random numbers. How would you go about actually drawing the sample? You can expect to get some numbers larger than 8260 and some repetitions. It is very unlikely that the first 120 orders selected would contain exactly 30 of each type.

19.28 Suppose that the survey of Exercise 19.26 yields the following results (times in days):

Type	Mean	Standard Deviation
A	3.21	.82
B	7.39	2.14
C	4.65	1.05
D	9.27	3.65

Assume that the four types of orders are equally represented in the population of all orders. Calculate a 95% confidence interval for the true mean time.

19.29 Assume that the population of orders in Exercise 19.28 consists of 2216 A's, 2715 B's, 1874 C's, and 1455 D's.

a. Recalculate the 95% confidence interval.

b. How sensitive is the interval to assumptions about the number of each type of order?

19.30 Assume that the population of orders is as given in Exercise 19.29. How large a sample is needed to estimate the mean time within $\pm .3$ day with 95% confidence? How should the sample be allocated among the four types?

19.31 An economic survey is designed to estimate the average amount spent on utilities for households in a city. Since no list of households is available, cluster sampling is used with divisions (wards) forming the clusters. A simple random sample of 20 wards is selected from the 60 wards of the city. Interviewers then obtain the cost of utilities from each household within the sampled wards. The total costs are tabulated below.

Ward	Number of Households	Total Amount Spent	Ward	Number of Households	Total Amount Spent
1	55	$2210	11	73	$2930
2	60	2390	12	64	2470
3	63	2430	13	69	2830
4	58	2380	14	58	2370
5	71	2760	15	63	2390
6	78	3110	16	75	2870
7	69	2780	17	78	3210
8	58	2370	18	51	2430
9	52	1990	19	67	2730
10	71	2810	20	70	2880

Estimate the average amount a household in the city spends on utilities and calculate a 95% confidence interval.

19.32 Refer to Exercise 19.31. Use the sample data to estimate τ, the total amount spent on utilities. Calculate a 95% confidence interval. The total number of households in the city is 3950.

19.33 The parts warehouse of a foreign-car importer contains 81,513 items. A complete inventory of all these items would obviously be very expensive; therefore sampling is employed. The parts are listed in nine categories. Random samples are taken of items in each category, and the dollar value of "shrinkage" in each sampled item is determined. Identify the sample survey design.

19.34 One inventory check for the importer of Exercise 19.33 yields the following data:

Category	Items Sampled	Items Included	Mean Shrinkage/Item	Variance
1	125	12,461	1.310	.125
2	84	8,402	.272	.092
3	10	6,977	.149	.011
4	92	9,206	.117	.021
5	118	11,817	1.970	.214
6	100	10,048	.132	.032
7	94	9,372	.098	.062
8	80	7,965	.216	.042
9	53	5,265	.020	.017

Calculate a 90% confidence interval for the total inventory shrinkage.

19.35 What total sample size is needed to make the confidence interval of Exercise 19.34 have a half-width of $300? If the sample size is allocated to categories in proportion to the number of items in the categories, how many items should be sampled in each category?

19.36 A large hospital surveys its accounts for the past year to see what proportion of accounts were not settled within 60 days. The accounts are stratified by size, and random samples are taken of all accounts. The following data are obtained:

Stratum	Account Size	Total Number of Accounts	Number of Accounts Sampled	Number of Accounts Not Settled in 60 Days
A	Under $100	8251	41	5
B	100–499	2917	29	6
C	500–999	843	42	8
D	1000–2499	487	49	15
E	2500 or over	202	40	23

Calculate a 95% confidence interval for the true proportion of accounts not settled in 60 days.

19.37 The formula for required stratified sample size given in Section 19.4 is based on the assumption that the sample size is allocated among strata in proportion to the relative sizes of the strata. This assumption does not hold for Exercise 19.36.

a. Use the formula to calculate the sample size required to give the half-width that was actually obtained in Exercise 19.36.

b. How much difference in required sample size is caused by the difference in allocation?

19.38 A campus survey is conducted to determine the average occupancy per car for staff and faculty members entering the university during the 7:30–9:00 A.M. rush-hour period for a representative mid-week day. There are four legal entrances to the university and it is assumed that data are collected from these checkpoints. The day chosen for study is a Wednesday in the middle of the semester. One person is stationed at each checkpoint and instructed to sample every second car. For each car included in the sample, the person first checks the sticker on the front of the car to verify that it is a faculty or staff member's car. Then the number of occupants (including the driver) is recorded. Assuming that the four checkpoints represent four strata and that the method of sampling simulates simple random sampling of each stratum, use the data in the following table to estimate the average occupancy per car. Give a 95% confidence interval.

Stratum (Entrance)	N_i	n_i	$\sum_i y_i$	s_i^2
North	294	147	185	.24
South	150	75	89	.26
East	230	115	144	.26
West	322	161	203	.18

19.39 Is simple random sampling an alternative to stratified random sampling for estimating μ in Exercise 19.38? Explain.

19.40 Refer to Exercise 19.38. Estimate τ, the total number of occupants entering the campus during the 7:30–9:00 A.M. period. Find a 95% confidence interval.

20

DATA MANAGEMENT AND REPORT PREPARATION

Over the past 19 chapters we've discussed particular statistical methods, how those methods are applied to specific data sets, and how findings from statistical analyses presented in the form of computer output are interpreted. We have not concentrated on the processing steps that one follows between the time the data are received and the time the data are available in computer-readable form for analysis, nor have we discussed the form and content of the report that summarizes the results of a statistical analysis. In this chapter we consider the data-processing steps and statistical report writing. The chapter is not a complete manual with all the tools required; rather, it is an overview—what a manager should know about these steps. As an example, the chapter reflects standard procedures in the pharmaceutical industry, which is highly regulated. Procedures differ somewhat in other industries.

20.1 PREPARING DATA FOR STATISTICAL ANALYSIS

We begin with a discussion of the steps involved in processing data from a study. In practice, these steps may consume 75% of the total effort from the receipt of the raw data to the presentation of results from the analysis. What are these steps, why are they so important, and why are they so time-consuming?

To answer these questions, let's list the major data-processing steps in the cycle, which begin with receipt of the data and end when the statistical analysis begins. Then we'll discuss each step separately.

Steps in Preparing Data for Analysis

1. Receiving the raw data source
2. Creating the data base from the raw data source
3. Editing the data base
4. Correcting and clarifying the raw data source
5. Finalizing the data base
6. Creating data files from the data base

raw data source

1. Receiving the raw data source. For each study that is to be summarized and analyzed, the data arrive in some form, which we'll refer to as the **raw data source**. For a clinical trial, the raw data source is usually case report forms, sheets of $8\frac{1}{2}''$ × 11″ paper, that have been used to record study data for each patient entered into the study. For other types of studies, the raw data source may be sheets of paper from a laboratory notebook, a magnetic tape (or any other form of machine-readable data), hand-tabulations, and so on.

data trail

It is important to retain the raw data source, since it is the beginning of the **data trail**, which leads from the raw data to the conclusions drawn from a study. Many consulting operations involved with the analysis and summarization of many different studies keep a log that contains vital information related to the study and raw data source. General information contained in a study log is shown below:

Log for Study Data

1. Date received, and from whom
2. Study investigator
3. Statistician (and others) assigned
4. Brief description of study
5. Treatments (compounds, preparations, etc.) studied
6. Raw data source
7. Response(s) measured
8. Reference number for study
9. Estimated (actual) completion date
10. Other pertinent information

Later, when the study has been analyzed and results have been communicated, additional information can be added to the log on how the study results were communicated, where these results are recorded, what data files have been saved, and where these files are stored.

2. Creating the data base from the raw data source. For most studies that are scheduled for a statistical analysis, a machine-readable data base is created. The steps taken to create the data base and the eventual form of the data base vary from one operation to another, depending on the software systems to be used in the statistical analysis. However, we can give a few guidelines based on the form of the entry system.

When the data are to be *key-entered* at a terminal, the raw data are first checked for legibility. Any illegible numbers or letters or other problems should be brought to the attention of the study coordinator. Then a coding guide that assigns column numbers and variable names to the data is filled out. Certain codes for missing values (e.g., not available) are also defined here. Also, it is helpful to give a brief description of each variable. The data file keyed in at the terminal is referred to as the **machine-readable data base**. A listing of the data base should be obtained and checked carefully against the raw data source. Any errors should be corrected at the terminal and verified against an updated listing.

machine readable data base

Sometimes data are received in machine-readable form. For these situations the magnetic tape or disk file is considered to be the data base. You must, however, have a coding guide to "read" the data base. Using the coding guide, obtain a listing of the data base and check it *carefully* to see that all numbers and characters look reasonable and that proper formats were used to create the file. Any problems that arise must be resolved before proceeding further.

Some data sets are so small that it is not necessary to create a machine-readable data file from the raw data source. Instead, calculations are performed by hand or the data are entered into an electronic calculator. For these situations check any calculations to see that they make sense. Don't believe everything you see; redoing the calculations is not a bad idea.

3. Editing the data base. The types of edits done and the completeness of the editing process really depend on the type of study and how concerned you are about the accuracy and completeness of the data prior to the analysis. For example, in using SAS files it is wise to examine the minimum, maximum, and frequency distribution for each variable to make certain nothing looks unreasonable.

Certain other checks should be made. Plot the data and look for problems. Also, certain **logic checks** should be done depending on the structure of the data. If, for example, data are recorded for patients at several different visits, then the data recorded for visit 2 can't be earlier than the data for visit 1; similarly, if a patient is lost to follow-up after visit 2, we can't have any data for that patient at later visits.

logic checks

For small data sets we can do these data edits by hand, but for large data sets the job may be too time consuming and tedious. If machine editing is required, look for a software system that allows the user to specify certain data edits. Even so, for more complicated edits and logic checks it may be necessary to have a customized edit program written in order to machine edit the data. This programming chore can be a time-consuming step; plan for this well in advance of the receipt of the data.

4. Correcting and clarifying the raw data source. Questions frequently arise concerning the legibility or accuracy of the raw data during any one of the steps from the receipt of the raw data to the communication of the results from the statistical analysis. We have found it helpful to keep a list of these problems or discrepancies in order to define the data trail for a study. If a correction (or clarification) is required to the raw data source, this should be indicated on the form and the appropriate change made to the raw data source. If no correction is required, this should be indicated on the form as well. Keep in mind that the machine-readable data base should be changed to reflect any changes made to the raw data source.

5. Finalizing the data base. You may have been led to believe that all data for a study arrive at one time. This, of course, is not always the case. For example, with a marketing survey, different geographic locations may be surveyed at different times and hence those responsible for data processing do not receive all the data at one time. All these subsets of data, however, must be processed through the cycles required to create, edit, and correct the data base. Eventually the study is declared complete and the data is processed into the data base. At this time the data base should be reviewed again and final corrections made before beginning the analysis. The reason for this is that, for large data sets, the analysis and summarization chores take considerable manpower and computer time. It's better to agree on a final data base analysis than to have to repeat all analyses on a changed data base at a later date.

6. Creating data files from the data base. Generally there are one or two sets of data files created from the machine-readable data base. The first set, referred to as **original files**, reflects the basic structure of the data base. A listing of the files is checked against the data base listing to verify that the variables have been read with correct formats and missing value codes have been retained. For some studies the original files are actually used for editing the data base.

original files

A second set of data files, called **work files**, may be created from the original files. Work files are designed to facilitate the analysis. They may require restructuring of the original files, a selection of important variables, or the creation or addition of new variables by insertion, computation, or transformation. A listing of the work files is checked against that of the original files to ensure proper restructuring and variable selection. Computed and transformed variables are checked by hand calculations to verify the program code.

work files

If original and work files are SAS data sets, you should utilize the documentation features provided by SAS. At the time an SAS data set is created, a descriptive label for the data set, of up to 40 characters, should be assigned. The label can be stored with the data set imprinted wherever the contents procedure is used to print the data set's contents. All variables can be given descriptive names, up to 8 characters in length, which are meaningful to those involved in the project. In addition, variable labels up to 40 characters in length can be used to provide additional information. Title statements can be included in the SAS code to identify the project and describe each job. For each file, a listing (proc print) and a dictionary (proc contents) can be retained.

For files created from the data base using other software packages, you should utilize the labeling and documentation features available in the computer program.

Even if appropriate statistical methods are applied to data, the conclusions drawn from the study are only as good as the data on which they are based. So you be the judge.

The amount of time spent on these data-processing chores before analysis really depends on the nature of the study, the quality of the raw data source, and how confident you want to be about the completeness and accuracy of the data.

20.2 GUIDELINES FOR A STATISTICAL ANALYSIS AND REPORT

In this section we briefly discuss a few guidelines for a statistical analysis and some important elements of a statistical report for communicating results. The statistical analysis of a large study can usually be broken down into three types of analyses: (1) preliminary analyses, (2) primary analyses, and (3) backup analyses.

preliminary analyses

The **preliminary analyses**, which are often descriptive or graphic, familiarize the statistician with the data and provide a foundation for all subsequent analyses. These analyses may include frequency distributions, histograms, descriptive statistics, an examination of comparability of the treatment groups, correlations, or univariate and bivariate plots.

primary analyses
backup analyses

Primary analyses are those used to address the objectives of the study and the analyses on which conclusions are drawn. **Backup analyses** include alternate methods for examining the data that confirm the results of the primary analyses; they may also include new statistical methods that are not as readily accepted as the more standard methods. Several guidelines for analyses are presented below:

Preliminary, Primary, and Backup Analyses

1. Analyses should be performed with software that has been extensively tested.
2. Computer output should be labeled to reflect which study is analyzed, what subjects (animals, patients, etc.) are used in the analysis, and a brief description of the analysis preferred. For example, TITLE statements in SAS are very helpful.
3. Variable labels and value labels (e.g., 0 = none, 1 = mild) should appear on the output.
4. A list of the data used in each analysis should be provided.
5. The output for all analyses should be checked *carefully*. Did the job run successfully? Are the sample sizes, means, and degrees of freedom correct? Other checks may be necessary as well.
6. All preliminary, primary, and backup analyses that provide the informational base from which study conclusions are drawn should be saved.

After the statistical analysis is completed, conclusions must be drawn and the results communicated to the intended audience. Sometimes it is necessary to communicate these results as a formal written statistical report. A general outline for a statistical report that we have found useful and informative is listed below:

General Outline for a Statistical Report

1. Summary
2. Introduction
3. Experimental design and study procedures
4. Descriptive statistics
5. Statistical methodology
6. Results and conclusions
7. Discussion
8. Data listings

20.3 DOCUMENTATION AND STORAGE OF RESULTS

The final part of this cycle of data processing, analysis, and summarization concerns the documentation and storage of results. For formal statistical analyses that are subject to careful scrutiny by others, it is important to provide detailed documentation for all data processing and the statistical analyses so that the data trail is clear and the data base or work files readily accessible. Then the reviewer can follow what has been done, redo it, or extend the analyses. The elements of a documentation and storage file depend on the particular setting in which you work. The contents for a general documentation storage file are shown below:

Study Documentation and Storage File

1. Statistical report
2. Study description
3. Random code (used to assign subjects to treatment groups)
4. Important correspondence
5. File creation information
6. Preliminary, primary, and backup analyses
7. Raw data source
8. A data management sheet, which includes the log as well as information on the storage of the data files

SUMMARY

We hope the information presented in this chapter has opened your eyes and broadened your perspective on data processing and statistical analysis. Most textbooks assume that the data are ready for analysis when received or displayed. In practice, however, much work is required to prepare the data for analysis and to write a report. The material presented here gives the flavor of what is being done in the pharmaceutical industry. Because many other businesses or experimental settings are less highly regulated, it may be possible to relax some of the steps that have been outlined here. The important point is that you should actively consider whether these steps are required for a study; don't just ignore them.

Besides discussing the steps required to prepare data for analysis, we have presented a few guidelines for a statistical analysis, a general outline for a statistical report, and the contents of a documentation and storage system. Remember, these are only examples of what can be done. Other variations are certainly reasonable and appropriate.

APPENDIX

Table 1	Binomial probabilities	763
Table 2	Poisson probabilities	769
Table 3	Normal curve areas	771
Table 4	Percentage points of the t distribution	772
Table 5	Percentage points of the χ^2 distribution	773
Table 6	Percentage points of the F distribution	775
Table 7	Critical values for the Wilcoxon signed-rank test	787
Table 8	Percentage points of the studentized range	788
Table 9	Random numbers	790

Table 1

Binomial probabilities (n between 2 and 6)

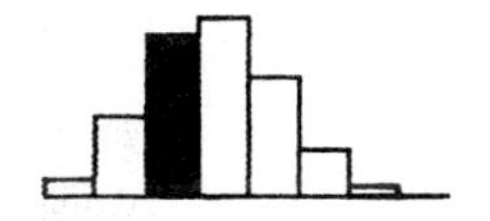

n=2

y↓	π .05	.10	.15	.20	.25	.30	.35	.40	.45	.50	
0	.9025	.8100	.7225	.6400	.5625	.4900	.4225	.3600	.3025	.2500	2
1	.0950	.1800	.2550	.3200	.3750	.4200	.4550	.4800	.4950	.5000	1
2	.0025	.0100	.0225	.0400	.0625	.0900	.1225	.1600	.2025	.2500	0
	.95	.90	.85	.80	.75	.70	.65	.60	.55	.50	y↑

n=3

y↓	π .05	.10	.15	.20	.25	.30	.35	.40	.45	.50	
0	.8574	.7290	.6141	.5120	.4219	.3430	.2746	.2160	.1664	.1250	3
1	.1354	.2430	.3251	.3840	.4219	.4410	.4436	.4320	.4084	.3750	2
2	.0071	.0270	.0574	.0960	.1406	.1890	.2389	.2880	.3341	.3750	1
3	.0001	.0010	.0034	.0080	.0156	.0270	.0429	.0640	.0911	.1250	0
	.95	.90	.85	.80	.75	.70	.65	.60	.55	.50	y↑

n=4

y↓	π .05	.10	.15	.20	.25	.30	.35	.40	.45	.50	
0	.8145	.6561	.5220	.4096	.3164	.2401	.1785	.1296	.0915	.0625	4
1	.1715	.2916	.3685	.4096	.4219	.4116	.3845	.3456	.2995	.2500	3
2	.0135	.0486	.0975	.1536	.2109	.2646	.3105	.3456	.3675	.3750	2
3	.0005	.0036	.0115	.0256	.0469	.0756	.1115	.1536	.2005	.2500	1
4	.0000	.0001	.0005	.0016	.0039	.0081	.0150	.0256	.0410	.0625	0
	.95	.90	.85	.80	.75	.70	.65	.60	.55	.50	y↑

n=5

y↓	π .05	.10	.15	.20	.25	.30	.35	.40	.45	.50	
0	.7738	.5905	.4437	.3277	.2373	.1681	.1160	.0778	.0503	.0313	5
1	.2036	.3281	.3915	.4096	.3955	.3602	.3124	.2592	.2059	.1563	4
2	.0214	.0729	.1382	.2048	.2637	.3087	.3364	.3456	.3369	.3125	3
3	.0011	.0081	.0244	.0512	.0879	.1323	.1811	.2304	.2757	.3125	2
4	.0000	.0005	.0022	.0064	.0146	.0284	.0488	.0768	.1128	.1563	1
5	.0000	.0000	.0001	.0003	.0010	.0024	.0053	.0102	.0185	.0313	0
	.95	.90	.85	.80	.75	.70	.65	.60	.55	.50	y↑

n=6

y↓	π .05	.10	.15	.20	.25	.30	.35	.40	.45	.50	
0	.7351	.5314	.3771	.2621	.1780	.1176	.0754	.0467	.0277	.0156	6
1	.2321	.3543	.3993	.3932	.3560	.3025	.2437	.1866	.1359	.0938	5
2	.0305	.0984	.1762	.2458	.2966	.3241	.3280	.3110	.2780	.2344	4
3	.0021	.0146	.0415	.0819	.1318	.1852	.2355	.2765	.3032	.3125	3
4	.0001	.0012	.0055	.0154	.0330	.0595	.0951	.1382	.1861	.2344	2
5	.0000	.0001	.0004	.0015	.0044	.0102	.0205	.0369	.0609	.0938	1
6	.0000	.0000	.0000	.0001	.0002	.0007	.0018	.0041	.0083	.0156	0
	.95	.90	.85	.80	.75	.70	.65	.60	.55	.50	y↑

Table 1 cont'd

Binomial probabilities (n between 7 and 10)

n=7

y ↓	.05	.10	.15	.20	.25	.30	.35	.40	.45	.50	
0	.6983	.4783	.3206	.2097	.1335	.0824	.0490	.0280	.0152	.0078	7
1	.2573	.3720	.3960	.3670	.3115	.2471	.1848	.1306	.0872	.0547	6
2	.0406	.1240	.2097	.2753	.3115	.3177	.2985	.2613	.2140	.1641	5
3	.0036	.0230	.0617	.1147	.1730	.2269	.2679	.2903	.2918	.2734	4
4	.0002	.0026	.0109	.0287	.0577	.0972	.1442	.1935	.2388	.2734	3
5	.0000	.0002	.0012	.0043	.0115	.0250	.0466	.0774	.1172	.1641	2
6	.0000	.0000	.0001	.0004	.0013	.0036	.0084	.0172	.0320	.0547	1
7	.0000	.0000	.0000	.0000	.0001	.0002	.0006	.0016	.0037	.0078	0
	.95	.90	.85	.80	.75	.70	.65	.60	.55	.50	y ↑

n=8

y ↓	.05	.10	.15	.20	.25	.30	.35	.40	.45	.50	
0	.6634	.4305	.2725	.1678	.1001	.0576	.0319	.0168	.0084	.0039	8
1	.2793	.3826	.3847	.3355	.2670	.1977	.1373	.0896	.0548	.0313	7
2	.0515	.1488	.2376	.2936	.3115	.2965	.2587	.2090	.1569	.1094	6
3	.0054	.0331	.0839	.1468	.2076	.2541	.2786	.2787	.2568	.2188	5
4	.0004	.0046	.0185	.0459	.0865	.1361	.1875	.2322	.2627	.2734	4
5	.0000	.0004	.0026	.0092	.0231	.0467	.0808	.1239	.1719	.2188	3
6	.0000	.0000	.0002	.0011	.0038	.0100	.0217	.0413	.0703	.1094	2
7	.0000	.0000	.0000	.0001	.0004	.0012	.0033	.0079	.0164	.0313	1
8	.0000	.0000	.0000	.0000	.0000	.0001	.0002	.0007	.0017	.0039	0
	.95	.90	.85	.80	.75	.70	.65	.60	.55	.50	y ↑

n=9

y ↓	.05	.10	.15	.20	.25	.30	.35	.40	.45	.50	
0	.6302	.3874	.2316	.1342	.0751	.0404	.0207	.0101	.0046	.0020	9
1	.2985	.3874	.3679	.3020	.2253	.1556	.1004	.0605	.0339	.0176	8
2	.0629	.1722	.2597	.3020	.3003	.2668	.2162	.1612	.1110	.0703	7
3	.0077	.0446	.1069	.1762	.2336	.2668	.2716	.2508	.2119	.1641	6
4	.0006	.0074	.0283	.0661	.1168	.1715	.2194	.2508	.2600	.2461	5
5	.0000	.0008	.0050	.0165	.0389	.0735	.1181	.1672	.2128	.2461	4
6	.0000	.0001	.0006	.0028	.0087	.0210	.0424	.0743	.1160	.1641	3
7	.0000	.0000	.0000	.0003	.0012	.0039	.0098	.0212	.0407	.0703	2
8	.0000	.0000	.0000	.0000	.0001	.0004	.0013	.0035	.0083	.0176	1
9	.0000	.0000	.0000	.0000	.0000	.0000	.0001	.0003	.0008	.0020	0
	.95	.90	.85	.80	.75	.70	.65	.60	.55	.50	y ↑

n=10

y ↓	.05	.10	.15	.20	.25	.30	.35	.40	.45	.50	
0	.5987	.3487	.1969	.1074	.0563	.0282	.0135	.0060	.0025	.0010	10
1	.3151	.3874	.3474	.2684	.1877	.1211	.0725	.0403	.0207	.0098	9
2	.0746	.1937	.2759	.3020	.2816	.2335	.1757	.1209	.0763	.0439	8
3	.0105	.0574	.1298	.2013	.2503	.2668	.2522	.2150	.1665	.1172	7
4	.0010	.0112	.0401	.0881	.1460	.2001	.2377	.2508	.2384	.2051	6
5	.0001	.0015	.0085	.0264	.0584	.1029	.1536	.2007	.2340	.2461	5
6	.0000	.0001	.0012	.0055	.0162	.0368	.0689	.1115	.1596	.2051	4
7	.0000	.0000	.0001	.0008	.0031	.0090	.0212	.0425	.0746	.1172	3
8	.0000	.0000	.0000	.0001	.0004	.0014	.0043	.0106	.0229	.0439	2
9	.0000	.0000	.0000	.0000	.0000	.0001	.0005	.0016	.0042	.0098	1
10	.0000	.0000	.0000	.0000	.0000	.0000	.0000	.0001	.0003	.0010	0
	.95	.90	.85	.80	.75	.70	.65	.60	.55	.50	y ↑

Table 1 cont'd

Binomial probabilities (n between 12 and 16)

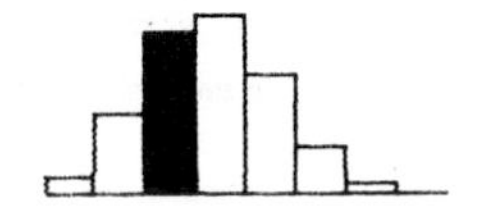

n=12

y↓	π .05	.10	.15	.20	.25	.30	.35	.40	.45	.50	
0	.5404	.2824	.1422	.0687	.0317	.0138	.0057	.0022	.0008	.0002	12
1	.3413	.3766	.3012	.2062	.1267	.0712	.0368	.0174	.0075	.0029	11
2	.0988	.2301	.2924	.2835	.2323	.1678	.1088	.0639	.0339	.0161	10
3	.0173	.0852	.1720	.2362	.2581	.2397	.1954	.1419	.0923	.0537	9
4	.0021	.0213	.0683	.1329	.1936	.2311	.2367	.2128	.1700	.1208	8
5	.0002	.0038	.0193	.0532	.1032	.1585	.2039	.2270	.2225	.1934	7
6	.0000	.0005	.0040	.0155	.0401	.0792	.1281	.1766	.2124	.2256	6
7	.0000	.0000	.0006	.0033	.0115	.0291	.0591	.1009	.1489	.1934	5
8	.0000	.0000	.0001	.0005	.0024	.0078	.0199	.0420	.0762	.1208	4
9	.0000	.0000	.0000	.0001	.0004	.0015	.0048	.0125	.0277	.0537	3
10	.0000	.0000	.0000	.0000	.0000	.0002	.0008	.0025	.0068	.0161	2
11	.0000	.0000	.0000	.0000	.0000	.0000	.0001	.0003	.0010	.0029	1
12	.0000	.0000	.0000	.0000	.0000	.0000	.0000	.0000	.0001	.0002	0
	.95	.90	.85	.80	.75	.70	.65	.60	.55	.50	y↑

n=14

y↓	π .05	.10	.15	.20	.25	.30	.35	.40	.45	.50	
0	.4877	.2288	.1028	.0440	.0178	.0068	.0024	.0008	.0002	.0001	14
1	.3593	.3559	.2539	.1539	.0832	.0407	.0181	.0073	.0027	.0009	13
2	.1229	.2570	.2912	.2501	.1802	.1134	.0634	.0317	.0141	.0056	12
3	.0259	.1142	.2056	.2501	.2402	.1943	.1366	.0845	.0462	.0222	11
4	.0037	.0349	.0998	.1720	.2202	.2290	.2022	.1549	.1040	.0611	10
5	.0004	.0078	.0352	.0860	.1468	.1963	.2178	.2066	.1701	.1222	9
6	.0000	.0013	.0093	.0322	.0734	.1262	.1759	.2066	.2088	.1833	8
7	.0000	.0002	.0019	.0092	.0280	.0618	.1082	.1574	.1952	.2095	7
8	.0000	.0000	.0003	.0020	.0082	.0232	.0510	.0918	.1398	.1833	6
9	.0000	.0000	.0000	.0003	.0018	.0066	.0183	.0408	.0762	.1222	5
10	.0000	.0000	.0000	.0000	.0003	.0014	.0049	.0136	.0312	.0611	4
11	.0000	.0000	.0000	.0000	.0000	.0002	.0010	.0033	.0093	.0222	3
12	.0000	.0000	.0000	.0000	.0000	.0000	.0001	.0005	.0019	.0056	2
13	.0000	.0000	.0000	.0000	.0000	.0000	.0000	.0001	.0002	.0009	1
14	.0000	.0000	.0000	.0000	.0000	.0000	.0000	.0000	.0000	.0001	0
	.95	.90	.85	.80	.75	.70	.65	.60	.55	.50	y↑

n=16

y↓	π .05	.10	.15	.20	.25	.30	.35	.40	.45	.50	
0	.4401	.1853	.0743	.0281	.0100	.0033	.0010	.0003	.0001	.0000	16
1	.3706	.3294	.2097	.1126	.0535	.0228	.0087	.0030	.0009	.0002	15
2	.1463	.2745	.2775	.2111	.1336	.0732	.0353	.0150	.0056	.0018	14
3	.0359	.1423	.2285	.2463	.2079	.1465	.0888	.0468	.0215	.0085	13
4	.0061	.0514	.1311	.2001	.2252	.2040	.1553	.1014	.0572	.0278	12
5	.0008	.0137	.0555	.1201	.1802	.2099	.2008	.1623	.1123	.0667	11
6	.0001	.0028	.0180	.0550	.1101	.1649	.1982	.1983	.1684	.1222	10
7	.0000	.0004	.0045	.0197	.0524	.1010	.1524	.1889	.1969	.1746	9
8	.0000	.0001	.0009	.0055	.0197	.0487	.0923	.1417	.1812	.1964	8
9	.0000	.0000	.0001	.0012	.0058	.0185	.0442	.0840	.1318	.1746	7
10	.0000	.0000	.0000	.0002	.0014	.0056	.0167	.0392	.0755	.1222	6
11	.0000	.0000	.0000	.0000	.0002	.0013	.0049	.0142	.0337	.0667	5
	.95	.90	.85	.80	.75	.70	.65	.60	.55	.50	y↑

n=16 continued on next page

Table 1 cont'd

Binomial probabilities (n between 16 and 20)

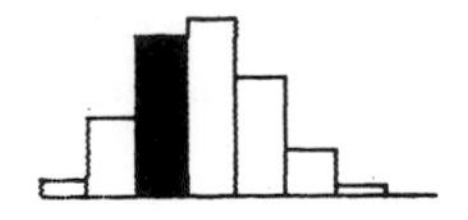

$n=16$ (continued from previous page)

					π						
y ↓	.05	.10	.15	.20	.25	.30	.35	.40	.45	.50	
12	.0000	.0000	.0000	.0000	.0000	.0002	.0011	.0040	.0115	.0278	4
13	.0000	.0000	.0000	.0000	.0000	.0000	.0002	.0008	.0029	.0085	3
14	.0000	.0000	.0000	.0000	.0000	.0000	.0000	.0001	.0005	.0018	2
15	.0000	.0000	.0000	.0000	.0000	.0000	.0000	.0000	.0001	.0002	1
	.95	.90	.85	.80	.75	.70	.65	.60	.55	.50	y ↑

$n=18$

					π						
y ↓	.05	.10	.15	.20	.25	.30	.35	.40	.45	.50	
0	.3972	.1501	.0536	.0180	.0056	.0016	.0004	.0001	.0000	.0000	18
1	.3763	.3002	.1704	.0811	.0338	.0126	.0042	.0012	.0003	.0001	17
2	.1683	.2835	.2556	.1723	.0958	.0458	.0190	.0069	.0022	.0006	16
3	.0473	.1680	.2406	.2297	.1704	.1046	.0547	.0246	.0095	.0031	15
4	.0093	.0700	.1592	.2153	.2130	.1681	.1104	.0614	.0291	.0117	14
5	.0014	.0218	.0787	.1507	.1988	.2017	.1664	.1146	.0666	.0327	13
6	.0002	.0052	.0301	.0816	.1436	.1873	.1941	.1655	.1181	.0708	12
7	.0000	.0010	.0091	.0350	.0820	.1376	.1792	.1892	.1657	.1214	11
8	.0000	.0002	.0022	.0120	.0376	.0811	.1327	.1734	.1864	.1669	10
9	.0000	.0000	.0004	.0033	.0139	.0386	.0794	.1284	.1694	.1855	9
10	.0000	.0000	.0001	.0008	.0042	.0149	.0385	.0771	.1248	.1669	8
11	.0000	.0000	.0000	.0001	.0010	.0046	.0151	.0374	.0742	.1214	7
12	.0000	.0000	.0000	.0000	.0002	.0012	.0047	.0145	.0354	.0708	6
13	.0000	.0000	.0000	.0000	.0000	.0002	.0012	.0045	.0134	.0327	5
14	.0000	.0000	.0000	.0000	.0000	.0000	.0002	.0011	.0039	.0117	4
15	.0000	.0000	.0000	.0000	.0000	.0000	.0000	.0002	.0009	.0031	3
16	.0000	.0000	.0000	.0000	.0000	.0000	.0000	.0000	.0001	.0006	2
17	.0000	.0000	.0000	.0000	.0000	.0000	.0000	.0000	.0000	.0001	1
	.95	.90	.85	.80	.75	.70	.65	.60	.55	.50	y ↑

$n=20$

					π						
y ↓	.05	.10	.15	.20	.25	.30	.35	.40	.45	.50	
0	.3585	.1216	.0388	.0115	.0032	.0008	.0002	.0000	.0000	.0000	20
1	.3774	.2702	.1368	.0576	.0211	.0068	.0020	.0005	.0001	.0000	19
2	.1887	.2852	.2293	.1369	.0669	.0278	.0100	.0031	.0008	.0002	18
3	.0596	.1901	.2428	.2054	.1339	.0716	.0323	.0123	.0040	.0011	17
4	.0133	.0898	.1821	.2182	.1897	.1304	.0738	.0350	.0139	.0046	16
5	.0022	.0319	.1028	.1746	.2023	.1789	.1272	.0746	.0365	.0148	15
6	.0003	.0089	.0454	.1091	.1686	.1916	.1712	.1244	.0746	.0370	14
7	.0000	.0020	.0160	.0545	.1124	.1643	.1844	.1659	.1221	.0739	13
8	.0000	.0004	.0046	.0222	.0609	.1144	.1614	.1797	.1623	.1201	12
9	.0000	.0001	.0011	.0074	.0271	.0654	.1158	.1597	.1771	.1602	11
10	.0000	.0000	.0002	.0020	.0099	.0308	.0686	.1171	.1593	.1762	10
11	.0000	.0000	.0000	.0005	.0030	.0120	.0336	.0710	.1185	.1602	9
12	.0000	.0000	.0000	.0001	.0008	.0039	.0136	.0355	.0727	.1201	8
13	.0000	.0000	.0000	.0000	.0002	.0010	.0045	.0146	.0366	.0739	7
14	.0000	.0000	.0000	.0000	.0000	.0002	.0012	.0049	.0150	.0370	6
15	.0000	.0000	.0000	.0000	.0000	.0000	.0003	.0013	.0049	.0148	5
16	.0000	.0000	.0000	.0000	.0000	.0000	.0000	.0003	.0013	.0046	4
17	.0000	.0000	.0000	.0000	.0000	.0000	.0000	.0000	.0002	.0011	3
18	.0000	.0000	.0000	.0000	.0000	.0000	.0000	.0000	.0000	.0002	2
	.95	.90	.85	.80	.75	.70	.65	.60	.55	.50	y ↑

Table 1 cont'd

Binomial probabilities (n=50 and 100)

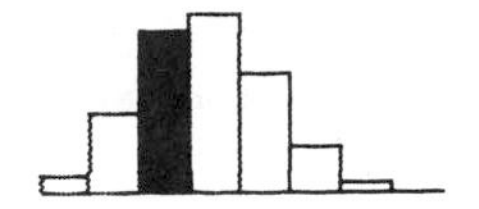

n=50

y ↓	.05	.10	.15	.20	.25	.30	.35	.40	.45	.50	
0	.0769	.0052	.0003	.0000	.0000	.0000	.0000	.0000	.0000	.0000	50
1	.2025	.0286	.0026	.0002	.0000	.0000	.0000	.0000	.0000	.0000	49
2	.2611	.0779	.0113	.0011	.0001	.0000	.0000	.0000	.0000	.0000	48
3	.2199	.1386	.0319	.0044	.0004	.0000	.0000	.0000	.0000	.0000	47
4	.1360	.1809	.0661	.0128	.0016	.0001	.0000	.0000	.0000	.0000	46
5	.0658	.1849	.1072	.0295	.0049	.0006	.0000	.0000	.0000	.0000	45
6	.0260	.1541	.1419	.0554	.0123	.0018	.0002	.0000	.0000	.0000	44
7	.0086	.1076	.1575	.0870	.0259	.0048	.0006	.0000	.0000	.0000	43
8	.0024	.0643	.1493	.1169	.0463	.0110	.0017	.0002	.0000	.0000	42
9	.0006	.0333	.1230	.1364	.0721	.0220	.0042	.0005	.0000	.0000	41
10	.0001	.0152	.0890	.1398	.0985	.0386	.0093	.0014	.0001	.0000	40
11	.0000	.0061	.0571	.1271	.1194	.0602	.0182	.0035	.0004	.0000	39
12	.0000	.0022	.0328	.1033	.1294	.0838	.0319	.0076	.0011	.0001	38
13	.0000	.0007	.0169	.0755	.1261	.1050	.0502	.0147	.0027	.0003	37
14	.0000	.0002	.0079	.0499	.1110	.1189	.0714	.0260	.0059	.0008	36
15	.0000	.0001	.0033	.0299	.0888	.1223	.0923	.0415	.0116	.0020	35
16	.0000	.0000	.0013	.0164	.0648	.1147	.1088	.0606	.0207	.0044	34
17	.0000	.0000	.0005	.0082	.0432	.0983	.1171	.0808	.0339	.0087	33
18	.0000	.0000	.0001	.0037	.0264	.0772	.1156	.0987	.0508	.0160	32
19	.0000	.0000	.0000	.0016	.0148	.0558	.1048	.1109	.0700	.0270	31
20	.0000	.0000	.0000	.0006	.0077	.0370	.0875	.1146	.0888	.0419	30
21	.0000	.0000	.0000	.0002	.0036	.0227	.0673	.1091	.1038	.0598	29
22	.0000	.0000	.0000	.0001	.0016	.0128	.0478	.0959	.1119	.0788	28
23	.0000	.0000	.0000	.0000	.0006	.0067	.0313	.0778	.1115	.0960	27
24	.0000	.0000	.0000	.0000	.0002	.0032	.0190	.0584	.1026	.1080	26
25	.0000	.0000	.0000	.0000	.0001	.0014	.0106	.0405	.0873	.1123	25
26	.0000	.0000	.0000	.0000	.0000	.0006	.0055	.0259	.0687	.1080	24
27	.0000	.0000	.0000	.0000	.0000	.0002	.0026	.0154	.0500	.0960	23
28	.0000	.0000	.0000	.0000	.0000	.0001	.0012	.0084	.0336	.0788	22
29	.0000	.0000	.0000	.0000	.0000	.0000	.0005	.0043	.0208	.0598	21
30	.0000	.0000	.0000	.0000	.0000	.0000	.0002	.0020	.0119	.0419	20
31	.0000	.0000	.0000	.0000	.0000	.0000	.0001	.0009	.0063	.0270	19
32	.0000	.0000	.0000	.0000	.0000	.0000	.0000	.0003	.0031	.0160	18
33	.0000	.0000	.0000	.0000	.0000	.0000	.0000	.0001	.0014	.0087	17
34	.0000	.0000	.0000	.0000	.0000	.0000	.0000	.0000	.0006	.0044	16
35	.0000	.0000	.0000	.0000	.0000	.0000	.0000	.0000	.0002	.0020	15
36	.0000	.0000	.0000	.0000	.0000	.0000	.0000	.0000	.0001	.0008	14
37	.0000	.0000	.0000	.0000	.0000	.0000	.0000	.0000	.0000	.0003	13
38	.0000	.0000	.0000	.0000	.0000	.0000	.0000	.0000	.0000	.0001	12
	.95	.90	.85	.80	.75	.70	.65	.60	.55	.50	y ↑

n=100

y ↓	.05	.10	.15	.20	.25	.30	.35	.40	.45	.50	
0	.0059	.0000	.0000	.0000	.0000	.0000	.0000	.0000	.0000	.0000	100
1	.0312	.0003	.0000	.0000	.0000	.0000	.0000	.0000	.0000	.0000	99
2	.0812	.0016	.0000	.0000	.0000	.0000	.0000	.0000	.0000	.0000	98
3	.1396	.0059	.0001	.0000	.0000	.0000	.0000	.0000	.0000	.0000	97
4	.1781	.0159	.0003	.0000	.0000	.0000	.0000	.0000	.0000	.0000	96
5	.1800	.0339	.0011	.0000	.0000	.0000	.0000	.0000	.0000	.0000	95
6	.1500	.0596	.0031	.0001	.0000	.0000	.0000	.0000	.0000	.0000	94
7	.1060	.0889	.0075	.0002	.0000	.0000	.0000	.0000	.0000	.0000	93
8	.0649	.1148	.0153	.0006	.0000	.0000	.0000	.0000	.0000	.0000	92
9	.0349	.1304	.0276	.0015	.0000	.0000	.0000	.0000	.0000	.0000	91
10	.0167	.1319	.0444	.0034	.0001	.0000	.0000	.0000	.0000	.0000	90
11	.0072	.1199	.0640	.0069	.0003	.0000	.0000	.0000	.0000	.0000	89
	.95	.90	.85	.80	.75	.70	.65	.60	.55	.50	y ↑

n=100 continued on next page

Table 1 cont'd

Binomial probabilities (n=100)

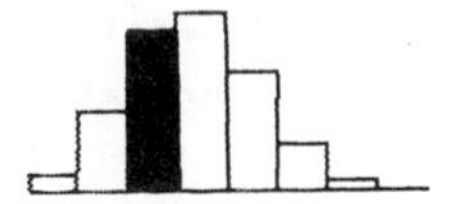

n=100 (continued from previous page)

π

y ↓	.05	.10	.15	.20	.25	.30	.35	.40	.45	.50	
12	.0028	.0988	.0838	.0128	.0006	.0000	.0000	.0000	.0000	.0000	88
13	.0010	.0743	.1001	.0216	.0014	.0000	.0000	.0000	.0000	.0000	87
14	.0003	.0513	.1098	.0335	.0030	.0001	.0000	.0000	.0000	.0000	86
15	.0001	.0327	.1111	.0481	.0057	.0002	.0000	.0000	.0000	.0000	85
16	.0000	.0193	.1041	.0638	.0100	.0006	.0000	.0000	.0000	.0000	84
17	.0000	.0106	.0908	.0789	.0165	.0012	.0000	.0000	.0000	.0000	83
18	.0000	.0054	.0739	.0909	.0254	.0024	.0001	.0000	.0000	.0000	82
19	.0000	.0026	.0563	.0981	.0365	.0044	.0002	.0000	.0000	.0000	81
20	.0000	.0012	.0402	.0993	.0493	.0076	.0004	.0000	.0000	.0000	80
21	.0000	.0005	.0270	.0946	.0626	.0124	.0009	.0000	.0000	.0000	79
22	.0000	.0002	.0171	.0849	.0749	.0190	.0017	.0001	.0000	.0000	78
23	.0000	.0001	.0103	.0720	.0847	.0277	.0032	.0001	.0000	.0000	77
24	.0000	.0000	.0058	.0577	.0906	.0380	.0055	.0003	.0000	.0000	76
25	.0000	.0000	.0031	.0439	.0918	.0496	.0090	.0006	.0000	.0000	75
26	.0000	.0000	.0016	.0316	.0883	.0613	.0140	.0012	.0000	.0000	74
27	.0000	.0000	.0008	.0217	.0806	.0720	.0207	.0022	.0001	.0000	73
28	.0000	.0000	.0004	.0141	.0701	.0804	.0290	.0038	.0002	.0000	72
29	.0000	.0000	.0002	.0088	.0580	.0856	.0388	.0063	.0004	.0000	71
30	.0000	.0000	.0001	.0052	.0458	.0868	.0494	.0100	.0008	.0000	70
31	.0000	.0000	.0000	.0029	.0344	.0840	.0601	.0151	.0014	.0001	69
32	.0000	.0000	.0000	.0016	.0248	.0776	.0698	.0217	.0025	.0001	68
33	.0000	.0000	.0000	.0008	.0170	.0685	.0774	.0297	.0043	.0002	67
34	.0000	.0000	.0000	.0004	.0112	.0579	.0821	.0391	.0069	.0005	66
35	.0000	.0000	.0000	.0002	.0070	.0468	.0834	.0491	.0106	.0009	65
36	.0000	.0000	.0000	.0001	.0042	.0362	.0811	.0591	.0157	.0016	64
37	.0000	.0000	.0000	.0000	.0024	.0268	.0755	.0682	.0222	.0027	63
38	.0000	.0000	.0000	.0000	.0013	.0191	.0674	.0754	.0301	.0045	62
39	.0000	.0000	.0000	.0000	.0007	.0130	.0577	.0799	.0391	.0071	61
40	.0000	.0000	.0000	.0000	.0004	.0085	.0474	.0812	.0488	.0108	60
41	.0000	.0000	.0000	.0000	.0002	.0053	.0373	.0792	.0584	.0159	59
42	.0000	.0000	.0000	.0000	.0001	.0032	.0282	.0742	.0672	.0223	58
43	.0000	.0000	.0000	.0000	.0000	.0019	.0205	.0667	.0741	.0301	57
44	.0000	.0000	.0000	.0000	.0000	.0010	.0143	.0576	.0786	.0390	56
45	.0000	.0000	.0000	.0000	.0000	.0005	.0096	.0478	.0800	.0485	55
46	.0000	.0000	.0000	.0000	.0000	.0003	.0062	.0381	.0782	.0580	54
47	.0000	.0000	.0000	.0000	.0000	.0001	.0038	.0292	.0736	.0666	53
48	.0000	.0000	.0000	.0000	.0000	.0001	.0023	.0215	.0665	.0735	52
49	.0000	.0000	.0000	.0000	.0000	.0000	.0013	.0152	.0577	.0780	51
50	.0000	.0000	.0000	.0000	.0000	.0000	.0007	.0103	.0482	.0796	50
51	.0000	.0000	.0000	.0000	.0000	.0000	.0004	.0068	.0386	.0780	49
52	.0000	.0000	.0000	.0000	.0000	.0000	.0002	.0042	.0298	.0735	48
53	.0000	.0000	.0000	.0000	.0000	.0000	.0001	.0026	.0221	.0666	47
54	.0000	.0000	.0000	.0000	.0000	.0000	.0000	.0015	.0157	.0580	46
55	.0000	.0000	.0000	.0000	.0000	.0000	.0000	.0008	.0108	.0485	45
56	.0000	.0000	.0000	.0000	.0000	.0000	.0000	.0004	.0071	.0390	44
57	.0000	.0000	.0000	.0000	.0000	.0000	.0000	.0002	.0045	.0301	43
58	.0000	.0000	.0000	.0000	.0000	.0000	.0000	.0001	.0027	.0223	42
59	.0000	.0000	.0000	.0000	.0000	.0000	.0000	.0001	.0016	.0159	41
60	.0000	.0000	.0000	.0000	.0000	.0000	.0000	.0000	.0009	.0108	40
61	.0000	.0000	.0000	.0000	.0000	.0000	.0000	.0000	.0005	.0071	39
62	.0000	.0000	.0000	.0000	.0000	.0000	.0000	.0000	.0002	.0045	38
63	.0000	.0000	.0000	.0000	.0000	.0000	.0000	.0000	.0001	.0027	37
64	.0000	.0000	.0000	.0000	.0000	.0000	.0000	.0000	.0001	.0016	36
65	.0000	.0000	.0000	.0000	.0000	.0000	.0000	.0000	.0000	.0009	35
66	.0000	.0000	.0000	.0000	.0000	.0000	.0000	.0000	.0000	.0005	34
67	.0000	.0000	.0000	.0000	.0000	.0000	.0000	.0000	.0000	.0002	33
68	.0000	.0000	.0000	.0000	.0000	.0000	.0000	.0000	.0000	.0001	32
69	.0000	.0000	.0000	.0000	.0000	.0000	.0000	.0000	.0000	.0001	31
	.95	.90	.85	.80	.75	.70	.65	.60	.55	.50	y ↑

Source: Computed by D. K. Hildebrand.

Table 2

Poisson probabilities
(mu between .1 and 5.0)

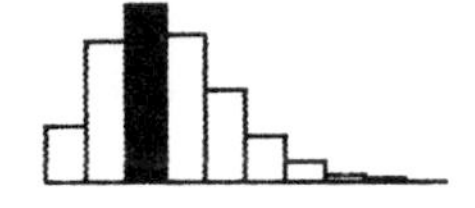

y	.1	.2	.3	.4	.5 (μ)	.6	.7	.8	.9	1.0
0	.9048	.8187	.7408	.6703	.6065	.5488	.4966	.4493	.4066	.3679
1	.0905	.1637	.2222	.2681	.3033	.3293	.3476	.3595	.3659	.3679
2	.0045	.0164	.0333	.0536	.0758	.0988	.1217	.1438	.1647	.1839
3	.0002	.0011	.0033	.0072	.0126	.0198	.0284	.0383	.0494	.0613
4	.0000	.0001	.0003	.0007	.0016	.0030	.0050	.0077	.0111	.0153
5	.0000	.0000	.0000	.0001	.0002	.0004	.0007	.0012	.0020	.0031
6	.0000	.0000	.0000	.0000	.0000	.0000	.0001	.0002	.0003	.0005

y	1.1	1.2	1.3	1.4	1.5 (μ)	1.6	1.7	1.8	1.9	2.0
0	.3329	.3012	.2725	.2466	.2231	.2019	.1827	.1653	.1496	.1353
1	.3662	.3614	.3543	.3452	.3347	.3230	.3106	.2975	.2842	.2707
2	.2014	.2169	.2303	.2417	.2510	.2584	.2640	.2678	.2700	.2707
3	.0738	.0867	.0998	.1128	.1255	.1378	.1496	.1607	.1710	.1804
4	.0203	.0260	.0324	.0395	.0471	.0551	.0636	.0723	.0812	.0902
5	.0045	.0062	.0084	.0111	.0141	.0176	.0216	.0260	.0309	.0361
6	.0008	.0012	.0018	.0026	.0035	.0047	.0061	.0078	.0098	.0120
7	.0001	.0002	.0003	.0005	.0008	.0011	.0015	.0020	.0027	.0034
8	.0000	.0000	.0001	.0001	.0001	.0002	.0003	.0005	.0006	.0009

y	2.1	2.2	2.3	2.4	2.5 (μ)	2.6	2.7	2.8	2.9	3.0
0	.1225	.1108	.1003	.0907	.0821	.0743	.0672	.0608	.0550	.0498
1	.2572	.2438	.2306	.2177	.2052	.1931	.1815	.1703	.1596	.1494
2	.2700	.2681	.2652	.2613	.2565	.2510	.2450	.2384	.2314	.2240
3	.1890	.1966	.2033	.2090	.2138	.2176	.2205	.2225	.2237	.2240
4	.0992	.1082	.1169	.1254	.1336	.1414	.1488	.1557	.1622	.1680
5	.0417	.0476	.0538	.0602	.0668	.0735	.0804	.0872	.0940	.1008
6	.0146	.0174	.0206	.0241	.0278	.0319	.0362	.0407	.0455	.0504
7	.0044	.0055	.0068	.0083	.0099	.0118	.0139	.0163	.0188	.0216
8	.0011	.0015	.0019	.0025	.0031	.0038	.0047	.0057	.0068	.0081
9	.0003	.0004	.0005	.0007	.0009	.0011	.0014	.0018	.0022	.0027
10	.0001	.0001	.0001	.0002	.0002	.0003	.0004	.0005	.0006	.0008
11	.0000	.0000	.0000	.0000	.0000	.0001	.0001	.0001	.0002	.0002

y	3.1	3.2	3.3	3.4	3.5 (μ)	3.6	3.7	3.8	3.9	4.0
0	.0450	.0408	.0369	.0334	.0302	.0273	.0247	.0224	.0202	.0183
1	.1397	.1304	.1217	.1135	.1057	.0984	.0915	.0850	.0789	.0733
2	.2165	.2087	.2008	.1929	.1850	.1771	.1692	.1615	.1539	.1465
3	.2237	.2226	.2209	.2186	.2158	.2125	.2087	.2046	.2001	.1954
4	.1733	.1781	.1823	.1858	.1888	.1912	.1931	.1944	.1951	.1954
5	.1075	.1140	.1203	.1264	.1322	.1377	.1429	.1477	.1522	.1563
6	.0555	.0608	.0662	.0716	.0771	.0826	.0881	.0936	.0989	.1042
7	.0246	.0278	.0312	.0348	.0385	.0425	.0466	.0508	.0551	.0595
8	.0095	.0111	.0129	.0148	.0169	.0191	.0215	.0241	.0269	.0298
9	.0033	.0040	.0047	.0056	.0066	.0076	.0089	.0102	.0116	.0132
10	.0010	.0013	.0016	.0019	.0023	.0028	.0033	.0039	.0045	.0053
11	.0003	.0004	.0005	.0006	.0007	.0009	.0011	.0013	.0016	.0019
12	.0001	.0001	.0001	.0002	.0002	.0003	.0003	.0004	.0005	.0006
13	.0000	.0000	.0000	.0000	.0001	.0001	.0001	.0001	.0002	.0002

y	4.1	4.2	4.3	4.4	4.5 (μ)	4.6	4.7	4.8	4.9	5.0
0	.0166	.0150	.0136	.0123	.0111	.0101	.0091	.0082	.0074	.0067
1	.0679	.0630	.0583	.0540	.0500	.0462	.0427	.0395	.0365	.0337
2	.1393	.1323	.1254	.1188	.1125	.1063	.1005	.0948	.0894	.0842
3	.1904	.1852	.1798	.1743	.1687	.1631	.1574	.1517	.1460	.1404
4	.1951	.1944	.1933	.1917	.1898	.1875	.1849	.1820	.1789	.1755
5	.1600	.1633	.1662	.1687	.1708	.1725	.1738	.1747	.1753	.1755
6	.1093	.1143	.1191	.1237	.1281	.1323	.1362	.1398	.1432	.1462
7	.0640	.0686	.0732	.0778	.0824	.0869	.0914	.0959	.1002	.1044
8	.0328	.0360	.0393	.0428	.0463	.0500	.0537	.0575	.0614	.0653
9	.0150	.0168	.0188	.0209	.0232	.0255	.0281	.0307	.0334	.0363
10	.0061	.0071	.0081	.0092	.0104	.0118	.0132	.0147	.0164	.0181
11	.0023	.0027	.0032	.0037	.0043	.0049	.0056	.0064	.0073	.0082
12	.0008	.0009	.0011	.0013	.0016	.0019	.0022	.0026	.0030	.0034
13	.0002	.0003	.0004	.0005	.0006	.0007	.0008	.0009	.0011	.0013
14	.0001	.0001	.0001	.0001	.0002	.0002	.0003	.0003	.0004	.0005
15	.0000	.0000	.0000	.0000	.0001	.0001	.0001	.0001	.0001	.0002

Table 2 cont'd

Poisson probabilities
(mu between 5.5 and 20.0)

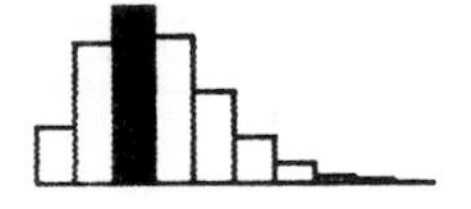

y	5.5	6.0	6.5	7.0	7.5 μ	8.0	8.5	9.0	9.5	10.0
0	.0041	.0025	.0015	.0009	.0006	.0003	.0002	.0001	.0001	.0000
1	.0225	.0149	.0098	.0064	.0041	.0027	.0017	.0011	.0007	.0005
2	.0618	.0446	.0318	.0223	.0156	.0107	.0074	.0050	.0034	.0023
3	.1133	.0892	.0688	.0521	.0389	.0286	.0208	.0150	.0107	.0076
4	.1558	.1339	.1118	.0912	.0729	.0573	.0443	.0337	.0254	.0189
5	.1714	.1606	.1454	.1277	.1094	.0916	.0752	.0607	.0483	.0378
6	.1571	.1606	.1575	.1490	.1367	.1221	.1066	.0911	.0764	.0631
7	.1234	.1377	.1462	.1490	.1465	.1396	.1294	.1171	.1037	.0901
8	.0849	.1033	.1188	.1304	.1373	.1396	.1375	.1318	.1232	.1126
9	.0519	.0688	.0858	.1014	.1144	.1241	.1299	.1318	.1300	.1251
10	.0285	.0413	.0558	.0710	.0858	.0993	.1104	.1186	.1235	.1251
11	.0143	.0225	.0330	.0452	.0585	.0722	.0853	.0970	.1067	.1137
12	.0065	.0113	.0179	.0263	.0366	.0481	.0604	.0728	.0844	.0948
13	.0028	.0052	.0089	.0142	.0211	.0296	.0395	.0504	.0617	.0729
14	.0011	.0022	.0041	.0071	.0113	.0169	.0240	.0324	.0419	.0521
15	.0004	.0009	.0018	.0033	.0057	.0090	.0136	.0194	.0265	.0347
16	.0001	.0003	.0007	.0014	.0026	.0045	.0072	.0109	.0157	.0217
17	.0000	.0001	.0003	.0006	.0012	.0021	.0036	.0058	.0088	.0128
18	.0000	.0000	.0001	.0002	.0005	.0009	.0017	.0029	.0046	.0071
19	.0000	.0000	.0000	.0001	.0002	.0004	.0008	.0014	.0023	.0037
20	.0000	.0000	.0000	.0000	.0001	.0002	.0003	.0006	.0011	.0019
21	.0000	.0000	.0000	.0000	.0000	.0001	.0001	.0003	.0005	.0009
22	.0000	.0000	.0000	.0000	.0000	.0000	.0001	.0001	.0002	.0004
23	.0000	.0000	.0000	.0000	.0000	.0000	.0000	.0000	.0001	.0002

y	11.0	12.0	13.0	14.0	15.0 μ	16.0	17.0	18.0	19.0	20.0
0	.0000	.0000	.0000	.0000	.0000	.0000	.0000	.0000	.0000	.0000
1	.0002	.0001	.0000	.0000	.0000	.0000	.0000	.0000	.0000	.0000
2	.0010	.0004	.0002	.0001	.0000	.0000	.0000	.0000	.0000	.0000
3	.0037	.0018	.0008	.0004	.0002	.0001	.0000	.0000	.0000	.0000
4	.0102	.0053	.0027	.0013	.0006	.0003	.0001	.0001	.0000	.0000
5	.0224	.0127	.0070	.0037	.0019	.0010	.0005	.0002	.0001	.0001
6	.0411	.0255	.0152	.0087	.0048	.0026	.0014	.0007	.0004	.0002
7	.0646	.0437	.0281	.0174	.0104	.0060	.0034	.0019	.0010	.0005
8	.0888	.0655	.0457	.0304	.0194	.0120	.0072	.0042	.0024	.0013
9	.1085	.0874	.0661	.0473	.0324	.0213	.0135	.0083	.0050	.0029
10	.1194	.1048	.0859	.0663	.0486	.0341	.0230	.0150	.0095	.0058
11	.1194	.1144	.1015	.0844	.0663	.0496	.0355	.0245	.0164	.0106
12	.1094	.1144	.1099	.0984	.0829	.0661	.0504	.0368	.0259	.0176
13	.0926	.1056	.1099	.1060	.0956	.0814	.0658	.0509	.0378	.0271
14	.0728	.0905	.1021	.1060	.1024	.0930	.0800	.0655	.0514	.0387
15	.0534	.0724	.0885	.0989	.1024	.0992	.0906	.0786	.0650	.0516
16	.0367	.0543	.0719	.0866	.0960	.0992	.0963	.0884	.0772	.0646
17	.0237	.0383	.0550	.0713	.0847	.0934	.0963	.0936	.0863	.0760
18	.0145	.0255	.0397	.0554	.0706	.0830	.0909	.0936	.0911	.0844
19	.0084	.0161	.0272	.0409	.0557	.0699	.0814	.0887	.0911	.0888
20	.0046	.0097	.0177	.0286	.0418	.0559	.0692	.0798	.0866	.0888
21	.0024	.0055	.0109	.0191	.0299	.0426	.0560	.0684	.0783	.0846
22	.0012	.0030	.0065	.0121	.0204	.0310	.0433	.0560	.0676	.0769
23	.0006	.0016	.0037	.0074	.0133	.0216	.0320	.0438	.0559	.0669
24	.0003	.0008	.0020	.0043	.0083	.0144	.0226	.0328	.0442	.0557
25	.0001	.0004	.0010	.0024	.0050	.0092	.0154	.0237	.0336	.0446
26	.0000	.0002	.0005	.0013	.0029	.0057	.0101	.0164	.0246	.0343
27	.0000	.0001	.0002	.0007	.0016	.0034	.0063	.0109	.0173	.0254
28	.0000	.0000	.0001	.0003	.0009	.0019	.0038	.0070	.0117	.0181
29	.0000	.0000	.0001	.0002	.0004	.0011	.0023	.0044	.0077	.0125
30	.0000	.0000	.0000	.0001	.0002	.0006	.0013	.0026	.0049	.0083
31	.0000	.0000	.0000	.0000	.0001	.0003	.0007	.0015	.0030	.0054
32	.0000	.0000	.0000	.0000	.0001	.0001	.0004	.0009	.0018	.0034
33	.0000	.0000	.0000	.0000	.0000	.0001	.0002	.0005	.0010	.0020

Source: Computed by D. K. Hildebrand.

Table 3
Normal curve areas

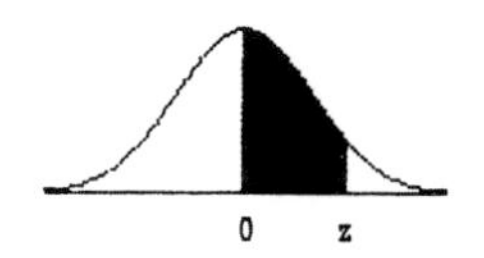

z	.00	.01	.02	.03	.04	.05	.06	.07	.08	.09
0.00	.0000	.0040	.0080	.0120	.0160	.0199	.0239	.0279	.0319	.0359
0.10	.0398	.0438	.0478	.0517	.0557	.0596	.0636	.0675	.0714	.0753
0.20	.0793	.0832	.0871	.0910	.0948	.0987	.1026	.1064	.1103	.1141
0.30	.1179	.1217	.1255	.1293	.1331	.1368	.1406	.1443	.1480	.1517
0.40	.1554	.1591	.1628	.1664	.1700	.1736	.1772	.1808	.1844	.1879
0.50	.1915	.1950	.1985	.2019	.2054	.2088	.2123	.2157	.2190	.2224
0.60	.2257	.2291	.2324	.2357	.2389	.2422	.2454	.2486	.2517	.2549
0.70	.2580	.2611	.2642	.2673	.2704	.2734	.2764	.2794	.2823	.2852
0.80	.2881	.2910	.2939	.2967	.2995	.3023	.3051	.3078	.3106	.3133
0.90	.3159	.3186	.3212	.3238	.3264	.3289	.3315	.3340	.3365	.3389
1.00	.3413	.3438	.3461	.3485	.3508	.3531	.3554	.3577	.3599	.3621
1.10	.3643	.3665	.3686	.3708	.3729	.3749	.3770	.3790	.3810	.3830
1.20	.3849	.3869	.3888	.3907	.3925	.3944	.3962	.3980	.3997	.4015
1.30	.4032	.4049	.4066	.4082	.4099	.4115	.4131	.4147	.4162	.4177
1.40	.4192	.4207	.4222	.4236	.4251	.4265	.4279	.4292	.4306	.4319
1.50	.4332	.4345	.4357	.4370	.4382	.4394	.4406	.4418	.4429	.4441
1.60	.4452	.4463	.4474	.4484	.4495	.4505	.4515	.4525	.4535	.4545
1.70	.4554	.4564	.4573	.4582	.4591	.4599	.4608	.4616	.4625	.4633
1.80	.4641	.4649	.4656	.4664	.4671	.4678	.4686	.4693	.4699	.4706
1.90	.4713	.4719	.4726	.4732	.4738	.4744	.4750	.4756	.4761	.4767
2.00	.4772	.4778	.4783	.4788	.4793	.4798	.4803	.4808	.4812	.4817
2.10	.4821	.4826	.4830	.4834	.4838	.4842	.4846	.4850	.4854	.4857
2.20	.4861	.4864	.4868	.4871	.4875	.4878	.4881	.4884	.4887	.4890
2.30	.4893	.4896	.4898	.4901	.4904	.4906	.4909	.4911	.4913	.4916
2.40	.4918	.4920	.4922	.4925	.4927	.4929	.4931	.4932	.4934	.4936
2.50	.4938	.4940	.4941	.4943	.4945	.4946	.4948	.4949	.4951	.4952
2.60	.4953	.4955	.4956	.4957	.4959	.4960	.4961	.4962	.4963	.4964
2.70	.4965	.4966	.4967	.4968	.4969	.4970	.4971	.4972	.4973	.4974
2.80	.4974	.4975	.4976	.4977	.4977	.4978	.4979	.4979	.4980	.4981
2.90	.4981	.4982	.4982	.4983	.4984	.4984	.4985	.4985	.4986	.4986
3.00	.4987	.4987	.4987	.4988	.4988	.4989	.4989	.4989	.4990	.4990

z	area
3.50	.49976737
4.00	.49996833
4.50	.49999660
5.00	.49999971

Source: Computed by P. J. Hildebrand.

Table 4

Percentage points of the t-distribution

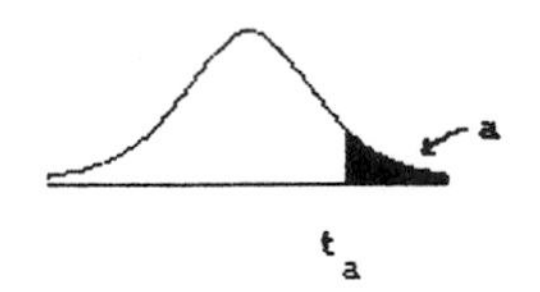

df	a=.1	a=.05	a=.025	a=.01	a=.005	a=.001
1	3.078	6.314	12.706	31.821	63.657	318.309
2	1.886	2.920	4.303	6.965	9.925	22.327
3	1.638	2.353	3.182	4.541	5.841	10.215
4	1.533	2.132	2.776	3.747	4.604	7.173
5	1.476	2.015	2.571	3.365	4.032	5.893
6	1.440	1.943	2.447	3.143	3.707	5.208
7	1.415	1.895	2.365	2.998	3.499	4.785
8	1.397	1.860	2.306	2.896	3.355	4.501
9	1.383	1.833	2.262	2.821	3.250	4.297
10	1.372	1.812	2.228	2.764	3.169	4.144
11	1.363	1.796	2.201	2.718	3.106	4.025
12	1.356	1.782	2.179	2.681	3.055	3.930
13	1.350	1.771	2.160	2.650	3.012	3.852
14	1.345	1.761	2.145	2.624	2.977	3.787
15	1.341	1.753	2.131	2.602	2.947	3.733
16	1.337	1.746	2.120	2.583	2.921	3.686
17	1.333	1.740	2.110	2.567	2.898	3.646
18	1.330	1.734	2.101	2.552	2.878	3.610
19	1.328	1.729	2.093	2.539	2.861	3.579
20	1.325	1.725	2.086	2.528	2.845	3.552
21	1.323	1.721	2.080	2.518	2.831	3.527
22	1.321	1.717	2.074	2.508	2.819	3.505
23	1.319	1.714	2.069	2.500	2.807	3.485
24	1.318	1.711	2.064	2.492	2.797	3.467
25	1.316	1.708	2.060	2.485	2.787	3.450
26	1.315	1.706	2.056	2.479	2.779	3.435
27	1.314	1.703	2.052	2.473	2.771	3.421
28	1.313	1.701	2.048	2.467	2.763	3.408
29	1.311	1.699	2.045	2.462	2.756	3.396
30	1.310	1.697	2.042	2.457	2.750	3.385
40	1.303	1.684	2.021	2.423	2.704	3.307
60	1.296	1.671	2.000	2.390	2.660	3.232
120	1.289	1.658	1.980	2.358	2.617	3.160
240	1.285	1.651	1.970	2.342	2.596	3.125
inf.	1.282	1.645	1.960	2.326	2.576	3.090

Source: Computed by P. J. Hildebrand.

Table 5

Percentage points of the chi-square distribution (a>.5)

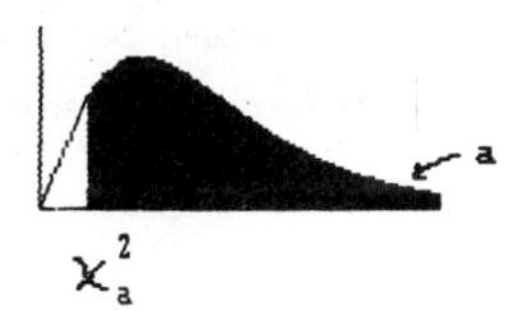

df	a = .999	a = .995	a = .99	a = .975	a = .95	a = .9
1	.000002	.000039	.000157	.000982	.003932	.01579
2	.002001	.01003	02010	.05064	.1026	.2107
3	.02430	.07172	.1148	.2158	.3518	.5844
4	.09080	.2070	.2971	.4844	.7107	1.064
5	.2102	.4117	.5543	.8312	1.145	1.610
6	.3811	.6757	.8721	1.237	1.635	2.204
7	.5985	.9893	1.239	1.690	2.167	2.833
8	.8571	1.344	1.646	2.180	2.733	3.490
9	1.152	1.735	2.088	2.700	3.325	4.168
10	1.479	2.156	2.558	3.247	3.940	4.865
11	1.834	2.603	3.053	3.816	4.575	5.578
12	2.214	3.074	3.571	4.404	5.226	6.304
13	2.617	3.565	4.107	5.009	5.892	7.042
14	3.041	4.075	4.660	5.629	6.571	7.790
15	3.483	4.601	5.229	6.262	7.261	8.547
16	3.942	5.142	5.812	6.908	7.962	9.312
17	4.416	5.697	6.408	7.564	8.672	10.09
18	4.905	6.265	7.015	8.231	9.390	10.86
19	5.407	6.844	7.633	8.907	10.12	11.65
20	5.921	7.434	8.260	9.591	10.85	12.44
21	6.447	8.034	8.897	10.28	11.59	13.24
22	6.983	8.643	9.542	10.98	12.34	14.04
23	7.529	9.260	10.20	11.69	13.09	14.85
24	8.085	9.886	10.86	12.40	13.85	15.66
25	8.649	10.52	11.52	13.12	14.61	16.47
26	9.222	11.16	12.20	13.84	15.38	17.29
27	9.803	11.81	12.88	14.57	16.15	18.11
28	10.39	12.46	13.56	15.31	16.93	18.94
29	10.99	13.12	14.26	16.05	17.71	19.77
30	11.59	13.79	14.95	16.79	18.49	20.60
40	17.92	20.71	22.16	24.43	26.51	29.05
50	24.67	27.99	29.71	32.36	34.76	37.69
60	31.74	35.53	37.48	40.48	43.19	46.46
70	39.04	43.28	45.44	48.76	51.74	55.33
80	46.52	51.17	53.54	57.15	60.39	64.28
90	54.16	59.20	61.75	65.65	69.13	73.29
100	61.92	67.33	70.06	74.22	77.93	82.36
120	77.76	83.85	86.92	91.57	95.70	100.62
240	177.95	187.32	191.99	198.98	205.14	212.39

Table 5 cont'd

Percentage points of the chi-square distribution (a<.5)

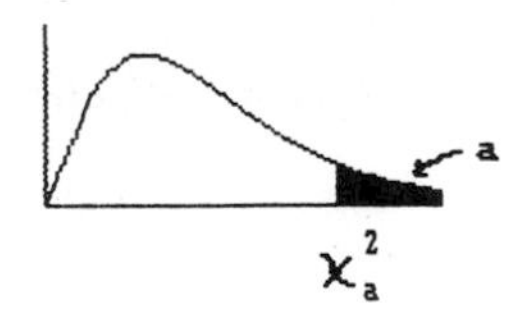

a=.1	a=.05	a=.025	a=.01	a=.005	a=.001	df
2.706	3.841	5.024	6.635	7.879	10.83	1
4.605	5.991	7.378	9.210	10.60	13.82	2
6.251	7.815	9.348	11.34	12.84	16.27	3
7.779	9.488	11.14	13.28	14.86	18.47	4
9.236	11.07	12.83	15.09	16.75	20.52	5
10.64	12.59	14.45	16.81	18.55	22.46	6
12.02	14.07	16.01	18.48	20.28	24.32	7
13.36	15.51	17.53	20.09	21.95	26.12	8
14.68	16.92	19.02	21.67	23.59	27.88	9
15.99	18.31	20.48	23.21	25.19	29.59	10
17.28	19.68	21.92	24.72	26.76	31.27	11
18.55	21.03	23.34	26.22	28.30	32.91	12
19.81	22.36	24.74	27.69	29.82	34.53	13
21.06	23.68	26.12	29.14	31.32	36.12	14
22.31	25.00	27.49	30.58	32.80	37.70	15
23.54	26.30	28.85	32.00	34.27	39.25	16
24.77	27.59	30.19	33.41	35.72	40.79	17
25.99	28.87	31.53	34.81	37.16	42.31	18
27.20	30.14	32.85	36.19	38.58	43.82	19
28.41	31.41	34.17	37.57	40.00	45.31	20
29.62	32.67	35.48	38.93	41.40	46.80	21
30.81	33.92	36.78	40.29	42.80	48.27	22
32.01	35.17	38.08	41.64	44.18	49.73	23
33.20	36.42	39.36	42.98	45.56	51.18	24
34.38	37.65	40.65	44.31	46.93	52.62	25
35.56	38.89	41.92	45.64	48.29	54.05	26
36.74	40.11	43.19	46.96	49.65	55.48	27
37.92	41.34	44.46	48.28	50.99	56.89	28
39.09	42.56	45.72	49.59	52.34	58.30	29
40.26	43.77	46.98	50.89	53.67	59.70	30
51.81	55.76	59.34	63.69	66.77	73.40	40
63.17	67.50	71.42	76.15	79.49	86.66	50
74.40	79.08	83.30	88.38	91.95	99.61	60
85.53	90.53	95.02	100.43	104.21	112.32	70
96.58	101.88	106.63	112.33	116.32	124.84	80
107.57	113.15	118.14	124.12	128.30	137.21	90
118.50	124.34	129.56	135.81	140.17	149.45	100
140.23	146.57	152.21	158.95	163.65	173.62	120
268.47	277.14	284.80	293.89	300.18	313.44	240

Source: Computed by P. J. Hildebrand.

Table 6

Percentage points of the F-distribution (df_2 between 1 and 6)

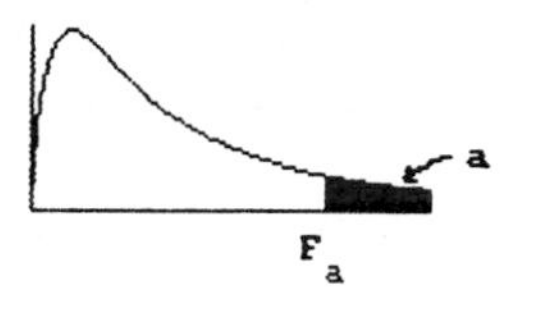

df_2	a	df_1 = 1	2	3	4	5	6	7	8	9	10
1	.25	5.83	7.50	8.20	8.58	8.82	8.98	9.10	9.19	9.26	9.32
	.10	39.86	49.50	53.59	55.83	57.24	58.20	58.91	59.44	59.86	60.19
	.05	161.4	199.5	215.7	224.6	230.2	234.0	236.8	238.9	240.5	241.9
	.025	647.8	799.5	864.2	899.6	921.8	937.1	948.2	956.7	963.3	968.6
	.01	4052	5000	5403	5625	5764	5859	5928	5981	6022	6056
2	.25	2.57	3.00	3.15	3.23	3.28	3.31	3.34	3.35	3.37	3.38
	.10	8.53	9.00	9.16	9.24	9.29	9.33	9.35	9.37	9.38	9.39
	.05	18.51	19.00	19.16	19.25	19.30	19.33	19.35	19.37	19.38	19.40
	.025	38.51	39.00	39.17	39.25	39.30	39.33	39.36	39.37	39.39	39.40
	.01	98.50	99.00	99.17	99.25	99.30	99.33	99.36	99.37	99.39	99.40
	.005	198.5	199.0	199.2	199.2	199.3	199.3	199.4	199.4	199.4	199.4
	.001	998.5	999.0	999.2	999.2	999.3	999.3	999.4	999.4	999.4	999.4
3	.25	2.02	2.28	2.36	2.39	2.41	2.42	2.43	2.44	2.44	2.44
	.10	5.54	5.46	5.39	5.34	5.31	5.28	5.27	5.25	5.24	5.23
	.05	10.13	9.55	9.28	9.12	9.01	8.94	8.89	8.85	8.81	8.79
	.025	17.44	16.04	15.44	15.10	14.88	14.73	14.62	14.54	14.47	14.42
	.01	34.12	30.82	29.46	28.71	28.24	27.91	27.67	27.49	27.35	27.23
	.005	55.55	49.80	47.47	46.19	45.39	44.84	44.43	44.13	43.88	43.69
	.001	167.0	148.5	141.1	137.1	134.6	132.8	131.6	130.6	129.9	129.2
4	.25	1.81	2.00	2.05	2.06	2.07	2.08	2.08	2.08	2.08	2.08
	.10	4.54	4.32	4.19	4.11	4.05	4.01	3.98	3.95	3.94	3.92
	.05	7.71	6.94	6.59	6.39	6.26	6.16	6.09	6.04	6.00	5.96
	.025	12.22	10.65	9.98	9.60	9.36	9.20	9.07	8.98	8.90	8.84
	.01	21.20	18.00	16.69	15.98	15.52	15.21	14.98	14.80	14.66	14.55
	.005	31.33	26.28	24.26	23.15	22.46	21.97	21.62	21.35	21.14	20.97
	.001	74.14	61.25	56.18	53.44	51.71	50.53	49.66	49.00	48.47	48.05
5	.25	1.69	1.85	1.88	1.89	1.89	1.89	1.89	1.89	1.89	1.89
	.10	4.06	3.78	3.62	3.52	3.45	3.40	3.37	3.34	3.32	3.30
	.05	6.61	5.79	5.41	5.19	5.05	4.95	4.88	4.82	4.77	4.74
	.025	10.01	8.43	7.76	7.39	7.15	6.98	6.85	6.76	6.68	6.62
	.01	16.26	13.27	12.06	11.39	10.97	10.67	10.46	10.29	10.16	10.05
	.005	22.78	18.31	16.53	15.56	14.94	14.51	14.20	13.96	13.77	13.62
	.001	47.18	37.12	33.20	31.09	29.75	28.83	28.16	27.65	27.24	26.92
6	.25	1.62	1.76	1.78	1.79	1.79	1.78	1.78	1.78	1.77	1.77
	.10	3.78	3.46	3.29	3.18	3.11	3.05	3.01	2.98	2.96	2.94
	.05	5.99	5.14	4.76	4.53	4.39	4.28	4.21	4.15	4.10	4.06
	.025	8.81	7.26	6.60	6.23	5.99	5.82	5.70	5.60	5.52	5.46
	.01	13.75	10.92	9.78	9.15	8.75	8.47	8.26	8.10	7.98	7.87
	.005	18.63	14.54	12.92	12.03	11.46	11.07	10.79	10.57	10.39	10.25
	.001	35.51	27.00	23.70	21.92	20.80	20.03	19.46	19.03	18.69	18.41

Table 6 cont'd

Percentage points of the F-distribution (df_2 between 1 and 6)

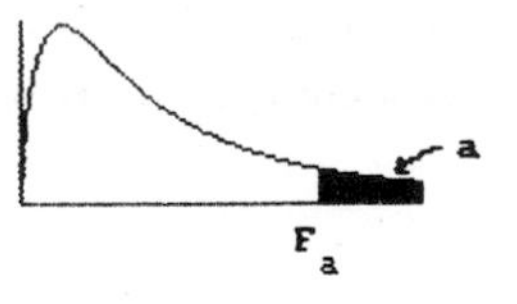

df_1

12	15	20	24	30	40	60	120	240	inf.	a	df_2
9.41	9.49	9.58	9.63	9.67	9.71	9.76	9.80	9.83	9.85	.25	1
60.71	61.22	61.74	62.00	62.26	62.53	62.79	63.06	63.19	63.33	.10	
243.9	245.9	248.0	249.1	250.1	251.1	252.2	253.3	253.8	254.3	.05	
976.7	984.9	993.1	997.2	1001	1006	1010	1014	1016	1018	.025	
6106	6157	6209	6235	6261	6287	6313	6339	6353	6366	.01	
3.39	3.41	3.43	3.43	3.44	3.45	3.46	3.47	3.47	3.48	.25	2
9.41	9.42	9.44	9.45	9.46	9.47	9.47	9.48	9.49	9.49	.10	
19.41	19.43	19.45	19.45	19.46	19.47	19.48	19.49	19.49	19.50	.05	
39.41	39.43	39.45	39.46	39.46	39.47	39.48	39.49	39.49	39.50	.025	
99.42	99.43	99.45	99.46	99.47	99.47	99.48	99.49	99.50	99.50	.01	
199.4	199.4	199.4	199.5	199.5	199.5	199.5	199.5	199.5	199.5	.005	
999.4	999.4	999.4	999.5	999.5	999.5	999.5	999.5	999.5	999.5	.001	
2.45	2.46	2.46	2.46	2.47	2.47	2.47	2.47	2.47	2.47	.25	3
5.22	5.20	5.18	5.18	5.17	5.16	5.15	5.14	5.14	5.13	.10	
8.74	8.70	8.66	8.64	8.62	8.59	8.57	8.55	8.54	8.53	.05	
14.34	14.25	14.17	14.12	14.08	14.04	13.99	13.95	13.92	13.90	.025	
27.05	26.87	26.69	26.60	26.50	26.41	26.32	26.22	26.17	26.13	.01	
43.39	43.08	42.78	42.62	42.47	42.31	42.15	41.99	41.91	41.83	.005	
128.3	127.4	126.4	125.9	125.4	125.0	124.5	124.0	123.7	123.5	.001	
2.08	2.08	2.08	2.08	2.08	2.08	2.08	2.08	2.08	2.08	.25	4
3.90	3.87	3.84	3.83	3.82	3.80	3.79	3.78	3.77	3.76	.10	
5.91	5.86	5.80	5.77	5.75	5.72	5.69	5.66	5.64	5.63	.05	
8.75	8.66	8.56	8.51	8.46	8.41	8.36	8.31	8.28	8.26	.025	
14.37	14.20	14.02	13.93	13.84	13.75	13.65	13.56	13.51	13.46	.01	
20.70	20.44	20.17	20.03	19.89	19.75	19.61	19.47	19.40	19.32	.005	
47.41	46.76	46.10	45.77	45.43	45.09	44.75	44.40	44.23	44.05	.001	
1.89	1.89	1.88	1.88	1.88	1.88	1.87	1.87	1.87	1.87	.25	5
3.27	3.24	3.21	3.19	3.17	3.16	3.14	3.12	3.11	3.10	.10	
4.68	4.62	4.56	4.53	4.50	4.46	4.43	4.40	4.38	4.36	.05	
6.52	6.43	6.33	6.28	6.23	6.18	6.12	6.07	6.04	6.02	.025	
9.89	9.72	9.55	9.47	9.38	9.29	9.20	9.11	9.07	9.02	.01	
13.38	13.15	12.90	12.78	12.66	12.53	12.40	12.27	12.21	12.14	.005	
26.42	25.91	25.39	25.13	24.87	24.60	24.33	24.06	23.92	23.79	.001	
1.77	1.76	1.76	1.75	1.75	1.75	1.74	1.74	1.74	1.74	.25	6
2.90	2.87	2.84	2.82	2.80	2.78	2.76	2.74	2.73	2.72	.10	
4.00	3.94	3.87	3.84	3.81	3.77	3.74	3.70	3.69	3.67	.05	
5.37	5.27	5.17	5.12	5.07	5.01	4.96	4.90	4.88	4.85	.025	
7.72	7.56	7.40	7.31	7.23	7.14	7.06	6.97	6.92	6.88	.01	
10.03	9.81	9.59	9.47	9.36	9.24	9.12	9.00	8.94	8.88	.005	
17.99	17.56	17.12	16.90	16.67	16.44	16.21	15.98	15.86	15.75	.001	

Table 6 cont'd

Percentage points of the F-distribution (df_2 between 7 and 12)

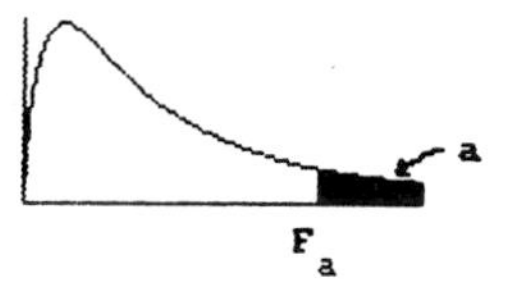

		df_1									
df_2	a	1	2	3	4	5	6	7	8	9	10
7	.25	1.57	1.70	1.72	1.72	1.71	1.71	1.70	1.70	1.69	1.69
	.10	3.59	3.26	3.07	2.96	2.88	2.83	2.78	2.75	2.72	2.70
	.05	5.59	4.74	4.35	4.12	3.97	3.87	3.79	3.73	3.68	3.64
	.025	8.07	6.54	5.89	5.52	5.29	5.12	4.99	4.90	4.82	4.76
	.01	12.25	9.55	8.45	7.85	7.46	7.19	6.99	6.84	6.72	6.62
	.005	16.24	12.40	10.88	10.05	9.52	9.16	8.89	8.68	8.51	8.38
	.001	29.25	21.69	18.77	17.20	16.21	15.52	15.02	14.63	14.33	14.08
8	.25	1.54	1.66	1.67	1.66	1.66	1.65	1.64	1.64	1.63	1.63
	.10	3.46	3.11	2.92	2.81	2.73	2.67	2.62	2.59	2.56	2.54
	.05	5.32	4.46	4.07	3.84	3.69	3.58	3.50	3.44	3.39	3.35
	.025	7.57	6.06	5.42	5.05	4.82	4.65	4.53	4.43	4.36	4.30
	.01	11.26	8.65	7.59	7.01	6.63	6.37	6.18	6.03	5.91	5.81
	.005	14.69	11.04	9.60	8.81	8.30	7.95	7.69	7.50	7.34	7.21
	.001	25.41	18.49	15.83	14.39	13.48	12.86	12.40	12.05	11.77	11.54
9	.25	1.51	1.62	1.63	1.63	1.62	1.61	1.60	1.60	1.59	1.59
	.10	3.36	3.01	2.81	2.69	2.61	2.55	2.51	2.47	2.44	2.42
	.05	5.12	4.26	3.86	3.63	3.48	3.37	3.29	3.23	3.18	3.14
	.025	7.21	5.71	5.08	4.72	4.48	4.32	4.20	4.10	4.03	3.96
	.01	10.56	8.02	6.99	6.42	6.06	5.80	5.61	5.47	5.35	5.26
	.005	13.61	10.11	8.72	7.96	7.47	7.13	6.88	6.69	6.54	6.42
	.001	22.86	16.39	13.90	12.56	11.71	11.13	10.70	10.37	10.11	9.89
10	.25	1.49	1.60	1.60	1.59	1.59	1.58	1.57	1.56	1.56	1.55
	.10	3.29	2.92	2.73	2.61	2.52	2.46	2.41	2.38	2.35	2.32
	.05	4.96	4.10	3.71	3.48	3.33	3.22	3.14	3.07	3.02	2.98
	.025	6.94	5.46	4.83	4.47	4.24	4.07	3.95	3.85	3.78	3.72
	.01	10.04	7.56	6.55	5.99	5.64	5.39	5.20	5.06	4.94	4.85
	.005	12.83	9.43	8.08	7.34	6.87	6.54	6.30	6.12	5.97	5.85
	.001	21.04	14.91	12.55	11.28	10.48	9.93	9.52	9.20	8.96	8.75
11	.25	1.47	1.58	1.58	1.57	1.56	1.55	1.54	1.53	1.53	1.52
	.10	3.23	2.86	2.66	2.54	2.45	2.39	2.34	2.30	2.27	2.25
	.05	4.84	3.98	3.59	3.36	3.20	3.09	3.01	2.95	2.90	2.85
	.025	6.72	5.26	4.63	4.28	4.04	3.88	3.76	3.66	3.59	3.53
	.01	9.65	7.21	6.22	5.67	5.32	5.07	4.89	4.74	4.63	4.54
	.005	12.23	8.91	7.60	6.88	6.42	6.10	5.86	5.68	5.54	5.42
	.001	19.69	13.81	11.56	10.35	9.58	9.05	8.66	8.35	8.12	7.92
12	.25	1.46	1.56	1.56	1.55	1.54	1.53	1.52	1.51	1.51	1.50
	.10	3.18	2.81	2.61	2.48	2.39	2.33	2.28	2.24	2.21	2.19
	.05	4.75	3.89	3.49	3.26	3.11	3.00	2.91	2.85	2.80	2.75
	.025	6.55	5.10	4.47	4.12	3.89	3.73	3.61	3.51	3.44	3.37
	.01	9.33	6.93	5.95	5.41	5.06	4.82	4.64	4.50	4.39	4.30
	.005	11.75	8.51	7.23	6.52	6.07	5.76	5.52	5.35	5.20	5.09
	.001	18.64	12.97	10.80	9.63	8.89	8.38	8.00	7.71	7.48	7.29

Table 6 cont'd

Percentage points of the F-distribution (df_2 between 7 and 12)

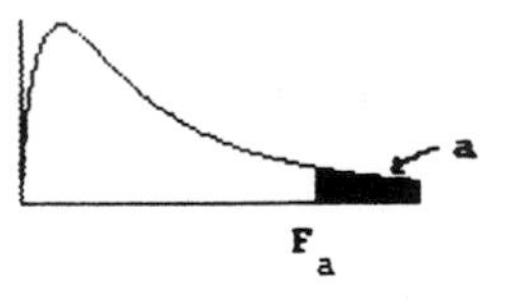

12	15	20	24	30	40	60	120	240	inf.	a	df_2
					df_1						
1.68	1.68	1.67	1.67	1.66	1.66	1.65	1.65	1.65	1.65	.25	7
2.67	2.63	2.59	2.58	2.56	2.54	2.51	2.49	2.48	2.47	.10	
3.57	3.51	3.44	3.41	3.38	3.34	3.30	3.27	3.25	3.23	.05	
4.67	4.57	4.47	4.41	4.36	4.31	4.25	4.20	4.17	4.14	.025	
6.47	6.31	6.16	6.07	5.99	5.91	5.82	5.74	5.69	5.65	.01	
8.18	7.97	7.75	7.64	7.53	7.42	7.31	7.19	7.13	7.08	.005	
13.71	13.32	12.93	12.73	12.53	12.33	12.12	11.91	11.80	11.70	.001	
1.62	1.62	1.61	1.60	1.60	1.59	1.59	1.58	1.58	1.58	.25	8
2.50	2.46	2.42	2.40	2.38	2.36	2.34	2.32	2.30	2.29	.10	
3.28	3.22	3.15	3.12	3.08	3.04	3.01	2.97	2.95	2.93	.05	
4.20	4.10	4.00	3.95	3.89	3.84	3.78	3.73	3.70	3.67	.025	
5.67	5.52	5.36	5.28	5.20	5.12	5.03	4.95	4.90	4.86	.01	
7.01	6.81	6.61	6.50	6.40	6.29	6.18	6.06	6.01	5.95	.005	
11.19	10.84	10.48	10.30	10.11	9.92	9.73	9.53	9.43	9.33	.001	
1.58	1.57	1.56	1.56	1.55	1.54	1.54	1.53	1.53	1.53	.25	9
2.38	2.34	2.30	2.28	2.25	2.23	2.21	2.18	2.17	2.16	.10	
3.07	3.01	2.94	2.90	2.86	2.83	2.79	2.75	2.73	2.71	.05	
3.87	3.77	3.67	3.61	3.56	3.51	3.45	3.39	3.36	3.33	.025	
5.11	4.96	4.81	4.73	4.65	4.57	4.48	4.40	4.35	4.31	.01	
6.23	6.03	5.83	5.73	5.62	5.52	5.41	5.30	5.24	5.19	.005	
9.57	9.24	8.90	8.72	8.55	8.37	8.19	8.00	7.91	7.81	.001	
1.54	1.53	1.52	1.52	1.51	1.51	1.50	1.49	1.49	1.48	.25	10
2.28	2.24	2.20	2.18	2.16	2.13	2.11	2.08	2.07	2.06	.10	
2.91	2.85	2.77	2.74	2.70	2.66	2.62	2.58	2.56	2.54	.05	
3.62	3.52	3.42	3.37	3.31	3.26	3.20	3.14	3.11	3.08	.025	
4.71	4.56	4.41	4.33	4.25	4.17	4.08	4.00	3.95	3.91	.01	
5.66	5.47	5.27	5.17	5.07	4.97	4.86	4.75	4.69	4.64	.005	
8.45	8.13	7.80	7.64	7.47	7.30	7.12	6.94	6.85	6.76	.001	
1.51	1.50	1.49	1.49	1.48	1.47	1.47	1.46	1.45	1.45	.25	11
2.21	2.17	2.12	2.10	2.08	2.05	2.03	2.00	1.99	1.97	.10	
2.79	2.72	2.65	2.61	2.57	2.53	2.49	2.45	2.43	2.40	.05	
3.43	3.33	3.23	3.17	3.12	3.06	3.00	2.94	2.91	2.88	.025	
4.40	4.25	4.10	4.02	3.94	3.86	3.78	3.69	3.65	3.60	.01	
5.24	5.05	4.86	4.76	4.65	4.55	4.45	4.34	4.28	4.23	.005	
7.63	7.32	7.01	6.85	6.68	6.52	6.35	6.18	6.09	6.00	.001	
1.49	1.48	1.47	1.46	1.45	1.45	1.44	1.43	1.43	1.42	.25	12
2.15	2.10	2.06	2.04	2.01	1.99	1.96	1.93	1.92	1.90	.10	
2.69	2.62	2.54	2.51	2.47	2.43	2.38	2.34	2.32	2.30	.05	
3.28	3.18	3.07	3.02	2.96	2.91	2.85	2.79	2.76	2.72	.025	
4.16	4.01	3.86	3.78	3.70	3.62	3.54	3.45	3.41	3.36	.01	
4.91	4.72	4.53	4.43	4.33	4.23	4.12	4.01	3.96	3.90	.005	
7.00	6.71	6.40	6.25	6.09	5.93	5.76	5.59	5.51	5.42	.001	

Table 6 cont'd

Percentage points of the F-distribution (df_2 between 13 and 18)

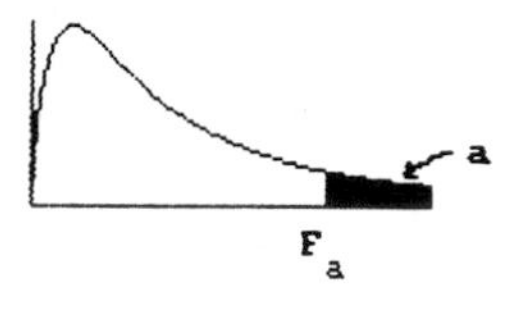

df_2	a	1	2	3	4	5	6	7	8	9	10
		df_1									
13	.25	1.45	1.55	1.55	1.53	1.52	1.51	1.50	1.49	1.49	1.48
	.10	3.14	2.76	2.56	2.43	2.35	2.28	2.23	2.20	2.16	2.14
	.05	4.67	3.81	3.41	3.18	3.03	2.92	2.83	2.77	2.71	2.67
	.025	6.41	4.97	4.35	4.00	3.77	3.60	3.48	3.39	3.31	3.25
	.01	9.07	6.70	5.74	5.21	4.86	4.62	4.44	4.30	4.19	4.10
	.005	11.37	8.19	6.93	6.23	5.79	5.48	5.25	5.08	4.94	4.82
	.001	17.82	12.31	10.21	9.07	8.35	7.86	7.49	7.21	6.98	6.80
14	.25	1.44	1.53	1.53	1.52	1.51	1.50	1.49	1.48	1.47	1.46
	.10	3.10	2.73	2.52	2.39	2.31	2.24	2.19	2.15	2.12	2.10
	.05	4.60	3.74	3.34	3.11	2.96	2.85	2.76	2.70	2.65	2.60
	.025	6.30	4.86	4.24	3.89	3.66	3.50	3.38	3.29	3.21	3.15
	.01	8.86	6.51	5.56	5.04	4.69	4.46	4.28	4.14	4.03	3.94
	.005	11.06	7.92	6.68	6.00	5.56	5.26	5.03	4.86	4.72	4.60
	.001	17.14	11.78	9.73	8.62	7.92	7.44	7.08	6.80	6.58	6.40
15	.25	1.43	1.52	1.52	1.51	1.49	1.48	1.47	1.46	1.46	1.45
	.10	3.07	2.70	2.49	2.36	2.27	2.21	2.16	2.12	2.09	2.06
	.05	4.54	3.68	3.29	3.06	2.90	2.79	2.71	2.64	2.59	2.54
	.025	6.20	4.77	4.15	3.80	3.58	3.41	3.29	3.20	3.12	3.06
	.01	8.68	6.36	5.42	4.89	4.56	4.32	4.14	4.00	3.89	3.80
	.005	10.80	7.70	6.48	5.80	5.37	5.07	4.85	4.67	4.54	4.42
	.001	16.59	11.34	9.34	8.25	7.57	7.09	6.74	6.47	6.26	6.08
16	.25	1.42	1.51	1.51	1.50	1.48	1.47	1.46	1.45	1.44	1.44
	.10	3.05	2.67	2.46	2.33	2.24	2.18	2.13	2.09	2.06	2.03
	.05	4.49	3.63	3.24	3.01	2.85	2.74	2.66	2.59	2.54	2.49
	.025	6.12	4.69	4.08	3.73	3.50	3.34	3.22	3.12	3.05	2.99
	.01	8.53	6.23	5.29	4.77	4.44	4.20	4.03	3.89	3.78	3.69
	.005	10.58	7.51	6.30	5.64	5.21	4.91	4.69	4.52	4.38	4.27
	.001	16.12	10.97	9.01	7.94	7.27	6.80	6.46	6.19	5.98	5.81
17	.25	1.42	1.51	1.50	1.49	1.47	1.46	1.45	1.44	1.43	1.43
	.10	3.03	2.64	2.44	2.31	2.22	2.15	2.10	2.06	2.03	2.00
	.05	4.45	3.59	3.20	2.96	2.81	2.70	2.61	2.55	2.49	2.45
	.025	6.04	4.62	4.01	3.66	3.44	3.28	3.16	3.06	2.98	2.92
	.01	8.40	6.11	5.18	4.67	4.34	4.10	3.93	3.79	3.68	3.59
	.005	10.38	7.35	6.16	5.50	5.07	4.78	4.56	4.39	4.25	4.14
	.001	15.72	10.66	8.73	7.68	7.02	6.56	6.22	5.96	5.75	5.58
18	.25	1.41	1.50	1.49	1.48	1.46	1.45	1.44	1.43	1.42	1.42
	.10	3.01	2.62	2.42	2.29	2.20	2.13	2.08	2.04	2.00	1.98
	.05	4.41	3.55	3.16	2.93	2.77	2.66	2.58	2.51	2.46	2.41
	.025	5.98	4.56	3.95	3.61	3.38	3.22	3.10	3.01	2.93	2.87
	.01	8.29	6.01	5.09	4.58	4.25	4.01	3.84	3.71	3.60	3.51
	.005	10.22	7.21	6.03	5.37	4.96	4.66	4.44	4.28	4.14	4.03
	.001	15.38	10.39	8.49	7.46	6.81	6.35	6.02	5.76	5.56	5.39

Table 6 cont'd

Percentage points of the F-distribution (df_2 between 13 and 18)

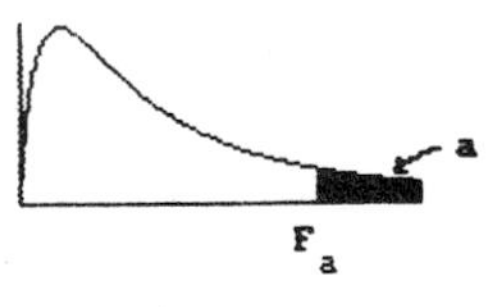

12	15	20	24	30	40	60	120	240	inf.	a	df_2
					df_1						
1.47	1.46	1.45	1.44	1.43	1.42	1.42	1.41	1.40	1.40	.25	13
2.10	2.05	2.01	1.98	1.96	1.93	1.90	1.88	1.86	1.85	.10	
2.60	2.53	2.46	2.42	2.38	2.34	2.30	2.25	2.23	2.21	.05	
3.15	3.05	2.95	2.89	2.84	2.78	2.72	2.66	2.63	2.60	.025	
3.96	3.82	3.66	3.59	3.51	3.43	3.34	3.25	3.21	3.17	.01	
4.64	4.46	4.27	4.17	4.07	3.97	3.87	3.76	3.70	3.65	.005	
6.52	6.23	5.93	5.78	5.63	5.47	5.30	5.14	5.05	4.97	.001	
1.45	1.44	1.43	1.42	1.41	1.41	1.40	1.39	1.38	1.38	.25	14
2.05	2.01	1.96	1.94	1.91	1.89	1.86	1.83	1.81	1.80	.10	
2.53	2.46	2.39	2.35	2.31	2.27	2.22	2.18	2.15	2.13	.05	
3.05	2.95	2.84	2.79	2.73	2.67	2.61	2.55	2.52	2.49	.025	
3.80	3.66	3.51	3.43	3.35	3.27	3.18	3.09	3.05	3.00	.01	
4.43	4.25	4.06	3.96	3.86	3.76	3.66	3.55	3.49	3.44	.005	
6.13	5.85	5.56	5.41	5.25	5.10	4.94	4.77	4.69	4.60	.001	
1.44	1.43	1.41	1.41	1.40	1.39	1.38	1.37	1.36	1.36	.25	15
2.02	1.97	1.92	1.90	1.87	1.85	1.82	1.79	1.77	1.76	.10	
2.48	2.40	2.33	2.29	2.25	2.20	2.16	2.11	2.09	2.07	.05	
2.96	2.86	2.76	2.70	2.64	2.59	2.52	2.46	2.43	2.40	.025	
3.67	3.52	3.37	3.29	3.21	3.13	3.05	2.96	2.91	2.87	.01	
4.25	4.07	3.88	3.79	3.69	3.58	3.48	3.37	3.32	3.26	.005	
5.81	5.54	5.25	5.10	4.95	4.80	4.64	4.47	4.39	4.31	.001	
1.43	1.41	1.40	1.39	1.38	1.37	1.36	1.35	1.35	1.34	.25	16
1.99	1.94	1.89	1.87	1.84	1.81	1.78	1.75	1.73	1.72	.10	
2.42	2.35	2.28	2.24	2.19	2.15	2.11	2.06	2.03	2.01	.05	
2.89	2.79	2.68	2.63	2.57	2.51	2.45	2.38	2.35	2.32	.025	
3.55	3.41	3.26	3.18	3.10	3.02	2.93	2.84	2.80	2.75	.01	
4.10	3.92	3.73	3.64	3.54	3.44	3.33	3.22	3.17	3.11	.005	
5.55	5.27	4.99	4.85	4.70	4.54	4.39	4.23	4.14	4.06	.001	
1.41	1.40	1.39	1.38	1.37	1.36	1.35	1.34	1.33	1.33	.25	17
1.96	1.91	1.86	1.84	1.81	1.78	1.75	1.72	1.70	1.69	.10	
2.38	2.31	2.23	2.19	2.15	2.10	2.06	2.01	1.99	1.96	.05	
2.82	2.72	2.62	2.56	2.50	2.44	2.38	2.32	2.28	2.25	.025	
3.46	3.31	3.16	3.08	3.00	2.92	2.83	2.75	2.70	2.65	.01	
3.97	3.79	3.61	3.51	3.41	3.31	3.21	3.10	3.04	2.98	.005	
5.32	5.05	4.78	4.63	4.48	4.33	4.18	4.02	3.93	3.85	.001	
1.40	1.39	1.38	1.37	1.36	1.35	1.34	1.33	1.32	1.32	.25	18
1.93	1.89	1.84	1.81	1.78	1.75	1.72	1.69	1.67	1.66	.10	
2.34	2.27	2.19	2.15	2.11	2.06	2.02	1.97	1.94	1.92	.05	
2.77	2.67	2.56	2.50	2.44	2.38	2.32	2.26	2.22	2.19	.025	
3.37	3.23	3.08	3.00	2.92	2.84	2.75	2.66	2.61	2.57	.01	
3.86	3.68	3.50	3.40	3.30	3.20	3.10	2.99	2.93	2.87	.005	
5.13	4.87	4.59	4.45	4.30	4.15	4.00	3.84	3.75	3.67	.001	

Table 6 cont'd

Percentage points of the F-distribution (df_2 between 19 and 24)

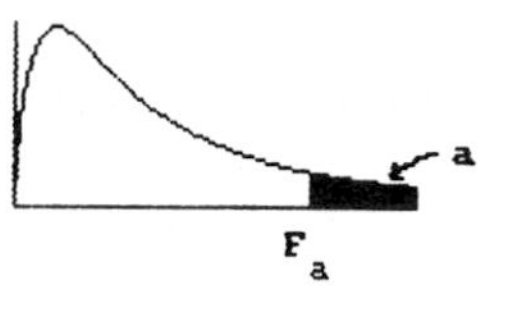

df_2	a	df_1 = 1	2	3	4	5	6	7	8	9	10
19	.25	1.41	1.49	1.49	1.47	1.46	1.44	1.43	1.42	1.41	1.41
	.10	2.99	2.61	2.40	2.27	2.18	2.11	2.06	2.02	1.98	1.96
	.05	4.38	3.52	3.13	2.90	2.74	2.63	2.54	2.48	2.42	2.38
	.025	5.92	4.51	3.90	3.56	3.33	3.17	3.05	2.96	2.88	2.82
	.01	8.18	5.93	5.01	4.50	4.17	3.94	3.77	3.63	3.52	3.43
	.005	10.07	7.09	5.92	5.27	4.85	4.56	4.34	4.18	4.04	3.93
	.001	15.08	10.16	8.28	7.27	6.62	6.18	5.85	5.59	5.39	5.22
20	.25	1.40	1.49	1.48	1.47	1.45	1.44	1.43	1.42	1.41	1.40
	.10	2.97	2.59	2.38	2.25	2.16	2.09	2.04	2.00	1.96	1.94
	.05	4.35	3.49	3.10	2.87	2.71	2.60	2.51	2.45	2.39	2.35
	.025	5.87	4.46	3.86	3.51	3.29	3.13	3.01	2.91	2.84	2.77
	.01	8.10	5.85	4.94	4.43	4.10	3.87	3.70	3.56	3.46	3.37
	.005	9.94	6.99	5.82	5.17	4.76	4.47	4.26	4.09	3.96	3.85
	.001	14.82	9.95	8.10	7.10	6.46	6.02	5.69	5.44	5.24	5.08
21	.25	1.40	1.48	1.48	1.46	1.44	1.43	1.42	1.41	1.40	1.39
	.10	2.96	2.57	2.36	2.23	2.14	2.08	2.02	1.98	1.95	1.92
	.05	4.32	3.47	3.07	2.84	2.68	2.57	2.49	2.42	2.37	2.32
	.025	5.83	4.42	3.82	3.48	3.25	3.09	2.97	2.87	2.80	2.73
	.01	8.02	5.78	4.87	4.37	4.04	3.81	3.64	3.51	3.40	3.31
	.005	9.83	6.89	5.73	5.09	4.68	4.39	4.18	4.01	3.88	3.77
	.001	14.59	9.77	7.94	6.95	6.32	5.88	5.56	5.31	5.11	4.95
22	.25	1.40	1.48	1.47	1.45	1.44	1.42	1.41	1.40	1.39	1.39
	.10	2.95	2.56	2.35	2.22	2.13	2.06	2.01	1.97	1.93	1.90
	.05	4.30	3.44	3.05	2.82	2.66	2.55	2.46	2.40	2.34	2.30
	.025	5.79	4.38	3.78	3.44	3.22	3.05	2.93	2.84	2.76	2.70
	.01	7.95	5.72	4.82	4.31	3.99	3.76	3.59	3.45	3.35	3.26
	.005	9.73	6.81	5.65	5.02	4.61	4.32	4.11	3.94	3.81	3.70
	.001	14.38	9.61	7.80	6.81	6.19	5.76	5.44	5.19	4.99	4.83
23	.25	1.39	1.47	1.47	1.45	1.43	1.42	1.41	1.40	1.39	1.38
	.10	2.94	2.55	2.34	2.21	2.11	2.05	1.99	1.95	1.92	1.89
	.05	4.28	3.42	3.03	2.80	2.64	2.53	2.44	2.37	2.32	2.27
	.025	5.75	4.35	3.75	3.41	3.18	3.02	2.90	2.81	2.73	2.67
	.01	7.88	5.66	4.76	4.26	3.94	3.71	3.54	3.41	3.30	3.21
	.005	9.63	6.73	5.58	4.95	4.54	4.26	4.05	3.88	3.75	3.64
	.001	14.20	9.47	7.67	6.70	6.08	5.65	5.33	5.09	4.89	4.73
24	.25	1.39	1.47	1.46	1.44	1.43	1.41	1.40	1.39	1.38	1.38
	.10	2.93	2.54	2.33	2.19	2.10	2.04	1.98	1.94	1.91	1.88
	.05	4.26	3.40	3.01	2.78	2.62	2.51	2.42	2.36	2.30	2.25
	.025	5.72	4.32	3.72	3.38	3.15	2.99	2.87	2.78	2.70	2.64
	.01	7.82	5.61	4.72	4.22	3.90	3.67	3.50	3.36	3.26	3.17
	.005	9.55	6.66	5.52	4.89	4.49	4.20	3.99	3.83	3.69	3.59
	.001	14.03	9.34	7.55	6.59	5.98	5.55	5.23	4.99	4.80	4.64

Table 6 cont'd

Percentage points of the F-distribution (df_2 between 19 and 24)

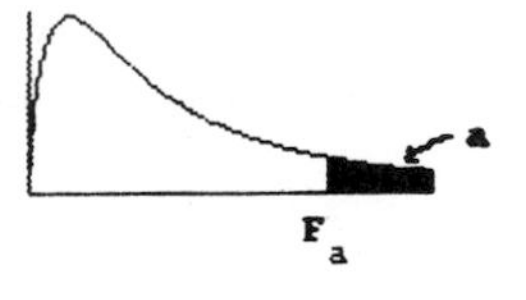

df_1											
12	**15**	**20**	**24**	**30**	**40**	**60**	**120**	**240**	**inf.**	**a**	**df_2**
1.40	1.38	1.37	1.36	1.35	1.34	1.33	1.32	1.31	1.30	.25	**19**
1.91	1.86	1.81	1.79	1.76	1.73	1.70	1.67	1.65	1.63	.10	
2.31	2.23	2.16	2.11	2.07	2.03	1.98	1.93	1.90	1.88	.05	
2.72	2.62	2.51	2.45	2.39	2.33	2.27	2.20	2.17	2.13	.025	
3.30	3.15	3.00	2.92	2.84	2.76	2.67	2.58	2.54	2.49	.01	
3.76	3.59	3.40	3.31	3.21	3.11	3.00	2.89	2.83	2.78	.005	
4.97	4.70	4.43	4.29	4.14	3.99	3.84	3.68	3.60	3.51	.001	
1.39	1.37	1.36	1.35	1.34	1.33	1.32	1.31	1.30	1.29	.25	**20**
1.89	1.84	1.79	1.77	1.74	1.71	1.68	1.64	1.63	1.61	.10	
2.28	2.20	2.12	2.08	2.04	1.99	1.95	1.90	1.87	1.84	.05	
2.68	2.57	2.46	2.41	2.35	2.29	2.22	2.16	2.12	2.09	.025	
3.23	3.09	2.94	2.86	2.78	2.69	2.61	2.52	2.47	2.42	.01	
3.68	3.50	3.32	3.22	3.12	3.02	2.92	2.81	2.75	2.69	.005	
4.82	4.56	4.29	4.15	4.00	3.86	3.70	3.54	3.46	3.38	.001	
1.38	1.37	1.35	1.34	1.33	1.32	1.31	1.30	1.29	1.28	.25	**21**
1.87	1.83	1.78	1.75	1.72	1.69	1.66	1.62	1.60	1.59	.10	
2.25	2.18	2.10	2.05	2.01	1.96	1.92	1.87	1.84	1.81	.05	
2.64	2.53	2.42	2.37	2.31	2.25	2.18	2.11	2.08	2.04	.025	
3.17	3.03	2.88	2.80	2.72	2.64	2.55	2.46	2.41	2.36	.01	
3.60	3.43	3.24	3.15	3.05	2.95	2.84	2.73	2.67	2.61	.005	
4.70	4.44	4.17	4.03	3.88	3.74	3.58	3.42	3.34	3.26	.001	
1.37	1.36	1.34	1.33	1.32	1.31	1.30	1.29	1.28	1.28	.25	**22**
1.86	1.81	1.76	1.73	1.70	1.67	1.64	1.60	1.59	1.57	.10	
2.23	2.15	2.07	2.03	1.98	1.94	1.89	1.84	1.81	1.78	.05	
2.60	2.50	2.39	2.33	2.27	2.21	2.14	2.08	2.04	2.00	.025	
3.12	2.98	2.83	2.75	2.67	2.58	2.50	2.40	2.35	2.31	.01	
3.54	3.36	3.18	3.08	2.98	2.88	2.77	2.66	2.60	2.55	.005	
4.58	4.33	4.06	3.92	3.78	3.63	3.48	3.32	3.23	3.15	.001	
1.37	1.35	1.34	1.33	1.32	1.31	1.30	1.28	1.28	1.27	.25	**23**
1.84	1.80	1.74	1.72	1.69	1.66	1.62	1.59	1.57	1.55	.10	
2.20	2.13	2.05	2.01	1.96	1.91	1.86	1.81	1.79	1.76	.05	
2.57	2.47	2.36	2.30	2.24	2.18	2.11	2.04	2.01	1.97	.025	
3.07	2.93	2.78	2.70	2.62	2.54	2.45	2.35	2.31	2.26	.01	
3.47	3.30	3.12	3.02	2.92	2.82	2.71	2.60	2.54	2.48	.005	
4.48	4.23	3.96	3.82	3.68	3.53	3.38	3.22	3.14	3.05	.001	
1.36	1.35	1.33	1.32	1.31	1.30	1.29	1.28	1.27	1.26	.25	**24**
1.83	1.78	1.73	1.70	1.67	1.64	1.61	1.57	1.55	1.53	.10	
2.18	2.11	2.03	1.98	1.94	1.89	1.84	1.79	1.76	1.73	.05	
2.54	2.44	2.33	2.27	2.21	2.15	2.08	2.01	1.97	1.94	.025	
3.03	2.89	2.74	2.66	2.58	2.49	2.40	2.31	2.26	2.21	.01	
3.42	3.25	3.06	2.97	2.87	2.77	2.66	2.55	2.49	2.43	.005	
4.39	4.14	3.87	3.74	3.59	3.45	3.29	3.14	3.05	2.97	.001	

Table 6 cont'd

Percentage points of the F-distribution (df_2 between 25 and 30)

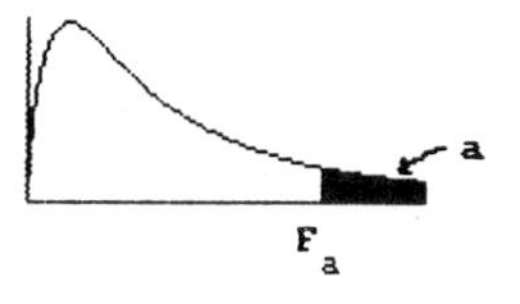

df_2	a	df_1 = 1	2	3	4	5	6	7	8	9	10
25	.25	1.39	1.47	1.46	1.44	1.42	1.41	1.40	1.39	1.38	1.37
	.10	2.92	2.53	2.32	2.18	2.09	2.02	1.97	1.93	1.89	1.87
	.05	4.24	3.39	2.99	2.76	2.60	2.49	2.40	2.34	2.28	2.24
	.025	5.69	4.29	3.69	3.35	3.13	2.97	2.85	2.75	2.68	2.61
	.01	7.77	5.57	4.68	4.18	3.85	3.63	3.46	3.32	3.22	3.13
	.005	9.48	6.60	5.46	4.84	4.43	4.15	3.94	3.78	3.64	3.54
	.001	13.88	9.22	7.45	6.49	5.89	5.46	5.15	4.91	4.71	4.56
26	.25	1.38	1.46	1.45	1.44	1.42	1.41	1.39	1.38	1.37	1.37
	.10	2.91	2.52	2.31	2.17	2.08	2.01	1.96	1.92	1.88	1.86
	.05	4.23	3.37	2.98	2.74	2.59	2.47	2.39	2.32	2.27	2.22
	.025	5.66	4.27	3.67	3.33	3.10	2.94	2.82	2.73	2.65	2.59
	.01	7.72	5.53	4.64	4.14	3.82	3.59	3.42	3.29	3.18	3.09
	.005	9.41	6.54	5.41	4.79	4.38	4.10	3.89	3.73	3.60	3.49
	.001	13.74	9.12	7.36	6.41	5.80	5.38	5.07	4.83	4.64	4.48
27	.25	1.38	1.46	1.45	1.43	1.42	1.40	1.39	1.38	1.37	1.36
	.10	2.90	2.51	2.30	2.17	2.07	2.00	1.95	1.91	1.87	1.85
	.05	4.21	3.35	2.96	2.73	2.57	2.46	2.37	2.31	2.25	2.20
	.025	5.63	4.24	3.65	3.31	3.08	2.92	2.80	2.71	2.63	2.57
	.01	7.68	5.49	4.60	4.11	3.78	3.56	3.39	3.26	3.15	3.06
	.005	9.34	6.49	5.36	4.74	4.34	4.06	3.85	3.69	3.56	3.45
	.001	13.61	9.02	7.27	6.33	5.73	5.31	5.00	4.76	4.57	4.41
28	.25	1.38	1.46	1.45	1.43	1.41	1.40	1.39	1.38	1.37	1.36
	.10	2.89	2.50	2.29	2.16	2.06	2.00	1.94	1.90	1.87	1.84
	.05	4.20	3.34	2.95	2.71	2.56	2.45	2.36	2.29	2.24	2.19
	.025	5.61	4.22	3.63	3.29	3.06	2.90	2.78	2.69	2.61	2.55
	.01	7.64	5.45	4.57	4.07	3.75	3.53	3.36	3.23	3.12	3.03
	.005	9.28	6.44	5.32	4.70	4.30	4.02	3.81	3.65	3.52	3.41
	.001	13.50	8.93	7.19	6.25	5.66	5.24	4.93	4.69	4.50	4.35
29	.25	1.38	1.45	1.45	1.43	1.41	1.40	1.38	1.37	1.36	1.35
	.10	2.89	2.50	2.28	2.15	2.06	1.99	1.93	1.89	1.86	1.83
	.05	4.18	3.33	2.93	2.70	2.55	2.43	2.35	2.28	2.22	2.18
	.025	5.59	4.20	3.61	3.27	3.04	2.88	2.76	2.67	2.59	2.53
	.01	7.60	5.42	4.54	4.04	3.73	3.50	3.33	3.20	3.09	3.00
	.005	9.23	6.40	5.28	4.66	4.26	3.98	3.77	3.61	3.48	3.38
	.001	13.39	8.85	7.12	6.19	5.59	5.18	4.87	4.64	4.45	4.29
30	.25	1.38	1.45	1.44	1.42	1.41	1.39	1.38	1.37	1.36	1.35
	.10	2.88	2.49	2.28	2.14	2.05	1.98	1.93	1.88	1.85	1.82
	.05	4.17	3.32	2.92	2.69	2.53	2.42	2.33	2.27	2.21	2.16
	.025	5.57	4.18	3.59	3.25	3.03	2.87	2.75	2.65	2.57	2.51
	.01	7.56	5.39	4.51	4.02	3.70	3.47	3.30	3.17	3.07	2.98
	.005	9.18	6.35	5.24	4.62	4.23	3.95	3.74	3.58	3.45	3.34
	.001	13.29	8.77	7.05	6.12	5.53	5.12	4.82	4.58	4.39	4.24

Table 6 cont'd

Percentage points of the F-distribution (df_2 between 25 and 30)

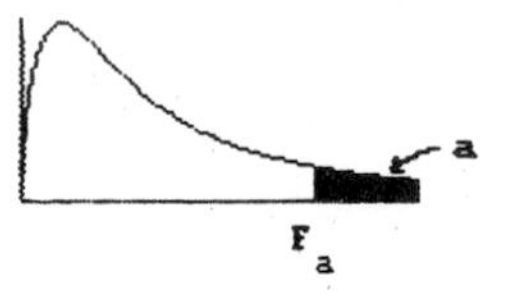

df_1: 12	15	20	24	30	40	60	120	240	inf.	a	df_2
1.36	1.34	1.33	1.32	1.31	1.29	1.28	1.27	1.26	1.25	.25	**25**
1.82	1.77	1.72	1.69	1.66	1.63	1.59	1.56	1.54	1.52	.10	
2.16	2.09	2.01	1.96	1.92	1.87	1.82	1.77	1.74	1.71	.05	
2.51	2.41	2.30	2.24	2.18	2.12	2.05	1.98	1.94	1.91	.025	
2.99	2.85	2.70	2.62	2.54	2.45	2.36	2.27	2.22	2.17	.01	
3.37	3.20	3.01	2.92	2.82	2.72	2.61	2.50	2.44	2.38	.005	
4.31	4.06	3.79	3.66	3.52	3.37	3.22	3.06	2.98	2.89	.001	
1.35	1.34	1.32	1.31	1.30	1.29	1.28	1.26	1.26	1.25	.25	**26**
1.81	1.76	1.71	1.68	1.65	1.61	1.58	1.54	1.52	1.50	.10	
2.15	2.07	1.99	1.95	1.90	1.85	1.80	1.75	1.72	1.69	.05	
2.49	2.39	2.28	2.22	2.16	2.09	2.03	1.95	1.92	1.88	.025	
2.96	2.81	2.66	2.58	2.50	2.42	2.33	2.23	2.18	2.13	.01	
3.33	3.15	2.97	2.87	2.77	2.67	2.56	2.45	2.39	2.33	.005	
4.24	3.99	3.72	3.59	3.44	3.30	3.15	2.99	2.90	2.82	.001	
1.35	1.33	1.32	1.31	1.30	1.28	1.27	1.26	1.25	1.24	.25	**27**
1.80	1.75	1.70	1.67	1.64	1.60	1.57	1.53	1.51	1.49	.10	
2.13	2.06	1.97	1.93	1.88	1.84	1.79	1.73	1.70	1.67	.05	
2.47	2.36	2.25	2.19	2.13	2.07	2.00	1.93	1.89	1.85	.025	
2.93	2.78	2.63	2.55	2.47	2.38	2.29	2.20	2.15	2.10	.01	
3.28	3.11	2.93	2.83	2.73	2.63	2.52	2.41	2.35	2.29	.005	
4.17	3.92	3.66	3.52	3.38	3.23	3.08	2.92	2.84	2.75	.001	
1.34	1.33	1.31	1.30	1.29	1.28	1.27	1.25	1.24	1.24	.25	**28**
1.79	1.74	1.69	1.66	1.63	1.59	1.56	1.52	1.50	1.48	.10	
2.12	2.04	1.96	1.91	1.87	1.82	1.77	1.71	1.68	1.65	.05	
2.45	2.34	2.23	2.17	2.11	2.05	1.98	1.91	1.87	1.83	.025	
2.90	2.75	2.60	2.52	2.44	2.35	2.26	2.17	2.12	2.06	.01	
3.25	3.07	2.89	2.79	2.69	2.59	2.48	2.37	2.31	2.25	.005	
4.11	3.86	3.60	3.46	3.32	3.18	3.02	2.86	2.78	2.69	.001	
1.34	1.32	1.31	1.30	1.29	1.27	1.26	1.25	1.24	1.23	.25	**29**
1.78	1.73	1.68	1.65	1.62	1.58	1.55	1.51	1.49	1.47	.10	
2.10	2.03	1.94	1.90	1.85	1.81	1.75	1.70	1.67	1.64	.05	
2.43	2.32	2.21	2.15	2.09	2.03	1.96	1.89	1.85	1.81	.025	
2.87	2.73	2.57	2.49	2.41	2.33	2.23	2.14	2.09	2.03	.01	
3.21	3.04	2.86	2.76	2.66	2.56	2.45	2.33	2.27	2.21	.005	
4.05	3.80	3.54	3.41	3.27	3.12	2.97	2.81	2.73	2.64	.001	
1.34	1.32	1.30	1.29	1.28	1.27	1.26	1.24	1.23	1.23	.25	**30**
1.77	1.72	1.67	1.64	1.61	1.57	1.54	1.50	1.48	1.46	.10	
2.09	2.01	1.93	1.89	1.84	1.79	1.74	1.68	1.65	1.62	.05	
2.41	2.31	2.20	2.14	2.07	2.01	1.94	1.87	1.83	1.79	.025	
2.84	2.70	2.55	2.47	2.39	2.30	2.21	2.11	2.06	2.01	.01	
3.18	3.01	2.82	2.73	2.63	2.52	2.42	2.30	2.24	2.18	.005	
4.00	3.75	3.49	3.36	3.22	3.07	2.92	2.76	2.68	2.59	.001	

Table 6 cont'd

Percentage points of the F-distribution (df_2 at least 40)

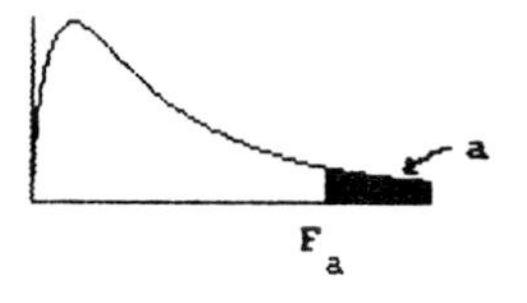

df_2	a	df_1 = 1	2	3	4	5	6	7	8	9	10
40	.25	1.36	1.44	1.42	1.40	1.39	1.37	1.36	1.35	1.34	1.33
	.10	2.84	2.44	2.23	2.09	2.00	1.93	1.87	1.83	1.79	1.76
	.05	4.08	3.23	2.84	2.61	2.45	2.34	2.25	2.18	2.12	2.08
	.025	5.42	4.05	3.46	3.13	2.90	2.74	2.62	2.53	2.45	2.39
	.01	7.31	5.18	4.31	3.83	3.51	3.29	3.12	2.99	2.89	2.80
	.005	8.83	6.07	4.98	4.37	3.99	3.71	3.51	3.35	3.22	3.12
	.001	12.61	8.25	6.59	5.70	5.13	4.73	4.44	4.21	4.02	3.87
60	.25	1.35	1.42	1.41	1.38	1.37	1.35	1.33	1.32	1.31	1.30
	.10	2.79	2.39	2.18	2.04	1.95	1.87	1.82	1.77	1.74	1.71
	.05	4.00	3.15	2.76	2.53	2.37	2.25	2.17	2.10	2.04	1.99
	.025	5.29	3.93	3.34	3.01	2.79	2.63	2.51	2.41	2.33	2.27
	.01	7.08	4.98	4.13	3.65	3.34	3.12	2.95	2.82	2.72	2.63
	.005	8.49	5.79	4.73	4.14	3.76	3.49	3.29	3.13	3.01	2.90
	.001	11.97	7.77	6.17	5.31	4.76	4.37	4.09	3.86	3.69	3.54
90	.25	1.34	1.41	1.39	1.37	1.35	1.33	1.32	1.31	1.30	1.29
	.10	2.76	2.36	2.15	2.01	1.91	1.84	1.78	1.74	1.70	1.67
	.05	3.95	3.10	2.71	2.47	2.32	2.20	2.11	2.04	1.99	1.94
	.025	5.20	3.84	3.26	2.93	2.71	2.55	2.43	2.34	2.26	2.19
	.01	6.93	4.85	4.01	3.53	3.23	3.01	2.84	2.72	2.61	2.52
	.005	8.28	5.62	4.57	3.99	3.62	3.35	3.15	3.00	2.87	2.77
	.001	11.57	7.47	5.91	5.06	4.53	4.15	3.87	3.65	3.48	3.34
120	.25	1.34	1.40	1.39	1.37	1.35	1.33	1.31	1.30	1.29	1.28
	.10	2.75	2.35	2.13	1.99	1.90	1.82	1.77	1.72	1.68	1.65
	.05	3.92	3.07	2.68	2.45	2.29	2.18	2.09	2.02	1.96	1.91
	.025	5.15	3.80	3.23	2.89	2.67	2.52	2.39	2.30	2.22	2.16
	.01	6.85	4.79	3.95	3.48	3.17	2.96	2.79	2.66	2.56	2.47
	.005	8.18	5.54	4.50	3.92	3.55	3.28	3.09	2.93	2.81	2.71
	.001	11.38	7.32	5.78	4.95	4.42	4.04	3.77	3.55	3.38	3.24
240	.25	1.33	1.39	1.38	1.36	1.34	1.32	1.30	1.29	1.27	1.27
	.10	2.73	2.32	2.10	1.97	1.87	1.80	1.74	1.70	1.65	1.63
	.05	3.88	3.03	2.64	2.41	2.25	2.14	2.04	1.98	1.92	1.87
	.025	5.09	3.75	3.17	2.84	2.62	2.46	2.34	2.25	2.17	2.10
	.01	6.74	4.69	3.86	3.40	3.09	2.88	2.71	2.59	2.48	2.40
	.005	8.03	5.42	4.38	3.82	3.45	3.19	2.99	2.84	2.71	2.61
	.001	11.10	7.11	5.60	4.78	4.25	3.89	3.62	3.41	3.24	3.09
inf.	.25	1.32	1.39	1.37	1.35	1.33	1.31	1.29	1.28	1.27	1.25
	.10	2.71	2.30	2.08	1.94	1.85	1.77	1.72	1.67	1.63	1.60
	.05	3.84	3.00	2.60	2.37	2.21	2.10	2.01	1.94	1.88	1.83
	.025	5.02	3.69	3.12	2.79	2.57	2.41	2.29	2.19	2.11	2.05
	.01	6.63	4.61	3.78	3.32	3.02	2.80	2.64	2.51	2.41	2.32
	.005	7.88	5.30	4.28	3.72	3.35	3.09	2.90	2.74	2.62	2.52
	.001	10.83	6.91	5.42	4.62	4.10	3.74	3.47	3.27	3.10	2.96

Table 6 cont'd

Percentage points of the F-distribution (df_2 at least 40)

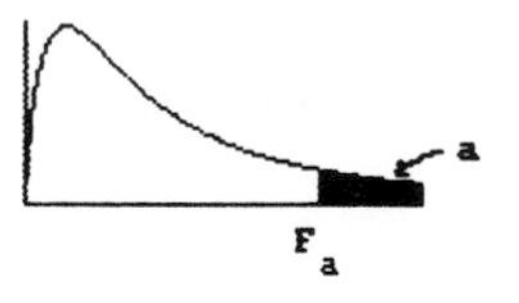

df_1 12	15	20	24	30	40	60	120	240	inf.	a	df_2
1.31	1.30	1.28	1.26	1.25	1.24	1.22	1.21	1.20	1.19	.25	**40**
1.71	1.66	1.61	1.57	1.54	1.51	1.47	1.42	1.40	1.38	.10	
2.00	1.92	1.84	1.79	1.74	1.69	1.64	1.58	1.54	1.51	.05	
2.29	2.18	2.07	2.01	1.94	1.88	1.80	1.72	1.68	1.64	.025	
2.66	2.52	2.37	2.29	2.20	2.11	2.02	1.92	1.86	1.80	.01	
2.95	2.78	2.60	2.50	2.40	2.30	2.18	2.06	2.00	1.93	.005	
3.64	3.40	3.14	3.01	2.87	2.73	2.57	2.41	2.32	2.23	.001	
1.29	1.27	1.25	1.24	1.22	1.21	1.19	1.17	1.16	1.15	.25	**60**
1.66	1.60	1.54	1.51	1.48	1.44	1.40	1.35	1.32	1.29	.10	
1.92	1.84	1.75	1.70	1.65	1.59	1.53	1.47	1.43	1.39	.05	
2.17	2.06	1.94	1.88	1.82	1.74	1.67	1.58	1.53	1.48	.025	
2.50	2.35	2.20	2.12	2.03	1.94	1.84	1.73	1.67	1.60	.01	
2.74	2.57	2.39	2.29	2.19	2.08	1.96	1.83	1.76	1.69	.005	
3.32	3.08	2.83	2.69	2.55	2.41	2.25	2.08	1.99	1.89	.001	
1.27	1.25	1.23	1.22	1.20	1.19	1.17	1.15	1.13	1.12	.25	**90**
1.62	1.56	1.50	1.47	1.43	1.39	1.35	1.29	1.26	1.23	.10	
1.86	1.78	1.69	1.64	1.59	1.53	1.46	1.39	1.35	1.30	.05	
2.09	1.98	1.86	1.80	1.73	1.66	1.58	1.48	1.43	1.37	.025	
2.39	2.24	2.09	2.00	1.92	1.82	1.72	1.60	1.53	1.46	.01	
2.61	2.44	2.25	2.15	2.05	1.94	1.82	1.68	1.61	1.52	.005	
3.11	2.88	2.63	2.50	2.36	2.21	2.05	1.87	1.77	1.66	.001	
1.26	1.24	1.22	1.21	1.19	1.18	1.16	1.13	1.12	1.10	.25	**120**
1.60	1.55	1.48	1.45	1.41	1.37	1.32	1.26	1.23	1.19	.10	
1.83	1.75	1.66	1.61	1.55	1.50	1.43	1.35	1.31	1.25	.05	
2.05	1.94	1.82	1.76	1.69	1.61	1.53	1.43	1.38	1.31	.025	
2.34	2.19	2.03	1.95	1.86	1.76	1.66	1.53	1.46	1.38	.01	
2.54	2.37	2.19	2.09	1.98	1.87	1.75	1.61	1.52	1.43	.005	
3.02	2.78	2.53	2.40	2.26	2.11	1.95	1.77	1.66	1.54	.001	
1.25	1.23	1.21	1.19	1.18	1.16	1.14	1.11	1.09	1.07	.25	**240**
1.57	1.52	1.45	1.42	1.38	1.33	1.28	1.22	1.18	1.13	.10	
1.79	1.71	1.61	1.56	1.51	1.44	1.37	1.29	1.24	1.17	.05	
2.00	1.89	1.77	1.70	1.63	1.55	1.46	1.35	1.29	1.21	.025	
2.26	2.11	1.96	1.87	1.78	1.68	1.57	1.43	1.35	1.25	.01	
2.45	2.28	2.09	1.99	1.89	1.77	1.64	1.49	1.40	1.28	.005	
2.88	2.65	2.40	2.26	2.12	1.97	1.80	1.61	1.49	1.35	.001	
1.24	1.22	1.19	1.18	1.16	1.14	1.12	1.08	1.06	1.00	.25	**inf.**
1.55	1.49	1.42	1.38	1.34	1.30	1.24	1.17	1.12	1.00	.10	
1.75	1.67	1.57	1.52	1.46	1.39	1.32	1.22	1.15	1.00	.05	
1.94	1.83	1.71	1.64	1.57	1.48	1.39	1.27	1.19	1.00	.025	
2.18	2.04	1.88	1.79	1.70	1.59	1.47	1.32	1.22	1.00	.01	
2.36	2.19	2.00	1.90	1.79	1.67	1.53	1.36	1.25	1.00	.005	
2.74	2.51	2.27	2.13	1.99	1.84	1.66	1.45	1.31	1.00	.001	

Source: Computed by P. J. Hildebrand.

Table 7

Critical values for the Wilcoxon signed rank test (n=5(1)54)

One-sided	Two-sided	n=5	n=6	n=7	n=8	n=9	n=10	n=11	n=12	n=13	n=14
p=.1	p=.2	2	3	5	8	10	14	17	21	26	31
p=.05	p=.1	0	2	3	5	8	10	13	17	21	25
p=.025	p=.05		0	2	3	5	8	10	13	17	21
p=.01	p=.02			0	1	3	5	7	9	12	15
p=.005	p=.01				0	1	3	5	7	9	12
p=.0025	p=.005					0	1	3	5	7	9
p=.001	p=.002						0	1	2	4	6

One-sided	Two-sided	n=15	n=16	n=17	n=18	n=19	n=20	n=21	n=22	n=23	n=24
p=.1	p=.2	36	42	48	55	62	69	77	86	94	104
p=.05	p=.1	30	35	41	47	53	60	67	75	83	91
p=.025	p=.05	25	29	34	40	46	52	58	65	73	81
p=.01	p=.02	19	23	27	32	37	43	49	55	62	69
p=.005	p=.01	15	19	23	27	32	37	42	48	54	61
p=.0025	p=.005	12	15	19	23	27	32	37	42	48	54
p=.001	p=.002	8	11	14	18	21	26	30	35	40	45

One-sided	Two-sided	n=25	n=26	n=27	n=28	n=29	n=30	n=31	n=32	n=33	n=34
p=.1	p=.2	113	124	134	145	157	169	181	194	207	221
p=.05	p=.1	100	110	119	130	140	151	163	175	187	200
p=.025	p=.05	89	98	107	116	126	137	147	159	170	182
p=.01	p=.02	76	84	92	101	110	120	130	140	151	162
p=.005	p=.01	68	75	83	91	100	109	118	128	138	148
p=.0025	p=.005	60	67	74	82	90	98	107	116	126	136
p=.001	p=.002	51	58	64	71	79	86	94	103	112	121

One-sided	Two-sided	n=35	n=36	n=37	n=38	n=39	n=40	n=41	n=42	n=43	n=44
p=.1	p=.2	235	250	265	281	297	313	330	348	365	384
p=.05	p=.1	213	227	241	256	271	286	302	319	336	353
p=.025	p=.05	195	208	221	235	249	264	279	294	310	327
p=.01	p=.02	173	185	198	211	224	238	252	266	281	296
p=.005	p=.01	159	171	182	194	207	220	233	247	261	276
p=.0025	p=.005	146	157	168	180	192	204	217	230	244	258
p=.001	p=.002	131	141	151	162	173	185	197	209	222	235

One-sided	Two-sided	n=45	n=46	n=47	n=48	n=49	n=50	n=51	n=52	n=53	n=54
p=.1	p=.2	402	422	441	462	482	503	525	547	569	592
p=.05	p=.1	371	389	407	426	446	466	486	507	529	550
p=.025	p=.05	343	361	378	396	415	434	453	473	494	514
p=.01	p=.02	312	328	345	362	379	397	416	434	454	473
p=.005	p=.01	291	307	322	339	355	373	390	408	427	445
p=.0025	p=.005	272	287	302	318	334	350	367	384	402	420
p=.001	p=.002	249	263	277	292	307	323	339	355	372	389

Source: Computed by P. J. Hildebrand.

TABLE 8 Percentage Points of the Studentized Range

Error d.f.	α	t = number of treatment means 2	3	4	5	6	7	8	9	10	11
5	.05	3.64	4.60	5.22	5.67	6.03	6.33	6.58	6.80	6.99	7.17
	.01	5.70	6.98	7.80	8.42	8.91	9.32	9.67	9.97	10.24	10.48
6	.05	3.46	4.34	4.90	5.30	5.63	5.90	6.12	6.32	6.49	6.65
	.01	5.24	6.33	7.03	7.56	7.97	8.32	8.61	8.87	9.10	9.30
7	.05	3.34	4.16	4.68	5.06	5.36	5.61	5.82	6.00	6.16	6.30
	.01	4.95	5.92	6.54	7.01	7.37	7.68	7.94	8.17	8.37	8.55
8	.05	3.26	4.04	4.53	4.89	5.17	5.40	5.60	5.77	5.92	6.05
	.01	4.75	5.64	6.20	6.62	6.96	7.24	7.47	7.68	7.86	8.03
9	.05	3.20	3.95	4.41	4.76	5.02	5.24	5.43	5.59	5.74	5.87
	.01	4.60	5.43	5.96	6.35	6.66	6.91	7.13	7.33	7.49	7.65
10	.05	3.15	3.88	4.33	4.65	4.91	5.12	5.30	5.46	5.60	5.72
	.01	4.48	5.27	5.77	6.14	6.43	6.67	6.87	7.05	7.21	7.36
11	.05	3.11	3.82	4.26	4.57	4.82	5.03	5.30	5.35	5.49	5.61
	.01	4.39	5.15	5.62	5.97	6.25	6.48	6.67	6.84	6.99	7.13
12	.05	3.08	3.77	4.20	4.52	4.75	4.95	5.12	5.27	5.39	5.51
	.01	4.32	5.05	5.50	5.84	6.10	6.32	6.51	6.67	6.81	6.94
13	.05	3.06	3.73	4.15	4.45	4.69	4.88	5.05	5.19	5.32	5.43
	.01	4.26	4.96	5.40	5.73	5.98	6.19	6.37	6.53	6.67	6.79
14	.05	3.03	3.70	4.11	4.41	4.64	4.83	4.99	5.13	5.25	5.36
	.01	4.21	4.89	5.32	5.63	5.88	6.08	6.26	6.41	6.54	6.66
15	.05	3.01	3.67	4.08	4.37	4.59	4.78	4.94	5.08	5.20	5.31
	.01	4.17	4.84	5.25	5.56	5.80	5.99	6.16	6.31	6.44	6.55
16	.05	3.00	3.65	4.05	4.33	4.56	4.74	4.90	5.03	5.15	5.26
	.01	4.13	4.79	5.19	5.49	5.72	5.92	6.08	6.22	6.35	6.46
17	.05	2.98	3.63	4.02	4.30	4.52	4.70	4.86	4.99	5.11	5.21
	.01	4.10	4.74	5.14	5.43	5.66	5.85	6.01	6.15	6.27	6.38
18	.05	2.97	3.61	4.00	4.28	4.49	4.67	4.82	4.96	5.07	5.17
	.01	4.07	4.70	5.09	5.38	5.60	5.79	5.94	6.08	6.20	6.31
19	.05	2.96	3.59	3.98	4.25	4.47	4.65	4.79	4.92	5.04	5.14
	.01	4.05	4.67	5.05	5.33	5.55	5.73	5.89	6.02	6.14	6.25
20	.05	2.95	3.58	3.96	4.23	4.45	4.62	4.77	4.90	5.01	5.11
	.01	4.02	4.64	5.02	5.29	5.51	5.69	5.84	5.97	6.09	6.19
24	.05	2.92	3.53	3.90	4.17	4.37	4.54	4.68	4.81	3.92	5.01
	.01	3.96	4.55	4.91	5.17	5.37	5.54	5.69	5.81	5.92	6.02
30	.05	2.89	3.49	3.85	4.10	4.30	4.46	4.60	4.72	4.82	4.92
	.01	3.89	4.45	4.80	5.05	5.24	5.40	5.54	5.65	5.76	5.85
40	.05	2.86	3.44	3.79	4.04	4.23	4.39	4.52	4.63	4.73	4.82
	.01	3.82	4.37	4.70	4.93	5.11	5.26	5.39	5.50	5.60	5.69
60	.05	2.83	3.40	3.74	3.98	4.16	4.31	4.44	4.55	4.65	4.73
	.01	3.76	4.28	4.59	4.82	4.99	5.13	5.25	5.36	5.45	5.53
120	.05	2.80	3.36	3.68	3.92	4.10	4.24	4.36	4.47	4.56	4.64
	.01	3.70	4.20	4.50	4.71	4.87	5.01	5.12	5.21	5.30	5.37
∞	.05	2.77	3.31	3.63	3.86	4.03	4.17	4.29	4.39	4.47	4.55
	.01	3.64	4.12	4.40	4.60	4.76	4.88	4.99	5.08	5.16	5.23

This table is abridged from E. S. Pearson and H. O. Hartley, eds., *Biometrika Tables for Statisticians*, Vol. 1, 2d ed. (New York: Cambridge University Press, 1958), Table 29. Reproduced with the kind permission of the editors and the trustees of *Biometrika*.

TABLE 8 (*continued*)

t = number of treatment means										
12	13	14	15	16	17	18	19	20	α	Error d.f.
7.32	7.47	7.60	7.72	7.83	7.93	8.03	8.12	8.21	.05	5
10.70	10.89	11.08	11.24	11.40	11.55	11.68	11.81	11.93	.01	
6.79	6.92	7.03	7.14	7.24	7.34	7.43	7.51	7.59	.05	6
9.48	9.65	9.81	9.95	10.08	10.21	10.32	10.43	10.54	.01	
6.43	6.55	6.66	6.76	6.85	6.94	7.02	7.10	7.17	.05	7
8.71	8.86	9.00	9.12	9.24	9.35	9.46	9.55	9.65	.01	
6.18	6.29	6.39	6.48	6.57	6.65	6.73	6.80	6.87	.05	8
8.18	8.31	8.44	8.55	8.66	8.76	8.85	8.94	9.03	.01	
5.98	6.09	6.19	6.28	6.36	6.44	6.51	6.58	6.64	.05	9
7.78	7.91	8.03	8.13	8.23	8.33	8.41	8.49	8.57	.01	
5.83	5.93	6.03	6.11	6.19	6.27	6.34	6.40	6.47	.05	10
7.49	7.60	7.71	7.81	7.91	7.99	8.08	8.15	8.23	.01	
5.71	5.81	5.90	5.98	6.06	6.13	6.20	6.27	6.33	.05	11
7.25	7.36	7.46	7.56	7.65	7.73	7.81	7.88	7.95	.01	
5.61	5.71	5.80	5.88	5.95	6.02	6.09	6.15	6.21	.05	12
7.06	7.17	7.26	7.36	7.44	7.52	7.59	7.66	7.73	.01	
5.53	5.63	5.71	5.79	5.86	5.93	5.99	6.05	6.11	.05	13
6.90	7.01	7.10	7.19	7.27	7.35	7.42	7.48	7.55	.01	
5.46	5.55	5.64	5.71	5.79	5.85	5.91	5.97	6.03	.05	14
6.77	6.87	6.96	7.05	7.13	7.20	7.27	7.33	7.39	.01	
5.40	5.49	5.57	5.65	5.72	5.78	5.85	5.90	5.96	.05	15
6.66	6.76	6.84	6.93	7.00	7.07	7.14	7.20	7.26	.01	
5.35	5.44	5.52	5.59	5.66	5.73	5.79	5.84	5.90	.05	16
6.56	6.66	6.74	6.82	6.90	6.97	7.03	7.09	7.15	.01	
5.31	5.39	5.47	5.54	5.61	5.67	5.73	5.79	5.84	.05	17
6.48	6.57	6.66	6.73	6.81	6.87	6.94	7.00	7.05	.01	
5.27	5.35	5.43	5.50	5.57	5.63	5.69	5.74	5.79	.05	18
6.41	6.50	6.58	6.65	6.73	6.79	6.85	6.91	6.97	.01	
5.23	5.31	5.39	5.46	5.53	5.59	5.65	5.70	5.75	.05	19
6.34	6.43	6.51	6.58	6.65	6.72	6.78	6.84	6.89	.01	
5.20	5.28	5.36	5.43	5.49	5.55	5.61	5.66	5.71	.05	20
6.28	6.37	6.45	6.52	6.59	6.65	6.71	6.77	6.82	.01	
5.10	5.18	5.25	5.32	5.38	5.44	5.49	5.55	5.59	.05	24
6.11	6.19	6.26	6.33	6.39	6.45	6.51	6.56	6.61	.01	
5.00	5.08	5.15	5.21	5.27	5.33	5.38	5.43	5.47	.05	30
5.93	6.01	6.08	6.14	6.20	6.26	6.31	6.36	6.41	.01	
4.90	4.98	5.04	5.11	5.16	5.22	5.27	5.31	5.36	.05	40
5.76	5.83	5.90	5.96	6.02	6.07	6.12	6.16	6.21	.01	
4.81	4.88	4.94	5.00	5.06	5.11	5.15	5.20	5.24	.05	60
5.60	5.67	5.73	5.78	5.84	5.89	5.93	5.97	6.01	.01	
4.71	4.78	4.84	4.90	4.95	5.00	5.04	5.09	5.13	.05	120
5.44	5.50	5.56	5.61	5.66	5.71	5.75	5.79	5.83	.01	
4.62	4.68	4.74	4.80	4.85	4.89	4.93	4.97	5.01	.05	∞
5.29	5.35	5.40	5.45	5.49	5.54	5.57	5.61	5.65	.01	

TABLE 9 Random Numbers

Line/Col.	(1)	(2)	(3)	(4)	(5)	(6)	(7)	(8)	(9)	(10)	(11)	(12)	(13)	(14)
1	10480	15011	01536	02011	81647	91646	69179	14194	62590	36207	20969	99570	91291	90700
2	22368	46573	25595	85393	30995	89198	27982	53402	93965	34095	52666	19174	39615	99505
3	24130	48360	22527	97265	76393	64809	15179	24830	49340	32081	30680	19655	63348	58629
4	42167	93093	06243	61680	07856	16376	39440	53537	71341	57004	00849	74917	97758	16379
5	37570	39975	81837	16656	06121	91782	60468	81305	49684	60672	14110	06927	01263	54613
6	77921	06907	11008	42751	27756	53498	18602	70659	90655	15053	21916	81825	44394	42880
7	99562	72905	56420	69994	98872	31016	71194	18738	44013	48840	63213	21069	10634	12952
8	96301	91977	05463	07972	18876	20922	94595	56869	69014	60045	18425	84903	42508	32307
9	89579	14342	63661	10281	17453	18103	57740	84378	25331	12566	58678	44947	05585	56941
10	85475	36857	53342	53988	53060	59533	38867	62300	08158	17983	16439	11458	18593	64952
11	28918	69578	88231	33276	70997	79936	56865	05859	90106	31595	01547	85590	91610	78188
12	63553	40961	48235	03427	49626	69445	18663	72695	52180	20847	12234	90511	33703	90322
13	09429	93969	52636	92737	88974	33488	36320	17617	30015	08272	84115	27156	30613	74952
14	10365	61129	87529	85689	48237	52267	67689	93394	01511	26358	85104	20285	29975	89868
15	07119	97336	71048	08178	77233	13916	47564	81056	97735	85977	29372	74461	28551	90707
16	51085	12765	51821	51259	77452	16308	60756	92144	49442	53900	70960	63990	75601	40719
17	02368	21382	52404	60268	89368	19885	55322	44819	01188	65255	64835	44919	05944	55157
18	01011	54092	33362	94904	31273	04146	18594	29852	71585	85030	51132	01915	92747	64951
19	52162	53916	46369	58586	23216	14513	83149	98736	23495	64350	94738	17752	35156	35749
20	07056	97628	33787	09998	42698	06691	76988	13602	51851	46104	88916	19509	25625	58104
21	48663	91245	85828	14346	09172	30168	90229	04734	59193	22178	30421	61666	99904	32812
22	54164	58492	22421	74103	47070	25306	76468	26384	58151	06646	21524	15227	96909	44592
23	32639	32363	05597	24200	13363	38005	94342	28728	35806	06912	17012	64161	18296	22851
24	29334	27001	87637	87308	58731	00256	45834	15398	46557	41135	10367	07684	36188	18510
25	02488	33062	28834	07351	19731	92420	60952	61280	50001	67658	32586	86679	50720	94953

REFERENCES

BOOKS AND ARTICLES

Beardsley, G., and Mansfield, E. (1978). "A Note on the Accuracy of Industrial Forecasts of the Profitability of New Products and Processes." *Journal of Business 51*:127–135.

Box, G. E. P., Hunter, W. G., and Hunter, J. S. (1978). *Statistics for Experimenters: An Introduction to Design, Data Analysis, and Model Building*. New York: Wiley.

Box, G. E. P., and Jenkins, G. (1970). *Time Series Analysis: Forecasting and Control.* San Francisco: Holden-Day.

Cochran, W. G. (1977). *Sampling Techniques*. New York: Wiley.

Cochran, W. G., and Cox, G. M. (1957). *Experimental Designs*, 2d ed. New York: Wiley.

Davis, S. M. (1979). "No Connection between Executive Age and Corporate Performance." *Harvard Business Review*, March/April, pp. 6–8.

DeGroot, M. H. (1970). *Optimal Statistical Decisions*. New York: McGraw-Hill.

Draper, N. R., and Smith, H. (1981). *Applied Regression Analysis*, 2d ed. New York: Wiley.

Greenberg, B. G., Kuebler, R. R., Abernathy, J. R., and Horvitz, D. G. (1971). "Application of Randomized Response Technique in Obtaining Quantitative Data." *Journal of the American Statistical Association* 66:245–250.

Hildebrand, D. K., Laing, J. D., and Rosenthal, H. (1977). *Prediction Analysis of Cross Classifications*. New York: Wiley.

Hollander, M., and Wolfe, D. (1973). *Nonparametric Statistical Methods*. New York: Wiley.

Johnston, J. (1977). *Econometric Methods*, 2d ed. New York: McGraw-Hill.

Kish, L. (1965). *Survey Sampling*, New York: Wiley.

Mallows, C. L. (1973) "Some Comments on C_p." *Technometrics* (15):661–675.

Mendenhall, W. (1968). *Introduction to Linear Models and the Design and Analysis of Experiments*. Belmont, Calif.: Wadsworth.

Mood, A. M., Graybill, F. A., and Boes, D. C. (1974) *Introduction to the Theory of Statistics*, 3d ed. New York: McGraw-Hill.

Nelson, C. R. (1977). *Applied Time Series Analysis for Managerial Forecasting*. San Francisco: Holden-Day.

Ott, L. (1984). *An Introduction to Statistical Methods and Data Analysis*, 2d ed. N. Scituate, Mass.: Duxbury Press.

Ott, L., Larson, R. F., and Mendenhall, W. (1987). *Statistics: A Tool for the Social Sciences*, 4th ed. Boston Mass.: Duxbury Press.

Pierce, D. A. (1980). "A Survey of Recent Developments in Seasonal Adjustment." *The American Statistician 34*:125–134.

Scheaffer, R. L., Mendenhall, W., and Ott, L. (1986). *Elementary Survey Sampling*, 3d ed. N. Scituate, Mass.: Duxbury Press.

Tukey, J. W. (1977). *Exploratory Data Analysis*. Reading, Mass.: Addison-Wesley.

von Neumann, J., and Morganstern, O. (1967). *Theory of Games and Economic Behavior*, 3d ed. New York: Wiley.

Warner, S. L. (1965). "Randomized Response: A Survey Technique for Eliminating Evasive Answer Bias." *Journal of the American Statistical Association* 60:63–69.

Welch, B. L. (1938). "The Significance of the Differences between Two Means When the Population Variances Are Unequal." *Biometrika* 29:350–362.

STATISTICAL PROGRAM PACKAGES

(BMDP)

Dixon, W. J. (1983). *BMDP Statistical Software*. Berkeley: University of California Press.

(IDA)

Ling, R. F., and Roberts, H. V. (1980). *User's Manual for IDA*. Palo Alto: Scientific Press.

(Minitab)

Ryan, T. A., Joiner, B. L., and Ryan, B. F. (1985). *Minitab Handbook*, N. Scituate, Mass.: Duxbury Press.

(SAS)

SAS Institute. (1983). *SAS Introductory Guide*. Cary, N.C.: SAS Institute.

(SPSS)

Norusis, Marija J. (1983). *SPSS^X Introductory Statistics Guide*. New York: McGraw-Hill.

ANSWERS TO EXERCISES

CHAPTER 2
SECTION 2.1, p. 14

2.1 a.

Class	Midpoint	Frequency
320–329	325	1
330–339	335	1
340–349	345	0
350–359	355	1
360–369	365	6
370–379	375	7
380–389	385	11
390–399	395	1

Note that this is just *one* of many possible frequency tables that can be constructed for these data. A histogram is simply a graphic display of the data as summarized in the table.

b.

Stem	Leaves
32	4
33	9
34	
35	9
36	6 0 6 3 7 4
37	5 5 9 4 7 1 9
38	5 0 4 3 6 7 4 6 5 1 5
39	0

2.2 a.

Class	Midpoint	Frequency
810–819	815	1
820–829	825	1
830–839	835	4
840–849	845	7
850–859	855	2
860–869	865	1
870–879	875	1

Note that this is just one possible summarization of the data. A histogram is simply a graphic display of the data as summarized in the table.

b.

Stem	Leaves
81	2
82	4
83	8 8 6 9
84	7 9 1 6 9 6 3
85	2 0
86	4
87	1

2.3 a.

Class	Midpoint	Frequency
11.80–11.89	11.85	1
11.90–11.99	11.95	2
12.00–12.09	12.05	6
12.10–12.19	12.15	4
12.20–12.29	12.25	6
12.30–12.39	12.35	1

Note that this is just one summarization of the data. A histogram is simply a graphic display of the data as summarized in the table above.

b.

Class	Midpoint	Frequency
11.80–11.89	11.85	1
11.90–11.99	11.95	2
12.00–12.09	12.05	5
12.10–12.19	12.15	1
12.20–12.29	12.25	1
12.30–12.39	12.35	0

MANUFACTURER E

Class	Midpoint	Frequency
11.80–11.89	11.85	0
11.90–11.99	11.95	0
12.00–12.09	12.05	1
12.10–12.19	12.15	3
12.20–12.29	12.25	5
12.30–12.39	12.35	1

MANUFACTURER S

Histograms are graphic displays of the data as summarized in each of the tables above.

c. The combined histogram shows 2 modal classes. The separate histograms show that each has one different modal class.

2.4 a. One possible summary, using a class width of 5:

Interval	Midpoint	Frequency
7.5–12.4	10	7
12.5–17.4	15	5
17.5–22.4	20	4
22.5–27.4	25	1
27.5–32.4	30	2
32.5–37.4	35	3
37.5–42.4	40	1
42.5–47.4	45	1
		Total 24

Divide by 24 to obtain relative frequencies.

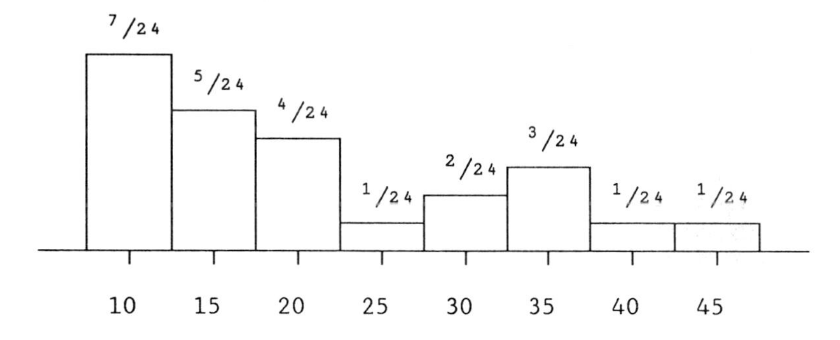

b. We could simply ignore the third digit, round off the numbers, or record the third digit as follows.

0	7.7 8.5 9.6 9.7
1	0.3 3.6 4.5 0.1 1.4 4.2 3.7
1	9.5 7.8 5.9
2	0.7 2.1
2	5.9 9.1
3	2.6 2.4
3	6.7 5.9
4	0.1
4	5.9

Both pictures show the right-skewness. The stem-and-leaf diagram shows the data.

SECTION 2.2, p. 18

2.5 a. $\bar{y} = 373.36$; median $= (377 + 379)/2 = 378$

b. Modal class is 380–389

c. Mean = 373.36, median = 378, mode = 385

2.6 mean = 15.1875, median = 14.5, mode = 18

2.7 a. One possibility is to use a class width of 300; several other choices would be reasonable.

Class	Midpoint	Frequency
150–449	300	3
450–749	600	3
750–1049	900	6
1050–1349	1200	3
1350–1649	1500	6
1650–1949	1800	0
1950–2249	2100	2
		Total 23

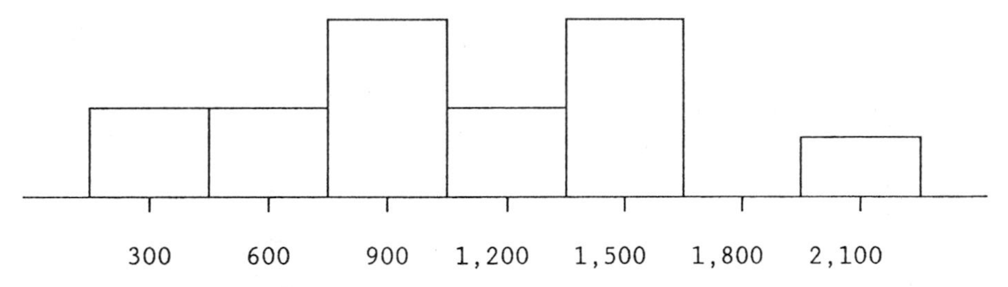

b. The histogram (a picture of these frequencies) is roughly symmetric around 1200 so $\bar{y}$ is roughly 1200.
c. median = 1039, mean = 1091.96
d. mean is slightly larger; slight right-skewness.

2.8 a. Not for quantitative data
b. median = 688
c. mean = 679.4
d. left-skewness

2.9

Stem	Leaves
54	7
⋮	
62	5
63	0
64	
65	6
66	4 7 7
67	7 9
68	8 8 8 8 8
69	1 4 7 9
70	0 1 2 3 3 3 8
71	1

Yes, data are left-skewed.

2.10

Number of members	a. mean	b. median
1	93.750	78.5
2	98.652	95.0
3	113.312	100.0
4	124.857	112.5
5	131.900	118.5

2.11 a. $\bar{y} = 108.72$
b. Yes, as a weighted average.
c. Median = 104
d. No, it can't.

SECTION 2.3, p. 26

2.12 a. $\bar{y} = 843.82$; $s = 13.44$ b. AD = 9.36; Range = 59
c. $\bar{y} \pm 2s$: 816.94 to 870.70
Actual: 15/17 or 88%
Empirical Rule: approximately 95%
Chebyshev's: at least 75%

2.13 a. Sample 1: Range = 10
Sample 2: Range = 12
b. Sample 1: $\bar{y} = 20$; $s^2 = 17.33$; $s = 4.16$
Sample 2: $\bar{y} = 20$; $s^2 = 8.50$; $s = 2.92$
c. Lineplots for both Sample 1 and Sample 2 data show that sample 1 has more variability.

2.14 $s^2 = 16.296$, $s = 4.037$

2.15 a. $s = 506.97$
b. 78.02 to 2105.90 includes 22/23, or 95.7%, of the data; this is very close to the Empirical Rule value.
c. Yes; the histogram is roughly mound shaped.

2.16 a. IQR = 748
b. Lower inner fence = −435.5, upper inner fence = 2556.5. No outliers.
c. The box goes from 686.5 to 1434.5 with a vertical bar at the median, 1039. The whiskers go down to 361 and up to 2197.

2.17 a. $\bar{y} = 679.4$, $s = 34.8$
b. 23/26, or 88.5%, of the data fall in this range. The Empirical Rule works poorly for these data.

2.18 a. median = 688, IQR = 34
b. Inner: 616 and 752
Outer: 565 and 803
The 547 value is a severe outlier.
c. The box goes from 667 to 701 with a line at 688. The whiskers go down to 625 and up to 711. The severe outlier, 547, is indicated by a o symbol.

2.19

				Inner Fences		Outer Fences		
Members	Median	25th	75th	Lower	Upper	Lower	Upper	Outliers
1	78.5	67.0	123.0	−17.0	207.0	−101.0	291.0	none
2	95.0	83.0	115.0	35.0	163.0	−13.0	211.0	none
3	100.0	85.0	135.0	11.25	209.25	−63.0	283.5	none
4	112.5	99.0	139.0	39.0	199.0	−21.0	259.0	251(*)
5+	128.5	111.0	136.0	73.5	173.5	36.0	211.0	206(*)

2.20

Members:	1	2	3	4	5+
s:	37.99	20.19	33.63	44.65	28.60

SECTION 2.4, p. 29

2.21 a. "SKEWNESS" = 1.766; "KURTOSIS" = 4.372
b. The histogram shows slight right-skewness, indicating that the mean is larger than the median. The skewness coefficient should be positive and it is. With roundoff error, skewness = 1.67.

2.22 a. KURTOSIS = 3.88
b. The coefficient indicates slightly more heavy-tailness than the normal distribution.

2.23 a. right-skewness b. slight left-skewness c. symmetric

2.24 a. Year 1: $\bar{y} = 2.083$, $s = 2.158$
Year 2: $\bar{y} = 2.308$, $s = 2.293$

b.

Year	$(\bar{y} - \text{median})/s$	Third-power Skewness
1	.525	1.048
2	.309	1.435

Strong right-skewness is indicated.

2.25 a. $\bar{y} = 2.196$, the average of the yearly means
b. $s = 2.181$, not an average

2.26 a. left-skewness
b. measure $= -.247$, indicating left-skewness

2.27 a. Nothing extreme.
b. measure $= .104$, moderate right-skewness

2.28 Kurtosis $= 2.467$, less than the 3.0 value for bell-shaped data. Data are slightly light-tailed.

SECTION 2.6, p. 34

2.29 a. median $= 58.35$
b. $\bar{y} = 57.88$

2.30 a. Literally, center rectangles at 27.4, 32.4, . . .
b. Mean slightly less than 37.4, maybe 36.5
c. $\bar{y} = 36.025$ (or 36.125 if using .5 midpoints). We don't have the actual salaries.

2.31 $s = 11.62$

2.32 $s^2 = 29.471$, $s = 5.421$

CHAPTER EXERCISES, p. 35

2.33 a. $\bar{y} = 4$; $s^2 = 52.85$; $s = 7.27$
b. $\bar{y} \pm 1s$: -3.27 to 11.27
Actual: 90%
Empirical Rule: 68%
The discrepancy is due to the right-skewness of the data. The Empirical Rule holds only for mound-shaped distributions.
c. Skewness $= +.344$, correctly indicating right-skewness.

2.35 a.

Class	Midpoint	Frequency
8–12	10	10
13–17	15	0
18–22	20	5
23–27	25	0
28–32	30	10

Note that this is just one summarization for these data. A histogram is simply a graphic display of the data as summarized in the table above. The data is symmetric (no skewness) and trimodal.
b. The data are distributed perfectly symmetrically about the value 20. Therefore, the mean and median are exactly the same, 20.
c. $s = 9.18$
Since the data are symmetrically distributed, the skewness coefficient must equal 0.

d. $\bar{y} \pm 1s$: 10.82 to 29.18
Actual: 36%
Empirical Rule: 68%

The preceding discrepancy is due to the trimodal distribution of the data. The Empirical Rule holds for mound-shaped distributions.

e. $\bar{y} \pm 2s$: 1.64 to 38.36
Actual: 100%
Chebyshev's: at least 75%

2.36 a. $\bar{y} = 0$; $s^2 = 117.43$; $s = 10.84$
b. skewness = $-.092$ using $(\bar{y} - \text{median})/s$
skewness = -2.49 using alternative formula
c. skewed right but with a left outlier
d. The outlier at -36 causes it.

2.37 a. Means are relevant for DIFF and CONDN.
b. MEAN DIFF = .775; MEAN CONDN = .425
c. Request means and standard deviations for each of the four truckers; this will allow us to compare the four trucking companies.

CHAPTER 3

SECTION 3, p. 44

3.1 a. If the editor calls on past experience as an editor, the long-run relative frequency interpretation could be used. If the editor is a new editor or if the textbook is one of a kind, the editor may simply guess that the probability is .8, thereby using the subjective interpretation.
b. Long-run relative frequency interpretation: over a huge number of trials, the gear failed to satisfy tolerances .2% of the time.
c. Classical interpretation: a random sample (all probabilities equal) of 100 employees is taken and the probability is obtained by counting.
d. If one is willing to assume that economic conditions in West Germany are going to be the same next year as in previous years, one could calculate the probability of next year's inflation rate exceeding 4% using the long-run interpretation. If one is not willing to assume the same economic conditions for the next year, one would have to guess the probability, thereby using the subjective interpretation.
e. Long-run relative frequency interpretation: if one observed over a long series of days that the demand for coronary care beds exceeding the normal capacity was .4% of the days observed, then one could use the long-run interpretation and state that the relevant probability is .004.

3.2 There are no "correct" answers to subjective probabilities.

SECTION 3.2, p. 47

3.3 The sample space, S, consists of 45 outcomes: (A12), (A13), (A14), (A15), (A16), (A23), (A24), (A25), (A26), (A34), (A35), (A36), (A45), (A46), (A56), (B12), (B13), (B14), (B15), (B16), (B23), (B24), (B25), (B26), (B34), (B35), (B36), (B45), (B46), (B56), (C12), (C13), (C14), (C15), (C16), (C23), (C24), (C25), (C26), (C34), (C35), (C36), (C45), (C46), (C56).

3.4 P(belong to the same division) = 3/45 = .067
P(belong to different divisions) = 12/45 = .267

3.5 An outcome would be a string of 100 numbers, each number being a 0 or a 1. It is unlikely that errors are as likely as correct accounts; the outcomes aren't equally likely.

3.6 $P(A) = .80$; $P(B) = .50$; P(A and B both occur) = .35; P(either A or B, or both) = .95

3.7 $P(A \text{ or } B) = .95 \neq P(A) + P(B) = 1.30$. A and B aren't mutually exclusive.

SECTION 3.3, p. 54

3.8 Let A be the event that generator 1 works properly, and B be the event that generator 2 works properly. Then $\overline{A}$ is the event that generator 1 fails. $(A \cup B)$ is the event that generator 1 works properly or generator 2 works properly or both generator 1 and 2 work properly.
$(A \cap B)$ is the event that both generator 1 and 2 work properly.
$(\overline{A} \cap \overline{B})$ is the event that both generator 1 and 2 fail.
The complement of $A \cup B$ is $\overline{A} \cap \overline{B}$.

3.9 $P(A \cap \overline{B}) = .03$; $P(\overline{A} \cap B) = .01$; $P(\overline{A} \cap \overline{B}) = .03$

3.10 $P(B \mid A) = .969$. Given that generator 1 works properly, the probability that 2 also works is .969.
$P(\overline{B} \mid A) = .031$. Given that generator 1 works properly, the probability that 2 fails is .031.
$P(B \mid \overline{A}) = .25$. Given that generator 1 fails, the probability that 2 works properly is .25.
$P(\overline{B} \mid \overline{A}) = .75$. Given that generator 1 fails, the probability that 2 also fails is .75.
$P(B \mid A) + P(\overline{B} \mid A) = 1$ but $P(B \mid A) + P(B \mid \overline{A}) \neq 1$

3.11 $P(A \cap B) \neq 0$; therefore, A and B are not mutually exclusive.
$P(A \cup B) = .97$

3.12 (A,B), (B,C) and (A,C) are not mutually exclusive pairs. $(A \cap B \cap C)$ is the event that the grant was ineligible and it was incomplete and received past the deadline.

3.13

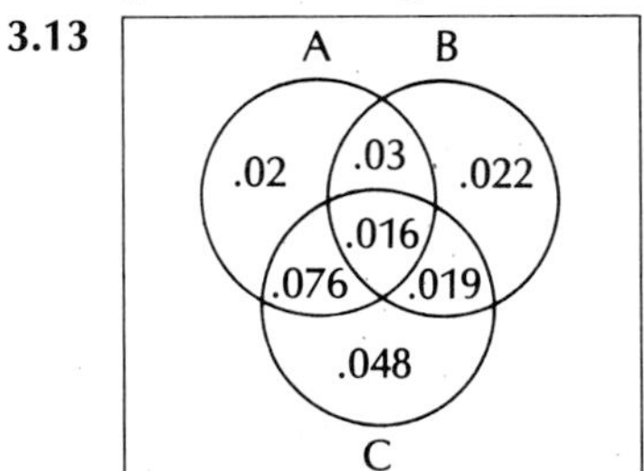

3.14 $P(\overline{A} \cap \overline{B} \cap \overline{C}) = .769$; $P(A \cup B \cup C) = .231$; $P(\overline{A} \cap \overline{B}) = .817$
$\overline{A} \cap \overline{B} \cap \overline{C}$ is "none of the events occur"; $A \cup B \cup C$ is "at least one occurs"; $\overline{A} \cap \overline{B}$ is "neither A nor B occurs."

3.15 a. P(defective and detected) = .09
b. P(defective and not detected) = .01

3.16 a. P(defective and not detected) = .002
b. P(defective and not detected) is lower when the original probability of defective drilling is lower, even if the probability of detection is only .4.

3.17 P(defective and never detected) = .002

SECTION 3.4, p. 58

3.18 a. $P(A) = .40$, $P(A \cap B) = .12$, $P(B \mid A) = .30$
b. $P(B \mid A) = P(B) = .30$. Therefore A and B are statistically independent.
c. Because A and B are independent, the test is not at all useful in predicting good sales achievement.

3.19 a. $P(A \cap \overline{B}) = .28$, $P(\overline{B} \mid A) = .70$
b. $P(\overline{B} \mid A) = P(\overline{B}) = .70$. Therefore A and $\overline{B}$ are statistically independent.

3.20 A is the event "worker comes from Plant 1." B is the event "response is poor."
a. $P(A) = .436$; $P(B) = .291$; $P(A \cap B) = .109$
b. $P(A \cap B) \neq P(A)P(B)$; therefore A and B are not independent.
c. $P(B \mid A) = .25 \neq P(B \mid \overline{A}) = .323$

3.21 Because $P(B \mid A) = P(B)$ and $P(B \mid \overline{A}) = P(B)$ by independence, $P(B \mid A) = P(B \mid \overline{A})$.

3.22 P(no substitute needed at any school) = .18 ?

3.23 The assumption of independent processes is not a realistic one. Take the case of a flu epidemic that affects an entire region.

SECTION 3.5, p. 65

3.24 a. Joint probability table:

		Retest: Major	Retest: Minor	Retest: None	
First Test	Major	.18	.30	.12	.60
	Minor	.03	.09	.18	.30
	None	.00	.02	.08	.10
		.21	.41	.38	1.00

b. P(Major at retest) = .21
c. P(Minor at retest) = .41; P(none at retest) = .38

3.25 P(Major at retest) = .21; P(Minor at retest) = .41; P(None at retest) = .38

3.26 b. P(program will show major bugs at all three tests) = .018
c. P(program will show major bugs at third test) = .021
d. P(program shows no bugs at second and third tests) = .896

3.27 a. P(first time bidder $\cap$ satisfactory service) = .06
b. P(satisfactory service) = .84
c. P(first time bidder|satisfactory service) = .071

CHAPTER EXERCISES, p. 67

3.28 A typical outcome of this experiment would be one with (say) 18 B's, 22 C's, and 20 D's, indicating that during that month, a random sample of 60 tickets indicated that airline A accepted 18 tickets from airline B, 22 tickets from airline C and 20 tickets from airline D. All outcomes should not be regarded as equally likely.

3.29 Let A be the event "MBA degree," and B be the event "undergraduate business degree."

a. Venn diagram

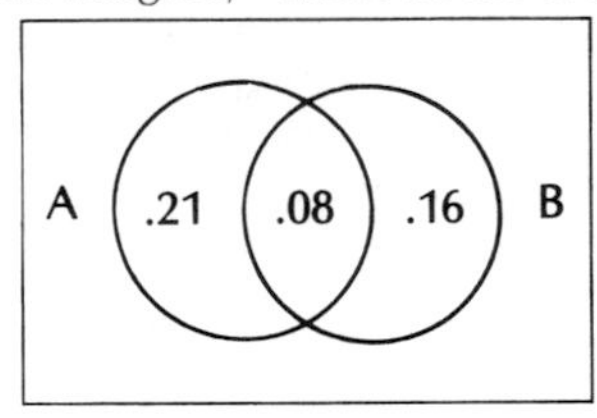

b. $P(A \cup B) = .45$ c. $P(\bar{A} \cap \bar{B}) = .55$

3.30 P(exactly one of A or B) = .37

3.31 Let A be the event "air-conditioning," and B be the event "power steering."

a. Venn diagram

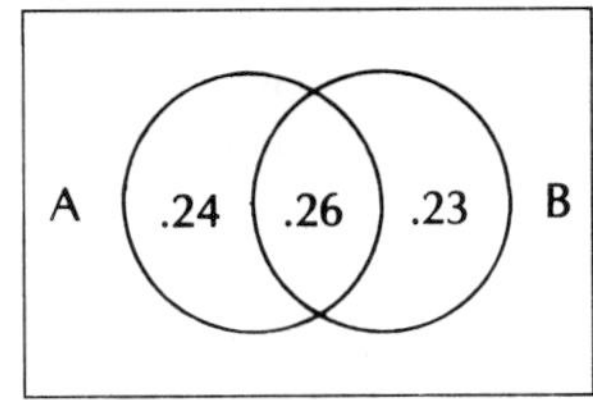

b. $P(A \cap \bar{B}) = .24$ c. $P(\bar{A} \cap \bar{B}) = .27$

3.32 Let A and B be events as defined in Exercise 3.31, and C be the event "automatic transmission."

a. Venn diagram

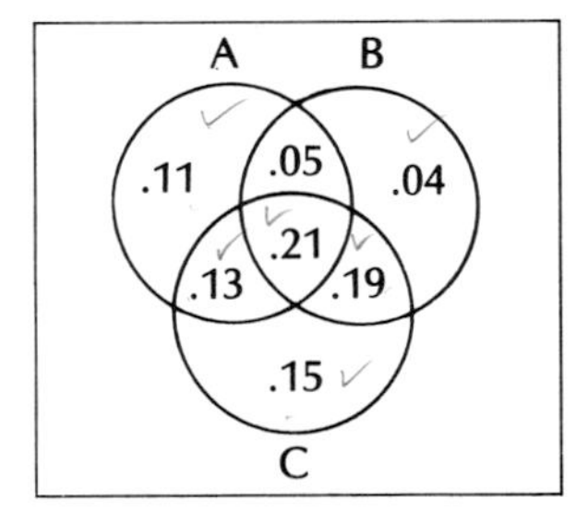

b. $P(A \cup B \cup C) = .88$ c. P(exactly 1 option ordered) = .30

3.33 $P(A \cap C) = P(A)P(C) = .34$. Therefore, A and C are statistically independent.

3.34 a. P(all 8 stocks beat the market) = $.5^8 = .00390625$

b. Assumptions: independence among stocks and even chance of the stock doing better or worse than the market.

3.35 a. P(no analyst gets 8 "winners") = $[1 - (.5)^8]^{100} = .676$

b. P(at least one analyst gets 8 "winners") = .324

3.36 a. Joint probability table

		Sales Level for Week 1				
		10	20	30	40	
Sales Level for Week 2	10	.16	.12	.08	.04	.40
	20	.12	.09	.06	.03	.30
	30	.08	.06	.04	.02	.20
	40	.04	.03	.02	.01	.10
		.40	.30	.20	.10	1.00

b. P(average sales level per week = 25) = .20

3.37 It is doubtful that the independence assumption is reasonable. If a recently released book makes the best seller list one week, sales the following week could be greatly affected.

3.38 a. $P(\text{received on time} \cap \text{specifications met}) = .60$

b. Joint probability table

	Received on Time	Received Late	
Specifications Met	.60	.15	.75
Specifications Not Met	.15	.10	.25
	.75	.25	1.00

c. $P(\text{specifications met}) = .75$

3.39 a. $P(\text{all 4 orders meet specifications}) = .316$

b. Assumption: the orders are independent of each other.

3.40 a. $P(\text{both pay}) = .25$

b. $P(\text{pay in full twice}) = .25$

c. Independence; unreasonable in b.

3.41 a. $P(\text{pay in full twice}) = .45$

b. $P(\text{pay in full neither month}) = .45$

c. $P(\text{exactly one month}) = .10$

3.42 $P(\text{first}|\text{second}) = .90$

3.43 a. $P(\text{no alteration}) = .52$

b. $P(\text{exactly one alteration}) = .26$

3.44 not independent

3.45 a. $P(\text{both jackets need alteration}) = .16$

b. Same $P(\text{alteration})$ and independence. Independence is dubious; an "odd-shaped" buyer of one suit would probably be "odd-shaped" for the other.

3.46 a. $P(\text{first key}) = .25$

b. $P(\text{one of first three}) = .643$

3.47 $P(\text{key 2 is master}|\text{door open on try 2}) = .5$

REVIEW EXERCISES CHAPTERS 2–3, p. 70

R1 a.

Supplier	Mean	Std. Dev.	Median	Skew = (mean − median) ≠ Std. Dev.
A	26.71	9.21	$\frac{22.6 + 23.9}{2} = 23.25$	.38
B	31.33	8.49	$\frac{28.0 + 28.4}{2} = 28.2$	.37
C	22.77	9.02	$\frac{19.7 + 19.9}{2} = 19.8$	.33
D	28.05	11.95	$\frac{23.5 + 24.5}{2} = 24.0$	.34

b. No, the data are severely right-skewed.

R2 $\bar{y} = 27.215$ is the average of supplier means.

$s^2 = 97.56$ is not an average.

R3 $P(\text{decrease}|\text{not correct}) = .160$

R4 a. The qualitative variable "type of bank" is excluded.

Variable	Deposits	Capital	Reserves	Bad debts
$\bar{y}$	8.683	2.617	2.475	1.794
s	9.112	3.229	2.647	1.333

b.

Variable	Deposits	Capital	Reserves	Bad Debts
25th %ile	3.67	.85	.92	.97
75th %ile	11.64	4.03	3.28	2.00
IQR	7.97	3.18	2.36	1.03
Lower inner fence	negative	negative	negative	negative
Upper inner fence	23.595	8.80	6.82	3.545
Lower outer fence	negative	negative	negative	negative
Upper outer fence	35.55	13.57	10.36	5.09
Outliers (* = mild, o = severe)	31.62(*)	10.63(*)	9.22(*)	3.97(*) 4.31(*)

R5 $P(\text{unacceptable}|\text{test shows excess}) = .0092/.0488 = .19$

R6 $P(\text{borderline}|\text{acceptable}) = P(\text{borderline}|\text{not acceptable}) = .05$

R7 a. $\bar{y} = 2.31$, $s = 1.38$

b. Meaningless; X_3 is qualitative.

R8

Variable	X_4	X_5	Y
Mean	32,938	40,977	8,038
St. dev.	21,561	26,433	6,256
Median	21,200	36,000	5,000

The mean of Y is (except for roundoff) equal to $\bar{x}_5 - \bar{x}_4$.
There is no such relation for medians or standard deviations.

R9 $(\bar{y} - \text{median})/s = .486$, more right-skewness than is suggested by a histogram.

R10 a. $P(\text{cancelled}) = .587$

b. $P(\text{bottom third}|\text{cancelled}) = .425/.587 = .724$

R11 No; $P(\text{bottom third}|\text{cancelled}) \neq P(\text{bottom third}) = .50$. Independence would mean that highly ranked shows were just as likely to be cancelled as poorly ranked ones. Not so!

R12 a. $P(\text{excellent} \cap \text{definite}) = .0432$

b. Independence, an unreasonable assumption. $P(\text{definite}|\text{excellent})$ should in fact be higher than .24, so the calculated probability should be too low.

R13 a. $\bar{y} = 2.477$

b. $s = 1.834$. As a population, $\sigma = 1.833$.

c. The range contains 1134/1533, or 74.0%, of the data, fairly close to 68%.

R14 a. $P(\text{at least } 1) = .869$

b. $P(\text{at most } 3|\text{at least } 1) = .738$

R15 a. The data are nearly symmetric; the two should be nearly equal. Median = 35.55; $\bar{y} = 35.41$.

b. Range for man = 11.3, range for women = 22.2; there are outliers in the data for women.

R16 For men, the box goes from 33.8 to 36.7 with a line at 35.55, whiskers down to 31.6 and up to 40.2, and a moderate outlier at 28.9. For women, the box goes from 30.5 to 33.6 with a line at 32.15, whiskers down to 29.8 and up to 38.1, a moderate outlier at 22.5, and a severe outlier at 44.7.

R17 $\bar{y} = 34.08$, a weighted average; median $= 34.0$, not directly related to the group medians.

R18 a. $P(\text{engine} \cap \text{not interior} \cap \text{not exterior}) = .32$
b. $P(\text{exactly one}) = .63$
c. No; $P(\text{engine} \cap \text{interior}) = .23 \neq P(\text{engine})P(\text{interior})$.

R19 a. Each plot is somewhat right-skewed.

b.

Base	Corn	Oats	Wheat
Mean	11.25	12.97	32.14
Median	10.75	12.25	31.0
n	20	18	29

In each case the mean is larger than the median, indicating right-skewness.

R20 $\bar{y} = 20.75$, median $= 15$. The combined data are bimodal, so neither is typical.

CHAPTER 4
SECTIONS 4.1 AND 4.2, p. 82

4.1 a. Let A through C represent the women recruiters, and D through H the men recruiters. The sample space consists of 56 outcomes: (AB), (AC), (AD), (AE), (AF), (AG), (AH), (BA), (BC), (BD), (BE), (BF), (BG), (BH), (CA), (CB), (CD), (CE), (CF), (CG), (CH), (DA), (DB), (DC), (DE), (DF), (DG), (DH), (EA), (EB), (EC), (ED), (EF), (EG), (EH), (FA), (FB), (FC), (FD), (FE), (FG), (FH), (GA), (GB), (GC), (GD), (GE), (GF), (GH), (HA), (HB), (HC), (HD), (HE), (HF), (HG).

b. The number of women selected for each outcome in the sample space is given in the order in which the outcomes are listed above: 2, 2, 1, 1, 1, 1, 1, 2, 2, 1, 1, 1, 1, 1, 2, 2, 1, 1, 1, 1, 1, 1, 1, 0, 0, 0, 0, 1, 1, 1, 0, 0, 0, 0, 1, 1, 1, 0, 0, 0, 0, 1, 1, 1, 0, 0, 0, 0, 1, 1, 1, 0, 0, 0, 0.

4.2

y	0	1	2
$P_Y(y)$	.357	.536	.107

4.3

y	0	1	2
$F_Y(y)$	.357	.893	1.00

4.4 b. $P(Y \leq 2) = .50$ c. $P(Y \geq 7) = .13$
d. $P(1 \leq Y \leq 5) = .71$

4.5

y	0	1	2	3	4	5	6	7	8	9	10
$F_Y(y)$	.10	.25	.50	.64	.73	.81	.87	.92	.96	.985	1.00

$P(Y \leq 2) = .50$ $P(Y \geq 7) = .13$
$P(1 \leq Y \leq 5) = .71$

4.6 a. $P(X \geq 3) = .64$
b. $P(2 \leq X \leq 6) = .60$
c. $P(X \geq 9) = .07$

4.7 a.

x	0	1	2	3	4	5	6	7	8	9	10
$F_X(x)$	.06	.20	.36	.50	.62	.72	.80	.87	.93	.97	1.00

b. $1 - F_X(2) = .64$, $F_X(6) - F_X(1) = .60$, $1 - F_X(8) = .07$

SECTION 4.3, p. 88

4.8 a. Table of $F_Y(y)$ values

y	$F_Y(y) = 5y^4 - 4y^5$	y	$F_Y(y) = 5y^4 - 4y^5$
0	0	.6	.33696
.1	.00046	.7	.52822
.2	.00672	.8	.73728
.3	.03078	.9	.91854
.4	.08704	1.0	1.00000
.5	.18750		

b. $P(Y \leq .8) = .73728$; $P(Y \geq .6) = .66304$; $P(.5 \leq Y \leq .9) = .73104$

4.9 a. Table of $F_Y(y)$ values

y	$F_Y(y) = 1 - e^{-.5y}$	y	$F_Y(y) = 1 - e^{-.5y}$
0	0	5	.9179
1	.3935	6	.9502
2	.6321	7	.9698
3	.7769	8	.9817
4	.8647		

b. $P(Y \leq .75) = .3127$; $P(Y \geq 4) = .1353$; $P(2 \leq Y \leq 3.5) = .1941$

4.10 a. $f_Y(y) = 5(4)y^3 - 4(5)y^4 \quad 0 \leq y \leq 1$

b. Integrate $f_Y(y)$; $P(.2 \leq Y \leq .7) = .5215$; $P(Y \leq .6) = .33696$; $P(Y \geq .5) = .8125$

c. $P(.2 \leq Y \leq .7) = .5215$; $P(Y \leq .6) = .33696$; $P(Y \geq .5) = .8125$

4.11 a. Table of $f_Y(y)$ values

y	$f_Y(y) = (4/27)(9y - 6y^2 + y^3)$	y	$f_Y(y) = (4/27)(9y - 6y^2 + y^3)$
0	0	1.50	.5000
.25	.2801	2.00	.2963
.50	.4629	2.50	.0926
.75	.5625	3.00	0
1.00	.5926		

b. $P(Y \leq 1.5) = .6875$; $P(Y \geq 2.0) = .1111$; $P(1 \leq Y \leq 2.5) = .5764$

c. Table of $F_Y(y)$ values

y	$F_Y(y) = \frac{4}{27}\left(\frac{9y^2}{2} - \frac{6y^3}{3} + \frac{y^4}{4}\right)$
0	0
.5	.1319
1.0	.4074
1.5	.6875
2.0	.8889
2.5	.9838
3.0	1.0000

4.12 $P(Y > 115) = .0430$

4.13 $F_Y(y) = .0009375\,[-10{,}400/3 + 40y - .1(y - 100)^3/3]$, $80 < y < 120$

4.14 a. $P(2 < Y < 4) = .1094$, regardless of the inclusion of $2.00000\cdots$ and $4.00000\cdots$

b. $P(0.5 < Y < 1.5) = P(1 < Y < 1.5) = .7037$, because $Y \leq 1$ is impossible.

4.15 a. $F_Y(y) = 1 - 1/y^3$, for $y > 1$

b. $1 - 1/y^3 = .99$ if $y = 4.642$

SECTIONS 4.4 AND 4.5, p. 95

4.16 b. $E(Y) = 4.32$

c. $E(Y)$ is pulled to the right, away from the bulk of the data because of the right-skewness of the data.

4.17 a. $\sigma_Y = 2.2798$ b. $\sigma_Y = \sqrt{23.86 - (4.32)^2} = 2.2798$

4.18 Actual probability: $P(2.0402 \leq Y \leq 6.5998) = .63$

Empirical Rule: $P(2.0402 \leq Y \leq 6.5998) = .68$

The discrepancy between the preceding probabilities is due to the right-skewness of the population distribution. The Empirical Rule holds for mound-shaped probability histograms.

4.19 a. Apartment #1: $E(Y) = 100$

Apartment #2: $E(Z) = 105$

b. Apartment #1: $\text{Var}(Y) = 2500$; $\sigma_Y = 50$

Apartment #2: $\text{Var}(Z) = 8475$; $\sigma_Z = 92.06$

4.20 a. Apartment #2 has higher return, but also higher risk.

b. For many people, the less risky apartment #1 would be preferred, despite its lower return.

4.21 a. $\mu_X = 3.97$

b. $\sigma_X^2 = 7.0891$

c. $\sigma_X^2 = 22.85 - (3.97)^2 = 7.0891$

4.22 $P(-1.35 < X < 9.29) = .97$, close to .95 (Empirical Rule) and well above .75 (Chebyshev's Inequality).

4.23 a. μ_Y should be 100, because the density is symmetric around 100.

b. $\mu_Y = 100$

c. $\sigma_Y^2 = 80.000$

4.24 Fairly well; the data are mound-ish. $P(91.056 < Y < 108.944) = .626$

4.25 a. $\mu_Y = 1.5$, the long-run average Y value

b. $\sigma_Y = .866$

4.26 a. The density is very right-skewed.

b. No; right-skewness. $P(.634 < Y < 2.366) = P(1 < Y < 2.366) = .924$, not close to .68.

4.27 a. $\mu_Y = 10$ by symmetry

b. $\sigma_Y = .039$

SECTION 4.6, p. 100

4.28 a. $P(X = 1, Y = 2) = .055$ b. $P(X \leq 2, Y \leq 2) = .21$

c.

x	1	2	3	4
$P_X(x)$	.23	.27	.27	.23

y	1	2	3	4
$P_Y(y)$	.23	.27	.27	.23

d. $P_{XY}(1, 1) = .03 \neq P_X(1)P_Y(1) = .0529$. Therefore, X and Y are not statistically independent.

4.29 Plug in x and y values.

4.30 a. Joint probability table

		y					
		0	1	2	3	4	5
	0	.01	.03	.025	.02	.01	.005
	1	.04	.12	.10	.08	.04	.02
x	2	.025	.075	.0625	.05	.025	.0125
	3	.02	.06	.05	.04	.02	.01
	4	.005	.015	.0125	.01	.005	.0025

b.

x	0	1	2	3	4	
$P_X(x)$	.10	.40	.25	.20	.05	
y	0	1	2	3	4	5
$P_Y(y)$	.10	.30	.25	.20	.10	.05

4.31 The assumption of independence is doubtful. A purchaser buying a speaker system on a given day most likely will also buy an amplifier.

4.32 a. $P_{XY}(2, 2) = .100$
b. $P(X = 2 \cap Y \leq 2) = .175$
c. $P(X \leq 2 \cap Y \leq 2) = .325$

4.33 a.

x	0	1	2	3	4
$P_X(x)$	.180	.195	.250	.195	.180

y	0	1	2	3	4
$P_Y(y)$	.150	.225	.250	.225	.150

No. For example, $P_{XY}(0, 0) = .010 \neq P_X(0)P_Y(0) = (.180)(.150) = .027$

4.34 $P_{Y|X}(y|x)$

		y				
		0	1	2	3	4
	0	.0556	.0833	.1667	.4167	.2778
	1	.1026	.1538	.2308	.3077	.2051
x	2	.1200	.1800	.4000	.1800	.1200
	3	.2051	.3077	.2308	.1538	.1026
	4	.2778	.4167	.1667	.0833	.0556

Not independent; the conditional probabilities change as x changes.

SECTIONS 4.7 AND 4.8, p. 108

4.35 a. $\mu_X = \mu_Y = 2$, by symmetry
b. $\sigma_X = 1.353$, $\sigma_Y = 1.285$

4.36 a. $\text{Cov}(X, Y) = -.70$
b. $\text{Corr}(X, Y) = -.403$. As X increases, Y tends to decrease; the two are dependent.

4.37

x	0	1	2	3	4
$\mu_{Y\mid X=x}$	2.778	2.359	2.000	1.641	1.222

The conditional expectation decreases as x increases.

4.38 a.

t	0	1	2	3	4	5	6	7	8
$P_T(t)$	.010	.035	.090	.205	.320	.205	.090	.035	.010

b. $\mu_T = 4$, $\sigma_T^2 = 2.08$
c. $\mu_T = 2 + 2$, $\sigma_T^2 = 1.83 + 1.65 + 2(-.70)$

4.39 a.

x	0	1	2
$P_X(x)$	.81	.18	.01

b. $\mu_X = .20$, $\sigma_X^2 = .18$
c. Independence; unreasonable if wear or misalignment tends to produce one defect after another, and if blocks were sampled one after another.

4.40 $P_{XY}(x, y)$

		y = 0	y = 1	y = 2	
x	0	.8100	0	0	.81
	1	.0180	.1620	0	.18
	2	.0001	.0018	.0081	.01
		.8281	.1638	.0081	

4.41 a. $\mu_Y = .1800$, $\sigma_Y = .4047$
b. $\text{Corr}(X, Y) = .943$
c. The more defectives, the more detected.

4.42 a.

y	0	1	2	3
$P_Y(y)$	.080	.314	.414	.192

b. $P(Y \geq 2) = .606$

c.

y	0	1	2	3
$F_Y(y)$	.080	.394	.808	1.000

$P(Y \geq 2) = 1 - .394$

4.43 $\mu_Y = 1.718$, $\sigma_Y = .8640$

4.44 a.

x	0	100	150	200	250	300	350	400
$P_X(x)$	.080	.020	.070	.224	.030	.096	.288	.192

b. $\mu_X = 280.8$, $\sigma_X = 128.69$

4.45 a. $\mu_T = 5054.4$
$\sigma_T = 298{,}104.48$
b. Independence is reasonable.

4.46 a. Density is left-skewed.
b. $P(.7 < Y < .9) = .4655$
c. $P(Y > .8) = .3446$

4.47 $\mu_Y = 0.7143$, $\sigma_Y = 0.1597$

4.48 a. $F_X(x) = (6/125)(2.5x^2 - x^3/3)$, for $0 < x < 5$
b. $P(X \geq 3) = 1 - F_X(3) = .352$
c. $\mu_X = 2.5$, $\sigma_X = 1.118$

4.49 Corr(X, Y) = 0 if X and Y are independent.

4.50 a. $f_Y(y) = .0625e^{-.5y}y^2$, for $y > 0$
b. $f_{X|Y=y}(x|y) = xe^{-x/y}/y^2$, for $x > 0$
c. Not independent; complex calls make both X and Y large. Note that $f_{X|Y}(x|y)$ depends on y, so X and Y are not independent.

4.51 $\mu_{X|Y=y} = 2y$

4.52 Cov(X, Y) = 96 − (6)(12) = 24

CHAPTER EXERCISES, p. 112

4.53 Let 1 through 3 represent the field engineers over 40
4 represent the field engineer under 40
5 and 6 represent the sales reps over 40
7 through 10 represent the sales reps under 40

a. The sample space consists of 60 possible outcomes for the three people chosen.
b. y = the number of persons over 40

y	0	1	2	3
$P_Y(y)$	.100	.433	.417	.050
$F_Y(y)$	.100	.533	.950	1.00

4.54 $E(Y) = 1.417$; $\text{Var}(Y) = .543$ $\sigma_Y = .737$

4.55 a. Table of $F_Y(y)$ values:

y	$F_Y(y)$
0	.02
1	.15
2	.35
3	.65
4	.84
5	.99
6	1.00

b. $E(Y) = 3$; $\sigma_Y = 1.33$

c. Actual probability: $P(1.67 \leq Y \leq 4.33) = .69$
Empirical Rule: $P(1.67 \leq Y \leq 4.33) = .68$

4.56 a. $E(Y) = .6666$; $\sigma_Y = .1785$
b. Actual probability: $P(.3096 \leq Y \leq 1.0236) = .9654$

4.57 a. Table of $F_Y(y)$ values

y	$F_Y(y) = 10y^3 - 15y^4 + 6y^5$
0	0
.25	.10352
.50	.5000
.75	.9058
1.00	1.000

b. $P(Y \leq .5) = .5$;
$P(.4 \leq Y \leq .6) = .36512$;
$P(Y \geq .7) = .16308$

4.58 a. $f_Y(y) = 30y^2 - 60y^3 + 30y^4$

b. Table of $f_Y(y)$ values

y	$f_Y(y) = 30y^2 - 60y^3 + 30y^4$
0	0
.25	1.054
.50	1.875
.75	1.054
1.00	0

c. $E(Y) = .5$;
$\text{Var}(Y) = .0357$;
$\sigma_Y = .1889$

d. $\mu_Y \pm \sigma_Y$ gives the interval (.3111, .6889).
$P(.3111 \leq Y \leq .6889) = .6438$; Empirical Rule probability is .68.

4.59 a. Table of $F_Y(y)$ values

y	$F_Y(y)$
0	.21
1	.59
2	.79
3	.90
4	.96
5	.99
6	1.00

b. $E(Y) = 1.56$
$\sigma_Y = 1.34$

4.60 a. Joint probability table

y_1 \ y_2	0	1	2	3	4	5	6
0	.0441	.0798	.0420	.0231	.0126	.0063	.0021
1	.0798	.1444	.0760	.0418	.0228	.0114	.0038
2	.0420	.0760	.0400	.0220	.0120	.0060	.0020
3	.0231	.0418	.0220	.0121	.0066	.0033	.0011
4	.0126	.0228	.0120	.0066	.0036	.0018	.0006
5	.0063	.0114	.0060	.0033	.0018	.0009	.0003
6	.0021	.0038	.0020	.0011	.0006	.0003	.0001

b. Table of $P_S(s)$ values

s	$P_S(s)$	s	$P_S(s)$
0	.0441	7	.0328
1	.1596	8	.0142
2	.2284	9	.0058
3	.1982	10	.0021
4	.1488	11	.0006
5	.1022	12	.0001
6	.0631		

c. $E(S) = 3.12$;
$\sigma_S = 1.9007$

4.61 a. $P(0 < X < .5 \cap 0 < Y < .5) = .0371$
b. $P(Y > 1) = .7688$

4.62 a. $f_X(x) = (3/320)(128/3 + 32x - 16x^2)$
b. $f_{Y|X}(y|x) = (48 - 12x^2 - 3y^2 + 12xy)/(128 + 96x - 48x^2)$

4.63 No; longer programs should have longer times. The conditional density of Y and $X = x$ varies with x, so there is dependence.

4.64 a. $P(X > .5 \cap Y > .5) = .0586$
b. $P(.1 < Y < .3) = .3903$

4.65 a. $f_X(x) = 12x(1 - x)^2$, $f_Y(y) = 20y(1 - y)^3$
b. $\text{Cov}(X, Y) = 0$, by independence

CHAPTER 5

SECTION 5.1, p. 123

5.1 $\binom{7}{3} = 35$ panels

5.2 $\binom{5}{2}\binom{2}{1} = 20$ panels with exactly 2

$\binom{5}{2}\binom{2}{1} + \binom{5}{3}\binom{2}{0} = 30$ panels with at least 2

5.3 $\binom{8}{4} = 70$ choices

5.4 $\binom{4}{3}\binom{4}{1} = 16$ choices

SECTION 5.2, p. 129

5.5 a. $\binom{10}{3}(.2)^3(.8)^7 = .2013$

b. $\binom{4}{2}(.4)^2(.6)^2 = .3456$

c. $\binom{16}{12}(.7)^{12}(.3)^4 = .2040$

5.6 a.

y	1	2	3
$P_Y(y)$	.35596	.29663	.13184
Table 1	.3560	.2966	.1318

b. The histogram is right-skewed.

c. $\mu_1 = 1.50$, $\sigma_Y = 1.0607$

5.7 a. $P(Y \geq 4) = .9840$

b. $P(Y > 4) = .9490$

c. $P(Y \leq 10) = .8723$

d. $P(Y \geq 16) = .0000$

5.8 $P(Y \leq 16) = .9840$ and $P(Y < 16) = .9490$
The events are logically identical, with renaming of success/failure.

5.9 a. Yes; each customer is a trial, the success probability shouldn't change, and the results should be independent.

b. $P(Y = 5) = .1789$

c. $P(Y \leq 5) = .4163$

d. mode = 6

5.10 $\mu_Y = 6$, $\sigma_Y = 2.05$

5.11 $P(3.95 \leq Y \leq 8.05) = .7796$, not overly close to .68

5.12 a. Independence may not hold, because successes may depend on storms, etc.

b. $P(Y \geq 85) = .9601$

5.13 $\mu_Y = 90$, $\sigma_Y = 3$

5.14 a. Taking $\pi = .10$, $P(Y = 0) = .3487$

b. $P(Y \geq 2) = .2639$

c. $\mu_Y = .80$

5.15 Histogram is very right-skewed.

SECTION 5.3, p. 132

5.16 a.

y	0	1	2	3
$P_Y(y)$	4/35	18/35	12/35	1/35

b. The distribution is slightly right-skewed.

5.17 $\mu_Y = 1.2857$, $\sigma_Y = 0.6999$

5.18 a. $P_Y(2) = .3$

b. $P_Y(2) = .4286$

c. $P_Y(2) = .5$

5.19 a. $P_Y(0) = .0776$

b. $P_Y(0) = 0$

5.20 a. $P_Y(4) = 1/70 = .014$

b.

y	0	1	2	3	4
$P_Y(y)$	1/70	16/70	36/70	16/70	1/70

5.21 $P(\text{at least } 1) = 1 - \binom{2375}{50} \Big/ \binom{2500}{50}$

5.22 $\mu_Y = 2.5$, $\sigma_Y^2 = 2.257$

5.23 $P(Y \geq 1) = .923$ with $n = 50$ and $\pi = .05$

$\mu_Y = 2.5$, $\sigma_Y^2 = 2.375$, close to the correct values.

5.24

y	1	2	3
$P_Y(y)$	5/35	20/35	10/35

SECTION 5.5, p. 138

5.25 a. From Table 2, $P_Y(1) = .2681$, .3476, and .0395, respectively.

b. Summing for $y = 0, 1, 2, 3$ in Table 2, $P(Y \leq 3) = .9211$ and .0817, respectively.

c. Summing for $y = 0, 1, \ldots, 10$ in Table 2, $P(Y \leq 10) = 1.0000$ and .5831, respectively.

5.26 The distribution is highly right-skewed.

5.27 a. $P(Y \leq 10) = .7060$ b. $P(Y \geq 7) = .7932$

c. $P(7 \leq Y \leq 11) = .5962$

5.28 $E(Y) = 9$; $\sigma_Y = 3$

5.29 If a fire burns several neighboring houses and two or more of them were insured by this firm, then the assumption that events happen one at a time doesn't hold.

5.30 a. $P(Y = 0) = .6703$ b. $P(Y \geq 2) = .0616$

5.31 $E(Y) = .4$; $\sigma_Y = .632$

5.32 a. $P(Y = 0) = .0003$ b. $P(Y \geq 5) = .900$

SECTION 5.6, p. 141

5.33 $P(60 < Y < 85) = 25/110$

5.34 $\mu_Y = 65$, $\sigma_Y = 31.75$

5.35 a. $P(10 < Y < 50) = 40/200 = .2$
b. $P(Y > 50) = 150/200 = .75$
c. $P(Y \leq 120) = 120/200 = .60$

5.36 a. $P(300 \leq Y \leq 1300) = 1000/10{,}000 = .1$
b. $\sigma_Y^2 = (10{,}000)^2/12 = 8{,}333{,}333.3$

5.37 a. $P(Y \geq 8) = (20 - 8)/20 = .60$
b. $\sigma_Y = \sqrt{(20)^2/12} = 5.774$

SECTION 5.7, p. 143

5.38 For $f_Y(y) = 0.4e^{-y/2.5}$ we have

y	0	.5	1.0	1.5	2.0
$f_Y(y)$	.4000	.3275	.2681	.2195	.1797

The density is highly right-skewed.

5.39 a. $P(Y > 2) = e^{-2/2} = .3679$
b. $P(Y > 1) = e^{-1/2} = .6065$
c. $P(1 < Y < 2) = e^{-1/2} - e^{-2/2} = .2386$
d. $P(1 \leq Y \leq 2) = P(1 < Y < 2)$ for a continuous random variable.

5.40 With $\mu = 1.25$, $P(Y > 1) = e^{-1/1.25} = .4493$ and $P(Y \geq 2) = e^{-2/1.25} = .2109$

5.41 a. $P_Y(0) = e^{-.8} = .4493$
b. $P_Y(0) = e^{-.8(2)} = .2109$
c. Same probabilities because same events.

5.42 a. $P(Y < 2.5) = 1 - e^{-2.5/5} = .3935$
b. $P(Y > 10) = e^{-10/5} = .1353$

5.43 a. $\mu = 1/5 = 0.2$
b. Poisson with $\mu = (0.2)(2.5) = 0.5$
$P(Y \geq 1) = 1 - P_Y(0) = .3935$
c. with $\mu = (0.2)(10) = 2.0$, $P(Y = 0) = .1353$

5.44 a. $P(20 < Y < 60) = .3834$
b. $\sigma_Y = 40$

5.45 The expected rate may not be constant.

5.46 a. $P(Y > 5) = .0163$
b. $P(\text{arrival time} < 3) = .4512$

5.47 With $\mu_Y = 12(1/4) = 3$, $P(Y \geq 6) = .0838$

5.48 a. $P(W > 7) = .2466$
b. $P(W > 14) = .0608$

5.49 With $\mu = 7(0.2) = 1.4$, $P(Y \geq 4) = .0538$

SECTION 5.8, p. 149

5.50 a. $P(0 \leq Z \leq 1.00) = .3413$
b. $P(0 \leq Z \leq 1.65) = .4505$
c. $P(-1.00 \leq Z \leq 0) = .3413$
d. $P(-1.28 \leq Z \leq 0) = .3997$
e. $P(-1.65 \leq Z \leq 1.65) = .9010$
f. $P(-1.28 \leq Z \leq 1.28) = .7994$
g. $P(-1.07 \leq Z \leq 2.33) = .8478$
h. $P(Z \geq 2.65) = .0040$
i. $P(Z \leq -2.42) = .0078$
j. $P(Z \geq 1.39 \text{ or } Z \leq -1.39) = .1646$

5.51 a. $k = 2.33$ b. $k = 2.33$ c. $k = 2.33$
d. $k = 1.00$ e. $k = 2.00$ f. $k = -1.645$

5.52 From Exercise 6.25 d. and e.:
$P(-1 \leq Z \leq 1) = .6826$ and $P(-2 \leq Z \leq 2) = .9544$
From the Empirical Rule:
$P(-1 \leq Z \leq 1) \cong .68$ and $P(-2 \leq Z \leq 2) \cong .95$
The normal curve probabilities are more precise statements of the Empirical Rule probabilities.

5.53 a. $P(Y \leq 130) = P(Z \leq 2)$
b. $P(Y \geq 82.5) = P(Z \geq -1.167)$
c. $P(Y \leq 130) = .9772$; $P(Y \geq 82.5) = .8790$
d. $P(Y > 106) = .3446$; $P(Y < 94) = .3446$; $P(94 < y < 106) = .3108$
e. $P(Y \leq 70) = .0228$; $P(Y \geq 130) = .0228$; $P(70 < Y < 130) = .9544$

5.54 a. $k = 24.75$ b. $k = 24.75$ c. $k = 112.6$
d. $k = 92.2$ e. $k = 112.6$ f. $k = 92.2$

5.55 a. $P(Y \geq 1000) = .3085$ b. $P(Y \leq 940) = .1587$
c. $P(960 \leq Y \leq 1060) = .6687$

5.56 a. $k = 928.8$ b. $k = 970$

5.57 a. $P(Y > 5.40) = .2266$ b. $P(4.70 < Y < 5.50) = .6826$
c. $P(Y > 3.90) = .9987$

SECTION 5.9, p. 153

5.58 a. Using binomial tables: $P(40 \leq Y \leq 60) = .965$
b. Using normal approximation: $P(40 \leq Y \leq 60) = .9544$
The normal approximation to the binomial is quite good. Note that $n\pi = 50$ and $n(1 - \pi) = 50$ both exceed 10.

5.59 Using a continuity-corrected normal approximation:
$P(39.5 \leq Y \leq 60.5) = .9642$, very close to the correct value.

5.60 Using binomial tables: $P(Y \geq 85) = .9601$
Using normal approximation: $P(Y \geq 85) = .9522$, quite close.

5.61 Using continuity-corrected normal approximation:
$P(Y \geq 84.5) = .9664$, closer.

5.62 We would expect the Poisson assumptions to be a reasonable approximation if the input errors occurred independently of each other during a given time period.

5.63 a. $E(Y) = 8$; $\sigma_Y = 2.828$ b. $P(5 \leq Y \leq 11) = .7884$

5.64 $P(4.5 \leq Y \leq 11.5) = .785$

CHAPTER EXERCISES, p. 157

5.65 a. Yes, the binomial assumptions are met.
b. $P(Y \geq 24) = .012$
c. Assuming the claim is true ($\pi = .15$), then $P(Y \geq 24) = .012$, a very small probability. Therefore, if in fact 24 of the 100 numbers turn out to be business phones, we would doubt the manufacturer's claim.

5.66 a. $E(Y) = 15$; $\sigma_Y^2 = 12.75$
b. $P(Y \geq 24) = .0059$; $P(Y \geq 23.5) = .0087$; fairly close.
5.67 a. $P(Y \leq 1) = .2794$ b. $P(Y \geq 4) = .2396$
5.68 a. $P(Y \leq 1) = .2873$ b. $P(Y \geq 4) = .2424$
The Poisson probabilities are a fairly good approximation to the binomial probabilities.
5.69 a. $\binom{30000}{100}$ b. $\binom{29700}{100}$ c. $\binom{29700}{100}/\binom{30000}{100}$
d. $\Sigma_{k=0}^{2} \binom{300}{k}\binom{29700}{100-k}/\binom{30000}{100}$
5.70 a. $P(Y = 0) = \binom{100}{0}(.01)^0(.99)^{100}$; $P(Y \leq 2) = \Sigma_{Y=0}^{2} \binom{100}{Y}(.01)^Y(.99)^{100-Y}$
b. $P(Y = 0) = .3678$; $P(Y \leq 2) = .9196$
5.71 a. $\binom{30}{14} = 145{,}422{,}675$ b. $\binom{6}{5}\binom{24}{9} = 7{,}845{,}024$ c. .059
5.72 a. $P(Y = 1) = .3614$; $P(Y \leq 1) = .6626$
b. $E(Y) = 1.2$; $\sigma_Y = 1.095$
5.73 A normal approximation would not be accurate since $\mu < 5$.
5.74 a. $P(Y \leq 72.8) = .6915$; $P(71.2 \leq Y \leq 72.8) = .3830$
b. $P(Y \geq 74) = .1056$ c. $k = 75.73$
5.75 a. $P(Y > 73) = .2659$ b. $P(Y = 3) = .055$
5.76 a. $P(Y > 5) = .0668$ b. $P(Y \leq 6) \approx 1$
5.77 a. $P(Y \geq 8) \approx 0$
b. Since $P(Y = 8)$ is nearly 0, the result conclusively indicates that the cutter was inefficient.
5.78 a. $P(Y < 8) = .8666$
b. We might expect the independent assumption not to hold; that is, it is likely that the occurrence of errors in transmission are dependent during a given time period.
5.79 $P(Y < 8) = .2203$
5.80 a. Binomial with $n = 1000$ and $\pi = .50$ under reasonable assumptions
b. $\mu_Y = 500$, $\sigma_Y = 15.811$
5.81 $P(Y \leq 460) \approx .0057$; accurate because $n\pi = n(1 - \pi) = 500$
5.82 a. $P(Y \leq 5) = \displaystyle\sum_{y=0}^{5} \binom{250}{y}(.01)^y(.99)^{250-y}$
b. Independence and constant probability seem to be reasonable.
5.83 a. $P(Y \leq 5) = .9441$
b. $P(Y \leq 5) = .9580$
c. Poisson, because $\mu_Y < 5$
5.84 a. $\mu_Y = 100$; $\sigma_Y = 99.499$
b. $P_Y(1) = .01 > P_Y(100) = .00370$
5.85 a. $P(Y \geq 400) = \displaystyle\sum_{y=400}^{\infty} \frac{(y-1)!}{(4-1)!(y-4)!}(.01)^4(.99)^{y-4}$
b. $\sigma_Y^2 = 39{,}600$
5.86 a. 120 choices
b. $P(Y \geq 2) = 110/455$
5.87 $\mu_Y = 1.0$, $\sigma_Y^2 = .4762$
5.88 a. $P(Y > 7) = .2466$
b. $P(Y > 14) = .0608$
5.89 $\mu_Y = 5$, $\sigma_Y = 5$
5.90 Nonclumping of crashes and independence of time periods. Independence would be violated.

5.91 a. $P(Y \leq 57.5) \approx P(Z < -2.81) = .0025$
b. Yes; $n\pi = 80$ and $n(1 - \pi) = 320$
c. Not plausible

5.92 a. $P(Y \leq 56) = \sum_{y=0}^{56} \binom{400}{y} (.2)^y (.8)^{400-y}$
b. $P(Y \leq 56) = .0013$ or $P(Y \leq 56.5) = .0016$
c. No; result is unlikely assuming $\pi = .20$.

5.93 Poisson ($\mu = 0.5$) probability $= .3935$

5.94 $P(\text{win}) = 1/3{,}838{,}380 = .0000002605$

5.95 a. $P(\text{exactly } 4) = .002192$
b. $P(\text{at least } 4) = .002246$

5.96 a. With slightly more numbers, $P(\text{win})$ should decrease a bit.
b. $P(\text{win}) = .0000001906$
c. Decreased quite substantially

5.97 a. Binomial with $n = 1{,}000{,}000$ and $\pi = 1/3{,}838{,}380$
b. $P_Y(0) = \binom{1{,}000{,}000}{0} (1/3{,}838{,}380)^0 (1 - 1/3{,}838{,}380)^{1{,}000{,}000}$
c. $P(Y \geq 2) = 1 - \sum_{y=0}^{1} \binom{1{,}000{,}000}{y} \pi^y (1 - \pi)^{1{,}000{,}000-y}$
where $\pi = 1/3{,}838{,}380$

5.98 a. $\mu_Y = 0.2605$, $\sigma_Y^2 = 0.2605$
b. $P(Y = 0) = e^{-0.2605} = .7707$

$$P(Y \geq 2) = 1 - e^{-0.2605} - e^{-0.2605} \frac{(.2605)^1}{1!} = .0286$$

Should be close; n is large and $n\pi < 1$.
c. $P(Y = 0) = .771$
$P(Y \geq 2) = .028$

5.99 No; $n\pi = 0.2605 < 5$

5.100 a. Geometric with $\pi = 1 - .771 = .229$
b. $\mu_X = 4.37$, $\sigma_X = 3.834$
c. $P_X(3) = (.771)^2(.229)$ for the probability of no winners in two periods and at least 1 winner in the third period.

5.101 The probability of success would increase (depending on the previous results).

5.102 a. $\mu = 3.5$
b. Poisson $P(Y \geq 4) = .4634$
c. Nonclumping and independence. Maybe there's an arsonist or a "take care" effect, but basically the assumptions seems sensible.

5.103 $P(W > 3) = .2231$, $\mu_W = 2$

5.104 a. $\mu_Y = 12$, $\sigma_Y = 3.464$
b. $P(Y > 8) = .8450$

5.105 $P(W \geq 2) = .0498$

5.106 a. No. Constant rate is irrelevant for Poisson probabilities.
b. No. Constant rate is irrelevant for Poisson probabilities.

CHAPTER 6
SECTION 6.1, p. 165

6.1 Select 375, 779 (ignore 995, 963, 895, and 854), 289, 635, 094, 103, 071, 510, 023, and 010.

6.2 Selection bias favoring those at home at those times.

6.3 a. Not all combinations of books have the same probability.
b. Select books from an inventory list, using random numbers.

6.4 Use random numbers between 0001 and 4256.

6.5 Size bias favoring holders of many seats.

6.6 No; there are possible biases. First strategy probably would check only the work of a few employees.

SECTION 6.2, p. 169

6.7 A sample of the 100 most recently repaired machines might be biased toward those machines that are in poorest condition.

6.8 One possible approach to collecting a sample of 100 times to breakdown is to obtain records for the population of all machines in the chain of laundromats, randomly choose 100 records, and record the time between the last 2 repairs for each of the chosen machines.

6.9 If I were an embezzler working in the backroom, I would make sure not to "tinker" with any transactions with serial numbers ending in 00.

6.10 An ideal way to randomly sample 1% of all transactions would be to collect all transaction numbers for each day and select 1% of them via a lottery or random number generator.

6.11 a. Both are symmetric; there is less variability for $n = 3$.
b. $P(\bar{Y}$ will be no more than \$.50 away from the mean$) = .60$

6.12 Sampling distribution, $n = 2$: $E(\bar{Y}) = 4$; $\sigma^2_{\bar{Y}} = .75$
Sampling distribution, $n = 3$: $E(\bar{Y}) = 4$; $\sigma^2_{\bar{Y}} = .3334$
The sample size does not affect the expected values but does affect the variances. As the sample size increases, the variance decreases.

6.13 b. $\mu = E(Y) = 7$ c. $E(\bar{Y}) = 7$; $\sigma^2_{\bar{Y}} = 1.575$
d. Probability $= .8401$

6.14 a. The histogram of the sampling distribution of $\bar{Y}$ shows smaller variance.
b. The histogram of the sampling distribution of $\bar{Y}$ shows less skewness.

6.15 b. The histogram looks very much like the theoretical probability histogram. The Monte Carlo simulation appears to be a good approximation to the sampling distribution of $\bar{Y}$.

SECTION 6.3, p. 172

6.16 $E(\bar{Y}) = \mu = 7$; $\sigma^2_{\bar{Y}} = 1.575$. The values agree.

6.17 a. $E(T) = 46{,}350$; $\sigma_T = 6158.9$ b. $E(\bar{Y}) = 927$; $\sigma_{\bar{Y}} = 123.2$

6.18 a. $E(\bar{Y}) = 28.2$; $\sigma_{\bar{Y}} = .218$ b. $E(\bar{Y}) = 28.2$; $\sigma_{\bar{Y}} = .109$

SECTION 6.4, p. 179

6.19 a. $E(T) = 3270$; $\sigma_T = 107.52$
b. $P(3150 \leq T \leq 3390) = .7372$
c. $\mu_{\bar{Y}} = 327$; $\sigma_{\bar{Y}} = 10.76$; $P(314 \leq \bar{Y} \leq 339) = .7555$

6.20 Range: 327 − 21 to 327 + 21 or 306 to 348

6.21 a. The population distribution is skewed right.

b.

Sample Size	$P(\mu - 2\sigma_{\bar{Y}} < \bar{Y} < \mu + 2\sigma_{\bar{Y}})$	Normal Approximation
2	.9664	.95
4	.9479	.95
8	.9681	.95
16	.9532	.95
32	.9533	.95

c.

Sample Size	$P(\mu - \sigma_{\bar{Y}} < \bar{Y} < \mu + \sigma_{\bar{Y}})$	Normal Approximation
2	.7840	.68
4	.6739	.68
8	.6946	.68
16	.6638	.68
32	.6984	.68

6.22 a. Sampling distribution of $\bar{Y}$ is approximately normal with $\mu_{\bar{Y}} = 927$ and $\sigma_{\bar{Y}} = 123.2$.
b. $P(\bar{Y} > 1100) = .0808$

6.23 The population might well be very skewed, so $n = 50$ may not be enough for a good approximation.

6.24 $P(\bar{Y} > 1100) = .0808$. There is only an 8% chance that the sample average takes on the value 1100 or greater. One might conclude that the repair claims this year will be a bit higher than in the past.

$P(\bar{Y} > 1000) = .2776$. We might conclude that the repair claims this year will be similar to those in the past.

SECTION 6.6, p. 188

6.25 a. Expected value and standard error of $\bar{Y}$
b. Close to $\mu_{\bar{Y}} = 50$ and $\sigma_{\bar{Y}} = 5$

6.26 Histogram is nearly normal. For a normal population, the theoretical distribution is exactly normal.

6.27 Expected value is close to 0 and standard error is close to .2.

6.28 Both histogram and normal plot indicate normal distribution, possibly with a very few mild outliers.

6.29 a. $\mu = 3$, $\sigma = 1.265$
b. $n = 10$: $\mu_{\bar{Y}} = 3$, $\sigma_{\bar{Y}} = .400$
$n = 30$: $\mu_{\bar{Y}} = 3$, $\sigma_{\bar{Y}} = .231$
c. Means very close, standard deviations a bit smaller.

6.30 a. Discreteness of means, from discreteness of population.
b. Yes; plot is nearly a straight line.

6.31 a. $\mu_{\bar{Y}} = 0$, $\sigma_{\bar{Y}} = .4772$
b. Both the average and the standard deviation are very close.

6.32 Both plots indicate a normal distribution.

CHAPTER EXERCISES, p. 196

6.33 a. Since the names of all Fortune 1000 firm directors are publicly available, we could simply sample from a list to obtain a sample of 200. There could be a problem of duplication, for directors of several firms.
b. No. No matter how large the sample size is, if the distribution of percentage income tax payments in the population is skewed, so will be the distribution of percentage income tax in the sample.

6.34 The classified ads are a collection of many contributions from many areas of interest. More than likely, each of the advertisers acts independently when deciding on the number of column-inches to buy. Therefore, by CLT, the individual Monday column-inch values can be expected to be normally distributed.

6.35 The union presented data from a committee of 22 unhappy workers. This is a biased sample.

6.36 a. The demand for a 5-pound sack of flour by an individual customer may be thought of as independent from every other customer. The weekly demand is the sum of these individual demands, and there are most likely a large number of customers for a supermarket. Therefore, by CLT, the weekly demand can be expected to be roughly normally distributed.
b. One way to select a random sample of size $n = 15$ would be to visit the store on a randomly selected day of the week at a randomly selected time of the day and observe the number of sacks of flour bought by the first 15 customers to go through the checkout lines.

6.37 a. $P(\bar{Y} > 73.0) = .0078$
b. 95% range for $\bar{Y}$: 71.1905 to 72.8095

6.38 a. We can obtain a listing of all of the 2571 sales categories and simply sample every 25th sales category or have a lottery with sales category codes.
b. A simple random sample may not be desirable. It's possible that there will not be equal representation from the various departments in the department store.
One possible sampling method alternative is to group the 2571 into various departments and randomly choose an equal number from within each department to make up the sample.

6.39 a. $E(\bar{Y}) = 2.2$; $\sigma_{\bar{Y}} = .16$ b. $E(\bar{Y}) = 2.2$; $\sigma_{\bar{Y}} = .1568$; very little difference
c. $P(\bar{Y} > 2.4) = .1056$

6.40 With a sample of size of 100, we would expect the normal approximation to be fairly good barring severe skewness. A plot of the sample data would indicate to us how much faith we would have in the approximation.

6.41 a. $\mu = 100$; $\sigma = 21.07$
b. The distribution is perfectly symmetric. There are two extreme (symmetric) outlier values.

6.42 a.

Sample Size	$P(\mu - \sigma_{\bar{Y}} < \bar{Y} < \mu + \sigma_{\bar{Y}})$	Normal Approximation
2	.8224	.6828
4	.8202	.6828
8	.7088	.6828

b. Approximation is good for $n = 8$.

c.

Sample Size	$P(\mu - 2\sigma_{\bar{Y}} < \bar{Y} < \mu + 2\sigma_{\bar{Y}})$	Normal Approximation
2	.9224	.95
4	.9024	.95
8	.9334	.95

Again, for $n = 8$, approximation is decent.

REVIEW EXERCISES, CHAPTERS 4–6, p. 198

R21 a. $P(Y > 210) = .4013$
b. $P(\bar{Y} > 210) = .1056$

R22 Part a. will be poorer; part b. is influenced by Central Limit Theorem.

R23 a. $P(Y \leq 3) = \sum_{y=0}^{3} \binom{50}{y}(.05)^y(.95)^{50-y}$
b. $P(Y \leq 3) = .7604$
c. Binomial assumptions; might have dependence, but assumptions seem fairly sensible.

R24 a. $\mu_X = 6.66667$, $\sigma_X = 2.687$
b. $P(X \geq 3) = .909$

R25 a. $\mu_T = 26.66668$, $\sigma_T^2 = 28.444$
b. Independence, only in variance calculation.

R26 $P(\bar{X} > 7) = .0401$; good approximation, using Central Limit Theorem, with $n = 200$.

R27 a. For a continuous random variable, probability is an area, not a sum.
b. $P(5 \leq Y \leq 8) = .5067$
c. $\mu_Y = 6$, $\sigma_Y = 2$

R28 $P(Y < 8|Y > 5) = .7370$

R29 Poisson; $P_Y(0) = .3012$

R30 Nonclumping and independence of impurities

R31 Not independent

		y						
		0	1	2	3	4	5	$P_X(x)$
	0	.0000	.2400	.1200	.0200	.0120	.0080	.40
	1	.0300	.1200	.0750	.0450	.0150	.0150	.30
x	2	.0075	.0450	.0600	.0150	.0120	.0105	.15
	3	.0030	.0150	.0220	.0300	.0200	.0100	.10
	4	.0005	.0020	.0075	.0150	.0200	.0050	.05
$P_Y(y)$		.0410	.4220	.2845	.1250	.0790	.0485	

R32 $\mu_Y = 1.9245$, $\sigma_Y = 1{,}2074$

R33 a. $P(Y \geq 2) = .2641$
b. Binomial assumptions

R34 No; not sampling a fixed number, and order is relevant.

R35 $\mu_Y = 2.07$, $\sigma_Y^2 = .2851$

R36 a. $P(0 < X < .3 \cap 0 < Y < .5) = .406875$
b. $f_X(x) = 2(1 - x)$, for $0 < x < 1$
c. Yes; $f_{XY}(x, y) = f_X(x)f_Y(y)$

R37 $\mu_Y = .25$, $\sigma_Y = .194$

R38 $\mu_{\overline{W}} = 33.33$, $\sigma_{\overline{W}}^2 = 103.90$

R39 $P(\overline{Y} > 265) = .0089$

R40 Used Central Limit Theorem; poor approximation if the distribution of individual values is highly skewed.

R41 $P(Y \leq 3) = .0281$

R42 Binomial assumption seem plausible.

R43 a. $P_{Y|X}(y|x) = (xy + 1)/(5 + 10x)$
b. No; as x gets larger, y (reasonably) tends to get larger.

R44 $\mu_Y = 2.743$, $\sigma_Y^2 = 1.448$

R45 Poisson; $P(Y \geq 10) = .2833$

R46 Nonclumping and independence appear plausible.

R47 $P(200 < X < 300 \cap 30 < Y < 50) = .0496$

R48 $f_{Y|X}(y|x) = .000006y(100 - y)$, for $0 < y < 100$
Independent

R49 $\mu_X = 170$, $\sigma_X^2 = 2500$
$\mu_Y = 50$, $\sigma_Y^2 = 500$
$\mu_T = 220$, $\sigma_T^2 = 3000$

R50 $\mu_{X'} = .170$, $\sigma_{X'}^2 = .002500$
$\mu_{Y'} = .050$, $\sigma_{Y'}^2 = .000500$
$\mu_{T'} = .220$, $\sigma_{T'}^2 = .003000$

R51 a. $P(\overline{X}' > .180) \approx .001$
b. Yes; $n = 250$ is large and the Central Limit Theorem applies.

CHAPTER 7
SECTION 7.1, p. 207

7.1 a. $\overline{y} = 23.985$, median $= 23.8$ b. $\overline{y}_t = 28.8438$

7.2 a. Frequency distribution table is shown below:

Class	Midpoint	Frequency	Relative Frequency
15.9–18.9	17.4	3	.15
19.0–22.0	20.5	2	.10
22.1–25.1	23.6	8	.40
25.2–28.2	26.7	5	.25
28.3–31.3	29.8	1	.05
31.4–34.4	32.9	1	.05

A histogram is a graphic display of the data as tabulated in the frequency distribution table above.

b. Data are fairly close to normally distributed.
c. The sample mean is most efficient for normal populations.

7.3 a. All estimators, the mean, median, and 20% trimmed mean, appear unbiased, as is to be expected given a normal population. (The averages are all very close to 100.)
b. The mean appears most efficient, since it has the smallest variance 26.52.

7.4 a. $\bar{y} = 236.4$, median $= 234.5$ b. $\bar{y}_t = 233.875$

7.5 a. A stem-and-leaf diagram of the data is shown below:

Stem	Leaves
17	1
18	5
19	3 9
20	4
21	0 6 8
22	1 3 5 8 8
23	0 4 5 7
24	0 1 3 5 9
25	1 4 7
26	2 3
27	1
28	0
29	
30	
31	
32	
33	
34	
35	
36	
37	9

The most conspicuous aspect of the display is the value 379, which is an obvious outlier.

b. No. Mean has been thrown off by the outlier. It will not be a good measure of central tendency.

7.6 a. Yes; $E(\hat{\mu}_1) = E(\hat{\mu}_2) = \mu$
b. $\text{Var}(\hat{\mu}_1) = .25\sigma^2 < \text{Var}(\hat{\mu}_2) = .26\sigma^2$

7.7 a. Yes; boxplots center at 0.
b. Median has narrowest boxplot.

7.8 Averages are essentially 0; median has smallest standard deviation.

7.9 a. Median again
b. Similar answers; efficiency doesn't appear to depend heavy on n.

7.10 Again, median is most efficient.

7.11 a. All should be unbiased, by symmetry.
b. Mean has smallest standard deviation.

7.12 Yes, unbiased and mean boxplot narrowest.

SECTION 7.2, p. 213

7.13 a. Sample without replacement.
b. Assuming sampling with replacement, $\sigma_{\bar{Y}} = .0422$
c. Assuming sampling without replacement, $\sigma_{\bar{Y}} = .0392$
d. The standard error of the mean when sampling without replacement is smaller by 3/1000.

7.14 Use the basic conclusion that the absolute sample size is more important in determining probable accuracy than is the fraction of the population that is sampled to explain the projections. Projections are based on absolute sample size, not percentages. For example, all projections may be based on samples of 50,000, which may be 1% of N.Y. population but 20% of the population of Wyoming.

SECTION 7.3, p. 217

7.15 a., b. At right is a table of likelihoods for specified value of θ.

c. MLE of θ is .5

θ	Likelihood $= \theta^4 e^{-8\theta}$
.1	.000045
.2	.000323
.3	.000735
.4	.001044
.5	.001145
.6	.001067
.7	.000888
.8	.000681
.9	.000490
1.0	.000335

7.16 Take derivative, set to zero, and solve.

7.17 a. Direct computation.

b. According to the table in part a., the MLE of θ is 1.5.

7.18 Take derivative, set to zero, and solve.

7.19 a. $\bar{y} = 5.4$, median $= 5.0$

b. $L(5.4) = .000007027$

$L(5.0) = .00001910$

c. At $\bar{y} = 5.4$, the likelihood is not as large as at 5.0; $\bar{Y}$ is not the maximum likelihood estimator.

CHAPTER EXERCISES, p. 219

7.20 a. Yes; averages are near 300, the population mean.

b. Trimmed mean has smallest variance.

7.21 a. No; for a skewed distribution, mean, median, and trimmed mean should differ and have different expected values.

b. Yes; the averages differ.

7.22 a. Yes; the boxplots are symmetric around 0.

b. No! The boxplot for the mean is much wider.

7.23 Yes. The averages are nearly 0; the standard deviation for the mean is much larger.

7.24 For outlier-prone data or (possibly) for highly skewed data.

7.25 No; proportion sampled is not critical for accuracy.

7.26 All are unbiased.

7.27 $\hat{\theta}_3$ has smallest variance; it gives most weight to Y_1, which has smaller variance.

7.28 a. $k = 3$

b.

π	.10	.11	.12	.13	.14	.15
1,000,000 $L(\pi)$	.0261	.0627	.136	.270	.498	.860

c. The maximum will occur at $\pi = .15$ or higher.

7.29 $\hat{\pi} = .3333$

7.30 a. $L(\theta) = \theta^3(25.36443)^{-(\theta+1)}$

b.

θ	1	2	3
$L(\theta)$	.00155	.00049	.000065

$\theta = 1$ is closest

7.31 $\hat{\theta} = 0.928 = n/\log_e[(1 + y_1)\cdots(1 + y_n)]$

7.32 a. MEAN 100.71 MEDIAN 100.10
b. Median; data indicate outliers.

7.33 a. S-shape indicates outliers.
b. Yes; the histogram showed outliers.

7.34 No; data appear to be roughly normal.

7.35 Plot is roughly a straight line, indicating near-normal data.

7.36 No; data are skewed.

7.37 Normal plot is curved, indicating skewness.

CHAPTER 8
SECTION 8.1, p. 230

8.1 a. 95% confidence interval for μ: $22.2 \leq \mu \leq 25.7$
b. 99% confidence interval for μ: $21.68 \leq \mu \leq 26.29$

8.2 We would expect that 95% of all confidence intervals calculated in this manner would include the parameter μ. "In this manner" means that if we sampled again and again (20 persons each time), calculated a sample mean each time and a 95% CI for the true mean each time, we would expect that 95% of them include μ.

8.3 Given the data, it seems reasonable to assume near-normality.

8.4 90% confidence interval for μ: $2.631 \leq \mu \leq 2.769$

8.5 95% confidence interval for μ: $4.4\% \leq \mu \leq 7.2\%$

8.6 The distribution of Y may not be normal, since shrinkage will not take on negative values. The CLT tells us that for n large and Y not severely skewed, the distribution of Y will be approximately normal. For $n = 36$ and possibly severe skewness, the approximation may not be good.

SECTION 8.2, p. 232

8.7 95% confidence interval for π: $.414 \leq \pi \leq .502$

8.8 We are 95% confident that the true long-run proportion of new product placements is in the interval (.414, .502). By 95% confident we mean that if the experiment is conducted again and again and a confidence interval calculated each time, we expect 95% of them to include the true long-run proportion, π.

8.9 90% confidence interval for π: $.60 \leq \pi \leq .74$

8.10 Yes. The sample size (125) is large enough and $n\hat{\pi}$ and $n(1 - \hat{\pi})$ are both greater than 5 ($n\hat{\pi} = 84$; $n(1 - \hat{\pi}) = 41$), so that the normal approximation to the binomial is accurate.

8.11 90% confidence interval for π: $.096 \leq \pi \leq .304$

SECTION 8.3, p. 235

8.12 a. For 90% CI with width of .50: $n \cong 693$
For 90% CI with width of .25: $n \cong 2771$
For 90% CI with width of .125: $n \cong 11{,}084$
b. In general, one must quadruple the sample size to cut the width of the confidence interval in half.

8.13 For 95% CI with width of $.3\sigma$: $n \cong 171$
For 95% CI with width of $.4\sigma$: $n \cong 96$

8.14 For 95% CI with width of \$50: $n \cong 983$

8.15 a. If $\sigma = 300$, $n \cong 553$. If $\sigma = 450$, $n \cong 1245$
b. If the sample size corresponding to $\sigma = 450$ is used ($n = 1245$) and σ is really 300, the confidence interval is $\bar{y} \pm 16.66$. This interval width is only about 33, much smaller than the desired width of 50.

8.16 Yes. The sample size ($n = 983$) is much larger than the usual values for application of the Central Limit Theorem in assuming normality of $\bar{Y}$.

8.17 a. For 95% confidence interval with width of .02 and using $\hat{\pi} = .5$: $n \cong 9604$
b. For 95% confidence interval with width of .02 and using $\hat{\pi} = .005$: $n \cong 191$
Using $\hat{\pi} = .08$: $n \cong 2827$
The sample size need only be 2827 for a width no larger than .02, given the assumption that $.005 \leq \pi \leq .08$.

SECTION 8.4, p. 240

8.18 a. $P(t > 1.638) = .1$
b. $P(t > 5.841) = .005$
c. $P(t < -2.353) = .05$
d. $P(-2.353 < t < 2.353) = .9$
e. $P(|t| > 3.182) = .05$
f. $P(|t| > 4.541) = .02$

8.19 $P(t > 1.638) = .10$; $P(z > 1.638) = .0507$
$P(|t| > 1.638) = .20$; $P(|z| > 1.638) = .1014$
The mistaken assumption of a normal distribution causes an understatement of the probabilities.

8.20 a. Table of simulated and theoretical relative frequencies:

Range	Actual Frequencies	Relative Frequencies	Theoretical Relative Frequencies
$t < -2.353$	44	.04	.05
$-2.353 < t < -1.638$	59	.0536	.05
$-1.638 < t < 1.638$	896	.8145	.80
$1.638 < t < 2.353$	47	.0427	.05
$2.353 < t$	54	.0491	.05
	1100	1.000	1.00

b. There is no evidence of a systematic departure from the theoretical frequencies.

8.21 $t_{.10,72} = 1.2946$ $t_{.05,72} = 1.6684$ $t_{.01,72} = 2.3836$

SECTION 8.5, p. 243

8.22 a. $\bar{y} = 56.87$; $s = 28.97$
b. 99% confidence interval for μ: $40 \leq \mu \leq 74$
8.23 For 90% confidence interval with width of 6 days: $n \cong 252$
8.24 $\bar{y} = 328.64$; $s = 15.49$
A 95% confidence interval for μ: $319.7 \leq \mu \leq 337.6$
8.25 a. Using the t tables with 19 d.f., $46.6 \leq \mu \leq 63.2$.
b. Data are outlier-prone; mean is inefficient.
8.26 a. $1.071 \leq \mu \leq 1.189$
b. $1.074 \leq \mu \leq 1.186$
c. z interval is (artificial and) narrower.
8.27 The 95% claim is slightly suspect for $n = 31$.

SECTIONS 8.6 AND 8.7, p. 250

8.28 a. Stem-and-leaf display of the data:

Stem	Leaves
29	8
30	7
31	9 6
32	6 0 9
33	1 5 5
34	1 7 6
35	1

b. The stem-and-leaf diagram shows no blatant skewness or extreme outliers, so there is no reason to doubt the approximate correctness of the 95% confidence interval.
8.29 $316 \leq$ median ≤ 346
8.30 $39 \leq$ median ≤ 57
8.31 a. $12.060 \leq \mu \leq 13.184$
b. $11.6 \leq$ median ≤ 13.6
8.32 a. The data appear nearly normal.
b. The more efficient mean has the narrower interval.
8.33 The normal plot is close to a straight line. There is no obvious nonnormality.

CHAPTER EXERCISES, p. 252

8.34 $3.99 \leq \mu \leq 6.41$
8.35 a. No; right-skewed
b. May be poor approximation with $n = 22$.
8.36 $n = 1537$
8.37 No. With $n = 1537$, the CLT certainly applies.
8.38 $.295 \leq \pi \leq .393$
8.39 a. $n = 1527$
b. $n = 1692$
8.40 $.498 \leq \pi \leq .622$
$.438 \leq \pi \leq .562$
$.398 \leq \pi \leq .522$

8.41 Nonsense, assuming a random sample. The sample fraction is not critical in determining accuracy.

8.42 $97.6 \leq \mu \leq 102.4$

8.43 a. No; the fraction isn't relevant to accuracy.
b. Yes; sample wasn't random. The machine may have cyclic behavior.

8.44 $.0168 \leq \mu \leq .0252$

8.45 Correction factor is .9989 and has no real effect.

8.46 Data severely right-skewed and n = only 22. Confidence level may be wrong.

8.47 a. Data are roughly normal, perhaps with slight outliers.
b. The mean interval should be narrower, barring outliers; $59.903 \leq \mu \leq 63.017$
c. $60 \leq \text{median} \leq 63$
Actually, the median interval is a trifle narrower.

8.48 There might be a slight S-shape, indicating a few moderate outliers.

CHAPTER 9
SECTION 9.1, p. 261

9.1 a. π = long-run proportion of bids resulting in takeovers
b. H_0: $\pi = .35$
c. H_a: $\pi < .35$
d. From Appendix Table 1 with $n = 20$, $P(Y \leq 3 | \pi = .35) = .0445$.

9.2 Yes. $y = 2$ is in the rejection region of part d, Exercise 9.1.

9.3 a. The proportion, π, of houses with swimming pools that must be purchased by the realtor is the relevant parameter.
b. H_a: $\pi > .05$ (one-sided)
c. H_0: $\pi \leq .05$ d. R.R.: If $y \geq 6$, reject H_0

9.4 Since $y = 7$, reject H_0 at $\alpha = .05$ level.

9.5 a. The population parameter, π, is the proportion of applications prepared by the new manager that are accepted for funding.
b. $\pi = .50$
c. H_a: $\pi \neq .5$, because a change in either direction would be relevant.
d. R.R.: If $y \leq 4$ or $y \geq 14$, reject H_0

9.6 Since $y = 7$, do not reject H_0 at $\alpha = .05$ level.

9.7 The assumptions for a binomial model are (a) fixed number of trials, (b) independence of trials, (c) constant probabilities. If these assumptions are not valid, use of binomial probabilities may not be appropriate. For example, if the decision to accept an application from the city is based on the number of applications already accepted, trials are not independent and probabilities are not constant.

9.8 a. $\pi = .40$
b. The selection method might decrease the rate of desirability of products, as well as increase it.
c. The intent is to improve desirability of products; H_a: $\pi > .40$

9.9 a. $\mu = 20(.40) = 8.0$
b. Reject H_0 if $y \leq 3$ or if $y \geq 13$
$P(Y \leq 3 \cup Y \geq 13 | \pi = .40) = .0370$

9.10 The binomial assumption of independence is violated.

SECTION 9.2, p. 264

9.11 a. $\beta = P(Y \geq 4|\pi = .25) = .7749$
b. power $= P(\text{reject } H_0|\pi = .25) = P(Y \leq 3|\pi = .25) = .2251$

9.12 As the population parameter gets farther from the H_0 value, β should decrease.

9.13 $\beta = P(Y \leq 5|\pi = .10) = .6161$

9.14 False positive (Type I) error

9.15 a. The probability that a false H_0 will not be rejected is β; all else equal, β decreases as n increases.
b. $\beta = P(Y \leq 9|\pi = .10) = .4513$, which is indeed smaller than $\beta = .6161$ as found in Exercise 9.13.

9.16 Type I (false positive): Claim a different marketing rate, when in fact the rate is the same.
Type II (false negative): Claim the same marketing rate, when in fact the rate is different.

9.17 $\beta = P(H_0 \text{ not rejected}) = P(4 \leq Y \leq 12)$ with $n = 20$.
Adding probabilities in Appendix Table 1 yields the following results.

π	.45	.50	.55	.65	.75	.80
β	.9370	.8671	.7477	.3990	.9019	.0322

The β curve decreases in an S-shape.

9.18 a. power$(\pi = .50) = 1 - \beta_{.50} = .1329$
b. As n increases, power increases.

SECTIONS 9.3 AND 9.4, p. 275

9.19 a. The parameter to be tested is the mean waiting time of nonemergency patients.
b. H_0: $\mu \leq 30$
H_a: $\mu > 30$
c. T.S. $z = \dfrac{\bar{y} - \mu_0}{\sigma/\sqrt{n}} = \dfrac{\bar{y} - 30}{10/\sqrt{22}}$
R.R. at $\alpha = .05$, reject H_0 if $z > 1.645$

9.20 T.S. $z = 3.7992$
Conclusion: Reject H_0 at $\alpha = .05$ level

9.21 $\beta = P(z < -.2311) = .4086$ for $\mu_a = 34$

Table of β probabilities:

μ_a	$\beta = P(H_0 \text{ not rejected}\|H_a \text{ true})$
32	.7602
34	.4086
36	.1212
38	.0175
40	.0012

9.22 All waiting times on a busy day would be long. Hence, independence would be a bad assumption.

9.23 H_0: $\mu = 1.5$
H_a: $\mu \neq 1.5$
T.S.: $z = -.5887$
RR: At $\alpha = .05$, reject H_0 if $z \geq 1.96$ or $z < -1.96$
Conclusion: Do not reject H_0.

9.24 Table of β probabilities:

μ_a	β
1.0	.7365
1.2	.8750
1.4	.9419
1.6	.9419
1.8	.8750
2.0	.7365

9.25 Plot of the data:

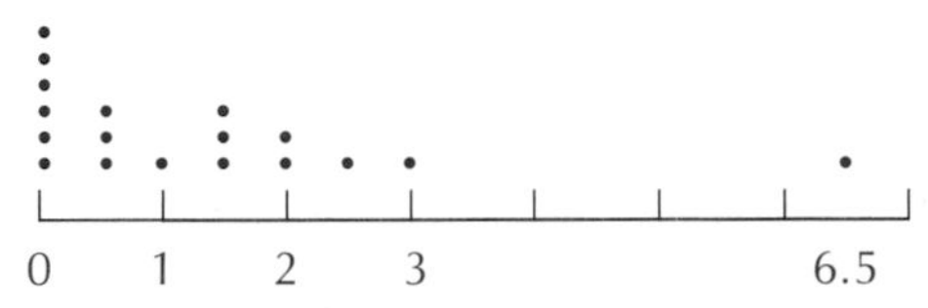

Since n is small (18) and the data are clearly skewed, the assumption of normality is not appropriate.

SECTION 9.5, p. 279

9.26 p-value $= P(z \geq 3.7992) \approx .0002$

9.27 p-value $= 2P(z \geq |-.5887|) = .555$

9.28 p-value $= P(Y \leq 4 | \pi = .15) = .1121$

9.29 p-value $= P(z \leq -1.39) = .0823$

9.30 a. "Z STATISTIC EQUALS": $z = 1.138$

b. "P VALUE": p-value $= .2549$

c. A two-tailed p-value is appropriate. The company is concerned with *any* deviation from the goal of 40%.

9.31 a. The p-value of .2549 exceeds the usual α values of .10, .05, and .01. Therefore, the sales manager is correct in concluding that the test was not statistically significant.

b. According to the results of the test, the probability of obtaining a sample mean 47.6 or larger or 40.4 or smaller if the true mean is 44 is .2549. This means that the result is not unlikely. From these results, we would not suspect that the situation was otherwise. However, the term "proved" is much too strong.

9.32 a. H_0: $\mu \geq 20{,}000$
H_a: $\mu < 20{,}000$

b. T.S. $z = -2.9524$
p-value $= P(z < -2.95) = .0016$

9.33 Since the p-value $\leq \alpha$-value for $\alpha = .10$, .05, and .01, the result is statistically significant at the usual α values. The result, however, is not what would be termed "practically significant." If the battery needs recharging after 19,695 calculations instead of 20,000, we seriously doubt that customers will return calculators for this reason.

SECTION 9.6, p. 285

9.34 H_0: $\mu \leq 1600$ T.S.: $t = 3.64$
H_a: $\mu > 1600$ RR: at $\alpha = .10$, reject H_0 if $t > 1.333$
Conclusion: Reject H_0

9.35 p-value $< .005$, and H_a is strongly supported.

9.36 a. H_0: $\mu = 0$ b. T.S.: $t = .3509$
H_a: $\mu > 0$ R.R.: At $\alpha = .05$, reject H_0 if $t > 1.943$
Conclusion: Do not reject H_0.

9.37 90% confidence interval for μ: $-2.3 \leq \mu_0 \leq 3.3$
Since $\mu = 0$ is contained in this interval, we cannot reject H_0: $\mu = 0$ at $\alpha = .10$.

9.38 No, the agency has not "conclusively established" that the device has *no* effect on mpg. The results indicate that no change in mpg is a possibility, but then so are the possible changes of 3 mpg or -1.5 mpg (according to the CI of 10.11). Since the confidence interval is quite wide, the β risk for alternative μ values may be large, and it is unwise to conclude "no change."

9.39 a. $\bar{d} = 1.8$; $s_d = 1.9889$
b. H_0: $\mu_d = 0$ T.S. $t = 2.8619$
H_a: $\mu_d > 0$ R.R.: At $\alpha = .01$, reject H_0 if $t > 2.821$
Conclusion: Reject H_0.

9.40 a. "T STATISTIC": $t = 2.86$
b. "P VALUE": $p = .0187$ two-sided p-value; one-sided p-value $= (1/2)(.0187) = .0098$.

SECTION 9.8, p. 291

9.41 a. The histogram or stem-and-leaf display shows no obvious nonnormality.
b. No; the data appear normal.

9.42 a. H_0: $\mu = 315$ T.S.: $t = 3.2955$
H_a: $\mu \neq 315$ RR: At $\alpha = .05$, reject H_0 if $t > 2.16$ or $t < -2.16$
Conclusion: Reject H_0.
p-value $< .01$
b. H_0: median $= 315$ T.S.: $y = 12$
H_a: median $\neq 315$ R.R.: at $\alpha = .05$, reject H_0 if $y \geq 12$ or $y \leq 2$
Conclusion: Reject H_0.
p-value $= .0130$
c. No, there is no serious difference in the conclusions of part a. and b.

9.43 95% confidence interval for true median: $316 \leq$ population median ≤ 346

9.44 a. H_0: $\mu \leq 45$ T.S.: $t = 1.965$
H_a: $\mu > 45$ R.R.: At $\alpha = .05$, reject H_0 if $t > 1.717$
Conclusion: Reject H_0.

b. A plot of the data:

Stem	Leaves
2	7
3	4 6 6 8 9 9 9
4	0 0 2 5 7
5	1 2 7
6	3
7	1 5
8	4
9	6
10	
11	0
12	
13	
14	7

Since the data are clearly skewed, the test based on the assumption of symmetry in part a. and its conclusion may not be appropriate.

9.45 a. H_0: median ≤ 45 T.S.: $y = 11$, $z = 0$
H_a: median > 45 R.R.: At $\alpha = .05$, reject H_0 if $z > 1.645$
Conclusion: Do not reject H_0.

b. Yes, there is a difference in the conclusions of 9.44(a) and 9.45(a). In 9.44(a), a one-sided test was performed on skewed data. The test, assuming symmetry, grossly underestimates probability in the one tail and overestimates the probability in the other. The conclusion is highly suspect. The test for the population median is designed to deal with this problem.

9.46 a. $\mu = 55$ but H_0: $\mu = 50$. Therefore, H_a is true and the simulation is evaluating P(reject $H_0|H_a$ true) = power.

b. The power is higher for the t test, for any α.

9.47 a. H_0: $\mu = 50$, and in fact $\mu = 50$. The simulation is evaluating $\alpha = P$(reject $H_0|H_0$ true).

b. The one-tailed probabilities are moderately wrong but get closer as n increases.

c. The two-tailed probabilities are reasonably accurate except when α is small. Again, the probabilities get closer as n increases.

9.48 a. Again $\mu = 50$ and H_0: $\mu = 50$. The simulation evaluates $\alpha = P$(reject $H_0|H_0$ true).

b. The probabilities appear close to correct for all n; perhaps they are slightly more conservative for large n.

SECTION 9.10, p. 298

9.49 95% confidence interval for μ: $.54 \leq \mu \leq 2.02$
In the test of Exercise 9.23, H_0: $\mu = 1.5$. The value $\mu = 1.5$ lies in the 95% confidence interval, so it seems plausible not to reject H_0. This is consistent with the result of the test at $\alpha = .05$.

9.50 90% confidence interval for μ: $42.4 \leq \mu \leq 52.8$
In the test of 9.30, H_0: $\mu = 44$, the value $\mu_0 = 44$ lies in the 90% confidence interval for μ, so it seems plausible not to reject H_0. This is consistent with the result of the test $\alpha = .10$.

9.51 99% confidence interval for μ: $19{,}429 \leq \mu \leq 19{,}961$
Testing H_0: $\mu = 20{,}000$ against H_a: $\mu \neq 20{,}000$: Since 20,000 does not lie in the 99% confidence interval for the true mean number of calculations, H_0 can be rejected at $\alpha = .01$.

SECTION 9.11, p. 300

9.52 a. H_0: $\pi \geq .135$
H_a: $\pi < .135$

b. Rejection of H_0 implies that the miller does not pay the full price of the grain to the seller. Not rejecting H_0 implies that the full price is paid.

c. If α is very small, it is more difficult to reject H_0. This would be disadvantageous to the miller if H_0 is false. The miller pays full price when it should not be paid and the finished product may not be of the quality desired and may lose money.

9.53 i. Implies that the cost of a Type I error is very high and as such α should be very small.
ii. Implies that the cost of a Type II error is not great and so the β value can be allowed to be fairly large.
iii. Makes the miller skeptical of any sample that indicates rejection of H_0, indicating that α should be small.

9.54 In the marketing of drugs, a Type I error is much more serious than a Type II error. As a result, the cost of a Type I error is much higher than the cost of a Type II error. For these reasons, the α value is set very low.

CHAPTER EXERCISES, p. 302

9.55 H_0: $\pi \leq .1$ T.S.: $y = 9$
H_a: $\pi > .1$ R.R.: At $\alpha = .05$, reject H_0 if $y \geq 10$
Conclusion: Do not reject H_0.

9.56 p-value $= P(Y \geq 9 | \pi = .1) = .0579$

9.57 One-sided alternative: consumers will benefit if the mpg is underestimated. Therefore, we should only be concerned if the mpg is overestimated and the consumer is misled by misadvertising.
Two-sided alternative: for technological reasons, a two-sided alternative may be desirable. For example, if the EPA set minimum standards to be attained by 1990, say, 30 mpg, a manufacturer may want to know how near it is to attaining the standard. Also, manufacturers may be interested in the effects of pollution control devices on mpg.

9.58 H_0: $\mu = 28.2$ T.S.: $z = -2.0203$
H_a: $\mu \neq 28.2$ R.R.: At $\alpha = .01$, reject H_0 if $z < -2.576$
Conclusion: Do not reject H_0.

9.59 p-value $= 2P(z \geq 2.0203) = .0434$

9.60 Although the result was not significant at the $\alpha = .01$ level, it *is* significant at the .05 and .10 levels, and one would seriously question the acceptance of 28.2 mpg as "truth."

9.61 99% confidence interval for μ: $24.8 \leq \mu \leq 28.6$
Reliably assume that the true mean is between 24.8 and 28.6 mpg.

9.62 a. 95% confidence interval for μ: $3.03 \leq \mu \leq 3.07$
b. H_0: $\mu = 3$ H_a: $\mu \neq 3$
Conclusion: Since 3 does not lie in the 95% confidence interval for μ, H_0 can be rejected at $\alpha = .05$.

9.63 a. Yes, the results were statistically significant.
b. Yes, this is misleading. The statement without any numerical clarification might tend to give the impression that business executives are completely without conscience, when in fact if given the numerical results, the deviation from "average" is so small that people would most probably ignore it.

9.64 p-value $= 2(z \geq 3.9648) \approx .00006$

REVIEW EXERCISES, CHAPTERS 7–9, p. 302

R52 a. $240.40 \leq \mu \leq 266.24$
b. A plot of the data shows a roughly normal shape without outliers. The sample mean should be reasonably efficient.

R53 Reject H_0; 230.2 is not in the interval.

R54 H_0: $\mu = 230.2$
H_a: $\mu \neq 230.2$
T.S. $z = \dfrac{\bar{y} - \mu_0}{\sigma/\sqrt{n}}$
R.R.: $|z| > 2.58$
Conclusion: $z = 4.62$; reject H_0.

R55 p-value $= 2p(z > 4.62) \approx 2P(z > 4.50) = .0000068$ (two-sided)

R56 Using $t_{.005,\ 49\ \text{d.f.}} \approx 2.684$ by interpolation, a 99% confidence interval is $239.62 \leq \mu \leq 267.02$; reject H_0: $\mu = 230.2$
Formal test T.S.: $t = 4.53 > 2.684$, reject H_0, p-value $< 2(.001)$
Same conclusions.

R57 $233 \leq \text{median} \leq 275$; reject H_0

R58 The median interval is wider; $\bar{y}$ is more efficient here.

R59 R.R.: Reject H_0 if $y \leq 36$; $P(Y \leq 36|\pi = .45) = .0429$

R60 R.R.: Reject H_0 if $y \leq 33$; $P(Y \leq 33|\pi = .45) = .0097$

R61 Yes; $y = 32 \leq 36$, so reject H_0.

R62 Yes; $y = 32 \leq 33$, so reject H_0 with p-value $< .01$.

R63 a. H_0: $\mu = 272.6$; H_a: $\mu > 272.6$ because higher yield is desired.
b. H_0: $\mu = 272.6$
H_a: $\mu > 272.6$
T.S.: $z = \dfrac{\bar{y} - 272.6}{67.3/\sqrt{25}}$
R.R.: Reject H_0 if $z > 1.645$

R64 $\beta = P(Z > .76) = .2236$

R65 The β probability is larger with μ_a closer to μ_0.

R66 $z = 3.09$; reject H_0

R67 p-value $= .001$

R68 Data are modestly left-skewed with possible outliers. A fences calculation shows that 135 and 185 are (oh, deer!) outliers. The mean may not be efficient.

R69

π	.2	.3	.4	.5
1,000,000 $L(\pi)$	.0792	.0164	.0009	.00001

$\hat{\pi} = .2$ or less

R70 a. $\hat{\pi} = 1/6 = .1667$
b. $\hat{\pi} = 1/(1 + \bar{x})$

R71 $\hat{\pi}$ is a (slightly) biased estimator.

R72 $.102 \leq \pi \leq .229$

R73 a. $n = 589.3$, rounded up to 590
b. $n = 1068$

R74 Yes; $n\hat{\pi} = 22$ and $n(1 - \hat{\pi}) = 111$ are much greater than 5.

R75 Biased in favor of large accounts.

R76 a. Using $t_{.025,240\ \text{d.f.}} = 1.970$, $5093 \leq \mu \leq 5669$
b. Skewness is a minor problem, given the CLT and $n = 241$. The major problem is the biased sampling.

R77 $78.91 \leq \mu \leq 81.83$

R78 Using a z approximation, $k = 8$; so $80.3 \leq \text{median} \leq 82.5$.

R79 The median interval is shorter, indicating that the sample median is more efficient. A plot shows left-skewness with outliers; a fence calculation shows that 67.6 and 68.5 are severe outliers. The median should be more efficient.

R80 $|t| = |-1.90|$, $.05 < p\text{-value} < .10$
R81 power $= 1 - P(Z > 2.49) = .9936$
R82 The given means were hypothetical population means: the p-value refers to a particular sample.
R83 $\hat{\theta} = 1.833$
R84 $\hat{\theta} = \left[\sum_{i=1}^{n} \log_e (1 + y_i)\right]/n$
R85 It is unbiased and most efficient.

CHAPTER 10

SECTION 10.1, p. 310

10.1 a. 95% confidence interval for true difference: $-2.6791 \le \mu_1 - \mu_2 \le 15.8791$
b. H_0: $\mu_1 - \mu_2 = 0$ T.S.: $z = 1.3941$
H_a: $\mu_1 - \mu_2 \ne 0$ R.R.: Reject H_0 if $|z| > 1.96$
Conclusion: Do not reject H_0.
Note also that 0 is included in the confidence interval.
10.2 p-value $= .165$
10.3 a. 90% confidence interval for true difference: $-.4782 \le \mu_1 - \mu_2 \le -.0218$
b. H_0: $\mu_1 - \mu_2 = 0$ T.S.: $z = -1.8025$
H_a: $\mu_2 - \mu_2 \ne 0$ R.R.: Reject H_0 if $|z| > 1.645$
Conclusion: Reject H_0.
10.4 p-value $= .0718$
10.5 It would seem that a large number of orders may take a relatively short amount of time to be processed and that a small number of orders may have a long processing time. It is believed that both data sets will be skewed right. Since both populations will be skewed in the same direction and since we are concerned with the difference in means, the distribution of $\bar{y}_1 - \bar{y}_2$ will most likely be more symmetric than the populations themselves and we can be confident in the z probabilities (especially since n_1 and n_2 are large).
10.6 a. $\mu_{\bar{K}-\bar{R}} = 78.28 - 75.79 = 2.49$
b. $\sigma_{\bar{K}-\bar{R}} = \sqrt{\dfrac{(9.63)^2}{8} + \dfrac{(11.25)^2}{12}} = 4.705$
10.7 μ would not change; $\sigma_{\bar{K}-\bar{R}} = 4.853$
10.8 a. $\mu_{\bar{Y}_1-\bar{Y}_2} = -4.21$
b. $\sigma^2_{\bar{Y}_1-\bar{Y}_2} = 2.31172$
10.9 The mean is still valid; mean calculations do not require independence, but the dependent sampling makes the variance wrong.

SECTION 10.2, p. 319

10.10 a. $s_p^2 = 1.1601$; $s_p = 1.077$
b. 95% confidence interval for true difference: $-1.2619 \le \mu_1 - \mu_2 \le .5619$
c. H_0: $\mu_1 - \mu_2 = 0$ T.S.: $t = -.7959$
H_a: $\mu_1 - \mu_2 \ne 0$ R.R.: Reject H_0 if $|t| > 2.074$
Conclusion: Retain H_0.
10.11 p-value: $p > .2$

10.12 a. The first concern is the correctness of nominal α probabilities. With unequal n's and variances, t' should give closer α probabilities.

b. Yes; the t' probabilities are very close to the nominal α probabilities, but the pooled variance t probabilities are not.

10.13

Source I	
2	1
2	5 9 8 7
3	2
3	8 5 6 7
4	1 3
4	

Source II	
1	8 3
2	2 5
3	3
4	0 7 3 2 6
5	1 4

The assumption of normality is mildly suspect. Both sets of data appear slightly bimodal. The conclusion of no difference in means seems appropriate given the mild problem.

10.14 a. $\bar{y}_1 = 10.37$; $\bar{y}_2 = 9.83$; $s_1 = .3234$; $s_2 = .2406$

b. $t = 4.2368$ c. $t' = 4.2368$ d. $t = t'$ because $n_1 = n_2$

10.15 a.

Sample 1

9.8–.9	x
10.0–.1	x
10.2–.3	x x x
10.4–.5	x
10.6–.7	x x x
10.8–.9	x

Sample 2

9.4–.5	x
9.6–.7	x x x
9.8–.9	x x x
10.0–.1	x x
10.2–.3	x

b. The assumption of normal populations does not appear seriously violated. And since $s_1^2 \approx 1.8\ s_2^2$ (not grossly unequal) and since $n_1 = n_2$, the assumption of equal variances is not inappropriate.

10.16 a. p-value: $p = .0005$, two-tailed

b. Since the p-values of both t tests are very small, the data give strong support to H_a that there is a decrease in mean potency.

10.17 a. $s_p^2 = \dfrac{9(10.46)^2 + 15(8.67)^2}{24} = 88.010$

b. With 24 d.f., $t_{.005} = 2.797$

$2.92 \le \mu_A - \mu_B \le 24.08$

c. Yes; $\mu_A - \mu_B = 0.00$ is not included in the interval.

10.18 $1.95 \le \mu_A - \mu_B \le 25.05$

By either method, 0.00 isn't included. The more believable t' confidence interval is somewhat wider.

10.19 a.

Sample	$\bar{y}$	s
I	13.43	5.638
II	15.21	6.174

b. $t = -.668$

c. $t' = -.668$

d. $t = t'$ whenever $n_1 = n_2$

10.20 a. No; $|t| < 2.101$

b. 2-tailed p-value $> .2$

c. d.f. ≈ 16

d. same conclusion

10.21 a. In principle, t' is better because $\sigma_1 \neq \sigma_2$. In practice, it doesn't matter much at all.
b. d.f. = 18 compared to average approx. d.f. = 13.1030

SECTION 10.3, p. 328

10.22 a. Source I sum of ranks = 133; Source II sum of ranks = 167
b. H_0: Populations are identical
H_a: Population 1 is shifted to right or left of Population 2
T.S.: $z = -.98$
R.R.: Reject H_0 if $|z| > 1.96$
Conclusion: Do not reject H_0.

10.23 a. The conclusions of Exercises 10.10 and 10.22 are the same—retain H_0.
b. Since the conclusions were the same and the data were nearly normal, one would accept the t result without serious question.

10.24 p-value: $p = .327$

10.25 a. Sample 1 rank sum = 146; Sample 2 rank sum = 64
b. $z = 3.0743$
c. H_0: Populations are identical.
H_a: Population 1 is shifted to the right of population 2.
d. R.R.: Reject H_0 if $z > 2.326$
Conclusion: Reject H_0. The research hypothesis is supported at $\alpha = .01$ level.

10.26 a. A two-tailed p-value is shown as .0021. A one-tailed p-value should be reported: $p = .00105$.
b. The conclusion of the rank sum test and the t test are the same—reject H_0.

10.27 a.

A:	39	41	42	44	49	53	56	57	63	71						
Rank:	11	13	14	16	18	21	23	24	25	26						
B:	27	28	29	30	31	33	35	36	37	38	40	43	45	50	52	54
Rank:	1	2	3	4	5	6	7	8	9	10	12	15	17	19	20	22

b. $T_1 = 191$, $T_2 = 160$
c. $z = 2.95 > 2.58$; yes, significant.

10.28 Stem-and-leaf displays are moderately right-skewed. There is no clear preference; both methods yield the same conclusion.

10.29 a. $|z| = |-1.66| < 1.96$; retain H_0
b. p-value = .097

10.30 Plots (stem-and-leaf or histogram) show that both samples are highly right-skewed. The rank sum test is preferred.

10.31 The claimed α values for both t tests are conservative (too large).

10.32 a. The means are unequal, so H_a is true; power = $P(\text{reject } H_0 | H_0 \text{ true})$.
b. The rank sum test

SECTION 10.4, p. 338

10.33 a. H_0: $\mu_d = 0$ T.S.: $t = 4.0602$
H_a: $\mu_d \neq 0$ R.R.: Reject H_0 if $|t| > 1.761$.
Conclusion: Reject H_0.
b. 90% confidence interval: $1.965 \leq \mu_d \leq 4.975$

10.34 a. H_0: The distribution of differences is symmetric around 0.
H_a: The differences tend to be larger than 0 or smaller than 0.
T.S.: $T = 7$
R.R.: Reject H_0 if $T \leq 26$
Conclusion: Reject H_0.
b. The conclusion of this test and the t test of Exercise 10.33 indicate that Display A has a better effect on sales than Display B.

10.35 Paired-sample t test: p-value $< .002$
Wilcoxon signed-rank test: p-value $< .002$

10.36 a. H_0: $\mu_1 - \mu_2 = 0$ T.S.: $t = .7600$
H_a: $\mu_1 - \mu_2 \neq 0$ R.R.: Reject H_0 if $|t| > 1.701$
Conclusion: Retain H_0.
b. The variability in sales volumes among the stores completely masks the variability due to the different displays when the data are treated as independent samples. The pairing process effectively eliminates this undesirable source of variation.

10.37 H_0: The median difference is zero
H_a: The median difference is not zero
T.S.: $y = 12$ R.R.: Reject H_0 if $y \leq 3$ or $y \geq 11$
Discard the 0 difference; take $n = 14$.
Conclusion: Reject H_0.

10.38 The signed-rank test has better power.

10.39 a. $t = 2.86$ b. $T = 6$

c.

t Test	Wilcoxon Signed-rank Test
H_0: $\mu_d = 0$	H_0: Differences are symmetric about 0
H_a: $\mu_d \neq 0$	H_a: Differences greater or less than 0
T.S.: $t = 2.86$	T.S.: $T = 6$
R.R.: Reject H_0 if $\|t\| > 2.262$	R.R.: Reject H_0 if $T \leq 8$
Conclusion: Reject H_0.	Conclusion: Reject H_0.

10.40 t test: p-value $= .0187$
Wilcoxon: p-value $= .0284$
Since the p-values for both tests are small, one would feel reasonably comfortable rejecting H_0.

10.41 Paired sampling is more effective because of the great variability in delivery times among the different destinations. To select destinations, choose as wide a variety as possible. The variability in delivery times among the destinations is virtually eliminated by the paired sample. In this way, one can study how the two companies compare in a number of different situations.

10.42 a. The same ZIP codes are used for both samples.
b. For A-B differences, $\bar{d} = -.375$, $s_d = .920$
c. $t = -1.41 > -1.796$; retain H_0
d. $.05 < p$-value $< .10$, one-tailed

10.43 a. $T = T_+ = 23 > 17$; retain H_0
b. p-value $> .1$
c. Yes, at $\alpha = .05$

10.44 No; H_0 is retained but not proved.

10.45 a. The data are paired by task (finding the same combinations).
b. $-48.1 \leq \mu_d \leq 5.5$, where d = Program 1 − Program 2
c. Retain H_0 at $\alpha = .05$; 0.0 is included.

10.46 $T = |T_-| = 9 > 5$; retain H_0

10.47 Plot the differences. Because of two outliers, the signed rank test is preferred. The same conclusion is found.

10.48 a. $-89.8 \leq \mu_1 - \mu_2 \leq 47.2$
b. The independent-sample interval is much wider; pairing was quite helpful in reducing variability.

10.49 90% confidence interval: $.0076 \leq \pi_1 - \pi_2 \leq .0304$

10.50 a. H_0: $\pi_1 - \pi_2 = 0$ T.S.: $z = 2.7536$
H_a: $\pi_1 - \pi_2 \neq 0$ R.R.: Reject H_0 if $|z| > 1.645$
Conclusion: Reject H_0.
b. Modified standard error $= \sigma_{\bar{\pi}} = .0075$
Modified z T.S.: $z = 2.5333$
Conclusion: Reject H_0.
The modification does not affect the conclusion.

10.51 z statistic: p-value $= .006$
Modified z statistic: p-value $= .0114$

10.52 a. 95% confidence interval for true difference: $.0785 \leq \pi_1 - \pi_2 \leq .1467$
b. H_0: $\pi_1 - \pi_2 = 0$ T.S.: $z = 6.4713$
H_a: $\pi_1 - \pi_2 > 0$ R.R.: Reject H_0 if $|z| > 1.645$
Conclusion: Reject H_0.

10.53 p-value $= P(Z > 6.4713) \approx 0$. The results of the study strongly support the research hypothesis.

10.54 Modified standard error: $\sigma_{\bar{\pi}} = .0180$
Modified z T.S.: $z = 6.2556$
p-value $= P(z > 6.2556) \approx 0$
Use of the modified standard error does not affect the conclusion.

10.55 a. $z = 2.31 > 1.645$; reject H_0
b. p-value $= .0208$, two-tailed

10.56 $z = 2.29$; essentially the same result

10.57 a. $.038 \leq \pi_1 - \pi_2 \leq .166$
b. Yes, more than adequate.
c. Yes; .000 is not included.

10.58 $z = 3.13$ (or $z = 3.11$ with the pooled proportion); reject H_0

10.59 a. $z = 1.64$ (or $z = 1.61$ using $\bar{\pi}$) < 1.96; retain H_0
b. No; H_0 is retained, not proved.

10.60 $-.039 \leq \pi_1 - \pi_2 \leq .439$
Supplier 1 may produce anywhere from 3.9% fewer reliable motors to 43.9% more reliable motors than supplier 2! The interval is uninformatively wide.

CHAPTER EXERCISES, p. 348

10.61 a. 95% confidence interval for true difference: $-.0241 \leq \pi_1 - \pi_2 \leq .0547$
b. H_0: $\pi_1 - \pi_2 = 0$ T.S.: $z = .7612$
H_a: $\pi_1 - \pi_2 \neq 0$ R.R.: Reject H_0 if $|z| > 1.96$
Conclusion: Retain H_0.

10.62 p-value $= .4472$

10.63 a. 90% confidence interval for true difference: $\$9.39 \leq \mu_1 - \mu_2 \leq \$28.83$
b. We are 90% confident that the mean error of those accounts in error at Bank B is at least \$9.39 larger than the mean error of those accounts in error at Bank A and at most \$28.83. The conclusion applies to the population of erroneous accounts at the banks.
c. H_0: $\mu_1 - \mu_2 = 0$ T.S.: $z = 3.2333$
H_a: $\mu_1 - \mu_2 > 0$ R.R.: Reject H_0 if $t > 1.96$
Conclusion: Reject H_0.

10.64 p-value $< .002$

10.65 a. H_0: $\mu_1 - \mu_2 = 0$ T.S.: $t = 3.2707$
H_a: $\mu_1 - \mu_2 \neq 0$ R.R.: Reject H_0 if $|t| > 1.994$ (interpolated)
Conclusion: Reject H_0.
b. H_0: $\mu_1 - \mu_2 = 0$ T.S.: $t' = 3.2333$
H_a: $\mu_1 - \mu_2 \neq 0$ R.R.: Reject H_0 if $|t'| > 2.0252$
Conclusion: Reject H_0.
Note that the t' statistic equals the large sample z statistic.

10.66 The t test should and does have higher power.

10.67 a. H_0: $\mu_1 - \mu_2 = 0$ T.S.: $t = .7999$
H_a: $\mu_1 - \mu_2 \neq 0$ R.R.: Reject H_0 if $|t| > 1.321$
Conclusion: Retain H_0.
b. H_0: The two populations are identical
H_a: Population 1 is shifted to the right of Population 2
T.S.: $z = .6928$
R.R.: Reject H_0 if $z > 1.282$
Conclusion: Retain H_0.

10.68

Variety R		Variety K	
3	71	3	24
4	73	4	86 90 90
5	32 40 87 98	5	08 74 99
6	11 16 42 80	6	00 19 21 67
7	11 46	7	20

Since the assumption of normality of the populations seems appropriate here, the t test is more powerful. However, since the conclusion of each of the tests is retention of H_0, choosing the more appropriate test is not critical.

10.69 t test: $.20 < p$-value
Wilcoxon rank sum test: p-value $\approx .25$

10.70 a. H_0: $\mu_d = 0$ T.S.: $t = 3.1542$
H_a: $\mu_d > 0$ R.R.: Reject H_0 if $t > 1.363$
Conclusion: Reject H_0.
b. H_0: The distribution of differences is symmetric around 0
H_a: The differences tend to be larger than 0
T.S.: $T = 8$
R.R.: Reject H_0 if $T \leq 17$
Conclusion: Reject H_0.

10.71

Stem	Leaves
−2.*	3
−1.	7
−0.	9
0.	
1.	1
2.	
3.	2
4.	2 3 6 7
5.	
6.	1
7.	9
8.	
9.	7

The distribution of differences appears heavy-tailed, although it has the central mode characteristic of normality. In the case of heavy tails, the signed rank test may be more powerful. In this situation, since the conclusions of both tests are the same, one test need not be chosen.

10.72 t test: $.001 < p\text{-value} < .005$ Wilcoxon: $.005 < p\text{-value} < .01$

10.73 Since the many plots on a farm may have a wide range in yields (due to fertility, irrigation, etc.) a paired sample experiment eliminates the effect of variability on the two varieties of trees with which we are concerned. In selecting plots, choose a wide range (rocky, smooth, fertile, poor mineral content, well irrigated, dry, shady, etc.) so that as many situations as possible are represented in the sample.

10.74 a. $t = 5.9126$ b. $t' = 4.4491$

c. H_0: $\mu_1 - \mu_2 = 0$ | H_0: $\mu_1 - \mu_2 = 0$
H_a: $\mu_1 - \mu_2 \neq 0$ | H_a: $\mu_1 - \mu_2 \neq 0$
T.S.: $t = 5.9126$ | T.S.: $t' = 4.4491$
R.R.: Reject H_0 if $|t| > 2.66$ | R.R.: Reject H_0 if $|t'| > 2.831$
Conclusion: Reject H_0. | Conclusion: Reject H_0.
p-value $= .0001$ | p-value $= .0002$

Since the H_a is strongly supported in both tests, choice of a proper statistic is not crucial here.

10.75 The t' statistic seems to be more reliable for this experiment. The sample sizes are unequal ($n_1 > 2n_2$). The sample variances are vastly different ($s_2^2 > 7s_1^2$). Also, the larger variance is associated with the smaller sample size. For these reasons, the t' statistic is believed more appropriate. The conclusions are the same.

10.76 a. T = sum of ranks for sample $R = 345$

b. $z = -4.29163$; p-value $= .0000$

c. H_0: the two populations are identical
H_a: Population 1 is shifted to the right of Population 2
T.S.: $z = -4.2963$
R.R.: Reject H_0 if $|z| > 2.5758$
Conclusion: Reject H_0.

CHAPTER 11
SECTION 11.1, p. 356

11.1 a. $P(Y > 46.96) = .01$
b. $P(Y > 18.11) = .90$
c. $P(Y < 12.88) = .01$
d. $P(12.88 < Y < 46.96) = .98$

11.2 a. $\chi^2_{.025} = 21.92$
b. $\chi^2_{.975} = 3.816$

11.3 a. $\chi^2_{.025} \approx 324.51$, $\chi^2_{.975} \approx 232.33$

11.4 a. .975 b. .95 c. .99

11.5 H_0: $\sigma^2 = .15^2 = .0225$ T.S.: $\chi^2 = 46.464$
H_a: $\sigma^2 < .0225$ R.R.: Reject H_0 if $\chi^2 < 43.19$
Conclusion: Do not reject H_0.

11.6 H_0: $\sigma^2 = .0225$ T.S.: $\chi^2 = 46.464$
H_a: $\sigma^2 > .0225$ R.R.: Reject H_0 if $\chi^2 > 79.08$
Conclusion: Retain H_0.
This choice of H_a is more generous.

11.7 a. $\bar{y} = 3.999$; $s = .0159$

b. H_0: $\sigma^2 = .011^2 = .0001$ T.S.: $\chi^2 = 63.2025$

H_a: $\sigma^2 > .0001$ R.R.: Reject H_0 if $\chi^2 > 37.65$

Conclusion: Reject H_0.

11.8 90% confidence interval: $.0002 < \sigma^2 < .0004$; $.0130 < \sigma < .0208$

11.9

```
3.95*  2
3.96
3.97   8  9
3.98   4  7
3.99   1  5  7  9  9  9
4.00   0  0  0  1  1  2  2  3  4  6  9
4.01   0  2
4.02
4.03   3
4.04   1
```

The distribution may be heavy-tailed, which may make the results suspect. However, the data seem to support H_a strongly so, although the probabilities may not be accurate, the conclusion seems appropriate.

11.10 H_0: $\sigma^2 = 4$ T.S.: $\chi^2 = 64.8$

H_a: $\sigma^2 < 4$ R.R.: Reject H_0 if $\chi^2 < 60.39$

Conclusion: Do not reject H_0.

11.11 p-value $\simeq .1$

SECTION 11.2, p. 361

11.12 H_0: $\sigma_1^2 = \sigma_2^2$ T.S.: $F = .2453$

H_a: $\sigma_1^2 \neq \sigma_2^2$ R.R.: At $\alpha = .05$, reject H_0 if $F > 3.47$ or $F < .288$

Conclusion: Reject H_0.

11.13 To place bounds on the p-value using more extensive tables:

1. In each table giving F statistics corresponding to a tail area $= \alpha$, locate the $F_{\alpha, 11, 11}$.
2. Find the two α values, α_1 and α_2, such that $1/F_{\alpha 1, 11, 11} < .2478 < 1/F_{\alpha 2, 11, 11}$.
3. Then $\alpha_1 < p$-value (one-sided) $< \alpha_2$, so the appropriate p-value for this two-sided test would be $2\alpha_1 < p < 2\alpha_2$.

11.14

```
Source I              Source II
2. | 1                1. | 8  3
2. | 5  9  8  7       2. | 2  5
3. | 2                3. | 3
3. | 8  5  6  7       4. | 0  7  3  2  6
4. | 1  3             5. | 1  4
4. |
```

The data of Source I are not centered about their mean. The distribution may be bimodal. The data of Source II appear mildly skewed. For these reasons, a more robust method may be preferred.

11.15 a. Since the patients' perceived benefit is a key factor in treatment, the times to onset of relief may be skewed. Those more skeptical may not perceive relief for a long while.

b. H_0: $\sigma_1^2 = \sigma_2^2$

H_a: $\sigma_1^2 \neq \sigma_2^2$

T.S.: $F = 1.5376$

R.R.: Reject H_0 if $F > 2.53$ or $F < .395$

Conclusion: Do not reject H_0.

11.16 95% confidence interval: $.607 < \sigma_1^2/\sigma_2^2 < 2.53$

11.17 Yes; the nominal α is much too small.

SECTION 11.3, p. 364

11.18 a. and b.

Number of Arrivals per Minute	Theoretical Proportions π_i	Expected Frequencies $E_i = n\pi_i$
0	.0224	44.8
1	.0850	179
2	.1615	323
3	.2046	409.2
4	.1944	388.8
5	.1477	295.4
6	.0936	187.2
7	.0508	101.6
8	.0241	48.2
9 or more	.0159	31.8

c. H_0: $\pi_i = \pi_{i,0}$ T.S.: $\chi^2 = 77.2595$
H_a: H_0 is not true R.R.: Reject H_0 if $\chi^2 > 21.67$
Conclusion: Reject H_0.

11.19 It appears that in the last four cells, six or more arrivals per minute, the Poisson model systematically overestimates the number of 1-minute periods that have arrivals falling in those categories.

11.20 a. The expected frequencies are 7.5, 11.25, and 6.25.
b. H_0: $\pi_i = \pi_{i,0}$ T.S.: $\chi^2 = 4.4933$
H_a: H_0 is not true R.R.: Reject H_0 if $\chi^2 > 5.991$
Conclusion: Retain H_0.

11.21 a. H_0: $\pi_i = \pi_{i,0}$ T.S.: $\chi^2 = 44.933$
H_a: H_0 is not true R.R.: Reject H_0 if $\chi^2 > 5.991$
Conclusion: Reject H_0.
b. This conclusion differs from the previous one. Here, the sample size is large and the expected cell frequencies are all much larger than 5. For these reasons, the probability of a Type II error is small compared to the previous situation, where it was very likely that H_0 would not be rejected when it is false.

SECTION 11.4, p. 369

11.22 a. $\chi^2 = 15.96704$ b. p-value $= .0139$
c. The null hypotheses of no relationship between school and rating can be rejected at the $\alpha = .05$ level. The presence of some relationship is indicated.
d. Yes. Not all the expected cell values are greater than or equal to five.

11.23

	Outstanding	Average	Poor
Most desirable	43.75%	52.08%	4.17%
Good	30.30	54.55	15.15
Adequate	16.00	56.00	28.00
Undesirable	17.65	47.06	35.29

The relation can be seen by the fact that the percentage of Poor ratings increases as the desirability of the school decreases.

11.24 a. Expected numbers under independence

	Under 30	30–39	40–49	Over 50	Total
Promoted	16	22.4	25.6	16	80
Not promoted	34	47.6	54.4	34	170
Total	50	70	80	50	250

b. d.f. = 3

c. H_0: Age and promotion are independent
H_a: Age and promotion are dependent
T.S.: $\chi^2 = 13.0252$
R.R.: Reject H_0 at $\alpha = .05$ if $\chi^2 > 7.815$
Conclusion: Reject H_0.

11.25 $.001 < p\text{-value} < .005$

11.26 a. Expected numbers under independence

	−39	40+	Total
Promoted	38.4	41.6	80
Not promoted	81.6	88.4	170
Total	120	130	250

b. H_0: Age and promotion are independent
H_a: Age and promotion are dependent
T.S.: $\chi^2 = .0118$
R.R.: Reject H_0 at $\alpha = .05$ if $\chi^2 > 3.841$
Conclusion: Retain H_0.

c. In part a., combining age group when there was no need to (all cells sufficiently larger than 5) had the effect of completely masking a relationship that was present, as indicated in Exercise 11.24.

CHAPTER EXERCISES, p. 377

11.27 a. Recall that we were considering two noise control systems for jet engines. The agency needs to consider the variability in noise levels as well as the mean in choosing between systems. Suppose one system has a lower mean noise level than the second. If this noise level is not stable (i.e., it has a larger variance) this could be disastrous for the community (when glass-shattered noise levels are reached). Although it has a higher mean noise level, the second system may be desirable if it is not much higher and it is stable.

b. H_0: $\sigma_1^2 = \sigma_2^2$
H_a: $\sigma_1^2 \neq \sigma_2^2$
T.S.: $F = .1336$
R.R.: At $\alpha = .05$, reject H_0 if $F > 3.10$ or $F < .383$
Conclusion: Reject H_0.

c. $p\text{-value} < .01$

11.28 System H

94–96	x x x
97–99	x x x x x x x x x x
100–102	x x x x x x x x x x x x x x x x x
103–105	x x x x x x x
106–108	x x
109–111	x

System R

79–84	x x
85 90	x x x x x x
91–96	x x x x x x
97–102	x x x x
103–108	x
109–114	x

The data plots show no extreme skewness or heavy-tailness. Therefore, there is no obvious reason to doubt the conclusion of 11.27.

11.29 H_0: $\sigma^2 = 4$ T.S.: $\chi^2 = 91.9059$
H_a: $\sigma^2 > 4$ R.R.: Reject H_0 at $\alpha = .05$ if $\chi^2 > 56.93$
Conclusion: Reject H_0.

11.30 a. $F = 1.42$
b. H_0: $\sigma_{\text{II}}^2 = \sigma_{\text{I}}^2$ T.S.: $F = 1.42$
H_a: $\sigma_{\text{II}}^2 > \sigma_{\text{I}}^2$ R.R.: Reject H_0 at $\alpha = .05$ if $F > 2.58$ (interpolated)
Conclusion: Retain H_0.
c. p-value $> .25$

11.31 H_0: $\sigma^2 = .08^2 = .0064$ T.S.: $\chi^2 = 53.77$
H_a: $\sigma^2 > .0064$ R.R.: Reject H_0 at $\alpha = .05$ if $\chi^2 > 22.36$
Conclusion: Reject H_0.

11.32 Process I

9.8	1 7 9
9.9	0 3 9 6
10.0	3 0
10.1	6 8
10.2	1 2
10.3	6

Process II

9.7	1
9.8	8 4 9 6 0
9.9	9 3
10.0	1
10.1	6 3
10.2	7 2
10.3	5

The data do not appear normal in either case. Since normality of the populations is an assumption underlying the F test, our conclusion may be suspect. However, looking at the spread of each data set, the conclusion that the variances may be equal appears reasonable.

11.33 a. The 25%, 40%, and 35% claim concerning opinions on union membership was made for industrial workers as a whole without regard to membership status. The relevant data are the column totals of those favoring, those indifferent, and those opposed industrial workers (i.e., 210, 240, and 150, respectively).

b. H_0: The proportions .25, .40, and .35 describe union preference among industrial laborers.
   H_a: The proportions are something else.
   T.S.: $\chi^2 = 41.1429$
   R.S.: Reject H_0 at $\alpha = .01$ if $\chi^2 > 9.21$
   Conclusion: Reject H_0 overwhelmingly.

11.34 p-value $< .001$

11.35 H_0: Union preference is independent of membership status.
H_a: Membership status and union preference are related.
T.S.: $\chi^2 = 162.795$
R.R.: Reject H_0 at $\alpha = .05$ if $\chi^2 > 5.99$
Conclusion: Reject H_0.

11.36

	Favor	Indifferent	Opposed
Members	70.0%	21.0%	9.0%
Nonmembers	17.5%	49.5%	33.0%

Far more members than nonmembers are in favor; far more nonmembers than members are opposed.

11.37 $\lambda = .272$, indicating a strong relation; recall that $\lambda > .3$ is rare.

11.38 a. Expected numbers under independence

		Opinion					
		1	2	3	4	5	
	A	42	107	78	34	39	300
Commercial	B	42	107	78	34	39	300
	C	42	107	78	34	39	300
		126	321	234	102	117	900

b. d.f. = 8

c. H_0: Opinion distribution is the same for each of three commercials.
   H_a: Opinion distribution and commercial viewed are somehow related.
   T.S.: $\chi^2 = 72.5208$
   R.R.: Reject H_0 at $\alpha = .01$ if $\chi^2 > 20.09$
   Conclusion: Reject H_0.

11.39 p-value $< .001$

11.40 For predicting opinion given commercial: $\lambda = .0069$
For predicting commercial given opinion: $\lambda = .1567$
Since both λ values are small, we believe that a weak relationship between commercial and opinion exists.

CHAPTER 12
SECTIONS 12.1 AND 12.2, p. 389

12.1 a. $\bar{y} = 1118.525$
b. SS(Between) = 105,529.85
c. SS(Within) = 38,244.08
d. SSB d.f. = 4; SSW d.f. = 35

12.2 See 12.1 b. and c.

12.3 H_0: $\mu_1 = \mu_2 = \mu_3 = \mu_4 = \mu_5$
H_a: Not all μ_i equal ($i = 1, \ldots, 5$)
T.S.: $F = 24.145$
R.R.: At $\alpha = .05$, reject H_0 if $F > 2.65$ (interpolated)
Conclusion: Reject H_0.
p-value $< .001$

12.4 The assumption of independence of samples is not a problem. The assumption of normality appears reasonable, as can be seen from plots of the data. The assumption of equal variances for the 5 distributions is the problem here; sample variance of Plan A is more than 5 times the sample variance of Plan E. Equal sample sizes may take care of this problem, and the result is very conclusive.

12.5 H_0: The distributions are identical
H_a: The distributions differ in location
T.S.: $H = 115.9976$
R.R.: At $\alpha = .05$, reject $H_0 > 9.49$
Conclusion: Reject H_0.
$p < .001$

12.6 The conclusions of the F test and the Kruskal-Wallis test agree; therefore, the choice is not crucial.

12.7 a. SS(Between) = 3478.17505669; SS(Within) = 7782.06984127
b. $F = 4.92$
c. H_0: $\mu_1 = \mu_2 = \mu_3 = \mu_4 = \mu_5$
H_a: Not all means are equal
T.S.: $F = 4.92$
R.R.: At $\alpha = .01$, reject H_0 if $F > 3.78$ (interpolated)
Conclusion: Reject H_0.

12.8 The assumptions of independent samples, normality of distributions, and equality of variances appear reasonable for the data.

12.9 a. $H = 16.10$
b. H_0: The distributions are identical.
H_a: The distributions differ in location.
T.S.: $H = 16.10$
R.R.: At $\alpha = .05$, reject H_0 if $H > 9.49$
Conclusion: Reject H_0.
c. Since the conclusions of the F test and K-W test are the same, choice is not crucial.

SECTION 12.3, p. 396

12.10 a. 95% Scheffé confidence intervals (could also use Tukey method):

$-112.051 < \mu_A - \mu_B < -4.429$
$-8.431 < \mu_A - \mu_C < 99.191$
$-23.311 < \mu_A - \mu_D < 84.311$
$-142.181 < \mu_A - \mu_E < -34.559$
$49.809 < \mu_B - \mu_C < 157.431$
$34.929 < \mu_B - \mu_D < 142.551$
$-83.941 < \mu_B - \mu_E < 23.681$
$-68.691 < \mu_C - \mu_D < 38.931$
$-187.561 < \mu_C - \mu_E < -79.939$
$-172.681 < \mu_D - \mu_E < -65.059$

b. The following differences are significant at $\alpha = .05$:

$\mu_A - \mu_B$, $\mu_B - \mu_D$, $\mu_A - \mu_E$, $\mu_C - \mu_E$, $\mu_B - \mu_C$, $\mu_D - \mu_E$

12.11 a. 99.5% confidence intervals:

$$
\begin{aligned}
-118.2476 &< \mu_A - \mu_B < 1.7946 \\
-14.8469 &< \mu_A - \mu_C < 105.6069 \\
-25.1842 &< \mu_A - \mu_D < 86.1842 \\
-141.4834 &< \mu_A - \mu_E < -35.2566 \\
53.6438 &< \mu_B - \mu_C < 153.5962 \\
44.3431 &< \mu_B - \mu_D < 133.1369 \\
-71.2564 &< \mu_B - \mu_E < 10.9964 \\
-59.5365 &< \mu_C - \mu_D < 29.7765 \\
-175.1565 &< \mu_C - \mu_E < -92.3435 \\
-153.3367 &< \mu_D - \mu_E < -84.4033
\end{aligned}
$$

b. If the use of the t statistic is correct for the data, the overall probability of error will be no more than $10(.005) = .05$. However, if all the assumptions for use of the t statistic are not met, the .005 probabilities of error may not be accurate and the overall error probability may be greater than .05.

12.12 All except three of the 99.5% confidence intervals calculated using a t statistic are smaller in width than the 95% confidence intervals calculated using Scheffé's Method.

12.13 The 99% Scheffé confidence intervals (for unequal n's):

$$
\begin{aligned}
-38.48 &< \mu_1 - \mu_2 < 3.74 & -27.30 &< \mu_2 - \mu_4 < 20.22 \\
-29.86 &< \mu_1 - \mu_3 < 15.42 & -11.02 &< \mu_2 - \mu_5 < 39.96 \\
-42.71 &< \mu_1 - \mu_4 < .89 & -38.82 &< \mu_3 - \mu_4 < 11.44 \\
-26.59 &< \mu_1 - \mu_5 < 20.79 & -22.45 &< \mu_3 - \mu_5 < 31.09 \\
-14.38 &< \mu_2 - \mu_3 < 34.68 & -8.05 &< \mu_4 - \mu_5 < 44.07
\end{aligned}
$$

According to the 99% CI limits, none of the pairs of means can be declared significantly different at $\alpha = .01$.

SECTIONS 12.4 AND 12.5, p. 409

12.14 a. $\bar{y}_{...} = 67$ is the overall average value, equally weighted over regions and products.

b.

Region	Region Effect
1	8
2	−8
3	0
4	4
5	−4

Product	Product Effect
A	−5
B	5
C	0

c. On the average, consumers rated Product A five points lower and Product B five points higher than the average rating of all three products over all 5 regions.

12.15 Table of interactions:

Product / Region	A	B	C
1	−2	0	2
2	1	0	−1
3	0	0	0
4	1	0	−1
5	0	0	0

12.16 The profile plot consists of five lines, each with an elbow. The lines all increase, then decrease, but they are not exactly parallel.

12.17 a. Yes, interaction is present since graph by regions is not parallel.
b. Yes, Product B is the favored product in every region.

12.18 a. $\bar{y}_{...} = 1.4$

b.

Row (Plan)	Row Effect
A	.3
B	.4
C	−.7

Column (Areas)	Column Effect
1	.2
2	0
3	−.2
4	0

c. $SS(Plan) = 80[(.3)^2 + (.4)^2 + (-.7)^2] = 59.20$

12.19 ANOVA Table:

Source	SS	d.f.	MS	F	p-value	Conclusion
Plan	59.2	2	29.6	24.67	$p < .001$	Reject H_0
Area	4.8	3	1.6	1.33	$p > .25$	Retain H_0
Interaction	2.4	6	.4	.33	$p > .25$	Retain H_0
Error	273.6	228	1.2			
Total	340.0	239				

12.20 b. A profile plot of the data suggests the possibility that some interaction is present since the graphs are not exactly parallel.
c. The F test indicates no interaction is present. Although the plot indicates some interaction, this departure from parallelism is not enough for statistical significance.

12.21 Tukey 95% confidence intervals:
$-.51 < \mu_1 - \mu_2 < .31$
$.59 < \mu_1 - \mu_3 < 1.41$ significantly different
$.69 < \mu_2 - \mu_3 < 1.51$ significantly different

12.22 a. SS(Tank type) = 3259.375; SS(Location) = 29,976.875
b. $F_{\text{Tank type}} = 13.138$; $F_{\text{Location}} = 51.785$
c. H_0: No significant differences among tank types
H_a: H_0 is false
T.S.: $F = 13.138$
R.R.: At $\alpha = .01$, reject H_0 if $F > 4.87$
Conclusion: Reject H_0.

12.23 p-value ≈ 0

12.24 If the data had been incorrectly analyzed as a one-factor ANOVA, the SS(Location) would become part of SS(Residual). The test is as follows.
H_0: No significant differences among tank types.
H_a: H_0 if false.
T.S.: $F = .96$
R.R.: At $\alpha = .01$, reject H_0 if $F > 4.57$
Conclusion: Do not reject H_0.
Here the differences among tank types are masked when location is not considered as a factor in ANOVA.

12.25 a. Mean yields by tank type: $\bar{y}_{1\cdot} = 136.375$; $\bar{y}_{2\cdot} = 128.625$; $\bar{y}_{3\cdot} = 112.125$; $\bar{y}_{4\cdot} = 114.125$
b. The following pairs of means are significantly different as determined by the Scheffé 99% confidence interval method: μ_1 and μ_3, and μ_1 and μ_4. Note that the Tukey method could be used also.

SECTION 12.6, p. 414

12.26 a.

System:	I	II	III	IV	
Mean:	189.833	173.833	211.000	164.333	

Task:	1	2	3	4	5	6
Mean:	54.00	326.75	195.25	86.75	38.25	407.50

Grand mean = 184.75

b. SS(System = 7505.5, SS(Tank) = 472,214.5, SS(Total) = 481,758.5, SS(Error) = 2038.5

c. $F = 18.41$; p-value $< .001$

12.27 For treatments (systems) $A_I = 1139$, $A_{II} = 1043$, $A_{III} = 1266$, $A_{IV} = 986$. $T = 4434$. SS(System) = 7505.5; similar calculations yield other SS.

12.28 Systems: All differences are significant except I vs. II and II vs. IV
Tasks: All differences are significant except 1 vs. 5.

12.29 a. Treatments are mixtures, blocks are investigators.
b. To control for possible variation among investigators.

12.30 a. MIXT is significant (p-value $= .0001 < .05$) but INVEST is not (p-value $= .2273 > .05$).
b. Mixture 2 has mean 2653.2; the average for the other three mixtures is 2400.6. The Scheffé interval is $\hat{l} \pm \sqrt{(I-1)F_\alpha}\sqrt{\text{MS(Error)}\Sigma c_j^2/n}$ where $\hat{l} = 2653.2 - 2400.6 = 252.6$, $I = 4$, $F_\alpha = 3.49$, MS(Error) = 68.858, $n = 5$, and the c_j are 1, $-1/3$, $-1/3$, $-1/3$. The interval doesn't include 0; the difference is significant.

12.31 No; F for INVEST isn't significant. If it had been, we would have evidence of variation among the investigators in their average measurements.

12.32 a. $F = 5.51$, $.01 < p\text{-value} < .025$
b. $F = 6.91$, $p < .001$; significant at $\alpha = .01$
Clearly, different people assign different average scores.

12.33 a. Tukey: $2.17 \leq \mu_1 - \mu_2 \leq 19.33$, $-.16 \leq \mu_1 - \mu_3 \leq 17.00$, $-10.91 \leq \mu_2 - \mu_3 \leq 6.25$
b. Only 1 vs. 2 is detectable (significant).

12.34 $\text{SS(FORM)} = 12[(57.667 - 51.278)^2 + (46.917 - 51.278)^2 + (49.250 - 51.278)^2] = 767.41$

SECTION 12.7, p. 419

12.35 a.

Source	F value
INSUQUAL	33.34
LAYOUT	24.72
INSUQUAL*LAYOUT	0.03
FAMTYPE	44.29
INSUQUAL*FAMTYPE	0.01
LAYOUT*FAMTYPE	24.21
INSUQUAL*LAYOUT*FAMTYPE	0.03

b. All the main factors, LAYOUT, INSULATION LEVEL, and FAMILY TYPE are significant at $\alpha = .01$. The LAYOUT/FAMILY TYPE interaction is the only significant interaction.

12.36 b. The profile plots agree with the conclusion of the corresponding F tests. The insulation quality-layout plot does not indicate the presence of interaction (the graphs are parallel) and the F test concludes no significant interaction. The layout family-type plot indicates interaction (the graphs intersect) and the F test concurs (p-value = .0001).

12.37 a. The Scheffé 95% confidence interval is $-8.11 < \mu_1 - \mu_2 < -3.89$. This confidence interval agrees with the F test that there is a significant difference in mean electricity usage for the two insulation qualities.

b. No, it does not make sense to compare the layout means ignoring family type, since the F test indicates the presence of interaction for the two factors.

c. All *but* the following pairs of means are significant as determined by the Tukey 95% confidence interval method: μ_3 and μ_6, μ_4 and μ_5, μ_4 and μ_6, and μ_5 and μ_6.

CHAPTER EXERCISES, p. 424

12.38 a. Design A: $\bar{y}_1 = 492.2,\ s_1^2 = 12351.29$
Design B: $\bar{y}_2 = 476.5,\ s_2^2 = 8360.06$
Design C: $\bar{y}_3 = 573.5,\ s_3^2 = 6258.94$

b., c. SS(Between) = 54,217.2667 and SS(Within) = 242,732.61

12.39 H_0: $\mu_1 = \mu_2 = \mu_3$

H_a: H_0 is false T.S.: $F = \dfrac{27{,}108.634}{8990.096} = 3.015$

R.R.: At $\alpha = .05$, reject H_0 if $F > 3.35$
Conclusion: Do not reject H_0.
$.05 < p\text{-value} < .10$

12.40 H_0: The distributions are identical.
H_a: The distributions differ in location.
T.S.: $H = 6.3277$
R.R.: At $\alpha = .05$, reject H_0 if $H > 5.99$
Conclusion: Reject H_0.
$.025 < p\text{-value} < .05$

12.41 Plots of data sets A, B, and C appear to be slightly skewed left. Since the sample sizes are small, the Central Limit Theorem may not have its effect and the Kruskal-Wallis conclusion may be preferred. Note also that the sample variances are unequal but the equality of sample sizes enable us to retain the assumption of equal variances. Note that 226 may be a bad outlier, casting doubt on the use of F.

12.42 The Scheffé 95% confidence intervals are
$-94.06 < \mu_1 - \mu_2 < 125.46$
$-191.06 < \mu_1 - \mu_3 < 28.46$
$-206.76 < \mu_2 - \mu_3 < 12.76$
None of the pairwise confidence intervals indicates significant differences among means.

12.43 a. $F = 23.10$

b. H_0: $\mu_1 = \mu_2 = \mu_3$
H_a: All means are not equal
T.S.: $F = 23.10$
R.R.: At $\alpha = .01$, reject H_0 if $F > 5.42$
Conclusion: Reject H_0.

c. p-value = .0001

12.44 H_0: The distributions are identical.
H_a: The distributions have different locations.
T.S.: $H = 18.87$
R.R.: At $\alpha = .01$, reject H_0 if $H > 9.21$
Conclusion: Reject H_0.

12.45 b. There appear to be serious violations of the ANOVA assumptions. None of the three data sets appears reasonably normal. Also, the three sample variances are not equal, and, since the sample sizes are unequal, the F statistic may be inaccurate.
c. Since both the F test and Kruskal-Wallis test show strong support of the research hypothesis, choice of the more appropriate test is not necessary here.

12.46 The Scheffé 99% confidence intervals are

$$-25.36 < \mu_1 - \mu_2 < -2.14$$
$$-36.26 < \mu_1 - \mu_3 < -12.59$$
$$-21.66 < \mu_2 - \mu_3 < .31$$

μ_1, and μ_2; μ_1 and μ_3 are significantly different.

12.47 a. The null hypothesis of equality of house prices is totally irrelevant here. Only if all the houses were built exactly alike by one contractor with similar locations would we expect house prices to be equal. The F test concurs; $F = 122.58$ has a p-value of 0.0001.
b. $F = 10.661$
c. The null hypothesis is rejected at all typical α levels; the p-value is .0001.

12.48 The Scheffé 95% confidence intervals are

$.515 < \mu_1 - \mu_2 < 8.819$	significantly different
$-5.99 < \mu_1 - \mu_3 < 2.319$	
$-6.735 < \mu_1 - \mu_4 < 1.569$	
$-10.652 < \mu_2 - \mu_3 < -2.348$	significantly different
$-11.402 < \mu_2 - \mu_4 < -3.098$	significantly different
$-4.902 < \mu_3 - \mu_4 < 3.402$	

12.49 a. $F = .34$; retain H_0.
b. It is very important to control for effects of house differences for this data. The variation in houses completely masks the variation in appraisers that we are attempting to study, as can be seen in the one-factor ANOVA. The two-factor ANOVA allowed us to view variation in appraisers by controlling for the effect of house differences.

12.50 b. H_0: No interaction between design and depth
H_a: H_0 is false
T.S.: $F = .93$
R.R.: At $\alpha = .05$, reject H_0 if $F > 3.89$
Conclusion: Do not reject H_0.
c. A profile plot of the data indicates no substantial interaction. This conclusion concurs with the result of the hypothesis test in part b.

12.51 a. H_0: no design effect
H_a: H_0 is false
T.S.: $F = 6.35$
R.R.: At $\alpha = .05$, reject H_0 if $F > 3.89$
Conclusion: There are significant effects of design.
p-value $= .0131$

b. H_0: No depth effect
H_a: H_0 is false.
T.S.: $F = 178.68$
R.R.: At $\alpha = .05$, reject H_0 if $F > 4.75$
Conclusion: Reject H_0, there are significant depth effects

12.52 The 95% confidence intervals, using Tukey's Method, are

$22.882 < \mu_A - \mu_B < 159.118$ significantly different
$-19.118 < \mu_A - \mu_C < 117.118$
$-110.118 < \mu_B - \mu_C < 26.118$

12.53 a. $F = .44$; this test statistic will cause nonrejection of the null hypothesis.
b. The graphs of the data are not parallel, indicating the presence of interaction. However, since the graphs are not seriously nonparallel, the interaction may not be substantial; it is not statistically significant.

12.54 Summary of F tests; all conclusions are based on $\alpha = .05$.

Null Hypothesis	F statistic	p-value	Conclusion
No three-factor interaction	.44	.6459	Retain H_0
No sweetness-acidity interaction	8.89	.0004	Reject H_0
No sweetness-color interaction	3.70	.0307	Reject H_0
No acidity-color interaction	2.92	.0927	Retain H_0
No sweetness effect	75.51	.0000	Reject H_0
No acidity effect	22.72	.0000	Reject H_0
No color effect	116.46	.0000	Reject H_0

12.55 Two means were omitted in the output; $\bar{y}_{H2N} = 35.16667$, $\bar{y}_{H2E} = 30.33333$. a. 95% confidence interval (Scheffé): $3.42 < \mu_{A1} - \mu_{A2} < 8.36$. The difference in average ratings is significant at $\alpha = .05$.
b. The following pairs of means are significantly different, by the Scheffé Method.

(μ_{L1N}, μ_{L1E})
(μ_{L1N}, μ_{L2E})
(μ_{L1N}, μ_{H1E})
(μ_{L1N}, μ_{H2N})
(μ_{L1N}, μ_{H2E})
(μ_{L1E}, μ_{M1N})
(μ_{L1E}, μ_{M2N})
(μ_{L2N}, μ_{L2E})
(μ_{L2N}, μ_{H1E})
(μ_{L2N}, μ_{H2N})
(μ_{L2N}, μ_{H2E})
(μ_{L2E}, μ_{M1N})
(μ_{L2E}, μ_{M2N})
(μ_{M1N}, μ_{M1E})
(μ_{M1N}, μ_{M2E})
(μ_{M1N}, μ_{H1N})
(μ_{M1N}, μ_{H1E})
(μ_{M1N}, μ_{H2N})
(μ_{M1N}, μ_{H2E})
(μ_{M1E}, μ_{H2N})
(μ_{M1E}, μ_{H2E})
(μ_{M1E}, μ_{M2N})
(μ_{M2N}, μ_{M2E})
(μ_{M2N}, μ_{H1E})
(μ_{M2N}, μ_{H2N})
(μ_{M2N}, μ_{H2E})
(μ_{M2E}, μ_{H2N})
(μ_{M2E}, μ_{H2E})
(μ_{H1N}, μ_{H2N})
(μ_{H1N}, μ_{H2E})

REVIEW EXERCISES, CHAPTERS 10–12, p. 430

R86 a. $-11.70 \le \mu_{old} - \mu_{new} \le 2.90$
b. Retain H_0; 0.00 is included in the interval.
c. Paired samples (same testers)

R87 a. $2.14 \le \mu_d \le 6.66$
b. Yes; 0.00 is not included.
c. Differences appear nearly normal. The 23 difference is not an outlier, by a fence calculation.

R88 Signed-rank test; discard the 0 difference, so $n = 39$. $T = |T_-| = 151.5 \leq 249$. Reject H_0; same conclusion.

R89 $t = 3.95$, p-value $< .002$; signed-rank p-value $< .002$ also.

R90 The paired-sample interval is much narrower; pairing was highly desirable.

R91 a. $F = 4.25 > F_{.01, 3, 60} = 4.13$; reject H_0. We have good evidence that at least one (sub)population mean differs from the others.
b. $.005 < p\text{-value} < .01$
c. Plots show some clear outliers; no other violations. Conclusion is still valid, but a rank test might be more effective.

R92 a. Using the Tukey method, $-17.06 \leq \mu_1 - \mu_2 \leq 2.70$, $-17.63 \leq \mu_1 - \mu_3 \leq 2.13$, $-20.25 \leq \mu_1 - \mu_4 \leq -0.49$, $-10.45 \leq \mu_2 - \mu_3 \leq 9.31$, $-13.07 \leq \mu_2 - \mu_4 \leq 6.69$, and $-12.50 \leq \mu_3 - \mu_4 \leq 7.26$.
b. Only the difference between means 1 and 4 is significant (either at $\alpha = .05$ or at $\alpha = .01$).

R93 a. $\hat{l} = \bar{y}_1 - (\bar{y}_2 + \bar{y}_3 + \bar{y}_4)/3 = -8.43$
b. Scheffé $S = 7.15$; $|\hat{l}| > S$ so it is significant.

R94 a. Yes; $H > \chi^2_{.01, 3\text{ d.f.}} = 11.34$
b. p-value $< .001$
c. The p-value for the Kruskal-Wallis test is (even) more conclusive, presumably because of the outliers in the data.

R95 a. $\chi^2 = 5.675 < 9.488$; not detectable (significant)
b. p-value $> .10$
c. Mildly. The expected values are small; 3 of the 9 are less than 5.

R96 Using degree of control to predict rating, $\lambda = .107$, indicating a modest relation.

R97 No. H_0 is retained, not proved, especially when n is small.

R98 Using $t_{.025, 45\text{ d.f.}} = 2.016$ by interpolation, $-115.5 \leq \mu_1 - \mu_2 \leq -51.5$. Reject H_0 at $\alpha = .05$ because 0 is not included.

R99 The sample sizes and sample variances are unequal (in the dangerous way—the larger n with the smaller s^2). The data are clearly left-skewed.

R100 Using t' with 19 d.f., $-122.8 \leq \mu_1 - \mu_2 \leq -44.2$. Again, reject H_0 at $\alpha = .05$.

R101 $z = -3.70$, $p < 2(.00024) = .00048$

R102 a. $F = 2.61$, $.01 < p\text{-value} < .025$ (assuming $F_{14, 31} \approx F_{14, 30}$). Retain H_0 at $\alpha = .01$.
b. Yes; the data are skewed to the left, and this F test is sensitive to nonnormality.

R103 a.

Program	A	B	C	(Grand Mean)
Mean	38.55	35.00	37.30	36.95

$\text{SS(Program)} = 20[(38.55 - 36.95)^2 + (35 - 36.95)^2 + (37.30 - 36.95)^2] = 1219.7$

$\text{SS(Error)} = 9[(5.190)^2 + (3.342)^2 + \cdots + (4.971)^2] = 1393.55$

b. $F = 2.51 < F_{.05, 2, 54} \approx 3.2$, not significant

R104 Using the Tukey method, the $\pm$ term is approximately 3.87. No program means differ by that much, so none of them is significant at $\alpha = .05$.

R105 The profile plots aren't close to parallel, so there is serious interaction. The program means are equally weighted averages, but in fact the service has fewer inexperienced people. The program means aren't a good indicator.

R106 Yes, the standard deviations for the inexperienced preparers are consistently quite a bit lower.

R107

Number of Cycles	1	2	3	4	5+
Expected	30.00	18.00	10.80	6.48	9.72
Observed	39	16	5	3	12

$\chi^2 = 8.441 < 9.488$; retain H_0. The frequencies are (barely) consistent with the theory.

R108 No; all *expected* frequencies are greater than 5.

R109 $.092 \leq \sigma \leq .144$

R110 $\chi^2 = 35.1232$, p-value $> .1$

R111 No. H_0 is retained, not accepted.

R112 The data are outlier-prone, and the test is sensitive to nonnormality.

R113 $t = t' = 4.027$, p-value $< .001$

R114 $T_1 = 254.5$, $z = -4.21$, p-value $< .0001$

R115 The rank test is preferred because of the outlier at 4.7. The conclusion is the same.

R116 $F = 4.598 > F_{.01, 19, 19} \approx 3.03$. Support H_a.

R117 Normality is crucial; the outlier makes the assumption shaky.

R118 a. $-.256 \leq \pi_1 - \pi_2 \leq .016$

b. Retain H_0: $\pi_1 - \pi_2 = 0$. We do not have enough evidence to conclude that there is a difference.

R119 Retain H_0; $|z| = |-1.73| < 1.96$

R120 Trials are not independent.

R121 $\chi^2 = 9.289$ with 3 d.f., $.025 < p$-value $< .05$

R122 None obvious; in particular, all $\hat{E}_{ij} > 5$.

R123 a. Paired-sample $t = 3.777 > t_{.025,\ 19\ \text{d.f.}} = 2.093$; reject H_0.

b. $p < .002$ (two-tailed); printer A clearly is slower.

R124 Signed-rank test; $T = |T_-| = 24 \leq 26$, so p-value $< .002$ as for the t test.

R125 s_d is much smaller than $\sqrt{s_A^2 + s_B^2} = 12.17$, indicating that a paired-sample method is much more effective.

R126 a. $3.96 \leq \sigma \leq 7.61$

b. The differences are left-skewed and -13 is beyond the lower outer fence. The χ^2 variance procedure is extremely sensitive to nonnormality.

R127 a. SS(Between) $= 12(21.17 - 22.718)^2 + 9(15.44 - 22.718)^2 + 18(27.39 - 22.718)^2 = 898.4$, as shown except for roundoff.

SS(Within) $= 11(8.26)^2 + 8(7.32)^2 + 17(8.15)^2 = 2{,}308.3$, again as shown except for roundoff.

b. $F = 7.00 > F_{.05, 2, 36} = 3.28$ (by rough interpolation); significant

R128 Kruskal-Wallis $H = 10.12 > 5.991$, significant at $\alpha = .05$; $.005 < p$-value $< .01$

R129 The F test assumes normality, so the Kruskal-Wallis is more appropriate.

R130 a. Because the n's differ, the Scheffé method is better.

b. Using $F_{.05} = 3.28$, the 95% Scheffé confidence intervals are

$$-3.31 \leq \mu_B - \mu_E \leq 14.77$$
$$-13.86 \leq \mu_B - \mu_L \leq 1.42$$
$$3.58 \leq \mu_E - \mu_L \leq 20.32$$

Only the Engineering and Liberal arts difference is significant.

R131 $\chi^2 = 16.61 > 9.488$; reject H_0

R132 Because 6 of the 9 $\hat{E}_{ij}$ values are less than 5, the probability may not be completely accurate. However, the χ^2 statistic is so far beyond the table value that the conclusion seems safe.

CHAPTER 13

SECTIONS 13.1 AND 13.2, p. 448

13.1 c. The plot of y vs. x' appears more nearly linear.

13.2 a. $\hat{y} = 14.29 + 1.475x$ b. $s_\epsilon = 1.36$

13.3 a. $\hat{y} = 14.876 + 10.523x'$ b. $s_\epsilon = 1.13$

13.4 The residual standard deviation in Exercise 13.3 is smaller. This result concurs with the choice of the model based on the plot of Exercise 13.1.

13.5 a. A plot of the data indicates that a linear equation is plausible.
b. $\hat{y} = 1.801 + .011x$ c. $s_\epsilon = .604$

13.6 Nothing blatant.

SECTION 13.3, p. 453

13.7 a. $.00934 < \beta < .01266$
b. The number of independent businesses (x) in sample zip-code areas has no effect in predicting the number of bank branches in these areas.
c. Generally, we have good reason to presume that the number of bank branches increases with the number of businesses in the same area. Therefore, the H_a might be stated as $H_a{:}\beta_1 > 0$
d. H_0: $\beta_1 = 0$ T.S.: $t = 12.029$
H_a: $\beta_1 > 0$ R.R.: At $\alpha = .05$, reject H_0 if $t > 1.812$
Conclusion: Reject H_0.

13.8 p-value $< .001$

13.9 $.4221 < \sigma_\epsilon < 1.059$

13.10 a. A plot of the data indicates that a linear equation relating y to x seems plausible. There is one very extreme point (runsize = 20.0, total cost = 1146).
b. $\hat{y} = 99.77704 + 51.91785x$; $s_\epsilon = 12.2068$
c. $50.717 < \beta_1 < 53.119$; the slope, β_1, could be interpreted as the variable cost per sticker, while the intercept, β_0, could be interpreted as the fixed cost.

13.11 a. $t = 88.527$ b. p-value = 0.0000 (two-tailed)

13.12 a. $F = 7835.946$; p-value = 0.0000
b. The conclusions of the F test and the t test are identical, since in simple regression models, $F = t^2$.

13.13 $10.05 < \sigma_\epsilon < 15.70$

SECTION 13.4, p. 458

13.14 a. $E(Y_{n+1}) = \hat{y}_{n+1} = 203.6$
b. $198.9 < E(Y_{n+1}) < 208.3$

13.15 Since $x_{x+1} = 2.0$ is close to $\bar{x} = 2.967$, the mean of the x-values used in determining the prediction equation, we would not expect major extrapolation in the prediction.

13.16 a. $\hat{y}_{n+1} = 203.6$ $178.2 < Y_{n+1} < 229.0$
b. Yes; $250 does not fall in the 95% prediction interval.

13.17 a. $\hat{y}_{n+1} = 15.738$; $13.810 < Y_{n+1} < 17.666$
b. As the x_{n+1} is very far away from the other x values, the prediction is a considerable extrapolation from the data.

13.18 a. $\hat{y}_{n+1} = 43.79$ b. $\hat{y}_{n+1} = 28.567$

c. The s_ϵ of predicting y with x' is 1.13; this is smaller than the s_ϵ of predicting y with x, which is 1.36. Therefore, it seems that prediction with x' is more reasonable.

13.19 $25.383 < Y_{n+1} < 31.751$

SECTIONS 13.5 AND 13.6, p. 467

13.20 $r_{yx} = .972$

13.21 a. H_0: $\rho_{yx} = 0$ T.S.: $t = 13.106$

H_a: $\rho_{yx} > 0$ R.R.: At $\alpha = .05$, reject H_0 if $t > 1.812$

Conclusion: Reject H_0.

b. In Exercise 13.7d., we rejected H_0: $\beta_1 = 0$ quite conclusively. That conclusion is coincident with this test of correlation. The t's are equal except for rounding.

13.22 a. $r_{yx}^2 = .9964$

b. Since $\hat{\beta}_1$ is positive, it follows that r_{yx} is positive.

13.23 With a smaller x (RUNSIZE) range, the correlation would probably be lower.

13.24 a. $r_{yx} = .9561$ b. $r_s = .9879$

c. A plot of the data indicates a nonlinear relation between y and x. It does appear, though, to be generally increasing.

CHAPTER EXERCISES, p. 470

13.25 b. $\hat{y} = 2.02 + 2.35x$ c. $\hat{y} = 51.37$

13.26 a. $s_\epsilon = 1.961$

b. $-3.922 < (y - \hat{y}) < 3.922$; all of the residuals for these data lie within this interval.

13.27 b. $\hat{y} = 5.859 + .015x$

c. H_0: $\beta_1 = 0$ T.S.: $t = 6.056$

H_a: $\beta_1 > 0$ R.R.: At $\alpha = .05$, reject H_0 if $t > 2.015$

Conclusion: Reject H_0.

p-value $< .001$

13.28 b. $\hat{y} = 1.008 + 5.312x'$

c. H_0: $\beta_1 = 0$ T.S.: $t = 7.019$

H_a: $\beta_1 > 0$ R.R.: At $\alpha = .05$, reject H_0 if $t > 2.015$

Conclusion: Reject H_0.

p-value $< .001$

13.29 The regression line in Exercise 13.28 appears to be the better fit. Comparison of the plots indicate y to be more linear in log x than in x. Also, the residual standard deviation in Exercise 13.28 is smaller than that of Exercise 13.27.

13.30 a. $1.988 < \beta_1 < 2.712$ b. $1.498 < \sigma_\epsilon^2 < 23.129$

13.31 a. $\hat{y} = 2.3548 + 42.3246\sqrt{x}$

b. $t = 15.1$ $p \ll .002$

13.32 $.564 < Y_{n+1} < 13.404$ (may contain roundoff error)

13.33 a. $\hat{y} = 149.056 + 50.118x$ b. $r_{yx}^2 = .823$

13.34 a. SS(Residual) = 87,5619; $s_\epsilon = 94.525$ b. $45.445 < \beta_1 < 54.791$

c. There is no point in testing H_0: $\beta_1 = 0$. Obviously, longer flights take more fuel; $\beta_1 > 0$.

13.35 a. $629.233 < E(Y_{n+1}) < 671.239$

b. $460.962 < Y_{n+1} < 839.510$; $Y_{n+1} = 570$ is contained in this interval. This value is not exceptionally low.

13.36 The value of β_1 is the fuel used for each hundred miles of air mileage. The value of β_0 might be interpreted as the fuel needed to take off.

13.37 a. $r_{yx} = .864$

b. H_0: $\rho = 0$ T.S.: $t = 5.427$
H_a: $\rho > 0$ R.R.: At $\alpha = .05$, reject H_0 if $t > 1.813$
Conclusion: Reject H_0.
p-value $< .001$

13.38 a. From the sample data, we can conclude the following:

1. The advertising expense accounts for $r_{yx}^2 = .746$ of the variability of sales.
2. From Exercise 13.37b., since we could reject H_0: $\rho = 0$ at any reasonable level, the correlation between advertising expense and sales is statistically significant.

b. Other uncontrolled factors that might affect sales are marketing strategy, competition, etc. These factors are represented in the regression model via the random error term.

c. The data hardly represent a random sample from the population of itnerest. The data have been collected over time, and therefore represent a time series.

13.39 $69.217 < Y_{n+1} < 137.933$

13.40 b. $y = 556.132 - 26.131x$
H_0: $\beta_1 = 0$ T.S.: $t = -2.823$
H_a: $\beta_1 \neq 0$ R.R.: At $\alpha = .05$, reject H_0 if $|t| > 2.228$
Conclusion: Reject H_0.

c. $210.891 < Y_{n+1} < 326.491$

13.41 a. $r_s = -.678$

b. H_0: $\rho_s = 0$ T.S.: $t = -2.917$
H_a: $\rho_s < 0$ R.R.: At $\alpha = .05$, reject H_0 if $t < -1.813$
Conclusion: Reject H_0.

c. From part b., we have significant evidence to conclude that the number of housing starts decreases with interest rates.

13.42 SS(Total) $= S_{yy} = 2508.625$, of which 2356.125, or 93.9%, is accounted for by SS(Between).

13.43 a. $\hat{y} = 68 + 5.75x_2$

b. H_0: $\beta_2 = 0$ T.S.: $t = .897$
H_a: $\beta_2 \neq 0$ R.R.: At $\alpha = .05$, reject H_0 if $|t| > 2.228$
Conclusion: Do not reject H_0.
p-value: $.20 < p$-value $.50$

13.44 a. $\hat{y} = 55.73 + 2.311x$ b. $s_\epsilon = 4.4445$

13.45 When $x_{n+1} = 21$, $\hat{y}_{n+1} = 104.268$; when $x_{n+1} = 40$, $\hat{y}_{n+1} = 148.183$. The $\pm$ term when $x_{n+1} = 21$ is 10.296; the $\pm$ term when $x_{n+1} = 40$ is 14.341. The plus-or-minus increases as one forecasts farther into the future.

13.46 $1.9488 < \beta_1 < 2.6732$; the value of β_1 is the monthly growth rate in sales.

13.47 a. $\hat{y} = 9.771 + .0501$ MILEAGE; $s_\epsilon = 2.202$

b. $.0486 < \beta_1 < .0516$

13.48 Examining a plot of the data, a linear equation relating y to x seems plausible. However, it appears that the variability of y increases with x; this would violate the constant variance assumption.

13.49 $22.32 < y_{n+1} < 31.30$; the extrapolation problem is not extremely serious. The value, $x_{n+1} = 340$, is relatively close to the mean of the x values, $\bar{x} = 395.04167$.

CHAPTER 14

SECTIONS 14.1 AND 14.2, p. 489

14.1 a. $\hat{y} = 326.39 + 136.10\text{PROMO} - 61.18\text{DEVEL} - 43.70\text{RESEARCH}$
b. $s_\epsilon = 25.63$ c. SS(Residual) = 13,136.23

14.2 $\hat{\beta}_1$ is the expected change in y(sales), per unit change in x, (promotion expenses), provided that all other x's (direct development expenditures and research effort) stay constant.

14.3 b. The plot of x vs. w indicates zero correlation.
c. $\hat{y} = 14.5 + 7x$ d. $s_\epsilon = 3.782$ e. $r_{yx}^2 = .733$

14.4 a. If the multiple regression equation $y = \beta_0 + \beta_1 x + \beta_2 w + \epsilon$ is used, the coefficient of x will not change from the value shown in Exercise 14.3 but the intercept value should change. Recall from Exercise 14.3 that the plot of x vs. w indicated a zero correlation between x and w; therefore, adding the w variable to the linear equation in x should in no way influence the effect of x and y. However, w presumably has an effect on y, so we expect the intercept to change in value.
b. $342 = 8.5(12) + 7.0(24) + 2.4(30)$
$740 = 8.5(24) + 7.0(56) + 2.4(60)$
$891 = 8.5(30) + 7.0(60) + 2.4(90)$
c. $s_\epsilon = 2.508$ d. $R_{y \cdot xw}^2 = .894$

14.5 a. If the multiple regression model $\hat{y} = \hat{\beta}_0 + \hat{\beta}_1 x + \hat{\beta}_2 v$ is used, we would expect both the coefficient of x and the intercept to change from the values shown in Exercise 14.3. Recall that the plot of x vs. v indicated a positive correlation between x and v; therefore, adding the v variable to the linear equation in x should influence the effect of x on y. Also, v presumably has an effect on y, so we would expect the intercept to change in value.
b. $342 = 17.0(12) - 3.0(24) + 5.0(42)$
$740 = 17.0(24) - 3.0(56) + 5.0(100)$
$1324 = 17.0(42) - 3.0(100) + 5.0(182)$
c. $s_\epsilon = 2.769$ d. $R_{y \cdot xv}^2 = .873$

14.6 a. $\hat{y} = 10 + 5x + 2w + 1v$; $s_\epsilon = 2.646$
b. $R_{y \cdot xwv}^2 = .895 > .873$ and $2.646 < 2.769$.

14.7 a. $y = 50.0195 + 6.64357143\text{ CAT1} + 7.3145\text{ CAT2} - 1.23142857\text{ CAT1SQ} - .7724\text{ CAT1CAT2} - 1.755\text{ CAT2SQ}$
b. SS(Residual) = 71.48900671; $s_\epsilon = 2.25972512$

14.8 $r_{y \cdot x_1 x_2 x_3 x_4 x_5}^2 = .862$

SECTION 14.3, p. 496

14.9 a. $F = 22.28$
b. H_0: $\beta_1 = \beta_2 = \beta_3 = 0$
H_a: At least one $\beta_j \neq 0$
T.S.: $F = 22.28$
R.R.: At $\alpha = .01$, reject H_0 if $F > 4.94$
Conclusion: Reject H_0.
c. $t = 4.84$
d. H_0: $\beta_1 = 0$ T.S.: $t = 4.84$
H_a: $\beta_1 \neq 0$ R.R.: At $\alpha = .05$, reject H_0 if $|t| > 2.086$
Conclusion: Reject H_0.

e. The conclusion of the test in part d. is that promotion has additional predictive value in predicting y over and above that contributed by the other independent variables, development expenditure, and research effort.

14.10 p-value = .0001, a two-tailed value.

14.11 DEVEL and RESEARCH each may have no additional predictive value in predicting y, as the "last predictor in." PROMO has additional predictive value in predicting y, as "last predictor in."

14.12 a. MS(Regression) = 159.67; MS(Residual) = 7.00
b. $F = 22.81$ c. p-value $< .001$
d. Since the p-value for the F test is extremely small, we would reject the null hypothesis, H_0: $\beta_2 = \beta_3 = \beta_4 = 0$ in favor of H_a: at least one $\beta_j \neq 0$.

14.13 a. $\hat{y} = 7.204 + 1.363\text{METAL} + .306\text{TEMP} + .0102\text{WATTS} - .00278\text{MET} \times \text{TEMP}$
b. Only WATTS has some proven additional predictive value in predicting y, as the "last predictor in," using $\alpha = .05$.

SECTION 14.4, p. 501

14.14 a. $R^2 = .769693$ b. $F = 22.28$
c. The F value on the output is the same as the value in part b.

14.15 a. $R^2 = .352298$ b. $F = 18.12$
c. H_0: $\beta_2 = \beta_3 = 0$; the independent variables promotion activities (x_2) and short-range research effort (x_3) have no predictive value once developmental expenditures (x_1) are included as predictor.
d. H_0: $\beta_2 = \beta_3 = 0$ T.S.: $F = 18.12$
H_a: H_0 is not true. R.R.: At $\alpha = .01$, reject H_0 if $F > 5.85$
Conclusion: Reject H_0.

14.16 a. $R^2 = .7444$ b. $R^2 = .7567$
c. $F = .430$; the null hypothesis tested is H_0: $\beta_3 = \beta_4 = 0$; that is, BUSIN and COMPET have no predictive value once INCOME is included as predictor.
d. H_0: $\beta_3 = \beta_4 = 0$ T.S.: $F = .721$
H_a: H_0 is not true. R.R.: At $\alpha = .05$, reject H_0 if $F > 6.11$
Conclusion: Retain H_0.

14.17 a. $\hat{y} = 50.0195 + 6.64357143\text{CAT1} + 7.3145\text{CAT2} - 1.23142857\text{CAT1SQ} - .7724\text{CAT1CAT2} - 1.1755\text{CAT2SQ}$
b. $\hat{y} = 70.31 - 2.676\text{CAT1} - .8802\text{CAT2}$
c. Complete model $R^2 = .862437$; reduced model $R^2 = .588452$
d. H_0: $\beta_3 = \beta_4 = \beta_5 = 0$
H_a: H_0 is not true.
T.S.: $F = 9.29$
R.R.: At $\alpha = .05$, reject H_0 if $F > 3.24$
Conclusion: Reject H_0; yes, there is evidence that the addition of the second-order terms x_3, x_4, and x_5 have improved the predictive value of the model.

SECTION 14.5, p. 504

14.18 a. For observation 21, $54.7081 < Y_{n+1} < 65.1438$
For observation 22, $57.0829 < Y_{n+1} < 67.6528$
b. For observation 21, $50.0281 < Y_{n+1} < 65.6985$
For observation 22, $51.0525 < Y_{n+1} < 66.4345$
c. Yes, the confidence limits for the model of part a. are tighter.

SECTION 14.6, p. 511

14.19 c. $\hat{\boldsymbol{\beta}} = \begin{bmatrix} 50.733 \\ 1.667 \\ 1.060 \end{bmatrix}$

14.20 a. $\mathbf{Y'Y} = 39341$ b. SS(Residual) = 118.7 c. $s_\epsilon = 3.145$

14.21 a. $F = 31.048$

b. H_0: $\beta_1 = \beta_2 = 0$
H_a: H_0 is not true.
T.S.: $F = 31.048$
R.R.: At $\alpha = .01$, reject H_0 if $F > 6.93$
Conclusion: Reject H_0.

14.22 H_0: $\beta_1 = 0$ R.R.: At $\alpha = .05$, reject H_0 if $|t| > 2.178$
H_a: $\beta_1 \neq 0$ Conclusion: Reject H_0.
T.S.: $t = 5.926$

H_0: $\beta_2 = 0$ R.R.: At $\alpha = .05$, reject H_0 if $|t| > 2.178$
H_a: $\beta_2 \neq 0$ Conclusion: Reject H_0.
T.S.: $t = 5.329$

14.23 b. $\hat{\boldsymbol{\beta}} = \begin{bmatrix} .200 \\ 1.264 \\ .740 \end{bmatrix}$ c. $s_\epsilon = 3.131$

14.24 a. $s_{\hat{\beta}_1} = 2.686$; $s_{\hat{\beta}_2} = 1.651$

b. $-3.493 < \beta_1 < 6.021$; $-2.184 < \beta_2 < 3.664$

c. H_0: $\beta_1 = 0$
H_a: $\beta_1 > 0$
T.S.: $t = .471$
R.R.: At $\alpha = .05$, reject H_0 if $t > 1.771$
Conclusion: Retain H_0.

14.25 a. SS(Residual) = 127.456; SS(Regression) = 224.554; SS(Total) = 352

b. $R^2_{y \cdot x_2 x_2} = .638$

c. H_0: $\beta_1 = \beta_2 = 0$
H_a: H_0 is not true.
T.S.: $F = 11.45$
R.R.: At $\alpha = .01$, reject H_0 if $F > 6.70$
Conclusion: Reject H_0.

14.26 a. $\hat{y}_{n+1} = 24.748$ b. $18.723 < Y_{n+1} < 30.773$
c. $\hat{y}_{n+1} = 20.308$ d. $2.456 < Y_{n+1} < 38.160$

CHAPTER EXERCISES, p. 514

14.27 a. $\hat{y}$ = −1.31949 + 5.54983 EDUC (2.70228) + .88514 INCOME (1.30846) + 1.92483 POP (1.37081) − 11.38928 FAMSIZE (6.66931)

b. $R^2 = .9625$, $s_\epsilon = 2.6863$

14.28 Collectively, the independent variables have at least some value in predicting DEMAND; $F = 44.891$ with a p-value that is essentially 0. The t tests all are not significant at $\alpha = .05$, and only t for EDUC is significant at $\alpha = .10$. There is little evidence as to which predictors have "last in" value.

14.29 a. $R^2 = .9423$
b. $F = 1.885 < 4.74$; retain H_0
We have no evidence that INCOME and POPN add predictive value.

14.30 a. $\hat{y} = -16.81979712 + 1.47018752x1 + .99477822x2 - .02400705x1x2 - .01031004x2SQ + .00024957x1x2SQ$; $s_\epsilon = 3.39011256$
b. The variable x_3 represents the interaction effect of number of subprograms (x_1) and average number of lines per subprogram (x_2) on the dependent variable, the number of writer days needed (y).
c. H_0: $\beta_3 = 0$ T.S.: $t = -1.01$
H_a: $\beta_3 \neq 0$ R.R.: At $\alpha = .05$, reject H_0 if $|t| > 2.087$
Conclusion: Retain H_0.

14.31 a. $\hat{y} = -16.81979712 + 1.47018752x1 + .99477822x2 - .02400705x1x2 - .01031004x2SQ + .00024957x1x2SQ$ (Complete Model);
$\hat{y} = .84008527 + 1.01583472x1 + .05582624x2$ (Reduced Model)
b. H_0: $\beta_3 = \beta_4 = \beta_5 = 0$
H_a: H_0 is not true.
T.S.: $F = .867$
R.R.: At $\alpha = .05$, reject H_0 if $F > 3.10$
Conclusion: Retain H_0.
p-value $> .25$

14.32 a. $\hat{y} = 9.85081550 + 2.49902738$POLINDEX
b. $s_\epsilon = 3.57034167$; $R^2 = .727886$
c. $\hat{y} = -2.67027671 + .72545042$POLLAG1
d. $s_\epsilon = 3.02277968$; $R^2 = .822896$

14.33 a. $y = 7.81896016 + 1.99503279$POLINDEX $+ .82254859$POLLAG1 $+ .26587278$POLLAG2
b. H_0: $\beta_{POLLAG1} = 0$
H_a: $\beta_{POLLAG1} \neq 0$
T.S.: $t = 2.56$
R.R.: At $\alpha = .05$, reject H_0 if p-value $< .05$
Conclusion: Reject H_0.

H_0: $\beta_{POLLAG2} = 0$
H_a: $\beta_{POLLAG2} \neq 0$
T.S.: $t = .99$
R.R.: At $\alpha = .05$, reject H_0 if p-value $< .05$
Conclusion: Retain H_0.

14.34 a. Shown in last part of output.
b. $s_\epsilon = 3.02277968$ c. $\hat{y}_{n+1} = 21.882$
d. $15.5779 < Y_{n+1} < 28.1861$

14.35 a. $\hat{y}_{n+1} = 28.98966$ b. $20.37166 < Y_{n+1} < 37.60766$

14.36 The prediction interval of Exercise 14.35 is wider than the prediction interval of Exercise 14.34 because the extrapolation penalty is larger.

14.37 a. $\hat{y} = .87272 + 2.54794$SIZE $+ .22028$PARKING $+ .58982$INCOME
(1.20083) (.15539) (.17806)
b. The intercept, .873, is predicted SALES with 0 floor space, 0 parking, and 0 income—a less-than-promising parlay. The coefficient of SIZE is the predicted increase in SALES from an additional 1000 square feet of floor space holding PARKING and INCOME constant. Similar interpretations hold for the other coefficients.
c. $R^2 = .7912$, $s_\epsilon = .7724$

14.38 $F = 15.158$ indicates that there's predictive value there somewhere. The t tests that INCOME definitely has incremental predictive value and that SIZE may have incremental value. There is no evidence that PARKING adds anything.

14.39 a. $\hat{y} = 102.7083 - .83333\text{PROTEIN} - 1.375\text{SUPPLEM} - 4.0\text{ANTIBIO}$
b. $s_\epsilon = 1.7096$ c. $R^2 = .9007$

14.40 a. $\hat{y}_{n+1} = 77.333$
b. The values of the independent variables don't represent a major extrapolation from the data; each of the given values (x_j) is equal to the mean of the sample x_j values, $\bar{x}_j$.
c. $76.469 < E(Y_{n+1}) < 78.197$

14.41 a. $\hat{y} = 89.83334 - .83333\text{PROTEIN}$ b. $R^2 = .5057$
c. H_0: $\beta_{\text{ANTIBIO}} = \beta_{\text{SUPPLEM}} = 0$
H_a: H_0 is not true.
T.S.: $F = 27.84$
R.R.: At $\alpha = .05$, reject H_0 if $F > 3.74$
Conclusion: Reject H_0.

14.42 a. $\hat{y} = 18.678 + .5420\text{SENIOR} + 1.2074\text{SEX} + 8.7779\text{RANKD1} + 4.4211\text{RANKKD2} + 2.7165\text{RANKD3} + .9225\text{DOCT}$
b. Given RANK, SENIOR, and DOCT, a male professor is predicted to have a salary 1.2074 thousand dollars higher than a comparable woman professor.
c. A full professor (of specified seniority, sex, and doctorate holding) is predicted to have a salary 8.7779 thousand dollars higher than a similar lecturer/instructor.

14.43 a. $t = 1.134 < 1.714$; retain H_0
b. Not proved; may be true, may be false.

14.44 a. $F = 64.646$
b. That none of the (obviously relevant) variables predicts salaries.
c. Yes; the p-value $= .0001$. A silly H_0 is rejected.

14.45 a. $R^2 = .9403$
b. $F < 1$; no evidence of incremental predictive value.

CHAPTER 15
SECTIONS 15.1 AND 15.2, p. 529

15.1 a. $r_{y_t x_t} = .635$ b. $r_{y_t x_{t-1}} = .974$
c. The "lag one" correlation is stronger. There is a delay between commission of a crime and sentencing to probation.

15.2 a. $y_t = 2.82 + 1.7308x_{t-1}$ b. $s_\epsilon = .1701$ c. $14.02 < y_{13} < 14.82$

15.3 With only two points, the correlation is either 1 or -1, which indicates the obvious fact that a straight line fits exactly through two points. With only one point, the correlation is undefined since the denominator of the formula is zero.

15.4 a. The dummy variables are DIV2 and DIV3; that is, define DIV2 = 1 if division = 2, DIV2 = 0, otherwise; define DIV3 = 1 if division = 3, DIV3 = 0, otherwise. If both DIV2 = 0 and DIV3 = 0, it follows by elimination that division = 1.
b. β_2 is the difference in expected ACTUAL sales between division 1 and division 2 for a fixed forecast. Similarly, β_3 is the difference in expected ACTUAL sales between division 1 and division 3 for a fixed forecast.
c. H_0: $\beta(\text{DIV2}) = 0$ T.S.: $t = -5.67$
H_a: $\beta(\text{DIV2}) \neq 0$ R.R.: At $\alpha = .05$, reject H_0 if p-value $< .05$
Conclusion: Reject H_0 since p-value $= .0001$.

H_0: $\beta(\text{DIV3}) = 0$ T.S.: $t = 1.25$
H_a: $\beta(\text{DIV3}) \neq 0$ R.R.: At $\alpha = .05$, reject H_0 if p-value $< .05$
Conclusion: Retain H_0 since p-value $= .2205$.

15.5 To complete a test of the null hypothesis that the dummy variables collectively have no predictive value, once the forecast has been included, we would have to run a simple regression model (using FORECAST as the only predictor variable), obtain computer printout so that we can calculate a R^2 reduced value required for the F test statistic.

15.6 a. H_0: $\beta_1 = \beta_2 = \beta_3 = 0$, which says that none of the variables in the multiple regression has any predictive value at all.

b. $F = 109.47$; at $\alpha = .01$, $F_{.01} = 4.60$, therefore reject H_0; p-value $= .0001$

15.7 a. $\hat{y}$ = FORECAST; the coefficient would be 1.000.

b. $1.184 < \beta_1 < 1.524$; the confidence interval doesn't include the value 1.

SECTION 15.3, p. 537

15.8 a. Pre-tax and post-tax income are highly correlated since post-tax income is roughly a percentage of pre-tax income.

Number of employees and gross sales are not necessarily highly correlated. It's plausible, though, that as the corporation gross sales increase, the number of employees needed increases.

b. A regression analysis based on profitability, size, and education level would be bedeviled by collinearity; it's likely that profitability in total dollars would be strongly correlated with size of firm. Some sort of ratio to size is needed.

15.9 a. Defining the "industry-type" variable as indicated would not be a good idea because a one-unit increase in INDUSTRY could mean either a change from the chemicals industry to the data-processing industry or a change from the data-processing industry to the electronics industry. There is no reason to assume that these two possible changes would predict the same change in any y.

b. An alternative approach for indicating these industries would be to define three dummy variables.

c. The factor of whether or not the firm matches employee contributions can be incorporated into the regression model by defining a dummy variable: 1 if matches, 0 if not.

15.10 a. A regression analysis based on CONTRIB, INCOME, and MATCHING would be bedeviled by collinearity and nonconstant variance. It's likely that INCOME and SIZE are highly correlated.

b. CONTRIB/INCOME represents the proportion of pre-tax income contributed by the corporation.

15.11 To test the suspicion, we could modify the regression model by introducing cross-product terms, involving SIZE and industry dummies.

15.12 a. To test this suspicion, a quadratic term $(\text{EDLEVEL})^2$ would be incorporated into the model.

b. If the consultant's suspicion is correct, the plot of residuals from a first-order regression model vs. EDLEVEL would show a parabolic pattern.

15.13 a. Each pair of the independent variables has correlation coefficient values of 0. Therefore, there is no collinearity present in the data.

b. Plots of the data indicate a zero correlation between MELT and CHILL as well as MELT and REPEL.

15.14 There is no obvious evidence of nonlinearity.

15.15 a. The difference in R^2 values is .913828 − .903581 = .010247
b. H_0: $\beta_{MELTSQ} = \beta_{CHILLSQ} = 0$
H_a: H_0 is not true.
T.S.: $F = 1.427$
R.R.: At $\alpha = .05$, reject H_0 if $F > 3.40$
Conclusion: Retain H_0.
c. Neither MELTSQ or KNIFESQ is a statistically significant predictor as "last predictor in."

SECTION 15.4, p. 547

15.16 a. REPEL, KNIFE, CHILL, MELT, SPEED (not entered)
b. REPEL, KNIFE, CHILL, MELT, SPEED
c. The ordering of the variables is the same.

15.17 H_0: $\beta_4 = \beta_5 = 0$
H_a: H_0 is not true.
T.S.: $F = 3.11$
R.R.: At $\alpha = .05$, reject H_0 if $F > 3.35$
Conclusion: Retain H_0, in a close call.

15.18 a. H_0: $\beta_1 = \cdots = \beta_7 = 0$
H_a: H_0 is not true.
T.S.: $F = 1.019$
Conclusion: The p-value > .25, therefore, retain H_0.
b. None of them.

15.19 a. The increment to R^2 is .192 − .060 = .132
b. H_0: $\beta_1 = \beta_2 = \beta_5 = 0$
H_a: H_0 is not true.
T.S.: $F = 1.805$
R.R.: At $\alpha = .05$, reject H_0 if $F > 2.92$
Conclusion: Retain H_0.
c. The simpler model is more sensible because the increment to R^2 obtained by using variables 1, 2, 5 in addition to 3, 4, 6, and 7 is not significant.

SECTIONS 15.5 AND 15.6, p. 566

15.20 a. There doesn't seem to be any strong suggestion of nonlinearity.
b. There doesn't seem to be a strongly suggested nonconstant variance problem.

15.21 a. $\hat{y}$ = 5759.7 + 230.79(MILAGE) − 8.4934(GASPRI) − 14.538(PREGAS) − 251.70(INTRAT) − 56.437(CARPRI)
b. 47.543

15.22 a. Yes, there appears to be a (positive) autocorrelation problem; $d = .7999$.
b. There is rather strong hint of a pattern in the sequence plot of residuals.

15.23 a. $\hat{y}$ = −.66169 − 89.662CPRIMI − 221.95CLAGPR − 109.50CINTRA + 5.8950CCARPR
b. $s_\epsilon = 37.339$
c. Not this time. Usually in the presence of autocorrelation, the residual standard deviation is biased downward, and differencing leads to a larger, more honest, estimate.

SECTION 15.7, p. 571

15.24 a. No; the average error is −2.0016 and the standard deviation of errors is 58.224.
b. It is 58.224, as compared to 37.339

15.25 a. In the full model, the error is always negative, clearly indicating a bias. There isn't a similar pattern for the simpler model.
b. In the simpler model, the standard deviation is actually smaller, at .0079447.
c. No; both bias and standard deviation results indicate that the model should be kept.

CHAPTER EXERCISES, p. 573

15.26 a. The plot of residuals use predicted values shows evidence of nonconstant variance.
b. If the dependent variable is redefined, we might well expect the problem of nonconstant variance to disappear.

15.27 a. The *X* correlations are 0.
b. The data are not a time series, and therefore we would not expect to find a problem of autocorrelation.
c. Given the nature of the problem, "heteroscedascity" is a possibility. A plot of residuals vs. predicted would be useful.

15.28 a. A plot of residuals against predicted values would allow us to determine if there were presence of either nonconstant variance or outliers. One could also detect an inappropriate model.
b. There doesn't seem to be any evidence that the problem of nonconstant variance exists. The point in the upper left corner is an outlier.

15.29 a. The lagged variables are number of installations in previous months and number of installations in four preceding months.
b. There is no obvious indication that any severe interaction could be expected.
c. There is an indication of possible need for nonlinear terms in the model. There was question as to whether support costs would increase in proportion to the number of installations.

15.30 a. Yes, there are two possible outliers.
b. Yes, there appears to be an autocorrelation problem.

15.31 There may be a slight nonconstant variance problem.

15.32 a. BCURRSQ, ACURRSQ, BPREV, APREV, BPREC4, BCURR, ACURR, APREV4—a rather strange sequence!
b. Increment to $R^2 = .0228 + .0095 + .0030 + .0030 = .0383$

15.33 a. Yes; $d = 1.34$, and there is a cyclic pattern in the plot of residuals and time.
b. Yes; in Figure 15.21a., "the plot thickens."

15.34 The sequence is very different.

15.35 a. No; there's no cyclic pattern.
b. Not really; the plot in Figure 15.22a. shows no pattern.

15.36 a. $\hat{y} = .520634 + 1.334738\text{DACURR} + 1.063336\text{DAPREV} + 2.025487\text{DBCURR} + .396343\text{DBPREV} - .020037\text{DBCURRSQ} + .119333\text{DAPREC4} + .009662\text{DACURRSQ} + .058331\text{DBPREC4}$
b. Yes; $F = 40.08$ with p-value .0000.
c. DACURR, DAPREV, DBCURR, DBPREV
d. $R^2 = .94129$; variation in all eight predictors accounts for 94.129% of the variation in *Y*.
e. $s_\epsilon = 3.35551$

15.37 a. Residuals -3.78, -2.03, 2.01, 4.94, 1.71, -1.78; $s = 3.25$
b. It's essentially the same.
c. Yes; ± 2 standard deviations is less than 7 units.

15.38 a. The SPECIALS variable is a dummy or indicator variable. SPECIALS should not be coded as 0, 1, and 2. Rather, two dummy variables should be used to define special packing.
b. β_8 is the difference in expected time between "no special packing" and "special packing at A" orders for a fixed NUMFREQ, NOMMOD, NOMRARE, NUMLOOSE, AVSZCAR, AVSZLB, and SKIDS. Similarly, β_9 is the difference in expected time for "no special packing" and "special packing at B" orders for a fixed NUMFREQ, NUMMOD, NUMRARE, NUMLOOSE, AVSZCAR, AVZZLB, and SKIDS.
c. NUMFREQ, NUMMOD, NUMRARE, and NUMLOOSE might be transformed because the supervisor expected travel time not to be directly proportional to the number of items.
d. Interaction terms might be useful because the supervisor expected assembly-station time to depend on NUM and AVSZ.

15.39 a. There is one very obvious outlier.
b. Nonconstant variance does not appear to be a problem.
c. The nature of this case study does not lend itself to a time series problem. Therefore, we would not expect to find a problem of autocorrelation.

15.40 Square roots is the more effective transformation.

15.41 a. Increased $R^2 = .0124$
b. H_0: $\beta_4 = \beta_5 = 0$
H_a: H_0 is not true.
T.S.: $F = 20.39$
R.R.: At $\alpha = .05$, reject H_0 if $F \geq 3.23$
Conclusion: Reject H_0.

15.42 a. $\hat{y} = .02436 + 1.80133$ (SQTNFPER) $+ 1.63110$ (SQTNMPER) $+ 2.73480$ (SQTNRPER) $+ 1.14912$ (NUMLPER) $+ .36002$ (ASZCRPER) $+ .21039$ (ASZLBPER) $+ 1.87525$ (SPECAPER) $+ 2.92493$ (SPECBPER) $+ 1.28820$ (SKIDSPER) Residual standard deviation $= .0492$
b. All the residual plots indicate no violation of assumptions.

15.43 a. Standard deviation of the regression prediction errors $= .08848$. Standard deviation of the superintendent's prediction errors $= .15273$.
b. The standard deviation in Exercise 15.42 was .0492. The regression standard deviation has increased to .08848.
c. Yes; the error should be cut in half, roughly.

REVIEW EXERCISES, CHAPTERS 13–15, p. 596

R133 a. $S_{xy} = 35{,}178{,}250$ and $S_{xx} = 1.010.887.8$
$\hat{y} = 1013.1027 + 34.79936x$
b. It is the predicted increase in direct cost per additional direct labor hour, the variable cost.
c. It is the fixed cost.
d. $s = 3396.71$

R134 $r_{xy} = .9031$; variation in direct labor hours accounts for $(.9031)^2 = .8156$, or 81.56%, of total direct cost

R135 a. $s_{\hat{\beta}_1} = 3.3784$
b. $27.82 \leq \beta_1 \leq 41.77$

R136 No; direct cost certainly should increase as direct labor hours increase. $t = 10.30$ is conclusive evidence to reject H_0.

R137 a. $\hat{y} = 31{,}985$; $25{,}481 \leq Y_{n+1} \leq 38{,}489$
b. $10{,}262 \leq Y_{n+1} \leq 22{,}108$
c. The interval in part a. is wider because of the extrapolation penalty.

R138 a. The constant-variance (homoscedasticity) assumption appears to be wrong.
b. The prediction intervals in R137 are most questionable.

R139 a. PURCH = −.744 + .0329 AGE + .00900 INCOME + .115 OWNER + .00818 EDUCN
b. For given income, ownership, and education, each one-year increase in age predicts a .0329 unit increase in purchases. The other slopes have similar interpretation.
c. Meaningless; there are extremely few 0-year-old, income-less, education-less card-holders.

R140 a. That are predictors were worthless
b. $F > 600$
c. AGE definitely adds predictive value, OWNER and EDUCN most likely add value, and INCOME may add value.

R141 Yes; in particular, AGE and INCOME are rather highly correlated.

R142 a. To test whether the effects of AGE, INCOME, and EDUCN on PURCH depend on OWNER.
b. $F = 1.52 < F_{.10,3,152} \approx 2.12$; retain H_0

R143 The variance is not constant.

R144 There is an evident outlier at the lower left of the plot.

R145 a. Yes; R^2 is slightly higher and s is lower.
b. Yes. Everything else is the same.
c. The t for LOGINC is larger.

R146 Yes; LEADING, RATE, and MONTH all are very highly correlated.

R147 a. $F = 321.310$ with p-value .0001
b. Yes for SUPPLY and RATE, no for LEADING.

R148 ROOT MSE $= .0359 = \sqrt{\text{MS(ERROR)}}$
This standard deviation is considerably lower than the standard deviation of SPREAD, indicating good predictive value.

R149 a. Definite cyclic pattern indicates autocorrelation.
b. $d = 0.807$
c. The claimed standard deviation and standard errors are too low, and R^2 is too high.

R150 a. They are quite similar.
b. No change; we would believe the results for the difference data.
c. Yes; $d = 2.003$

R151 No particular pattern; there shouldn't be a cyclic pattern with $d = 2.003$.

R152 The correlations among independent variables are much lower.

R153 a. $\hat{y} = -3.333 + .0794\text{GAS} + 7.861\text{TRAFFIC} - .132\text{ AVGINC}$
b. No; Gas = 0, TRAFFIC = 0, and AVGINC = 0 are all impossible.

R154 a. SS(Regression) = 480.17195
b. GAS accounts for 457.42848, TRAFFIC adds an additional 20.61406, and AVGINC adds 2.12941.
c. AVGINC seems to be of little value.
d. Delete AVGINC and perhaps TRAFFIC.

R155 a. Not extreme; highest correlation of independent variables is .566.
b. d isn't shown, but the data aren't a time series.

R156 One outlier (upper right); no other pattern evident.

R157 a. The coefficient of TRAFFIC has decreased considerably.
b. $s = 3.26$ compared to previous $s = 4.34$
c. $R^2 = .513$ compared to previous $R^2 = .600$

R158 $F = 0.54$, not close to significant.

R159 a. The plot is perfectly rectangular.
b. $r_{TEMP,PRES} = 0$
c. No collinearity

R160 a. YIELD = $-2.41 + .0262$TEMP $+ 5.33$PRES
b. $\pm .962$

R161 $F = 82.14$, t's = 9.09 and 9.04; TEMP and PRES definitely have predictive value, both jointly and incrementally

R162 The model is wrong; there's a clear curve in the plot.

R163 a. Not at all
b. Only modestly; s decreases slightly and R^2 increases slightly. The t statistic is not highly significant.

R164 a. Yes; $t = 8.33$
b. $R^2 = .759$

R165 Autocorrelation is shown by Durbin-Watson statistic = 0.43. R^2 is too high; s and the standard errors are too low.

R166 a. Yes; $t = 18.16$
b. $R^2 = .940$
c. Surprisingly, both values are larger

R167 Not entirely; $d = 1.43$

R168 a. That pressure difference didn't affect YIELD.
b. $F = 414.234$, $t = 20.353$

R169 s = ROOT MSE = 2.435 = $\sqrt{\text{MS(Error)}}$
Most (95%) prediction errors will be within $\pm 2(2.435) = \pm 4.870$

R170 a. Yes, there's a clear curve in the plot.
b. No obvious outliers

R171 $R^2 = .9437$ is larger than the previous R^2 of .9367.

CHAPTER 16

SECTION 16.1, p. 619

16.1 a. $I_2 = 109.8$; $I_3 = 120.8$ b. $I_2 = 109.2$; $I_3 = 120.2$

16.2 a. $I_2 = 109.2$; $I_3 = 118.5$; $I_3 = 128.6$
b. $I_2 = 105.7$; $I_3 = 112.8$; $I_4 = 119.0$
c. $I_2 = 107.9$; $I_3 = 112.3$; $I_4 = 112.7$

16.3 a. The weighted index value is consistently lower than the simple aggregate index.
b. The difference is due to the change in relative weights.

16.4 a. $I_2 = 102.9$; $I_3 = 113.2$; $I_4 = 129.1$; $I_5 = 133.3$
b. $I_2 = 107.5$; $I_3 = 115.8$; $I_4 = 128.9$; $I_5 = 139.2$
c. Any differences can be attributed to the relatively large steel weights.

SECTION 16.2, p. 628

16.5 From the plot of the raw data, we see that there is an obvious upward trend. There is also a seasonal pattern to the data; the sales peak in September. If there is a longer-term cyclic effect, it is not very obvious.

16.6 $\hat{y}_t = 100.513 + .5963t$

16.7 From the plot of the detrended values, we see a seasonal pattern. The sales peak occurs in September.

16.8 a.

Month	J	F	M	A	M	J
Value	.985	.983	.981	.987	.988	.983

Month	J	A	S	O	N	D
Value	.988	.989	.987	.987	.989	.987

b.

t	61	62	63	64	65	66
forecast	137.1611	124.1477	124.6861	125.2246	139.4118	125.8819

t	67	68	69	70	71	72
forecast	140.4651	154.8854	169.8476	156.0526	142.4217	143.0163

16.9 There is a clear decreasing trend in the values. Years 1 and 2 are all positive, whereas Years 4 and 5 are all negative.

16.10 b. The upward trend is clear; linear trend should be adequate.

c. There doesn't seem to be any strong visual evidence of a cyclical or seasonal pattern.

16.11 $\log \hat{y}_t = 4.56295 + .0197t$

16.12 b. There doesn't seem to be any strong visual evidence of a seasonal or cyclical effects.

16.13

t	33	34	35	36
forecast	183.6533	187.3072	191.0337	194.8344

SECTION 16.3, p. 635

16.14 a.

Month	3-Month	5-Month	Month	3-Month	5-Month
1	*	*	13	114.7	106.0
2	*	*	14	95.0	105.0
3	95.7	*	15	91.0	104.2
4	122.0	*	16	95.7	97.2
5	129.7	115.0	17	109.3	101.6
6	131.7	118.6	18	115.0	104.4
7	109.0	119.2	19	109.3	106.2
8	101.0	118.2	20	99.3	106.6
9	95.0	102.4	21	92.3	102.4
10	94.0	97.8	22	92.3	96.0
11	109.7	105.0	23	95.7	95.0
12	115.0	106.0	24	102.7	98.4

c. Average squared-error 3-month moving average = 627.613; for 5-month moving average = 337.6589

16.15 a.

Month	3-Month	5-Month	Month	3-Month	5-Month
1	*	*	13	105	97
2	*	*	14	96	97
3	97	*	15	93	96
4	101	*	16	93	96
5	120	101	17	110	96
6	120	107	18	110	110
7	107	107	19	110	110
8	100	107	20	95	110
9	96	100	21	93	95
10	96	97	22	93	93
11	97	97	23	93	93
12	105	97	24	105	95

c. Average squared-error or 3-month running median = 557.8571; for 5-month running median = 291.0526

16.16 a.

Month	Forecast	Month	Forecast
1	89.00	13	110.71
2	71.20	14	98.94
3	91.84	15	86.99
4	99.17	16	91.79
5	154.23	17	106.36
6	126.85	18	121.27
7	110.97	19	112.25
8	102.19	20	96.85
9	97.24	21	95.37
10	90.65	22	90.27
11	95.73	23	92.45
12	133.55	24	102.49

b. Average squared error = 510.901

16.17 The 5-month running median is best.

16.18 b. The 3-week moving average

16.19 b. The 3-week running median

16.20 Yes. The 3-week running median forecast method is best on both criteria.

16.21 c. Perhaps $\alpha = .8$ is overresponding.

16.22 $\alpha = .8$

SECTION 16.4, p. 647

16.23 a. There is a long-term upward trend.
b. There does not seem to be any evidence of a linear trend in the changes.

16.24 a. $r_1 = .693$, by calculation.
b. The lag correlations taper off, rather than dropping to 0.

16.25 It appears that the model yields useful forecasts of changes in sales.

16.26 The estimated lag correlations for the data of Exercise 16.24 are close to but higher than the estimated lag correlations for the model of Exercise 16.25.

16.27 a. There is a suggestion of a sharp drop between lags 1 and 2.
b. An autoregressive parameter is needed as suggested by the gradual tapering of autocorrelations. The suggestion of a drop at lag 2 indicates that at least one moving average term could be used.

16.28 No

16.29 No

CHAPTER EXERCISES, p. 652

16.31 a. $\hat{y}_t = 429.30 + 4.892t$

16.32 a.

Month	J	F	M	A	M	J
Index	.975	.995	.990	.9925	1.0075	1.0075

Month	J	A	S	O	N	D
Index	1.015	1.0025	.9975	1.005	1.0075	1.005

b.

Month	J	A	S	O	N	D
Index	.985	1.017	1.023	1.005	.983	.975

16.33 There is some curvature; try an exponential trend.

16.34 $\log \hat{y}_t = 1.4363 + .0215t$

16.35 a. There appears to be a seasonal component in the data. The data values peak in the 3rd quarter.

b.

Quarter	1	2	3	4
Index	.89525	1.042375	1.272625	.842375

	Quarter			
Year	1	2	3	4
1	.891	.445	.856	.853
2	.872	.950	1.119	1.042
3	.975	1.010	1.142	.989
4	.939	.991	1.126	.983
5	.938	.998	1.126	.982
6	.957	.996	1.129	.995
7	.945	.992	1.131	.985
8	.923	.983	1.143	.992

16.36 b. Quarter 3 is consistently underestimated. Since the $\alpha = .05$ is fairly small, the previous observations are "remembered" more. Thus, the lower values in Quarters 1 and 2 keep the 3rd Quarter estimate too low.

16.37 a.

Quarter	5-quarter Moving Average	Quarter	5-quarter Moving Average
1	***	17	.9852
2	***	18	1.0188
3	***	19	1.0614
4	***	20	.9766
5	.9898	21	.9952
6	1.0172	22	1.0234
7	1.0646	23	1.0644
8	.9788	24	.9854
9	.9900	25	.9972
10	1.0208	26	1.0182
11	1.0702	27	1.0656
12	.98	28	.9806
13	.9906	29	.9808
14	1.0176	30	1.0104
15	1.0608	31	1.0634
16	.9738	32	.9772

b.

Quarter	5-quarter Running Median	Quarter	5-quarter Running Median
1	***	17	.893
2	***	18	1.042
3	***	19	1.058
4	***	20	.893
5	.891	21	.933
6	1.028	22	1.034
7	1.028	23	1.034
8	.891	24	.933
9	.907	25	.933
10	1.028	26	1.034
11	1.045	27	1.038
12	.907	28	.896
13	.907	29	.896
14	1.042	30	1.038
15	1.042	31	1.044
16	.890	32	.861

c. No. The figures are clearly not good predictors.

d. Work with a seasonal adjustment.

16.38 a. $I_1 = 107.6$; $I_2 = 118.8$; $I_3 = 133.3$; $I_4 = 147.8$; $I_5 = 169.5$; $I_6 = 186.5$; $I_7 = 207.9$

b. $I_1 = 107.8$; $I_2 = 119.6$; $I_3 = 134.2$; $I_4 = 148.9$; $I_5 = 170.9$; $I_6 = 188.0$; $I_7 = 209.5$

c. The difference is very small.

16.39 a. A positive linear trend equation seems appropriate.

b. $\hat{y}_t = 135.204 + 2.559t$

16.40 b. The lag correlations decrease smoothly.
c. AR[1], AR[2]

16.41 a. $\hat{y}_t = 1.401 + 409y_{t-1}$
$\hat{y}_t = 1.205 + .378y_{t-1} + .108y_{t-2}$
b. No c. The AR(1) model seems adequate.

CHAPTER 17

SECTION 17.1, p. 661

17.1 a. $A - \{40, 41, \ldots, 100\}$
b. $\Theta = \{$very light, light, moderate, heavy, very heavy$\}$

17.2 a. Ordering 50 cars is an unreasonable action.
b. Ordering 100 cars is also clearly unreasonable.

17.3 a. $A = \{1, 2, \ldots, 30\}$ b. $\Theta = \{1000, 2000, \ldots, 30{,}000\}$

17.4 There are no inadmissable actions.

17.5 a. $A = \{100\%, 75\%, 50\%, 25\%, 0\%\}$
b. $\Theta = \{$(law not passed), (law passed and rental level rises), (law passed and rental level stays the same)$\}$
c. Payoff matrix

	θ_1	θ_2	θ_3
A_1	100,000	40,000	−20,000
A_2	75,000	30,000	−15,000
A_3	50,000	20,000	−10,000
A_4	25,000	10,000	−5,000
A_5	0	0	0

17.6 a. $A = \{$settle, contest$\}$; $\Theta = \{$(settled), (found not guilty), (found guilty by Judge A), found guilty by Judge B)$\}$
b. Payoff matrix

	θ_1	θ_2	θ_3
A_1	−80,000	−80,000	−80,000
A_2	−20,000	−150,000	−180,000

SECTION 17.2, p. 666

17.7 "Full production" has the highest payoff.

17.8 a. "Full production" has the highest payoff.
b. "Full production" has the highest payoff.
c. "Limited production" has the highest payoff.
d. The action "full production" remains optimal if the payoffs and probabilities are varied one at a time. But if we vary both the payoffs and probabilities, the action "full production" becomes second best to "limited production."

17.9 "Moderate" has the highest payoff.

17.10 a. To perform a sensitivity analysis, add .5 million to the profits for "low" and "high" prices.
b. "High" has the highest payoff.

17.11 a. "Moderate" has the highest payoff.
b. "High" has the highest payoff.
c. The action "moderate" remains optimal if the probabilities are varied. But if we vary the payoffs, either alone or in combination with varied probabilities, the action "moderate" becomes a second best decision to "high."

17.12 "Go to trial" has the lower expected cost.

17.13 "Settlement" has the lower expected cost if the probability is increased to .6. "Go to trial" has the lower expected cost if the probability is decreased to .3.

SECTION 17.3, p. 673

17.14 a. "90%" has the highest expected profit.
b. "80%" has the lowest standard deviation.
c. Yes, there is a conflict.

17.15 a. "90%" has the highest expected profit.
b. If our goal was to maximize expected return, there would be no reason to modify the choice of the optimal action found in Exercise 17.14. However, if our goal was to minimize risk, we would change our initial choice "80%" to "90%."

17.16 a. "90%" has the highest expected return.
b. "80%" has the lowest risk.
c. No, the optimal actions do not change.

17.17 a. Cov. $(Y_1Y_2) = -84$ b. $\rho_{Y_1Y_2} = -.374$

17.18 a. Expected return = 500 b. Variance = 1992
c. Expected return = 500; variance = 1842

17.19 a. Expected return = 1,091,333.332; variance = 14,444,444,000
b. Expected return = 1,091,500; variance = 17,000,000,000
Expected return = 1,091,000; variance = 22,000,000,000

17.20 a. Expected return table

		$R(a_i)$
Order	50	$30(.20) + 40(.20) + 40(.20) + 40(.20) + 40(.20) = 38$
	60	$35(.20) + 50(.20) + 55(.20) + 55(.20) + 55(.20) = 50$
	70	$20(.20) + 40(.20) + 60(.20) + 70(.20) + 70(.20) = 52$
	80	$25(.20) + 30(.20) + 55(.20) + 80(.20) + 85(.20) = 55$
	90	$-10(.20) + 20(.20) + 50(.20) + 80(.20) + 110(.20) = 50$
	100	$-25(.20) + 10(.20) + 45(.20) + 80(.20) + 105(.20) = 43$

b. Risk Table

		Risk, Standard Deviation
Order	50	$[(30)^2(.20) + (40)^2(.20) + (40)^2(.20) + (40)^2(.20) + (40)^2(.20) - (38)^2]^{1/2} = 4$
	60	$[(35)^2(.20) + (50)^2(.20) + (55)^2(.20) + (55)^2(.20) + (55)^2(.20) - (50)^2]^{1/2} = 7.745$
	70	$[(20)^2(.20) + (40)^2(.20) + (60)^2(.20) + (70)^2(.20) + (70)^2(.20) - (52)^2]^{1/2} = 19.39$
	80	$[(25)^2(.20) + (30)^2(.20) + (55)^2(.20) + (80)^2(.20) + (85)^2(.20) - (55)^2]^{1/2} = 24.69$
	90	$[(-10)^2(.20) + (20)^2(.20) + (50)^2(.20) + (80)^2(.20) + (110)^2(.20) - (50)^2]^{1/2} = 42.43$
	100	$[(-25)^2(.20) + (10)^2(.20) + (45)^2(.20) + (80)^2(.20) + (105)^2(.20) - (43)^2]^{1/2} = 46.75$

c. No. The order quantity with the highest expected return is "80," whereas the order quantity with the lowest risk is "50."

17.21 These figures alone do not show the inferiority of ordering "50" cards. Even though the quantity "50" has a lower expected return, it also has the lower risk.

17.22 a. Expected return = 3269 b. Variance = 554,000

17.23 Expected return = 3271.50; variance = 787,500. If an investor desires low risk, he would prefer the portfolio of Exercise 17.22.

SECTION 17.4, p. 678

17.24 a. "Expansion" has a higher expected payoff but also a higher standard deviation.
b. "Expansion" has a higher expected utility.

17.25 The revised prior probabilities have affected the conclusions of Exercise 17.24. The optimal choice changes from "expansion" to "no expansion."

17.26 "Expansion" has the higher expected utility. The revised utility function has not affected the conclusion of Exercise 17.24.

17.27 a. Expected return table

		$R(a_i)$
Strategy	1	$270(.10) + 370(.20) + 470(.40) + 570(.30) = 460$
	2	$310(.10) + 385(.20) + 460(.40) + 535(.30) = 452.5$
	3	$350(.10) + 400(.20) + 450(.40) + 500(.30) = 445$
	4	$390(.10) + 415(.20) + 440(.40) + 465(.30) = 437.5$
	5	$480(.10) + 430(.20) + 430(.40) + 430(.30) = 430$

The expected return is highest for "strategy 1" and declines to lowest for "strategy 5."

b. Variance table

		Var (a_i)
Strategy	1	$[(270)^2(.10) + (370)^2(.20) + (470)^2(.40) + (570)^2(.30) - (460)^2] = 8900$
	2	$[(310)^2(.10) + (385)^2(.20) + (460)^2(.40) + (535)^2(.30) - (452.5)^2] = 5006.25$
	3	$[(350)^2(.10) + (400)^2(.20) + (450)^2(.40) + (500)^2(.30) - (445)^2] = 2225$
	4	$[(390)^2(.10) + (415)^2(.20) + (440)^2(.40) + (465)^2(.30) - (437.5)^2] = 556.25$
	5	$[(430)^2(.10) + (430)^2(.20) + (430)^2(.40) + (430)^2(.30) - (430)^2] = 0$

The variance is highest for "strategy 1" and declines to lowest for "strategy 5."

17.28 Investment "strategy 1" has the highest expected utility.

CHAPTER EXERCISES, p. 683

17.29 a. A = {(promotion 1), (promotion 2)}
b. Θ = {(Both competitors, A and B), (Competitor A only), (Competitor B only), (Neither A or B)}

17.30 a. "Promotion 2" has the higher expected payoff.
b. "Promotion 1" has the lower risk.

17.31 a. A plot of the utility function indicates risk aversion.
b. "Promotion 1" has the higher expected utility.
c. The preferred promotion, "Promotion 1," has also lower risk but not higher expected payoff.

17.32 "Promotion 2" has the higher expected payoff and "Promotion 1" has the lower risk if the probability is decreased to .5. "Promotion 2" has the higher expected payoff and "Promotion 1" has the lower risk if the probability is increased to .7. The conclusions of Exercise 17.31 do not change.

17.33 a. "Replace" has the highest expected payoff.
b. "No change" has the lowest variance.

17.34 "Replace" has the highest expected utility.

17.35 Revision 1 – "replace" and "upgrade" have the highest expected payoff.
Revision 2 – "upgrade" has the highest expected payoff.
Revision 3 – "replace" has the highest expected payoff.
Revision 4 – "replace" has the highest expected payoff.

17.36 a. $\text{Cov}(X, Y) = -.000108$; $\rho_{xy} = -.779$
b. The covariance is negative because there is a tendency for Project 2 net returns to decrease as Project 1 net returns increase.

17.37 a. Expected return = 100; variance = 7176
b. Expected return = 100; variance = 18800
c. The risk is lower because if we invest in both projects, even if the net return from project 2 is low, the net return from project 1 is likely to be high (due to the negative covariance).

17.38 a. "Preferred" has the highest expected payoff.
b. "Preferred" has the lowest variance.

17.39 "Common" has the highest expected payoff; "preferred" has the lowest variance. The revised probabilities have affected the decision about the action with the highest expected payoff.

17.40 a. "Preferred" has the highest expected utility.
b. "Common" has the highest expected utility.

17.41 a. Expected return = 13,080; variance = 7,360,000; standard deviation = 2712.93
b. Expected return = 13,068; variance = 2,484,000; standard deviation = 1576.07
c. Assuming a somewhat risk-averse investor, the portfolio consisting of $3000 invested in each of the four investments is preferred.

17.42 The portfolio consisting of $3000 invested in each of the four investments has the higher expected utility.

CHAPTER 18
SECTION 18.1, p. 691

18.1 a. $P(\text{Defaulted}|\text{Poor}) = .029$ b. $P(\text{Dafaulted}|\text{Poor}) = .029$

18.2 $P(\text{Defaulted}|\text{Fair}) = P(\text{Defaulted}) = .01$, by independence

18.3 $P(A|\text{Defective}) = .5$, higher than the prior probability, 1/3.

18.4 $P(A|Y = 0) = .228$

18.5 $P(\text{Type 1}|\text{Fire}) = .40$

18.6 $P(\text{Unsuccess}|\text{Good}) = .714 > P(\text{Unsuccess}) = .40$

18.7 $P(A_2|B_1) = .443$; $P(A_2|B_2) = .226$; $P(A_2|B_3) = .513$; $P(A_2|B_4) = .276$

SECTION 18.2, p. 699

18.8 a. "Medium" has the highest expected return.
b. EVPI = 1.24. The pilot project would be desirable.

18.9 a. $P(\text{Good Yield}|\text{Favorable}) = .581$; $P(\text{Fair Yield}|\text{Favorable}) = .322$; $P(\text{Poor Yield}|\text{Favorable}) = .097$
b. "Large" has the best expected value.
c. $P(\text{Favorable}) = .31$

d. P(Good Yield|Mediocre) = .047; P(Fair Yield|Mediocre) = .814; P(Poor Yield| Mediocre) = .139; "Medium" has the best expected value; P(Mediocre) = .43. P(Good Yield|Unfavorable) = 0; P(Fair Yield|Unfavorable) = .192; P(Poor Yield| Unfavorable) = .808; "No scale" has the best expected value; P(Unfavorable) = .26

18.10 EVSI = .768; the pilot project would be desirable.

18.11 a. The "no display" option is always worse than the "15 ft., consolidated" option.
b. "30 ft., scattered" has the highest expected profit.
c. EVPI = .27, the survey is not worth its cost, 2.4.

18.12 a. P(Little) = .075
b. P(25%|Little) = .667; P(50%|Little) = .333; P(75%|Little) = 0; P(100%|Little) = 0
c. "30 ft., scattered" has the highest expected profit, given a "little" result
d. P(Some) = .22; P(25%|Some) = .682; P(50%|Some) = .227; P(75%|Some) = .091; P(100%|Some) = 0; "30 ft., scattered" has the highest expected profit, given a "some" result. P(Half) = .245; P(25%|Half) = .612; P(50%|Half) = .306; P(75%|Half) = .082; P(100%|Half) = 0; "30 ft., scattered" has the highest expected profit, given a "half" result. P(Most) = .28; P(100%|Most) = .018; "30 ft., scattered" has the highest expected profit, given a "most" result. P(All) = .18; P(25% All) = .278; P(50%|All) = .139; P(75%|All) = .333; P(100%|All) = .250; "30 ft. consolidated" has the highest expected profit, given an "all" result.

18.13 EVSI = .0873; the survey is not worthwhile, at a cost of .4.

SECTION 18.3, p. 707

18.14 a. P(no change) = .37; P(Up 10%) = .32; P(Up 20%) = .31
Posterior probabilities

	θ_i		
	Steady	Small Increase	Large Increase
No change	.757	.162	.081
Up 10%	.250	.563	.187
Up 20%	.129	.194	.677

b.

Result:	No change	Up 10%	Up 20%
Optimal Action:	No expansion	Expansion	Expansion

c. Yes.

18.15 Buy

18.16 Buy

18.17 Buy

18.18 P(Little) = .07; P(Some) = .20; P(Half) = .23; P(Most) = .28; P(All) = .22
Posterior probabilities

	θ_i			
	25%	50%	75%	100%
Little	.571	.429	0	0
Some	.600	.300	.100	0
Half	.522	.391	.087	0
Most	.286	.321	.357	.036
All	.182	.136	.273	.409

No changes in optimal decisions.

SECTION 18.4, p. 714

18.19 a. Loss table

		Outcome θ_j				
		Severe	Cold	Normal	Mild	Warm
	Full	0	0	20	40	70
Order	90%	40	0	0	0	10
a_i	80%	70	20	5	0	5
	70%	90	50	30	10	0
Probability		.10	.20	.40	.20	.10

b. The quantity that minimizes expected loss is "90%"; this is the same optimal quantity found in Exercise 17.14.

18.20 a. Payoff table

		Actual Proportions, θ								
		.10	.20	.30	.40	.50	.60	.70	.80	.90
	.10	.10	.30	.50	.70	.90	1.10	1.30	1.50	1.70
	.20	.20	.20	.40	.60	.80	1.00	1.20	1.40	1.60
	.30	.30	.30	.30	.50	.70	.90	1.10	1.30	1.50
Estimates	.40	.40	.40	.40	.40	.60	.80	1.00	1.20	1.40
a	.50	.50	.50	.50	.50	.50	.70	.90	1.10	1.30
	.60	.60	.60	.60	.60	.60	.60	.80	1.00	1.20
	.70	.70	.70	.70	.70	.70	.70	.70	.90	1.10
	.80	.80	.80	.80	.80	.80	.80	.80	.80	1.00
	.90	.90	.90	.90	.90	.90	.90	.90	.90	.90

b. Loss table

		.10	.20	.30	.40	.50	.60	.70	.80	.90
	.10	.00	.10	.20	.30	.40	.50	.60	.70	.80
	.20	.10	.00	.10	.20	.30	.40	.50	.60	.70
	.30	.20	.10	.00	.10	.20	.30	.40	.50	.60
Estimates	.40	.30	.20	.10	.00	.10	.20	.30	.40	.50
a_i	.50	.40	.30	.20	.10	.00	.10	.20	.30	.40
	.60	.50	.40	.30	.20	.10	.00	.10	.20	.30
	.70	.60	.50	.40	.30	.20	.10	.00	.10	.20
	.80	.70	.60	.50	.40	.30	.20	.10	.00	.10
	.90	.80	.70	.60	.50	.40	.30	.20	.10	.00

c. The loss function is an absolute error loss function with $c = 100{,}000$ (or $c = 1\ (100{,}000)$). For example, $L(.20, .10) = 1|.20 - .10| = .10$.

18.21 a. The given prior density is a beta density with $\alpha = 2$ and $\beta = 2$.

b. The mean = median = .5

c. $a^* = .5$

18.22 a. The posterior density of θ is a beta density with $\alpha' = 42$ and $\beta' = 62$.

b. For large values of α' and β', the density is approximately normal with mean, .4038, and variance, .0023; the median = .4038.

c. $a^* = .4038$

d. Because of the large sample size ($n = 100$), the sample result $p = 40/100$ carries more weight than the prior density estimate of $p = .50$.

18.23 a. Payoff matrix

Stock a	Demand θ 5	6	7	8	9	10	. . .	50
5	5	5.70	6.40	7.10	7.80	8.50	. . .	
6	4.90	6	6.70	7.40	8.10	8.80	. . .	
7	4.80	5.90	7	7.70	8.40	9.10	. . .	
8	4.70	5.80	6.90	8	8.70	9.40	. . .	
9	4.60	5.70	6.80	8.90	9	9.70	. . .	
10	4.50	5.60	6.70	8.80	8.90	10.00		
.	.	.	.	.	.	.	.	
.	.	.	.	.	.	.	.	
.	.	.	.	.	.	.	.	
50								50

b. Loss table

Stock a	Demand θ 5	6	7	8	9	10	. . .	50
5	0	.30	.60	.90	1.2	1.5	. . .	
6	.10	0	.30	.60	.90	1.2	. . .	
7	.20	.10	0	.30	.60	.90	. . .	
8	.30	.20	.10	0	.30	.60	. . .	
9	.40	.30	.20	.10	0	.30	. . .	
10	.50	.40	.30	.20	.10	0		
.	.	.	.	.	.	.		
.	.	.	.	.	.	.		
.	.	.	.	.	.	.		
50								0

c. The loss function is an angle shaped loss function with $c_1 = .1$ and $c_2 = .3$. For example,
$L(5, 7) = .3(7 - 5) = .60$
$L(7, 5) = .1(7 - 5) = .20$

18.24 $a^* = 35.36$ (thousands)

18.25 a. The posterior distribution of θ is normal with mean, 33.37, and standard deviation, 3.179 (thousands).
b. $a^* = 35.499$ (thousands)

18.26 a. $a^* = 2000$ b. $E[L(a, \theta)] = 22{,}500{,}000$

18.27 a. The posterior distribution of μ is normal with mean, 1898.78, and variance, 1355.01.
b. $a^* = 1898.78$ c. $E[L(a, \theta)] = 1{,}355{,}010$

18.28 The posterior distribution of μ is normal with mean, 1900.88, and variance, 1805.05; $a^* = 1900.88$; $E(L(a, \theta)] = 1{,}805{,}050$.

CHAPTER EXERCISES, p. 718

18.29 a. "Method 1" has the highest expected return.
b. Perfect information is worth $53,000.

18.30 a. $P(\text{Down sharply}) = .12$; $P(\text{Down slightly}) = .25$; $P(\text{Constant}) = .38$; $P(\text{Up slightly}) = .19$; $P(\text{Up sharply}) = .06$

Posterior probabilities

	θ_i		
	Down	No Change	Up 10%
Down sharply	.750	.250	0
Down slightly	.480	.480	.040
Constant	.158	.789	.053
Up slightly	.158	.632	.210
Up sharply	0	.500	.500

b.

Opinion:	Down sharply	Down slightly	Constant	Up slightly	Up sharply
Optimal Method:	1	1	1	1	2

c. EVSI = 3.8968; the company should not hire the consultant.

18.31 a. The alternative likelihoods indicate greater value to the consultant's opinion.

b. Posterior probabilities

	Outcome θ_i		
	Down	No Change	Up 10%
Down sharply	1	0	0
Down slightly	.625	.375	0
Constant	.065	.913	.022
Up slightly	0	.643	.357
Up sharply	0	0	1

$P(\text{Down sharply}) = .12$; $P(\text{Down slightly}) = .24$; $P(\text{Constant}) = .46$; $P(\text{Up slightly}) = .14$; $P(\text{Up sharply}) = .05$

The action that has the highest expected value given each of the opinions DOWN SHARPLY, DOWN SLIGHTLY, and CONSTANT is "Method 1." The action that has the highest expected value given the opinions UP SLIGHTLY and UP SHARPLY is "Method 2."

The overall decision changes from Exercise 18.30. The decision in this exercise is to hire the consultant.

18.32 a. The given prior density is a beta density with $\alpha = 4$ and $\beta = 2$.

b.

θ	0	.2	.4	.6	.8	1.0
$f(\theta)$	0	.128	.768	1.728	2.048	0

c. $P(\theta < .6) = .33696$

18.33 a. The posterior density of θ is a beta density with $\alpha' = 74$ and $\beta' = 60$.

b. For large α' and β', the density is approximately normal with mean, .552, and variance, .00183; $P(\theta < .6) = .8686$.

18.34 H_0: $\theta \geq .60$

H_a: $\theta < .60$

T.S.: $z = -1.227$

R.R.: At $\alpha = .05$, reject if $z < -1.645$

Conclusion: Retain H_0.

From Exercise 18.33, we see that there is a fairly high probability of obtaining a proportion value less than .60. Yet, the preceding test leads us not to conclude $\theta < .60$.

18.35 $P(\mu < 2) = .9332$

18.36 a. The posterior distribution of μ is normal with mean, 1.378, and variance, .111.

b. $P(\mu < 2) = 1.00$

18.37 a. H_0: $\mu = 2$

H_a: $\mu < 2$

T.S.: $z - -1.116$

R.R.: At $\alpha = .05$, reject H_0 if $z < -1.645$; at $\alpha = .10$, reject H_0 if $z < -1.28$

Conclusion: Retain H_0.

b. The variance in the decision theory approach is smaller using the prior distribution approach rather than the single random sampling hypothesis testing approach.

18.38 a. "New car buyers" has the highest expected payoff.

b. "New car buyers" has the lowest expected loss.

18.39 $P(\text{Down } 10\%) = .15$; $P(\text{Down } 5\%) = .215$; $P(\text{Steady}) = .26$; $P(\text{Up } 5\%) = .22$; $P(\text{Up } 10\%) = .155$.

a. Posterior probabilities

	Down 10%	Down 5%	Steady	Up 5%	Up 10%
Down 10%	.200	.267	.200	.300	.033
Down 5%	.139	.233	.279	.279	.070
Steady	.077	.154	.462	.230	.077
Up 5%	.068	.182	.273	.341	.136
Up 10%	.032	.194	.194	.387	.194

b. The optimal aim for each possible value of summer sales is "new car."

18.40 The company should defer its choice at a cost of $4000.

18.41 a. Loss table

		Outcome θ_j (Proportion Demanded)					
		.05	.10	.15	.20	.25	.30
	.05	0	10	20	30	40	50
	.10	15	0	10	20	30	40
a_i	.15	30	15	0	10	20	30
Proportion	.20	45	30	15	0	10	20
Ordered	.25	60	45	30	15	0	10
	.30	75	60	45	30	15	0

b. The loss function is angle shaped with $c_1 = 300$ and $c_2 = 200$. For example,

$L(.05, .10) = 200(.10 - .05) = 10$

$L(.10, .05) = 300(.10 - .05) = 15$

c. $a^* =$ 40th percentile of $f(\theta)$

CHAPTER 19

SECTION 19.1, p. 727

19.1 A minor selection bias would be the homeowners who do not own telephones. A more serious bias would be the omission of those who have nonpublished phone numbers.

19.2 A possible selection bias would be the omission of those wheelchair-bound people who are not members of the statewide association.

19.3 $.36 < \pi < .52$

19.4 A possible selection bias would be the omission of those owners not available for contact during the month of March or the omission of those owners not available for contact during the specified times for call-ups.

19.5 a. $.311 < \pi < .441$ b. $3.83 < \mu < 4.57$

19.6 Large sales have a tendency to push the running totals over a $10,000 interval, and thereby would be more likely to be in the sample.

19.7 $\$101.27 < \mu < \$123.21$

19.8 The bias to larger sales being included in the sample will probably give an inflated estimate of the mean. Also, the inclusion of these larger sales most likely increases the estimate of the variability.

SECTION 19.2, p. 733

19.9 $192.998 < \mu < 195.782$

19.10 $95.55 < \mu < 104.05$ (in thousands of dollars); stratification has helped.

19.11 a. $10.297 < 10.743$ b. $9.824 < \mu < 10.976$

19.12 $10.297 < \mu < 10.783$; there is no substantial shift in the confidence interval because the means and standard deviations for each strata are very similar.

19.13 $.1535 < \pi < .2461$

19.14 $.153 < \pi < .247$; the individual strata proportions are very much alike, so that the stratified calculation is very close to the simple random sampling calculation.

SECTION 19.3, p. 739

19.15 a. $9.302 < \mu < 11.498$

b. A simple random sample would have the survey team riding all over town searching for individual houses. It would also require a complete list of household addresses. Cluster sampling merely requires the identification for city blocks.

19.16 a. $9.772 < \mu < 11.078$

b. The width of the interval assuming a simple random sample is smaller.

19.17 a. $\bar{y}_c = 100.32$ b. $99.56 < \mu < 101.08$

19.18 a. $\bar{y}_c = .1048$ b. $.0977 < \pi < .1119$ (in thousands)

c. $.1986 < \pi < .2158$

SECTION 19.4, p. 744

19.19 $n = 602$; $n_i = 61$ ($i = 1, \ldots, 10$)

19.20 a. $n = 1135$ b. Yes. The within-stratum variance is lower.

19.21 $n = 4$

19.22 $n = 4$

CHAPTER EXERCISES, p. 751

19.24 a. Cluster sampling

b. Cluster sampling is preferred, as in this case, when the elements of the population are located over a large area.

19.25 a. $-.84 < \mu < 9.487$ b. $3.728 < \mu < 4.932$
c. $n \approx 366$

19.26 a. Stratified random sampling
b. Stratification by order type would yield data that should be more homogeneous within each stratum than in the population as a whole. Confidence intervals should be smaller than those constructed assuming simple random sample.

19.27 The simplest procedure would be to continue drawing random numbers until there are 30 samples of each type.

19.28 $5.736 < \mu < 6.524$

19.29 a. $5.623 < \mu < 6.337$
b. Assuming equal representation: $5.736 < \mu < 6.524$
Assuming unequal representation: $5.623 < \mu < 6.337$, not drastically different

19.30 $n = 187$; $n_1 = 50$; $n_2 = 62$; $n_3 = 42$; $n_4 = 33$

19.31 $\$39.54 < \mu < \$40.80$

19.32 $\$156{,}192.51 < \mu < \$161{,}150.49$

19.33 Stratified sampling

19.34 $46{,}735.48 < \mu < 49{,}417.26$

19.35 $n = 13{,}252$; $n_1 = 2026$; $n_2 = 1366$; $n_3 = 1134$; $n_4 = 1497$; $n_5 = 1921$; $n_6 = 1921$; $n_6 = 1633$; $n_7 = 1524$; $n_8 = 1295$; $n_9 = 856$

19.36 $.086 < \pi < .234$

19.37 a. $n = 94$
b. The effect of using proportional allocation is to more than halve the sampling size.

19.38 $1.401 < \mu < 1.459$

19.39 Simple random sampling is not advisable. If two (or more) riders in the same carpool were sampled, a car's total occupancy would be counted more than once.

19.40 $1395 < \tau < 1453.56$

INDEX

Absolute error, as loss function, 709–710
Actions:
in decision theory, 658
listing all possible, 659
AD (*see* Average deviation)
Addition law:
general, 50
for mutually exclusive events, 49
Addition of outcome probabilities, 46
Additive effects, 480
Additive model, in time series, 625
Aggregate price index:
simple, 617
weighted, 618
Alpha:
choice of, 260
computed at boundary value of null hypothesis, 260
and costs of error, 299
effect on beta, 264
probability of Type I error, 259
Alternative formulas, in analysis of variance, 384
Alternative hypothesis (*see* Research hypothesis)
Analysis of variance:
assumptions of, 382
notation for, 381
for testing equality of several means, 383
Analysis of variance table:
in randomized block designs, 413
in two-factor studies, 405
Angle-shaped error, as loss function, 709-710
ANOVA (*see* Analysis of variance)
Antiderivative, 119
AR model (*see* Autoregressive model)
ARMA model (*see* Autoregressive, moving average model)
Association, strength of, 368
Assumptions:
of analysis of variance, 386
for confidence interval, 245–246
in goodness-of-fit test, 363
for inference about correlation, 461
in multiple regression, 478
for paired-sample tests, 338
of rank sum test, 326
of regression analysis, 439
for two-sample methods, 307
Attained significance level (*see* p-value)
Autocorrelation(s):
coefficients in time series, 639
Durbin-Watson test for, 559
effect of, 558–559
for general Box-Jenkins model, 644
for moving average model, 642
for theoretical autoregressive model, 640
in time-series regression, 557
Autoregressive model:
correlogram of, 644
and forecasting, 646
higher-order, 641
in regression, 563
theoretical autocorrelations of, 640
of time series, 637
Autoregressive, moving average models, 644
Average (*see also* Mean, Median, Mode):
moving (*see* Moving average)
weighted, 17–18
Average deviation, 20
Axioms, of probability, 46

Backup analyses, 759
Backward elimination, stepwise regression, 544–545
Balanced design:
and equal-variance assumption, 387
in two-factor analysis of variance, 399
Bar chart, 10
Bayes' Theorem, 686, 689
for random variables, 691
Bernoulli trials, 124–125
assumptions, 126
Beta:
calculation in binomial test, 263
and computer simulation, 284
and costs of error, 299
factors influencing, 264
as function of population parameter, 264
and sample size, 275
for *t* test of a mean, 283
and width of confidence interval, 298
for *z* test, 272–275
Beta distribution, as prior probability distribution, 713
Between-sample variability, 730
Bias:
in estimating correlation, 461
non-response, 724
selection, 162, 723
sensitive-question, 724
size, 724
from stepwise regression, 545

Binomial experiment, 125–126
Binomial probability distribution, 126–128
  approximation by Poisson, 138
  mean and variance, 129
  normal approximation of, 151
Block factor, in randomized block designs, 412
Boundary of null hypothesis, and computation of alpha, 260
Boundary value, of null hypothesis, 267–268
Box-and-whiskers plot (*see* Box plot)
Box-Jenkins:
  general model, 644
  models, of cyclic behavior, 638
  seasonal models, 646
  time series models, 637–647
Box plot, 25

Cancelling variation, by pairing, 332
Central Limit Theorem:
  and analysis of variance, 386–387
  in distribution, 714
  effect in two-sample methods, 315
  misinterpretations, 181
  for other statistics, 181
  and population distributions, 180
  and portfolio analysis, 673
  sample size for, 175
  for sums and means, 174
  and test for mean, 266
Central tendency, measures of, 17
Chain rule, 117
Chebyshev's Inequality:
  definition of, 22
  for random variables, 93
Chi-square:
  approximation in goodness-of-fit test, 363
  distribution of sample variance, 255, 352
  tables, 352–353
Chi-square distribution:
  definition of, 254
  for large d.f., 355
  properties of, 255
Chi-square goodness-of-fit test, 362, 363
Chi-square test:
  goodness-of-fit, 362
  of independence, 367
  for variance and standard deviation, 355
Class boundaries, 9
Classes:
  choice of, 8–9
  width of, 8
Classical interpretation:
  of probability, 42
  and random sampling, 121
  and sampling distribution, 167
Cluster sample, 722
Cluster sampling, 734
  two-stage, 746
Cobb-Douglas production function, estimation of, 533
Cochran-Orcutt method, with autocorrelation, 564
Cochran's Theorem:
  in analysis of variance, 383
  for two-factor analysis of variance, 405
Coefficient of determination (*see* R-squared)
Coefficients (*see* Slope, Intercept, Partial slope)
Collinearity:
  and addition of R-squares, 488–489
  and choice of independent variables, 522
  definition of, 488, 522
  effect on standard error of partial slope, 494
  and extrapolation, 504
  and *F* and *t* tests in multiple regression, 496
  of lagged variables, 527
  and stepwise regression, 543
Combinations, number of, 122–123
Comparison, general definition of, 394
Complement, of an event, 48
Complement law, 51
Complete model, 498
Components, of time series (*see* Trend, Cycles, Seasonal)
Computer packages, for simple regression, 452
Computers, use of, 30–31
Concave utility functions, 676–677
Conclusiveness, and *p*-value, 277
Conditional density, of continuous random variables, 107
Conditional distribution, 99
  and independence, 99
Conditional probability, 51–52
  and independence, 55
  in probability trees, 63
Confidence interval(s) (*see also* Prediction interval):
  and beta, 298
  for difference of means (sigmas known), 308
  for difference of means (unknown, equal sigmas), 312
  for difference of proportions, 342
  and hypothesis tests, 296–298
  interpretation of, 229
  for mean, cluster sampling, 735
  for mean, sigma known, 228
  for mean, stratified sampling, 729
  for mean in multiple regression, 455
  for median, 247
  one-sided, 297
  for paired samples, 335
  for partial slope, 494
  for proportion, 231
  for proportion, cluster sampling, 737
  for proportion, simple random sampling, 726
  for proportion, stratified sampling, 731
  for ratio of variances, 360
  for row means in two-factor studies, 408
  Scheffé method for all pairs of means, 393
  Scheffé method for comparison of several means, 395
  for slope in simple regression, 449
  t for mean, 242
  for total, cluster sampling, 735
  for total, simple random sampling, 725
  for total, stratified sampling, 729
  Tukey method for all pairs of means, 391
  for two-factor means using Tukey method, 407
  for variance and standard

Confidence interval(s) (Continued)
deviation, 354
for variance in simple regression, 452
Confidence level, 232–233
Confounded effects, 411
Constant, smoothing (*see* Smoothing constant)
Constant probability:
assumption in goodness-of-fit tests, 363
assumption of Bernoulli trials, 124
Constant variance, assumption in regression, 439 (*see also* Homoscedasticity)
Constants, effect on mean and variance, 113–114
Continuity correction, for normal approximation, 151
Continuous random variable, 77
Correlation:
assumptions for inference about, 461
calculation of, 459
coefficient of, 459
of discrete random variables, 102
of independent variables, 488 (*see also* Collinearity)
of independent variables, and partial slope, 485
interpretation of, 460
and nonlinear relation, 460
and nonlinearity, 465–466
and range of independent variable, 461–462
rank, 466
relation to slope, 475
and SS, 459
Correlogram:
for general model, 644–645
standard error for, 643
for theoretical autoregressive model, 640–641
of time series data, 639
Covariance:
of discrete random variables, 102
and independence, 103
and variance of a sum, 115
C_p statistic:
in multiple regression, 545
random error of, 546–547
Cross-product terms:
for interaction, 535
for interaction in multiple
Cross-product terms (Continued)
regression, 481–482
Cumulative distribution function, 78
for continuous random variables, 83
as integral of density, 87
and probability tables, 80
Cycles:
in additive model of time series, 625
and Box-Jenkins models, 638
in deseasonalized data, 627
in multiplicative model, 626
and spectral analysis, 627
in time series, 621

Data, as numerical values, 1
Data base, creation of, 756
Data files, creation of, 758
Data trail, 756
Decisions, and probability theory, 42
Decision trees, 700–707
Definite integral, 119
Degrees of freedom:
for analysis of variance, 383
for chi-square, 353
for chi-square distribution, 254–255
for chi-square independence test, 367
for comparison of row means, 408
for confidence interval for regression variance, 452
for confidence interval of partial slope, 494
for error in multiple regression, 527
of *F* statistic, 358
for *F* test in multiple regression, 492, 497
for *F* test in simple regression, 451
for goodness-of-fit test, 362, 363
for Kruskal-Wallis test, 388
for paired samples, 335
in randomized block designs, 413
for regression, 446
for Scheffé comparison of means, 393, 395
for SS (Residual) in multiple regression, 486
Degrees of freedom (Continued)
for *t* distribution, 256
for *t* statistic, 237–238
for *t* test of a mean, 281
for *t* test of correlation, 463
for test of regression variance, 457
for test of several partial slopes, 499
for test of slope in simple regression, 449
for Tukey method for all pairs of means, 391
for two independent samples, 313
for two-factor analysis of variance, 405
for two-sample *t′* test, 315
Density, probability (*see* Probability density)
Dependent variable, 372
in regression, 436
Derivative(s):
of a function, 116
partial, 118
table of, 117
Deseasonalized data, 627
Detectable (*see* Significant)
Detrending, in time series, 625
Deviations from the mean, 20
d.f. (*see* Degrees of freedom):
Differences, with autocorrelation, 563–564
Discrete random variable, 77
Distribution, probability (*see* Probability distribution)
Distribution function, cumulative (*see* Cumulative distribution function)
Documentation, of study, 760
Double integrals, 120
Dummy variables:
for many-category variable, 524–525
for two-category variable, 523–524
Durbin-Watson test, 559

Effects, additive, 480
Efficiency:
and confidence intervals, 246
and median interval, 247
Efficient estimator, 205
Elasticity of demand, estimation of, 533–534

Empirical Rule:
and Central Limit Theorem, 181
definition of, 22
for random variables, 93
Equally likely outcomes, and classical probability, 42
Equal variance (*see* Homoscedasticity)
Equal variances:
in analysis of variance, 387
test of, 359
Error, in regression studies, 438 (*see also* Type I error, Type II error)
Estimate, sample statistic as, 166 (*see also* Estimator)
Estimated expected values, in independence test, 366
Estimator:
definition of, 203
efficient, 205
interval, 227
maximum likelihood, 215
robust, 206
unbiased, 204
Event, 42, 45
EVPI, 693–694
EVSI, 695–698
Expected frequencies, and chi-square approximation, 368
Expected numbers:
and chi-square approximation, 363
in goodness-of-fit test, 362
Expected return, of action, 663
Expected utility, in decision theory, 675–678
Expected value:
and adding a constant, 113
of beta distribution, 713
of continuous random variable, 94
of difference of proportions, 342
of discrete random variable, 89
estimated, 366
as generalization of a mean, 90–91
as long-run average, 90
and multiplying by a constant, 114
of perfect information (*see* EVPI)
of portfolio, 670
of a proportion, 231
of rank sum, 323
Expected value (Continued)
of sample information (*see* EVSI)
of sample mean, 171
of sample sum, 170
of sum, 115
Expected-value tree, 664
Experiment, in probability theory, 44–45
Experiments, and quality control, 417
Exponential distribution, assumptions for, 141–142
Exponential probability density, 142
Exponential smoothing:
choice of constant, 634
computation of, 633
definition of, 633
Exponential trend, 623
in time series, 622
Extrapolation:
effect on standard error of intercept, 445
in multiple regression, 502–503, 504
Extrapolation penalty, 455

F distribution, definition and tables, 358
F statistic, definition of, 358
F test(s):
analysis of variance, 383
of equality of variances, 359
in factorial experiments, 417
interpretation of result in multiple regression, 492
in multiple regression, 492
in randomized block designs, 414
and R-squared, 497
for several partial slopes, 498–499
for slope in simple regression, 451
in two-factor analysis of variance, 406
Factorial experiment, 417
False negative, as Type II error, 262
False positive, as Type I error, 262
Fences, and outlier check, 24
Finite population correction factor, 132, 725
First-order model, 479
Fit, to trend equation, 624
Fixed independent variables, in regression, 438
Forecasting, with Box-Jenkins models, 646–647
Forward selection, in stepwise regression, 543
Frequency:
relative, 8
table of, 7
Function, definition of, 116
Fundamental Theorem of Calculus, 119

Gambles, 680
General linear model, 479
Geometric probability distribution, 133
mean and variance, 134
Good fit, conclusion of test, 363
Goodness-of-fit test, for several proportions, 361, 362
Grouped data, 8
summarization of, 31–33
Groups, summations over, 39–40 (*see also* Classes)

Harmonic mean, use with unbalanced design, 392
Heaviness of tails, 28 (*see also* Outliers)
Heteroscedasticity, effect in regression, 553 (*see also* Homoscedasticity)
Higher-order terms, in multiple regression, 532
Hinges (*see* IQR)
Histogram:
definition of, 10–11
probability, 78
relative frequency, 11
sample, 168
Homogeneity, test of (*see* Chi-square test, of independence)
Homoscedasticity, assumption in regression, 552–553
Hypergeometric:
mean, 132
probability distribution, 130–131
relation to binomial, 131
variance, 132

Hypothesis (*see* Null hypothesis, Research hypothesis)
Hypothesis test(s):
- chi-square of independence, 367
- and confidence interval(s), 296–298
- and decisions, 298–300
- for difference of means (sigmas known), 310
- for difference of means (unknown, equal sigmas), 314
- for difference of proportions, 343
- of equality of variances, 359
- *F* for several partial slopes, 498–499
- *F* in analysis of variance, 383
- *F* in factorial experiments, 417
- *F* in multiple regression, 492
- *F* in randomized block designs, 414
- *F* in simple regression, 451
- *F* in two-factor analysis of variance, 406
- *F* using R-squared, 497
- goodness-of-fit, 362
- of homogeneity (*see* Chi-square test, of independence)
- Kruskal-Wallis, 388
- for mean (with known standard deviation), 267
- for paired samples, 335
- in randomized block design, 414
- rank sum, 323–324
- relation of *F* and *t* in multiple regression, 495–496
- *t* for correlation, 463
- *t* for partial slope, 495
- *t* for rank correlation, 467
- *t* for slope in simple regression, 449
- *t* and stepwise regression, 544
- *t′* for two samples, unequal variances, 315
- Tukey method for all pairs of means, 392
- for variance and standard deviation, 355

Hypothesis testing:
- basic strategy, 258
- and decision theory, 716
- effect of prior probabilities, 299
- five steps of, 261

Independence:
- in analysis of variance, 387
- as assumption, 56
- assumption for confidence interval, 245
- assumption in goodness-of-fit tests, 363
- assumption in regression, 557 (*see also* Autocorrelation)
- assumption in regression analysis, 438
- assumption in two-sample test, 314
- assumption of Bernoulli trials, 124
- chi-square test of, 367
- of continuous random variables, 107
- and covariance, 103
- of events, 55
- hypothesis of, 365
- and multiplication law, 56
- in probability trees, 63
- of random variables, 99
- of samples, 308
- of several events, 57
- and variance of a sum, 115

Independent processes, 57
Independent variable(s), 373
- fixed in regression, 438
- ideal choice of, 522–523
- in regression, 436
- selection of, 522–526

Index:
- price vs. quantity, 616
- seasonal, 626

Index numbers, price (*see* Price index)
Indicator variables (*see* Dummy variables)
Inference, using probability theory, 41–42, 66
Inner fences, 24
Integrals, definite, 119
Integration, to find probability, 86
Interaction:
- and additive effects model, 480
- concept of, 397–398
- and cross-product terms, 535
- detected by profile plot, 398–399
- *F* test for, 406
- in factorial experiments, 417
- ignored in randomized block designs, 412

Interaction (Continued)
- and interpretation of main effects, 406–407
- in multiple regression, 480–482, 534–536
- residual plots for, 531–532
- SS in analysis of variance, 402
- test for in multiple regression, 535–536
- in two-factor model, 398

Intercept:
- estimate of, 441
- estimate of, in multiple regression, 483
- least squares estimate of, 475
- in regression analysis, 437
- standard error of in multiple regression, 494

Interpretation, of dummy variable coefficients, 524
Interquartile range (*see* IQR)
Intersection, of events, 48
Interval estimates, 226
Interval estimator, 227
IQR, 23–24
- and outliers, 24

Jackknife method, in regression, 551
Joint probability, 53
Joint probability density, for continuous random variables, 105
Joint probability distribution, 97

Kruskal-Wallis test, of several distributions, 387
Kurtosis:
- definition of, 28
- sensitivity to outliers, 28–29

Lagged variables, 526–529
- choice of lags, 527
- for forecasting, 526–527

Lambda, measure of relation, 373
Large-sample method, for difference of means, 308
Large-sample test of mean, using z, 271
Laspeyres index, definition of, 619
Last predictor in, interpretation of multiple regression *t* test, 495
Least squares:
- definition of, 440
- effect of outliers, 550

Least squares (Continued)
estimate of Box-Jenkins coefficients, 645–646
estimate of trend, 623–624
estimates in multiple regression, 483
estimates of slope and intercept, 475
estimation of parameters, 441
weighted, 553
Likelihood, in Bayes' Theorem, 687
Likelihood function, 214
Linearity assumption:
in multiple regression, 530–531
in regression analysis, 437–438
Linear trend, 623
in time series, 622
Line plot, 10
Location, measures of, 17
Log, of study data, 756
Logarithmic transformation, 533
Logic checks, 756
Loss, in decision theory, 708–709
Loss functions, in decision theory, 709

Main effects:
F tests for, 406
in the presence of interaction, 406–407
in two-factor analysis of variance, 401
Mallows' C_p (*see* C_p statistic)
Marginal probabilities, in probability trees, 63
Marginal probability, 53
Marginal probability distribution, 97
Matched samples (*see* Paired samples)
Maximum and minimum:
by derivatives, 117–118
by partial derivatives, 118
Maximum likelihood, 213–215
Mean (*see also* Expected value):
of binomial distribution, 129
Central Limit Theorem for, 174
confidence interval for, cluster sampling, 735
confidence interval for, stratified sampling, 729
confidence interval with sigma known, 228
definition of, 16
Mean (Continued)
expected value of, 171
of exponential density, 142
of geometric distribution, 134
of grouped data, 32–33
of hypergeometric distribution, 132
negative binomial, 135
of Poisson distribution, 138
population, 17
in profile plot, 398–399
sample, 17
sample size for, 233
standard error of, 197–198
t interval for, 242
test for (with known standard deviation), 265–271
trimmed, 17
of uniform distribution, 140
as weighted average of group means, 17–18
Mean (sample), expected value and standard error of, 171
Mean square (MS):
in analysis of variance, 383
in randomized block designs, 413
in two-factor analysis of variance, 405
Median:
confidence interval for, 247
definition of, 16
as estimator, 203
of grouped data, 32
robustness of, 290
running, choice of time period, 633
running, in time series, 631
Median test, as binomial test, 291
Midpoints, of classes, 9
Modal prediction, and lambda, 373
Mode, definition of, 15
Model selection, of Box-Jenkins models, 644–646
Monte Carlo method (*see* Simulation)
Moving average:
autocorrelations for, 642
choice of time period, 633
summary of time series, 630
Moving average model(s):
Box-Jenkins, in time series, 641
correlogram of, 644–645
and forecasting, 646
MS (*see* Mean square)
MS (Residual), and residual standard deviation, 486
Multicollinearity (*see* Collinearity)
Multinomial assumptions, 363
Multiple regression:
assumptions, 478
definition of, 477
model for, 478
Multiple *t* tests, objection in analysis of variance, 391
Multiplication law, 53
for independent events, 56
Multiplicative model, in time series, 624
Mutually exclusive events, 48
addition law for, 49

Negative binomial, mean and variance, 135
Negative binomial distribution, 134
Noise, white (*see* White noise)
Non-response bias, 724
Nonlinearity, and correlation, 465–466
Nonnormality:
effect in analysis of variance, 386–387
effect on *t* test of a mean, 286–290
and individual predictions in regression, 549
and Kruskal-Wallis rank test, 387
and variance inference, 356
and variance ratio methods, 360
Nonparametric methods (*see* Rank sum test, Signed-rank test, Kruskal-Wallis test)
Normal, probability density, 144
Normal approximation:
to a binomial, 231
to binomial, 151, 295
Normal approximation to binomial, by Central Limit Theorem, 180
Normal curve, 145
Normal distribution:
and Empirical Rule, 23
as prior probability distribution, 711–712
as sampling distribution, 173
Normal equations, in multiple regression, 483

Normality:
  assumption in *t* test of difference of means, 315
  not assumed in rank test, 326
  and paired-sample methods, 338
Normal population:
  assumption for confidence interval, 245
  and Central Limit Theorem, 180
  and *z* interval, 228
Normal probability plot, 184–186
  and outliers, 249
  and sampling distributions, 187–188
Notation, for two-factor analysis of variance, 401
Null hypothesis, definition of, 258
Numerical payoffs, in decision theory, 659

One-sided research hypothesis, 258, 268
Optimal action, for specified loss function, 710
Order statistics, and confidence intervals, 247
Ordinal data, and rank sum test, 326
Outcome(s), 45
  in decision theory, 658
  of an experiment, 42
  listing all possible, 659
Outer fences, 24
Outliers:
  and analysis of variance, 386–387
  definition of, 17
  detection in multiple regression, 551
  effect on Central Limit Theorem, 175
  effect in multiple regression, 550–551
  effect on range, 20
  and paired-sample methods, 338
  and rank sum vs. *t* tests, 327
  test for, 24
  and trimmed means, 17
  and variance inference, 356
  and variance ratio methods, 360
Overall alpha risk, in analysis of variance, 381

p-value:
  definition of, 276
  interpretation of, 277
  for a *t* test of a mean, 282
  two-tailed, 277
  and Universal Rejection Region, 278
  of *z* test, 277
Paasche index, definition of, 619
Paired samples, 308
  for control of variability, 332
Paired-sample methods, extended to randomized block designs, 412
Partial derivatives, 118
Partial slope(s), 478
  confidence interval for, 494
  estimate of, 483
  least squares estimates using matrix notation, 507
  relation to simple regression slope, 485
  standard error of, in multiple regression, 494
  test for all in multiple regression, 492
  test for several, 498–499
Payoff, expected, 662–663
Payoff function:
  in decision theory, 659
  numerical, in decision theory, 659
Percentage analysis, 371–372
Perfect information, expected value of (*see* EVPI)
Permutations, number of, 122
Personal probabilities (*see* Subjective probabilities)
Point estimation, 203
Poisson probability distribution, 137
  approximation to binomial, 138
  assumptions for, 136
  mean and variance, 138
Pooled proportion, in *z* test, 344
Pooled variance, 313
  generalized to MS (Within), 391
Portfolio, variance of, 669–670
Posterior distribution:
  beta, 713
  normal, 712
Posterior probabilities, in Bayes' Theorem, 687
Power, of a *t* test, 284 (*see also* Beta)

Power of a test, definition of, 263
PRE, interpretation of lambda, 373
Predictability, and relation, 373
Prediction, and residual standard deviation, 487
Prediction interval, for individual value in regression, 457
Preliminary analyses, 759
Price index:
  definition of, 616
  effect of choice of weights, 618–619
  simple aggregate, 617
  use of, 616
  weighted aggregate, 618
  with constant price increases, 616
Primary analyses, 759
Prior probabilities, in Bayes' Theorem, 687
Probability:
  as area, 86
  conditional (*see* Conditional probability)
  joint, 53
  marginal, 53
  role in decision theory, 661
Probability axioms, 46
Probability density, 88
  as derivative of cdf, 87
  exponential, 142
  normal, 144
  uniform, 140
Probability distribution:
  binomial, 128
  continuous, 86
  geometric, 133
  hypergeometric, 130–131
  negative binomial, 134
  Poisson, 137
  properties of, 78
Probability histogram, 78
Probability tables, 60
Probability tree, 61–63
  for Bayes' Theorem, 687
Probable range, for a parameter, 227
Profile plot, 398–399
Proportion:
  confidence interval for, 231
  confidence interval for, cluster sampling, 737
  confidence interval for, simple random sampling, 726
  confidence interval for, stratified

Proportion (Continued)
    sampling, 731
  sample size for, 234
  z test for, 296
Proportionate reduction in error (*see* PRE)
Proportions, two-sample procedures for, 342–343

Quadratic transformation, 534
Qualitative data, 11
  means by computer, 31
Qualitative random variable, 75
Quality control, and experimental design, 417–419
Quantitative data, 11
Quantitative random variable, 75
Quantity index, 616
Quartiles, 24

R-squared:
  and F in multiple regression, 497
  in multiple regression, 487–489
  as proportionate reduction in squared error, 460
  and SS (Regression), 461, 476
Random effect, in time series, 621
Random error term, in regression, 438
Randomized block designs, 412
Randomized-response technique, 746–747
Random numbers, use of table, 164
Random sample, simple, 722
Random sampling, and classical probability, 42
Random variable:
  continuous, 77
  definition of, 75
  discrete, 77
  notation for, 76
Range, 20
Range of independent variable, effect on correlation, 461–462
Rank correlation, 466
  hypothesis test for, 467
Rank sum test, 323–324
  and Kruskal-Wallis test, 387 (*see also* Kruskal-Wallis test)
Reduced model, 498
Reexpression (*see* Transformation)
Regression:
  multiple (*see* Multiple regression)
  robust, 552
Regression analysis, concept of, 436
Regret (*see* Loss)
Rejection region:
  definition of, 259
  and Type I error, 259
  using sample mean, 266
  using z statistic, 267
Relative frequency, 8
  interpretation of probability, 42–43
  interpretation of sampling distribution, 167
Replacement, and variance of mean, 211
Report, format, 760
Research hypothesis:
  definition of, 258
  one-sided, 258, 268
  one-sided vs. two-sided, 258
Residual plots:
  for detecting outliers, 551
  detection of autocorrelation, 559
  for heteroscedasticity, 553–554
  for nonlinearity, 531–533
Residuals:
  and autocorrelation, 559
  in regression, 446
Residual standard deviation:
  effect of autocorrelation, 559
  in multiple regression, 486
  and probable error of prediction, 487
  in regression, 446
Return, expected, 663
Risk:
  ignored by expected value criterion, 664
  and return, 671–672
  and variance, 668
Risk aversion, and concave utility function, 677–678
Robust estimator, 206
Robustness:
  and confidence intervals, 246
  of median, 290
Robust regression, 552
Running medians (*see* Median, running)
Sample fraction, vs. sample size, 212
Sample histogram, 168
Sample information, expected value of (*see* EVSI)
Sample mean, sampling distribution of, 173–174 (*see also* Central Limit Theorem)
Sample size:
  for Central Limit Theorem, 175
  for chi-square independence test, 368
  for confidence interval, 233
  for confidence interval, all sampling methods, 741
  for difference of proportions, 343
  effect on beta, 264
  for large-sample test of mean, 271
  to obtain desired beta, 275
  for proportion, 234
  and proportion .5, 235
  vs. sample fraction, 212
  and standard error of mean, 172
  for t interval for mean, 242–243
  for z test of a proportion, 296
Sample space, 45
Sampling:
  random, 162–164
  systematic, 745–746
Sampling distribution:
  definition of, 166
  interpretation of, 167
  not observed, 167
  simulation of, 183–184
Sampling frame, 164, 723
Scatter plot, definition of, 440
Scheffé method:
  compared to Tukey method, 394
  for comparison of row means in two-factor studies, 408
Seasonal:
  in additive model of time series, 625
  Box-Jenkins models, 646
  in multiplicative model, 626
  pattern in time series, 621
Seasonal index, in time series, 626
Second-order model, 479
Selection bias, 162, 723
Sensitive-question bias, 724
Sensitivity analysis, 661, 665–666

Separate variance t test, 316 (*see also* t test with unequal variances in two samples)
Sequences:
  number of, 122
  of symbols, 122
Shocks, in time series models, 641–642
Shortcut formula, for variance of a random variable, 92
Shortcut formulas:
  in analysis of variance, 384
  in randomized block designs, 414
  for SS in two-factor analysis of variance, 403
Signed-rank test, 336
Significance:
  compared to strength of relation, 464
  interpretation of, 279
Significant, definition of, 279
Sign test:
  for paired samples, 336
  simulation of, 288
Simple aggregate price index, 617
Simple random sample, 722
Simulation:
  of Central Limit Theorem, 176–177
  of chi-square variance test, 356
  of confidence interval, 229
  of estimators, 206
  vs. mathematical derivation, 186
  of power, 284
  of power of t, 288
  of rank sum vs. t tests, 326, 327, 330, 331
  of sampling distribution, 168
  of sampling distributions, 181–183
  of t and signed-rank tests, 339
  of t test for outlier-prone population, 287
  of t test under normality, 292
  of t test under skewness, 289, 293
  of t test with discrete population, 294
  of t vs. signed-rank test, 349
  of two-sample t tests, 318, 319, 320
  of two-sample t tests, 322
Simulation (Continued)
  of variance-ratio test with outliers, 361
Size bias, 724
Skewness:
  and analysis of variance, 386–387
  effect in regression, 549–550
  effect on Central Limit Theorem, 175
  effect on confidence intervals, 245–246
  effect on mean, 17
  effect on rank-sum vs. t tests, 326
  effect on risk measures, 672–673
  effect on t test of a mean, 288–289
  effect on two-sample t test, 315
  mean vs. median, 27
  measures of, 27–28
  and paired-sample methods, 338
  and variance inference, 356
  and variance ratio methods, 360
Slope:
  estimate of, 441
  least squares estimate of, 475
  partial, 478
  in regression analysis, 437
  relation to correlation, 475
Small-sample, interval for mean, 242
Small-sample test (*see* t test)
Smoothing:
  exponential (*see* Exponential smoothing)
  of time series, 630
Smoothing constant, choice of, 633
Spectral analysis, 627
Spectrum, of a time series, 627
Squared error, as loss function, 709–710
SS(Between), 382
  of ranks in Kruskal-Wallis test, 387
  shortcut (alternative) formulas, 384
SS(Block):
  in randomized block designs, 413
  shortcut formula in randomized
SS(Block) (Continued)
    block design, 414
SS(Columns):
  shortcut formula for, 403
  in two-factor analysis of variance, 402
SS(Error), in randomized block designs, 413
SS(Interaction):
  shortcut formula for, 403
  in two-factor analysis of variance, 402
SS(Regression):
  in multiple regression, 492
  and R-squared, 461, 476
  using matrix notation, 508
SS(Residual):
  in multiple regression, 486
  and R-squared, 461
  using matrix notation, 508
SS(Rows):
  shortcut formula for, 403
  in two-factor analysis of variance, 402
SS(Total), 413
  algebraic analysis of, 429–430
  in analysis of variance, 381
  in multiple regression, 487
  shortcut formula, 384, 403, 414
  using matrix notation, 508
SS(Treatment):
  in randomized block designs, 413
  shortcut formula in randomized block designs, 414
SS(Within), 382
  shortcut (alternative) formulas, 384
  shortcut formula for, 403
Standard deviation (*see also* Variance):
  confidence interval for, 354
  definition of, 21
  of a discrete random variable, 91
  of grouped data, 33
  hypothesis test for, 355
  interpretation of (*see* Empirical Rule, Chebyshev's Inequality)
  around regression line, 446
  residual, 446
  of residuals in multiple regression, 486
Standard error (*see also* Standard deviation):

for Box-Jenkins models, 643
definition of, 171
of difference of means (independent samples), 307
of difference of proportions, 342
effect of autocorrelation, 559
of efficient estimator, 205
estimated, 230
for forecasting mean, 454
for individual prediction in multiple regression, 503, 510
of intercept, 441
of intercept, effect of extrapolation, 445
of partial slope using matrix notation, 509
for predicting mean in multiple regression, 510
of a proportion, 231
of sample mean, 171, 197–198
of sample sum, 170
of slope in simple regression analysis, 444, 449
of slope, effect of x variability, 444–445
of slopes and intercept in multiple regression, 494
Standard error of estimate (*see* Residual standard deviation)
Standard normal distribution, 145
States of nature (*see* Outcome)
Stationary time series, 638
Statistic:
sample, 166
sampling distribution of, 166
Statistical independence (*see* Independence)
Statistically detectable (*see* Significant)
Statistically significant (*see* Significant)
Statistical test (*see* Hypothesis test)
Stem-and-leaf diagram, definition of, 12–13
Steps, multiple regression modelling, 521
Stepwise regression, 543–545
bias of, 545
Stopping rules, for stepwise regression, 543–544
Strategy, of hypothesis testing, 258
Stratified sample, 722
Stratified sampling, 728
Strength of relation:
compared to statistical
Strength of relation (Continued)
significance, 464
and correlation, 459
Student's t (*see* t statistic)
Subjective probabilities, 43
Subsets:
number of, 122–123
of symbols, 122
Sum:
Central Limit Theorem for, 174
expected value and standard error of, 170
Summation sign, definition of, 39
Systematic sampling, 745–746

t confidence interval: for difference of means, 312; for mean, 242
t distribution:
definition of, 256
properties of, 237
table of, 238
and z distribution, 239–240
t statistic, 237
definition of, 256
t test:
of a mean, 281
of correlation, 463
for difference of means, 314
for paired samples, 335
for partial slope, 495
of slope in simple regression, 449
and stepwise regression, 544
with unequal variances in two samples, 315
t test of a mean, and p-value, 282
Table:
of normal distribution, 145
probability, 60
Target population, 723
Test statistic, definition of, 258
Ties, in rank sum test, 323
Time series, stationary, 638
Time series analysis, classical, 620
Tolerable width, of a confidence interval, 233
Total:
confidence interval for, cluster sampling, 735
confidence interval for, simple random sampling, 725
confidence interval for, stratified sampling, 729
Transformation:
to achieve constant variance, 554–555
of dependent variable, 536
of independent variables, 532–536
logarithmic, 533
quadratic, 534
Treatment factor, in randomized block designs, 412
Tree:
decision, 700–707
probability, 61–63
Trend:
in additive model of time series, 625
exponential, 622, 623
forms of in time series, 622
Gompertz, 622
linear, 622, 623
logistic, 622
in multiplicative model, 626
S-shaped, 624
in time series, 621
Trimmed mean, 17
and confidence intervals, 246
as estimator, 203
Tukey method:
compared to Scheffé method, 394
for pairs of cell means in two-factor studies, 407
for row means in two-factor studies, 408
Two-sided research hypothesis, 258
Two-stage cluster sampling, 746
Two-tailed p-value, 277
Two-tailed test, using z, 269 (*see also* Two-sided research hypothesis)
Type I error, 263
definition of, 259
as false positive, 262
overall risk of, 381
Type II error, 263 (*see also* Beta)
definition of, 262
as false negative, 262
in goodness-of-fit tests, 363

Unbiased estimator, 204
of difference of means, 307
Unequal variance (*see* Heteroscedasticity)
Uniform probability density, 140

Uniform probability distribution, 82
Union, of events, 48
Universal Rejection Region, 278
Utility function:
- concavity and risk, 676–677
- construction of, 681–682
- definition, 675

Validation:
- methods for, 570
- need for, 569

Variability:
- controlled by pairing, 332
- need for measures of, 19

Variables, dependent and independent, 372
Variance(s):
- and adding a constant, 113
- of beta distribution, 713
- of binomial distribution, 129
- chi-square distribution of, 255
- confidence interval for, 354
- confidence interval for ratio of, 360
- of continuous random variable, 95
- definition of, 21
- of difference of means (independent samples), 307
- of difference of proportions, 342

Variance(s) (Continued)
- of discrete random variable, 91
- distribution of, 352
- equal (assumption), 312 (*see also* Homoscedasticity)
- of exponential density, 142
- of geometric distribution, 134
- of grouped data, 33
- of hypergeometric distribution, 132
- hypothesis test for, 355
- and multiplying by a constant, 114
- negative binomial, 135
- of Poisson distribution, 138
- pooled, 313
- population, 21
- and portfolio analysis, 669
- of rank sum, 323
- as risk measure, 668
- of sample mean, 171
- of sample sum, 170
- short cut formula for, 21, 92
- of sum, 115
- test of equality of, 359
- of uniform distribution, 140
- with and without replacement, 211

Variances not equal, effect on two-sample t test, 315
Venn diagrams, 48

Weighted aggregate, price index, 618
Weighted average, 17–18
Weighted index, choice of weights, 618–619
Weighted least squares, 553
- with heteroscedasticity, 554

White noise:
- in moving average model, 641
- in time series, 637

Wilcoxon rank sum test (*see* Rank sum test)
Wilcoxon signed-rank test (*see* Signed-rank test)
Within-sample variability, 730
Work files, 758

z interval, for a mean, 228
z score, 145
z statistic, for testing mean, 266
z table, approximation to chi-square table, 355
z test:
- for difference of proportions, 343
- and p-value, 277
- for proportion, 296

Zero differences, in signed-rank test, 336

GLOSSARY OF SYMBOLS: Roman Letters

Symbol	Meaning	Chapter(s)
S_{xx}	Sum of squared X deviations	13
S_{xy}	Sum of products of X and Y deviations	13
S	Plus-or-minus for Scheffé type comparisons	12
S	Sample space	3
T	Sum of ranks	10
T_+	Sum of positive rank	10
T_-	Sum of negative rank	11
t_a	Tabled t value, right-tail area a	8–15
t	A t statistic	8–15
T.S.	Test statistic	9–15
U	Number of errors with unknown independent variable	11
U	Utility	17
$v(a_i, \theta_j)$	Payoff of action a_i for state of nature θ_j	17, 18
Var(X)	Variance of random variable X	all
w	Interval width for grouped data	2
X	Random variable	4–6
X	Independent variable	13–15
x_i	ith value of X	13–15
x_{n+1}	Value of X for which prediction is made	13
X	Matrix of independent variables' values	14
$\overline{X}, \overline{x}$	Average of x values	15, 19
Y	Random variable	4ff
Y	Dependent variable	13–15
Y_{n+1}	Actual Y value at x_{n+1}	13, 14
$\hat{Y}, \hat{y}$	Predicted Y value	13–15
$\overline{Y}, \overline{y}$	Average of Y values	2–16
z	Standardized normal statistic	5ff
z_a	Tabled standardized normal value, right-tail area a	5ff